DATE DUE

Demco

Handbook of Epigenetics

Handbook of Epigenetics

The New Molecular and Medical Genetics

Edited by

Trygve Tollefsbol

Department of Biology
University of Alabama at Birmingham
Birmingham, AL, 35294-1170, USA

Amsterdam • Boston • Heidelberg • London • New York • Oxford
Paris • San Diego • San Francisco • Singapore • Sydney • Tokyo

Academic Press is an imprint of Elsevier

Academic Press is an imprint of Elsevier
32 Jamestown Road, London NW1 7BY, UK
30 Corporate Drive, Suite 400, Burlington, MA 01803, USA
525 B Street, Suite 1800, San Diego, CA 92101-4495, USA

First edition 2011

Notice
No responsibility is assumed by the publisher for any injury and/or damage to persons
or property as a matter of products liability, negligence or otherwise, or from any use or
operation of any methods, products, instructions or ideas contained in the material herein.
Because of rapid advances in the medical sciences, in particular, independent verification of
diagnoses and drug dosages should be made

British Library Cataloguing-in-Publication Data
A catalogue record for this book is available from the British Library

Library of Congress Cataloging-in-Publication Data
A catalog record for this book is available from the Library of Congress

ISBN: 978-0-12-375709-8

For information on all Academic Press publications
visit our website at www.elsevierdirect.com

Typeset by MPS Limited, a Macmillan Company, Chennai, India
www.macmillansolutions.com

Printed and bound in United States of America

10 11 12 13   10 9 8 7 6 5 4 3 2 1

CONTENTS

v

CONTENTS

CONTRIBUTORS

Masahiko Abematsu
Laboratory of Molecular Neuroscience
Graduate School of Biological Sciences
Nara Institute of Science and Technology
Nara 630-0101, Japan

David W. Anderson
Department of Pathology
St. Luke's Hospital and Health Network
801 Ostrum Street
Bethlehem, PA 18015, USA

Roberta H. Andronikos
Murdoch Childrens Research Institute
Flemington Road
Parkville, Victoria 3052, Australia

Adriana Arita
Nelson Institute of Environmental Medicine
New York University School of Medicine
57 Old Forge Road, New York, NY 10987, USA

Zoya Avramova
School of Biological Sciences, UNL
E-300 Beadle Center
P.O. Box 88066
Lincoln, NE 68588-0666, USA

Autumn J. Bernal
Department of Radiation Oncology
Box 3433
Duke University Medical Center
Durham, NC 27710, USA

Javier Campión
Department of Nutrition and Food Sciences,
    Physiology and Toxicology
University of Navarra, 31080 Pamplona
Navarra, Spain

Vince Castranova
PPRB/NIOSH
Basic Pharmaceutical Sciences
West Virginia University
1095 Willowdale Road
Morgantown, WV 26505, USA

Frances A. Champagne
Department of Psychology
Columbia University
Room 406, Schermerhorn Hall
1190 Amsterdam Avenue
New York, NY 10027, USA

Wendy Chao
Department of Ophthalmology
Schepens Eye Research Institute
Harvard Medical School
20 Staniford Street
Boston, MA 02114, USA

Fei Chen
PPRB/NIOSH
Basic Pharmaceutical Sciences
West Virginia University
1095 Willowdale Road
Morgantown, WV 26505, USA

Xiaowei Sylvia Chen
Department of Biochemistry
University of Otago
Dunedin 9054, New Zealand

Xiaodong Cheng
Department of Biochemistry
Emory University School of Medicine
1510 Clifton Road
Atlanta, Georgia 30322, USA

Lesley Collins
Massey Genome Service
Massey University
Palmerston North, New Zealand

Max Costa
Nelson Institute of Environmental Medicine
New York University School of Medicine
57 Old Forge Road, New York, NY 10987, USA

James P. Curley
Department of Psychology
Columbia University
Room 406, Schermerhorn Hall
1190 Amsterdam Avenue
New York, NY 10027, USA

Walter Doerfler
Institute for Virology
Erlangen University Medical School
Schlossgarten 4
D-91054 Erlangen, Germany

Tamar Dvash
Department of Human Genetics
The Eli and Edythe Broad Center for Regenerative
    Medicine and Stem Cell Research

David Geffen School of Medicine
University of California Los Angeles
695 Charles E. Young Drive South
Los Angeles, CA 90095, USA

Thomas Eggermann
Institute of Human Genetics
University Hospital
RWTH Aachen, Pauwelsstrasse 30, DE-52074
Aachen, Germany

Guoping Fan
Department of Human Genetics
The Eli and Edythe Broad Center for Regenerative
  Medicine and Stem Cell Research
David Geffen School of Medicine
University of California Los Angeles
695 Charles E. Young Drive South
Los Angeles, CA 90095, USA

Tamara B. Franklin
Brain Research Institute
Medical Faculty of the University of Zürich and
  Department of Biology
Swiss Federal Institute of Technology
Winterthurerstrasse 190
CH-8057 Zürich, Switzerland

Steffen Gay
Center of Experimental Rheumatology
University Hospital and Zurich Center of
  Integrative Human Physiology (ZIHP)
University Hospital Zürich,
8091, Zürich, Switzerland

Eric Gilson
Laboratory of Biology and Pathology of Genomes
  CNRS UMR6267/INSERM U998/UNS
28 Avenue Valombrose, Faculte de Medecine
06107 Nice Cedex 2, France

Johannes Gräff
Brain Research Institute
Medical Faculty of the University of Zürich
  and Department of Biology
Swiss Federal Institute of Technology
Winterthurerstrasse 190
CH-8057 Zürich, Switzerland

Hideharu Hashimoto
Department of Biochemistry
Emory University School of Medicine
1510 Clifton Road
Atlanta, GA 30322, USA

Naoko Hattori
Carcinogenesis Division
National Cancer Center Research Institute
1-1 Tsukiji 5-chome
Chuo-ku, Tokyo 104-0045, Japan

Zdenko Herceg
Epigenetics Group
International Agency for Research on
  Cancer (IARC)
150 cours Albert Thomas
69372 Lyon cedex 08, France

Norbert Hochstein
QIAGEN
Qiagen Strasse 1
40724, Hilden, Germany

John R. Horton
Department of Biochemistry
Emory University School of Medicine
1510 Clifton Road
Atlanta, GA 30322, USA

Simon H. House
22 Stanley Street
Southsea, Portsmouth, Hants
PO5 2DS, UK

Karolina Janitz
The Ramaciotti Centre for Gene Function Analysis
The University of New South Wales, Sydney
NSW 2052, Australia

Michal Janitz
School of Biotechnology and Biomolecular
  Sciences
The University of New South Wales, Sydney
NSW 2052, Australia

Randy L. Jirtle
Department of Radiation Oncology
Box 3433
Duke University Medical Center
Durham, NC 27710, USA

Ji-Hoon E. Joo
Murdoch Children's Research Institute
Flemington Road
Parkville, Victoria 3052, Australia

Berry Juliandi
Laboratory of Molecular Neuroscience
Graduate School of Biological Sciences
Nara Institute of Science and Technology
Nara 630-0101, Japan

Astrid Jungel
Center of Experimental Rheumatology
University Hospital and Zurich Center of
  Integrative Human Physiology (ZIHP)
University Hospital Zürich,
8091, Zürich, Switzerland

Hong Kan
PPRB/NIOSH
Basic Pharmaceutical Sciences
West Virginia University
1095 Willowdale Road
Morgantown, WV 26505, USA

Tae Hoon Kim
Yale University School of Medicine
333 Cedar Street SHM I-142B
New Haven, CT 06520-8005, USA

Yoon Jung Kim
Yale University School of Medicine
333 Cedar Street SHM I-142B
New Haven, CT 06520-8005, USA

Hervé Lalucque
Institut de Genetique et Microbiologie
UMR CNRS 8621
Univ Paris Sud 11
91405 Orsay cedex, France

Sheila C.S. Lima
Epigenetics Group
International Agency for Research on Cancer (IARC)
150 Cours Albert Thomas
Rhone-Alpes, 69008 Lyon, France

Sally A. Litherland
Metabolic Signaling and Disease Program
Diabetes & Obesity Center
Sanford-Burnham Medical Research Center
   at Lake Nona
6400 Sanger Road
Orlando, FL 32827, USA

John C. Lucchesi
Department of Biology
Emory University
1510 Clifton Road, Atlanta, GA 30322, USA

Hanna Maciejewska-Rodrigues
Center of Experimental Rheumatology
University Hospital and Zurich Center of
   Integrative Human Physiology (ZIHP)
University Hospital Zürich,
8091, Zürich, Switzerland

Frédérique Magdinier
Laboratoire de Génétique Médicale et
   Génomique Fonctionnelle
INSERM, UMR S_910/Université de la
   Méditerannée
Faculté de Médecine de la Timone
27 Bd Jean Moulin 13385
   Marseille cedex05, France

Fabienne Malagnac
Institut de Genetique et Microbiologie
UMR CNRS 8621
Univ Paris Sud 11
91405 Orsay cedex, France

Isabelle M. Mansuy
Brain Research Institute
Medical Faculty of the University of Zürich
   and Department of Biology
Swiss Federal Institute of Technology
Winterthurerstrasse 190, CH-8057 Zürich
Switzerland

J. Alfredo Martinez
Department of Nutrition and Food Sciences,
   Physiology and Toxicology
University of Navarra, 31080 Pamplona
Navarra, Spain

Rahia Mashoodh
Department of Psychology
Columbia University
Room 406, Schermerhorn Hall
1190 Amsterdam Avenue
New York, NY 10027, USA

Fermin I. Milagro
Department of Nutrition and Food Sciences,
   Physiology and Toxicology
University of Navarra, 31080 Pamplona
Navarra, Spain

Ashleigh Miller
Department of Medicine
Division of Hematology and Oncology
University of California-San Francisco
San Francisco, CA 94143, USA

Pamela Munster
Department of Medicine
Division of Hematology and Oncology
University of California-San Francisco
San Francisco, CA 94143, USA

Susan K. Murphy
Department Gynecologic Oncology
Box 91092
Duke University Medical Center
Durham, NC 27710, USA

Rabih Murr
Epigenetics Group
International Agency for Research on
   Cancer (IARC)
150 cours Albert Thomas
69372 Lyon cedex 08, France

CONTRIBUTORS

Kinichi Nakashima
Laboratory of Molecular Neuroscience
Graduate School of Biological Sciences
Nara Institute of Science and Technology
Nara 630-0101, Japan

Anja Naumann
Institute for Virology
Erlangen University Medical School
Schlossgarten 4
D-91054 Erlangen, Germany

Alexandre Ottaviani
Laboratoire de Biologie Moléculaire de la Cellule
CNRS UMR5239, Ecole Normale Supérieure
    de Lyon
UCBL1, IFR128, 46 allée d'Italie
69364 Lyon Cedex 07, France

Jacob Peedicayil
Department of Pharmacology and Clinical
    Pharmacology
Christian Medical College
Vellore, India

Gerd P. Pfeifer
Division of Biology
Beckman Research Institute of the
    City of Hope
1500 East Duarte Road
Duarte, CA 91010, USA

Luis Felipe Ribeiro Pinto
Divisão de Genética — Coordenação de Pesquisa
Instituto Nacional de Câncer
Rua André Cavalcanti, 37, 40 Andar
CEP: 20231-050
Riode Janiero, RJ, Brazil

Vincenzo Pirrotta
Department of Molecular Biology and
    Biochemistry
Rutgers University
604 Allison Road
Piscataway, NJ 08854, USA

Pier Lorenzo Puri
Istituto Dulbecco Telethon at Istituto
    Di Ricovero e Cura a Carattere Scientifico
Santa Lucia Fondazione and European Brain
    Research Institute
64 Via del Fosso di Fiorano
00143 Rome, Italy
The Sanford/Burnham Institute for
    Medical Research
10901 North Torrey Pines Road
La Jolla, CA 92037-1062, USA

Tibor A. Rauch
Section of Molecular Medicine
Department of Orthopedic Surgery
Rush University Medical Center
Chicago, IL 60612, USA

Lee B. Riley
Department of Oncology,
St. Luke's Hospital and Health Network,
801 Ostrum Street,
Bethlehem, PA 18015, USA

Eric D. Roth
Department of Neurobiology
1825 University Blvd, SHEL 1070
Birmingham, AL 35294, USA

Tania L. Roth
Department of Neurobiology
1825 University Blvd, SHEL 1070
Birmingham, AL 35294, USA

Richard Saffery
Murdoch Childrens Research Institute
Flemington Road
Parkville, Victoria 3052, Australia

Vittorio Sartorelli
Laboratory of Muscle Stem Cells and
    Gene Regulation
National Institute of Arthritis, Musculoskeletal
    and Skin Diseases
National Institutes of Health
Bethesda, MD, USA

Caroline Schluth-Bolard
Laboratoire de Biologie Moléculaire de la Cellule
CNRS UMR5239, Ecole Normale Supérieure
    de Lyon
UCBL1, IFR128, 46 allée d'Italie
69364 Lyon Cedex 07, France

Barbara Schönfeld
Allan Wilson Centre for Molecular Ecology
    and Evolution
Massey University
Palmerston North, New Zealand

Axel Schumacher
Centre for Addiction and Mental Health,
Toronto, ON M5T 1R8, Canada

Philippe Silar
Institut de Genetique et Microbiologie
UMR CNRS 8621
Univ Paris Sud 11
91405 Orsay cedex, France

Kjetil Søreide
Department of Surgery
Stavanger University Hospital
Armaver Hansenvei 20
POB 8100, N-4068
Stavanger, Norway

J. David Sweatt
Department of Neurobiology
1825 University Blvd, SHEL 1070
Birmingham, AL 35294, USA

Scott Thomas
Department of Medicine
Division of Hematology and Oncology
University of California-San Francisco
San Francisco, CA 94143, USA

Kenneth T. Thurn
Department of Medicine
Division of Hematology and Oncology
University of California-San Francisco
San Francisco, CA 94143, USA

Trygve O. Tollefsbol
Department of Biology
175 Campbell Hall
1300 University Boulevard
University of Alabama at Birmingham
Birmingham, AL 35294-1170, USA

Toshikazu Ushijima
Carcinogenesis Division
National Cancer Center Research Institute
1-1 Tsukiji 5-chome
Chuo-ku, Tokyo 104-0045, Japan

David A. Wacker
Yale University School of Medicine
333 Cedar Street SHM I-142B
New Haven, CT 06520, USA

Stefanie Weber
Institute for Virology
Erlangen University Medical School
Schlossgarten 4
D-91054 Erlangen, Germany

Xing Zhang
Department of Biochemistry
Emory University School of Medicine
1510 Clifton Road
Atlanta, GA 30322, USA

Epigenetics is considered by many to be the "new genetics" because many biological processes are controlled not through gene mutations, but rather through reversible and heritable epigenetic phenomena ranging from DNA methylation to histone modifications to prions. Epigenetic processes occur in diverse organisms and control a vast array of biological functions, such as tissue/organ regeneration, X-chromosome inactivation, stem cell differentiation, genomic imprinting, and aging. Epigenetic aberrations underlie many diseases, including cancer and disorders of the immune, endocrine, and nervous systems; clinical intervention is already in place for some of these disorders and many novel epigenetic therapies are likely on the horizon.

Handbook of Epigenetics: The New Molecular and Medical Genetics is the first comprehensive analysis of epigenetics, and summarizes recent advances in this intriguing field of study. This book will interest students and researchers in both academics and industry by illuminating the evolution of epigenetics, the epigenetic basis of normal and pathological processes, and the practical applications of epigenetics in research and therapeutics.

CHAPTER 1

Epigenetics: The New Science of Genetics

Trygve O. Tollefsbol[1,2,3,4,5]
[1]Department of Biology, University of Alabama at Birmingham, AL 35294
[2]Center for Aging, University of Alabama at Birmingham, AL 35294
[3]Comprehensive Cancer Center, University of Alabama at Birmingham, AL 35294
[4]Nutrition Obesity Research Center, University of Alabama at Birmingham, AL 35294
[5]Comprehensive Diabetes Center, University of Alabama at Birmingham, AL 35294, USA

INTRODUCTION

The term epigenetics was first introduced in 1942 by Conrad Waddington and was defined as the causal interactions between genes and their products that allow for phenotypic expression [1]. This term has now been somewhat redefined and although there are many variants of the definition of this term today, a consensus definition is that epigenetics is the collective heritable changes in phenotype due to processes that arise independent of primary DNA sequence. This heritability of epigenetic information was for many years thought to be limited to cellular divisions. However, it is now apparent that epigenetic processes can be transferred in organisms from one generation to another [2–3]. This phenomenon was first described in plants [4] and has been expanded to include yeast, *Drosophila*, mouse and, possibly, humans [5–7].

THE BASICS OF DNA METHYLATION AND HISTONE MODIFICATIONS

In most eukaryotes DNA methylation, the most studied of epigenetic processes, consists of transfer of a methyl moiety from S-adenosylmethionine (SAM) to the 5-position of cytosines in certain CpG dinucleotides. This important transfer reaction is catalyzed by the DNA methyltransferases (DNMTs). The three major DNMTs are DNMT1, 3A and 3B and DNMT1 catalyzes what is referred to as maintenance methylation that occurs during each cellular replication as the DNA is duplicated. The other major DNMTs, 3A and 3B, are known more for their relatively higher *de novo* methylation activity where new 5-methylcytosines are introduced in the genome at sites that were not previously methylated. The most significant aspect of DNA methylation, which can also influence such processes as X chromosome inactivation and cellular differentiation, is its effects on gene expression. In general, the more methylated a gene regulatory region, the more likely it is that the gene activity will become down-regulated and vice versa although there are some notable exceptions to this dogma [8]. Chapter 2 of this book reviews the mechanisms of DNA methylation, methyl-CpG recognition and demethylation in mammals. Recent advances have highlighted important roles of UHRF1 and DNMT3L that are required for maintenance and *de novo* methylation,

1

Handbook of Epigenetics: The New Molecular and Medical Genetics. DOI: 10.1016/B978-0-12-375709-8.00001-0

respectively, and the potential inclusion of 5-hydroxymethylcytosine with 5-methylcytosine in expressing the impact of DNA methylation on the genome.

Chromatin changes are another central epigenetic process that have an impact not only on gene expression, but also many other biological processes. Posttranslational modifications of histones such as acetylation and methylation occur in a site-specific manner that influences the binding and activities of other proteins that influence gene regulation. The histone acetyltransferases (HATs) catalyze histone acetylation and the histone deacetylases (HDACs) result in removal of acetyl groups from key histones that comprise the chromatin. These modifications can occur at numerous sties in the histones and are most common in the amino terminal regions of these proteins as reviewed in Chapter 3. In general, increased histone acetylation is associated with greater gene activity and vice versa. By contrast, methylation of histones has variable effects on gene activity where lysine 4 (K4) methylation of histone H3 is often associated with increasing gene activity whereas methylation of lysine 9 (K9) of histone H3 may lead to transcriptional repression. There is also considerable crosstalk between DNA methylation and histone modifications [9] such that cytosine methylation may increase the likelihood of H3-K9 methylation and H3-K9 methylation may promote cytosine methylation.

ADDITIONAL EPIGENETIC PROCESSES

Among the most exciting advances in epigenetics have been the discoveries that many other processes besides DNA methylation and histone modifications impact the epigenetic behavior of cells. For instance, non-coding RNA (Chapter 4) including both short and long forms, often share protein and RNA components with the RNA interference (RNAi) pathway and they may also influence more traditional aspects of epigenetics such as DNA methylation and chromatin marking. These effects appear to be widespread and occur in organisms ranging from protists to humans. Prions are fascinating in that they can influence epigenetic processes independent of DNA and chromatin. In Chapter 5 it is shown that structural heredity also is important in epigenetic expression where alternative states of macromolecular complexes or regulatory networks can have a major effect on phenotypic expression independent of changes in DNA sequences. The prion proteins are able to switch their structure in an autocatalytic manner that can not only influence epigenetic expression, but also lead to human disease. The position of a gene in a given chromosome can also greatly influence its expression (Chapter 6). Upon rearrangement, a gene may be relocated to a heterochromatic region of the genome leading to gene silencing and many other gene position effects have been described, some of which may also lead to various human diseases. Polycomb mechanisms are another relatively new aspect of epigenetics that control all of the major cellular differentiation pathways and are also involved in cell fate. Polycomb repression is very dynamic and can be easily reversed by activators and they also raise the threshold of the signals or activators required for transcriptional activation which places these fascinating proteins within the realm of epigenetic processes (Chapter 7). Therefore, although DNA methylation and histone modifications are mainstays of epigenetics, recent advances have greatly expanded the epigenetic world to include many other processes such as non-coding RNA, prions, chromosome position effects and Polycomb mechanisms.

EPIGENETIC TECHNOLOGY

Many of the advances in epigenetics that have driven this field for the past two decades can be traced back to the technological breakthroughs that have made the many discoveries possible. We now have a wealth of information about key gene-specific epigenetic changes that occur in a myriad of biological processes. In Chapter 8, gene-specific techniques for determining DNA methylation are reviewed. These methods include bisulfite sequencing, methylation-specific PCR (MSP) and quantitative MSP. These techniques can be applied not only to mechanisms of epigenetic gene control, but to diagnostic processes as well. In addition, there have been important breakthroughs in analyses of the methylome at high resolution. Microarray platforms and high-throughput sequencing have made possible new techniques to analyze genome-wide features of epigenetics that are based on uses of methylation-sensitive restriction enzymes, sodium bisulfite conversion and affinity capture with antibodies or proteins that select methylated DNA sequences. Techniques such as restriction landmark genomic sequencing (RLGS), methylation-sensitive restriction fingerprinting, methylation-specific digital karyotyping, targeted and whole genome bisulfite sequencing, methylated DNA immunoprecipitation (MeDIP) and the methylated-CpG island recovery assay are reviewed in Chapter 9. Mechanisms for lysine 9 methylation of histone H3 are reviewed in Chapter 10 and chromatin immunoprecipitation (ChIP) and chromosome conformation capture (3C) are covered in Chapter 11. The 3C-based method allows analyses of the spatial proximity of distant functional genomic sites to render a three dimensional view of the genome within the nucleus itself. Since there has been much information derived from epigenomic approaches, methods to analyze data from ChIP-on-chip and ChIP-seq, for example, are becoming increasingly important and are delineated in Chapter 12. There is no question that developments in the tools for assessing epigenetic information have been and will continue to be important factors in advancing epigenetics.

MODEL ORGANISMS OF EPIGENETICS

Epigenetic processes are widespread and much of our extant knowledge about epigenetics has been derived from model systems, both typical and unique. The ease of manipulation of eukaryotic microbes has facilitated discoveries in the molecular mechanisms of basic epigenetic processes (Chapter 13). In these cases epigenetics may play a key role in genomic protection from invasive DNA elements and in identifying the importance of gene silencing mechanisms in evolution. *Drosophila* is a mainstay model in biology in general and the epigenetics field is not an exception in this regard. For example, Chapter 14 offers a number of examples of transgenerational inheritance in *Drosophila* and this model system also shows promise in unraveling the evolutionary aspects of epigenetics. Probably the most useful model system in epigenetics to date is the mouse model (Chapter 15). Randy Jirtle and colleagues review numerous different mouse models that are important in many epigenetic processes such as transgenerational epigenetics and imprinting and these models have potential in illuminating human diseases such as diabetes, neurological disorders and cancer. Plant models (Chapter 16) are of great importance in epigenetics due in part to their plasticity and their ability to silence transposable elements. RNAi silencing in plants has been at the forefront of epigenetics and plant models will likely lead the way in several other epigenetic processes in the future. Thus, model development, like the advances in techniques, have made many of the most exciting discoveries in epigenetics possible for a number of years.

METABOLISM AND EPIGENETICS

Epigenetics is intricately linked to changes in the metabolism of organisms and these two processes cannot be fully understood separately. S-adenosylmethionine (SAM) is a universal methyl donor and drives many epigenetic processes (Chapter 17) and the importance of SAM in epigenetic mechanisms is vast. Metabolic functions can also influence the chromatin which is a major mediator of epigenetic processes (Chapter 18). It is now apparent that various environmental influences and metabolic compounds can regulate the many enzymes that modify histones in mammals. Thus, metabolic processes impact DNA methylation and chromatin remodeling, the two major epigenetic mediators, and it is likely that this relatively new field will continue to advance in an exponential manner.

FUNCTIONS OF EPIGENETICS

The functions of epigenetics are indeed numerous and it would be next to impossible to do complete justice in one book to this ever-expanding field. However, Chapters 19–25 illustrate a few of the many different functions that epigenetics mediates. Stem cells rely in part on signals from the environment and epigenetic mechanisms such as DNA methylation, histone modification, and microRNA (miRNA) have central roles in how stem cells respond to environmental influences (Chapter 19). Regenerative medicine is dependent upon stem cells and skeletal muscle regeneration (Chapter 20) involves key changes in the epigenome that regulate gene expression in muscle progenitors through chromatin as well as microRNA epigenetic changes. It has been known that epigenetics is important in X chromosome inactivation for quite some time although advances in this area are continuing to move rapidly. It is now apparent that X chromosome inactivation is regulated not only through the genes *Tsix* and *Xist*, but also pluripotency factors that affect *Xist* expression (Chapter 21). Genomic imprinting, likewise, has been known to be epigenetic-based for many years, but discoveries in this area of epigenetics continue to move at a rapid pace. Genomic imprinting is not limited to mammals but also occurs through analogous processes in plants and invertebrates and it can occur in specific tissues or during critical developmental stages (Chapter 22). Profound new discoveries have recently occurred in the area of the epigenetics of memory processes. Recent exciting discoveries have shown that gene regulation through epigenetic mechanisms is necessary for changes in adult brain function and behavior based on life experiences (Chapter 23). Moreover, new drugs that impact epigenetic mechanisms may have future uses in treating or alleviating cognitive dysfunction. Transgenerational inheritance (Chapter 24) is also a form of memory based in part on epigenetics in that early life experiences that impact epigenetic markers can greatly influence adult health and risk for diseases. In addition, the aging process is a form of epigenetic memory and experience, in that our genes are epigenetically modified from our parents and also during our entire life spans, that can significantly impact the longevity of humans as well as our risk for the numerous age-related diseases, many of which are also epigenetically-based (Chapter 25). It is therefore apparent that epigenetics influences a number of different functions and it is highly likely that many additional functions of epigenetics will be discovered in the future.

EVOLUTIONARY EPIGENETICS

Although many think of epigenetic processes as being inherent and static to a specific organism, it is apparent that epigenetics has been a major force behind the evolutionary creation of new species. Chapter 26 reveals that epigenetic mechanisms have a major influence on mutations. The evolutionary impact of epigenetics is in full force even today with the ever-changing environment that can modulate gene expression through epigenetic processes. For example, rapid changes in diet and the modern lifestyle as well as environmental pollution are undoubtedly impacting not only the human epigenome, but also the evolution of many of the more primitive species that in turn greatly affect the environment.

EPIGENETIC EPIDEMIOLOGY

Dietary factors are highly variable not only between individuals, but also among human populations and various nonhuman species. Many studies have shown that diet has a profound effect on the epigenetic expression of the genome and therefore on the phenotype. DNA methylation is the epigenetic process that has been most often associated with diet and changes in the diet may not only induce varying epigenetic expressions, but, paradoxically, a changed diet may also transfix epigenetic changes that can then be transferred to the next generation in a stable manner (Chapter 27). Environmental agents other than diet also impact the epigenome. For example, Chapter 28 reviews the many environmental agents that can lead to alterations in the epigenome thereby inducing toxicity or carcinogenesis. Moreover, invasion by foreign agents can influence the epigenome (Chapter 29). Viruses and bacteria, for example, play a major role in altering the epigenetic expression of the genome and these processes may lead to human diseases such as cancer. Chapter 30 by Walter Doerfler and colleagues illustrates details of the role of adenovirus type 12 (Ad12) in reshaping the hamster genome and they also provide analyses of the human *FMR1* promoter that is impacted by DNA methylation in the fragile X syndrome. Drugs also reshape the epigenome, which has opened the new field of pharmacoepigenomics. It is clear that certain populations respond differently to drugs and much of this variation may be explained by epigenetic factors (Chapter 31). Thus, epidemiological factors have great importance in epigenetics and this is influenced by diet, environmental agents, infections, drugs and likely many other factors as well.

EPIGENETICS AND HUMAN DISEASES

For the medical community, a major interest in epigenetics stems from the role of epigenetic changes in the etiology, progression and diagnosis of human diseases. Cancer has long been associated with epigenetic alterations, and DNA methylation, chromatin modifications, and RNA-dependent regulation have all been shown to affect the incidence and severity of cancer (Chapter 32). Many immune disorders such as systemic lupus erythematosus (SLE) and rheumatoid arthritis as well as autoimmune disorders such as multiple sclerosis have been associated with epigenetic aberrations (Chapter 33) and epigenetic processes have also been linked to brain disorders (Chapter 34). In the latter case, the Rett syndrome, Alzheimer's disease, Huntington's disease and even autism to name a few have been associated in at least some way with epigenetic alterations. Even schizophrenia and depression may have an epigenetic basis in their expression. System metabolic disorders may also be related to epigenetic aberrations. For example, obesity, gestational diabetes and hypertension can influence the fetal chromatin and lead to an increased incidence in adult disease later in life (Chapter 35). Since genomic imprinting is based on epigenetic mechanisms, it may come as no surprise that defects in imprinting can lead to a number of human diseases (Chapter 36). The Prader–Willi syndrome, Angelman syndrome, Silver Russell syndrome and many other imprinting disorders such as transient neonatal diabetes mellitus are due to imprinting disorders that are based on epigenetic defects. Therefore, the number of diseases impacted by epigenetic processes is large and advances in the treatment of these disorders will likely depend in part on breakthroughs in epigenetic therapy.

EPIGENETIC THERAPY

Although there are many epigenetic therapies that are in use and on the horizon, histone modifying drugs have probably received the most attention in the clinics. Chief among these are the histone deacetylase (HDAC) inhibitors. Vorinostat (Zolinza), for example, has been approved by the Food and Drug Administration for use in the treatment of patients with cutaneous T-cell lymphoma (Chapter 37). Many different HDAC inhibitors have been developed and it is likely that significant improvements will occur for HDAC inhibitors as well as many other drugs that can normalize aberrations in not only histone modifications, but also DNA methylation and perhaps some of the many other epigenetic processes that have been discovered.

CONCLUSION

Advances in understanding the basic mechanisms of DNA methylation and histone modifications have raised the field of epigenetics well beyond original expectations. This area of research has also significantly expanded horizontally with the illumination of other epigenetic processes such as non-coding RNA, prion changes and Polycomb mechanisms and it is likely that additional epigenetic processes will be discovered in the not too distant future. A major driving force in epigenetics has been the outstanding development of new technology that has not only served to stimulate new discoveries, but has also expanded the field by allowing for novel discoveries possible only through the use of these new tools. Advances in new model organisms for understanding epigenetic processes have also greatly stimulated this field of study. We now know that epigenetics is not only intricately associated with metabolism, but also functions in stem cell behavior, X chromosome inactivation, tissue regeneration, genomic imprinting, the transfer of information through generations, neurological memory processes and even the aging of organisms. Epigenetics has also played roles in evolution and has served as a molecular driver of mutations. Moreover, the changing environment is currently reshaping the evolution of many organisms through plastic epigenetic processes. Epidemiological factors such as diet, environmental exposure, microbial infections and drugs are also influencing our daily lives through epigenetics. Diseases that have been associated with epigenetic processes range from schizophrenia to cancer and the list of these diseases is rapidly expanding. Fortunately, the field of epigenetic therapy is also expanding and the hope is that the future will see many novel treatments for the numerous diseases that are derived from epigenetic defects.

References

1. Waddington CH. The epigenotype. Endeavour 1942;1:18–20.
2. Liu L, Li Y, Tollefsbol TO. Gene-environment interactions and epigenetic basis of human diseases. Curr Issues Mol Biol 2008;10:25–36.
3. Chong S, Whitelaw E. Epigenetic germline inheritance. Curr Opin Genet Dev 2004;14:692–6.
4. Brink RA, Styles ED, Axtell JD. Paramutation: directed genetic change. Paramutation occurs in somatic cells and heritably alters the functional state of a locus. Science 1968;159:161–70.
5. Cavalli G, Paro R. Epigenetic inheritance of active chromatin after removal of the main transactivator. Science 1999;286:955–8.
6. Grewal SI, Klar AJ. Chromosomal inheritance of epigenetic states in fission yeast during mitosis and meiosis. Cell 1996;86:95–101.
7. Rakyan VK, Blewitt ME, Druker R, Preis JI, Whitelaw E. Metastable epialleles in mammals. Trends Genet 2002;18:348–51.
8. Lai SR, Phipps SM, Liu L, Andrews LG, Tollefsbol TO. Epigenetic control of telomerase and modes of telomere maintenance in aging and abnormal systems. *Front Biosci* 2005;**10**:1779–96.
9. Fuks F, Burgers WA, Brehm A, Hughes-Davies L, Kouzarides T. DNA methyltransferase Dnmt1 associates with histone deacetylase activity. *Nat Genet* 2000;**24**:88–91.

Molecular Mechanisms of Epigenetics

Mechanisms of DNA Methylation, Methyl-CpG Recognition, and Demethylation in Mammals

Xiaodong Cheng, Hideharu Hashimoto, John R. Horton, and Xing Zhang
Department of Biochemistry, Emory University School of Medicine, Atlanta, GA 30322, USA

INTRODUCTION

The control of transcription initiation in mammalian cells can be very broadly divided into three categories: intrinsic promoter strength and availability of core transcription machinery [1–3], the actions of promoter- or regulon-specific transcription factors (positive and negative) [4–6], and the control of DNA accessibility by altering chromatin structure [6–8]. This latter category, including posttranslational modifications to histones and postreplicational modification of DNA, is the focus of recent extensive studies. Nucleosomes are the fundamental building blocks of eukaryotic chromatin, and consist of ~146 base pairs of DNA wrapped twice around a histone octamer [9]. A variety of protein-modifying enzymes (including methyltransferases, MTases] is responsible for histone modification, primarily at their flexible N-termini [10–12]. Here, we summarize the most recent structural and biochemical advances in the study of mammalian DNA MTases and their associated protein factor(s), and will touch on the functional links between histone modification and that of DNA.

In mammals and other vertebrates, DNA methylation occurs at the C5 position of cytosine (5mC), mostly within CpG dinucleotides (Fig. 2.1A), with the Dnmt enzymes using a conserved mechanism [13] that has been studied best in the bacterial 5mC MTase M.HhaI [14–18]. Briefly, this mechanism involves MTase binding to the DNA, eversion of the target nucleotide so that it projects out of the double helix ("base flipping"), covalent attack of a conserved Cys nucleophile on cytosine C6, transfer of the methyl group from S-adenosyl-L-methionine (AdoMet) to the activated cytosine C5, and the various release steps. This methylation, together with histone modifications, plays an important role in modulating chromatin structure, thus controlling gene expression and many other chromatin-dependent processes [19]. The resulting epigenetic effects maintain the various patterns of gene expression in different cell types [20]. Epigenetic processes include genomic imprinting [21], gene silencing [22,23], X chromosome inactivation [24], reprogramming in transferred nuclei [25,26], and some elements of carcinogenesis [27].

9

Handbook of Epigenetics: The New Molecular and Medical Genetics. DOI: 10.1016/B978-0-12-375709-8.00002-2

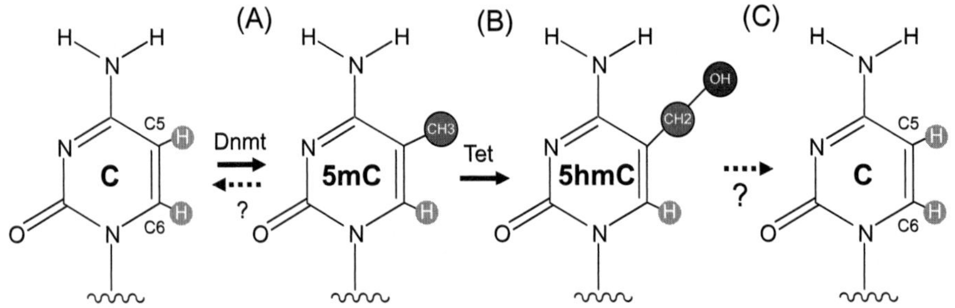

FIGURE 2.1
DNA cytosine methylation, hydroxylation, and demethylation. (A) The question mark indicates possible activity of DNA demethylases [150–155]. (B) Conversion of 5 mC to 5 hmC in mammalian DNA by the MLL fusion partner TET1 [129]. (C) It is currently unknown whether 5hmC is an end product or an intermediate in active DNA demethylation. The question mark indicates a possible MTase-assisted removal of the C5-bound hydroxymethyl group [131] (Please refer to color plate section)

DNA methylation is also associated with phenomena such as DNA repair [28], initiation of sexual dimorphism [29], progression through cell division checkpoints [30], and suppression of the huge number of transposable and retroviral elements in the mammalian genome [31–33].

MAMMALIAN DNA MTases

In mammals, Dnmts include three members, in two families that are structurally and functionally distinct (Fig. 2.2A). The Dnmt3a and Dnmt3b establish the initial CpG methylation pattern *de novo*, while Dnmt1 maintains this pattern during chromosome replication [34] and repair [35] (Fig. 2.2B). As befits a maintenance MTase, Dnmt1 has a 30–40-fold preference for hemimethylated sites [discussed in refs 36 and 37]. However, this division of labor is not absolute, as Dnmt1 activity is required for *de novo* methylation at non-CpG cytosines [38], and perhaps to an extent even in CpG islands [39,40].

The Dnmt3 family includes two active *de novo* Dnmts, Dnmt3a and Dnmt3b, and one regulatory factor, Dnmt3-Like protein (Dnmt3L) [41] (Fig. 2.2A). Dnmt3a and Dnmt3b have similar domain arrangements: both contain a variable region at the N-terminus, followed by a PWWP domain that may be involved in nonspecific DNA binding [42,43], a Cys-rich 3-Zn binding domain (comprising six CXXC motifs), and a C-terminal catalytic domain. The amino acid sequence of Dnmt3L is very similar to that of Dnmt3a and Dnmt3b in the Cys-rich 3-Zn binding domain, but it lacks the conserved residues required for DNA MTase activity in the C-terminal domain.

Dnmt3L IS A REGULATORY FACTOR FOR *DE NOVO* DNA METHYLATION

The phenotype of Dnmt3L knockout mice is indistinguishable from that of a Dnmt3a germ-cell-specific conditional knockout, with both having altered sex-specific *de novo* methylation of DNA sequences in germ cells and dispersed retrotransposons [44–47]. These results indicate that Dnmt3a and Dnmt3L are both required for the methylation of most imprinted loci in germ cells. Dnmt3L co-localizes and co-immunoprecipitates with both Dnmt3a and Dnmt3b [48], and enhances *de novo* methylation by both of these MTases [49–53]. The minimal regions required for interaction between Dnmt3L and Dnmt3a (or Dnmt3b), and for stimulated activity, are in the C-terminal domains of both proteins [50–54], as illustrated by the structure of the complex between C-terminal domains of Dnmt3a and Dnmt3L [55] (Fig. 2.3A).

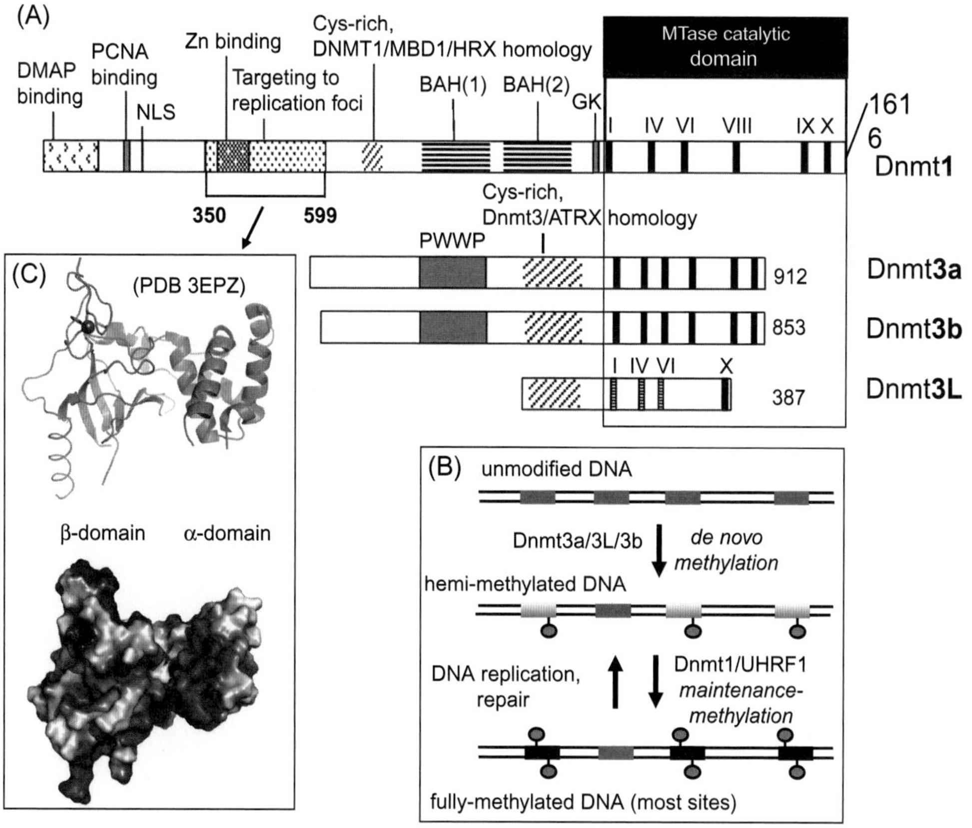

FIGURE 2.2

Schematic representation of Dnmt1 and Dnmt3. (A) Roman numerals refer to conserved motifs of DNA MTases [156]; motif IV includes the Cys nucleophile that forms a transient covalent bond to C6 of the target cytosine. (B) Maintenance vs. *de novo* methylation. The rectangular segments are substrate sequences (usually CpG), and the small ball shapes represent methyl groups on the cytosines. Following replication or repair, the duplex is methylated on one strand only. (C) The first domain structure of Dnmt1 (residues 350–599; PDB 3EPZ) [89] contains targeting sequence association with replication foci [88]. (Please refer to color plate section)

Both Dnmt3a and Dnmt3L C-terminal domains have the characteristic fold of Class I AdoMet-dependent MTases [56]. However, the methylation reaction product S-adenosyl-L-homocysteine (AdoHcy) was found only in Dnmt3a and not in Dnmt3L. This is consistent with Dnmt3a being the catalytic component of the complex, while Dnmt3L is inactive and unable to bind AdoMet [52,53]. The overall Dnmt3a/Dnmt3L C-terminal complex is ~16 nm long, which is greater than the diameter of a 11-nm core nucleosome (Fig. 2.3A). This complex contains two monomers of Dnmt3a and two of Dnmt3L, forming a tetramer with two 3L-3a interfaces and one 3a-3a interface (3L-3a-3a-3L). Substituting key non-catalytic residues at the Dnmt3a-3L or Dnmt3a-3a interfaces eliminates enzymatic activity, indicating that both interfaces are essential for catalysis [55].

DIMERIC Dnmt3a SUGGESTS *DE NOVO* DNA METHYLATION DEPENDS ON CpG SPACING

Among known active DNA MTases, Dnmt3a and Dnmt3b have the smallest DNA binding domain (though it is absent altogether in Dnmt3L). However, dimerization via the 3a-3a interface brings two active sites together and effectively doubles the DNA-binding surface. Superimposing the Dnmt3a structure, onto that of M.HhaI complexed with a short oligonucleotide [14], yielded a model such that the two active sites are located in the DNA major groove and dimeric Dnmt3a could methylate two CpGs

11

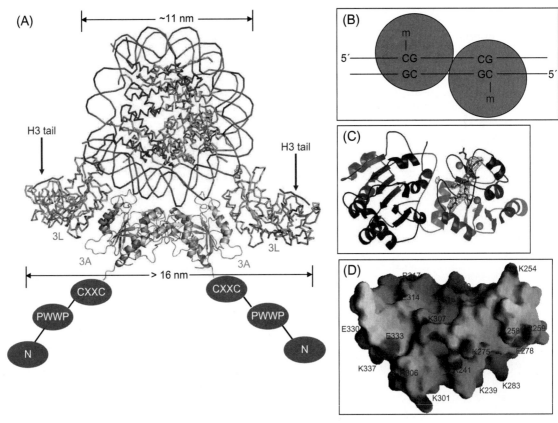

FIGURE 2.3

A model of interactions between Dnmt3a-3L tetramer and a nucleosome. (A) A nucleosome is shown, docked to a Dnmt3L-3a-3a-3L tetramer (3a-C in green; 3L full length in gray). The position of a peptide derived from the sequence of the histone H3 amino terminus (purple) is shown, taken from a co-crystal structure with this peptide bound to Dnmt3L [71]. Wrapping the tetramer around the nucleosome, the two Dnmt3L molecules could bind both histone tails from one nucleosome. The amino-proximal portion of Dnmt3a is labeled as N (for N-terminal domain), PWWP domain, and CXXC domain. By analogy to Dnmt3L, the CXXC domain of Dnmt3a might interact with histone tails from neighboring nucleosomes. (B) The Dnmt3a dimer could in theory methylate two CpGs separated by one helical turn in one binding event. (C) Structure of Dnmt3L with a bound histone H3 N-terminal tail (orange) [71]. (D) The PWWP domain structure of mouse Dnmt3b, rich in basic residues [42]. (Please refer to color plate section)

separated by one helical turn in one binding event (Fig. 2.3B). A periodicity in the activity of Dnmt3a on long DNA substrates revealed a correlation of methylated CpG sites at distances of 8–10 base pairs, and the structural model of oligomeric Dnmt3a docked to DNA may explain this pattern [55]. Similar periodicity is observed for the frequency of CpG sites in the differentially-methylated regions of 12 maternally-imprinted mouse genes [55]. These results suggest a basis for the recognition and methylation of differentially-methylated regions in imprinted genes, involving detection of both CpG spacing and nucleosome modification (see next section). Zhang et al. (2009) analyzed the methylation status of a large number of CpG sites (total of 580,427) of chromosome 21 and found that CpG DNA methylation patterns are correlated with the CpG periodicity of nine base pairs [57]. More recently, an 8–10 base-pair periodicity has also been evident for non-CpG methylation in embryonic stem cells [58]. Non-CG methylation disappeared upon induced differentiation of the embryonic stem cells, and was restored in induced pluripotent stem cells. Similarly, a 10-bp correlation of non-CpG DNA methylation by *Arabidopsis thaliana* DRM2 (which is related to mammalian Dnmt3a) has been observed [59].

Dnmt3L CONNECTS UNMETHYLATED HISTONE H3 LYSINE 4 TO *DE NOVO* DNA METHYLATION

DNA methylation and histone modifications are intricately connected with each other [60–62]. In fact, genome-scale DNA methylation profiles suggest that DNA methylation is better correlated with histone methylation patterns than with the underlying genome sequence context [62]. Specifically, DNA methylation is correlated with the absence of H3K4 methylation and the presence of H3K9 methylation. Methylation of histone H3 lysine 4 (H3K4) [63] has been suggested to protect gene promoters from *de novo* DNA methylation in somatic cells [64,65]. There have been reports of an inverse relationship between H3K4 methylation and allele-specific DNA methylation at differentially methylated regions [57,62,66–69]. More recently, AOF1 (amine-oxidase flavin-containing domain 1), a homolog of histone H3 lysine 4 demethylase (LSD1), has been shown to be required for *de novo* DNA methylation of imprinted genes in oocytes [70], suggesting that demethylation of H3K4 is critical for establishing the DNA methylation imprints during oogenesis.

The mammalian *de novo* DNA methylation Dnmt3L-Dnmt3a machinery could translate patterns of H3K4 methylation, which are not known to be themselves preserved during chromosome replication, into heritable patterns of DNA methylation that mediate transcriptional silencing of the affected sequences [71]. Dnmt3a is fully active on nucleosomal DNA *in vitro* [72]. Dnmt3a2 is a shorter isoform of Dnmt3a, predominant in embryonic stem cells and embryonal carcinoma cells and detectable in testis, ovary, thymus, and spleen, that is also required for genomic imprinting [73]. Dnmt3a2 and Dnmt3b, along with the four core histones, were identified as the main *in vivo* interaction partners of epitope-tagged Dnmt3L [71]. Peptide interaction assays showed that Dnmt3L specifically interacts with the extreme amino terminus of histone H3; this interaction was strongly inhibited by H3K4 methylation, but was insensitive to modifications at other positions [71]. Co-crystallization of Dnmt3L with the amino tail of H3 showed this tail bound to the Cys-rich 3-Zn binding domain of Dnmt3L (Fig. 2.3C), and substitution of key residues in the binding site eliminated the H3-Dnmt3L interaction. These data suggest that Dnmt3L is a probe of H3K4 methylation, and if the methylation is absent then Dnmt3L induces *de novo* DNA methylation by docking activated Dnmt3a2 to the nucleosome.

Mouse ES cells that lack the H3 lysine 9 (H3K9) MTases Suv39h1 and Suv39h2 show slight demethylation of satellite DNA [74]. G9a and GLP (G9a-like protein) – two related euchromatin-associated H3K9 methyltransferases [75] – have been implicated in DNA methylation at various loci, including imprinting center [76,77], retrotransposons and satellite repeats [78], a G9a/GLP target promoter [79], and a set of embryonic genes [80]. In filamentous fungi *Neurospora*, the H3K9 methyltransferase DIM-5 is required for DNA methylation [81–84], whereas in *Arabidopsis* the H3K9 methyltransferase KRYPTONITE is required for DNA methylation [85]. This suggests an evolutionarily-conserved silencing pathway in which H3K9 methylation correlates with DNA methylation. However, how H3K9 methylation contributes to DNA methylation is not clear, particularly in mammalian cells. G9a interacts directly with Dnmt1 during replication [86]. In addition, the G9a ankyrin repeat domain has been suggested to interact with Dnmt3a [80,87], a possible way for G9a to induce *de novo* DNA methylation [78].

A STRUCTURAL FRAGMENT OF Dnmt1

At the time of this writing (August, 2009), one domain structure is available (Fig. 2.2C) for part of the large 183 kDa Dnmt1 protein. The region (residues 350–599 of human Dnmt1) was initially identified as a novel targeting sequence association with replication foci [88]. This sequence has the properties expected of a targeting sequence in that it is not required for enzymatic activity, prevents proper targeting when deleted, and, when fused to β-galactosidase, causes the fusion protein to associate with replication foci in a cell cycle-dependent manner. The domain structure, solved by the Structural Genomics Consortium at Toronto (PDB 3EPZ) [89], adopts a mainly β structure in the N-terminal half and a helix bundle in the C-terminal half (Fig. 2.2C).

Dnmt1 itself is subject to posttranslational modifications, including phosphorylation (Ser515 in mouse Dnmt1) [90,91] and methylation (Lys142 in human Dnmt1) [92]. Methylation of Dnmt1 at Lys142, mediated by Set7 (a protein lysine methyltransferase), resulted in its decreased stability [92]. Reciprocally, enhanced Dnmt1 methylation in the background of total deletion of LSD1 (a protein lysine demethylase) correlates with reduced Dnmt1 stability *in vivo* and progressive loss of DNA methylation [93]. Furthermore, it was hypothesized that polymers present on PARP-1 (PARylated poly(ADP-ribose) polymerase 1) interact noncovalently with Dnmt1, preventing Dnmt1 enzymatic activity. In the absence of poly(ADP-ribosyl)ation of PARP-1, Dnmt1 is free to methylate DNA; if, in contrast, high levels of PARylated PARP-1 persist, Dnmt1 will be stably inhibited, preventing DNA methylation [94].

THE SRA DOMAIN OF UHRF1 FLIPS 5-METHYLCYTOSINE OUT OF THE DNA HELIX

An accessory protein UHRF1 (ubiquitin-like, containing PHD and RING finger domains 1) targets Dnmt1 to hemimethylated replication forks (and presumably repair sites) [95–97]. The murine ortholog of this protein is also known as NP95 (nuclear protein of 95 kDa) [98–100]; the human ortholog is called ICBP90 (inverted CCAAT binding protein of 90 kDa) [101].

The crystal structure of the SET and RING associated (SRA) domain of UHRF1 in complex with DNA containing a hemimethylated CpG site was recently determined [102–104]. They reveal that the SRA domain flips the 5-methylcytosine (5mC) completely out of the DNA helix (Fig. 2.4A,B) and is positioned in a binding pocket with planar stacking contacts, and both Watson-Crick polar hydrogen bonds and van der Waals interactions specific for 5mC that distinguishes 5mC from cytosine. The structure also suggests an explanation for the preference for hemimethylated sites. In the major groove side, a backbone carbonyl oxygen is close to the C5 ring carbon of the unmethylated cytosine, forming a C=O⋯H-C hydrogen bond. The addition of a methyl group to C5 of the unmethylated cytosine would cause a steric clash between the methyl group and SRA.

BASE FLIPPING MECHANISM

Base flipping is a conserved mechanism that is widely used by nucleotide modifying enzymes, including DNA MTases [13,14], DNA repair enzymes [105–108], and RNA modification enzymes [109]. This mechanism, first discovered in the bacterial 5mC MTase M.HhaI [14], involves enzyme binding to the DNA and eversion of the target nucleotide so that it projects out of the double helix and into the active-site pocket. The SRA domain is the first-discovered non-enzymatic sequence-specific DNA binding protein domain that uses the base flipping mechanism in its interaction with DNA.

There is no apparent sequence or structural similarity between the SRA and the DNA MTase domain (or of DNA repair enzymes). However, the phosphodiester backbone pinching [110] due to extensive protein-phosphate contacts surrounding the flipped nucleotide, the use of two loops to approach DNA from the major and minor grooves simultaneously, and the binding of the flipped base in a concave pocket are analogous to the DNA MTases (Fig. 2.5A,B) [111]. Furthermore, enzymes use base flipping to gain access to a DNA base to perform chemistry on it, but the SRA domain probably uses base flipping to increase its protein-DNA interface and to prevent the SRA domain from linear diffusion away from the

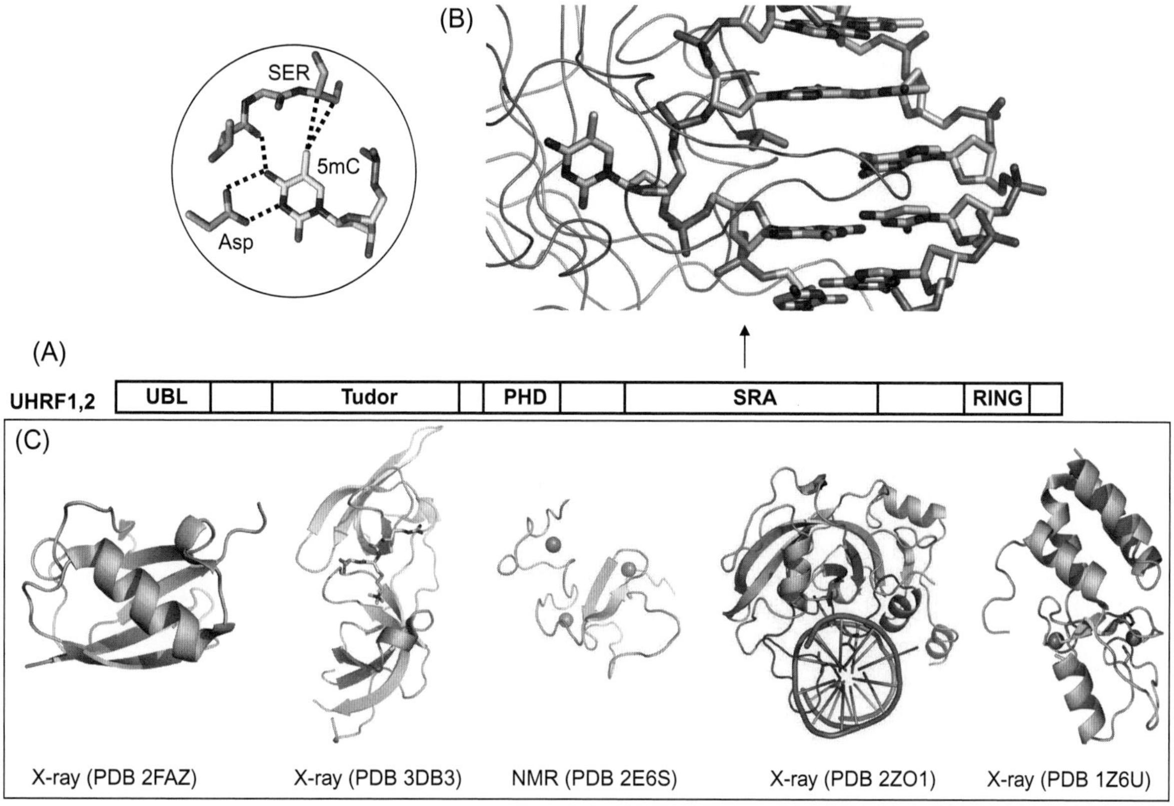

FIGURE 2.4

UHRF1 – a multi-domain protein. (A) Schematic representation of UHRF1 and its homolog UHRF2. (B) Structure of SRA-DNA complex. The 5mC flips out and is bound in a cage-like pocket. (C) Five domain structures are currently available. (Please refer to color plate section)

site on the DNA. This may be particularly important for the SRA domain, as its recognition sequence is only two base pairs. The surface area buried at the SRA-DNA interface is increased approximately 70% from what is buried at the MBD1-DNA interface [104] that does not involve base flipping [112,113] (Fig. 2.5C).

The 5mC base flipping by the SRA domain might also provide a more general mechanism to distinguish the methylated parental strand from the unmethylated daughter strand, an ability particularly important for mismatch repair if an error occurs during DNA replication. Supporting this hypothesis, the expression of ICBP90 (the human ortholog of UHRF1) is deregulated in cancer cells [114], and mouse UHRF1-null cells are more sensitive to DNA damaging agents and DNA replication arrest [99]. We therefore suggest that the SRA-DNA interaction (through recognition and flipping of the 5mC) serves as an anchor to keep UHRF1 at hemimethylated CpG site where it recruits Dnmt1 for maintenance methylation, and perhaps other proteins such as DNA repair enzymes for mismatch repair.

UHRF1-HISTONE INTERACTIONS

Besides the SRA domain, the Structural Genomics Consortium at Toronto has solved three additional domain structures by X-ray crystallography (Fig. 2.4C): the N-terminal ubiquitin-like domain (PDB 2FAZ [115]) and the tandem tudor domain with and without bound histone H3K9me3 (PDB 3DB3 and 3DB4 [116] for human UHRF1), and the C-terminal RING domain (PDB 1Z6U [117] of human UHRF2), whereas the RIKEN Structural Genomics/Proteomics Initiative at Japan solved an NMR structure for the PHD domain of human UHRF2 (PDB 2E6S [118]). Individual domain functions have been

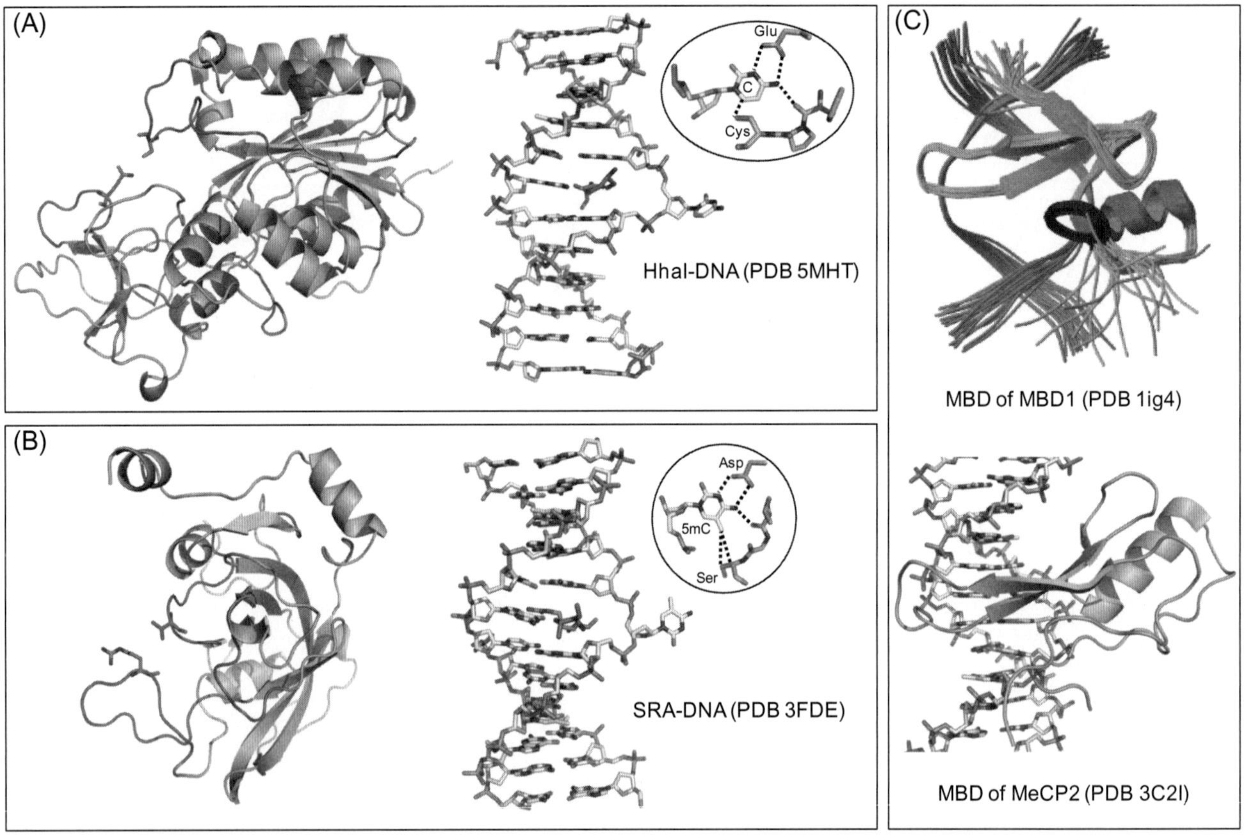

FIGURE 2.5
Comparison of base flipping by the SRA domain and HhaI methyltransferase. (A,B) DNA structures bound by HhaI (A) or SRA (B) show a flipped nucleotide. The intercalating amino acids are shown in each case. Structures of HhaI (A) and SRA (B) show the two opposite-side DNA-approaching loops. (C) NMR structure of MBD1-DNA (top) and X-ray structure of MeCP2-DNA (bottom) showed MBD domain inserts a beta-hairpin through the DNA major groove. The methyl-binding domains of MBD1 [112] and MeCP2 [113], instead of using a base-flipping mechanism, recognize changes in hydration of the major groove of a fully methylated CpG rather than detecting methyl groups directly. (Please refer to color plate section)

suggested for the PHD that may be involved in histone H3 tail binding [98,101], and for the RING domain that may confer E3 ubiquitin ligase activity on histones [98]. The new structure of the tandem tudor domain bound with histone H3 tail contains three structured H3 residues, H3R8-H3K9me3-H3S10 (PDB 3DB3).

REPLICATION-COUPLED CROSSTALK BETWEEN DNA METHYLATION AND HISTONE MODIFICATIONS

The fact that UHRF1 contains modules, within the same polypeptide, recognizing both DNA silencing marks (via the SRA) and histone silencing marks (via the tudor and/or PHD) suggests it may be a key component to couple the preservation of histone-modification through the cell cycle with maintenance DNA methylation (Fig. 2.6). We hypothesize that UHRF1 brings the two components (histones and DNA) carrying appropriate markers (on the tails of H3 and hemimethylated CpG sites) ready to be assembled into a nucleosome after replication. In this context the missing key experiment is whether these domains act independently or in a cooperative fashion. For example, does binding of UHRF1 SRA domain to hemi-methylated DNA improve the binding of UHRF1 tudor and/or PHD domains to histone tail? Furthermore, the E3 ubiquitin ligase activity [98], residing in the C-terminal RING domain, may add an additional level of modification, such as H2A

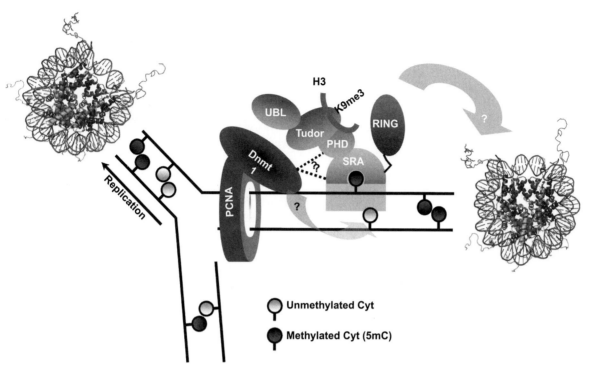

FIGURE 2.6

Hypothetical model of UHRF1-mediated replication-coupled crosstalk between DNA methylation and histone modifications. Existence of both silencing mark readers recognizing DNA (via the SRA) and histone (via the tudor and/or PHD) facilitates the idea of maintenance and conversion of epigenetic silencing marks on both DNA and histone modifications.

17

ubiquitylation that is enriched with inactive genes [119]. In a separate note, a recent study suggested Polycomb proteins remain bound to chromatin and DNA during DNA replication *in vitro* [120]. Retention of Polycomb proteins through DNA replication may contribute to maintenance of transcriptional silencing through cell division.

DNA DEMETHYLATION VIA HYDROXYLATION?

5-hydroxymethylcytosine (5hmC) has long been noted in bacterial phage DNA [121–125], and its presence in mammalian cells [126] was believed to be a by-product of oxidative DNA damage [127]. Recently, using isolated relatively homogeneous populations of Purkinje and granule neuronal nuclei of adult mouse brains, Kriaucionis and Heintz found that significant fractions (~40%) of cytosine nucleotides correspond to 5hmCs, the amount of which inversely correlates with 5mC and nuclear heterochromatin in neurons [128]. Even more fascinating, a conserved mammalian-specific family of TET (ten-eleven translocation) proteins that converts 5mC to 5hmC (Fig. 2.1B) was identified [129]. One of these proteins, TET1, is fused to the MLL protein in a subset of acute myeloid leukemia patients. Overproduction of TET1 in human cells led to the appearance of 5hmC. A concomitant reduction in DNA 5mC indicated that 5hmC is an oxidation product of 5mC. 5hmC was detected in ES cells and this percentage decreased with RNAi knockdown of TET1. The surprising finding of a 5mC oxidation pathway raises numerous questions, such as whether oxidation of 5mC is an important epigenetic modification, either as an end product or as an intermediate in active DNA demethylation, as supported by the presence of 5hmC DNA excision repair glycosylase [130]. It is intriguing to note that bacterial 5mC MTase M.HhaI can promote the reverse reaction *in vitro* – the removal of formaldehyde from 5hmC to yield the unmodified cytosine [131] (Fig. 2.1C). New lines of research will likely be catalyzed by the presence of 5hmC in mammalian DNA.

DNA DEMETHYLATION VIA GLYCOSYLATION?

A recent study reported that MBD4 [132], a protein containing an N-terminal methyl-CpG-binding domain (MBD) and a C-terminal glycosylase domain [133], is phosphorylated via protein kinase C (PKC) by parathyroid hormone stimulation [134]. Phosphorylated MBD4 promotes incision of methylated DNA through glycosylase activity, and a base-excision repair process seems to complete DNA demethylation in the MBD4-bound promoter. Such parathyroid-hormone-induced MBD4 phosphorylation and subsequent DNA demethylation and transcriptional derepression are impaired in Mbd4(–/–) mice.

Dnmt2, AN ENIGMATIC DNA METHYLTRANSFERASE HOMOLOG

No review on mammalian DNA methylation will be complete without mention of Dnmt2, a small protein of 391 residues in human, initially identified based on its conservation of sequence motifs to known DNA cytosine-C5 methyltransferases [135,136]. Targeted deletion of Dnmt2 gene indicated Dnmt2 is not required for global *de novo* or maintenance methylation of DNA in embryonic stem cells [137]. Baculovirus-expressed Dnmt2 protein failed to methylate DNA *in vitro* [137], whereas bacteria-expressed Dnmt2 showed no detectable activity in one study [138] or residual activity towards DNA in another study [139]. Structure of human Dnmt2 is closely related in overall structure to M.HhaI [138], a bacterial DNA methyltransferase. In 2006, Goll et al. revealed that human Dnmt2 methylates cytosine 38 in the anticodon loop of aspartic acid transfer RNA (tRNA[Asp]) [140]. The function of tRNA[Asp] methylation by Dnmt2 is highly conserved, as human Dnmt2 protein restored methylation *in vitro* to tRNA[Asp] from Dnmt2-deficient strains of mouse, *Arabidopsis thaliana*, and *Drosophila melanogaster* [140]. Not surprisingly, human Dnmt2 methylates RNA cytosine-C5 using a DNA methyltransferase-like catalytic mechanism [141], because of high sequence and structure conservation as well as identical chemistry between DNA cytosine and RNA cytosine.

In parallel, study in *Drosophila* revealed that depletion of dDnmt2 had no detectable effect on embryonic development, whereas overexpression of dDnmt2 resulted in significant genomic DNA hypermethylation at CpT and CpA dinucleotides [142]. A weak but significant activity for Dnmt2 was detected in a non-CpG dinucleotide context in flies overexpressing mouse Dnmt2 [143]. More recently, Phalke et al. showed that dDnmt2 controls retrotransposon silencing in *Drosophila* somatic cells [144]. Loss of dDnmt2 eliminates histone H4 lysine 20 (H4K20) trimethylation (mediated by Suv4-20) at retrotransposons and impairs maintenance of retrotransposon silencing. The new study uncovers a previously unappreciated role of dDnmt2 in DNA methylation in retrotransposon silencing and telomere integrity in *Drosophila* [144] and will help to resolve the Dnmt2 enigma [145] and whether two substrates are better than one [146].

CONCLUSION AND PERSPECTIVES

The experimental characterizations of Dnmts and their associated protein factors are providing a rapidly and convergent picture of the kinetic mechanisms (activities of oligomers [147]), binding partners (UHRF1-Dnmt1 [95,96] and Dnmt3L-Dnmt3a [55]), chromatin recognition [62] (histone binding such as H3K4me0 [71] and H3K9me3 [74]), RNA-directed DNA methylation [148,149], methylation-dependent regulation (by protein lysine methyltransferases G9a/GLP [78–80] and by protein lysine demethylase LSD1 [93]), and the discovery of 5hmC in mammalian genome [128,129]. However, understanding the basis for establishing, maintaining, and disturbing DNA methylation patterns will require a much better understanding of the union between structure and function in the Dnmts and their associated protein factors than we currently possess. Without understanding the interactions and spatial relationships between their modular domains, or whether inter-domain interactions contribute to target specificity, it is not possible to construct a temporal sequence of events or causal relationships in gene silencing.

ACKNOWLEDGEMENTS

We thank most warmly our colleagues and coworkers whose hard work was responsible for much of the DNA and histone methylation work cited in this review. Work in the authors' laboratory is supported in part by grants (US National Institutes of Health GM049245, GM068680, and DK082678) and the Georgia Research Alliance. This chapter is an updated version of our previous articles "Mammalian DNA methyltransferases: A structural perspective" (originally published in *Structure* **16**, March 2008, pp. 341–350; with permission from Elsevier) and "UHRF1, a modular multi-domain protein, regulates replication-coupled crosstalk between DNA methylation and histone modifications" (originally published in *Epigenetics* **4**, pp. 8–14; January 2009, with permission from Landes Bioscience).

References

1. Dvir A, Conaway JW, Conaway RC. Mechanism of transcription initiation and promoter escape by RNA polymerase II. *Curr Opin Genet Dev* 2001;**11**:209–14.

2. Sandelin A, Carninci P, Lenhard B, Ponjavic J, Hayashizaki Y, Hume DA. Mammalian RNA polymerase II core promoters: insights from genome-wide studies. *Nat Rev Genet* 2007;**8**:424–36.

3. Tran K, Gralla JD. Control of the timing of promoter escape and RNA catalysis by the transcription factor IIb fingertip. *J Biol Chem* 2008;**283**:15665–71.

4. Malik S, Roeder RG. Dynamic regulation of pol II transcription by the mammalian Mediator complex. *Trends Biochem Sci* 2005;**30**:256–63.

5. Hoffmann A, Natoli G, Ghosh G. Transcriptional regulation via the NF-kappaB signaling module. *Oncogene* 2006;**25**:6706–16.

6. Carrera I, Treisman JE. Message in a nucleus: signaling to the transcriptional machinery. *Curr Opin Genet Dev* 2008;**18**:397–403.

7. Li B, Carey M, Workman JL. The role of chromatin during transcription. *Cell* 2007;**128**:707–19.

8. Berger SL. The complex language of chromatin regulation during transcription. *Nature* 2007;**447**:407–12.

9. Luger K, Mader AW, Richmond RK, Sargent DF, Richmond TJ. Crystal structure of the nucleosome core particle at 2.8 A resolution. *Nature* 1997;**389**:251–60.

10. Shilatifard A. Chromatin modifications by methylation and ubiquitination: implications in the regulation of gene expression. *Annu Rev Biochem* 2006;**75**:243–69.

11. Shi Y. Histone lysine demethylases: emerging roles in development, physiology and disease. *Nat Rev Genet* 2007;**8**:829–33.

12. Bhaumik SR, Smith E, Shilatifard A. Covalent modifications of histones during development and disease pathogenesis. *Nat Struct Mol Biol* 2007;**14**:1008–16.

13. Cheng X, Roberts RJ. AdoMet-dependent methylation, DNA methyltransferases and base flipping. *Nucleic Acids Res* 2001;**29**:3784–95.

14. Klimasauskas S, Kumar S, Roberts RJ, Cheng X. HhaI methyltransferase flips its target base out of the DNA helix. *Cell* 1994;**76**:357–69.

15. Sheikhnejad G, Brank A, Christman JK, Goddard A, Alvarez E, Ford Jr H, et al. Mechanism of inhibition of DNA (cytosine C5)-methyltransferases by oligodeoxyribonucleotides containing 5,6-dihydro-5-azacytosine. *J Mol Biol* 1999;**285**:2021–34.

16. Wu JC, Santi DV. Kinetic and catalytic mechanism of HhaI methyltransferase. *J Biol Chem* 1987;**262**:4778–86.

17. Youngblood B, Shieh FK, Buller F, Bullock T, Reich NO. S-adenosyl-L-methionine-dependent methyl transfer: observable precatalytic intermediates during DNA cytosine methylation. *Biochem* 2007;**46**:8766–75.

18. Zhang X, Bruice TC. The mechanism of M.HhaI DNA C5 cytosine methyltransferase enzyme: a quantum mechanics/molecular mechanics approach. *Proc Natl Acad Sci USA* 2006;**103**:6148–53.

19. Kouzarides T. Chromatin modifications and their function. *Cell* 2007;**128**:693–705.

20. Turner BM. Defining an epigenetic code. *Nat Cell Biol* 2007;**9**:2–6.

21. Hore TA, Rapkins RW, Graves JA. Construction and evolution of imprinted loci in mammals. *Trends Genet* 2007;**23**:440–8.

22. Miranda TB, Jones PA. DNA methylation: the nuts and bolts of repression. *J Cell Physiol* 2007;**213**:384–90.

23. Lande-Diner L, Zhang J, Ben-Porath I, Amariglio N, Keshet I, Hecht M, et al. Role of DNA methylation in stable gene repression. *J Biol Chem* 2007;**282**:12194–200.

24. Yen ZC, Meyer IM, Karalic S, Brown CJ. A cross-species comparison of X-chromosome inactivation in Eutheria. *Genomics* 2007;**90**:453–63.

25. Yang X, Smith SL, Tian XC, Lewin HA, Renard JP, Wakayama T. Nuclear reprogramming of cloned embryos and its implications for therapeutic cloning. *Nat Genet* 2007;**39**:295–302.

26. Reik W. Stability and flexibility of epigenetic gene regulation in mammalian development. *Nature* 2007;**447**:425–32.

27. Gronbaek K, Hother C, Jones PA. Epigenetic changes in cancer: *APMIS* 2007;**115**:1039–59.

28. Walsh CP, Xu GL. Cytosine methylation and DNA repair. *Curr Top Microbiol Immunol* 2006;**301**:283–315.

29. Schaefer CB, Ooi SK, Bestor TH, Bourc'his D. Epigenetic decisions in mammalian germ cells. *Science* 2007;**316**:398–9.

30. Unterberger A, Andrews SD, Weaver IC, Szyf M. DNA methyltransferase 1 knockdown activates a replication stress checkpoint. *Mol Cell Biol* 2006;**26**:7575–86.

31. Bird A. Does DNA methylation control transposition of selfish elements in the germline? *Trends Genet* 1997;**13**:469–72.

32. Howard G, Eiges R, Gaudet F, Jaenisch R, Eden A. Activation and transposition of endogenous retroviral elements in hypomethylation induced tumors in mice. *Oncogene* 2008;**27**:404–8.

33. Yoder JA, Walsh CP, Bestor TH. Cytosine methylation and the ecology of intragenomic parasites. *Trends Genet* 1997;**13**:335–440.

34. Chen T, Li E. Establishment and maintenance of DNA methylation patterns in mammals. *Curr Top Microbiol Immunol* 2006;**301**:179–201.

35. Mortusewicz O, Schermelleh L, Walter J, Cardoso MC, Leonhardt H. Recruitment of DNA methyltransferase I to DNA repair sites. *Proc Natl Acad Sci USA* 2005;**102**:8905–9.

36. Jeltsch A. On the enzymatic properties of Dnmt1: specificity, processivity, mechanism of linear diffusion and allosteric regulation of the enzyme. *Epigenetics* 2006;**1**:63–6.

37. Ooi SK, Bestor TH. Cytosine methylation: remaining faithful. *Curr Biol* 2008;**18**:R174–6.

38. Grandjean V, Yaman R, Cuzin F, Rassoulzadegan M. Inheritance of an epigenetic mark: The CpG DNA methyltransferase 1 is required for de novo establishment of a complex pattern of non-CpG methylation. *PLoS ONE* 2007;**2**:e1136.

39. Feltus FA, Lee EK, Costello JF, Plass C, Vertino PM. Predicting aberrant CpG island methylation. *Proc Natl Acad Sci USA* 2003;**100**:12253–8.

40. Jair KW, Bachman KE, Suzuki H, Ting AH, Rhee I, Yen RW, et al. De novo CpG island methylation in human cancer cells. *Cancer Res* 2006;**66**:682–92.

41. Bestor TH. The DNA methyltransferases of mammals. *Hum Mol Genet* 2000;**9**:2395–402.

42. Qiu C, Sawada K, Zhang X, Cheng X. The PWWP domain of mammalian DNA methyltransferase Dnmt3b defines a new family of DNA-binding folds. *Nat Struct Biol* 2002;**9**:217–24.

43. Lukasik SM, Cierpicki T, Borloz M, Grembecka J, Everett A, Bushweller JH. High resolution structure of the HDGF PWWP domain: a potential DNA binding domain. *Protein Sci* 2006;**15**:314–23.

44. Bourc'his D, Bestor TH. Meiotic catastrophe and retrotransposon reactivation in male germ cells lacking Dnmt3L. *Nature* 2004;**431**:96–99.

45. Bourc'his D, Xu GL, Lin CS, Bollman B, Bestor TH. Dnmt3L and the establishment of maternal genomic imprints. *Science* 2001;**294**:2536–9.

46. Kaneda M, Okano M, Hata K, Sado T, Tsujimoto N, Li E, et al. Essential role for de novo DNA methyltransferase Dnmt3a in paternal and maternal imprinting. *Nature* 2004;**429**:900–3.

47. Webster KE, O'Bryan MK, Fletcher S, Crewther PE, Aapola U, Craig J, et al. Meiotic and epigenetic defects in Dnmt3L-knockout mouse spermatogenesis. *Proc Natl Acad Sci USA* 2005;**102**:4068–73.

48. Hata K, Okano M, Lei H, Li E. Dnmt3L cooperates with the Dnmt3 family of de novo DNA methyltransferases to establish maternal imprints in mice. *Development* 2002;**129**:1983–93.

49. Chedin F, Lieber MR, Hsieh CL. The DNA methyltransferase-like protein DNMT3L stimulates de novo methylation by Dnmt3a. *Proc Natl Acad Sci USA* 2002;**99**:16916–21.

50. Suetake I, Shinozaki F, Miyagawa J, Takeshima H, Tajima S. DNMT3L stimulates the DNA methylation activity of Dnmt3a and Dnmt3b through a direct interaction. *J Biol Chem* 2004;**279**:27816–23.

51. Chen ZX, Mann JR, Hsieh CL, Riggs AD, Chedin F. Physical and functional interactions between the human DNMT3L protein and members of the de novo methyltransferase family. *J Cell Biochem* 2005;**95**:902–17.

52. Gowher H, Liebert K, Hermann A, Xu G, Jeltsch A. Mechanism of stimulation of catalytic activity of Dnmt3A and Dnmt3B DNA-(cytosine-C5)-methyltransferases by Dnmt3L. *J Biol Chem* 2005;**280**:13341–8.

53. Kareta MS, Botello ZM, Ennis JJ, Chou C, Chedin F. Reconstitution and mechanism of the stimulation of de novo methylation by human DNMT3L. *J Biol Chem* 2006;**281**:25893–902.

54. Margot JB, Ehrenhofer-Murray AE, Leonhardt H. Interactions within the mammalian DNA methyltransferase family. *BMC Mol Biol* 2003;**4**:7.

55. Jia D, Jurkowska RZ, Zhang X, Jeltsch A, Cheng X. Structure of Dnmt3a bound to Dnmt3L suggests a model for de novo DNA methylation. *Nature* 2007;**449**:248–51.

56. Schubert HL, Blumenthal RM, Cheng X. Many paths to methyltransfer: a chronicle of convergence. *Trends Biochem Sci* 2003;**28**:329–35.

57. Zhang Y, Rohde C, Tierling S, Jurkowski TP, Bock C, Santacruz D, et al. DNA methylation analysis of chromosome 21 gene promoters at single base pair and single allele resolution. *PLoS Genet* 2009;**5**:e1000438.

58. Lister R, Pelizzola M, Dowen RH, Hawkins RD, Hon G, Tonti-Filippini J, et al. Human DNA methylomes at base resolution show widespread epigenomic differences. *Nature* 2009;**462**:315–22.

59. Cokus SJ, Feng S, Zhang X, Chen Z, Merriman B, Haudenschild CD, et al. Shotgun bisulphite sequencing of the Arabidopsis genome reveals DNA methylation patterning. *Nature* 2008;**452**:215–19.

60. Vaissiere T, Sawan C, Herceg Z. Epigenetic interplay between histone modifications and DNA methylation in gene silencing. *Mutat Res* 2008;**659**:40–8.

61. Wu J, Wang SH, Potter D, Liu JC, Smith LT, Wu YZ, et al. Diverse histone modifications on histone 3 lysine 9 and their relation to DNA methylation in specifying gene silencing. *BMC Genomics* 2007;**8**:131.

62. Meissner A, Mikkelsen TS, Gu H, Wernig M, Hanna J, Sivachenko A, et al. Genome-scale DNA methylation maps of pluripotent and differentiated cells. *Nature* 2008;**454**:766–70.

63. Shilatifard A. Molecular implementation and physiological roles for histone H3 lysine 4 (H3K4) methylation. *Curr Opin Cell Biol* 2008;**20**:341–8.

64. Weber M, Hellmann I, Stadler MB, Ramos L, Paabo S, Rebhan M, et al. Distribution, silencing potential and evolutionary impact of promoter DNA methylation in the human genome. *Nat Genet* 2007;**39**:457–66.

65. Appanah R, Dickerson DR, Goyal P, Groudine M, Lorincz MC. An unmethylated 3′ promoter-proximal region is required for efficient transcription initiation. *PLoS Genet* 2007;**3**:e27.

66. Fournier C, Goto Y, Ballestar E, Delaval K, Hever AM, Esteller M, et al. Allele-specific histone lysine methylation marks regulatory regions at imprinted mouse genes. *EMBO J* 2002;**21**:6560–70.

67. Vu TH, Li T, Hoffman AR. Promoter-restricted histone code, not the differentially methylated DNA regions or antisense transcripts, marks the imprinting status of IGF2R in human and mouse. *Hum Mol Genet* 2004;**13**:2233–45.

68. Yamasaki Y, Kayashima T, Soejima H, Kinoshita A, Yoshiura K, Matsumoto N, et al. Neuron-specific relaxation of Igf2r imprinting is associated with neuron-specific histone modifications and lack of its antisense transcript Air. *Hum Mol Genet* 2005;**14**:2511–20.

69. Delaval K, Govin J, Cerqueira F, Rousseaux S, Khochbin S, Feil R. Differential histone modifications mark mouse imprinting control regions during spermatogenesis. *EMBO J* 2007;**26**:720–9.

70. Ciccone DN, Su H, Hevi S, Gay F, Lei H, Bajko J, et al. KDM1B is a histone H3K4 demethylase required to establish maternal genomic imprints. *Nature* 2009;**461**:415–18.

71. Ooi SKT, Qiu C, Bernstein E, Li K, Jia D, Yang Z, et al. DNMT3L connects unmethylated lysine 4 of histone H3 to de novo methylation of DNA. *Nature* 2007;**448**:714–7.

72. Gowher H, Stockdale CJ, Goyal R, Ferreira H, Owen-Hughes T, Jeltsch A. De novo methylation of nucleosomal DNA by the mammalian Dnmt1 and Dnmt3A DNA methyltransferases. *Biochem* 2005;**44**:9899–904.

73. Chen T, Ueda Y, Xie S, Li E. A novel Dnmt3a isoform produced from an alternative promoter localizes to euchromatin and its expression correlates with active de novo methylation. *J Biol Chem* 2002;**277**:38746–54.

74. Lehnertz B, Ueda Y, Derijck AA, Braunschweig U, Perez-Burgos L, Kubicek S, et al. Suv39h-mediated histone H3 lysine 9 methylation directs DNA methylation to major satellite repeats at pericentric heterochromatin. *Curr Biol* 2003;**13**:1192–200.

75. Tachibana M, Ueda J, Fukuda M, Takeda N, Ohta T, Iwanari H, et al. Histone methyltransferases G9a and GLP form heteromeric complexes and are both crucial for methylation of euchromatin at H3-K9. *Genes Dev* 2005;**19**:815–26.

76. Xin Z, Tachibana M, Guggiari M, Heard E, Shinkai Y, Wagstaff J. Role of histone methyltransferase G9a in CpG methylation of the Prader-Willi syndrome imprinting center. *J Biol Chem* 2003;**278**:14996–5000.

77. Wagschal A, Sutherland HG, Woodfine K, Henckel A, Chebli K, Schulz R, et al. G9a histone methyltransferase contributes to imprinting in the mouse placenta. *Mol Cell Biol* 2008;**28**:1104–13.

78. Dong KB, Maksakova IA, Mohn F, Leung D, Appanah R, Lee S, et al. DNA methylation in ES cells requires the lysine methyltransferase G9a but not its catalytic activity. *EMBO J* 2008;**27**:2691–701.

79. Tachibana M, Matsumura Y, Fukuda M, Kimura H, Shinkai Y. G9a/GLP complexes independently mediate H3K9 and DNA methylation to silence transcription. *Embo J* 2008;**27**:2681–90.

80. Epsztejn-Litman S, Feldman N, Abu-Remaileh M, Shufaro Y, Gerson A, Ueda J, et al. De novo DNA methylation promoted by G9a prevents reprogramming of embryonically silenced genes. *Nat Struct Mol Biol* 2008;**15**:1176–83.

81. Tamaru H, Selker EU. A histone H3 methyltransferase controls DNA methylation in Neurospora crassa. *Nature* 2001;**414**:277–83.

82. Tamaru H, Zhang X, McMillen D, Singh PB, Nakayama J, Grewal SI, et al. Trimethylated lysine 9 of histone H3 is a mark for DNA methylation in Neurospora crassa. *Nat Genet* 2003;**34**:75–9.

83. Honda S, Selker EU. Direct interaction between DNA methyltransferase DIM-2 and HP1 is required for DNA methylation in Neurospora crassa. *Mol Cell Biol* 2008;**28**:6044–55.

84. Adhvaryu KK, Selker EU. Protein phosphatase PP1 is required for normal DNA methylation in Neurospora. *Genes Dev* 2008;**22**:3391–6.

85. Jackson JP, Lindroth AM, Cao X, Jacobsen SE. Control of CpNpG DNA methylation by the KRYPTONITE histone H3 methyltransferase. *Nature* 2002;**416**:556–60.

86. Esteve PO, Chin HG, Smallwood A, Feehery GR, Gangisetty O, Karpf AR, et al. Direct interaction between DNMT1 and G9a coordinates DNA and histone methylation during replication. *Genes Dev* 2006;**20**:3089–103.

87. Cedar H, Bergman Y. Linking DNA methylation and histone modification: patterns and paradigms. *Nat Rev Genet* 2009;**10**:295–304.

88. Leonhardt H, Page AW, Weier HU, Bestor TH. A targeting sequence directs DNA methyltransferase to sites of DNA replication in mammalian nuclei. *Cell* 1992;**71**:865–73.

89. Walker JR, Avvakumov GV, Xue S, Li Y, Bountra C, Weigelt J, et al. Structure of the replication foci-targeting sequence of human DNA cytosine methyltransferase DNMT1; 2008. *Structural Genomics Consortium* DOI:10.2210/pdb3epz/pdb.

90. Goyal R, Rathert P, Laser H, Gowher H, Jeltsch A. Phosphorylation of serine-515 activates the mammalian maintenance methyltransferase Dnmt1. *Epigenetics* 2007;**2**:155–60.

91. Glickman JF, Pavlovich JG, Reich NO. Peptide mapping of the murine DNA methyltransferase reveals a major phosphorylation site and the start of translation. *J Biol Chem* 1997;**272**:17851–7.

92. Esteve PO, Chin HG, Benner J, Feehery GR, Samaranayake M, Horwitz GA, et al. Regulation of DNMT1 stability through SET7-mediated lysine methylation in mammalian cells. *Proc Natl Acad Sci USA* 2009;**106**:5076–81.

93. Wang J, Hevi S, Kurash JK, Lei H, Gay F, Bajko J, et al. The lysine demethylase LSD1 (KDM1) is required for maintenance of global DNA methylation. *Nat Genet* 2009;**41**:125–9.

94. Caiafa P, Guastafierro T, Zampieri M. Epigenetics: poly(ADP-ribosyl)ation of PARP-1 regulates genomic methylation patterns. *FASEB J* 2009;**23**:672–8.

95. Sharif J, Muto M, Takebayashi S, Suetake I, Iwamatsu A, Endo TA, et al. The SRA protein Np95 mediates epigenetic inheritance by recruiting Dnmt1 to methylated DNA. *Nature* 2007;**450**:908–12.

96. Bostick M, Kim JK, Esteve PO, Clark A, Pradhan S, Jacobsen SE. UHRF1 plays a role in maintaining DNA methylation in mammalian cells. *Science* 2007;**317**:1760–4.

97. Achour M, Jacq X, Ronde P, Alhosin M, Charlot C, Chataigneau T, et al. The interaction of the SRA domain of ICBP90 with a novel domain of DNMT1 is involved in the regulation of VEGF gene expression. *Oncogene* 2008;**27**:2187–97.

98. Citterio E, Papait R, Nicassio F, Vecchi M, Gomiero P, Mantovani R, et al. Np95 is a histone-binding protein endowed with ubiquitin ligase activity. *Mol Cell Biol* 2004;**24**:2526–35.

99. Muto M, Kanari Y, Kubo E, Takabe T, Kurihara T, Fujimori A, et al. Targeted disruption of Np95 gene renders murine embryonic stem cells hypersensitive to DNA damaging agents and DNA replication blocks. *J Biol Chem* 2002;**277**:34549–55.

100. Papait R, Pistore C, Negri D, Pecoraro D, Cantarini L, Bonapace IM. Np95 is implicated in pericentromeric heterochromatin replication and in major satellite silencing. *Mol Biol Cell* 2007;**18**:1098–106.

101. Karagianni P, Amazit L, Qin J, Wong J. ICBP90, a novel methyl K9 H3 binding protein linking protein ubiquitination with heterochromatin formation. *Mol Cell Biol* 2008;**28**:705–17.

102. Arita K, Ariyoshi M, Tochio H, Nakamura Y, Shirakawa M. Recognition of hemi-methylated DNA by the SRA protein UHRF1 by a base-flipping mechanism. *Nature* 2008;**455**:818–21.

103. Avvakumov GV, Walker JR, Xue S, Li Y, Duan S, Bronner C, et al. Structural basis for recognition of hemi-methylated DNA by the SRA domain of human UHRF1. *Nature* 2008;**455**:822–5.

104. Hashimoto H, Horton JR, Zhang X, Bostick M, Jacobsen SE, Cheng X. The SRA domain of UHRF1 flips 5-methylcytosine out of the DNA helix. *Nature* 2008;**455**:826–9.

105. Yang CG, Yi C, Duguid EM, Sullivan CT, Jian X, Rice PA, et al. Crystal structures of DNA/RNA repair enzymes AlkB and ABH2 bound to dsDNA. *Nature* 2008;**452**:961–5.

106. Min JH, Pavletich NP. Recognition of DNA damage by the Rad4 nucleotide excision repair protein. *Nature* 2007;**449**:570–5.

107. Parker JB, Bianchet MA, Krosky DJ, Friedman JI, Amzel LM, Stivers JT. Enzymatic capture of an extrahelical thymine in the search for uracil in DNA. *Nature* 2007;**449**:433–7.

108. Scrima A, Konickova R, Czyzewski BK, Kawasaki Y, Jeffrey PD, Groisman R, et al. Structural basis of UV DNA-damage recognition by the DDB1-DDB2 complex. *Cell* 2008;**135**:1213–23.

109. Lee TT, Agarwalla S, Stroud RM. A unique RNA Fold in the RumA-RNA-cofactor ternary complex contributes to substrate selectivity and enzymatic function. *Cell* 2005;**120**:599–611.

110. Werner RM, Jiang YL, Gordley RG, Jagadeesh GJ, Ladner JE, Xiao G, et al. Stressing-out DNA? The contribution of serine-phosphodiester interactions in catalysis by uracil DNA glycosylase. *Biochem* 2000;**39**:12585–94.

111. Cheng X, Blumenthal RM. Finding a basis for flipping bases. *Structure* 1996;**4**:639–45.

112. Ohki I, Shimotake N, Fujita N, Jee J, Ikegami T, Nakao M, et al. Solution structure of the methyl-CpG binding domain of human MBD1 in complex with methylated DNA. *Cell* 2001;**105**:487–97.

113. Ho KL, McNae IW, Schmiedeberg L, Klose RJ, Bird AP, Walkinshaw MD. MeCP2 binding to DNA depends upon hydration at methyl-CpG. *Mol Cell* 2008;**29**:525–31.

114. Mousli M, Hopfner R, Abbady AQ, Monte D, Jeanblanc M, Oudet P, et al. ICBP90 belongs to a new family of proteins with an expression that is deregulated in cancer cells. *Br J Cancer* 2003;**89**:120–7.

115. Walker JR, Wybenga-Groot L, Doherty RS, Finerty Jr PJ, Newman E, Mackenzie FM, et al. Ubiquitin-like domain of human nuclear zinc finger protein NP95; 2005. DOI: 10.2210/pdb2faz/pdb

116. Walker JR, Avvakumov GV, Xue S, Dong A, Li Y, Bountra C, et al. Cryptic tandem tudor domains in UHRF1 interact with H3K9ME and are important for pericentric heterochromatin replication; 2008. DOI: 10.2210/pdb3db3/pdb and DOI: 10.2210/pdb3db4/pdb

117. Walker JR, Avvakumov GV, Xue S, Newman EM, Mackenzie F, Sundstrom M, et al. 2.1 Angstrom crystal structure of the human ubiquitin Liagse NIRF; 2005. DOI: 10.2210/pdb1z60/pdb

118. Kadirvel S, He F, Muto Y, Inoue M, Kigawa T, Shirouzu M, et al. Solution structure of the PHD domain in RING finger protein. 2006;107. DOI: 10.2210/pdb2e65/pdb

119. Lee N, Zhang Y. Chemical answers to epigenetic crosstalk. *Nat Chem Biol* 2008;**4**:335–7.

120. Francis NJ, Follmer NE, Simon MD, Aghia G, Butler JD. Polycomb proteins remain bound to chromatin and DNA during DNA replication in vitro. *Cell* 2009;**137**:110–22.

121. Wyatt GR, Cohen SS. The bases of the nucleic acids of some bacterial and animal viruses: the occurrence of 5-hydroxymethylcytosine. *Biochem J* 1953;**55**:774–82.

122. Wiberg JS. Amber mutants of bacteriophage T4 defective in deoxycytidine diphosphatase and deoxycytidine triphosphatase. On the role of 5-hydroxymethylcytosine in bacteriophage deoxyribonucleic acid. *J Biol Chem* 1967;**242**:5824–9.

123. Anisymova NI, Gabrilovich IM, Soshina NV, Cherenkevich SN. 5-Hydroxymethylcytosine-containing Klebsiella bacteriophage. *Biochim Biophys Acta* 1969;**190**:225–7.

124. Fleischman RA, Campbell JL, Richardson CC. Modification and restriction of T-even bacteriophages. In vitro degradation of deoxyribonucleic acid containing 5-hydroxymethylctosine. *J Biol Chem* 1976;**251**:1561–70.

125. Warren RA. Modified bases in bacteriophage DNAs. *Annu Rev Microbiol* 1980;**34**:137–58.

126. Penn NW, Suwalski R, O'Riley C, Bojanowski K, Yura R. The presence of 5-hydroxymethylcytosine in animal deoxyribonucleic acid. *Biochem J* 1972;**126**:781–90.

127. Castro GD, Diaz Gomez MI, Castro JA. 5-Methylcytosine attack by hydroxyl free radicals and during carbon tetrachloride promoted liver microsomal lipid peroxidation: structure of reaction products. *Chem Biol Interact* 1996;**99**:289–99.

128. Kriaucionis S, Heintz N. The nuclear DNA base 5-hydroxymethylcytosine is present in Purkinje neurons and the brain. *Science* 2009;**324**:929–30.

129. Tahiliani M, Koh KP, Shen Y, Pastor WA, Bandukwala H, Brudno Y, et al. Conversion of 5-methylcytosine to 5-hydroxymethylcytosine in mammalian DNA by the MLL fusion partner TET1. *Science* 1988;**324**:930–5.

130. Cannon SV, Cummings A, Teebor GW. 5-Hydroxymethylcytosine DNA glycosylase activity in mammalian tissue. *Biochem Biophys Res Commun* 1988;**151**:1173–9.

131. Liutkeviciute Z, Lukinavicius G, Masevicius V, Daujotyte D, Klimasauskas S. Cytosine-5-methyltransferases add aldehydes to DNA. *Nat Chem Biol* 2009;**5**:400–2.

132. Hendrich B, Hardeland U, Ng HH, Jiricny J, Bird A. The thymine glycosylase MBD4 can bind to the product of deamination at methylated CpG sites. *Nature* 1999;**401**:301–4.

133. Wu P, Qiu C, Sohail A, Zhang X, Bhagwat AS, Cheng X. Mismatch repair in methylated DNA. Structure and activity of the mismatch-specific thymine glycosylase domain of methyl-CpG-binding protein MBD4. *J Biol Chem* 2003;**278**:5285–91.

134. Kim MS, Kondo T, Takada I, Youn MY, Yamamoto Y, Takahashi S, et al. DNA demethylation in hormone-induced transcriptional derepression. *Nature* 2009;**461**:1007–12.

135. Vilain A, Apiou F, Dutrillaux B, Malfoy B. Assignment of candidate DNA methyltransferase gene (DNMT2) to human chromosome band 10p15.1 by in situ hybridization. *Cytogenet Cell Genet* 1998;**82**:120.

136. Yoder JA, Bestor TH. A candidate mammalian DNA methyltransferase related to pmt1p of fission yeast. *Hum Mol Genet* 1998;**7**:279–84.

137. Okano M, Xie S, Li E. Dnmt2 is not required for de novo and maintenance methylation of viral DNA in embryonic stem cells. *Nucleic Acids Res* 1998;**26**:2536–40.

138. Dong A, Yoder JA, Zhang X, Zhou L, Bestor TH, Cheng X. Structure of human DNMT2, an enigmatic DNA methyltransferase homolog that displays denaturant-resistant binding to DNA. *Nucleic Acids Res* 2001;**29**:439–48.

139. Hermann A, Schmitt S, Jeltsch A. The human Dnmt2 has residual DNA-(cytosine-C5) methyltransferase activity. *J Biol Chem* 2003;**278**:31717–21.

140. Goll MG, Kirpekar F, Maggert KA, Yoder JA, Hsieh CL, Zhang X, et al. Methylation of tRNAAsp by the DNA methyltransferase homolog Dnmt2. *Science* 2006;**311**:395–8.

141. Jurkowski TP, Meusburger M, Phalke S, Helm M, Nellen W, Reuter G, et al. Human DNMT2 methylates tRNA(Asp) molecules using a DNA methyltransferase-like catalytic mechanism. *RNA* 2008;**14**:1663–970.

142. Kunert N, Marhold J, Stanke J, Stach D, Lyko F. A Dnmt2-like protein mediates DNA methylation in Drosophila. *Development* 2003;**130**:5083–90.

143. Mund C, Musch T, Strodicke M, Assmann B, Li E, Lyko F. Comparative analysis of DNA methylation patterns in transgenic Drosophila overexpressing mouse DNA methyltransferases. *Biochem J* 2004;**378**:763–8.

144. Phalke S, Nickel O, Walluscheck D, Hortig F, Onorati MC, Reuter G. Retrotransposon silencing and telomere integrity in somatic cells of Drosophila depends on the cytosine-5 methyltransferase DNMT2. *Nat Genet* 2009;**41**:696–702.

145. Schaefer M, Lyko F. Solving the Dnmt2 enigma. *Chromosoma* 2009.

146. Jeltsch A, Nellen W, Lyko F. Two substrates are better than one: dual specificities for Dnmt2 methyltransferases. *Trends Biochem Sci* 2006;**31**:306–8.

147. Jurkowska RZ, Anspach N, Urbanke C, Jia D, Reinhardt R, Nellen W, et al. Formation of nucleoprotein filaments by mammalian DNA methyltransferase Dnmt3a in complex with regulator Dnmt3L. *Nucleic Acids Res* 2008.

148. Kanno T, Bucher E, Daxinger L, Huettel B, Bohmdorfer G, Gregor W, et al. A structural-maintenance-of-chromosomes hinge domain-containing protein is required for RNA-directed DNA methylation. *Nat Genet* 2008;**40**:670–5.

149. Aravin AA, Sachidanandam R, Bourc'his D, Schaefer C, Pezic D, Toth KF, et al. A piRNA pathway primed by individual transposons is linked to de novo DNA methylation in mice. *Mol Cell* 2008;**31**:785–99.

150. Kress C, Thomassin H, Grange T. Active cytosine demethylation triggered by a nuclear receptor involves DNA strand breaks. *Proc Natl Acad Sci USA* 2006;**103**:11112–7.

151. Vairapandi M. Characterization of DNA demethylation in normal and cancerous cell lines and the regulatory role of cell cycle proteins in human DNA demethylase activity. *J Cell Biochem* 2004;**91**:572–83.

152. Kangaspeska S, Stride B, Metivier R, Polycarpou-Schwarz M, Ibberson D, Carmouche RP, et al. Transient cyclical methylation of promoter DNA. *Nature* 2008;**452**:112–15.

153. Metivier R, Gallais R, Tiffoche C, Le Peron C, Jurkowska RZ, Carmouche RP, et al. Cyclical DNA methylation of a transcriptionally active promoter. *Nature* 2008;**452**:45–50.

154. Ooi SK, Bestor TH. The colorful history of active DNA demethylation. *Cell* 2008;**133**:1145–8.

155. Rai K, Huggins IJ, James SR, Karpf AR, Jones DA, Cairns BR. DNA demethylation in zebrafish involves the coupling of a deaminase, a glycosylase, and gadd45. *Cell* 2008;**135**:1201–12.

156. Kumar S, Cheng X, Klimasauskas S, Mi S, Posfai J, Roberts RJ, et al. The DNA (cytosine-5) methyltransferases. *Nucleic Acids Res* 1994;**22**:1–10.

CHAPTER 3

Mechanisms of Histone Modifications

Zdenko Herceg[1] and Rabih Murr[1,2]
[1]Epigenetics Group. International Agency for Research on Cancer (IARC), Lyon, France
[2]Friedrich Miescher Institute for Biomedical Research, Basel, Switzerland

INTRODUCTION

The term "epigenetic" was first introduced by Conrad Waddington in 1942 to describe "The interactions of genes with their environment that bring the phenotype into being". Currently, it includes all features such as chromatin and DNA modifications that are heritable and stable over rounds of cell division, but do not alter the nucleotide sequence within the underlying DNA [1]. Over the years, a wide variety of products and events have been lumped into epigenetics. These include paramutation, bookmarking, imprinting, gene silencing, X chromosome inactivation, position effect variegation, reprogramming, transvection, infection agents like prions, maternal conditioning, RNA interference, non coding RNA, small RNAs, DNA methylation and chromatin modifications. In this chapter, we will focus on epigenetic mechanisms involving histone modifications and recent development establishing a link between chromatin modifications (with an emphasis on acetylation and methylation) and cellular processes such as transcription and DNA repair.

HISTONE MODIFICATIONS

In all eukaryotes, chromatin is a highly condensed structure that forms the scaffold of fundamental nuclear processes such as transcription, replication and DNA repair [2]. Chromatin exists in at least two conceptually distinct functional forms: a condensed form during mitosis and meiosis that generally lacks DNA regulatory activity, called heterochromatin; and a looser decondensed form, which provides the environment for DNA regulatory processes, called euchromatin. Nucleosomes are the building blocks of chromatin and they represent two turns of genomic DNA (147 base pairs) wrapped around an octamer of two subunits of each of the core histones H2A, H2B, H3, and H4. The amino-terminal portion of the core histone proteins contains a flexible and highly basic tail region, which is conserved across various species and is subject to various post-translational modifications (Fig. 3.1). The structure of chromatin fulfils essential functions, not only by condensing and protecting DNA, but also in preserving genetic information and controlling gene expression [3]. However, given its compacted structure, chromatin hinders several important cellular processes including, transcription, replication, and the detection/repair of DNA breaks [4,5]. Therefore, chromatin

Handbook of Epigenetics: The New Molecular and Medical Genetics. DOI: 10.1016/B978-0-12-375709-8.00003-4

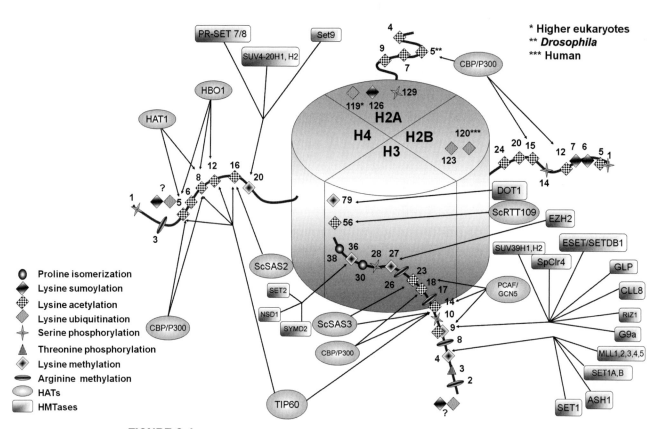

FIGURE 3.1

Schematic representation of a nucleosome and major histone modifications. Post-translational modifications of histones occur primarily on N-terminal tails of core histones (H2A, H2B, H3, and H4) and include acetylation, methylation, phosphorylation, and ubiquitination. Note that several lysines (e.g. Lys 9) can be either acetylated or methylated.

must be first made relaxed to allow access of cellular machineries to chromatin DNA. This leads to one of the most fundamental questions in biology – how is chromatin remodeled? A part of the answer resides in the fact that cells have evolved cellular mechanisms that alter the structure of chromatin. These activities include ATP-dependent nucleosome mobilization (chromatin remodeling) and post-translational histone modifications.

Chromatin modifications can occur through covalent additions to histones. Histone amino-terminal tails are most frequently targets of modifications. There are at least 60 different residues on histones where modifications have been detected and it is likely that this number is underestimated; emergence of new techniques will, without any doubt, help identify new target residues and new modifications. There are, to date, at least eight different types of histone modification: acetylation, methylation, phosphorylation, ubiquitination, sumoylation, ADP ribosylation, deimination, and proline isomerization [6–8]. Traditionally, two mechanisms are thought to govern the function of these modifications. First, these different marks could affect the nucleosome-nucleosome or DNA-nucleosome interactions through the addition of physical entities or by changing histone charges. Second, different marks could represent a docking site for the recruitment of specific proteins which could result in different cellular outcomes. Additionally, numerous reports raised the possibility that all these modifications are combinatorial and interdependent and therefore may form the "histone code", which means that combination of different modifications may result in distinct and consistent cellular outcomes [9,10]. The molecular mechanisms, the role, and the interdependence of these modifications will be discussed in the following paragraphs.

Proline Isomerization

Isomerization is defined as the transformation of a molecule into a different isomere. Isomerization of proteins was first described in 1968 [11] and it was shown to dramatically affect protein conformation by disrupting the secondary structure of polypeptides. It can adopt two distinct conformations: *cis* or *trans*. Isomerization occurs spontaneously, but enzymes called proline isomerases have evolved in order to accelerate switching between different conformations (*cis-trans*). The first evidence that histones can be isomerized was reported in 2006 [12] when Frp4 was identified as a histone isomerase of prolines 30 and 38 (P30 and P38) on histone H3 tail [Fig. 3.1]. The conformational status of P38 is necessary for the induction of lysine 36 of histone H3 (H3K36) methylation and its isomerization appears to inhibit the ability of Set2 to methylate H3K36.

Sumoylation

Sumoylation consists in the addition of a "Small Ubiquitin-related MOdifier protein" (SUMO) of ~100 amino acids. Similar to ubiquitination, SUMO is always covalently attached to other proteins through the activities of members of an enzymatic cascade (E1-E2-E3). Histone sumoylation was first reported in 2003, when Shiio et al. found that H4 can be modified by SUMO and they suggested that this modification leads to the repression of transcriptional activity through the recruitment of HDACs and HP1 proteins [13]. Recently, it was demonstrated that all four core histones can be sumoylated in yeast [Fig. 3.1]. The putative sumoylation sites were identified as K6/7 and to a lesser extent K16/17 of H2B, K126 of H2A, and all four lysines in the N-terminal tail of H4. Histone sumoylation has a role in transcription repression by opposing other active marks such as acetylation and ubiquitination [14].

Ubiquitination

Ubiquitin is a 76 amino acid protein highly conserved in eukaryotes. Ubiquitination (or ubiquitylation) refers to the post-translational modification of the ε-amino group of a lysine residue by the covalent attachment of one (monoubiquitination) or more (polyubiquitination) ubiquitin monomers. Typically, polyubiquitination marks a protein to be degraded via the 26S proteasome, whereas monoubiquitination modifies protein function.

Histone H2A was the first histone identified to be ubiquitinated [15]. Later on, H2B (K119, K120 (K123 in yeast) and K143), H3 and H1 were also reported to be ubiquitinated [16] [Fig. 3.1]. Histones appear to be mostly monoubiquitinated, although H2A and H2B may be polyubiquitinated [17,18]. As for non-histone proteins, histones ubiquitination consists of formation of an isopeptide bond between the C-terminus of ubiquitin and a lysine side chain of histone by sequential actions of E1 activating, E2 conjugating, and E3 ligase enzymes. E2 and E3 play a crucial role in specifying the protein to be ubiquitinated. E3-ligases mostly belong to HECT (Homologous to E6AP-C Terminus) or RING (Really Interesting New Gene) protein families. H2BK123 specific E2 in yeast, Rad6 was the first histone E2 to be identified at the beginning of the century [19]. Rad6 activity is combined to the RING finger E3 ligase, Bre1. Homologous of Rad6, Rhp6 in *drosophila* and HR6A/B in humans, as well as homologous of Bre1, Brl1 in *Drosophila* and RNF20 in humans were also shown to be involved in H2B ubiquitination [20–24]. Histone H2A ubiquitination is dependent on the Polycomb repressive complex 1 (PRC1); PRC2 set up the H3K27me3 marks that are recognized by the PRC1 complex which would ubiquitinate H2A and silence gene expression. At least two members of the PRC1 complex, RING1b (also known as Rnf2) and BMI1, were found to form a heterodimer that ubiquitinates H2A [25–29]. The PRC1 complex is formed of four core proteins that include the PcG proteins Polycomb, Polyhomeotic, Posterior sex combs (PsC) and RING (also known as sex comb extra), in addition to many other proteins

[30–32]. Ortholog of PRC1 and RING1b also exist in complexes distinct from PRC1 and these are the RING-associated factor (dRAF) in *D. melanogaster* and bCL6 corepressor (bCoR) in mammals. These PRC1-like complexes can also ubiquitinate H2A [26,33]. The addition of the ubiquitin moiety to histones is reversible through the activity of deubiquitinating enzymes that consist of ubiquitin C-terminal hydrolases and ubiquitin-specific processing proteases (UBPs). Sixteen UBPs have been identified so far in yeast. These UBPs differ by the length of their amino-terminal part that confers their specificity [34,35]. The best studied UBPs are UBP8 and UBP10 that are specific for H2B. UBP8 belongs to the SAGA complex [36–38] and orthologs were found in *Drosophila* (Nonstop) and in human (USP22) and they were both involved in H2B deubiquitination in the context of the SAGA complex [39–42]. UBP10 activity is SAGA-independent but SIR-dependent and its orthologs in higher eukaryotes are still to be defined [43–45].

Ubiquitinated H2A at K119 (uH2A) was shown to be important for transcriptional activation and several active genes were shown to contain a high percentage of uH2A [46,47,48]. Surprisingly, uH2A was also linked to transcription inhibition [48,49,50,51]. Similarly, ubiquitination of H2B was linked to both activation and inhibition of transcription [35,52]. The ambiguity on the role of the mono-ubiquitinated H2B (uH2B) in transcription regulation was mostly due to the lack of specific antibodies. However, the elaboration of a suitable anti-uH2B monoclonal antibody by using a branched peptide as an antigen partially clarified this matter and opted for a positive correlation between uH2B-K120 and gene expression. Indeed, this antibody was used in ChIP-Chip experiments on tiling arrays and the results showed a preferential association of uH2BK120 with the transcribed regions of highly expressed genes [53]. As this mark is not associated to distal gene promoters but rather to transcription start site (TSS) and further to gene bodies of active genes, it was suggested that it is linked to transcription elongation rather than initiation. Further proof of the correlation of uH2B with active transcription came from an elegant biochemical study performed in Tom Muir's laboratory. In this study, the authors used two traceless orthogonal expressed protein ligation (EPL) reactions to chemically and specifically ubiquitinate H2BK120 incorporated into chemically defined nucleosomes. The results showed a direct activation of H3-K79 methylation by hDot1L, a mark related to gene activation [54].

Histone ubiquitination may affect other histone modifications. For example, histone deacetylase 6 (HDAC6) was shown to bind to ubiquitin through its zinc-finger domain. H3K4 and H3K79 methylation was shown to be dependent on Rad6-mediated H2BK123 ubiquitination [54–58]. The effect of ubiquitination on histone methylation can explain its role in both activation and inhibition of transcription. For instance, it has been proposed that ubiquitination of H2B occurs mostly in euchromatin leading to H3K4 and H3K79 methylation, which would prevent Sir proteins from association with active euchromatic regions, thereby restricting Sir proteins to heterochromatic regions to mediate silencing [59]. At the same time in euchromatin, the ubiquitination would activate the transcription by methylating H3K4 and by facilitating the transcriptional elongation [60,61].

ADP-ribosylation

ADP-ribosylation is a post-translational modification defined by the addition of an ADP-ribose moiety onto a protein using NAD+ as a substrate. If the transfer takes place on an amino-acid acceptor, it is referred to as mono- or poly-(ADP-ribosyl)ation (PARation) and if it occurs on an acetyl group it is called O-acetyl-ADP-ribosylation. Mono(ADP-ribosyl)ation is mediated by ADP ribosyl transferases (ART) and the enzymes responsible for the PARation are the poly(ADP-ribose) polymerases (PARPs) [62,63]. All core histones and linker histone H1 are subjects to mono(ADP-ribosyl)ation either in response to genotoxic stress or in physiological conditions depending on the cell cycle stage, proliferation activity, and degree of terminal differentiation [64,65]. PARation can also be detected on the majority of histone types. There seems to be some specificity in the activity of PARP proteins on histones; for example PARP1 seems to preferentially poly(ADP-ribosyl)ate the linker histone H1 whereas PARP2 prefers core histones. In response to single strand break (SSB), PARP1 and PARP2 poly(ADP-ribosyl)ate the C- and N-terminus of histones H1 and H2B leading to the relaxation of the chromatin structure facilitating the access of single strand break repair (SSBR)/base excision repair (BER) factors to the site of damage. This has been explained, at least in part, by the fact that PARation of histones leads to their removal from the chromatin. The removal of histones results in the opening of the chromatin structure (the same mechanism leads to transcriptional activation). Moreover, PAR is used for tagging the region affected by DNA damage allowing adequate response of the cell according to the extent of damage signaled by the presence of PAR moieties. On the other hand, recent studies indicated that PARP-dependent ribosylation in response to DNA damage may induce local chromatin condensation rather than relaxation. Indeed, PAR moieties are recognized by the macrodomain of the histone variant macroH2A1.1. This would lead to a transient condensation or looping, increase phosphorylation of H2AX at the sites of break and reduced recruitment of Ku70/80 leading to altered DNA damage response [66]. It is difficult to reconcile this last study with the rest of the literature on the role of PARP in DNA repair. The obvious explanation is that Poly(ADP-ribosyl)ation leads to a quick and transient compaction of the chromatin that would protect the DNA from additional damage and this is rapidly reversed allowing DNA repair to take place. This hypothesis is supported by the quick and transient nature of the PAR-macroH2A1.1 interaction-dependent chromatin condensation that is gradually lost after the reduction of Poly(ADP-ribosyl)ation levels.

The role of mono- and poly(ADP-ribosyl)ation in DNA repair and transcription may be explained by their interaction with other chromatin modifications in the context of the "histone code". For example, Mono-ADP-ribosylation on H4 seems to occur preferentially when H4 is hyperacetylated and mono-ADP-ribosylation of histone H1.3 on arginine 33 (R33) may reduce cyclic AMP-dependent phosphorylation of Serine 36 (S36) [67].

Phosphorylation

Protein phosphorylation represents the addition of a phosphate (PO_4) group to a protein molecule. Phosphorylation is catalyzed by various specific protein kinases, whereas phosphatases mediate removal of the phosphate group. Histones can also get phosphorylated and the most studied sites of histone phosphorylation are the serine 10 of histone H3 (H3S10) that is deposited by the Aurora-B kinase during mitosis (Fig. 3.1) and S139 (129 in yeast) of H2A variant, H2AX, DNA damage-dependent phosphorylation by ATM and ATR. H2AX could be additionally phosphorylated on tyrosine 142. H4 (S1) and linker histone H1 (S18, S173, S189, T11, T138, and T155) were also shown to be phosphorylated by the CK2 and DNA-PK respectively.

ROLE OF HISTONE PHOSPHORYLATION IN TRANSCRIPTION REGULATION

The relationship between histone phosphorylation and gene expression is far from being totally understood. The phosphorylation of H3S10 (H3S10P) was initially linked to chromosome condensation and segregation during mitosis and meiosis [68,69]. The role of this mark during mitosis was investigated in detail in a study performed in the laboratory of David Allis [70]. The authors followed the status of HP1-α, -β and -γ during mitosis. These proteins are recruited via interaction with H3K9me3 and lead to heterochromatinization. However, during mitosis, these marks are ejected from chromatin even though H3K9me3 marks are preserved. The study demonstrated that the addition of the H3S10P mark is responsible for the eviction of these proteins. It was proposed that this would lead to better recruitment of players necessary for proper condensation and segregation of chromosomes, although this hypothesis awaits experimental confirmation. The role of H3S10P in chromatin condensation suggests that it should be involved in transcriptional repression;

however, evidence has accumulated indicating that this mark has rather an important role in transcriptional activation of genes in various organisms. For instance, induction of heat-shock genes in *Drosophila* is concomitant to a high increase of H3S10P [71,72]. On the other hand, dephosphorylation of H3S10 is carried out by a phosphatase called PP2A (Protein Phosphatase 2A) and results in the inhibition of transcription [72]. Additionally, H3S10P was shown to be important for the activation of NFKB-regulated genes but also "immediate early" genes such as c-fos and c-jun. This phosphorylation is proposed to induce the accumulation of phosphor-binding protein 14-3-3 [73]. Genome-wide ChIP-Chip analysis in budding yeast has shown that several kinases are not only present in the cytoplasm but also on the chromatin of specific genes [74], which suggests that the kinase signal transduction cascade could have direct effect on gene expression by phosphorylating the histones of specific genes or gene promoters [8].

ROLE OF HISTONE PHOSPHORYLATION IN DNA REPAIR

Beside its role in chromosomal condensation and transcription, phosphorylation of histones, in particular phosphorylation of H2AX, has a role in DNA damage response and DNA repair. Rapid phosphorylation of H2AX, at serine 129 (γH2AX) by the PI3K kinases at double strand break (DSB) sites, is one of the first and most easily detectable DNA damage signaling post-translational events. It anticorrelates to the phosphorylation of Tyrosine (Y) 142 of H2AX. Indeed, recent studies have shown that H2AX-Y142 is constitutively phosphorylated under physiological conditions by the activity of WSTF (Williams–Beuren syndrome transcription factor) and is dephosphorylated in response to DNA damage via the activity of Eya tyrosine phosphatase in correlation to the increase in serine phosphorylation. The kinase activity of WSTF as well as the phosphatase activity of Eya shown to be important for the early recruitment of phospho-ATM and MDC1 to sites of DNA damage, thus privileging DNA repair over apoptosis [75,76]. γH2AX can be detected over kilobases (in yeast) or megabases (in mammalian cells) from sites of DSBs [77,78] and is required for the retention/accumulation of repair proteins [79,80,81]. γH2AX also plays a role in cohesion binding to a large region around DSB, an event thought to be important for sister chromatid cohesion in post-replicative repair [82,83]. Interestingly, γH2AX is required for the recruitment of the NuA4 histone acetyl-transferase (HAT) complex (yeast homolog of mammalian TIP60) to sites of DNA DSBs induced by HO endonuclease [80]. The recruitment of this HAT complex to γH2AX is mediated by Arp4 and leads to acetylation of chromatin surrounding the break site, thereby facilitating efficient repair of DNA damage [80]. As well as being a component of NuA4 HAT complex, Arp4 is a subunit of the ATP-dependent chromatin remodeling complex INO80/SWR1. It was shown that INO80/SWR1 is also recruited to γH2AX around DNA breaks and its remodeling activity seems to be required for the repair of DNA DSBs [80,84,85]. Hence it would appear that cells can utilize the activities of both histone modifying and remodeling complexes in order to facilitate DNA repair.

The precise role of γH2AX in DSBs is still under debate. Originally, it was suggested that the phosphorylation of H2AX is essential for the recruitment of DNA repair enzymes [86] through their BRCT (BRCA1 COOH Terminal) domain [87]. However, a study by Celeste et al., changed our understanding on the role of γH2AX by demonstrating that DNA repair proteins, including Brca1 and Nbs1, are recruited to DNA breaks even in the absence of γH2AX. On the other hand, the presence of γH2AX is essential for the formation of IRIF (Irradiation-Induced Foci) [79], indicating that the role of H2AX phosphorylation may be dispensable for the original recruitment of DNA repair factors but indispensable for the accumulation/retention of these factors at DNA break sites. DNA break-associated histone phosphorylation can also occur on H4S1 through the activity of Casein Kinase II (CK2) in response to DNA damage and this facilitates double strand break (DSB) repair via nonhomologous end joining (NHEJ) [88]. A study from Côté's laboratory demonstrated that this phosphorylation coincides with a decrease in acetylation, suggesting that it occurs after H4 deacetylation and that these events may regulate chromatin restoration after repair is completed [89]. Finally, linker histone H1 is found to be phosphorylated by PI3K family member DNA-PKcs, and this phosphorylation is required for efficient DNA repair by NHEJ [90–92].

Methylation

Protein methylation is a covalent modification that represents the addition of a methyl group from the donor S-adenosylmethionine (SAM) on carboxyl groups of glutamate, leucine, and isoprenylated cysteine, or on the side-chain nitrogen atoms of lysine, arginine, and histidine residues [93]. However, histone methylation occurs only on arginines and lysines. Arginines can be mono- or dimethylated whereas lysines can be mono-, di- or trimethylated [8]. Arginine methylation can be either symmetrical or asymmetrical.

The enzymes responsible for histone methylation are grouped into three different classes: [1] the lysine-specific SET domain-containing histone methyltransferases (HMT) involved in methylation of lysines 4, 9, 27, and 36 of histone H3 and lysine 20 of histone H4; [2] non-SET domain-containing lysine methyltransferases involved in methylating lysine 79 of histone H3; and [3] arginine methyltransferases involved in methylating arginines 2, 17, and 26 of histone H3 as well as arginine 3 of histone H4 [Fig. 3.1].

Whereas most covalent histone modifications are reversible, until recently it was unknown whether methyl groups could be actively removed from histones. The first histone demethylase to be discovered was LSD1, which mainly demethylates H3K4 but could also demethylate H3K9, when it is present in a complex with the androgen receptor [94,95]. A flow of other related enzymes were subsequently discovered and classified into two families of histone lysine demethylases: JMD2 and JARID1 families. The JMD2 (Jumonji C (JmcC)-domain containing proteins) family includes JHDM3A (Jumonji C (JmjC)-domain-containing histone demethylase 3A; also known as JMJD2A); JMJD2C/GASC1 [96], which can both demethylate H3K9 and H3K36 [97–99]; and JMJD2B [100] and JMJD2D [101] which demethylate H3K9, JHDM1 (JmjC domain-containing histone demethylase 1) (demethylates H3K36) and UTX [97,98,102–104]. JARID1 proteins include RBP2, PLU1, SMCY/Jarid1ds and SMCX [105,106]. Histone arginine methylation marks were shown to be reversible. The first report about arginine demethylation suggested that methylated arginines on histones H3 (R3) and H4 (R17) are converted into citrulline via the activity of the human "peptidylarginine deiminase 4" protein, Pad4; this process was called "demethylimination or deimination" because the methyl group is removed along with the imine group of arginine [107,108]. Pad4 can deiminate multiple arginine sites on histones H3 (R2, R8, R17, and R26) and H4 (R3) [107]. Beside its function as an intermediate in the histone demethylation process, deimination has been involved in the estrogen signaling pathway [109]. On the other hand, a direct histone arginine demethylase, namely JMJD6, was recently identified and was found to belong to the JMD2 family [110].

ROLE OF HISTONE METHYLATION IN TRANSCRIPTION REGULATION

The methylation mark on histones could be related to activation, elongation, or repression of gene expression, For example, H3K4me, me2 and me3 have been found on active promoters and linked to transcription initiation and elongation [60,61,111–113] whereas HK36me2/me3 have been correlated to transcription elongation [10,60,114–116]. To obtain a more detailed picture on histone methylation distribution along genes, ChIP-Chip and ChIP-seq experiments were performed and showed that H3K4me3 peaks at 5′ ends and at promoter proximal regions of active genes, H3K4me2 peaks at active gene bodies whereas H3K4me is enriched at the 5′ end of active genes. On the other hand H3K36me2/3 marks are enriched in active gene bodies and mostly at the 3′ end of active genes [114,117–123]. The precise underlying mechanism of H3K4me-dependent transcription regulation is still not clear. One possibility would be that histone modifying complexes or chromatin remodeling factors, such as Taf3 [124], could recognize and bind to the methylation mark through their PHD (Plant Homeo Domain), thus activating the transcription. H3K36 methylation marks are then recognized by the chromodomain of the Eaf3 subunit of The Rpd3S HDAC [120,125–128]. This would lead to the deacetylation of gene bodies and thus prevent, transcription initiation at cryptic sites of gene bodies. Methylation of H3K79 was also implicated in transcription activation and elongation [129]; however, this should be taken carefully, because it is based mainly on the fact that this mark was related to the activation of HOXA9 and that it limits the spreading of heterochromatin by preventing Sir2 and Sir3 from spreading into euchromatin. Moreover, detailed genome-wide study showed that while both H3K79me2 and H3K79me3 are enriched in gene bodies in yeast and *Drosophila*, only *Drosophila* H3K79me2 correlates with active transcription [122,130].

Three lysine methylation sites are connected to transcriptional repression: H3K9, H3K27, and H4K20. Very little is known regarding the repression functions of H4K20 methylation compared to a large number of studies on the two other repressive marks. Methylation of H3K9 is carried out by SUV39H1 and SUV39H2 in humans (its homolog Clr4, cryptic locus regulator 4, in *Schizosaccharomyces pombe*, and Su(var)3–9, suppressor of position-effect variegation, in *Drosophila*). These HMT have been shown to contain an SET domain. SET domain usually contains 130–140 amino acids and is a common feature of Trithorax (Thx) and Polycomb (PcG) group proteins which are involved in transcriptional activation and repression, respectively. Su(var)3–9 and its homologs were shown to be important for proper heterochromatin formation. These findings suggest a role for H3K9 methylation in gene silencing through correct heterochromatin folding [131,132]. It is now well established that HP1 recognizes methylated H3K9 through its chromodomain, contributing in part to the formation of heterochromatin. How are H3K9me2/3 and subsequent heterochromatinization initially targeted to DNA sequences? Two mechanisms could serve as the initial trigger for H3K9me: DNA-binding factors such as transcription factors or RNAi. Evidence for direct targeting of H3K9me by RNAi came first from studies on the core RNAi machinery which includes Dicer (Dcr), Argonaute (Ago) and RNA-dependent RNA polymerase (RdRP) in *S. pombe* [133]. Later, several studies on different organisms demonstrated the role of RNAi in heterochromatin establishment [134–144]. The involvement of transcription factors such as Atf1, PCR1, and Taz1 in targeting heterochromatinization was also reported [145–149]. Although H3K9me was traditionally linked to repression, a recent study showed that H3K9me3 could be located in the gene bodies of active genes along with HP1 [150]. This observation led to the currently used model where H3K9me within the coding regions is activator whereas H3K9me in the promoters is repressive. H3K27 methylation has been implicated in the silencing of HOX gene expression. There are also indications that the same mark would be involved in the inactivation of the X chromosome and silencing during genomic imprinting. Interestingly, ChIP-Chip and ChIP-seq studies in ES cells indicated that some of the genes which are not expressed in ES cells have both repressive (H3K27me3) and active (H3K4me3) marks at their promoters, forming the so-called "bivalent domains". Along differentiation, bivalent domain genes are resolved into monovalent by getting rid of one associated mark and thus get either stably activated or stably repressed. Therefore, these bivalent domains were thought to keep genes repressed at a certain developmental window but poised for activation in another subsequent developmental stage [111,151,152].

Arginine methylation was thought to be an activation mark as suggested by the fact that protein arginine methyltransferases are recruited to promoters by transcription factors [153]. One example of promoters regulated by histone arginine methylation is the pS2 promoter, a downstream target in the ER (Estrogen Receptor) pathway. Indeed, Metivier and colleagues showed that the transcription of this gene goes ON/OFF in a very controlled and specific fashion forming cycles of activation [154]. The activation of pS2 transcription correlated with the recruitment of protein arginine methyltransferase 1 (PRMT1) and Cofactor Associated Arginine Methyltransferase 1 (CARM1) recruitment. However, recent studies from the same group as well as another group found that these activation cycles could actually result from cycles of DNA methylation/demethylation and were probably not due to arginine methylation marks [155,156]. We now know that the arginine methylation effect on transcription could be either activating or repressive and depends on the type of arginine methyltransferase (RMT) involved [157]. For example, type I RMT, which include CARM1, PRMT1, and PRMT2 and generate monomethyl-arginine and asymmetric dimethyl-arginine derivatives, are involved in activation, while type II arginine methyltransferase PRMT5, which generates monomethyl-arginine and symmetric dimethyl-arginine derivatives, is involved in repression [158–164]. On the other hand, PRMT5 was shown to be associated with transcriptional repression. It associates with mSin3/HDAC and Brg1/hBrm and it is recruited to genes involved in control of cell proliferation (e.g. c-Myc target gene: cad and tumor suppressors: ST7 and NM23) in correlation with their repression [165,166].

ROLE OF HISTONE METHYLATION IN DNA REPAIR

The role of histone methylation in the DNA damage response and DNA repair is less clear than the role of histone acetylation and phosphorylation; however, the involvement of lysine methylation, in processes other than transcriptional regulation, has recently received considerable attention. Methylation of H4 by Set9 histone lysine methyltransferase functions to localize Crb2, a DNA damage sensor and checkpoint protein in *Schizosaccharomyces pombe*, to sites of DNA damage hence increasing cellular survival following genotoxic stress [167]. Crb2 recruitment to DNA repair foci is dependent on the recognition of methylated H4K20 via the double tudor domains of Crb2 [168]. Subsequently, ionizing radiation-induced DNA damage generates nuclear foci at sites of DNA repair, which contain methylated H4K20 and the cell-cycle checkpoint protein Crb2 [167]. Similarly, the mammalian homolog of Crb2, 53BP1, also binds methylated H3 at sites of DNA DSBs [168,169]. Interestingly, Crb2 and 53BP1 do not recognize the trimethyl form of K20, which may indicate a different role of this modification in response to DNA damage.

Acetylation

Acetylation describes a reaction that introduces an acetyl functional group into an organic compound. Both histones and non-histone proteins can be acetylated. Histone acetylation consists in the transfer of an acetyl group from acetyl-CoA to the lysine ε-amino groups on the N-terminal tails of histones. This enzymatic activity is catalyzed by enzymes called histone acetyltransferases (HATs). The acetyl-CoA recognizes a specific domain within HATs called AT domain, Arg/Gln-X-X-Gly-X-gly/Ala. HAT enzymes often exist in multisubunit complexes that count one HAT catalytic subunit, adapter proteins, several other molecules of unknown function and, in many cases, a large scaffold protein called TRRAP. Acetylation can occur on specific lysines in all four histones (H3, H4, H2B, and H2A) [Fig. 3.1]. Hyperacetylation of histones is considered as a hallmark of transcriptionally active regions. Studies also revealed that the role of acetylation is not exclusively related to transcription, but can affect other DNA-based cellular processes such as DNA repair and replication.

CLASSIFICATION OF HATs

Two classifications can be used to separate HATs.

First Classification

This divides HAT complexes into two big classes based on their suspected cellular origin and functions: A-type and B-type HATs. A-type HATs are nuclear enzymes that catalyze acetylation on already deposited histones in the context of the chromatin. B-type HATs are cytoplasmic enzymes and thought to be responsible for the acetylation of newly synthesized histones leading to their transport from the cytoplasm to the nucleus where they are deposited onto newly replicated DNA [170,171].

Second Classification

Instead of classifying HATs based on their cellular localization, modern classification uses structural criteria such as the presence or absence of chromodomains, bromodomains, and zinc finger domains. This classification separates the HATs into two major families: Gcn5-related acetyltransferases (GNATs) and the MYST (for 'MOZ, Ybf2/Sas3, Sas2, and Tip60)-related HATs. To these families, we can add p300/CBP HATs, the general transcription factor HATs which include the TFIID subunit $TAF_{II}250$, and the nuclear hormone-related HATs: SRC1 and ACTR (SRC3). The classification of these different families is not based on functional criteria. Due to space restriction, only complexes based on the second classification will be detailed in the following.

GNAT SUPERFAMILY

All the GNAT superfamily members share structural and sequence similarity to Gcn5. This superfamily is characterized by four regions with different degrees of conservation (labeled A to D) spanning over 100 residues. These regions were first defined by the comparison between Gcn5 and B-type Hat1. Motif A, also called AT domain, contains an Arg/Gln-X-XGly-X-Gly/Ala sequence and is shared with other HAT families. It is the most highly conserved and is important for acetyl-CoA recognition and binding. Tridimensional structure of this motif is highly conserved in all 15 GNAT proteins crystallized so far [172]. The C motif is found in most of the GNAT family acetyltransferases but not in the majority of known HATs. The GNAT superfamily contains over 10,000 members distributed in all kingdoms of life including histone acetyltransferases (HATs) but also nonhistone AT (see http://supfam.mrc-lmb.cam.ac.uk/SUPERFAMILY). The most relevant HATs of this family are: Gcn5, PCAF, Hat1, Elp3, and Hpa2.

MYST SUPERFAMILY

The MYST family was named after its founding members: MOZ, Ybf2/Sas3, Sas2, and Tip60 [173]. These proteins are grouped together on the basis of their close sequence similarities and their possession of a particular acetyltransferase homology region (part of motif A of the GNAT superfamily) that binds Acetyl-CoA [174], in addition to a zinc finger domain called C2HC (C-X2-C-X13-H-X-C), and a "E-R" motif (Esa1-Rpd3), both needed for the enzymatic activity and for the substrate recognition [175]. Recently, additional members of this family were identified including Esa1 in yeast, MOF in *Drosophila*, and HBO1 and MORF in mammals. Despite their structural similarities, the members of this superfamily have various functions in various organisms. They resemble those of the GNAT family as both have an AT domain [176], but differ by the fact that they have different C- and N-termin, leading to different substrates. In addition, MYST family members possess either a chromodomain or an additional zinc finger domain termed the PHD domain [177,178].

p300/CBP

p300 and CBP are often referred to as a single entity, since the two proteins are considered as structural and functional homologs and both proteins were subsequently shown be functionally interchangeable. But the two proteins diverge in several functional and structural properties. Indeed, some studies identified phosphorylation residues which are specific for each of the two proteins [179]. Another difference is that, in response to ionizing radiation (IR), p300, but not CBP, is important for apoptosis induction (probably through the activation of p53) [180]. In addition, whereas both proteins are necessary for apoptosis and G1 arrest of F9 embryo carcinoma cells, differentiation and induction of the cell cycle inhibitor p21/Cip1 critically depends on p300, while induction of p27/Kip1 requires CBP [181]. The most striking divergence comes from loss of function studies showing that individual knockouts of each of the two proteins resulted in two different phenotypes [182,183]. The specificity of the acetyltransferase activity of both proteins may explain, at least in part, the different functions between the two proteins. For instance, CBP was recently shown to have a preference for acetylating K12 on histone H4, while p300 preferentially acetylates K8 on histone H4 *in vivo* [184]. However, both proteins were shown to be able to acetylate H3K56 in collaboration with ASF1A histone chaperone. Another histone chaperone, CAF1, is needed for the incorporation of H3K56Ac into the chromatin, notably in response to DNA damage [185]. Other studies showed that H3K56 can be acetylated in the cytoplasm by Gcn5-containing HAT complex called HatB3.1 prior to its transport to the nucleus [186,187], which makes Gcn5 both nuclear and cytoplasmic. Another HAT-like complex in yeast, called Rtt109p, was also defined as responsible for H3K56 acetylation in the cytoplasm [188]. P300/CBP are large proteins (~300 kDa) containing more than 2400 residues. Four interaction domains have been characterized throughout their sequence. These include a bromodomain motif [189,190], which is also found in several other HATs such as Gcn5 and PCAF. P300/CBP have homologs in most metazoas but not in inferior eukaryotes including yeast. They were first identified as transcriptional adaptors for many different transcription factors that directly contact DNA-bound activators. *In vitro* studies seem to indicate that p300/CBP preferentially acetylate K12 and K15 of H2B, K14 K18 and K56 of H3K5, and K8 of H4 [191]. HAT proteins have also been directly implicated in transcriptional activation brought about by hormone signals. The HAT activity of human coactivators ACTR, SRC-1, and TIF2, which interact with nuclear hormone receptors, confirms the involvement of acetylation in yet another system of transcriptional regulation and defines a unique family of HATs. The members of this family share several similarities including HAT domain in the C-terminus, and an N-terminal, basic helix-loop-helix/PAS region [192], as well as receptor and coactivator interaction domains.

HAT COMPLEXES

Most HAT enzymes, alone, are not able to acetylate histones in the context of nucleosomes. However, when present in multisubunit complexes, these enzymes become more stable and more histone-type specific. Furthermore, the substrate of the HAT enzyme may change according to the HAT complex to which it belongs. This modification in specificity is further confirmed by the fact that distinct HAT complexes having distinct substrate specificity may share common subunits. For instance, TRRAP is shared between several HAT complexes, and STAGA and TFTC complexes share all their subunits except some high molecular weight TAFs which are not part of STAGA. HAT complexes were purified in both humans and yeast, were functionally equivalent in the two organisms, and were divided into several families.

Between all HAT complexes, GNAT or SAGA-like HAT complexes (SAGA, SLIK, PCAF, STAGA, TFTC) are unique by the fact that they contain TAFs. The enzymatic subunit of these complexes can be represented by Gcn5 or PCAF. To date, two complexes belonging to this group have been discovered in yeast (SAGA and SALSA/SLIK) and three in humans (PCAF, STAGA, TFTC). It is important to note that GNAT or SAGA-like complexes may exist in flies and in mice [193,194]. Subunits of these complexes include Ada proteins, Spt proteins, TAFs, SAP130, and TRRAP. NuA3 (nucleosomal acetyltransferase of histone H3) is one of the yeast HAT complexes identified in the study carried out by Grant et al. [195]. It is a 500-kDa complex and it exclusively acetylates histone H3 in nucleosomes. Peptide sequencing of proteins from the purified NuA3 complex identified Sas3, a MYST protein involved in silencing, as the catalytic HAT subunit of the complex. NuA3 also contains the TBP-associated factor, yTAF(II)30. In addition, Yng1 was identified as a subunit of NuA3: it belongs to PHD finger-containing proteins and was recently found to interact with H3K4me3. The interaction between Yng1 and H3K4me3 seems to promote NuA3 HAT activity at K14 of H3 and transcription at a subset of targeted ORFs [196]. *In vitro* studies on NuA3, like those on Ada, indicated that both complexes failed to interact with activation domains or to activate transcription in a specific way [197,198]. NuA4 (nucleosomal acetyltransferase of histone H4)/TIP60 complex is another yeast HAT complex identified by Grant et al. [195] (complex 2), at the same time as SAGA, NuA3, and ADA. Its human homolog is called TIP60 [199]. As with Gcn5, NuA4 and TIP60 enzymes are able to acetylate histones H4, H3, and H2A when in free form but they are not capable of acetylating histones folded into nucleosomes and their activities seem to depend on the presence of other proteins in the context of multisubunit complexes. These complexes also seem to restrict their activities to histones H4 and H2A [199]. A homolog of these complexes was also recently identified in *Drosophila* (dmTIP60) [193,200].

Further studies identified three new complexes that share several subunits of the TIP60/NuA4 complex. The first two complexes have been identified in humans *in vivo* and they are very similar to TIP60: these are the p400 complex [201] and another complex that contains TRRAP-BAF53-TIP48-TIP49 [202]. P400 is deprived of any HAT activity but can hydrolyze ATP [193]. The second complex has a HAT activity; however, the enzyme responsible for this activity is still not defined. The third complex was identified in yeast, and it represents a sort of "mini" NuA4 complex containing only three subunits (Tip60p/NuA4-Ing3-Epc1). This complex is termed "Piccolo NuA4" and its homolog also exists in humans and seems to represent the catalytic core of TIP60 [203].

HISTONE DEACETYLASES

There are three distinct families of histone deacetylases: the class I and class II histone deacetylases, and the class III NAD-dependent enzymes of the Sir family. They are involved in multiple signaling pathways and are present in numerous repressive chromatin complexes. Similarly to HATs, these enzymes do not appear to show much specificity for a particular acetyl group. However, yeast enzyme Hda1 seems to have higher specificity for H3 and H2B whereas Hos2 is specific for H3 and H4. The fission yeast class III deacetylase Sir2 and its human homolog SirT2 preferentially deacetylate H4K16ac [204]. Recent reports indicated that Sir2/SirT2 is also able to deacetylate H3K56 [185].

ROLE OF HISTONE ACETYLATION IN TRANSCRIPTION REGULATION

The "traditional" role of histone acetylation is transcription regulation. The first evidence of the involvement of HATs in transcription dates back to 1964, when it was observed that chromatin regions of actively transcribed genes tend to have hyperacetylated histones [6]. The addition of acetyl groups to histone tails was proposed to neutralize the histone charge, which weakens histone-DNA interaction, relaxing the chromatin structure and facilitating the access of transcription machinery [205]. For example, work from Craig Peterson's laboratory demonstrated that the incorporation of H4K16Ac into nucleosomal arrays impedes the formation of compacted chromatin fibers and prevents the ATP-mediated chromatin remodeling factors from mediating nucleosome sliding [206,207]. In addition, two other mechanisms by which histone acetylation facilitates transcription have been proposed. First, there is evidence that histone acetylation may serve as a specific docking site for the recruitment of transcription regulators [208–211]. Second, histone acetylation may also act in combination with other histone modifications (methylation, phosphorylation, and ubiquitination) to form the "histone code" that dictates biological outcomes including gene transcription [9,212]. HAT complexes from both GNAT and MYST families were shown to be recruited to activator-bound nucleosomes resulting in transcriptional activation [89,213,214]. The recruitment of SAGA leads to acetylation of promoter-proximal H3, whereas recruitment of NuA4 results in a broader domain of H4 acetylation (>3 kbp) [214]. This hyperacetylation of histones was linked to transcription activation [215], and NuA4-dependent acetylation of histone H4 was shown to affect transcription of specific genes such as His4, Lys2 [216], ribosomal proteins, and heat-shock proteins [217]. Arabi and colleagues have shown that TRRAP (a subunit of many HAT complexes) is recruited by c-Myc to the promoters of Pol I transcribed genes. The recruitment of TRRAP leads to increased histone acetylation, followed by recruitment of RNA polymerase I and activation of rRNA transcription [218]. Interaction between several activators and Tra1 (yeast homologs of TRRAP) cofactor was demonstrated in yeast and this interaction is essential for efficient transcriptional activation [219]. For example, c-Myc binding correlates with regions of acetylated histones [220]. The effect of Myc oncoproteins on chromatin structure was studied in more details by Knoepfler and coworkers, who found that c-Myc and N-Myc are involved in the widespread maintenance of active chromatin, probably through upregulation of GCN5 [221]. In mammals, TRRAP has also been implicated in the regulation of transcription. For instance, TRRAP activates the transcription of target genes through the recruitment of Tip60 and Gcn5/PCAF to their promoters, thus acetylating histones H4 and H3, respectively [222,223]. H3K56Ac was also implicated in transcriptional activation. H3K56 residue is facing the major groove of the DNA within the nucleosome, so it is in a particularly good position to affect histone/DNA interactions when acetylated [224–226].

ROLE OF HISTONE ACETYLATION IN DNA REPAIR

While the role of HAT enzymes in transcriptional regulation is well established [227–229], a plethora of recent reports has also implicated HATs and histone acetylation in DNA damage detection and DNA repair. TATA box-binding protein-free TAFII (TFTC), a complex containing Gcn5 HAT, appears to preferentially acetylate histone H3 in nucleosomes containing UV-damaged DNA in mammalian cells [230], whereas STAGA (SPT3-TAFII31-GCN5L acetyltransferase), another Gcn5 containing HAT complex, associates with UV-damage-binding factor [231]. Yeast strains with mutations in the N-terminal tail of histone H4, a subject for acetylation, were shown to be deficient in both DNA DSB repair and replication-coupled repair, and Esa1 (catalytic component of the yeast NuA4) was found to be responsible for this acetylation. Tip60 (mammalian homolog of Esa1) was also shown to be important in DNA DSB repair following genotoxic stress [199]. In addition, mutations in Yng2, a component of the yeast NuA4 HAT complex, results in hypersensitivity to and inefficient repair of DNA damage caused by genotoxic agents that induce replication fork stall [232]. Finally, mutations in either specific lysine residues in histone H3 or the yeast acetyltransferase HAT1 result in hypersensitivity to DNA DSB-inducing agents [233].

Mechanistic data for the role of acetylation in DNA repair has arisen from several recent reports. Binding of the NuA4 HAT complex at sites of DNA damage and site-specific histone H4 acetylation were found to occur concomitantly with histone H2A phosphorylation after induction of DSBs [80,90]. Additionally, histone H3 acetylation is an abundant modification of newly synthesized histones and defects in this acetylation result in sensitivity to DNA damaging agents that cause DNA breaks during replication [224]. Furthermore, localized histone H3 and H4 acetylation and deacetylation is triggered by homology directed repair of DSBs. Consistent with this finding, Gcn5 and Esa1 HATs are recruited to chromatin around a DSB induced by HO endonuclease in yeast [234]. Alongside histone modification on the amino-terminal tails of histones, histone core modifications also play a role in DNA repair. This is exemplified by the role of H3K56 acetylation in response to DNA damage. In budding yeast acetylation of H3K56 is deposited on newly synthesized histones during S phase and disappears in G2. However, in the presence of DNA damage the deacetylases for H3K56, Hst3, and Hst4 (two paralogs of Sir2) are downregulated and the modification persists [235,236]. The Rtt109 enzyme, which acetylates H3K56, has recently been implicated in genome stability and DNA replication [225,237,238]. Furthermore, recent evidence has revealed that histone acetylation by TRRAP/TIP60 HAT is important for recruitment/loading of repair proteins to sites of DNA DSBs and homology-directed DNA repair [239]. These findings lead to a model in which induction of DSBs leads to the recruitment of the TIP60/NuA4 complex to DSBs and concomitant acetylation of H4 N-terminal tails [4,240].

SUMMARY AND PERSPECTIVES

Research on chromatin modifications is a newly emerging field that holds the promise of further advancing our understanding of tumorigenesis and facilitating the development of novel strategies to prevent, diagnose, and treat cancer. Chromatin modifications act in a coordinated and orderly fashion to regulate cellular processes such as transcription, DNA replication, and DNA repair. These processes may be regulated by TRRAP/HAT and there is an intimate and self-reinforcing cross-talk and interdependence between histone-modifying complexes and other histone-modifying activities such as acetylation, phosphorylation, and methylation.

Consistent with the critical function of histone modifications in key cellular processes, a large body of evidence has suggested that these complexes are intimately linked to human pathologies. Most notably, recent genetic and molecular studies have directly implicated histone modifications and histone-modifying complexes in human cancer. The fact that epigenetic alterations are, in contrast to genetic changes, reversible, has important implications for human cancer treatment as aberrant histone modifications are potential molecular targets for therapeutic intervention in human malignancies.

References

1. Bird A. DNA methylation patterns and epigenetic memory. *Genes Dev* 2002;**16**:6–21.
2. Marmorstein R. Protein modules that manipulate histone tails for chromatin regulation. *Nat Rev Mol Cell Biol* 2001;**2**:422–32.
3. Fyodorov DV, Kadonaga JT. The many faces of chromatin remodeling: Switching beyond transcription. *Cell* 2001;**106**:523–5.
4. Loizou JI, Murr R, Finkbeiner MG, Sawan C, Wang ZQ, Herceg Z. Epigenetic information in chromatin: the code of entry for DNA repair. *Cell Cycle* 2006;**5**:696–701.
5. Peterson CL, Cote J. Cellular machineries for chromosomal DNA repair. *Genes Dev* 2004;**18**:602–16.
6. Allfrey VG, Faulkner R, Mirsky AE. Acetylation and methylation of histones and their possible role in the regulation of RNA synthesis. *Proc Natl Acad Sci USA* 1964;**51**:786–94.
7. Allfrey VG, Mirsky AE. Structural modifications of histones and their possible role in the regulation of RNA synthesis. *Science* 1964;**144**:559.
8. Kouzarides T. Chromatin modifications and their function. *Cell* 2007;**128**:693–705.
9. Jenuwein T, Allis CD. Translating the histone code. *Science* 2001;**293**:1074–80.
10. Strahl BD, Allis CD. The language of covalent histone modifications. *Nature* 2000;**403**:41–5.
11. Ramachandran GN, Venkatachalam CM. Stereochemical criteria for polypeptides and proteins. IV. Standard dimensions for the cis-peptide unit and conformation of cis-polypeptides. *Biopolymers* 1968;**6**:1255–62.
12. Nelson CJ, Santos-Rosa H, Kouzarides T. Proline isomerization of histone H3 regulates lysine methylation and gene expression. *Cell* 2006;**126**:905–16.
13. Shiio Y, Eisenman RN. Histone sumoylation is associated with transcriptional repression. *Proc Natl Acad Sci USA* 2003;**100**:13225–30.
14. Nathan D, Ingvarsdottir K, Sterner DE, Bylebyl GR, Dokmanovic M, Dorsey JA, et al. Histone sumoylation is a negative regulator in Saccharomyces cerevisiae and shows dynamic interplay with positive-acting histone modifications. *Genes Dev* 2006;**20**:966–76.
15. Goldknopf IL, Taylor CW, Baum RM, Yeoman LC, Olson MO, Prestayko AW, et al. Isolation and characterization of protein A24, a "histone-like" non-histone chromosomal protein. *J Biol Chem* 1975;**250**:7182–7.
16. West MH, Bonner WM. Histone 2B can be modified by the attachment of ubiquitin. *Nucleic Acids Res* 1980;**8**:4671–80.
17. Geng F, Tansey WP. Polyubiquitylation of histone H2B. *Mol Biol Cell* 2008;**19**:3616–24.
18. Zhou W, Wang X, Rosenfeld MG. Histone H2A ubiquitination in transcriptional regulation and DNA damage repair. *Int J Biochem Cell Biol* 2009;**41**:12–15.

19. Robzyk K, Recht J, Osley MA. Rad6-dependent ubiquitination of histone H2B in yeast. *Science* 2000;**287**:501–4.

20. Kim J, Hake SB, Roeder RG. The human homolog of yeast BRE1 functions as a transcriptional coactivator through direct activator interactions. *Mol Cell* 2005;**20**:759–70.

21. Koken MH, Reynolds P, Jaspers-Dekker I, Prakash L, Prakash S, Bootsma D, et al. Structural and functional conservation of two human homologs of the yeast DNA repair gene RAD6. *Proc Natl Acad Sci USA* 1991;**88**:8865–9.

22. Roest HP, van Klaveren J, de Wit J, van Gurp CG, Koken MH, Vermey M, et al. Inactivation of the HR6B ubiquitin-conjugating DNA repair enzyme in mice causes male sterility associated with chromatin modification. *Cell* 1996;**86**:799–810.

23. Tanny JC, Erdjument-Bromage H, Tempst P, Allis CD. Ubiquitylation of histone H2B controls RNA polymerase II transcription elongation independently of histone H3 methylation. *Genes Dev* 2007;**21**:835–47.

24. Zhu B, Zheng Y, Pham AD, Mandal SS, Erdjument-Bromage H, Tempst P, et al. Monoubiquitination of human histone H2B: the factors involved and their roles in HOX gene regulation. *Mol Cell* 2005;**20**:601–11.

25. Buchwald G, van der Stoop P, Weichenrieder O, Perrakis A, van Lohuizen M, Sixma TK. Structure and E3-ligase activity of the Ring-Ring complex of polycomb proteins Bmi1 and Ring1b. *EMBO J* 2006;**25**:2465–74.

26. Cao R, Tsukada Y, Zhang Y. Role of Bmi-1 and Ring1A in H2A ubiquitylation and Hox gene silencing. *Mol Cell* 2005;**20**:845–54.

27. Li Z, Cao R, Wang M, Myers MP, Zhang Y, Xu RM. Structure of a Bmi-1-Ring1B polycomb group ubiquitin ligase complex. *J Biol Chem* 2006;**281**:20643–9.

28. Simon JA, Kingston RE. Mechanisms of polycomb gene silencing: knowns and unknowns. *Nat Rev Mol Cell Biol* 2009;**10**:697–708.

29. Wang H, Wang L, Erdjument-Bromage H, Vidal M, Tempst P, Jones RS, et al. Role of histone H2A ubiquitination in polycomb silencing. *Nature* 2004;**431**:873–8.

30. Francis NJ, Saurin AJ, Shao Z, Kingston RE. Reconstitution of a functional core polycomb repressive complex. *Mol Cell* 2001;**8**:545–56.

31. Mohd-Sarip A, Cleard F, Mishra RK, Karch F, Verrijzer CP. Synergistic recognition of an epigenetic DNA element by Pleiohomeotic and a polycomb core complex. *Genes Dev* 2005;**19**:1755–60.

32. Shao Z, Raible F, Mollaaghababa R, Guyon JR, Wu CT, Bender W, et al. Stabilization of chromatin structure by PRC1, a polycomb complex. *Cell* 1999;**98**:37–46.

33. Lagarou A, Mohd-Sarip A, Moshkin YM, Chalkley GE, Bezstarosti K, Demmers JA, et al. dKDM2 couples histone H2A ubiquitylation to histone H3 demethylation during polycomb group silencing. *Genes Dev* 2008;**22**:2799–810.

34. Hurley JH, Lee S, Prag G. Ubiquitin-binding domains. *Biochem J* 2006;**399**:361–72.

35. Zhang Y. Transcriptional regulation by histone ubiquitination and deubiquitination. *Genes Dev* 2003;**17**:2733–40.

36. Daniel JA, Torok MS, Sun ZW, Schieltz D, Allis CD, Yates JR 3rd, et al. Deubiquitination of histone H2B by a yeast acetyltransferase complex regulates transcription. *J Biol Chem* 2004;**279**:1867–71.

37. Henry KW, Wyce A, Lo WS, Duggan LJ, Emre NC, Kao CF, et al. Transcriptional activation via sequential histone H2B ubiquitylation and deubiquitylation, mediated by SAGA-associated Ubp8. *Genes Dev* 2003;**17**:2648–63.

38. Shukla A, Stanojevic N, Duan Z, Sen P, Bhaumik SR. Ubp8p, a histone deubiquitinase whose association with SAGA is mediated by Sgf11p, differentially regulates lysine 4 methylation of histone H3 in vivo. *Mol Cell Biol* 2006;**26**:3339–52.

39. Weake VM, Lee KK, Guelman S, Lin CH, Seidel C, Abmayr SM, et al. SAGA-mediated H2B deubiquitination controls the development of neuronal connectivity in the Drosophila visual system. *EMBO J* 2008;**27**:394–405.

40. Zhang XY, Pfeiffer HK, Thorne AW, McMahon SB. USP22, an hSAGA subunit and potential cancer stem cell marker, reverses the polycomb-catalyzed ubiquitylation of histone H2A. *Cell Cycle* 2008;**7**:1522–4.

41. Zhang XY, Varthi M, Sykes SM, Phillips C, Warzecha C, Zhu W, et al. The putative cancer stem cell marker USP22 is a subunit of the human SAGA complex required for activated transcription and cell-cycle progression. *Mol Cell* 2008;**29**:102–11.

42. Zhao Y, Lang G, Ito S, Bonnet J, Metzger E, Sawatsubashi S, et al. A TFTC/STAGA module mediates histone H2A and H2B deubiquitination, coactivates nuclear receptors, and counteracts heterochromatin silencing. *Mol Cell* 2008;**29**:92–101.

43. Emre NC, Ingvarsdottir K, Wyce A, Wood A, Krogan NJ, Henry KW, et al. Maintenance of low histone ubiquitylation by Ubp10 correlates with telomere-proximal Sir2 association and gene silencing. *Mol Cell* 2005;**17**:585–94.

44. Gardner RG, Nelson ZW, Gottschling DE. Ubp10/Dot4p regulates the persistence of ubiquitinated histone H2B: distinct roles in telomeric silencing and general chromatin. *Mol Cell Biol* 2005;**25**:6123–39.

45. Osley MA. Regulation of histone H2A and H2B ubiquitylation. *Brief Funct Genomic Proteomic* 2006;**5**:179–89.

46. Barsoum J, Varshavsky A. Preferential localization of variant nucleosomes near the 5'-end of the mouse dihydrofolate reductase gene. *J Biol Chem* 1985;**260**:7688–97.

47. Levinger L, Varshavsky A. Selective arrangement of ubiquitinated and D1 protein-containing nucleosomes within the Drosophila genome. *Cell* 1982;**28**:375–85.

48. Nickel BE, Allis CD, Davie JR. Ubiquitinated histone H2B is preferentially located in transcriptionally active chromatin. *Biochemistry* 1989;**28**:958–63.

49. Dawson BA, Herman T, Haas AL, Lough J. Affinity isolation of active murine erythroleukemia cell chromatin: uniform distribution of ubiquitinated histone H2A between active and inactive fractions. *J Cell Biochem* 1991;**46**:166–73.

50. Huang SY, Barnard MB, Xu M, Matsui S, Rose SM, Garrard WT. The active immunoglobulin kappa chain gene is packaged by non-ubiquitin-conjugated nucleosomes. *Proc Natl Acad Sci USA* 1986;**83**:3738–42.

51. Parlow MH, Haas AL, Lough J. Enrichment of ubiquitinated histone H2A in a low salt extract of micrococcal nuclease-digested myotube nuclei. *J Biol Chem* 1990;**265**:7507–12.

52. Laribee RN, Fuchs SM, Strahl BD. H2B ubiquitylation in transcriptional control: a FACT-finding mission. *Genes Dev* 2007;**21**:737–43.

53. Minsky N, Shema E, Field Y, Schuster M, Segal E, Oren M. Monoubiquitinated H2B is associated with the transcribed region of highly expressed genes in human cells. *Nat Cell Biol* 2008;**10**:483–8.

54. McGinty RK, Kim J, Chatterjee C, Roeder RG, Muir TW. Chemically ubiquitylated histone H2B stimulates hDot1L-mediated intranucleosomal methylation. *Nature* 2008;**453**:812–16.

55. Briggs SD, Xiao T, Sun ZW, Caldwell JA, Shabanowitz J, Hunt DF, et al. Gene silencing: trans-histone regulatory pathway in chromatin. *Nature* 2002;**418**:498.

56. Dover J, Schneider J, Tawiah-Boateng MA, Wood A, Dean K, Johnston M, et al. Methylation of histone H3 by COMPASS requires ubiquitination of histone H2B by Rad6. *J Biol Chem* 2002;**277**:28368–71.

57. Ng HH, Xu RM, Zhang Y, Struhl K. Ubiquitination of histone H2B by Rad6 is required for efficient Dot1-mediated methylation of histone H3 lysine 79. *J Biol Chem* 2002;**277**:34655–7.

58. Sun ZW, Allis CD. Ubiquitination of histone H2B regulates H3 methylation and gene silencing in yeast. *Nature* 2002;**418**:104–8.

59. van Leeuwen F, Gafken PR, Gottschling DE. Dot1p modulates silencing in yeast by methylation of the nucleosome core. *Cell* 2002;**109**:745–56.

60. Krogan NJ, Kim M, Tong A, Golshani A, Cagney G, Canadien V, et al. Methylation of histone H3 by Set2 in *Saccharomyces cerevisiae* is linked to transcriptional elongation by RNA polymerase II. *Mol Cell Biol* 2003;**23**:4207–18.

61. Ng HH, Robert F, Young RA, Struhl K. Targeted recruitment of Set1 histone methylase by elongating Pol II provides a localized mark and memory of recent transcriptional activity. *Mol Cell* 2003;**11**:709–19.

62. Hassa PO, Haenni SS, Elser M, Hottiger MO. Nuclear ADP-ribosylation reactions in mammalian cells: where are we today and where are we going? *Microbiol Mol Biol Rev* 2006;**70**:789–829.

63. Schreiber V, Dantzer F, Ame JC, de Murcia G. Poly(ADP-ribose): novel functions for an old molecule. *Nat Rev Mol Cell Biol* 2006;**7**:517–28.

64. Kreimeyer A, Adamietz P, Hilz H. Alkylation-induced mono(ADP-ribosyl)-histones H1 and H2B. Hydroxylamine-resistant linkage in hepatoma cells. *Biol Chem Hoppe Seyler* 1985;**366**:537–44.

65. Kreimeyer A, Wielckens K, Adamietz P, Hilz H. DNA repair-associated ADP-ribosylation in vivo. Modification of histone H1 differs from that of the principal acceptor proteins. *J Biol Chem* 1984;**259**:890–6.

66. Timinszky G, Till S, Hassa PO, Hothorn M, Kustatscher G, Nijmeijer B, et al. A macrodomain-containing histone rearranges chromatin upon sensing PARP1 activation. *Nat Struct Mol Biol* 2009;**16**:923–9.

67. Ushiroyama T, Tanigawa Y, Tsuchiya M, Matsuura R, Ueki M, Sugimoto O, et al. Amino acid sequence of histone H1 at the ADP-ribose-accepting site and ADP-ribose X histone-H1 adduct as an inhibitor of cyclic-AMP-dependent phosphorylation. *Eur J Biochem* 1985;**151**:173–7.

68. Gurley LR, D'Anna JA, Barham SS, Deaven LL, Tobey RA. Histone phosphorylation and chromatin structure during mitosis in Chinese hamster cells. *Eur J Biochem* 1978;**84**:1–15.

69. Wei Y, Mizzen CA, Cook RG, Gorovsky MA, Allis CD. Phosphorylation of histone H3 at serine 10 is correlated with chromosome condensation during mitosis and meiosis in Tetrahymena. *Proc Natl Acad Sci USA* 1998;**95**:7480–4.

70. Fischle W, Tseng BS, Dormann HL, Ueberheide BM, Garcia BA, Shabanowitz J, et al. Regulation of HP1-chromatin binding by histone H3 methylation and phosphorylation. *Nature* 2005;**438**:1116–22.

71. Nowak SJ, Corces VG. Phosphorylation of histone H3 correlates with transcriptionally active loci. *Genes Dev* 2000;**14**:3003–13.

72. Nowak SJ, Corces VG. Phosphorylation of histone H3: a balancing act between chromosome condensation and transcriptional activation. *Trends Genet* 2004;**20**:214–20.

73. Macdonald N, Welburn JP, Noble ME, Nguyen A, Yaffe MB, Clynes D, et al. Molecular basis for the recognition of phosphorylated and phosphoacetylated histone h3 by 14-3-3. *Mol Cell* 2005;**20**:199–211.

74. Pokholok DK, Zeitlinger J, Hannett NM, Reynolds DB, Young RA. Activated signal transduction kinases frequently occupy target genes. *Science* 2006;**313**:533–6.

75. Cook PJ, Ju BG, Telese F, Wang X, Glass CK, Rosenfeld MG. Tyrosine dephosphorylation of H2AX modulates apoptosis and survival decisions. *Nature* 2009;**458**:591–6.

76. Xiao A, Li H, Shechter D, Ahn SH, Fabrizio LA, Erdjument-Bromage H, et al. WSTF regulates the H2A.X DNA damage response via a novel tyrosine kinase activity. *Nature* 2009;**457**:57–62.

77. Redon C, Pilch D, Rogakou E, Sedelnikova O, Newrock K, Bonner W. Histone H2A variants H2AX and H2AZ. *Curr Opin Genet Dev* 2002;**12**:162–9.

78. Shroff R, Arbel-Eden A, Pilch D, Ira G, Bonner WM, Petrini JH, et al. Distribution and dynamics of chromatin modification induced by a defined DNA double-strand break. *Curr Biol* 2004;**14**:1703–11.

79. Celeste A, Fernandez-Capetillo O, Kruhlak MJ, Pilch DR, Staudt DW, Lee A, et al. Histone H2AX phosphorylation is dispensable for the initial recognition of DNA breaks. *Nat Cell Biol* 2003;**5**:675–9.

80. Downs JA, Allard S, Jobin-Robitaille O, Javaheri A, Auger A, Bouchard N, et al. Binding of chromatin-modifying activities to phosphorylated histone H2A at DNA damage sites. *Mol Cell* 2004;**16**:979–90.

81. Stucki M, Clapperton JA, Mohammad D, Yaffe MB, Smerdon SJ, Jackson SP. MDC1 directly binds phosphorylated histone H2AX to regulate cellular responses to DNA double-strand breaks. *Cell* 2005;**123**:1213–26.

82. Strom L, Lindroos HB, Shirahige K, Sjogren C. Postreplicative recruitment of cohesin to double-strand breaks is required for DNA repair. *Mol Cell* 2004;**16**:1003–15.

83. Unal E, Arbel-Eden A, Sattler U, Shroff R, Lichten M, Haber JE, et al. DNA damage response pathway uses histone modification to assemble a double-strand break-specific cohesin domain. *Mol Cell* 2004;**16**:991–1002.

84. Morrison AJ, Highland J, Krogan NJ, Arbel-Eden A, Greenblatt JF, Haber JE, et al. INO80 and gamma-H2AX interaction links ATP-dependent chromatin remodeling to DNA damage repair. *Cell* 2004;**119**:767–75.

85. van Attikum H, Fritsch O, Hohn B, Gasser SM. Recruitment of the INO80 complex by H2A phosphorylation links ATP-dependent chromatin remodeling with DNA double-strand break repair. *Cell* 2004;**119**:777–88.

86. Rogakou EP, Boon C, Redon C, Bonner WM. Megabase chromatin domains involved in DNA double-strand breaks in vivo. *J Cell Biol* 1999;**146**:905–16.

87. Kobayashi J, Tauchi H, Sakamoto S, Nakamura A, Morishima K, Matsuura S, et al. NBS1 localizes to gamma-H2AX foci through interaction with the FHA/BRCT domain. *Curr Biol* 2002;**12**:1846–51.

88. Cheung WL, Turner FB, Krishnamoorthy T, Wolner B, Ahn SH, Foley M, et al. Phosphorylation of histone H4 serine 1 during DNA damage requires casein kinase II in *S. cerevisiae. Curr Biol* 2005;**15**:656–60.

89. Utley RT, Lacoste N, Jobin-Robitaille O, Allard S, Cote J. Regulation of NuA4 histone acetyltransferase activity in transcription and DNA repair by phosphorylation of histone H4. *Mol Cell Biol* 2005;**25**:8179–90.

90. Downs JA, Cote J. Dynamics of chromatin during the repair of DNA double-strand breaks. *Cell Cycle* 2005;**4**:1373–6.

91. Harvey AC, Downs JA. What functions do linker histones provide? *Mol Microbiol* 2004;**53**:771–5.

92. Kysela B, Chovanec M, Jeggo PA. Phosphorylation of linker histones by DNA-dependent protein kinase is required for DNA ligase IV-dependent ligation in the presence of histone H1. *Proc Natl Acad Sci USA* 2005;**102**:1877–82.

93. Clarke S. Protein methylation. *Curr Opin Cell Biol* 1993;**5**:977–83.

94. Metzger E, Wissmann M, Yin N, Muller JM, Schneider R, Peters AH, et al. LSD1 demethylates repressive histone marks to promote androgen-receptor-dependent transcription. *Nature* 2005;**437**:436–9.

95. Shi Y, Lan F, Matson C, Mulligan P, Whetstine JR, Cole PA, et al. Histone demethylation mediated by the nuclear amine oxidase homolog LSD1. *Cell* 2004;**119**:941–53.

96. Cloos PA, Christensen J, Agger K, Maiolica A, Rappsilber J, Antal T, et al. The putative oncogene GASC1 demethylates tri- and dimethylated lysine 9 on histone H3. *Nature* 2006;**442**:307–11.

97. Tsukada Y, Fang J, Erdjument-Bromage H, Warren ME, Borchers CH, Tempst P, et al. Histone demethylation by a family of JmjC domain-containing proteins. *Nature* 2006;**439**:811–16.

98. Whetstine JR, Nottke A, Lan F, Huarte M, Smolikov S, Chen Z, et al. Reversal of histone lysine trimethylation by the JMJD2 family of histone demethylases. *Cell* 2006;**125**:467–81.

99. Yamane K, Toumazou C, Tsukada Y, Erdjument-Bromage H, Tempst P, Wong J, et al. JHDM2A, a JmjC-containing H3K9 demethylase, facilitates transcription activation by androgen receptor. *Cell* 2006;**125**:483–95.

100. Fodor BD, Kubicek S, Yonezawa M, O'Sullivan RJ, Sengupta R, Perez-Burgos L, et al. Jmjd2b antagonizes H3K9 trimethylation at pericentric heterochromatin in mammalian cells. *Genes Dev* 2006;**20**:1557–62.

101. Shin S, Janknecht R. Diversity within the JMJD2 histone demethylase family. *Biochem Biophys Res Commun* 2007;**353**:973–7.

102. Chen Z, Zang J, Whetstine J, Hong X, Davrazou F, Kutateladze TG, et al. Structural insights into histone demethylation by JMJD2 family members. *Cell* 2006;**125**:691–702.

103. Klose RJ, Yamane K, Bae Y, Zhang D, Erdjument-Bromage H, Tempst P, et al. The transcriptional repressor JHDM3A demethylates trimethyl histone H3 lysine 9 and lysine 36. *Nature* 2006;**442**:312–16.

104. Lee MG, Villa R, Trojer P, Norman J, Yan KP, Reinberg D, et al. Demethylation of H3K27 regulates polycomb recruitment and H2A ubiquitination. *Science* 2007;**318**:447–50.

105. Christensen J, Agger K, Cloos PA, Pasini D, Rose S, Sennels L, et al. RBP2 belongs to a family of demethylases, specific for tri- and dimethylated lysine 4 on histone 3. *Cell* 2007;**128**:1063–76.

106. Iwase S, Lan F, Bayliss P, de la Torre-Ubieta L, Huarte M, Qi HH, et al. The X-linked mental retardation gene SMCX/JARID1C defines a family of histone H3 lysine 4 demethylases. *Cell* 2007;**128**:1077–88.

107. Cuthbert GL, Daujat S, Snowden AW, Erdjument-Bromage H, Hagiwara T, Yamada M, et al. Histone deimination antagonizes arginine methylation. *Cell* 2004;**118**:545–53.

108. Wang Y, Wysocka J, Sayegh J, Lee YH, Perlin JR, Leonelli L, et al. Human PAD4 regulates histone arginine methylation levels via demethylimination. *Science* 2004;**306**:279–83.

109. Denis H, Deplus R, Putmans P, Yamada M, Metivier R, Fuks F. Functional connection between deimination and deacetylation of histones. *Mol Cell Biol* 2009;**29**:4982–93.

110. Chang B, Chen Y, Zhao Y, Bruick RK. JMJD6 is a histone arginine demethylase. *Science* 2007;**318**:444–7.

111. Bernstein BE, Humphrey EL, Erlich RL, Schneider R, Bouman P, Liu JS, et al. Methylation of histone H3 Lys 4 in coding regions of active genes. *Proc Natl Acad Sci USA* 2002;**99**:8695–700.

41

112. Briggs SD, Bryk M, Strahl BD, Cheung WL, Davie JK, Dent SY, et al. Histone H3 lysine 4 methylation is mediated by Set1 and required for cell growth and rDNA silencing in *Saccharomyces cerevisiae. Genes Dev* 2001;**15**:3286–95.

113. Santos-Rosa H, Schneider R, Bannister AJ, Sherriff J, Bernstein BE, Emre NC, et al. Active genes are tri-methylated at K4 of histone H3. *Nature* 2002;**419**:407–11.

114. Kizer KO, Phatnani HP, Shibata Y, Hall H, Greenleaf AL, Strahl BD. A novel domain in Set2 mediates RNA polymerase II interaction and couples histone H3 K36 methylation with transcript elongation. *Mol Cell Biol* 2005;**25**:3305–16.

115. Li B, Howe L, Anderson S, Yates JR 3rd, Workman JL. The Set2 histone methyltransferase functions through the phosphorylated carboxyl-terminal domain of RNA polymerase II. *J Biol Chem* 2003;**278**:8897–903.

116. Xiao B, Wilson JR, Gamblin SJ. SET domains and histone methylation. *Curr Opin Struct Biol* 2003;**13**:699–705.

117. Bannister AJ, Schneider R, Myers FA, Thorne AW, Crane-Robinson C, Kouzarides T. Spatial distribution of di- and tri-methyl lysine 36 of histone H3 at active genes. *J Biol Chem* 2005;**280**:17732–6.

118. Barski A, Cuddapah S, Cui K, Roh TY, Schones DE, Wang Z, et al. High-resolution profiling of histone methylations in the human genome. *Cell* 2007;**129**:823–37.

119. Bell O, Wirbelauer C, Hild M, Scharf AN, Schwaiger M, MacAlpine DM, et al. Localized H3K36 methylation states define histone H4K16 acetylation during transcriptional elongation in Drosophila. *EMBO J* 2007;**26**:4974–84.

120. Lee JS, Shilatifard A. A site to remember: H3K36 methylation a mark for histone deacetylation. *Mutat Res* 2007;**618**:130–4.

121. Mikkelsen TS, Ku M, Jaffe DB, Issac B, Lieberman E, Giannoukos G, et al. Genome-wide maps of chromatin state in pluripotent and lineage-committed cells. *Nature* 2007;**448**:553–60.

122. Pokholok DK, Harbison CT, Levine S, Cole M, Hannett NM, Lee TI, et al. Genome-wide map of nucleosome acetylation and methylation in yeast. *Cell* 2005;**122**:517–27.

123. Rao B, Shibata Y, Strahl BD, Lieb JD. Dimethylation of histone H3 at lysine 36 demarcates regulatory and nonregulatory chromatin genome-wide. *Mol Cell Biol* 2005;**25**:9447–59.

124. Vermeulen M, Mulder KW, Denissov S, Pijnappel WW, van Schaik FM, Varier RA, et al. Selective anchoring of TFIID to nucleosomes by trimethylation of histone H3 lysine 4. *Cell* 2007;**131**:58–69.

125. Carrozza MJ, Li B, Florens L, Suganuma T, Swanson SK, Lee KK, et al. Histone H3 methylation by Set2 directs deacetylation of coding regions by Rpd3S to suppress spurious intragenic transcription. *Cell* 2005;**123**:581–92.

126. Joshi AA, Struhl K. Eaf3 chromodomain interaction with methylated H3-K36 links histone deacetylation to Pol II elongation. *Mol Cell* 2005;**20**:971–8.

127. Keogh MC, Kurdistani SK, Morris SA, Ahn SH, Podolny V, Collins SR, et al. Cotranscriptional set2 methylation of histone H3 lysine 36 recruits a repressive Rpd3 complex. *Cell* 2005;**123**:593–605.

128. Li B, Gogol M, Carey M, Lee D, Seidel C, Workman JL. Combined action of PHD and chromo domains directs the Rpd3S HDAC to transcribed chromatin. *Science* 2007;**316**:1050–4.

129. Steger DJ, Lefterova MI, Ying L, Stonestrom AJ, Schupp M, Zhuo D, et al. DOT1L/KMT4 recruitment and H3K79 methylation are ubiquitously coupled with gene transcription in mammalian cells. *Mol Cell Biol* 2008;**28**:2825–39.

130. Schubeler D, MacAlpine DM, Scalzo D, Wirbelauer C, Kooperberg C, van Leeuwen F, et al. The histone modification pattern of active genes revealed through genome-wide chromatin analysis of a higher eukaryote. *Genes Dev* 2004;**18**:1263–71.

131. Richards EJ, Elgin SC. Epigenetic codes for heterochromatin formation and silencing: rounding up the usual suspects. *Cell* 2002;**108**:489–500.

132. Tschiersch B, Hofmann A, Krauss V, Dorn R, Korge G, Reuter G. The protein encoded by the Drosophila position-effect variegation suppressor gene Su(var)3-9 combines domains of antagonistic regulators of homeotic gene complexes. *EMBO J* 1994;**13**:3822–31.

133. Allshire RC, Nimmo ER, Ekwall K, Javerzat JP, Cranston G. Mutations derepressing silent centromeric domains in fission yeast disrupt chromosome segregation. *Genes Dev* 1995;**9**:218–33.

134. Cam HP, Sugiyama T, Chen ES, Chen X, FitzGerald PC, Grewal SI. Comprehensive analysis of heterochromatin- and RNAi-mediated epigenetic control of the fission yeast genome. *Nat Genet* 2005;**37**:809–19.

135. Grewal SI, Elgin SC. Transcription and RNA interference in the formation of heterochromatin. *Nature* 2007;**447**:399–406.

136. Grewal SI, Jia S. Heterochromatin revisited. *Nat Rev Genet* 2007;**8**:35–46.

137. Hall IM, Shankaranarayana GD, Noma K, Ayoub N, Cohen A, Grewal SI. Establishment and maintenance of a heterochromatin domain. *Science* 2002;**297**:2232–7.

138. Kato H, Goto DB, Martienssen RA, Urano T, Furukawa K, Murakami Y. RNA polymerase II is required for RNAi-dependent heterochromatin assembly. *Science* 2005;**309**:467–9.

139. Matzke MA, Birchler JA. RNAi-mediated pathways in the nucleus. *Nat Rev Genet* 2005;**6**:24–35.

140. Mochizuki K, Fine NA, Fujisawa T, Gorovsky MA. Analysis of a piwi-related gene implicates small RNAs in genome rearrangement in tetrahymena. *Cell* 2002;**110**:689–99.

141. Pal-Bhadra M, Leibovitch BA, Gandhi SG, Rao M, Bhadra U, Birchler JA, et al. Heterochromatic silencing and HP1 localization in Drosophila are dependent on the RNAi machinery. *Science* 2004;**303**:669–72.

142. Reinhart BJ, Bartel DP. Small RNAs correspond to centromere heterochromatic repeats. *Science* 2002;**297**:1831.

143. Taverna SD, Coyne RS, Allis CD. Methylation of histone h3 at lysine 9 targets programmed DNA elimination in tetrahymena. *Cell* 2002;**110**:701–11.

144. Volpe TA, Kidner C, Hall IM, Teng G, Grewal SI, Martienssen RA. Regulation of heterochromatic silencing and histone H3 lysine-9 methylation by RNAi. *Science* 2002;**297**:1833–7.

145. Hansen KR, Ibarra PT, Thon G. Evolutionary-conserved telomere-linked helicase genes of fission yeast are repressed by silencing factors, RNAi components and the telomere-binding protein Taz1. *Nucleic Acids Res* 2006;**34**:78–88.

146. Jia S, Noma K, Grewal SI. RNAi-independent heterochromatin nucleation by the stress-activated ATF/CREB family proteins. *Science* 2004;**304**:1971–6.

147. Kanoh J, Sadaie M, Urano T, Ishikawa F. Telomere binding protein Taz1 establishes Swi6 heterochromatin independently of RNAi at telomeres. *Curr Biol* 2005;**15**:1808–19.

148. Kim HS, Choi ES, Shin JA, Jang YK, Park SD. Regulation of Swi6/HP1-dependent heterochromatin assembly by cooperation of components of the mitogen-activated protein kinase pathway and a histone deacetylase Clr6. *J Biol Chem* 2004;**279**:42850–9.

149. Yamada T, Fischle W, Sugiyama T, Allis CD, Grewal SI. The nucleation and maintenance of heterochromatin by a histone deacetylase in fission yeast. *Mol Cell* 2005;**20**:173–85.

150. Vakoc CR, Mandat SA, Olenchock BA, Blobel GA. Histone H3 lysine 9 methylation and HP1gamma are associated with transcription elongation through mammalian chromatin. *Mol Cell* 2005;**19**:381–91.

151. Pan G, Tian S, Nie J, Yang C, Ruotti V, Wei H, et al. Whole-genome analysis of histone H3 lysine 4 and lysine 27 methylation in human embryonic stem cells. *Cell Stem Cell* 2007;**1**:299–312.

152. Zhao XD, Han X, Chew JL, Liu J, Chiu KP, Choo A, et al. Whole-genome mapping of histone H3 Lys4 and 27 trimethylations reveals distinct genomic compartments in human embryonic stem cells. *Cell Stem Cell* 2007;**1**:286–98.

153. Lee DY, Teyssier C, Strahl BD, Stallcup MR. Role of protein methylation in regulation of transcription. *Endocr Rev* 2005;**26**:147–70.

154. Metivier R, Penot G, Hubner MR, Reid G, Brand H, Kos M, et al. Estrogen receptor-alpha directs ordered, cyclical, and combinatorial recruitment of cofactors on a natural target promoter. *Cell* 2003;**115**:751–63.

155. Kangaspeska S, Stride B, Metivier R, Polycarpou-Schwarz M, Ibberson D, Carmouche RP, et al. Transient cyclical methylation of promoter DNA. *Nature* 2008;**452**:112–15.

156. Metivier R, Gallais R, Tiffoche C, Le Peron C, Jurkowska RZ, Carmouche RP, et al. Cyclical DNA methylation of a transcriptionally active promoter. *Nature* 2008;**452**:45–50.

157. Wysocka J, Allis CD, Coonrod S. Histone arginine methylation and its dynamic regulation. *Front Biosci* 2006;**11**:344–55.

158. Chen D, Huang SM, Stallcup MR. Synergistic, p160 coactivator-dependent enhancement of estrogen receptor function by CARM1 and p300. *J Biol Chem* 2000;**275**:40810–16.

159. Koh SS, Chen D, Lee YH, Stallcup MR. Synergistic enhancement of nuclear receptor function by p160 coactivators and two coactivators with protein methyltransferase activities. *J Biol Chem* 2001;**276**:1089–98.

160. Koh SS, Li H, Lee YH, Widelitz RB, Chuong CM, Stallcup MR. Synergistic coactivator function by coactivator-associated arginine methyltransferase (CARM) 1 and beta-catenin with two different classes of DNA-binding transcriptional activators. *J Biol Chem* 2002;**277**:26031–5.

161. Lee YH, Koh SS, Zhang X, Cheng X, Stallcup MR. Synergy among nuclear receptor coactivators: selective requirement for protein methyltransferase and acetyltransferase activities. *Mol Cell Biol* 2002;**22**:3621–32.

162. Qi C, Chang J, Zhu Y, Yeldandi AV, Rao SM, Zhu YJ. Identification of protein arginine methyltransferase 2 as a coactivator for estrogen receptor alpha. *J Biol Chem* 2002;**277**:28624–30.

163. Rezai-Zadeh N, Zhang X, Namour F, Fejer G, Wen YD, Yao YL, et al. Targeted recruitment of a histone H4-specific methyltransferase by the transcription factor YY1. *Genes Dev* 2003;**17**:1019–29.

164. Xu W, Cho H, Kadam S, Banayo EM, Anderson S, Yates JR 3rd, et al. A methylation-mediator complex in hormone signaling. *Genes Dev* 2004;**18**:144–56.

165. Cosgrove MS, Boeke JD, Wolberger C. Regulated nucleosome mobility and the histone code. *Nat Struct Mol Biol* 2004;**11**:1037–43.

166. Pal S, Vishwanath SN, Erdjument-Bromage H, Tempst P, Sif S. Human SWI/SNF-associated PRMT5 methylates histone H3 arginine 8 and negatively regulates expression of ST7 and NM23 tumor suppressor genes. *Mol Cell Biol* 2004;**24**:9630–45.

167. Sanders SL, Portoso M, Mata J, Bahler J, Allshire RC, Kouzarides T. Methylation of histone H4 lysine 20 controls recruitment of Crb2 to sites of DNA damage. *Cell* 2004;**119**:603–14.

168. Botuyan MV, Lee J, Ward IM, Kim JE, Thompson JR, Chen J, et al. Structural basis for the methylation state-specific recognition of histone H4-K20 by 53BP1 and Crb2 in DNA repair. *Cell* 2006;**127**:1361–73.

169. Huyen Y, Zgheib O, Ditullio RA Jr., Gorgoulis VG, Zacharatos P, Petty TJ, et al. Methylated lysine 79 of histone H3 targets 53BP1 to DNA double-strand breaks. *Nature* 2004;**432**:406–11.

170. Allis CD, Chicoine LG, Richman R, Schulman IG. Deposition-related histone acetylation in micronuclei of conjugating Tetrahymena. *Proc Natl Acad Sci USA* 1985;**82**:8048–52.

171. Ruiz-Carrillo A, Wangh LJ, Allfrey VG. Processing of newly synthesized histone molecules. *Science* 1975;**190**:117–28.

172. Vetting MW, LP SdC, Yu M, Hegde SS, Magnet S, Roderick SL, et al. Structure and functions of the GNAT superfamily of acetyltransferases. *Arch Biochem Biophys* 2005;**433**:212–26.

173. Borrow J, Stanton Jr VP, Andresen JM, Becher R, Behm FG, Chaganti RS, et al. The translocation t(8;16)(p11;p13) of acute myeloid leukaemia fuses a putative acetyltransferase to the CREB-binding protein. *Nat Genet* 1996;**14**:33–41.

43

174. Neuwald AF, Landsman D. GCN5-related histone N-acetyltransferases belong to a diverse superfamily that includes the yeast SPT10 protein. *Trends Biochem Sci* 1997;**22**:154–5.

175. Akhtar A, Becker PB. The histone H4 acetyltransferase MOF uses a C2HC zinc finger for substrate recognition. *EMBO Rep* 2001;**2**:113–18.

176. Yan Y, Barlev NA, Haley RH, Berger SL, Marmorstein R. Crystal structure of yeast Esa1 suggests a unified mechanism for catalysis and substrate binding by histone acetyltransferases. *Mol Cell* 2000;**6**:1195–205.

177. Doyon Y, Cote J. The highly conserved and multifunctional NuA4 HAT complex. *Curr Opin Genet Dev* 2004;**14**:147–54.

178. Doyon Y, Selleck W, Lane WS, Tan S, Cote J. Structural and functional conservation of the NuA4 histone acetyltransferase complex from yeast to humans. *Mol Cell Biol* 2004;**24**:1884–96.

179. Kalkhoven E. CBP and p300: HATs for different occasions. *Biochem Pharmacol* 2004;**68**:1145–55.

180. Yuan ZM, Huang Y, Ishiko T, Nakada S, Utsugisawa T, Shioya H, et al. Function for p300 and not CBP in the apoptotic response to DNA damage. *Oncogene* 1999;**18**:5714–17.

181. Kawasaki H, Eckner R, Yao TP, Taira K, Chiu R, Livingston DM, et al. Distinct roles of the co-activators p300 and CBP in retinoic-acid-induced F9-cell differentiation. *Nature* 1998;**393**:284–9.

182. Kung AL, Rebel VI, Bronson RT, Ch'ng LE, Sieff CA, Livingston DM, et al. Gene dose-dependent control of hematopoiesis and hematologic tumor suppression by CBP. *Genes Dev* 2000;**14**:272–7.

183. Yao TP, Oh SP, Fuchs M, Zhou ND, Ch'ng LE, Newsome D, et al. Gene dosage-dependent embryonic development and proliferation defects in mice lacking the transcriptional integrator p300. *Cell* 1998;**93**:361–72.

184. McManus KJ, Hendzel MJ. Quantitative analysis of CBP- and P300-induced histone acetylations in vivo using native chromatin. *Mol Cell Biol* 2003;**23**:7611–27.

185. Das C, Lucia MS, Hansen KC, Tyler JK. CBP/p300-mediated acetylation of histone H3 on lysine 56. *Nature* 2009;**459**:113–17.

186. Adkins MW, Carson JJ, English CM, Ramey CJ, Tyler JK. The histone chaperone anti-silencing function 1 stimulates the acetylation of newly synthesized histone H3 in S-phase. *J Biol Chem* 2007;**282**:1334–40.

187. Sklenar AR, Parthun MR. Characterization of yeast histone H3-specific type B histone acetyltransferases identifies an ADA2-independent Gcn5p activity. *BMC Biochem* 2004;**5**:11.

188. Tsubota T, Berndsen CE, Erkmann JA, Smith CL, Yang L, Freitas MA, et al. Histone H3-K56 acetylation is catalyzed by histone chaperone-dependent complexes. *Mol Cell* 2007;**25**:703–12.

189. Haynes SR, Dollard C, Winston F, Beck S, Trowsdale J, Dawid IB. The bromodomain: a conserved sequence found in human, Drosophila and yeast proteins. *Nucleic Acids Res* 1992;**20**:2603.

190. Jeanmougin F, Wurtz JM, Le Douarin B, Chambon P, Losson R. The bromodomain revisited. *Trends Biochem Sci* 1997;**22**:151–3.

191. Schiltz RL, Mizzen CA, Vassilev A, Cook RG, Allis CD, Nakatani Y. Overlapping but distinct patterns of histone acetylation by the human coactivators p300 and PCAF within nucleosomal substrates. *J Biol Chem* 1999;**274**:1189–92.

192. Taylor BL, Zhulin IB. PAS domains: internal sensors of oxygen, redox potential, and light. *Microbiol Mol Biol Rev* 1999;**63**:479–506.

193. Kusch T, Florens L, Macdonald WH, Swanson SK, Glaser RL, Yates JR 3rd, et al. Acetylation by Tip60 is required for selective histone variant exchange at DNA lesions. *Science* 2004;**306**:2084–7.

194. Muratoglu S, Georgieva S, Papai G, Scheer E, Enunlu I, Komonyi O, et al. Two different Drosophila ADA2 homologues are present in distinct GCN5 histone acetyltransferase-containing complexes. *Mol Cell Biol* 2003;**23**:306–21.

195. Grant PA, Duggan L, Cote J, Roberts SM, Brownell JE, Candau R, et al. Yeast Gcn5 functions in two multisubunit complexes to acetylate nucleosomal histones: characterization of an Ada complex and the SAGA (Spt/Ada) complex. *Genes Dev* 1997;**11**:1640–50.

196. Taverna SD, Ilin S, Rogers RS, Tanny JC, Lavender H, Li H, et al. Yng1 PHD finger binding to H3 trimethylated at K4 promotes NuA3 HAT activity at K14 of H3 and transcription at a subset of targeted ORFs. *Mol Cell* 2006;**24**:785–96.

197. Utley RT, Ikeda K, Grant PA, Cote J, Steger DJ, Eberharter A, et al. Transcriptional activators direct histone acetyltransferase complexes to nucleosomes. *Nature* 1998;**394**:498–502.

198. Wallberg AE, Neely KE, Gustafsson JA, Workman JL, Wright AP, Grant PA. Histone acetyltransferase complexes can mediate transcriptional activation by the major glucocorticoid receptor activation domain. *Mol Cell Biol* 1999;**19**:5952–9.

199. Ikura T, Ogryzko VV, Grigoriev M, Groisman R, Wang J, Horikoshi M, et al. Involvement of the TIP60 histone acetylase complex in DNA repair and apoptosis. *Cell* 2000;**102**:463–73.

200. Sapountzi V, Logan IR, Robson CN. Cellular functions of TIP60. *Int J Biochem Cell Biol* 2006;**38**:1496–509.

201. Fuchs M, Gerber J, Drapkin R, Sif S, Ikura T, Ogryzko V, et al. The p400 complex is an essential E1A transformation target. *Cell* 2001;**106**:297–307.

202. Park J, Wood MA, Cole MD. BAF53 forms distinct nuclear complexes and functions as a critical c-Myc-interacting nuclear cofactor for oncogenic transformation. *Mol Cell Biol* 2002;**22**:1307–16.

203. Boudreault AA, Cronier D, Selleck W, Lacoste N, Utley RT, Allard S, et al. Yeast enhancer of polycomb defines global Esa1-dependent acetylation of chromatin. *Genes Dev* 2003;**17**:1415–28.

204. Vaquero A, Loyola A, Reinberg D. The constantly changing face of chromatin. *Sci Aging Knowledge Environ* 2003;**2003**:RE4.

205. Workman JL, Kingston RE. Alteration of nucleosome structure as a mechanism of transcriptional regulation. *Annu Rev Biochem* 1998;**67**:545–79.

206. Shogren-Knaak M, Ishii H, Sun JM, Pazin MJ, Davie JR, Peterson CL. Histone H4-K16 acetylation controls chromatin structure and protein interactions. *Science* 2006;**311**:844–7.

207. Shogren-Knaak M, Peterson CL. Switching on chromatin: mechanistic role of histone H4-K16 acetylation. *Cell Cycle* 2006;**5**:1361–5.

208. de la Cruz X, Lois S, Sanchez-Molina S, Martinez-Balbas MA. Do protein motifs read the histone code? *Bioessays* 2005;**27**:164–75.

209. Dhalluin C, Carlson JE, Zeng L, He C, Aggarwal AK, Zhou MM. Structure and ligand of a histone acetyltransferase bromodomain. *Nature* 1999;**399**:491–6.

210. Hassan AH, Prochasson P, Neely KE, Galasinski SC, Chandy M, Carrozza MJ, et al. Function and selectivity of bromodomains in anchoring chromatin-modifying complexes to promoter nucleosomes. *Cell* 2002;**111**:369–79.

211. Owen DJ, Ornaghi P, Yang JC, Lowe N, Evans PR, Ballario P, et al. The structural basis for the recognition of acetylated histone H4 by the bromodomain of histone acetyltransferase gcn5p. *EMBO J* 2000;**19**:6141–9.

212. Rea S, Eisenhaber F, O'Carroll D, Strahl BD, Sun ZW, Schmid M, et al. Regulation of chromatin structure by site-specific histone H3 methyltransferases. *Nature* 2000;**406**:593–9.

213. Brownell JE, Allis CD. Special HATs for special occasions: linking histone acetylation to chromatin assembly and gene activation. *Curr Opin Genet Dev* 1996;**6**:176–84.

214. Vignali M, Steger DJ, Neely KE, Workman JL. Distribution of acetylated histones resulting from Gal4-VP16 recruitment of SAGA and NuA4 complexes. *EMBO J* 2000;**19**:2629–40.

215. Nourani A, Doyon Y, Utley RT, Allard S, Lane WS, Cote J. Role of an ING1 growth regulator in transcriptional activation and targeted histone acetylation by the NuA4 complex. *Mol Cell Biol* 2001;**21**:7629–40.

216. Galarneau L, Nourani A, Boudreault AA, Zhang Y, Heliot L, Allard S, et al. Multiple links between the NuA4 histone acetyltransferase complex and epigenetic control of transcription. *Mol Cell* 2000;**5**:927–7.

217. Reid JL, Iyer VR, Brown PO, Struhl K. Coordinate regulation of yeast ribosomal protein genes is associated with targeted recruitment of Esa1 histone acetylase. *Mol Cell* 2000;**6**:1297–307.

218. Arabi A, Wu S, Ridderstrale K, Bierhoff H, Shiue C, Fatyol K, et al. c-Myc associates with ribosomal DNA and activates RNA polymerase I transcription. *Nat Cell Biol* 2005;**7**:303–10.

219. Brown CE, Howe L, Sousa K, Alley SC, Carrozza MJ, Tan S, et al. Recruitment of HAT complexes by direct activator interactions with the ATM-related Tra1 subunit. *Science* 2001;**292**:2333–7.

220. Fernandez PC, Frank SR, Wang L, Schroeder M, Liu S, Greene J, et al. Genomic targets of the human c-Myc protein. *Genes Dev* 2003;**17**:1115–29.

221. Knoepfler PS, Zhang XY, Cheng PF, Gafken PR, McMahon SB, Eisenman RN. Myc influences global chromatin structure. *EMBO J* 2006;**25**:2723–34.

222. Herceg Z, Li H, Cuenin C, Shukla V, Radolf M, Steinlein P, et al. Genome-wide analysis of gene expression regulated by the HAT cofactor Trrap in conditional knockout cells. *Nucleic Acids Res* 2003;**31**:7011–23.

223. Li H, Cuenin C, Murr R, Wang ZQ, Herceg Z. HAT cofactor Trrap regulates the mitotic checkpoint by modulation of Mad1 and Mad2 expression. *EMBO J* 2004;**23**:4824–34.

224. Masumoto H, Hawke D, Kobayashi R, Verreault A. A role for cell-cycle-regulated histone H3 lysine 56 acetylation in the DNA damage response. *Nature* 2005;**436**:294–8.

225. Schneider J, Bajwa P, Johnson FC, Bhaumik SR, Shilatifard A. Rtt109 is required for proper H3K56 acetylation: a chromatin mark associated with the elongating RNA polymerase II. *J Biol Chem* 2006;**281**:37270–4.

226. Xu F, Zhang K, Grunstein M. Acetylation in histone H3 globular domain regulates gene expression in yeast. *Cell* 2005;**121**:375–85.

227. Brown CE, Lechner T, Howe L, Workman JL. The many HATs of transcription coactivators. *Trends Biochem Sci* 2000;**25**:15–19.

228. Carrozza MJ, Utley RT, Workman JL, Cote J. The diverse functions of histone acetyltransferase complexes. *Trends Genet* 2003;**19**:321–9.

229. Herceg Z, Wang ZQ. Rendez-vous at mitosis: TRRAPed in the chromatin. *Cell Cycle* 2005;**4**:383–7.

230. Brand M, Moggs JG, Oulad-Abdelghani M, Lejeune F, Dilworth FJ, Stevenin J, et al. UV-damaged DNA-binding protein in the TFTC complex links DNA damage recognition to nucleosome acetylation. *EMBO J* 2001;**20**:3187–96.

231. Martinez E, Palhan VB, Tjernberg A, Lymar ES, Gamper AM, Kundu TK, et al. Human STAGA complex is a chromatin-acetylating transcription coactivator that interacts with pre-mRNA splicing and DNA damage-binding factors in vivo. *Mol Cell Biol* 2001;**21**:6782–95.

232. Choy JS, Kron SJ. NuA4 subunit Yng2 function in intra-S-phase DNA damage response. *Mol Cell Biol* 2002;**22**:8215–25.

233. Qin S, Parthun MR. Histone H3 and the histone acetyltransferase Hat1p contribute to DNA double-strand break repair. *Mol Cell Biol* 2002;**22**:8353–65.

234. Tamburini BA, Tyler JK. Localized histone acetylation and deacetylation triggered by the homologous recombination pathway of double-strand DNA repair. *Mol Cell Biol* 2005;**25**:4903–13.

235. Celic I, Masumoto H, Griffith WP, Meluh P, Cotter RJ, Boeke JD, et al. The sirtuins hst3 and Hst4p preserve genome integrity by controlling histone h3 lysine 56 deacetylation. *Curr Biol* 2006;**16**:1280–9.

236. Maas NL, Miller KM, DeFazio LG, Toczyski DP. Cell cycle and checkpoint regulation of histone H3 K56 acetylation by Hst3 and Hst4. *Mol Cell* 2006;**23**:109–19.

237. Driscoll R, Hudson A, Jackson SP. Yeast Rtt109 promotes genome stability by acetylating histone H3 on lysine 56. *Science* 2007;**315**:649–52.

238. Han J, Zhou H, Horazdovsky B, Zhang K, Xu RM, Zhang Z. Rtt109 acetylates histone H3 lysine 56 and functions in DNA replication. *Science* 2007;**315**:653–5.

239. Murr R, Loizou JI, Yang YG, Cuenin C, Li H, Wang ZQ, et al. Histone acetylation by Trrap-Tip60 modulates loading of repair proteins and repair of DNA double-strand breaks. *Nat Cell Biol* 2006;**8**:91–9.

240. Murr R, Vaissiere T, Sawan C, Shukla V, Herceg Z. Orchestration of chromatin-based processes: mind the TRRAP. *Oncogene* 2007;**26**:5358–72.

Additional Epigenetic Processes

CHAPTER 4

The Epigenetics of Non-coding RNA

Lesley J. Collins[1], Barbara Schönfeld[2,3], and Xiaowei Sylvia Chen[4]
[1]Massey Genome Service, Massey University, Palmerston North, New Zealand
[2]Allan Wilson Centre for Molecular Ecology and Evolution, Massey University, Palmerston North, New Zealand
[3]Institute of Molecular BioSciences, Massey University, Palmerston North, New Zealand
[4]Department of Biochemistry, University of Otago, Dunedin 9054, New Zealand

INTRODUCTION

Non-protein-coding RNAs (ncRNAs) are RNAs that are transcribed from DNA but are not translated into proteins. Many are functional and are involved in the processing and regulation of other RNAs such as mRNA, tRNA, and rRNA. Processing-type ncRNAs include small nuclear RNAs (snRNAs) involved in splicing, small nucleolar RNAs (snoRNAs) that modify nucleotides in rRNAs and other RNAs, and RNase P that cleaves pre-tRNAs. Other small ncRNAs such as microRNAs (miRNAs) and short interfering RNAs (siRNAs) are involved in the regulation of target mRNAs and chromatin. Although many of these latter ncRNA classes are grouped under the term RNA interference (RNAi), it has become clear that there are many different ways that ncRNAs can interact with genes to up-regulate or down-regulate expression, to silence translation, or guide methylation [1–3]. Adding to these classes are long ncRNAs (typically >200 nt) that have also been implicated in gene regulation [4]. All of these ncRNAs form a network of processes, the RNA-infrastructure [2] that spans the cell not only spatially as RNAs move across the cell, but also temporally as the RNAs regulate gene processes during the cell cycle. Thus, the regulation of RNA processes may not only be transcriptional or translational, but also from their biogenesis and processing pathways [2]. However, when talking about gene regulation, it is RNAi that immediately comes to mind (especially in multicellular organisms) and it appears that RNAi-based ncRNAs and some longer ncRNAs have roles in epigenetic processes [5]. Some of these roles have been known for some time (e.g. X-chromosome inactivation [6] and gene imprinting [7]) but other roles in non-developmental mechanisms and cancer are only just coming to light.

We can cover only some of these mechanisms here but further reviews are available [5,7–10]. Although work in this area has clearly concentrated on mammalian examples there are many interesting mechanisms coming to light from non-mammalian species which we will cover to a small extent here. Presently we can divide the epigenetic-related classes of ncRNAs into two main groups; the long ncRNAs, and short ncRNAs including miRNAs, siRNAs, and Piwi-interacting RNAs (piRNAs). This chapter reviews both the long and short classes of ncRNAs involved in epigenetic regulation: those that generally act as *cis*-acting silencers, but also as *trans*-acting regulators of site specific modification and imprinted gene-silencing (Table 4.1). As the examples in the following sections will show, we are still very much in the early days

Handbook of Epigenetics: The New Molecular and Medical Genetics. DOI: 10.1016/B978-0-12-375709-8.00004-6

TABLE 4.1 ncRNAs Discussed in This Chapter and Their Abbreviations

ncRNA		Length	Short Description	Suggested Reviews and Examples
miRNA	Micro RNA	21–23 nt	ssRNA folds into dsRNA structure; after processing and binding to RISC complex they target mRNAs to regulate translation.	Reviews [12,13,15,25] Figure 4.1A Rtl1 [16–19] Figure 4.2 miR-290 [20,21]
siRNA	Short interfering or silencing RNA	20–25 nt	Regulate a specific gene using complementary sequence. Post-Transcriptional Gene Silencing (PTGS) and Transcriptional Gene Silencing (TGS) pathways. Plants also use RNA-directed DNA methylation (RdDM).	Reviews [9,25–28] Figure 4.1B FLC gene [9,31,32]
piRNA	Piwi-interacting RNA	27–30 nt	Interact with PIWI proteins for chromatin regulation and transposon silencing. Scan RNAs (scnRNAs) are a type of piRNA.	Reviews [33–36] Figure 4.1C
XiRNAs	XCI inactivation linked small RNAs	24–42 nt	Produced from Xist and Tsix long ncRNAs, required for controlling methylation of the future inactive X chromosome and of the Xist promoter region on the future active X chromosome.	XiRNAs in XCI [45,51,52] Figure 4.3
Long ncRNAs		>200 nt	Many have specific targets and are critical for X chromosome inactivation in mammals (XCI), meiotic sex chromosome inactivation (MSCI), RoX (RNA on X) +system in insects, and Hox gene regulation.	XCI [40,41,45, 49,52] Figure 4.3 MSCI [54,55] RoX [56–58] HOX [63,65] Figure 4.4

of investigating how many characterized ncRNAs work to regulate processes such as RNA editing and methylation.

SHORT ncRNAs AND EPIGENETICS

RNAi is a mechanism by which short double-stranded RNAs (dsRNA) are used for sequence-specific regulation of gene expression, where some of the nucleotides on the ncRNA bind to either the coding or promoter region of an mRNA. This binding interferes with normal mRNA processing and consequently silences the expression of the mRNA. The three major classes are microRNA (miRNA), short interfering RNA (siRNA) and Piwi-interacting RNA (piRNA) which differ in their biogenesis and modes of target regulation [11] (Fig. 4.1). Although best known for roles in regulating mRNA transcripts, these short ncRNAs are also directly involved in other cellular processes including chromatin-mediated gene silencing and DNA rearrangements [2,12]. We will go through each class in turn highlighting how they are different, and review recent studies that indicate their use in epigenetics.

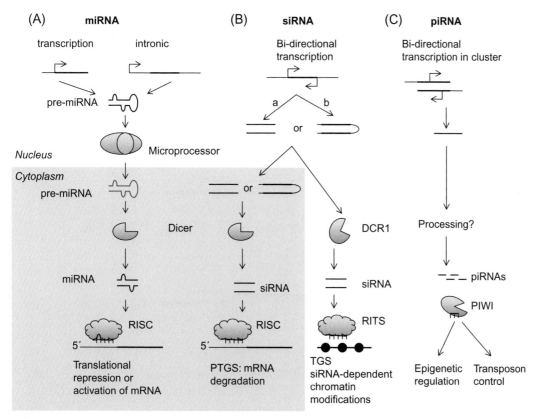

FIGURE 4.1
Processing pathways of small regulatory ncRNAs. (A) miRNAs are initially single-stranded RNAs (ssRNAs) produced via transcription or through splicing, which fold into stem-loop structures to form imperfect double-stranded RNA molecules (dsRNAs). These are then processed by the RNase III endoribonuclease (generally Dicer) before being denatured. One of the RNA strands (usually the less stable of the two) binds to the RNA-induced silencing complex (RISC), which then binds to a specific target mRNA that contains sequence complementary to the miRNA, to induce either cleavage or degradation, or block translation. (B) siRNAs are produced as dsRNAs, and can enter the Post-Transcriptional Gene Silencing (PTGS) pathway, which leads to mRNA degradation in the cytoplasm, or the Transcriptional Gene Silencing (TGS) pathway involved in chromatin modification. (C) piRNAs are ssRNAs produced in clusters and cleaved to individual units through an as yet undefined processing mechanism. They then bind to PIWI proteins to induce epigenetic regulation and transposon control.

51

miRNAs are perhaps the most well known of the regulatory ncRNA classes, and the general miRNA processing pathway is shown in Figure 4.1. Yet we cannot assume that all miRNAs within one species will regulate their genes in the same way in another species. Studies have shown that not only can a single miRNA down-regulate expression of hundreds of its target genes [13], but some miRNAs use alternative methods of down-regulation, such as accelerated deadenylation of the polyA tail [14]. Other studies revealed that animal miRNAs can induce translational up-regulation, and that some plant miRNAs can function as translational inhibitors contrary to their original functional descriptions [reviewed in Ref. 15]. However, miRNAs are not merely regulating mRNA targets, but are also involved in intricate mechanisms that involve feedback, self-regulation and in some cases methylation.

An example (Fig. 4.2) comes from the mouse Dlk1-Dio3 region in which three protein genes, i.e. delta-like 1 (Dlk1), retrotransposon gene (Rtl1), and Dio3, are expressed exclusively from the paternal chromosome [16]. On the maternal chromosome these protein-coding genes are normally repressed, and several other transcripts are produced including one antisense to the Rtl1 gene. Regional imprinting of Rtl1 is predetermined by the methylation status of the nearby intergenic differentially methylated region (IG-DMR), which is methylated in the paternal chromosome, but not in the maternal. The maternally inherited unmethylated

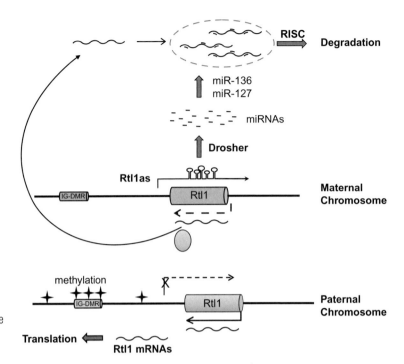

FIGURE 4.2

miRNAs and imprinting.
Methylation of the IG-DMR region on the paternal chromosome represses the expression of the Rtl1a antisense transcript, allowing expression of the Rtl1 transcript. Without this methylation, as on the maternal strand, the Rtl1a is processed to produce miRNAs including miR-136 and miR-127, which complementarily bind to the Rtl1 transcript and induce degradation.

state of IG-DMR is essential for maintaining the repression of the protein-coding genes and for the expression of the antisense transcript [16,17]. The Rtl1as (antisense) transcript (also known as antiPeg11) forms hairpin structures from which after processing, miRNAs are released including miR-127 and miR-136 [18]. These miRNAs are located near 2 CpG islands in the Rtl1 transcript, and regulate the expression of Rtl1 *in trans* by guiding RISC-mediated cleavage of any maternal transcript. Aberrant epigenetic reprogramming of miR-127, miR-136, or Rtl1 result in late-fetal and/or neonatal lethality [19].

miRNAs have also been shown to be important in stem-cell self-renewal and differentiation (reviewed in Ref. 12). There are two types of stem cell, tissue stem cells (which include somatic and germline cells which develop, maintain, and repair tissues in developing and adult organisms), and embryonic stem cells (ES) which develop from an embryo to give rise to the fetus. Self-renewal (or self-replication) in tissue cells results in asymmetrical division, whereby one daughter cell retains the stem-cell properties, and the other daughter cell is committed to a differentiated function. This behavior is controlled inter-cellularly (between cells by cell signalling), as well as intra-cellularly through epigenetic, transcriptional, translational, and post-translational mechanisms. Recently, miRNAs have been found to be important players in controlling stem-cell fate and behavior. One example is the mouse miR-290–295 miRNA cluster, a group of miRNAs that share a 5′ proximal AAGUGC motif [20]. The expression of this cluster increases during pre-implantation development and remains high in undifferentiated ES cells, but then decreases after ES cell differentiation [20]. The miR-290 miRNAs act as post-transcriptional regulators of retinoblastoma-like 2 (Rbl2), which in turn acts as a transcriptional repressor of DNA methyl transferases (DNMTs), Dnmt3a and Dnmt3b. DNMTs epigenetically silence OCT4, a key transcription factor of ES cell renewal and differentiation [20,21]. Repression of Dnmt3a and Dnmt3b results in hypomethylation of the genome and especially the telomeres, leading to the appearance of long telomeres and increased telomere recombination. Alternatively, if Dicer is knocked out, miRNAs are depleted and the methylation of the Oct4 promoter is severely impaired during differentiation [20]. Many other candidate targets of the AAGUGC seed-containing miRNAs have been identified as well as many indirectly regulated targets [20], but it remains to be seen how other aspects of self-renewal and differentiation are affected by the miR-290 cluster.

This is only one example of many that show how miRNAs are directly or indirectly regulating key self-renewal or differentiating genes by either directly or indirectly affecting methylation processes. It is also possible for a miRNA to regulate another miRNA. An example of this action is miR-184, which negatively regulates miR-205 in human epithelial cells. Interfering with miR-205 dampens the Akt signaling pathway and is associated with a marked increase in keratinocyte apoptosis and cell death [22]. Current research (e.g. 22–24) is finding that more and more of such miRNAs are subjected to feedback from their target genes, and serve as a warning that what may appear at first to be "simple" regulation of an mRNA by an miRNA, may in fact have hidden features only revealed upon a detailed investigation of a mechanism.

As with miRNAs there are many subclasses of siRNAs that can be processed either as sense–antisense pairs (e.g. bidirectional promoter produced; Fig. 4.1 – siRNA pathway A), or as double-stranded transcripts which are subsequently cleaved by Dicer (Fig. 4.1 – siRNA pathway B) [25]. siRNA-based mechanisms have been well-studied in plants and fission yeast [9,26]. Although at first siRNAs in animals were mostly considered to be from foreign DNA or RNA (i.e. viral-induced exo-siRNAs), recent studies have characterized many more endogenously encoded siRNAs (endo-siRNAs) that appear to have a role in transposon control [reviewed in Ref. 27]. However, concentrating on the more well-known mechanisms in plants, one group of endo-siRNAs are the RDR2-dependent siRNAs which are preferentially associated with transposons, retroelements, and repetitive DNA, but also appear to guide methylation of specific DNA regions [26]. In plants, fission yeast, and to a small extent in mammals [28], both the transcriptional (TGS) as well as post-transcriptional gene silencing (PTGS) pathways are activated by dsRNAs. With the PTGS pathway, siRNAs direct mRNA degradation in the cytoplasm with no epigenetic incidence. However, TGS acts in the nucleus and is associated with chromatin modifications that silence transcription, and are maintained throughout the phases of the cell cycle [9].

In fission yeast more complicated models have been characterized. During TGS, the RITS (RNAi-Induced Transcriptional Silencing) complex is similar to RISC in containing Argonaute, but (unlike RISC) RITS localizes exclusively to the nucleus and contains at least one chromatin-binding module called a chromodomain [9]. Bound to a siRNA it mediates sequence specific heterochromatin formation and histone methylation. Recent models propose that RITS and RDRC (RNA-directed RNA polymerase Complex) are recruited to the site of intended heterochromatin formation when their associated siRNAs bind to a nascent RNA being transcribed at that site [29]. Thus, the binding of RITS to chromatin initiates heterochromatin formation which in turn results in TGS. Assembly of heterochromatin at a given genomic site comes with a heritable silencing of transcription. In fission yeast this mechanism is widely used to regulate heterochromatin formation, and a positive-feedback loop involving RDRP couples siRNA production to chromatin modifications [9].

In plants, siRNAs are involved in RNA-directed DNA Methylation (RdDM), which was first observed in viroid infected tobacco plants where sequences similar in sequence to the viral genes became methylated [30]. The exact mechanism for RdDM has not yet been characterized but in a general model [9], the plant specific RNA Polymerase IV is (somehow) recruited to a target genomic site; once there it synthesizes an ssRNA which RDR2 uses as a template to construct dsRNA that is processed by DCL3 (plant dicer) into siRNAs that bind AGO4 proteins. An AGO4 protein bound to an siRNA is thought to form a complex with PolIVb and DRM2 to guide DNA and chromatin methylation at the target genomic region [9,26]. One example is the FLC gene (Flowering locus gene C), a key MADS box transcription factor with key cell differentiation roles similar to that of HOX in animals. FLC gene expression is low during flowering in *Arabidopsis thaliana*, maintained by the Polycomb group of silencing proteins. A transposon in an intron of FLC is believed to nucleate formation of silent chromatin by attracting DNA and H3K9 methylation [9,31]. siRNAs complementary to the 3′ end of the FLC gene have now been detected and their accumulation requires DCL2, RDR2, and PolIVa [32]. However, the siRNAs do not depend on the transposon but instead on antisense transcription of FLC 3′ UTR by a mechanism that is not yet clear [9].

The use of piRNAs (PIWI-interacting RNAs) in epigenetic processes is (like the siRNAs) only just coming under detailed investigation. Although also found in mammals and some ciliates piRNAs have been studied in greater detail in *Drosophila melanogaster*, in both germline and, more recently, soma cells [33,34], where they play critical roles in transposon "control" (i.e. preventing transposon activation and hence keeping the levels of transposons interrupting genes to a minimum) [35]. *Drosophila* piRNAs reside in clusters usually within heterochromatin or at heterochromatin–euchromatin boundaries. These piRNA clusters are repeat-rich regions composed of ancient fragmented transposon copies representing all major classes and element families [35]. Unlike miRNAs and siRNAs, piRNAs are not produced by "Dicing" (Fig. 4.1), but mainly by bi-directional promoters and what is known as the "ping pong" cycle of biogenesis and amplification. This cycle is initiated by primary piRNAs arising from piRNA clusters. Those piRNAs that are antisense to expressed transposons identify and cleave their targets, resulting in a set of new sense piRNAs in an AGO3 complex termed secondary piRNA. The AGO3-bound piRNA targets any transposon target that contains antisense transposon sequences. This cleavage then generates additional antisense piRNAs and the cycle can continue. This forms an effective small ncRNA-based transposon immune system.

piRNAs are now being proposed as possible vectors for carrying epigenetic inheritance [36]. An example comes from *Drosophila* strains that differ in the presence of a specific transposon, where crosses produce sterile progeny (hybrid dysgenesis), but only if the transposon is paternally inherited. Maternally inherited piRNAs are thought to play a role in this transposon silencing [36]. Both PIWI and Aubergine (Aub) proteins are deposited into developing oocytes and accumulate in the pole plasm suggesting a mechanism of transfer of maternal piRNAs into the germ lines of their progeny [36]. piRNA clusters alone have been shown to be insufficient to inactivate some transposons within a single generation. Instead maternally inherited siRNAs appear to prime the "resistance"-type control system at each generation to achieve full immunity. It is also thought that, since environment can influence the content of maternal small RNA populations, these RNAs could epigenetically alter the phenotype of progeny [35].

In mammals, transposon control by TGS occurs using PIWI-type proteins Milli and Miwi2 (Line-1 non-LTR), and IAP (LTR) retrotransposons, along with DNA methylation during embryogenesis in male germ cells (prospermatogonia) [35]. Like AGO3, Mili binds preferentially to piRNAs corresponding to transposon sense strands while Miwi2 contains mainly antisense piRNAs. piRNAs in prospermatogonia are derived from transposon rich piRNA clusters. There is evidence for a ping-pong amplification cycle as seen in *Drosophila*, but as yet its involvement in epigenetic inheritance is not characterized. Although the transmission of phenotype via piRNAs has only been demonstrated to date in *Drosophila*, the accumulation of small RNAs in the oocytes of other species is known and opens the way for this phenomenon to be more widespread [35].

LONG ncRNAs AND EPIGENETICS

During the last few years, evidence of complex, long ncRNA mediated epigenetic control systems has increased dramatically [3,37,38]. In a famous example, X chromosome inactivation (XCI) studied largely in mice, ensures only one of the two X chromosomes in XX females is expressed during development, and involves two long ncRNAs: Xist (17 kb) and its antisense transcript Tsix (40 kb) (reviewed in Refs 6,39,40). Xist RNA is expressed at a low level in both females and males before differentiation [41], but upon cell differentiation, Xist RNA coats the future inactive X chromosome (Xi) triggering extensive histone methylation

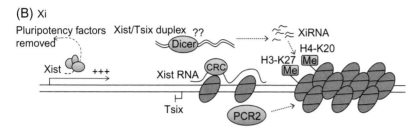

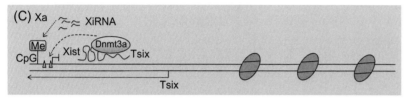

FIGURE 4.3

A general model of placental mammal X Chromosome Inactivation (XCI). (A) Prior to XCI, Tsix is expressed at a high level and triggers H3-K4 dimethylation in itself and the Xist gene, leading to active transcription of Xist and Tsix. This results in an equal chance for transcription and ensures random initiation of XCI. (B) During XCI, expression of Xist is elevated upon removal of the pluripotency factors from the first intron on Xist. Xist RNA then coats the future Xi *in cis* and recruits the chromatin repressive complex (CRC) to Xi. Xist RNA also forms RNA duplex with Tsix RNA and is processed into 24 to 42 nt XiRNAs through the possible action of Dicer. XiRNA then directs H3-K27 trimethylation and H4-K20 monomethylation on the future Xi. The Xi status is maintained by the Polycomb repressive complex PCR2. (C) On the future Xa, Tsix is associated with methyltransferase Dnm3a and directs methylation on the Xist promoter to ensure expression of X-linked genes through repression of Xist. XiRNA is also involved by directing methylation of Xist on CpG islands.

55

[42], whereas Tsix appears to restrict Xist activity on the future active X chromosome (Xa) [43]. Recent studies, especially in mice, have revealed a more complex regulatory network of XCI which involves the interaction of long and short ncRNAs (Fig. 4.3).

To explain in more detail, in mice pre-XCI embryonic stem cells (ES) (Fig. 4.3A), Tsix is transcribed at a much higher level than Xist and triggers cytosine methylation within both *Tsix* and *Xist* genes, resulting in epigenetically equal competency for transcription and random X-inactivation [44]. The transcriptional level of Xist is elevated when the major pluripotency factors Nanog, Oct3/4, and Sox2 dissociate from intron 1 within *Xist* initiating XCI [44] (Fig. 4.3B and C). The coating of Xist on the future inactive X chromosome (Xi) (Fig. 4.3B) forms a silent chromatin compartment where X-linked genes become "localized" through Xist binding [45]. Xist RNA is required for chromosome-wide methylation in undifferentiated ES cells during the onset of X inactivation; however, once established, the maintenance of the heterochromatic state is independent of Xist RNA [46]. In contrast, the Polycomb repressive complex PCR2 is recruited by the RepA (a 1.6 kb ncRNA within Xist), and responsible for the maintenance of Xi [47,48]. On the future active X chromosome Xa (Fig. 4.3C), the level of *Xist* expression is largely controlled by its antisense transcript Tsix. The expression of Tsix is restricted to Xa [49] and associates with the DNA methyltransferase *Dnmt3a* to direct methylation on *Xist* promoter [50]. However, this methylation event is transient and does not play a role during the initiation of XCI [50]. Besides directing histone modification, the Tsix RNA can also down-regulate *Xist* expression through antisense binding. It has been shown that Tsix transcription across the Xist promoter is crucial for Xist regulation [44].

In addition small ncRNAs are also involved in XCI. Dicer-dependent XiRNAs are produced from both the Xist and Tsix ncRNA transcripts [44] and are required for methylation along the future inactive X chromosome, and methylation of the CpG island of the Xist promoter region in the future active X chromosome Xa [51]. Although XiRNAs are produced with Dicer, RNAi is not directly involved in X chromosome inactivation; instead it appears to maintain the steady-state level of the Xist RNA [52]. Adding to this increasingly complex network, RepA has been found to mediate the heterochromatic configuration of the Xist promoter through recruiting PRC2 [53].

Although most of the studies on XCI have been in placental mammals (and especially mice), recent work in marsupials has shown that a very similar mechanism exists although marsupials do not have the Xist RNA [54]. Here it is thought that male meiotic sex chromosome inactivation (MSCI) plays a greater role in dosage compensation. In mice, (reviewed in Ref. 55), MSCI silencing of the X chromosome genes is initiated during male meiosis, but unlike XCI, MSCI is transient, occurring during each round of spermatogenesis with some X-linked genes reactivating, then subsequently becoming silenced in the female. In marsupials, this can be demonstrated since XCI appears not to result from inheriting an X chromosome already inactivated by MSCI, but instead the inactivation takes place in the female (although the exact timing is not as yet known) [54]. It is also suggested [55] that some X-linked miRNAs escape MSCI and may contribute to the mechanisms regulating MSCI in an RNAi-like manner. Commonalities between the placental mammal and marsupial models (including enrichment of H3K27 trimethylation on the Xi and association of the Xi with the nucleolus [54]), indicate that aspects of the XCI system may be more conserved than originally thought [54].

In insects however, dosage compensation is achieved not by silencing but by a 2-fold increase of X-linked genes in males, relative to females [56–58]. In this mechanism the male-specific-lethal (MSL) complex (consisting of MSL1, MSL2, MSL3, MOF (*males absent on first*), and MLE (*maleless*)) binds to genes along the male X chromosome. Associated with this complex are two long ncRNAs, roX1 and roX2 (*RNA on X*), that direct activation, rather than silencing, of their target genes [56,59]. roX1 and roX2 transcripts spread along the X chromosome recruiting the histone deacetylation protein complex, which generates an open chromatin conformation to facilitate active transcription [60,61]. How roX RNA regulates changes in the localization and activity of the MSL complex, is still poorly understood [62], and likewise how the MSL complex achieves dosage compensation [58]. Studies are beginning to indicate that target genes are enriched at the 3′ end and not at promoter sites leading to a model that the MSL complex affects elongation, resulting perhaps in hyper-transcription of the targeted genes or chromatin looping [58]. What is clear is that the RoX RNAs are a key part of the insect dosage compensation mechanism, and more study is needed to uncover the finer details [58].

Recent studies have also revealed long non-coding RNAs regulating the *Hox* gene cluster in insects and vertebrates (reviewed in Ref. 63). First found in *Drosophila*, the Hox family of proteins are critical determinants of correct patterning of the axis during embryonic development [64]. A large number of non-coding transcripts have been identified within the *Hox* gene cluster [63], the majority of which are found as antisense transcripts from intergenic regions, and are coordinately induced with their 3′-end *Hox* genes [65]. In *Drosophila*, the Bithorax *Hox* gene cluster (BX-C) regulation is extremely complex, with the regulatory region containing enhancers, silencers, maintenance elements, boundary elements, and possibly other elements not yet characterized [63].

Included in this regulation are the long ncRNAs, bxd RNAs, and iab RNAs [64], involved in regulation of their downstream Hox genes, *Ultrabithorax* (*Ubx*), *Abdominal-A* (*abd-A*), and *Abdominal-B* (*abd-B*) [63,66]. Bxd RNAs are expressed in different cells and germ layers, consistent with each bxd ncRNA having a unique role [63]. This spatial regulation may account for the observed mosaic expression pattern of the Hox genes in early embryos

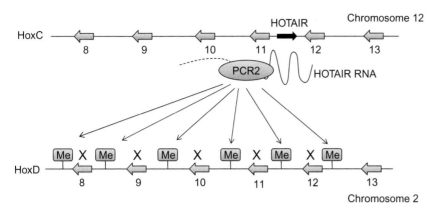

FIGURE 4.4

Long ncRNA regulation of Hox genes. Human HOTAIR RNA is expressed on the antisense strand within the HoxC gene cluster on chromosome 12. The HOTAIR RNA associates with Polycomb repressive complex PCR2 which triggers methylation along the HoxD gene cluster on chromosome 2, leading to silencing of the HoxD genes.

[63,66]. In particular there has been some attention focused on one of these ncRNAs, HOTAIR, identified as regulating chromatin silencing of the adjacent *Hox* locus [65]. Figure 4.4 illustrates the mechanism by which the HOTAIR RNA regulates expression of *HOX* gene clusters through epigenetic control. HOTAIR, a 2158 nt spliced and polyadenylated long ncRNA is transcribed as a single copy on the antisense strand of the *HOXC* gene [65]. siRNA knockdown of HOTAIR results in transcriptional activation of the *HOXD* gene locus spanning four genes on a different chromosome. The HOTAIR RNA is physically associated with the Polycomb Repressive Complex 2 (PCR2), and is required for H3K27me3 modification and transcriptional silencing at HoxD [48,65]. Interestingly, HOTAIR transcription is linked to Polycomb group protein deposition and HOXD silencing on a different chromosome, demonstrating the action of an ncRNA *in trans* [48,65]. This action, but *in cis*, is seen in other long ncRNAs such as RepA, Kcnq1ot1, and AIR [48].

A number of models have been studied to reveal the roles of long ncRNAs in imprinted gene clusters. Short ncRNAs clearly have a role in epigenetic imprinting. In mouse, for example, 80 genes are grouped into clusters [7,67] and in many cases, one or more ncRNAs expressed from within a gene cluster play a crucial role in regulating the expression the gene cluster [68]. This regulation directs chromatin modification forming an "epigenetic memory" within the same cell lineage [69]. Expression of genes in an imprinted cluster is generally controlled by a *cis*-regulatory region, the Imprint Control Element (ICE), which carries parental information in the form of DNA methylation [70]. Several gene clusters controlled by ICE are insulin-like growth factor 2 (*Igf2*), insulin-like growth factor 2 receptor (*Igf2r*), potassium voltage-gated channel (*Kcnq1*), and guanine nucleotide binding protein α stimulating factor (*Gnas*). Each of these clusters carries one ncRNA gene on the parental chromosome with unmethylated ICE [70]. The paternally imprinted *Igf2* cluster contains a 2.5 kb spliced long ncRNA H19, which correlates with the methylation silencing of *Igf2* genes [71], despite not having a direct role in maintaining silencing of the *Igf2* cluster [72]. Although not necessary for the silencing of the *Igf2* cluster, transgenic studies have revealed that H19 expression is sufficient for acquiring paternal-allele-specific methylation of the *Igf2* genes [73]. It is possible that some of these long imprinting ncRNAs are in fact miRNA precursors [7], but as with many of the ncRNAs discussed here, details will emerge as further investigations proceed.

CONCLUSIONS

The investigation of ncRNA-related epigenetic mechanisms is at this point relatively new, but the rise of new sequencing technologies has already revealed epigenetic regulation at the

57

genome level. Deep-sequencing technologies (also known as Next Generation Sequencing or NGS) have not only enabled analysis of histone modifications and methylation sites across entire genomes [74], but are enabling the detection of ncRNAs important in the regulation of these modifications. An example is Wang et al. (2009) [75] where maize organ specific distributions of canonical miRNAs and endogenous siRNAs have been linked to epigenetic modifications, H3K27me3, and DNA methylation [75]. This provides an important link between the epigenome and the transcriptome.

An interesting titbit is that the role of ncRNA in epigenetics has even been investigated in space. Spaceflight is a unique environment comprising of cosmic irradiation, microgravity, and space magnetic fields. A study of rice plants germinated from seed subjected to spaceflight showed altered methylation patterns and gene expression in six transposable elements and 11 cellular genes including siRNA related proteins Ago1 and Ago4 [76]. All of the detected alterations in the cellular genes were hypermethylation events occurring at CNG sites. This is consistent with the idea that plant CNG methylation is more prone to perturbation by environmental stresses [76,77].

While most studies have been conducted on major model organisms, there is now some information on ncRNA-based epigenetic mechanisms in protist lineages. piRNA-type RNAs (scan RNAs or scnRNAs) from ciliates are produced during the reorganization of the macronucleus during sexual development when some exons can become "scrambled" [35]. In *Tetrahymena thermophila* ~6000 IES internal eliminated sequences consisting of transposon-like and other repeats are targeted for removal by RNA-directed heterochromatin marking by scnRNAs [78]. Although the actual molecular mechanism is not as yet known, scnRNAs pair with either DNA or RNA from the parental macronucleus to be sorted, and then "selected" transcripts are moved to the newly developed macronucleus where they induce heterochromatin formation on the IES prior to elimination.

Recently, long ncRNAs have been found in the malaria parasite *Plasmodium falciparum*, where sterile sense and antisense RNAs are transcribed from the *var* virulence gene family and coat chromatin in a similar way to the animal Xist RNA and the *Drosophila* roX RNAs [79]. The regulation of the VSP genes involved in antigenic switching in the Diplomonad *Giardia lamblia* is also thought to be epigenetically regulated [80]. Subsequently there has been the identification of key RNAi proteins [81,82], some miRNAs [83,84], and a little on the regulatory mechanism of the VSP genes [85]; however, nothing is known as yet about chromatin modifications in *Giardia*. Further studies on protists are essential if we are to understand how ncRNAs in general regulate epigenetics and to understand how such mechanisms evolved.

Despite the individual variances in these pathways, miRNAs, siRNAs, and piRNAs all share several key protein components including Argonaute, PIWI, RDRP, and Dicer. Many RNA-directed epigenetic regulation events thus appear to be sharing protein and RNA components with the RNAi pathway if not dependent on the latter. We also note that evolution of ncRNAs by duplication could allow epigenetic states (e.g. methylation and imprinting) between the two copies to differ [1,86]. Since there are instances where a single *trans*-acting siRNA may have ~2300 predicted gene targets [7], this type of duplication could possibly result in a significant change in phenotype [1]. There is no doubt that the next few years will see more a greater understanding of ncRNA-related epigenetic mechanisms and perhaps then we can move on to constructive evolutionary analysis.

ACKNOWLEDGEMENTS

Many thanks go to David Penny, Peter Lockhart, and all, at the Allan Wilson Centre for Molecular Ecology and Evolution in Palmerston North, for support and encouragement. Funding for this project was from the Allan Wilson Centre, with funding also supplied to LJC by the Health Research Council of New Zealand.

References

1. Collins LJ, Chen XS. Ancestral RNA: the RNA biology of the eukaryotic ancestor. *RNA Biology* 2009;**6**:1–8.

2. Collins LJ, Penny D. The RNA infrastructure: dark matter of the eukaryotic cell? *Trends Genet* 2009;**25**:120–8.

3. Amaral PP, Dinger ME, Mercer TR, Mattick JS. The eukaryotic genome as an RNA machine. *Science* 2008;**319**:1787–9.

4. Mattick JS. The genetic signatures of noncoding RNAs. *PLoS Genet* 2009;**5**:e1000459.

5. Kurokawa R, Rosenfeld MG, Glass CK. Transcriptional regulation through noncoding RNAs and epigenetic modifications. *RNA Biol* 2009;**6**.

6. Chow J, Heard E. X inactivation and the complexities of silencing a sex chromosome. *Curr Opin Cell Biol* 2009;**21**:359–66.

7. Royo H, Cavaille J. Non-coding RNAs in imprinted gene clusters. *Biol Cell* 2008;**100**:149–66.

8. Whitehead J, Pandey GK, Kanduri C. Regulation of the mammalian epigenome by long noncoding RNAs. *Biochim Biophys Acta* 2009;**1790**:936–47.

9. Verdel A, Vavasseur A, Le Gorrec M, Touat-Todeschini L. Common themes in siRNA-mediated epigenetic silencing pathways. *Int J Dev Biol* 2009;**53**:245–57.

10. Costa FF. Non-coding RNAs, epigenetics and complexity. *Gene* 2008;**410**:9–17.

11. Shrey K, Suchit A, Nishant M, Vibha R. RNA interference: emerging diagnostics and therapeutics tool. *Biochem Biophys Res Commun* 2009;**386**:273–7.

12. Gangaraju VK, Lin H. microRNAs: key regulators of stem cells. *Nat Rev Mol Cell Biol* 2009;**10**:116–25.

13. Lim LP, Lau NC, Garrett-Engele P, Grimson A, Schelter JM, Castle J, et al. Microarray analysis shows that some microRNAs downregulate large numbers of target mRNAs. *Nature* 2005;**433**:769–73.

14. Eulalio A, Huntzinger E, Nishihara T, Rehwinkel J, Fauser M, Izaurralde E. Deadenylation is a widespread effect of miRNA regulation. *RNA* 2009;**15**:21–32.

15. Zhang R, Su B. Small but influential: the role of microRNAs on gene regulatory network and 3′UTR evolution. *J Genet Genomics* 2009;**36**:1–6.

16. Lin SP, Coan P, da Rocha ST, Seitz H, Cavaille J, Teng PW, et al. Differential regulation of imprinting in the murine embryo and placenta by the Dlk1-Dio3 imprinting control region. *Development* 2007;**134**:417–26.

17. Kagami M, Sekita Y, Nishimura G, Irie M, Kato F, Okada M, et al. Deletions and epimutations affecting the human 14q32.2 imprinted region in individuals with paternal and maternal upd(14)-like phenotypes. *Nat Genet* 2008;**40**:237–42.

18. Seitz H, Youngson N, Lin SP, Dalbert S, Paulsen M, Bachellerie JP, et al. Imprinted microRNA genes transcribed antisense to a reciprocally imprinted retrotransposon-like gene. *Nat Genet* 2003;**34**:261–2.

19. Cui XS, Zhang DX, Ko YG, Kim NH. Aberrant epigenetic reprogramming of imprinted microRNA-127 and Rtl1 in cloned mouse embryos. *Biochem Biophys Res Commun* 2009;**379**:390–4.

20. Sinkkonen L, Hugenschmidt T, Berninger P, Gaidatzis D, Mohn F, Artus-Revel CG, et al. microRNAs control de novo DNA methylation through regulation of transcriptional repressors in mouse embryonic stem cells. *Nat Struct Mol Biol* 2008;**15**:259–67.

21. Benetti R, Gonzalo S, Jaco I, Munoz P, Gonzalez S, Schoeftner S, et al. A mammalian microRNA cluster controls DNA methylation and telomere recombination via Rbl2-dependent regulation of DNA methyltransferases. *Nat Struct Mol Biol* 2008;**15**:268–79.

22. Yu J, Ryan DG, Getsios S, Oliveira-Fernandes M, Fatima A, Lavker RM. MicroRNA-184 antagonizes microRNA-205 to maintain SHIP2 levels in epithelia. *Proc Natl Acad Sci USA* 2008;**105**:19300–5.

23. Yang X, Feng M, Jiang X, Wu Z, Li Z, Aau M, et al. miR-449a and miR-449b are direct transcriptional targets of E2F1 and negatively regulate pRb-E2F1 activity through a feedback loop by targeting CDK6 and CDC25A. *Genes Dev* 2009;**23**:2388–93.

24. Zhao C, Sun G, Li S, Shi Y. A feedback regulatory loop involving microRNA-9 and nuclear receptor TLX in neural stem cell fate determination. *Nat Struct Mol Biol* 2009;**16**:365–71.

25. Carthew RW, Sontheimer EJ. Origins and mechanisms of miRNAs and siRNAs. *Cell* 2009;**136**:642–55.

26. Henderson IR, Jacobsen SE. Epigenetic inheritance in plants. *Nature* 2007;**447**:418–24.

27. Golden DE, Gerbasi VR, Sontheimer EJ. An inside job for siRNAs. *Mol Cell* 2008;**31**:309–12.

28. Kim DH, Saetrom P, Snove O Jr., Rossi JJ. microRNA-directed transcriptional gene silencing in mammalian cells. *Proc Natl Acad Sci USA* 2008;**105**:16230–5.

29. Motamedi MR, Verdel A, Colmenares SU, Gerber SA, Gygi SP, Moazed D. Two RNAi complexes, RITS and RDRC, physically interact and localize to noncoding centromeric RNAs. *Cell* 2004;**119**:789–802.

30. Wassenegger M, Heimes S, Riedel L, Sanger HL. RNA-directed de novo methylation of genomic sequences in plants. *Cell* 1994;**76**:567–76.

59

31. Liu J, He Y, Amasino R, Chen X. siRNAs targeting an intronic transposon in the regulation of natural flowering behavior in *Arabidopsis*. *Genes Dev* 2004;**18**:2873–8.

32. Swiezewski S, Crevillen P, Liu F, Ecker JR, Jerzmanowski A, Dean C. Small RNA-mediated chromatin silencing directed to the 3′ region of the Arabidopsis gene encoding the developmental regulator, FLC. *Proc Natl Acad Sci USA* 2007;**104**:3633–8.

33. Li C, Vagin VV, Lee S, Xu J, Ma S, Xi H, et al. Collapse of germline piRNAs in the absence of Argonaute3 reveals somatic piRNAs in flies. *Cell* 2009;**137**:509–21.

34. Malone CD, Brennecke J, Dus M, Stark A, McCombie WR, Sachidanandam R, et al. Specialized piRNA pathways act in germline and somatic tissues of the *Drosophila* ovary. *Cell* 2009;**137**:522–35.

35. Malone CD, Hannon GJ. Small RNAs as guardians of the genome. *Cell* 2009;**136**:656–68.

36. Brennecke J, Malone CD, Aravin AA, Sachidanandam R, Stark A, Hannon GJ. An epigenetic role for maternally inherited piRNAs in transposon silencing. *Science* 2008;**322**:1387–92.

37. Guttman M, Amit I, Garber M, French C, Lin MF, Feldser D, et al. Chromatin signature reveals over a thousand highly conserved large non-coding RNAs in mammals. *Nature* 2009;**458**:223–7.

38. Kapranov P, Willingham AT, Gingeras TR. Genome-wide transcription and the implications for genomic organization. *Nat Rev Genet* 2007;**8**:413–23.

39. Lee JT. Lessons from X-chromosome inactivation: long ncRNA as guides and tethers to the epigenome. *Genes Dev* 2009;**23**:1831–42.

40. Erwin JA, Lee JT. New twists in X-chromosome inactivation. *Curr Opin Cell Biol* 2008;**20**:349–55.

41. Lee JT, Davidow LS, Warshawsky D. Tsix, a gene antisense to Xist at the X-inactivation centre. *Nat Genet* 1999;**21**:400–4.

42. Heard E, Rougeulle C, Arnaud D, Avner P, Allis CD, Spector DL. Methylation of histone H3 at Lys-9 is an early mark on the X chromosome during X inactivation. *Cell* 2001;**107**:727–38.

43. Stavropoulos N, Lu N, Lee JT. A functional role for Tsix transcription in blocking Xist RNA accumulation but not in X-chromosome choice. *Proc Natl Acad Sci USA* 2001;**98**:10232–7.

44. Navarro P, Pichard S, Ciaudo C, Avner P, Rougeulle C. Tsix transcription across the Xist gene alters chromatin conformation without affecting Xist transcription: implications for X-chromosome inactivation. *Genes Dev* 2005;**19**:1474–84.

45. Chaumeil J, Le Baccon P, Wutz A, Heard E. A novel role for Xist RNA in the formation of a repressive nuclear compartment into which genes are recruited when silenced. *Genes Dev* 2006;**20**:2223–7.

46. Kohlmaier A, Savarese F, Lachner M, Martens J, Jenuwein T, Wutz A. A chromosomal memory triggered by Xist regulates histone methylation in X inactivation. *PLoS Biol* 2004;**2**:E171.

47. Wang J, Mager J, Chen Y, Schneider E, Cross JC, Nagy A, et al. Imprinted X inactivation maintained by a mouse polycomb group gene. *Nat Genet* 2001;**28**:371–5.

48. Gieni RS, Hendzel MJ. Polycomb group protein gene silencing, non-coding RNA, stem cells, and cancer. *Biochem Cell Biol* 2009;**87**:711–46.

49. Lee JT, Lu N. Targeted mutagenesis of Tsix leads to nonrandom X inactivation. *Cell* 1999;**99**:47–57.

50. Sun BK, Deaton AM, Lee JT. A transient heterochromatic state in Xist preempts X inactivation choice without RNA stabilization. *Mol Cell* 2006;**21**:617–28.

51. Ogawa Y, Sun BK, Lee JT. Intersection of the RNA interference and X-inactivation pathways. *Science* 2008;**320**:1336–41.

52. Kanellopoulou C, Muljo SA, Dimitrov SD, Chen X, Colin C, Plath K, et al. X chromosome inactivation in the absence of dicer. *Proc Natl Acad Sci USA* 2009;**106**:1122–27.

53. Zhao J, Sun BK, Erwin JA, Song JJ, Lee JT. Polycomb proteins targeted by a short repeat RNA to the mouse X chromosome. *Science* 2008;**322**:750–6.

54. Mahadevaiah SK, Royo H, Vandeberg JL, McCarrey JR, Mackay S, Turner JM. Key features of the X inactivation process are conserved between marsupials and eutherians. *Curr Biol* 2009;**19**:1478–84.

55. Yan W, McCarrey JR. Sex chromosome inactivation in the male. *Epigenetics* 2009;**4**:452–6.

56. Bai X, Larschan E, Kwon SY, Badenhorst P, Kuroda MI. Regional control of chromatin organization by noncoding roX RNAs and the NURF remodeling complex in *Drosophila melanogaster*. *Genetics* 2007;**176**:1491–9.

57. Franke A, Baker BS. The rox1 and rox2 RNAs are essential components of the compensasome, which mediates dosage compensation in *Drosophila*. *Mol Cell* 1999;**4**:117–22.

58. Ilik I, Akhtar A. roX RNAs: non-coding regulators of the male X chromosome in flies. *RNA Biol* 2009;**6**:113–21.

59. Deng X, Meller VH. Molecularly severe roX1 mutations contribute to dosage compensation in *Drosophila*. *Genesis* 2009;**47**:49–54.

60. Park Y, Kelley RL, Oh H, Kuroda MI, Meller VH. Extent of chromatin spreading determined by roX RNA recruitment of MSL proteins. *Science* 2002;**298**:1620–3.

61. Oh H, Park Y, Kuroda MI. Local spreading of MSL complexes from roX genes on the *Drosophila* X chromosome. *Genes Dev* 2003;**17**:1334–9.

62. Menon DU, Meller VH. Imprinting of the Y chromosome influences dosage compensation in roX1 roX2 *Drosophila melanogaster*. *Genetics* 2009;**183**:811–20.

63. Brock HW, Hodgson JW, Petruk S, Mazo A. Regulatory noncoding RNAs at Hox loci. *Biochem Cell Biol* 2009;**87**:27–34.

64. Lipshitz HD, Peattie DA, Hogness DS. Novel transcripts from the Ultrabithorax domain of the bithorax complex. *Genes Dev* 1987;**1**:307–22.

65. Rinn JL, Kertesz M, Wang JK, Squazzo SL, Xu X, Brugmann SA, et al. Functional demarcation of active and silent chromatin domains in human HOX loci by noncoding RNAs. *Cell* 2007;**129**:1311–23.

66. Petruk S, Sedkov Y, Brock HW, Mazo A. A model for initiation of mosaic HOX gene expression patterns by non-coding RNAs in early embryos. *RNA Biol* 2007;**4**:1–6.

67. Verona RI, Mann MR, Bartolomei MS. Genomic imprinting: intricacies of epigenetic regulation in clusters. *Annu Rev Cell Dev Biol* 2003;**19**:237–59.

68. O'Neill MJ. The influence of non-coding RNAs on allele-specific gene expression in mammals. *Hum Mol Genet* 2005;**14**(Spec No 1):R113–20.

69. Martin C, Zhang Y. Mechanisms of epigenetic inheritance. *Curr Opin Cell Biol* 2007;**19**:266–72.

70. Lewis A, Reik W. How imprinting centres work. *Cytogenet Genome Res* 2006;**113**:81–9.

71. Feil R, Walter J, Allen ND, Reik W. Developmental control of allelic methylation in the imprinted mouse Igf2 and H19 genes. *Development* 1994;**120**:2933–43.

72. Thorvaldsen JL, Fedoriw AM, Nguyen S, Bartolomei MS. Developmental profile of H19 differentially methylated domain (DMD) deletion alleles reveals multiple roles of the DMD in regulating allelic expression and DNA methylation at the imprinted H19/Igf2 locus. *Mol Cell Biol* 2006;**26**:1245–58.

73. Cranston MJ, Spinka TL, Elson DA, Bartolomei MS. Elucidation of the minimal sequence required to imprint H19 transgenes. *Genomics* 2001;**73**:98–107.

74. Park PJ. Epigenetics meets next-generation sequencing. *Epigenetics* 2008;**3**:318–21.

75. Wang X, Elling AA, Li X, Li N, Peng Z, He G, et al. Genome-wide and organ-specific landscapes of epigenetic modifications and their relationships to mRNA and small RNA transcriptomes in maize. *Plant Cell* 2009;**21**:1053–69.

76. Ou X, Long L, Zhang Y, Xue Y, Liu J, Lin X, et al. Spaceflight induces both transient and heritable alterations in DNA methylation and gene expression in rice (*Oryza sativa* L.). *Mutat Res* 2009;**662**:44–53.

77. Boyko A, Kathiria P, Zemp FJ, Yao Y, Pogribny I, Kovalchuk I. Transgenerational changes in the genome stability and methylation in pathogen-infected plants: (virus-induced plant genome instability). *Nucleic Acids Res* 2007;**35**:1714–25.

78. Kurth HM, Mochizuki K. Non-coding RNA: a bridge between small RNA and DNA. *RNA Biol* 2009;**6**:138–40.

79. Epp C, Li F, Howitt CA, Chookajorn T, Deitsch KW. Chromatin associated sense and antisense noncoding RNAs are transcribed from the *var* gene family of virulence genes of the malaria parasite *Plasmodium falciparum*. *RNA* 2009;**15**:116–27.

80. Kulakova L, Singer SM, Conrad J, Nash TE. Epigenetic mechanisms are involved in the control of *Giardia lamblia* antigenic variation. *Mol Microbiol* 2006;**61**:1533–42.

81. Iyer LM, Anantharaman V, Wolf MY, Aravind L. Comparative genomics of transcription factors and chromatin proteins in parasitic protists and other eukaryotes. *Int J Parasitol* 2008;**38**:1–31.

82. Macrae IJ, Li F, Zhou K, Cande WZ, Doudna JA. Structure of Dicer and mechanistic implications for RNAi. *Cold Spring Harb Symp Quant Biol* 2006;**71**:73–80.

83. Chen XS, Collins LJ, Biggs PJ, Penny D. High throughput genome-wide survey of small RNAs from the parasitic protists *Giardia intestinalis* and *Trichomonas vaginalis*. *Genome Biol Evol* 2009;**1**:165–75.

84. Chen XS, White WT, Collins LJ, Penny D. Computational identification of four spliceosomal snRNAs from the deep-branching eukaryote *Giardia intestinalis*. *PLoS ONE* 2008;**3**:e3106.

85. Prucca CG, Slavin I, Quiroga R, Elias EV, Rivero FD, Saura A, et al. Antigenic variation in *Giardia lamblia* is regulated by RNA interference. *Nature* 2008;**456**:750–4.

86. Gu X, Su Z, Huang Y. Simultaneous expansions of microRNAs and protein-coding genes by gene/genome duplications in early vertebrates. *J Exp Zool B Mol Dev Evol* 2009;**312**:164–70.

61

CHAPTER 5

Prions and Prion-like Phenomena in Epigenetic Inheritance

Hervé Lalucque, Fabienne Malagnac, and Philippe Silar
UFR des Sciences du Vivant, Univ Paris 7 Denis Diderot, 75013 Paris; Institut de Génétique et Microbiologie, UMR CNRS 8621, Univ Paris Sud 11, 91405 Orsay Cedex, France

"This discussion is based on the idea of two types of cellular regulatory systems, both capable of maintaining persistent cellular characteristics but achieving homeostasis by different means. The current concept of a primary genetic material (DNA), replicating by a template mechanism, is opposed to a homeostatic system operating by, perhaps, self-regulating metabolic patterns."

D.L. Nanney, 1958 [1]

63

In the early period of the 20th century, almost as soon as the laws of Mendel were rediscovered, characters that would not follow the rules of classical Mendelian segregation were discovered. Most of these cases of non-mendelian heredity are presently accounted for by mutations in eukaryotic organelle genomes (plastes and mitochondria), by cytoplasmic symbionts, or by viruses and virus-like particles. However, a subset of these phenomena cannot be explained by the presence of nucleic acid-bearing entities in the cytoplasm. Theoretical considerations, first made by Max Delbrück [2], proposed that negatively-interacting metabolic networks could generate alternative states, stable enough to be passed on during cell division. Similarly, as early as 1961, a model based on structure inheritance was proposed by Marcou and Rizet to account for a case of non-mendelian inheritance in the fungus *Podospora anserina* [3]. Early experiments with the lactose operon of *Escherichia coli* [4] proved that indeed metabolic networks could generate inheritable alternative metabolic states, and studies with paramecia showed that complex sub-cellular structures, e.g. cilia, which direct their own assembly in a template-assisted fashion [5], could create alternative states that were inheritable during cell division and even sexual reproduction. This led to the definition of two kinds of epigenetic inheritance 6]: the structural inheritance, based on the transmission of alternative structures of macromolecules and macromolecular complexes, and the regulatory inheritance, based on the alternative states adopted by metabolic or regulatory pathways. As we shall see below, such a clear-cut difference may not be made in some instances, which are clearly a mix of the two types.

Handbook of Epigenetics: The New Molecular and Medical Genetics. DOI: 10.1016/B978-0-12-375709-8.00005-8

It is interesting to note that the acceptance that DNA, and not proteins, is the genetic material took a long time. Purification of DNA associated with genetic transformation were the key experiments that permitted final recognition. The discovery in the mid-1960s of mitochondrial and plastid genomic DNAs [7,8] has then eclipsed for three decades the studies on many "genes" with unorthodox segregation that could be due to structural or regulatory inheritance. Their analysis was re-ignited by the proposal of R. Wickner in 1994 [9] that two of them, the [PSI⁺] and [URE3] elements of the yeast *Saccharomyees cerevisiae*, could be due to inheritable changes in protein structure. At that time, the concern with the mad-cow disease, whose etiologic agent appears to be composed only of proteins, made the scientific community more receptive to unorthodox ideas regarding inheritance. As we shall see, we now can transform cells to alternative "states" with purified prions, a feat that would have postponed the recognition of DNA as the genetic material, and reinforced the hypothesis that proteins were the genetic material, would one of the unorthodox "genes" have been chosen in the pioneering transformation experiments!

Here, we will review only a few cases of structural and regulatory inheritance, due to space constraint. Indeed, an ever increasing array of phenomena is now attributed to prions and prion-like elements. Only those prototypic are discussed below.

STRUCTURAL HEREDITY

At the present time, two different kinds of structural heredity have been clearly demonstrated, that of prions, which is based on the structural changes in a single polypeptide, and the cytotaxis, in which a large macromolecular complex is concerned.

Prions of *S. cerevisiae* and *P. anserina*

The term "prion" for *pr*oteinaceous *in*fectious particle was first proposed by Prusiner [10] to characterize the etiologic agents of some, at the time, bizarre diseases of mammals called transmissible spongiform encephalopathies (TSEs), including scrapie in sheep, as well as Kuru and Creutzfeldt–Jakob disease in Human. A basic definition of this term is the following: a protein able to adopt two distinct conformations, one of which can convert the other one. Usually, prion proteins may adopt monomeric or oligomeric states. The protein can change spontaneously from the monomeric to the oligomeric forms with a low frequency. Importantly, oligomers trigger the switch of the monomers towards the oligomeric and infectious form. This auto-catalytic process leads thus to the depletion of the monomers and to the accumulation of the oligomers. TSEs are probably caused by an aberrant folding of the PrP^c protein into the infectious PrP^{Sc} form [11].

Afterwards, this concept was successfully extended to explain the peculiar features of two non-mendelian elements of *Saccharomyees cerevisiae*, [PSI⁺] and [URE3] [9]. Now, the term prion is no longer restricted to TSE agents, but refers to any protein able to adopt an infectious conformation. The prion transition alters the function of the protein and consequently the phenotype of the cell. The aggregated state, as well as the associated phenotype, is infectious and stably transmitted from generations to generations both by mitosis and meiosis. Thus, yeast prions act as protein-based genetic element corresponding to an elegant epigenetic heredity. Several genetic criteria have been retained to suggest a prion behavior for a cellular protein [12]: (1) a prion can be cured, but it can reappear in the cured strain with a constant frequency because the protein able to change to an infectious form is still present; (2) overproduction of a protein capable of becoming a prion increases the frequency of the prion arising *de novo*; and (3) if the prion phenotype is due to the absence of the normal form of the protein, then the phenotype of null mutant of the gene for the protein is the same as that of the strain containing the aggregated prion. Also this gene is required for the prion to propagate.

TABLE 5.1 Summary of the Prion Proteins in Different Organisms

Protein	Organism	Prion	Infectious Prion Conformation*	Reference
PrP	Mammals	TSE agent	Suspected	[11]
HET-s	P. anserina	HET-s	Yes	[14]
Sup35	S. cerevisiae	[PSI+]	Yes	[39,110,111]
Ure2	S. cerevisiae	[URE3]	Yes	[112]
Rnq1	S. cerevisiae	[PIN+]	Yes	[113]
New1	S. cerevisiae	[NU+]	?	[29]
Swi1	S. cerevisiae	[SWI+]	?	[114]
Cyc8	S. cerevisiae	[OCT+]	?	[115]
Mca1	S. cerevisiae	[MCA]	?	[25]
Pma1 (?)	S. cerevisiae	[GAR+]	?	[116]
CPEB	A. californica	–	?	[17]

*Indicate whether the synthetic prion protein transforms prion-free cells to prion-containing cells. Two studies suggest that a dozen additional yeast proteins, not indicated in the table, may also be prion [29,117]

The first two yeast proteins obeying the above criteria and showed to be true prions in yeast were eRF3 and Ure2p. [PSI+], the prion of the release factor eRF3, also called Sup35p, affects the efficiency of translation termination. This may result in significant morphological or physiological switch when the transition is made [13]. [URE3], the prion of the protein Ure2p alters nitrogen catabolism. Although the genetic, biochemical, and cell biological analysis of these two prions, especially [PSI+], has boosted the comprehension of prion properties, the definitive demonstration that a protein is infectious based on cell transformation was obtained in the ascomycete fungus P. anserina [14]. This organism contains a true prion that displays the expected properties, the HET-s protein involved in heterokaryon incompatibility [15]. Prion aggregates of HET-s obtained from recombinant protein made in E. coli were introduced by ballistic transformation into prion-free cells of P. anserina and were shown to induce a phenotypic conversion towards the prion-containing cells [14]. This demonstration is formally equivalent to transformation experiments conducted by Avery et al. [16] proving that DNA is the support of genetic information. To date, this kind of transformation demonstration has been performed for only few additional examples in S. cerevisiae but an ever-growing number of proteins are strongly suspected to be prions (Table 5.1). For most prions, only the monomeric form has a biological function. To date, the only exception to this rule is HET-s, which appears active only in the oligomeric and infectious conformation [15]. This is also suspected for Aplysia CPEB [17], a neuronal mRNA translation regulator. This case is somewhat special, as it was shown to have prion properties in yeast, but as yet not in animals.

Domain analysis showed that the yeast and P. anserina prions contain a modular prion domain, dispensable for the cellular function but required and sufficient for the prion properties [18,19]. Most proteins fused to it behave like a prion. For example, in cells expressing a fusion protein composed of a prion domain and the GFP, two populations of cells may be observed, one with a homogeneous cytoplasmic fluorescence corresponding to the monomeric form and the other displaying intense punctuated foci due to oligomerization of the fusion protein [19]. Studies have focused on these domains to detect important features. In S. cerevisiae, an essential feature is richness in glutamines (Q) and asparagines (N), since all known yeast prion have such a domain. The sequence has probably no importance, because a prion domain with a sequence randomly shuffled is still able to form prion [20]. A bioinformatic analysis of the yeast genome performed to select gene coding a protein with a Q/N rich domain [21] permitted the identification of the Rnq1 protein (Rich in N and Q) that has prion properties, confirming the importance of this feature. However, it may not be universal, since the prion domains of Het and PrP are not Q/N rich and exhibit no obvious bias in their amino-acid composition. Despite their variation in primary sequences, all prions appear to adopt a similar conformation [see Ref. 22 for a review]. The monomer is soluble, rich in α-helix, and protease sensitive. The infectious form is oligomeric, rich in β-sheet, and partially protease resistant. The oligomers of all prions are abled to form amyloid fibers. These fibrous aggregates, identified by a birefringence when stained with congo red, are characteristically composed of proteins rich in β-sheet structures. These kinds of fiber are detected in prion disease in mammals in the brain of affected individuals and also *in vitro* for recombinant yeast prion proteins. Note that these fibers are not restricted to prion proteins; indeed some native amyloïds exist like the curlin protein in *E. coli* and even in the silk of some spiders. Progress is being are made in the understanding of the structural basis of infectivity [23].

As expected for a mechanism based on protein conformations, chaperones are implicated in the stability of prions. Hsp104, Hsp70, and Hsp40 families of chaperon, but also Hsp70 or Hsp40 activating proteins, are involved in this regulation. Hsp104 breaks aggregates into smaller ones; Hsp40 and Hsp70 help to refold the proteins into their native conformation. Hsp104p was the first chaperone identified to modulate prion [24]. Surprisingly, both over-expression and depletion of Hsp104 led to the loss of [*PSI*⁺]. Hsp104 is required to break large aggregates of prion producing new oligomers efficient for polymerization or seeds. The normal level of Hsp104 permits keeping seeds large enough to not be refolded by Hsp40/Hsp70, as well as to have an adequate number of seeds for efficient segregation during cellular division. Among chaperones, Hsp104 appears to be the major player as it is required for all the yeast prions, except [MCA] [25]. Moreover, the first compound destabilizing [*PSI*⁺], guanidium hydrochloride, is an inhibitor of Hsp104 [26]. However, Hsp104 over-expression affects only [*PSI*⁺] and [MCA], indicating a complex role for Hsp104 with a general effect on yeast prions and another activity specific to [*PSI*⁺] and [MCA]. The effect of over-expression/depletion of members of the families Hsp40 and Hsp70 on the prion stability is more complex than that for Hsp104 (see Ref. 27 for a review).

The search for others players modulating prion stability provided one of the most astonishing results in prion studies: the appearance of [*PSI*⁺] is itself controlled by another prion [28]. In the absence of this prion, called [*PIN*⁺] for PSI-inducible in yeast cells, PSI is unable to appear. This PIN element is usually due to the prion conformation of Rnq1, but experimentally several other prions including [URE3] permit the formation of [*PSI*⁺] independently of Rnq1 [29]. The effect of [*PIN*⁺] is not limited to [*PSI*⁺], as it also influences the appearance of [URE3] [30]. These kinds of interaction seems to be not restricted to positive ones as it is suggested that [*PSI*⁺] and [URE3] antagonize each other's propagation and *de novo* appearance [31]. Destabilizing interactions between [*PSI*⁺] and [*PIN*⁺] have also been described [32]. These data exemplify interactions between prions in a manner similar to interactions between alleles of different genes, emphasizing the similarity between true genes and "prion genes".

The similarity of prions to true genes is even more pronounced, since "allelic variants" called strains have been discovered for several prions [30,33–36]. Indeed, the observation that cells with the prion [*PSI*⁺] may present different stabilities of [*PSI*⁺] during cell division and different expressivity of the phenotype led to the proposal that strains of [*PSI*⁺] exist [33]. The same observation was also made with TSE diseases that would appear to be caused by various strains of infectious agents [35,37,38]. The question is how a protein able to switch to an inactive and infection state may be connected to several phenotypes. The answer lies in the ability of a unique prion protein to adopt many distinct infectious conformations [36,39]. As the first aggregated proteins appear spontaneously, one conformation is adopted among many possibilities. Then, this conformation acts as a template for further aggregation and is accurately transmitted to the successive cycles of conversion. Each conformation gives birth to a strain or variant of prion presenting distinct properties: stability, structure of the amyloid fibers, number of seeds, but also proportion of the prion protein in the aggregated form versus the soluble form [36,39,40]. At the level of the organism, these different strains trigger phenotypes with a more or less pronounced severity or effects. Interestingly, some are able to act as template on proteins with a different primary structure. This enables for example, some of the conformers from one species, but not all of them, to transmit its conformation to the homologous prion protein from a different species, a process known as barrier species crossing [41].

Prions are clearly endowed with the ability to transmit information from one cell generation to the next one and importantly the aggregated form can be purified and used in transformation experiments. Some are even responsible for transmitting diseases in mammals, clearly demonstrating their stability outside cells or organisms. They should thus be considered as true "hereditary units" in their ability to carry genetic information. At the present time, their actual role in cell physiology is unclear. Mammalian prions are clearly detrimental infectious agents. The *P. anserina* prion may be beneficial [15], while, in yeast, prions are regarded either as enabling adaptation [13] or diseases [42].

Self-Driven Assembly of Hsp60 Mitochondrial Chaperonin

Unlike prions, which can be viewed as abnormal proteins, the *S. cerevisiae* Hsp60 chaperonin provides a clear example of a structure catalyzing its own folding. It demonstrates the necessity for correctly folded pre-existing oligomers to ensure the correct folding of further monomers [43]. Some proteins, which are imported from the cytosol into mitochondria, cross the mitochondrial membranes in an unfolded conformation and then are folded in the matrix by Hsp60. Monomers of Hsp60 form a complex, arranged as two stacked 7-mer rings. Once assembled in the matrix of mitochondria, these 14-mer complexes bind unfolded proteins to catalyze their proper folding in an ATP-dependent manner. But Hsp60 proteins are also encoded by a nuclear gene and translated in the cytosol as precursors, which are then translocated into the mitochondrial matrix. So, how could they assemble themselves without pre-existing 14-mer complexes to fold them?

67

To address this question, Cheng et al. took advantage of a temperature-sensitive lethal mutation in the *hsp60* gene [43]. At 23°C, the *hsp60*[ts] mutant cells grew normally but when the temperature was shifted to 37°C, the mutant cells stopped growing within one generation, because the impaired Hsp60 complex fails in folding and assembly of imported mitochondrial proteins. An attempt to rescue the growth deficient phenotype of *hsp60*[ts] mutant strains was set up with a high copy plasmid, containing the coding sequence of the wild-type Hsp60 precursor, driven by the inducible galactose promoter. Cultures were first shifted from 23°C to 37°C and two hours later, expression of wild-type *hsp60* was induced by addition of galactose. In these conditions, the growth deficient phenotype of the mutant strain was not rescued. But strikingly, when expression of wild-type Hsp60 subunits was induced by addition of galactose for two hours before the temperature shift to 37°C, the mutant cells could grow. This means that wild-type hsp60 complexes can rescue the mutant phenotype at restrictive temperature only when expressed at permissive temperature, indicating that Hsp60 is required for its own assembly. In another words, newly Hsp60 imported subunits can be assembled only if pre-existing Hsp60 complexes are present in the matrix of mitochondria. More generally, this study strongly suggests that biogenesis of organelles such as mitochondria are probably not a *de novo* process, but rather relies on pre-existing structures, acting as a template. If this template is lost along the path, although the protein subunits are produced, no functional organelles would be made.

Cytotaxis of Cilia and Other Complex Structures

Prions and Hsp60 are homopolymers of a single protein. However, in the cells, most structures are built up from several different polypeptides and additional molecules, such as RNAs and cofactors. While in many cases it has been shown that these complex structures are able to correctly fold themselves spontaneously, often with the help of chaperones, in other cases the pre-existence of some structural information is necessary to obtain a correct organization. This was first shown by Beisson and Sonneborn on the orientation of cilia in *paramecium* [5]. Ciliates, like *paramecium*, are large cells that display a complex organization. In particular, their cortex is endowed with cilia that are all oriented in the same direction, permitting efficient swimming. Beisson and Sonneborn "grafted" in an inverted orientation rows of cilia in *Paramecium aurelia*, producing "variants" with abnormal swimming behavior [5]. These variants could be maintained over 800 mitotic generations and maintained in sexual crosses. This heritability is due to the fact that pre-existing cilia direct the correct insertion and orientation of newly formed cilia. This process whereby an old cellular structure orders a new one was called cytotaxis [44]. Cytotaxis of cilia/flagella have also been described in *Tetrahymena* [45] *Chlamydomonas* [46], and *Trypanosoma* [47]. Additional examples for other structures have been described in *Paramecium*, including handedness [48,49] and doublets [see Ref. 50 for a review]. Even in *S. cerevisiae*, cortical inheritance has been described [51]. The mechanism involved in cilium insertion in *Paramecium* has been analyzed at the ultra-structural level [52] and mutant searches [50] have uncovered nuclear *Tetrahymena* and *Paramecium* mutants with altered cortical elements, which should enable us to fully understand, at the molecular level, how the old structure directs the construction of the new one.

MIXED HEREDITY: A PRION THAT PROPAGATES BY COVALENT AUTO-ACTIVATION

In this section, we will present one process that stands at the border between the structural and the regulatory inheritance. It is based on the inheritable auto-catalytic cleavage of a protease [53]. The yeast protease B, PrB, is a subtilisin/furin class serine protease derived from a larger, catalytically inactive pro-form encoded by the gene *PRB1*. The final steps in the maturation of the pro-enzyme PrB are sequential truncations occurring in the lysosome-like yeast vacuole, catalyzed by protease A, PrA, and finally PrB itself. Mature PrB protease activates other vacuolar hydrolases such as carboxipeptidase Y (CpY), whose activity can be easily assayed.

Deletion of *PEP4*, encoding the PrA precursor, leads to accumulation of the immature form of PrB and therefore loss of its activity, as seen by lack of CpY activity. However, the disappearance of mature PrB after deletion of *PEP4* is progressive. CpY activity can be detected in *pep4Δ* strains for more than 20 mitotic generations. This hysteresis of PrB activity is referred to as "phenotypic lag" and it is believed to reflect dilution during growth of PEP4 mRNA and PrA protease. The phenotypic lag was initially observed during growth on dextrose medium because dextrose represses *PRB1* transcription. Roberts and Wickner [53] tested if this lag might be prolonged after transfer onto glycerol medium, which does not repress *PRB1* expression. The authors sporulated a diploid heterozygous for *PEP4* deletion and germinated the meiotic products on glycerol medium. They found that *CpY* remained active indefinitely, even in the colonies derived from the *pep4Δ* spores. However, when transferred onto dextrose medium, which represses *PRB1* transcription, CpY activity of these *pep4Δ* cells was progressively lost and not restored by a return to glycerol medium. The "PrB$^+$" state expressing the CpY hydrolase is infectious during cytoduction experiments (i.e. cytoplasmic mixing without karyogamy), even when both donor and recipient *pep4Δ* strains are grown on glycerol. This was the demonstration that the PrB$^+$ state is triggered by a cytoplasmic and infectious factor called [β]. They further showed that the cells mutated for PrB do not contain [β] and that the over-expression of *PRB1* increases the frequency of [β] appearance in *pep4Δ* cells that have been previously cured by extended growth on dextrose medium. In this system, PrpA is only needed for the initial conversion of PrB in the absence of [β]. To the authors, the fairly unconventional behavior of the PrB$^+$ state is reminiscent of that of structure-based prion, indicating that any enzyme could be a prion, provided that its activity depends on self-modification *in trans* and that there is a mechanism by which it can be transmitted from individual to individual [54]. As we will see below, an additional example of such behavior exhibited by kinases involved in signaling indicates that this is indeed the case. However, because these kinases are involved in regulation, we will discuss them in the next section dealing with regulatory inheritance.

REGULATORY INHERITANCE

There is now a large body of literature dealing with the behavior of regulatory networks, especially their ability to generate emergent properties enabling cells to finely tune their response to various environmental changes [see Refs 55–57 for recent reviews]. In some cases, these properties result in the generation of bistable states that are inheritable in a more or less faithful fashion [58–63]. Below we will discuss three examples of such regulatory inheritance.

The Lactose Operon and its Positive Feedback Loop

In their seminal studies of 1956 [4], Novick and Weiner, and later on Cohn and Horibata [64–66], showed that under defined conditions, it is possible to obtain an epigenetic inheritability of the activation status of the lactose operon in *E. coli*. Indeed, when grown on a gratuitous inducer at low concentrations, *E. coli* are either not induced for their lactose operon or fully induced, and are never found in an intermediate state. Non-induced cells can change spontaneously towards the induced state with a constant probability. Thus, when transferred from a medium lacking the inducer towards a medium containing the inducer, the population accumulates more and more cells that have made the transition towards the active state. These cells do not invade the population since the induced cells grow more slowly than the non-induced ones, permitting a dynamic equilibrium in the population. That both the on and off states are inheritable was demonstrated by diluting an equilibrated population in new medium that contained the inducer at an inoculum of one cell per new culture. Two kinds of culture where obtained, one composed of fully induced cells that originated from a bacterium that was already induced and the other which accumulated induced cells at the level of the parental culture before dilution. The cause of this behavior is the presence of a self-positive regulatory loop in the lactose regulation, whereby the entrance of enough inducer inside the cell activates the operon and especially the production of permease, which in turn allows more inducer to enter the cell. It was recently demonstrated that the stochastic complete dissociation of the lactose repressor, which binds as a tetramer, triggers the initial burst of production of the permease [67]. Other information gained from the study of the lactose operon is that the behavior of the cells is strongly dependent upon the inducer and glucose concentrations [68]. Glucose represses the operon while the inducer activates it. Three main regulatory behaviors can be adopted by the cells: (i) the monostable state induced at low glucose and high inducer concentrations; (ii) the monostable state uninduced at high glucose and low inducer lever; and (iii) the bistable state described above at intermediate concentrations. Importantly, cells that are placed in these intermediate concentrations will behave differently if they originate from the monostable induced or monostable uninduced conditions.

This lactose operon is prototypic of systems with a positive auto-regulatory loop, and mathematic models describing their properties are available [68–70]. More complex inheritable units can be envisioned. They are all based on positive auto-regulatory loops or derivative the thereof, the reciprocal double negative loop – the one present in the C1/Cro interaction of the lambda phage. That these transcription factors that negatively regulate other are able to produce each an epigenetic inheritable switch has been known for a long time [62,71]. Readers interested in the emergent properties of regulatory networks, including epigenetic inheritance, can refer to a number of excellent recent reviews [55–57]. Most of the research carried out today is performed on man-made regulatory networks. We discuss below two examples of regulatory inheritance encountered in wild organisms: the sectors of filamentous fungi and yeast phenotypic switches.

69

Crippled Growth, a Self-Sustained and Mitotically Inheritable Signaling Pathway in the Filamentous Fungus *Podospora anserina*

P. anserina is a saprophytic filamentous fungus that has been used as a model organism for decades. In the 1990s, Silar and his colleagues noticed that sectors of altered growth could be seen on *P. anserina* growing thalli [72]. This cell degeneration phenomenon, called Crippled Growth (CG), was easily visible macroscopically, displaying highly pigmented, flat, and female-sterile mycelium as opposed to normal growth (NG). Curiously, the development of these sectors occurred only in special genetic or environmental conditions [72,73]. The switch is controlled in both directions by environmental stimuli [72]. It was rapidly demonstrated that no nucleic acid was involved in the genesis of these sectors and that the presence of *C*, a cytoplasmic and infectious factor, was associated with CG. The mycelium of *P. anserina* can thus exhibit a bistability at the morphological level. Similar phenomena were previously described and reported to be very frequent in filamentous ascomycetes (see Ref. 74 for a review). They were generally due to the presence in the cell of cytoplasmic and infectious factors, whose properties appear strikingly similar to prions. Apart from CG, only the "secteur" phenomenon of *Nectria haematococca* has been studied [75], but in this instance no clear model on how the infectious factor is generated is presently available. In the case of CG, a genetic analysis [73] permitted the retrieval of numerous genes, which are required to produce *C*. Some of these "*IDC*" genes (Impaired in the Development of Crippled growth) were cloned and shown to encode a MAP kinase kinase kinase (MAPKKK) [76] and a MAP kinase kinase (MAPKK) [77]. These two proteins are members of a large family of kinases present in all eukaryotes. They act in a sequential manner, i.e. the MAPKKK activates by phosphorylation the MAPKK, which in turn activates by phosphorylation a MAP kinase (or MAPK). These *IDC* mutants unable to produce *C* were null mutants of either the MAPKKK or MAPKK genes. Further genetic inactivation of the gene coding the downstream MAP kinase also showed it to be a key element in the genesis of *C* [77]. Moreover, over-expression of the MAPKKK and MAPK was shown to facilitate the development of CG [76,77]. (i) Presence of a cytoplasmic and infectious factor, (ii) necessity of a gene for its propagation, and (iii) increased frequency of appearance of the infectious factor when the gene is over-expressed are properties exhibited by genes coding for prions. Here, the three genes coding the MAPKKK, MAPKK and MAPK display these properties. A model related to that of prions but based on an autocatalytic activation loop in the MAPK cascade has thus been proposed to account for the *C* element [76,77]. Some element(s) downstream of the MAPK would be able to activate directly or indirectly the upstream MAPKKK. In this model, the *C* element corresponds to components of the cascade in the active state, which are able to activate *in trans* other molecules that are in the inactive state. This results in the complete conversion of the inactive factors to their active form. This strikingly resembles the ability of prions to promote their own aggregation or that of the [β] "prion" in *S. cerevisiae* to promote its own maturation. This model is supported by experiments in *Xenopus* eggs, in which it was demonstrated that the presence of a positive self-regulation in the p42 MAP kinase cascade entails the presence of only two states: one in which no active MAPK is present and the other in which all MAPK molecules are active, the intermediate states being transient [78]. It was also shown that transfer of cytoplasm from an activated egg to an inactivated one results in complete activation of the MAPK cascade in the recipient [78]. This property is conserved over three transfers, and therefore in conditions where no cytoplasm originating from the first egg is present. In essence, this is strikingly similar to the cytoplasmic and infectious factors detected in CG and related phenomena.

This regulatory inheritance has many properties in common with prions. It however displays several differences. First, it relies on many proteins (the whole signaling cascade or at least a subset of the cascade). This implies that the genetic basis is more complex than for prion [73]. In the case of CG, additional factors have been identified and one of them has been shown to be necessary for producing the *C* element, but likely not to be present in the regulatory loop [79]. Many genes restricting the spread of *C* have also been identified, adding another level of complexity [73]. Second, the development and/or spreading of *C* are highly dependent upon the environmental conditions, a property also exhibited by the [β] factor. Because regulatory networks can adopt complex behavior, depending upon the level of expression of the key "flexibility loci" [80], it is not surprising that the determinism of CG is quite complex and depends upon numerous genetic and environmental factors.

The White/Opaque Switch of *Candida albicans*, an Epigenetic Switch at the Transcription Level

Many species of yeast have the ability to switch at various frequencies between different states [13,81–86]. These switches may be caused by classical transcriptional gene silencing (see Chapter 13 by Malagnac and Silar in this book, for a review) or prions [13]. However, regulatory inheritance may also be involved. The most studied of the switches is the white/opaque transition exhibited by *C. albicans* [87]. This transition is present in this diploid fungus only when homozygous for the mating type [88,89]. The cells may then adopt two morphologies: roundish cells forming white colonies and bigger, more elongated cell forming colonies that are more translucent. In fact, the two types of cell differ by an impressive array of differences [90–92]. *C. albicans* causes mycosis in humans and the switch likely enables the fungus to adapt to the various niches it will encounter in the human body [93,94].

To understand how the transition is controlled, genes down-regulated by the a1-α2 heterodimers encoded at the *C. albicans* mating type were searched, since the white/opaque switch is specific of strains homozygous for their mating-type [88]. Transcriptomic or chromatin-immunoprecipitation approaches identified the WOR1/TOS9 transcription factor as specifically expressed in opaque cells [95–97]. Gene inactivation of WOR1 showed that the cells were locked in the white state. Ectopic expression of WOR1, even as a single pulse or in cells heterozygous for the mating type, was sufficient to convert the whole cell population to the opaque state. Finally, it was shown that WOR1 binds the promoter of its own gene and thereby activates its own expression [97]. Overall, these data permitted the formulation of a model for the white/opaque transition based on the self-activation of the WOR1 transcription factor [95–97]. WOR1 is absent in white cells. Random fluctuation in the transcription of the WOR1 locus permits the expression of a few molecules of WOR1, resulting in further transcription of the WOR1 gene, locking the cell in a state with a high concentration of WOR1. The other targets of WOR1 are then regulated, promoting the physiological and morphological changes to the opaque state. Although it has not yet been formally demonstrated that the opaque state is infectious towards the white state, the similarity of this bistable system with the ones created by classical, β-type, and C-type infectious prions is evident.

Recent studies showed that the WOR1 positive regulatory loop is embedded in a complex network of transcription factors with positive and negative feedback loops [98]. The multiplicity of the feedbacks appears to ensure a faithful transmission of the white and opaque states through numerous cell generations and permits us to explain the previously known roles of various transcription factors and chromatin remodeling factors in the control of the transition [99–102]. As with the Crippled Growth of *P. anserina*, the environment controls the switch in both directions. While high temperature triggers the opaque to white switch [87,103], numerous factors are able to trigger the white to opaque transition [104,105], and a recent study has shown that slowing the cell cycle by many means is sufficient to increase the white to opaque switch frequency [106].

CONCLUSION

The various examples presented above show that epigenetic states can be conferred in many ways (Fig. 5.1), provided that an auto-regulatory loop (or a double repression) is present. The presence of this loop ensures that two mutually exclusive states may be exhibited by cells with an identical genome and grown in the same conditions. In the case of the structural

71

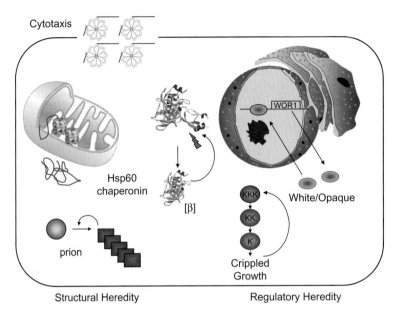

Cytotaxis

Hsp60 chaperonin

[β]

prion

WOR1

White/Opaque

KKK

KK

K

Crippled Growth

Structural Heredity

Regulatory Heredity

FIGURE 5.1
Schematic diagram of the prions and prion-like elements of eukaryotes discussed in this chapter.
(Please refer to color plate section)

inheritance that we have presented, the loops ensure the faithful reproduction of a structure made of proteins (prions, Hsp60, or cilia). There is, however, the suggestion that another component of the cell, the membranes, may adopt alternative states [107]. In general, the influence of the environment on this kind of inheritance is moderate. On the contrary, the regulatory inheritance is usually greatly influenced by the environment, since any modification in the concentration of key factors [80] under the influence of external stimuli may drastically alter the behavior of the pathway. Importantly, in the case of regulatory inheritance the phenotype of the cells will depend upon their history – however, not in a directed fashion as in a Lamarckian inheritance. Often, this inheritance results from the emergent properties of complex networks or structural properties of domains that adopt particular structures. It is thus not easy to know whether the inheritable behavior is a by-product of these, or whether it participates in a process essential to the life cycle. If the white/opaque transition appears to confer a selective advantage to *C. albicans*, what of yeast prions and Crippled Growth? Certainly in some cases, prions and the prion-like element are involved in differentiation processes [108]. However, human prions clearly cause severe diseases and it has been proposed that regulatory inheritance may be involved in cancer formation [109]. This regulatory inheritance is presently known in both eukaryotes and prokaryotes, and occurs at all levels of gene regulation (transcription, signal transduction, carbon metabolism...). As yet the full scope of this kind of inheritance based on prion and prion-like phenomena is unknown. We here have presented data obtained with a few model microorganisms, since they are more easily tractable than the multicellular animals and plants. However, we propose that the same epigenetic mechanisms are prevalent in all organisms.

References

1. Nanney DL. Epigenetic control systems. *Proc Natl Acad Sci USA* 1958;**44**:712–17.

2. Delbrück M. Comment of an article. In: Sonneborn TM, Beale GH, editors. *Unités biologiques douées de continuité génétique*. Paris France: CNRS; 1949. pp. 33–4.

3. Marcou D. Notion de longévité et nature cytoplasmique du déterminant de la sénescence chez quelques champignons. *Ann Sci Nat Bot Sér* 1961;**12**(2):653–764.

4. Novick A, Weiner M. Enzyme induction as an all-or-none phenomenon. *Proc Natl Acad Sci USA* 1957;**43**:553–66.

5. Beisson J, Sonneborn TM. Cytoplasmic inheritance of the organization of the cell cortex in *Paramecium aurelia*. *Proc Natl Acad Sci USA* 1965;**53**:275–82.

6. Beisson J. *Non-nucleic acid inheritance and epigenetic phenomena*. Cell Biology. San Francisco USA: Academic Press Inc.; 1977. pp. 375–421.

7. Kirk JTO. Roots: The discovery of chloroplast DNA. *BioEssays* 1986;**4**:36–8.

8. Mounolou JC, Lacroute F. Mitochondrial DNA: an advance in eukaryotic cell biology in the 1960s. *Biol Cell* 2005;**97**:743–8.

9. Wickner RB. [URE3] as an altered URE2 protein: evidence for a prion analog in *Saccharomyces cerevisiae*. *Science* 1994;**264**:566–9.

10. Prusiner SB. Novel proteinaceous infectious particles cause scrapie. *Science* 1982;**216**:136–44.

11. Legname G, Baskakov IV, Nguyen HO, Riesner D, Cohen FE, DeArmond SJ, et al. Synthetic mammalian prions. *Science* 2004;**305**:673–6.

12. Wickner RB, Shewmaker F, Kryndushkin D, Edskes HK. Protein inheritance (prions) based on parallel in-register beta-sheet amyloid structures. *Bioessays* 2008;**30**:955–64.

13. True HL, Berlin I, Lindquist SL. Epigenetic regulation of translation reveals hidden genetic variation to produce complex traits. *Nature* 2004;**431**:184–7.

14. Maddelein ML, Dos Reis S, Duvezin-Caubet S, Coulary-Salin B, Saupe SJ. Amyloid aggregates of the HET-s prion protein are infectious. *Proc Natl Acad Sci USA* 2002;**99**:7402–7.

15. Coustou V, Deleu C, Saupe S, Begueret J. The protein product of the het-s heterokaryon incompatibility gene of the fungus *Podospora anserina* behaves as a prion analog. *Proc Natl Acad Sci USA* 1997;**94**:9773–8.

16. Avery OT, McLeod CM, McCarty M. Studies on the chemical nature of the substance inducing transformation of pneumococcal types. *J Exp Med* 1944;**79**:137–58.

17. Si K, Lindquist S, Kandel ER. A neuronal isoform of the aplysia CPEB has prion-like properties. *Cell* 2003;**115**:879–91.

18. Li L, Lindquist S. Creating a protein-based element of inheritance. *Science* 2000;**287**:661–4.

19. Patino MM, Liu JJ, Glover JR, Lindquist S. Support for the prion hypothesis for inheritance of a phenotypic trait in yeast. *Science* 1996;**273**:622–6.

20. Ross ED, Baxa U, Wickner RB. Scrambled prion domains form prions and amyloid. *Mol Cell Biol* 2004;**24**:7206–13.

21. Sondheimer N, Lindquist S. Rnq1: an epigenetic modifier of protein function in yeast. *Mol Cell* 2000;**5**:163–72.

22. Baxa U, Cassese T, Kajava AV, Steven AC. Structure, function, and amyloidogenesis of fungal prions: filament polymorphism and prion variants. *Adv Protein Chem* 2006;**73**:125–80.

23. Tessier PM, Lindquist S. Unraveling infectious structures, strain variants and species barriers for the yeast prion [PSI+]. *Nat Struct Mol Biol* 2009;**16**:598–605.

24. Chernoff YO, Lindquist SL, Ono B, Inge-Vechtomov SG, Liebman SW. Role of the chaperone protein Hsp104 in propagation of the yeast prion-like factor [psi+]. *Science* 1995;**268**:880–4.

25. Nemecek J, Nakayashiki T, Wickner RB. A prion of yeast metacaspase homolog (Mca1p) detected by a genetic screen. *Proc Natl Acad Sci USA* 2009;**106**:1892–6.

26. Jung G, Masison DC. Guanidine hydrochloride inhibits Hsp104 activity in vivo: a possible explanation for its effect in curing yeast prions. *Curr Microbiol* 2001;**43**:7–10.

27. Rikhvanov EG, Romanova NV, Chernoff YO. Chaperone effects on prion and nonprion aggregates. *Prion* 2007;**1**:217–22.

28. Derkatch IL, Bradley ME, Zhou P, Chernoff YO, Liebman SW. Genetic and environmental factors affecting the de novo appearance of the [PSI+] prion in *Saccharomyces cerevisiae*. *Genetics* 1997;**147**:507–19.

29. Derkatch IL, Bradley ME, Hong JY, Liebman SW. Prions affect the appearance of other prions: the story of [PIN(+)]. *Cell* 2001;**106**:171–82.

30. Bradley ME, Edskes HK, Hong JY, Wickner RB, Liebman SW. Interactions among prions and prion "strains" in yeast. *Proc Natl Acad Sci USA* 2002;**99**(Suppl 4):16392–9.

31. Schwimmer C, Masison DC. Antagonistic interactions between yeast [PSI(+)] and [URE3] prions and curing of [URE3] by Hsp70 protein chaperone Ssa1p but not by Ssa2p. *Mol Cell Biol* 2002;**22**:3590–8.

32. Bradley ME, Liebman SW. Destabilizing interactions among [PSI(+)] and [PIN(+)] yeast prion variants. *Genetics* 2003;**165**:1675–85.

33. Derkatch IL, Chernoff YO, Kushnirov VV, Inge-Vechtomov SG, Liebman SW. Genesis and variability of [PSI] prion factors in *Saccharomyces cerevisiae*. *Genetics* 1996;**144**:1375–86.

34. Schlumpberger M, Wille H, Baldwin MA, Butler DA, Herskowitz I, Prusiner SB. The prion domain of yeast Ure2p induces autocatalytic formation of amyloid fibers by a recombinant fusion protein. *Protein Sci* 2000;**9**:440–51.

73

35. Morales R, Abid K, Soto C. The prion strain phenomenon: molecular basis and unprecedented features. *Biochim Biophys Acta* 2007;**1772**:681–91.

36. Tanaka M, Chien P, Naber N, Cooke R, Weissman JS. Conformational variations in an infectious protein determine prion strain differences. *Nature* 2004;**428**:323–8.

37. Pattison IH, Millson GC. Further experimental observations on scrapie. *J Comp Pathol* 1961;**71**:350–9.

38. Pattison IH, Millson GC. Scrapie produced experimentally in goats with special reference to the clinical syndrome. *J Comp Pathol* 1961;**71**:101–9.

39. King CY, Diaz-Avalos R. Protein-only transmission of three yeast prion strains. *Nature* 2004;**428**:319–23.

40. Tanaka M, Collins SR, Toyama BH, Weissman JS. The physical basis of how prion conformations determine strain phenotypes. *Nature* 2006;**442**:585–9.

41. Tanaka M, Chien P, Yonekura K, Weissman JS. Mechanism of cross-species prion transmission: an infectious conformation compatible with two highly divergent yeast prion proteins. *Cell* 2005;**121**:49–62.

42. Nakayashiki T, Kurtzman CP, Edskes HK, Wickner RB. Yeast prions [URE3] and [PSI+] are diseases. *Proc Natl Acad Sci USA* 2005;**102**:10575–80.

43. Cheng MY, Hartl FU, Horwich AL. The mitochondrial chaperonin hsp60 is required for its own assembly. *Nature* 1990;**348**:455–8.

44. Sonneborn TM. The determinants and evolution of life. The differentiation of cells. *Proc Natl Acad Sci USA* 1964;**51**:915–29.

45. Ng SF, Frankel J. 180 degrees rotation of ciliary rows and its morphogenetic implications in *Tetrahymena pyriformis*. *Proc Natl Acad Sci USA* 1977;**74**:1115–19.

46. Feldman JL, Geimer S, Marshall WF. The mother centriole plays an instructive role in defining cell geometry. *PLoS Biol* 2007;**5**:e149.

47. Moreira-Leite FF, Sherwin T, Kohl L, Gull K. A trypanosome structure involved in transmitting cytoplasmic information during cell division. *Science* 2001;**294**:610–12.

48. Nelsen EM, Frankel J. Maintenance and regulation of cellular handedness in *Tetrahymena*. *Development* 1989;**105**:457–71.

49. Nelsen EM, Frankel J, Jenkins LM. Non-genic inheritance of cellular handedness. *Development* 1989;**105**:447–56.

50. Frankel J. What do genic mutations tell us about the structural patterning of a complex single-celled organism? *Eukaryot Cell* 2008;**7**:1617–39.

51. Chen T, Hiroko T, Chaudhuri A, Inose F, Lord M, Tanaka S, et al. Multigenerational cortical inheritance of the Rax2 protein in orienting polarity and division in yeast. *Science* 2000;**290**:1975–8.

52. Iftode F, Fleury-Aubusson A. Structural inheritance in Paramecium: ultrastructural evidence for basal body and associated rootlets polarity transmission through binary fission. *Biol Cell* 2003;**95**:39–51.

53. Roberts BT, Wickner RB. Heritable activity: a prion that propagates by covalent autoactivation. *Genes Dev* 2003;**17**:2083–7.

54. Roberts BT, Wickner RB. A new kind of prion: a modified protein necessary for its own modification. *Cell Cycle* 2004;**3**:100–3.

55. Brandman O, Meyer T. Feedback loops shape cellular signals in space and time. *Science* 2008;**322**:390–5.

56. Drubin DA, Way JC, Silver PA. Designing biological systems. *Genes Dev* 2007;**21**:242–54.

57. Sprinzak D, Elowitz MB. Reconstruction of genetic circuits. *Nature* 2005;**438**:443–8.

58. Becskei A, Seraphin B, Serrano L. Positive feedback in eukaryotic gene networks: cell differentiation by graded to binary response conversion. *EMBO J* 2001;**20**:2528–35.

59. Gardner TS, Cantor CR, Collins JJ. Construction of a genetic toggle switch in *Escherichia coli*. *Nature* 2000;**403**:339–42.

60. Kobayashi H, Kaern M, Araki M, Chung K, Gardner TS, Cantor CR, et al. Programmable cells: interfacing natural and engineered gene networks. *Proc Natl Acad Sci USA* 2004;**101**:8414–19.

61. Tchuraev RN, Stupak IV, Tropynina TS, Stupak EE. Epigenes: design and construction of new hereditary units. *FEBS Lett* 2000;**486**:200–2.

62. Toman Z, Dambly C, Radman M. Induction of a stable, heritable epigenetic change by mutagenic carcinogens: a new test system. *IARC Sci Publ* 1980:243–55.

63. Ajo-Franklin CM, Drubin DA, Eskin JA, Gee EP, Landgraf D, Phillips I, et al. Rational design of memory in eukaryotic cells. *Genes Dev* 2007;**21**:2271–6.

64. Cohn M, Horibata K. Analysis of the differentiation and of the heterogeneity within a population of *Escherichia coli* undergoing induced beta-galactosidase synthesis. *J Bacteriol* 1959;**78**:613–23.

65. Cohn M, Horibata K. Inhibition by glucose of the induced synthesis of the beta-galactoside-enzyme system of *Escherichia coli*. Analysis of maintenance. *J Bacteriol* 1959;**78**:601–12.

66. Cohn M, Horibata K. Physiology of the inhibition by glucose of the induced synthesis of the beta-galactoside enzyme system of *Escherichia coli*. *J Bacteriol* 1959;**78**:624–35.

67. Choi PJ, Cai L, Frieda K, Xie XS. A stochastic single-molecule event triggers phenotype switching of a bacterial cell. *Science* 2008;**322**:442–6.

68. Ozbudak EM, Thattai M, Lim HN, Shraiman BI, Van Oudenaarden A. Multistability in the lactose utilization network of *Escherichia coli*. *Nature* 2004;**427**:737–40.

69. Laurent M, Kellershohn N. Multistability: a major means of differentiation and evolution in biological systems. *Trends Biochem Sci* 1999;**24**:418–22.

70. Santillan M, Mackey MC. Quantitative approaches to the study of bistability in the lac operon of *Escherichia coli*. *J R Soc Interface* 2008;**5**(Suppl 1):S29–39.

71. Toman Z, Dambly-Chaudiere C, Tenenbaum L, Radman M. A system for detection of genetic and epigenetic alterations in *Escherichia coli* induced by DNA-damaging agents. *J Mol Biol* 1985;**186**:97–105.

72. Silar P, Haedens V, Rossignol M, Lalucque H. Propagation of a novel cytoplasmic, infectious and deleterious determinant is controlled by translational accuracy in *Podospora anserina*. *Genetics* 1999;**151**:87–95.

73. Haedens V, Malagnac F, Silar P. Genetic control of an epigenetic cell degeneration syndrome in *Podospora anserina*. *Fungal Genet Biol* 2005;**42**:564–77.

74. Silar P, Daboussi MJ. Non-conventional infectious elements in filamentous fungi. *Trends Genet* 1999;**15**:141–5.

75. Graziani S, Silar P, Daboussi MJ. Bistability and hysteresis of the 'Secteur' differentiation are controlled by a two-gene locus in *Nectria haematococca*. *BMC Biol* 2004;**2**:18.

76. Kicka S, Silar P. PaASK1, a mitogen-activated protein kinase kinase kinase that controls cell degeneration and cell differentiation in *Podospora anserina*. *Genetics* 2004;**166**:1241–52.

77. Kicka S, Bonnet C, Sobering AK, Ganesan LP, Silar P. A mitotically inheritable unit containing a MAP kinase module. *Proc Natl Acad Sci USA* 2006;**103**:13445–50.

78. Bagowski CP, Ferrell JE Jr.. Bistability in the JNK cascade. *Curr Biol* 2001;**11**:1176–82.

79. Jamet-Vierny C, Debuchy R, Prigent M, Silar P. IDC1, a Pezizomycotina-specific gene that belongs to the PaMpk1 MAP kinase transduction cascade of the filamentous fungus *Podospora anserina*. *Fungal Genet Biol* 2007;**44**:1219–30.

80. Bhalla US, Ram PT, Iyengar R. MAP kinase phosphatase as a locus of flexibility in a mitogen-activated protein kinase signaling network. *Science* 2002;**297**:1018–23.

81. Clemons KV, Hanson LC, Stevens DA. Colony phenotype switching in clinical and non-clinical isolates of *Saccharomyces cerevisiae*. *J Med Vet Mycol* 1996;**34**:259–64.

82. Lachke SA, Joly S, Daniels K, Soll DR. Phenotypic switching and filamentation in *Candida glabrata*. *Microbiology* 2002;**148**:2661–74.

83. Lachke SA, Srikantha T, Tsai LK, Daniels K, Soll DR. Phenotypic switching in *Candida glabrata* involves phase-specific regulation of the metallothionein gene MT-II and the newly discovered hemolysin gene HLP. *Infect Immun* 2000;**68**:884–95.

84. Fries BC, Goldman DL, Casadevall A. Phenotypic switching in *Cryptococcus neoformans*. *Microbes Infect* 2002;**4**:1345–52.

85. Enger L, Joly S, Pujol C, Simonson P, Pfaller M, Soll DR. Cloning and characterization of a complex DNA fingerprinting probe for *Candida parapsilosis*. *J Clin Microbiol* 2001;**39**:658–69.

86. Joly S, Pujol C, Schroppel K, Soll DR. Development of two species-specific fingerprinting probes for broad computer-assisted epidemiological studies of *Candida tropicalis*. *J Clin Microbiol* 1996;**34**:3063–71.

87. Slutsky B, Staebell M, Anderson J, Risen L, Pfaller M, Soll DR. "White-opaque transition": a second high-frequency switching system in *Candida albicans*. *J Bacteriol* 1987;**169**:189–97.

88. Miller MG, Johnson AD. White-opaque switching in *Candida albicans* is controlled by mating-type locus homeodomain proteins and allows efficient mating. *Cell* 2002;**110**:293–302.

89. Lockhart SR, Pujol C, Daniels KJ, Miller MG, Johnson AD, Pfaller MA, et al. In *Candida albicans*, white-opaque switchers are homozygous for mating type. *Genetics* 2002;**162**:737–45.

90. Soll DR. High-frequency switching in *Candida albicans*. *Clin Microbiol Rev* 1992;**5**:183–203.

91. Soll DR. Gene regulation during high-frequency switching in *Candida albicans*. *Microbiology* 1997;**143** (Pt 2):279–88.

92. Lan CY, Newport G, Murillo LA, Jones T, Scherer S, Davis RW, et al. Metabolic specialization associated with phenotypic switching in *Candida albicans*. *Proc Natl Acad Sci USA* 2002;**99**:14907–12.

93. Vargas K, Messer SA, Pfaller M, Lockhart SR, Stapleton JT, Hellstein J, et al. Elevated phenotypic switching and drug resistance of *Candida albicans* from human immunodeficiency virus-positive individuals prior to first thrush episode. *J Clin Microbiol* 2000;**38**:3595–607.

94. Lohse MB, Johnson AD. Differential phagocytosis of white versus opaque *Candida albicans* by *Drosophila* and mouse phagocytes. *PLoS One* 2008;**3**:e1473.

75

95. Huang G, Wang H, Chou S, Nie X, Chen J, Liu H. Bistable expression of WOR1, a master regulator of white-opaque switching in *Candida albicans*. *Proc Natl Acad Sci USA* 2006;**103**:12813–18.

96. Srikantha T, Borneman AR, Daniels KJ, Pujol C, Wu W, Seringhaus MR, et al. TOS9 regulates white-opaque switching in *Candida albicans*. *Eukaryot Cell* 2006;**5**:1674–87.

97. Zordan RE, Galgoczy DJ, Johnson AD. Epigenetic properties of white-opaque switching in *Candida albicans* are based on a self-sustaining transcriptional feedback loop. *Proc Natl Acad Sci USA* 2006;**103**:12807–12.

98. Zordan RE, Miller MG, Galgoczy DJ, Tuch BB, Johnson AD. Interlocking transcriptional feedback loops control white-opaque switching in *Candida albicans*. *PLoS Biol* 2007;**5**:e256.

99. Sonneborn A, Tebarth B, Ernst JF. Control of white-opaque phenotypic switching in *Candida albicans* by the Efg1p morphogenetic regulator. *Infect Immun* 1999;**67**:4655–60.

100. Perez-Martin J, Uria JA, Johnson AD. Phenotypic switching in *Candida albicans* is controlled by a SIR2 gene. *EMBO J* 1999;**18**:2580–92.

101. Klar AJ, Srikantha T, Soll DR. A histone deacetylation inhibitor and mutant promote colony-type switching of the human pathogen *Candida albicans*. *Genetics* 2001;**158**:919–24.

102. Srikantha T, Tsai L, Daniels K, Klar AJ, Soll DR. The histone deacetylase genes HDA1 and RPD3 play distinct roles in regulation of high-frequency phenotypic switching in *Candida albicans*. *J Bacteriol* 2001;**183**:4614–25.

103. Rikkerink EH, Magee BB, Magee PT. Opaque-white phenotype transition: a programmed morphological transition in *Candida albicans*. *J Bacteriol* 1988;**170**:895–9.

104. Huang G, Srikantha T, Sahni N, Yi S, Soll DR. CO(2) regulates white-to-opaque switching in *Candida albicans*. *Curr Biol* 2009;**19**:330–4.

105. Ramirez-Zavala B, Reuss O, Park YN, Ohlsen K, Morschhauser J. Environmental induction of white-opaque switching in *Candida albicans*. *PLoS Pathog* 2008;**4**:e1000089.

106. Alby K, Bennett R. Stress-induced phenotypic switching in *Candida albicans*. *Mol Biol Cell* 2009;**20**:3178–91.

107. Lockshon D. A heritable structural alteration of the yeast mitochondrion. *Genetics* 2002;**161**:1425–35.

108. Malagnac F, Silar P. Regulation, cell differentiation and protein-based inheritance. *Cell Cycle* 2006;**5**:2584–7.

109. Blagosklonny MV. Molecular theory of cancer. *Cancer Biol Ther* 2005;**4**:621–7.

110. Sparrer HE, Santoso A, Szoka FC Jr., Weissman JS. Evidence for the prion hypothesis: induction of the yeast [PSI+] factor by in vitro-converted Sup35 protein. *Science* 2000;**289**:595–9.

111. Tanaka M, Chien P, Naber N, Cooke R, Weissman JS. Conformational variations in an infectious protein determine prion strain differences. *Nature* 2004;**428**:323–8.

112. Brachmann A, Baxa U, Wickner RB. Prion generation in vitro: amyloid of Ure2p is infectious. *EMBO J* 2005;**24**:3082–92.

113. Patel BK, Liebman SW. "Prion-proof" for [PIN+]: infection with in vitro-made amyloid aggregates of Rnq1p-(132–405) induces [PIN+]. *J Mol Biol* 2007;**365**:773–82.

114. Du Z, Park KW, Yu H, Fan Q, Li L. Newly identified prion linked to the chromatin-remodeling factor Swi1 in *Saccharomyces cerevisiae*. *Nat Genet* 2008;**40**:460–5.

115. Patel BK, Gavin-Smyth J, Liebman SW. The yeast global transcriptional co-repressor protein Cyc8 can propagate as a prion. *Nat Cell Biol* 2009;**11**:344–9.

116. Brown JC, Lindquist S. A heritable switch in carbon source utilization driven by an unusual yeast prion. *Genes Dev* 2009;**23**:2320–32.

117. Alberti S, Halfmann R, King O, Kapila A, Lindquist S. A systematic survey identifies prions and illuminates sequence features of prionogenic proteins. *Cell* 2009;**137**:146–58.

CHAPTER 6

Chromosomal Position Effects and Gene Variegation: Impact in Pathologies

Caroline Schluth-Bolard[1,*], **Alexandre Ottaviani**[1,*], **Eric Gilson**[1,2,3], and **Frédérique Magdinier**[1,4]

[1]Laboratoire de Biologie Moléculaire de la Cellule, Ecole Normale Supérieure de Lyon, 69364 Lyon Cedex 07
[2]Department of Medical Genetics, Archet 2 Hopital, CHU de Nice
[3]Laboratory of Biology and Pathology of Genomes, University of Nice-Sophia Antipolis, CNRS UMR 6267/INSERM U998, Faculty of Medicine, Nice, France
[4]Laboratoire de Génétique Médicale et Génomique Fonctionnelle, Faculté de Médecine de la Timone, 13385 Marseille Cedex 5, France

INTRODUCTION

The genome of higher eukaryotes is composed of thousands of genes and even more interspersed non-coding sequences. Constraining up to tens of billions of bases within a nucleus of a few microns in diameter requires a high level of DNA compaction that must also exhibit high plasticity in order to allow efficient realization of cellular functions such as replication, transcription, or repair. DNA sequence is the first determinant of chromatin organization and cross-talks between the DNA sequence, the protein complexes involved in the chromatin architecture and the structural components of the nucleus provide a proper subnuclear environment that ensures correct spatial and temporal gene expression.

Each chromatin state can be defined by its level of compaction, its topological state, the positioning and the spacing of nucleosomes, its histone code predicting how the posttranslational modifications of specific amino acids of the core histones (H2A, H2B, H3, and H4) are translated into distinct information [1], the presence of histone variants, the covalent modification of the underlying DNA, its composition in non-histone binding factors, the spatial localization within the nucleoplasm, and its dynamics during cell cycle. Generally, open chromatin, where most of the transcription occurs, is referred to as "euchromatin" whereas condensed chromatin, where transcription is generally inhibited, is referred to as "heterochromatin" although various types of chromatin structure are evoked under these denominations. Heterochromatin was originally described as a portion of the genome deeply stained from metaphase to interphase associated with the pericentric

*Contributed equally to this work.

Handbook of Epigenetics: The New Molecular and Medical Genetics. DOI: 10.1016/B978-0-12-375709-8.00006-X

regions, telomeres, and some interstitial domains. In higher eukaryotes, this constitutive heterochromatin is hypoacetylated, lacks methylation of H3K4 but is enriched in methylated DNA, histone H3K9 di and tri-methylation, H3K27 and H4K20 methylation, HP1 binding, and can spread over flanking regions thereby inducing transcriptional silencing [2,3].

Genes, regulatory elements, and repetitive DNA are interspersed resulting, at the chromosome level, in a mosaic of condensed and open regions. The proximity of different types of chromatin can influence gene expression either positively (enhancer proximity) or negatively (silencer proximity) [4–6]. Furthermore, changes in chromosomal structure can influence DNA transactions by triggering the delocalization of specific chromatin factors. In addition, upon rearrangement, a gene relocated in the vicinity of heterochromatin can become silent in a subset of cells, leading to a characteristic variegated pattern of expression as a consequence of a position effect (position effect variegation or PEV). The different types of position effect are globally referred herein as Chromosomal Position Effect, or CPE. Among them, telomere proximity resembles classical CPE by triggering the silencing of genes located at their proximity, defining the Telomeric Position Effect (or TPE) [7].

To an extent, the identity of chromatin domains is maintained by different factors such as *cis*-regulators, "fuzzy boundaries" or insulators that limit the influence of one region on to the adjacent one [8–10]. Many short sequences can act as *cis*-regulators of gene transcription while "fuzzy boundaries" define regions at the limit between euchromatin and heterochromatin, which are not precisely defined and may change with time. We will develop later in the text the involvement of insulators – specialized elements – which can limit these position effects and separate distinct domains.

The goal of this chapter is not to provide a detailed review of all the experimental work that has been published on CPE, PEV, or TPE in different cellular or animal models but rather to describe the main features of these epigenetic changes and the consequences of CPE or CPE-like mechanisms in human pathologies.

CPE AND TPE, LESSONS FROM MODEL ORGANISMS

Chromosomal position effect was originally discovered in flies in a study of X-ray-induced chromosomal rearrangements or P element insertions that placed euchromatic genes into heterochromatic regions and rearrangements that positioned euchromatin domains into heterochromatin or *vice versa*.

In the 1920s, Sturtevant first described surprising phenotypical changes linked to duplication of the *Bar* locus by observing the number of facets in the eyes of flies [11]. Normal females have two copies of the locus on each X chromosome but *Bar* mutant females have four copies of the gene and display changes in the number of facets. Through the characterization of several mutants with an increasing number of copies of the *Bar* gene, Sturtevant concluded that rearrangement and duplication of the *Bar* locus had an influence on *Bar* gene expression. Counter-intuitively, increasing the number of *Bar* genes decreases the size of the eye suggesting a repressive effect on gene expression upon increase in gene copy number, possibly due to a complex mechanism of control that he named "position effect" [4,11].

Later on, Muller showed that an inversion of the X chromosome leading to the relocalization of the *white* gene close to pericentromeric heterochromatin was associated with a "mottled" phenotype with each eye having some white (mutant) and red (wild-type) regions with variation from eye to eye among the flies carrying such mutations. This phenomenon was dubbed Position Effect Variegation (PEV) [12,13].

The observations from Sturtevant and Muller led to the description of two types of mechanism that can be defined as stable position effect and position effect variegation, respectively. The mechanism associated with the repression of the *Bar* gene is reminiscent of what was later observed upon multimerization of transgenes where the repetition of the sequences induced silencing (i.e. repeat induced silencing) [14,15] or of what occurs at some loci with copy number variations (see below). On the other hand, PEV is produced by translocation that moves euchromatic genes into heterochromatin or inserts heterochromatin-prone sequences into euchromatin domains [16–18] (Fig. 6.1) as observed experimentally in the past in numerous attempts at non-targeted transgenesis.

Indeed, in eukaryotes, heterochromatin is an integral part of the genome and establishment of constitutive heterochromatin plays key functions in the stability and maintenance of the genome during mitosis and meiosis. In *Drosophila*, the archetypal model of PEV, silencing and phenotypical variegation is observed when a gene is brought close to pericentric chromatin. In such cases, gene silencing by heterochromatinization is clonally initiated in a certain number of cells, and is inheritable, and the variegated phenotype results from the spreading of heterochromatic features into euchromatin regions, thereby inactivating flanking genes. Gene variegation results from the length of spreading which might be variable from cell to cell leading to the characteristic mosaic appearance. Generally, the strength of PEV is inversely proportional to the distance from the breakpoint. This model has been exemplified by studies carried on the white gene responsible for the red pigmentation of the eye. Upon translocation close to pericentric heterochromatin, the expression of the white gene is decreased or silenced in a subset of cells leading to a patchy pattern with facets of different colors ranging from white to different shades of yellow, orange, or red [19,20].

The genetic dissection of this process has been performed by means of dominant suppressor (Su(var), suppressor of variegation) and enhancer (E(var), enhancer of variegation) mutations. Since the discovery of PEV, 50 to 150 loci that impact on PEV have been described in flies [4,20] and most of the modifiers studied so far in *Drosophila* or other model

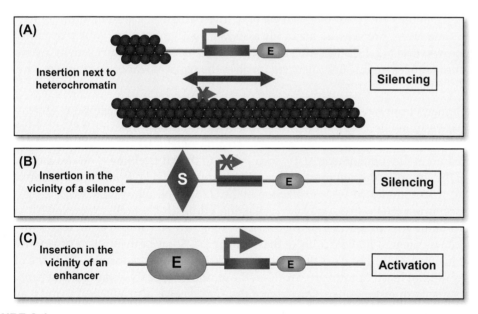

FIGURE 6.1
Chromatin configurations leading to the differential expression of an integrated transgene. Expression of integrated transgenes, even if they contain their own enhancers (E) can be modulated by the chromatin context at the site of integration. (A) Condensed chromatin (constitutive heterochromatin, repetitive DNA, telomeric DNA for instance) can induce the spreading of heterochromatin marks, the repositioning of nucleosomes, the recruitment of chromatin remodeling complexes, the silencing of the transgene, and the appearance of a variegated expression pattern from cell to cell. This phenomenon is called PEV (Position Effect Variegation) and has been observed in all eukaryotic cells. Alternatively, a transgene can be misregulated by the proximity of *cis*-regulating elements such as silencer (B) or enhancers (C) at the integration site, which might influence the expression of the gene of interest. (Please refer to color plate section)

organisms are components of heterochromatin, enzymes that modify histones, non-histone proteins, or nuclear architectural proteins.

For instance, loss of function of the Su(var) genes causes suppression of PEV whereas additional gene copies enhance silencing. Su(var)2-5 encodes the heterochromatin protein 1, HP1 [18,21] and Su(var)3-9, the histone H3 K9 methyltransferase (HMTase) [22,23]. HP1 and Su(var)3-9 work together with the Zn finger protein Su(var)3-7 [24,25] to maintain stable silencing. The suppressor effect of Su(var)3-9 dominates almost all enhancer mutations indicating a pivotal role for H3K9 trimethylation in heterochromatin spreading. Several mutants which antagonize Su(var)3-9-mediated PEV have been identified such as the H3K9 deacetylase [26] or the JIL mutants controlling the H3 serine 10 dephosphorylation [27]. Also, HP1 recruits the Suv4-20 histone methyltransferase and mutations in the *Suv4-20* gene affect PEV [28]. Recently heterogenous nuclear ribonucleoproteins (hnRNPs) bound to chromatin and associated with HP1 have been identified as dominant suppressors of PEV [29].

Consistent with the existence of a dynamic balance between HMTases and demethylases that controls the antagonistic H3K4 and H3K9 methylation marks at the border between eu- and heterochromatin, the *Drosophila* homolog of the mammalian LSD1 amine oxidase that demethylates H3K4me2 and H3K4me1 restricts the extend of H3K4me2, a mark associated with open chromatin, at heterochromatic sequences and facilitates H3K9 methylation and PEV [30]. Other modulators of the chromatin state or transcriptional coactivators such as histone acetyl transferases (HATs), components of the TIP60 complex, or Rsf (remodeling and spacing factor) are also required for PEV in Drosophila by mediating histone variant exchanges [31–37]. Another example involves Try, a member of the trithorax group, which encodes the GAGA factor that interacts with the *Drosophila* counterpart of the FACT complex [38–40]. Mutation in Trl enhances PEV by blocking heterochromatin spreading after replacement of the methylated H3K9 by the H3.3 histone variant, which is a preferred target of the permissive H3K4 methylation [41]. Interestingly, nucleosomes containing the H3.3 and H2A.Z variants are massively associated with promoters, enhancers, and insulator elements [42,43] suggesting that in different species the barrier activity against heterochromatin spreading and silencing involves the replacement of nucleosome core histones by histone variants.

Most telomeres from *Saccharomyces cerevisiae* to *Homo sapiens* silence neighboring genes. Telomeric Position Effect (TPE) in yeast was first demonstrated by insertion of a construct containing a *URA3* auxotrophic marker next to an array of telomeric repeats. Integration of this construct at the subtelomeric *ADH4* locus, close to the VII-L telomere, deletes the terminal 15 kb of the chromosome and positions the *URA3* promoter 1.1 kb from the newly-formed telomere, termed truncated VII-L. Expression of the *URA3* gene allows growth of the cells on plates lacking uracil. However, on plates containing a drug toxic for cells expressing *URA3* (5-fluoro-orotic acid, or 5-FOA), 20 to 60% of the cells were still able to grow, suggesting that the *URA3* was silenced in the vicinity of the telomere [7]. Some of the features of TPE were concomitantly described, such as the stochastic reversibility, since the same cells plated onto a medium without uracil can still grow without the amino acid; or the promoter independence and variegated expression, since expression of the *ADE2* gene is also repressed in the same context and colonies obtained present white (ade2$^+$) and red (ade2$^-$) sectors.

Unlike many organisms, *Drosophila* species lack short repeats-based telomeres and telomerase but maintain their telomeres by the transposition of the *HeTA* and *TART* retrotransposons to chromosome ends [44], and, proximal to the terminal transposon array, *Drosophila* telomeres carry several kilobases of conserved complex satellites termed Telomere Associated Sequences (TAS) [reviewed in Ref. 45]. Despite these structural differences, *Drosophila melanogaster* exhibits telomeric silencing as observed when reporter genes are inserted at a telomeric position [46] and shares many of the features of telomeres in other species.

Genetic modifiers of position effect variegation (PEV) display little or no effect on TPE, suggesting the existence of specialized mechanisms. Experiments in *S. cerevisiae* allowed the identification of more than 50 proteins that can modulate TPE [7,47,48]. However, among deletion mutants of these different proteins, only a few exhibit a specific and complete suppression of telomeric silencing [49]. Among them, Sir-complex proteins (Sir2p, Sir3p, and Sir4p for Silent Information Regulators) [50], Ku heterodimer components (yKu70p and yKu80p) [51,52], and C-terminal domain of Rap1p [53] are absolutely required [5,49]. In other organisms, most of the few factors mediating TPE identified so far are functional orthologs of *S. cerevisiae* proteins. For instance, in *S. pombe*, the telomeric repeats-binding protein Taz1p recruits spRap1 (ortholog of Rap1p in *S. cerevisiae* [54,55]) at the telomere, and both proteins are required for TPE [56,57]. Nevertheless in this model, no link could be established between Ku and TPE. Although they have no ortholog in *S. cerevisiae*, HP1 proteins are involved in TPE in *Drosophila* (on chromosome 4 [58]) and fission yeast [55,59] where they could play similar roles of the budding yeast Sir3p and Sir4p in the spreading of heterochromatin. Interestingly, in *S. cerevisiae*, the Sir proteins are in limited amount and concentrated at the telomere clusters located near the nuclear periphery [60,61]. Therefore, the establishment of silencing requires the localization of the *cis*-acting silencer at these Sir-rich regions of the nucleoplasm [61]. Thus, the nuclear positioning constitutes regulatory information, eventually transmitted to the progeny. Such type of transmission is also found in *Drosophila* for the *Brown^D* locus [62] and in mammals [63].

In higher eukaryotes, both telomeres and subtelomeres contain nucleosomes [64–66] that are enriched in chromatin marks found at constitutive heterochromatin regions [67–71]. However, the capacity of mammalian telomeres to induce position effect has been controversial for many years. The first evidence *in vivo* came from the analysis of replication timing of human chromosome 22 carrying a chromosomal abnormality frequently observed in pathologies such as cancer or genetic diseases [72]. However, other studies implied that human telomeres neither modulate the expression of nearby genes nor affect the homeostasis of telomeres [73,74]. More recently, evidence for transcriptional silencing in the vicinity of human telomeres was provided experimentally by a telomere seeding procedure where natural telomeric regions have been replaced by artificial ones containing a reporter gene [75,76]. This powerful experimental system revealed that telomere length and architecture together with chromatin remodeling factors are involved in TPE. The involvement of TPE in human pathology and health is yet poorly understood but increasing lines of evidence suggest that it might be implicated in both physiological and pathological conditions.

How can spreading of silencing be accomplished? Several mechanisms have been proposed elsewhere [77] and the existence of long-range but also discontinuous [78–80] or *trans* silencing [61,81,82] led to conclude that multiple modes of action have evolved to silence gene at a distance.

SETTING THE FRONTIERS OF CHROMATIN DOMAINS

Active and inactive chromatin domains are often juxtaposed but their respective identity is maintained by specialized elements [8–10]. The first insulator elements, *scs* and *scs'*, were described in *Drosophila* in 1985 by Udvardy and colleagues in a study of the flanking elements at the heat shock locus [83]. Many more insulators have been described since and they are now classified in two non-exclusive subclasses defined by their properties: enhancer-blocking insulators and barrier insulators [9] (Fig. 6.2B,C). Enhancer-blocking insulators can disrupt the communication between a promoter and a *cis*-regulatory element when placed in-between, without preventing them from interacting with other genetic elements (Fig. 6.2C). Barrier insulators protect genes or regions from the spreading of heterochromatin (Fig. 6.2B). If some insulators harbor both properties, these two functions are often uncoupled [84].

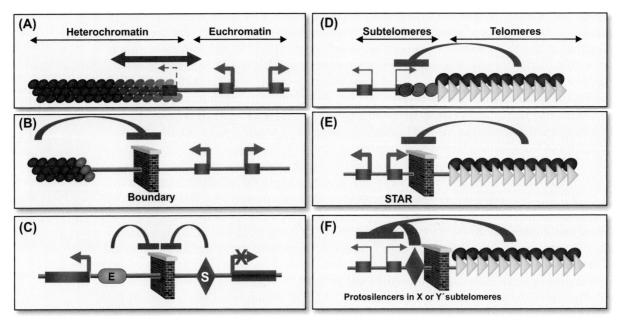

FIGURE 6.2

Modulation of position effect variegation by boundary elements. Two types of boundary element able to separate functional domains have been described in the literature [9,10]. (A) Regions named "fuzzy boundaries" depend on a dynamic equilibrium of chromatin proteins and are the more prone to variegation. (B) DNA sequences named "insulator elements" recruit specific factors to a precise location and define strict borders between chromatin regions. Two types of insulator element exist depending on their specificity (see text for details and [9,10]) and can be classified as boundary elements (B) or enhancer blocking insulators (C). (D) In *S. cerevisiae*, telomeres silence proximal gene through a mechanism named Telomeric Position Effect involving the nucleation of heterochromatin and the spreading of chromatin modifications from the telomere to the subtelomeric region [5,7]. (E) However, each of the 32 yeast chromosomes has a different composition in subtelomeric elements that can modulate TPE. Strong boundary elements dubbed STAR for SubTelomeric Antisilencing Regions consisting of binding sites for Tbf1p and Reb1p [78] are located in some subtelomeres and prevent the spreading of TPE and shelter gene expression. (F) Nevertheless, some elements in X or Y′ subtelomeric repeats can resume and reinforce TPE, bypassing the protective effect of STARs. The core X sequence behaves as a protosilencer, i.e. it does not act as a silencer by itself but reinforces silencing when located in the proximity of a master silencer [245,246]. Different combinations of STAR and protosilencer at native telomeres are likely to contribute to their respective behaviors with regard to TPE. (Please refer to color plate section)

Insulator elements bind specific proteins. In *Drosophila*, five proteins with enhancer-blocking activity have been identified: Zw5, BEAF-32 [85,86], the GAGA factor [87,88], Su(Hw) [89], and an ortholog of the CCCTC-binding factor CTCF [90,91]. CTCF is the only protein conserved and displaying such activity in vertebrates [92].

There are two non-exclusive mechanistic models explaining how these sequences can block the spreading of repressive marks (Fig. 6.2).

The first model is based on competition between opposite histone modification activities (Fig. 6.2A). In *S. cerevisiae*, the main component of the right border at the HMR mating type locus is a transfer RNA gene and transcription by the RNA polymerase III is essential for barrier activity [93]. Sas2p and Gcn5p acetyltransferases are also important for this function and their artificial targeting is sufficient to obtain barrier activity [94]. Protection against TPE can be brought by the recruitment of histone acetyltransferases by Reb1p and Tbf1p at STARs [78] (Fig. 6.2E). Binding of bromodomain proteins, such as Bdf1 in budding yeast, to acetylated histones, could compete against the activity of deacetylases, such as Sir2p [95].

In vertebrates, at the *β-globin* locus, barrier activity of the *5'HS4* insulator depends on the binding of the USF1 and USF2 transcription factors. These two proteins then recruit the SET7/9 histone methyltransferase that methylates H3K4, and the PCAF or p300 acetyltransferases that can acetylate H3 and H4 histones [96], suggesting a dynamic model based on competition between euchromatin- and heterochromatin-remodeling complexes (Fig. 6.2A). Based on these observations, frontiers between open and silenced chromatin could be considered either as "hot spots" for the recruitment of histone modifying enzymes, or undefined boundaries resulting from an equilibrium between different levels of chromatin condensation and depending upon the dosage of different factors. These "fuzzy boundaries" could be more subject to environmental variations and thus lead to variegated gene expression (reviewed in Refs 5,8).

The second model considers barrier activity as a physical block to the spreading of chromatin modifications and is also supported by several lines of evidence (Fig. 6.2B). In budding yeast, components of chromatin remodeling complexes are also directly involved in insulator activity at the *HMR* right boundary, or can harbor barrier activity when artificially tethered [97,98]. Another strategy that stops the spreading of heterochromatin relies on the properties of various sequences to induce a local nucleosome exclusion [99] or the replacement of core histones by histone variants [38,42,43], suggesting that preventing locally the interaction of remodeling complexes, by the protection of substrates or nucleosome positioning, stops the spreading of condensed chromatin conformation.

Since insulators separate independent entities, the mapping of binding sites for insulator proteins would provide a global picture of functional genomic domains. In *Drosophila*, the Su(Hw) protein can be detected on polytene chromosomes at band/interband transitions, at *gypsy* retrotransposons and at numerous other sites [100], suggesting an important role for this protein in genome partitioning [101,102].

In human cells, recent genome-wide studies in different cell types identified up to 39,000 putative sites for CTCF that binds to a wide range of sequences possibly through the combinatorial use of its eleven zinc fingers. Strikingly, distribution of these sites along chromosomes follows the distribution of genes, as for transcription factors, suggesting a key role across the genome [103–105]. Regions depleted in CTCF sites often correspond to clusters of co-expressed genes while regions enriched in those sites generally contain genes with multiple alternative promoters. Interestingly, many of these sites overlap with binding sites for cohesins [106–108] or correspond to boundaries between internal and peripheral sequences based on their interaction with B-type Lamin (LAD) [109], a component of the inner nuclear membrane.

All these results further support a genome-wide role of CTCF in shaping metazoan transcriptional maps but also in the partitioning of functional domains through the interaction with major components of the genome architecture. The main functions of CTCF in enhancer blocking activities or chromatin organization and looping have been described recently in several reviews [110,111] but recent examples relevant to regulation of the position effect in human health will be detailed below.

Very little is known on the dysregulation of insulator elements in human diseases, however recent pieces of evidence suggest a role for CTCF in pathologies. For instance, microdeletion or microduplications of the CTCF sites at the *IGF2/H19* locus are associated with some cases of non-syndromic Wilms' tumors [112]. CTCF also flanks CTG/CAG trinucleotide repeats at several disease-associated loci, such as the *DM1* locus implicated in myotonic dystrophy (or Steinert disease, MIM 160900) [113,114] and other diseases linked to triplet expansion. Myotonic dystrophy is an autosomal dominant multisystemic disorder characterized by myotonia, muscular dystrophy, cataracts, hypogonadism, cardiac conduction anomaly, and diabetes mellitus. This muscular dystrophy is caused by a CTG repeat expansion in the 3'UTR region of the *DMPK* gene. This repeat is normally present in 5 to 34 copies in the general population and is increased up to 50–1000 copies in DM1 patients. The physiopathology is not fully elicited and may be linked to a complex mechanism involving chromatin changes and non-coding RNAs [115–117]. CTG repeats are flanked by CTCF binding-sites [114]. In the congenital form of DM1, hypermethylation of CpG dinucleotides [118], a more compact chromatin [119], and loss of CTCF binding [114] have been observed, suggesting that chromatin modifications could play a role in, at least, the more severe forms of DM1.

The presence of CTCF restricts the extent of antisense transcription and constrains the spreading of heterochromatin from the trinucleotide repeats [113,114]. However, since CTCF interacts also with the RNA polymerase II large subunit and initiates the transcription of a reporter gene [113], the vertebrate insulator might have a bivalent role around the triplets by initiating, activating and restricting the bidirectional transcription, which in turn both enhances and limits heterochromatin formation at the DM1 locus. Similar mechanisms might also be implicated in other triplet expansion diseases [120].

Non-coding RNA and CTCF may also be associated with another muscular dystrophy named FSHD (Facio-Scapulo-Humeral Dystrophy) [121,122]. This autosomal dominant pathology is linked to the shortening of an array of repeated macrosatellite elements at the distal end of the q arm of chromosome 4 [123,124]. Normal 4q35 ends carry from 11 up to 150 copies of this element while this number is reduced to 1 to 10 repeats in most patients [125,126] and the hypothesis of position effect mechanisms has been proposed for many years to explain the physiopathology of this complex disease. D4Z4 acts as a potent CTCF- and A-type Lamins-dependent insulator in cells from patients but not in control cells [121] and is transcribed into several overlapping sense and antisense transcripts [122].

Although the link between CTCF, non coding RNA, heterochromatin formation, and position effects at the FSHD locus has not been established [121,122], a recent study on arrays of DXZ4 macrosatellite repeats localized at the Xq23-24 locus which display differential chromatin structures and CTCF binding [127] suggests that similar mechanisms exist in the repetitive DNA complement of the human genome. The CTCF binding site in DXZ4 is close to a bidirectional promoter and has a unidirectional enhancer-blocking property. CTCF binding is associated with absence of CpG methylation and bidirectional transcription of the full repeats on the inactive X chromosome (Xi). On the active X (Xa), absence of CTCF binding correlates with generation of four small antisense transcripts that correspond to sites of increased H3K9me3. Albeit the complete mechanisms surrounding this macrosatellite gain of function are not yet fully understood, it suggests, once again, an involvement of the RNAi machinery and a role for the CTCF protein or repeated elements in organizing specific domains in the genome.

Moreover, these examples illustrate the importance of borders and insulators in blocking the propagation of chromatin modifications for proper gene regulation but also in the onset of pathologies linked to mechanisms reminiscent of position effect. Thus, although never investigated in the light of CTCF function, one can speculate that hypomorphic mutations in factors associated with CTCF binding sites such as cohesins [106–108] or Lamins [109] might also alter insulation activity and have consequences on the regulation of CPE in human cells. For instance, cohesin genes are mutated in the Cornelia de Lange syndrome, a heterogeneous developmental disorder characterized by multiple abnormalities (abnormal facies, prenatal and postnatal growth retardation, mental retardation and, in many cases, upper limb anomalies) while mutations in the gene encoding A-type Lamins give rise to at least ten distinct and heterogeneous genetic diseases called laminopathies [128] including an autosomal dominant form of Emery–Dreifuss muscular dystrophy (EDMD) [129], a rare dominantly inherited muscular dystrophy (limb girdle muscular dystrophy type 1B, LGMD1B [130], the Hutchinson–Gilford progeria, a rare syndrome of rapid aging [131,132], or familial partial lipodystrophy (FPLD). Therefore, disrupting the integrity of the complexes mediating the partitioning of functional chromatin domains may cause a wide range of defects by affecting insulators and chromatin boundaries that could lead or contribute to these multi-symptomatic syndromes.

CPE AND GENE TRANSFER

As we described previously, a transgene located in constitutive heterochromatin adopts the compact nucleosomal structure of the insertion site [133]. Moreover, if some transgenes do not variegate when repeated in tandem, certain ones are prone to silencing in such configuration, challenging the use of transgenesis for experimental or therapeutic purposes [14,15].

Viral vector-mediated gene transfer has become a promising and potent tool for the treatment of life threatening diseases. A variety of integrating vectors for gene delivery exists. Some of them have integration preferences while some exhibit random genomic integration. However, a significant risk of insertional mutagenesis emerged from early therapeutic assays [134–137], linked to genotoxic effects after mutagenesis of critical cellular genes (knock-out of the gene, changes in the spatio-temporal expression pattern, truncation of the gene product). In other gene therapy trials and animal models, transgenes have been susceptible to a substantial reduction and variegation in gene expression attributable to changes in the chromatin structure, and efficacy of gene transfer *in vivo* has been compromised in many cases [138–141].

One possible issue regarding the "position effect" problem is the use of insulator elements that would block the effects of neighboring *cis*-regulators and the spreading of heterochomatin towards the transgene (but also potentially, protect the sequences at the site of integration). Thus, flanking transgenes with insulators has been attempted to minimize position effect and insertional mutagenesis in several species. Well-known insulators include the 5′*HS4* element from the chicken *β globin* insulator [142,143], the *scs* and *scs'* from the Drosophila *HSP70* locus, the *gypsy* retroviral sequence or the upstream sequence of the sea urchin *ARS* gene [144]. Although insulators can decrease expression variability in some contexts, they often give only partial protection [145–147]. Hence, position effect still represents an obstacle to gene transfer, especially to the goal of cellular therapy.

85

CHROMOSOMAL POSITION EFFECT IN HUMAN PATHOLOGIES

Mechanisms through which CPE can cause human diseases are diverse: separation of the transcription unit from an essential distant regulatory element, juxtaposition of the gene with the enhancer element of another gene, competition for the same regulatory element, and classical position effect variegation in which a gene is moved to a new chromatin environment [148]. CPE-associated pathogenesis has been described in cancers as well as in constitutional pathologies, essentially in the context of chromosomal rearrangements (translocations, deletions, inversions).

For instance, malignant hemopathies are characterized by acquired chromosomal rearrangements, mainly translocations, which are clonal, non-random, recurrent, and often specific for a tumor type. These translocations have two main consequences: the formation of a chimerical gene encoding a new fusion protein or the combination between the coding region of a gene and the promoter/enhancer region of another one leading to inappropriate overexpression of the former gene [149]. A well-known example is the t(8;14)(q24;q32) translocation observed in Burkitt's lymphoma, an aggressive B cell neoplasm. This translocation places the *c-MYC* gene (chr. 8) near the enhancer of the heavy chain of immunoglobulin on chromosome 14 resulting in overexpression of *c-MYC* in B cells [150]. Other translocations of genes like *BCL2* (follicular lymphoma) [151], *BCL6* (diffuse large B cell lymphoma) [152], and the *TMPRSS2-ETS* fusion gene involved in prostate cancer [153] can also be cited as examples. Possibly also, the translocation of a heterochromatic region might lead to long-range repression of genes in the rearranged chromosome. This was suggested to occur for rearrangements of the heterochromatic 1q12 region frequently found in hematologic malignancies [154].

The involvement of the chromosomal position effect in constitutional pathologies appears in most cases more difficult to prove. Indeed, although this mechanism is often put forward in accounting for balanced reciprocal translocation with abnormal phenotype in which no gene is disrupted [155,156], a firm demonstration is rarely found in the literature [157], which can be explained giving several reasons.

First, in developmental syndromes, the causative gene often shows a specific spatio-temporal pattern of expression and access to the tissue from the patient is often impossible, challenging the analysis of gene expression and regulation. Second, regulatory elements are often located far from the gene of interest, sometimes included in another neighboring gene, and can regulate more than one gene. Finally, if the position effect seems evident when a chromosomal breakpoint occurs in the vicinity of a gene for which the associated phenotype is well known, it is less obvious when the phenotype is not related to the gene or when the gene function is unknown. Pathologies in which the chromosomal position effect have been suspected or proved are summarized in Table 6.1. Some examples will be further detailed.

PAX6 and Aniridia: A Well-Characterized Position Effect Pathology

Aniridia (MIM #106210) is an autosomal dominant panocular disorder characterized by complete or partial iris and foveal hypoplasia resulting in reduced visual acuity and nystagmus presenting in early infancy together with ocular abnormalities such as cataract, glaucoma, corneal opacification, and vascularization. Aniridia can be isolated or part of the WAGR syndrome (Wilms' tumor, Aniridia, Genitourinary anomalies and mental Retardation, 11p13 microdeletion). It is caused by haploinsufficiency of *PAX6*, a member of the paired box family that encodes a transcriptional regulator involved in oculogenesis and other developmental processes. In some aniridia patients carrying 11p13 rearrangements (deletion or translocation), the breakpoint occurred downstream of the *PAX6* transcript unit [158] in the final intron of the *ELP4* gene highlighting the putative existence of unknown elements and factors controlling gene expression. Using the Small eye mouse model, *DNase* hypersensitivity mapping and reporter gene systems, Kleinjan et al. [157] demonstrated that a regulatory region is located 130 kb downstream of *PAX6* polyadenylation site and that this region contains a tissue-specific enhancer. Chromosomal rearrangements involving this downstream region are responsible for cases of aniridia associated with a position effect. This case illustrates how the study of pathological cases can advance our knowledge of genome organization and the identification of distant regulatory *cis*-elements.

Split-Hand/Foot Malformation Type 1: Position Effect Involving Several Genes?

The split-hand/foot malformation (MIM %183600) (SHFM1) is an autosomal dominant congenital limb defect with incomplete penetrance and variable expressivity. It involves the central ray of the hand and feet and affects randomly one to all of the four limbs and is characterized by a median cleft of the hand or foot (ectrodactyly) resulting in an aspect of "lobster claw". SHFM1 is caused by chromosomal rearrangements of the 7q21q22 region (deletions, translocations, inversions) [159–162]. The minimal deleted interval has been narrowed to a 900 kb region [163] that contains three candidate genes: *DLX5*, *DLX6*, and *DSS1*. *DLX5* and *DLX6* are *distal-less*-related homeogenes expressed in mouse embryo skeletal structures [164]. *DSS1* encodes a highly conserved acidic protein expressed during limb, craniofacial, skin, and genitourinary mouse development [159]. In mouse, the concomitant deletion of *Dlx5* and *Dlx6* is necessary to cause a limb phenotype [165]. In humans, no mutation has been identified in these candidate genes to date [159] and none of these genes is directly disrupted by rearrangement breakpoints. So, it has been proposed that SHFM1 is due to a position effect by disruption of a long-range regulatory sequence controlling *DLX5*, *DLX6*, and *DSS1* genes [166] but the regulatory region has not been identified yet.

TABLE 6.1 Position Effect in Human Pathology: Non-exhaustive Literature Review of Genes for Which a Position Effect had been Proved or Put Forward in Constitutional Diseases

Gene	Chromosome	Phenotype	Regulation Element	Modification	Reference
APC	5q22.2	Adenomatous polyposis		Disruption	[247]
DLX5/DLX6/DSS 1	7q22	Split-Hand/Foot Malformation type I	5'/3'	Disruption	[159]
FOXC1	6p25	Primary Congenital Glaucoma	5'	Disruption	[144]
FOXC2	16q24.3	Lymphedema Distichiasis	3'	Disruption	[249]
FOXL2	3q23	Blepharophimosis Ptosis Epicanthus inversus Syndrome	5'/3'	Disruption/deletion	[250,251]
FOXP2	7q31	Speech and language disorder	3'	Disruption	[252]
GLI3	7p13	Greig Cephalopolysyndactyly syndrome	3'	Disruption	[253]
HBA	16p13.3	Alpha-Thalassemia	3'	Deletion	[179]
HBB	11p15.5	Gamma/Beta-Thalassemia	5'	Disruption/deletion	[254,255]
HOXB	17q21.3	Mental retardation, hexadactyly	5'	Disruption	[256]
HOXD	2q31	Limb malformations	5'/3'	Disruption	[257,258]
LCT	2q21	Adult-type hypolactasia	5'	Mutation	[259]
MAF	16q23	Cataract, anterior segment dysgenesis, microphthalmia	5'	Disruption	[260]
PAX6	11p13	Aniridia	3'	Disruption	[261]
PITX2	4q25	Rieger syndrome	5'	Disruption	[262]
PLP1	Xq22	Pelizaeus–Merzbacher syndrome/Spastic Paraplegia, neuropathy	5'/3'	Disruption/Duplication	[176,177]
POU3F4	Xq21.1	X-linked Deafness	5'	Deletion	[263]
REEP3	10q21.3	Autism	5'	Disruption	[264]
RUNX2	6p21	Cleido Cranial Dysplasia	5'	Disruption	[265]
SALL1	16q12.1	Townes-Brocks syndrome	3'	Disruption	[266]
SDC2	8q22	Autism, Multiple Exostoses	3'	Disruption	[267]
SHH	7q36	Holoprosencephaly	5'	Disruption	[173]
SHH	7q36	Pre-axial Polydactyly	5'	Disruption	[174]
SHOX	Xp22.3	Leri-Weill Dyschondrosteosis	3'	Deletion/Mutation	[268] [269]
SIX3	2p21	Holoprosencephaly	5'	Disruption	[270]
SOST	17q21	Van Buchem disease	3'	Deletion	[271]
SOX3	Xq27.1	Hypoparathyroidism	3'	Disruption	[272]
SOX9	17q24.3	Campomelic Dysplasia/Pierre Robin Sequence	5'/3'	Disruption/deletion/mutation	[170,172]
SRY	Yp11.3	Sex Reversal	3'	Deletion	[273]
TGFB2	1q41	Peters anomaly	3'	Disruption	[274]
TRPS1	8q23.3	Ambras syndrome	3'/5'	Disruption/deletion	[275]
TWIST	7p21.1	Saethre-Chotzen syndrome	3'	Disruption	[31,216]

SOX9: The Phenotypic Spectrum of Campomelic Dysplasia

SOX9 is a transcription factor of the HMG box family that regulates chondrogenesis and testis development. Its haploinsufficiency is responsible for campomelic dysplasia (MIM #114290), a rare autosomal dominant disorder involving shortening and bowing of long bones, skeletal malformations, Pierre Robin sequence (PRS, associating micrognathia, glossoptosis, and cleft palate), hypoplastic lung, and male-to-female sex-reversal in two thirds of cases. Besides classical mutations, translocation or inversion breakpoints sparing the *SOX9* sequence have been described [167–171] and allowed the identification of a long-range regulatory region located in a gene desert up to 1 Mb from the 5′ end of the *SOX9* gene [170]. Recently, another *SOX9* regulatory element has been identified at a distance of 1.5 Mb from *SOX9* in patients with PRS carrying translocation, deletion, and also point mutation of this element [172]. *In vitro*, the mutation abrogates gene activation and the presence of a developmentally regulated enhancer was postulated. In the developing mouse mandible, this region presents features of decompacted chromatin (H3K4 methylation) in cells expressing *SOX9* suggesting that this sequence contributes to the tissue-specific regulation of the gene. Interestingly also, transcriptional activation correlates with the presence of CTCF and p300 suggesting the existence of an insulator involved in the transcriptional activation and chromatin remodeling of the locus. Upon deletion of the enhancer, position effects may alter the tissue-specific *SOX9* regulation leading to PRS. This example illustrates, once again, the strength of long-distance gene regulation and suggests that a continuum of severity depends in some cases on the distance between the regulatory element and the transcript unit.

SHH: One Gene, Two Phenotypes

SonicHedgehog (*SHH*) encodes a secreted signaling protein that establishes cell fate during development. *SHH* is expressed in notochord, footplate of neural tube, posterior limb buds, and gut in human embryo. Mutations of *SHH* have been identified in patients with isolated holoprosencephaly (HPE), an anomaly of the brain in which the developing forebrain fails to divide into two separate hemispheres (MIM #142945). The severity of HPE is extremely variable even within the same family and the disease is often associated with facial anomaly, mental retardation, and seizures. HPE patients with translocation breakpoint lying 15 to 265 kb upstream from *SHH* have been reported [173] and the phenotype was attributed to position effect. Surprisingly, a t(5;7)(q11;q36) translocation was reported in a patient with preaxial polydactyly (PPD), a common limb malformation characterized by an additional digit on the side of the thumb or great toe (MIM #174500), reminiscent of the Sasquach mouse, a model of PPD, due to a transgenic insertion in the syntenic region. The genetic lesions in both human and mouse lie within the intron 5 of the *LMBR1/lmbr1* gene, located 1 Mb upstream of *SHH* suggesting that the mutation interrupts a regulator of the SonicHedgehog gene [174]. Further studies revealed the existence of a regulatory sequence that acts as a specific enhancer controlling the spatio-temporal expression of *SHH* during limb development. Moreover, point mutations of this regulatory sequence have been identified in patients with PPD, confirming the role for this enhancer in limb development [175]. In this example, the phenotype depends on the type of regulatory sequence affected.

PLP1 and Pelizaeus–Merzbacher Disease: When Position Effect Rescues the Phenotype

PLP1-related diseases are disorders of myelin formation in the central nervous system (CNS) and include Pelizaeus–Merzbacher disease (PMD) (MIM #312080) and type 2 spastic paraplegia (MIM #312920). PMD is an X-linked recessive disease that begins in infancy and manifests by nystagmus, hypotonia, and cognitive impairment progressing to severe spasticity and ataxia. It is caused by duplications, point mutations and deletions of the *PLP1* gene. This gene is sensitive to dosage since an extra copy of the gene affects the development of oligodendrocytes and is associated with abnormal CNS myelinization.

Although some cases of PMD resulting from position effect linked to small rearrangements in the region surrounding the coding sequence have been reported [176,177], Inoue and colleagues described an interesting familial case in which position effect rescues the phenotype [178] (Fig. 6.3). The mother carried a balanced insertion of a small segment of chromosome X including the *PLP1* gene into the terminal region of the long arm of chromosome 19 (ins(19;X) (13.4;q22.2q22.2)). She displayed some features of PMD with late-onset. One of her sons, showing a mild form of PMD, inherited the derivative X and carried no *PLP1* copy. The second son was healthy although he inherited the derivative chromosome 19, and carried two copies of *PLP1*. However, the *PLP1* copy inserted on chromosome 19 may be silenced by the chromatin context and the influence of telomere proximity in the proband's brother, rescuing thereby the phenotype usually associated with abnormal *PLP1* dosage.

Alpha-Thalassemia

Alpha-thalassemia (MIM +141800) is a heritable form of anemia characterized by a decreased level of alpha-globin chain that composes hemoglobin. Alpha-globin is expressed from four genes (*HBA1* and *HBA2* on each chromosome 16) that are under the control of a 5′ regulatory element, 5′-HS40. Usually, alpha-thalassemia is linked to the deletion of

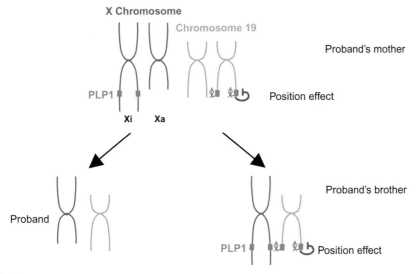

FIGURE 6.3
When the position effect rescues the phenotype. In the majority of patients with Pelizaeus–Merzbacher disease (PMD), duplication of the *PLP1* gene is responsible for the abnormal development of oligodendrocytes and the demyelinization of central nervous system neurons. In one family described in the literature [178], *PLP1* deletion resulted from a maternal balanced submicroscopic insertional translocation of the entire *PLP1* gene to the telomere of chromosome 19. The proband was a 10-year-old boy who showed early motor development and mental retardation. He had a healthy male sibling. The mother presented mild features of PMD such as difficulty in walking at the third decade, changes in her personality, and progressive mental deterioration. The patient carries a deletion of the *PLP1* gene on the X chromosome and further analysis of the family revealed that the translocation was inherited from his mother. The latter carried a translocation of the *PLP1* gene on the distal end of chromosome 19. The patient inherited the translocated X chromosome but did not inherit the derivative chromosome 19 containing *PLP1* while his healthy brother inherited this extra copy of the gene but does not carry the deleted X chromosome. However, the gene on chromosome 19 was silenced by the position effect and *PLP1* gene dosage was normal. Interestingly, in the mother, skewed X inactivation allowed the expression of the *PLP1* gene present on the inactive X chromosome suggesting that a low but detectable PLP1 level was associated with the late onset of the clinical signs. (Please refer to color plate section)

either the alpha-globin genes or the 5′ regulatory element. Recently, Barbour et al. reported an alpha-thalassemia patient presenting a 18 kb deletion downstream of the alpha-globin cluster that did not affect either the gene or the 5′HS40 [179]. However, the gene was hypermethylated at the DNA level and *HBA2* gene expression was abolished while the production of an antisense RNA was detected [180]. In this example, the deletion within the *alpha globin* locus placed a truncated copy of *LUC7L*, a widely expressed gene transcribed from the opposite strand, next to the polyadenylation site of *HBA2*. This rearrangement extended the transcription of the *LUC7L* across the *HBA2* gene, leading to the production of an antisense RNA, which in turn, mediated *HBA2* silencing.

DISORDERS OF UNSTABLE REPEAT EXPANSIONS

Repeat expansion diseases are caused by an increase in the number of trinucleotide repeats and share some common features. These short repeats are polymorphic in the general population. Beyond a certain number of repeats, they become unstable and tend to increase at the next generation (dynamic mutation). The disease appears when the number of triplets exceeds a certain critical threshold. Trinucleotide repeats are either found in coding region resulting in the production of a protein with altered function (i.e. Huntington disease), or in non-coding regions resulting in an altered transcription and loss of protein synthesis (Fragile-X syndrome, Friedreich's ataxia, myotonic dystrophy) [181].

In non-coding repeat expansion diseases, the expansion is transcribed but not translated. Several reports have suggested that the pathophysiology of these diseases may implicate epigenetic changes and heterochromatin formation since heterochromatin marks have been found at expanded loci [17].

Fragile X syndrome (FXS) (MIM #300624) is the most common form of inherited mental retardation also characterized by delayed language and autistic-like behavior. FXS is caused by the expansion of a CGG trinucleotide in the 5′ UTR of the *FMR1* gene. In the normal population, the number of repeats varies from 6 to 54. Between 55 to 200 CGG, the repeats become meiotically unstable (premutation) while the full mutation corresponds to more than 200 repeats and is accompanied by a hypermethylation of the repeats and neighboring sequences [182]. These changes in the chromatin structure result in the silencing of the *FMR1* gene and the absence of FMRP protein, a regulator of translation in brain neurons [183]. The *FMR1* hypermethylation could act directly by preventing the binding of transcription factors. However, features of inactive chromatin have also been observed on mutated allele like lysine hypoacetylation of histone H3 and H4 and methylation of H3K9 residues [184,185] suggesting that the expansion of the repeats and, in particular, of methyl-sensitive sites, directly contributes to the phenotype.

Another interesting disease linked to triplet expansion is the Friedreich's ataxia (FRDA) (MIM #2293000), an autosomal recessive neurodegenerative disease that begins in childhood and is characterized by difficulties to coordinate movements, dysarthria, loss of reflexes, pes cavus, scoliosis, cardiomyopathy, and diabetes mellitus. The causative gene, *FXN*, encodes the frataxin, a protein involved in the assembly and transport of iron-sulfur proteins of the mitochondrial respiratory chain. FRDA is caused by the expansion of a GAA trinucleotide in the first intron of *FXN*. Repeats range from 6 to 34 in the general population and there are over 66 in patients. Expansion is associated with a decrease in *FXN* transcription level [186]. Different mechanisms could be implicated in gene silencing. First, it has been shown that GAA repeats adopt an unusual conformation able to inhibit transcription *in vitro* [187]. Then, a role for heterochromatin formation has been proposed based on the observation of patients' samples [188,189] and on an elegant mouse model using the PEV-sensitive human *CD2* gene as a reporter [190]. This system revealed that the inclusion of triplet repeats (GAA found in Friedreich's ataxia and CTG repeats found in myotonic dystrophy) confers variegation of expression, independently of the chromosomal site of integration, by inducing a more dense nucleosomal organization of the transgene promoter. Moreover, in FRDA patients, specific CpG sites are hypermethylated in the *FXN* intron 1 compared to control, and hypoacetylation of histones H3 and H4, and hypermethylation of H3K9, are thought to modulate promoter activity [188,189].

In favor of a role for chromatin remodeling and heterochromatinization in the pathophysiology of triplet expansion diseases, reactivation of *FMR1* and *FXN* transcription is induced with DNA hypomethylating (5-aza-z'deoxycytidine) or histone hyperacetylating agents [191,192]. However, the mechanisms of chromatin remodeling are not fully elucidated yet. But, consistent with what is known in other species in the spreading of heterochromatin and PEV, a role for non-coding RNA is now emerging as a keystone in the pathogenesis of these diseases [193].

COPY NUMBER VARIANTS: A NEW CHALLENGE IN THE DECIPHERING OF CPE

In addition to short nucleotide polymorphisms (SNPs), DNA copy number variation (CNV) represents a considerable source of human genetic diversity. Copy number variants are usually considered as DNA sequences ≥1 kb in length, present in a variable number of copies [194,195]. Nearly 6225 CNV loci, which represent approximately 28.8% of a human genome and 1447 different regions, have been documented (http://projects.tcag. ca/variation) but this list is increasing continually. The majority of CNVs are biallelic polymorphisms (deletions, insertions, or duplications larger than 1 kb and up to several megabases) but different regions with segmental duplications have been observed, likely due to non-allelic homologous recombination [196]. Another process at the origin of CNVs, fork stalling and template switching (FoSTeS) involving DNA replication rather than recombination, has been recently proposed for the complex rearrangement of the *PLP1* gene involved in the Pelizaeus–Merzbacher syndrome (PMD) [197] and might account for some of the variation existing in the human genome.

91

The recent advent of CNVs as a major source of diversity in human genetics has also laid new ground for the understanding of inherited diseases, cancer genetics, inter-individual variations, and mosaicism. As discussed earlier in the text, having two copies of a gene on two separate chromosomes does not necessarily have the same impact as carrying these two copies at the same locus. With regard to CPE, not all gene CNVs result in changes in gene expression levels but a negative correlation between copy number and expression level was found in 5–15% of CNVs [176,198–200].

For instance, in the Williams–Beuren syndrome, CNVs and position effect mechanisms might be involved in the modulation of the phenotype. This syndrome (MIM #194050) is a recurrent autosomal dominant genomic disorder caused by 7q11.2 microdeletions. Clinically, it is characterized by mental retardation, typical facial features, supravalvular aortic stenosis and other cardiovascular anomalies, over-friendliness, visuospatial impairment, and hypercalcemia. The classical Williams–Beuren deletion includes 28 genes. Recently, the level of expression of these genes but also of genes surrounding the deletion has been studied [201] and consistently, the aneuploid genes showed an expression level that was half that of the control. Surprisingly, nonhemizygous genes located several megabases away from the deletion harbored also significantly reduced expression level (*ASL*, *KCTD7*, *HIP1*, *POR*, *MDH2*) leading the authors of this finding to propose that the effect could be mediated by disturbances of the copy number of long-range *cis*-regulatory elements and that genes not included in the deletion may also participate in the phenotype of Williams–Beuren syndrome.

Thus, the actual challenge is now to identify the regulatory networks and epigenetic changes underlying the regulation of CNVs, especially in pathologies. However, since CNVs can alter expression of genes located *in cis*, at a long distance of the variation, or *in trans* [200], the mechanism involved in CPE or TPE discussed throughout this chapter might also in the near future become an important milestone in uncovering the regulation of CNVs. In addition, a better and more precise knowledge of the higher-order chromatin structure induced by copy number changes would likely bring important insights for understanding the complexity of the regulation of human genome variations and how it relates to phenotypical differences and susceptibility to diseases.

TELOMERIC POSITION EFFECT IN HUMAN PATHOLOGIES

The most distal unique regions of the chromosomes and telomeres are separated by different types of subtelomeric repeats varying in size from 10 to up to hundreds of kb in human cells. Subtelomeres are gene rich regions and RNAs produced by these regions include transcripts from multi-copy protein-encoding gene families, single genes, and a large variety of non-coding RNAs (reviewed in Ref. 45). In the human population, the subtelomeric regions are highly polymorphic and the rate of recombination at chromosome ends is higher than in the rest of the genome. The nature of telomeric chromatin differs from that of global constitutive heterochromatin due to its DNA sequences, the binding of specific factors, and the particular structure of its nucleosomal fiber. Human telomeres influence the maintenance and recombination of the adjacent subtelomeric regions. However, if the 46 human telomeres share the same repeated (TTAGGG)n sequence, the unique composition of each individual subtelomere concurs with the propensity of telomeres to regulate gene expression but also to the replication and maintenance of chromosome ends. Subtelomeres may buffer or facilitate the spreading of silencing that emanates from the telomere as observed in other species. Large variations of subtelomeric DNA were among the first examples of CNVs in human cells and are detected at almost every subtelomere. Thus these large disparities might affect the abundance of telomeric transcripts.

These regions are associated with genome evolution but also human disorders and the existence of TPE has been evoked in many cases of patients with truncated chromosomes ends that have been repaired by the process of telomeric healing (*de novo* addition of telomeric repeats by telomerase at breaks devoid of exact telomeric repeat sequence match), telomeric capture (resulting from a break-induced replication event between a truncated chromosome and the distal arm of another chromosome), or formation of ring chromosomes. However, the molecular pathogeneses associated with these rearrangements have never been investigated.

Subtelomeric imbalances that include deletions, duplications, unbalanced translocations, and complex rearrangements [202] are terminal as well as interstitial and extremely variable in size [203]. Fifty percent of them are inherited. Telomeric polymorphisms are found in at least 1% of the population. Some telomeric polymorphisms and transmitted subtelomeric imbalances are benign and not associated with any phenotypical manifestation [204,205]. These latter have been detected at 24 of the 41 telomeres [206]. Several mechanisms may explain the absence of the abnormal phenotype in carriers of the subtelomeric imbalances: variable expressivity, unmasking of a recessive allele, somatic mosaicism in the normal parent, and/or epigenetic modifications.

The importance of subtelomeric rearrangements affecting all chromosomes with the exception of the short arms of acrocentric chromosomes is well established in 5 to 10% of idiopathic mental retardation [207–209]. The refinement of diagnostic techniques allowed the identification of a number of genes, but many cases remain poorly characterized [210–212] and rather a few deleterious subtelomeric imbalances are associated with a distinct, recognizable phenotype. Among well-delineated syndromes, genotype–phenotype correlation studies have uncovered that a single gene (*SHANK3*) may be responsible for the phenotype of the 22qter deletion syndrome (global development delay, hypotonia, autistic-like behavior, normal to accelerated growth, absent to severely delayed speech) [213,214]. In other syndromes, critical regions have been delineated. For example, cri-du-chat syndrome (5pter deletion) is characterized by microcephaly, facial dysmorphism, high-pitched cat-like cry, severe mental retardation, and speech delay [215]. Although three critical regions corresponding to cry (5p15.31), speech delay (5p15.32–15.33) and facial dysmorphism (5p15.31–15.2) [216] are known, no gene has been identified for each specific feature yet. Finally, no candidate gene could be identified in some cases. The 1p36 monosomy syndrome is the most commonly found subtelomeric microdeletion syndrome with a frequency of 1/5000. It is characterized by mental retardation, developmental delay, hearing impairment, seizures, growth impairment, hypotonia, heart defect, and distinctive dysmorphic features [217]. Attempts have been made to demonstrate that monosomy 1p36 is a contiguous gene syndrome. However, no correlation between the deletion size and the number of clinical features could be observed [217]. There is neither common breakpoint nor common deletion interval in monosomy 1p36 patients. Moreover, two of six patients presenting with very similar features were investigated using tiling path array Comparative Genomic Hydridization (CGH) but no overlapping 1p36 deletions could be detected [218] leading to the hypothesis that the 1p36 monosomy syndrome might be due to a positional effect rather than haploinsufficiency of contiguous genes. Moreover, gene dosage alone cannot account for the phenotype in many cases and alternative mechanisms, in particular, epigenetic modifications, may be involved. Among the hundreds of patients analyzed, the size of the subtelomeric region disrupted may be accompanied by variable degrees of chromatin condensation and explain the penetrance of the clinical manifestations [212].

Besides subtelomeric rearrangements, hundreds of patients have been reported with various combinations of malformations, minor abnormalities, and growth retardation usually associated with mental retardation linked to the formation of a ring chromosome [219,220]. Ring chromosomes are thought to be formed by deletion near the end(s) of chromosomes followed by fusion at breakage points and have been described for all human chromosomes. The resulting phenotypes vary greatly depending on the size and the nature of the deleted segments. Most ring chromosomes are formed by fusion of the deleted ends of both chromosome arms coupled with the loss of genetic material. However, in a few cases, the rings are formed by telomere–telomere fusion with little or no loss of chromosomal material and have intact subtelomeric and telomeric sequences suggesting that the "ring syndrome" might be associated with the silencing of genes in the vicinity of a longer telomere. Intact rings putatively causing telomeric position effects have been reported for the different autosomes [221–223].

Because of the paucity of cases reported and the variability in the size of the terminal deletion, genotype–phenotype correlations are not clearly established for most syndromes involving either subtelomeric imbalance or the formation of rings. Gene abnormalities and dosage cannot always account for the phenotype and from one case to another, the size of the deleted region could be associated with variable degrees of chromatin condensation and explain the penetrance of phenotypes. Epigenetic modifications including DNA methylation, post-translational modification of histones, and non-coding RNAs provide a mechanism for modulation of gene expression or chromatin accessibility that can be influenced by exposure to environmental factors and parental effects. Thus, in diseases involving subtelomeric imbalance, it is conceivable that genes residing in close proximity to a rearranged region or a healed telomere become epigenetically inactivated contributing to the phenotype. However, characterization of the rearrangement's effect on gene expression is still needed to prove that the pathological manifestation is caused by modification of the chromatin architecture at telomeric and subtelomeric loci.

POSITION EFFECTS AND AGING

Aging is characterized as a progressive and generalized impairment of cellular functions resulting in an increasing susceptibility to environmental stress and a wide range of diseases.

At the cellular level, aging, usually named senescence, involves highly dynamic series of modifications that further lead to widespread changes at the level of the whole body and increasing evidence supports a link between modification of chromatin, senescence, and aging. Causality between CPE, TPE, and aging is not clearly established but the large-scale shift of the constitutive heterochromatin compartment [224], the emergence of heterochromatin foci [225–227] concomitant to telomere shortening (replicative senescence), which might delocalize components of the telomeric heterochromatin to non-telomeric sites [61], raises the intriguing possibility of a global reprogramming of gene expression, which could result from position effect mechanisms.

Besides heterochromatinization, the chromosomal position effect in aging tissues might also be affected by the age-related loss of epigenetic silencing of repeated sequences observed for example in heart and in neurons [228–230]. An active skewing of X inactivation has also been reported with increased age [231–233] and might be associated with X-linked disorders in elderly women such as X-linked hyper IgM syndrome, combined immunodeficiency, and chronic granulomatous disease [234–236].

Also, lymphopenia [237–241] in older individuals together with the increased rate of condensed chromatin in these cells as observed for other tissues [242,243] suggests that the expansion of silenced regions through PEV or TPE might also affect the defense against pathogens and may explain why older subjects are more susceptible than younger ones to infections [244].

Thus, the cytogenetically visible lesions such as translocations, insertions, and dicentric and acentric fragments inherent to the aging process might lead to altered expression of pro- or anti-aging genes by CPE. In addition, telomere shortening, which can lead to cellular senescence and appears to contribute to some aging processes in mammals, might alleviate the repression of pro-aging genes, which otherwise would have succumbed to TPE.

CONCLUSIONS

The extent by which epigenetic changes and position effects contribute to the diversity of human phenotypes is increasingly being recognized, especially when genome-wide association fails to establish a clear genotype–/phenotype correlation. Through this review, our aim was to emphasize that besides genetic changes linked to inherited pathologies, alterations of chromatin and disruption of the equilibrium maintaining epigenetic information results in variegated phenotypes, which might be associated with the onset, susceptibility or penetrance of various diseases.

Unraveling the complexity of human CPE and TPE in the context of health and diseases is contingent upon our knowledge of epigenetic regulations implicated in pathogenic pathways and decades of extensive research on model organisms should help to validate the hypothetical mechanisms involved. Also, the emergence of micro RNAs or non-coding RNAs in the regulation of chromatin architecture, lays new ground for the deciphering of the events involved in the pathogenesis of these various cases.

In the future, a better understanding of human PEV and the identification of proteins and pathways involved in its regulation depending on the genomic context might thus be considered to hold the promise of modifying the clinical severity of a wide range of pathologies either by modulating the chromatin changes affecting the locus of interest or, indirectly, by facilitating gene transfer strategies.

Together with epigenetic regulation in general, deciphering the mechanisms involved in gene variegation is thus a major challenge of the post-genomic era for the understanding and cure of a wide range of human diseases.

ACKNOWLEDGEMENTS

The work at the Gilson lab is supported by the Ligue Nationale contre le Cancer (Equipe labellisée) and by the Association Française contre les Myopathies (AFM).

References

1. Strahl BD, Allis CD. The language of covalent histone modifications. *Nature* 2000;**403**:41–5.

2. Peng JC, Karpen GH. Epigenetic regulation of heterochromatic DNA stability. *Curr Opin Genet Dev* 2008;**18**:204–11.

3. Trojer P, Reinberg D. Facultative heterochromatin: is there a distinctive molecular signature? *Mol Cell* 2007;**28**:1–13.

4. Girton JR, Johansen KM. Chromatin structure and the regulation of gene expression: the lessons of PEV in *Drosophila*. *Adv Genet* 2008;**61**:1–43.

5. Ottaviani A, Gilson E, Magdinier F. Telomeric position effect: from the yeast paradigm to human pathologies? *Biochimie* 2008;**90**:93–107.

6. Rabbitts TH, Forster A, Baer R, Hamlyn PH. Transcription enhancer identified near the human C mu immunoglobulin heavy chain gene is unavailable to the translocated c-myc gene in a Burkitt lymphoma. *Nature* 1983;**306**:806–9.

7. Gottschling DE, Aparicio OM, Billington BL, Zakian VA. Position effect at *S. cerevisiae* telomeres: reversible repression of pol II transcription. *Cell* 1990;**63**:751–62.

8. Fourel G, Magdinier F, Gilson E. Insulator dynamics and the setting of chromatin domains. *Bioessays* 2004;**26**:523–32.

9. Gaszner M, Felsenfeld G. Insulators: exploiting transcriptional and epigenetic mechanisms. *Nat Rev Genet* 2006;**7**:703–13.

10. Valenzuela L, Kamakaka RT. Chromatin insulators. *Annu Rev Genet* 2006;**40**:107–38.

11. Sturtevant A. The effect of unequal crossing over the bar locus in *Drosophila*. *Genetics* 1925;**10**:117–47.

12. Muller. Types of visible variations induced by X-rays in *Drosophila*. *J Genet* 1930;**22**:254–99.

13. Muller. Further studies on the nature and causes of gene mutations. *Proc Sixth Int Congr Genet* 1932;**1**:213–55.

14. Dorer DR, Henikoff S. Transgene repeat arrays interact with distant heterochromatin and cause silencing in cis and trans. *Genetics* 1997;**147**:1181–90.

15. Garrick D, Fiering S, Martin DI, Whitelaw E. Repeat-induced gene silencing in mammals. *Nat Genet* 1998;**18**:56–9.

16. Eissenberg JC, James TC, Foster-Hartnett DM, Hartnett T, Ngan V, Elgin SC. Mutation in a heterochromatin-specific chromosomal protein is associated with suppression of position-effect variegation in *Drosophila melanogaster*. *Proc Natl Acad Sci USA* 1990;**87**:9923–7.

17. Festenstein R, Sharghi-Namini S, Fox M, Roderick K, Tolaini M, Norton T, et al. Heterochromatin protein 1 modifies mammalian PEV in a dose- and chromosomal-context-dependent manner. *Nat Genet* 1999;**23**:457–61.

18. James TC, Elgin SC. Identification of a nonhistone chromosomal protein associated with heterochromatin in *Drosophila melanogaster* and its gene. *Mol Cell Biol* 1986;**6**:3862–72.

19. Schotta G, Ebert A, Dorn R, Reuter G. Position-effect variegation and the genetic dissection of chromatin regulation in *Drosophila*. *Semin Cell Dev Biol* 2003;**14**:67–75.

20. Schulze SR, Wallrath LL. Gene regulation by chromatin structure: paradigms established in *Drosophila melanogaster*. *Annu Rev Entomol* 2007;**52**:171–92.

21. Eissenberg JC, Morris GD, Reuter G, Hartnett T. The heterochromatin-associated protein HP-1 is an essential protein in Drosophila with dosage-dependent effects on position-effect variegation. *Genetics* 1992;**131**:345–52.

22. Rea S, Eisenhaber F, O'Carroll D, Strahl BD, Sun ZW, Schmid M, et al. Regulation of chromatin structure by site-specific histone H3 methyltransferases. *Nature* 2000;**406**:593–9.

23. Schotta G, Ebert A, Krauss V, Fischer A, Hoffmann J, Rea S, et al. Central role of Drosophila SU(VAR)3-9 in histone H3-K9 methylation and heterochromatic gene silencing. *EMBO J* 2002;**21**:1121–31.

24. Ebert A, Lein S, Schotta G, Reuter G. Histone modification and the control of heterochromatic gene silencing in *Drosophila*. *Chromosome Res* 2006;**14**:377–92.

25. Jaquet Y, Delattre M, Spierer A, Spierer P. Functional dissection of the *Drosophila* modifier of variegation Su(var)3-7. *Development* 2002;**129**:3975–82.

26. Czermin B, Schotta G, Hulsmann BB, Brehm A, Becker PB, Reuter G, et al. Physical and functional association of SU(VAR)3-9 and HDAC1 in *Drosophila*. *EMBO Rep* 2001;**2**:915–19.

95

27. Ebert A, Schotta G, Lein S, Kubicek S, Krauss V, Jenuwein T, et al. Su(var) genes regulate the balance between euchromatin and heterochromatin in Drosophila. *Genes Dev* 2004;**18**:2973–83.

28. Schotta G, Lachner M, Sarma K, Ebert A, Sengupta R, Reuter G, et al. A silencing pathway to induce H3-K9 and H4-K20 trimethylation at constitutive heterochromatin. *Genes Dev* 2004;**18**:1251–62.

29. Piacentini L, Fanti L, Negri R, Del Vescovo V, Fatica A, Altieri F, et al. Heterochromatin protein 1 (HP1a) positively regulates euchromatic gene expression through RNA transcript association and interaction with hnRNPs in Drosophila. *PLoS Genet* 2009;**5**:e1000670.

30. Rudolph T, Yonezawa M, Lein S, Heidrich K, Kubicek S, Schafer C, et al. Heterochromatin formation in Drosophila is initiated through active removal of H3K4 methylation by the LSD1 homolog SU(VAR)3-3. *Mol Cell* 2007;**26**:103–15.

31. Cai Y, Jin J, Tomomori-Sato C, Sato S, Sorokina I, Parmely TJ, et al. Identification of new subunits of the multiprotein mammalian TRRAP/TIP60-containing histone acetyltransferase complex. *J Biol Chem* 2003;**278**:42733–6.

32. Doyon Y, Selleck W, Lane WS, Tan S, Cote J. Structural and functional conservation of the NuA4 histone acetyltransferase complex from yeast to humans. *Mol Cell Biol* 2004;**24**:1884–96.

33. Hanai K, Furuhashi H, Yamamoto T, Akasaka K, Hirose S. RSF governs silent chromatin formation via histone H2Av replacement. *PLoS Genet* 2008;**4**:e1000011.

34. Ikura T, Ogryzko VV, Grigoriev M, Groisman R, Wang J, Horikoshi M, et al. Involvement of the TIP60 histone acetylase complex in DNA repair and apoptosis. *Cell* 2000;**102**:463–73.

35. Kusch T, Florens L, Macdonald WH, Swanson SK, Glaser RL, Yates JR 3rd, et al. Acetylation by TIP60 is required for selective histone variant exchange at DNA lesions. *Science* 2004;**306**:2084–7.

36. Qi D, Jin H, Lilja T, Mannervik M. Drosophila Reptin and other TIP60 complex components promote generation of silent chromatin. *Genetics* 2006;**174**:241–51.

37. Swaminathan J, Baxter EM, Corces VG. The role of histone H2Av variant replacement and histone H4 acetylation in the establishment of Drosophila heterochromatin. *Genes Dev* 2005;**19**:65–76.

38. Nakayama T, Nishioka K, Dong YX, Shimojima T, Hirose S. Drosophila GAGA factor directs histone H3.3 replacement that prevents the heterochromatin spreading. *Genes Dev* 2007;**21**:552–61.

39. Orphanides G, LeRoy G, Chang CH, Luse DS, Reinberg D. FACT, a factor that facilitates transcript elongation through nucleosomes. *Cell* 1998;**92**:105–16.

40. Orphanides G, Wu WH, Lane WS, Hampsey M, Reinberg D. The chromatin-specific transcription elongation factor FACT comprises human SPT16 and SSRP1 proteins. *Nature* 1999;**400**:284–8.

41. Farkas G, Gausz J, Galloni M, Reuter G, Gyurkovics H, Karch F. The trithorax-like gene encodes the Drosophila GAGA factor. *Nature* 1994;**371**:806–8.

42. Jin C, Felsenfeld G. Nucleosome stability mediated by histone variants H3.3 and H2A.Z. *Genes Dev* 2007;**21**:1519–29.

43. Jin C, Zang C, Wei G, Cui K, Peng W, Zhao K, et al. H3.3/H2A.Z double variant-containing nucleosomes mark 'nucleosome-free regions' of active promoters and other regulatory regions. *Nat Genet* 2009;**41**:941–5.

44. Mason JM, Biessmann H. The unusual telomeres of Drosophila. *Trends Genet* 1995;**11**:58–62.

45. Mefford HC, Trask BJ. The complex structure and dynamic evolution of human subtelomeres. *Nat Rev Genet* 2002;**3**:91–102.

46. Gehring WJ, Klemenz R, Weber U, Kloter U. Functional analysis of the white gene of Drosophila by P-factor-mediated transformation. *Embo J* 1984;**3**:2077–85.

47. Craven RJ, Petes TD. Involvement of the checkpoint protein Mec1p in silencing of gene expression at telomeres in *Saccharomyces cerevisiae*. *Mol Cell Biol* 2000;**20**:2378–84.

48. Pryde FE, Louis EJ. Limitations of silencing at native yeast telomeres. *EMBO J* 1999;**18**:2538–50.

49. Mondoux MA, Zakian VA. Telomere position effect: Silencing near the end. In: De Lange T, Lundblad V, Blackburn EH, editors. *Telomeres*. 2nd ed. Cold Spring Harbor, N.Y.: Cold Spring Harbor Laboratory Press; 2006. pp. 261–316.

50. Aparicio OM, Billington BL, Gottschling DE. Modifiers of position effect are shared between telomeric and silent mating-type loci in *S. cerevisiae*. *Cell* 1991;**66**:1279–87.

51. Boulton SJ, Jackson SP. Components of the Ku-dependent non-homologous end-joining pathway are involved in telomeric length maintenance and telomeric silencing. *EMBO J* 1998;**17**:1819–28.

52. Laroche T, Martin SG, Gotta M, Gorham HC, Pryde FE, Louis EJ, et al. Mutation of yeast Ku genes disrupts the subnuclear organization of telomeres. *Curr Biol* 1998;**8**:653–6.

53. Kyrion G, Boakye KA, Lustig AJ. C-terminal truncation of RAP1 results in the deregulation of telomere size, stability, and function in *Saccharomyces cerevisiae*. *Mol Cell Biol* 1992;**12**:5159–73.

54. Cooper JP, Nimmo ER, Allshire RC, Cech TR. Regulation of telomere length and function by a Myb-domain protein in fission yeast. *Nature* 1997;**385**:744–7.

55. Kanoh J, Ishikawa F. spRap1 and spRif1, recruited to telomeres by Taz1, are essential for telomere function in fission yeast. *Curr Biol* 2001;**11**:1624–30.

56. Nimmo ER, Pidoux AL, Perry PE, Allshire RC. Defective meiosis in telomere-silencing mutants of *Schizosaccharomyces pombe*. *Nature* 1998;**392**:825–8.

57. Park MJ, Jang YK, Choi ES, Kim HS, Park SD. Fission yeast Rap1 homolog is a telomere-specific silencing factor and interacts with Taz1p. *Mol Cells* 2002;**13**:327–33.

58. Wallrath LL, Elgin SC. Position effect variegation in Drosophila is associated with an altered chromatin structure. *Genes Dev* 1995;**9**:1263–77.

59. Ekwall K, Javerzat JP, Lorentz A, Schmidt H, Cranston G, Allshire R. The chromodomain protein Swi6: a key component at fission yeast centromeres. *Science* 1995;**269**:1429–31.

60. Gotta M, Laroche T, Formenton A, Maillet L, Scherthan H, Gasser SM. The clustering of telomeres and colocalization with Rap1, Sir3, and Sir4 proteins in wild-type *Saccharomyces cerevisiae*. *J Cell Biol* 1996;**134**:1349–63.

61. Maillet L, Boscheron C, Gotta M, Marcand S, Gilson E, Gasser SM. Evidence for silencing compartments within the yeast nucleus: a role for telomere proximity and Sir protein concentration in silencer-mediated repression. *Genes Dev* 1996;**10**:1796–811.

62. Csink AK, Henikoff S. Genetic modification of heterochromatic association and nuclear organization in Drosophila. *Nature* 1996;**381**:529–31.

63. Brown KE, Guest SS, Smale ST, Hahm K, Merkenschlager M, Fisher AG. Association of transcriptionally silent genes with ikaros complexes at centromeric heterochromatin. *Cell* 1997;**91**:845–54.

64. de Lange T, Shiue L, Myers RM, Cox DR, Naylor SL, Killery AM, et al. Structure and variability of human chromosome ends. *Mol Cell Biol* 1990;**10**:518–27.

65. Makarov VL, Lejnine S, Bedoyan J, Langmore JP. Nucleosomal organization of telomere-specific chromatin in rat. *Cell* 1993;**73**:775–87.

66. Pisano S, Galati A, Cacchione S. Telomeric nucleosomes: forgotten players at chromosome ends. *Cell Mol Life Sci* 2008;**65**(22):3553–63.

67. Benetti R, Garcia-Cao M, Blasco MA. Telomere length regulates the epigenetic status of mammalian telomeres and subtelomeres. *Nat Genet* 2007;**39**:243–50.

68. Benetti R, Gonzalo S, Jaco I, Schotta G, Klatt P, Jenuwein T, et al. Suv4-20h deficiency results in telomere elongation and derepression of telomere recombination. *J Cell Biol* 2007;**178**:925–36.

69. Garcia-Cao M, O'Sullivan R, Peters AH, Jenuwein T, Blasco MA. Epigenetic regulation of telomere length in mammalian cells by the Suv39h1 and Suv39h2 histone methyltransferases. *Nat Genet* 2004;**36**:94–9.

70. Gonzalo S, Garcia-Cao M, Fraga MF, Schotta G, Peters AH, Cotter SE, et al. Role of the RB1 family in stabilizing histone methylation at constitutive heterochromatin. *Nat Cell Biol* 2005;**7**:420–8.

71. Gonzalo S, Jaco I, Fraga MF, Chen T, Li E, Esteller M, et al. DNA methyltransferases control telomere length and telomere recombination in mammalian cells. *Nat Cell Biol* 2006;**8**:416–24.

72. Ofir R, Wong AC, McDermid HE, Skorecki KL, Selig S. Position effect of human telomeric repeats on replication timing. *Proc Natl Acad Sci USA* 1999;**96**:11434–9.

73. Bayne RA, Broccoli D, Taggart MH, Thomson EJ, Farr CJ, Cooke HJ. Sandwiching of a gene within 12 kb of a functional telomere and alpha satellite does not result in silencing. *Hum Mol Genet* 1994;**3**:539–46.

74. Sprung CN, Sabatier L, Murnane JP. Effect of telomere length on telomeric gene expression. *Nucleic Acids Res* 1996;**24**:4336–40.

75. Baur JA, Zou Y, Shay JW, Wright WE. Telomere position effect in human cells. *Science* 2001;**292**:2075–7.

76. Koering CE, Pollice A, Zibella MP, Bauwens S, Puisieux A, Brunori M, et al. Human telomeric position effect is determined by chromosomal context and telomeric chromatin integrity. *EMBO Rep* 2002;**3**:1055–61.

77. Talbert PB, Henikoff S. Spreading of silent chromatin: inaction at a distance. *Nat Rev Genet* 2006;**7**:793–803.

78. Fourel G, Revardel E, Koering CE, Gilson E. Cohabitation of insulators and silencing elements in yeast subtelomeric regions. *EMBO J* 1999;**18**:2522–37.

79. Kmita M, Kondo T, Duboule D. Targeted inversion of a polar silencer within the HoxD complex re-allocates domains of enhancer sharing. *Nat Genet* 2000;**26**:451–4.

80. Talbert PB, Henikoff S. A reexamination of spreading of position-effect variegation in the white-roughest region of *Drosophila melanogaster*. *Genetics* 2000;**154**:259–72.

81. Henikoff S, Dreesen TD. Trans-inactivation of the Drosophila brown gene: evidence for transcriptional repression and somatic pairing dependence. *Proc Natl Acad Sci USA* 1989;**86**:6704–8.

82. Sass GL, Henikoff S. Pairing-dependent mislocalization of a Drosophila brown gene reporter to a heterochromatic environment. *Genetics* 1999;**152**:595–604.

83. Udvardy A, Maine E, Schedl P. The 87A7 chromomere. Identification of novel chromatin structures flanking the heat shock locus that may define the boundaries of higher order domains. *J Mol Biol* 1985;**185**:341–58.

84. Recillas-Targa F, Pikaart MJ, Burgess-Beusse B, Bell AC, Litt MD, West AG, et al. Position-effect protection and enhancer blocking by the chicken beta-globin insulator are separable activities. *Proc Natl Acad Sci USA* 2002;**99**:6883–8.

85. Gaszner M, Vazquez J, Schedl P. The Zw5 protein, a component of the scs chromatin domain boundary, is able to block enhancer-promoter interaction. *Genes Dev* 1999;**13**:2098–107.

86. Zhao K, Hart CM, Laemmli UK. Visualization of chromosomal domains with boundary element-associated factor BEAF-32. *Cell* 1995;**81**:879–89.

87. Belozerov VE, Majumder P, Shen P, Cai HN. A novel boundary element may facilitate independent gene regulation in the antennapedia complex of Drosophila. *EMBO J* 2003;**22**:3113–21.

88. Ohtsuki S, Levine M. GAGA mediates the enhancer blocking activity of the eve promoter in the Drosophila embryo. *Genes Dev* 1998;**12**:3325–30.

89. Gerasimova TI, Gdula DA, Gerasimov DV, Simonova O, Corces VG. A Drosophila protein that imparts directionality on a chromatin insulator is an enhancer of position-effect variegation. *Cell* 1995;**82**:587–97.

90. Gerasimova TI, Lei EP, Bushey AM, Corces VG. Coordinated control of dCTCF and gypsy chromatin insulators in Drosophila. *Mol Cell* 2007;**28**:761–72.

91. Moon H, Filippova G, Loukinov D, Pugacheva E, Chen Q, Smith ST, et al. CTCF is conserved from Drosophila to humans and confers enhancer blocking of the Fab-8 insulator. *EMBO Rep* 2005;**6**:165–70.

92. Bell AC, West AG, Felsenfeld G. The protein CTCF is required for the enhancer blocking activity of vertebrate insulators. *Cell* 1999;**98**:387–96.

93. Donze D, Adams CR, Rine J, Kamakaka RT. The boundaries of the silenced HMR domain in *Saccharomyces cerevisiae*. *Genes Dev* 1999;**13**:698–708.

94. Donze D, Kamakaka RT. RNA polymerase III and RNA polymerase II promoter complexes are heterochromatin barriers in *Saccharomyces cerevisiae*. *EMBO J* 2001;**20**:520–31.

95. Ladurner AG, Inouye C, Jain R, Tjian R. Bromodomains mediate an acetyl-histone encoded antisilencing function at heterochromatin boundaries. *Mol Cell* 2003;**11**:365–76.

96. West AG, Huang S, Gaszner M, Litt MD, Felsenfeld G. Recruitment of histone modifications by USF proteins at a vertebrate barrier element. *Mol Cell* 2004;**16**:453–63.

97. Jambunathan N, Martinez AW, Robert EC, Agochukwu NB, Ibos ME, Dugas SL, et al. Multiple bromodomain genes are involved in restricting the spread of heterochromatic silencing at the *Saccharomyces cerevisiae* HMR-tRNA boundary. *Genetics* 2005;**171**:913–22.

98. Oki M, Valenzuela L, Chiba T, Ito T, Kamakaka RT. Barrier proteins remodel and modify chromatin to restrict silenced domains. *Mol Cell Biol* 2004;**24**:1956–67.

99. Bi X, Yu Q, Sandmeier JJ, Zou Y. Formation of boundaries of transcriptionally silent chromatin by nucleosome-excluding structures. *Mol Cell Biol* 2004;**24**:2118–31.

100. Gerasimova TI, Corces VG. Polycomb and trithorax group proteins mediate the function of a chromatin insulator. *Cell* 1998;**92**:511–21.

101. Parnell TJ, Kuhn EJ, Gilmore BL, Helou C, Wold MS, Geyer PK. Identification of genomic sites that bind the Drosophila suppressor of Hairy-wing insulator protein. *Mol Cell Biol* 2006;**26**:5983–93.

102. Ramos E, Ghosh D, Baxter E, Corces VG. Genomic organization of gypsy chromatin insulators in *Drosophila melanogaster*. *Genetics* 2006;**172**:2337–49.

103. Barski A, Cuddapah S, Cui K, Roh TY, Schones DE, Wang Z, et al. High-resolution profiling of histone methylations in the human genome. *Cell* 2007;**129**:823–37.

104. Kim TH, Abdullaev ZK, Smith AD, Ching KA, Loukinov DI, Green RD, et al. Analysis of the vertebrate insulator protein CTCF-binding sites in the human genome. *Cell* 2007;**128**:1231–45.

105. Xie X, Mikkelsen TS, Gnirke A, Lindblad-Toh K, Kellis M, Lander ES. Systematic discovery of regulatory motifs in conserved regions of the human genome, including thousands of CTCF insulator sites. *Proc Natl Acad Sci USA* 2007;**104**:7145–50.

106. Parelho V, Hadjur S, Spivakov M, Leleu M, Sauer S, Gregson HC, et al. Cohesins functionally associate with CTCF on mammalian chromosome arms. *Cell* 2008;**132**:422–33.

107. Stedman W, Kang H, Lin S, Kissil JL, Bartolomei MS, Lieberman PM. Cohesins localize with CTCF at the KSHV latency control region and at cellular c-myc and H19/Igf2 insulators. *EMBO J* 2008;**27**:654–66.

108. Wendt KS, Yoshida K, Itoh T, Bando M, Koch B, Schirghuber E, et al. Cohesin mediates transcriptional insulation by CCCTC-binding factor. *Nature* 2008;**451**:796–801.

109. Guelen L, Pagie L, Brasset E, Meuleman W, Faza MB, Talhout W, et al. Domain organization of human chromosomes revealed by mapping of nuclear lamina interactions. *Nature* 2008;**453**:948–51.

110. Filippova GN. Genetics and epigenetics of the multifunctional protein CTCF. *Curr Top Dev Biol* 2008;**80**:337–60.

111. Phillips JE, Corces VG. CTCF: master weaver of the genome. *Cell* 2009;**137**:1194–211.

112. Scott RH, Douglas J, Baskcomb L, Huxter N, Barker K, Hanks S, et al. Constitutional 11p15 abnormalities, including heritable imprinting center mutations, cause nonsyndromic Wilms tumor. *Nat Genet* 2008;**40**:1329–34.

113. Cho DH, Thienes CP, Mahoney SE, Analau E, Filippova GN, Tapscott SJ. Antisense transcription and heterochromatin at the DM1 CTG repeats are constrained by CTCF. *Mol Cell* 2005;**20**:483–9.

114. Filippova GN, Thienes CP, Penn BH, Cho DH, Hu YJ, Moore JM, et al. CTCF-binding sites flank CTG/CAG repeats and form a methylation-sensitive insulator at the DM1 locus. *Nat Genet* 2001;**28**:335–43.

115. Jansen G, Groenen PJ, Bachner D, Jap PH, Coerwinkel M, Oerlemans F, et al. Abnormal myotonic dystrophy protein kinase levels produce only mild myopathy in mice. *Nat Genet* 1996;**13**:316–24.

116. Mankodi A, Logigian E, Callahan L, McClain C, White R, Henderson D, et al. Myotonic dystrophy in transgenic mice expressing an expanded CUG repeat. *Science* 2000;**289**:1769–73.

117. Reddy S, Smith DB, Rich MM, Leferovich JM, Reilly P, Davis BM, et al. Mice lacking the myotonic dystrophy protein kinase develop a late onset progressive myopathy. *Nat Genet* 1996;**13**:325–35.

118. Steinbach P, Glaser D, Vogel W, Wolf M, Schwemmle S. The DMPK gene of severely affected myotonic dystrophy patients is hypermethylated proximal to the largely expanded CTG repeat. *Am J Hum Genet* 1998;**62**:278–85.

119. Otten AD, Tapscott SJ. Triplet repeat expansion in myotonic dystrophy alters the adjacent chromatin structure. *Proc Natl Acad Sci USA* 1995;**92**:5465–9.

120. Libby RT, Hagerman KA, Pineda VV, Lau R, Cho DH, Baccam SL, et al. CTCF cis-regulates trinucleotide repeat instability in an epigenetic manner: a novel basis for mutational hot spot determination. *PLoS Genet* 2008;**4**:e1000257.

121. Ottaviani A, Rival-Gervier S, Boussouar A, Foerster AM, Rondier D, Sacconi S, et al. The D4Z4 macrosatellite repeat acts as a CTCF and A-type lamins-dependent insulator in facio-scapulo-humeral dystrophy. *PLoS Genet* 2009;**5**:e1000394.

122. Snider L, Asawachaicharn A, Tyler AE, Geng LN, Petek LM, Maves L, et al. RNA transcripts, miRNA-sized fragments and proteins produced from D4Z4 units: new candidates for the pathophysiology of facioscapulohumeral dystrophy. *Hum Mol Genet* 2009;**18**:2414–30.

123. Hewitt JE, Lyle R, Clark LN, Valleley EM, Wright TJ, Wijmenga C, et al. Analysis of the tandem repeat locus D4Z4 associated with facioscapulohumeral muscular dystrophy. *Hum Mol Genet* 1994;**3**:1287–95.

124. Lyle R, Wright TJ, Clark LN, Hewitt JE. The FSHD-associated repeat, D4Z4, is a member of a dispersed family of homeobox-containing repeats, subsets of which are clustered on the short arms of the acrocentric chromosomes. *Genomics* 1995;**28**:389–97.

125. van Deutekom JC, Wijmenga C, van Tienhoven EA, Gruter AM, Hewitt JE, Padberg GW, et al. FSHD associated DNA rearrangements are due to deletions of integral copies of a 3.2 kb tandemly repeated unit. *Hum Mol Genet* 1993;**2**:2037–42.

126. Winokur ST, Bengtsson U, Feddersen J, Mathews KD, Weiffenbach B, Bailey H, et al. The DNA rearrangement associated with facioscapulohumeral muscular dystrophy involves a heterochromatin-associated repetitive element: implications for a role of chromatin structure in the pathogenesis of the disease. *Chromosome Res* 1994;**2**:225–34.

127. Chadwick BP. DXZ4 chromatin adopts an opposing conformation to that of the surrounding chromosome and acquires a novel inactive X specific role involving CTCF and anti-sense transcripts. *Genome Res* 2008;**18**:1259–69.

128. Worman HJ, Bonne G. "Laminopathies": a wide spectrum of human diseases. *Exp Cell Res* 2007;**313**:2121–33.

129. Bonne G, Di Barletta MR, Varnous S, Becane HM, Hammouda EH, Merlini L, et al. Mutations in the gene encoding lamin A/C cause autosomal dominant Emery-Dreifuss muscular dystrophy. *Nat Genet* 1999;**21**:285–8.

130. Muchir A, Bonne G, van der Kooi AJ, van Meegen M, Baas F, Bolhuis PA, et al. Identification of mutations in the gene encoding lamins A/C in autosomal dominant limb girdle muscular dystrophy with atrioventricular conduction disturbances (LGMD1B). *Hum Mol Genet* 2000;**9**:1453–9.

131. De Sandre-Giovannoli A, Bernard R, Cau P, Navarro C, Amiel J, Boccaccio I, et al. Lamin a truncation in Hutchinson-Gilford progeria. *Science* 2003;**300**:2055.

132. Eriksson M, Brown WT, Gordon LB, Glynn MW, Singer J, Scott L, et al. Recurrent de novo point mutations in lamin A cause Hutchinson–Gilford progeria syndrome. *Nature* 2003;**423**:293–8.

133. Sun FL, Cuaycong MH, Elgin SC. Long-range nucleosome ordering is associated with gene silencing in *Drosophila melanogaster* pericentric heterochromatin. *Mol Cell Biol* 2001;**21**:2867–79.

134. Hacein-Bey-Abina S, Le Deist F, Carlier F, Bouneaud C, Hue C, De Villartay JP, et al. Sustained correction of X-linked severe combined immunodeficiency by ex vivo gene therapy. *N Engl J Med* 2002;**346**:1185–93.

135. Hacein-Bey-Abina S, Von Kalle C, Schmidt M, McCormack MP, Wulffraat N, Lebouloch P, et al. LMO2-associated clonal T cell proliferation in two patients after gene therapy for SCID-X1. *Science* 2003;**302**:415–19.

136. Modlich U, Kustikova OS, Schmidt M, Rudolph C, Meyer J, Li Z, et al. Leukemias following retroviral transfer of multidrug resistance 1 (MDR1) are driven by combinatorial insertional mutagenesis. *Blood* 2005;**105**:4235–46.

137. Wu X, Li Y, Crise B, Burgess SM. Transcription start regions in the human genome are favored targets for MLV integration. *Science* 2003;**300**:1749–51.

138. Ellis J. Silencing and variegation of gammaretrovirus and lentivirus vectors. *Hum Gene Ther* 2005;**16**:1241–6.

139. Ishii N, Asao H, Kimura Y, Takeshita T, Nakamura M, Tsuchiya S, et al. Impairment of ligand binding and growth signaling of mutant IL-2 receptor gamma-chains in patients with X-linked severe combined immunodeficiency. *J Immunol* 1994;**153**:1310–17.

140. Rosenqvist N, Hard Af Segerstad C, Samuelsson C, Johansen J, Lundberg C. Activation of silenced transgene expression in neural precursor cell lines by inhibitors of histone deacetylation. *J Gene Med* 2002;**4**:248–57.

141. Sinn PL, Sauter SL, McCray Jr PB. Gene therapy progress and prospects: development of improved lentiviral and retroviral vectors – design, biosafety, and production. *Gene Ther* 2005;**12**:1089–98.

142. Chung JH, Whiteley M, Felsenfeld G. A 5′ element of the chicken beta-globin domain serves as an insulator in human erythroid cells and protects against position effect in Drosophila. *Cell* 1993;**74**:505–14.

143. Potts W, Tucker D, Wood H, Martin C. Chicken beta-globin 5′HS4 insulators function to reduce variability in transgenic founder mice. *Biochem Biophys Res Commun* 2000;**273**:1015–18.

144. Akasaka K, Nishimura A, Takata K, Mitsunaga K, Mibuka F, Ueda H, et al. Upstream element of the sea urchin arylsulfatase gene serves as an insulator. *Cell Mol Biol (Noisy-le-grand)* 1999;**45**:555–65.

145. Giraldo P, Rival-Gervier S, Houdebine LM, Montoliu L. The potential benefits of insulators on heterologous constructs in transgenic animals. *Transgenic Res* 2003;**12**:751–5.

146. Jakobsson J, Rosenqvist N, Thompson L, Barraud P, Lundberg C. Dynamics of transgene expression in a neural stem cell line transduced with lentiviral vectors incorporating the cHS4 insulator. *Exp Cell Res* 2004;**298**:611–23.

147. Yao S, Sukonnik T, Kean T, Bharadwaj RR, Pasceri P, Ellis J. Retrovirus silencing, variegation, extinction, and memory are controlled by a dynamic interplay of multiple epigenetic modifications. *Mol Ther* 2004;**10**:27–36.

148. Kleinjan DJ, van Heyningen V. Position effect in human genetic disease. *Hum Mol Genet* 1998;**7**:1611–18.

149. Rabbitts TH. Chromosomal translocations in human cancer. *Nature* 1994;**372**:143–9.

150. Rabbitts TH. Translocations, master genes, and differences between the origins of acute and chronic leukemias. *Cell* 1991;**67**:641–4.

151. Korsmeyer SJ. Chromosomal translocations in lymphoid malignancies reveal novel proto-oncogenes. *Annu Rev Immunol* 1992;**10**:785–807.

152. Lossos IS, Akasaka T, Martinez-Climent JA, Siebert R, Levy R. The BCL6 gene in B-cell lymphomas with 3q27 translocations is expressed mainly from the rearranged allele irrespective of the partner gene. *Leukemia* 2003;**17**:1390–7.

153. Kumar-Sinha C, Tomlins SA, Chinnaiyan AM. Recurrent gene fusions in prostate cancer. *Nat Rev Cancer* 2008;**8**:497–511.

154. Barki-Celli L, Lefebvre C, Le Baccon P, Nadeau G, Bonnefoix T, Usson Y, et al. Differences in nuclear positioning of 1q12 pericentric heterochromatin in normal and tumor B lymphocytes with 1q rearrangements. *Genes Chromosomes Cancer* 2005;**43**:339–49.

155. State MW, Greally JM, Cuker A, Bowers PN, Henegariu O, Morgan TM, et al. Epigenetic abnormalities associated with a chromosome 18(q21-q22) inversion and a Gilles de la Tourette syndrome phenotype. *Proc Natl Acad Sci USA* 2003;**100**:4684–9.

156. Wieacker P, Apeshiotis N, Jakubiczka S, Volleth M, Wieland I. Familial translocation t(1;9) associated with macromastia: molecular cloning of the breakpoints. *Sex Dev* 2007;**1**:35–41.

157. Kleinjan DA, Seawright A, Schedl A, Quinlan RA, Danes S, van Heyningen V. Aniridia-associated translocations, DNase hypersensitivity, sequence comparison and transgenic analysis redefine the functional domain of PAX6. *Hum Mol Genet* 2001;**10**:2049–59.

158. Lauderdale JD, Wilensky JS, Oliver ER, Walton DS, Glaser T. 3′ deletions cause aniridia by preventing PAX6 gene expression. *Proc Natl Acad Sci USA* 2000;**97**:13755–9.

159. Crackower MA, Scherer SW, Rommens JM, Hui CC, Poorkaj P, Soder S, et al. Characterization of the split hand/split foot malformation locus SHFM1 at 7q21.3-q22.1 and analysis of a candidate gene for its expression during limb development. *Hum Mol Genet* 1996;**5**:571–9.

160. Ignatius J, Knuutila S, Scherer SW, Trask B, Kere J. Split hand/split foot malformation, deafness, and mental retardation with a complex cytogenetic rearrangement involving 7q21.3. *J Med Genet* 1996;**33**:507–10.

161. Scherer SW, Poorkaj P, Allen T, Kim J, Geshuri D, Nunes M, et al. Fine mapping of the autosomal dominant split hand/split foot locus on chromosome 7, band q21.3-q22.1. *Am J Hum Genet* 1994;**55**:12–20.

162. van Silfhout AT, van den Akker PC, Dijkhuizen T, Verheij JB, Olderode-Berends MJ, Kok K, et al. Split hand/foot malformation due to chromosome 7q aberrations(SHFM1): additional support for functional haploinsufficiency as the causative mechanism. *Eur J Hum Genet* 2009;**17**(11):1432–8.

163. Wieland I, Muschke P, Jakubiczka S, Volleth M, Freigang B, Wieacker PF. Refinement of the deletion in 7q21.3 associated with split hand/foot malformation type 1 and Mondini dysplasia. *J Med Genet* 2004;**41**:e54.

164. Simeone A, Acampora D, Pannese M, D'Esposito M, Stornaiuolo A, Gulisano M, et al. Cloning and characterization of two members of the vertebrate Dlx gene family. *Proc Natl Acad Sci USA* 1994;**91**:2250–4.

165. Merlo GR, Paleari L, Mantero S, Genova F, Beverdam A, Palmisano GL, et al. Mouse model of split hand/foot malformation type I. *Genesis* 2002;**33**:97–101.

166. Scherer SW, Cheung J, MacDonald JR, Osborne LR, Nakabayashi K, Herbrick JA, et al. Human chromosome 7: DNA sequence and biology. *Science* 2003;**300**:767–72.

167. Foster JW, Dominguez-Steglich MA, Guioli S, Kwok C, Weller PA, Stevanovic M, et al. Campomelic dysplasia and autosomal sex reversal caused by mutations in an SRY-related gene. *Nature* 1994;**372**:525–30.

168. Pfeifer D, Kist R, Dewar K, Devon K, Lander ES, Birren B, et al. Campomelic dysplasia translocation breakpoints are scattered over 1 Mb proximal to SOX9: evidence for an extended control region. *Am J Hum Genet* 1999;**65**:111–24.

169. Pop R, Conz C, Lindenberg KS, Blesson S, Schmalenberger B, Briault S, et al. Screening of the 1 Mb SOX9 5′ control region by array CGH identifies a large deletion in a case of campomelic dysplasia with XY sex reversal. *J Med Genet* 2004;**41**:e47.

170. Velagaleti GV, Bien-Willner GA, Northup JK, Lockhart LH, Hawkins JC, Jalal SM, et al. Position effects due to chromosome breakpoints that map approximately 900 Kb upstream and approximately 1.3 Mb downstream of SOX9 in two patients with campomelic dysplasia. *Am J Hum Genet* 2005;**76**:652–62.

171. Wagner T, Wirth J, Meyer J, Zabel B, Held M, Zimmer J, et al. Autosomal sex reversal and campomelic dysplasia are caused by mutations in and around the SRY-related gene SOX9. *Cell* 1994;**79**:1111–20.

172. Benko S, Fantes JA, Amiel J, Kleinjan DJ, Thomas S, Ramsay J, et al. Highly conserved non-coding elements on either side of SOX9 associated with Pierre Robin sequence. *Nat Genet* 2009;**41**:359–64.

173. Roessler E, Ward DE, Gaudenz K, Belloni E, Scherer SW, Donnai D, et al. Cytogenetic rearrangements involving the loss of the Sonic Hedgehog gene at 7q36 cause holoprosencephaly. *Hum Genet* 1997;**100**:172–81.

174. Lettice LA, Horikoshi T, Heaney SJ, van Baren MJ, van der Linde HC, Breedveld GJ, et al. Disruption of a long-range cis-acting regulator for Shh causes preaxial polydactyly. *Proc Natl Acad Sci USA* 2002;**99**:7548–53.

175. Lettice LA, Heaney SJ, Purdie LA, Li L, de Beer P, Oostra BA, et al. A long-range Shh enhancer regulates expression in the developing limb and fin and is associated with preaxial polydactyly. *Hum Mol Genet* 2003;**12**:1725–35.

176. Lee JA, Madrid RE, Sperle K, Ritterson CM, Hobson GM, Garbern J, et al. Spastic paraplegia type 2 associated with axonal neuropathy and apparent PLP1 position effect. *Ann Neurol* 2006;**59**:398–403.

177. Muncke N, Wogatzky BS, Breuning M, Sistermans EA, Endris V, Ross M, et al. Position effect on PLP1 may cause a subset of Pelizaeus-Merzbacher disease symptoms. *J Med Genet* 2004;**41**:e121.

178. Inoue K, Osaka H, Thurston VC, Clarke JT, Yoneyama A, Rosenbarker L, et al. Genomic rearrangements resulting in PLP1 deletion occur by nonhomologous end joining and cause different dysmyelinating phenotypes in males and females. *Am J Hum Genet* 2002;**71**:838–53.

179. Barbour VM, Tufarelli C, Sharpe JA, Smith ZE, Ayyub H, Heinlein CA, et al. Alpha-thalassemia resulting from a negative chromosomal position effect. *Blood* 2000;**96**:800–7.

180. Tufarelli C, Stanley JA, Garrick D, Sharpe JA, Ayyub H, Wood WG, et al. Transcription of antisense RNA leading to gene silencing and methylation as a novel cause of human genetic disease. *Nat Genet* 2003;**34**:157–65.

181. Brouwer JR, Willemsen R, Oostra BA. Microsatellite repeat instability and neurological disease. *Bioessays* 2009;**31**:71–83.

182. Oberle I, Rousseau F, Heitz D, Kretz C, Devys D, Hanauer A, et al. Instability of a 550-base pair DNA segment and abnormal methylation in Fragile X syndrome. *Science* 1991;**252**:1097–102.

183. Pieretti M, Zhang FP, Fu YH, Warren ST, Oostra BA, Caskey CT, et al. Absence of expression of the FMR-1 gene in Fragile X syndrome. *Cell* 1991;**66**:817–22.

184. Coffee B, Zhang F, Ceman S, Warren ST, Reines D. Histone modifications depict an aberrantly heterochromatinized FMR1 gene in Fragile X syndrome. *Am J Hum Genet* 2002;**71**:923–32.

185. Coffee B, Zhang F, Warren ST, Reines D. Acetylated histones are associated with FMR1 in normal but not Fragile X-syndrome cells. *Nat Genet* 1999;**22**:98–101.

186. Sakamoto N, Ohshima K, Montermini L, Pandolfo M, Wells RD. Sticky DNA, a self-associated complex formed at long GAA*TTC repeats in intron 1 of the frataxin gene, inhibits transcription. *J Biol Chem* 2001;**276**:27171–7.

187. Bidichandani SI, Ashizawa T, Patel PI. The GAA triplet-repeat expansion in Friedreich ataxia interferes with transcription and may be associated with an unusual DNA structure. *Am J Hum Genet* 1998;**62**:111–21.

188. Al-Mahdawi S, Pinto RM, Ismail O, Varshney D, Lymperi S, Sandi C, et al. The Friedreich ataxia GAA repeat expansion mutation induces comparable epigenetic changes in human and transgenic mouse brain and heart tissues. *Hum Mol Genet* 2008;**17**:735–46.

189. Greene E, Mahishi L, Entezam A, Kumari D, Usdin K. Repeat-induced epigenetic changes in intron 1 of the frataxin gene and its consequences in Friedreich ataxia. *Nucleic Acids Res* 2007;**35**:3383–90.

101

190. Saveliev A, Everett C, Sharpe T, Webster Z, Festenstein R. DNA triplet repeats mediate heterochromatin-protein-1-sensitive variegated gene silencing. *Nature* 2003;**422**:909–13.

191. Chiurazzi P, Pomponi MG, Pietrobono R, Bakker CE, Neri G, Oostra BA. Synergistic effect of histone hyperacetylation and DNA demethylation in the reactivation of the FMR1 gene. *Hum Mol Genet* 1999;**8**:2317–23.

192. Herman D, Jenssen K, Burnett R, Soragni E, Perlman SL, Gottesfeld JM. Histone deacetylase inhibitors reverse gene silencing in Friedreich's ataxia. *Nat Chem Biol* 2006;**2**:551–8.

193. Kumari D, Usdin K. Chromatin remodeling in the noncoding repeat expansion diseases. *J Biol Chem* 2009;**284**:7413–7.

194. Redon R, Ishikawa S, Fitch KR, Feuk L, Perry GH, Andrews TD, et al. Global variation in copy number in the human genome. *Nature* 2006;**444**:444–54.

195. Scherer SW, Lee C, Birney E, Altshuler DM, Eichler EE, Carter NP, et al. Challenges and standards in integrating surveys of structural variation. *Nat Genet* 2007;**39**:S7–15.

196. Goidts V, Cooper DN, Armengol L, Schempp W, Conroy J, Estivill X, et al. Complex patterns of copy number variation at sites of segmental duplications: an important category of structural variation in the human genome. *Hum Genet* 2006;**120**:270–84.

197. Lee JA, Carvalho CM, Lupski JR. A DNA replication mechanism for generating nonrecurrent rearrangements associated with genomic disorders. *Cell* 2007;**131**:1235–47.

198. Aldred PM, Hollox EJ, Armour JA. Copy number polymorphism and expression level variation of the human alpha-defensin genes DEFA1 and DEFA3. *Hum Mol Genet* 2005;**14**:2045–52.

199. McCarroll SA, Hadnott TN, Perry GH, Sabeti PC, Zody MC, Barrett JC, et al. Common deletion polymorphisms in the human genome. *Nat Genet* 2006;**38**:86–92.

200. Stranger BE, Forrest MS, Dunning M, Ingle CE, Beazley C, Thorne N, et al. Relative impact of nucleotide and copy number variation on gene expression phenotypes. *Science* 2007;**315**:848–53.

201. Merla G, Howald C, Henrichsen CN, Lyle R, Wyss C, Zabot MT, et al. Submicroscopic deletion in patients with Williams-Beuren syndrome influences expression levels of the nonhemizygous flanking genes. *Am J Hum Genet* 2006;**79**:332–41.

202. Shao L, Shaw CA, Lu XY, Sahoo T, Bacino CA, Lalani SR, et al. Identification of chromosome abnormalities in subtelomeric regions by microarray analysis: a study of 5,380 cases. *Am J Med Genet A* 2008;**146A**:2242–51.

203. Ballif BC, Sulpizio SG, Lloyd RM, Minier SL, Theisen A, Bejjani BA, et al. The clinical utility of enhanced subtelomeric coverage in array CGH. *Am J Med Genet A* 2007;**143A**:1850–7.

204. Barber JC. Terminal 3p deletions: phenotypic variability, chromosomal non-penetrance, or gene modification? *Am J Med Genet A* 2008;**146A**:1899–901.

205. Ledbetter DH, Martin CL. Cryptic telomere imbalance: a 15-year update. *Am J Med Genet C Semin Med Genet* 2007;**145C**:327–34.

206. Balikova I, Menten B, de Ravel T, Le Caignec C, Thienpont B, Urbina M, et al. Subtelomeric imbalances in phenotypically normal individuals. *Hum Mutat* 2007;**28**:958–67.

207. Flint J, Wilkie AO, Buckle VJ, Winter RM, Holland AJ, McDermid HE. The detection of subtelomeric chromosomal rearrangements in idiopathic mental retardation. *Nat Genet* 1995;**9**:132–40.

208. Giraudeau F, Aubert D, Young I, Horsley S, Knight S, Kearney L, et al. Molecular-cytogenetic detection of a deletion of 1p36.3. *J Med Genet* 1997;**34**:314–17.

209. Horsley SW, Knight SJ, Nixon J, Huson S, Fitchett M, Boone RA, et al. Del(18p) shown to be a cryptic translocation using a multiprobe FISH assay for subtelomeric chromosome rearrangements. *J Med Genet* 1998;**35**:722–6.

210. Kleefstra T, Brunner HG, Amiel J, Oudakker AR, Nillesen WM, Magee A, et al. Loss-of-function mutations in euchromatin histone methyl transferase 1 (EHMT1) cause the 9q34 subtelomeric deletion syndrome. *Am J Hum Genet* 2006;**79**:370–7.

211. Lamb J, Harris PC, Wilkie AO, Wood WG, Dauwerse JG, Higgs DR. De novo truncation of chromosome 16p and healing with (TTAGGG)n in the alpha-thalassemia/mental retardation syndrome (ATR-16). *Am J Hum Genet* 1993;**52**:668–76.

212. Walter S, Sandig K, Hinkel GK, Mitulla B, Ounap K, Sims G, et al. Subtelomere FISH in 50 children with mental retardation and minor anomalies, identified by a checklist, detects 10 rearrangements including a de novo balanced translocation of chromosomes 17p13.3 and 20q13.33. *Am J Med Genet A* 2004;**128**:364–73.

213. Bonaglia MC, Giorda R, Mani E, Aceti G, Anderlid BM, Baroncini A, et al. Identification of a recurrent breakpoint within the SHANK3 gene in the 22q13.3 deletion syndrome. *J Med Genet* 2006;**43**:822–8.

214. Wilson HL, Wong AC, Shaw SR, Tse WY, Stapleton GA, Phelan MC, et al. Molecular characterisation of the 22q13 deletion syndrome supports the role of haploinsufficiency of SHANK3/PROSAP2 in the major neurological symptoms. *J Med Genet* 2003;**40**:575–84.

102

215. Cerruti Mainardi P. Cri du Chat syndrome. *Orphanet J Rare Dis* 2006;**1**:33.

216. Zhang X, Snijders A, Segraves R, Niebuhr A, Albertson D, Yang H, et al. High-resolution mapping of genotype-phenotype relationships in cri du chat syndrome using array comparative genomic hybridization. *Am J Hum Genet* 2005;**76**:312–26.

217. Gajecka M, Mackay KL, Shaffer LG. Monosomy 1p36 deletion syndrome. *Am J Med Genet C Semin Med Genet* 2007;**145C**:346–56.

218. Redon R, Rio M, Gregory SG, Cooper RA, Fiegler H, Sanlaville D, et al. Tiling path resolution mapping of constitutional 1p36 deletions by array-CGH: contiguous gene deletion or "deletion with positional effect" syndrome? *J Med Genet* 2005;**42**:166–71.

219. Cote GB, Katsantoni A, Deligeorgis D. The cytogenetic and clinical implications of a ring chromosome 2. *Ann Genet* 1981;**24**:231–5.

220. Kosztolanyi G. Does "ring syndrome" exist? an analysis of 207 case reports on patients with a ring autosome. *Hum Genet* 1987;**75**:174–9.

221. Pezzolo A, Gimelli G, Cohen A, Lavaggetto A, Romano C, Fogu G, et al. Presence of telomeric and subtelomeric sequences at the fusion points of ring chromosomes indicates that the ring syndrome is caused by ring instability. *Hum Genet* 1993;**92**:23–7.

222. Sigurdardottir S, Goodman BK, Rutberg J, Thomas GH, Jabs EW, Geraghty MT. Clinical, cytogenetic, and fluorescence in situ hybridization findings in two cases of "complete ring" syndrome. *Am J Med Genet* 1999;**87**:384–90.

223. Vermeesch JR, Baten E, Fryns JP, Devriendt K. Ring syndrome caused by ring chromosome 7 without loss of subtelomeric sequences. *Clin Genet* 2002;**62**:415–17.

224. Sarg B, Koutzamani E, Helliger W, Rundquist I, Lindner HH. Postsynthetic trimethylation of histone H4 at lysine 20 in mammalian tissues is associated with aging. *J Biol Chem* 2002;**277**:39195–201.

225. Funayama R, Saito M, Tanobe H, Ishikawa F. Loss of linker histone H1 in cellular senescence. *J Cell Biol* 2006;**175**:869–80.

226. Narita M, Nunez S, Heard E, Lin AW, Hearn SA, Spector DL, et al. Rb-mediated heterochromatin formation and silencing of E2F target genes during cellular senescence. *Cell* 2003;**113**:703–16.

227. Zhang R, Poustovoitov MV, Ye X, Santos HA, Chen W, Daganzo SM, et al. Formation of MacroH2A-containing senescence-associated heterochromatin foci and senescence driven by ASF1a and HIRA. *Dev Cell* 2005;**8**:19–30.

228. Gaubatz JW, Cutler RG. Mouse satellite DNA is transcribed in senescent cardiac muscle. *J Biol Chem* 1990;**265**:17753–8.

229. Gaubatz JW, Flores SC. Tissue-specific and age-related variations in repetitive sequences of mouse extrachromosomal circular DNAs. *Mutat Res* 1990;**237**:29–36.

230. Shen S, Liu A, Li J, Wolubah C, Casaccia-Bonnefil P. Epigenetic memory loss in aging oligodendrocytes in the corpus callosum. *Neurobiol Aging* 2008;**29**:452–63.

231. Busque L, Mio R, Mattioli J, Brais E, Blais N, Lalonde Y, et al. Nonrandom X-inactivation patterns in normal females: lyonization ratios vary with age. *Blood* 1996;**88**:59–65.

232. Gale RE, Fielding AK, Harrison CN, Linch DC. Acquired skewing of X-chromosome inactivation patterns in myeloid cells of the elderly suggests stochastic clonal loss with age. *Br J Haematol* 1997;**98**:512–19.

233. Sharp A, Robinson D, Jacobs P. Age- and tissue-specific variation of X chromosome inactivation ratios in normal women. *Hum Genet* 2000;**107**:343–9.

234. Au WY, Ma ES, Lam VM, Chan JL, Pang A, Kwong YL. Glucose 6-phosphate dehydrogenase (G6PD) deficiency in elderly Chinese women heterozygous for G6PD variants. *Am J Med Genet A* 2004;**129A**:208–11.

235. Cazzola M, May A, Bergamaschi G, Cerani P, Rosti V, Bishop DF. Familial-skewed X-chromosome inactivation as a predisposing factor for late-onset X-linked sideroblastic anemia in carrier females. *Blood* 2000;**96**:4363–5.

236. Invernizzi P, Pasini S, Selmi C, Miozzo M, Podda M. Skewing of X chromosome inactivation in autoimmunity. *Autoimmunity* 2008;**41**:272–7.

237. Allman D, Miller JP. B cell development and receptor diversity during aging. *Curr Opin Immunol* 2005;**17**:463–7.

238. Allman D, Miller JP. The aging of early B-cell precursors. *Immunol Rev* 2005;**205**:18–29.

239. Lombardi G, Di Somma C, Rota F, Colao A. Associated hormonal decline in aging: is there a role for GH therapy in aging men? *J Endocrinol Invest* 2005;**28**:99–108.

240. Pifer J, Stephan RP, Lill-Elghanian DA, Le PT, Witte PL. Role of stromal cells and their products in protecting young and aged B-lineage precursors from dexamethasone-induced apoptosis. *Mech Ageing Dev* 2003;**124**:207–18.

241. Stephan RP, Lill-Elghanian DA, Witte PL. Development of B cells in aged mice: decline in the ability of pro-B cells to respond to IL-7 but not to other growth factors. *J Immunol* 1997;**158**:1598–609.

242. Lezhava T. Chromosome and aging: genetic conception of aging. *Biogerontology* 2001;**2**:253–60.

243. Lezhava T, Jokhadze T. Activation of pericentromeric and telomeric heterochromatin in cultured lymphocytes from old individuals. *Ann N Y Acad Sci* 2007;**1100**:387–99.

244. Krabbe KS, Pedersen M, Bruunsgaard H. Inflammatory mediators in the elderly. *Exp Gerontol* 2004;**39**:687–99.

245. Boscheron C, Maillet L, Marcand S, Tsai-Pflugfelder M, Gasser SM, Gilson E. Cooperation at a distance between silencers and proto-silencers at the yeast HML locus. *Embo J* 1996;**15**:2184–95.

246. Lebrun E, Revardel E, Boscheron C, Li R, Gilson E, Fourel G. Protosilencers in *Saccharomyces cerevisiae* subtelomeric regions. *Genetics* 2001;**158**:167–76.

247. de Chadarevian JP, Dunn S, Malatack JJ, Ganguly A, Blecker U, Punnett HH. Chromosome rearrangement with no apparent gene mutation in familial adenomatous polyposis and hepatocellular neoplasia. *Pediatr Dev Pathol* 2002;**5**:69–75.

248. Nishimura DY, Searby CC, Alward WL, Walton D, Craig JE, Mackey DA, et al. A spectrum of FOXC1 mutations suggests gene dosage as a mechanism for developmental defects of the anterior chamber of the eye. *Am J Hum Genet* 2001;**68**:364–72.

249. Fang J, Dagenais SL, Erickson RP, Avlt MF, Glynn MW, Gorski JL, et al. Mutations in FOXC2 (MFH-1), a forkhead family transcription factor, are responsible for the hereditary lymphadema-distichiasis syndrome. *AM J Hom Genet* 2000;**67**(6):1382–8.

250. Crisponi L, Deiana M, Loi A, Chiappe F, Uda M, Amati P, et al. The putative forkhead transcription factor FOXL2 is mutated in blepharophimosis/ptosis/epicanthus inversus syndrome. *Nat Genet* 2001;**27**:159–66.

251. Beysen D, Raes J, Leroy BP, Lucassen A, Yates JR, Clayton-Smith J, et al. Deletions involving long-range conserved nongenic sequences upstream and downstream of FOXL2 as a novel disease-causing mechanism in blepharophimosis syndrome. *Am J Hum Genet* 2005;**77**:205–18.

252. Kosho T, Sakazume S, Kawame H, Wakui K, Wada T, Okoshi Y, et al. De-novo balanced translocation between 7q31 and 10p14 in a girl with central precocious puberty, moderate mental retardation, and severe speech impairment. *Clin Dysmorphol* 2008;**17**:31–4.

253. Vortkamp A, Gessler M, Grzeschik KH. GLI3 zinc-finger gene interrupted by translocations in Greig syndrome families. *Nature* 1991;**352**:539–40.

254. Kioussis D, Vanin E, deLange T, Flavell RA, Grosveld FG. Beta-globin gene inactivation by DNA translocation in gamma beta-thalassaemia. *Nature* 1983;**306**:662–6.

255. Driscoll MC, Dobkin CS, Alter BP. Gamma delta beta-thalassemia due to a de novo mutation deleting the 5′ beta-globin gene activation-region hypersensitive sites. *Proc Natl Acad Sci USA* 1989;**86**:7470–4.

256. Yue Y, Farcas R, Thiel G, Bommer C, Grossmann B, Galetzka D, et al. Haaf T. De novo t(12;17)(p13.3;q21.3) translocation with a breakpoint near the 5′ end of the HOXB gene cluster in a patient with developmental delay and skeletal malformations. *Eur J Hum Genet* 2007;**15**:570–7.

257. Spitz F, Montavon T, Monso-Hinard C, Morris M, Ventruto ML, Antonarakis S, et al. t(2;8) balanced translocation with breakpoints near the human HOXD complex causes mesomelic dysplasia and vertebral defects. *Genomics* 2002;**79**:493–8.

258. Dlugaszewska B, Silahtaroglu A, Menzel C, Kubart S, Cohen M, Mundlos S, et al. Breakpoints around the HOXD cluster result in various limb malformations. *J Med Genet* 2006;**43**:111–18.

259. Enattah NS, Sahi T, Savilahti E, Terwilliger JD, Peltonen L, Jarvela I. Identification of a variant associated with adult-type hypolactasia. *Nat Genet* 2002;**30**:233–7.

260. Jamieson RV, Perveen R, Kerr B, Carette M, Yardley J, Heon E, et al. Domain disruption and mutation of the bZIP transcription factor, MAF, associated with cataract, ocular anterior segment dysgenesis and coloboma. *Hum Mol Genet* 2002;**11**:33–42.

261. Fantes J, Redeker B, Breen M, Boyle S, Brown J, Fletcher J, et al. Aniridia-associated cytogenetic rearrangements suggest that a position effect may cause the mutant phenotype. *Hum Mol Genet* 1995;**4**:415–22.

262. Flomen RH, Gorman PA, Vatcheva R, Groet J, Barisic I, Ligutic I, et al. Rieger syndrome locus: a new reciprocal translocation t(4;12)(q25;q15) and a deletion del(4)(q25q27) both break between markers D4S2945 and D4S193. *J Med Genet* 1997;**34**:191–5.

263. de Kok YJ, Vossenaar ER, Cremers CW, Dahl N, Laporte J, Hu LJ, et al. Identification of a hot spot for microdeletions in patients with X-linked deafness type 3 (DFN3) 900 kb proximal to the DFN3 gene POU3F4. *Hum Mol Genet* 1996;**5**:1229–35.

264. Castermans D, Vermeesch JR, Fryns JP, Steyaert JG, Van de Ven WJ, Creemers JW, et al. Identification and characterization of the TRIP8 and REEP3 genes on chromosome 10q21.3 as novel candidate genes for autism. *Eur J Hum Genet* 2007;**15**:422–31.

265. Fernandez BA, Siegel-Bartelt J, Herbrick JA, Teshima I, Scherer SW. Holoprosencephaly and cleidocranial dysplasia in a patient due to two position-effect mutations: case report and review of the literature. *Clin Genet* 2005;**68**:349–59.

266. Marlin S, Blanchard S, Slim R, Lacombe D, Denoyelle F, Alessandri JL, et al. Townes-Brocks syndrome: detection of a SALL1 mutation hot spot and evidence for a position effect in one patient. *Hum Mutat* 1999;**14**:377–86.

267. Ishikawa-Brush Y, Powell JF, Bolton P, Miller AP, Francis F, Willard HF, et al. Autism and multiple exostoses associated with an X;8 translocation occurring within the GRPR gene and 3' to the SDC2 gene. *Hum Mol Genet* 1997;**6**:1241–50.

268. Benito-Sanz S, Thomas NS, Huber C, Gorbenko del Blanco D, Aza-Carmona M, Crolla JA, et al. A novel class of Pseudoautosomal region 1 deletions downstream of SHOX is associated with Leri–Weill dyschondrosteosis. *Am J Hum Genet* 2005;**77**:533–44.

269. Sabherwal N, Bangs F, Roth R, Weiss B, Jantz K, Tiecke E, et al. Long-range conserved non-coding SHOX sequences regulate expression in developing chicken limb and are associated with short stature phenotypes in human patients. *Hum Mol Genet* 2007;**16**:210–22.

270. Wallis DE, Roessler E, Hehr U, Nanni L, Wiltshire T, Richieri-Costa A, et al. Mutations in the homeodomain of the human SIX3 gene cause holoprosencephaly. *Nat Genet* 1999;**22**(2):196–8.

271. Balemans W, Patel N, Ebeling M, Van Hul E, Wuyts W, Lacza C, et al. Identification of a 52 kb deletion downstream of the SOST gene in patients with van Buchem disease. *J Med Genet* 2002;**39**:91–7.

272. Bowl MR, Nesbit MA, Harding B, Levy E, Jefferson A, Volpi E, et al. An interstitial deletion-insertion involving chromosomes 2p25.3 and Xq27.1, near SOX3, causes X-linked recessive hypoparathyroidism. *J Clin Invest* 2005;**115**:2822–31.

273. McElreavey K, Vilain E, Barbaux S, Fuqua JS, Fechner PY, Souleyreau N, et al. Loss of sequences 3' to the testis-determining gene, SRY, including the Y pseudoautosomal boundary associated with partial testicular determination. *Proc Natl Acad Sci USA* 1996;**93**:8590–4.

274. David D, Cardoso J, Marques B, Marques R, Silva ED, Santos H, et al. Molecular characterization of a familial translocation implicates disruption of HDAC9 and possible position effect on TGFbeta2 in the pathogenesis of Peters' anomaly. *Genomics* 2003;**81**:489–503.

275. Fantauzzo KA, Tadin-Strapps M, You Y, Mentzer SE, Baumeister FA, Cianfarani S, et al. A position effect on TRPS1 is associated with Ambras syndrome in humans and the Koala phenotype in mice. *Hum Mol Genet* 2008;**17**:3539–51.

276. Rose CS, Patel P, Reardon W, Malcolm S, Winter RM. The TWIST gene, although not disrupted in Saethre–Chotzen patients with apparently balanced translocations of 7p21, is mutated in familial and sporadic cases. *Hum Mol Genet* 1997;**6**:1369–73.

CHAPTER 7

Polycomb Mechanisms and Epigenetic Control of Gene Activity

Vincenzo Pirrotta
Rutgers University, Department of Molecular Biology and Biochemistry, Piscataway, NJ 08854, USA

INTRODUCTION

The Polycomb Group (PcG) proteins were first discovered in the study of *Drosophila* homeotic genes and their regulation. They derive their name from the fact that the first sign of a decrease in PcG function is often a homeotic transformation of posterior legs towards anterior legs, which have a characteristic comb-like set of bristles. Studies in flies, mammals, and other metazoans have now shown that they are part of a genome-wide mechanism for controlling genomic priorities that governs the ability of genes to respond to signals, and that channels differentiation pathways and ultimately determines the developmental fate and identities of cells. PcG mechanisms do this by targeting most if not all of the key genes in the genome that govern cellular decisions from cell cycle progression to apoptosis and including in particular the effectors controlling the variety of differentiation programs.

PcG complexes are found in the nuclei of most cells but the mechanisms by which they are targeted, which genes they affect, how they repress expression, and how the repressed state is remembered from one cell cycle to the next remain important research questions. In this article, I have not tried to cover all aspects of PcG mechanisms but have tried to give the basics and to give an overview of the current understanding of the role played by these mechanisms, using *Drosophila* as a model to gain insight into their biological function in mammals. PcG mechanisms have been the subject of a number of recent reviews that treat many aspects in greater detail and to which the reader is referred for additional information [1–6].

THE HARDWARE

The essential PcG machinery consists of two multiprotein complexes, Polycomb Repressive Complex 1 and 2, or PRC1 and PRC2.

PRC1

The core of this complex is a quartet of proteins. In *Drosophila*, these are the Polycomb protein itself (PC), Posterior Sex Combs (PSC), Polyhomeotic (PH), and dRING [7,8].

Handbook of Epigenetics: The New Molecular and Medical Genetics. DOI: 10.1016/B978-0-12-375709-8.00007-1

PC contains a chromodomain, a structure that binds methyl-lysine, in the case of PC – specifically the trimethylated form of lysine 27 of histone H3 (H3K27me3) [9]. PSC and dRING are RING domain proteins that together mediate the ubiquitylation of histone H2A [11,12]. PH contains a protein–protein interaction domain called SPM, which interacts with a similar domain in another PcG protein Sex Comb on Midleg (SCM), which appears to be more loosely associated with PRC1 [12,13]. In *Drosophila*, SU(Z)2 provides an alternative to PSC, while two distinct genes – *ph-p* and *ph-d* – provide the PH function. A corresponding PRC1 complex in mammals contains the mammalian homologs of the *Drosophila* proteins [14] except that, as is often the case, each component has one or more relatives that appear to be used in different circumstances. Thus there are several PC homologs (now called CBX2, CBX4, CBX6, CBX7, and CBX8) [15], at least two PSC homologs (Bmi1 and Mel18) [16], two RING homologs (Ring1A and Ring1B) [17], and three PH homologs (Phc1, 2, 3) [18].

PRC2

The catalytic component of this complex is Enhancer of Zeste [E(Z)], a SET domain histone methyltransferase [19–23]. Enzymatic activity requires, however, the other components, Suppressor of Zeste-12 [SU(Z)12], Extra Sex Combs (ESC) or close homolog ESCL, and the histone chaperone RbAp48. *In vivo*, the target of this complex is histone H3K27, and PRC2 is responsible for all mono-, di- and tri-methylation at this position [24]. The mammalian PRC2 is entirely homologous, including the proteins Suz12, Eed, and the E(Z) homolog Ezh2 (or its more specialized homolog Ezh1). Both in flies and in mammals, variants of this complex have been found, containing additional factors or, in mammals, different isoforms of Eed that modulate its specificity [25–27].

While PRC1 and PRC2 are the heart of the PcG mechanisms, other proteins play important roles at many target genes. Among these is Pleiohomeotic (PHO) and its close relative PHOL, which are homologs of the mammalian YY1 protein [28,29]. PHO/YY1 is a DNA-binding protein that is thought to play an important role in recruiting PcG machinery to target genes. PHO is associated with dSFMBT, a homolog of mammalian SFMBT containing multiple MBT methyl-lysine binding domains, and both are necessary for the effective repression of *Drosophila* homeotic genes [29]. Classical genetics, including one recent systematic screen [31], has identified a number of other genes that make some contribution to PcG silencing of homeotic genes and it is likely that most if not all of the major players have now been identified. To the extent that they have been characterized, they seem to be less central to the PcG mechanisms but their molecular analysis may yet hold some surprises and some valuable hints. The molecular characterization of one of these, SXC, may be such a hint. SXC was found to be the *Drosophila* homolog of OGT, a highly conserved N-acetylglucosamine (GlcNAc) transferase responsible for O-linked GlcNAcylation of many proteins [32,33]. O-GlcNAcylation is enriched at *Drosophila* PREs and, in fact, at least one protein, the PH component of PRC1, was found to be glycosylated by OGT in the fly as well as in mammals [32,34]. Many proteins and many transcription factors are O-glycosylated, including GAF, a DNA-binding protein frequently found at PREs, but it is not clear what the precise role of this modification might be. The loss of OGT function does not affect the binding of PRC1 or the H3K27 methylation but it strongly reduces the repressive effects. It is unclear at this point whether OGT itself binds to PcG target genes. It is interesting in this respect that the CTD of RNA pol II is also a target for O-glycosylation by OGT and that this modification prevents the Serine-2 phosphorylation that is essential for transcription elongation [35]. It remains possible therefore that the glycosylation of PH is not the essential role of OGT and, in particular, if OGT is itself associated with PREs, that the glycosylation of the RNA pol II CTD is the key to PcG repression of transcription.

Non-canonical PRC1 complexes have also been reported. One, called dRAF, lacks PC and PH but includes dRING, PSC, and the F-box protein dKDM2, homologous to the mammalian KDM2 histone demethylase that targets H3K36 [36]. It is not clear at present what the functional relationship of dRAF to the PRC1 complex might be and whether it might represent a stage in the functional cycle of a single structure. Genetic analysis shows that dKDM2 contributes to homeotic gene silencing and antagonizes the effects of Trithorax and ASH1. Interestingly, F-box proteins are implicated in ubiquitylation, and depletion of dKDM2 also causes global loss of H2Aub to the same extent as knockdown of dRING or PSC while PC knockdown has no effect. These results suggest that the H2A ubiquitylation activity attributed to PRC1 might instead be largely due to the dRAF complex. A global mapping of H2Aub in the genome has not yet been reported and will clearly be necessary to evaluate the role of this important modification in PcG mechanisms and other processes.

THE POLYCOMB RESPONSE ELEMENT

DNA fragments from PcG target genes that confer the ability to bind and respond to PcG mechanisms were identified in *Drosophila* by genetic and functional analyses [37–39]. These Polycomb Response Elements (PREs) are regions of several hundred base pairs that can recruit both PRC1 and PRC2 complexes and produce PcG-dependent repression of neighboring genes [40]. PREs have unusual properties. In *Drosophila* transgenic constructs and in the genome, a PRE can repress several genes, often at distances of many tens of kb from their promoters. PRE-containing transgenes often display a pairing effect: the repressive effect becomes much stronger when the transgene is made homozygous [37,41]. This is attributed to the homologous pairing of chromosomes that, in *Drosophila*, brings the two copies of the transgene in close proximity. This pairing effect suggests that the physical proximity of two or more PREs enhances or stabilizes the repressive effect. The ability to act over a distance is strikingly illustrated by the fact that if the PRE is deleted from one copy of the transgene, the remaining PRE on the homologous chromosome can repress both copies [42].

Analyses of PcG protein distribution using chromatin immunoprecipitation (ChIP) or DamID showed that known PREs were indeed binding sites of PcG proteins and that similar binding sites were found associated with hundreds of other genes [43–45]. The binding profiles showed that PcG proteins are for the most part sharply localized to narrow chromatin regions corresponding to known or presumptive PREs. This is particularly true for E(Z) and PSC proteins while PC often tends to spread beyond the PRE region. Although often situated near the promoters of the gene they control, PREs can also be found at distances of tens of kb, sometimes in intronic regions or downstream of the gene. Some genes, particularly genes with complex regulatory regions or multiple promoters, have more than one PRE. How the functioning of multiple PREs at the same gene can be integrated is not entirely clear but, as an example, extensive genetic and functional analysis at the *Drosophila Abd-B* gene has shown that different regulatory modules, each with its own PRE, are separated from one another by insulator or boundary elements that in some way do not interfere with the ability of each module to act upon the *Abd-B* gene [reviewed in Ref. 46].

RECRUITERS

The *Drosophila* PREs as a group lack well defined sequence similarities. However, they often contain recognizable motifs corresponding to consensus binding sequences for certain DNA binding proteins. Foremost among these is the Pleiohomeotic (PHO) protein [28,29,47]. PHO or its closely related paralog PHO-like (PHOL), is associated with dSFMBT and both are necessary for the effective repression of the homeotic genes in *Drosophila* [30]. dSFMBT contains MBT domains that can bind to methyl-lysines but the interactions reported are with mono- and di-methylated H3K9 and H4K20, modifications that are not generally associated with PcG target sites. *In vitro*, PHO cooperates with PRC1 in binding specifically to a PRE [48].

At least three other DNA binding proteins are often associated with PREs: DSP1 [49], GAF/TRL [41,50,51], and Pipsqueak [52,53]. With the genomic repertoire of PcG binding

109

sites now available in *Drosophila*, it seems clear that not all these proteins are present at all PREs and that there are at least some PcG binding sites that lack all of these. The most probable interpretation of these observations is that a set of proteins cooperate in recruiting PcG complexes to target genes, that no single protein is likely to be sufficient for effective recruitment, and that there are probably additional proteins, not yet identified, that can participate in the recruitment. It cannot be excluded that some DNA binding factors may be required for *de novo* PcG recruitment but not for continued maintenance of PcG repression. Similarly, it is likely that recruitment of PcG complexes involves the participation of nucleosome remodeling activities to make the PRE region accessible, as suggested by the transient presence of histone H3.3 [see Ref. 54].

The situation in mammalian genomes is far less clear. Despite much effort over the past fifteen years, mammalian PREs have proved much more elusive and, until recently, no clear evidence of a sequence-based recruiting element had been produced. Binding profiles produced by ChIP/chip or ChIP/seq methods show no sharp localization as in *Drosophila*. While many target genes show a broad profile of PcG binding peaking near the promoter, others are associated with very broad binding domains for all the PcG proteins tested, which do not betray the possible existence of underlying PRE-like recruiting elements. Does this mean that PcG complexes are recruited by entirely different methods in mammals? A recent report has now brightened the outlook. Sing et al. [55], guided by the effects of a genetic inversion at the mouse *kreisler* gene, identified a 450 bp region that is 92% conserved between humans and mouse and contains PHO/YY1 consensus binding sites as well as GAGAG motifs recognized by the *Drosophila* GAF. This fragment was shown to bind PcG proteins and mediate PcG-dependent repression of a reporter gene in *Drosophila*. More to the point, PcG protein binding and PcG-mediated repression of a reporter gene could be demonstrated also in mouse. This does not necessarily mean that hidden under all the broad peaks of PcG binding in mammalian genes there lurks a well defined PRE. In mammals and in *Drosophila*, different genes may well use different mechanisms to recruit PcG complexes.

THE H3K27 METHYL MARK

The PRC2 complex is responsible for all H3K27 methylation. In mammals, the monomethylation is not abolished by knockout of *Ezh1* and *Ezh2* but is still dependent on Eed [56]. The lower degrees of H3K27 methylation remain a puzzle. While H3K27me3 is the characteristic mark associated with PcG repression, a quantitatively much more important product of PRC2 is H3K27 dimethylation. This constitutes more than 50% of genomic H3 [24] and is present virtually everywhere in the genome except in regions rich in H3K27me3 and in actively transcribed regions, where it is underrepresented. The function of this pervasive methylation is unknown but it is clearly dependent on PRC2 in both flies and mammals and is lost when PRC2 components are mutated. The H3K27me3 mark is specifically found at PcG target genes, in association with the binding of PRC2. The fact that at these sites PRC2 is converted to a trimethyl transferase is not just due to the stable binding of PRC2 but to the association of the complex with additional factors. A component that is required for trimethylation is Polycomb-like (PCL), whose mammalian homolog is PHF1. Loss of PCL/PHF1 causes the loss of H3K27me3 without affecting the level of H3K27me2 in the genome [57–59]. In *Drosophila*, PCL is associated with PcG target sites, presumably at the PRE, and PCL/PHF has been purified biochemically as a component of an alternative form of the PRC2 complex [24,56,57].

How is the H3K27me3 distributed? Genome-wide ChIP analysis shows that in *Drosophila* and in mammals, the chromatin region that becomes trimethylated is very broad, generally including the regulatory region, and most or all of the transcription unit and often well beyond [43,44,60]. In *Drosophila*, the H3K27me3 domain includes the PRE region but not the PRE itself, which appears to be generally depleted of nucleosomes, probably because they are displaced by DNA-binding proteins [61,62]. In contrast, the PRC1 and PRC2 complexes are localized at the PRE. To account for the much broader reach of the H3K27me3 domain, a looping model was proposed, according to which the PRE region with its bound complexes loops to interact with flanking nucleosomes [61,63]. Such looping could be mediated by the PC chromodomain and could extend over a large region or until prevented by chromatin insulators or antagonistic histone modifications. In this view, the H3K27 methylation domain serves not only to stabilize the binding of PcG complexes but also as an access path for PRC1 to contact the promoter region and antagonize transcription initiation. Insulator elements have been shown in fact to prevent the spread of the methylation domain and to block the repressive action of the PRE on a distal promoter [42,61].

RECRUITMENT BY H3K27me3

The discovery that the PC chromodomain has a specific affinity for H3K27me3, the methyl mark deposited by the PRC2 complex, led inexorably to the attractive conclusion that PcG complex recruitment was in some way explained by PRC2 laying the mark and PRC1 then binding to this mark, a view that is still widely held. A sequential recruitment process was proposed by Wang et al. [64], according to which, the DNA-binding protein PHO initiates the process by recruiting PRC2, which then methylates surrounding nucleosomes at H3K27. PRC1would then be recruited by the affinity of the PC chromodomain for the H3K27me3. The facts do not support this scenario. First, the observed *in vitro* affinities of the PC chromodomain for nucleosomes containing H3K27me3 are in the micromolar range [9], orders of magnitude weaker than the affinities of DNA-binding transcription factors, and unlikely to lead to substantial binding under physiological conditions. The inability of the H3K27me3 mark to recruit PRC1 is manifested by the fact that while PcG target genes are embedded in a large methylation domain, the PRC1 and PRC2 complexes are generally much more narrowly localized, at least in *Drosophila* [43]. In many mammalian genes, it is true, the distribution of the PcG proteins often is nearly as broad as that of the methyl mark but even in mammals the PcG proteins do not bind to all sites containing H3K27me3 [65–68]. The loss of H3K27 methylation during development in *Drosophila E(z)* or *esc escl* mutants does not lead to loss of PRC1 binding for several days during larval development [69]. It remains true, however, that, in the long term, PRC1 binding is lost from many genomic sites in the absence of PRC2 function. Therefore, it is likely that H3K27me3 does play a role in stabilizing the binding of PRC1 after its recruitment. It should also be noted that experiments in which the PC protein is targeted to a reporter gene by fusion to a LexA or GAL4 DNA binding domain show that, in the early *Drosophila* embryo, this fusion protein by itself can target repression of a reporter gene containing the corresponding binding site [70–72]. This repression requires PRC2 activity, implying that PRC1 can recruit PRC2. Conversely, targeted PRC2 can recruit PRC1. Immunoprecipitation experiments showed that in the early embryo PRC1 may interact with PRC2 and PHO. In view of the evidence that presumed DNA-binding recruiters such as PHO/YY1 have been shown to interact with both PRC1 and PRC2 [72,73], a sequential model for recruitment based on H3K27 methylation is not compelling at this point.

It is important to bear in mind that PcG binding sites in the genome are not all the same. The level of binding, the binding profile, the degree and distribution of H3K27 trimethylation, and the relative amounts of different proteins, in particular the relationship between PRC1 and PRC2, vary greatly from site to site. Some sites bind very little or no PRC2 and lack appreciable H3K27me3 although the associated gene is clearly regulated by PcG mechanisms: an example of this is the *Drosophila ph* locus, encoding the PcG protein PH [43]. Similarly, the genome-wide studies have shown a few sites that apparently bind E(Z) but lack PRC1 components.

Finally, in at least some cases, PcG complexes can be recruited by binding to certain non-coding RNAs. This is particularly clear in the case of mammalian X inactivation where the

111

cis-acting Xist RNA is the key element that initiates inactivation of one X chromosome. It has been known for some time that an early step in this process is the recruitment of both PRC1 and PRC2 to the X chromosome that is undergoing inactivation. A recent report shows that a short non-coding RNA called RepA, within the Xist region, is produced early in the process of inactivation, directly recruits PRC2 through the Ezh2 component and is necessary for the up-regulation of Xist expression and subsequent spread of PcG complex binding to the inactive X chromosome [74]. That RNA involvement may not be limited to X inactivation is suggested by the report that a non-coding RNA is implicated in the PcG repression of mammalian *Hox* genes [75]. It is not clear at this point whether these represent exceptional cases, examples of one class of PcG recruiting mechanisms, or a glimpse of a much more widespread involvement of ncRNAs in PcG repression.

HOW DO PcG MECHANISMS REPRESS?

Early concepts of the repression mechanism derived by analogy from the then current ideas about heterochromatic silencing. The idea was that heterochromatin proteins packaged chromatin into a highly condensed structure that rendered the DNA inaccessible to transcription factors and RNA pol II. In this view, supported by some though not all enzymatic accessibility studies, PcG-repressed genes would not be "open for business" whether or not transcriptional activators were present [76,77]. Furthermore, *in vitro* experiments indicated that the PRC1 complex bound to a non-specific nucleosomal template inhibits chromatin remodeling by the SWI/SNF complex [7]. *In vitro*, the addition of sufficient PRC1 complex to a nucleosome array causes a degree of clumping detected by electron microscopy [78]. However, placing a strong PRE next to a heat shock gene does not prevent the binding of RNA pol II or TFIID to the promoter, nor does it block the access of heat shock factor upon heat induction [79]. Nevertheless, the RNA polymerase is unable to initiate transcription, suggesting that the repressive mechanism interferes with the pre-initiation complex. It is possible, as discussed above, that PRC1-dependent ubiquitylation of histone H2A interferes with transcription initiation. It also possible that repression is effected by multiple mechanisms, for example by methylating or ubiquitylating non-histone components of the transcriptional machinery. Chromatin condensation might then be a consequence rather than a cause of transcriptional repression. Alternatively, a degree of aggregation of the nucleosomes might be one additional factor contributing to transcriptional interference. Furthermore, work using imaging methods has shown that PcG protein binding, as well as that of heterochromatin proteins, is highly dynamic and cannot be interpreted as resulting in stable nucleosome condensation [80–82]. Although research has concentrated on the effect of PcG mechanisms on binding of transcription factors and on transcription initiation, the realization that the repression is often only partial and dynamic (see below) makes it very likely that other stages in the transcription process might also be targeted. PcG mechanisms might, for example, affect transcription elongation, or splicing and processing of the RNA.

EPIGENETIC MAINTENANCE OR CELLULAR MEMORY

In any one cell, some of the potential PcG targets bind PcG proteins and are repressed while others are active. What determines which genes will be repressed? The early answer for the *Drosophila Hox* genes, the prototypical PcG targets, was that if the gene had been active in the early embryo, it remained potentially unrepressed by PcG mechanisms and could function at later stages, provided the appropriate activators were present [83]. In cells in which the gene was not active in the early embryo, that gene bound PcG proteins and was repressed. In other words, once the PcG proteins established repression, this repressed state was remembered in succeeding cell cycles and was re-established in the cellular progeny. Conversely, if the gene had been active in the early embryo, this activity somehow set a protective barrier that prevented PcG repression at later stages. Genetic evidence showed that the functions responsible for this kind of memory of the active state were associated with the *trithorax* and *ash1* genes (see below).

The close association of H3K27 trimethylation with PcG repression immediately suggests that this methyl mark might be responsible for the epigenetic memory of the repressed state and this has been widely assumed. In order for such a histone mark to be responsible for the epigenetic memory it must be able to perform two functions. One is to facilitate or stabilize the binding of PcG complexes and the other is to trigger the self-renewal of the mark every cell cycle. As discussed above, the methyl mark does appear to stabilize the binding of PRC1 to target genes and, although the mechanism is not entirely clear, it provides an advantage over transcriptional reactivation. It has been argued that the PRC2 complex associates with the replication fork and might recognize the H3K27me3 mark [84]. Recent work has now shown how this recognition functions and allows the methyl mark to direct its own self-renewal [85]. During chromatin replication, old nucleosomes are disrupted but the core H3/H4 tetramer is thought to be randomly redistributed to the two daughter DNA molecules [reviewed in Ref. 86]. The newly replicated DNA therefore has half the complement of nucleosomes and new nucleosomes must be deposited to fill the gaps. As a result, the methyl marks present on the old chromatin become diluted two-fold. To maintain the mark therefore, it is essential to restore full H3K27 trimethylation rapidly and efficiently. Structural studies of the Eed protein have revealed that residues from different WD40 repeats combine to form an aromatic cage that binds repressive histone methyl marks like H3K27me3. Like the chromodomain binding, this is not a very tight interaction but sufficient to provide the PRC2 complex with a way to sense the presence of H3K27me3 marks in a newly replicated region that still bears one half of the old nucleosomes. A second discovery was that, while recombinant PRC2 complex normally has only a modest methyltransferase activity on reconstituted nucleosomes, the presence of H3K27me3 peptides in the reaction greatly stimulates the catalytic activity on an unmethylated target nucleosome. As a consequence, the presence of old, methylated nucleosomes in newly replicated chromatin would greatly enhance the methylation of the newly deposited nucleosomes, thus restoring the full density of the methyl mark. The importance of this mechanism was demonstrated by the fact that aromatic cage mutations that abolish the H3K27me3 binding ability of the *Drosophila* Eed homolog ESC result in lethality, derepression of the homeotic genes, and dramatic depletion of both H3K27me3 and H3K27me2 [85].

MAINTENANCE OF THE NON-REPRESSED STATE

What prevents PcG complexes from acting on target genes that should be in an active mode? Overexpression of an activator trumps PcG repression and can turn on a repressed gene [87–89] However, the genetic evidence in *Drosophila* shows that the maintenance of a program of expression of PcG target genes requires the *trithorax* (*trx*) and *ash1* genes. These functions do not activate or derepress target genes but they cooperate with activators to stimulate transcription, antagonize PcG repression, and preserve an epigenetic memory of the derepressed state. Interestingly, however, these are necessary only in the presence of PcG complexes. That is, the genetic evidence in *Drosophila* shows that when the PcG repressive mechanisms are disabled, *trx* and *ash1* functions can be dispensed with. This strongly implies that their main function is to antagonize PcG repression and, in fact, loss of *trx* causes the PcG repression of homeotic genes in domains where they are normally active [90,91]. TRX and ASH1 are both SET domain proteins. While TRX has been shown to have a H3K4 methyltransferase activity [92–94], the *in vivo* target of ASH1 has been much less clear [95–97]. Most likely, however, the most recent report is correct in proposing that it methylates H3K36. TRX binds to PREs both in the active and in the PcG repressed state [61,98] and is in some way necessary for the recruitment of ASH1 when the target gene is in the active mode. In vivo evidence suggests in fact that TRX and ASH1 cooperate [99,100] in binding to target genes but their molecular action is still obscure. The mammalian homologs of TRX are MLL1 and 2 (Mixed Lineage Leukemia), which, like TRX, are needed for the effective expression of *Hox* genes. MLL1/2 form complexes similar to those formed by the yeast Set1 methyltransferase, responsible for H3K4 methylation associated with transcriptional activity and have been often considered to be the corresponding mammalian functions [101,102]. This cannot be strictly true: mammals (and *Drosophila*) possess true homologs of yeast Set1, and Set1A and B, that are necessary for effective transcriptional elongation [103,104]. Unlike Set1 mutations, which cause depletion of H3K4me3, loss of MLL1/2 has little effect on H3K4me3, indicating that, like the fly TRX, they are specialized functions that affect a specific subset of genes [105]. How H3K4 and/or H3K36 methylation antagonize or prevent PcG repression is far from clear but both TRX and ASH1 have been reported to act in conjunction with the CBP histone acetylase [106–108]. A recent report found that the derepressed state of PcG target genes involves CBP-dependent H3K27 acetylation, also requires TRX function, and is antagonistic to PcG repression [109]. This is highly suggestive since H3K27 acetylation is clearly incompatible with H3K27 methylation. Recent results have shown that the derepressed state of PcG target genes is in fact associated with a broad domain of H3K27ac that replaces the domain of H3K27 trimethylation present in the repressed state [123]. Additional work will be necessary to explain how TRX and ASH1 produce a derepressed state that can be maintained epigenetically.

PcG BINDING DOES NOT NECESSARILY RESULT IN SILENCING

The early studies of the role of PcG mechanisms in *Hox* gene expression indicated that repression of these target genes dependent on early decisions, was essentially a complete silencing and was stably maintained through later development. The realization that PcG mechanisms target a far larger number of genes, many of these important for later developmental decisions and response to signaling events, strongly suggested that this all-or-nothing silencing could not be the general rule. Many genes need to be repressed for a time and then be reactivated. More importantly, evidence now shows that PcG binding can coexist with some level of transcriptional activity. Compelling examples of this are the *Drosophila ph* and *Psc* genes, which encode the PH and PSC core components of the PRC1 complex. Both genes bind PcG complexes but are necessarily also transcriptionally active [43,110]. Although it remains possible that they cycle on and off depending on the level of PcG proteins or on the cell cycle, it is clear that, in this case at least. PcG mechanisms do not permanently silence. Studies in cultured cells show a subset of genes that bind both PcG proteins and RNA pol II and that possess both the repressive H3K27me3 mark and the active H3K4me3 mark in the promoter region. The level of expression of these genes is modulated when the relative levels of PC and TRX are altered [123]. The PcG mechanisms in these cases act in a highly dynamic way rather than producing permanent repression. Similar conclusions were reached by Oktaba et al. [111] studying the effect of PcG loss during *Drosophila* development. These observations echo the well documented case of "bivalent" chromatin states in mammalian stem cells.

THE BIVALENT STATE

Embryonic stem (ES) cells are pluripotent: they are able to enter any differentiation pathway and to contribute to any tissue if introduced into an early embryo. Therefore, the genes controlling all pathways are potentially available for expression. However, these genes must not be expressed if the cells are to remain pluripotent. The critical regulatory genes are targets of a complex regulatory circuitry involving the pluripotency factors, *oct4*, *sox2*, and *nanog* [reviewed in Ref. 112]. At the same time, these genes show the presence of the PcG-associated H3K27me3 mark and, at the same time, of H3K4me3, a mark usually associated with transcriptional activity. This condition has been called "bivalent" and thought to correspond to a poised state ready to switch into a fully active mode or into a fully repressed mode [68]. Upon differentiation, the bivalent state is resolved towards active transcription or long term repression and, correspondingly, either the active mark or the repressive mark is exclusively retained. A detailed study of genes in such a bivalent state showed that they are not in fact transcriptionally repressed in the usual sense [113]. Their promoter regions are acetylated and bind RNA pol II whose CTD is phosphorylated at Serine-5 but not Serine-2. These genes are not only poised but active, producing low levels of transcripts relative to the amount of polymerase present at the promoter. Loss of both Ring1A and Ring1B causes the progressive loss of H2A ubiquitylation without affecting the distributions of PRC2 or H3K27me3. The loss of ubiquitylation is accompanied by the transcriptional up-regulation of the bivalent genes and onset of differentiation. The simplest interpretation of these results is that H2A ubiquitylation interferes with some stage of the transcription process. Consistent with this is the report that H2A ubiquitylation prevents recruitment of the FACT complex and therefore interferes with effective transcriptional elongation [114]. It remains to be explained why, should the CTD of pol II fail to be phosphorylated at Ser-2, how nevertheless methyltransferase complexes are recruited to methylate H3K4.

The presence of a stalled RNA pol II at many key genes in *Drosophila* embryos has been reported [115]. Among these genes are included the heat shock genes, as has been known for a long time, but the majority are important regulatory genes that control developmental decisions, genes that must be repressed in some embryonic domains but expressed in others. It is not surprising therefore that most of these genes correspond to those found to be PcG targets in cultured cells [43]. Interestingly, most of these genes do not bind RNA pol II in the cultured cells (Y. Schwartz, T. Kahn and V.P., unpublished), indicating that the stalled state is not a constitutive feature, as in the heat shock genes, but is a specific feature of these genes in the early embryo.

Perhaps a good way to envision this state is suggested by the role of TRX and ASH1 seen in *Drosophila*. According to this, the bivalent state would correspond to a dynamic equilibrium between TRX/MLL and ASH1L (the mammalian ASH1 homolog) on one hand, and PcG mechanisms on the other, presumably driven by subliminal amounts of activators and resulting in low levels of transcription, possibly frequently aborted or incompletely spliced. In the absence of any activators, PcG target genes would be simply repressed. When subliminal levels of activators are available (in the pluripotent state), the bivalent state would be reached with extensive binding of TRX/MLL and ASH1L. The role of PcG complexes would thus be to raise the threshold for activators that, abetted by TRX and ASH1, would be required for productive transcriptional activation. Decreased levels of the pluripotency factors and increased levels of specific activators or repressors would tilt the equilibrium towards either full activation or full repression.

DIFFERENTIATION

The elevated levels of PcG proteins in ES cells and the binding to a multitude of target genes controlling differentiation pathways suggested that PcG repression might be essential for the maintenance of the pluripotent state. Surprisingly, it has been possible to generate ES cells lacking at least PRC2 function [56,116,117]. In these cells, the expression of many PcG target genes is elevated but the cells remain pluripotent though less stable. What is defective in these cells is the ability to differentiate correctly. If differentiation is induced, the pluripotency genes and genes specific for other differentiation pathways are not properly turned off, the appropriate differentiation genes are not correctly activated, and the differentiation program fails to be carried out. A recent study analyzing the role of PcG mechanisms in epidermal differentiation used a conditional knockout of *Ezh2* to show that, once differentiation is under way [118], PcG repression is no longer needed to keep other differentiation pathways turned off. Other mechanisms, probably initiated by G9a methylation of H3K9 and recruitment of DNA methyltransferases, have already established long-term silencing [119,120]. The lack of PRC2 function, however, disrupts the normal process of differentiation, resulting in the premature expression of the differentiated functions and exit from the cell cycle before the appropriate precursor cells have been produced. PcG complexes regulate *Ink4A* and *B*, two important inhibitors of cyclin-dependent kinases. In normal epidermal differentiation, the expression of PRC2 components gradually winds down through a poorly specified mechanism, allowing the gradual derepression of *Ink4A/B* and eventual exit from the cell cycle as the epidermal differentiation program unfolds. Ebbing levels of PcG proteins are not the only thing that determines the reactivation of *Ink4A/B*. As intimated above, balance between repression and expression depends on the relative levels of PcG proteins, TRX, the relevant transcriptional activators, and the chromatin remodeling complexes that refashion the chromatin landscape to allow transcriptional activity. In certain cell types, the levels of the SWI/SNF chromatin remodeler are a determining factor in the decision whether to derepress *Ink4A/B*, inducing exit from the cell cycle and cellular senescence [121]. The SWI/SNF complex facilitates eviction of the PcG complexes and the action of MLL/TRX.

This kind of program must be run in reverse to produce induced pluripotent stem (iPS) cells [122]. Reprogramming differentiated cells involves not only the reactivation of the pluripotency gene circuits but, critically, the re-establishment of PcG repression at the *Ink4A* and *B* genes, thus permitting cell cycle progression and proliferation.

CONCLUSIONS

If PcG repression is so plastic, dynamic, and easily reversed by the presence of activators, what is epigenetic about it and how does it differ from any other negative regulator? One way to view the role of PcG mechanisms is that they raise the threshold of the signals or activators required to turn on a gene. Furthermore, this higher threshold is transmitted to the cellular progeny. As a result, these mechanisms can differentiate the ability of cells to respond to transcriptional signals depending on the history of the cell lineage. This, as developmental biologist have known for a long time, is essential for pattern formation, morphogenesis, and organogenesis. As differentiation proceeds, alternative pathways have to be turned off more or less permanently. In mammals, this does not seem to be done by PcG mechanisms but by DNA methylation. Although phylogenetically ancient and present in other insects, DNA methylation is not used in *Drosophila* development and PcG mechanisms have adapted to take over that function in ways that are adequate at least for the short lifespan and rapid developmental strategies of the fruitfly.

How PcG mechanisms are modulated in response to metabolic states, intercellular signaling, stress response mechanisms, and many other factors in development, aging, homeostasis, and disease, are questions that are now being actively studied with potential for many exciting advances. In this article, I have used PcG mechanisms in the plural to emphasize the realization that there is no single account that describes all the ways in which these versatile proteins are used in the genome or in different cells and different developmental contexts. Only some of these are discussed here and, doubtless, many more are yet to be discovered.

References

1. Schwartz YB, Pirrotta V. Polycomb silencing mechanisms and the management of genomic programmes. *Nat Rev Genet* 2007;**8**:9–22.

2. Schuettengruber B, Chourrout D, Vervoort M, Leblanc B, Cavalli G. Genome regulation by Polycomb and Trithorax proteins. *Cell* 2007;**128**:735–45.

3. Ringrose L, Paro R. Polycomb/Trithorax response elements and epigenetic memory of cell identity. *Development* 2007;**134**:223–32.

4. Simon JA, Lange CA. Roles of the Ezh2 histone methyltransferase in cancer epigenetics. *Mutation Res.* 2008;**647**:21–9.

5. Müller J, Verrijzer P. Biochemical mechanisms of gene regulation by Polycomb group protein complexes. *Curr Opin Genet Dev* 2009;**19**:150–8.

6. Niessen HEC, Demmers JAA, Voncken JW. Talking to chromatin: post-translational modulation of Polycomb group function. *Epigenetics & Chromatin* 2009;**2**:10.

7. Shao Z, Raible F, Mollaaghababa R, Guyon JR, Wu C-t, Bender W, et al. Stabilization of chromatin structure by PRC1, a Polycomb complex. *Cell* 1999;**98**:37–46.

8. Saurin AJ, Shao Z, Erdjument-Bromage H, Tempst P, Kingston RE. A *Drosophila* Polycomb group complex includes Zeste and dTAFII proteins. *Nature* 2001;**412**:655–60.

9. Fischle W, Wang Y, Jacobs SA, Kim Y, Allis CD, Khorasanizadeh S. Molecular basis for the discrimination of repressive methyl-lysine marks in histone H3 by Polycomb and HP1 chromodomains. *Genes Dev* 2003;**17**:1870–81.

10. de Napoles M, Mermoud JE, Wakao R, Tang YA, Endoh M, Appanah R, et al. Polycomb group proteins Ring1A/B link ubiquitylation of histone H2A to heritable gene silencing and X inactivation. *Dev Cell* 2004;**7**:663–76.

11. Wang H, Wang L, Erdjument-Bromage H, Vidal M, Tempst P, Jones RS, et al. Role of histone H2A ubiquitination in Polycomb silencing. *Nature* 2004;**431**:873–8.

12. Kim CA, Gingery M, Pilpa RM, Bowie JU. The SAM domain of polyhomeotic forms a helical polymer. *Nat Struct Biol* 2002;**9**:453–6.

13. Peterson AJ, Mallin DR, Francis NJ, Ketel CS, Stamm J, Voeller RK, et al. Requirement for sex comb on midleg protein interactions in *Drosophila* Polycomb group repression. *Genetics* 2004;**167**:1225–39.

14. Levine SS, Weiss A, Erdjument-Bromage H, Shao Z, Tempst P, Kingston RE. The core of the Polycomb repressive complex is compositionally and functionally conserved in flies and humans. *Mol Cell Biol* 2002;**22**:6070–8.

15. Tajul-Arifinm K, Teasdale R, Ravasi T, Hume DA, Mattick JS. Identification and analysis of chromodomain-containing proteins encoded in the mouse transcriptome. *Genome Res* 2003;**13**:1416–29.

16. Akasaka T, van Lohuizen M, van der Lugt N, Mizutani-Koseki Y, Kanno M, Taniguchi M, et al. Mice doubly deficient for the Polycomb Group genes Mel18 and Bmi1 reveal synergy and requirement for maintenance but not initiation of Hox gene expression. *Development* 2001;**128**:1587–97.

17. Voncken JW, Roelen BAJ, Roefs M, de Vries S, Marino S, Deschamps J, et al. Rnf2 (Ring1b) deficiency causes gastrulation arrest and cell cycle inhibition. *Proc Natl Acad Sci USA* 2003;**100**:2468–73.

18. Isono K-i, Fujimura Y-i, Shinga J, Yamaki M, O-Wang J, Takihara Y, et al. Mammalian polyhomeotic homologues Phc2 and Phc1 act in synergy to mediate Polycomb repression of Hox Genes. *Mol Cell Biol* 2005;**25**:6694–706.

19. Tie F, Furuyama T, Prasad-Sinha J, Jane E, Harte PJ. The *Drosophila* Polycomb Group proteins ESC and E(Z) are present in a complex containing the histone-binding protein p55 and the histone deacetylase RPD3. *Development* 2001;**128**:275–86.

20. Cao R, Wang L, Wang H, Xin L, Erdjument-Bromage H, Tempst P, et al. Role of histone H3 lysine 27 methylation in Polycomb-Group silencing. *Science* 2002;**298**:1039–43.

21. Czermin B, Melfi. R, McCabe D, Seitz V, Imhof A, Pirrotta V. *Drosophila* Enhancer of Zeste/ESC complexes have a histone H3 methyltransferase activity that marks chromosomal Polycomb sites. *Cell* 2002;**111**:185–96.

22. Müller J, Hart CM, Francis NJ, Vargas ML, Sengupta A, Wild B, et al. Histone methyltransferase activity of a *Drosophila* Polycomb Group repressor complex. *Cell* 2002;**111**:197–208.

23. Kuzmichev A, Nishioka K, Erdjument-Bromage H, Tempst P, Reinberg D. Histone methyltransferase activity associated with a human multiprotein complex containing the Enhancer of Zeste protein. *Genes Dev* 2002;**22**:2893–905.

24. Ebert A, Schotta G, Lein S, Kubicek S, Krauss V, Jenuwein T, et al. Su(var) genes regulate the balance between euchromatin and heterochromatin in *Drosophila*. *Genes Dev* 2004;**18**:2973–83.

25. Tie F, Prasad-Sinha J, Birve A, Rasmuson-Lestander Å, Harte PJ. A 1-megadalton ESC/E(Z) complex from *Drosophila* that contains Polycomb-like and RPD3. *Mol Cell Biol* 2003;**23**:3352–62.

26. Furuyama T, Banerjee R, Breen TR, Harte PJ. SIR2 is required for Polycomb silencing and is associated with an E(z) histone methyltransferase complex. *Curr Biol* 2004;**14**:1812–21.

27. Kuzmichev A, Jenuwein T, Tempst P, Reinberg D. Different Ezh2-containing complexes target methylation of histone H1 or nucleosomal histone H3. *Mol Cell* 2004;**14**:183–93.

28. Brown JL, Mucci D, Whiteley MML, Kassis JA. The *Drosophila* Polycomb group gene pleiohomeotic encodes a DNA binding protein with homology to the transcription factor YY1. *Mol Cell* 1998;**4**:1057–106.

29. Brown JL, Fritsch C, Mueller J, Kassis JA. The *Drosophila* pho-like gene encodes a YY1-related DNA binding protein that is redundant with pleiohomeotic in homeotic gene silencing. *Development* 2003;**130**:285–94.

30. Klymenko T, Papp B, Fischle W, Kocher T, Schelder M, Fritsch C, et al. A Polycomb group protein complex with sequence-specific DNA-binding and selective methyl-lysine-binding activities. *Genes Dev* 2006;**20**:1110–22.

31. de Ayala Alonso AG, Gutierrez L, Fritsch C, Papp B, Beuchle D, Muller J. A genetic screen identifies novel Polycomb group genes in *Drosophila*. *Genetics* 2007;**176**:2099–108.

32. Gambetta MC, Oktaba K, Muller J. Essential role of the glycosyltransferase Sxc/Ogt in Polycomb repression. *Science* 2009;**325**:93–6.

33. Sinclair DAR, Syrzycka M, Macauley MS, Rastgardani T, Komljenovic I, Vocadlo DJ, et al. *Drosophila* O-GlcNAc transferase (OGT) is encoded by the Polycomb group (PcG) gene, super sex combs (sxc). *Proc Natl Acad Sci USA* 2009;**106**:13427–32.

34. Chalkley RJ, Thalhammer A, Schoepfer R, Burlingame AL. Identification of protein O-GlcNAcylation sites using electron transfer dissociation mass spectrometry on native peptides. *Proc Natl Acad Sci USA* 2009;**106**:8894–9.

35. Phatnani HP, Greenleaf AL. Phosphorylation and functions of the RNA polymerase II CTD. *Genes Dev* 2006;**20**:2922–36.

36. Lagarou A, Mohd-Sarip A, Moshkin YM, Chalkley GE, Bezstarosti K, Demmers JAA, et al. dKDM2 couples histone H2A ubiquitylation to histone H3 demethylation during Polycomb group silencing. *Genes Dev* 2008;**22**:2799–810.

37. Fauvarque M-O, Dura J-M. Polyhomeotic regulatory sequences induce developmental regulator-dependent variegation and targeted P-element insertions in *Drosophila*. *Genes Dev* 1993;**7**:1508–20.

38. Chan C-S, Rastelli L, Pirrotta V. A Polycomb response element in the Ubx gene that determines an epigenetically inherited state of repression. *EMBO J* 1994;**13**:2553–64.

39. Kassis JA. Unusual properties of regulatory DNA from the *Drosophila* engrailed gene: three "pairing-sensitive" sites within a 1.6 kb region. *Genetics* 1994;**136**:1025–38.

40. Americo G, Whiteley M, Brown JL, Fujioka M, Jaynes JB, Kassis JA. A complex array of DNA-binding complexes required for pairing-sensitive silencing by a Polycomb Group response element from the *Drosophila* engrailed gene. *Genetics* 2002;**160**:1561–71.

41. Horard B, Tatout C, Poux S, Pirrotta V. Structure of a Polycomb response element and in vitro binding of Polycomb Group complexes containing GAGA factor. *Mol Cell Biol* 2000;**20**:3187–97.

42. Sigrist CJA, Pirrotta V. Chromatin insulator elements block the silencing of a target gene by the *Drosophila* Polycomb Response Element (PRE) but allow trans interactions between PREs on different chromosomes. *Genetics* 1997;**147**:209–21.

43. Schwartz YB, Kahn TG, Nix DA, Li X-Y, Bourgon R, Biggin M, et al. Genome-wide analysis of Polycomb targets in *Drosophila melanogaster*. *Nat Genet* 2006;**38**:700–5.

44. Négre N, Hennetin J, Sun LV, Lavrov S, Bellis M, White KP, et al. Chromosomal distribution of PcG proteins during *Drosophila* development. *PLoS Biology* 2006;**4**:e170.

45. Tolhuis B, Muljrers I, de Wit E, Teunissen H, Talhout W, van Steensel B, et al. Genome-wide profiling of PRC1 and PRC2 Polycomb chromatin binding in *Drosophila melanogaster*. *Nat Genet* 2006;**38**:694–9.

46. Maeda RK, Karch F. The ABC of the BX-C: the bithorax complex explained. *Development* 2006;**133**:11413–22.

47. Fritsch C, Brown JL, Kassis JA, Müller J. The DNA-binding Polycomb group protein pleiohomeotic mediates silencing of a *Drosophila* homeotic gene. *Development* 1999;**126**:3905–13.

48. Mohd-Sarip A, Cleard F, Mishra RK, Karch F, Verrijzer CP. Synergistic recognition of an epigenetic DNA element by pleiohomeotic and a Polycomb core complex. *Genes Dev* 2005;**19**:1755–60.

49. Déjardin J, Rappailles A, Cuvier O, Grimaud C, Decoville M, Locker D, et al. Recruitment of *Drosophila* Polycomb group proteins to chromatin by DSP1. *Nature* 2005;**434**:533–8.

50. Hagstrom K, Müller M, Schedl P. A Polycomb and GAGA dependent silencer adjoins the Fab7 boundary in the *Drosophila* bithorax complex. *Genetics* 1997;**146**:1365–80.

51. Mahmoudi T, Zuijderduijn LM, Mohd-Sarip A, Verrijzer CP. GAGA facilitates binding of pleiohomeotic to a chromatinized Polycomb response element. *Nucleic Acids Res* 2003;**31**:4147–56.

52. Huang D-H, Chang Y-L, Yang C, Pan I-C, King B. Pipsqueak encodes a factor essential for sequence-specific targeting of a Polycomb group protein complex. *Mol Cell Biol* 2002;**22**:6261–71.

53. Hodgson JW, Argiropoulos B, Brock HW. Site-specific recognition of a 70-base-pair element containing d(GA)(n) repeats mediates bithoraxoid Polycomb group response element-specific silencing. *Mol Cell Biol* 2001;**21**:4528–43.

54. Mito Y, Henikoff JG, Henikoff S. Histone replacement marks the boundaries of cis-regulatory domains. *Science* 2007;**315**:1408–11.

55. Sing A, Pannell D, Karaiskakis A, Sturgeon K, Djabali M, Ellis J, et al. A vertebrate Polycomb response element governs segmentation of the posterior hindbrain. *Cell* 2009;**138**:885–97.

56. Montgomery ND, Yee D, Chen A, Kalantry S, Chamberlain SJ, Otte AP, et al. The murine Polycomb Group protein Eed is required for global histone H3 lysine-27 methylation. *Curr Biol* 2005;**15**:942–7.

57. Nekrasov M, Klymenko T, Fraterman S, Papp B, Oktaba K, Köcher T, et al. Pcl-PRC2 is needed to generate high levels of H3-K27 trimethylation at Polycomb target genes. *EMBO J* 2007;**26**:4078–88.

58. Sarma K, Margueron R, Ivanov A, Pirrotta V, Reinberg D. Ezh2 requires PHF1 to efficiently catalyze H3 lysine 27 trimethylation in vivo. *Mol Cell Biol* 2008;**28**:2718–31.

59. Cao R, Wang H, He J, Erdjument-Bromage H, Tempst P, Zhang Y. Role of hPHF1 in H3K27 methylation and Hox gene silencing. *Mol Cell Biol* 2008;**28**:1862–72.

60. Barski A, Cuddapah S, Cui K, Roh T-Y, Schones DE, Wang Z, et al. High-resolution profiling of histone methylations in the human genome. *Cell* 2007;**129**:823–37.

61. Kahn TG, Schwartz YB, Dellino GI, Pirrotta V. Polycomb complexes and the propagation of the methylation mark at the *Drosophila* Ubx gene. *J Biol Chem* 2006;**281**:29064–75.

62. Mohd-Sarip A, van der Knaap JA, Wyman C, Kanaar R, Schedl P, Verrijzer CP. Architecture of a Polycomb nucleoprotein complex. *Mol Cell* 2006;**24**:91–100.

63. Papp B, Muller J. Histone trimethylation and the maintenance of transcriptional ON and OFF states by trxG and PcG proteins. *Genes Dev* 2006;**20**:2041–54.

64. Wang L, Brown JL, Cao R, Zhang Y, Kassis JA, Jones RS. Hierarchical recruitment of Polycomb group silencing complexes. *Mol Cell* 2004;**14**:637–46.

65. Lee TI, Jenner RG, Boyer LA, Guenther MG, Levine SS, Kumar RM, et al. Control of developmental regulators by Polycomb in human embryonic stem cells. *Cell* 2006;**125**:301–13.

66. Boyer LA, Plath K, Zeitlinger J, Brambrink T, Medeiros LA, Lee TI, et al. Polycomb complexes repress developmental regulators in murine embryonic stem cells. *Nature* 2006;**441**:349–53.

67. Bracken AP, Dietrich N, Pasini D, Hansen KH, Helin K. Genome-wide mapping of Polycomb target genes unravels their roles in cell fate transitions. *Genes Dev* 2006;**20**:1123–36.

68. Bernstein BE, Mikkelsen TS, Xie X, Kamal M, Huebert DJ, Cuff J, et al. A bivalent chromatin structure marks key developmental genes in embryonic stem cells. *Cell* 2006;**125**:315–26.

69. Ohno K, McCabe D, Czermin B, Imhof A, Pirrotta V. ESC, ESCL and their roles in Polycomb Group mechanisms. *Mechanisms of Development* 2008;**125**:527–41.

70. Müller J. Transcriptional silencing by the Polycomb protein in *Drosophila* embryos. *EMBO J* 1995;**14**:1209–20.

71. Poux S, McCabe D, Pirrotta V. Recruitment of components of Polycomb Group chromatin complexes in *Drosophila*. *Development* 2001;**128**:75–85.

72. Poux S, Melfi R, Pirrotta V. Establishment of Polycomb silencing requires a transient interaction between PC and ESC. *Genes Dev* 2001;**15**:2509–14.

73. Kim SY, Paylor SW, Magnuson T, Schumacher A. Juxtaposed Polycomb complexes co-regulate vertebral identity. *Development* 2006;**133**:4957–68.

74. Zhao J, Sun BK, Erwin JA, Song J-J, Lee JT. Polycomb proteins targeted by a short repeat RNA to the mouse X chromosome. *Science* 2008;**322**:750–6.

75. Rinn JL, Kertesz M, Wang JK, Squazzo SL, Xu X, Brugmann SA, et al. Functional demarcation of active and silent chromatin domains in human HOX loci by noncoding RNAs. *Cell* 2007;**129**:1311–23.

76. McCall K, Bender W. Probes for chromatin accessibility in the Drosophila bithorax complex respond differently to Polycomb-mediated repression. *EMBO J* 1996;**15**:569–80.

77. Boivin A, Dura J-M. In vivo chromatin accessibility correlates with gene silencing in Drosophila. *Genetics* 1998;**150**:1539–49.

78. Francis NJ, Kingston RE, Woodcock CL. Chromatin compaction by a Polycomb Group protein complex. *Science* 2004;**306**:1574–7.

79. Dellino GI, Schwartz YB, Farkas G, McCabe D, Elgin SCR, Pirrotta V. Polycomb silencing blocks transcription initiation. *Mol Cell* 2004;**13**:887–93.

80. Ficz G, Heintzmann R, Arndt-Jovin DJ. Polycomb group protein complexes exchange rapidly in living Drosophila. *Development* 2005;**132**:3963–76.

81. Cheutin T, McNairn A, Jenuwein T, Gilbert DM, Singh PB, Misteli T. Maintenance of stable heterochromatin domains by dynamic HP1 binding. *Science* 2003;**299**:721–5.

82. Chen D, Dundr M, Wang C, Leung A, Lamond A, Misteli T, et al. Condensed mitotic chromatin is accessible to transcription factors and chromatin structural proteins. *J Cell Biol* 2005;**168**:41–54.

83. Poux S, Kostic C, Pirrotta V. Hunchback-independent silencing of late Ubx enhancers by a Polycomb Group Response Element. *EMBO J* 1996;**15**:4713–22.

84. Hansen KH, Bracken AP, Pasini D, Dietrich N, Gehani SS, Monrad A, et al. A model for transmission of the H3K27me3 epigenetic mark. *Nat Cell Biol* 2008;**10**:1291–300.

85. Margueron R, Justin N, Ohno K, Sharpe ML, Son J, Drury WJ III, et al. Role of the Polycomb protein EED in the propagation of repressive histone marks. *Nature* 2009 (in press).

86. Groth A, Rocha W, Verreault A, Almouzni G. Chromatin challenges during DNA replication and repair. *Cell* 2007;**128**:721–33.

87. Zink D, Paro R. *Drosophila* Polycomb-group regulated chromatin inhibits the accessibility of a trans-activator to its target DNA. *EMBO J* 1995;**14**:5660–71.

88. Cavalli G, Paro R. The *Drosophila* Fab-7 chromosomal element conveys epigenetic inheritance during mitosis and meiosis. *Cell* 1998;**93**:505–18.

89. Cavalli G, Paro R. Epigenetic inheritance of active chromatin after removal of the main transactivator. *Science* 1999;**286**:955–8.

90. Poux S, Horard B, Sigrist CJA, Pirrotta V. The *Drosophila* trithorax protein is a coactivator required to prevent re-establishment of Polycomb silencing. *Development* 2002;**129**:2483–93.

91. Klymenko T, Müller J. The histone methyltransferases trithorax and Ash1 prevent transcriptional silencing by Polycomb group proteins. *EMBO Reports* 2004;**5**:373–7.

92. Milne TA, Briggs SD, Brock HW, Martin ME, Gibbs D, Allis CD, et al. MLL targets SET domain methyltransferase activity to Hox gene promoters. *Mol Ciell* 2002;**10**:1107–17.

93. Nakamura T, Mori T, Tada S, Krajewski S, Rozovskaia T, Wassell R, et al. ALL-1 is a histone methyltransferase that assembles a supercomplex of proteins involved in transcriptional regulation. *Mol Cell* 2002;**10**:1119–28.

94. Smith ST, Petruk S, Sedkov Y, Cho E, Tillib S, Canaani E, et al. Modulation of heat shock gene expression by the TAC1 chromatin-modifying complex. *Nat Cell Biol* 2004;**6**:162–7.

95. Beisel C, Imhof A, Greene J, Kremmer E, Sauer F. Histone methylation by the *Drosophila* epigenetic transcriptional regulator Ash1. *Nature* 2002;**419**:857–62.

96. Byrd KN, Shearn A. ASH1, a Drosophila trithorax group protein, is required for methylation of lysine 4 residues on histone H3. *Proc Natl Acad Sci USA* 2003;**100**:11535–40.

97. Tanaka Y, Katagiri Z, Kawahashi K, Kioussis D, Kitajima S. Trithorax-group protein ASH1 methylates histone H3 lysine 36. *Gene* 2007;**397**:161–8.

98. Orlando V, Jane EP, Chinwalla V, Harte PJ, Paro R. Binding of trithorax and Polycomb proteins to the bithorax complex: dynamic changes during early *Drosophila* embryogenesis. *EMBO J* 1998;**17**:5141–50.

99. Srinivasan S, Dorighi KM, Tamkun JW. *Drosophila* Kismet regulates histone H3 lysine 27 methylation and early elongation by RNA polymerase II. *PLoS Genetics* 2008;**4**:e1000217.

100. Petruk S, Smith ST, Sedkov Y, Mazo A. Association of trxG and PcG proteins with the bxd maintenance element depends on transcriptional activity. *Development* 2008;**135**:2383–90.

101. Hughes CM, Rozenblatt-Rosen O, Milne TA, Copeland TD, Levine SS, Lee JC, et al. Menin associates with a Trithorax family histone methyltransferase complex and with the *Hoxc8* locus. *Mol Cell* 2004;**13**:587–97.

102. Steward MM, Lee J-S, O'Donovan A, Wyatt M, Bernstein BE, Shilatifard A. Molecular regulation of H3K4 trimethylation by ASH2L, a shared subunit of MLL complexes. *Nat Struct Mol Biol* 2006;**13**:852–4.

103. Lee J-H, Tate CM, You J-S, Skalnik DG. Identification and characterization of the human Set1B histone H3-Lys4 methyltransferase complex. *J Biol Chem* 2007;**282**:13419–28.

104. Lee J-H, Skalnik DG. Wdr82 Is a C-terminal domain-binding protein that recruits the Setd1A histone H3-Lys4 methyltransferase complex to transcription start sites of transcribed human genes. *Mol Cell Biol* 2008;**28**:609–18.

105. Wu M, Wang PF, Lee JS, Martin-Brown S, Florens L, Washburn M, et al. Molecular regulation of H3K4 Trimethylation by Wdr82, a component of human Set1/COMPASS. *Mol Cell Biol* 2008;**28**:7337–44.

106. Petruk S, Sedkov Y, Smith S, Tillib S, Kraevski V, Nakamura T, et al. Trithorax and dCBP acting in a complex to maintain expression of a homeotic gene. *Science* 2001;**294**:1331–4.

107. Bantignies F, Goodman RH, Smolik SM. Functional interaction between the coactivator *Drosophila* CREB-binding protein and ASH1, a member of the trithorax group of chromatin modifiers. *Mol Cell Biol* 2000;**20**:9317–30.

108. Ernst P, Wang J, Huang M, Goodman RH, Korsmayer SJ. MLL and CREB bind cooperatively to the nuclear coactivator CREB-binding protein. *Mol Cell Biol* 2001;**21**:2249–58.

109. Tie F, Banerjee R, Stratton CA, Prasad-Sinha J, Stepanik V, Zlobin A, et al. CBP-mediated acetylation of histone H3 lysine 27 antagonizes *Drosophila* Polycomb silencing. *Development* 2009;**136**:3131–41.

110. Bloyer S, Cavalli G, Brock HW, Dura J-M. Identification and characterization of polyhomeotic PREs and TREs. *Dev Biol* 2003;**261**:426–42.

111. Oktaba K, Gutiérrez L, Gagneur J, Girardot C, Sengupta AK, Furlong EEM, et al. Dynamic regulation by Polycomb group protein complexes controls pattern formation and the cell cycle in *Drosophila*. *Developmental Cell* 2008;**15**:877–89.

112. Niwa H. How is pluripotency determined and maintained? *Development* 2007;**134**:635–46.

113. Stock JK, Giadrossi S, Casanova M, Brookes E, Vidal M, Koseki H, et al. Ring1-mediated ubiquitination of H2A restrains poised RNA polymerase II at bivalent genes in mouse ES cells. *Nat Cell Biol* 2007;**9**:1428–35.

114. Zhou W, Zhu P, Wang JK, Pascual G, Ohgi KA, Lozach J, et al. Histone H2A monoubiquitination represses transcription by inhibiting RNA polymerase II transcriptional elongation. *Mol Cell* 2008;**29**:69–80.

115. Zeitlinger J, Stark A, Kellis M, Hong J-W, Nechaev S, Adelman K, et al. RNA polymerase stalling at developmental control genes in the *Drosophila melanogaster* embryo. *Nat Genet* 2007;**39**:1512–16.

116. Pasini D, Bracken AP, Hansen JB, Capillo M, Helin K. The Polycomb group protein Suz12 is required for embryonic stem cell differentiation. *Mol Cell Biol* 2007;**27**:3769–79.

117. Chamberlain SJ, Yee D, Magnuson T. Polycomb repressive complex 2 is dispensable for maintenance of embryonic stem cell pluripotency. *Stem Cells* 2008;**26**:1496–505.

118. Ezhkova E, Pasolli HA, Parker JS, Stokes N, Su IH, Hannon G, et al. Ezh2 Orchestrates gene expression for the stepwise differentiation of tissue-specific stem cells. *Cell* 2009;**136**:1122–35.

119. Feldman N, Gerson A, Fang J, Li E, Zhang Y, Sihinkai Y, et al. G9a-mediated irreversible epigenetic inactivation of Oct3/4 during early embryogenesis. *Nat Cell Biol* 2006;**8**:188–94.

120. Epsztejn-Litman S, Feldman N, Abu-Remaileh M, Shufaro Y, Gerson A, Ueda J, et al. De novo DNA methylation promoted by G9a prevents reprogramming of embryonically silenced genes. *Nat Struct Mol Biol* 2008;**15**:1176–83.

121. Kheradmand Kia S, Gorski MM, Giannakopoulos S, Verrijzer CP. SWI/SNF Mediates Polycomb eviction and epigenetic reprogramming of the INK4b-ARF-INK4a locus. *Mol Cell Biol* 2008;**28**:3457–64.

122. Jaenisch R, Young R. Stem cells, the molecular circuitry of pluripotency and nuclear reprogramming. *Cell* 2008;**132**:567–82.

123. Schwartz YB, Kahn G, Stenberg P, Ohno K, Bourgon R, Pirrotta V. Alternative epigenetic chromatin states of Polycomb target genes. *PLoSgenes* 2010;**6**:e1000805.

SECTION III

Epigenetic Technology

CHAPTER 8

Analysis of Gene-specific DNA Methylation

Naoko Hattori and Toshikazu Ushijima
National Cancer Center Research Institute, Tokyo 104-0045, Japan

INTRODUCTION

Gene- or region-specific DNA methylation analysis is necessary in various situations, and a variety of methods are available. It is important to become familiar with the characteristics of each technique, including the required amount of DNA, flexibility in selection of CpG sites to analyze, how quantitative the technique is, technical complexity, and the cost (Table 8.1). For example, if one wants to analyze DNA methylation as a cause of gene silencing, a specific region that controls gene expression should be analyzed [1], and a method with flexibility in selecting a region to analyze should be used. If one aims for diagnostic applications, a method that is highly accurate should be adopted.

In this chapter, we first introduce principles of DNA methylation analysis, and then summarize characteristics of individual methods. Finally, we will provide tips necessary to perform bisulfite sequencing, methylation-specific PCR (MSP), and quantitative MSP.

PRINCIPLES OF DNA METHYLATION ANALYSIS

DNA methylation can be analyzed based on several principles that differentially recognize 5-methylcytosine (C^m) from cytosine (C). The first principle depends upon methylation-sensitive restriction enzymes whose activity is affected by the presence of a methyl group on a cytosine at a CpG site(s) within restriction sites (Fig. 8.1A). The vast majority of methylation-sensitive restriction enzymes, such as *Hpa*II and *Sma*I, are inactive on methylated CpG sites, but a unique methylation-sensitive restriction enzyme, *Mcr*BC, is inactive on unmethylated CpG sites. Differential cleavage can be detected by Southern-blot hybridization.

The second principle depends on bisulfite-mediated DNA conversion. This treatment converts unmethylated C into uracil (U) very rapidly, whereas it converts methylated C extremely slowly [2]. Under optimized conditions, a difference in methylation status of a CpG site can be converted into a difference of sequence, UpG or CpG. Once a difference of methylation status is converted into a difference of DNA sequence, it can be detected by various techniques, such as bisulfite sequencing, methylation-specific PCR (MSP), real-time MSP, combined bisulfite restriction analysis (COBRA), pyrosequencing, and MassARRAY® analysis (Table 8.1).

Third, methylated cytosines can be specifically recognized by an anti-methylcytidine antibody or a methylated DNA binding (MBD) protein. After appropriate shearing of DNA,

Handbook of Epigenetics: The New Molecular and Medical Genetics. DOI: 10.1016/B978-0-12-375709-8.00008-3

TABLE 8.1 Characteristics of Methods for Gene-specific Methylation Analysis

	Amount of DNA Required	Flexibility in Selection of a Region Analyzed	Quantification	Ease of Use	Cost	Application
Southern-blot hybridization	Large	Low	No	Intermediate	Low	Detection of methylation/ unmethylation at specific CpG sites
Bisulfite sequencing	Small	High	No	Intermediate	Low	Analysis of methylation pattern on individual DNA molecules
COBRA	Small	Low	Yes	Easy	Low	Detection of DNA molecules methylated/unmethylated at a specific CpG site
MSP	Small	High	No	Easy	Low	Detection of DNA molecules methylated/unmethylated at a specific region
Real-time MSP using SYBR Green I	Small	High	Yes	Easy	Low	Quantitative analysis of DNA molecules methylated/ unmethylated at a specific region
MethyLight	Small	Very high	Yes	Easy	Intermediate	Quantitative analysis of DNA molecules methylated/ unmethylated at a specific region
Pyrosequencing	Small	High	Yes	Intermediate	High	Quantitative methylation analysis of multiple CpG sites
MassARRAY®	Small	High	Yes	Easy	High	Quantitative methylation analysis of multiple CpG sites

(A)

(B)

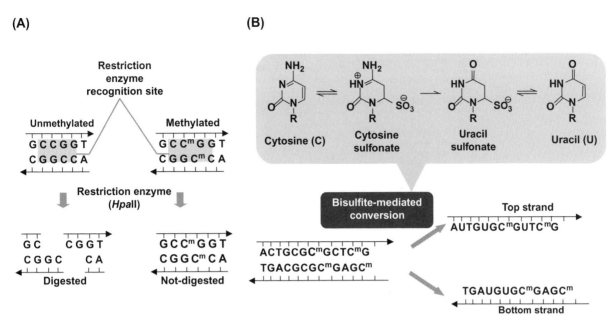

FIGURE 8.1

(A) Methods of DNA methylation detection. Detection by methylation-sensitive restriction enzymes. Genomic DNA is digested with a methylation-sensitive restriction enzyme (*Hpa*II in this figure) when its restriction site (CCGG) is unmethylated, but not digested when the site is methylated. Whether genomic DNA is digested or not represents the methylation status in the original DNA. C^m stands for methylated cytosine. (B) Detection by bisulfite-mediated DNA conversion. Unmethylated cytosines are converted very rapidly into uracil by deamination whereas methylated cytosines are converted extremely slowly. Therefore, a difference in methylation status of a CpG site can be converted into a difference of sequence, UpG or CpG. After bisulfite-mediated DNA conversion, the upper and lower strands are no longer complementary.

methylated DNA can be collected using these affinity methods. This principle is mainly used for genome-wide screening techniques [3].

Fourth, the fraction of methylcytosine in the entire genomic DNA can be measured by HPLC or mass spectrometry [4]. Since this method does not contain sequence information, this can be used solely to measure global methylation levels.

CHARACTERISTICS OF INDIVIDUAL TECHNIQUES
Southern-blot Hybridization

Southern-blot hybridization for DNA methylation analysis is based on DNA digestion by a methylation-sensitive restriction enzyme and subsequent hybridization using a probe for a specific genomic region [5]. The methylation status of a restriction recognition site can be detected by monitoring the band positions of DNA fragments flanking the restriction sites. The advantage of this technique is its quantitative results reflecting the amounts of digested and undigested DNA molecules. Southern blot analysis is especially useful for analysis of repetitive sequences because multiple similar sequences in the genome can be analyzed by a single probe. On the other hand, this technique analyzes only a limited number of CpG sites located within restriction recognition sites, and requires a large amount of high-quality DNA. Although this technique was frequently used before bisulfite conversion-based techniques became popular, it has recently been used only occasionally.

Bisulfite Sequencing

Bisulfite-converted DNA is amplified by PCR using primers located in genomic regions lacking CpG sites. The PCR product is then sequenced, usually after cloning of the PCR product, and CpG sites within the amplified region are interrogated (Fig. 8.2A) [6]. Cytosine

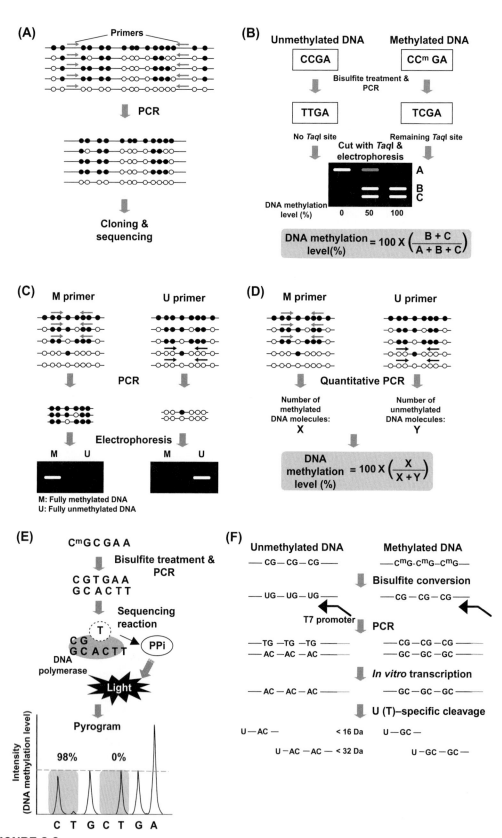

FIGURE 8.2

Principles of individual techniques for DNA methylation analysis. Methylated and unmethylated CpG sites are shown by closed and open circles, respectively. (A) Bisulfite sequencing. Bisulfite-converted DNA is amplified by PCR with primers covering no CpG sites. The PCR product is cloned, and individual clones are sequenced. This technique

(C) and thymine (T) at a CpG site in the converted DNA show methylated and unmethylated C, respectively, in the original DNA. This technique enables us to investigate the methylation status of every single CpG site between the primers, and how multiple CpG sites in a single DNA molecule are methylated. DNA methylation of almost any region can be analyzed using this method. A possible disadvantage is that this technique is labor-intensive, requiring that at least 10 clones per single sample be sequenced. There are also some technical pitfalls that will be described later.

Combined Bisulfite Restriction Analysis (COBRA)

The COBRA technique is based on the appearance or disappearance of a restriction enzyme recognition site after bisulfite conversion (Fig. 8.2B) [7]. By quantifying the ratio of digested and undigested PCR products, the ratio of methylated and unmethylated DNA molecules can be quantified. This technique is suitable for detecting the methylation level of a CpG site quantitatively, and has the advantage of ease of procedure. Since multiple CpG sites within a small genomic region are coordinately methylated or unmethylated [4,8], analysis of a single CpG site can predict the methylation status of the surrounding region. A disadvantage is that CpG sites that can be analyzed by COBRA are limited.

Recently, a modified protocol for COBRA, Bio-COBRA, was developed [9]. Bio-COBRA incorporates an electrophoresis step of the digested PCR product in a microfluidics chip, such as Bioanalyzer (Agilent), and provides rapid and quantitative assessment of DNA methylation statuses in a large sample set.

Methylation-specific PCR (MSP)

This technique interrogates methylation statuses of several CpGs at primer sites by performing PCR with primers specific to methylated or unmethylated sequences and observing the presence or absence of a PCR product (Fig. 8.2C) [10]. If both forward and reverse primer regions are methylated, intervening CpG sites are also likely to be methylated. DNA molecules with mosaic methylation patterns at primer sites are not amplified. This technique has high flexibility in selecting a genomic region to analyze because PCR primers can be designed at arbitrary positions, even if the region to be analyzed is CpG-rich, and it is technically simple. At the same time, MSP can easily produce false positive and false negative results. Therefore, it is critically important to use the optimal number of PCR cycles and annealing temperatures with appropriate negative controls, which will be described in the third section of this chapter.

Real-time MSP and MethyLight

Real-time MSP is performed by real-time detection of MSP products. By comparing amplification of test samples with standard samples that contain known numbers of DNA

129

provides a methylation pattern of individual DNA molecules at single CpG resolution. (B) COBRA. Bisulfite-converted DNA is amplified by PCR with primers covering no CpG sites, and the PCR product is digested with a restriction enzyme (*Taq*I in this figure). In the COBRA assay shown here, if the cytosine in the CpG site is methylated, the restriction site will remain. On the other hand, if the site is unmethylated, the restriction site will disappear. Quantitative analysis of methylation levels is achieved by subsequent gel electrophoresis and measurement of cleaved and uncleaved bands. (C) MSP. Methylation statuses at several CpGs within primer sequences are interrogated by performing PCR with primers specific to methylated or unmethylated templates and monitoring the presence or absence of a PCR product. PCR conditions are optimized using fully methylated DNA and fully unmethylated DNA. (D) Real-time MSP. The numbers of methylated and unmethylated DNA molecules are quantified by real-time MSP. (E) Pyrosequencing. C/T polymorphisms in the PCR product are investigated by measuring pyrophosphate released at individual sites. The amount of pyrophosphate is converted into a light signal, and then shown as a pyrogram. (F) MassARRAY®. The PCR product amplified from bisulfite-converted DNA is transcribed *in vitro*, and cleaved by RNase A. The difference in the mass of a product with C and that with T (16 Da) is detected by MALDI-TOF mass spectrometry.

molecules, numbers of methylated and unmethylated DNA molecules can be quantified (Fig. 8.2D). A methylation level can be calculated based on these numbers of DNA molecules. PCR products can be detected by an intercalating dye like SYBR® Green I (real-time MSP), or by a TaqMan probe (MethyLight) [11]. Since a TaqMan probe anneals only to a specific sequence (methylated or unmethylated sequence), MethyLight has higher specificity than quantitative MSP although a TaqMan probe is costly. Intercalating dye can detect even non-specifically amplified DNA and primer dimers, and confirmation of specific amplification by melting analysis of the PCR product is essential. It is reported that the use of a new fluorescent dye, such as SYTO-82, can produce more accurate melting results [12]. The real-time MSP and MethyLight techniques have a lot of flexibility in selecting a genomic region to analyze, as does MSP, and are accurate and sensitive in quantifying DNA methylation levels. The high accuracy and sensitivity of these techniques make them suitable for analysis of a large number of clinical samples.

Pyrosequencing

Pyrosequencing detects methylation levels of individual CpG sites in a PCR product obtained by primers common to methylated and unmethylated sequences after bisulfite conversion. The amounts of C and T at individual sites are converted into the amounts of pyrophosphates released using the primer extension method, and their amounts are accurately quantified bioluminometrically using the Pyrosequencer system (QIAGEN) (Fig. 8.2E). The advantages of pyrosequencing are its accurate quantitative results and ease of daily procedure. However, design of suitable primers is difficult, depending upon the local sequence, and an instrument specifically designed for this analysis is unavoidably necessary.

MassARRAY®

MassARRAY® also detects methylation levels of individual CpG sites in a PCR product using primers common to methylated and unmethylated sequences after bisulfite conversion. In this technique, the PCR is performed with a reverse primer coupled with a T7 promoter tag. The PCR product is transcribed *in vitro* using a single dNTP analog, which can be substituted for its rNTP. The *in vitro* transcript is then cleaved by RNase A, which digests at pyrimidine bases, in a base-specific manner (Fig. 8.2F). If dCTP was used during the *in vitro* transcription, the RNase A will cleave at every uracil. A difference in the mass of product with C and that with T (16 Da) is detected by matrix-assisted laser desorption/ionization time-of-flight (MALDI-TOF) mass spectrometry. MassARRAY® is a powerful technique to quantitatively investigate DNA methylation statuses of multiple CpG sites in a large number of samples, but has a disadvantage in the cost of the instrument.

TIPS FOR BISULFITE SEQUENCING

Bisulfite sequencing is capable of analyzing detailed DNA methylation patterns of individual DNA molecules in given regions of the genome. It also provides quantitative information on the ratio of methylated and unmethylated DNA molecules. At the same time, although this technique is generally considered as technically simple, caution must be exercised to obtain unbiased results.

PCR Conditions for Unbiased Amplification

It is well known that, depending upon PCR conditions, there can be a PCR bias that leads to preferential amplification of either unmethylated or methylated DNA [13,14]. In most cases, unmethylated DNA is preferentially amplified, but methylated DNA can be preferentially amplified with specific primers [13]. To avoid this PCR bias, a PCR condition that equally amplifies fully methylated and fully unmethylated DNA controls should be established by selecting an optimal primer set and an optimal annealing temperature (Fig. 8.3A) [14].

(A)

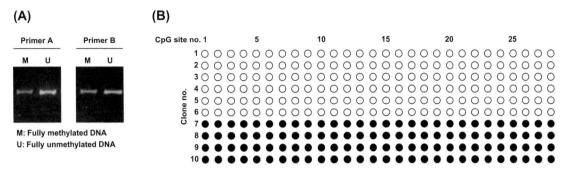

(B)

FIGURE 8.3
Optimization of bisulfite sequencing. (A) Comparison of two primer sets for bisulfite sequencing. The influence of primers on PCR efficiency was examined using fully methylated DNA (S) and fully unmethylated DNA (G). Primer A predominantly amplified unmethylated DNA, whereas primer B equally amplified both methylated and unmethylated DNA. (B) Confirmation of unbiased amplification. Methylated and unmethylated cytosines are shown by closed and open circles, respectively. The proportion of methylated clones was 40%, indicating appropriate PCR conditions and unbiased amplification were achieved.

Fully methylated DNA can be prepared by treatment of DNA with *SssI* methylase (*SssI*), and fully unmethylated DNA can be prepared by amplifying normal DNA with a GenomiPhi DNA amplification kit (GenomiPhi). When accurate estimation of the ratio of methylated and unmethylated DNA is necessary, control DNA containing an equal number of fully methylated and unmethylated DNA molecules should be prepared by mixing such DNA, and simultaneously analyzed to obtain a ratio of 40% to 60% (Fig. 8.3B).

PCR Cycles to Avoid Artifacts

Even if optimal PCR conditions are used, PCR cycles should be minimized as long as a sufficient amount of a PCR product for cloning is obtained. Excessive PCR cycles cause denaturation of the PCR product in the absence of Taq polymerase activity, and produce the amplification of chimeric products and even PCR products that were not present in the template DNA. Excessive PCR cycles also exaggerate the difference in PCR efficiency between methylated and unmethylated DNA.

TIPS FOR MSP AND QUANTITATIVE MSP

MSP is flexible in selecting regions for analysis and can be performed with ease and at a low cost. Real-time MSP provides accurate, sensitive, and quantitative assessment of DNA methylation levels. Under good conditions, DNA methylation levels obtained by real-time MSP have a variation ≤20% of the mean methylation level. To maximize these advantages, there are some tips for conducting MSP and real-time MSP.

Primer Design

A genomic region should be carefully selected as in other analyses, and primers specific to methylated or unmethylated DNA should be designed in the same region. The 3' end of a primer should be located at a polymorphic C/T site, and multiple CpG sites should be located near the 3' end (Fig. 8.4A). Difficulty in designing primers specific to unmethylated DNA is frequently encountered, and use of the other DNA strand (bottom strand) is often helpful.

PCR Conditions for Specific Amplification

The annealing temperature and magnesium concentration should be optimized using the fully methylated and fully unmethylated DNA controls. A good condition for primers specific to methylated DNA shows ample amplification of fully methylated DNA and no

131

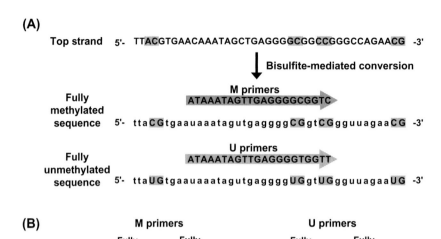

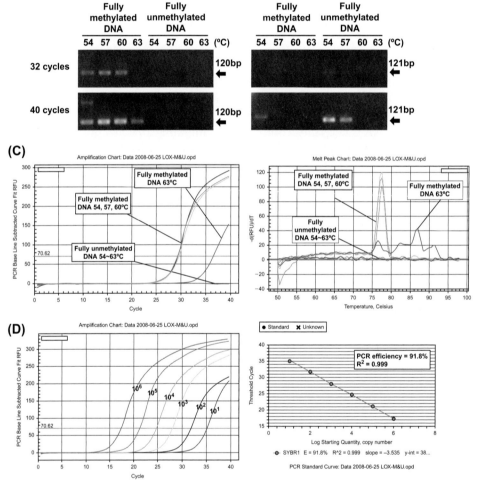

FIGURE 8.4

Optimization of real-time PCR conditions. (A) Primer design for MSP and real-time MSP. Primers specific to methylated and unmethylated DNA (M and U primers, respectively) should contain multiple CpG sites near and at their 3′ ends, and are desirably located in the same region. (B) Optimization of the annealing temperature for MSP. For the M primer, annealing temperatures of 57 and 60°C did not amplify fully unmethylated DNA, but amplified fully methylated DNA with good efficiency. For the U primer, only an annealing temperature of 57°C yielded specific and efficient amplification. (C) Optimization of the annealing temperature for real-time MSP. The real-time PCR amplification curve showed high PCR efficiency under annealing temperatures of 54, 57, and 60°C. The melting curve showed a single peak, thus specific amplification, under annealing temperatures of 54, 57, and 60°C. If multiple good annealing temperatures are available, a higher temperature is preferable for specificity. Optimal conditions in real-time MSP are occasionally different from those in MSP even if the same primer set is used. (D) Real-time MSP using standard DNA. Correlation using multiple standard DNA (R2) was >0.98, and PCR efficiency was >80%.

amplification of fully unmethylated DNA (Fig. 8.4B). A good condition for primers specific to unmethylated DNA amplifies fully unmethylated DNA, but not fully methylated DNA.

In the case of real-time MSP, the best conditions can be determined by the amplification curve and the melting curve (Fig. 8.4C). The amplification curve under good conditions shows a steep rise at an early PCR cycle, and a flat plateau. The melting curve under the best PCR conditions shows a single sharp peak.

Preparation of Standard DNA

To quantify DNA methylation levels by real-time MSP, standard DNA with known numbers of DNA molecules is necessary. This can be prepared in two ways. First, the PCR product can be purified by a gel-filtration column to remove unused nucleotides and primers. Second, the PCR product of MSP is cloned into a plasmid, and the plasmid is linearized by a restriction enzyme. Since the molecular weight of the PCR product or the plasmid with the insert can be calculated, the number of DNA molecules in a measured weight of solution can be calculated. Preparation of standard DNA by cloning a PCR product has the advantage of accuracy and availability of a large amount of standard DNA, but has the disadvantage of being a complex procedure.

Quantity of Template DNA

Both MSP and real-time MSP can achieve high sensitivity, such as detecting one methylated DNA molecule among 1000 molecules. However, substantial loss in the number of DNA molecules that can serve as a PCR template takes place during bisulfite-mediated conversion. Namely, although the weight of DNA decreases only slightly, the number of template DNA molecules measured by quantitative PCR decreases down to 5 to 10% of DNA before the treatment [15]. Therefore, caution must be exercised as to how many copies of template DNA are present in a PCR solution. Supposing that one human haploid genome weighs 3.6 pg and that 10% of DNA molecules are recovered as a template for PCR after bisulfite-mediated conversion, only 28 molecules are available for PCR of a single target sequence in a DNA sample that originated from 1 ng of genomic DNA before bisulfite treatment. If one wants to have a sensitivity of 1%, 1000 molecules (10 methylated molecules) in a PCR solution will be necessary, and this corresponds to 36 ng DNA in a reaction.

133

EPILOGUE

Regional DNA methylation analysis is applied not only for basic research but also for diagnostic purposes. Selecting an appropriate technique and conducting experiments under good conditions are required to obtain reliable data. We hope that this chapter will help investigators to select appropriate techniques.

References

1. Ushijima T. Detection and interpretation of altered methylation patterns in cancer cells. *Nat Rev Cancer* 2005;**5**:223–31.
2. Hayatsu H, Wataya Y, Kazushige K. The addition of sodium bisulfite to uracil and to cytosine. *J Am Chem Soc* 1970;**92**:724–6.
3. Rauch T, Pfeifer GP. Methylated-CpG island recovery assay: a new technique for the rapid detection of methylated-CpG islands in cancer. *Lab Invest* 2005;**85**:1172–80.
4. Kaneda A, Tsukamoto T, Takamura-Enya T, Watanabe N, Kaminishi M, Sugimura T, et al. Frequent hypomethylation in multiple promoter CpG islands is associated with global hypomethylation, but not with frequent promoter hypermethylation. *Cancer Sci* 2004;**95**:58–64.
5. Bird AP, Southern EM. Use of restriction enzymes to study eukaryotic DNA methylation: I. The methylation pattern in ribosomal DNA from *Xenopus laevis*. *J Mol Biol* 1978;**118**:27–47.
6. Clark SJ, Harrison J, Paul CL, Frommer M. High sensitivity mapping of methylated cytosines. *Nucleic Acids Res* 1994;**22**:2990–7.

7. Xiong Z, Laird PW. COBRA: a sensitive and quantitative DNA methylation assay. *Nucleic Acids Res* 1997;**25**:2532–4.

8. Kaneda A, Kaminishi M, Yanagihara K, Sugimura T, Ushijima T. Identification of silencing of nine genes in human gastric cancers. *Cancer Res* 2002;**62**:6645–50.

9. Brena RM, Auer H, Kornacker K, Plass C. Quantification of DNA methylation in electrofluidics chips (Bio-COBRA). *Nat Protoc* 2006;**1**:52–8.

10. Herman JG, Graff JR, Myohanen S, Nelkin BD, Baylin SB. Methylation-specific PCR: a novel PCR assay for methylation status of CpG islands. *Proc Natl Acad Sci USA* 1996;**93**:9821–6.

11. Eads CA, Danenberg KD, Kawakami K, Saltz LB, Blake C, Shibata D, et al. MethyLight: a high-throughput assay to measure DNA methylation. *Nucleic Acids Res* 2000;**28**:E32.

12. Gudnason H, Dufva M, Bang DD, Wolff A. Comparison of multiple DNA dyes for real-time PCR: effects of dye concentration and sequence composition on DNA amplification and melting temperature. *Nucleic Acids Res* 2007;**35**:e127.

13. Warnecke PM, Stirzaker C, Melki JR, Millar DS, Paul CL, Clark SJ. Detection and measurement of PCR bias in quantitative methylation analysis of bisulphite-treated DNA. *Nucleic Acids Res* 1997;**25**:4422–6.

14. Warnecke PM, Stirzaker C, Song J, Grunau C, Melki JR, Clark SJ. Identification and resolution of artifacts in bisulfite sequencing. *Methods* 2002;**27**:101–7.

15. Munson K, Clark J, Lamparska-Kupsik K, Smith SS. Recovery of bisulfite-converted genomic sequences in the methylation-sensitive QPCR. *Nucleic Acids Res* 2007;**35**:2893–903.

CHAPTER 9

Methods for Assessing Genome-wide DNA Methylation

Tibor A. Rauch[1] and Gerd P. Pfeifer[2]
[1]Section of Molecular Medicine, Department of Orthopedic Surgery, Rush University Medical Center, Chicago, IL 60612, USA
[2]Division of Biology, Beckman Research Institute of the City of Hope, Duarte, CA 91010, USA

INTRODUCTION

Historically, DNA methylation was the first epigenetic modification to be discovered [1]. The processes governing DNA methylation as well as the biological properties of this modification are quite well understood. A functional role of DNA methylation in epigenetic control, gene regulation, X chromosome inactivation, and cell differentiation was first proposed in 1975 [2,3]. Today, the connections between DNA methylation and other epigenetic marks such as histone modifications are being studied. Aberrations in DNA methylation patterns as a cause or consequence of diseases, such as cancer, are analyzed and described by many laboratories.

To explore DNA methylation profiles at genome scale, a wide range of approaches have been developed. Most of the methods were originally used for detecting methylation changes at the single gene level but by coupling them with extensive cloning and sequencing work or combining them with microarray platforms, genome-wide analysis tools have been developed. On microarray platforms, promoter or CpG island arrays are often used to analyze important regulatory regions. Tiling arrays can be used to investigate segments of specific chromosomes or the entire genome. Most methylation analysis methods can be categorized into several well-characterized groups on the basis of their principles (Fig. 9.1).

Several methods are based on restriction endonucleases that possess altered sensitivity towards methylated cytosine residues present in the cleavage site. In this way the restriction endonuclease digestion pattern depends on the methylation status of the cleavage sites and ultimately reflects methylation profiles of the given chromosomal region. Other techniques are based on antibodies or proteins that bind to methylated DNA. Resolution at the single nucleotide level often requires bisulfite sequencing approaches for which high throughput techniques are currently being developed. In this review, we will describe several of the more commonly used genome-scale methods for DNA methylation analysis in some detail.

Handbook of Epigenetics: The New Molecular and Medical Genetics. DOI: 10.1016/B978-0-12-375709-8.00009-5

Genome-wide DNA methylation profiling approaches		
Methylation sensitive endonuclease-based methods	**Sodium bisulfite treatment-based methods**	**Biological affinity-based methods**
Restriction landmark genomic scanning (RLGS)	MALDI-TOF MS	Methylated DNA immuno-precipitation (MeDIP)
Methylation sensitive fingerprinting (MSRF)	Bisulfite sequencing (targeted and whole genome)	MBD-affinity column (MAC)
Methylated-CpG island amplification coupled to microarray (MCAM)	Golden Gate and Infinium assays	Methylated-CpG island recovery assay (MIRA)
Methylation-sensitive representational difference analysis (MS-RDA)	Pyrosequencing	
Differential methylation hybridization (DMH)		
HELP assay		
Methylation-specific digital karyotyping (MSDK)		
McrBC-based methods		

FIGURE 9.1

Frequently used methods for large-scale analysis of DNA methylation profiles.

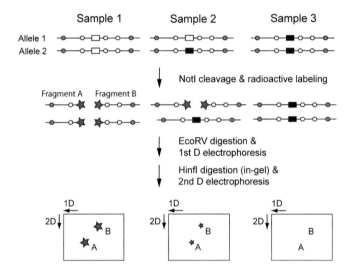

FIGURE 9.2

Restriction landmark genomic scanning (RLGS). A more detailed description of the main steps of RLGS is provided in the text. Open and filled rectangles label unmethylated and methylated NotI sites, respectively. Circles represent EcoRV and HinfI cleavage sites. (Please refer to color plate section)

RESTRICTION LANDMARK GENOMIC SCANNING (RLGS)

RLGS (Fig. 9.2) was originally applied to identify genes involved in genomic imprinting and was later adopted for mapping genome-wide DNA methylation changes [4,5]. Isolated genomic DNA is digested with NotI enzyme that cuts unmethylated CG-rich sequences (5′-GCGGCCGC-3′) and the cleaved ends are subsequently radioactively labeled. Next, the labeled genomic DNA fraction is further digested with a more frequently cutting enzyme (e.g. EcoRV: 5′-GATATC-3′) and separated on an agarose gel (1D electrophoresis). The double-digested and separated DNA fragments are subjected to a third in-gel digestion by a frequently cutting enzyme (e.g. HinfI: 5′-GANTC-3′) and separated on a polyacrylamide gel (2D electrophoresis). Other combinations of restriction enzymes have been used as well [6,7]. RLGS provides excellent and reproducible results but it is a very laborious technique. Identification of the differentially methylated regions from radioactive spots used to be a tedious task but recently established human and mouse RLGS databases (http://genome.gsc.riken.go.jp/RLGS/RLGShome.html) and the constantly improving bioinformatics tools (e.g. virtual spot analysis) are of great help in identifying the critical differentially methylated regions [8].

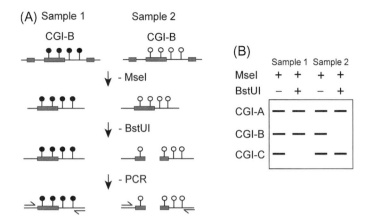

FIGURE 9.3
Methylation sensitive restriction fingerprinting (MSRF). (A) Principle of MSRF. Details can be found in the text. Filled and open lollipops mark methylated and unmethylated CpG sites, respectively. Rectangles refer to MseI and BstUI cleavage sites. Arrows show the locations of the arbitrary primers' annealing sites. (B) Theoretical outcome of an MSRF experiment (autoradiogram). CGI-A: there is no difference between samples 1 and 2. CGI-B: hypermethylation in sample 1 (this case is outlined in panel A). CGI-C: hypomethylation in sample 1. (Please refer to color plate section)

TECHNIQUES BASED ON METHYLATION-SENSITIVE RESTRICTION ENDONUCLEASES AND PCR

Methylation-sensitive restriction fingerprinting (MSRF) (Fig. 9.3) was one of the first methods that allowed researchers to identify multiple de novo methylated or demethylated sequences occurring in different tissue samples [9,10]. For example, DNA samples isolated from tissues are first digested with a restriction enzyme that rarely cuts within CpG islands (e.g. MseI: 5′-TTAA-3′). The MseI-digested genomic DNA is then divided into two aliquots and one of them is cut with an enzyme that frequently cleaves CpG-containing sequences (e.g. BstUI: 5′-CGCG-3′ or HpaII: 5′-CCGG-3′) in a methylation-sensitive manner, i.e. the methylated sequences are resistant to cutting. Subsequently, the BstUI or HpaII cut and uncut (control) samples are subjected to PCR amplification with short arbitrary primer pairs in the presence of radioactive dNTPs. Radioactively labeled PCR fragments are separated on a gel and visualized by autoradiography (Fig. 9.3). Fragments showing disease-specific or tissue-specific patterns are excised from the dried gel and reamplified. After cloning and sequencing of the amplified bands the necessary information is gained for data base searches and target gene identification. Because of the requirement for gel electrophoresis, the application of MSRF is limited nowadays.

Methylation-sensitive representational difference analysis (MS-RDA) is a genome-wide analysis method that isolates DNA fragments differentially methylated between two samples [11]. One important characteristic of MS-RDA is that it enriches the unmethylated CpG-rich fraction of the genome (e.g. by digestion with HpaII enzyme). In MS-RDA, genomic DNA is digested with a methylation-sensitive restriction enzyme, and DNA fragments are amplified by PCR using a universal adaptor primer. DNA fragments differentially methylated between samples will be present in one amplicon, but not in the other. Genomic subtraction is performed using the two amplicons and the resulting fragments are cloned and sequenced.

Another example of restriction enzyme-based approaches is differential methylation hybridization (DMH), which allows the simultaneous determination of the methylation status of a large number of CpG-island loci [12]. CpG islands containing DNA fragments are gridded on high-density arrays, genomic DNA from the tissues of interest is digested with methylation-sensitive enzymes, and digestion products are used as templates for PCR after ligation of linkers. The resulting fragments are used as probes to screen for hypermethylated

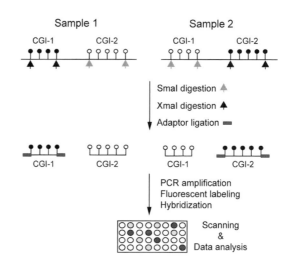

Sample 1   Sample 2

FIGURE 9.4
The main steps of methylated CpG island amplification coupled microarray (MCAM). Details are in the text. Arrows point to SmaI and XmaI cleavage sites. (Please refer to color plate section)

sequences, for example in cancer tissues. A recently developed related method is the HELP assay (HpaII tiny fragment enrichment by ligation-mediated PCR), which involves co-hybridization of the DNA samples to a genomic DNA microarray after cutting with a methylation-sensitive restriction enzyme or its methylation-insensitive isoschizomer [13]. A variation of this approach can be used in combination with high-throughput DNA sequencing [14].

Methylated CpG island amplification coupled microarray (MCAM) also is based on methylation-sensitive restriction enzymes and PCR using flanking primers followed by sequence analysis or microarray probing. MCAM is a powerful approach for simultaneous identification of differentially-methylated genomic regions [15,16] (Fig. 9.4). Initially, the genomic DNA is cleaved with a methylation-sensitive enzyme (e.g. SmaI: 5′-CCCGGG-3′) that creates blunt ends and in this way contributes to the elimination of unmethylated sequences from the subsequent adaptor ligation step. The second enzyme XmaI recognizes the same sequence independently of methylation status and creates sticky ends that make the appropriate adaptor ligation possible. The adaptor-ligated methylated fraction is PCR-amplified, labeled with fluorescent dyes, and hybridized onto a microarray platform. Performing MCAM on normal and disease-derived tissue can easily reveal a number of disease-associated DNA methylation changes (Fig. 9.4).

Another variation of current methylation microarray approaches includes the use of the methylation-dependent restriction enzyme McrBC, which cleaves methylated DNA only. Although the exact cleavage position of this enzyme does not coincide directly with the methylation site, the technique can effectively be used for genome-scale DNA methylation analysis [17,18].

METHYLATION-SPECIFIC DIGITAL KARYOTYPING (MSDK)

The principle of MSDK roots back to the serial analysis of gene expression (SAGE) approach that is based on the ultimate reads of concatemerized sequence tags [19] (Fig. 9.5). The obtained short sequence reads can provide information that can be traced back to the original genomic region. As a first step, the purified genomic DNA is fragmentized by cutting with the methylation-sensitive AscI enzyme (5′-GGCGCGCC-3′), and subsequently biotinylated linkers are ligated to the ends. As the linker-ligated fragments can be quite long, a frequently cutting enzyme (NlaIII: 5′-ACGT-3′) is used to eliminate the central part of the fragments. The successfully ligated and shortened fragments are captured by using streptavidin-coated beads

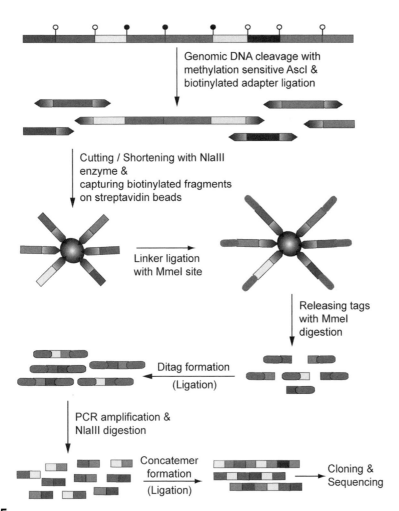

FIGURE 9.5
Methylation specific digital karyotyping (MSDK). A step-by-step introduction to the main steps can be found in the text. Filled and open lollipops mark methylated and unmethylated CpG sites, respectively. (Please refer to color plate section)

and divided into two aliquots. In the subsequent reaction, two different adaptors are ligated to free NlaIII-digested ends. The ligated adaptors harbor an MmeI recognition site. The enzyme cuts 17 bp away from its recognition site and in this way just a short 17 bp-long sequence tag is retained from the original genomic region. After MmeI digestion, the captured sequences are released from the beads and the two aliquots are combined and ligated. Tags having complementary overhanging ends form ditags, and in the subsequent step they are PCR-amplified. The adaptors are removed from the amplified ditags by NlaIII digestion and ligated together in order to form several hundred bp long concatemers. These concatemers are cloned and sequenced, and by applying proper bioinformatics analysis, the sequence information is extracted and aligned to the genome. MSDK is reliable but is a time-consuming method. Generation of MSDK libraries can be completed in 7–10 days, whereas sequencing and data analysis require an additional 3–4 weeks.

Numerous variations (e.g. application of different enzyme pairs, neoschizomers) of the outlined restriction enzyme-based approaches have been used but all of them are based on similar principles. The main drawback of the methylation-sensitive cleavage-dependent methods is that they can provide only limited methylation profile analysis since there is no restriction enzyme that cleaves appropriately within all CpG islands. The use of multiple restriction enzymes may address some of these limitations [20]. The second set of techniques to be described next makes use of the differential sensitivity of cytosine and 5-methylcytosine towards chemical modification and/or cleavage. Under prolonged and harsh incubation with sodium bisulfite, cytosines are deaminated and converted into uracils. Methylated cytosines are resistant to bisulfite modification and, therefore, any cytosine that remains unchanged in bisulfite-treated DNA is inferred to be methylated. This methodology allows single base resolution but is technically challenging when applied at the whole genome level.

TARGETED AND WHOLE GENOME BISULFITE SEQUENCING

The basis of sodium bisulfite sequencing is the fact that cytosine is deaminated to uracil by sodium bisulfite but 5-methylcytosine is resistant to bisulfite-induced deamination [21–23]. Bisulfite sequencing provides single base resolution for analysis of DNA methylation patterns [24]. It is based on sequencing of PCR products of bisulfite-treated DNA to profile DNA methylation at specific loci or along entire segments of chromosomes as recently reported [25]. The latter project analyzed DNA methylation patterns across human chromosomes 6, 20, and 22 in 12 human tissues, covering 22% of the CpGs on these chromosomes.

Ideally, one would like to apply this approach to whole genome methylation profiling. The recently developed approaches for massively parallel sequence analysis, for example on the Roche 454, Illumina/Solexa, and ABI/SOLiD platforms, have provided technology for unprecedented large scale analysis on a rapid time scale and for a reasonable and still declining cost. Whole genome bisulfite sequencing has been accomplished for the relatively small genome of the plant *Arabidopsis thaliana* [26] and, more recently, for the genome of two human cell types [27]. However, for mammalian genomes, there are still significant challenges. The treatment of DNA with sodium bisulfite effectively converts the genome into three bases (A, G, U/T) with the exception of the few methylated cytosines that remain. Thus, it is extremely difficult to align short sequence reads unambiguously to their original genomic location. The development of technology with longer read length may overcome this problem. In the meantime, researchers have developed approaches that allow limited and targeted bisulfite sequence analysis of critical parts of the genome. If the DNA is pre-digested with an enzyme that targets CpG islands (e.g. MspI: 5'-CCGG-3'), reduced representation methylation maps can be derived [28]. One disadvantage of this approach is that most (>90%) of the sequence reads come from unmethylated DNA (i.e. MspI cuts both methylated and unmethylated 5'CCGG sequences) and it depends, of course on the presence of MspI sites. Another published approach is based on the enrichment of specific genomic targets using padlock probes. The method specifically captures a subset of genomic targets for single-molecule bisulfite sequencing [29,30]. For example, a set of ~30,000 padlock probes was designed to assess methylation of ~66,000 CpG sites within 2020 CpG islands on human chromosome 12, chromosome 20, and 34 additional selected regions [30]. Other genomic enrichment approaches for targeted bisulfite sequencing are currently being developed [31].

OTHER SODIUM BISULFITE BASED APPROACHES

After conversion of genomic DNA with sodium bisulfite, several medium- to high-throughput approaches have been developed to analyze methylation profiles of a number of genes simultaneously.

In the mass spectrometry-based MALDI-TOF MS (Sequenom approach), bisulfite-pretreated genomic DNA is used as a template for PCR amplification of specific genes [32]. One of the PCR primers is designed in such a way that a T7 RNA polymerase promoter is attached to the 5' end. By using this promoter the amplicon is transcribed *in vitro* into single stranded RNA and subsequently digested with ribonuclease A at uracil residues. MALDI-TOF mass spectrometry of the cleavage products can provide quantitative data regarding the methylation status of most CpG dinucleotides in the tested region because (i) the endonuclease cleavage occurs quantitatively so that the potential fragment variants are predictable; (ii) the methylation state-related sequence alteration from C to T (or G to A on the opposite strand) yields a 16-Da mass difference (Fig. 9.6). MALDI-TOF MS is a medium-throughput approach that provides direct quantitative methylation data for many genes simultaneously. The analysis is usually conducted in multi-well format and is limited by the number of gene targets that are amplified and analyzed in each run.

Pyrosequencing is a sequencing-by-synthesis based technology that makes use of the luminometric detection of pyrophosphate released upon nucleotide incorporation. Using bisulfite-converted DNA as a template, the feasibility and reliability of this technology for the quantification of methylation at single CpGs within PCR fragments has been shown [33,34]. A target region of up to 350 bp is amplified by PCR using primers complementary to the bisulfite-converted DNA sequence. One of the two amplification primers carries a biotin residue at its 5′-terminus for purification of single-stranded templates. An internal pyrosequencing primer complementary to the single-stranded template is hybridized to the template, and the pyrosequencing reactions are performed by sequential addition of single nucleotides in a predefined order. This method is quantitative but limitations exist with regard to the requirement to place the primers into regions that are free of CpG sites.

The Illumina GoldenGate assay has been designed for high-throughput quantitative measurements of DNA methylation at a large number of genes [35]. It can analyze up to 1536 targeted CpG sites in 96 samples simultaneously. An updated version of this technology called Infinium assay allows researchers to interrogate 27,578 informative CpG sites per sample covering more than 14,000 genes. The method is based on gene-specific

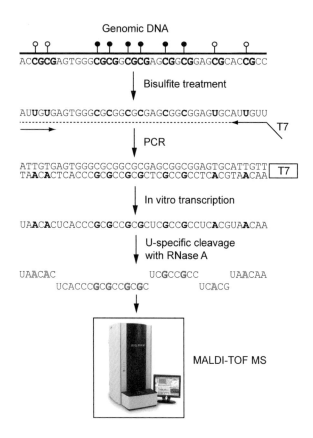

FIGURE 9.6
MALDI-TOF MS ("Sequenom" approach). The principle of MALDI-TOF MS methylation detection, which is based on bisulfite conversion, is outlined in the text. Filled and open lollipops mark methylated and unmethylated CpG sites, respectively. Boldfaced and colored letters help in following the nucleotide changes through the procedure. (Please refer to color plate section)

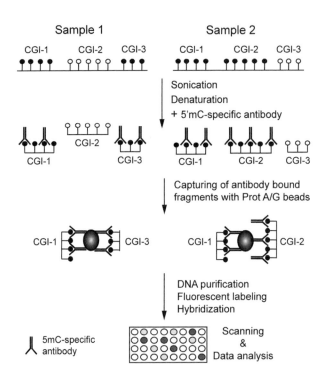

FIGURE 9.7
Methylated DNA immunoprecipitation (MeDIP). Delineation of MeDIP is given in the text. Filled and open lollipops mark methylated and unmethylated CpG sites, respectively. (Please refer to color plate section)

and methylation-dependent single nucleotide primer extension on bisulfite-converted DNA using primers that distinguish between unmethylated and methylated CpGs after bisulfite conversion. Single-base extension of the probes incorporates a labeled ddNTP, which is subsequently stained with a fluorescence reagent. The level of methylation for the interrogated locus can be determined by calculating the ratio of the fluorescent signals from the methylated versus the unmethylated sites. Only a few CpG sites are assayed per locus.

Affinity purification is the principle of the third group of techniques applied for genome-wide DNA methylation mapping. Methylated DNA fragments can be affinity purified by using either a 5mC-specific antibody or proteins that specifically bind to methylated DNA sequences [36–39]. Conversely, DNA fragments containing unmethylated CpG sites can be enriched by using protein domains, e.g. the CXXC domains that have a selective affinity for unmethylated CpG-containing DNA [40].

METHYLATED DNA IMMUNOPRECIPITATION (MeDIP)

A 5-methylcytosine-specific antibody has been raised and used for immunoprecipitation of densely methylated DNA sequences [37] (Fig. 9.7). In this method, the fragments generated by sonication of genomic DNA are incubated with anti-5mC antibody and the captured methylated fraction is extracted from the reaction by using Protein A/G beads. Isolated DNA fragments are deproteinized, fluorescently labeled, and hybridized onto different microarray platforms or analyzed by massively parallel sequencing. The main limitation of the MeDIP method is the quality of 5mC antibodies used in the procedure and the requirement of single-stranded (denatured) DNA for analysis. Methylation profiles can be generated with a resolution of approximately 100 base pairs.

MBD-AFFINITY COLUMN (MAC)

The proteins MBD1, MBD2, MBD3, MBD4, and MeCP2 comprise a small family of nuclear proteins that share a common methyl-CpG binding domain (MBD). Each of these proteins,

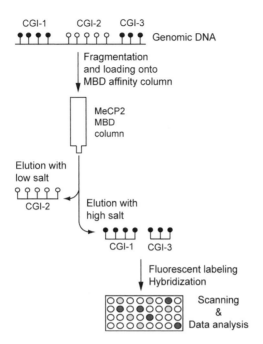

FIGURE 9.8
MBD-affinity column (MAC). Description of MAC is found in the text. Filled and open lollipops mark methylated and unmethylated CpG sites, respectively. (Please refer to color plate section)

with the exception of human MBD3, is capable of binding specifically to methylated DNA. An affinity column-based method was created that employs the methyl-CpG binding domain of MeCP2 [41,42]. Fragmentized (sonicated or endonuclease-digested) genomic DNA is loaded onto the column, which strongly retains those sequences that are highly methylated. The elution of the methylation-enriched fraction can be achieved by high salt containing buffer. The eluted fraction is labeled with fluorescent dye and hybridized onto a microarray platform [Fig. 9.8]. The main drawback of MAC is the high amount of starting genomic DNA required for column purification.

METHYLATED-CpG ISLAND RECOVERY ASSAY (MIRA)

Among the MBD proteins, MBD2b has the strongest affinity to methylated DNA and can form heterodimers with other MBD proteins via its C-terminal coiled-coil domain. MBD3L1, a related protein and member of the MBD2/3 sub-family, has no DNA binding domain itself but it can be a binding partner of MBD2b via heterodimer formation. The MBD2b/MBD3L1 protein complex has higher affinity to methylated DNA than MBD2b alone [39]. In the MIRA procedure, the fragmentized (restriction enzyme-cut or sonicated) genomic DNA is incubated with the bacterially-expressed and purified GST-MBD2b and His-MBD3L1 proteins. The high affinity MBD2b/MBD3L1 complex specifically binds to the methylated genomic DNA fragments and, since MBD2b is GST-tagged, it is easy to purify the complex containing methylated DNA by applying glutathione-coated magnetic beads [39,43–45] (Fig. 9.9). MIRA does not depend on restriction enzyme recognition sites or sodium bisulfite conversion of the DNA and it works on double-stranded DNA. There is no dependence on DNA sequences other than that it requires a minimum of two methylated CpGs in the captured fragment. MIRA can reliably be performed with a few hundred nanograms of genomic DNA. Recently, the MIRA technique has been used to profile DNA methylation patterns at a resolution of 100 base pairs in the entire genome of normal human B lymphocytes [46]. MIRA analysis is compatible with different microarray platforms [43,44] and with high-throughput DNA sequencing on the Illumina Genome Analyzer (T.A. Rauch, X. Zhong, H. Gao, and G.P. Pfeifer, unpublished data).

143

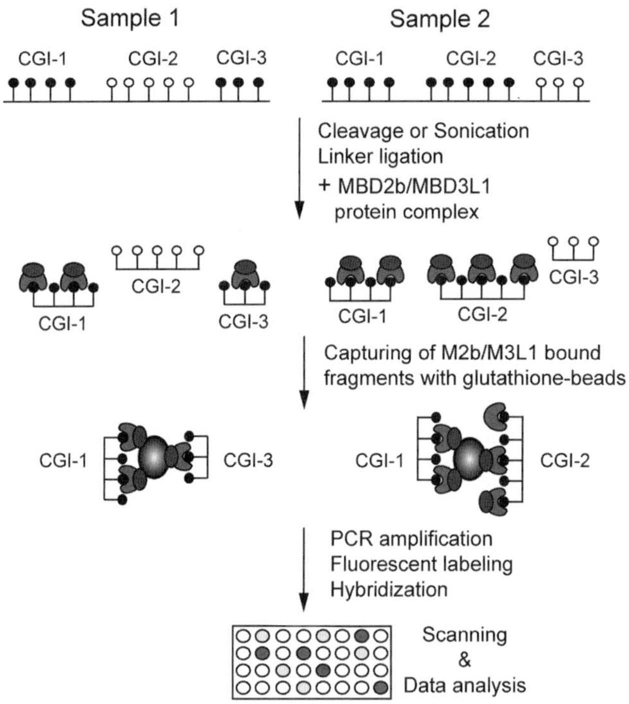

FIGURE 9.9
Methylated-CpG island recovery assay (MIRA). The steps of the MIRA procedure are described in the text. Filled and open lollipops mark methylated and unmethylated CpG sites, respectively. (Please refer to color plate section)

144

FUTURE DIRECTIONS AND CHALLENGES

The many different DNA methylation profiling techniques described in this review will continue to be applied in various settings of molecular analysis and probably will at some point move into the clinical arena for disease diagnosis and treatment stratification. The particular method of choice depends on many parameters including the scope of analysis that is desired, the exact questions to be pursued, and sample size, as well as cost. It is expected that the technological development of high-throughput approaches will continue at an accelerated pace allowing cost effective analysis of large sample series. The current challenges surrounding whole genome bisulfite sequencing will probably be solved in the near future. However, it may not always be desirable or necessary to have information on the methylation status of each and every one of the ~28 million CpG sites in the haploid human genome. For determining the number of methylated genes, in cancer for example, a lower resolution approach will be fully sufficient, and for disease diagnosis, only a subset of specific genes will most likely be useful.

A recent technological challenge affecting almost all types of DNA methylation analysis has emerged from the discovery of 5-hydroxymethylcytosine in mammalian DNA [47,48]. This base modification, hitherto thought to be only a product of oxidative DNA damage, is in fact produced by an enzymatic pathway involving the TET1 hydroxylase, appears to be quite abundant in certain mammalian tissues, and may have regulatory roles distinct from that of 5-methylcytosine. It should be noted that 5-hydroxymethylcytosine prevents cleavage by certain (and probably most) methylation-sensitive restriction endonucleases making it indistinguishable from 5-methylcytosine [49]. Moreover, it is likely that 5-hydroxymethylcytosine in DNA is resistant to bisulfite-induced deamination similarly to 5-methylcytosine. In fact, it has been shown that 5-hydroxymethylcytosine reacts with bisulfite and, instead of leading to deamination, this reaction gives rise to cytosine 5-methylenesulfonate as the product [50]. Cytosine 5-methylenesulfonate was only very slowly deaminated by treatment with bisulfite. 5-hydroxymethylcytosine may or may not be recognized by anti-5-methylcytosine antibodies and by proteins that bind to methylated CpGs [51]. Using existing bisulfite-based technology, it is practically impossible to distinguish these two base modifications in mammalian DNA [51].

Entirely new technologies are on the horizon. Single molecule nanopore DNA sequencing is a promising new technique in which 5-methylcytosine can be distinguished from the other four standard DNA bases [52]. The data suggest that this system may be compatible with an approach where individual nucleotides from a DNA fragment are released continuously by an exonuclease and identified directly without fluorescent labeling.

References

1. Wyatt GR. Occurrence of 5-methylcytosine in nucleic acids. *Nature* 1950;**166**:237–8.

2. Holliday R, Pugh JE. DNA modification mechanisms and gene activity during development. *Science* 1975;**187**:226–32.

3. Riggs AD. X inactivation, differentiation, and DNA methylation. *Cytogenet Cell Genet* 1975;**14**:9–25.

4. Hatada I, Hayashizaki Y, Hirotsune S, Komatsubara H, Mukai T. A genomic scanning method for higher organisms using restriction sites as landmarks. *Proc Natl Acad Sci USA* 1991;**88**:9523–7.

5. Costello JF, Fruhwald MC, Smiraglia DJ, Rush LJ, Robertson GP, Gao X, et al. Aberrant CpG-island methylation has non-random and tumour-type-specific patterns. *Nat Genet* 2000;**24**:132–8.

6. Dai Z, Weichenhan D, Wu YZ, Hall JL, Rush LJ, Smith LT, et al. An AscI boundary library for the studies of genetic and epigenetic alterations in CpG islands. *Genome Res* 2002;**12**:1591–8.

7. Ando Y, Hayashizaki Y. Restriction landmark genomic scanning. *Nat Protoc* 2006;**1**:2774–83.

8. Smiraglia DJ, Kazhiyur-Mannar R, Oakes CC, Wu YZ, Liang P, Ansari T, et al. Restriction landmark genomic scanning (RLGS) spot identification by second generation virtual RLGS in multiple genomes with multiple enzyme combinations. *BMC Genomics* 2007;**8**:446.

9. Gonzalgo ML, Liang G, Spruck CH 3rd, Zingg JM, Rideout WM 3rd, Jones PA. Identification and characterization of differentially methylated regions of genomic DNA by methylation-sensitive arbitrarily primed PCR. *Cancer Res* 1997;**57**:594–9.

10. Huang TH, Laux DE, Hamlin BC, Tran P, Tran H, Lubahn DB. Identification of DNA methylation markers for human breast carcinomas using the methylation-sensitive restriction fingerprinting technique. *Cancer Res* 1997;**57**:1030–4.

11. Ushijima T, Morimura K, Hosoya Y, Okonogi H, Tatematsu M, Sugimura T, et al. Establishment of methylation-sensitive-representational difference analysis and isolation of hypo- and hypermethylated genomic fragments in mouse liver tumors. *Proc Natl Acad Sci USA* 1997;**94**:2284–9.

12. Huang TH, Perry MR, Laux DE. Methylation profiling of CpG islands in human breast cancer cells. *Hum Mol Genet* 1999;**8**:459–70.

13. Khulan B, Thompson RF, Ye K, Fazzari MJ, Suzuki M, Stasiek E, et al. Comparative isoschizomer profiling of cytosine methylation: the HELP assay. *Genome Res* 2006;**16**:1046–55.

14. Oda M, Glass JL, Thompson RF, Mo Y, Olivier EN, Figueroa ME, et al. High-resolution genome-wide cytosine methylation profiling with simultaneous copy number analysis and optimization for limited cell numbers. *Nucleic Acids Res* 2009;**37**:3829–39.

15. Toyota M, Ho C, Ahuja N, Jair KW, Li Q, Ohe-Toyota M, et al. Identification of differentially methylated sequences in colorectal cancer by methylated CpG island amplification. *Cancer Res* 1999;**59**:2307–12.

16. Estecio MR, Yan PS, Ibrahim AE, Tellez CS, Shen L, Huang TH, et al. High-throughput methylation profiling by MCA coupled to CpG island microarray. *Genome Res* 2007;**17**:1529–36.

17. Nouzova M, Holtan N, Oshiro MM, Isett RB, Munoz-Rodriguez JL, List AF, et al. Epigenomic changes during leukemia cell differentiation: analysis of histone acetylation and cytosine methylation using CpG island microarrays. *J Pharmacol Exp Ther* 2004;**311**:968–81.

18. Lippman Z, Gendrel AV, Colot V, Martienssen R. Profiling DNA methylation patterns using genomic tiling microarrays. *Nat Methods* 2005;**2**:219–24.

19. Hu M, Yao J, Cai L, Bachman KE, van den Brule F, Velculescu V, et al. Distinct epigenetic changes in the stromal cells of breast cancers. *Nat Genet* 2005;**37**:899–905.

20. Schumacher A, Kapranov P, Kaminsky Z, Flanagan J, Assadzadeh A, Yau P, et al. Microarray-based DNA methylation profiling: technology and applications. *Nucleic Acids Res* 2006;**34**:528–42.

21. Hayatsu H, Wataya Y, Kazushige K. The addition of sodium bisulfite to uracil and to cytosine. *J Am Chem Soc* 1970;**92**:724–6.

22. Clark SJ, Harrison J, Paul CL, Frommer M. High sensitivity mapping of methylated cytosines. *Nucleic Acids Res* 1994;**22**:2990–7.

23. Hayatsu H. Discovery of bisulfite-mediated cytosine conversion to uracil, the key reaction for DNA methylation analysis – a personal account. *Proc Jpn Acad Ser B Phys Biol Sci* 2008;**84**:321–30.

24. Frommer M, McDonald LE, Millar DS, Collis CM, Watt F, Grigg GW, et al. A genomic sequencing protocol that yields a positive display of 5-methylcytosine residues in individual DNA strands. *Proc Natl Acad Sci USA* 1992;**89**:1827–31.

25. Eckhardt F, Lewin J, Cortese R, Rakyan VK, Attwood J, Burger M, et al. DNA methylation profiling of human chromosomes 6, 20 and 22. *Nat Genet* 2006;**38**:1378–85.

26. Cokus SJ, Feng S, Zhang X, Chen Z, Merriman B, Haudenschild CD, et al. Shotgun bisulphite sequencing of the Arabidopsis genome reveals DNA methylation patterning. *Nature* 2008;**452**:215–19.

27. Lister R, Pelizzola M, Dowen RH, Hawkins RD, Hon G, Tonti-Filippini J, et al. Human DNA methylomes at base resolution show widespread epigenomic differences. *Nature* 2009;**462**:315–22.

28. Meissner A, Mikkelsen TS, Gu H, Wernig M, Hanna J, Sivachenko A, et al. Genome-scale DNA methylation maps of pluripotent and differentiated cells. *Nature* 2008;**454**:766–70.

29. Ball MP, Li JB, Gao Y, Lee JH, LeProust EM, Park IH, et al. Targeted and genome-scale strategies reveal gene-body methylation signatures in human cells. *Nat Biotechnol* 2009;**27**:361–8.

30. Deng J, Shoemaker R, Xie B, Gore A, LeProust EM, Antosiewicz-Bourget J, et al. Targeted bisulfite sequencing reveals changes in DNA methylation associated with nuclear reprogramming. *Nat Biotechnol* 2009;**27**:353–60.

31. Lister R, Ecker JR. Finding the fifth base: genome-wide sequencing of cytosine methylation. *Genome Res* 2009;**19**:959–66.

32. Ehrich M, Nelson MR, Stanssens P, Zabeau M, Liloglou T, Xinarianos G, et al. Quantitative high-throughput analysis of DNA methylation patterns by base-specific cleavage and mass spectrometry. *Proc Natl Acad Sci USA* 2005;**102**:15785–90.

33. Tost J, Gut IG. DNA methylation analysis by pyrosequencing. *Nat Protoc* 2007;**2**:2265–75.

34. Vaissiere T, Cuenin C, Paliwal A, Vineis P, Hoek G, Krzyzanowski M, et al. Quantitative analysis of DNA methylation after whole bisulfitome amplification of a minute amount of DNA from body fluids. *Epigenetics* 2009;**4**:221–30.

35. Bibikova M, Fan JB. GoldenGate assay for DNA methylation profiling. *Methods Mol Biol* 2009;**507**:149–63.

36. Rauch T, Pfeifer GP. Methylated-CpG island recovery assay: a new technique for the rapid detection of methylated-CpG islands in cancer. *Lab Invest* 2005;**85**:1172–80.

37. Weber M, Davies JJ, Wittig D, Oakeley EJ, Haase M, Lam WL, et al. Chromosome-wide and promoter-specific analyses identify sites of differential DNA methylation in normal and transformed human cells. *Nat Genet* 2005;**37**:853–62.

38. Gebhard C, Schwarzfischer L, Pham TH, Schilling E, Klug M, Andreesen R, et al. Genome-wide profiling of CpG methylation identifies novel targets of aberrant hypermethylation in myeloid leukemia. *Cancer Res* 2006;**66**:6118–28.

39. Rauch T, Li H, Wu X, Pfeifer GP. MIRA-assisted microarray analysis, a new technology for the determination of DNA methylation patterns, identifies frequent methylation of homeodomain-containing genes in lung cancer cells. *Cancer Res* 2006;**66**:7939–47.

40. Illingworth R, Kerr A, Desousa D, Jorgensen H, Ellis P, Stalker J, et al. A novel CpG island set identifies tissue-specific methylation at developmental gene loci. *PLoS Biol* 2008;**6**:e22.

41. Cross SH, Charlton JA, Nan X, Bird AP. Purification of CpG islands using a methylated DNA binding column. *Nat Genet* 1994;**6**:236–44.

42. Zhang X, Yazaki J, Sundaresan A, Cokus S, Chan SW, Chen H, et al. Genome-wide high-resolution mapping and functional analysis of DNA methylation in *Arabidopsis*. *Cell* 2006;**126**:1189–201.

43. Rauch T, Wang Z, Zhang X, Zhong X, Wu X, Lau SK, et al. Homeobox gene methylation in lung cancer studied by genome-wide analysis with a microarray-based methylated CpG island recovery assay. *Proc Natl Acad Sci USA* 2007;**104**:5527–32.

44. Rauch TA, Zhong X, Wu X, Wang M, Kernstine KH, Wang Z, et al. High-resolution mapping of DNA hypermethylation and hypomethylation in lung cancer. *Proc Natl Acad Sci USA* 2008;**105**:252–7.

45. Rauch TA, Pfeifer GP. The MIRA method for DNA methylation analysis. *Methods Mol Biol* 2009;**507**:65–75.

46. Rauch TA, Wu X, Zhong X, Riggs AD, Pfeifer GP. A human B cell methylome at 100-base pair resolution. *Proc Natl Acad Sci USA* 2009;**106**:671–8.

47. Kriaucionis S, Heintz N. The nuclear DNA base 5-hydroxymethylcytosine is present in Purkinje neurons and the brain. *Science* 2009;**324**:929–30.

48. Tahiliani M, Koh KP, Shen Y, Pastor WA, Bandukwala H, Brudno Y, et al. Conversion of 5-methylcytosine to 5-hydroxymethylcytosine in mammalian DNA by MLL partner TET1. *Science* 2009;**324**:930–5.

49. Tardy-Planechaud S, Fujimoto J, Lin SS, Sowers LC. Solid phase synthesis and restriction endonuclease cleavage of oligodeoxynucleotides containing 5-(hydroxymethyl)-cytosine. *Nucleic Acids Res* 1997;**25**:553–9.

50. Hayatsu H, Shiragami M. Reaction of bisulfite with the 5-hydroxymethyl group in pyrimidines and in phage DNAs. *Biochemistry* 1979;**18**:632–7.

51. Jin S.-G, Kadam S, Pfeifer GP. Examination of the specificity of DNA methylation profiling techniques towards 5-methylcytosine and 5-hydroxymethylcytosine. *Nucleic Acids Res* 2010; in press.

52. Clarke J, Wu HC, Jayasinghe L, Patel A, Reid S, Bayley H. Continuous base identification for single-molecule nanopore DNA sequencing. *Nat Nanotechnol* 2009;**4**:265–70.

Methylation of Lysine 9 of Histone H3: Role of Heterochromatin Modulation and Tumorigenesis[†]

Fei Chen, Hong Kan, and Vince Castranova
Laboratory of Cancer Signaling and Epigenetics, Health Effects Laboratory Division, Pathology and Physiology Research Branch, National Institute for Occupational Safety and Health, Morgantown, WV 26505, USA

149

INTRODUCTION

The basic building blocks of the mammalian genome are the nucleosomal core particles composed of 146 to 147 bp of DNA wrapped around a histone octamer containing two copies each of histone H2A, H2B, H3, and H4. The histone H3 and H4 hold highly dynamic N-terminal ends rich in basic amino acids that are subjected to epigenetic modifications including reversible methylation, acetylation, ubiquitination, and phosphorylation [1]. These modifications may act individually, sequentially, or in combination to form a "histone code" that regulates the assembly of nucleosome beads into higher order chromatin structures through interaction with other regulatory proteins.

Based on the degree of chromatin condensation, the nuclear chromatin can be generally categorized into two different types, heterochromatin and euchromatin [2]. Heterochromatin is highly condensed and silent in transcription. It is also relatively stable throughout different cell cycle phases [3]. In contrast, euchromatin is decondensed in G1, S, and G2 phases of the cell cycle. The genes in the euchromatin region are actively transcribed either constitutively or inducibly. Heterochromatin can be subdivided into constitutive heterochromatin and facultative heterochromatin. It is unclear whether there is a third type of heterochromatin, intermediate or transient heterochromatin, because of lack of detailed structural and biochemical information. The initiation, propagation and maintenance of heterochromatin is largely controlled by the formation of trimethylation of the lysine 9 on

[†]**Disclaimer:** The opinions expressed in this manuscript are those of the authors and do not necessarily represent the views of the National Institute for Occupational Safety and Health, Center for Disease Control and Prevention of the USA.

Handbook of Epigenetics: The New Molecular and Medical Genetics. DOI: 10.1016/B978-0-12-375709-8.00010-1

histone H3 (H3K9me3), along with some other synergistic epigenetic modifications [4]. Considering the importance in controlling the packing and usage of genetic information, it has been recently recognized that alteration in chromatin structure due to epigenetic changes may be an event essential for cancer development [5].

SUBTYPES OF HETEROCHROMATIN
Constitutive Heterochromatin

The constitutive heterochromatin primarily encompasses the regions containing a high density of repetitive DNA elements, such as clusters of satellite sequences and transposable elements at centromeres, pericentric foci, and telomeres [3]. The transposable elements (transponsons), which are abundant in the human genome, are highly mutagenic because of their ability to target protein-coding genes for insertion, causing chromosome breakage and promoting illegitimate genome rearrangement [6]. Thus, the constitutive heterochromatin maintained by H3K9me3 is pivotal for genomic integrity by preventing abnormal chromosome segregation, recombination, and DNA replication.

Facultative Heterochromatin

The facultative heterochromatin, on the other hand, is found mainly at the developmentally regulated loci, rRNA gene, or, under certain circumstances, at the transcriptionally active region of euchromatin, where the chromatin state can change in response to cellular signals [3]. Through interplay with DNA methylation, facultative heterochromatin is the key for normal cell lineage development and cell differentiation by somatic methylation and inactivation of the germline-specific genes [7]. In addition, the facultative heterochromatin appears to be responsible for the allelic exclusion, genomic imprinting, or inactivation of the X chromosome and the gene loci of immunoglobulins (Igh/Igκ) and T-cell receptor-α and -β [8–10]. The allelic exclusion of Igh and Igκ was attributed to their reposition to and associate with centromeric constitutive heterochromatin in the nucleus [11]. It is mechanistically plausible, thus, that the constitutive heterochromatin in the centromeric region might be able to instruct the formation of the facultative heterochromatin on either the same chromatin fiber (*cis*) or a different chromatin fiber (*trans*). Such instructive process may be made through nuclear compartmentalization and intra- and inter-chromosomal interactions, leading to silencing of the long-distance or intergenetic genes, such as immunoglobulins and T-cell receptors [12–14].

In general, both types of heterochromatin initiated and maintained by H3K9me3 are repressive for gene transcription by antagonizing lysine 9 acetylation, serine 10 phosphorylation, and tri-methylation of lysine 4 on the histone H3 (H3K4me3), the markers for active gene transcription. Accordingly, any measure that impedes H3K9me3 will change the overall nucleosome dynamics and result in genomic instability, de-differentiation of the cells, aberrant gene expression, and, consequently, the tumorigenic transformation.

Intermediate or Transient Heterochromatin

When both constitutive heterochromatin and facultative heterochromatin occur in a large region of the chromosomes and are cytologically stable, an addition type of heterochromatin, intermediate or transient heterochromatin, has been mentioned in a few reports recently [15]. Unlike the constitutive heterochromatin and facultative heterochromatin, this type of heterochromatin largely occurs within the actively transcribing euchromatin region or a single gene locus. It is not surprising, therefore, that intermediate heterochromatin has features of both the inhibiting chromatin markers, such as H3K9me3, H3K9me2, and H3K27me3, and the active chromatin markers including H3K4me3 and acetylations of H3K9. Functionally, this type of heterochromatin may fine-tune the expression of the given genes through inducible epigenetic changes on a few nucleosomes.

Methylation of Lysine 9 of Histone H3: Role of Heterochromatin Modulation and Tumorigenesis

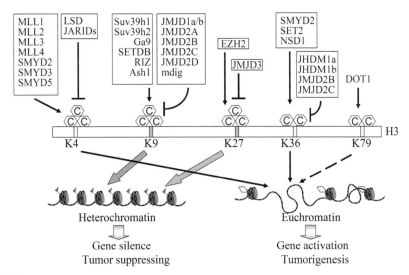

FIGURE 10.1

Regulation of histone H3 methylation and chromatin structure. The lysine (K) residues 4, 9, 27, 36, and 79 can be methylated by the addition of methyl groups, CH3 (C), by methyltransferases (enclosed in rectangle box and indicated by arrows). Several histone demethylases (enclosed in oval box and indicated by the T bars) relatively specific for the individual methylated lysine are depicted also. Tri-methylation of the lysine 9 and/or lysine 27 facilitates formation of the heterochromatin to suppress gene expression and maintain genomic stability. In contrast, tri-methylation of lysines 4, 36, and possibly 79 induces formation of the euchromatin structure for active transcription of the genes (see text for the details).

HISTONE H3 METHYLATION AND HETEROCHROMATIN

Post-translational modifications of the key lysine residues on histone proteins, most importantly histone H3, determine the formation of different chromatin structures. Among the well-documented histone modifications, it is fair to say that methylation of the N-terminal lysine residues of histone H3 is the most critical in this regard. As mentioned earlier, lysines 4, 9, 27, 36, and 79 of histone H3 can be modified through methyltransferase and demethylases to form mono-, di-, or tri-methylated lysines. Although both mono- and di-methylations are attributed to the chromatin structure and the transcription status of the genes, the most extensively studied histone methylation and epigenetic regulation is the tri-methylation on lysines 4, 9, 27, 36, and 79 of histone H3 (H3K4me3, H3K9me3, H3K27me3, H3K36me3, and H3K79me3) (Fig. 10.1). Accumulating evidence supports the notion that H3K4me3, H3K36me3, and possibly H3K79me3 facilitate the opening of the chromatin configuration to form euchromatin, which is also associated with serine 10 phosphorylation and lysine 9 acetylation of the histone H3 for active transcription of genes [1]. In contrast, H3K9me3 and H3K27me3 are mainly involved in the initiation, propagation, and maintenance of the highly compacted heterochromatin to silence gene expression (Fig. 10.1) [2].

TRI-METHYL LYSINE 9 OF HISTONE H3 (H3K9me3) AND CELL CYCLE

Methylation of lysine 9 on histone H3 (H3K9) is critically regulated by both methyltransferase and demethylases (Fig. 10.1). The most characterized H3K9 methytransferases include G9a for mono- and di-methylation and SUV39h1/h2 for di- and tri-methylation of H3K9. Some other H3K9 methyltransferase, such as SETDB, RIZ1, and Ash1, have also been identified but with less clear biochemical characterization [4]. The methylated H3K9, whether it is H3K9me2 or H3K9me3, can be demethylated by a family of relatively specific H3K9 demethylases, such as JMJD1a/b, JMJD2a/b/c/d, and the recently identified Mdig protein [16,17]. An additional H3K9me3 demethylase is LSD1, a histone demethylase originally identified as a bona fide H3K4, rather than H3K9 demethylase [18]. Demethylation of H3K4me3 by LSD1 will inactivate transcription of genes in the actively transcribed euchromatin region. However, there is evidence suggesting that, when complexed with androgen receptors, LSD1 acts as an H3K9me2 or H3K9me3 demethylase to activate gene transcription [19].

The most characterized function of H3K9me3 is its role in stabilizing constitutive heterochromatin and facultative heterochromatin primarily encompassing the regions abundant in repetitive DNA in chromosomes. However, several recent studies revealed a critical role of H3K9me3 in mediating transcriptional repressor- and miRNA-induced short-term silencing of the actively expressed genes or ectopically expressed exogenous gene [17,20]. Furthermore, it is believed that H3K9me3 determines specific remodeling status of a single nucleosome through altering the epigenetic landscape, possibly by initial nucleation and spreading of the heterochromatin, in an active gene locus. Enhancement of H3K9me3 in this case is repressive for gene transcription. In contrast, attenuation of H3K9me3 will ultimately lead to transcriptional competence, as exemplified by the mitogenic signal-induced HIV transcription in microglial cells and TNFα expression in monocytes [21,22]. In addition to the gene promoter of rRNA as reported by Martens et al. [23] in differentiated embryonic stem cells, repression of S-phase gene expression, mainly the E2F-targeting genes including B-myb, cyclin A, cyclin D1, cyclin E, DHFR, and PCNA, by H3K9me3 has been observed in skeletal myoblasts, HeLa cells, 3T3 cells, and human diploid fibroblasts [24–27]. The association of H3K9me3 with the promoters of the E2F-targeted S-phase genes was considered as an essential step for the cell cycle exit of the cells undergoing differentiation and senescence. An intact Rb pathway is required for the formation of the silent chromatin at the promoters of the E2F-targeted S phase genes, since Rb is able to recruit histone methyltransferase SUV39H1 that induces H3K9me3 [25,27]. Genetic deficiency in either Rb or SUV39H1 is highly tumorigenic due to the loss of the silent chromatin at the promoters of these S phase genes [25,28]. More recent biochemical evidence also indicates that H3K9me3 is able to repress expression of the androgen receptor-targeting genes [29]. An intriguing finding is that H3K9me3 mediated by SUV39h1 is able to attenuate the Tax-induced transactivation of HTLV-1 LTR [20]. In the case of the cell cycle regulatory genes, although it seems unlikely that the S phase genes are repressed in an actively growing cell population, the H3K9me3-based short-term suppression may function as a natural brake to prevent excessive expression of the S-phase genes that otherwise will be tumorigenic. It has been demonstrated that the H3K9me3 is an important mediator for Rb-induced cell cycle exit or differentiation when the cells are committed to specialized lineage development or in earlier G1 phase. It remains unclear, however, whether H3K9me3 is capable of regulating these genes for cell cycle transition in developmentally mature cells, transformed cells, and cells at later G1 phase, G1/S boundary, and S phase in response to either Rb signals or non-Rb signals.

Whether H3K9me3 delays the cell cycle progression or induces irreversible cell cycle exit, such as terminal differentiation and senescence, may be determined by the extent of histone methylation on the nucleosome in the entire genomic region of these S phase or E2F targeting genes. This possible feature of H3K9me3 is reminiscent of the established function of the activating E2F proteins that are needed for cell proliferation as well as being involved in inducing cell differentiation and apoptosis [30]. It is conceivable that in a certain cellular context, the level of H3K9me3 may be lowered below the threshold necessary for cell cycle exit but remain high enough to execute other functions, such as antagonizing the H3K4me3 that is important for the active transcription of the E2F-targeting genes [31]. The key genes regulating S phase include those for DNA synthesis (DNA polymerase α or DHFR) and cell cycle control (cyclin E and cyclin D). Thus, it is very likely that H3K9me3 in the promoter region of these genes prevents excessive expression of these genes and, therefore, delays cell cycle transition and/or proliferation.

H3K9me3 AND TRANSCRIPTIONAL REPRESSORS

Other evidence supporting the role of H3K9me3 in short-term silencing at specific genes in different tissues arises from its capability to recruit transcriptional repressors at the actively transcribed gene loci [32]. It has been demonstrated that the formation of H3K9me3 in any actively expressed gene loci can cause transient repression of the related genes by recruiting some transcriptional repressor complexes, such as Repressor Element 1 Silencing Transcription factor (REST) and CoREST. Both REST and CoREST contain histone deacetylases (HDACs) and the H3K4me3 demethylases, LSD1 and, possibly, Rbp4 [33]. As a result of REST and CoREST recruitment by H3K9me3, the formation of the active gene transcription marker, H3K4me3, and acetylation on lysines 9 and 14 of histone H3 will be blocked, leading to suppression of the gene transcription. The short-term silencing of the actively expressed genes can also be achieved by the microRNA (miRNA)-induced formation and spreading of the silent chromatin, in which the H3K9me3 is an essential mediator for the interaction of Argonaute (Ago) protein in the miRNA machinery and the nucleosome or chromatin fiber [34].

The short-term silencing of active genes by H3K9me3 is partially supported by the most recent genome annotation of the *Drosophila* heterochromatin region, which suggested that a considerable number of genes encoding proteins for membrane cation transporters, DNA binding, protein binding, and protein kinase regulators are located in the silent chromatin regions [35]. Additional evidence suggesting the transient inhibition of H3K9me3 on actively transcribed genes is the fact that both H3K9me2 and H3K9me3 are essential for the binding of the heterochromatin protein 1 (HP1). Tethering of HP1α and HP1β to H3K9me2 or H3K9me3 immediately upstream of a promoter triggers silencing of the gene due to the recruitment of DNA methyltransferases and additional histone methylases that leads to a localized chromatin condensation. Supporting evidence in this regard is reported by Smallwood et al. [36] that indicates transcriptional repression of the anti-apoptotic gene, Survivin, and the cell cycle regulatory genes, including CDC2 and CDC25C, by promoter occupancy of H3K9me2 and HP1 proteins. This observation was further supported by the fact that H3K9me3 and HP1 proteins impair the assembly of the preinitiation complex [37]. Thus, H3K9me3 can be functioning as a natural brake to prevent unnecessary over-transcription of the actively expressed genes. Attenuation of the H3K9me3, by either over-expression of demethylases or deficiency in H3K9 methyltransferases, therefore, will foster sustained expression of the genes involved in either cell cycle transition or proliferation.

H3K9me3 AND CANCER

Certain genetic abnormalities or polymorphisms have been found to be associated with a greater risk of cancer development in humans [38]. It has been widely acknowledged that carcinogenesis is largely determined through gene–environment interactions. The best example for this assumption is tobacco smoking-induced lung cancer. It has been known for a long time that tobacco smoking is a dominant risk factor for lung cancer. However, only a limited percentage, about 11%, of tobacco smokers develop lung cancer [39]. Clearly the genetic signature, which is largely affected by epigenetic modification of the genome, for a given individual, determines the outcome of the gene-environment interaction through affecting the expression and function of the oncogenes, tumor suppressors, checkpoint proteins, and enzymes responsible for the metabolism of carcinogens. A number of genome-wide loss-of-heterozygosity studies have identified activating mutations in K-Ras, c-myc, EGFR, cyclin D1, telomerase, and BCL2, and mutations that compromise the function of p53, p16INK/p14ARF, fragile histidine triad (FHIT), RASFF1A, SEMA3B, and Rb-pathway in cancer patients. In addition, chromosomal comparative genomic hybridization and high-resolution genomic profiles of cancers, especially lung cancer, reveal a high number of recurrent chromosomal aberrations, particularly amplifications and deletions in certain chromosome regions, including 1q36, 3q11.2-12.3, 3q25-27, and 5p13-14, where some oncogenes are localized [40]. Furthermore, various case-control and cohort studies indicate the presence of a rare autosomal dominant gene predisposing to some types of cancers in members of families with cancer history [38,41].

Epigenetic modification of the histone proteins, especially the methylation of the histone H3, in cancer development has been an area of emerging research in recent years. It is now generally accepted that in addition to genetic aberrations, cancer can be viewed as a disease that is initiated by epigenetic changes in which gene expression is altered without genomic abnormalities under many circumstances. These epigenetic changes include loss or gain of methylations in both DNA and histone proteins. Many epigenetic studies of cancers focused on hyper- or hypo-methylation of DNA in the promoter region of tumor suppressors, such as adenomatous polyposis coli (APC), retinoic acid receptor β-2, and H-cadherin [42], whereas investigations on the alternations of histone methylation in cancers are just beginning.

Histone methylation has traditionally been viewed as a permanent mark and an irreversible process. Since the first discovery of histone lysine specific demethylases-1 (LSD1) in 2004 [18], however, such a view has changed recently. Since 2006, several groups independently have identified a JmjC family of histone demethylases that demethylate di- or tri-methylated lysines 4, 9, and 36 on histone H3 [29,43–46]. Although histone methylation has been considered as a key event for gene transcription, it is plausible to speculate that such modification can also regulate DNA replication, recombination, and damage repair [47]. The addition of a methyl group to lysine is unable to change the charges that affect the chromatin structure due to the nature of the small molecular weight of the methyl group. It is very likely, thus, that methylations on the side chains of lysine residues create binding code for regulatory proteins.

There is no compelling evidence suggesting that cancers develop purely by aberrations in histone methylation or its signaling pathways. However, global down-regulation of H3K9me3 has been observed in several types of human cancer, such as colorectal cancer, ovarian cancer, and lung cancer, which arises from either the deficiency of H3K9 methyltransferases or elevated activity or expression of H3K9 demethylases [48,49]. In experimental animal studies, mice with gene deficiency of SUV39, a methytransferase catalyzing tri-methylation of lysine 9 on histone H3, suffer dramatic genome instability and develop B-cell lymphomas due to a substantial loss of H3K9me3 [50]. An increased lung cancer risk has been recently reported in people carrying 1624G > C polymorphism of SUV39h2 [51]. Since this polymorphism occurred in the 3′-UTR region, it is very likely that this polymorphism may reduce the mRNA stability or translational potential of SUV39h2 and, therefore, compromise the level of H3K9me3. Reduced expression of RIZ1, another methyltransferase for H3K9, has been frequently observed in lung cancer, breast cancer, hepatocellular carcinoma, colon cancer, neuroblastoma, and melanoma [52]. The gene encoding RIZ1 is mapped in a chromosomal region that is frequently deleted in human cancer, 1p36 [52]. It is very likely, therefore, that RIZ1 has some tumor suppressor-like activity. In human chronic myelogenous leukemia (CML), the magnitude of RIZ1 reduction is positively correlated with the disease progress from the chronic phase to blast crisis associated with increased cell proliferation, apoptosis resistance, and poor differentiation [53].

In addition to the observed decrease in activity or expression of the H3K9 methyltransferases, several types of human cancers also exhibit over-expression of H3K9 demethylases that reduce the level of H3K9me3 to promote tumorigenesis. The first evidence showing over-expression of H3K9 demethylase in cancer is from the observed amplification of DNA copy number of Gene Amplified in Squamous Cell Carcinoma 1 (GASC1) at chromosome 9p23-24 in cell lines derived from esophageal squamous cell carcinomas (ESCs) [54]. GASC1 was later re-named JMJD2C and confirmed as a H3K9 demethylase to demethylate di- and tri-methylated H3K9 [44,55]. Most recently, over-expression of JMJD2C due to amplification of the DNA copy number has also been demonstrated in human acute myeloid leukemia [56].

The recent studies on histone demethylation and lung cancer provide other evidence showing that increased expression of H3K9 demethylase is possibly responsible for the reduction of H3K9me3 in cancer tissues or cells [17]. The development of human lung cancer caused by tobacco smoking or other environmental hazards involves a complex sequence of interdependent events that include DNA damage, genomic mutations, chromosomal translocation, and re-arrangement [57]. Despite a number of reports suggesting epigenetic abnormalities of the lung cancer cells, mostly focusing on DNA CpG methylation, the contribution of altered histone methylation to cancer development is limited. We and others previously identified a JmjC domain-containing protein, Mineral Dust-Induced Gene (Mdig), from human alveolar macrophages, which is also named Myc-Induced Nuclear Antigen 53 (Mina53) and Nucleolar Protein 52 (NO52) [17,58–60]. In the latest studies, we provided biochemical and clinical evidence showing that Mdig is a potential H3K9me3 demethylase that is over-expressed in lung cancer [17]. Increased levels of Mdig protein were detected in 16 out of 19 human lung cancer tissues as compared to the case-matched noncancerous lung tissues, which correlate with a decreased level of H3K9me3. Other evidence clearly implicated Mdig protein as a member of the JmjC domain-containing histone demethylase family that is capable of reducing the level of H3K9me3 after over-expression in the cells derived from the lung or bronchus. Furthermore, Mdig protein and mRNA can be induced in lung cells by components of tobacco smoking and occupational mineral dust (Lu et al., unpublished observation). Taken together, all of these findings highlight the critical role of H3K9me3 demethylation in the development of lung cancer.

PERSPECTIVES AND CONCLUSION

As discussed above, the functional hallmarks of H3K9me3 include establishment and maintenance of constitutive and facultative heterochromatins, recruitment of the transcriptional repressor complexes at the active gene loci, and contribution to the microRNA-induced silent chromatin formation over a normally expressed euchromatic locus. Accordingly, it is mechanistically plausible to assume that any improper regulation of H3K9me3 will alter the chromatin structure, the stability of the transposons in the genome and the accessibility of the genome by the environmental DNA damaging or mutagenic agents. Because of its importance in the overall dynamics of the genome, abnormalities of the regulation of H3K9me3 will certainly make a causative contribution to both early and subsequent stages of cancer development. An intriguing question to be asked is whether changes in H3K9me3 can serve as a biomarker for diagnosis and prognosis of cancers. Many challenges still remain in understanding its function and regulation in malignant transformation of cells, carcinogenesis of the tissue, and tumorigenesis. Furthermore, there is an urgent need to assess the potential of targeting H3K9me3 or its regulatory signaling in cancer therapy.

References

1. Shi Y. Histone lysine demethylases: emerging roles in development, physiology and disease. *Nat Rev Genet* 2007;**8**:829–33.

2. Shilatifard A. Chromatin modifications by methylation and ubiquitination: implications in the regulation of gene expression. *Annu Rev Biochem* 2006;**75**:243–69.

3. Grewal SI, Jia S. Heterochromatin revisited. *Nat Rev Genet* 2007;**8**:35–46.

4. Klose RJ, Zhang Y. Regulation of histone methylation by demethylimination and demethylation. *Nat Rev Mol Cell Biol* 2007;**8**:307–18.

5. Zhang K, Dent SY. Histone modifying enzymes and cancer: going beyond histones. *J Cell Biochem* 2005;**96**:1137–48.

6. Henikoff S. Heterochromatin function in complex genomes. *Biochim Biophys Acta* 2000;**1470**:1–8.

7. Feldman N, Gerson A, Fang J, Li E, Zhang Y, Shinkai Y, et al. G9a-mediated irreversible epigenetic inactivation of Oct-3/4 during early embryogenesis. *Nat Cell Biol* 2006;**8**:188–94.

8. Heard E. Delving into the diversity of facultative heterochromatin: the epigenetics of the inactive X chromosome. *Curr Opin Genet Dev* 2005;**15**:482–9.

9. Corcoran AE. Immunoglobulin locus silencing and allelic exclusion. *Semin Immunol* 2005;**17**:141–54.

10. Skok JA, Gisler R, Novatchkova M, Farmer D, de Laat W, Busslinger M. Reversible contraction by looping of the Tcra and Tcrb loci in rearranging thymocytes. *Nat Immunol* 2007;**8**:378–87.

11. Roldan E, Fuxa M, Chong W, Martinez D, Novatchkova M, Busslinger M, et al. Locus "decontraction" and centromeric recruitment contribute to allelic exclusion of the immunoglobulin heavy-chain gene. *Nat Immunol* 2005;**6**:31–41.

12. Fraser P, Bickmore W. Nuclear organization of the genome and the potential for gene regulation. *Nature* 2007;**447**:413–17.

13. Talbert PB, Henikoff S. Spreading of silent chromatin: inaction at a distance. *Nat Rev Genet* 2006;**7**:793–803.

14. Lanctot C, Cheutin T, Cremer M, Cavalli G, Cremer T. Dynamic genome architecture in the nuclear space: regulation of gene expression in three dimensions. *Nat Rev Genet* 2007;**8**:104–15.

15. Nishimura T, Paszkowski J. Epigenetic transitions in plants not associated with changes in DNA or histone modification. *Biochim Biophys Acta* 2007;**1769**:393–8.

16. Agger K, Christensen J, Cloos PA, Helin K. The emerging functions of histone demethylases. *Curr Opin Genet Dev* 2008;**18**:159–68.

17. Lu Y, Chang Q, Zhang Y, Beezhold K, Rojanasakul Y, Zhao H, et al. Lung cancer-associated JmjC domain protein mdig suppresses formation of tri-methyl lysine 9 of histone H3. *Cell Cycle* 2009;**8**:2101–9.

18. Shi Y, Lan F, Matson C, Mulligan P, Whetstine JR, Cole PA, et al. Histone demethylation mediated by the nuclear amine oxidase homolog LSD1. *Cell* 2004;**119**:941–53.

19. Metzger E, Wissmann M, Yin N, Muller JM, Schneider R, Peters AH, et al. LSD1 demethylates repressive histone marks to promote androgen-receptor-dependent transcription. *Nature* 2005;**437**:436–9.

20. Kamoi K, Yamamoto K, Misawa A, Miyake A, Ishida T, Tanaka Y, et al. SUV39H1 interacts with HTLV-1 Tax and abrogates Tax transactivation of HTLV-1 LTR. *Retrovirology* 2006;**3**:5.

21. Marban C, Suzanne S, Dequiedt F, de Walque S, Redel L, Van Lint C, et al. Recruitment of chromatin-modifying enzymes by CTIP2 promotes HIV-1 transcriptional silencing. *EMBO J* 2007;**26**:412–23.

22. Sullivan KE, Reddy AB, Dietzmann K, Suriano AR, Kocieda VP, Stewart M, et al. Epigenetic regulation of tumor necrosis factor alpha. *Mol Cell Biol* 2007;**27**:5147–60.

23. Martens JH, O'Sullivan RJ, Braunschweig U, Opravil S, Radolf M, Steinlein P, et al. The profile of repeat-associated histone lysine methylation states in the mouse epigenome. *EMBO J* 2005;**24**:800–12.

24. Ait-Si-Ali S, Guasconi V, Fritsch L, Yahi H, Sekhri R, Naguibneva I, et al. A Suv39h-dependent mechanism for silencing S-phase genes in differentiating but not in cycling cells. *EMBO J* 2004;**23**:605–15.

25. Narita M, Nunez S, Heard E, Narita M, Lin AW, Hearn SA, et al. Rb-mediated heterochromatin formation and silencing of E2F target genes during cellular senescence. *Cell* 2003;**113**:703–16.

26. Nicolas E, Roumillac C, Trouche D. Balance between acetylation and methylation of histone H3 lysine 9 on the E2F-responsive dihydrofolate reductase promoter. *Mol Cell Biol* 2003;**23**:1614–22.

27. Nielsen SJ, Schneider R, Bauer UM, Bannister AJ, Morrison A, O'Carroll D, et al. Rb targets histone H3 methylation and HP1 to promoters. *Nature* 2001;**412**:561–5.

28. Braig M, Lee S, Loddenkemper C, Rudolph C, Peters AH, Schlegelberger B, et al. Oncogene-induced senescence as an initial barrier in lymphoma development. *Nature* 2005;**436**:660–5.

29. Yamane K, Toumazou C, Tsukada Y, Erdjument-Bromage H, Tempst P, Wong J, et al. JHDM2A, a JmjC-containing H3K9 demethylase, facilitates transcription activation by androgen receptor. *Cell* 2006;**125**:483–95.

30. Stanelle J, Putzer BM. E2F1-induced apoptosis: turning killers into therapeutics. *Trends Mol Med* 2006;**12**:177–85.

31. Tyagi S, Chabes AL, Wysocka J, Herr W. E2F activation of S phase promoters via association with HCF-1 and the MLL family of histone H3K4 methyltransferases. *Mol Cell* 2007;**27**:107–19.

32. Ooi L, Wood IC. Chromatin crosstalk in development and disease: lessons from REST. *Nat Rev Genet* 2007;**8**:544–54.

33. Yang M, Gocke CB, Luo X, Borek D, Tomchick DR, Machius M, et al. Structural basis for CoREST-dependent demethylation of nucleosomes by the human LSD1 histone demethylase. *Mol Cell* 2006;**23**:377–87.

34. Grewal SI, Elgin SC. Transcription and RNA interference in the formation of heterochromatin. *Nature* 2007;**447**:399–406.

35. Smith CD, Shu S, Mungall CJ, Karpen GH. The release 5.1 annotation of *Drosophila melanogaster* heterochromatin. *Science* 2007;**316**:1586–91.

36. Smallwood A, Esteve PO, Pradhan S, Carey M. Functional cooperation between HP1 and DNMT1 mediates gene silencing. *Genes Dev* 2007;**21**:1169–78.

37. Smallwood A, Black JC, Tanese N, Pradhan S, Carey M. HP1-mediated silencing targets Pol II coactivator complexes. *Nat Struct Mol Biol* 2008;**15**:318–20.

38. Matakidou A, Eisen T, Houlston RS. Systematic review of the relationship between family history and lung cancer risk. *Br J Cancer* 2005;**93**:825–33.

39. Lippman SM, Spitz MR. Lung cancer chemoprevention: an integrated approach. *J Clin Oncol* 2001;**19**:S74–82.

40. Tonon G, Wong KK, Maulik G, Brennan C, Feng B, Zhang Y, et al. High-resolution genomic profiles of human lung cancer. *Proc Natl Acad Sci USA* 2005;**102**:9625–30.

41. Minna JD, Roth JA, Gazdar AF. Focus on lung cancer. *Cancer Cell* 2002;**1**:49–52.

42. Zochbauer-Muller S, Gazdar AF, Minna JD. Molecular pathogenesis of lung cancer. *Annu Rev Physiol* 2002;**64**:681–708.

43. Whetstine JR, Nottke A, Lan F, Huarte M, Smolikov S, Chen Z, et al. Reversal of histone lysine trimethylation by the JMJD2 family of histone demethylases. *Cell* 2006;**125**:467–81.

44. Cloos PA, Christensen J, Agger K, Maiolica A, Rappsilber J, Antal T, et al. The putative oncogene GASC1 demethylates tri- and dimethylated lysine 9 on histone H3. *Nature* 2006;**442**:307–11.

45. Fodor BD, Kubicek S, Yonezawa M, O'Sullivan RJ, Sengupta R, Perez-Burgos L, et al. Jmjd2b antagonizes H3K9 trimethylation at pericentric heterochromatin in mammalian cells. *Genes Dev* 2006;**20**:1557–62.

46. Tsukada Y, Fang J, Erdjument-Bromage H, Warren ME, Borchers CH, Tempst P, et al. Histone demethylation by a family of JmjC domain-containing proteins. *Nature* 2006;**439**:811–16.

47. Huyen Y, Zgheib O, Ditullio Jr RA, Gorgoulis VG, Zacharatos P, Petty TJ, et al. Methylated lysine 79 of histone H3 targets 53BP1 to DNA double-strand breaks. *Nature* 2004;**432**:406–11.

48. Espino PS, Drobic B, Dunn KL, Davie JR. Histone modifications as a platform for cancer therapy. *J Cell Biochem* 2005;**94**:1088–102.

49. Hamamoto R, Furukawa Y, Morita M, Iimura Y, Silva FP, Li M, et al. SMYD3 encodes a histone methyltransferase involved in the proliferation of cancer cells. *Nat Cell Biol* 2004;**6**:731–40.

50. Peters AH, O'Carroll D, Scherthan H, Mechtler K, Sauer S, Schofer C, et al. Loss of the Suv39h histone methyltransferases impairs mammalian heterochromatin and genome stability. *Cell* 2001;**107**:323–37.

51. Yoon KA, Hwangbo B, Kim IJ, Park S, Kim HS, Kee HJ, et al. Novel polymorphisms in the SUV39H2 histone methyltransferase and the risk of lung cancer. *Carcinogenesis* 2006;**27**:2217–22.

52. Gibbons RJ. Histone modifying and chromatin remodelling enzymes in cancer and dysplastic syndromes. *Hum Mol Genet* 2005;**14**(Spec No 1):R85–92.

53. Lakshmikuttyamma A, Takahashi N, Pastural E, Torlakovic E, Amin HM, Garcia-Manero G, et al. RIZ1 is potential CML tumor suppressor that is down-regulated during disease progression. *J Hematol Oncol* 2009;**2**:28.

54. Yang ZQ, Imoto I, Fukuda Y, Pimkhaokham A, Shimada Y, Imamura M, et al. Identification of a novel gene, GASC1, within an amplicon at 9p23-24 frequently detected in esophageal cancer cell lines. *Cancer Res* 2000;**60**:4735–9.

55. Katoh M, Katoh M. Identification and characterization of JMJD2 family genes in silico. *Int J Oncol* 2004;**24**:1623–8.

56. Helias C, Struski S, Gervais C, Leymarie V, Mauvieux L, Herbrecht R, et al. Polycythemia vera transforming to acute myeloid leukemia and complex abnormalities including 9p homogeneously staining region with amplification of MLLT3, JMJD2C, JAK2, and SMARCA2. *Cancer Genet Cytogenet* 2008;**180**:51–5.

57. Futreal PA, Kasprzyk A, Birney E, Mullikin JC, Wooster R, Stratton MR, et al. Cancer and genomics. *Nature* 2001;**409**:850–2.

58. Zhang Y, Lu Y, Yuan BZ, Castranova V, Shi X, Stauffer JL, et al. The Human mineral dust-induced gene, mdig, is a cell growth regulating gene associated with lung cancer. *Oncogene* 2005;**24**:4873–82.

59. Tsuneoka M, Koda Y, Soejima M, Teye K, Kimura H. A novel myc target gene, mina53, that is involved in cell proliferation. *J Biol Chem* 2002;**277**:35450–9.

60. Eilbracht J, Kneissel S, Hofmann A, Schmidt-Zachmann MS. Protein NO52 – a constitutive nucleolar component sharing high sequence homologies to protein NO66. *Eur J Cell Biol* 2005;**84**:279–94.

157

Chromatin Modifications Distinguish Genomic Features and Physical Organization of the Nucleus

David A. Wacker[1], Yoon Jung Kim[2], and Tae Hoon Kim[2]
[1]Yale University, School of Medicine, New Haven, CT 06510
[2]Yale University, School of Medicine, Department of Genetics, New Haven, CT 06510, USA

INTRODUCTION

Chromatin immunoprecipitation (ChIP) initially developed by Gilmour and Lis [1,3] and Solomon and Varshavsky [4] involves incubating living cells with formaldehyde to fix proteins to their DNA substrates inside the cells (Fig. 11.1). The crosslinked chromosomes are extracted and fragmented by physical shearing or enzymatic digestion. The specific DNA sequences associated with a particular protein complex are then isolated by immuno-affinity purification using a specific antibody against the protein. The purified DNA fragments are assayed by a variety of molecular techniques, such as Southern blot or polymerase chain reaction (PCR), to determine association of particular DNA sequences with the protein of interest [5,6]. A method to detect protein-DNA interaction sites in the yeast genome using DNA microarrays (ChIP-chip) was introduced in 2000 [7,8]. The commercial availability of high-density oligonucleotide arrays representing the entire human genome have facilitated comprehensive mapping of protein DNA interaction sites by ChIP-chip [9]. This genome-wide approach to investigating protein-DNA interactions was extended by the adaptation of Serial Analysis of Gene Expression (SAGE) [10] technique to analysis of ChIP DNA [11,12]. Subsequently, technological advances in high throughput sequencing [13] (ChIP-seq) have opened a new chapter in ChIP-based analysis of gene control and epigenomics, allowing potentially improved cost effectiveness, and surveillance of the entire genome without the bias introduced by a pre-designed chip. Direct sequencing of ChIP DNA has also made it possible to interrogate a significant fraction of repeat elements in the genome that was technically inaccessible using DNA microarrays.

Chromosome conformation capture (3C), initially pioneered and developed by Dekker and co-workers, allows determination of interaction between any pair of loci across the genome inside intact cells [2,14]. Like ChIP, 3C utilizes formaldehyde crosslinking of intact cells to preserve genomic interaction for analysis (Fig. 11.1). Crosslinked chromatin is digested with a restriction enzyme and digested genomic DNA is ligated under dilute conditions

Handbook of Epigenetics: The New Molecular and Medical Genetics. DOI: 10.1016/B978-0-12-375709-8.00011-3

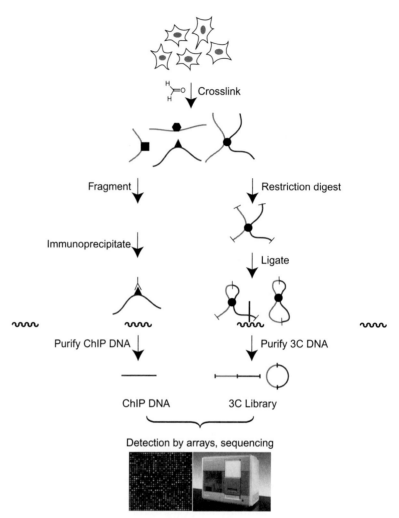

FIGURE 11.1

Overview of ChIP and 3C techniques. Cultured or primary cells are treated with formaldehyde to chemically crosslink DNA-binding proteins to their binding loci *in situ*. The crosslinked nuclei are processed for ChIP as outlined in the left panels or for 3C as outlined in right panels. Nuclear DNA is fragmented to yield DNA segments with associated proteins (squares, triangles, and circles, here) still intact. Proteins of interest (black triangle, here) are selected for by immuno-affinity precipitation, and crosslinks are reversed. The resulting co-precipitated DNA is then analyzed by whole-genome microarray (ChIP-chip) or high-throughput sequencing (ChIP-seq). For analyzing three-dimensional folding and looping, 3C employs crosslinked chromatin that has been restriction digested and ligated under dilute conditions for intramolecular ligation of DNA ends. The ligated DNA junctions are purified and analyzed by PCR, DNA microarrays, or sequencing. (Please refer to color plate section)

that favor intra-molecular ligation of crosslinked chromatin segments. Crosslinks are then reversed and the ligation mixture is purified. 3C yields a genome-wide ligation product library in which each ligation product corresponds to a specific interaction between the two corresponding loci that occur in the nuclei. Several extensions of 3C methods allow genome-wide interrogation of a particular sequence against the entire genome (4C) [15] and comprehensive region-wide mapping of looping interactions (5C) [16]. Combined with imaging techniques, these approaches allow reconstruction of folding and packaging of the genome in the nucleus that influences gene activity and regulation.

FEATURES OF THE EPIGENOME

Various aspects of chromatin function and regulation across the entire genome have been analyzed by ChIP-based methods, and these approaches promise to yield more insights

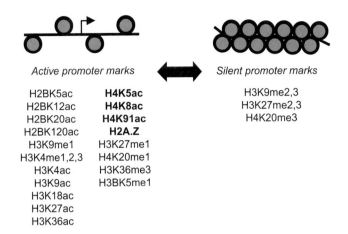

FIGURE 11.2
Enrichment of active and silent promoters for specific histone marks. Outline of the distribution of histone marks to active (left) versus silent (right) promoters. The 17 "backbone" active chromatin marks are shown in bold. me1, me2, me3 = mono-, di-, tri-methyl, respectively. Of note, these represent three separate, independent modifications of a given residue. ac = acetyl. Based on data from Barski et al. [20] and Wang et al. [19].

regarding the dynamic structure and function of the epigenome. Applications have included basal transcription machinery, notably transcription factor IID (TFIID) [9]; a plethora of gene-specific regulators such as estrogen receptor [17], p53 [12], and neuron-restrictive silencing factor (NRSF) [13]; and elements regulating chromatin structure such as CCCTC-binding factor [CTCF] [18]. In addition, these techniques have successfully been applied to map histone modifications and to define epigenetic states across the genome [19,20].

An especially complete map of histone marks in human primary CD4[+] T-cells has been provided by two related studies [19,20] (Fig. 11.2). Because of their global nature, these studies were able to uncover many novel and unexpected functions of histone modifications. While H3K27 and K9 di- and trimethylation (H3K27-me2 and -me3 and H3K9-me2 and -me3) were confirmed as repressive marks, monomethylation of these residues proved to be associated with active promoters. Additionally, H2BK5 monomethylation, a previously unexplored mark, was found to associate downstream of active transcription start sites. Exploration of CTCF binding sites and known active enhancers revealed enrichment for H3K4 methylation and H2A.Z at these elements [20], a finding which has also been demonstrated in CD133[+] hematopoetic progenitor cells [21]. As expected, several of the findings of these studies confirm on a whole-genome scale what was already known or suspected; for example di- and trimethylation of H3K4 are often found within 1 kb of the transcription start site of active genes [20].

Combining the data from these two studies, the authors identify 17 chromatin modifications that concurrently appear at 3286 active promoters (Fig. 11.2) [19]. While the significance of this "backbone" of modifications remains obscure, the elevated expression levels observed of backbone-modified genes, and the relative absence of promoters with only 16 of the 17 modifications, suggests that these form a single functional unit which establishes a transcription-friendly environment. While the discovery of the "backbone" modifications is important, additional experiments, such as sequential ChIP, are needed to confirm and explore the significance of this concordance of modifications that define active euchromatin.

Histone modifications occurring at non-genic stretches of DNA such as pericentromeres, subtelomeres and gene deserts (stretches which are not near centromeres or telomeres, but nonetheless are depleted of genes) have also recently been scrutinized with somewhat unexpected results [22]. While inactivating marks H3K9-me2 and -me3 were abundant in gene deserts, surprisingly all other marks associated with dormant chromatin such as

H3K27me3 were not over-represented. H3K9 di- and tri-methyl marks were also found at pericentromeric DNA as well as centromere-specific mark H4K20me3. H4R3me2 also localized to pericentromeres, which is unexpected in that H4R3 methylation has been identified as an activating mark, although no differentiation has yet been made between mono-, di-, and tri-methylation. Similarly, activating marks such as acetylation of H2AK5 (H2AK5ac) and H3K14 (H3K14ac) were found at subtelomeres, as well as silencing mark H3K27me2/3. A comparison of non-genic regions with regions near weakly, moderately, and strongly expressed genes was also very revealing. As expected, H3K9-me2 and -me3 were most strongly associated with non-genic stretches, and abundance of these modifications decreased with increased expression in genic areas. Indeed, trimethylation of this residue was virtually unique to non-genic regions. Conversely, H3K27me3 was most abundant at silent genic regions, with abundance at non-genic regions approximately equivalent to that found at moderately-expressed genes. This suggests that in this setting, H3K27me3 may be a temporary silencing mark, easily undone upon gene activation, whereas H3K9me3 may play a role in more permanent silencing. It should be noted, though, that in other settings, such as X-chromosome inactivation, H3K27me3 is associated with, though not sufficient for, more permanent silencing [23].

Because histone modifications play a pivotal role in dictating gene expression, cells must have a reliable mechanism for segregating domains with active marks from those with inactive marks. CTCF is a ubiquitously expressed nuclear protein which under various contexts has been shown to act as a transcriptional activator [24], repressor [25], and insulator [26]. Surveying CTCF binding in primary lung fibroblasts (IMR90) by ChIP-chip revealed that half of almost 14,000 binding sites found were located near or in genes, while the other half were dispersed to inter-genic DNA. As CTCF functions as an insulator, there was a paucity of binding sites in regions containing clusters of co-regulated genes, but significant enrichment of CTCF at genes with multiple alternative promoters, where multiple activators may need to be separated from inappropriate targets. Furthermore, a fundamental containment pattern is revealed by analysis of the layout of CTCF sites relative to nearby genes. Essentially, one or more genes (average 2.5) were isolated from neighboring regions by a CTCF binding site on either side.

A number of distinct properties have been ascribed to CTCF, including enhancer blocking and chromatin barrier activity. Chromatin barrier activity of CTCF across the genome was explored using ChIP-seq data generated from HeLa, Jurkat, and CD4$^+$ T-cells [27]. Localization maps for H3K27me3 (repressive) and H2AK5ac (activating) histone modifications were aligned, and the boundaries between functional domains were identified. Incidence of CTCF binding at these boundaries was then assessed, with the finding that 4% (CD4$^+$ T-cell) and 2% (HeLa) of H3K27me3 domains had at least one edge within one kilobase of a CTCF binding site, and ~5% of CTCF binding sites in both cell types were within one kilobase of a domain boundary. Importantly, boundaries of these large extended heterochromatin domains can be missed if the number of mapped sequence tags in the ChIP-seq data set are not enough to saturate the heterochromatin domains, suggesting that perhaps an even greater percentage of CTCF binding sites correspond with boundaries. Nonetheless, no consensus DNA sequence was found to be unique to this limited number of barrier CTCF binding sites, and many of these CTCF barrier binding sites were cell-type specific (58% occurring in only one cell type). These observations suggest that the insulation activity of CTCF might be defined by the genomic context such as activities of neighboring genes and their associated *cis*-regulatory (i.e. from the same chromosome) elements proximal to the insulator elements.

It has been thought that cell type-specific programming is a result of differential engagement of regulatory elements, such as enhancers and silencers, with the promoters. However, discriminating between promoters and enhancers has been difficult due to their similar molecular and epigenetic features (nucleosome-free regions, DNase hypersensitivity, mono-, di-, or tri-methylated H3K4) [28,29]. Several attempts have been made to map generalized enhancers by sequence motif, but these have met with variable success [30,31]. Recently several global studies have been performed using ChIP-chip to map putative enhancers in the ENCODE regions of the genome without prior knowledge of the transcription factors involved. The histone acetyltransferase p300/CBP was initially identified as one of several factors recruited to the enhancer region upon gene activation [28,32]. Bing Ren and co-workers [33] analyzed p300 locations in HeLa cells before and after treatment with interferon gamma (IFN-γ). They found that the p300 binding sites resembled the general features of enhancers and these sites show enhancer activity in reporter assays. In addition, they were able to discriminate between enhancers and promoters by absence of H3K4me3 at p300 sites, but a strong presence of mono-methylated H3K4. Extending this strategy of mapping transcriptional coactivators to identify enhancers, Visel et al. [34] mapped p300 binding sites in mouse embryo forebrain, midbrain, and limb using ChIP-seq and evaluated their tissue-specific activity by reporter assay. They found the vast majority of identified enhancers were tissue-specific, demonstrating the feasibility of mapping regulatory controls using ChIP for particular tissues throughout multiple stages of development. Extending the enhancer identification strategy using differentiating histone marks, Bing Ren and co-workers [35] performed ChIP-chip on chromatin marks in a different set of five cell lines to identify enhancers across the entire human genome. They observed that most active enhancers were specific to one cell type. The enhancer maps thus far generated will be instrumental in identifying aberrant changes in programming in cancer cells as the disease state progresses. Computational and bioinformatic analyses of the comprehensive enhancer maps will also enable reconstruction of *cis*-regulatory network across the entire genome, to classify and refine various *cis*-regulatory modules, and to investigate the developmental and evolutionary landscape of these control elements.

EPIGENOME DURING DIFFERENTIATION

As the regulatory functions of many histone modifications are well defined, their roles in controlling cell differentiation and de-differentiation programs are beginning to be investigated with a greater interest and intensity. A number of recent genome-wide studies have analyzed the role of histone and DNA modifications in controlling the transcription of genes promoting or inhibiting differentiation in human embryonic stem (ES) cells [36–41]. These studies have unmasked global trends in the control of several classes of genes in these cells and their differentiated progeny.

The roles of histone H3K4 and H3K27 trimethylation in regulating expression and potential expression of genes have been recently investigated in mouse ES cells (mES) using ChIP-seq [41], human hES3 cells using ChIP-PET [40], and human H1 ES cells using ChIP-chip [39]. Those promoters associated with only H3K4me3 demonstrated avid expression of their products, with approximately 80% of genes in this group being actively expressed in hES3 cells [40]. Conversely, association with H3K27me3, or with neither K4 nor K27 trimethylation ("neither"), was repressive. Genes in this category showed very low expression, with less than 1% of "neither" genes expressing in hES3 cells [40]. An unexpected product of these studies was the demonstration of a "bivalent" state, in which both H3K4 and H3K27 are trimethylated at the same promoter [42,43]. This pairing of marks occurs throughout the stem cell genome, and is associated with a moderately-expressed state which is poised to resolve quickly to frank activation (H3K4me3-only) or deeper, more permanent silencing (H3K27me3-only or neither mark) [39–41]. In a genome-wide survey of bivalently marked promoters in mES cells upon differentiation to neural progenitor cells, only 8% retained a bivalent mark, while 46% were promoted to H3K4me3 only, and 40% were left with either H3K27me3-only or neither mark. The function of each gene was most deterministic of its fate: genes with functions specific to the new cell type were activated, while unrelated genes were suppressed [39,41]. Because of the global nature of these studies, a breakdown of each promoter class by ontological family was possible. Promoters marked by only H3K4me3 tended to be housekeeping genes such as those associated with proliferation or metabolism [37,39–41]. In contrast, promoters with neither methyl mark enriched for genes expressed only in highly differentiated cell types such as olfactory receptors and immune response factors [39,40]. Bivalent promoters were most often found associated with genes for developmental transcription factors and morphogens [39–41], poised to be activated if cell-type appropriate or more permanently silenced if not. Finally, it should be noted that the exact nature of bivalent markings cannot be fully elucidated by ChIP-based techniques as they are unable to distinguish between two co-existing marks at a single promoter in the same cell and two exclusive marks at a single promoter each arising alone, but in parallel sub-populations of cells.

Complementary to histone modification, methylation of CpG sites on DNA represents another side of the gene-regulation coin, and two recent studies have provided insight into global trends of this modification in ES and differentiated cells. The first sequenced genome-wide in mouse ES cells by bisulphate sequencing [36]; the second employed antibodies against methylated DNA to survey mouse ES cell promoters in a ChIP-chip assay [38]. Two classes of promoters were defined, those with high CpG density (HCP) which often contained a CpG island, and those with low CpG density (LCP). The likelihood of methylation at a specific CpG was inversely correlated with CpG density: at HCP promoters incidence of methylation was less than 1%, whereas ~90% of CpGs at LCPs carried this mark [36]. A similar trend of hypermethylation at CpG-poor promoters and hypomethylation at CpG-islands was also observed in human somatic tissues; however, methylation status of the promoter was not a strict indicator of gene expression levels [44]. Ontologically, greater than 50% of the unmethylated genes coded housekeeping genes, while methylated genes were enriched for specialized products such as sensory receptors and cell signaling molecules [38]. Strikingly, methylation of DNA and histones seems to be anti-correlated: of H3K4me3-modified genes (alone or bivalent), only 40–50% have DNA methylation; conversely 87% of genes lacking both histone marks show CpG methylation [38,41]. Furthermore, the presence of H3K4me2 seems to be a strong predictor of decreased DNA methylation levels [36,44]. From these observations, a global model becomes apparent, in which histone and DNA modifications work synergistically to drive or repress transcription in a developmentally-appropriate way. As HCPs are generally unmethylated, histone modifications play a prominent role in their regulation. Upon differentiation, however, more permanent silencing by DNA methylation occurs where appropriate. Concurrently, at DNA methylation-prone LCPs, regulation is less dependent on histone methylation, and more so on dynamic methylation and demethylation of DNA [36].

A separate study of promoter CpG and H3K27 methylation during differentiation of mouse ES cells to glutamatergic pyramidal neurons further reinforces this model [45]. Specific attention was given in this study to the fate of bivalently-modified, CpG island-containing ES cell promoters. Such promoters were significantly more likely to acquire DNA methylation during differentiation than other promoters, and the likelihood of DNA methylation was linked to histone modification fate: promoters resolving to H3K27me3 only were more likely to gain CpG methylation than those promoters resolving to H3K4me3 only. Furthermore, a great deal of plasticity was observed with regard to gain or loss of H3K27 methylation during differentiation, and in some cases this modification seemed to be used for temporary de-activation of genes. This was particularly true of several genes specific to glutamatergic pyramidal neurons, which acquired H3K27 methylation upon differentiation from ES cell to neuronal progenitor, but then were again demethylated when these progenitors differentiated to their terminal form. These findings again suggest that DNA methylation and histone modification are used in concert to carry out the complicated genetic programming of differentiation.

In addition to their role regulating stem cell differentiation, histone modifications also play an integral part in regulating differentiation further down the developmental tree, for example in hematopoetic stem and progenitor cells. Cui et al. [21] have investigated dynamic patterns of histone modification during differentiation of CD133$^+$ progenitor cells into CD36$^+$ erythrocyte precursors. They found that a significant number of bivalent genes in the progenitor cells retained their bivalent state upon differentiation (24%, whereas in ES cells only 8% remained bivalent), but nonetheless 53% lost the H3K4me3 mark (potentially deactivating) and 19% lost the H3K27me3 mark (potentially activating). Among the up-regulated group were factors important for oxygen transport, while the down-regulated group included factors related to immune response and cellular defense such as IRF1, STAT1, and STAT2. Of note, there were also significant changes to other developmental genes such as HOXA5, 7, 9, and 10 which lost activating marks and acquired repressive H3K27me3, and HOXB5 and 6 which exchanged H3K4me3 for H3K27me3.

In addition to bivalent histone trimethylation, there also seems to be a role for H3K4 dimethylation (H3K4me2) in poising genes for expression in mouse hematopoetic progenitor cell lines (U-EML cells) [37]. As would be expected, H3K4me3- and H3K4me2-associated promoters drove gene expression at a high rate (79% expression). However, genes associated with only H3K4me2 were significantly under-expressed by comparison (21% expression). Of note, many genes related to hematopoiesis fell into this latter group. Upon differentiation to an erythroid line (E-EML), H3K4me2$^+$/H3K4me3$^-$ genes demonstrated a bimodal behavior, with erythroid-specific genes gaining H3K4 trimethylation, and those specific to myeloid and lymphoid cell types remaining H3K4me2$^+$/H3K4me3$^-$ or losing H3K4 methylation altogether. Thus, H3K4 dimethylation provides another mechanism by which genes can be "poised" in multipotent progenitor cells. Interestingly, mouse ES cells lack H3K4me2$^+$/H3K4me3$^-$associated promoters, indicating that these genes must acquire their poised state at some point during the differentiation from stem cells to hematopoietic progenitor cells.

Even after a hematologic cell's lineage has been determined, histone modifications continue to play a role in differentiation of specific subtype. The maturation of naïve CD8$^+$ T-cells to memory T-cells involves increased H3K4 trimethylation at 486 genes, and increased H3K27 trimethylation at 271 [46]. Of those genes gaining H3K4me3, 434 show an immediate up-regulation, but the other 52 remain poised, only increasing in expression once the resting memory T-cell is activated. There is also a role for histone modification in CD4$^+$ cell differentiation, but this role seems more limited. An analysis of modifications in naïve CD4$^+$ cells in comparison with Th1, Th2, Th17, and Treg CD4$^+$ cells reveals that most H3K4 and H3K27 trimethylation islands are conserved among all Th cell types [47]. Only T regulatory cells contained a significant number of unique modification islands. Furthermore, although expression of type-specific signature genes (for example Il17a in Th17 cells) was associated with H3K4me3 modification, the H3K27me3 mark was not generally associated with repression of signature genes specific to other cell types.

CANCER EPIGENOME

More recently, genome-wide ChIP technologies have been used to study somatic cell de-differentiation into neoplastic cells. Much of this effort has been focused on leukemia, as has recently been reviewed [48]. A combined survey of CpG methylation using HELP (*Hpa*II-tiny fragment Enrichment by Ligation-mediated PCR), H3K9 acetylation using ChIP-chip, and global gene expression in 2 ALL (acute lymphoblastic leukemia) strains and 3 AML (acute myeloid leukemia) strains has suggested significant variation in gene expression patterns between these two diseases [49]. One-hundred ninety genes showed different methylation patterns, 374 had differential H3K9 acetylation, and over 1300 had differential expression levels between ALL and AML cells.

In another study, the role of the leukemia-associated MLL-AF4 fusion protein was studied in SEM cells, a B-cell ALL cell line which is known to express this fusion protein due to a t(4,11) translocation [50]. Two-hundred twenty-six binding sites were detected, including 126 previously unrecognized binding sites in gene-associated areas, and including sites near PROM1 (CD133), RUNX1, ETV6 HoxA9, and FLT3 all of which are known hematopoetic transcription factors. The functional effect of this binding was assessed by expression analysis in 132 ALL strains. Of those expressing MLL fusions, over 66% showed elevated expression of these hematopoetic genes, whereas a significantly lower percentage of non-MLL fusion strains over-expressed these genes. Finally, as MLL is known to be an H3K4 methyltransferase associated with transcription elongation, localization of the elongation factor ENL and elongation-associated marks H3K79me2 and H3K4me3 was carried out in SEM cells. ENL was found to be mis-targeted to MLL-fusion binding sites including the HOXA locus, and the histone modifications aberrantly co-localized with 98% of MLL-AF4 binding sites.

Meanwhile, Ke et al. [51] have investigated the role of H3K4 and H3K27 trimethylation patterns in metastatic prostate cancer. Using ChIP-chip with promoter arrays, they identified an approximately similar number of H3K27me3 sites in metastatic prostate cancer line PC3 and benign prostate line EP156T (36,922 and 34,480). Of note, though, while the number of H3K27me3 sites was similar, there was little overlap in the actual locations of this modification. In contrast, the number of H3K4me3 sites in PC3 cells was significantly less than in the benign cells, but those remaining largely overlapped between the two cell types. To demonstrate the effect of modification changes, the expression of affected genes was assayed by microarray. Changes from H3K4me3-only in benign cells to H3K27me3-only in malignant cells or vice versa were the most infrequent changes observed, but were also associated with the greatest change in expression level. In contrast changes from H3K27me3-only to neither mark or H3K4me3-only to both marks were most common, but had only small effects on expression. Interestingly, genes coding cell-to-cell contact machinery were among those with a significant increase in H3K27me3 and decrease in H3K4me3 with corresponding decrease in transcription in metastatic cells. This group includes desmosome-associated products plakophilins, desmoplasmin, PERP, desmocollin, and desmoglein, and gap junction-associated GJB family genes. As these factors function to maintain cohesion between cells, their down-regulation may be necessary for metastatic cells to disjoin from the primary lesion. Finally, as may be expected, various developmental loci also showed changes in expression and histone modification during oncogenesis, including HOXA (down-regulated), and HOXC (up-regulated) [51].

Thus, it seems that genetic programs for differentiation and similarly de-differentiation are at a very fundamental level carried out by epigenetic modifications and de-modifications of DNA and histones, to effect expression or silencing of developmental and proliferative factors. As we learn more about the effects of individual modifications and their exact role in developmental programs, we will be able to more effectively counter malfunctions in the programs, with positive impact on the diseases that these malfunctions cause.

PHYSICAL ORGANIZATION OF THE GENOME

Chromosomes are folded and packaged within the limiting nuclear volume, and their spatial arrangement within the nucleus plays roles in gene regulation, chromosome segregation, genome stability, and other aspects of chromosome behavior. Packaging of the chromosomes into heterochromatin and euchromatin affects not only the activity of the underlying sequence but its location within the nucleus. Heterochromatin is often localized to the periphery of the nucleus and/or in distinct structures known as Polycomb bodies. In contrast, euchromatin is found in the nuclear interior and associates with transcriptional factories and chromatin hubs [52–54]. This general positioning of the chromatin within the nucleus can also be regulated developmentally to generate an inverted orientation of euchromatin at the nuclear periphery and heterochromatin in the nuclear interior [55].

Understanding the mechanisms of packaging and segregating of distinct chromatin domains into large organized centers of transcriptional repression and activation is the next frontier of epigenomics. Technical advances in cellular visualization methods such as fluorescence in situ hybridization (FISH) and immunofluorescence-FISH (immuno-FISH) have enabled the scanning of sequences or regions that are anchored by heterochromatin or euchromatin-forming proteins [56]. Development of chromosome conformation capture (3C) and related technologies has enabled molecular views of nuclear packaging and location [14,57].

The interphase nucleus is well compartmentalized, with chromosomes occupying discrete territories and various regulatory proteins present in specific nuclear bodies [58]. Gene-rich chromosome regions are presented deeper within the nucleus than gene-poor regions, which are closer to the nuclear laminae which line the inner surface of the nuclear envelop [54]. Recently, nuclear lamina associated genomic regions were investigated with DamID (DNA adenine methyltransferase identification), which detects potential protein interaction sites for a protein by analyzing DNA adenine methylation in cells that express the protein of interest fused to the *E. coli* DNA adenine methyltransferase [59]. Consistent with cytological observations, DamID analysis of the nuclear membrane protein Lamin B1 and its associated emerin indicated that a majority of the lamina-associated chromatin is gene-poor, repeat-rich, and enriched in the H3K27me3 silencing mark [60]. Additionally, the genes in these lamina-associated domains were less likely to recruit RNA Polymerase II or activating histone marks, indicating a low transcriptional activity within these domains. In addition, knock-down of Lamin B1 resulted in the gain of AcH3 and almost complete loss of H3K9me2 in these normally heterchromatic regions [61]. These results are consistent with previous observations that genes proximal to the nuclear lamina are often, but not exclusively, transcriptionally repressed [62–64]. Similar to the radial arrangement of gene dense and gene poor chromatin in the nuclei, 3D architectural and spatial interrelationships of different histone lysine methylation sites in various human cell types have been investigated. Trimethyl-H3K4 and monomethyl-H4K20 were largely restricted to active chromatin-containing nuclear zones, while trimethyl-H3K27 and trimethyl-H3K9 were arranged distinct from the nuclear zones, showing clear separation of these regions. [65–67].

Polycomb group (PcG) proteins accumulate in the pericentric heterochromatin structures within the nucleus known as Polycomb bodies, which have already been mentioned. Genes located within these bodies are stably silenced [68,69]. ChIP analyses using antibodies for repressive histone marker H3K27me3 and PcG proteins have determined that their post translational modifications are essential for both the localization to nuclear bodies *in vivo* and transcriptional repression of target genes during development [70–72]. Promyelocytic leukemia protein nuclear bodies (PML-NBs) are discrete nuclear structures analogous to PcG bodies. Recent evidence suggests that PML-NBs are sites of multiple levels of transcriptional regulation [73]. The transcription factors and regulators dynamically localize to these bodies [74,75]. PML-NBs are non-randomly associated with gene-rich and transcriptionally active loci. For example, active genes on the X chromosome interact with PML-NBs more often than their homologs on the inactivated chromosome [76]. PML-NBs are also associated with transcriptional repression. Transcriptional repressors and heterochromatin proteins (HP1) co-localize with these PML-NBs [77,78]. In addition, PML-NBs are found near centromeric regions and are involved in establishing condensed heterochromatin [79–81]. Furthermore, recent studies of the Major Histocompatibility Complex (MHC) locus indicates that PML-NBs can regulate transcription by organizing the locus into different high-order chromatin-loop structures that are actively transcribed [82].

Active RNA polymerases are anchored to a nuclear substructure to form transcription factories and serve as an important organizing center in the nucleus [56,83]. These transcription factories can be visualized with antibodies for the phosphorylated form of RNA polymerase II, which marks the sites of active transcriptional elongation [84]. Recent studies using visual data from FISH and chromosome conformation capture (3C) have

167

shown that tissue-specific genes that are widely separated in the genome can localize to the same transcriptional factory [85,86]. Osborne et al. showed that hybridization signals from β-globin transcription co-localize with signals from Eraf, another erythroid-specific gene, via a *cis* interaction; that is, they are located on the same chromosome, but are separated by 25 Mb. Similarly long-range interactions and co-localizations can also occur between different chromosomes *in trans*.

Differential association and interaction across the genome within specific subnuclear structures are regulated events that influence gene activity and regulation and are exemplified by developmental expression of the murine Th2 cytokine gene and the interferon-γ (IFNγ) gene which is coincident with regulated co-localization of these loci on chromosome 11 and 10 [86]. Another example of regulated co-localization of co-regulated genes involve the proto-oncogene Myc and *immunoglobin heavy chain* (*Igh*) gene. During immediate early (IE) gene induction in mouse B lymphocytes, the *Myc* gene on chromosome 15 in the mouse genome rapidly relocates to the same transcription factory that is transcribing its oncogenic translocation partner *Igh* gene on chromosome 12. This observation suggests that nuclear proximity might be a factor in recurrent genomic rearrangements in cancers [87–89]. These studies highlight various mechanisms that mediate differential localization and long range interactions that underlie transcriptional regulation.

CONCLUSION

The spatial organization of genomic DNA in the nucleus thus represents the next frontier in functional genomics and epigenomics. Extension of 3C analysis to the entire human genome coupled with advances in imaging and FISH techniques promises to provide us with a global understanding of how the entire genome is packaged into the nucleus and how the resulting three dimensional topology of the genome influences its activity. Combined with detailed genome-wide maps of heterochromatin and euchromatin, an unprecedented molecular view of gene activity, epigenetic structure, and 3D organization of the genome will be possible.

ACKNOWLEDGEMENTS

Work in the laboratory is supported by the Rita Allen Foundation, Sidney Kimmel Foundation for Cancer Research, Yale Comprehensive Cancer Center, and Yale University School of Medicine. We thank Spiro Razis for his comments on the manuscript. We apologize to the numerous researchers whose work couldn't be mentioned in this chapter.

References

1. Cecchini KR, Raja Banerjee A, Kim TH. Towards a genome-wide reconstruction of cis-regulatory networks in the human genome. *Semin Cell Dev Biol* 2009;**20**:842–8.

2. Dekker J. Gene regulation in the third dimension. *Science* 2008;**319**(5871):1793–4.

3. Gilmour DS, Lis JT. Detecting protein-DNA interactions in vivo: distribution of RNA polymerase on specific bacterial genes. *Proc Natl Acad Sci USA* 1984;**81**(14):4275–9.

4. Solomon MJ, Varshavsky A. Formaldehyde-mediated DNA-protein crosslinking: a probe for in vivo chromatin structures. *Proc Natl Acad Sci USA* 1985;**82**(19):6470–4.

5. Orlando V, Paro R. Mapping Polycomb-repressed domains in the bithorax complex using in vivo formaldehyde cross-linked chromatin. *Cell* 1993;**75**(6):1187–98.

6. Hecht A, Strahl-Bolsinger S, Grunstein M. Spreading of transcriptional repressor SIR3 from telomeric heterochromatin. *Nature* 1996;**383**(6595):92–6.

7. Ren B, Robert F, Wyrick JJ, Aparicio O, Jennings EG, Simon I, et al. Genome-wide location and function of DNA binding proteins. *Science* 2000;**290**(5500):2306–9.

8. Iyer VR, Horak CE, Scafe CS, Botstein D, Snyder M, Brown PO. Genomic binding sites of the yeast cell-cycle transcription factors SBF and MBF. *Nature* 2001;**409**(6819):533–8.

9. Kim TH, Barrera LO, Zheng M, Qu C, Singer MA, Richmond TA, et al. A high-resolution map of active promoters in the human genome. *Nature* 2005;**436**(7052):876–80.

10. Velculescu VE, Zhang L, Vogelstein B, Kinzler KW. Serial analysis of gene expression. *Science* 1995;**270**(5235):484–7.

11. Roh TY, Ngau WC, Cui K, Landsman D, Zhao K. High-resolution genome-wide mapping of histone modifications. *Nat Biotechnol* 2004;**22**(8):1013–16.

12. Wei CL, Wu Q, Vega VB, Chiu KP, Ng P, Zhang T, et al. A global map of p53 transcription-factor binding sites in the human genome. *Cell* 2006;**124**(1):207–19.

13. Johnson DS, Mortazavi A, Myers RM, Wold B. Genome-wide mapping of in vivo protein-DNA interactions. *Science* 2007;**316**(5830):1497–502.

14. Dekker J, Rippe K, Dekker M, Kleckner N. Capturing chromosome conformation. *Science* 2002;**295**(5558):1306–11.

15. Simonis M, Klous P, Splinter E, Moshkin Y, Willemsen R, de Wit E, et al. Nuclear organization of active and inactive chromatin domains uncovered by chromosome conformation capture-on-chip (4C). *Nat Genet* 2006;**38**(11):1348–54.

16. Dostie J, Richmond TA, Arnaout RA, Selzer RR, Lee WL, Honan TA, et al. Chromosome Conformation Capture Carbon Copy (5C): a massively parallel solution for mapping interactions between genomic elements. *Genome Res* 2006;**16**(10):1299–309.

17. Carroll JS, Meyer CA, Song J, Li W, Geistlinger TR, Eeckhoute J, et al. Genome-wide analysis of estrogen receptor binding sites. *Nat Genet* 2006;**38**(11):1289–97.

18. Kim TH, Abdullaev ZK, Smith AD, Ching KA, Loukinov DI, Green RD, et al. Analysis of the vertebrate insulator protein CTCF-binding sites in the human genome. *Cell* 2007;**128**(6):1231–45.

19. Wang Z, Zang C, Rosenfeld JA, Schones DE, Barski A, Cuddapah S, et al. Combinatorial patterns of histone acetylations and methylations in the human genome. *Nat Genet* 2008;**40**(7):897–903.

20. Barski A, Cuddapah S, Cui K, Roh TY, Schones DE, Wang Z, et al. High-resolution profiling of histone methylations in the human genome. *Cell* 2007;**129**(4):823–37.

21. Cui K, Zang C, Roh TY, Schones DE, Childs RW, Peng W, et al. Chromatin signatures in multipotent human hematopoietic stem cells indicate the fate of bivalent genes during differentiation. *Cell Stem Cell* 2009;**4**(1):80–93.

22. Rosenfeld JA, Wang Z, Schones DE, Zhao K, DeSalle R, Zhang MQ. Determination of enriched histone modifications in non-genic portions of the human genome. *BMC Genomics* 2009;**10**:143.

23. Plath K, Fang J, Mlynarczyk-Evans SK, Cao R, Worringer KA, Wang H, et al. Role of histone H3 lysine 27 methylation in X inactivation. *Science* 2003;**300**(5616):131–5.

24. Vostrov AA, Quitschke WW. The zinc finger protein CTCF binds to the APBbeta domain of the amyloid beta-protein precursor promoter. Evidence for a role in transcriptional activation. *J Biol Chem* 1997;**272**(52):33353–9.

25. Lobanenkov VV, Nicolas RH, Adler VV, Paterson H, Klenova EM, Polotskaja AV, et al. A novel sequence-specific DNA binding protein which interacts with three regularly spaced direct repeats of the CCCTC-motif in the 5′-flanking sequence of the chicken c-myc gene. *Oncogene* 1990;**5**(12):1743–53.

26. Bell AC, West AG, Felsenfeld G. The protein CTCF is required for the enhancer blocking activity of vertebrate insulators. *Cell* 1999;**98**(3):387–96.

27. Cuddapah S, Jothi R, Schones DE, Roh TY, Cui K, Zhao K. Global analysis of the insulator binding protein CTCF in chromatin barrier regions reveals demarcation of active and repressive domains. *Genome Res* 2009;**19**(1):24–32.

28. Hatzis P, Talianidis I. Dynamics of enhancer-promoter communication during differentiation-induced gene activation. *Mol Cell* 2002;**10**(6):1467–77.

29. Bernstein BE, Kamal M, Lindblad-Toh K, Bekiranov S, Bailey DK, Huebert DJ, et al. Genomic maps and comparative analysis of histone modifications in human and mouse. *Cell* 2005;**120**(2):169–81.

30. Prabhakar S, Poulin F, Shoukry M, Afzal V, Rubin EM, Couronne O, et al. Close sequence comparisons are sufficient to identify human cis-regulatory elements. *Genome Res* 2006;**16**(7):855–63.

31. Pennacchio LA, Loots GG, Nobrega MA, Ovcharenko I. Predicting tissue-specific enhancers in the human genome. *Genome Res* 2007;**17**(2):201–11.

32. Merika M, Williams AJ, Chen G, Collins T, Thanos D. Recruitment of CBP/p300 by the IFN beta enhanceosome is required for synergistic activation of transcription. *Mol Cell* 1998;**1**(2):277–87.

33. Heintzman ND, Stuart RK, Hon G, Fu Y, Ching CW, Hawkins RD, et al. Distinct and predictive chromatin signatures of transcriptional promoters and enhancers in the human genome. 2007;**39**(3):311–318.

34. Visel A, Blow MJ, Li Z, Zhang T, Akiyama JA, Holt A, et al. ChIP-seq accurately predicts tissue-specific activity of enhancers. *Nature* 2009;**457**(7231):854–8.

35. Heintzman ND, Hon GC, Hawkins RD, Kheradpour P, Stark A, Harp LF, et al. Histone modifications at human enhancers reflect global cell-type-specific gene expression. *Nature* 2009;**459**(7243):108–12.

36. Meissner A, Mikkelsen TS, Gu H, Wernig M, Hanna J, Sivachenko A, et al. Genome-scale DNA methylation maps of pluripotent and differentiated cells. *Nature* 2008;**454**(7205):766–70.

37. Orford K, Kharchenko P, Lai W, Dao MC, Worhunsky DJ, Ferro A, et al. Differential H3K4 methylation identifies developmentally poised hematopoietic genes. *Dev Cell* 2008;**14**(5):798–809.

38. Fouse SD, Shen Y, Pellegrini M, Cole S, Meissner A, Van Neste L, et al. Promoter CpG methylation contributes to ES cell gene regulation in parallel with Oct4/Nanog, PcG complex, and histone H3 K4/K27 trimethylation. *Cell Stem Cell* 2008;**2**(2):160–9.

39. Pan G, Tian S, Nie J, Yang C, Ruotti V, Wei H, et al. Whole-genome analysis of histone H3 lysine 4 and lysine 27 methylation in human embryonic stem cells. *Cell Stem Cell* 2007;**1**(3):299–312.

40. Zhao XD, Han X, Chew JL, Liu J, Chiu KP, Choo A, et al. Whole-genome mapping of histone H3 Lys4 and 27 trimethylations reveals distinct genomic compartments in human embryonic stem cells. *Cell Stem Cell* 2007;**1**(3):286–98.

41. Mikkelsen TS, Ku M, Jaffe DB, Issac B, Lieberman E, Giannoukos G, et al. Genome-wide maps of chromatin state in pluripotent and lineage-committed cells. *Nature* 2007;**448**(7153):553–60.

42. Bernstein BE, Mikkelsen TS, Xie X, Kamal M, Huebert DJ, Cuff J, et al. A bivalent chromatin structure marks key developmental genes in embryonic stem cells. *Cell* 2006;**125**(2):315–26.

43. Azuara V, Perry P, Sauer S, Spivakov M, Jorgensen HF, John RM, et al. Chromatin signatures of pluripotent cell lines. *Nat Cell Biol* 2006;**8**(5):532–8.

44. Weber M, Hellmann I, Stadler MB, Ramos L, Paabo S, Rebhan M, et al. Distribution, silencing potential and evolutionary impact of promoter DNA methylation in the human genome. *Nat Genet* 2007;**39**(4):457–66.

45. Mohn F, Weber M, Rebhan M, Roloff TC, Richter J, Stadler MB, et al. Lineage-specific polycomb targets and de novo DNA methylation define restriction and potential of neuronal progenitors. *Mol Cell* 2008;**30**(6):755–66.

46. Araki Y, Wang Z, Zang C, Wood WH 3rd, Schones D, Cui K, et al. Genome-wide analysis of histone methylation reveals chromatin state-based regulation of gene transcription and function of memory CD8+ T cells. *Immunity* 2009;**30**(6):912–25.

47. Wei G, Wei L, Zhu J, Zang C, Hu-Li J, Yao Z, et al. Global mapping of H3K4me3 and H3K27me3 reveals specificity and plasticity in lineage fate determination of differentiating CD4+ T cells. *Immunity* 2009;**30**(1):155–67.

48. Neff T, Armstrong SA. Chromatin maps, histone modifications and leukemia. *Leukemia* 2009;**23**(7):1243–51.

49. Figueroa ME, Reimers M, Thompson RF, Ye K, Li Y, Selzer RR, et al. An integrative genomic and epigenomic approach for the study of transcriptional regulation. *PLoS One* 2008;**3**(3):e1882.

50. Guenther MG, Lawton LN, Rozovskaia T, Frampton GM, Levine SS, Volkert TL, et al. Aberrant chromatin at genes encoding stem cell regulators in human mixed-lineage leukemia. *Genes Dev* 2008;**22**(24):3403–8.

51. Ke XS, Qu Y, Rostad K, Li WC, Lin B, Halvorsen OJ, et al. Genome-wide profiling of histone h3 lysine 4 and lysine 27 trimethylation reveals an epigenetic signature in prostate carcinogenesis. *PLoS One* 2009;**4**(3):e4687.

52. Kosak ST, Scalzo D, Alworth SV, Li F, Palmer S, Enver T, et al. Coordinate gene regulation during hematopoiesis is related to genomic organization. *PLoS Biol* 2007;**5**(11):e309.

53. Misteli T. Beyond the sequence: cellular organization of genome function. *Cell* 2007;**128**(4):787–800.

54. Cremer T, Cremer C. Chromosome territories, nuclear architecture and gene regulation in mammalian cells. *Nat Rev Genet* 2001;**2**(4):292–301.

55. Solovei I, Kreysing M, Lanctot C, Kosem S, Peichl L, Cremer T, et al. Nuclear architecture of rod photoreceptor cells adapts to vision in mammalian evolution. *Cell* 2009;**137**(2):356–68.

56. Sutherland H, Bickmore WA. Transcription factories: gene expression in unions? *Nat Rev Genet* 2009;**10**(7):457–66.

57. Miele A, Dekker J. Mapping cis- and trans- chromatin interaction networks using Chromosome Conformation Capture (3C). *Methods Mol Biol* 2009;**464**:105–21.

58. Spector DL. The dynamics of chromosome organization and gene regulation. *Annu Rev Biochem* 2003;**72**:573–608.

59. van Steensel B, Henikoff S. Identification of in vivo DNA targets of chromatin proteins using tethered dam methyltransferase. *Nat Biotechnol* 2000;**18**(4):424–8.

60. Guelen L, Pagie L, Brasset E, Meuleman W, Faza MB, Talhout W, et al. Domain organization of human chromosomes revealed by mapping of nuclear lamina interactions. *Nature* 2008;**453**(7197):948–51.

61. Shimi T, Pfleghaar K, Kojima S, Pack CG, Solovei I, Goldman AE, et al. The A- and B-type nuclear lamin networks: microdomains involved in chromatin organization and transcription. *Genes Dev* 2008;**22**(24):3409–21.

62. Finlan LE, Sproul D, Thomson I, Boyle S, Kerr E, Perry P, et al. Recruitment to the nuclear periphery can alter expression of genes in human cells. *PLoS Genet* 2008;**4**(3):e1000039.

63. Kumaran RI, Spector DL. A genetic locus targeted to the nuclear periphery in living cells maintains its transcriptional competence. *J Cell Biol* 2008;**180**(1):51–65.

170

64. Reddy KL, Zullo JM, Bertolino E, Singh H. Transcriptional repression mediated by repositioning of genes to the nuclear lamina. *Nature* 2008;**452**(7184):243–7.

65. Cremer M, Zinner R, Stein S, Albiez H, Wagler B, Cremer C, et al. Three dimensional analysis of histone methylation patterns in normal and tumor cell nuclei. *Eur J Histochem* 2004;**48**(1):15–28.

66. Zinner R, Albiez H, Walter J, Peters AH, Cremer T, Cremer M. Histone lysine methylation patterns in human cell types are arranged in distinct three-dimensional nuclear zones. *Histochem Cell Biol* 2006;**125**(1–2):3–19.

67. Bartova E, Harnicarova A, Krejci J, Strasak L, Kozubek S. Single-cell c-myc gene expression in relationship to nuclear domains. *Chromosome Res* 2008;**16**(2):325–43.

68. Saurin AJ, Shiels C, Williamson J, Satijn DP, Otte AP, Sheer D, et al. The human polycomb group complex associates with pericentromeric heterochromatin to form a novel nuclear domain. *J Cell Biol* 1998;**142**(4):887–98.

69. Voncken JW, Roelen BA, Roefs M, de Vries S, Verhoeven E, Marino S, et al. Rnf2 (Ring1b) deficiency causes gastrulation arrest and cell cycle inhibition. *Proc Natl Acad Sci USA* 2003;**100**(5):2468–73.

70. Zhang H, Christoforou A, Aravind L, Emmons SW, van den Heuvel S, Haber DA. The *C. elegans* Polycomb gene SOP-2 encodes an RNA binding protein. *Mol Cell* 2004;**14**(6):841–7.

71. Gambetta MC, Oktaba K, Muller J. Essential role of the glycosyltransferase sxc/Ogt in polycomb repression. *Science* 2009;**325**(5936):93–6.

72. Oktaba K, Gutierrez L, Gagneur J, Girardot C, Sengupta AK, Furlong EE, et al. Dynamic regulation by polycomb group protein complexes controls pattern formation and the cell cycle in *Drosophila*. *Dev Cell* 2008;**15**(6):877–89.

73. Bernardi R, Pandolfi PP. Structure, dynamics and functions of promyelocytic leukaemia nuclear bodies. *Nat Rev Mol Cell Biol* 2007;**8**(12):1006–16.

74. Boisvert FM, Hendzel MJ, Bazett-Jones DP. Promyelocytic leukemia (PML) nuclear bodies are protein structures that do not accumulate RNA. *J Cell Biol* 2000;**148**(2):283–92.

75. Zhong S, Salomoni P, Pandolfi PP. The transcriptional role of PML and the nuclear body. *Nat Cell Biol* 2000;**2**(5):E85–90.

76. Wang J, Shiels C, Sasieni P, Wu PJ, Islam SA, Freemont PS, et al. Promyelocytic leukemia nuclear bodies associate with transcriptionally active genomic regions. *J Cell Biol* 2004;**164**(4):515–26.

77. Seeler JS, Marchio A, Sitterlin D, Transy C, Dejean A. Interaction of SP100 with HP1 proteins: a link between the promyelocytic leukemia-associated nuclear bodies and the chromatin compartment. *Proc Natl Acad Sci USA* 1998;**95**(13):7316–21.

78. Tashiro S, Muto A, Tanimoto K, Tsuchiya H, Suzuki H, Hoshino H, et al. Repression of PML nuclear body-associated transcription by oxidative stress-activated Bach2. *Mol Cell Biol* 2004;**24**(8):3473–84.

79. Luciani JJ, Depetris D, Usson Y, Metzler-Guillemain C, Mignon-Ravix C, Mitchell MJ, et al. PML nuclear bodies are highly organised DNA-protein structures with a function in heterochromatin remodelling at the G2 phase. *J Cell Sci* 2006;**119**(12):2518–31.

80. Zhang R, Poustovoitov MV, Ye X, Santos HA, Chen W, Daganzo SM, et al. Formation of MacroH2A-containing senescence-associated heterochromatin foci and senescence driven by ASF1a and HIRA. *Dev Cell* 2005;**8**(1):19–30.

81. Ye X, Zerlanko B, Zhang R, Somaiah N, Lipinski M, Salomoni P, et al. Definition of pRB- and p53-dependent and -independent steps in HIRA/ASF1a-mediated formation of senescence-associated heterochromatin foci. *Mol Cell Biol* 2007;**27**(7):2452–65.

82. Kumar PP, Bischof O, Purbey PK, Notani D, Urlaub H, Dejean A, et al. Functional interaction between PML and SATB1 regulates chromatin-loop architecture and transcription of the MHC class I locus. *Nat Cell Biol* 2007;**9**(1):45–56.

83. Iborra FJ, Pombo A, Jackson DA, Cook PR. Active RNA polymerases are localized within discrete transcription "factories" in human nuclei. *J Cell Sci* 1996;**109**(6):1427–36.

84. Phatnani HP, Greenleaf AL. Phosphorylation and functions of the RNA polymerase II CTD. *Genes Dev* 2006;**20**(21):2922–36.

85. Osborne CS, Chakalova L, Brown KE, Carter D, Horton A, Debrand E, et al. Active genes dynamically colocalize to shared sites of ongoing transcription. *Nat Genet* 2004;**36**(10):1065–71.

86. Lee GR, Spilianakis CG, Flavell RA. Hypersensitive site 7 of the TH2 locus control region is essential for expressing TH2 cytokine genes and for long-range intrachromosomal interactions. *Nat Immunol* 2005;**6**(1):42–8.

87. Osborne CS, Chakalova L, Mitchell JA, Horton A, Wood AL, Bolland DJ, et al. Myc dynamically and preferentially relocates to a transcription factory occupied by Igh. *PLoS Biol* 2007;**5**(8):e192.

88. Kozubek S, Lukasova E, Ryznar L, Kozubek M, Liskova A, Govorun RD, et al. Distribution of ABL and BCR genes in cell nuclei of normal and irradiated lymphocytes. *Blood* 1997;**89**(12):4537–45.

89. Roix JJ, McQueen PG, Munson PJ, Parada LA, Misteli T. Spatial proximity of translocation-prone gene loci in human lymphomas. *Nat Genet* 2003;**34**(3):287–91.

CHAPTER 12

Assessing Epigenetic Information

Karolina Janitz[1] and Michal Janitz[2]
[1]Ramaciotti Centre for Gene Function Analysis, The University of New South Wales, Sydney, NSW 2052
[2]School of Biotechnology and Biomolecular Sciences, The University of New South Wales, Sydney, NSW 2052, Australia

INTRODUCTION

Epigenetics is currently one of the most rapidly developing areas of molecular biology. Remarkable technological progress has enabled genome-scale analysis of epigenetic mechanisms. The analysis of epigenetic data has challenged computational bioscience, and computational methods have already been involved in the analysis of epigenetic questions. Nucleotide sequence shows how the genetic program is read. Epigenetic information is encoded as chemical modifications of cytosine bases and histone proteins that assist genome condensation. These modifications regulate the way the genome is converted at different stages of cell growth in various tissues and morbidity conditions by influencing DNA availability via the structure of the chromatin [1–3]. In recent years, enormous progress in the ability to characterize widespread epigenetic modifications has taken place, and novel patterns of genome regulation have begun to emerge.

EPIGENETIC DATA GENERATION AND ANALYSIS

Chromatin immunoprecipitation (ChIP) is a powerful tool used to study protein–DNA interaction and chromatin changes associated with gene expression. As such, it can be used for the collection of epigenetic data that cover the entire genome. This technique enables enrichment of DNA fragments bound to specific proteins and the identification of the specific sequences of these fragments using microarrays that encompass the genome (tiling arrays) or next-generation DNA sequencing (Fig. 12.1). A third technique, bisulfite sequencing, is limited to the detection of DNA methylation, and we recommend the original research of Hajkova et al. [4] and the review reported by Zilberman and Henikoff [5] for further details of this technique.

ChIP-on-chip Technique

ChIP-on-chip employs chromatin immunoprecipitation (ChIP), which is used to enrich for specific DNA fragments, and genome tiling microarrays to discover differences between tested and control DNA. To start, cells are treated with formaldehyde to crosslink proteins and their bound DNA. Next, chromatin is extracted and cut into smaller fragments of

Handbook of Epigenetics: The New Molecular and Medical Genetics. DOI: 10.1016/B978-0-12-375709-8.00012-5

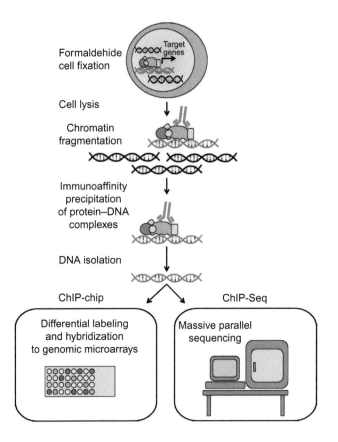

FIGURE 12.1
The identification of protein–DNA interactions on a genome-wide level can be performed by chromatin immunoprecipitation (ChIP) based on microarrays (ChIP-on-chip) or next-generation sequencing technology platforms (ChIP-Seq). *Adapted by permission from Macmillan Publishers Ltd: Leukemia 2009.*

about 500 base pairs. The size of the fragments limits the resolution of the method. Certain fragments are enriched using antibodies against modifications of histones or chromatin proteins. Subsequently, DNA is released from these fragments and hybridized to genome tiling microarrays. Regions that are significantly over-represented in DNA immunoprecipitates compared to control DNA are presumed to be epigenetically modified or bound to proteins in accord with the type of antibody used. In a ChIP-on-chip variant called methyl-DNA immunoprecipitation (MeDIP) [6], purified DNA is subjected to immunoprecipitation using antibodies against methylated cytosine. This technique allows for the creation of genome-scale DNA methylation maps. Although these methods are used in many laboratories around the world, the quality of the antibodies used requires a thorough examination, a condition not always fulfilled. Moreover, background noise resulting from cross hybridization and varying affinities of different oligomers requires consideration during data processing.

ChIP-on-chip Data Processing

A difficult statistical problem is to differentiate the signal from the noise within data generated by ChIP-on-chip. Although numerous techniques have been developed that employ averages calculated with sliding window methods [7] and hidden Markov models [8,9], these techniques are all characterized by a significant frequency of errors. They are also not adjusted to properties typical of the wide domains characteristic of many epigenetic markers. The lack of reproducibility of genome-wide ChIP-on-chip experiments may be partially of technical origin [10,11] and depends to a significant degree on the selection of proper statistical models and enrichment-separation filters. Another important factor is that ChIP analysis permits the detection of rarely occurring protein–DNA interactions that may not be physiologically significant. A recent study [12] has resulted in the observation that most tightly bound sites are a much better foundation for predicting gene expression patterns than regions that are less tightly bound. This suggests that ChIP experiments analyzing the interactions of a single factor may be an inefficient way of characterizing regulatory functions of the binding protein. However, inclusion of ChIP data for another binding protein, such as a cofactor or related epigenetic mark, may enhance the power of prediction.

An important element in the interpretation of ChIP results is the level of correlation between the binding site or epigenetic mark and the biological context. The context can be defined as the dominant cell cycle phase of a given population of cells, the differentiation and viability status of the cell, tissue or organism, and the evolutionary diversity between species. When binding regions or epigenetic marks are very dynamic and easily succumb to the influence of these factors, the formulation of general principles becomes much more difficult. Some of these principles were determined before the sequencing of the human genome, for example the idea that transcription is controlled mainly by regulatory sequence fragments usually located upstream of the target sequence. This idea contributed to the development of the promoter arrays used in ChIP-on-chip experiments. It also had an impact on subsequent data analysis, especially the correlation between transcription factor binding sites and modifications of gene expression. A good example comes from the studies of Guttmann et al., who identified large non-coding RNAs by determining the location of H3K4me3 and histone H3 trimethylated at lysine 36 (H3K36me3) [13]. In another study, Mikkelsen and co-workers used genome-wide chromatin-state maps of mouse embryonic stem cells, neural progenitor cells, and embryonic fibroblasts to observe chromatin changes during the development of immature cells into adult tissue [14].

Recent studies carried out by the Encyclopedia of DNA Elements (ENCODE) Project focusing on 1% of the human genome (30 Mb) have yielded surprising new insights into the regulatory mechanisms of gene expression [15]. ENCODE focused on a set of 44 genome regions with the aim of identifying functional elements in the human genome. This enabled the identification of promoters of previously known transcripts as well as novel ones. Due to its non-tendentious multi-group approach, ENCODE questioned certain dogmas that were the basis of previous interpretations of ChIP data [15]. Most bases in the human genome are associated with at least one primary transcript, and transcripts from completely separate regions may link to form a molecule that encodes a protein. Moreover, chromatin accessibility and histone modifications are positive factors that enable the prediction of transcription start sites and their activity. Additionally, regulatory sequences surrounding the transcription start site are located symmetrically around it and have a decreased tendency to be in upstream regions. These observations illustrated that most functional DNA sequence elements are not actively influenced by evolution. These elements are neutral, and even though their biological activity yields no specific advantages to the organism, they may provide raw material for subsequent stages of natural selection and the emergence of new species.

A prominent example of the comprehensive toolbox for microarray data analysis, including ChIP data, is Bioconductor (www.bioconductor.org) [16]. The Bioconductor initiative is devoted to the development of software and algorithms for statistical analysis and visualization of high-throughput experimental data. The programs are written in R and are an open source that is freely available to the scientific community. Bioconductor consists of so-called containers and workflows for the initial processing and subsequent analysis of high-throughput data. Containers have been defined for the processing and management of metadata in the Minimum Information about a Microarray Experiment (MIAME) data model [17]. Moreover, containers enable the management and analysis of sample-level data and custom metadata about samples. The mapping of large quantities of annotations between proprietary probe identifiers and public database or ontology identifiers can also be performed.

ChIP-Seq Methodology

ChIP-Seq [18–20], which unites ChIP with next-generation sequencing technologies, may replace ChIP-on-chip and become the predominant means of identifying genome-wide protein-DNA interactions in humans. The range of applications, high resolution, and cost-effectiveness of ChIP-Seq together with its ability to sequence millions of bases within a few days will allow the mapping of protein-DNA interactions on a genome-wide scale.

DNA fragments obtained by ChIP are sequenced directly in ChIP-Seq using next-generation DNA sequencers, such as the Illumina Genome Analyzer. Although the length of the analyzed DNA may range from 200 bp to 1 kb, only the first 36–100 nucleotides at the end of the DNA fragment are typically sequenced. The short reads are aligned to the reference genome and only those reads (tags) that match are used for further analysis. Typically, genome tags obtained with high frequency are considered as transcription factor binding sites [18–20].

Although this approach aids in the correct identification of binding sites, the short length of the reads presents a challenge in terms of the ability to localize the binding sites to a specific genomic location. Moreover, the resolution of the identified binding sites may even be equal to the length of the tested DNA, if not longer. Additionally, transcription factor binding sites are often clustered in critical regulatory regions and are thus located close to one another. To understand the structure of regulatory elements and to describe the influence of each binding site or transcription factor, it is necessary to develop sensitive and precise methods for the identification of protein–DNA binding sites. Additionally, such methods should be both reliable and adaptable to enable the control of elements such as antibody specificity and sequencing errors that influence data quality. Nevertheless, the enhanced spatial resolution of next-generation sequencing has greatly facilitated the genome-wide identification of binding sites for transcription factors and their sequence motifs. This elevated resolution becomes especially relevant for studies of the nucleosome. As a result, a number of studies have attempted to systematically investigate patterns of nucleosome positioning, histone variants, and modifications [19,21–24].

Calculation Algorithms for Epigenetic Data Extraction from ChIP-Seq Experiments

Many unsolved problems limiting the widespread application of ChIP-Seq are related to the manipulation and interpretation of the data obtained from next-generation DNA sequencing devices. The tools usually used for sequence alignment are based on the Smith-Waterman alignment algorithm, ensuring a precise and optimal solution, and on the Basic Local Alignment Search Tool (BLAST), which guarantees a fast and almost optimal solution [25,26]. Even though the BLAST algorithm undergoes continual modifications [27,28], the strategy on which this algorithm is based is unsuitable for fast alignment of short fragments to the genome of origin. Although these methods offer a wide range of capabilities, the amount of time required to process millions of fragments is significant, and thus these methods are not suitable for highly efficient sequence identification. Fortunately, a new generation of programs optimized for aligning short reads was created for this purpose.

ELAND (Efficient Local Alignment of Nucleotide Data) by Illumina is the most popular tool employed for alignment. It can quickly identify the starting point of a short sequence if there is unequivocal alignment with a maximum of two differing bases. If this condition is not fulfilled and there are a few equally probable starting points for a given fragment, a result is not presented. In the case of many mammalian DNA datasets, this procedure enables the identification of only 50–65% of fragments [18,20]. This identification frequency results

TABLE 12. 1 Selected Short-read Analysis Software

Program	Website	Open Source?	Handles ABI Color Space?	Maximum Read Length
Bowtie	http://bowtie.cbcb.umd.edu	Yes	No	None
BWA	http://maq.sourceforge.net/bwa-man.shtml	Yes	Yes	None
Maq	http://maq.sourceforge.net/bwa-man.shtml	Yes	Yes	127
Mosaik	http://bioinformatics.bc.edu/marthlab/Mosaik	No	Yes	None
Novoalign	http://www.novocraft.com	No	No	None
SOAP2	http://soap.genomics.org.cn	No	No	60
ZOOM	http://www.bioinfor.com	No	Yes	240

Adapted by permission from Macmillan Publishers Ltd: Nat Biotechnol 2009.

from the inability of ELAND to align sequences containing insertions or deletions (so-called *gapped alignment*), and the algorithm limits the number of mis-calls (SNPs or sequencing errors) that may be accepted in an alignment. It is expected that the ratio of the number of mapped fragments to the number of all sequenced fragments will improve as new methods for short read alignment are developed.

Many competing software packages, summarized in Table 12.1, have already been offered. For example, Maq and Bowtie use a computational strategy known as "indexing" in order to accelerate their mapping algorithms. Maq utilizes spaced seed indexing and builds assemblies by mapping shotgun short reads to a reference genome, using quality scores to derive genotype calls of the consensus sequence of a diploid genome. Maq makes full use of mate-pair information and estimates the error probability of each read alignment [29]. Bowtie indexes the reference genome using a scheme based on the Burrows–Wheeler transform and the FM Index [30]. The program aligns a read one character at a time to the Burrows–Wheeler-transformed genome. Each successively aligned new character allows Bowtie to winnow the list of positions to which the read might map. Each of these programs has certain advantages, and many can perform more sophisticated functions, including mapping one fragment to multiple starting points and encompassing insertions and deletions. For these reasons, such programs are gaining dominance over the ELAND software for reading short sequences.

Another major problem in the use of short reads occurs during data interpretation. A method has been invented [20] in which a minimum threshold of sequence tags based on ROC analysis [31] must be present within 100 base pairs. Another method takes advantage of so-called peaks, where examined sequences should widen to the average length of the precipitated DNA fragments and the number of observations of each base in each widened sequence is displayed in histogram form in which peaks are identified [18]. Peaks that are above the false detection threshold are preserved. Both methods allow for a quick scan of the genome for enriched regions representing binding sites. An advantage of the latter method is the simplicity of the graphic presentation of the peaks or sequenced reads, which makes it relatively easy to visually interpret the results. However, in situations where binding site sequences overlap or when secondary interactions occur, the peaks may become more complex, which makes the identification of the actual binding site of a given protein more challenging. Moreover, as sequencing depth increases, fragments derived from non-specific binding start to gather on the limbs of each peak and impede the identification of peak boundaries.

The shortness (at present 36–100 bp) of the sequences obtained with the Illumina technology constitutes the next challenge for data interpretation. Short reads often result in an insufficient number of sequences capable of predicting the actual genomic starting point of the DNA fragment. This insufficiency is due to ambiguity deriving from sequences that are repeated in the genome. Similarly, entire regions that do not exist in a genome may be sequenced, or an obtained sequence may contain so many mutations that it has insufficient similarity to the analogous site in the genome of interest. Each of these situations causes data loss due to an inability to interpret certain subsets of results obtained during data analysis. This issue will probably be solved when the Illumina technology improves or competitive technologies enabling the use of longer sequence reads are introduced to the market. The use of the paired-end read protocol during DNA sequencing is another potential solution to the problem that can deliver additional fragments and allow better identification of the origin of DNA fragments.

Recently, ShortRead, as an extension of the Bioconductor suite for initial analysis of short-read DNA sequences, has been released [32]. The main features of ShortRead include data input, quality assessment, data transformation, and access to downstream Bioconductor analysis options. The package allows input of diverse sequence-related files into R and output of common data formats. There are also quality assessment tools and an HTML-based report-generating feature.

EPIGENETICS OF CpG ISLANDS

CpG islands (CGIs) are regions of the genome that contain a large number of CpG dinucleotide repeats. In mammalian genomes, CpG islands usually extend for 300–3000 base pairs. They are located within and close to sites of about 40% of mammalian gene promoters. It is estimated that in mammalian genomes about 80% of CpG dinucleotides are methylated. However, CpG dinucleotides in regions abundant in GC pairs, such as CpG clusters and CpG islands (CGIs), are usually unmethylated, and this is an important feature of gene promoters and gene expression control [33]. Although most CGIs linked to promoters are non-methylated, recent studies have revealed that the majority of CGIs may be completely methylated in normal cells [34,35].

Numerous algorithms have been formulated [36–40] to identify CGIs. These algorithms may be divided into two groups: (i) traditional algorithms based on three parameters (length, number of GC pairs, and the ratio of observed to expected number of CpG dinucleotides) and (ii) algorithms based on the statistical properties of sequences that do not employ the three criteria used by traditional algorithms.

The first traditional algorithm was postulated by Gardiner-Garden and Frommer [37], and its criteria include: length over 200 base pairs, over 50% GC pairs, and a ratio of observed to expected number of CpG dinucleotides over 0.60. However, many repetitive elements that commonly occur in genomes (such as *Alu* repeats) also meet these criteria. To avoid this problem, non-repeating parts of the genome were exclusively used to search for CGIs [41]. This algorithm was subsequently modified with more stringent criteria, such as length over 500 base pairs, over 55% GC pairs, and a ratio of observed to expected number of CpG dinucleotides over 0.65 [36]. Lately, two algorithms have been formulated (CpGcluster and CG clusters) that use statistical sequence properties and do not impose *a priori* conditions regarding sequence length. CpGcluster detects CpG clusters through statistical significance based on the physical distance between neighboring CpG dinucleotides in a chromosome. This program assumes that the distances between neighboring CpGs are different in a CGI than in other regions [40]. CG clusters are defined as fragments rich in CG detected on the basis of empirical reference points specific for different species. Recently, progress has been made in research on DNA methylation in chromosomes and whole genomes due to the rapid advancement of sequencing technologies [42]. By means of experimentally verified data concerning DNA methylation, researchers predicted methylation status based on sequence features in the vicinity of CGIs, such as transcription factor binding sites, DNA sequence motifs, and repetitive elements [43–45]. This research significantly improved our comprehension of the relationship between CGIs, DNA structure, and methylation status.

CGIs constitute a field of interest with regard to methylation profile detection in large-scale experiments using various platforms. In most studies on CGI annotation, the Gardiner-Garden and Frommer algorithms were employed [46–48]. Although a resolution of one base provides the most useful information, it has been recently suggested that, with regard to the currently available microarray-based platforms, it might be sufficient to determine the average level of methylation in sites rich in CpG dinucleotides [49].

CONCLUSIONS

The continually increasing efficiency and declining cost of next-generation sequencing will soon enable researchers to study the epigenomes of hundreds of individuals and thus enable the study of epigenetic variability in the human population. Such investigations will provide a challenge for bioinformatics analysis, as large sets of data will have to be processed in comparative studies, including data regarding haploid gene expression variability. Until now, epigenetic research has focused mainly on modifications of chromatin structure through changes in interactions between proteins and DNA. Another interesting point is the growing need to integrate epigenetic data derived from different platforms. A good example of an efficient integrative approach was presented by Mathur and co-workers [50], who combined ChIP-on-chip and ChIP-PET (for a review, please refer to Ng et al. [51]) to study regulatory networks in mouse embryonic stem cells. Therefore, the growth of cost-effective genomic and proteomic technologies and the rapid development of integrative and comparative approaches in computational epigenetics stand to illuminate the epigenomic landscape at an unprecedented level of detail.

References

1. Margueron R, Trojer P, Reinberg D. The key to development: interpreting the histone code? *Curr Opin Genet Dev* 2005;**15**:163–76.

2. Bird A. DNA methylation patterns and epigenetic memory. *Genes Dev* 2002;**16**:6–21.

3. Goll MG, Bestor TH. Eukaryotic cytosine methyltransferases. *Annu Rev Biochem* 2005;**74**:481–514.

4. Hajkova P, el-Maarri O, Engemann S, Oswald J, Olek A, Walter J. DNA-methylation analysis by the bisulfite-assisted genomic sequencing method. *Methods Mol Biol* 2002;**200**:143–54.

5. Zilberman D, Henikoff S. Genome-wide analysis of DNA methylation patterns. *Development* 2007;**134**:3959–65.

6. Sørensen AL, Collas P. Immunoprecipation of methylated DNA. *Methods Mol Biol* 2009;**567**:249–62.

7. Jothi R, Cuddapah S, Barski A, Cui K, Zhao K. Genome-wide identification of in vivo protein-DNA binding sites from ChIP-Seq data. *Nucleic Acids Res* 2008;**36**:5221–31.

8. Du J, Rozowsky JS, Korbel JO, Zhang ZD, Royce TE, Schultz MH, et al. A supervised hidden Markov model framework for efficiently segmenting tiling array data in transcriptional and chIP-chip experiments: systematically incorporating validated biological knowledge. *Bioinformatics* 2006;**22**:3016–24.

9. Ji H, Wong WH. TileMap: create chromosomal map of tiling array hybridizations. *Bioinformatics* 2005;**21**:3629–36.

10. Zecchini V, Mills IG. Putting chromatin immunoprecipitation into context. *J Cell Biochem* 2009;**107**:19–29.

11. Johnson DS, Li W, Gordon DB, Bhattacharjee A, Curry B, Ghosh J, et al. Systematic evaluation of variability in ChIP-chip experiments using predefined DNA targets. *Genome Res* 2008;**18**:393–403.

12. Li XY, MacArthur S, Bourgon R, Nix D, Pollard DA, Iyer VN, et al. Transcription factors bind thousands of active and inactive regions in the *Drosophila* blastoderm. *PLoS Biol* 2008;**6**:e27.

13. Guttman M, Amit I, Garber M, French C, Lin MF, Feldser D, et al. Chromatin signature reveals over a thousand highly conserved large non-coding RNAs in mammals. *Nature* 2009;**458**:223–7.

14. Mikkelsen TS, Ku M, Jaffe DB, Issac B, Lieberman E, Giannoukos G, et al. Genome-wide maps of chromatin state in pluripotent and lineage-committed cells. *Nature* 2007;**448**:553–60.

15. Birney E, Stamatoyannopoulos JA, Dutta A, Guigo R, Gingeras TR, Margulies EH, et al. Identification and analysis of functional elements in 1% of the human genome by the ENCODE pilot project. *Nature* 2007;**447**:799–816.

16. Reimers M, Carey VJ. Bioconductor: an open source framework for bioinformatics and computational biology. *Methods Enzymol* 2006;**411**:119–34.

17. Brazma A, Hingamp P, Quackenbush J, Sherlock G, Spellman P, Stoeckert C, et al. Minimum information about a microarray experiment (MIAME): toward standards for microarray data. *Nature Genet* 2001;**29**:365–71.

18. Robertson G, Hirst M, Bainbridge M, Bilenky M, Zhao Y, Zeng T, et al. Genome-wide profiles of STAT1 DNA association using chromatin immunoprecipitation and massively parallel sequencing. *Nat Methods* 2007;**4**:651–57.

19. Barski A, Cuddapah S, Cui K, Roh TY, Schones DE, Wang Z, et al. High-resolution profiling of histone methylations in the human genome. *Cell* 2007;**129**:823–37.

20. Johnson DS, Mortazavi A, Myers RM, Wold B. Genome-wide mapping of in vivo protein-DNA interactions. *Science* 2007;**316**:1497–502.

21. Wang Z, Zang C, Rosenfeld JA, Schones DE, Barski A, Cuddapah S, et al. Combinatorial patterns of histone acetylations and methylations in the human genome. *Nat Genet* 2008;**40**:897–903.

22. Bernstein BE, Mikkelsen TS, Xie X, Kamal M, Huebert DJ, Cuff J, et al. A bivalent chromatin structure marks key developmental genes in embryonic stem cells. *Cell* 2006;**125**:315–26.

23. Albert I, Mavrich TN, Tomsho LP, Qi J, Zanton SJ, Schuster SC, et al. Translational and rotational settings of H2A.Z nucleosomes across the *Saccharomyces cerevisiae* genome. *Nature* 2007;**446**:572–6.

24. Mavrich TN, Jiang C, Ioshikhes IP, Li X, Venters BJ, Zanton SJ, et al. Nucleosome organization in the *Drosophila* genome. *Nature* 2008;**453**:358–62.

25. Smith TF, Waterman MS. Identification of common molecular subsequences. *J Mol Biol* 1981;**147**:195–7.

26. Altschul SF, Gish W, Miller W, Myers EW, Lipman DJ. Basic local alignment search tool. *J Mol Biol* 1990;**215**:403–10.

27. Gertz EM, Yu YK, Agarwala R, Schaffer AA, Altschul SF. Composition-based statistics and translated nucleotide searches: improving the TBLASTN module of BLAST. *BMC Biol* 2006;**4**:41.

28. Schaffer AA, Aravind L, Madden TL, Shavirin S, Spouge JL, Wolf YI, et al. Improving the accuracy of PSI-BLAST protein database searches with composition-based statistics and other refinements. *Nucleic Acids Res* 2001;**29**:2994–3005.

29. Li H, Ruan J, Durbin R. Mapping short DNA sequencing reads and calling variants using mapping quality scores. *Genome Res* 2008;**18**:1851–8.

30. Langmead B, Trapnell C, Pop M, Salzberg SL. Ultrafast and memory-efficient alignment of short DNA sequences to the human genome. *Genome Biol* 2009;**10**:R25.

31. Lasko TA, Bhagwat JG, Zou KH, Ohno-Machado L. The use of receiver operating characteristic curves in biomedical informatics. *J Biomed Inform* 2005;**38**:404–15.

32. Morgan M, Anders S, Lawrence M, Aboyoun P, Pagès H, Gentleman R. ShortRead: a bioconductor package for input, quality assessment and exploration of high-throughput sequence data. *Bioinformatics* 2009;**25**:2607–8.

33. Antequera F. Structure, function and evolution of CpG island promoters. *Cell Mol Life Sci* 2003;**60**:1647–58.

34. Illingworth R, Kerr A, Desousa D, Jorgensen H, Ellis P, Stalker J, et al. A novel CpG island set identifies tissue-specific methylation at developmental gene loci. *PLoS Biol* 2008;**6**:e22.

35. Eckhardt F, Lewin J, Cortese R, Rakyan VK, Attwood J, Burger M, et al. DNA methylation profiling of human chromosomes 6, 20 and 22. *Nat Genet* 2006;**38**:1378–85.

36. Takai D, Jones PA. Comprehensive analysis of CpG islands in human chromosomes 21 and 22. *Proc Natl Acad Sci USA* 2002;**99**:3740–5.

37. Gardiner-Garden M, Frommer M. CpG islands in vertebrate genomes. *J Mol Biol* 1987;**196**:261–82.

38. Ponger L, Mouchiroud D. CpGProD: identifying CpG islands associated with transcription start sites in large genomic mammalian sequences. *Bioinformatics* 2002;**18**:631–3.

39. Glass JL, Thompson RF, Khulan B, Figueroa ME, Olivier EN, Oakley EJ, et al. CG dinucleotide clustering is a species-specific property of the genome. *Nucleic Acids Res* 2007;**35**:6798–807.

40. Hackenberg M, Previti C, Luque-Escamilla PL, Carpena P, Martinez-Aroza J, Oliver JL. CpGcluster: a distance-based algorithm for CpG-island detection. *BMC Bioinformatics* 2006;**7**:446.

41. Lander ES, Linton LM, Birren B, Nusbaum C, Zody MC, Baldwin J, et al. Initial sequencing and analysis of the human genome. *Nature* 2001;**409**:860–921.

42. Mardis ER. The impact of next-generation sequencing technology on genetics. *Trends Genet* 2008;**24**:133–41.

43. Das R, Dimitrova N, Xuan Z, Rollins RA, Haghighi F, Edwards JR, et al. Computational prediction of methylation status in human genomic sequences. *Proc Natl Acad Sci USA* 2006;**103**:10713–16.

44. Feltus FA, Lee EK, Costello JF, Plass C, Vertino PM. Predicting aberrant CpG island methylation. *Proc Natl Acad Sci USA* 2003;**100**:12253–8.

45. Fang F, Fan S, Zhang X, Zhang MQ. Predicting methylation status of CpG islands in the human brain. *Bioinformatics* 2006;**22**:2204–9.

46. Meissner A, Mikkelsen TS, Gu H, Wernig M, Hanna J, Sivachenko A, et al. Genome-scale DNA methylation maps of pluripotent and differentiated cells. *Nature* 2008;**454**:766–70.

47. Shann YJ, Cheng C, Chiao CH, Chen DT, Li PH, Hsu MT. Genome-wide mapping and characterization of hypomethylated sites in human tissues and breast cancer cell lines. *Genome Res* 2008;**18**:791–801.

48. Rakyan VK, Down TA, Thorne NP, Flicek P, Kulesha E, Graf S, et al. An integrated resource for genome-wide identification and analysis of human tissue-specific differentially methylated regions (tDMRs). *Genome Res* 2008;**18**:1518–29.

49. Bock C, Walter J, Paulsen M, Lengauer T. Inter-individual variation of DNA methylation and its implications for large-scale epigenome mapping. *Nucleic Acids Res* 2008;**36**:e55.

50. Mathur D, Danford TW, Boyer LA, Young RA, Gifford DK, Jaenisch R. Analysis of the mouse embryonic stem cell regulatory networks obtained by ChIP-chip and ChIP-PET. *Genome Biol* 2008;**9**:R126.

51. Ng P, Wei CL, Ruan Y. Paired-end diTagging for transcriptome and genome analysis. *Curr Protoc Mol Biol* 2007. Chapter 21:Unit 21.12.

Model Organisms of Epigenetics

CHAPTER 13

Epigenetics of Eukaryotic Microbes

Fabienne Malagnac and Philippe Silar
UFR des Sciences du Vivant, Univ Paris 7 Denis Diderot, 75013 Paris; Institut de Génétique et Microbiologie, UMR CNRS 8621, Univ Paris Sud 11, 91405 Orsay Cedex, France

INTRODUCTION

Eukaryotic microbes encompass the vast majority of the eukaryotic diversity [1]. Although some have been used as laboratory models for decades and other are important plagues to humankind, their biology is often less well-known than that of animals and plants. This applies to the knowledge of epigenetic processes, which is scarce in most microbial eukaryotes. Nevertheless, key discoveries regarding the molecular processes involved in epigenetic inheritance have been made with these organisms, especially model fungi, such as *Schizosaccharomyces pombe* and *Neurospora crassa*. Here, we will not attempt to provide an exhaustive overview of epigenetic phenomena in eukaryotic microbes, due to space constraint, but rather give a close look at the major contributions brought by these model organisms, especially filamentous fungi. Readers interested in other epigenetic phenomena such as prions and related phenomena in eukaryotic microbes are invited to read Chapter 5 by Lalucque et al. in this book.

To date, silencing phenomenon can be divided into two categories, transcriptional gene silencing (TGS), when no transcript of the targeted gene is produced, and post transcriptional gene silencing (PTGS), when transcripts are produced but specifically degraded before translation could occur. The latter, PTGS, is known as RNA interference (in animals) or co-suppression (in plants). But, PTGS and TGS also exist in fungi and protists, as will be illustrated below.

POST-TRANSCRIPTIONAL GENE SILENCING

Phylogenetic surveys of proteins involved in PTGS have shown that they are present in all lineages of eukaryotes [2,3], and thus that the ancestors of the eukaryotes were likely endowed with some primitive PTGS mechanisms. However, some organisms lack the PTGS machinery (see below), indicating that PTGS is not mandatory for efficient survival. In these early eukaryotes, PTGS could either degrade mRNA with the help of small guide RNA (e.g. siRNA and related molecules) or modify histones leading to transcriptional gene silencing, two functions that are nowadays widely conserved among eukaryotic microbes [2]. In the RNAi world, a tremendous body of work has been accomplished by taking advantage of the nematode *Caenorhabolitis elegans* [4–8]. This outstanding scientific adventure made Greg Mello and Andrew Fire Nobel Price laureates in 2006. But fungi, especially the bread mould *N. crassa*, although less famous have been instrumental in deciphering PTGS at the molecular

Handbook of Epigenetics: The New Molecular and Medical Genetics. DOI: 10.1016/B978-0-12-375709-8.00013-7

185

level. For years, this species has been a great contributor to research in many scientific fields, but as regards homologous-based control of gene expression, it shows outstanding features. In *N. crassa*, two PTGS mechanisms have been extensively studied so far: quelling and meiotic silencing of unpaired DNA. But *N. crassa* also presents a TGS mechanism, the repeat induced point mutation (RIP) phenomenon, which will be discussed at the end of this chapter.

QUELLING IN *N. crassa*

Quelling, as designated by Romano and Macino in 1992 [9], was the first truly reversible homology dependent gene silencing process, discovered in fungi. Indeed, these authors showed that endogenous expression of the *al-1* gene, involved in *N. crassa* carotenoid biosynthesis, could be silenced after transformation with homologous *al-1* sequences. This silencing was easily detected as transgenic lines ranged from wild-type orange color to light yellow and even pure white, the latter being the phenotype of *al-1* null mutant strains. But, upon vegetative growth, silenced genes were reactivated at high frequency, which often correlated with genomic rearrangements leading to partial losses of the transgenic repeats. Since then, it has been demonstrated that quelling, which is triggered during the vegetative phase, affects expression of both transgenic and endogenous homologous copies. Heterokaryons made from *al-1* silenced transgenic nuclei mixed together with wild-type nuclei revealed that quelling is dominant [10]. At the time, this latter feature strongly suggested that quelling relies on diffusible molecule(s), acting *in trans*, rather than on a DNA-DNA pairing mechanism. When the transcriptional status of the silenced loci was investigated, initiation appeared normal but no accumulation of transcripts could be detected [10]. Although DNA methylation is often detected on repeats, this epigenetic modification is not required for quelling, since silencing is fully efficient in *dim-2* mutant strains that show no DNA methylation [10]. However, methylation of lysine 9 of histones H3 (H3K9me), which is also a common epigenetic modification of chromatin, has an indirect effect on quelling [11]. Mutants defective for *dim-5*, a gene encoding a H3K9 methyltransferase [12], were unable to properly maintain quelling, because of the frequent loss of transgenes in tandem.

To further characterize the molecular bases of quelling, a mutant screen was set up by Cogoni and Macino, generating a series of quelling-deficient mutants (qde) [13]. The *qde-1* mutant was defective in an RNA-dependent RNA polymerase (RdRP) [14]. This gave the first clue that RNA components were involved in quelling. Afterwards, it was demonstrated that *Arabidopsis thaliana* and *C. elegans* homologous genes [15,16], both encoding RdRP, are required for PTGS and RNAi, respectively, indicating that the silencing machinery is evolutionarily conserved. The RNA mediated silencing model was further supported by the identification of the second gene, *qde-2*, as encoding a protein with a piwi-Paz domain that is also found in the Argonaute protein family previously characterized in plants [17]. Again, the Argonaute proteins, through the RNA-induced silencing complex (RISC) are now known to be essential for the RNA silencing pathway in numerous eukaryotes. The last of the *qde* mutants, *qde-3*, was impaired in a gene coding for a RecQ DNA helicase, suggesting the involvement of a nucleic acid pairing step [18]. Later on, DCL-1 and DCL-2, two *N. crassa* Rnase III dicer-like proteins partially redundant, were reported to be both involved in quelling by producing siRNAs of 21–25 nucleotides [19]. Biochemical purification of QDE-2 led to the identification of the exonuclease QIP [20]. QIP is thought to degrade the passenger strand of siRNA duplexes, and strains deleted for the corresponding gene are deficient for quelling. Looking for proteins that physically interact with the QDE-1 RdRP led to the discovery of the replication protein A (RPA) [21]. This finding is the first link that has been established between RNA silencing and DNA replication and opens a new field of investigations.

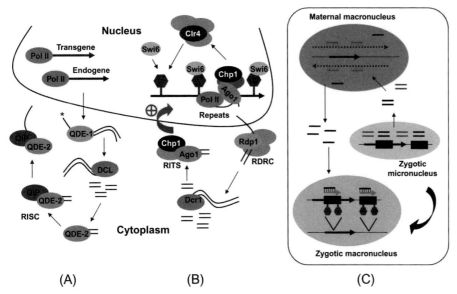

FIGURE 13.1
Models of RNAi in various eukaryotic microbes. (A) Quelling in *N. crassa.* Aberrant RNAs (*) are produced at loci that present repeats
in large tandem arrays. Features of these aberrant RNAs are unknown, but they must be recognized by the RdRP QDE-1 and then
convert into double stranded RNA molecules (dsRNA) [147]. dsRNA molecules are the typical substrate of the Dicer-like proteins
DCL-1 and DCL-2 that chop them into siRNAs of 21–25 nucleotides. These siRNAs are integrated into the RISC complex, along
with the Argonaute QDE-2 protein. They are then processed by the QIP nuclease and used as specific guides to target homologous
mRNAs, which, once trapped, are most likely degraded by QDE-2. (B) RNAi silencing in *S. pombe.* The nascent transcript model
proposes that RNA pol II continuously generates non-coding transcripts (*) from reverse promoter of heterochromatic repeats
[36]. These aberrant RNAs are first cleaved by Ago1 and then recruited by the RNA-directed RNA polymerase complex (RDRC)
to be converted into dsRNA by Rdp1 [148]. Using these dsRNAs as substrate, Dcr1 produces siRNA, which then bind to RNA-
induced transcriptional silencing (RITS) complex, by means of Ago1 [149,150]. While RISC complexes target and degrade
cytoplasmic mRNA, the RITS complex is tethered to chromatin through protein-protein interactions established between the
chromodomain protein Chip1 and the H3K9me nucleosomes [35] (hexagons). The close association of the RITS complex and
chromatin allows base-paring interactions between siRNA loaded on Ago1 and the nascent non-coding transcript soon to be
cleaved by this protein. This amplification step of siRNA is likely to form a positive-feedback loop (plus arrow), which is believed
to ensure the heterochromatin inheritance through cell divisions. As long as siRNA from a specific genomic region are produced,
they continuously target the Clr4 histone methyltransferase complex (CLRC) to nucleosomes [151,152]. Thus, using H3K9me
as signposts, heterochromatin spreads to large genomic territories in a sequence-independent but Swi6-dependent manner.
As a result, transcription of the forward strand is silenced as in classical TGS systems. Gray ovals: known additional effectors.
(C) Genome-scanning model in *Paramecium.* Because the micronucleus genome is unrearranged (rectangles represent IESs), it
produces both IES-homologous (black) and non-IES-homologous (gray) scnRNAs. These diffusible molecules would enter and scan
the IES-free maternal macronucleus. As a result of pairing with the maternal ncRNAs (dotted arrows), the non-IES-homologous
scnRNAs would be sequestered. The remaining pool of scnRNA, highly enriched with IES-homologous scnRNAs, would be free to
reach the developing zygotic macronucleus and pair with the nascent transcripts. At the IES targeted loci, chromatin shows H3K9
methylation [153] (hexagons), suggesting that this excision mechanism might have a TGS component. As for *S. pombe,* chromatin
modifications could be used as signposts to direct an endonuclease towards the IESs to be excised. The curved arrow indicates that
zygotic micronuclei develop into zygotic macronuclei throughout the course of the sexual phase. (Please refer to color plate section)

With the quelling mutants, *N. crassa* led the way to establish the molecular bases of RNAi-
mediated gene silencing (Fig. 13.1A), that we now know is widely conserved among
eukaryotes. However, to date, how genomic repetitive elements are identified as quelling
targets is still unclear.

MEIOTIC SILENCING BY UNPAIRED DNA IN *N. crassa*

Besides quelling, *N. crassa* presents a second PTGS mechanism, specifically active during
meiosis. First described as "meiotic transvection" (regulation dependent on pairing of
alleles) [22], it causes unpaired DNA to silence all the genes homologous to it, whether or

not they are themselves paired [23]. To better characterize this fascinating process, a gene fusion between histone H1 and the green flourescent protein (GFP) was transformed into *N. crassa* [24]. The transgenic strains were then crossed. When both parental strains harbored the hH1-GFP construct at the same locus, hH1-GFP was expressed all along meiosis. However, when a wild-type strain, carrying no hHP1-GFP transgene, was crossed to a hH1-GFP strain, the transgene was silenced during meiosis since no green fluorescence could be detected. But, once sexual reproduction was over, 12 to 24 hours after spore formation, the expression of the silenced hH1-GFP transgene gradually resumed. Thus, meiotic silencing operates in a limited period of the *N. crassa* life cycle, and with respect to timing, it seems to be the opposite of quelling. Nonetheless, as with quelling, meiotic silencing affects not only the unpaired copies but any additional copy sharing homology with them. This suggested that a mobile *trans*-acting signal is involved in meiotic silencing.

Once more, genetic screens set up to select suppressors of meiotic silencing allowed Metzenberg and collaborators to clarify the links between DNA pairing and this new RNA silencing-related mechanism [23]. One of the mutant strains, *sad-1*, uncovered the first gene involved in meiotic silencing. It encodes an RdRP similar to QDE-1. The *sms-2* (suppressor of meiotic silencing-2) and *sms-3* (suppressor of meiotic silencing-3) mutants are affected in genes encoding paralogs of QDE-2 and DCL-2, respectively [25]. Characterization of the *sad-2* mutant strains unraveled a protein of an unknown function, not yet identified as a component of the RNA-based silencing pathways [24]. SAD-2 and SAD-1 likely interact together, since the perinuclear localization of SAD-1 depends on the presence of SAD-2. Altogether, these findings tell us that, although different sets of proteins are required to operate quelling or meiotic silencing, the general machinery, by itself, is very similar [26]. Interestingly enough, *sad-1* mutant strains can perform interspecific crosses, which are otherwise barren when done with wild-type strains, suggesting that meiotic silencing could be one of the mechanisms by which genetic barrier is built between species, given that interspecific crosses might display unpaired DNA due to chromosomal variation. Genes encoding SAD-1-like protein can be found in a large number of fungal genomes, but to date, meiotic silencing has been described only in *N. crassa*, and is either absent or substantially reduced in the closely related species *Neurospora tetrasperma* [27].

PTGS IN OTHER FILAMENTOUS FUNGI

More generally, in filamentous fungi other than *N. crassa*, involvement of typical RNA silencing proteins such as Dicer in homology-based silencing phenomena is known at least in *Aspergillus nidulans* [28] and *Magnaporthe grisea* [29,30]. Production of siRNAs was detected in *A. nidulans* [28], *M. grisea* [29], and *Mucor circinelloides* [31]. Recent availability of numerous fungal genomes in public databases enables searches for the typical RNA silencing components by *in silico* approaches. The discovery of homologs of genes required for PTGS shows that an ever growing number of fungi are endowed with the RNAi machinery. As a matter of fact, genes seemingly involved in PTGS can be found in the four major groups of Eumycota: Ascomycota, Basidiomycota, Zygomycota, and Chytridiomycota, although this last group appears to lack QDE-1 RdRP.

Strikingly, only a very narrow subset of species, including the basidiomycete *Ustilago maydis* and the ascomycetous yeasts, both pre-whole genome duplication (WGD) species (*Ashbya gossypii*, *Kluyveromyces lactis*, *Kluyveromyces waltii*) and post-WGD species (*Saccharomyces cerevisiae*, *Saccharomyces bayanus*, *Candida glabrata*, *Candida guilliermondii*, and *Candida lusitaniae*) lack the complete set of typical RNAi proteins [32]. This finding suggests that PTGS has been recently and repetitively lost during budding-yeast evolution and therefore might not be essential for fungal survival over long periods of time. But some other budding-yeasts including the pre-WGD species *Candida albicans* and the post-WGD *Saccharomyces castellii* and

Kluyveromyces polysporus display Argonaute proteins but no canonical dicer [32]. Recently, it has been discovered that these species are in fact endowed with Dicer proteins which present a RNAseIII domain but no helicase or PAZ domains [33]. Nevertheless, these atypical Dicers produce siRNA, which are mostly targeted to transposable elements and subtelomeric repeats [33]. By introducing *S. castellii* Dicer and Argonaute genes into *S. cerevisiae*, Drinnenberg and his colleagues were even able to obtained RNAi silenced genes [33]!

The actual role of PTGS in fungi is somewhat unclear. In *N. crassa*, it has been hypothesized that quelling and meiotic silencing would protect the genome from incoming selfish genetic elements. It is also possible that some regions of the genome need PTGS for proper structuration. Indeed, data obtained with *S. pombe* have uncovered a connection between TGS and PTGS.

SILENCING IN *S. pombe*, WHEN PTGS MEETS TGS

S. pombe is the yeast of choice to study heterochromatin assembly, partly because its genome contains a large array of heterochromatic regions (pericentric and subtelomeric regions, rDNA, and silent mating-type loci). By contrast to euchromatin, the chromatin of these regions shows enrichment for Swi6 (the *S. pombe* HP1 homolog), Clr4 (the *S. pombe* homolog of Su(var)39 histone methyltransferase), and hypoacetylated H3K9me. Reporter genes inserted into these heterochromatic regions are silenced. Evidence for a functional link between RNAi machinery and heterochromatic gene-silencing assembly first came from deletion mutants of RNAi components. Indeed, deletions of Argonaute (*Ago1*), Dicer (*Dcr1*), or RNA-dependent RNA polymerase (*Rdp1*) genes impair epigenetic silencing at centromeres and the initiation of heterochromatin assembly at the mat locus, resulting in a loss of H3K9 methylation and Swi6 localization from these loci [34–36]. These findings were somehow puzzling since RNAi requires transcription while heterochromatin assembly results in TGS, as shown by silenced reporter genes. Nonetheless, small RNAs [20–22 nt) sharing homology with repeats present in the pericentric region could be detected [37]. Soon after, it was demonstrated that the RNA Pol II subunit Rpb7, contrary to other Pol II subunits, promotes pre-siRNA transcription of the so-called aberrant RNA required for RNAi-directed chromatin silencing [38,39]. Schematic representation of RNAi silencing and heterochromatin assembly in *S. pombe* is given in Figure 13.1B. The proposed model postulates that RNAi-mediated heterochromatin assembly in fission yeast appears to require initial nucleation sites that are then used as platforms to spread, but this spreading is *cis*-restricted. Boundary elements, such as that of inverted repeat (IR) of the mating-type region, prevent heterochromatin from invading the neighboring euchromatic regions [40]. This *cis* restriction is under the control of the ribonuclease Eri1, presumably by local degradation of excess siRNA [41].

To date, despite a good understanding of the involvement of the RNAi pathway in heterochromatin assembly, how histone-modifying activities, such as methylation and deacetylation, are localized in the first place remains to be determined. Addressing this question will help to understand the partition at a whole genome scale of heterochromatic regions versus euchromatic regions.

PTGS IN PROTISTS

Among protozoa and algae, PTGS has been demonstrated to be functional in alveolata (ciliates), discicristata (trypanosomes [42] and possibly *Leishmania* [43]), and unicellular green algae [44]. In many instances, the discovery of PTGS processes has led to their utilization in gene knockdown [45–47], with little study on the molecular modalities of gene silencing, some exceptions being *Trypanosoma brucei* [48,49] and *Chlamydomonas reinhardtii* [44,50–52]. PTGS pathways have been most extensively studied in ciliates in which, as in the worm *C. elegans*, silencing of gene expression can be obtained after either (i) transformation of the somatic nucleus with transgenes critically lacking a 3′ untranslated region leading to the production of dsRNA (23–24-nt siRNA) and subsequent degradation of homologous mRNA [53]; (ii) direct injection of dsRNA; or (iii) even feeding with bacteria expressing ciliate homologous dsRNA [54]. In *Tetrahymena thermophila*, production of the 23–24-nt siRNA has been shown to be dependent upon Dicer Dcr2 and the RNA-dependent RNA polymerase Rdr1 [55], as canonical RNAi pathways are. However, it is from studies focused on sexual development of this organism and *Paramecium tetraurelia* that a fascinating genome editing system, mediated by a second and distinct small RNA pathway, has been discovered.

RNA MEDIATED DEVELOPMENT IN CILIATES

P. tetraurelia is a unicellular eukaryote that contains two functionally distinct nuclei, namely germline micronuclei and somatic macronuclei. The diploid germline micronuclei, which undergo meiosis, are transcriptionally inactive during vegetative growth, whereas the highly polyploid somatic macronuclei ($\sim$800 n) are responsible for gene expression all along the life cycle, but are lost after fertilization. The mating process of *P. tetraurelia* is also very peculiar. Indeed, right after meiosis, three of the four haploid nuclei degenerate. In each conjugating partner, the remaining nucleus is then duplicated through a mitotic division. This duplication allows a reciprocal exchange of haploid nuclei between the mating paramecia. Once karyogamy has occurred, the resulting zygotes present a diploid micronucleus and deliquescent macronuclei. Therefore, brand new zygotic macronuclei have to be built up. This is achieved, after two micronucleus divisions, by massive endoreplication and extensive rearrangements of two of the four nuclei, the ones that lie at the posterior side of the cell. Chromosomes are heavily fragmented into shorter molecules capped by *de novo* telomere addition [56], but the most striking feature of those rearrangements is the precise excision of tens of thousands of single-copy short non-coding internal sequences (IESs) [57,58], which makes the macronuclei an expurgated version of the micronuclei. How can such an astonishing editing effort be performed?

First hints of an epigenetic compound implicated in that genome-wide rearrangement process came from transformation experiments on *P. tetraurelia* [59–61]. When an IES sequence is integrated into vegetative macronuclei, excision of the corresponding IES in the new macronuclei of sexual progeny is specifically inhibited [62]. The IES retention, which makes it present in all macromolecular copies is then maternally (cytoplasmically) inherited in the following sexual generations. This was clearly reminiscent of an epigenetic homology-based mechanism. Later on, in *T. thermophila*, developmental rearrangements were shown to depend on the *TWI1* gene, which encodes a protein homologous to Piwi-like proteins [63], on the *DCL1* gene, encoding a Dicer-like protein [64] and on Ema1p a putative RNA helicase [65]. In *P. tetraurelia*, identification of the Nowa1 and Nowa2, two RNA binding proteins required to remove the IESs from the developing macronuclei [66], further indicated that the cross-talk between nuclei at work during genome rearrangements is related to an RNAi pathway. Thus, unlike the canonical RNAi pathway, this second homology-dependent silencing system is restricted to sexual development, precisely when germline DNA rearrangements take place. It produces a specific class of 25-nt siRNA, called "scan RNAs" (scnRNAs) [67,68]. Microinjection of a 25-nt synthetic RNA duplex mimicking the structure of scnRNAs was shown to actually promote excision of the homologous IESs in the developing zygotic macronuclei [69]. Furthermore, in *Paramecium*, non-protein-coding transcripts (ncRNAs) produced from the somatic maternal macronucleus (devoid of IESs) are essential for IES excision in the developing zygotic macronucleus [69]. From this set of data, a whole "genome-scanning" model [70,71] has been proposed (Fig. 13.1C). According to this model, the epigenetic developmental program resulting in massive but precise DNA elimination would be based on a genomic subtraction between deletion-inducing scnRNAs and protective non-coding transcripts.

Studying sexual development in *P. tetraurelia* and *T. thermophila* has brought a lot to epigenetic fields, especially by giving intriguing new insights of how diverse homology-dependent mechanisms can be. The recruitment of the PTGS machinery in ciliates to help shape a new somatic genome free of selfish DNA elements is reminiscent of the roles attributed to PTGS in protecting the filamentous fungus genomes and in defining genomic heterochromatin territories of *S. pombe*. In other protists, such as in *T. brucei*, it was shown that transposons are reactivated in PTGS deficient mutants, confirming a role of PTGS in defending the genome against expression, and possibly expansion, of junk DNA [72].

TRANSCRIPTIONAL GENE SILENCING

Chromatin and chromatin-based gene regulation is present in many eukaryotes [73]. Again, a phylogenetic survey of chromatin proteins show that they are widely conserved [74], an argument in favor of an ancient origin of chromatin-based gene silencing. Yet, some eukaryotes have lost all chromatin, arguing that, like PTGS mechanisms, TGS pathways are not mandatory for survival. The best known of these organisms lacking typical chromatin are dinoflagellates. Indeed, these highly successful protists are considered to be one of the three major constituents of the phytoplankton. They have no nucleosomes [75] and have huge genomes condensed in the liquid crystal state [76,77]. For eukaryotic organisms that have lost canonical histones, this liquid crystal state of DNA may be the only option for retaining the necessary chromosomal compactness with segregation capability.

TGS modulates gene expression for various purposes, including antigen variability, mating type switching, protection against transposons and, possibly, development. As with PTGS mechanisms, fungi have greatly contributed to elucidating TGS mechanisms. Due to lack of space we are not able to discuss gene extinction in *S. cerevisiae*, where TGS is known to regulate silencing at mating-type cassettes, variegation in expression of telomere-located genes and recombination at the rDNA repeats. Importantly, *S. cerevisiae* lacks the HP1 protein, involved in the other eukaryotes in packaging heterochromatin. The production of heterochromatin in this yeast relies on a different set of proteins. Readers interested in *S. cerevisiae* TGS can refer to recent reviews [78–81].

We will discuss two TGS mechanisms of filamentous ascomycetes (Pezizomycotina), *A. immersus* and *N. crassa*. Although the *N. crassa* RIP process [82] was discovered before the Methylation Induced Premeiotically (MIP) process [83] of *A. immersus*, the latter will be dealt with first as it is truly a TGS system, but it is important to note that much of what was discovered about MIP was aided by the prior discovery of RIP. *N. crassa* and *A. immersus* are haploid during their vegetative growth phase. But when two haploid strains of compatible mating type encounter each other, sexual reproduction takes place. This first results in the formation of a transient dikaryotic cell. This feature, where two haploid nuclei are brought together, after mating, during an extended period within the same cell, is unique to higher fungi, the Dikaryomycota. Once karyogamy occurs a diploid cell is formed that undergoes meiosis immediately, which is then followed by post-meiotic mitosis generating asci with eight haploid ascospores. TGS in both *A. immersus* and *N. crassa* has been detected in the progeny after meiosis and affects genes present in two copies or more, in the same nucleus, during the dikaryotic phase.

METHYLATION INDUCED PREMEIOTICALY IN *A. immersus*

DNA methylation is a common epigenetic modification that can be detected in eubacteria, protists, fungi, plants, and animals. In eukaryotes, DNA methylation is restricted to the cytosine residues, either to any cytosine residues in plant and fungal genomes, or cytosine located within CpG dinucleotides in genomes of animals. As documented in other chapters of this book, DNA methylation has a strong impact on gene expression. Namely, in association with chromatin remodeling factors, it acts as a switch that can reversibly turn ON and OFF gene transcription. Methylation as a regulator of gene expression has been especially well studied in *A. immersus*. In this fungus, genes present in more than one copy, the so-called repeats obtained after integrative transformation, frequently lost their expression after the first round of sexual reproduction [83–85]. In the 1990s, Rossignol and his co-workers were able to demonstrate that this spontaneous inactivation, clearly triggered by repeats, was not due to mutations but rather to epimutations, since systematic DNA sequencing showed no mutation in the inactivated strains [86]. The observed silencing of gene expression was faithfully maintained throughout numerous mitotic and meiotic divisions, even if the repeats had segregated away from each other, but was proved to be reversible under selective pressure. With no exception, the silenced repeated genes were found heavily methylated. Most of their cytosine residues were modified. Furthermore, in all cases, the methylation pattern was strictly co-extensive with the length of the duplication. Since the repeats have to be present in the same haploid nucleus for the silencing to occur (a single copy present in the other nucleus was not inactivated), it was inferred that this inactivation process takes place in the dikaryotic nuclei, in a period between fertilization and karyogamy during which the two haploid nuclei involved in the cross are both present in the same cell but have not yet fused.

Tandem repeats as short as 400 pb, and ectopic duplications of 600 pb in length, can be efficiently targeted by MIP [87]. In addition to *de novo* methylation of the cytosine residues within the MIPed alleles, silencing was accompanied by either the absence of transcripts or the presence of truncated transcripts [88]. This was indicative of a TGS type of silencing mechanism. Sequencing and mapping of truncated transcripts made it clear that, once initiated, transcription can progress up to the boundary of the adjacent duplicated and methylated region, but reaching this point, the transcription elongation stops abruptly, leading to the production of unusual shorten transcripts. Thus, even though TGS is a conserved process among eukaryotes, effects on transcription are quite different between fungi on one hand, and plants and animals on the other hand. In plants and mammals, methylation of promoter regions correlates with lack of transcription initiation. In *A. immersus*, methylation of promoters does not prevent initiation of transcription, but methylation in the body of a duplicated gene inhibits transcription elongation from both copies. To date, no explanation has been found to account for this discrepancy. The chromatin states of the MIPed alleles was investigated [89]. Partial micrococcal nuclease digestion evidenced that the sensitive sites present along the unmethylated regions are no longer observed along the MIPed ones. Hence MIP is able to change the chromatin compaction of its genomic targets. Again, the extent of methylation and chromatin remodeling are alike. What role DNA methylation plays in these changes remains to be determined. In addition, these chromatin changes are associated with an increase in dimethylation on H3K9, and a decrease in dimethylation on H3K4 [89]. Contrary to the case with other organisms that display TGS, such as plants, no decrease in acetylation of histones H4 was observed. Is this why transcription initiation in *A. immersus* seems independent of the chromatin states and methylation status of the promoters? Or is it because promoters are not as well defined in filamentous fungi as in plants and animals?

Because it was so easy to get portions of DNA methylated through MIP, transfer of methylation between alleles was investigated in the *A. immersus* genome. This transfer was shown to be as frequent and polarized as gene conversion is [90]. This was a first indication that methylation transfer and recombination might be mechanistically related. A second clue came when crossing-over frequency was measured between two markers flanking an *A. immersus* spore color gene [91]. When the two homologs were methylated, the crossing-over frequency was reduced several hundredfold. This demonstrates that DNA methylation strongly inhibits homologous recombination. This also supports, on experimental bases, the hypothesis that methylation prevents homologous recombination between dispersed DNA repeats and therefore contributes to genome integrity.

The only MIP mutant that has been characterized is impaired in a gene, *masc1*, encoding a protein that bears all motifs of the catalytic domain of eukaryotic C5-DNA-methyltransferases

(DMT) [92]. However, despite its canonical DMT structure, no enzymatic activity was ever detected in standard *in vitro* assays. Although methylation was fully maintained on previously MIPed alleles, the *masc1* mutation prevents the *de novo* methylation of newly formed DNA repeats through MIP. Interestingly, crosses involving *masc1* mutant strains of the compatible mating types were arrested at an early stage of sexual reproduction and therefore barren. This indicates that the Masc1 protein, in addition to being required for the MIP process, plays a crucial role in sexual development. Curiously, DmtA, the Masc1 ortholog of *A. nidulans*, a fungus thought to have no DNA methylation and no TGS system, is also essential for early sexual development [93]. Is there a class of DMT-like proteins involved in early steps of fungal sexual reproduction? Is MIP a mechanism that evolved to protect the spreading of repeats across the *A. immersus* genome, in order not to have to deal with unpaired DNA during meiosis, as meiotic silencing does in *N. crassa*? To date, these questions remain to be addressed.

REPEAT INDUCED POINT MUTATION IN *N. crassa*

N. crassa, in addition to quelling and meiotic silencing, also displays a TGS-related mechanism, RIP. It was first discovered by Selker and collaborators in 1987 [82]. Like MIP, this premeiotic silencing process takes place at the dikaryotic stage of the sexual cycle. DNA repeats longer than 400 pb [94] that share a nucleotide identity greater than 80% are irreversibly mutagenized via C:G to T:A transitions. As an outcome of RIP, the *Neurospora* genome reveals a complete absence of intact mobile elements [26,95] and natural repeats display an AT-rich content. Interestingly, while the bulk of the *N. crassa* genome is unmethylated, RIPed repeats are heavily methylated. Furthermore it was shown that these AT-rich regions are by themselves a positive signal that promotes DNA methylation [96–98]. Whether DNA methylation is installed before the mutagenesis as the first step of RIP or only after the cross, in vegetative cells, is still not elucidated. Nonetheless, DNA methylation is associated with most of the sequences affected by RIP, and methylated cytosines are not limited to CpG dinucleotides [99]. If the RIPed sequences encompass genes, their expression is silenced, due to a strong reduction in transcription [100]. Run-on experiments have demonstrated that transcripts are initiated, even from methylated promoters but that elongation is blocked when the RNA polymerase II stalls in methylated regions lying in the body of the RIPed genes. However, DNA methylation alone is not sufficient to block transcription, which strongly suggests that others factors, likely linked to chromatin remodeling, might turn the RIPed region into silent heterochromatin. Altogether, these features define a two component system. Before meiosis, RIP introduces true mutations in the *N. crassa* genome and is therefore non-reversible. Reversibility is a property exhibited by most of the proper epigenetic phenomena, MIP included. But during vegetative life, DNA and H3K9 methylation [12], two genuine epigenetic modifications, maintain the transcriptional silencing of the RIPed alleles.

Mechanistically, majors questions remain to be answered [101]. One of them is how repeats identify each other. Since none or all the copies of repeated DNA are RIPed, the idea that this silencing mechanism can involve a DNA–DNA pairing step has been proposed. Moreover, the fact that RIP cannot be transmitted from one nucleus to the other in the dikaryotic cells suggests that it may not work through a diffusible signal [102]. Indeed, the *N. crassa qde* mutants impaired in the RNAi machinery – see the earlier text on "quelling" – can establish and maintain DNA or H3K9 methylation very well [103]. Thus, it is very unlikely that RNA intermediates can participate in the RIP homologously-based gene expression silencing. Another crucial question is how RIP mutations occur. It has been proposed that methylated cytosines are prone to be spontaneously deaminated at high frequency which would result in a cytosine to thymidine conversion. Alternatively, a DNA-cytidine deaminase might directly perform the conversion [101]. But so far, no experimental clue has arisen to confirm any of these hypotheses.

TABLE 13.1 Repeat Induced Point Mutation (RIP) in Fungi

Organism	Evidence	Reference
Neurospora crassa	Experimental	[82,101]
Podospora anserina	Experimental	[131,132]
Leptospheria maculans	Experimental	[110,133]
Magnaporthe grisea	Experimental	[134,135]
Magnaporthe oryzae	*In silico*	[136]
Aspergillus fumigatus	*In silico*	[137]
Aspergillus nidulans	*In silico*	[138]
Aspergillus niger	*In silico*	[139]
Fusarium oxysporum	*In silico*	[140–142]
Fusarium graminearum	Experimental	[143]
Nectria haematococca	Experimental	[144]
Microbotryum violaceum	*In silico*	[145]
Penicillium chrysogenum	*In silico*	[139]
Stagonospora nodorum	*In silico*	[146]

Experimental: functional RIP has been evidenced by experimental methods. *In silico* signatures: genomic sequences show typical C:G to T:A transitions, mostly by sequencing DNA repeats such as transposons; there is no experimental proof of functional RIP.

To date, only one RIP defective mutant has been characterized, whereas several mutations that impair DNA methylation with no effect on RIP are known [12,103–106]. Mutation in the *rid-1* gene encoding a putative DNA methyltransferase protein results in fertile but RIP defective strains [107]. As for *Ascobolus* Masc1 protein, *in vitro* assays did not reveal any DNA methyltransferase activity. Again, function of this DMT-like protein remains mysterious.

RIP/TGS IN OTHER FILAMENTOUS FUNGI

RIP has also been observed in numerous filamentous ascomycetes (Table 13.1). Although a common feature, this silencing system appears less efficient in other fungi than in *N. crassa* and has still no clear physiological role besides its impact upon genomic plasticity. Indeed, on one hand, RIP counters selfish DNA and therefore protect genomes from expansion of junk DNA, but on the other hand, it has significant cost on genome evolution by preventing the appearance of paralogs, as illustrated by the *N. crassa* genome, where creation of new genes through duplication is almost impossible [26]. By contrast, in *P. anserina*, where RIP is weak, numerous segmental duplications are detected [108]. The fact that large genes may duplicate is not contradictory to the presence of RIP, since, when moderately efficient, it can accelerate gene divergence as described for the het-D/E family [109]. Interestingly, in a field population of *L. maculans*, multiple independent RIP events were shown to be responsible for evolution of the AvrLm6 locus toward virulence, within a single season [110].

CHROMATIN-BASED REGULATION OF SECONDARY METABOLITE GENE CLUSTER EXPRESSION

Aspergilli are fungi of particular importance both as pathogens (human and plants) and as industrial organisms used in a wide range of productions. Synthesis of an amazing number of secondary metabolites, some of economic value, others poisonous, is one of the most remarkable properties of these fungi. The genes encoding secondary metabolites are generally grouped into clusters. It is difficult to monitor the production of these compounds since some clusters may be silenced [111]. Deletion of the *A. nidulans laeA* gene encoding an O-methyltransferase blocks the expression of the sterigmatocystin, penicillin, and lovastatin gene clusters [112,113]. Conversely, overexpression of *laeA* leads to increased penicillin and lovastatin gene transcription [112]. Recently, mutants of *A. nidulans* impaired for hdaA [114], a histone deacetylase, and CclA [115] involved in H3K9 methylation showed activation of several otherwise cryptic secondary metabolite clusters. These results led to the hypothesis that epigenetic mechanisms based on histone modifications might be crucial regulators for secondary metabolite clusters and provide a framework to attempt to control their expression.

TGS IN PROTISTS

Data on TGS in protozoa and algae are scarce. Some are available for the green algae *C. reinhardtii* [52,116,117]. However, most of them come from studies on *Plasmodium falciparum*, the malaria parasite (an apicomplexan), and on *T. brucei*, the agent of sleeping sickness (a discicristatan). A common fascinating property of these evolutionary very divergent intracellular parasites is their ability to perform antigenic variation. The multigenic VAR family of *P. falciparum* and VSG family of *T. brucei* both encode glycoproteins that coat the surface of the cells. The VSG genes and the VAR genes are localized at subtelomeric loci [118]. Only one gene of the family is expressed at a time. Moreover, using a periodic switch of the expressed gene, parasites can alter their antigenic signature and thus escape the immune system of the host [119]. Antigenic variation is the main reason that makes malaria or sleeping sickness chronic diseases. But to establish such an unusual mono-allelic expression, the parasites must dispose of a mechanism that tightly regulates *in situ* the switching and the mutually exclusive transcription of the VAR and VSG genes. Among others, epigenetic regulation has been postulated [120]. Notably, while the available apicomplexa genomes [121] show very few DNA-binding factors, it seems that numerous non-coding RNA are expressed in these parasites [122]. Another uncommon feature is that the VSG and VAR families are transcribed by Pol I, a polymerase exclusively involved in ribosomal DNA transcription in other eukaryotes.

In *T. brucei*, RNAi mediated knock-down of *ISWI*, a gene encoding a chromatin remodeling factor, results in derepression of the silenced VSG genes [123]. In the same organism, deletion of *DOT1B*, a gene encoding an enzyme responsible for trimethylation of H3K76, also leads to tenfold derepression of silent VSG genes [124]. But the link between TGS-based telomeric silencing and VSG regulation of expression is not so straightforward. Indeed, mutants impaired in the gene encoding SIR2rp1, a sirtuin, show activation of Pol I reporter constructs, but not of the endogenous VSG genes [125].

In *P. falciparum*, activation and silencing of VAR genes correlate with specific histone tail marks: H3K9 acetylation and H3K4 methylation have been shown to be associated with VAR gene activation [126], whereas tri-methylation of H3K9 is associated with VAR gene silencing [127]. Conversely to *T. brucei*, *P. falciparum* homologs of the histone deacetylase Sir2 are involved in the regulation of antigenic variation, in both mutual exclusion and silencing [128–130]. Further characterization of TGS pathways in such parasites might provide therapeutic prospects.

CONCLUSION

Although still patchy, the available data concerning gene silencing show that, in many eukaryotic microbes, both PTGS and TGS occur with modalities similar to those described in animals and plants. However, differences may occur as exemplified by the complete loss of the PTGS machinery in some fungi, the lack of HP1 in *S. cerevisiae*, and the lack of true chromatin in dinoflagellates. Gene silencing is involved in a variety of unrelated physiological processes in the form of clonal regulation of gene expression (antigen variation in parasites), genome defense (RIP and MIP), and genome structuration (macronuclei formation in ciliates and PTGS in *S. pombe*). Interestingly, the PTGS phenomenon of meiotic silencing could participate in the formation of species by an original mechanism. We expect that the exploration of these mechanisms both in well-tracked models and in more exotic species is likely to provide further original modalities and roles for both TGS and PTGS.

References

1. Cavalier-Smith T. Megaphylogeny, cell body plans, adaptive zones: causes and timing of eukaryote basal radiations. *J Eukaryot Microbiol* 2009;**56**:26–33.

2. Cerutti H, Casas-Mollano JA. On the origin and functions of RNA-mediated silencing: from protists to man. *Curr Genet* 2006;**50**:81–99.

3. Shabalina SA, Koonin EV. Origins and evolution of eukaryotic RNA interference. *Trends Ecol Evol* 2008;**23**:578–87.

4. Fire A, Xu S, Montgomery MK, Kostas SA, Driver SE, Mello CC. Potent and specific genetic interference by double-stranded RNA in *Caenorhabditis elegans*. *Nature* 1998;**391**:806–11.

5. Guo S, Kemphues KJ. par-1, a gene required for establishing polarity in *C. elegans* embryos, encodes a putative Ser/Thr kinase that is asymmetrically distributed. *Cell* 1995;**81**:611–20.

6. Timmons L, Fire A. Specific interference by ingested dsRNA. *Nature* 1998;**395**:854.

7. Boisvert ME, Simard MJ. RNAi pathway in *C. elegans*: the argonautes and collaborators. *Curr Top Microbiol Immunol* 2008;**320**:21–36.

8. Hunter CP, Winston WM, Molodowitch C, Feinberg EH, Shih J, Sutherlin M, et al. Systemic RNAi in *Caenorhabditis elegans*. *Cold Spring Harb Symp Quant Biol* 2006;**71**:95–100.

9. Romano N, Macino G. Quelling: transient inactivation of gene expression in *Neurospora crassa* by transformation with homologous sequences. *Mol Microbiol* 1992;**6**:3343–53.

10. Cogoni C, Irelan JT, Schumacher M, Schmidhauser TJ, Selker EU, Macino G. Transgene silencing of the al-1 gene in vegetative cells of *Neurospora* is mediated by a cytoplasmic effector and does not depend on DNA-DNA interactions or DNA methylation. *EMBO J* 1996;**15**:3153–63.

11. Chicas A, Forrest EC, Sepich S, Cogoni C, Macino G. Small interfering RNAs that trigger posttranscriptional gene silencing are not required for the histone H3 Lys9 methylation necessary for transgenic tandem repeat stabilization in *Neurospora crassa*. *Mol Cell Biol* 2005;**25**:3793–801.

12. Tamaru H, Selker EU. A histone H3 methyltransferase controls DNA methylation in *Neurospora crassa*. *Nature* 2001;**414**:277–83.

13. Cogoni C, Macino G. Isolation of quelling-defective (qde) mutants impaired in posttranscriptional transgene-induced gene silencing in *Neurospora crassa*. *Proc Natl Acad Sci USA* 1997;**94**:10233–8.

14. Cogoni C, Macino G. Gene silencing in *Neurospora crassa* requires a protein homologous to RNA-dependent RNA polymerase. *Nature* 1999;**399**:166–9.

15. Mourrain P, Beclin C, Elmayan T, Feuerbach F, Godon C, Morel JB, et al. Arabidopsis SGS2 and SGS3 genes are required for posttranscriptional gene silencing and natural virus resistance. *Cell* 2000;**101**:533–42.

16. Smardon A, Spoerke JM, Stacey SC, Klein ME, Mackin N, Maine EM. EGO-1 is related to RNA-directed RNA polymerase and functions in germ-line development and RNA interference in *C. elegans*. *Curr Biol* 2000;**10**:169–78.

17. Catalanotto C, Azzalin G, Macino G, Cogoni C. Gene silencing in worms and fungi. *Nature* 2000;**404**:245.

18. Cogoni C, Macino G. Posttranscriptional gene silencing in *Neurospora* by a RecQ DNA helicase. *Science* 1999;**286**:2342–4.

19. Catalanotto C, Pallotta M, ReFalo P, Sachs MS, Vayssie L, Macino G, et al. Redundancy of the two dicer genes in transgene-induced posttranscriptional gene silencing in *Neurospora crassa*. *Mol Cell Biol* 2004;**24**:2536–45.

20. Maiti M, Lee HC, Liu Y. QIP, a putative exonuclease, interacts with the *Neurospora* Argonaute protein and facilitates conversion of duplex siRNA into single strands. *Genes Dev* 2007;**21**:590–600.

21. Nolan T, Cecere G, Mancone C, Alonzi T, Tripodi M, Catalanotto C, et al. The RNA-dependent RNA polymerase essential for post-transcriptional gene silencing in *Neurospora crassa* interacts with replication protein A. *Nucleic Acids Res* 2008;**36**:532–8.

22. Aramayo R, Metzenberg RL. Meiotic transvection in fungi. *Cell* 1996;**86**:103–13.

23. Shiu PK, Raju NB, Zickler D, Metzenberg RL. Meiotic silencing by unpaired DNA. *Cell* 2001;**107**:905–16.

24. Shiu PK, Zickler D, Raju NB, Ruprich-Robert G, Metzenberg RL. SAD-2 is required for meiotic silencing by unpaired DNA and perinuclear localization of SAD-1 RNA-directed RNA polymerase. *Proc Natl Acad Sci USA* 2006;**103**:2243–8.

25. Lee DW, Pratt RJ, McLaughlin M, Aramayo R. An argonaute-like protein is required for meiotic silencing. *Genetics* 2003;**164**:821–8.

26. Galagan JE, Calvo SE, Borkovich KA, Selker EU, Read ND, Jaffe D, et al. The genome sequence of the filamentous fungus *Neurospora crassa*. *Nature* 2003;**422**:859–68.

27. Jacobson DJ, Raju NB, Freitag M. Evidence for the absence of meiotic silencing by unpaired DNA in *Neurospora tetrasperma*. *Fungal Genet Biol* 2008;**45**:351–62.

28. Hammond TM, Keller NP. RNA silencing in *Aspergillus nidulans* is independent of RNA-dependent RNA polymerases. *Genetics* 2005;**169**:607–17.

29. Kadotani N, Nakayashiki H, Tosa Y, Mayama S. RNA silencing in the phytopathogenic fungus *Magnaporthe oryzae*. *Mol Plant Microbe Interact* 2003;**16**:769–76.

30. Kadotani N, Nakayashiki H, Tosa Y, Mayama S. One of the two Dicer-like proteins in the filamentous fungi *Magnaporthe oryzae* genome is responsible for hairpin RNA-triggered RNA silencing and related small interfering RNA accumulation. *J Biol Chem* 2004;**279**:44467–74.

31. Nicolas FE, Torres-Martinez S, Ruiz-Vazquez RM. Two classes of small antisense RNAs in fungal RNA silencing triggered by non-integrative transgenes. *Embo J* 2003;**22**:3983–91.

32. Scannell DR, Frank AC, Conant GC, Byrne KP, Woolfit M, Wolfe KH. Independent sorting-out of thousands of duplicated gene pairs in two yeast species descended from a whole-genome duplication. *Proc Natl Acad Sci USA* 2007;**104**:8397–402.

33. Drinnenberg IA, Weinberg DE, Xie KT, Mower JP, Wolfe KH, Fink GR, et al. RNAi in budding yeast. *Science* 2009;**326**:544–50.

34. Hall IM, Shankaranarayana GD, Noma K, Ayoub N, Cohen A, Grewal SI. Establishment and maintenance of a heterochromatin domain. *Science* 2002;**297**:2232–37.

35. Verdel A, Jia S, Gerber S, Sugiyama T, Gygi S, Grewal SI, et al. RNAi-mediated targeting of heterochromatin by the RITS complex. *Science* 2004;**303**:672–6.

36. Volpe TA, Kidner C, Hall IM, Teng G, Grewal SI, Martienssen RA. Regulation of heterochromatic silencing and histone H3 lysine-9 methylation by RNAi. *Science* 2002;**297**:1833–7.

37. Reinhart BJ, Bartel DP. Small RNAs correspond to centromere heterochromatic repeats. *Science* 2002;**297**:1831.

38. Djupedal I, Portoso M, Spahr H, Bonilla C, Gustafsson CM, Allshire RC, et al. RNA Pol II subunit Rpb7 promotes centromeric transcription and RNAi-directed chromatin silencing. *Genes Dev* 2005;**19**:2301–6.

39. Kato H, Goto DB, Martienssen RA, Urano T, Furukawa K, Murakami Y. RNA polymerase II is required for RNAi-dependent heterochromatin assembly. *Science* 2005;**309**:467–9.

40. Noma K, Allis CD, Grewal SI. Transitions in distinct histone H3 methylation patterns at the heterochromatin domain boundaries. *Science* 2001;**293**:1150–5.

41. Buhler M, Verdel A, Moazed D. Tethering RITS to a nascent transcript initiates RNAi- and heterochromatin-dependent gene silencing. *Cell* 2006;**125**:873–86.

42. Balana-Fouce R, Reguera RM. RNA interference in *Trypanosoma brucei*: a high-throughput engine for functional genomics in trypanosomatids? *Trends Parasitol* 2007;**23**:348–51.

43. Peacock CS, Seeger K, Harris D, Murphy L, Ruiz JC, Quail MA, et al. Comparative genomic analysis of three *Leishmania* species that cause diverse human disease. *Nat Genet* 2007;**39**:839–47.

44. Zhao T, Li G, Mi S, Li S, Hannon GJ, Wang XJ, et al. A complex system of small RNAs in the unicellular green alga *Chlamydomonas reinhardtii*. *Genes Dev* 2007;**21**:1190–203.

45. Bellofatto V, Palenchar JB. RNA interference as a genetic tool in trypanosomes. *Methods Mol Biol* 2008;**442**:83–94.

46. Meissner M, Breinich MS, Gilson PR, Crabb BS. Molecular genetic tools in *Toxoplasma* and *Plasmodium*: achievements and future needs. *Curr Opin Microbiol* 2007;**10**:349–56.

47. Militello KT, Refour P, Comeaux CA, Duraisingh MT. Antisense RNA and RNAi in protozoan parasites: working hard or hardly working? *Mol Biochem Parasitol* 2008;**157**:117–26.

48. Best A, Handoko L, Schluter E, Goringer HU. In vitro synthesized small interfering RNAs elicit RNA interference in African trypanosomes: an *in vitro* and *in vivo* analysis. *J Biol Chem* 2005;**280**:20573–9.

49. Shi H, Tschudi C, Ullu E. An unusual Dicer-like1 protein fuels the RNA interference pathway in *Trypanosoma brucei*. *RNA* 2006;**12**:2063–72.

50. Yamasaki T, Miyasaka H, Ohama T. Unstable RNAi effects through epigenetic silencing of an inverted repeat transgene in *Chlamydomonas reinhardtii*. *Genetics* 2008;**180**:1927–44.

51. Rohr J, Sarkar N, Balenger S, Jeong BR, Cerutti H. Tandem inverted repeat system for selection of effective transgenic RNAi strains in *Chlamydomonas*. *Plant J* 2004;**40**:611–21.

52. Casas-Mollano JA, van Dijk K, Eisenhart J, Cerutti H. SET3p monomethylates histone H3 on lysine 9 and is required for the silencing of tandemly repeated transgenes in *Chlamydomonas*. *Nucleic Acids Res* 2007;**35**:939–50.

53. Ruiz F, Vayssie L, Klotz C, Sperling L, Madeddu L. Homology-dependent gene silencing in *Paramecium*. *Mol Biol Cell* 1998;**9**:931–43.

54. Galvani A, Sperling L. RNA interference by feeding in *Paramecium*. *Trends Genet* 2002;**18**:11–12.

55. Lee SR, Collins K. Physical and functional coupling of RNA-dependent RNA polymerase and Dicer in the biogenesis of endogenous siRNAs. *Nat Struct Mol Biol* 2007;**14**:604–10.

56. Aury JM, Jaillon O, Duret L, Noel B, Jubin C, Porcel BM, et al. Global trends of whole-genome duplications revealed by the ciliate *Paramecium tetraurelia*. *Nature* 2006;**444**:171–8.

57. Klobutcher LA, Herrick G. Developmental genome reorganization in ciliated protozoa: the transposon link. *Prog Nucleic Acid Res Mol Biol* 1997;**56**:1–62.

58. Meyer E, Garnier O. Non-Mendelian inheritance and homology-dependent effects in ciliates. *Adv Genet* 2002;**46**:305–37.

59. Koizumi S, Kobayashi S. Microinjection of plasmid DNA encoding the A surface antigen of *Paramecium tetraurelia* restores the ability to regenerate a wild-type macronucleus. *Mol Cell Biol* 1989;**9**:4398–401.

60. Jessop-Murray H, Martin LD, Gilley D, Preer JR Jr., Polisky B. Permanent rescue of a non-Mendelian mutation of *Paramecium* by microinjection of specific DNA sequences. *Genetics* 1991;**129**:727–34.

61. You Y, Aufderheide K, Morand J, Rodkey K, Forney J. Macronuclear transformation with specific DNA fragments controls the content of the new macronuclear genome in *Paramecium tetraurelia. Mol Cell Biol* 1991;**11**:1133–7.

62. Duharcourt S, Butler A, Meyer E. Epigenetic self-regulation of developmental excision of an internal eliminated sequence on *Paramecium tetraurelia. Genes Dev* 1995;**9**:2065–77.

63. Mochizuki K, Fine NA, Fujisawa T, Gorovsky MA. Analysis of a piwi-related gene implicates small RNAs in genome rearrangement in *Tetrahymena. Cell* 2002;**110**:689–99.

64. Mochizuki K, Gorovsky MA. A Dicer-like protein in *Tetrahymena* has distinct functions in genome rearrangement, chromosome segregation, and meiotic prophase. *Genes Dev* 2005;**19**:77–89.

65. Aronica L, Bednenko J, Noto T, DeSouza LV, Siu KW, Loidl J, et al. Study of an RNA helicase implicates small RNA-noncoding RNA interactions in programmed DNA elimination in *Tetrahymena. Genes Dev* 2008;**22**:2228–41.

66. Nowacki M, Zagorski-Ostoja W, Meyer E. Nowa1p and Nowa2 p: novel putative RNA binding proteins involved in trans-nuclear crosstalk in *Paramecium tetraurelia. Curr Biol* 2005;**15**:1616–28.

67. Lee SR, Collins K. Two classes of endogenous small RNAs in *Tetrahymena thermophila. Genes Dev* 2006;**20**:28–33.

68. Lepere G, Nowacki M, Serrano V, Gout JF, Guglielmi G, Duharcourt S, et al. Silencing-associated and meiosis-specific small RNA pathways in *Paramecium tetraurelia. Nucleic Acids Res* 2009;**37**:903–15.

69. Lepere G, Betermier M, Meyer E, Duharcourt S. Maternal noncoding transcripts antagonize the targeting of DNA elimination by scanRNAs in *Paramecium tetraurelia. Genes Dev* 2008;**22**:1501–12.

70. Duharcourt S, Lepere G, Meyer E. Developmental genome rearrangements in ciliates: a natural genomic subtraction mediated by non-coding transcripts. *Trends Genet* 2009;**25**:344–50.

71. Malone CD, Hannon GJ. Small RNAs as guardians of the genome. *Cell* 2009;**136**:656–68.

72. Shi H, Djikeng A, Tschudi C, Ullu E. Argonaute protein in the early divergent eukaryote *Trypanosoma brucei*: control of small interfering RNA accumulation and retroposon transcript abundance. *Mol Cell Biol* 2004;**24**:420–7.

73. Grewal SI, Jia S. Heterochromatin revisited. *Nat Rev Genet* 2007;**8**:35–46.

74. Iyer LM, Anantharaman V, Wolf MY, Aravind L. Comparative genomics of transcription factors and chromatin proteins in parasitic protists and other eukaryotes. *Intl J Parasitol* 2008;**38**:1–31.

75. Moreno Diaz de la Espina S, Alverca E, Cuadrado A, Franca S. Organization of the genome and gene expression in a nuclear environment lacking histones and nucleosomes: the amazing dinoflagellates. *Eur J Cell Biol* 2005;**84**:137–49.

76. Livolant F, Maestre MF. Circular dichroism microscopy of compact forms of DNA and chromatin in vivo and in vitro: cholesteric liquid-crystalline phases of DNA and single dinoflagellate nuclei. *Biochemistry* 1988;**27**:3056–68.

77. Rill RL, Livolant F, Aldrich HC, Davidson MW. Electron microscopy of liquid crystalline DNA: direct evidence for cholesteric-like organization of DNA in dinoflagellate chromosomes. *Chromosoma* 1989;**98**:280–6.

78. Brickner JH. Transcriptional memory at the nuclear periphery. *Curr Opin Cell Biol* 2009;**21**:127–33.

79. Gao L, Gross DS. Using genomics and proteomics to investigate mechanisms of transcriptional silencing in *Saccharomyces cerevisiae. Brief Funct Genomic Proteomic* 2006;**5**:280–8.

80. Huang Y. Transcriptional silencing in *Saccharomyces cerevisiae* and *Schizosaccharomyces pombe. Nucleic Acids Res* 2002;**30**:1465–82.

81. Rusche LN, Kirchmaier AL, Rine J. The establishment, inheritance, and function of silenced chromatin in *Saccharomyces cerevisiae. Annu Rev Biochem* 2003;**72**:481–516.

82. Selker EU, Jensen BC, Richardson GA. A portable signal causing faithful DNA methylation de novo in *Neurospora crassa. Science* 1987;**238**:48–53.

83. Goyon C, Faugeron G. Targeted transformation of *Ascobolus immersus* and de novo methylation of the resulting duplicated DNA sequences. *Mol Cell Biol* 1989;**9**:2818–27.

84. Faugeron G, Rhounim L, Rossignol JL. How does the cell count the number of ectopic copies of a gene in the premeiotic inactivation process acting in *Ascobolus immersus? Genetics* 1990;**124**:585–91.

85. Goyon C, Faugeron G, Rossignol JL. Molecular cloning and characterization of the met2 gene from *Ascobolus immersus. Gene* 1988;**63**:297–308.

86. Rhounim L, Rossignol JL, Faugeron G. Epimutation of repeated genes in *Ascobolus immersus. EMBO J* 1992;**11**:4451–57.

87. Goyon C, Barry C, Gregoire A, Faugeron G, Rossignol JL. Methylation of DNA repeats of decreasing sizes in *Ascobolus immersus*. *Mol Cell Biol* 1996;**16**:3054–65.

88. Barry C, Faugeron G, Rossignol JL. Methylation induced premeiotically in *Ascobolus*: coextension with DNA repeat lengths and effect on transcript elongation. *Proc Natl Acad Sci USA* 1993;**90**:4557–61.

89. Barra JL, Holmes AM, Gregoire A, Rossignol JL, Faugeron G. Novel relationships among DNA methylation, histone modifications and gene expression in *Ascobolus*. *Mol Microbiol* 2005;**57**:180–95.

90. Colot V, Maloisel L, Rossignol JL. Interchromosomal transfer of epigenetic states in *Ascobolus*: transfer of DNA methylation is mechanistically related to homologous recombination. *Cell* 1996;**86**:855–64.

91. Maloisel L, Rossignol JL. Suppression of crossing-over by DNA methylation in *Ascobolus*. *Genes Dev* 1998;**12**:1381–9.

92. Malagnac F, Wendel B, Goyon C, Faugeron G, Zickler D, Rossignol JL, et al. A gene essential for de novo methylation and development in *Ascobolus* reveals a novel type of eukaryotic DNA methyltransferase structure. *Cell* 1997;**91**:281–90.

93. Lee DW, Freitag M, Selker EU, Aramayo R. A cytosine methyltransferase homologue is essential for sexual development in *Aspergillus nidulans*. *PLoS One* 2008;**3**:e2531.

94. Watters MK, Randall TA, Margolin BS, Selker EU, Stadler DR. Action of repeat-induced point mutation on both strands of a duplex and on tandem duplications of various sizes in *Neurospora*. *Genetics* 1999;**153**:705–14.

95. Kinsey JA, Garrett-Engele PW, Cambareri EB, Selker EU. The *Neurospora* transposon Tad is sensitive to repeat-induced point mutation (RIP). *Genetics* 1994;**138**:657–64.

96. Selker EU, Stevens JN. Signal for DNA methylation associated with tandem duplication in *Neurospora crassa*. *Mol Cell Biol* 1987;**7**:1032–8.

97. Miao VP, Freitag M, Selker EU. Short TpA-rich segments of the zeta-eta region induce DNA methylation in *Neurospora crassa*. *J Mol Biol* 2000;**300**:249–73.

98. Tamaru H, Selker EU. Synthesis of signals for de novo DNA methylation in *Neurospora crassa*. *Mol Cell Biol* 2003;**23**:2379–94.

99. Singer MJ, Marcotte BA, Selker EU. DNA methylation associated with repeat-induced point mutation in *Neurospora crassa*. *Mol Cell Biol* 1995;**15**:5586–97.

100. Rountree MR, Selker EU. DNA methylation inhibits elongation but not initiation of transcription in *Neurospora crassa*. *Genes Dev* 1997;**11**:2383–95.

101. Selker EU. Premeiotic instability of repeated sequences in *Neurospora crassa*. *Annu Rev Genet* 1990;**24**:579–613.

102. Selker EU. Epigenetic phenomena in filamentous fungi: useful paradigms or repeat-induced confusion? *Trends Genet* 1997;**13**:296–301.

103. Freitag M, Lee DW, Kothe GO, Pratt RJ, Aramayo R, Selker EU. DNA methylation is independent of RNA interference in *Neurospora*. *Science* 2004;**304**:1939.

104. Adhvaryu KK, Selker EU. Protein phosphatase PP1 is required for normal DNA methylation in *Neurospora*. *Genes Dev* 2008;**22**:3391–6.

105. Freitag M, Hickey PC, Khlafallah TK, Read ND, Selker EU. HP1 is essential for DNA methylation in *Neurospora*. *Mol Cell* 2004;**13**:427–34.

106. Kouzminova E, Selker EU. dim-2 encodes a DNA methyltransferase responsible for all known cytosine methylation in *Neurospora*. *EMBO J* 2001;**20**:4309–23.

107. Freitag M, Williams RL, Kothe GO, Selker EU. A cytosine methyltransferase homologue is essential for repeat-induced point mutation in *Neurospora crassa*. *Proc Natl Acad Sci USA* 2002;**99**:8802–7.

108. Espagne E, Lespinet O, Malagnac F, Da Silva C, Jaillon O, Porcel BM, et al. The genome sequence of the model ascomycete fungus *Podospora anserina*. *Genome Biol* 2008;**9**:R77.

109. Paoletti M, Saupe SJ, Clave C. Genesis of a fungal non-self recognition repertoire. *PLoS ONE* 2007;**2**:e283.

110. Fudal I, Ross S, Brun H, Besnard AL, Ermel M, Kuhn ML, et al. Repeat-induced point mutation (RIP) as an alternative mechanism of evolution toward virulence in *Leptosphaeria maculans*. *Mol Plant Microbe Interact* 2009;**22**:932–41.

111. Yu JH, Keller N. Regulation of secondary metabolism in filamentous fungi. *Annu Rev Phytopathol* 2005;**43**:437–58.

112. Bok JW, Keller NP. LaeA, a regulator of secondary metabolism in *Aspergillus* spp. *Eukaryot Cell* 2004;**3**:527–35.

113. Stack D, Neville C, Doyle S. Nonribosomal peptide synthesis in *Aspergillus fumigatus* and other fungi. *Microbiology* 2007;**153**:1297–306.

114. Shwab EK, Bok JW, Tribus M, Galehr J, Graessle S, Keller NP. Histone deacetylase activity regulates chemical diversity in *Aspergillus*. *Eukaryot Cell* 2007;**6**:1656–64.

115. Bok JW, Chiang YM, Szewczyk E, Reyes-Dominguez Y, Davidson AD, Sanchez JF, et al. Chromatin-level regulation of biosynthetic gene clusters. *Nat Chem Biol* 2009;**5**:462–4.

116. van Dijk K, Marley KE, Jeong BR, Xu J, Hesson J, Cerny RL, et al. Monomethyl histone H3 lysine 4 as an epigenetic mark for silenced euchromatin in *Chlamydomonas*. *Plant Cell* 2005;**17**:2439–53.

117. Casas-Mollano JA, Jeong BR, Xu J, Moriyama H, Cerutti H. The MUT9p kinase phosphorylates histone H3 threonine 3 and is necessary for heritable epigenetic silencing in *Chlamydomonas*. *Proc Natl Acad Sci USA* 2008;**105**:6486–91.

118. Hertz-Fowler C, Figueiredo LM, Quail MA, Becker M, Jackson A, Bason N, et al. Telomeric expression sites are highly conserved in *Trypanosoma brucei*. *PLoS ONE* 2008;**3**:e3527.

119. Kyes S, Horrocks P, Newbold C. Antigenic variation at the infected red cell surface in malaria. *Annu Rev Microbiol* 2001;**55**:673–707.

120. Ralph SA, Scherf A. The epigenetic control of antigenic variation in *Plasmodium falciparum*. *Curr Opin Microbiol* 2005;**8**:434–40.

121. Templeton TJ, Iyer LM, Anantharaman V, Enomoto S, Abrahante JE, Subramanian GM, et al. Comparative analysis of apicomplexa and genomic diversity in eukaryotes. *Genome Res* 2004;**14**:1686–95.

122. Upadhyay R, Bawankar P, Malhotra D, Patankar S. A screen for conserved sequences with biased base composition identifies noncoding RNAs in the A-T rich genome of *Plasmodium falciparum*. *Mol Biochem Parasitol* 2005;**144**:149–58.

123. Hughes K, Wand M, Foulston L, Young R, Harley K, Terry S, et al. A novel ISWI is involved in VSG expression site downregulation in African trypanosomes. *EMBO J* 2007;**26**:2400–10.

124. Figueiredo LM, Janzen CJ, Cross GA. A histone methyltransferase modulates antigenic variation in African trypanosomes. *PLoS Biol* 2008;**6**:e161.

125. Alsford S, Kawahara T, Isamah C, Horn D. A sirtuin in the African trypanosome is involved in both DNA repair and telomeric gene silencing but is not required for antigenic variation. *Mol Microbiol* 2007;**63**:724–36.

126. Lopez-Rubio JJ, Gontijo AM, Nunes MC, Issar N, Hernandez Rivas R, Scherf A. 5' flanking region of var genes nucleate histone modification patterns linked to phenotypic inheritance of virulence traits in malaria parasites. *Mol Microbiol* 2007;**66**:1296–305.

127. Chookajorn T, Dzikowski R, Frank M, Li F, Jiwani AZ, Hartl DL, et al. Epigenetic memory at malaria virulence genes. *Proc Natl Acad Sci USA* 2007;**104**:899–902.

128. Duraisingh MT, Voss TS, Marty AJ, Duffy MF, Good RT, Thompson JK, et al. Heterochromatin silencing and locus repositioning linked to regulation of virulence genes in *Plasmodium falciparum*. *Cell* 2005;**121**:13–24.

129. Freitas-Junior LH, Hernandez-Rivas R, Ralph SA, Montiel-Condado D, Ruvalcaba-Salazar OK, Rojas-Meza AP, et al. Telomeric heterochromatin propagation and histone acetylation control mutually exclusive expression of antigenic variation genes in malaria parasites. *Cell* 2005;**121**:25–36.

130. Tonkin CJ, Carret CK, Duraisingh MT, Voss TS, Ralph SA, Hommel M, et al. Sir2 paralogues cooperate to regulate virulence genes and antigenic variation in *Plasmodium falciparum*. *PLoS Biol* 2009;**7**:e84.

131. Graia F, Lespinet O, Rimbault B, Dequard-Chablat M, Coppin E, Picard M. Genome quality control: RIP (repeat-induced point mutation) comes to *Podospora*. *Mol Microbiol* 2001;**40**:586–95.

132. Coppin E, Silar P. Identification of PaPKS1, a polyketide synthase involved in melanin formation and its use as a genetic tool in *Podospora anserina*. *Mycol Res* 2007;**111**:901–8.

133. Idnurm A, Howlett BJ. Analysis of loss of pathogenicity mutants reveals that repeat-induced point mutations can occur in the Dothideomycete *Leptosphaeria maculans*. *Fungal Genet Biol* 2003;**39**:31–7.

134. Ikeda K, Nakayashiki H, Kataoka T, Tamba H, Hashimoto Y, Tosa Y, et al. Repeat-induced point mutation (RIP) in *Magnaporthe grisea*: implications for its sexual cycle in the natural field context. *Mol Microbiol* 2002;**45**:1355–64.

135. Nakayashiki H, Nishimoto N, Ikeda K, Tosa Y, Mayama S. Degenerate MAGGY elements in a subgroup of *Pyricularia grisea*: a possible example of successful capture of a genetic invader by a fungal genome. *Mol Gen Genet* 1999;**261**:958–66.

136. Farman ML. Telomeres in the rice blast fungus *Magnaporthe oryzae*: the world of the end as we know it. *FEMS Microbiol Lett* 2007;**273**:125–32.

137. Neuveglise C, Sarfati J, Latge JP, Paris S. Afut1, a retrotransposon-like element from *Aspergillus fumigatus*. *Nucleic Acids Res* 1996;**24**:1428–34.

138. Nielsen ML, Hermansen TD, Aleksenko A. A family of DNA repeats in *Aspergillus nidulans* has assimilated degenerated retrotransposons. *Mol Genet Genomics* 2001;**265**:883–7.

139. Braumann I, van den Berg M, Kempken F. Repeat induced point mutation in two asexual fungi, *Aspergillus niger* and *Penicillium chrysogenum*. *Curr Genet* 2008;**53**:287–97.

140. Hua-Van A, Hericourt F, Capy P, Daboussi MJ, Langin T. Three highly divergent subfamilies of the impala transposable element coexist in the genome of the fungus *Fusarium oxysporum*. *Mol Gen Genet* 1998;**259**:354–62.

141. Hua-Van A, Langin T, Daboussi MJ. Evolutionary history of the impala transposon in *Fusarium oxysporum*. *Mol Biol Evol* 2001;**18**:1959–69.

142. Julien J, Poirier-Hamon S, Brygoo Y. Foret1, a reverse transcriptase-like sequence in the filamentous fungus *Fusarium oxysporum*. *Nucleic Acids Res* 1992;**20**:3933–7.

143. Cuomo CA, Guldener U, Xu JR, Trail F, Turgeon BG, Di Pietro A, et al. The *Fusarium graminearum* genome reveals a link between localized polymorphism and pathogen specialization. *Science* 2007;**317**:1400–2.

144. Coleman JJ, Rounsley SD, Rodriguez-Carres M, Kuo A, Wasmann CC, Grimwood J, et al. The genome of *Nectria haematococca*: contribution of supernumerary chromosomes to gene expansion. *PLoS Genet* 2009;**5**:e1000618.

145. Hood ME, Katawczik M, Giraud T. Repeat-induced point mutation and the population structure of transposable elements in *Microbotryum violaceum*. *Genetics* 2005;**170**:1081–9.

146. Hane JK, Oliver RP. RIPCAL: a tool for alignment-based analysis of repeat-induced point mutations in fungal genomic sequences. *BMC Bioinformatics* 2008;**9**:478.

147. Salgado PS, Koivunen MR, Makeyev EV, Bamford DH, Stuart DI, Grimes JM. The structure of an RNAi polymerase links RNA silencing and transcription. *PLoS Biol* 2006;**4**:e434.

148. Sugiyama T, Cam H, Verdel A, Moazed D, Grewal SI. RNA-dependent RNA polymerase is an essential component of a self-enforcing loop coupling heterochromatin assembly to siRNA production. *Proc Natl Acad Sci USA* 2005;**102**:152–7.

149. Buker SM, Iida T, Buhler M, Villen J, Gygi SP, Nakayama J, et al. Two different Argonaute complexes are required for siRNA generation and heterochromatin assembly in fission yeast. *Nat Struct Mol Biol* 2007;**14**:200–7.

150. Motamedi MR, Verdel A, Colmenares SU, Gerber SA, Gygi SP, Moazed D. Two RNAi complexes, RITS and RDRC, physically interact and localize to noncoding centromeric RNAs. *Cell* 2004;**119**:789–802.

151. Iida T, Nakayama J, Moazed D. siRNA-mediated heterochromatin establishment requires HP1 and is associated with antisense transcription. *Mol Cell* 2008;**31**:178–89.

152. Zhang K, Mosch K, Fischle W, Grewal SI. Roles of the Clr4 methyltransferase complex in nucleation, spreading and maintenance of heterochromatin. *Nat Struct Mol Biol* 2008;**15**:381–8.

153. Liu Y, Mochizuki K, Gorovsky MA. Histone H3 lysine 9 methylation is required for DNA elimination in developing macronuclei in *Tetrahymena*. *Proc Natl Acad Sci USA* 2004;**101**:1679–84.

201

CHAPTER 14

Drosophila Epigenetics

John C. Lucchesi
Department of Biology, Emory University, Atlanta, GA 30322, USA

INTRODUCTION

The term "epigenetics" was coined by Conrad Waddington, who introduced it in print
in 1957. Since Waddington was a *Drosophila* developmental geneticist, it is appropriate
to begin a chapter on *Drosophila* epigenetics with a consideration of the evolution of
the concept and of the term. To Waddington, epigenesis was the sum-total of all of the
regulatory events that are responsible for the development of a fertilized egg into an adult
organism. Today, our understanding of this phenomenon stems from the realization that
all of the cells of a developing embryo are identical with respect to their genetic material
and that differentiation is achieved by the differential expression of genes in regions of
the embryo during the course of development. These cell or tissue-specific expression
forms of the genome can be referred to as epigenomes, each of which is characteristic of a
particular modality of cellular differentiation. Epigenomes arise through the influence of a
wide range of environmental factors that include gradients of maternal morphogens, and
intercellular signals as well as extra-embryonic environmental factors. These stimuli result
in covalent modifications of the DNA that do not alter in any way the nucleotide-based
genetic code, and of many of the DNA-associated proteins; they also result in a change in
the architecture of the association. This set of particulars addresses the responsiveness of
epigenetic modifications. A number of these modifications are heritable, as is evidenced at
the cellular level by the fact that following mitosis the daughters of a somatic cell exhibit the
same pattern of gene expression as does the parent cell. An intriguing and largely unexpected
aspect of some epigenetic modifications is their transgenerational transmission – in other
words, their heritability through the germ line.

The primary goal of this chapter is to highlight the unique contributions of *Drosophila* to our
current understanding of epigenetic inheritance. The biochemical nature and the molecular
characterization of the covalent modification of DNA and histones, and the mechanisms
that underlie the remodeling of nucleosome organization, as well as instances of epigenetic
regulation that were discovered or are more productively studied in other organisms, will
not be discussed, even when similar regulation has been described in *Drosophila*. Exceptions
to this general approach will be made if a parallel study in fruit flies has or would provide
easier or quicker progress. The main topics that will be discussed are the formation of
heterochromatin, gene silencing and activation, epigenetic memory and the nature and
function of boundary elements.

The birth of the field of study of epigenetic phenomena can be traced to the discovery
of position effect variegation in *Drosophila* by Herman J. Muller, who was the first to use
X-rays to induce mutations. In 1941, Muller reported the occurrence of chromosome

Handbook of Epigenetics: The New Molecular and Medical Genetics. DOI: 10.1016/B978-0-12-375709-8.00014-9

203

rearrangements that resulted in the mutant expression of certain genes, but only in some sectors of the adult fly; in other sectors, the genes had normal, wild type expression. Muller first called these occurrences "eversporting displacements" and later "position effect variegation". In 1950, Jack Schultz discovered that the extent of the variegation was affected by different genetic factors. These results initiated the discovery of a large number of genes that, when mutated, either enhanced or diminished the variegated phenotype. Several of these genes encode structural proteins or enzymatic factors that are directly involved in the formation of heterochromatin in all multicellular eukaryotes investigated to date. The landmark characterization of Polycomb (*Pc*) by Ed Lewis in 1978 opened the way for the identification of the Polycomb group (PcG) of genes. Twenty years later, Peter Ingham described the founding member of the trithorax group (trxG) of genes. Members of these two gene groups are primarily involved in the regulation of normal development, from yeast to humans. Another broad area of investigation initiated in *Drosophila* was the discovery that mutations induced by the insertion of the Gypsy transposable element could be suppressed by the loss of function of an unlinked gene, leading to the discovery of insulators or boundary elements. Lastly, the first evidence for the transmission of epigenetic programs across cell generations was obtained while studying the function of PcG and trxG proteins.

THE NATURE AND FUNCTIONAL CHARACTERISTICS OF HETEROCHROMATIN

Heterochromatin is that fraction of chromatin that is highly condensed during the interphase of the cell cycle (G0, G1, S, and G2 phases) and that is generally associated with the absence of genes or with the repression of gene activity. Operationally, it is often useful to distinguish between constitutive and facultative hetrochromatin. Facultative hetrochromatin is found in regions of the genome that are condensed and inactive in some cell lineages, although they are uncondensed and are active in others. Here, one can distinguish more global regions involving whole chromosomes (for example, the classic mammalian X chromosome inactivation) or involving whole sets of chromosomes (the paternal genome in some beetles) from more localized regions involving small groups of genes or individual gene domains (for example, homeotic genes in those regions of the developing embryo where their expression is inappropriate). Constitutive hetrochromatin refers to particular regions of chromatin that are always condensed and never expressed.

In *Drosophila*, constitutive heterochromatin is found around the centromeres of all the chromosomes. In addition, the Y chromosome is entirely heterochromatic in all somatic tissues. High-resolution *in situ* hybridization reveals that these regions consist of blocks of transposable elements embedded within segments of repetitive DNA sequences [1]. Heterochromatin is relatively gene-poor, replicates in late S phase, and does not allow meiotic recombination.

Formation of Heterochromatin

THE STUDY OF POSITION EFFECT VARIEGATION HAS LED TO THE DISCOVERY OF FACTORS AND STRUCTURAL PROTEINS THAT CONSTITUTE HETEROCHROMATIN

Following the first occurrence of mutations leading to a mottled phenotype discovered by H.J. Muller in the early 1930s, it soon became apparent that in all cases, newly induced chromosomal rearrangements had relocated the affected genes within or near a region of constitutive heterochromatin (Fig. 14.1); the Russian geneticists N.B. Dubinin, B.N. Sidorov, and I.B. Panshin showed that the genes, themselves, were not altered in any way as they retained their wild type expression following their relocation by crossing over to a euchromatic domain or when excising them from their heterochromatic environment as extrachromosomal circles [2]. These observations led to the hypothesis that encroachment of heterochromatin into the newly adjacent euchromatic segment was responsible for

(A) Normal X chromosome   (B) Inverted X chromosome

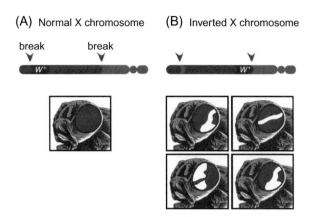

FIGURE 14.1
Diagram of the In(1)w[m4] inversion that relocates the w[+] gene next to centromeric heterochromatin. The resulting clonal silencing of the w[+] gene is evidenced by pigmentless sectors in the inversion-bearing flies. (Please refer to color plate section)

the inactivation of the genes in question. This contention was supported by the very early observation that the addition of heterochromatin, in the form of a Y chromosome in females or an extra Y chromosome in males carrying a variegating rearrangement, diminished or suppressed the presence and extent of the mutant sectors. Deletion of heterochromatin, for example by removing the Y chromosome from the genome of males, had the opposite effect [3]. In addition, there were several reports describing the influence of genetic background on the extent of variegation on a gene caused by a particular rearrangement. Approximately thirty years later, a mutation identifying the first gene that contributed to the occurrence of variegation was discovered and was named Suppressor-of-Variegation (*Su-V*) locus [4]. This finding initiated a number of genetic screens for mutations that affect position effect variegation and a large number of genes were identified by their loss-of-function mutations that either suppressed variegation (fewer or no sectors of somatic inactivation of the variegating reporter gene) or enhanced it (more and larger sectors) [5,6]. In many instances, these modifiers of position effect exhibited a dosage-dependent response: one dose of the wild type allele of the modifier locus (i.e. heterozygosity for the suppressor of variegation (Su(var) mutant allele) reduced variegation while three doses of the wild type gene enhanced it. The reverse was true for genes identified by loss-of-function mutations that enhanced variegation ([E(var)]. Not surprisingly, the molecular characterization of several of the position effect modifiers revealed that they function to provide the covalent modifications of nucleosomal histones necessary for the onset of heterochromatin formation or that they encode structural protein components of heterochromatin [7].

The products of the following Su(var) genes were identified as histone modifying enzymes. *Su(var)3-9* encodes a histone H3 methyltransferase responsible for lysine 9 di- and tri-methylation: H3K9me2 and H3K9me3 [8,9]. The dimethylated form is the major mark present in constitutive heterochromatin [10]. The Su(var)3-9 protein is associated with the histone deacetylase HDAC1/Rpd3 [9] that is encoded by *Su(var)3-26* [11]. Another histone methyl transferase is encoded by *Suv4-20* [12]; it dimethylates the majority of histone H4 at lysine 20 (H4K20me2). The trimethylated form (H3K20me3) depends on the presence of HP1 (defined below) suggesting that it is enriched in heterochromatin [13]. Note that although the effect of Suv4-20 on position effect variegation [12] could not be confirmed by another laboratory [14] it has been firmly established by subsequent experiments [15]. *Pr-Set7* encodes a histone methyl transferase that monomethylates histone H4 at lysine 20 (H4K20me). *PR-Set7* mutants suppress variegation [16]. *Su(var)3-3* encodes an amine oxidase that demethylates Histone H3 mono- and dimethylated at lysine 4 (H3K4me and H3K4me2) and is the *Drosophila* homolog of mammalian LSD1 [17]. *Little imaginal discs* (*lid*) encodes a histone demethylase that targets H3K4me3 [18,19]. Mutations in the gene *chameau* (*chm*) are dominant suppressors of position effect variegation [20]; the product of this gene is a histone H4 acetyl transferase [21]. *Su(var)2-1* encodes a protein correlated with histone deacetylation [22]. A mutant allele of *Su(var)3-1* was subsequently identified as a dominant negative allele of *JIL-1* [10], a gene that encodes a tandem kinase responsible for the phosphorylation of serine 10 on histone H3 (H3S10ph) in euchromatin [23]. Suppression of position effect variegation was observed in flies heterozygous for *Su(var)3-6*, a gene that encodes one of four different catalytic subunits of Protein Phosphatase 1 (PP1), a protein serine/threonine phosphatase [24].

Mutations that suppress position effect variegation also identified several genes responsible for structural proteins associated with heterochromatin. *Su(var)2-5* encodes the protein HP1 that occurs in most organisms, from fission yeast to humans. In *Drosophila*, HP1 is found in the heterochromatin of centromeres and telomeres, and on many sites throughout the euchromatic arms of the chromosomes [25,26]. *Su(var)3-7* encodes a protein with 7 Zinc fingers. Its distribution is identical to that of HP1 and it co-immunoprecipitates with HP1 [27–29]. HP2, a second protein that associates with HP1 in the heterochromatin of centromeres and telomeres, is produced by *Su(var)2-HP2* [30]. *Caravaggio* encodes the HOAP (HP1 Origin recognition complex-Associated) protein. Together with HP1, it is involved in the capping complex at telomeres and it is present in the heterochromatin of centromeres. Loss-of-function mutations behave as suppressors of heterochromatic silencing [31,32]. *Modulo* mutants are dominant suppressors of variegation [33]. MOD is a DNA-binding protein that is found in centromeric heterochromatin and very prominently in the nucleolus; in polytene chromosomes, it is also present throughout the euchromatic chromosomal arms on almost all bands but not in puffs or in interbands [34].

An additional set of structural proteins have been found to bind directly to DNA or RNA in centromeric heterochromatin regions. D1 is a protein with multiple AT-hook DNA binding domains that associates with AT-rich tandemly repeated (satellite) DNA sequences that are present in centromeric heterochromatin [35]. Decreased levels of D1 protein in flies heterozygous for a loss-of-function mutation lead to the suppression of position-effect variegation of a gene relocated to the centromeric region where the AT-rich satellites occur [36]. DDP1 is a protein with multiple KH single-stranded nucleic acid binding domains that associates with centromeric heterochromatin. Mutants that decrease the level of DDP1 behave as dominant suppressors of variegation. In homozygous mutant individuals, the presence of HP1 and the level of H3K9me2 are substantially reduced [37].

A number of additional genes have been implicated in the process of heterochromatin formation and gene silencing because their loss-of-function mutations enhance (rather than suppress) position effect variegation. For example, *E(var)3-93D*, also known as *modifier of mdg4* [*mod(mdg4)*], produces a protein that associates with the Su(Hw) protein (discussed below); mutations in *mod(mdg4)* act as classical enhancers of position-effect variegation [38]. *E(var)3-9* encodes a Zinc finger protein of novel sequence [39]. Some of the enhancers of variegation gene products are transcription factors and, although it may be expected that a decrease in their level may result in a concomitant increase in silencing, the role that they play is complex and not easily explained. As intuitively expected, increasing the level of the GAL4 activator counteracts the silencing of a gene inserted near heterochromatin and driven by the GAL4 promoter [40]. Concordant with this line of reasoning, the *E(var)3-93E* gene encodes the E2F1 protein, a member of the *Drosophila* E2F class of transcription factors that is required for transcription in the G1 and S phases and for cell proliferation [41].

Much less clear-cut conclusions can be drawn regarding the effect of factors in the JAK/STAT pathway. Loss of function mutations of *Su(var)2-10* act as dominant suppressors of position effect variegation; the protein product of this gene is highly homologous to members of the

Protein Inhibitor of Activated STAT (PIAS) group of proteins that down-regulate the action of STAT transcription factors [42]. In *Drosophila*, there is a single Janus kinase (JAK) produced by the gene *hopscotch (hop)* that activates a single STAT protein (STAT92E). A loss-of-function mutation of *hop* enhances position effect variegation while a gain-of-function allele encoding a hyperactive JAK kinase suppresses variegation [43]. Yet, loss of STAT92E function has the same effect as the overactive hop mutation and over-expression has the opposite effect on position effect variegation [44]. These results could be explained by suggesting that the unphosphorylated, inactive form of STAT92E, which associates with HP1, is required for stabilizing HP1 at heterochromatic sites. Another contradictory example is provided by *Domina (Dom)* that also encodes a product that belongs to the FKH/WH transcription factor protein family; yet, loss-of-function mutations are dominant suppressors of position effect variegation [45].

Before ending this section, mention should be made that additional proteins homologous to HP1 (more accurately referred to as HP1a) exist in *Drosophila*. HP1b is found in heterochromatin as well as in euchromatic sites while HP1c is found exclusively in euchromatic regions [46]. In these regions, HP1c co-localizes with activated but still poised RNAPII and H3K4me3; its binding to chromatin depends on two specific putative transcription factors with which it collaborates to enhance gene transcription [47]. Two additional paralogs – HP1d and HP1e – are expressed predominantly in ovaries [48] and testes [49], respectively. Although their function is unknown, their restricted presence suggests that there may be significant differences in chromatin organization and function between the soma and the germ line.

HETEROCHROMATIN FORMATION INVOLVES A SEQUENCE OF HISTONE MODIFICATIONS

Histone modifications attract various structural proteins, nucleosome assembly factors, and chromatin remodeling complexes and can act as either repressive or active marks. The acetylation of lysine residues on histone tails generally regulates transcription in a positive manner by inhibiting chromatin compaction as well as recruiting factors and complexes that promote gene activity. Histone lysine methylation can be correlated with transcriptional activation (methylation of H3-K4, H3-K36, and H3-K79) or repression (methylation of H3-K9, H3-K27, and H4-K20).

As can be surmised by the long (albeit partial) list of enzymatic and non-enzymatic proteins listed in the previous section, the formation of heterochromatin is a particularly complex process that is still poorly understood. It is possible, nevertheless, to order some of the factors involved in a reasonably sequential pathway [12,17]. Heterochromatin formation is initiated with the demethylation of H3K4me2 by the SU(VAR)3-3 LSD1 demethylase, followed by the deacetylation of H3K9 by HDAC1/RPD3. These modifications are essential for the di- and trimethylation of H3K9 by the SU(VAR)3-9 methyl transferase. This histone mark is recognized by the chromo domain of the HP1a protein that is targeted to the incipient heterochromatin together with its associated proteins HP2, and SU(VAR)3-7. HP1a was found to interact with several proteins in the two-hybrid–based protein-interaction map of the fly proteome [50]. Four of these proteins (HP3 to HP6) bind to the same genomic sites as HP1 and are redistributed in its absence, and mutant alleles of HP4, 5, and 6 act as dominant suppressors of variegation [51]. HP1a also recruits the *Suv4-20* histone methyl transferase to di- and trimethylate histone H4K20. It may also interact with the DDP1 protein. A second domain in HP1a, named the chromo shadow domain because of its similarity to canonical chromo domains, is thought to be involved in HP1a self-association as well as in interactions with other proteins.

HP2 co-immunoprecipitates with the Nucleosome Remodeling Factor NURF and the histone chaperone Nap-1, suggesting that these chromatin-remodeling factors are involved in heterochromatin formation [52]. This contention is reinforced by the discovery that loss-of-function alleles of *Anti-silencing factor 1* (*Asf1*) are dominant suppressors of position effect variegation [53]. The ASF1 protein bound to histones H3 and H4 constitutes the replication-coupling assembly factor RCAF that was found to facilitate the CAF-1 complex-mediated assembly of nucleosomes onto newly-replicated DNA *in vitro* [54].

RNA INTERFERENCE PLAYS A CRITICAL ROLE IN HETEROCHROMATIN FORMATION

The genomes of most multicellular eukaryotes contain a large number of copies of various transposable elements that are either vestigial or active. Silencing by RNA interference has been implicated as a major defense against the deleterious effects of transposition [55]. The major RNA interference defenses against transposable elements involve the recognition and cleavage of the elements' transcripts. Approximately 30% of the *Drosophila* genome consists of heterochromatin that is largely made up of DNA transposons, retroviruses, and tandemly repeated simple sequences. In retrospect, it is not surprising that the silencing mechanisms that evolved to combat the mutagenic effects of transposition would be subsumed to regulate the organism's own genetic pathways. Following the seminal observation in fission yeast [48], abundant evidence has been garnered that the RNA interference pathway is involved in the regulation of heterochromatin formation, in a variety of model organisms including *Drosophila* [56].

Interfering RNAs are of three general types: siRNAs derived from long double-stranded RNAs thought to function as a viral defense mechanism (the pathway that generates siRNAs is used in RNAi knock-down experiments); miRNAs that are transcribed by RNA Polymerase II from non-coding regions present in the euchromatic portion of the genome; piRNAs that can be produced from different sources including sense and antisense transcripts. The piRNAs of interest to the present discussion are referred to as repeat-associated small interfering RNAs (rasiRNAs) or heterochromatic RNAs (hcRNAs); they are derived from long double-stranded RNA precursors that are transcribed from specific genomic repeat regions composed primarily of copies of transposable elements termed rasiRNA clusters. These clusters are located in the centromeric heterochromatin, the fourth chromosome, and at telomeres [57,58]. In *Drosophila*, a large number of small RNAs that contain sequences from all known forms of repetitive elements, such as retrotransposons, DNA transposons, satellite, and microsatellite DNA sequences, have been identified [59]. RasiRNAs occur primarily in ovaries, testes, and in early embryos. Their presence depends upon two Argonaute family members *piwi* and *aubergine* (*aub*), as well as two putative RNA helicases, *spn-E* (aka *hls*) and *armitage* (*armi*) [60–62].

Mutations that interfere with the synthesis of rasiRNAs that are homologous to particular transposable elements result in increased expression and mobilization of these elements in the germline. As transposons are generally found in the heterochromatic regions of the genome, their silencing could result from a direct role of rasiRNAs in the formation of hetrochromatin. Of course, silencing could be achieved by a post-transcriptional mechanism. A direct role of rasiRNAs in the formation of heterochromatin is suggested by the observation that certain mutations in the rasiRNA pathway affect the chromatin organization of transposable elements [63]. This remodeling is not seen in somatic cells suggesting that, once heterochromatin has formed, it can be preserved in the absence of RNA interference. Yet, the possibility that rasiRNAs do play some role in heterochromatin formation in the soma is indicated by the observation that mutations in their synthesis act as weak suppressors of position-effect variegation [56]. SiRNAs have also been linked to the formation of heterochromatin by the observation that loss-of-function mutations in the pathway by which they achieve their interference result in mislocalization of HP1 and abnormal centromeric heterochromatin formation [64].

A unifying explanation for all of these experimental results would have the rasiRNA pathway play the major role in regulating the expression and mobilization of transposable elements in the germ line via a post-transcriptional mechanism or by inducing the formation of heterochromatin; in the early embryo, both the rasiRNA and the siRNA pathways would play a role in the continued inhibition of transposable elements and in heterochromatin formation. The classical hallmarks of heterochromatin – HP1a and methylated H3K9 – would function as end-points for either pathway [65]. Although RNA interference is most likely responsible for the formation of centromeric heterochromatin, it is clear that the targeting of HP1 can occur in an RNAi-independent manner, for example at the telomeres (discussed below).

HETEROCHROMATIN FORMATION IS LIMITED TO SPECIFIC CHROMOSOME DOMAINS

In nuclei with structurally normal chromosomes, the separation of constitutive heterochromatin from the adjacent euchromatic region is fixed, presumably by specific boundary elements. Although nothing is known of the molecular nature of these elements, their existence is supported by the fact that the over-expression of heterochromatin components that were found to enhance position effect variegation fails to extend the normal heterochromatin domains. In contrast, when chromosome rearrangements position euchromatic regions in contact with heterochromatin, silencing of the genes that have been relocated occurs because heterochromatin formation extends into these genes' domain. In these cases, the frequency of silencing is increased by the over-expression of heterochromatin components, indicating that the boundary element that normally contained the spread of the heterochromatic region has been relocated elsewhere allowing heterochromatin to invade regions where it normally is not present. Consistent with this invasive model were the observations that the inactivation of the gene closest to the heterochromatin/euchromatin breakpoint was often accompanied by the inactivation (i.e. by the variegation) of the next gene, and sometimes, by the concomitant inactivation of additional more distal genes. In these early studies, inactivation of the distal genes never occurred without the inactivation of the more proximal genes inferring that heterochromatin spread continuously from the breakpoint. Recent molecular data would support this hypothesis: chromatin immunoprecipitation experiments have shown that the level of H3K9me2 is highest in the euchromatic segment immediately adjacent to the breakpoint of a variegating rearrangement and decreases as one proceeds distally [17,66]. An observation that appeared to contradict this general conclusion derived from the cytological examination in polytene chromosomes of the heterochromatin present in the euchromatic region adjacent to the breakpoint of position-effect rearrangements: in some cases, rather than a single compacted block obliterating the banded appearance of the euchromatic region, the breakpoint region exhibited a discontinuous compaction with zones of heterochromatin separated by morphologically normal regions with distinct bands and interbands [66]. More recently, some examples of inactivation being able to skip over the most proximal gene and affect a gene farther away from the heterochromatin/euchromatin junction have been reported [67].

In *Drosophila*, homologous chromosomes are paired throughout most of the cell cycle (from G0 to G2) in somatic cells. This can allow the spreading of heterochromatin formation caused by a position effect rearrangement on one chromosome to induce heterochromatin formation on the normal-sequence homolog with which it is paired [68].

Telomeric Heterochromatin

Drosophila telomeres contain a region of repeated sequences followed by an array of specific retrotransposons capped by a protein complex [32] that binds to the end of the array, regardless of the terminal DNA sequence [69]. HP1 is an integral part of the cap (its absence causes multiple telomere fusions) and it binds directly to DNA; it also binds in the usual

209

H3K9me3-dependent manner to the repeated regions of the telomeres where it contributes to the formation of repressive telomeric heterochromatin [70].

Telomeres can generate position effect variegation of eukaryotic genes inserted into their domain. Many of the standard Su(var) mutations that modify position effect variegation in rearrangements involving centromeric heterochromatin have no effect on telomere-induced position effects. Screens of the genome for modifiers of the latter [71,72] have led to the discovery of several modifiers, only a subset of which also affect centromeric position effect variegation. A number of these modifiers are known regulators of chromatin architecture and euchromatic gene function leading to the conclusion that the telomeric silencing mechanism may be widely used [72].

Intercalary Heterochromatin

In the large, polytenic chromosomes produced by endoreplication in larval salivary glands, the bulk of centromeric heterochromatin is under-replicated and, therefore, under-represented. Dispersed within the euchromatic arms of the chromosomes are distinct regions that have been termed intercalary heterochromatin. Originally identified as sites that associate with each other or with centromeric heterochromatin, forming intra- or interchromosomal contacts (ectopic pairing), these regions share a number of characteristics with centromeric heterochromatin: late replication, under-replication in polytenic chromosomes, increased frequency of irradiation-induced chromosome breakage, and some histone modifications and factors. The under-replication of heterochromatic regions in polytene chromosomes is controlled by the *Suppressor of Underreplication* (SuUR) gene [73]. The product of this gene is an AT hook-containing protein that has no homolog in other model organisms (although it may be present in other dipterans). Based on the criterion of late-replication determined cytologically, there appear to be around 240 intercalary chromatin sites in the polytene chromosomes of *Drosophila* [74].

While the under-replication feature of specific regions of the genome (centromeric and intercalary) is clearly a characteristic of polytene chromosomes, late replication sites occur in other tissues and have been mapped using DNA-DNA microarrays in diploid cells [75]. Using similar microarray techniques, a number of these sites were found to coincide with under-replicated sites in polytene chromosomes. In contrast to the general characteristic of centromeric heterochromatin regions, the intercalary under-replicated sites include clusters of unique genes, many of which are co-expressed [76]. Although the presence of highly repetitive sequences has not been determined, HP1 is often found in these regions as well as members of the Polycomb Group of proteins (discussed below).

The Unique Heterochromatin Attributes of Chromosome IV

The small dot-like chromosome (IV in *D. melanogaster*, element F in interspecific comparisons) of the *Drosophila* karyotype is approximately 4.5 megabases in length. This chromosome contains a short arm that is fully heterochromatic. Beyond the heterochromatin block of the centromeric region, the long arm (1.2 megabases) is made up of regions of heterochromatin interspersed with segments containing canonical genes. These "euchromatic" segments contain an unusually high level of repetitive DNA consisting predominantly of transposable-element sequences and some simple repeats [77]. Perhaps for this reason, the chromosome is late-replicating [78] and normally does not undergo meiotic crossing over [79]. Nevertheless, this entire region is amplified normally in polytene chromosomes.

As expected, some of the proteins and factors associated with centromeric heterochromatin are present on chromosome IV, with some notable differences in the 1.2 Mb segment. HP1 and HP2 are present [30] and so are SU(VAR)3-7 [28], HOAP [31,32], and SU(VAR)3-9.

Surprisingly, the H3K9me2 present in this segment is not the product of SU(VAR)3-9 but of a different histone methyl transferase, DmSETDB1, produced by the *eggless* (*egg*) gene [80,81].

Another unique feature of the fourth chromosome is the presence of a protein encoded by the *Painting of fourth* (*Pof*) gene [82]. POF is a putative RNA-binding protein that associates exclusively with chromosome IV in *D. melanogaster* males and females, while in some other species of *Drosophila* it also associates with the X chromosome in males [83]. Presence of the POF protein is necessary for chromosome IV genes to transcribe at normal levels. On this chromosome, POF binds at the very same sites where HP1 is found and their respective presence on the chromosome are interdependent. Furthermore, this binding seems to be targeted preferentially to the exons of active genes, rather than to transposable elements. Of some interest is the observation that knock-down of HP1, which should lead to a decrease in POF association, results in a measure of up-regulation. These interactions suggest that POF and HP1 constitute a balancing mechanism for chromosome IV gene regulation [84,85].

Genes that Reside in Heterochromatin

The paradoxical existence of genes within heterochromatic domains was first suspected many years ago with the discovery that the Y chromosome was required for male fertility leading to the expectation that it must contain genes necessary for this function. The first gene residing in centromeric heterochromatin – the *light* (*lt*) gene – was found to respond in opposite fashion to all the euchromatic genes that had been studied with respect to position effect variegation: while the expression of these genes relocated near heterochromatin was diminished or suppressed in males by the removal of the Y chromosome, the expression of the *light* gene, present in the heterochromatic segment of the rearrangement, was enhanced; furthermore, an additional Y chromosome, instead of enhancing expression, as it does for variegating euchromatic genes, led to greater frequency of the light gene inactivation. These early observations highlighted the existence of genes that *depended* on their heterochromatic context for proper function. Additional genes residing in the constitutive heterochromatin regions of the *Drosophila* genome were discovered, and when several of these genes were relocated near euchromatin by chromosomal rearrangements, their expression was found to be enhanced by known Su(var) mutations and suppressed by some of the known E(var) mutations [86–88].

With the relatively recent sequencing of heterochromatic regions, the number of functional genes present in heterochromatin has increased from a few dozen discovered by conventional mutagenesis to a minimum of 230 to 254 protein coding genes [89]. The introns of these genes are much longer, on average, than those of euchromatic genes and consist almost exclusively of repeated transposable element sequences. In addition, a dozen genes that transcribe non-coding RNAs and a number of pseudogenes have been identified. Perhaps as might be expected, the distribution of histone H3 methylated at lysine 9 is different in its relation with heterochromatic genes than with euchromatic genes. While the 5′ regions of heterochromatic genes are enriched in H3K4me2 and H3K9ac – marks of euchromatic gene activity – the remainder of the transcribed regions contains H3K9me2 [90]. SU(VAR)3-9 and dmSETDB1 are not responsible for this methylation.

Modulation of Heterochromatin During Euchromatic Gene Transcription

In addition to their heterochromatin location, HP1a and SU(VAR)3-9 are present together at euchromatic sites that contain repetitive sequences; they are also frequently found independently of one another on many expressed genes where their binding is inversely correlated with the levels of gene expression. These results suggest that these two proteins can be present in different complexes [51]. HP1a is found in developmental and induced puffs on polytene chromosomes; although these puffs form in the absence of functional HP1, the steady-state level of transcript of the Hsp70 locus, which was studied in detail, is proportional to the level of HP1 present in the nucleus [91]. Microarray analysis of control and HP1a-depleteted synchronized cultured embryonic cells showed that a significant number of genes that regulate the cell cycle components, including DNA replication and chromosome segregation, depend on HP1 for their transcriptional activity [92]. Using a similar approach, hundreds of genes were found to be positively or negatively regulated by HP1 in larvae [93]. High resolution mapping of HP1 reveals that it associates with active euchromatic genes where it favors the coding regions over the promoters with a distinct preference towards the 3′ end. These genes display the chromatin characteristics of active genes: presence of H3K4me2 in the promoter region and histone H3.3 replacement of H3 (discussed below). This histone variant is present in the upstream portion of the coding region where HP1 is absent or present at a lower level suggesting that H3.3 and HP1 are mutually exclusive features of many active genes [94].

As mentioned above, HP1c displays a distribution that is exclusively euchromatic. HP1c is found together with several functionally unrelated transcription factors as well as co-activators or co-repressors at many genomic sites referred to as "co-localization spots". These spots are in actively transcribed regions and are enriched in RNAPII as well as the histone variant H3.3. Since the number of transcription factors in *Drosophila* is much larger than the number of the factors included in this study, it is likely that the co-localization spots include many additional proteins [95]. One cannot help wondering if these findings are in any way related to the "transcription factories" detected in mouse cells [96].

REGULATION OF GENE EXPRESSION
Developmental Regulation by Polycomb and Trithorax Proteins

Drosophila development begins with the establishment of the anterior-posterior and dorso-ventral axes by gene products that are deposited in the egg cytoplasm during oogenesis. Some of these products are transcription factors that regulate the expression of segmentation genes responsible for subdividing the embryo into a series of paired segments, each with its own anterior-posterior polarity. The differentiation of these segments into specific body regions is achieved by the homeotic genes that in turn, regulate the expression of realization genes. The homeotic genes share a characteristic, specific DNA-binding domain, the homeobox, and for this reason they are often referred to as HOX genes. There are two clusters of HOX genes in *Drosophila melanogaster*: the Antennapedia complex (ANT-C) that includes *labial* (*lab*), *proboscipedia* (*pb*), *Deformed* (*Dfd*), *Sex combs reduced* (*Scr*), and *Antennapedia* (*Antp*), and the Bithorax complex (BX-C) that includes *Ultrabithorax* (*Ubx*), *abdominalA* (*abdA*), and *AbdominalB* (*AbdB*). Specific HOX genes must be activated and remain active in particular regions of the embryo while others must be kept in a permanently repressed state. This is the responsibility of two groups of proteins with antagonistic functions – the Polycomb group (PcG) and the trithorax group (trxG). The genes encoding these proteins were discovered because loss-of-function mutations affected the expression of HOX genes and resulted in homeiotic transformations [97]. Some of these genes were found to be structurally similar to genes that affected position effect variegation suggesting a commonality in the regulatory mechanisms [98]. In addition to prescribing the activity of HOX genes, PcG and trxG genes regulate the expression of numerous other genes during development.

PcG REGULATORY PROTEINS
PcG Proteins form Regulatory Complexes

PcG gene products form three different types of multi-protein complexes. The Polycomb Repressive Complex 2 (PRC2) contains the histone methyl transferase Enhancer of zeste [E(Z)], Extra sex combs (ESC), Suppressor of zeste 12 [SU(Z)12], and Nucleosome remodeling factor p55 (Nurf-55) [99]. The Polycomb Repressive Complex 1 (PRC1) contains

Polycomb (PC), Polyhomeiotic (PH), Posterior Sex Combs (PSC) and Sex Combs Extra (SCE/dRing); it also contains other components, including Tata-Box Protein (TBP)-associated factors [100]. More recently, a third complex has been identified, the so-called PHORC complex that includes two proteins: Scm-related-gene-containing-Four-Malignant-Brain-Tumor-domains (SFMBT) and Pleiohomeiotic (PHO) that has a DNA-binding sequence [101].

The different PcG complexes appear to exist in several forms. For example, PRC2 complexes of 600 KD [102] and of 1MD that contains the polycomb-like (PLC) protein [103] have been purified. Adding to this complexity, several PcG genes appear to have closely related orthologs within the group with which they share similar functions. Examples of such pairs include PSC and SU(Z)2, ESC and ESCL (Extra Sex Comb-Like), and PHO and PHOL (Pleiohomeiotic-Like).

PcG Complexes Associate with the Genome at Specific DNA Binding Sites

The PcG complexes bind to sequence elements called Polycomb Response Elements (PRE). These elements consist of several hundred base-pairs that include some short motifs targeted by DNA-binding factors such as the GAGA factor (GAF), Pipsqueak (PSQ), the Dorsal Switch Protein 1 (DSP1), and the Sp1/KLF family of proteins [104–106]. Using an algorithm based on GAF binding sites and on the binding site of PHO and of the Zeste (Z) protein that has been found in several PcG binding sites, over a hundred and fifty presumed PREs were found throughout the euchromatic genome [107]. The complementary approach of mapping the location of PcG proteins along chromosomes by chromatin immunoprecipitation revealed that the algorithm had missed many PREs and that many of the PREs that it had identified were not occupied by PcG complexes [108,109]. These observations suggest that through their own binding, other proteins may attract PcG complexes to PREs (Fig. 14.2).

The level of product of a reporter gene in transgenic flies is usually greater in the homozygous than in the heterozygous condition; surprisingly, the reverse is the case, i.e. less gene product in homozygotes than in heterozygotes, if the transgenic construct contains a PRE. This observation highlights a characteristic of at least some PREs: pairing-dependent or pairing-sensitive silencing [110,111].

The Binding of PcG Complexes Induces and Depends on Post-Translational Chromatin Modifications

As should be evident from the previous statements, the targeting of PcG complexes to PREs to achieve the desired gene silencing is a complex and still poorly understood process. Nevertheless, certain parameters of a general nature have been identified [112,113]. The ESC and ESCL proteins are critical for the initial association, which is thought to involve PRC2 and to result in the trimethylation of H3K27 by E(Z). The latter does not occur at the PREs themselves since they appear to be nucleosome-free; rather it occurs on neighboring nucleosomes and may be involved in the spreading of repression throughout the targeted transcriptional units. Although initially thought to recruit the PRC1 complex via the chromodomain of its PC subunit, the role of H3K27me3 has been revised: *in vitro*, PC has a very modest binding affinity for this chromatin mark [114]; furthermore, while the presence of H3K27me3 extends through entire transcription unit domains, PRC1 proteins are found exclusively at the site of the adjacent PRE [115].

As mentioned above, the PHORC complex has DNA binding activity via its PHO subunit. The SFMBT protein binds to mono- and dimethylated H3K9 and H4K20, both of which are marks of repressed chromatin [101]. This specificity is not well understood, given that silenced HOX genes such as Ubx are decorated with trimethylated H3K9, H3K27, and H4K20.

213

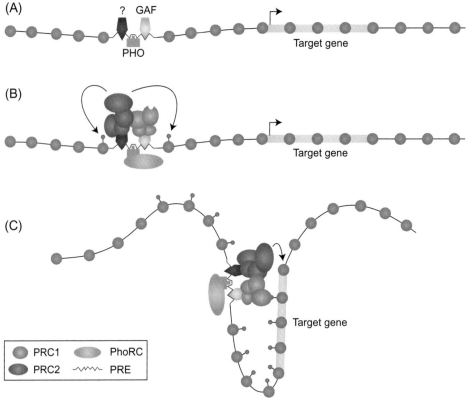

FIGURE 14.2
A model depicting the binding of PHO and GAF to a PRE, the recruitment of PRC1 and PRC2 resulting in methylation of nucleosomes, and the extension of the methylated domain by loop formation *(from [113], reprinted with permission from Macmillan Publishers Ltd).* (Please refer to color plate section)

The dRING protein is an E3 histone ubiquitine ligase responsible for the majority of the histone H2A ubiquitination [116]. In addition to its presence in PRC1, dRING is also found in another complex with PSC (Posterior Sex Combs) but not Polycomb (PC) or Polyhomeiotic (PH). Among other known proteins, this complex includes the *Drosophila* homolog of KDM2 – a H3K36me2 demethylase. In addition to this activity, KDM2 stimulates the ubiquitinase activity of dRING [117].

The PcG genes include *grappa* (*gpp*) which encodes a histone methyltransferase that is required for the methylation of H3K79 [118]. Loss-of-function mutations suppress telomere-mediated but not centromere-mediated position effect variegation. These mutations interact genetically with mutations in PcG genes and suppress the pairing-sensitive silencing of PRE-bearing transgenes. There is no information regarding the presence of the GPP gene product in the known PcG complexes.

trxG REGULATORY PROTEINS

The body of experimental information on the trxG genes is less voluminous than for the PcG genes and is mostly associated with a characterization of the interactions between these two gene groups (discussed below). All of the HOX genes that have been activated in the early embryo and that must be kept active during segment differentiation are targeted by gene products of members of the trithorax group, preventing their default-silencing by PcG complexes [119,120]. Transcription through the non-coding region of PREs is also correlated with a failure of PcG-mediated silencing [121].

TrxG Proteins Form Regulatory Complexes

Many of the genes that encode the trxG proteins were first identified in genetic screens for suppressors of mutations of *Pc* or of the HOX gene *Antennapedia* (*Antp*) [122]. Two of the trxG proteins are histone methyl transferases: *trithorax* (*trx*) that is H3K4-specific and *absent, small, or homeotic discs 1* (*ash1*). Loss of ASH1 results in a dramatic loss of H3K4 methylation on larval polytene chromosomes suggesting that it is the essential H3K4 methyl transferase; no changes are observed in either H4K20 or H3K36 methylation [123]. Perhaps not surprisingly, the specificity of Ash1 is different when a recombinant enzyme is assayed *in vitro* [124]. Surprisingly, the trxG genes include *little imaginal discs* (*lid*), a JmjC + N demethylase specific for H3K4me3 [18,19]. In spite of this usually repressive enzymatic activity, LID appears to contribute to the expression of the HOX gene *Ultrabithorax* (*Ubx*) in cultured cells [19] and in imaginal discs [125]. An explanation for this contradiction is suggested by the recent evidence that LID is a subunit of a multiprotein complex that includes the histone deacetylase RPD3 normally associated with repression of gene activity [126]; in this complex, RPD3 is inactive.

The TRX protein is present in a complex – Trithorax Acetylation Complex (TAC1) – that includes the histone acetyl transferase CBP and SBF1, the *Drosophila* ortholog of the human anti-phosphatase protein Set-Binding Factor 1 [127]. TAC1 has been shown to regulate the HOX gene *Ubx*'s expression.

TrxG proteins and their complexes bind to specific DNA sequences called TREs. Because they often overlap, PREs and TREs are sometimes referred to as MEs or maintenance elements [128].

TrxG Proteins Interact with Chaperones and Cohesins to Carry Out Their Functions

215

A relationship between trxG proteins and different molecular chaperones [129,130] has led to the discovery that *trithorax* requires Hsp90 to sustain the active state of HOX genes [131]. HSP90 is a ubiquitous chaperone involved in numerous regulatory pathways, under heat-shock stress or at normal temperature; partial loss of function of the protein leads to morphological abnormalities that resemble homeotic mutations. These phenotypes are explained by the functional interaction of HSP90 with trxG proteins.

TrxG proteins also interact with members of the cohesin complex. Cohesins were initially characterized as complexes that maintain the intimate association of sister chromatids. A more fine-grained analysis revealed that cohesin proteins and the Nipped-B complex that loads them onto chromatids co-localize extensively along chromosomes where they bind to a subset of transcribed regions [132]; as expected, they are absent from genes that are silenced by PcG complexes. In fact, RAD21, a major component of the cohesin complex encoded by the *verthandi* (*vtd*) gene has been added to the trxG list because of the suppressor-of-Pc activity of loss-of-function alleles [133].

INTERACTIONS BETWEEN PcG AND trxG FACTORS

By chromatin immunoprecipitation (ChIP) of their respective subunits, all three PcG complexes and TRX were found to bind to the two PREs flanking the *Ubx* gene, whether this gene was on or off [134]. When the gene is inactive, the entire Ubx region is enriched in trimethylated H3K9, H3K27, and H4K20. When *Ubx* was active, there was a strong reduction of PCG and TRX binding at the downstream PRE and ASH1 was observed bound in the coding region of the gene; the histone modifications present throughout the region were replaced in the promoter and coding regions by H4K20me1 and H3K4me3. Generally similar results were obtained by a ChIP-chip analysis of the distribution of several members of the PCG and trxG proteins and of histone modifications on the ANT-C and BX-C HOX gene complexes. Both PHO and PRC1 bind the regulatory sequences of silent HOX genes and TRX is always found at these sites [135]. The results of this analysis revealed that although the chromatin characteristics of two different HOX genes are similar when they are both inactive, in the active state these characteristics are different suggesting that maintenance of activity can be achieved by somewhat different means. Different conclusions were reached with experiments that made use of transgenes with a reporter gene under the control of a HOX gene maintenance element (ME) [136]. Using immunofluorescence to examine the presence of different trxG and PcG proteins on the transgenes in individual cells of larval salivary glands, the binding of TRX and PRC2 or PRC2 was found to be mutually exclusive.

A direct interaction at the level of histone modification occurs between the TRX-dependent CBP acetylation of H3K27, its deacetylation by RPD3 and its trimethylation by the PcG methyl transferase E(Z). These interactions provide a mechanism by which TRX antagonizes or impedes PcG silencing [137].

Chromosome-Level Epigenetic Regulation of Transcription: Dosage Compensation

Epigenetic modifications control the expression of the genome at the level of individual genes, of large genomic domains (locus control regions), or of entire chromosomes. Dosage compensation is an example of chromosomal-level epigenetic regulation.

THE MSL COMPLEX IS RESPONSIBLE FOR DOSAGE COMPENSATION

Dosage compensation refers to the equalization of most X-linked gene products between males, which have one X chromosome and a single dose of X-linked genes, and females, which have two X's and two doses of such genes. It is mediated by the MSL complex consisting of a core of five protein subunits encoded by *male-specific lethal 1, 2, and 3* (*msl1, msl2, msl3*), *males absent on the first* (*mof*), and *maleless* (*mle*), as well as one of two non-coding RNAs [*RNA on the X1 and 2* (*roX1* and *roX2*)] [138,139]. These RNAs are very different in size (4.1 to 4.3 kb and 0.6 kb, respectively); nevertheless, they share a hairpin secondary structure and several short stretches of sequence [140,141], and they are completely interchangeable with respect to the function of the MSL complex in dosage compensation (Fig. 14.3B). The complex preferentially associates with numerous sites on

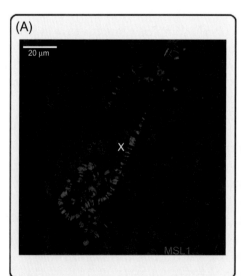
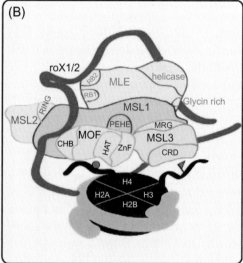

FIGURE 14.3
(A) The MSL complex is distributed along the male X chromosome as evidenced by indirect immunostaining of one of its protein components. (B) Subunit components of the MSL complex *(from Ref. 139, reprinted with permission from Springer Science + Business Media). (Please refer to color plate section)*

216

the X chromosome in somatic cells of males but not of females (Fig. 14.3A). It is responsible for an enhancement of the transcriptional rate of a substantial number of X-linked genes, thereby compensating for the difference in dosage of these genes between males and females. Although all of the *msl* genes are transcribed in females, the complex is absent because the female-specific RNA-binding Sex-lethal (SXL) protein that is responsible for female differentiation prevents the translation of the *msl2* gene transcript [142,143].

In males, the complex is believed to assemble at the locus of the two *roX* genes and then spread to numerous additional sites along the X chromosome for which it has a complete range of affinity levels [144,145]. Using ChIP-chip or ChIP followed by sequencing, from 130 to 150 X-chromosome sites predominantly enriched for a GA repeated sequence for which the MSL complex has particular affinity have been identified [146,147]. From these sites, the complex is attracted to activated genes [148,149] where it displays substantial affinity to the H3K36me3 mark characteristic of transcription.

HISTONE ACETYLATION IS CORRELATED WITH THE ENHANCEMENT OF X-LINKED GENE TRANSCRIPTION IN MALES

The presence of the MSL complex on the male X chromosome is correlated with a significant increase of histone H4 acetylated at lysine 16 [H4K16ac; [150]]. This acetylation is the result of the activity of MOF – a histone acetyltransferase of the MYST family [151]. Rather than targeting promoters, the acetylation occurs throughout transcriptional units with a significant increase towards the 3′ end [152–154]. For this reason, the function of the MSL complex is unlikely to be the initiation of gene activity; rather the enhancement in the level of gene expression that leads to dosage compensation may be the result of an enhancement in the rate of transcription elongation.

The X-ray crystal structure of nucleosomes suggests the occurrence of an inter-nucleosomal interaction involving the acidic patch formed by an H2A/H2B dimer in one nucleosome and the basic tail of histone H4 from a neighboring nucleosome [155,156]. Recently, the acetylation of H4 at lysine 16 was shown to prevent the conversion of reconstituted nucleosomal arrays into 30 nanometer fibers that are thought to represent a level of compaction of native chromatin unfavorable to transcription [157]. These observations suggest that the acetylation of H4K16 opens the X-chromosome chromatin and that dosage compensation is achieved by facilitating nucleosome eviction ahead of the processing RNP II and increasing the rate of elongation.

THE MSL COMPLEX INCLUDES AN RNA/DNA HELICASE

In addition to this enzyme, the MSL complex of *Drosophila* includes an ATP-dependent DEXH-box RNA/DNA helicase (MLE) that prefers double-stranded RNA or RNA/DNA hybrid substrates with a short 3′ overhang [158]. MLE is related to the ATPases present in complexes that remodel chromatin by altering the positioning or the architectural relationship between histone octamers and DNA [159]. The ATPase activity is required for MLE's role in the transcriptional enhancement of a targeted gene while the helicase activity is necessary for the spreading of the complex along the X chromosome [160].

INTERACTION WITH HETEROCHROMATIN PROTEINS

Over-expression of Su(var)3-7 results in morphological effects in the larval salivary gland polytene chromosomes of both males and females, but the male X is most affected as it assumes a very small and highly compacted shape [161]. In these chromosomes, the distribution of the MSL complex is abnormal [162]. Loss of Su(var)3-7 results in a polytene X chromosome in males that is shorter, bloated and with a banding sequence that is much less distinct. The bloated appearance requires the presence of an active MSL complex [163]. Loss of HP1 has the same bloating effect on the X chromosome morphology in males.

HP1 binding is moderately elevated along the whole male X chromosome in comparison to the autosomes. Females do not exhibit such an X-specific enrichment [94]. In contrast, loss of HP2, a protein that interacts with HP1 [52], does not have a specific effect on the X chromosome in males [164].

Different results were obtained RNAi knock-down of HP1. Sex-biased defective chromosome segregation, alterations in histone modifications, specific changes in transcription, and a skewed sex ratio in surviving progeny were observed. But the morphology of the X chromosome and the global level and distribution of H4K16ac seemed unaffected in male HP1 knock-down progeny, suggesting that the sex-biased effect on male viability was not caused by misregulation of dosage compensation [165]. High-resolution mapping of HP1 in autosomal regions revealed that HP1 target-genes are also marked by the histone variant H3.3 (discussed below) and H3K4me2 that are characteristic of active chromatin [165,166].

The Jil-1 kinase, responsible for the bulk of H3S10 phosphorylation during interphase, localizes to interband regions in all polytene chromosomes but is substantially more abundant on the X chromosome in males [23,167]. Loss-of-function alleles result in global changes in the morphology of polytene chromosomes with the male X, once again, shorter, fatter and without any evidence of banding [168]. JIL-1 loss-of-function alleles allow the spreading of H3K9me2 and HP1, suggesting that JIL-1 normally marks and preserves the limits of euchromatic domains [169]. Recently, Kristen Johansen and collaborators have provided direct evidence that the H3S10 phosphorylation mediated by JIL-1 induces an open chromatin state [170]. It is tempting to interpret these interactions in terms of the relative domain boundaries between euchromatin and intercalary heterochromatin on the X chromosome in males and their effect on transcription enhancement.

INTERACTION WITH THE GENERAL CHROMATIN REMODELING COMPLEXES NURF, ACF, CHRAC AND ATAC

The X chromosome in males responds dramatically to the loss-of-function of the general chromatin assembly complexes ACF and CHRAC and the nucleosome repositioning complex NURF. Loss-of-function mutations in ISWI (Imitation Switch) protein, the ATPase common to all three complexes, transform the male X chromosome in salivary gland polytene chromosome preparations into a chromatin mass that has lost all morphological features [171]. Loss-of-function mutations in a subunit unique to the NURF complex have the same effect on the X-chromosome morphology; this effect can be rescued in males by preventing the occurrence of H4K16ac. In a mutant *nurf* background, loss-of-function mutations in either *roX1* or *roX2* lead to a more normal appearance of the polytenic X in the general region of the mutation; conversely, a wild type *roX* transgene relocated to an ectopic autosomal location nucleated a region of disorganization at its site of insertion [172].

Mutations in *dAda3* gene cause a defect in the banding organization of polytene chromosomes in both males and females; once again, the X chromosome in males is more severely affected [173]. Mutations in the histone acetyl transferase Gcn5 and the ATAC (Ada two A containing) complex component Ada2a induce a specific decondensation of the X chromosome in mutant males [174]. In contrast, the *Drosophila* RSF (Remodeling and Spacing Factor) complex consisting of dRsf1 and ISWI does not affect the appearance of polytene chromosomes in either sex [175].

Histone Replacement

HISTONE VARIANTS AND TRANSCRIPTION

H3.3 Replaces H3 in the Domain of Actively Transcribed Genes

Canonical histones are synthesized during the S phase and the bulk of nucleosome assembly occurs behind the replication fork. In contrast, histone variants such as *Drosophila* H3.3

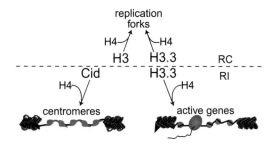

FIGURE 14.4

Diagram of the incorporation pathways of histones. H3 uses a replication-coupled (RC) mechanism and H3.3 and Cid a replication independent (RI) mechanism. Note that some degree of H3.3 incorporation can occur in an replication-coupled manner. *(from Ref. 217, reprinted with permission from The National Academy of Sciences, USA).* (Please refer to color plate section)

are synthesized in all phases of the cell cycle [176]. Histone H3.3 is used in a replication-independent (RI) nucleosome assembly process that occurs at transcriptionally active loci throughout the cell cycle [177]. Although H3.3 differs from H3 in only four amino acid residues, the difference is sufficient to target this variant to the RI pathway. Histone H3 replacement is initiated immediately upon induction of a gene, occurs throughout the length of the transcriptional unit, and is retained long after the gene is repressed [178,179].

The presence of H3.3 is greater on the hypertranscribing X-chromosome of males than in females or on the autosomes of either sex. In contrast to the distribution of H4K16ac, which is skewed towards the 3′ end, histone H3.3 is more enriched at the 5′ end of transcribing genes [166]. A fine-grained study of H3.3 distribution along a broad region that includes the ANT-C and BX-C complexes reveals that, although the two complexes exhibit a relatively low ratio of H3.3 to H3 histones, prominent peaks of H3.3 are found at the regulatory domains, at some of the PREs and promoters [180]. Many of the regulatory regions include hypersensitive sites and the enrichment in H3.3 at these sites indicates that a continuous process of nucleosome disruption and replacement allows high levels of H3.3 incorporation while exposing the DNA to nuclease digestion (Fig. 14.4).

Given the facts just discussed, the report that H3.3 is dispensable (in the soma although not in the germline) was surprising. As expected, the histone H3 covalent modifications that are characteristic of active gene function are found predominantly in H3.3 [181]. Deletion of the two genes that encode H3.3, or replacement of H3.3 with a form where H3K4 was mutated to H3A4 preventing methylation, has no effect on the output of major developmental signaling pathways or on viability; furthermore, the total levels of H3K4me3 remained unchanged suggesting that in the absence of H3.3, this modification is switched to canonical H3 [182].

H2Av Participates in Gene Silencing

This *Drosophila* homolog of the H2A variant H2AZ.A is present in centromeric heterochromatin and throughout the chromosome arms [183,184]. The gene that encodes this variant (*His2Av*) is classified as a PcG gene on the basis that a reduction in its product enhances the phenotype of mutations in *Pc* [185]. H2Av plays an important role in recruiting PC to PRE sites but does not affect E(Z) recruitment or H3K27me3 levels, suggesting that the latter precedes the former. In addition reducing H2Av levels results in a reduction in H3K9me2 and HP1 and a loss of H4K12ac in centromeric heterochromatin and, not surprisingly, suppresses position effect variegation [185].

RSF-1 is a protein encoded by the *dRsf-1* gene that interacts with the ATPase ISWI (imitation switch) protein to deposit nucleosomes. Mutations in *dRSF-1* act as suppressors of variegation and reduce the level of H3K9me2 with a concomitant decrease in H2Av [175]. RSF1 and ISWI co-immunoprecipitate with subunits of the histone acetyl tranferase Tip60 complex that had been shown to facilitate silenced chromatin. These results identify another ISWI-containing complex that mediates H2A exchange for H2Av [175].

HISTONE VARIANTS AND THE REPAIR OF DNA DAMAGE

H2AX is a histone H2A variant that is targeted to the site of DNA damage. In *Drosophila*, this H2A variant (H2Av) is a hybrid molecule that has the H2Av globular portion and a H2AX C-terminal tail. It is present ubiquitously and randomly throughout the chromatin. During the cellular response to DNA double-strand break damage, a phosphorylated form (S139ph) is exchanged for the unphosphorylated H2Av at the site of double-strand breaks. The Tip60 complex catalyzes this process and acetylates H2AvS139ph prior to the exchange [186]. Marking the sites of double-strand breaks in this fashion is thought to target them for the factors involved in non-homologous end joining or homologous recombination pathways.

CENTROMERE-SPECIFIC HISTONE VARIANTS

The nucleosomes in the centromere chromatin of most eukaryotic organisms contain a histone H3 variant that in *Drosophila* is referred to as CID or centromere identifier [187]. The presence of CID represents the epigenetic mark that identifies a chromosomal region as a centromere; its presence is required for all aspects of centromere assembly and function. When chromosomes replicate, existing nucleosomes are transmitted randomly to the daughter chromatids and new nucleosomes including those containing CID are deposited by chromatin assembly chaperones such as ASF1 (Antisilencing Factor 1) and complexes such as CAF-1 (Chromatin Assembly Factor 1). The timing within the cell cycle of centromere DNA replication has not been unanimously established: some experimental results suggest that it occurs early in S phase, while the pericentric heterochromatin replicates late [188]; other experiments place the time of replication late in the S phase [189]. Surprisingly, CID incorporation into centromeres occurs during early anaphase and is independent of DNA replication [190]. Although a large number of proteins associated with centromeres have been identified, the parameters that regulate the incorporation of CID into the nucleosomes and the factor that target them to the particular DNA sequence that gives rise to a centromere are poorly understood. Using a genome-wide RNAi screen for loss of CID, several gene products were identified that are required for CID localization: CENP-C (originally identified as a human autoantigen) and CAL1 (a novel constitutive centromere protein) are essential for CID incorporation; all three proteins are mutually dependent for localization to the centromere. In addition, cyclin A and APC (anaphase-promoting complex) regulate centromere assembly and propagation; the latter is regulated by RCA1 which inhibits the function of Fzr/Cdh1, a positive regulator of the APC complex [191].

Centromeric chromatin is embedded in pericentric heterochromatin and consists of blocks of CID-containing nucleosomes interspersed with blocks of canonical H3-containing nucleosomes. Yet the histone modifications of centromeric nucleosomes are different from those that characterize hetero- or euchromatin: H3K9me2/me3, H3K9ac, and H4K5/K8/K12/K16ac are absent; H3K4me2 (but not me3) is present [192]. The epigenetic character of centromeres is further evidenced by the observation that centromeres can form on chromosome fragments that are completely free of the functional centromeric DNA that is present in the chromosomes from which they are derived; in addition, these neocentromeres not only insure the recovery of the fragments after their induction by ionizing radiation but also exhibit transgenerational propagation [193]. Dicentric chromosomes usually fail to be transmitted because they result in breakage-fusion-bridge cycles. In exceptional cases, one of the two centromeres becomes epigenetically inactivated; which one remains active appears to be a random process but, once established, it is transmitted clonally to daughter cells [194].

CHROMATIN INSULATORS

Insulators have traditionally been defined as sequences along the DNA that attract particular proteins for the purpose of preventing distal enhancers from activating or silencing genes. This definition was extended to include sequence elements that block the inappropriate spread of heterochromatin (also referred to as boundary elements). Evidence from the study of enhancers in *Drosophila* indicates that their most important function may be to organize the genome into functional clusters.

Types of Insulator

Scs and *scs'* (specialized chromatin structures) were the first insulators to be discovered; they flank the region that includes the two *heat shock protein 70* (*hsp70A/B*) genes at 87A on the cytological map [195] and interact with the BEAF-32A and -32B (Boundary Associated Element 32) and ZW5 (Zeste-White region gene 5) proteins. These proteins are present at hundreds of sites throughout the genome [196].

The *gypsy insulator* exerts its action through the binding of the Suppressor of Hairy wing [Su(Hw)] protein and two additional proteins: Modifier of mdg4 [Mod(mdg4)] and Centrosomal Protein 190 (CP190). These proteins co-localize at hundreds of sites in the genome that do not correspond to the *gypsy* retrotransposon insertion sites.

The *Fab* insulators are present in the Bx-C region with its three homeotic genes and nine regulatory sites to insure the specific and exclusive activation of genes in different regions during development. Given the evolutionary conservation of the HOX genes, it may not be surprising that most of these insulators (*Fab-2, -3, -4, -6,* and *-8* but not *Fab-7*) are bound by CTCF, the protein associated with the major type of insulator known in vertebrates [197]. *Mcp*, an additional insulator in the Bx-C region, is also bound by CTCF. The sites that are bound by this protein are also bound by CP190 [198,199]. The *Fab-7* insulator may be targeted to chromatin by the GAGA factor (GAF) protein [200].

Insulator Functions

INSULATORS ESTABLISH THE BOUNDARIES OF REGULATORY DOMAINS

An insulator placed between two enhancers upstream of a gene will block the distal but not the proximal enhancer from regulating transcription. It does not prevent the distal enhancer from regulating the transcription of another promoter positioned directly upstream. Furthermore, an insulator separating a promoter from a distal enhancer followed by a more proximal silencer once again blocks the enhancer but has no effect on the silencer [201].

Early during development, a group of segmentation genes initially activated by maternal factors subdivide the embryo into 14 parasegments which later give rise to the segments of the larva and the adult. As mentioned in a previous section, the differentiation of the segments into the adult anatomy is the responsibility of the two HOX gene clusters, ANT-C and Bx-C. In order to achieve the orderly, differential expression of the eight genes within these clusters, each parasegment-specific regulatory domain should be flanked by domain boundary elements [202]. At the time of this writing, only two Bx-C insulators were known, *Fab-7* and *Mcp*. The discovery of many additional insulators as well as PREs and TREs associated with the HOX gene complexes suggests that the regulation of these complexes is parsed into independently regulated domains. A key parameter of this model is the relationship of these regulatory elements with one another. During the activation or repression of the domains, the spreading effect of the PREs must be limited by domain boundaries. In fact, available data indicate that boundary elements and PREs are in close proximity [197].

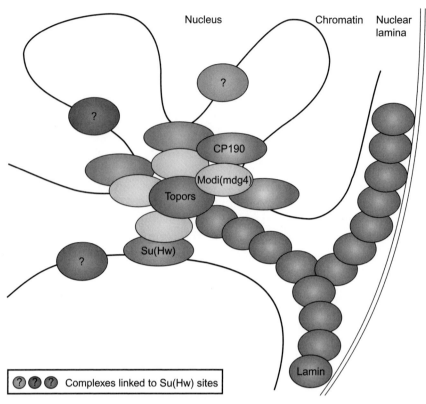

FIGURE 14.5

Detailed diagram of the association of Su(Hw) complexes with each other and with the nuclear matrix *(from Ref. 218, reprinted with permission from Macmillan Publishers Ltd).* (Please refer to color plate section)

INSULATORS MODIFY CHROMATIN STRUCTURE

Depletion of the BEAF insulator protein leads to a significant increase in the level and distribution of H3K9me3 and a concomitant repression of genes adjacent to its binding sites [203]. These results indicate that BEAF blocks H3K9 methylation and thereby preserves the transcription of adjacent genes. Consistent with these results is the observation that a dominant-negative suppressor of BEAF acts as an enhancer of position effect variegation, presumably by allowing the spread of H3K9me3 beyond the junction of the euchromatin-heterochromatin rearrangement [204].

INSULATORS AFFECT NUCLEAR ORGANIZATION

The initial realization of this effect was the observation that although insulator proteins such as Su(Hw) and Mod(mdg4) occupy hundreds of sites on larval salivary gland polytene chromosomes, they are present in a very limited number of locations (20 to 25) in interphase nuclei. The insulator property that provides the explanation for this surprising aspect of nuclear architecture was the tendency to form aggregates (Fig. 14.5). Multiple *gypsy*-like insulators (sites that are identified by the presence of Su(Hw) and Mod(mdg4) but are not associated with *gypsy* retrotransposon insertions) interact to form rosette-like structures in the nucleus [204,205]; the proteins of the *scs* and *scs'* insulators associate with each other juxtaposing these two insulators and forming a DNA loop between them [206].

The presence of an ectopic *gypsy* insulator causes the region where it is inserted to move from its normal location inside the nucleus to the nuclear periphery [207]. This observation suggested that the aggregation of insulator proteins and insulator sequences, referred to as insulator bodies, are associated with the nuclear matrix. This was demonstrated to be the

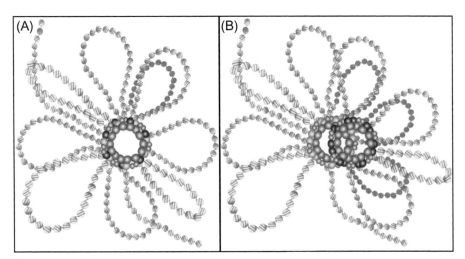

FIGURE 14.6
Models of the formation of DNA rosettes by the association of Su(Hw) (blue balls), Mod(mdg4) (green balls), CTCF (purple balls) and CP190 (pink balls). (A) Su(Hw) and CTCF insulators bind distinct but interspersed DNA sequences and, therefore, cluster together in the same insulator body. (B) Each insulator type forms its own insulator body but different bodies cluster together in particular nuclear regions *(from Ref. 211, reprinted with permission from Elsevier Limited).* (Please refer to color plate section)

case [208]. The Topoisomerase I-interacting protein dTopors, a ubiquitin ligase, interacts with Su(Hw) and Mod(mdg4) as well as with Lamin Dm0 is involved in recruitment to the matrix [209]. Components of the RNAi machinery may be involved as well. Using immunoaffinity purification, a protein in the double-stranded RNA interference pathway – the RNA helicase Rm62 – was found to associate with CP190. Loss of function of this protein improves insulator activity of a gypsy transposon while loss of function of Argonaute (proteins that either degrade or prevent the translation of target RNAs) reduces this activity [210]. In the matrix, the insulator bodies formed by Su(Hw) enhancers overlap with those formed by CTCF enhancers, suggesting an association between them mediated by their common protein CP190 [211].

The architectural organization of much of the genome into rosettes of DNA loops attached to the matrix by insulator-body hubs must have direct implications for gene regulation (Fig. 14.6). Although a specific demonstration of this contention has not yet been realized, a concordant correlative observation consists of the fact that the distributions of different insulator proteins (Su(Hw), BEAF, CP190, and CTCF) are different in different cell types [199].

DNA METHYLATION

Given the importance of this epigenetic modification in fungi, plants, and vertebrates, it is appropriate to devote a major section to its discussion. In spite of this consideration, the discussion will be very brief because until recently [212], DNA methylation was thought to be absent in *Drosophila*. The DNA methyl transferase DNMT2 identified in humans is present in *Drosophila* (dDNMT2). Similarly to the human enzyme, dDNMT2 methylates a small specific RNA: aspartic acid transfer RNA (tRNA^{Asp}). Loss of function mutations in the *Mt2* gene have no effect on the viability or fertility of these mutant individuals [213]. In contrast to human DNMT2 that does not methylate DNA, the *Drosophila* enzyme methylates cytosine nucleotides in the DNA of very early (1 to 5 hour-old) embryos [15]. While the absence of dDNMT2 activity had no effect on position effect variegation of genes relocated within or near centromeric heterochromatin, the variegation of reporter transgenes that landed within the domain of retrotransposons was strongly suppressed due to their enhanced transcription. In addition, the absence of dDNMT2 activity led to the loss of subtelomeric DNA repeats.

223

These results indicate that DNA methylation during an early developmental window is responsible for maintaining retrotransposons in a repressed state and is necessary for the integrity of telomeres. How these parameters are maintained after dDNMT2 is no longer active is an unanswered question.

PERSPECTIVE

Epigenetic chromatin modifications are responsible for the differential expression of the genome in time and space. They enable the rapid response to extra-cellular stimuli as well as the transmission of gene-expression states through cell divisions (cellular memory). Epigenetic changes can be perpetuated in the absence of the stimuli that caused them and they can be transmitted through meiosis (transgenerational memory). These facts validate the study of epigenetics as a central endeavor in biomedical research, in general; the increasing list of examples of transgenerational inheritance in *Drosophila* [214] validates the continued use of this model organism for this purpose.

Because all the factors that mediate epigenetic regulation are encoded in the genome, natural selection cannot operate on epigenetic variants without simultaneously operating on the underlying genetic variations. Yet, epigenetic phenotypes can exhibit rapid variation since they can result from responses to external stimuli. In contrast, genetic phenotypes resulting from mutations generally vary at a much smaller rate. These considerations raise the question whether epigenetic traits represent the preferred "fodder" of natural selection. Here again, the track record established with *Drosophila* for the experimental testing of evolutionary forces (see for example [215–217] bodes well.

References

1. Pimpinelli S, Berloco M, Fanti L, Dimitri P, Bonaccorsi S, Marchetti E, et al. Transposable elements are stable structural components of *Drosophila* melanogaster heterochromatin. *Proc Natl Acad Sci USA* 1995;**92**:3804–8.

2. Ahmad K, Golic KG. Somatic reversion of chromosomal position effects in *Drosophila melanogaster*. *Genetics* 1996;**144**:657–70.

3. Gowen JW, Gay EH. Chromosome constitution and behavior in eversporting and mottling in *Drosophila melanogaster*. *Genetics* 1934;**19**:189–208.

4. Spofford JB. Single-locus modification of position-effect variegation in *Drosophila melanogaster*. I. White variegation. *Genetics* 1967;**57**:751–66.

5. Reuter G, Wolff I. Isolation of dominant suppressor mutations for position-effect variegation in *Drosophila melanogaster*. *Mol Gen Genet* 1981;**182**:516–19.

6. Sinclair DA, Lloyd VK, Grigliatti TA. Characterization of mutations that enhance position-effect variegation in *Drosophila melanogaster*. *Mol Gen Genet* 1989;**216**:328–33.

7. Girton JR, Johansen KM. Chromatin structure and the regulation of gene expression: the lessons of PEV in *Drosophila*. *Adv Genet* 2008;**61**:1–43.

8. Tschiersch B, Hofmann A, Krauss V, Dorn R, Korge G, Reuter G. The protein encoded by the *Drosophila* position-effect variegation suppressor gene Su(var)3-9 combines domains of antagonistic regulators of homeotic gene complexes. *EMBO J* 1994;**13**:3822–31.

9. Czermin B, Schotta G, Hulsmann BB, Brehm A, Becker PB, Reuter G, et al. Physical and functional association of SU(VAR)3-9 and HDAC1 in *Drosophila*. *EMBO Rep* 2001;**2**:915–19.

10. Ebert A, Schotta G, Lein S, Kubicek S, Krauss V, Jenuwein T, et al. Su(var) genes regulate the balance between euchromatin and heterochromatin in *Drosophila*. *Genes Dev* 2004;**18**:2973–83.

11. Mottus R, Sobel RE, Grigliatti TA. Mutational analysis of a histone deacetylase in *Drosophila melanogaster*: missense mutations suppress gene silencing associated with position effect variegation. *Genetics* 2000;**154**:657–68.

12. Schotta G, Lachner M, Sarma K, Ebert A, Sengupta R, Reuter G, et al. A silencing pathway to induce H3-K9 and H4-K20 trimethylation at constitutive heterochromatin. *Genes Dev* 2004;**18**:1251–62.

13. Yang H, Pesavento JJ, Starnes TW, Cryderman DE, Wallrath LL, Kelleher NL, et al. Preferential dimethylation of histone H4 lysine 20 by Suv4-20. *J Biol Chem* 2008;**283**:12085–92.

14. Sakaguchi A, Karachentsev D, Seth-Pasricha M, Druzhinina M, Steward R. Functional characterization of the *Drosophila* Hmt4-20/Suv4-20 histone methyltransferase. *Genetics* 2008;**179**:317–22.

15. Phalke S, Nickel O, Walluscheck D, Hortig F, Onorati MC, Reuter G. Retrotransposon silencing and telomere integrity in somatic cells of *Drosophila* depends on the cytosine-5 methyltransferase DNMT2. *Nat Genet* 2009;**41**:696–702.

16. Karachentsev D, Sarma K, Reinberg D, Steward R. PR-Set7-dependent methylation of histone H4 Lys 20 functions in repression of gene expression and is essential for mitosis. *Genes Dev* 2005;**19**:431–5.

17. Rudolph T, Yonezawa M, Lein S, Heidrich K, Kubicek S, Schafer C, et al. Heterochromatin formation in *Drosophila* is initiated through active removal of H3K4 methylation by the LSD1 homolog SU(VAR)3-3. *Mol Cell* 2007;**26**:103–15.

18. Eissenberg JC, Lee MG, Schneider J, Ilvarsonn A, Shiekhattar R, Shilatifard A. The trithorax-group gene in *Drosophila* little imaginal discs encodes a trimethylated histone H3 Lys4 demethylase. *Nat Struct Mol Biol* 2007;**14**:344–6.

19. Lee N, Zhang J, Klose RJ, Erdjument-Bromage H, Tempst P, Jones RS, et al. The trithorax-group protein Lid is a histone H3 trimethyl-Lys4 demethylase. *Nat Struct Mol Biol* 2007;**14**:341–3.

20. Grienenberger A, Miotto B, Sagnier T, Cavalli G, Schramke V, Geli V, et al. The MYST domain acetyltransferase Chameau functions in epigenetic mechanisms of transcriptional repression. *Curr Biol* 2002;**12**:762–6.

21. Miotto B, Sagnier T, Berenger H, Bohmann D, Pradel J, Graba Y. Chameau HAT and DRpd3 HDAC function as antagonistic cofactors of JNK/AP-1-dependent transcription during *Drosophila* metamorphosis. *Genes Dev* 2006;**20**:101–12.

22. Dorn R, Heymann S, Lindigkeit R, Reuter G. Suppressor mutation of position-effect variegation in *Drosophila-melanogaster* affecting chromatin properties. *Chromosoma* 1986;**93**:398–403.

23. Jin Y, Wang Y, Walker DL, Dong H, Conley C, Johansen J, et al. JIL-1: a novel chromosomal tandem kinase implicated in transcriptional regulation in *Drosophila*. *Mol Cell* 1999;**4**:129–35.

24. Dombradi V, Cohen PT. Protein phosphorylation is involved in the regulation of chromatin condensation during interphase. *FEBS Lett* 1992;**312**:21–6.

25. James TC, Elgin SC. Identification of a nonhistone chromosomal protein associated with heterochromatin in *Drosophila* melanogaster and its gene. *Mol Cell Biol* 1986;**6**:3862–72.

26. James TC, Eissenberg JC, Craig C, Dietrich V, Hobson A, Elgin SC. Distribution patterns of HP1, a heterochromatin-associated nonhistone chromosomal protein of *Drosophila*. *Eur J Cell Biol* 1989;**50**:170–80.

27. Reuter G, Giarre M, Farah J, Gausz J, Spierer A, Spierer P. Dependence of position-effect variegation in *Drosophila* on dose of a gene encoding an unusual zinc-finger protein. *Nature* 1990;**344**:219–23.

28. Cleard F, Delattre M, Spierer P. SU(VAR)3-7, a *Drosophila* heterochromatin-associated protein and companion of HP1 in the genomic silencing of position-effect variegation. *EMBO J* 1997;**16**:5280–8.

29. Delattre M, Spierer A, Tonka CH, Spierer P. The genomic silencing of position-effect variegation in *Drosophila* *melanogaster*: interaction between the heterochromatin-associated proteins Su(var)3-7 and HP1. *J Cell Sci* 2000;**113(Pt 23)**:4253–61.

30. Shaffer CD, Stephens GE, Thompson BA, Funches L, Bernat JA, Craig CA, et al. Heterochromatin protein 2 (HP2), a partner of HP1 in *Drosophila* heterochromatin. *Proc Natl Acad Sci USA* 2002;**99**:14332–7.

31. Badugu R, Shareef MM, Kellum R. Novel *Drosophila* heterochromatin protein 1 (HP1)/origin recognition complex-associated protein (HOAP) repeat motif in HP1/HOAP interactions and chromocenter associations. *J Biol Chem* 2003;**278**:34491–8.

32. Cenci G, Siriaco G, Raffa GD, Kellum R, Gatti M. The *Drosophila* HOAP protein is required for telomere capping. *Nat Cell Biol* 2003;**5**:82–4.

33. Garzino V, Pereira A, Laurenti P, Graba Y, Levis RW, Le Parco Y, et al. Cell lineage-specific expression of modulo, a dose-dependent modifier of variegation in *Drosophila*. *EMBO J* 1992;**11**:4471–9.

34. Perrin L, Demakova O, Fanti L, Kallenbach S, Saingery S, Mal'ceva NI, et al. Dynamics of the sub-nuclear distribution of Modulo and the regulation of position-effect variegation by nucleolus in *Drosophila*. *J Cell Sci* 1998;**111(Pt 18)**:2753–61.

35. Rodriguez Alfageme C, Rudkin GT, Cohen LH. Isolation, properties and cellular distribution of D1, a chromosomal protein of *Drosophila*. *Chromosoma* 1980;**78**:1–31.

36. Aulner N, Monod C, Mandicourt G, Jullien D, Cuvier O, Sall A, et al. The AT-hook protein D1 is essential for *Drosophila melanogaster* development and is implicated in position-effect variegation. *Mol Cell Biol* 2002;**22**:1218–32.

37. Huertas D, Cortes A, Casanova J, Azorin F. *Drosophila* DDP1, a multi-KH-domain protein, contributes to centromeric silencing and chromosome segregation. *Curr Biol* 2004;**14**:1611–20.

38. Gerasimova TI, Gdula DA, Gerasimov DV, Simonova O, Corces VG. A *Drosophila* protein that imparts directionality on a chromatin insulator is an enhancer of position-effect variegation. *Cell* 1995;**82**:587–97.

39. Weiler KS. E(var)3-9 of *Drosophila melanogaster* encodes a zinc finger protein. *Genetics* 2007;**177**:167–78.

40. Ahmad K, Henikoff S. Modulation of a transcription factor counteracts heterochromatic gene silencing in *Drosophila*. *Cell* 2001;**104**:839–47.

41. Seum C, Spierer A, Pauli D, Szidonya J, Reuter G, Spierer P. Position-effect variegation in *Drosophila* depends on dose of the gene encoding the E2F transcriptional activator and cell cycle regulator. *Development* 1996;**122**:1949–56.

42. Hari KL, Cook KR, Karpen GH. The *Drosophila* Su(var)2-10 locus regulates chromosome structure and function and encodes a member of the PIAS protein family. *Genes Dev* 2001;**15**:1334–48.

43. Shi S, Calhoun HC, Xia F, Li J, Le L, Li WX. JAK signaling globally counteracts heterochromatic gene silencing. *Nat Genet* 2006;**38**:1071–6.

44. Shi S, Larson K, Guo D, Lim SJ, Dutta P, Yan SJ, et al. *Drosophila* STAT is required for directly maintaining HP1 localization and heterochromatin stability. *Nat Cell Biol* 2008;**10**:489–96.

45. Strodicke M, Karberg S, Korge G. Domina (Dom), a new *Drosophila* member of the FKH/WH gene family, affects morphogenesis and is a suppressor of position-effect variegation. *Mech Dev* 2000;**96**:67–78.

46. Smothers JF, Henikoff S. The hinge and chromo shadow domain impart distinct targeting of HP1-like proteins. *Mol Cell Biol* 2001;**21**:2555–69.

47. Font-Burgada J, Rossell D, Auer H, Azorin F. *Drosophila* HP1c isoform interacts with the zinc-finger proteins WOC and Relative-of-WOC to regulate gene expression. *Genes Dev* 2008;**22**:3007–23.

48. Volpe AM, Horowitz H, Grafer CM, Jackson SM, Berg CA. *Drosophila* rhino encodes a female-specific chromo-domain protein that affects chromosome structure and egg polarity. *Genetics* 2001;**159**:1117–34.

49. Vermaak D, Henikoff S, Malik HS. Positive selection drives the evolution of rhino, a member of the heterochromatin protein 1 family in *Drosophila*. *PLoS Genet* 2005;**1**:96–108.

50. Giot L, Bader JS, Brouwer C, Chaudhuri A, Kuang B, Li Y, et al. A protein interaction map of *Drosophila melanogaster*. *Science* 2003;**302**:1727–36.

51. Greil F, van der Kraan I, Delrow J, Smothers JF, de Wit E, Bussemaker HJ, et al. Distinct HP1 and Su(var)3-9 complexes bind to sets of developmentally coexpressed genes depending on chromosomal location. *Genes Dev* 2003;**17**:2825–38.

52. Stephens GE, Xiao H, Lankenau DH, Wu C, Elgin SC. Heterochromatin protein 2 interacts with Nap-1 and NURF: a link between heterochromatin-induced gene silencing and the chromatin remodeling machinery in *Drosophila*. *Biochemistry* 2006;**45**:14990–9.

53. Moshkin YM, Armstrong JA, Maeda RK, Tamkun JW, Verrijzer P, Kennison JA, et al. Histone chaperone ASF1 cooperates with the Brahma chromatin-remodelling machinery. *Genes Dev* 2002;**16**:2621–6.

54. Tyler JK, Adams CR, Chen SR, Kobayashi R, Kamakaka RT, Kadonaga JT. The RCAF complex mediates chromatin assembly during DNA replication and repair. *Nature* 1999;**402**:555–60.

55. Siomi H, Siomi MC. Interactions between transposable elements and Argonautes have (probably) been shaping the *Drosophila* genome throughout evolution. *Curr Opin Genet Dev* 2008;**18**:181–7.

56. Pal-Bhadra M, Leibovitch BA, Gandhi SG, Rao M, Bhadra U, Birchler JA, et al. Heterochromatic silencing and HP1 localization in *Drosophila* are dependent on the RNAi machinery. *Science* 2004;**303**:669–72.

57. Brennecke J, Aravin AA, Stark A, Dus M, Kellis M, Sachidanandam R, et al. Discrete small RNA-generating loci as master regulators of transposon activity in *Drosophila*. *Cell* 2007;**128**:1089–103.

58. Yin H, Lin H. An epigenetic activation role of Piwi and a Piwi-associated piRNA in *Drosophila melanogaster*. *Nature* 2007;**450**:304–8.

59. Aravin AA, Lagos-Quintana M, Yalcin A, Zavolan M, Marks D, Snyder B, et al. The small RNA profile during *Drosophila melanogaster* development. *Dev Cell* 2003;**5**:337–50.

60. Gillespie DE, Berg CA. Homeless is required for RNA localization in *Drosophila* oogenesis and encodes a new member of the DE-H family of RNA-dependent ATPases. *Genes Dev* 1995;**9**:2495–508.

61. Cook HA, Koppetsch BS, Wu J, Theurkauf WE. The *Drosophila* SDE3 homolog armitage is required for oskar mRNA silencing and embryonic axis specification. *Cell* 2004;**116**:817–29.

62. Vagin VV, Sigova A, Li C, Seitz H, Gvozdev V, Zamore PD. A distinct small RNA pathway silences selfish genetic elements in the germline. *Science* 2006;**313**:320–4.

63. Klenov MS, Lavrov SA, Stolyarenko AD, Ryazansky SS, Aravin AA, Tuschl T, et al. Repeat-associated siRNAs cause chromatin silencing of retrotransposons in the *Drosophila melanogaster* germline. *Nucleic Acids Res* 2007;**35**:5430–8.

64. Deshpande G, Calhoun G, Schedl P. *Drosophila* argonaute-2 is required early in embryogenesis for the assembly of centric/centromeric heterochromatin, nuclear division, nuclear migration, and germ-cell formation. *Genes Dev* 2005;**19**:1680–5.

65. Huisinga KL, Elgin SC. Small RNA-directed heterochromatin formation in the context of development: what flies might learn from fission yeast. *Biochim Biophys Acta* 2009;**1789**:3–16.

66. Belyaeva ES, Zhimulev IF. Cytogenetic and molecular aspects of position effect variegation in *Drosophila*. III. Continuous and discontinuous compaction of chromosomal material as a result of position effect variegation. *Chromosoma* 1991;**100**:453–66.

67. Talbert PB, Henikoff S. A reexamination of spreading of position-effect variegation in the white-roughest region of *Drosophila melanogaster*. *Genetics* 2000;**154**:259–72.

68. Henikoff S, Dreesen TD. Trans-inactivation of the *Drosophila* brown gene: evidence for transcriptional repression and somatic pairing dependence. *Proc Natl Acad Sci USA* 1989;**86**:6704–8.

69. Capkova Frydrychova R, Biessmann H, Mason JM. Regulation of telomere length in *Drosophila*. *Cytogenet Genome Res* 2008;**122**:356–64.

70. Perrini B, Piacentini L, Fanti L, Altieri F, Chichiarelli S, Berloco M, et al. HP1 controls telomere capping, telomere elongation, and telomere silencing by two different mechanisms in *Drosophila*. *Mol Cell* 2004;**15**:467–76.

71. Mason JM, Ransom J, Konev AY. A deficiency screen for dominant suppressors of telomeric silencing in *Drosophila*. *Genetics* 2004;**168**:1353–70.

72. Doheny JG, Mottus R, Grigliatti TA. Telomeric position effect – a third silencing mechanism in eukaryotes. *PLoS ONE* 2008;**3**:e3864.

73. Belyaeva ES, Zhimulev IF, Volkova EI, Alekseyenko AA, Moshkin YM, Koryakov DE. Su(UR)ES: a gene suppressing DNA underreplication in intercalary and pericentric heterochromatin of *Drosophila melanogaster* polytene chromosomes. *Proc Natl Acad Sci USA* 1998;**95**:7532–7.

74. Zhimulev IF, Belyaeva ES. Intercalary heterochromatin and genetic silencing. *Bioessays* 2003;**25**:1040–51.

75. Schubeler D, Scalzo D, Kooperberg C, van Steensel B, Delrow J, Groudine M. Genome-wide DNA replication profile for *Drosophila melanogaster*: a link between transcription and replication timing. *Nat Genet* 2002;**32**:438–42.

76. Belyakin SN, Christophides GK, Alekseyenko AA, Kriventseva EV, Belyaeva ES, Nanayev RA, et al. Genomic analysis of *Drosophila* chromosome underreplication reveals a link between replication control and transcriptional territories. *Proc Natl Acad Sci USA* 2005;**102**:8269–74.

77. Slawson EE, Shaffer CD, Malone CD, Leung W, Kellmann E, Shevchek RB, et al. Comparison of dot chromosome sequences from *D. melanogaster* and *D. virilis* reveals an enrichment of DNA transposon sequences in heterochromatic domains. *Genome Biol* 2006;**7**:R15.

78. Zhimulev IF, Belyaeva ES, Makunin IV, Pirrotta V, Volkova EI, Alekseyenko AA, et al. Influence of the SuUR gene on intercalary heterochromatin in *Drosophila melanogaster* polytene chromosomes. *Chromosoma* 2003;**111**:377–98.

79. Sandler L, Szauter P. The effect of recombination-defective meiotic mutants on fourth-chromosome crossing over in *Drosophila melanogaster*. *Genetics* 1978;**90**:699–712.

80. Seum C, Reo E, Peng H, Rauscher FJ 3rd, Spierer P, Bontron S. *Drosophila* SETDB1 is required for chromosome 4 silencing. *PLoS Genet* 2007;**3**:e76.

81. Tzeng TY, Lee CH, Chan LW, Shen CK. Epigenetic regulation of the *Drosophila* chromosome 4 by the histone H3K9 methyltransferase dSETDB1. *Proc Natl Acad Sci USA* 2007;**104**:12691–6.

82. Larsson J, Chen JD, Rasheva V, Rasmuson-Lestander A, Pirrotta V. Painting of fourth, a chromosome-specific protein in *Drosophila*. *Proc Natl Acad Sci USA* 2001;**98**:6273–8.

83. Larsson J, Svensson MJ, Stenberg P, Makitalo M. Painting of fourth in genus *Drosophila* suggests autosome-specific gene regulation. *Proc Natl Acad Sci USA* 2004;**101**:9728–33.

84. Johansson AM, Stenberg P, Pettersson F, Larsson J. POF and HP1 bind expressed exons, suggesting a balancing mechanism for gene regulation. *PLoS Genet* 2007;**3**:e209.

85. Johansson AM, Stenberg P, Bernhardsson C, Larsson J. Painting of fourth and chromosome-wide regulation of the 4th chromosome in *Drosophila melanogaster*. *EMBO J* 2007;**26**:2307–16.

86. Hearn MG, Hedrick A, Grigliatti TA, Wakimoto BT. The effect of modifiers of position-effect variegation on the variegation of heterochromatic genes of *Drosophila melanogaster*. *Genetics* 1991;**128**:785–97.

87. Eberl DF, Duyf BJ, Hilliker AJ. The role of heterochromatin in the expression of a heterochromatic gene, the rolled locus of *Drosophila melanogaster*. *Genetics* 1993;**134**:277–92.

88. Weiler KS, Wakimoto BT. Suppression of heterochromatic gene variegation can be used to distinguish and characterize E(var) genes potentially important for chromosome structure in *Drosophila melanogaster*. *Mol Genet Genomics* 2002;**266**:922–32.

89. Smith CD, Shu S, Mungall CJ, Karpen GH. The Release 5.1 annotation of *Drosophila melanogaster* heterochromatin. *Science* 2007;**316**:1586–91.

90. Yasuhara JC, Wakimoto BT. Molecular landscape of modified histones in *Drosophila* heterochromatic genes and euchromatin-heterochromatin transition zones. *PLoS Genet* 2008;**4**:e16.

91. Piacentini L, Fanti L, Berloco M, Perrini B, Pimpinelli S. Heterochromatin protein 1 (HP1) is associated with induced gene expression in *Drosophila* euchromatin. *J Cell Biol* 2003;**161**:707–14.

92. De Lucia F, Ni JQ, Vaillant C, Sun FL. HP1 modulates the transcription of cell-cycle regulators in *Drosophila melanogaster*. *Nucleic Acids Res* 2005;**33**:2852–8.

93. Cryderman DE, Grade SK, Li Y, Fanti L, Pimpinelli S, Wallrath LL. Role of *Drosophila* HP1 in euchromatic gene expression. *Dev Dyn* 2005;**232**:767–74.

94. de Wit E, Greil F, van Steensel B. Genome-wide HP1 binding in *Drosophila*: developmental plasticity and genomic targeting signals. *Genome Res* 2005;**15**:1265–73.

95. Moorman C, Sun LV, Wang J, de Wit E, Talhout W, Ward LD, et al. Hotspots of transcription factor colocalization in the genome of *Drosophila melanogaster*. *Proc Natl Acad Sci USA* 2006;**103**:12027–32.

96. Osborne CS, Chakalova L, Brown KE, Carter D, Horton A, Debrand E, et al. Active genes dynamically colocalize to shared sites of ongoing transcription. *Nat Genet* 2004;**36**:1065–71.

97. Lewis EB. A gene complex controlling segmentation in *Drosophila*. *Nature* 1978;**276**:565–70.

98. Laible G, Wolf A, Dorn R, Reuter G, Nislow C, Lebersorger A, et al. Mammalian homologues of the Polycomb-group gene Enhancer of zeste mediate gene silencing in *Drosophila* heterochromatin and at *S. cerevisiae* telomeres. *EMBO J* 1997;**16**:3219–32.

99. Czermin B, Melfi R, McCabe D, Seitz V, Imhof A, Pirrotta V. *Drosophila* enhancer of Zeste/ESC complexes have a histone H3 methyltransferase activity that marks chromosomal Polycomb sites. *Cell* 2002;**111**:185–96.

100. Saurin AJ, Shao Z, Erdjument-Bromage H, Tempst P, Kingston RE. A *Drosophila* Polycomb group complex includes Zeste and dTAFII proteins. *Nature* 2001;**412**:655–60.

101. Klymenko T, Papp B, Fischle W, Kocher T, Schelder M, Fritsch C, et al. A Polycomb group protein complex with sequence-specific DNA-binding and selective methyl-lysine-binding activities. *Genes Dev* 2006;**20**:1110–22.

102. Ng J, Hart CM, Morgan K, Simon JA. A *Drosophila* ESC-E(Z) protein complex is distinct from other polycomb group complexes and contains covalently modified ESC. *Mol Cell Biol* 2000;**20**:3069–78.

103. Tie F, Prasad-Sinha J, Birve A, Rasmuson-Lestander A, Harte PJ. A 1-megadalton ESC/E(Z) complex from *Drosophila* that contains polycomblike and RPD3. *Mol Cell Biol* 2003;**23**:3352–62.

104. Horard B, Tatout C, Poux S, Pirrotta V. Structure of a polycomb response element and in vitro binding of polycomb group complexes containing GAGA factor. *Mol Cell Biol* 2000;**20**:3187–97.

105. Dejardin J, Rappailles A, Cuvier O, Grimaud C, Decoville M, Locker D, et al. Recruitment of *Drosophila* Polycomb group proteins to chromatin by DSP1. *Nature* 2005;**434**:533–8.

106. Brown JL, Grau DJ, DeVido SK, Kassis JA. An Sp1/KLF binding site is important for the activity of a Polycomb group response element from the *Drosophila* engrailed gene. *Nucleic Acids Res* 2005;**33**:5181–9.

107. Ringrose L, Rehmsmeier M, Dura JM, Paro R. Genome-wide prediction of Polycomb/Trithorax response elements in *Drosophila melanogaster*. *Dev Cell* 2003;**5**:759–71.

108. Tolhuis B, de Wit E, Muijrers I, Teunissen H, Talhout W, van Steensel B, et al. Genome-wide profiling of PRC1 and PRC2 Polycomb chromatin binding in *Drosophila melanogaster*. *Nat Genet* 2006;**38**:694–9.

109. Negre N, Hennetin J, Sun LV, Lavrov S, Bellis M, White KP, et al. Chromosomal distribution of PcG proteins during *Drosophila* development. *PLoS Biol* 2006;**4**:e170.

110. Kassis JA, VanSickle EP, Sensabaugh SM. A fragment of engrailed regulatory DNA can mediate transvection of the white gene in *Drosophila*. *Genetics* 1991;**128**:751–61.

111. Gindhart JG Jr., Kaufman TC. Identification of Polycomb and trithorax group responsive elements in the regulatory region of the *Drosophila* homeotic gene Sex combs reduced. *Genetics* 1995;**139**:797–814.

112. Wang L, Brown JL, Cao R, Zhang Y, Kassis JA, Jones RS. Hierarchical recruitment of polycomb group silencing complexes. *Mol Cell* 2004;**14**:637–46.

113. Schwartz YB, Pirrotta V. Polycomb silencing mechanisms and the management of genomic programmes. *Nat Rev Genet* 2007;**8**:9–22.

114. Ringrose L, Ehret H, Paro R. Distinct contributions of histone H3 lysine 9 and 27 methylation to locus-specific stability of polycomb complexes. *Mol Cell* 2004;**16**:641–53.

115. Schwartz YB, Kahn TG, Nix DA, Li XY, Bourgon R, Biggin M, et al. Genome-wide analysis of Polycomb targets in *Drosophila melanogaster*. *Nat Genet* 2006;**38**:700–5.

116. Wang H, Wang L, Erdjument-Bromage H, Vidal M, Tempst P, Jones RS, et al. Role of histone H2A ubiquitination in Polycomb silencing. *Nature* 2004;**431**:873–8.

117. Lagarou A, Mohd-Sarip A, Moshkin YM, Chalkley GE, Bezstarosti K, Demmers JA, et al. dKDM2 couples histone H2A ubiquitylation to histone H3 demethylation during Polycomb group silencing. *Genes Dev* 2008;**22**:2799–810.

118. Shanower GA, Muller M, Blanton JL, Honti V, Gyurkovics H, Schedl P. Characterization of the grappa gene, the *Drosophila* histone H3 lysine 79 methyltransferase. *Genetics* 2005;**169**:173–84.

119. Poux S, Horard B, Sigrist CJ, Pirrotta V. The *Drosophila* trithorax protein is a coactivator required to prevent re-establishment of polycomb silencing. *Development* 2002;**129**:2483–93.

120. Klymenko T, Muller J. The histone methyltransferases Trithorax and Ash1 prevent transcriptional silencing by Polycomb group proteins. *EMBO Rep* 2004;**5**:373–7.

121. Schmitt S. Prestel M, Paro R. Intergenic transcription through a polycomb group response element counteracts silencing. *Genes Dev* 2005;**19**:697–708.

122. Gildea JJ, Lopez R, Shearn A. A screen for new trithorax group genes identified little imaginal discs, the *Drosophila melanogaster* homologue of human retinoblastoma binding protein 2. *Genetics* 2000;**156**:645–63.

123. Byrd KN, Shearn A. ASH1, a *Drosophila* trithorax group protein, is required for methylation of lysine 4 residues on histone H3. *Proc Natl Acad Sci USA* 2003;**100**:11535–40.

124. Tanaka Y, Katagiri Z, Kawahashi K, Kioussis D, Kitajima S. Trithorax-group protein ASH1 methylates histone H3 lysine 36. *Gene* 2007;**397**:161–8.

125. Lloret-Llinares M, Carre C, Vaquero A, de Olano N, Azorin F. Characterization of *Drosophila melanogaster* JmjC+N histone demethylases. *Nucleic Acids Res* 2008;**36**:2852–63.

126. Lee N, Erdjument-Bromage H, Tempst P, Jones RS, Zhang Y. The H3K4 demethylase lid associates with and inhibits histone deacetylase Rpd3. *Mol Cell Biol* 2009;**29**:1401–10.

127. Petruk S, Sedkov Y, Smith S, Tillib S, Kraevski V, Nakamura T, et al. Trithorax and dCBP acting in a complex to maintain expression of a homeotic gene. *Science* 2001;**294**:1331–4.

128. Fujioka M, Yusibova GL, Zhou J, Jaynes JB. The DNA-binding Polycomb-group protein Pleiohomeotic maintains both active and repressed transcriptional states through a single site. *Development* 2008;**135**:4131–9.

129. Mollaaghababa R, Sipos L, Tiong SY, Papoulas O, Armstrong JA, Tamkun JW, et al. Mutations in *Drosophila* heat shock cognate 4 are enhancers of Polycomb. *Proc Natl Acad Sci USA* 2001;**98**:3958–63.

130. Wang YJ, Brock HW. Polyhomeotic stably associates with molecular chaperones Hsc4 and Droj2 in *Drosophila* Kc1 cells. *Dev Biol* 2003;**262**:350–60.

131. Tariq M, Nussbaumer U, Chen Y, Beisel C, Paro R. Trithorax requires Hsp90 for maintenance of active chromatin at sites of gene expression. *Proc Natl Acad Sci USA* 2009;**106**:1157–62.

132. Misulovin Z, Schwartz YB, Li XY, Kahn TG, Gause M, MacArthur S, et al. Association of cohesin and Nipped-B with transcriptionally active regions of the *Drosophila melanogaster* genome. *Chromosoma* 2008;**117**:89–102.

133. Hallson G, Syrzycka M, Beck SA, Kennison JA, Dorsett D, Page SL, et al. The *Drosophila* cohesin subunit Rad21 is a trithorax group (trxG) protein. *Proc Natl Acad Sci USA* 2008;**105**:12405–10.

134. Papp B, Muller J. Histone trimethylation and the maintenance of transcriptional ON and OFF states by trxG and PcG proteins. *Genes Dev* 2006;**20**:2041–54.

135. Beisel C, Buness A, Roustan-Espinosa IM, Koch B, Schmitt S, Haas SA, et al. Comparing active and repressed expression states of genes controlled by the Polycomb/Trithorax group proteins. *Proc Natl Acad Sci USA* 2007;**104**:16615–20.

136. Petruk S, Smith ST, Sedkov Y, Mazo A. Association of trxG and PcG proteins with the bxd maintenance element depends on transcriptional activity. *Development* 2008;**135**:2383–90.

137. Tie F, Banerjee R, Stratton CA, Prasad-Sinha J, Stepanik V, Zlobin A, et al. CBP-mediated acetylation of histone H3 lysine 27 antagonizes *Drosophila* Polycomb silencing. *Development* 2009;**136**:3131–41.

138. Gelbart ME, Kuroda MI. *Drosophila* dosage compensation: a complex voyage to the X chromosome. *Development* 2009;**136**:1399–410.

139. Hallacli E, Akhtar A. X chromosomal regulation in flies: when less is more. *Chromosome Res* 2009;**17**:603–19.

140. Park SW, Kang Y, Sypula JG, Choi J, Oh H, Park Y. An evolutionarily conserved domain of roX2 RNA is sufficient for induction of H4-Lys16 acetylation on the *Drosophila* X chromosome. *Genetics* 2007;**177**:1429–37.

141. Park SW, Kuroda MI, Park Y. Regulation of histone H4 Lys16 acetylation by predicted alternative secondary structures in roX noncoding RNAs. *Mol Cell Biol* 2008;**28**:4952–62.

142. Bashaw GJ, Baker BS. The regulation of the *Drosophila* msl-2 gene reveals a function for Sex-lethal in translational control. *Cell* 1997;**89**:789–98.

143. Kelley RL, Wang J, Bell L, Kuroda MI. Sex lethal controls dosage compensation in *Drosophila* by a non-splicing mechanism. *Nature* 1997;**387**:195–9.

144. Fagegaltier D, Baker BS. X chromosome sites autonomously recruit the dosage compensation complex in *Drosophila* males. *PLoS Biol* 2004;**2**:e341.

145. Oh H, Bone JR, Kuroda MI. Multiple classes of MSL binding sites target dosage compensation to the X chromosome of *Drosophila*. *Curr Biol* 2004;**14**:481–7.

146. Alekseyenko AA, Peng S, Larschan E, Gorchakov AA, Lee OK, Kharchenko P, et al. A sequence motif within chromatin entry sites directs MSL establishment on the *Drosophila* X chromosome. *Cell* 2008;**134**:599–609.

147. Straub T, Grimaud C, Gilfillan GD, Mitterweger A, Becker PB. The chromosomal high-affinity binding sites for the *Drosophila* dosage compensation complex. *PLoS Genet* 2008;**4**:e1000302.

148. Sass GL, Pannuti A, Lucchesi JC. Male-specific lethal complex of *Drosophila* targets activated regions of the X chromosome for chromatin remodeling. *Proc Natl Acad Sci USA* 2003;**100**:8287–91.

149. Kind J, Akhtar A. Cotranscriptional recruitment of the dosage compensation complex to X-linked target genes. *Genes Dev* 2007;**21**:2030–40.

150. Bone JR, Lavender J, Richman R, Palmer MJ, Turner BM, Kuroda MI. Acetylated histone H4 on the male X chromosome is associated with dosage compensation in *Drosophila*. *Genes Dev* 1994;**8**:96–104.

151. Hilfiker A, Hilfiker-Kleiner D, Pannuti A, Lucchesi JC. mof, a putative acetyl transferase gene related to the Tip60 and MOZ human genes and to the SAS genes of yeast, is required for dosage compensation in *Drosophila*. *EMBO J* 1997;**16**:2054–60.

152. Smith ER, Allis CD, Lucchesi JC. Linking global histone acetylation to the transcription enhancement of X-chromosomal genes in *Drosophila* males. *J Biol Chem* 2001;**276**:31483–6.

153. Alekseyenko AA, Larschan E, Lai WR, Park PJ, Kuroda MI. High-resolution ChIP-chip analysis reveals that the *Drosophila* MSL complex selectively identifies active genes on the male X chromosome. *Genes Dev* 2006;**20**:848–57.

154. Gilfillan GD, Straub T, de Wit E, Greil F, Lamm R, van Steensel B, et al. Chromosome-wide gene-specific targeting of the *Drosophila* dosage compensation complex. *Genes Dev* 2006;**20**:858–70.

155. Luger K, Mader AW, Richmond RK, Sargent DF, Richmond TJ. Crystal structure of the nucleosome core particle at 2.8 Å resolution. *Nature* 1997;**389**:251–60.

156. Davey CA, Sargent DF, Luger K, Maeder AW, Richmond TJ. Solvent mediated interactions in the structure of the nucleosome core particle at 1.9 Å resolution. *J Mol Biol* 2002;**319**:1097–113.

157. Shogren-Knaak M, Ishii H, Sun JM, Pazin MJ, Davie JR, Peterson CL. Histone H4-K16 acetylation controls chromatin structure and protein interactions. *Science* 2006;**311**:844–7.

158. Lee CG, Chang KA, Kuroda MI, Hurwitz J. The NTPase/helicase activities of Drosophila maleless, an essential factor in dosage compensation. *EMBO J* 1997;**16**:2671–81.

159. Hogan C, Varga-Weisz P. The regulation of ATP-dependent nucleosome remodelling factors. *Mutat Res* 2007;**618**:41–51.

160. Morra R, Smith ER, Yokoyama R, Lucchesi JC. The MLE subunit of the *Drosophila* MSL complex uses its ATPase activity for dosage compensation and its helicase activity for targeting. *Mol Cell Biol* 2008;**28**:958–66.

161. Delattre M, Spierer A, Jaquet Y, Spierer P. Increased expression of *Drosophila* Su(var)3-7 triggers Su(var)3-9-dependent heterochromatin formation. *J Cell Sci* 2004;**117**:6239–47.

162. Spierer A, Begeot F, Spierer P, Delattre M. SU(VAR)3-7 links heterochromatin and dosage compensation in *Drosophila*. *PLoS Genet* 2008;**4**:e1000066.

163. Spierer A, Seum C, Delattre M, Spierer P. Loss of the modifiers of variegation Su(var)3-7 or HP1 impacts male X polytene chromosome morphology and dosage compensation. *J Cell Sci* 2005;**118**:5047–57.

164. Shaffer CD, Cenci G, Thompson B, Stephens GE, Slawson EE, Adu-Wusu K, et al. The large isoform of *Drosophila melanogaster* heterochromatin protein 2 plays a critical role in gene silencing and chromosome structure. *Genetics* 2006;**174**:1189–204.

165. Liu LP, Ni JQ, Shi YD, Oakeley EJ, Sun FL. Sex-specific role of *Drosophila melanogaster* HP1 in regulating chromatin structure and gene transcription. *Nat Genet* 2005;**37**:1361–6.

166. Mito Y, Henikoff JG, Henikoff S. Genome-scale profiling of histone H3.3 replacement patterns. *Nat Genet* 2005;**37**:1090–7.

167. Jin Y, Wang Y, Johansen J, Johansen KM. JIL-1, a chromosomal kinase implicated in regulation of chromatin structure, associates with the male specific lethal (MSL) dosage compensation complex. *J Cell Biol* 2000;**149**:1005–10.

168. Deng H, Zhang W, Bao X, Martin JN, Girton J, Johansen J, et al. The JIL-1 kinase regulates the structure of *Drosophila* polytene chromosomes. *Chromosoma* 2005;**114**:173–82.

169. Zhang W, Deng H, Bao X, Lerach S, Girton J, Johansen J, et al. The JIL-1 histone H3S10 kinase regulates dimethyl H3K9 modifications and heterochromatic spreading in *Drosophila*. *Development* 2006;**133**:229–35.

170. Deng H, Bao X, Cai W, Blacketer MJ, Belmont AS, Girton J, et al. Ectopic histone H3S10 phosphorylation causes chromatin structure remodeling in *Drosophila*. *Development* 2008;**135**:699–705.

171. Deuring R, Fanti L, Armstrong JA, Sarte M, Papoulas O, Prestel M, et al. The ISWI chromatin-remodeling protein is required for gene expression and the maintenance of higher order chromatin structure in vivo. *Mol Cell* 2000;**5**:355–65.

172. Bai X, Larschan E, Kwon SY, Badenhorst P, Kuroda MI. Regional control of chromatin organization by noncoding roX RNAs and the NURF remodeling complex in *Drosophila melanogaster*. *Genetics* 2007;**176**:1491–9.

173. Grau B, Popescu C, Torroja L, Ortuno-Sahagun D, Boros I, Ferrus A. Transcriptional adaptor ADA3 of *Drosophila melanogaster* is required for histone modification, position effect variegation, and transcription. *Mol Cell Biol* 2008;**28**:376–85.

174. Carre C, Ciurciu A, Komonyi O, Jacquier C, Fagegaltier D, Pidoux J, et al. The *Drosophila* NURF remodelling and the ATAC histone acetylase complexes functionally interact and are required for global chromosome organization. *EMBO Rep* 2008;**9**:187–92.

175. Hanai K, Furuhashi H, Yamamoto T, Akasaka K, Hirose S. RSF governs silent chromatin formation via histone H2Av replacement. *PLoS Genet* 2008;**4**:e1000011.

176. Fretzin S, Allan BD, van Daal A, Elgin SC. A *Drosophila melanogaster* H3.3 cDNA encodes a histone variant identical with the vertebrate H3.3. *Gene* 1991;**107**:341–2.

177. Ahmad K, Henikoff S. The histone variant H3.3 marks active chromatin by replication-independent nucleosome assembly. *Mol Cell* 2002;**9**:1191–200.

178. Schwartz BE, Ahmad K. Transcriptional activation triggers deposition and removal of the histone variant H3.3. *Genes Dev* 2005;**19**:804–14.

179. Wirbelauer C, Bell O, Schubeler D. Variant histone H3.3 is deposited at sites of nucleosomal displacement throughout transcribed genes while active histone modifications show a promoter-proximal bias. *Genes Dev* 2005;**19**:1761–6.

180. Mito Y, Henikoff JG, Henikoff S. Histone replacement marks the boundaries of cis-regulatory domains. *Science* 2007;**315**:1408–11.

181. McKittrick E, Gafken PR, Ahmad K, Henikoff S. Histone H3.3 is enriched in covalent modifications associated with active chromatin. *Proc Natl Acad Sci USA* 2004;**101**:1525–30.

182. Hodl M, Basler K. Transcription in the absence of histone H3.3. *Curr Biol* 2009;**19**:1221–6.

183. van Daal A, Elgin SC. A histone variant, H2AvD, is essential in *Drosophila melanogaster*. *Mol Biol Cell* 1992;**3**:593–602.

184. Leach TJ, Mazzeo M, Chotkowski HL, Madigan JP, Wotring MG, Glaser RL. Histone H2A.Z is widely but nonrandomly distributed in chromosomes of *Drosophila melanogaster*. *J Biol Chem* 2000;**275**:23267–72.

185. Swaminathan J, Baxter EM, Corces VG. The role of histone H2Av variant replacement and histone H4 acetylation in the establishment of *Drosophila* heterochromatin. *Genes Dev* 2005;**19**:65–76.

186. Kusch T, Florens L, Macdonald WH, Swanson SK, Glaser RL, Yates JR 3rd, et al. Acetylation by Tip60 is required for selective histone variant exchange at DNA lesions. *Science* 2004;**306**:2084–7.

187. Henikoff S, Ahmad K, Platero JS, van Steensel B. Heterochromatic deposition of centromeric histone H3-like proteins. *Proc Natl Acad Sci USA* 2000;**97**:716–21.

188. Ahmad K, Henikoff S. Centromeres are specialized replication domains in heterochromatin. *J Cell Biol* 2001;**153**:101–10.

189. Sullivan B, Karpen G. Centromere identity in *Drosophila* is not determined in vivo by replication timing. *J Cell Biol* 2001;**154**:683–90.

190. Schuh M, Lehner CF, Heidmann S. Incorporation of *Drosophila* CID/CENP-A and CENP-C into centromeres during early embryonic anaphase. *Curr Biol* 2007;**17**:237–43.

191. Erhardt S, Mellone BG, Betts CM, Zhang W, Karpen GH, Straight AF. Genome-wide analysis reveals a cell cycle-dependent mechanism controlling centromere propagation. *J Cell Biol* 2008;**183**:805–18.

192. Sullivan BA, Karpen GH. Centromeric chromatin exhibits a histone modification pattern that is distinct from both euchromatin and heterochromatin. *Nat Struct Mol Biol* 2004;**11**:1076–83.

193. Williams BC, Murphy TD, Goldberg ML, Karpen GH. Neocentromere activity of structurally acentric mini-chromosomes in *Drosophila*. *Nat Genet* 1998;**18**:30–7.

194. Agudo M, Abad JP, Molina I, Losada A, Ripoll P, Villasante A. A dicentric chromosome of *Drosophila melanogaster* showing alternate centromere inactivation. *Chromosoma* 2000;**109**:190–6.

195. Udvardy A, Maine E, Schedl P. The 87A7 chromomere. Identification of novel chromatin structures flanking the heat shock locus that may define the boundaries of higher order domains. *J Mol Biol* 1985;**185**:341–58.

196. Zhao K, Hart CM, Laemmli UK. Visualization of chromosomal domains with boundary element-associated factor BEAF-32. *Cell* 1995;**81**:879–89.

197. Holohan EE, Kwong C, Adryan B, Bartkuhn M, Herold M, Renkawitz R, et al. CTCF genomic binding sites in *Drosophila* and the organisation of the bithorax complex. *PLoS Genet* 2007;**3**:e112.

198. Bartkuhn M, Straub T, Herold M, Herrmann M, Rathke C, Saumweber H, et al. Active promoters and insulators are marked by the centrosomal protein 190. *EMBO J* 2009;**28**:877–88.

199. Bushey AM, Ramos E, Corces VG. Three subclasses of a *Drosophila* insulator show distinct and cell type-specific genomic distributions. *Genes Dev* 2009;**23**:1338–50.

200. Schweinsberg S, Hagstrom K, Gohl D, Schedl P, Kumar RP, Mishra R, et al. The enhancer-blocking activity of the Fab-7 boundary from the *Drosophila* bithorax complex requires GAGA-factor-binding sites. *Genetics* 2004;**168**:1371–84.

201. Cai H, Levine M. Modulation of enhancer-promoter interactions by insulators in the *Drosophila* embryo. *Nature* 1995;**376**:533–6.

231

202. Mihaly J, Hogga I, Barges S, Galloni M, Mishra RK, Hagstrom K, et al. Chromatin domain boundaries in the Bithorax complex. *Cell Mol Life Sci* 1998;**54**:60–70.

203. Emberly E, Blattes R, Schuettengruber B, Hennion M, Jiang N, Hart CM, et al. BEAF regulates cell-cycle genes through the controlled deposition of H3K9 methylation marks into its conserved dual-core binding sites. *PLoS Biol* 2008;**6**:2896–910.

204. Gilbert MK, Tan YY, Hart CM. The *Drosophila* boundary element-associated factors BEAF-32A and BEAF-32B affect chromatin structure. *Genetics* 2006;**173**:1365–75.

205. Ghosh D, Gerasimova TI, Corces VG. Interactions between the Su(Hw) and Mod(mdg4) proteins required for gypsy insulator function. *EMBO J* 2001;**20**:2518–27.

206. Blanton J, Gaszner M, Schedl P. Protein:protein interactions and the pairing of boundary elements in vivo. *Genes Dev* 2003;**17**:664–75.

207. Gerasimova TI, Byrd K, Corces VG. A chromatin insulator determines the nuclear localization of DNA. *Mol Cell* 2000;**6**:1025–35.

208. Byrd K, Corces VG. Visualization of chromatin domains created by the gypsy insulator of *Drosophila*. *J Cell Biol* 2003;**162**:565–74.

209. Capelson M, Corces VG. The ubiquitin ligase dTopors directs the nuclear organization of a chromatin insulator. *Mol Cell* 2005;**20**:105–16.

210. Lei EP, Corces VG. RNA interference machinery influences the nuclear organization of a chromatin insulator. *Nat Genet* 2006;**38**:936–41.

211. Gerasimova TI, Lei EP, Bushey AM, Corces VG. Coordinated control of dCTCF and gypsy chromatin insulators in *Drosophila*. *Mol Cell* 2007;**28**:761–72.

212. Lyko F, Ramsahoye BH, Jaenisch R. DNA methylation in *Drosophila melanogaster*. *Nature* 2000;**408**:538–40.

213. Goll MG, Kirpekar F, Maggert KA, Yoder JA, Hsieh CL, Zhang X, et al. Methylation of tRNAAsp by the DNA methyltransferase homolog Dnmt2. *Science* 2006;**311**:395–8.

214. Ruden DM, Lu X. Hsp90 affecting chromatin remodeling might explain transgenerational epigenetic inheritance in *Drosophila*. *Curr Genomics* 2008;**9**:500–8.

215. Rice WR. Sexually antagonistic genes: experimental evidence. *Science* 1992;**256**:1436–9.

216. Rice WR, Chippindale AK. Sexual recombination and the power of natural selection. *Science* 2001;**294**:555–9.

217. McGraw LA, Gibson G, Clark AG, Wolfner MF. Strain-dependent differences in several reproductive traits are not accompanied by early postmating transcriptome changes in female *Drosophila melanogaster*. *Genetics* 2009;**181**:1273–80.

218. Ahmad K, Henikoff S. Histone H3 variants specify modes of chromatin assembly. *Proc Natl Acad Sci USA* 2002;**99**(Suppl 4):16477–84.

219. Gaszner M, Felsenfeld G. Insulators: exploiting transcriptional and epigenetic mechanisms. *Nat Rev Genet* 2006;**7**:703–13.

CHAPTER 15

Mouse Models of Epigenetic Inheritance

Autumn J. Bernal[1], Susan K. Murphy[2], and Randy L. Jirtle[3]
[1]Department of Radiation Oncology, Duke University Medical Center, University Program in Genetics and Genomics, Integrated Toxicology and Environmental Health Program, Duke University, Durham, North Carolina 27710
[2]Department of Obstetrics and Gynecology, Duke University Medical Center, Division of Gynecologic Oncology, Durham, North Carolina 27708
[3]Department of Radiation Oncology, Duke University Medical Center, Durham, North Carolina 27710, USA

EARLY EXPOSURES AND ADULT DISEASE: THE MAMMALIAN MODEL ADVANTAGE

During WWII, heavy fighting, blocked railways, ruined bridges, and a German-issued food embargo left the Dutch population consuming, per capita, less than 1000 calories per day. Researchers later found that the famine not only affected the exposed cohort, it resulted in lasting consequences on the population's health for several generations. Individuals exposed to famine during gestation had increases in coronary heart disease, breast cancer, obesity, stress responsiveness, disturbed blood coagulation, obstructive airway disease, and microalbuminuria [1]. The outcomes resulting from the Dutch Hunger Winter exemplify the "Developmental Origins of Disease" hypothesis, which was first proposed by David J. P. Barker in the 1990s. This hypothesis simply states that early developmental exposures can lead to diseases later in life [2].

Although mechanistically still largely unexplained, increasing evidence implicates epigenetic changes as a primary impetus in disease development [3,4]. In the Dutch population, DNA methylation levels were decreased at the imprinted *Insulin-Like Growth Factor 2 (IGF2)* locus up to six decades following exposure to the famine [5]. As an adaptive response to stimuli, such as nutritional deprivation during development, epigenetic regulation alters gene expression to promote compensatory adjustments in metabolism. Unfortunately, these early adaptive epigenetic changes have consequences later in life when the metabolic changes no longer coincide with the external environment, resulting in pathologies such as coronary heart disease and obesity.

The Dutch Famine studies support the theory that early environmental exposures affect the inheritance of epigenetic marks, disrupt epigenetic profiles, and lead to disease progression later in life [1,3,4,6–14]. The inability to experimentally manipulate such exposures in humans leads to the necessity for mammalian models to further explore the molecular mechanisms behind this process. Many currently employed mammalian models provide

Handbook of Epigenetics: The New Molecular and Medical Genetics. DOI: 10.1016/B978-0-12-375709-8.00015-0

TABLE 15.1 Model Features Useful for the Study of Epigenetic Inheritance

Feature	Mice	Plants	Lower Organisms
DNA methylation	+	+	−[*]
Gene-specific imprinting	+	+	−
Metastable epialleles	+	−	−
Developing embryo analysis	+	−[**]	−
Transgenerational exposure studies	+	−	−

[*]Certain types of yeast possess DNA methylation. Other lower organisms, such as *C. elegans* do not. *Drosophila* exhibits very low levels of DNA methylation.
[**]Analysis of the developing embryo is possible in plants, but difficult due to the fact that it is surrounded by endosperm and embedded in the tissue of the parent plant.

researchers with the ability to mimic human *in utero* exposures by modifying maternal nutritional, chemical, or psychological status before, during, and after pregnancy [15]. Mammalian models are also useful for mechanistic analysis of unique forms of epigenetic regulation and inheritance such as metastable epialleles and gene-specific imprinting. Although plants also contain these forms of regulation, the ability to study them in the developing embryo is limited [16] (Table 15.1).

Metastable epialleles are alleles found to be variably expressed in response to intrinsic or extrinsic influences on CpG methylation and are valuable biosensors for the effects of such exposures. Three metastable epialleles are described in detail below. The mouse models that carry these alleles have been utilized to determine the establishment and disruption of methylation marks during development. Following a detailed description of metastable epialleles, genomic imprinting is briefly described. Genomic imprinting is a form of epigenetic gene regulation that is disrupted in human disease. The regulation of imprinted genes and metastable epialleles is contingent upon the correct inheritance, establishment, and maintenance of DNA methylation marks during early development and throughout life. The established mouse models described below are tools that could be used to analyze the regulation of imprinted genes. Other epigenetic marks, such as histone modifications and regulatory small RNAs undoubtedly contribute to the establishment and maintenance of methylation marks. The potential for these models to address inheritance of these additional epigenetic marks will also be briefly discussed.

METASTABLE EPIALLELES

Animal models have long been used to elucidate the effects of developmental exposures on adult disease, but certain models have proven to be invaluable for examining the role that epigenetic modifications play in development [15]. Mice that contain metastable epialleles are especially useful models for studying the inheritance of epigenetic marks and the developmental origins of adult disease. Metastable epialleles are identical alleles that are variably expressed in genetically identical individuals due to epigenetic modifications that are established during early development [17]. The creation of the epigenotype at each allele is stochastic and is associated with widely varying phenotypes. Stochastic epigenotype establishment can be brought about through several mechanisms. First, metastable epiallele epigenetic marks are not always completely reset during genome-wide epigenetic reprogramming that occurs during gametogenesis [18,19]. Incomplete erasure and reestablishment leads to mitotic inheritance of the residual epigenetic marks in the somatic tissues and to their passage to the next generation in the gametes through transgenerational epigenetic inheritance. Second, metastable epialleles may be subject to environmentally induced lability both during development and later in life [11,20–22].

Metastable epialleles are often associated with the presence of retrotransposable elements. Three murine metastable epialleles (A^{vy}, $Axin^{Fu}$, $Cabp^{IAP}$) are associated with intracisternal

A-particle (IAP) insertions. IAP retrotransposons are prevalent in the mouse genome and consist of elements up to 7 kb in full length [23,24]. In the mouse, approximately 1000 copies of IAP retrotransposons are present in each cell [23]. These elements, along with several others, comprise Class II endogenous retroviruses, which make up 3% of the mouse genome [25].

The long terminal repeats (LTRs) flanking IAPs carry promoters that initiate transcription of the IAP and in some cases adjacent host sequences. Methylation status at LTRs influences expression of the IAP and, in some cases, the nearby genes [26–28]. In fact, IAP hypomethylation at the 5' LTR has been attributed to constitutive expression of IAP transcripts in many mouse tumors [23,28]. Inactivation of DNA methyltransferase 1 (Dnmt1) results in demethylation of IAP LTRs in the preimplantation embryo [29]. Likewise, de novo methyltransferase Dnmt3b knockouts causes slight demethylation and embryonic death, while Dnmt3a knockouts have normal methylation of endogenous retroviral elements, but die at 4 weeks of age [30]. IAP RNA transcripts have been identified in most murine tissues, and these RNAs are increased dramatically in DNA methyltransferase knockout mice [23,31].

The ability of retrotransposons, such as IAPs, to dramatically affect gene expression has led researchers to dub them the "wild cards" of the epigenome [32]. Humans have 10-fold fewer IAP retrotransposons, but about 9 percent of the human genome still consists of other classes of retrotransposable elements [33]. In order to understand their effects more fully, global mapping techniques are currently being employed to further examine the integration sites of IAP retrotransposons, which should provide valuable information regarding the groups of genes most likely to be influenced by their presence [34,35].

The *Agouti Viable Yellow* (*A^vy*) Mouse Model

The murine *Agouti* gene encodes a paracrine signaling molecule that promotes follicular melanocytes to produce yellow phaeomelanin pigment instead of black eumelanin pigment. Transcription is initiated from a hair cycle-specific promoter in exon 2 of the *Agouti* (*A*) allele. Transcription of the *A* allele normally occurs only in the skin, where transient *A* expression in hair follicles during a specific stage of hair growth results in a sub-apical yellow band on each black hair, causing the brown (agouti) coat color of wild-type mice [36].

The *Agouti viable yellow* (*A^vy*) allele was first described in the early 1960s and resulted from the insertion of an IAP retrotransposon upstream of the transcription start site of the *Agouti* gene [21,36,37] (Fig. 15.1A). A cryptic promoter in the proximal end of the *A^vy* IAP promotes constitutive ectopic *Agouti* transcription, leading to yellow fur, obesity, and carcinogenesis [18,38]. CpG methylation in the *A^vy* IAP correlates inversely with ectopic *Agouti* expression. The degree of methylation varies dramatically among individual isogenic *A^vy/a* mice, causing coat color to range from yellow (unmethylated) to pseudoagouti (methylated) [18] (Fig. 15.1B and C).

In addition to the visible coat color change, extensive signaling from the binding of agouti to the melanocortin receptor in all tissues including the hypothalamus of *A^vy/a* mice makes them more prone to obesity and cancer [39]. These physiological effects are positively correlated to ectopic *Agouti* expression, as seen in the week 15 isogenic *A^vy/a* littermates shown in Figure 15.1C. This litany of possible phenotypes makes the *A^vy* mouse model a unique biosensor for studying transgenerational inheritance of epigenetic marks and the disruption of these marks by developmental exposures.

The *Axin Fused* (*Axin^Fu*) Mouse Model

In 2002, a pair of monozygotic human twins was found to be discordant for a caudal duplication syndrome; one twin had a duplication of the distal spine and a tumor,

235

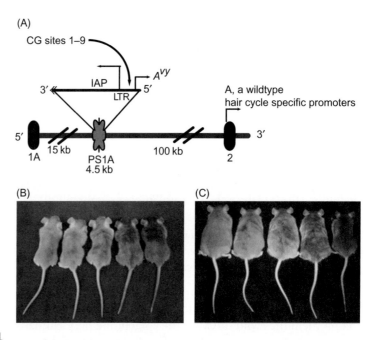

(A)

CG sites 1–9

A^{vy}

IAP

3′   5′

LTR

A, a wildtype
hair cycle specific promoters

5′   3′

1A   15 kb   PS1A   100 kb   2
                        4.5 kb

(B)   (C)

FIGURE 15.1

(A) The A^{vy} metastable epiallele contains a contraoriented intracisternal A particle insertion within pseudoexon 1A, a duplication of exon 1A. Normal transcription occurs from a hair cycle specific promoter in exon 2. The IAP insertion upstream of the wild type promoter leads to constitutive expression of *Agouti* from the IAP cryptic promoter. Methylation of the nine CpG sites upstream of the cryptic promoter inversely correlates with A^{vy} expression. (B) Three-week-old, genetically identical, A^{vy} mice with varying coat colors. Yellow mice (left) are hypomethylated upstream of the A^{vy} promoter while pseudoagouti mice (right) are hypermethylated at these CpG sites. (C) Increasing levels of ectopic expression of *Agouti* in 15-week-old A^{vy} mice (from right to left) leads to obesity, tumorigenesis, and diabetes.

while the other remained healthy. A mutation in the human *AXIN1* gene was thought responsible, yet no mutation was found upon sequencing [10]. Later, in 2006, Oates and colleagues discovered that methylation at the *AXIN1* gene promoter region was significantly increased in the affected versus unaffected twin, demonstrating the potential for epigenetic mechanisms to alter phenotype of genetically identical humans [40].

In mice, the wild type *Axin* gene encodes the Axin protein, which inhibits *Wnt* signaling and is therefore involved in mammalian embryonic axis formation [41]. The *Axin* gene is ubiquitously expressed during embryonic development and in adulthood [41]. The $Axin^{Fu}$ allele, first discovered in 1937, contains an IAP insertion within intron 6 of the gene. This insertion results in expression of a truncated but biologically active *Axin* transcript that results in axial duplications and tail kinks forming during development [42–44] (Fig. 15.2A and B). $Axin^{Fu}$ mice have kinked tails of varying severity. Moreover, the extent of the kink in the tail is inversely related to the degree of IAP methylation at the $Axin^{Fu}$ locus [19]. Like A^{vy}, the $Axin^{Fu}$ model also provides a powerful tool for analyzing the ability of developmental exposures to affect genomic methylation and phenotype.

The Mouse CDK5 Activator Binding Protein (Cabp^{IAP}) Metastable Epiallele

In 2004, a novel metastable epiallele called $Cabp^{IAP}$ was identified by Druker and colleagues through searching murine C57BL/6J cDNA databases for sequences homologous to IAP LTRs. They discovered a novel sequence within intron 6 of a previously unidentified gene in the mouse [45]. The gene shows sequence homology to the rat *CDK5 activator binding*

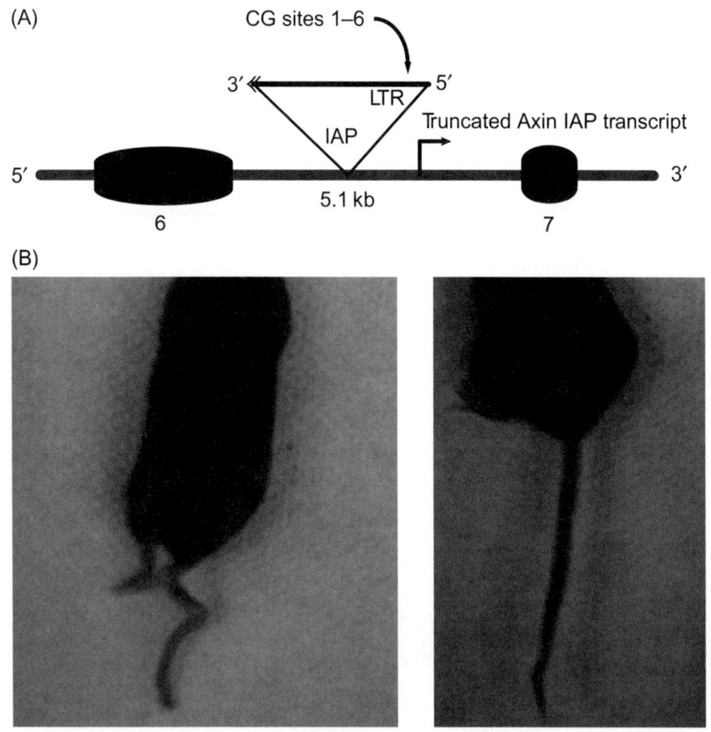

FIGURE 15.2
(A) The *Axin^Fu* metastable epiallele contains a contraoriented intracisternal A particle insertion within intron 6 of the *Axin* gene. The IAP insertion can lead to transcriptional activation of a downstream cryptic promoter, resulting in a truncated *Axin* transcript. Methylation of the six CpG sites in the IAP upstream of the cryptic promoter inversely correlates with *Axin^Fu* expression and extent of tail kink. (B) Examples of *Axin^Fu* mice with varying severities of tail kink. The left mouse exhibits severe tail kink resulting from increased *Axin^Fu* expression and hypomethylation at the allele. The right mouse exhibits only slight tail kink due to lower levels of *Axin^Fu* expression and increased methylation of the IAP.

protein (*Cabp*) responsible for CDK5 kinase inhibition. Murine *Cabp* is located on mouse chromosome 2, contains 14 exons, and normally produces a 2 kb transcript.

Like the *Agouti* and *Axin* genes, *Cabp^IAP* contains a contraoriented IAP retrotransposon insert. This IAP insert is specific to the C57BL/6J mouse strain, indicating a recent retrotransposition. *Cabp^IAP* is associated with an aberrant 1.3 kb transcript initiated from a cryptic promoter in the 5′ LTR, which is variably expressed in genetically identical littermates. Like *A^vy* and *Axin^Fu*, variable expressivity is inversely correlated to cytosine methylation in the 5′ LTR of the IAP element. Unlike *A^vy* and *Axin^Fu*, mice that are hypomethylated at the 5′ LTR also produce a number of short *Cabp* transcripts that start at the normal 5′ promoter, but end prematurely, immediately 5′ of the IAP element (Fig. 15.3). An additional short transcript resulting from the IAP insertion originates within intron 6. Therefore, *Cabp^IAP* is the first metastable epiallele for which both upstream and downstream effects on gene transcription have been demonstrated.

The existence of a second metastable epiallele within the congenic *A^vy* and *Axin^Fu* mice allows for examination of the epigenetic inheritance and the effects of environmental exposures at more than one metastable locus within a single animal. Consequently, researchers can now determine if retrotransposable elements within metastable epialleles share functional epigenetic hallmarks or if there are differences in structural features of the *A^vy*, *Axin^Fu*, and *Cabp^IAP* loci that confer variation in epigenetic mark inheritance and susceptibility to environmental exposures (Table 15.2). Future studies should aim to incorporate the analysis of the *Cabp^IAP* metastable epiallele in the above mouse models.

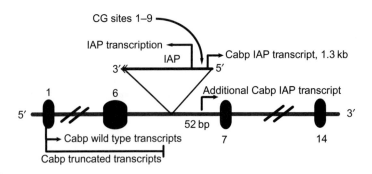

FIGURE 15.3

The *Cabp^IAP* metastable epiallele contains a contraoriented intracisternal A particle insertion within intron 6. The IAP insertion induces a cryptic promoter at the 5′ end of the insertion as well as a downstream cryptic promoter within intron 6, resulting in production of truncated *Cabp* transcripts. In addition to these transcripts, the insertion of the IAP also leads to production of truncated transcript from the wild type exon 1 promoter when the CG sites within the IAP are hypomethylated. Methylation of the nine CpG sites in the IAP upstream of the cryptic promoter suppresses transcription of *Cabp^IAP* transcripts as well as the truncated transcripts from the wild type promoter, leading to normal *Cabp* transcription.

TABLE 15.2 Defining Features of Three Mouse Metastable Epialleles

Metastable Epiallele	Gene	Location	Methylation	Hypomethylation	Epigenetic Inheritance
A^vy	*Agouti*	Chromosome 2; pseudoexon 1A	Brown coat color (pseudoagouti)	Yellow coat color, obesity, diabetes, carcinogenesis	Maternal Transmission
Axin^Fu	*Axin*	Chromosome 17; Intron 6	Straight tail	Kinked tail	Maternal and paternal transmission
Cabp^IAP	*Cabp*	Chromosome 2; Intron 6	Inactive IAP. No transcripts upstream or downstream	1.3 kb aberrant transcript, short cabp transcripts	Phenotype not yet identified due to aberrant transcript

INHERITANCE OF EPIGENETIC MARKS AT METASTABLE EPIALLELES

In the late 1970s, breeding studies involving both *A^vy* and *Axin^Fu* revealed inheritance of coat color or tail kink phenotype [46,47]. For example, pseudoagouti *A^vy* mothers produced more pseudoagouti offspring, and penetrant *Axin^Fu* mothers and fathers produced more offspring with tail kinks [19]. These observations suggested that epigenetic modifications are heritable across generations through inefficient erasure of methylation marks during gametogenesis; however, further developmental investigation indicates the presence of additional regulatory marks and/or mechanisms for epigenetic inheritance.

Epigenetic Inheritance in *A^vy* Mice

Using embryo transfer experiments in inbred mouse strains, Whitelaw and colleagues determined that the observed inheritance patterns of the *A^vy* allele occurred due to incomplete epigenetic reprogramming [18,19]. Furthermore, they observed different patterns of inheritance between paternally and maternally transmitted marks. If the *A^vy* allele was transmitted maternally from a yellow or pseudoagouti dam, the offspring were more likely to be yellow or pseudoagouti, respectively; however, paternal transmission of the *A^vy* allele did

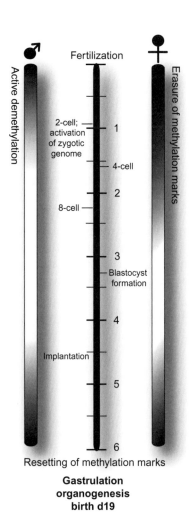

FIGURE 15.4

The erasure and establishment of methylation marks at the A^{vy} allele differs in the case of male transmission or female transmission. Active demethylation occurs very quickly following fertilization and the male-derived A^{vy} allele is completely demethylated by the 4-cell stage (represented by loss of shading). In contrast, methylation of the female-derived allele is slowly erased, reaching complete demethylation by the blastocyst stage (loss of shading). Reestablishment of methylation occurs for both male and female transmitted alleles following implantation (represented by increased shading).

not result in phenotypic inheritance. This discovery was hypothesized to be due to complete erasure of epigenetic marks in the male germ line, but incomplete erasure in the female germ line [48].

Blewitt, Whitelaw, and colleagues' later findings elaborated on the previous discovery, showing that the paternally derived A^{vy} allele is subject to rapid demethylation immediately after fertilization, while the maternally derived A^{vy} allele undergoes methylation erasure later, before implantation [49]. In zygotes, a dramatic decrease in methylation occurs at paternally derived A^{vy} alleles, as compared to methylation observed in mature sperm. This demethylation is due to the active demethylation process that is indeed seen across the paternal genome. The methylation remains low in two-cell embryos (day 1) and blastocysts (day 3) and is thought to reset between implantation (day 4.5) and midgestation (day 9) (Fig. 15.4).

In contrast, oocytes and zygotes showed higher methylation patterns at the maternally derived A^{vy} locus in pseudoagouti versus yellow mice. Yet, surprisingly, the maternal allele derived from pseudoagouti dams is completely unmethylated by the time it reaches the blastocyst stage (day 3) (see Fig. 15.4). Incomplete erasure of the methylation marks was initially hypothesized to contribute to the transgenerational inheritance of the maternally transmitted allele. Yet, these finding indicate that the marks are completely cleared in the blastocyst. Thus these observation are contrary to the initial hypothesis and question the role that methylation plays in the inheritance of phenotype seen from pseudoagouti dams.

To further explore the possibility that methylation may not be the inherited mark at the A^{vy} allele, Whitelaw's group utilized a *Mel18* knockout, which encodes for a mammalian Polycomb group protein. Unexpectedly, they discovered that epigenetic inheritance of the A^{vy} allele through the paternal germline was established in *Mel18* haplo-insufficient offspring. The abnormal paternal epigenetic inheritance indicates that another epigenetic mark aside from DNA methylation is incompletely cleared in the heterozygous knockout offspring and leads to epigenetic inheritance and phenotypic change. Moreover, the events occur after zygotic genome activation and complete demethylation of paternally-derived DNA, further supporting the role of another inherited mark, such as histone modifications, in transgenerational epigenetic inheritance at the A^{vy} allele.

Epigenetic Inheritance in *Axin^Fu* Mice

Contrary to the A^{vy} mice, the *Axin^Fu* mice display both maternal and paternal transmission of the epiallele and consequent inheritance of the phenotype [19]. In addition, the methylation state of the *Axin^Fu* adult sperm was consistent with the methylation state seen in somatic tissues of the animal, indicating that epigenetic reprogramming does not occur at this locus during gametogenesis. The allele is hypothesized to be resistant to demethylation during gametogenesis and the epigenetic marks are also thought to be incompletely cleared post-fertilization. Stochastic reestablishment of methylation marks ensures that some marks will remain and the memory of the parental alleles will influence the inheritance of the parental phenotype in the offspring [19].

Future Studies in Epigenetic Inheritance

Although the mechanisms behind the inheritance of epigenetic marks at the A^{vy} allele are incompletely defined, the data suggest further analysis of additional marks, such as histone modifications, for their role in maintaining epigenetic memory of the maternal allele. Likewise, the *Axin^Fu* model should also be examined for other marks that contribute to epigenetic inheritance. Despite unknown elements, phenotypic changes in these models are undoubtedly linked to epigenetic inheritance. Thus, these models serve as excellent biosensors for the effects of knockouts of other proteins that may be involved in epigenetic memory. With increased research and time, the complement of epigenetic marks should be entirely defined.

MOUSE MODELS AS A RESEARCHER'S TOOL

The A^{vy} and *Axin^Fu* mouse models, along with the *Cabp^IAP* metastable epiallele, have become valuable tools to study the effects of nutrition and chemical exposure on epigenetic programming. Persistent epigenetic adaptations that occur early in development in response to maternal nutrition have been associated with alterations in methylation, coat color, and tail kink phenotypes among genetically identical littermates [11,21,50]. These epigenetic adaptations are becoming increasingly quantifiable and may also play an important role in developmental plasticity and disease susceptibility [51,52].

The A^{vy} and *Axin^Fu* Models and Methyl Donors

Maternal epigenetics and methyl supplements affect *Agouti* and *Axin* gene expression in A^{vy} and *Axin^Fu* mice [53]. Diet-derived methyl donors and co-factors are necessary for the synthesis of *S*-adenosylmethionine (SAM) that provides the methyl groups required for

DNA methylation. Environmental factors that alter early nutrition or SAM synthesis can potentially influence adult phenotype via alterations in CpG methylation at critically important, epigenetically labile regions in the genome.

Using the A^{vy} mouse model, Cooney et al. and Waterland et al. demonstrated that maternal dietary methyl supplementation with folic acid, vitamin B12, choline, and betaine shifts the coat color distribution of the offspring towards the pseudoagouti phenotype [50,54] (Table 15.3). Moreover, it was also demonstrated that the shift in coat color distribution was caused by increased methylation at seven A^{vy} CpG sites [54]. Methylation profiles at the seven CpG sites were highly correlated in tissues from ectodermal (brain and tail), endodermal (liver) and mesodermal (kidney) origins, indicating that methylation profiles at the A^{vy} locus are established at the totipotent stem cell stage of embryogenesis. In addition, methylation in postnatal day (PND) 21 tissues was correlated with methylation in PND100 tissues, demonstrating that this pattern of DNA methylation is maintained over time.

The $Axin^{Fu}$ mice have also been utilized to analyze the effects of maternal methyl donor supplementation. Waterland et al. found that methyl donor supplementation before and during gestation significantly increased methylation and decreased the tail kink in the offspring. Furthermore, this increase in methylation was specific to the tail region, providing evidence that the timing of the effect in this study occurred around midgestation [55].

To analyze whether these methyl supplement-induced marks are inherited transgenerationally, Cropley and colleagues exposed only A^{vy} pseudoagouti F0 mothers to methyl donor or control diets [56]. They then analyzed the F2 generation and found that pseudoagouti mothers that had been exposed to methyl donors *in utero* had darker offspring. The authors concluded that the methyl supplements induced a germ-line epigenetic modification in the murine A^{vy} allele [56].

Waterland et al. later found, however, that the epigenetic changes spurred by methyl supplementation are not inherited by the next generation through new germ line epigenetic modifications [57]. In this study, the researchers weaned A^{vy}/a mice onto either control or

TABLE 15.3 Summary of Developmental Exposures Shown to Alter Metastable Epialleles

Metastable Epiallele	Preconception and/or Gestational Treatment				
	Methyl Donors	**Genistein**	**Bisphenol A**	***In vitro* Culture**	**Heterozygous Paternal Mutation of Epigenetic Modifier Genes**
A^{vy}	Hypermethylation [47], Pseudoagouti coat color shift [51], Prevents trans-generational amplification of obesity [55], May maintain epigenetic marks in germline as it is reset in the next generation [54]	Hyper-methylation [64], Pseudo-agouti coat color shift [64], Reduced adult-onset obesity in offspring [64]	Hypomethylation [67], Shift towards yellow coat color [67]	Shift in offspring coat color towards yellow [69]	Shift in offspring coat color towards yellow [70]
$Axin^{Fu}$	Hypermethylation [52] Less tail kink [52]				
$Cabp^{IAP}$			Hypomethylation [67]		

methyl supplemented diets. These F0 dams were then mated and F1 and F2 subsequent generations were weaned onto the same diet as their mother's, stopping at the F3 generation. The animals were then rated for coat color, an indicator of methylation. As expected, the supplemented mice had darker coat colors, indicating higher levels of methylation compared to controls; however, this increase was stable across generations and did not increase as would be expected if the effects of maternal diet on offspring A^{vy} methylation had been transmitted to the next generation. This study provides insight into Cropley's initial findings, suggesting that environmental influences maintain epigenetic information at the A^{vy} germ line as it is reset in the next generation, but does not confer new germ line epigenetic information (see Table 15.3).

Recently, Waterland and colleagues demonstrated that methyl donor supplementation also prevents transgenerational amplification of obesity [58]. In this study, obesity was shown to amplify through three successive generations of yellow, obese A^{vy} mothers. This transgenerational amplification of obesity in the third generation was prevented by a diet supplemented with methyl donors in prior generations. Furthermore, the methyl supplementation to yellow, obese A^{vy} mothers did not significantly alter transgenerational epigenetic changes at the A^{vy} allele, as no significant alteration in coat color was observed in the final generation. Additionally, the authors found no association between body weight and offspring coat color phenotype, suggesting again that A^{vy} methylation was not significantly altered and indicating that the A^{vy} allele was not substantially involved in this effect. Thus, the authors concluded that the methyl donors acted to prevent the amplification of obesity by directing epigenetic change at alternative locations that may have direct relevance to human obesity (see Table 15.3).

The A^{vy} Model and Isoflavones

The A^{vy} model has also been used to examine the effects of isoflavones on the epigenome. Isoflavones represent a class of phytoestrogens present in soy and soy products that are active in multiple biological systems, including estrogen receptor and non-estrogen receptor mediated signaling pathways [59,60]. Genistein, the major isoflavone in soy, exhibits mixed estrogen agonist and antagonist properties, inhibits tyrosine kinase activity, and scavenges free radicals depending on timing, dose, and the tissue type [61–63]. In addition, genistein and other isoflavones can interact with the estrogen receptor to enhance histone acetylation [64].

To determine if maternal genistein affects offspring by altering the epigenome *in utero*, Dolinoy et al. assessed coat color, DNA methylation, and body weight in heterozygous viable yellow Agouti (A^{vy}/a) offspring [20,65]. The results show that genistein induces CpG hypermethylation of six sites in A^{vy}, shifts coat color distribution towards pseudoagouti, and decreases the incidence of adult-onset obesity in the offspring [66]. This is the first evidence that early *in utero* exposure to genistein results in decreased adult chronic disease susceptibility by producing permanent alterations in the epigenome (see Table 15.3).

Dolinoy's methylation analysis serves as an example for future studies employing A^{vy} mice and other metastable epialleles. In addition to assessing average methylation over the region, Dolinoy determined the methylation level at each of the nine A^{vy} CpG sites for all animals using quantitative bisulfite sequencing [67]. This method provided insight into site-specific methylation and its importance to phenotypic change. The analysis revealed significantly different methylation profiles between the unsupplemented and genistein-supplemented diet groups at CpG dinucleotides 4 to 9. Moreover, the statistical significance of CpG site 4 was an order of magnitude greater than that for CpG sites 5 to 9. The enhanced significance of CpG site 4, coupled with the general increase in methylation closer to the cryptic A^{vy} promoter, suggests that CpG site 4 represents a boundary to methylation spreading, and may be particularly important in determining the epigenetically regulated mosaicism in A^{vy} mouse coat color.

Just as in the methyl supplementation studies performed by Waterland and Cooney, average methylation in PND21 tail tissues from a subset of genistein-supplemented animals was highly correlated with average methylation in PND150 tissues derived from the ectoderm (brain and tail), mesoderm (kidney), and endoderm (liver). The low variability in methylation across tissues derived from different germ layers relative to high variability between individual animals indicates that the establishment of epigenotype at A^{vy} occurs at the totipotent stem cell stage of embryonic development. Furthermore, the concordance between A^{vy} methylation in PND21 tail and that in the various tissues of the same animal at PND150 demonstrates that genistein-induced epigenetic changes persist to adulthood.

Methyl Donors, Bisphenol A, and the A^{vy} Mouse

One of the most interesting studies to come from the use of the A^{vy} mice in nutritional developmental research was Dolinoy et al.'s bisphenol A (BPA) study. This study showed that genistein and other methyl donors can counteract the hypomethylating effects of BPA, a compound found in baby bottles, food containers, and a variety of other plastics [68]. BPA supplementation induced hypomethylation of the A^{vy} allele, leading to an increased proportion of yellow mice in the population. This hypomethylating effect was significantly counteracted by addition of methyl donors to the diet, restoring the exposed population to similar coat color frequencies that are seen in the control population. Additional analysis of the Cab^{IAP} allele in these same mice showed significant methylation changes that coincided with the alterations at the A^{vy} locus. This study not only demonstrated the potential for the use of the A^{vy} mouse as a toxicological biosensor, but also shed light on the capabilities of the mouse to be utilized for the examination of therapeutic agents meant to counteract negative epigenetic modifications. The above studies represent a systematic method for examining the effects of exposures at the A^{vy} or $Axin^{Fu}$ alleles and are outlined in Figure 15.5.

The A^{vy} Mouse Model and *in vitro* Fertilization

The A^{vy} mice are useful for more than just maternal nutritional and chemical exposure studies. They can also be used to analyze the epigenetic mechanisms behind development, and those mechanisms that become disrupted during *in vitro* fertilization (IVF). Children born through IVF, which involves human embryo culture, have a small but increased incidence of loss of imprinting and consequent elevated risk of Beckwith–Wiedeman and Angelman syndromes and certain cancers [69]. Although this phenomenon can be difficult to study in humans, the A^{vy} mice provide a unique opportunity for examining the effects of an alternative preimplantation environment. Using the A^{vy} mice, Morgan and colleagues demonstrated that the culture of zygotes to the blastocyst stage changes the postnatal expression of the epigenetically labile A^{vy} allele [70]. In this study, A^{vy}/a males were mated to ovulation-induced females; zygotes were collected and grown in the same commercial IVF media used for human preimplantation embryo growth and then transplanted to pseudo-pregnant recipient females. Alternatively, blastocysts were transferred without embryo culture, or pregnancy was allowed to occur naturally. The authors found that the offspring produced following exposure to *in vitro* culture were, by proportion, significantly more yellow than those conceived without culture or with natural pregnancy. This study successfully demonstrated that the preimplantation environment is important for regulation of the A^{vy} allele and supports the notion that the environment encountered during this sensitive period of development has the potential to affect changes in the epigenetic regulation of the human genome.

The A^{vy} Model and Paternal Effect Genes

Further demonstrating the versatility of the A^{vy} model, Chong et al. employed A^{vy} mice to analyze genes that show paternal effects in the mouse [71]. Paternal effects occur when

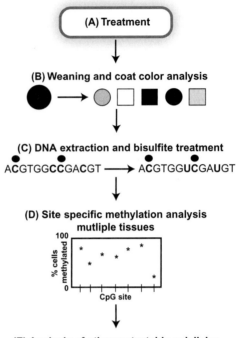

FIGURE 15.5

Utilizing A^{vy} mice for determining the effects of developmental exposures on the epigenome. (A) Mice are exposed to treatment during gestation and/or lactation. (B) Mice are weaned 21 days postnatal, coat colors are analyzed, and significance is assessed with proper statistical tests. (C) DNA is extracted from multiple tissues to determine the exposure's effects on the epigenome. DNA is modified with sodium bisulfite to allow for examination of DNA methylation. (D) Methylation is analyzed in a site-specific manner in order to correlate methylation at each site with phenotype. (E) Additional analysis of other metastable epialleles and imprinted genes aids in determining how exposure affects other parts of the epigenome, including regions intricately tied to human disease.

heterozygous mutations affect the phenotype of the offspring that do not inherit the mutated allele. Chong et al. showed that disruption of the epigenetic state in A^{vy} mouse germ cells could affect the epigenetic state of the following generation, regardless of inheritance of the mutation. The group accomplished this by analyzing the ability for wild type male mice, with heterozygous mutations in genes involved in epigenetic regulation, to show paternal effects on the maternally inherited A^{vy} allele in wild type offspring that do not inherit the mutant allele. Among the genes they found to show paternal effects are *Smarca5*, which encodes the chromatin remodeler Snf2h, and *Dnmt1*, which encodes the DNA methyltransferase enzyme that maintains somatic DNA methylation profiles during DNA replication. A shift towards yellow in offspring coat color was observed in the A^{vy} mice from both epigenetic programming mutations. Thus, the A^{vy} mice, carrying the reporter allele responsible for coat color, were essential in this study to show that the phenotype of the offspring is influenced by the untransmitted genotype of the male parent through epigenetic alterations originally established in the gametes. Although the offspring do not carry the mutant genotype, the parental gametic changes were thought to shift ratios of sperm heterochromatin and euchromotin and affect methylation of sensitive alleles [71].

GENOMIC IMPRINTING

Just as these mouse models have been employed for their metastable epialleles, they are also beneficial for studying the regulation of imprinted genes shared with humans. Imprinting

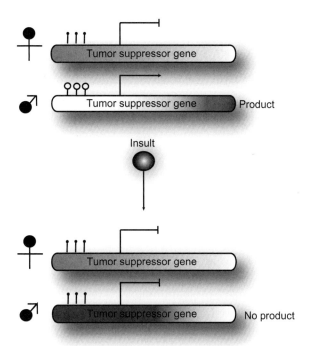

FIGURE 15.6
Imprinted genes are normally expressed in a parent of origin manner, resulting in apparent haploid expression. Environmental insults occurring early in development, during imprint mark establishment, can alter expression of the active allele, for example via epigenetic alterations such as increased methylation (filled lollipops) and lead to deregulation, as depicted for the maternally imprinted tumor suppressor gene shown here.

is a form of gene regulation in which inherited epigenetic modifications drive differential gene expression in a parent-of-origin dependent manner. In mammals, imprinting is exclusive to therian species [72]. Since the identification of the first imprinted genes in 1991, approximately 80 imprinted genes have been identified in mice and humans, with 29 of these genes showing imprinted expression in both species [73,74].

Imprinted genes are particularly relevant for study in disease occurrence due to their importance in development and growth combined with their normal haploid state and consequent vulnerability to deregulation. Since imprinted genes are functionally haploid, they are denied the protection from recessive mutations that diploidy normally provides (Fig. 15.6). Imprinted gene deregulation is associated with a number of neurodevelopmental and psychological disorders as well as increased susceptibility to cancer [7,75].

DNA methylation is a heritable epigenetic imprint mark that distinguishes the otherwise genetically identical parental alleles. For many imprinted genes, the methylation status of the two parental alleles is divergent. These regions are termed differentially methylated regions (DMRs). DNA methylation is also thought to interact with histone modifications and chromatin packaging to regulate imprinted gene expression. In fact, Waterland and colleagues have shown that developmental exposures can influence imprinting by altering methylation levels of imprinted gene DMRs. In their study, they discovered that post-weaning diet alters genomic imprinting at the *insulin-like growth factor 2* (*Igf2*) locus and methylation at the *Igf2* DMR2 region [76]. *Igf2* is a potent mitogenic growth factor [77–79]. Their findings suggest that imprinted genes may be susceptible to other developmental exposures and in fact may be responsive to such exposures in a manner similar to that observed at metastable epialleles.

The mouse models described above provide an avenue to look at the influence of early exposures on altered imprinted gene regulation and the potential for transgenerational inheritance of such defects. These models also provide the means to analyze these regions in detail along with the epigenetic mechanisms necessary for establishment and maintenance of imprinted gene expression. Future studies should take advantage of common imprinted genes between humans and mice in order to analyze the effects of exposures on regulation of these important genes.

MOUSE MODELS, HISTONE MODIFICATIONS, AND SMALL REGULATORY RNAs

Although the most widely understood inherited mark is DNA methylation, histone modifications and regulatory small RNAs undoubtedly contribute to the establishment and maintenance of methylation marks. Indeed, previous studies examining the inheritance of phenotype in the A^{vy} mouse indicate that methylation is not alone in controlling IAP expression [49]. Given that DNA is tightly bound around histone proteins, chromatin must also be restructured during epigenetic programming. As a result, it is not surprising that DNA methylation changes during development have been linked to chromatin structure and histone modifications [80]. Histone acetylation and methylation maintain chromatin in a transcriptionally active (acetylated) or silenced (deacetylated) state and are necessary for marking DNA sequences for methylation, although the details remain undetermined.

The particulars about how methyltransferase and histone interactions drive methylation are still being actively researched. Cofactors and methyl-CpG binding proteins, which are not enzymatically active, are thought to regulate the association and interaction with histones to direct CpG methylation (reviewed in Ref. 81). Several other factors and specific histone modifications have been recognized for their role in DNA methylation from loss of function and biochemical studies. For example, small RNAs have been found to work upstream of methyltransferase machinery to mediate methylation of male germ cell long interspersed nucleotide element-1 (LINE1) and Intracisternal A Particle (IAP) retrotransposons [82,83].

Studies that utilize the A^{vy} allele as a biosensor, such as Chong et al.'s [71], will prove useful for analyzing the inheritance of histone modifications and regulatory small RNAs. The group demonstrated that mutations in *Smarca5*, which encodes the chromatin remodeler Snf2h, and *Dnmt1*, which encodes the DNA methyltransferase enzyme that maintains somatic DNA methylation profiles during DNA replication, lead to hypomethylation at the A^{vy} allele. Additional loss-of-function studies can help elucidate the interactions between DNA methylation and other epigenetic alterations that influence epigenetic inheritance.

THE FUTURE OF MOUSE MODELS IN EPIGENETIC RESEARCH

As researchers continue to unfold the mechanisms behind the developmental origins of adult diseases, such as obesity, cancer, diabetes, and other metabolic syndromes, epigenetic regulation continually finds precedence for study. The A^{vy} and $Axin^{Fu}$ mouse models have become powerful tools for examining transgenerational inheritance of epigenetic marks as well as unique biosensors for early developmental exposures to nutritional supplements and chemical contaminants that disrupt epigenetic programming.

With the increased utility of these models, our understanding of epigenetic mark inheritance at metastable epialleles and imprinted genes, and across the genome, will expand. Further research will provide information about additional inherited epigenetic marks. As the model is more completely defined, the possibilities for their expanded use to address a broad array of research questions involved in complex human disease such as diabetes, neurological disorders, and cancer increase.

References

1. Roseboom T, de Rooij S, Painter R. The Dutch famine and its long-term consequences for adult health. *Early Hum Dev* 2006;**82**:485–91.

2. Barker DJ, Clark PM. Fetal undernutrition and disease in later life. *Rev Reprod* 1997;**2**:105–12.

3. Liu L, Li Y, Tollefsbol TO. Gene–environment interactions and epigenetic basis of human diseases. *Curr Issues Mol Biol* 2008;**10**:25–36.

4. Szyf M. The early life environment and the epigenome. *Biochim Biophys Acta* 2009;**1790**:878–85.

5. Heijmans BT, Tobi EW, Stein AD, Putter H, Blauw GJ, Susser ES, et al. Persistent epigenetic differences associated with prenatal exposure to famine in humans. *Proc Natl Acad Sci USA* 2008;**105**:17046–9.

6. Abel KM, Allin MP, Jirtle RL. Schizophrenia, cancer and imprinting: early nutritional influences. *Br J Psychiatry* 2006;**188**:394.

7. Falls JG, Pulford DJ, Wylie AA, Jirtle RL. Genomic imprinting: implications for human disease. *Am J Pathol* 1999;**154**:635–47.

8. Jirtle RL. Genomic imprinting and cancer. *Exp Cell Res* 1999;**248**:18–24.

9. Jirtle RL, Skinner MK. Environmental epigenomics and disease susceptibility. *Nat Rev Genet* 2007;**8**:253–62.

10. Kroes HY, Takahashi M, Zijlstra RJ, Baert JA, Kooi KA, Hofstra RM, et al. Two cases of the caudal duplication anomaly including a discordant monozygotic twin. *Am J Med Genet* 2002;**112**:390–3.

11. Waterland RA. Does nutrition during infancy and early childhood contribute to later obesity via metabolic imprinting of epigenetic gene regulatory mechanisms? *Nestle Nutr Workshop Ser Pediatr Program* 2005;**56**: 157–71. Discussion 171–54.

12. Waterland RA. Epigenetic mechanisms and gastrointestinal development. *J Pediatr* 2006;**149**:S137–42.

13. Weidman JR, Dolinoy DC, Murphy SK, Jirtle RL. Cancer susceptibility: epigenetic manifestation of environmental exposures. *Cancer J* 2007;**13**:9–16.

14. Manipalviratn S, DeCherney A, Segars J. Imprinting disorders and assisted reproductive technology. *Fertil Steril* 2009;**91**:305–15.

15. Nathanielsz PW. Animal models that elucidate basic principles of the developmental origins of adult diseases. *ILAR J* 2006;**47**:73–82.

16. Finnegan EJ, Whitelaw E. Leaving the past behind. *PLoS Genet* 2008;**4**:e1000248.

17. Rakyan VK BM, Druker R, Preis JI, Whitelaw E. Metastable epialleles in mammals. *Trends Genet* 2002;**18**:348–51.

18. Morgan HD, Sutherland HG, Martin DI, Whitelaw E. Epigenetic inheritance at the agouti locus in the mouse. *Nat Genet* 1999;**23**:314–18.

19. Rakyan VK, Chong S, Champ ME, Cuthbert PC, Morgan HD, Luu KV, et al. Transgenerational inheritance of epigenetic states at the murine Axin(Fu) allele occurs after maternal and paternal transmission. *Proc Natl Acad Sci USA* 2003;**100**:2538–43.

20. Dolinoy DC, Weidman JR, Waterland RA, Jirtle RL. Maternal genistein alters coat color and protects Avy mouse offspring from obesity by modifying the fetal epigenome. *Environ Health Perspect* 2006;**114**:567–72.

21. Waterland RA, Jirtle RL. Transposable elements: targets for early nutritional effects on epigenetic gene regulation. *Mol Cell Biol* 2003;**23**:5293–300.

22. Waterland RA, Jirtle RL. Early nutrition, epigenetic changes at transposons and imprinted genes, and enhanced susceptibility to adult chronic diseases. *Nutrition* 2004;**20**:63–8.

23. Kuff EL, Lueders KK. The intracisternal A-particle gene family: structure and functional aspects. *Adv Cancer Res* 1988;**51**:183–276.

24. Maksakova IA, Romanish MT, Gagnier L, Dunn CA, van de Lagemaat LN, Mager DL. Retroviral elements and their hosts: insertional mutagenesis in the mouse germ line. *PLoS Genet* 2006;**2**:e2.

25. Consortium MGS. Initial sequencing and comparative analysis of the mouse genome. *Nature* 2002;**420**:520–62.

26. Lewin B. *Genes*. New York: Oxford University Press; 2000.

27. Falzon M, Kuff EL. Multiple protein-binding sites in an intracisternal A particle long terminal repeat. *J Virol* 1988;**62**:4070–7.

28. Mietz JA, Kuff EL. Tissue and strain-specific patterns of endogenous proviral hypomethylation analyzed by two-dimensional gel electrophoresis. *PNAS* 1990;**87**:2269–73.

29. Kurihara Y, Kawamura Y, Uchijima Y, Amamo T, Kobayashi H, Asano T, et al. Maintenance of genomic methylation patterns during preimplantation development requires the somatic form of DNA methyltransferase I. *Dev Biol* 2008;**313**:335–46.

30. Okano M, Bell DW, Haber DA, Li E. DNA methyltransferases Dnmt3a and Dnmt3b are essential for de novo methylation and mammalian development. *Cell* 1999;**99**:247–57.

31. Walsh CP, Chaillet JR, Bestor TH. Transcription of IAP endogenous retroviruses is constrained by cytosine methylation. *Nat Genet* 1998;**20**:116–17.

32. Cropley JE, Martin DI. Controlling elements are wild cards in the epigenomic deck. *Proc Natl Acad Sci USA* 2007;**104**:18879–80.

33. International Human Genome Sequencing Consortium 2001.

34. Takabatake T, Ishihara H, Ohmachi Y, Tanaka I, Nakamura MM, Fujikawa K, et al. Microarray-based global mapping of integration sites for the retrotransposon, intracisternal A-particle, in the mouse genome. *Nucleic Acids Res* 2008;**36**:e59.

35. Zhang Y, Maksakova IA, Gagnier L, van de Lagemaat LN, Mager DL. Genome-wide assessments reveal extremely high levels of polymorphism of two active families of mouse endogenous retroviral elements. *PLoS Genet* 2008;**4**:e1000007.

36. Duhl DM, Stevens ME, Vrieling H, Saxon PJ, Miller MW, Epstein CJ, et al. Pleiotropic effects of the mouse lethal yellow (Ay) mutation explained by deletion of a maternally expressed gene and the simultaneous production of agouti fusion RNAs. *Development* 1994;**120**:1695–708.

37. Dickies MM. A new viable yellow mutation in the house mouse. *J Hered* 1962;**53**:84–6.

38. Miltenberger RJ, Mynatt RL, Wilkinson JE, Woychik RP. The role of the agouti gene in the yellow obese syndrome. *J Nutr* 1997;**127**:1902S–1907S.

39. Pan W, Kastin AJ. Mahogany, blood-brain barrier, and fat mass surge in Avy mice. *Int J Obes* 2007;**31**:1030–2.

40. Oates NA, van Vliet J, Duffy DL, Kroes HY, Martin NG, Boomsma M, et al. Increased DNA methylation at the AXIN1 gene in a monozygotic twin from a pair discordant for a caudal duplication anomoly. *Am J Hum Genet* 2006;**79**:155–62.

41. Zeng L, Fagotto F, Zhang T, Hsu W, Vasicek TJ, Perry WL 3rd, et al. The mouse Fused locus encodes Axin, an inhibitor of the Wnt signaling pathway that regulates embryonic axis formation. *Cell* 1997;**90**:181–92.

42. Flood WD, Ruvinsky A. Alternative splicing and expressivity of the Axin(Fu) allele in mice. *Heredity* 2001;**87**:146–52.

43. Vasicek TJ, Zeng L, Guan XJ, Zhang T, Costantini F, Tilghman SM. Two dominant mutations in the mouse fused gene are the result of transposon insertions. *Genetics* 1997;**147**:777–86.

44. Reed SC. The Inheritance and Expression of Fused, a New Mutation in the House Mouse. *Genetics* 1937;**22**:1–13.

45. Druker R, Bruxner TJ, Lehrbach NJ, Whitelaw E. Complex patterns of transcription at the insertion site of a retrotransposon in the mouse. *Nucleic Acids Res* 2004;**32**:5800–8.

46. Wolff GL. Influence of maternal phenotype on metabolic differentiation of agouti locus mutants in the mouse. *Genetics* 1978;**88**:529–39.

47. Belyaev DK, Ruvinsky AO, Borodin PM. Inheritance of alternative states of the fused gene in mice. *J Hered* 1981;**72**:107–12.

48. Morgan Hugh D, David I.K. Martin HGES, Whitelaw E. Epigenetic inheritance at the agouti locus in the mouse. *Nat Genet* 1999;**23**:314–18.

49. Blewitt ME, Vickaryous NK, Paldi A, Koseki H, Whitelaw E. Dynamic reprogramming of DNA methylation at an epigenetically sensitive allele in mice. *PLoS Genet* 2006;**2**:e49.

50. Cooney CA, Dave AA, Wolff GL. Maternal methyl supplements in mice affect epigenetic variation and DNA methylation of offspring. *J Nutr* 2002;**132**:2393S–400S.

51. Ounpraseuth S, Rafferty TM, McDonald-Phillips RE, Gammill WM, Siegel ER, Wheeler KL, et al. A method to quantify mouse coat-color proportions. *PLoS ONE* 2009;**4**:e5414.

52. Bateson P, Barker D, Clutton-Brock T, Deb D, D'Udine B, Foley RA, et al. Developmental plasticity and human health. *Nature* 2004;**430**:419–21.

53. Wolff GL, Kodell RL, Moore SR, Cooney CA. Maternal epigenetics and methyl supplements affect agouti gene expression in Avy/a mice. *FASEB J* 1998;**12**:949–57.

54. Waterland RA. Do maternal methyl supplements in mice affect DNA methylation of offspring? *J Nutr* 2003;**133**:239.

55. Waterland RA, Dolinoy DC, Lin JR, Smith CA, Shi X, Tahiliani KG. Maternal methyl supplements increase offspring DNA methylation at Axin fused. *Genesis* 2006;**44**:401–6.

56. Cropley JE, Suter CM, Martin DI. Methyl donors change the germline epigenetic state of the A(vy) allele. *FASEB J* 2007;**21**:3021–2.

57. Waterland RA, Travisano M, Tahiliani KG. Diet-induced hypermethylation at agouti viable yellow is not inherited transgenerationally through the female. *FASEB J* 2007;**21**:3380–5.

58. Waterland RA, Travisano M, Tahiliani KG, Rached MT, Mirza S. Methyl donor supplementation prevents transgenerational amplification of obesity. *Int J Obes (Lond)* 2008;**32**:1373–9.

59. Lamartiniere CA, Cotroneo MS, Fritz WA, Wang J, Mentor-Marcel R, Elgavish A. Genistein chemoprevention: timing and mechanisms of action in murine mammary and prostate. *J Nutr* 2002;**132**:552S–8S.

60. Valachovicova T, Slivova V, Sliva D. Cellular and physiological effects of soy flavonoids. *Mini Rev Med Chem* 2004;**4**:881–7.

61. Price KR, Fenwick GR. Naturally occurring oestrogens in foods–a r eview. *Food Addit Contam* 1985;**2**:73–106.

62. Akiyama T, Ishida J, Nakagawa S, Ogawara H, Watanabe S, Itoh N, et al. Genistein, a specific inhibitor of tyrosine-specific protein kinases. *J Biol Chem* 1987;**262**:5592–5.

63. Wei H, Wei L, Frenkel K, Bowen R, Barnes S. Inhibition of tumor promoter-induced hydrogen peroxide formation in vitro and in vivo by genistein. *Nutr Cancer* 1993;**20**:1–12.

64. Hong T, Nakagawa T, Pan W, Kim MY, Kraus WL, Ikehara T, et al. Isoflavones stimulate estrogen receptor-mediated core histone acetylation. *Biochem Biophys Res Commun* 2004;**317**:259–64.

65. Ziegler RG, Hoover RN, Pike MC, Hildesheim A, Nomura AM, West DW, et al. Migration patterns and breast cancer risk in Asian-American women. *J Natl Cancer Inst* 1993;**85**:1819–27.

66. Fritz WA, Wang J, Eltoum IE, Lamartiniere CA. Dietary genistein down-regulates androgen and estrogen receptor expression in the rat prostate. *Mol Cell Endocrinol* 2002;**186**:89–99.

67. Grunau C, Clark SJ, Rosenthal A. Bisulfite genomic sequencing: systematic investigation of critical experimental parameters. *Nucleic Acids Res* 2001;**29**:E65.

68. Dolinoy DC, Huang D, Jirtle RL. Maternal nutrient supplementation counteracts bisphenol A-induced DNA hypomethylation in early development. *Proc Natl Acad Sci USA* 2007;**104**:13056–61.

69. Paoloni-Giacobino A, Chaillet JR. Genomic imprinting and assisted reproduction. *Reprod Health* 2004;**1**:6.

70. Morgan HD, Jin XL, Li A, Whitelaw E, O'Neill C. The culture of zygotes to the blastocyst stage changes the postnatal expression of an epigentically labile allele, agouti viable yellow, in mice. *Biol Reprod* 2008;**79**:618–23.

71. Chong S, Vickaryous N, Ashe A, Zamudio N, Youngson N, Hemley S, et al. Modifiers of epigenetic reprogramming show paternal effects in the mouse. *Nat Genet* 2007;**39**:614–22.

72. Killian JK, Nolan CM, Stewart N, Munday BL, Andersen NA, Nicol S, et al. Monotreme IGF2 expression and ancestral origin of genomic imprinting. *J Exp Zool* 2001;**291**:205–12.

73. Luedi PP, Hartemink AJ, Jirtle RL. Genome-wide prediction of imprinted murine genes. *Genome Res* 2005;**15**:875–84.

74. Morison IM, Ramsay JP, Spencer HG. A census of mammalian imprinting. *Trends Genet* 2005;**21**:457–65.

75. Murphy SK, Jirtle RL. Imprinting evolution and the price of silence. *Bioessays* 2003;**25**:577–88.

76. Waterland RA, Lin JR, Smith CA, Jirtle RL. Post-weaning diet affects genomic imprinting at the insulin-like growth factor 2 (Igf2) locus. *Hum Mol Genet* 2006;**15**:705–16.

77. DeChiara TM, Robertson EJ, Efstratiadis A. Parental imprinting of the mouse insulin-like growth factor II gene. *Cell* 1991;**64**:849–59.

78. Barlow DP, Stoger R, Herrmann BG, Saito K, Schweifer N. The mouse insulin-like growth factor type-2 receptor is imprinted and closely linked to the Tme locus. *Nature* 1991;**349**:84–7.

79. Ghosh P, Dahms NM, Kornfeld S. Mannose 6-phosphate receptors: new twists in the tale. *Nat Rev Mol Cell Biol* 2003;**4**:202–12.

80. Weaver JR, Susiarjo M. Imprinting and epigenetic changes in the early embryo. *Mamm Genome* 2009. DOI 10.10007/s00335-009-9225-2.

81. Ooi SK, O'Donnell AH, Bestor TH. Mammalian cytosine methylation at a glance. *J Cell Sci* 2009;**122**:2787–91.

82. Aravin AA, Bourc'his D. Small RNA guides for de novo DNA methylation in mammalian germ cells. *Genes Dev* 2008;**22**:970–5.

83. Aravin AA, Sachidanandam R, Bourc'his D, Schaefer C, Pezic D, Toth KF, et al. A piRNA pathway primed by individual transposons is linked to de novo DNA methylation in mice. *Mol Cell* 2008;**31**:785–99.

CHAPTER 16

Epigenetic Regulatory Mechanisms in Plants

Zoya Avramova
School of Biological Sciences, UNL, E-300 Beadle Center, P.O. Box 88066, Lincoln, NE
68588-0666, USA

THE PLANT EPIGENOME

Genomes are defined by their primary sequence, which provides the genetic blueprint
of a species. Eukaryotic DNA functions within the context of chromatin, which provides
additional layers of gene regulation referred to as "epigenetic". The commonly found
definition of epigenetics is that of a "study of heritable changes in genome function that
occur without a change in DNA sequence" (Ref. 1 and ref. therein). However, evidence
that neuronal gene-expression states are also regulated by epigenetic mechanisms, despite
evidence that neuronal cells do not divide, has opened space for a broader unifying
definition that keeps "the sense of prevailing usage but avoids constraints imposed by
stringently required heritability" [1].

251

Epigenetic mechanisms regulate developmental programs, stress responses and adaptation,
senescence, disease, and various patterns of non-Mendelian inheritance. The totipotency
of plant cells, in addition to the ability of plants to withstand biotic, abiotic, and genome
stresses, such as changes in chromosome number and massive presence of transposable
elements, reflects the plasticity of plant genomes and makes them an excellent system to
study epigenetic phenomena. Genome plasticity is determined by the EPIGENOME. DNA
methylation and histone modification profiles define epigenomes of animals and plants.
The main molecular mechanisms operating in epigenetic phenomena are DNA methylation,
histone modifications, and RNA-based mechanisms, often referred to as "the three pillars of
epigenetics" [2]. Recent advances in genome research technologies, deep sequencing analysis
in particular, have led to an explosion of studies and novel results that are re-shaping our
views. Non-coding RNAs (ncRNAs) are emerging as central players responsible for the
establishment, maintenance, and regulation of plant genome epigenetic structure [3].

At the molecular level, a unifying view of epigenetics postulates that DNA methylation and
histone modification patterns provide "information" instructing genome function. Following
this information, the chromatin remodelers (the ATPase-containing machines) re-position
the nucleosomes modulating thus the access of Polymerase II (Pol II) to genes. NcRNAs
(small silencing RNAs, in particular) are the molecular mechanism integrating numerous
seemingly disparate cellular events [4] (Fig. 16.1). Long-standing questions about the
molecular basis of pluripotency, tumorigenesis, apoptosis, position effect variegation (PEV),
paramutation, imprinting, and cell identity are finding answers in small RNAs.

Handbook of Epigenetics: The New Molecular and Medical Genetics. DOI: 10.1016/B978-0-12-375709-8.00016-2

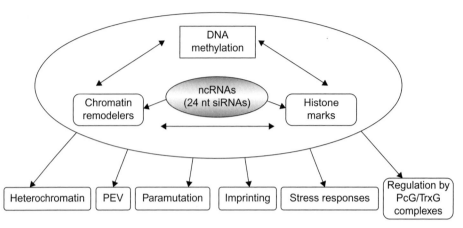

FIGURE 16.1

The main factors involved in epigenetic mechanisms regulating various plant processes. The noncoding small 24 nt siRNAs are the integrating function capable of mediating the activities of the chromatin remodeling machines, the histone modifying enzymes, and the DNA methylating mechanisms, as suggested by Costa [3]

Plants respond to internal and external stresses by altering expression of specific genes involved in the response. A significant fraction of plant genomes is made of repeated DNA sequences and transposable elements (TE) which, if activated, may cause genome malfunction and instability. NcRNAs are involved in coordinating genome function and in keeping TEs silent. RNA-mediated silencing is an evolutionarily conserved mechanism through which double-stranded RNAs (dsRNAs) induce inactivation of cognate sequences. Once established, silent chromatin states can be propagated even in the absence of the initial cues. High-resolution analysis of the *Arabidopsis* exosome revealed an astonishing world of ncRNAs including a novel class of plant RNAs matching the 5′-end of Pol II transcripts (upstream noncoding transcripts) resembling the CUTs (cryptic untranslated transcripts) in yeast and the P-associated short transcripts in mammals [5]. Among the transcripts, many represented precursors for the ~24-nt siRNAs. The latter, referred to as heterochromatic siRNAs, guide RNA-dependent DNA methylation (RdDM) and histone H3 Lysine9 (H3K9) methylation of transposons and heterochromatin-associated repetitive sequences in *Arabidopsis* [6–10].

A number of recent reviews have focused on specific aspects of ncRNAs biogenesis, and on the similarities, differences, and the crosstalk between cellular pathways involving ncRNAs, as well as on their kingdom-specific roles in organismal functions and epigenetic phenomena [4,10–12]. Here, a few plant-specific epigenetic mechanisms including silencing of TEs, heterochromatin formation, and genome re-programming, as well as the phenomena of paramutation and imprinting, will be overviewed through the prism of the small heterochromatic 24 nt siRNAs (Fig. 16.1). The involvement of the PcG/TrxG-related activities in plant development and the presence of dual silencing/activating (H3K27me3/H3K4me3) marks at developmental genes in *Arabidopsis* will be briefly discussed. Due to space limitations, the newly emerging topic of the epigenetic regulation of plant responses to biotic and abiotic stresses will not be covered. However, recently published results and reviews provide insights into epigenetic mechanisms operating in plants under stress [13–16], suggesting that environmental factors may increase genomic flexibility even in successive, untreated generations, increasing, thus, the potential for adaptation [16].

THE SMALL RNAs IN EPIGENETIC REGULATION OF SILENCING IN PLANTS

About a decade ago Fire and Mello (1998) established that in *C. elegans*, dsRNA (termed RNA interference, RNAi) can trigger gene silencing [17]. The next year Hamilton and

Baulcombe (1999) discovered that a critical step in the dsRNA silencing function is its conversion into small interfering RNAs (siRNAs) by the nuclease activity of a dsRNA-specific (RNase III family) ribonuclease (Dicer) [18]. These findings had an enormous impact on the entire field of molecular biology as they outlined unifying features for seemingly disparate processes, like homologous gene silencing, co-suppression, defense against viral infections, transposon-silencing, DNA methylation, heterochromatin formation, paramutation, nucleolar dominance, and imprinting. Small RNA mediated silencing mechanisms may operate at different levels: at the transcriptional level (through chromatin) and at the posttranscriptional and translation levels (through mRNA degradation). It is important to emphasize that the minimal machinery executing the various types of silencing is built by similar, albeit highly specific, activities.

The steps common for all RNAi-involving pathways constitute: (i) formation of a dsRNA; (ii) its processing by a Dicer (DCL) enzyme to shorter (20–30 nucleotides) dsRNA duplexes (bearing a 5′ phosphate and 3′ hydroxyl group with two-nucleotide overhangs at the 3′ ends); (iii) binding of the small RNA duplexes to a protein from the Argonaute (AGO) family; and (iv) targeting of the RNA-induced complex to mRNA (or DNA) guided by the strand complementary to the small dsRNA, called the guide.

Upon their formation, the two-nucleotide 3′ overhangs of the liberated small RNA duplex are methylated by a specific methyltransferase HUA ENHANCER1 (HEN1) protecting the small RNA from polyuridylation and degradation [19,20]. Methylation of the 24-nt siRNAs generated by the RDR2–DCL3–AGO4 pathway (see further below) might be taking place in the Cajal bodies. Co-localization of the 24-nt siRNAs with AGO4, Pol IVa, RDR2, and DCL3 in the Cajal bodies indicated that multiple steps in siRNA biogenesis were coupled *in vivo* [21].

The small RNAs are grouped in two categories based on the mode of their biogenesis: microRNAs, miRNAs, and small interfering RNAs, siRNAs. miRNAs are generated from single-stranded RNA transcripts (transcribed from *MIR* genes) and have the ability to fold back onto themselves to produce imperfectly double-stranded stem loop precursor structures [see Refs 22–25]. siRNAs are processed from long, perfectly double-stranded RNA. siRNAs can be both exogenous and endogenous in origin and provide an epigenetic component of chromatin structure, gene silencing, and resistance against biotic and abiotic stresses.

In plants, several classes of siRNAs derived from distinct loci are: *cis*-acting siRNAs (*casi*RNAs), representing the most abundant endogenously produced siRNAs in plants; *trans*-acting siRNAs (*tasi*RNAs), generated by the convergence of the miRNA and siRNA pathways in plants; and natural antisense transcript-derived siRNAs (*natsi*RNAs), produced in response to stress. *natsi*RNAs are generated from a pair of convergently transcribed RNAs: typically, one transcript is expressed constitutively, whereas the complementary RNA is transcribed only when the plant is subjected to stress [11–14,26,27].

CHROMATIN-BASED EPIGENETIC MECHANISMS OPERATING IN TRANSCRIPTIONAL GENE SILENCING (TGS)

Despite the fundamental similarities found in all eukaryotes using small RNA regulatory mechanisms, kingdom and species-specific characteristics have evolved to satisfy unique needs. For example, plant cells have evolved pathways upstream of Dicer and downstream of AGO to recognize repeated DNAs and methylated sequences. Instead of cleaving mRNA, AGO4-siRNA complexes recruit chromatin-modifying enzymes. Some plant-exclusive features of chromatin-based epigenetic mechanisms operating at transcriptional gene silencing (TGS) will be discussed. The main components of these mechanisms were identified by forward genetic screen analysis of *Arabidopsis* mutants impaired in TGS. These include DCL3 (DICER3), AGO4 (ARGONAUTE4), Pol IV (RNA-POLYMERASE IV), Pol V, RDR2 (RNA DEPENDENT RNA POLYMERASE2), DRD1 (DEFECTIVE IN RNA-DIRECTED

DNA METHYLATION1) and DRM2 (DOMAINS REARRANGED METHYLTRANSFERASE2). Together, these activities control the accumulation of endogenous 24-nt siRNAs [28]. Several of the proteins involved in the biogenesis of the 24-nt siRNAs are genetically redundant, whereas others have specialized roles.

The Dicers

The sources for dsRNA are variable and dsRNAs can serve as precursors of different classes of small RNAs. Nonetheless, the specific enzyme activity degrading dsRNA, DCL (Dicer), is an ancestrally conserved function of the RNAi machinery [23]. Most animals encode a single Dicer (insects encode two) but in *Arabidopsis* the *DICER-LIKE* genes have proliferated to four (*DCL1–DCL4*). The founding member (*DCL1*) of the plant-specific RNase III-like endonuclease family was identified in a mutant line, carpel factory (caf), displaying floral meristem and organ morphogenesis defects [29]. DCL1 cleaves endogenous dsRNAs to produce both siRNAs and miRNAs; DCL2 and DCL4 process dsRNA precursor into 21- and 22-nt siRNAs and upon combining with AGO proteins guide degradation of homologous RNA in posttranscriptional gene silencing (PTGS) [30–32]. DCL3 is the nuclease involved in all known chromatin-dependent TGS events in plants producing the 24-nt siRNAs (heterochromatic siRNAs) that bind AGO4.

The ARGONAUTE (AGO) proteins were named after the characteristic squid-like phenotype of *ago1* mutant *Arabidopsis* plants. A founding member of the *AGO* gene family, *AGO1* plays pleiotropic roles in embryonic development, cell differentiation, maintenance of stem identity, and organ polarity [33]. ARGONAUTE (AGO) proteins are the integral players in all known small RNA-directed regulatory pathways. The AGO family members segregate into three sister groups based on their phylogenetic relatedness and capacity to bind a particular class of small RNAs. Members of Group 1 (called the AGO proteins) bind to both miRNAs and siRNAs; Group 2 members (the PIWI proteins) interact with piRNAs, and Group 3 members (described only in *C. elegans*) bind to secondary siRNAs. Plants encode only AGO (Group 1) proteins [23,34].

All ARGONAUTE proteins carry conserved PAZ, MID, and PIWI domains at the C-terminal and variable N-terminal domains. The PAZ domain recognizes and binds the 3′ end of small RNAs, the MID domain binds to the 5′ phosphate of small RNAs, and the PIWI domain adopts a folded structure similar to that of RNaseH enzymes exhibiting endonuclease (slicer) activity [35]. The PIWI domains specifically interact with GW (glycine-tryptophan) repeat-containing partner proteins [36].

Among the ten *Arabidopsis* AGO proteins [23,34] slicer activity has been demonstrated for AGO1, AGO4, and AGO7. Only AGO4 and AGO6 operate in the DCL3-siRNAs TGS pathway and may be partially redundant [9,37,38]. The roles of the other AGO proteins are less clear. Although *AGO9* and *AGO8* belong in the same sister group as AGO4, mutations in *AGO9* and *AGO8* did not display obvious developmental defects or aberrant small RNA levels [34,39].

Some processes upstream of DCL and downstream of AGO involve unique plant specific proteins. For example, RNA-dependent polymerases (RDRs), Chromomethylase3 (CMT3), and RNA polymerases, Pol IV and Pol V, participate in the process of RNA-directed DNA methylation (RdDM).

RNA-Dependent RNA Polymerases (RDRs)

These enzymes produce dsRNAs used as substrates for DCLs in various small RNA pathways. Studied initially in plant antiviral defense (reviewed in Ref. 4) the RDRs participate in a number of endogenous functions beyond cellular defense. Heterochromatin structure, gene expression, and silencing of transposable element involve RDRs. These enzymes may initiate the RNAi pathway by producing the trigger dsRNA or may enhance the RNAi response by amplifying the amount of dsRNA.

In *Arabidopsis*, RDR2 generates dsRNA from single-stranded transcripts either by de novo second-strand synthesis from "aberrant" RNA templates (presumably lacking a 5′ cap or a polyA tail) or by using siRNAs as primers to synthesize RNA complementary to the target mRNA. DCL3 cleaves the dsRNAs to generate the 24-nt siRNAs. Hundreds of thousands of RDR2-DCL3-dependent 24-nt siRNAs mapping to heterochromatic regions containing DNA repeats, transposons, or silent euchromatin (the heterochromatic siRNAs) have been identified in *Arabidopsis thaliana*, rice, and tomato [40].

Many species outside plants and fungi do not have an RDR despite utilizing the RNAi machinery [23]. Animals, with the notable exception of *C. elegans*, do not have *RDR* genes and *S. cerevisiae* also does not carry *RDR* genes; it is noted that this species is devoid of the RNAi mechanism altogether. Absence of RDR activity in these species indicates that long dsRNA can derive from various sources, such as simultaneous sense and antisense transcription by Pol II or single-stranded RNA transcribed by RNA Pol II from inverted repeats and can form double-stranded hairpin RNAs after mono-directional transcription. *Arabidopsis* and rice have six identifiable RDRs, three of which (RDR3a, RDR3b, and RDR3c) form a distinct phylogenic clade for which no function has been established. The other three, RDR1, RDR2, and RDR6 have direct orthologs in many plant species and contain the catalytic DLDGD motif; all three function upstream of DCL closely linked to both DCL and AGO [41]. Transcriptional silencing of transposons and repeats in the nucleus typically involves DCL3 and AGO4 downstream of RDR2 [8–10]. The accumulation of RDR2-dependent siRNAs is linked to DNA methylation in RNA-directed DNA methylation (RdDM).

RNA-Directed DNA Methylation (RdDM) of Plant Genomes

The first indication that RNA could direct methylation of DNA came from observations that viroid RNA injected in plant cells caused *de novo* cytosine methylation of homologous genomic DNA sequences [42]. In plants, the RNA-directed DNA methylation (RdDM) pathway controls the establishment of DNA methylation at three sequence contexts (CG, CHG, and CHH) [43]. Three DNA methyltransferases cooperate to establish the genome methylation profile: CHROMOMETHYLASE3 (CMT3) and DOMAINS REARRANGED METHYLTRANFERASE2 (DRM2) produce *de novo* cytosine methylation, whereas the maintenance methylase, MET1, controls the symmetrical CG methylation on both DNA strands [44].

The *CMT3*-like genes, specific to the plant kingdom, encode methyltransferase proteins containing a chromodomain [45]. Through the chromodomain CMT3 binds dimethylated lysine 9 on histone H3 (H3K9me2) and together with SUVH4 (the activity that establishes the H3K9me2 mark, known also as KRYPTONITE, KYP) CMT3 generates a feedforward loop maintaining CHG methylation [46]. siRNAs guide CMT3 to sequences targeted for non-CG methylations and loss of *CMT3* function causes a large decrease in CHG methylation (and to a lesser extent in CHH) [47–50].

The *DRM* genes share homology with the mammalian *Dnmt3* genes encoding *de novo* methyltransferases [51]. At CG sites, the *DRM* genes are required for the establishment, but not for the maintenance, of preexisting CG methylation. DRM is guided to the targeted sequences by siRNAs and may act redundantly with CMT3 to establish and maintain CHG and CHH methylations (reviewed in Ref. 46).

Symmetric (CG) methylation is achieved by MET1 with the help of DECREASED DNA METHYLATION1 (DDM1) and VARIANT IN METHYLATION1 (VIM1). DDM1 is a chromatin-remodeling factor from the SNF2-family of ATPases. VIM1 is an unconventional methylcytosine-binding protein that is enriched at methylated genomic loci and at chromocenters. It binds to hemimethylated CG through the SRA (SET- and RING-associated) domain. Binding of SRA stabilizes the interaction and prevents sliding [52]. VIM1 can also bind to histones and it was suggested that VIM1 participates in methylated DNA-nucleosome interactions to maintain centromeric heterochromatin [53]. Loss of *MET1* or *DDM1* causes massive genome demethylation, transposon reactivation, and stochastic developmental defects [54,55]. Some results suggested that, once lost, CG methylation in plants could not be restored with fidelity [56]. However, the robust and specific restoration of the CG-methylation observed for the *Arabidopsis* centromeric repeats and transposons mediated by RNAi reported recently [57] challenged this view (see further below).

Although sounding paradoxical, siRNAs may also guide DNA demethylating activities [58,59]. The DNA glycosylase-lyase protein REPRESSOR OF SILENCING 1 (ROS1) can remove methylated residues and screens for suppressors of *ros1* mutations have identified RdDM factors; the protein ROS3 may guide demethylation by ROS1 [60]. The interplay between siRNA-directed DNA methylation and demethylation pathways might be required for the balance between the two epigenetic states.

High-resolution mapping of cytosine methylation in *Arabidopsis* confirmed previous reports that DNA in the centromeric regions and in repeat sequences was highly methylated [55,61,62]. In addition, it revealed unexpected patterns in the coding regions: less than 5% of genes are methylated at their promoters but about 30% are methylated in their open reading frames [63,64]. Unlike their mammalian counterparts, plant introns are almost completely devoid of TEs and clusters of dense CG methylation are accumulated at exons, but deficient in introns [65]. These patterns were found in highly transcribed and in constitutively expressed genes, whereas genes displaying lower level and tissue-specific expression patterns had methylated promoters [62]. These DNA methylation profiles contrast with the distribution of methylated cytosines in mammalian genomes where the CG islands in gene promoters are hypomethylated [66].

The gene body methylation in plants is almost exclusively restricted to CG, in marked contrast to the methylation of CG, CHG, and CHH sites typically seen at repeated sequences. Gene body methylation may result from two conflicting activities: one imposing it at CG sites, and one preventing extension to CHG sites. Importantly, the latter activity is not targeted towards silent transposable elements and is likely coupled to transcription elongation, suggesting that CHG methylation hinders this step [67]. According to a model, transcription of genes by Pol II attracts in its wake the maintenance DNA methyltransferase MET1 as well as a H3K9 methyltransferase activity. Gene transcription could also recruit the JmjC-domain containing histone demethylase, IBM1, which by demethylating H3K9 would prevent its recognition by the chromodomain CHG methyltransferase CMT3. Thus, targeting of DNA methylation seems to differ significantly for genes and TEs, despite the fact that many factors are shared by these two processes [68].

Pol IV and Pol V

In *Arabidopsis*, the RdDM machinery involves two plant-specific RNA polymerases, Pol IV and Pol V. Their largest subunits (NRPD1 and NRPE1, respectively) are related to the largest subunit of Pol II (RPB1) but Pol IV and Pol V function exclusively in the RNA-driven silencing pathway.

Pol V can generate uncapped and nonpolyadenylated transcripts from noncoding sequences that are targeted by RdDM. Pol V transcripts originate from intergenic noncoding regions triggering the siRNA-pathway. The subsequent chromatin modifications established *via* the siRNA-directed machinery impede transcription of adjacent regions by Pol II and Pol III [69]. In an *nrpe1* mutant, Pol V-generated transcripts disappear and methylation is lost, allowing uni- and/or bi-directional transcription by Pol II and Pol III. These findings suggest a unique mode for chromatin-based gene silencing based on Pol V generated transcripts [70] (see Fig. 16.2). The model is supported by the pervasive intergenic transcription found in eukaryotic genomes [71].

Pol IV uses the genomic DNA as a template to produce a single-stranded RNA transcript, which is then converted to dsRNA (by RDR2) to be used as a substrate by DCL3. Endogenous loci producing the 24-nt class of chromatin-targeting RNAs are dependent on Pol IV and on RDR2 [72,73]. Pol IV may directly transcribe a methylated DNA template, producing an aberrant (improperly processed or terminated) RNA that is copied by RDR2 to dsRNA precursors of siRNAs that trigger methylation [74,75] (Fig. 16.2A).

Subunits and Partners of Pol IV and Pol V

The N-terminal portions of NRPD1, NRPE1, and RPB1 containing the catalytic domains are conserved in Pol II, Pol IV, and Pol V. By contrast, their C-terminal domains (CTDs) differ in a very significant way: the CTD of RPB1 contains a reiterated heptapeptide recruiting proteins that process nascent RNA and catalyze histone modifications associated with Pol II transcription [76]. The NRPE1 of Pol V contains the WG/GW repeats, which can specifically interact with AGO4 [38]. Thus, the unique CTDs in the largest subunits of Pol II and Pol V attract different factors for different transcription functions [75].

Pol IV and Pol V share the same second largest subunit (NRPD2/NRPE2) and a smaller subunit, RDM2, similar in sequence to the Pol II subunit RPB4 [77]. Despite similarity in

(A)

de novo methylation

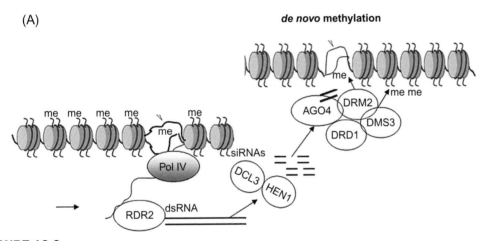

257

FIGURE 16.2

AGO4-siRNAs complexes involved in chromatin modifications. (A) A model for *de novo* DNA methylation involving Pol IV transcription as suggested by Matzke et al. [80]. The role of Pol IV is to produce single stranded RNA transcripts to be used as substrates by RDR2. Pol IV may transcribe from methylated DNA (as illustrated): DRM2 establishes new methyl groups at DNA sequences complementary to the small RNA loaded onto AGO4. The single stranded RNA produced from methylated DNA by Pol IV is used as a template for a dsRNA synthesis by RDR2 triggering the 24 nt siRNA pathway. dsRNA is processed by DCL3 and HEN1 into small 3′-end methylated siRNAs. The 24 nt siRNAs guide the AGO4 complex containing DRM2/DRD1/DMS3 to homologous genomic sequences. DRD1, a putative SNF2-like chromatin remodeler, and DMS3, an SMC-hinge domain-containing protein are accessory subunits of the complex [80]. (B) A model for spreading of silent chromatin and inhibiting Pol II activity through Pol V transcription, according to Wierzbicki et al. [85]. siRNAs and Pol V transcripts are produced by two independent pathways that collaborate to silence genes and to block Pol II activity. Pol V transcribes noncoding sequences enabled by DRD1 and DMS3. AGO4-siRNA complexes originated in a separate pathway recognize target loci by pairing with Pol V generated transcripts (see text). AGO4 recruits also DNA and histone modifiers (see panel C) to generate heterochromatin. The mechanism of recruiting chromatin modifiers is not clear. (C) AGO4-siRNA complexes in histone modifications and in DNA methylation establishing and propagating silenced chromatin. Once at a target locus, AGO4 and siRNA complexes might recruit several different chromatin-modifying enzymes to effect gene silencing. The order of action of these chromatin-modifying enzymes is not known, and their relative importance for gene silencing might be locus-specific. (I) Establishing the silencing H3K9me2 mark: SUVH4/KYP cooperates with the AGO4 complex to establish H3K9me2 according to [8–10,37,93]. (II) Removal of activating marks: LDL enzyme brought about by the AGO4-siRNA complex demethylates H3K4me3; de-ubiquitination of ubiquitinated H2B (H2Bubi) by the ubiquitinase SUP32 recruited and targeted by AGO4-siRNA [94]. (III) Establishing the CNG methylation: guided by homologous RNAs, AGO4 recruits the DNA methyltransferase CMT3 to produce CNG methylation at target loci [43,46,48]. (Please refer to color plate section)

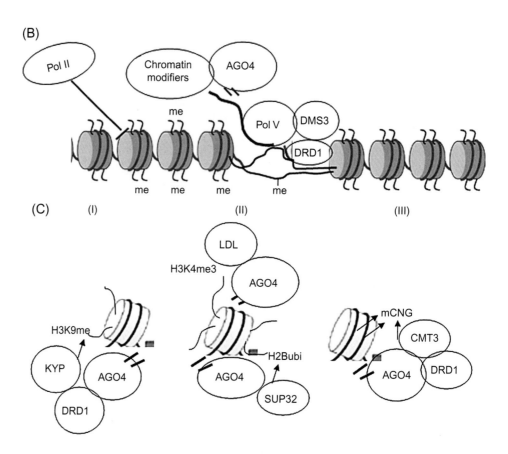

FIGURE 16.2
(*Continued*)

amino acid sequence, RDM2 has evolved as a Pol IV and Pol V specific subunit that does not function in Pol II complexes. Furthermore, the subunit differences have been localized to regions important for template entry and RNA exit points [78] suggesting that Pol IV and Pol V are evolutionarily derivatives of Pol II specialized for generating and/or using noncoding RNAs for chromatin-based gene silencing [70,75,79–81]. As a result, plants have remarkably diversified their transcriptional machinery enhancing their ability to modulate the epigenetic states of their genomes during transcription.

The SNF2-like ATPase nucleosome-remodeling factor DEFECTIVE IN RNA-DIRECTED DNA METHYLATION (DRD1) functions with Pol IV and Pol V complexes [75,82]. The cloning of a maize SNF2 ATPase protein (RMR1) related to DRD1 showed that a chromatin remodeling activity was required for paramutation at the *purple plant* (*pl1*) gene locus [83] (see further below).

DEFECTIVE IN MERISTEM SILENCING 3 (DMS3) is a protein consisting solely of the hinge domain region found in the structural maintenance of chromosomes (SMC) proteins, known to be involved in chromosome architecture [84]. DMS3 and DRD1 are involved in the assembly of Pol V initiation complex [85].

How are AGO and siRNAs Targeted to Specific Chromatin Sites?

How repeated sequences are selected for siRNA production and how the complex is recruited to the DNA to be silenced is still unclear. Lack of biochemically-tractable initiation-of-silencing functional assays does not allow following the AGO-siRNA targeting steps in the context of intact chromatin. However, recent studies have provided groundbreaking insights into the sequence of the events. In *Arabidopsis*, targeting of DRM2 depends on AGO4-bound siRNAs as a guide [6,9,39]. An RdDM effector, KTF1, with similarity to the transcription elongation factor SPT5, contains a C-terminal extension rich in GW/WG repeats. KTF1 co-localizes with ARGONAUTE 4 (AGO4) in nuclear foci and binds AGO4 and RNA transcripts. Thus, KTF1 acts as an adaptor binding transcripts generated by Pol V and as a recruiter for AGO4-bound siRNAs to form an RdDM effector complex [86]. Chromatin-targeting siRNAs might base pair directly with unwound genomic DNA or could bind to a locus by RNA–RNA interaction with a nascent single-stranded transcript. Base pairing between AGO4-associated siRNAs and nascent Pol V transcripts has been observed indeed, supporting a recent model according to which AGO4 is recruited to target loci by Pol V transcription [85]. Furthermore, siRNAs and Pol V transcripts are produced by two independent pathways that collaborate to promote heterochromatin formation and gene silencing [85]. In one pathway, Pol IV, RDR2, and DCL3 produce 24-nt siRNAs that associate with AGO4; in a separate pathway, DRD1 and DMS3 enable Pol V transcription of noncoding sequences at target loci. siRNA–AGO4 complexes are guided to target loci by interacting with Pol V transcripts. The proposed direct interaction between AGO4 and Pol V [70] has not been detected *in vivo* suggesting that AGO4 recruitment to chromatin is primarily an RNA mediated process although protein–protein interactions are not excluded [85]. Interestingly, in *S. pombe*, heterochromatin formation is achieved by the RITS complex guided to chromatin *via* associations with nascent Pol II transcripts [85a], suggesting that plants and yeast use similar RNA guidance mechanisms for recruiting Argonaute-containing transcriptional silencing complexes to target loci. Apparently, the unique RNA polymerase, Pol V, has evolved in plants for the specialized role of generating noncoding RNAs that can serve as scaffolds for Argonaute recruitment [85] (Fig. 16.2B).

Furthermore, AGO4 can function through two separable mechanisms: by recruiting components that signal DNA methylation independently of its catalytic activity and/or by the catalytic activity required for the generation of secondary siRNAs reinforcing its repressive effects [87].

The *SINE*-related tandem repeat in the promoter of *FLOWERING WAGENINGEN* (*FWA*) gene provides an example for the involvement of repeated sequence in the production of siRNA. The siRNAs, then, recruit RdDM to an unmethylated *FWA* promoter *in trans* to silence *FWA* in vegetative tissues. However, the tandem repeat is dispensable as an *FWA* locus without repeats could also recruit siRNA-producing proteins [88,89]. Likewise, many transposons in heterochromatin do not contain tandem repeats suggesting that additional signals guide RdDM to silent heterochromatin. Furthermore, repeat-independent siRNA production could result also from aberrant RNA processing of very highly transcribed transgenes (a phenomenon termed sense post-transcriptional gene silencing (S-PTGS) [90]. Additional signals include DNA methylation, modified histone marks, and proteins that "read" the marks and recruit the siRNA complex by interaction with its components. For example, the two SRA domain factors (SUVH9 and SUVH2) bind methylated cytosines and are essential both for *de novo* and for maintenance DNA methylations: SUVH9 preferentially binds asymmetric, while SUVH2 preferentially binds symmetrically, CG methylated sites [91]. The methylcytosine binding domain proteins MBD6 and MBD10 act in conjunction with RdDM to effect large-scale silencing of ribosomal DNA loci in the phenomenon of nucleolar dominance [92].

Non-CG methylation of developmental genes can be readily restored after it is lost, suggesting that targeting signals exist and remain in cells in the absence of DNA methyltransferase [44]. For the CNG methylation at the *SUP* locus controlled by CMT3, this signal may come from H3K9me2 (established by SUVH4/KYP) [50,93]; at other loci, only the H3K9me2 or a combination of pathways involving siRNAs may be used, whereas at the *Ta3* locus, CMT3 can propagate CNG methylation without siRNAs or DRD1 [44,88]. Histone H3 lysine4 (H3K4) demethylation helps *de novo* DNA methylation of an *FWA* transgene and histone H2B de-ubiquitination regulates gene silencing *via* siRNAs [94] (Fig. 16.2C).

TRANSPOSABLE ELEMENTS (TEs) AND HETEROCHROMATIN

DNA methylation is an epigenetic mark associated with silencing of TEs constituting about 45% and 75% of the human and maize genomes, respectively [95]. Their potential to transpose may cause significant damage to the host genome. Consequently, eukaryotes have evolved mechanisms, including epigenetic RNAi mediated chromatin modifications, to respond to the genome invaders and to suppress their activity. Among the earliest recognized mechanisms for keeping TEs silent is their sequestration in domains of constitutive heterochromatin. DNA methylation, histone modifications associated with silenced states, and condensed chromatin packing are features characteristic of heterochromatin found at the centromeres and the telomeres of eukaryotic chromosomes. The accumulation of TEs in regions essential for genome integrity suggests that silencing of TEs activity is critical for proper chromosome function. On the other hand, employment of TEs in essential cellular functions indicates that eukaryotes have "learnt" to not only silence, but also to take advantage of their presence. The DNA of centromeric heterochromatin in its condensed state is essential for the recruitment of the cohesin complex mediating sister-chromatid cohesion [96]. TEs are also a source of genetic and epigenetic material that can be utilized by the host to control chromatin structure, gene activity, phenotypic diversity, paramutation and imprinting.

Presence of endogenous centromeric repeats is important but not sufficient to guarantee kinetochore formation [reviewed in Ref. 97]. The establishment and maintenance of centromeric chromatin (characterized by the centromere-specific histone H3 variant, CENP-A, as the key determinant of centromere identity and the location of kinetochores) is epigenetically regulated. In the fission yeast *S. pombe*, the centromeric heterochromatin is maintained by non-coding transcripts from the outer (pericentromeric) repeats. These transcripts are processed into small interfering RNAs (siRNA) targeted to homologous sequences; together with recruited heterochromatin proteins, siRNAs are essential for the establishment of CENP-A centromeric heterochromatin. The Argonaute-associated proteins, Chp1 and Tas3, and the Suv39 and HP1 homologs, Clr4 and Swi6HP1, are required for establishing the centromeric heterochromatin [98]. Once assembled, CENP-A chromatin is propagated by epigenetic means in the absence of heterochromatin. These studies identified an important, potentially conserved, role for RNAi in directing centromere and kinetochore formation [98].

In *Arabidopsis*, the heterochromatin is located mainly at the centromeres, represented by retrotransposons interspersed among arrays of satellite repeats, at the pericentromeric regions composed mainly of DNA transposons, and in the knobs representing jumbled TE islands triggering RNAi-mediated silencing, perhaps through read-through transcription (AGI, 2000). The 180 base pairs centromeric methylated satellite repeats are found in thousands of tandem copies. However, in mutants lacking either MET1, or the histone deacetylase HDA6/SIL1, or the chromatin remodeling ATPase DDM1, the silenced TEs and satellite repeats were reactivated causing de-condensation of centromeres. Silencing lost in *met1* or *hda6* was reestablished in backcrosses to wild type, but silencing lost in RNAi mutants and *ddm1* was not. 24-nt siRNAs corresponding to the centromeric sequences were retained in *met1* and *hda6*, but not in *ddm1*, suggesting that siRNAs are guiding DDM1 for the silencing of centromeric DNA [7,46,99]. It should be noted that *S. pombe* lacks DNA methylation and plants do not have homologs of Argonaute-associated proteins Chp1 and Tas3 suggesting a convergent evolution of the siRNAs chromatin targeting pathways in *S. pombe* and in plants.

In grasses, a *Ty3/gypsy* class of centromere-specific retrotransposons are conserved and highly enriched in domains associated with CENP-A and with the flanking centromeric satellite DNAs [100]. These elements are actively transcribed in maize suggesting that deposition of centromeric histones might be a transcription-coupled event [101]. Transcribed repeats silence the retrotransposons and transcripts from retrotransposons help silencing the repeats suggesting roles for siRNAs in the evolution of centromeres [102].

In addition to DNA methylation, constitutive heterochromatin in *Arabidopsis* is marked by "repressive" histone modifications, including histone H3 dimethylation at Lys9 (H3K9me2) and methylation at Lys27 (H3K27me). H3K9me2 overlaps almost exclusively with transposons and other repeats, while H3K27me3 is associated mostly with inactive euchromatin. Nontranscribed genes may be associated with H3K27me3, H3K4me1, and H3K4me2 [55,103–108]. H3K9me2 and H3K27me1 are mediated by SUVH2, SUVH4 (KYP), SUVH5, and SUVH6, and *ARABIDOPSIS* TRITHORAX-RELATED PROTEIN 5 (ATXR5) and ATXR6 [109,110] (for a review on plant heterochromatin proteins see Ref. 110 and ref. therein).

Little is known about histone methylation in large-genome plants, which make up the bulk of the angiosperms. Combining high cytological resolution of maize pachytene chromosomes, three-dimensional light microscopy, and the ability to quantify staining patterns relative to cytological features, Shi and Dawe [111] reported that each methylation state identified different regions of the epigenome: H3K27me2 marks classical heterochromatin, H3K4me2 is limited to areas clearly demarcating the euchromatic gene space, while H3K9me2 and H3K27me3 occur in euchromatic domains; H3K9me3 is associated with centromeres and H4K20me2/3 is nearly or completely absent in maize. H3K9me2 appears excluded from repeats and associating with genes but does not overlap with either H3K27me3 or H3K4me2 [111]. Apparently, the presumed epigenetic code has the capacity to evolve along with changes in genome structure.

CG methylation provides distinct and direct information for a specific subset of histone methylation marks illustrating a crosstalk between DNA methylation and histone modifications (Fig. 16.1). CG methylation characteristic of heterochromatin specifically prevented H3K27 trimethylation but H3K27 mono- and dimethylation label silent heterochromatin independently of DNA methylation [108].

POSITION EFFECT VARIEGATION (PEV)

Heterochromatin can epigenetically influence the expression of nearby genes causing variegated phenotypes in genetically identical cells. The PEV phenomenon, described by H. Muller in 1938, is illustrated by the *Drosophila* gene, *white*, which shows unstable expression when moved in close proximity to heterochromatin. In plants, variegated gene expression has been reported in *Oenothera blandina* after X-ray chromosomal disruptions and translocations [112,113]. A phenomenon similar to PEV is observed when transgenes are inserted into heterochromatic regions in fission yeast [114] and proximity to TEs might be responsible for the silencing effects. Furthermore, gene screens for suppressors/enhancers of PEV have identified the same factors as those involved in TE silencing and heterochromatin structure, including the RNAi-mediated silencing implicating siRNAs in PEV [115].

Heterochromatin can spread linearly to about 10 kb until it encounters a boundary. TEs can function as nucleation centers for facultative heterochromatin, as well as insulators [116]. RdDM spreading is associated with the production of secondary siRNAs, which originate outside the region targeted by primary siRNAs [74,80]. An interesting example is that the spreading of methylation and siRNAs from a *LINE* element into the adjacent gene (*BONSAI*) is dependent on the chromatin remodeler DDM1 and a histone demethylase (INCREASE IN BONSAI METHYLATION1, IBM1) [54,117] but is not mediated by DRM2 or other components of the RNAi-dependent *de novo* methylation machinery [67].

EPIGENETIC VARIATIONS AND PLANT EVOLUTION

Because of their polymorphic locations and abilities to spread epigenetic marks, TE can influence transcription of nearby genes or cause readthrough, which would be subjected to silencing. Thus, TEs can produce phenotypic variability by forming epialleles that are metastable in nature and have variegated expression that resembles PEV. Epialleles are formed due to the proximity of a gene to a TE, and are regulated by the epigenetic mechanisms that the TE recruits. Natural epigenetic variation can originate from polymorphisms in transposon insertions and repeats, as illustrated by the siRNA-producing loci and DNA methylation in different *Arabidopsis* species and ecotypes of *A. thaliana* [118]. A Mutator element in the first intron of the floral repressor FLOWERING LOCUS C (FLC) in the ecotype Landsberg erecta (Ler) but not Columbia (Col) is a target of siRNAs that induce histone modifications and flowering time differences [119]. A HAT transposon in the promoter of *FLC* produces abundant siRNAs in Ler but not in Col, resulting in hypermethylation of the promoter only in Ler [120]. These results illustrated the role of the siRNAs-silencing pathway on the evolution of flowering time strategies and speciation. *FWA* epigenetic mutants cause a heritable late-flowering phenotype due to ectopic expression of the *FWA* gene in vegetative tissue. The expression level is heritable but variable within the examined 21 *Arabidopsis* accessions that have two direct repeats at the promoter [89].

A naturally-occurring mutant of *Linaria vulgaris* displaying a strikingly different flower phenotype with radial symmetry instead of bilateral, results from a different expression of the *Lcyc* gene. The *Lcyc* gene controlling flower symmetry is extensively methylated and transcriptionally silent in the radial flower mutant. The modification co-segregates with the mutant phenotype, is heritable, and occasionally reverts phenotypically during somatic development, correlating with demethylation of *Lcyc* and restoration of gene expression [121]. A spontaneous epimutation was identified at the *Colorless non-ripening, Cnr,* locus, a component in the regulatory network controlling tomato fruit ripening [122]. Mounting evidence is suggesting that epialleles and epigenetic mutations might play a more significant role in plant developmental processes, in the generation of natural variation and plant evolution, than has hitherto been suspected. The consequences of transgenerational epigenetic effects driven by *cis-* and *trans-*acting effects, chromatin modifications, RNA-mediated pathways, and regulatory networks modulating differential expression of homologous genes in polyploids might also facilitate adaptive evolution of polyploid plants and domestication of crops [16,123,124]. Some epialleles may undergo paramutation, a *trans-*interaction between alleles that induce heritable expression change in one allele.

Paramutation

Paramutation is one of the best-known examples of non-Mendelian inheritance arising from allelic interactions that lead to meiotically heritable gene silencing. Because changes in gene expression are not associated with changes in DNA sequence, paramutation is a classical example of an epigenetic phenomenon. Among the best-studied examples are the four maize loci, *Pl1, p1, r1,* and *b1,* involved in the anthocyanin synthesis pathway. *B′* and *Pl′* display strong paramutation strengths, while *r1* and *p1* are variable [125,126]. All encode transcription factors that increase pigmentation in the plant and a paramutation event at any of these loci is reflected by a decrease in coloration, providing an easily tractable system to study the phenomenon [127]. The allele that is capable of affecting (silencing) expression from the homologous locus is referred to as the paramutagenic allele and is usually expressed at low levels; by contrast, the affected (paramutable) allele is highly expressed. The paramutable *B-I* allele is transcribed at about a 20-fold higher level (providing for the strong red coloration of plants) than the paramutagenic *B′* allele found in colorless plants [128]. Combining the two alleles (in *B-I/B′* heterozygote) results in silencing of *B-I* transcription with 100% penetrance. The most remarkable feature, however, is that the silenced *B-I* allele acquires paramutagenic capability itself, effectively silencing transcription from other homologous alleles in subsequent generations – secondary paramutation ("the vampire" effect).

The molecular basis of paramutation has been a long-standing enigma but recent studies have provided breakthrough insights. Fine structural analysis revealed that an array of seven direct tandem 853-base repeats located ~100 kb upstream of *b1* are necessary for the paramutagenicity of *B'* and the ability of *B-I* to undergo paramutation [129,130]. The sequences are present as a single copy in neutral *b1* alleles, while alleles with three repeats show partial paramutational ability [129–132]. Thereby, the presence and the numbers, as well as the organization of the repeats (as observed at the *p1* locus, [133]), are essential for paramutation.

A critical question is how the silencing information is communicated. Some maize mutants deficient in the establishment and/or maintenance of paramutation at distinct loci are defective in genes encoding RdDM factors: a genetic suppressor of paramutation, *Mediator of Paramutation1* (*MOP1*), encodes the maize ortholog of RDR2; *Required to Maintain Repression*, *RMR1*, encodes an SNF2-like putative chromatin-remodeling factor related to DRD1 and CLSY1, and RMR6 is the largest subunit of Pol IV (RPD1). All of these factors are required for siRNA production, for DNA methylation at the silent epialleles, and for paramutation at the *purple plant* (*pl1*) locus [83,134]. MOP1 is needed for paramutations at all four loci [126]. The *mop1* gene is also required for silencing transgenes and *Mutator*-like transposons in maize [134]. The role of the chromatin remodeler RMR1 is not exactly known but it may act as a co-factor for Pol IV and, thus, be involved in interactions between siRNAs and chromatin [80].

RMR6, the largest subunit of the maize Pol IV (ZmRPD1), is required for both paramutation and for normal maize development [135]. ZmRPD1 is essential for accumulating the majority of 24-nt siRNAs indicating that it operates at repetitive DNAs. However, the biochemical function of ZmRPD1 remains unclear as it failed to produce detectable RNA transcripts for genomic regions represented by those siRNAs [133]. Furthermore, the primary polymerase for repetitive DNAs, including hypermethylated and repressed transposons in maize, seems to be Pol II [83] and not Pol IV, as in *Arabidopsis*. Transposon sequences proximal to Pol II templates could interfere with Pol II-dependent RNA synthesis, resulting in the production of abnormal Pol II transcripts, which could trigger the siRNA pathway [133].

Although essential for paramutation, the RNAi machinery is not the only factor. For example, the 853 nt repeats are transcribed from both *B'* and *B-I* loci indicating that siRNAs, alone, are not sufficient to establish paramutation. Recruitment of siRNA machinery to a locus is not always sufficient for the RNA-directed DNA methylation either. For example, differences between the silent (*FWA*) and the unmethylated (*fwa-1*) epialleles in *Arabidopsis* were not accounted for by siRNA production: the repeat-derived siRNAs accumulate equally in plants with wild-type *FWA*, and those with *fwa-1* and an introduced transgene can silence an unmethylated *fwa-1* endogenous gene [88,89].

How interallelic transfer of epigenetic information is achieved remains to be determined. Some models propose *trans*-communication between epialleles, including physical pairing of alleles and transfer of silencing complexes, histone modifiers, nucleosome re-positioning factors, and DNA-methylating activities on the paramutable locus [126]. In some cases, acquisition of DNA methylation accompanies *trans*-inactivation of paramutable alleles as found for the maize *r1*, *p1*, the petunia *A* loci, and *FWA* [126,136,137]. However, cytosine methylation is not the causative factor for establishing the *B'* silent state, despite *B-I* and *B'* alleles having different methylation profiles [129,130]. Thereby, other factors contribute to the epigenetic states and to the ability of certain epialleles to influence homologous sequences both *in cis* and *in trans*. Such factors could be chromosomal location (PEV), ploidy, environmental factors, and histone modifications. For example, the tomato *sulf* locus mapping near heterochromatin experiences silencing effects dependent on ploidy [138] and inactivation of an active transgene *HPT* locus in *Arabidopsis* is observed in a tetraploid but not diploid background [139].

263

IMPRINTING, GENOME RE-SETTING, AND REPROGRAMMING
Imprinting

Differential expression of alleles of the same gene depending on the parent-of-origin (gene imprinting) is thought to have evolved independently in mammals and in flowering plants [140]. Imprinting occurs in the nourishing tissues, the placenta of mammals, and the endosperm of plants. The formation of the endosperm, a process involving a double fertilization of the central cell (CC) by the pollen, is a defining feature of flowering plants. Pollen contains three nuclei: the vegetative nucleus (VN) is in the cell developing into the pollen tube; as it grows, the pollen tube transports the two sperm nuclei to the ovules. One of the sperm nuclei fertilizes the egg, and the third nucleus fertilizes the diploid central cell to form the triploid endosperm. After fertilization, the proliferation of the endosperm ensures nutritional resources for the embryo. In humans, mutations of imprinted genes are associated with developmental disorders and diseases; mutations in plant imprinted genes lead to defective reproduction and loss of viability [142].

Ten imprinted genes are currently recognized in *Arabidopsis*: two encode Polycomb group proteins (*MEDEA, MEA,* and *FERTILIZATION INDEPENDENT SEED2, FIS2*), an RNA-binding protein (*MATERNALLY EXPRESSED PAB C-TERMINAL, MPC*), two encode transcription factors (*FWA* and *PHERES1, PHE1*), and four genes encode class IV homeodomain transcription factors [141–144]. Although imprinted in the endosperm, the *PHE1* gene is set apart from *MEA, FIS2, FWA,* and *MPC* because *PHE1* is expressed from the paternal allele [145,146]. Assuming that genes with endosperm-preferred expression are less methylated at 5′ sequences in the endosperm than in the embryo and that less methylated genes in the endosperm exhibit endosperm-preferred expression, it is estimated that there are ~50 imprinted genes in *Arabidopsis* encoding mainly transcription factors and chromatin-related functions [144].

In mammals, imprinting is reflected by differential methylation of specific sequences in the gametes [147]. In *Arabidopsis*, imprinting is usually due to differences in the epigenetic marks (histone and DNA methylation) on alleles in the central cell, which are maintained in the endosperm [148]. The 5-methylcytosine DNA glycosylase *DEMETER (DME)* is expressed only in the CC before fertilization and demethylates the maternal alleles of imprinted genes establishing methylation asymmetry between embryo and endosperm [144,148]. A subset of Pol IV-dependent siRNAs specifically expressed from the maternal chromosomes was shown to accumulate in the maternal gametophyte and to persist during seed development linking genomic imprinting with RNA silencing mechanisms [149].

Bulk methylation in wild-type endosperm is lower in all sequence contexts compared with the embryo [150]. Genome-wide maize endosperm has 13% less 5-methylcytosine than embryos or leaves [151] and an imprinted gene is less methylated in the CC than in the egg cell or sperm [152]. Transposable elements are more heavily methylated than protein-coding genes, and genes are more methylated within their bodies than at their 5′ and 3′ ends. The reduced CG methylation at repeats and gene-bodies in *Arabidopsis* wild type endosperm was partially restored to levels found in other tissues in the *dme* endosperm, indicating that the CG demethylation is specific to maternal sequences [144]. In contrast to CG, methylations of CHG and CHH were reduced in *dme* endosperm suggesting that DME activity is necessary for up-regulating RNAi-mediated methylation in endosperm and for activating the TEs by demethylating them.

Importantly, the genome-wide CG demethylation of the maternal endosperm genome is accompanied by similarly extensive CHH hypermethylation of the TEs in the embryo revealing that siRNA accumulation in the CC contributes to enhanced methylation and silencing of elements in the egg cell (and later in the embryo). This could happen through siRNA transport, which could be the original force behind the evolution of the central cell demethylation. Thus, RNAi drives a substantial reconfiguration of the methylation landscape in the seed suggesting that imprinting in plants evolved from targeted methylation of TEs to reinforce transposon silencing in the embryo [144,148].

Genome Re-setting and Reprogramming in the Male Gametophyte

A strikingly similar process occurs in the *Arabidopsis* male gametophyte where reactivation of the TEs in the pollen vegetative nucleus (VN) reinforces silencing of the TEs in reproductive sperm cells [153]. Apparently similar mechanisms operate in germ lines for detecting aberrant RNAs and for silencing TE in the embryo [148,149,153]. However, in pollen, TEs are reactivated and transpose, but only in the VN, which accompanies the sperm cells but does not provide DNA to the fertilized zygote. In the gametes, the mutagenic activity of TEs is epigenetically suppressed by siRNAs preventing transmission to the next generation. The expression of the small RNAs coincides with down-regulation of the heterochromatin remodeler *DDM1* and of many TE siRNAs. An unknown DNA demethylase active only in the VN may act to selectively remove DNA methylation from some TEs.

The TEs are transiently activated in a coordinated fashion and down-regulation of the genes responsible for TEs silencing is confined to the VN of pollen. A silenced TE is transiently reactivated in maize pollen as well [153] and TE expression has been noted in the pollen of rice [154] indicating that the reactivation of silenced TEs in pollen is conserved among flowering plants. In contrast, the TEs in terminally differentiated senescing leaf cells are not coordinately reactivated suggesting that TE activation in the pollen VN represents a cell type-specific epigenetic reprogramming that has evolved for a function. Changes in histone H4 acetylation and in histone variants consistent with reactivation of TEs and loss of heterochromatin observed in the pollen from both *Lilium* and *Arabidopsis* [155–157] support a genome-wide reprogramming taking place in pollen.

To answer the question of why epigenetic reactivation of TEs is needed in the VN of pollen Slotkin and co-authors [153] suggest that the relative position of the VN next to the sperm cells is important for the silencing of TEs in the next generation. Interestingly, transposon-related siRNAs (21 nt long from the *Athila* retrotransposon family) are generated and accumulate in pollen and sperm [153]. The authors propose that these 21-nucleotide siRNAs, originating in the VN, travel to the adjacent sperm cells to reinforce silencing. Thus, only those transposons with the potential to be expressed (because they were expressed in the vegetative nucleus) would be targeted by siRNAs in sperm nuclei. Although new transposition events occur in pollen they are not inherited because the transposon activation occurs in the VN, which does not contribute DNA to the zygote akin to the TE reactivation that takes place in the endosperm. During reprogramming in *Drosophila* and mouse germlines, epigenetic marks are first lost and then robustly reset each generation resulting in transient TE expression [158,159]. Subsequent remethylation and silencing in sperm depends on the sperm-specific piRNA [160]. Movement of signals (small RNAs) from germline companion cells into germ cells conserved in insect nurse cells is consistent with speculation that the evolution of the sperm-companion vegetative cells interactions has promoted TE silencing in angiosperms. In contrast to the model suggesting that imprinted genes in the endosperm have evolved from TE-silencing mechanisms (see above), the authors conclude that the molecular mechanisms involved in the permanent silencing of foreign DNA have evolved from mechanisms required for the successful development of an embryo.

Genome Reprogramming During Flowering

Plant developmental and environmental responses involve reprogramming at specific genome loci so that the normal program of plant development is reiterated in each generation. Epigenetic repression of *FLOWERING LOCUS C (FLC)* in winter-annual ecotypes of *Arabidopsis* by prolonged cold (vernalization) ensures that plants flower in spring and not during winter. Flowering is induced by the photoperiod (day length) and/or temperature, which stimulate *FLOWERING LOCUS T (FT)*. FLC represses the activity of *FT* to prevent flowering [161].

The activities of both paternally and maternally derived *FLC* reporter genes are reset after vernalization, but the timing of their initial expression differs. The paternal gene copy is active during early gametogenesis and in the single-celled zygote, whereas the maternal copy is not expressed until the early multicellular embryo stage. In the progeny, the paternally derived *FLC* gene is expressed in the single-celled zygote and through embryo development, but not in the fertilized CC, which generates the endosperm. *FLC* activity during late embryo development is a prerequisite for the repressive action of *FLC* on the flowering of the adult plant [162].

Positioned at the convergence node of at least four distinct pathways that block transition from the vegetative to the reproductive stage, *FLC* gene is repressed by low temperature through changes in *FLC* chromatin. Chromatin based mechanisms involve histone modifications [163–166], replacement by histone variants [167–169], and a Pol IV-dependent antisense RNA initiating in the 3'- region of the gene [170].

Transcription from the *AP1* locus, required for the initiation of flowering and the reprogramming of the *AP1* locus to an actively transcribed state, is accompanied by a change of the chromatin structure at the *AP1* promoter. It involves removal of a nucleosome from the transcription start site and dynamic re-positioning of the TSS-nucleosome in a process that is developmentally regulated [165].

Restoring Lost Methylation at Heterochromatin

Given the importance of TEs for the formation of heterochromatin for centromere function and the role of DNA methylation in keeping the TEs silent, it is critical that cells maintain TE methylation levels throughout cell divisions and developmental transitions. It may be expected then that mechanisms guarding cells against accidental loss of heterochromatic CG methylation exist. Indeed, a novel corrective mechanism for restoring lost methylation from regions that need to remain silent was recently revealed [57]. Loss of CG DNA methylation during gametophyte generations was restored through the RNAi machinery in a robust and specific CG-re-methylation of the *Arabidopsis* centromeric repeats and transposons. Methylation was RDR2 dependent, was guided by 24-nt siRNAs corresponding to transposons and repetitive DNA sequences, and did not spread to adjacent sequences. Furthermore, the RNAi machinery is crucial for discriminating remethylatable from non-remethylatable sequences and is re-silencing only transposons activated in the *ddm1* mutants. Clearly, ability to re-methylate plays a protective role against the deleterious effects of active transposable elements. Apparently, it is advantageous for cells to maintain ongoing production of siRNAs from repeated DNAs, either as a backup for CG DNA methylation or to ensure rapid initiation of silencing at new transposon insertions [171].

POLYCOMB GROUP (PcG) AND TRITHORAX GROUP (TrxG) COMPLEXES IN PLANTS

In animals, expression states of homeotic genes (active or silent) are maintained and faithfully propagated throughout development by the counteracting activities of the PcG/TrxG complexes (Ref. 172 and ref. therein). In contrast to animals, plant organs (leaves, flowers) and seeds originate from the same undifferentiated meristem active throughout the life cycle. Although differentiation and organogenesis are not fixed in embryogenesis, PcG/TrxG homologs play roles in plant development as well. In plants, as in animals, development of a wrong organ at a wrong place (homeosis) is a consequence of a mutation of a homeotic gene. Unlike their animal counterparts, plant homeotic genes are not clustered and belong to the MADS-box family of transcription factors but nonetheless, the PcG/TrxG complexes similarly regulate their expression by modifying their nucleosomes. Like their metazoan counterparts, the *Arabidopsis* PcG complexes establish H3K27me3 through the biochemical activity of Enhancer of zeste (EZ)-related proteins [173–175], while Trithorax family members specifically tri-methylate histone H3K4 [176–178]. In both plant and metazoan chromatins, the H3K27me3 and H3K4me3 modifications are associated with transcriptionally silenced and active gene states, respectively.

At animal genes, two classes of PcG repressor complexes, PRC2 and PRC1, participate in the transcription-resistant chromatin structure [172]. PRC2 catalyzes the H3K27 tri-methylation resulting in the recruitment of PRC1, which maintains suppression by catalyzing H2A monoubiquitination [179].

In *Arabidopsis,* PRC2 complexes are conserved both structurally and functionally and, like their animal counterparts, form 600 kDa complexes involved in development, flowering, and imprinting [174,177]. The *Arabidopsis* H3K27me3 modifying mechanisms have proliferated to three *EZ* homologs (CURLY LEAVES, CLF, SWINGER, SWN, and MEDEA, MEA) forming at least three distinct PRC2 complexes containing proteins homologous to the metazoan PRC2 components: EMBRYONIC FLOWER2 (EMF2), VERNALIZATION2 (VRN2), FERTILIZATION INDEPENDENT SEED2 (FIS2) corresponding to SU(Z)12, and two WD-40 proteins (FERTILIZATION INDEPENDENT ENDOSPERM (FIE) and MULTICOPY SUPPRESSOR OF IRA1 (MSI1)) corresponding to ESC and P55, respectively. The different PRC2 complexes may have distinct functions as MEA expression is limited to the female gametophyte and embryo development, while CLF and SWN are expressed in adult plants [180].

Genes for proteins homologous to the PRC1 complex have not been identified in *Arabidopsis* but a functionally similar complex might be formed by the LIKE-HETEROCHROMATIN PROTEIN1 (LHP1), two RING-domain containing proteins, and a plant-specific protein, EMBRYONIC FLOWER1 (EMF1) [181]. LHP1 localizes at chromatin domains rich in H3K27me3 [103,105] and EMF1 is involved in the H3K27 tri-methylation [182].

The H3K4me3 mark of actively transcribed genes is established by the methyl transferase activity of the trithorax protein, which functions in a complex (COMPASS) conserved in yeast, *Drosophila*, and humans [183]. Genetic, biochemical, and molecular characteristics of the *ARABIDOPSIS HOMOLOG OF TRITHORAX1, ATX1,* have defined it as the plant counterpart of animal trithorax factors [106,176] but a plant COMPASS complex has not been identified yet. A WDR5-related protein capable of binding ATX1 was reported recently [184]. The evolution and function of *Trithorax*-related genes in plants, as well as the role of ATX1 in maintaining normal levels of homeotic gene expression during plant development and transition to flowering, for organ identity, and for biotic and abiotic response mechanisms in *Arabidopsis,* was recently reviewed [178].

Antagonistic PcG/TrxG Functions in *Arabidopsis*

The flower homeotic gene *AGAMUS, AG,* is silent in young seedlings and in vegetative tissues but its correct expression during flowering is critical for flower organ development and identity. Expression of *AG* is suppressed by the *Arabidopsis* homolog of EZ, CLF, and up-regulated by the homolog of trithorax, ATX1 [173,176], supporting the idea that CLF and ATX1 have counteracting activities at the *AG* locus. Interestingly, loss of both ATX1 and CLF functions in *atx1⁻/⁻ clf⁻/⁻* mutants rescued the single-mutant phenotypes suggesting that the Trx-like and the EZ-like plant homologs counterbalance each other at the shared locus [107]. Partial normalization of axial–skeletal transformations in mice was also observed when *Mll* (a human homolog of trithorax) and *BMI*-1 (a PcG component) were simultaneously deleted [185]. The molecular basis of this remarkable shift towards wild type phenotypes in the double *atx1/clf* mutants was related to the partial restoration of the histone marks on the *AG* nucleosomes that were lost in the single mutants. Restored patterns, however, were not identical with the initial patterns, an observation that could account for the variability and instability of phenotypes often seen in epigenetic mutants. At the molecular level, both H3K4me3 and H3K27me3 marks were required for the normal suppression of *AG* in leaves [107], perhaps establishing a chromatin state similar to the bivalent states of embryonic stem cell chromatin [186]. Contrary to the expectation that absent *ATX1* and *CLF* functions would erase the H3K4me3 and H3K27me3 marks, there was a partial restoration of the marks on the *AG*-nucleosomes in the double-mutant chromatin. The results suggested that in the absence of both ATX1 and CLF their roles could be undertaken by a different pair of modifiers supporting a model in which the PcG and TrxG-complexes form specific pairs to generate simultaneously present H3K4me3 and H3K27me3 marks. ATX1 and CLF physically interact providing a mechanistic basis for the observed effects [107].

Bivalent Chromatin Marks

Simultaneously present H3K4me3 and H3K27me3 marks found at silent genes in embryonic stem cells has suggested that co-existing "activating" and "silencing" nucleosomal modifications establish a bivalent chromatin state at loci "poised" for transcription later in development [186]. In addition to pluripotent cells, K4me3-K27me3 co-localization is functional in more differentiated cells as well [187]. Dual H3K4me3-H3K27me3 marks labeling the non-expressing *AG* locus in young seedling chromatin might similarly reflect a bivalent chromatin state to be expressed at a later developmental stage [107]. Coexisting H3K4me3 and H3K27me3 marks were found at two other loci (*FLC* and *AP1*) involved in the transition to flowering [165] suggesting that bivalent chromatin states might be a general characteristic of developmentally regulated genes in plants. The distribution patterns of the two antagonistic marks at promoters and at downstream gene-body nucleosomes are different. Presence of H3K4me3 and H3K27me3 on downstream nucleosomes remained unchanged throughout developmental transitions, independently of the transcriptional activity of *AG*, *AP1*, or *FLC*. In stark contrast, the H3K4me3/H3K27me3 profile at 5'-TSS nucleosomes changed dynamically, reflecting changes in transcription [165]. "Activating" H3K4me3 and "silencing" H3K27me3 modifications co-exist at 5'-end nucleosomes of both the transcriptionally active *FLC*- and the silent *AG* genes, while highly transcribed *AP1* displays neither of the two marks suggesting that distinct mechanisms "read" and operate at each locus even for genes belonging to the same MADS-box family.

Regulation of Imprinted Genes by PcG

PcG proteins directly regulate the silencing of the paternal allele in the male gametophyte of the imprinted gene *MEA* [188] and reduce the level of bi-parental *MEA* expression in vegetative tissues [189]. Thus, the PcG component, MEA, participates in both maternally and paternally derived PcG complexes to silence the paternal *MEA* allele in the endosperm. Polycomb group proteins and DNA methylation regulate also the *PHE1* gene imprinting. The paternal copy of *PHE1* is preferentially expressed while the maternal allele is silenced in the *Arabidopsis* endosperm [145] where the Polycomb group complex (MEA, FIE, FIS2) silences the maternal *PHE1* allele after fertilization. Loss of DNA methylation at a site 3' of the maternal *PHE1* allele is required for its silencing by the PcG proteins suggesting that DME might demethylate the maternal *PHE1* allele in the central cell [146] triggering the siRNA silencing pathway.

CHROMATIN REMODELING

Ultimately, it is the structure of chromatin that provides the permissive or restrictive environment for the transcriptional machinery exercising, thus, epigenetic control over a gene's expression. How chromatin structure (chromatin remodeling, involving assembly/disassembly, or re-positioning of nucleosomes) drives or responds to changes in gene expression is a subject of intensive studies. Revealing how chromatin remodeling, epigenetic marks (DNA methylation and histone modifications), and small RNAs are related mechanistically to achieve coordinated genome-wide control is among the most complex matters. Studies in animal systems are leading the way and have provided significant insights into the crosstalk of these factors at the molecular level. The nucleosome chaperones (CAF-1, NAP1, and HIRA) and chromatin remodelers (the SWI/SNF, RSC, ISWI, INO80, SWR1, and Mi-2/CHD) are specialized multi-protein machines regulating access to nucleosomal DNA by altering the structure, composition, and positioning of nucleosomes. ATP-dependent activities can remodel chromatin by either mobilizing nucleosomes on DNA or by exchanging one histone variant for another, within the nucleosome [190].

Components of the remodeling machinery are conserved throughout eukaryotes. In *Arabidopsis*, homologs of individual components of the nucleosome chaperone complexes, CAF-1, NAP1, and HIRA [191–194], of the CHD-type proteins PICKLE (PKL) and MOM1 [195,196], and of the ATP-dependent SWI/SNF remodelers [197–201] have been identified and shown to influence gene expression and plant development [200–204]. Most of the plant Snf2 proteins carry similar function as their yeast and animal homologs but some have been adapted for functions occurring only in plants. Forty-one members of the *Arabidopsis* Snf2 family fall into 19 distinct subfamilies reflecting the expansion of the SWI/SNF ATPase regulatory repertoire, while preserving essential ancestral functions [200,204]. DDM1, CLSY, and DRD1 are plant-specific ATPase activities involved in genome-wide DNA methylation, transposon silencing, and Pol IV-Pol V functions tightly coupled and guided by the 24-nt siRNAs, as discussed above. In contrast to yeast, *Drosophila*, and mammals, isolation and biochemical characterization of a full-size complex of plant origin has yet to be achieved for any of the chromatin-remodeling activities. However, homologs for the core subunits of the SWR1 chromatin remodeling complex have been characterized in *Arabidopsis* and shown to be functionally related to those described in yeast and human [205]. The SWR1-like chromatin-remodeling complex contains also a plant-specific protein, SEF, which genetically and physically interacts with the ATPase subunit counterpart, PIE1, and together with the ARP6 homolog (ESD) control gene expression at the chromatin level [206].

269

CONCLUSIONS AND PERSPECTIVES

- TEs are the major component of heterochromatin at the centromeres and telomeres. Regulation of the TEs activity is required for proper chromosome function, and epigenetic mechanisms in plants are largely oriented towards repressing TEs. Histone and DNA modifications are common epigenetic tools but they may be used in kingdom and species-specific ways. For example, in *Arabidopsis* CG, CHG and CHH methylations are present mainly in repeats, whereas the gene bodies are CG methylated [62–65]. It remains to be seen how general these DNA methylation patterns would be for other plants. In plants, the activities establishing the epigenetic chromatin marks are largely dependent on guidance by the small heterochromatic 24-nt siRNAs.

- Findings of siRNAs in all three eukaryotic kingdoms indicate that the siRNA machinery was present in the last common ancestor of plants, animals, and fungi [23]. Although the machinery might be ancient, the siRNA have diversified over time to acquire specialized roles. Unique plant specific proteins function upstream of DCL3 and downstream of AGO4 to produce 24-nt siRNAs that guide DNA methylation and heterochromatin formation. Instead of cleaving mRNA, AGO4-siRNA complexes recruit chromatin-modifying enzymes. Some of the proteins involved in the biogenesis of the 24-nt siRNAs are genetically redundant, whereas others have specialized roles. The diversification of RNA silencing pathways in plants reflects the intricate ways evolved by the sessile organisms to cope with biotic, abiotic, and genome stresses.

- Pol IV and Pol V transcription complexes have evolved as evolutionary derivatives of the Pol II mechanism specialized for generating and/or using noncoding RNAs for chromatin-based gene silencing. Elucidation of their function helped solve the paradox that transcription of DNA sequences that are silent at the chromatin level is required for the assembly of chromatin in a silent state nonpermissive for transcription by Pol II and Pol III. Thus, plants have enhanced their ability to modulate the epigenetic states of their genomes by remarkably diversifying their transcriptional machinery [85].

- As the ancestral role of chromatin-targeted siRNAs is the genome-wide suppression of repeated DNA, the number of endogenous genes that are controlled by this system might be small in *Arabidopsis*. However, in species with large genomes, like maize, the epigenetic control of TEs by chromatin-targeted RNAi has a much more important role regulating developmental genes [133,135].

- In addition to silencing TEs, flowering plants have evolved intricate ways to implement siRNA pathways in the regulation of pollen and embryo development through gene imprinting. In a highly specific RNAi-targeted process, transposon activation and siRNA accumulation in the central cell contribute to enhanced methylation and silencing of repetitive elements in the egg (and later the embryo). The model viewing imprinted genes not as exceptional sequences specifically targeted for demethylation in the central cell but rather as part of a process that reshapes DNA methylation of the entire maternal genome in the endosperm [148,149] is consistent with the hypothesis that imprinting arose as a byproduct of silencing the invading foreign DNA [207]. Transposon-silencing mechanisms might have been co-opted for the regulation of ribosome biogenesis and nucleolar dominance in interspecies crosses as well [208].

- Transient TE reactivation occurring in the pollen is limited to the VN and is signaling TE silencing in the neighboring sperm preventing, *via* siRNAs, transposon activation in the embryo. It was proposed that the molecular mechanisms involved in the permanent silencing of foreign DNA have evolved from mechanisms required for the successful development of an embryo [153]. This model proposes that epigenetic silencing of TEs has evolved from a developmental process in stark contrast to the models for the origin of gene imprinting in the endosperm and the nucleolar dominance which suggested that the TEs silencing mechanism has been co-opted for developmental and nucleolar functions.

- In addition to organismal development, epigenetic control of TEs has had a role in genome evolution. Epialleles are formed due to the proximity of a gene to a TE, and the regulatory complexes that the TE recruits. Natural epigenetic variation can originate from polymorphisms in transposon insertions and repeats, as illustrated by different *Arabidopsis* species and ecotypes of *A. thaliana* [95,118–120]. The consequences of transgenerational epigenetic effects for speciation and adaptive evolution are increasingly attracting attention [16,120–124].

- As in animals, plant developmental processes are regulated by antagonistic PcG/TrxG-related activities. Dual histone methylations (H3K27me3 and H3K4me3) mark silent genes in animal stem cells and non-differentiated cells establishing a bivalent chromatin state at loci poised for transcription later in development. Dual activating/silencing marks found at developmentally regulated *Arabidopsis* genes illustrate features of the epigenetic "code" conserved in animal and plant kingdoms despite differences in the developmental patterns. Co-existing H3K4me3 and H3K27me3 may form a distinct bi-modular "syllable" in the histone "code" that conveys specific "meaning" at different genes [165].

- DNA and histone modifications are linked with chromatin remodeling and nucleosome positioning. Although individual components of the remodeling machinery are conserved throughout eukaryotes, still very little is known about this mechanism in plants as full-size remodeling complexes have not been isolated and characterized biochemically. Given the existence of plant-specific histone modifications and DNA methylations, one might expect that interactions with the nucleosome remodeling machinery might be plant-specific as well.

- Uncovering features of plant-specific "dialects" in the epigenetic language "written" by the histone and DNA modifications, finding plant-unique ways of employing the enormously complex small RNAs mechanisms, and establishing correlations between chromatin-driven genome reprogramming processes in plants, would continue to be among the most challenging, fascinating, and revealing endeavors of contemporary molecular biology research.

References

1. Bird A. Perceptions of epigenetics. *Nature* 2007;**447**:396–8.

2. Grant-Downton RT, Dickinson HG. Epigenetics and its implications for plant biology. The epigenetic network in plants. *Ann Bot (Lond)* 2005;**96**:1143–64.

3. Costa F. Non-coding RNAs, epigenetics and complexity. *Gene* 2008;**410**:9–17.

4. Mlotshwa S, Pruss G, Vance V. Small RNAs in viral infection and host defence. *Trens Plant Sci* 2008;**13**:375–82.

5. Chekanova JA, Gregory BD, Reverdatto SV, Chen H, Kumar R, Hooker T, et al. Genome-wide high-resolution mapping of exosome substrates reveals hidden features in the *Arabidopsis* transcriptome. *Cell* 2007;**131**:1340–53.

6. Cao X, Aufsatz W, Zilberman D, Mette MF, Huang MS, Matzke M, et al. Role of the DRM and CMT3 methyltransferases in RNA-directed DNA methylation. *Curr Biol* 2003;**13**:2212–17.

7. Lippman Z, Martienssen R. The role of RNA interference in heterochromatic silencing. *Nature* 2004;**431**:364–70.

8. Xie Z, Johansen LK, Gustafson AM, Kasschau KD, Lellis AD, Zilberman D, et al. Genetic and functional diversification of small RNA pathways in plants. *PLoS Biol* 2004;**2**:e104.

9. Zilberman D, Cao X, Johansen LK, Xie Z, Carrington JC, Jacobsen SE. Role of *Arabidopsis* ARGONAUTE4 in RNA-directed DNA methylation triggered by inverted repeats. *Curr Biol* 2004;**14**:1214–20.

10. Zhang X. The epigenetic landscape of plants. *Science* 2008;**320**:489–92.

11. Eamens A, Wang M, Neil B, Smith A, Waterhouse PM. RNA silencing in plants: yesterday, today, and tomorrow. *Plant Physiol* 2008;**147**:456–68.

12. Ghildiyal M, Zamore PD. Small silencing RNAs: an expanding universe. *Nat Rev Genet* 2009;**10**:94–108.

13. Chinnusamy V, Gong Z, Zhu JK. Abscisic acid-mediated epigenetic processes in plant development and stress responses. *J Integr Plant Biol* 2008;**50**:1187–95.

14. Boyko A, Kovalchuk I. Epigenetic control of plant stress response genes. *Env Mol Mut* 2008;**49**:61–72.

15. Alvarez-Venegas R, Al-Abdallat A, Guo M, Alfano J, Avramova Z. Epigenetic control of a transcription factor at the node of convergence of two signaling pathways. *Epigenetics* 2007;**2**:106–13.

16. Molinier J, Ries G, Zipfel C, Honh B. Transgeneration memory of stress in plants. *Nature* 2006;**442**:1046–9.

17. Fire A, Xu S, Montgomery MK, Kostas SA, Driver SE, Mello CC. Potent and specific genetic interference by double-stranded RNA in *Caenorhabditis elegans*. *Nature* 1998;**391**:806–11.

18. Hamilton AJ, Baulcombe DC. A species of small antisense RNA in posttranscriptional gene silencing in plants. *Science* 1999;**286**:950–2.

19. Chen X, Liu J, Cheng Y, Jia D. *HEN1* functions pleiotropically in *Arabidopsis* development and acts in C function in the flower. *Development* 2002;**129**:1085–94.

20. Yu B, Yang Z, Li J, Minakhina S, Yang M, Padgett RW, et al. Methylation as a crucial step in plant microRNA biogenesis. *Science* 2005;**307**:932–5.

21. Li CF, Pontes O, El-Shami M, Henderson IR, Bernatavichute YV, Chan SW, et al. An ARGONAUTE4-containing nuclear processing center colocalized with Cajal bodies in *Arabidopsis thaliana*. *Cell* 2006;**126**:93–106.

22. Brodersen P, Sakvarelidze-Achard L, Bruun-Rasmussen M, Dunoyer P, Yamamoto YY, Sieburth L, et al. Widespread translational inhibition by plant miRNAs and siRNAs. *Science* 2008;**320**:1185–90.

23. Cerutti H, Casas-Mollano JA. On the origin and functions of RNA-mediated silencing: from protists to man. *Curr Genet* 2006;**50**:81–99.

24. Axtell MJ, Synder JA, Bartel DP. Common functions for diverse small RNAs of land plants. *Plant Cell* 2007;**19**:1750–69.

25. Voinnet O. Use, tolerance and avoidance of amplified RNA silencing by plants. *Trends Plant Sci* 2008;**13**:317–28.

26. Katiyar-Agarwal S, Morgan R, Dahlbeck D, Borsani O, Villegas A Jr., Zhu JK, et al. A pathogen-inducible endogenous siRNA in plant immunity. *Proc Natl Acad Sci USA* 2006;**103**:18002–7.

27. Brosnan CA, Mitter N, Christie M, Smith NA, Waterhouse PM, Carroll BJ. Nuclear gene silencing directs reception of long-distance mRNA silencing in *Arabidopsis*. *Proc Natl Acad Sci USA* 2007;**104**:14741–6.

28. Pontes O, Li CF, Nunes PC, Haag J, Ream T, Vitins A, et al. The *Arabidopsis* chromatin-modifying nuclear siRNA pathway involves a nucleolar RNA processing center. *Cell* 2006;**126**:79–92.

29. Jacobsen SE, Running MP, Meyerowitz EM. Disruption of an RNA helicase/RNase III gene in *Arabidopsis* causes unregulated cell division in floral meristems. *Development* 1999;**126**:5231–43.

30. Bernstein E, Caudy AA, Hammond SM, Hannon GJ. Role for a bidentate ribonuclease in the initiation step of RNA interference. *Nature* 2001;**409**:363–6.

31. Bouche N, Lauressergues D, Gasciolli V, Vaucheret H. An antagonistic function for *Arabidopsis* DCL2 in development and a new function for DCL4 in generating viral siRNAs. *EMBO J* 2006;**25**:3347–56.

32. Deleris A, Gallego-Bartolome J, Bao J, Kasschau KD, Carrington JC, Voinnet O. Hierarchical action and inhibition of plant Dicer-like proteins in antiviral defense. *Science* 2006;**313**:68–71.

33. Bohmert K, Camus I, Bellini C, Bouchez D, Caboche M, Benning C. AGO1 defines a novel locus of *Arabidopsis* controlling leaf development. *EMBO J* 1998;**17**:170–80.

34. Vaucheret H. Plant ARGONAUTES. *Trends Plant Sci* 2008;**13**:350–8.

35. Liu J, Carmell MA, Rivas FV, Marsden CG, Thomson JM, Song JJ, et al. Argonaute2 is the catalytic engine of mammalian RNAi. *Science* 2004;**305**:1437–41.

36. El-Shami M, Pontier D, Lahmy S, Braun L, Picart C, Vega D, et al. Reiterated WG/GW motifs form functionally and evolutionarily conserved ARGONAUTE-binding platforms in RNAi-related components. *Genes Dev* 2007;**21**:2539–44.

37. Zilberman D, Cao X, Jacobsen SE. ARGONAUTE4 control of locus-specific siRNA accumulation and DNA and histone methylation. *Science* 2003;**299**:716–19.

38. Zheng X, Zhu J, Kapoor A, Zhu JK. Role of *Arabidopsis* AGO6 in siRNA accumulation, DNA methylation and transcriptional gene silencing. *EMBO J* 2007;**26**:1691–701.

39. Takeda A, Iwasaki S, Watanabe T, Utsumi M, Watanabe Y. The mechanism selecting the guide strand from small RNA duplexes is different among Argonaute proteins. *Plant Cell Physiol* 2008;**49**:493–500.

40. Backman TW, Sullivan CM, Cumbie JS, Miller ZA, Chapman EJ, Fahlgren N, et al. Update of ASRP: the *Arabidopsis* Small RNA Project database. *Nucleic Acids Res* **36**(Database issue) 2008:D982–5.

41. Wassenegger M, Krczal G. Nomenclature and functions of RNA-directed RNA polymerases. *Trends Plant Sci* 2006;**11**:142–51.

42. Wassenegger M, Heimes S, Riedel L, Sanger HL. RNA-directed de novo methylation of genomic sequences in plants. *Cell* 1994;**76**:567–76.

43. Chan SW, Zilberman D, Xie Z, Johansen LK, Carrington JC. RNA silencing genes control de novo DNA methylation. *Science* 2004;**303**:1336.

44. Chan SW, Henderson IR, Jacobsen SE. Gardening the genome: DNA methylation in *Arabidopsis thaliana*. *Nat Rev Genet* 2005;**6**:351–60.

45. Henikoff S, Comai L. A DNA methyltransferase homolog with a chromodomain exists in multiple polymorphic forms in *Arabidopsis*. *Genetics* 1998;**149**:307–18.

46. Henderson IR, Jacobsen SE. Epigenetic inheritance in plants. *Nature* 2007;**447**:418–24.

47. Cao X, Jacobsen SE. Role of the *Arabidopsis* DRM methyltransferases in de novo DNA methylation and gene silencing. *Curr Biol* 2002;**12**:1138–44.

48. Cao X, Jacobsen SE. Locus-specific control of asymmetric and CpNpG methylation by the DRM and CMT3 methyltransferase genes. *Proc Natl Acad Sci USA* 2002;**99**:16491–8.

49. Lindroth AM, Cao X, Jackson JP, Zilberman D, McCallum CM, Henikoff S, et al. Requirement of *CHROMOMETHYLASE3* for maintenance of CpXpG methylation. *Science* 2001;**292**:2077–80.

50. Bartee L, Malagnac F, Bender J. *Arabidopsis* cmt3 chromomethylase mutations block non-CG methylation and silencing of an endogenous gene. *Genes Dev* 2001;**15**:1753–8.

51. Cao X, Springer NM, Muszynski MG, Phillips RL, Kaeppler S. Conserved plant genes with similarity to mammalian de novo DNA methyltransferases. *Proc Natl Acad Sci USA* 2000;**97**:4979–84.

52. Hashimoto H, Horton JR, Zhang X, Bostick M, Jacobsen SE, Cheng X. The SRA domain of UHRF1 flips 5-methylcytosine out of the DNA helix. *Nature* 2008;**455**:826–9.

53. Woo HR, Pontes O, Pikaard CS, Richards EJ. VIM1, a methylcytosine-binding protein required for centromeric heterochromatinization. *Genes Dev* 2007;**21**:267–77.

54. Saze H, Kakutani T. Heritable epigenetic mutation of a transposon-flanked *Arabidopsis* gene due to lack of the chromatin remodelling factor DDM1. *EMBO J* 2007;**26**:3641–52.

55. Lippman Z, Gendrel AV, Black M, Vaughn MW, Dedhia N, McCombie WR, et al. Role of transposable elements in heterochromatin and epigenetic control. *Nature* 2004;**430**:471–6.

56. Mathieu O, Reinders J, Caikovski M, Smathajitt C, Paszkowski A. Transgenerational stability of the *Arabidopsis* epigenome is coordinated by CG methylation. *Cell* 2007;**130**:851–62.

57. Teixeira FK, Heredia F, Sarazin A, Roudier F, Boccara M, Ciaudo C, et al. A role for RNAi in the selective correction of DNA methylation defects. *Science* 2009;**323**:1600–4.

58. Huettel B, Tatsuo Kanno, Daxinger L, Aufsatz W, Matzke AJ, Matzke M. Endogenous targets of RNA-directed DNA methylation and Pol IV in *Arabidopsis*. *EMBO J* 2006;**25**:2828–36.

59. Penterman J, Uzawa R, Fischer RL. Genetic interactions between DNA demethylation and methylation in *Arabidopsis*. *Plant Physiol* 2007;**145**:1549–57.

60. Zheng X, Pontes O, Zhu J, Miki D, Zhang F, Li WX, et al. ROS3 is an RNA-binding protein required for DNA demethylation in *Arabidopsis*. *Nature* 2008;**455**:1259–62.

61. Zhang X, Yazaki J, Sundaresan A, Cokus S, Chan SW, Chen H, et al. Genome-wide high-resolution mapping and functional analysis of DNA methylation in *Arabidopsis*. *Cell* 2006;**126**:1189–201.

62. Zilberman D, Gehring M, Tran RK, Ballinger T, Henikoff S. Genome-wide analysis of *Arabidopsis thaliana* DNA methylation uncovers an interdependence between methylation and transcription. *Nat Genet* 2007;**39**:61–9.

63. Cokus SJ, Feng S, Zhang X, Chen Z, Merriman B, Haudenschild CD, et al. Shotgun bisulphite sequencing of the *Arabidopsis* genome reveals DNA methylation patterning. *Nature* 2008;**452**:215–19.

64. Lister R, O'Malley RC, Tonti-Filippini J, Gregory BD, Berry CC, Millar AH, et al. Highly integrated single-base resolution maps of the epigenome in *Arabidopsis*. *Cell* 2008;**133**:523–36.

65. Tran RK, Zilberman D, de Bustos C, Ditt RF, Henikoff JG, Lindroth AM, et al. Chromatin and siRNA pathways cooperate to maintain DNA methylation of small transposable elements in *Arabidopsis*. *Genome Biol* 2005;**6**:R90.

66. Bernstein BE, Meissner A, Lander ES. The mammalian epigenome. *Cell* 2007;**128**:669–81.

67. Miura A, Nakamura M, Inagaki S, Kobayashi A, Saze H, Kakutani T. An *Arabidopsis* jmjC-domain protein protects transcribed genes from DNA methylation at CHG sites. *EMBO J* 2009;**28**:1078–86.

68. Teixeira FK, Colot V. Gene body DNA methylation in plants: a means to an end or an end to a means? *EMBO J* 2009;**28**:997–8.

69. Wierzbicki AT, Haag JR, Pikaard CS. Noncoding transcription by RNA polymerase IVb/Pol V mediates transcriptional silencing of overlapping and adjacent genes. *Cell* 2008;**135**:635–48.

70. Daxinger L, Kanno T, Matzke M. Pol V transcribes to silence. *Cell* 2008;**135**:592–4.

71. Kapronov P, Willingham AT, Gingeras TR. Genome-wide transcription and the implications for genomic organization. *Nat Rev Genet* 2007;**8**:413–23.

72. Kasschau KD, Fahlgren N, Chapman EJ, Sullivan CM, Cumbie J,S, Givan SA, et al. Genome-wide profiling and analysis of *Arabidopsis* siRNAs. *PLoS Biol* 2007;**5**:e57.

73. Zhang X, Henderson IR, Lu C, Green PJ, Jacobsen SE. Role of RNA polymerase IV in plant small RNA metabolism. *Proc Natl Acad Sci USA* 2007;**104**:4536–41.

74. Daxinger L, Kanno T, Bucher E, van der Winden J, Naumann U, Matzke AJM, et al. A stepwise pathway for biogenesis of 24-nt secondary siRNAs and spreading of DNA methylation. *EMBO J* 2009;**28**:48–57.

75. Pikaard CS, Haag JR, Ream T, Wierzbicki AT. Roles of RNA polymerase IV in gene silencing. *Trends Plant Sci* 2008;**13**:390–7.

76. Egloff S, Murphy S. Cracking the RNA polymerase II CTD code. *Trends Genet* 2008;**24**:280–8.

77. He XJ, Hsu YF, Pontes O, Zhu J, Lu J, Bressan RA, et al. NRPD4, a protein similar to the RPB4 subunit of RNA polymerase II, is a component of RNA polymerases IV and V and is required for siRNA production, RNA-directed DNA methylation, and transcriptional gene silencing. *Genes Dev* 2009;**23**:318–30.

78. Ream TS, Haag JR, Wierzbicki A, Nicora CD, Norbeck A, Zhu JK, et al. Subunit compositions of the RNA silencing enzymes Pol IV and Pol V reveal their origins as specialized forms of RNA polymerase II. *Mol Cell* 2009;**33**:192–203.

79. Huang L, Jones AM, Searle I, Patel K, Vogler H, Hubner NC, et al. An atypical RNA polymerase involved in RNA silencing shares small subunits with RNA polymerase II. *Nat Struct Mol Biol* 2009;**16**:91–3.

80. Matzke M, Kanno T, Daxinger L, Huettel B, Matzke AJ. RNA-mediated chromatin-based silencing in plants. *Curr Opin Cell Biol* 2009;**21**:367–76.

81. Mosher RA, Schwach F, Studholme D, Baulcombe DC. PolIVb influences RNA-directed DNA methylation independently of its role in siRNA biogenesis. *Proc Natl Acad Sci USA* 2008;**105**:3145–50.

82. Kanno T, Mette MF, Kreil DP, Aufsatz W, Matzke M, Matzke AJ. Involvement of putative SNF2 chromatin remodeling protein DRD1 in RNA-directed DNA methylation. *Curr Biol* 2004;**14**:801–5.

83. Stonaker Hale CJ, J. L. Gross SM, Hollick JB. A novel Snf2 protein maintains trans-generational regulatory states established by paramutation in maize. *PLoS Biol* 2007;**5**:2156–65.

84. Kanno T, Bucher E, Daxinger L, Huettel B, Bohmdorfer G, Gregor W, et al. A structural maintenance of chromosomes hinge domain-containing protein is required for RNA-directed DNA methylation. *Nat Genet* 2008;**40**:670–5.

85. Wierzbicki AT, Ream TS, Haag JR, Pikaard CS. RNA polymerase V transcription guides ARGONAUTE4 to chromatin. *Nat Genet* 2009;**41**:630–4.

85a. Buhler M, Moazed D. Transcription and RNAi in heterochromatic gene silencing. *Nat Struct Mol Biol* 2007;**14**:1041–8.

86. He XJ, Hsu YF, Wierzbicki AT, Pontes O, Pikaard CS, Liu HL, et al. An effector of RNA-directed DNA methylation in *Arabidopsis* is an ARGONAUTE 4- and RNA-binding protein. *Cell* 2009;**137**:498–508.

87. Qi Y, He X, Wang X-J, Kohany O, Jurka J, Hannon GJ. Distinct catalytic and non-catalytic roles of ARGONAUTE4 in RNA-directed DNA methylation. *Nature* 2006;**443**:1008–12.

88. Chan SW, Zhang X, Bernatavichute YV, Jacobsen SE. Two-step recruitment of RNA-directed DNA methylation to tandem repeats. *PLoS Biol* 2006;**4**:e363.

89. Fujimoto R, Kinoshita Y, Kawabe A, Kinoshita T, Takashima K, Nordborg M, et al. Evolution and control of imprinted FWA genes in the genus *Arabidopsis*. *PLoS Genetics* 2008;**4**:e:1000048.

90. Beclin C, Boutet S, Waterhouse P, Vaucheret H. A branched pathway for transgene-induced RNA silencing in plants. *Curr Biol* 2002;**12**:684–8.

91. Johnson LM, Law JA, Khattar A, Henderson IR, Jacobsen SE. SRA-domain proteins required for DRM2-mediated de novo DNA methylation. *PLoS Genet* 2008;**4**(11):e1000280.

92. Preuss SB, Costa-Nunes P, Tucker S, Pontes O, Lawrence RJ, Mosher R, et al. Multimegabase silencing in nucleolar dominance involves siRNA-directed DNA methylation and specific methylcytosine-binding proteins. *Mol Cell* 2008;**32**:673–84.

93. Jackson JP, Lindroth AM, Cao X, Jacobsen SE. Control of CpNpG DNA methylation by the KRYPTONITE histone H3 methyltransferase. *Nature* 2002;**416**:556–60.

94. Sridhar VV, Kapoor A, Zhang K, Zhu J, Zhou T, Hasegawa PM, et al. Control of DNA methylation and heterochromatic silencing by histone H2B deubiquitination. *Nature* 2007;**447**:735–8.

95. Kazazian HH Jr. Mobile elements: drivers of genome evolution. *Science* 2004;**303**:1626–32.

96. Kloc A, Zaratiegui M, Nora E, Martienssen R. RNA interference guides histone modification during the S phase of chromosomal replication. *Curr Biol* 2008;**18**:490–5.

97. Durand-Dubief M, Ekwall K. Heterochromatin tells CENP-A where to go. *BioEssays* 2008;**30**:526–9.

98. Folco HD, Pidoux AP, Urano T, Allshire RC. Heterochromatin and RNAi are required to establish CENP-A chromatin at centromeres. *Science* 2008;**319**:94–7.

99. Matzke MA, Birchler JA. RNAi-mediated pathways in the nucleus. *Nat Rev Genet* 2005;**6**:24–35.

100. Ma J, Wing RA, Bennetzen JL, Jackson SA. Plant centromere organization: a dynamic structure with conserved functions. *Trends Genet* 2007;**23**:134–9.

101. Jiang J, Birchler JA, Parrott WA, Dawe RK. A molecular view of plant centromeres. *Trends Plant Sci* 2003;**8**:570–5.

102. May BP, Lippman ZB, Fang Y, Spector DL, Martienssen RA. Differential regulation of strand-specific transcripts from *Arabidopsis* centromeric satellite repeats. *PLoS Genet* 2005;**1**(6):e79.

103. Turck F, Roudier F, Farrona S, Martin-Magniette ML, Guillaume E, Buisine N, et al. *Arabidopsis* TFL2/LHP1 specifically associates with genes marked by trimethylation of histone H3 lysine 27. *PLoS Genet* 2007;**3**(6):e86.

104. Mathieu O, Probst AV, Paszkowski J. Distinct regulation of histone H3 methylation at lysines 27 and 9 by CpG methylation in *Arabidopsis*. *EMBO J* 2005;**24**:2783–91.

105. Zhang X, Clarenz O, Cokus S, Bernatavichute YV, Pellegrini M, Goodrich J, et al. Whole-genome analysis of histone H3 lysine 27 trimethylation in *Arabidopsis*. *PLoS Biol* 2007;**5**(5):e129.

106. Alvarez-Venegas R, Avramova Z. Methylation Patterns of Histone H3 Lys 4, Lys 9 and Lys 27 in transcriptionally active and inactive *Arabidopsis* genes and in *atx1* mutants. *Nucleic Acid Res* 2005;**33**:5199–520.

107. Saleh A, Al-Abdallat A, Ndamukong I, Alvarez-Venegas R, Avramova Z. The *Arabidopsis* homologs of *trithorax* (ATX1) and *enhancer of zeste* (CLF) establish "bivalent chromatin marks" at the silent *AGAMOUS* locus. *Nucleic Acids Res* 2007;**35**:6290–6.

108. Zhang X, Bernatavichute YV, Cokus S, Pellegrini M, Jacobsen SE. Genome-wide analysis of mono-, di- and trimethylation of histone H3 lysine 4 in *Arabidopsis thaliana*. *Genome Biol* 2009;**10**(6):R62.

109. Fischer A, Hofmann I, Naumann K, Reuter G. Heterochromatin proteins and the control of heterochromatic gene silencing in *Arabidopsis*. *J Plant Physiol* 2006;**163**:358–68.

110. Jacob Y, Feng S, LeBlanc CA, Bernatavichute YV, Stroud H, Cokus S, et al. ATXR5 and ATXR6 are H3K27 monomethyl-transferases required for chromatin structure and gene silencing. *Nat Struct Mol Biol* 2009;**16**:763–8.

111. Shi J, Dawe RK. Partitioning of the maize epigenome by the number of methyl groups on histone H3 lysines 9 and 27. *Genetics* 2006;**173**:1571–83.

112. Catcheside DG. A position effect in *Oenothera*. *J Genet* 1938;**38**:345–52.

113. Catcheside DG. A position effect in *Oenothera*. *J Genet* 1949;**48**:31–42.

114. Grewal SI, Klar AJ. Chromosomal inheritance of epigenetic states in fission yeast during mitosis and meiosis. *Cell* 1996;**86**:95–101.

115. Haynes KA, Caudy AA, Collins L, Elgin SC. Element 1360 and RNAi components contribute to HP1-dependent silencing of a pericentric reporter. *Curr Biol* 2006;**16**:2222–7.

116. Slotkin RK, Martienssen R. Transposable elements and the epigenetic regulation of the genome. *Nat Rev Genet* 2007;**8**:272–85.

117. Saze H, Kakutani T. Control of genic DNA methylation by a jmjC domain-containing protein in *Arabidopsis thaliana*. *Science* 2008;**319**:462–5.

118. Vaughn MW, Tanurdži M, Lippman Z, Jiang H, Carrasquillo R, Rabinowicz PD, et al. Epigenetic Natural Variation in *Arabidopsis thaliana*. *PLoS Biol* 2007;**5**(7):e174.

119. Liu J, He Y, Amasino R, Chen X. siRNAs targeting an intronic transposon in the regulation of natural flowering behavior in *Arabidopsis*. *Genes and Dev* 2004;**18**:2873–8.

120. Zhai J, Lui J, Liu B, Li P, Meyers BC, Chen X, et al. Small RNA-directed epigenetic natural variation in *Arabidopsis thaliana*. *PLoS Genet* 2008;**4**(4):e1000056.

121. Cubas P, Vincent C, Coen E. An epigenetic mutation responsible for natural variation in floral symmetry. *Nature* 1999;**401**:157–61.

122. Manning K, Tör M, Poole M, Hong Y, Thompson AJ, King GJ, et al. A naturally occurring epigenetic mutation in a gene encoding an SBP-box transcription factor inhibits tomato fruit ripening. *Nat Genet* 2006;**38**:948–52.

123. Chen ZJ. Genetic and epigenetic mechanisms for gene expression and phenotypic variation in plant polyploids. *Annu Rev Plant Biol* 2007;**58**:377–406.

124. Johansen F, Colot V, Jansen RC. Epigenome dynamics: a quantitative genetics perspective. *Nat Rev Genet* 2008;**9**:883–90.

125. Chandler V, Alleman M. Paramutation: epigenetic instructions passed across generations. *Genetics* 2008;**178**:1839–44.

126. Stam M. Paramutation: a heritable change in gene expression by allelic interactions in trans. *Mol Plant* 2009;**2**:578–88.

127. Chandler VL, Eggleston WB, Dorweiler JE. Paramutation in maize. *Plant Mol Biol* 2000;**43**:121–45.

128. Patterson GI, Chandler VL. Paramutation in maize and related allelic interactions. *Curr Top Microbiol Immunol* 1995;**197**:121–41.

129. Stam M, Belele C, Ramakrishna W, Dorweiler JE, Bennetzen JL, Chandler VL. The regulatory regions required for B′ paramutation and expression are located far upstream of the maize b1 transcribed sequences. *Genetics* 2002;**162**:917–30.

130. Stam M, Belele C, Dorweiler JE, Chandler VL. Differential chromatin structure within a tandem array 100 kb upstream of the maize b1 locus is associated with paramutation. *Genes Dev* 2002;**16**:1906–18.

131. Kermicle JL, Eggleston WB, Alleman M. Organization of paramutagenicity in R-stippled maize. *Genetics* 1995;**141**:361–72.

132. Sidorenko L, Chandler V. RNA-dependent RNA polymerase is required for enhancer-mediated transcriptional silencing associated with paramutation at the maize p1 gene. *Genetics* 2008;**180**:1983–93.

133. Erhard KF Jr., Stonaker JL, Parkinson SE, Lim JP, Hale C, Hollick JB. RNA polymerase IV functions in paramutation in *Zea mays*. *Science* 2009;**323**:1201–5.

134. Lisch D, Carey CC, Dorweiler JE, Chandler VL. A mutation that prevents paramutation in maize also reverses *Mutator* transposon methylation and silencing. *Proc Natl Acad Sci USA* 2002;**99**:6130–5.

135. Parkinson SE, Gross SM, Hollick JB. Maize sex determination and abaxial leaf fates are canalized by a factor that maintains repressed epigenetic states. *Dev Biol* 2007;**308**:462–73.

136. Sidorenko LV, Peterson T. Transgene-induced silencing identifies sequences involved in the establishment of paramutation of the maize p1 gene. *Plant Cell* 2001;**13**:319–35.

137. Meyer P, Heidmann I, Niedenhof I. Differences in DNA-methylation are associated with a paramutation phenomenon in transgenic petunia. *Plant J* 1993;**4**:89–100.

138. Hagemann R, Berg W. Paramutation at the sulfurea locus of Lycopersicon esculentum Mill. Determination of the time of occurrence of paramutation by the quantitative evaluation of the variegation. *Theoret Appl Genet* 1978;**53**:113–23.

139. Scheid M, Afsar K, Paszkowski J. Formation of stable epialleles and their paramutation-like interaction in tetraploid *Arabidopsis thaliana*. *Nat Genet* 2003;**34**:450–4.

140. Feil R, Berger F. Convergent evolution of genomic imprinting in plants and mammals. *Trends Genet* 2007;**23**:192–9.

141. Huh JH, Bauer MJ, Hsieh TF, Fischer R. Endosperm gene imprinting and seed development. *Curr Opin Genet Dev* 2007;**17**:480–5.

275

142. Tiwari S, Schulz R, Ikeda Y, Dytham L, Bravo J, Mathers L, et al. MATERNALLY EXPRESSED PAB C-TERMINAL, a novel imprinted gene in *Arabidopsis*, encodes the conserved C-terminal domain of polyadenylate binding proteins. *Plant Cell* 2008;**20**:2387–98.

143. Day RC, Herridge RP, Ambrose BA, Macknight RC. Transcriptome analysis of proliferating *Arabidopsis* endosperm reveals biological implications for the control of syncytial division, cytokinin signaling, and gene expression regulation. *Plant Physiol* 2008;**148**:1964–84.

144. Gehring M, Bubb KL, Henikoff S. Extensive demethylation of repetitive elements during seed development underlies gene imprinting. *Science* 2009;**324**:1447–51.

145. Kohler C, Page DR, Gagliardini V, Grossniklaus U. The *Arabidopsis thaliana* MEDEA Polycomb group protein controls expression of PHERES1 by parental imprinting. *Nat Genet* 2005;**37**:28–30.

146. Makarevich G, Villar CB, Erilova A, Köhler C. Mechanism of PHERES1 imprinting in *Arabidopsis*. *J Cell Sci* 2008;**121**:906–12.

147. Munshi A, Duvvuri S. Genomic imprinting: the story of the other half and the conflicts of silencing. *J Genet Genomics* 2007;**34**:93–103.

148. Hsieh TF, Ibarra CA, Silva P, Zemach A, Eshed-Williams L, Fischer RL, et al. Genome-wide demethylation of *Arabidopsis* endosperm. *Science* 2009;**324**:1451–4.

149. Mosher RA, Melnyk CW, Kelly KA, Dunn RM, Studholme DJ, Baulcombe DC. Uniparental expression of PolIV-dependent siRNAs in developing endosperm of *Arabidopsis*. *Nature* 2009;**460**:283–6.

150. Choi Y, Gehring M, Johnson L, Hannon M, Harada JJ, Goldberg RB, et al. DEMETER, a DNA glycosylase domain protein, is required for endosperm gene imprinting and seed viability in *Arabidopsis*. *Cell* 2002;**110**:33–42.

151. Lauria M, Rupe M, Guo M, Kranz E, Pirona R, Viotti A, et al. Extensive maternal DNA hypomethylation in the endosperm of *Zea mays*. *Plant Cell* 2004;**16**:510–22.

152. Gutierrez-Marcos JF, Costa LM, Dal Prà M, Scholten S, Kranz E, Perez P, et al. Epigenetic asymmetry of imprinted genes in plant gametes. *Nat Genet* 2006;**38**:876.

153. Slotkin RK, Vaughn M, Borges F, Tanurdzić M, Becker JD, Feijó JA, et al. Epigenetic reprogramming and small RNA silencing of transposable elements in pollen. *Cell* 2009;**136**:461–72.

154. Nobuta K, Venu RC, Lu C, Belo A, Vemaraju K, Kulkarni K, et al. An expression atlas of rice mRNAs and small RNAs. *Nat Biotechnol* 2007;**25**:473–7.

155. Janousek B, Zluvova J, Vyskot B. Histone H4 acetylation and DNA methylation dynamics during pollen development. *Protoplasma* 2000;**211**:116–22.

156. Ingouff M, Hamamura Y, Gourgues M, Higashiyama T, Berger F. Distinct dynamics of HISTONE3 variants between the two fertilization products in plants. *Curr Biol* 2007;**17**:1032–7.

157. Okada T, Singh MB, Bhalla PL. Histone H3 variants in male gametic cells of lily and H3 methylation in mature pollen. *Plant Mol Biol* 2006;**62**:503–12.

158. Dupressoir A, Heidmann T. Germ line-specific expression of intracisternal A-particle retrotransposons in transgenic mice. *Mol Cell Biol* 1996;**16**:4495–503.

159. Pasyukova E, Nuzhdin S, Li W, Flavell AJ. Germ line transposition of the *copia* retrotransposon in *Drosophila melanogaster* is restricted to males by tissue-specific control of *copia* RNA levels. *Mol Gen Genet* 1997;**255**:115–24.

160. Carmell MA, Girard A, van de Kant HJ, Bourc'his D, Bestor TH, de Rooij DG, et al. MIWI2 is essential for spermatogenesis and repression of transposons in the mouse male germline. *Dev Cell* 2007;**12**:503–14.

161. Helliwell CA, Wood CC, Robertson M, Peacock WJ, Dennis ES. The *Arabidopsis* FLC protein interacts directly *in vivo* with *SOC1* and *FT* chromatin and is part of a high-molecular-weight protein complex. *Plant J* 2006;**46**:183–92.

162. Sheldon CC, Hills MG, Lister C, Dean C, Dennis ED, Peacock WJ. Resetting of *FLOWERING LOCUS C* expression after epigenetic repression by vernalization. *Proc Natl Acad Sci* 2008;**105**:2214–19.

163. He Y, Amasino RM. Role of chromatin modification in flowering-time control. *Trends Plant Sci* 2005;**10**:30–5.

164. Pien S, Fleury D, Mylne JS, Crevillen P, Inzé D, Avramova Z, et al. *ARABIDOPSIS* TRITHORAX1 dynamically regulates FLOWERING LOCUS C activation via histone 3 lysine 4 trimethylation. *Plant Cell* 2008;**20**:580–8.

165. Saleh A, Alvarez-Venegas R, Avramova Z. Dynamic and stable histone H3 methylation patterns at the *Arabidopsis* FLC and AP1 loci. *Gene* 2008;**423**:43–7.

166. Liu F, Quesada V, Crevillén P, Bäurle I, Swiezewski S, Dean C. The *Arabidopsis* RNA-binding protein FCA requires a lysine-specific demethylase 1 homolog to downregulate FLC. *Mol Cell* 2007;**28**:398–407.

167. Deal RB, Topp CN, McKinney EC, Meagher RB. Repression of flowering in *Arabidopsis* requires activation of FLOWERING LOCUS C expression by the histone variant H2A.Z. *Plant Cell* 2007;**19**:74–83.

168. March-Díaz R, Reyes JC. The beauty of being a variant: H2A.Z and the SWR1 complex in plants. *Mol Plant* 2009;**2**:565–77.

169. Choi K, Park C, Lee J, Oh M, Noh B, Lee I. *Arabidopsis* homologs of components of the SWR1 complex regulate flowering and plant development. *Development* 2007;**134**:1931–41.

170. Swiezewski S, Crevillen P, Liu F, Ecker JR, Jerzmanowski A, Dean C. Small RNA-mediated chromatin silencing directed to the 3′ region of the *Arabidopsis* gene encoding the developmental regulator, *FLC. Proc Natl Acad Sci USA* 2007;**104**:3633–8.

171. Law JA, Jacobsen SE. Dynamic DNA methylation. *Science* 2009;**323**:1568–9.

172. Schuettengruber B, Chourrout D, Vervoort M, Leblanc B, Cavalli G. Genome regulation by polycomb and trithorax proteins. *Cell* 2007;**128**:735–45.

173. Goodrich J, Puangsomlee P, Martin M, Long D, Meyerowitz EM, Coupland G. A Polycomb-group gene regulates homeotic gene expression in *Arabidopsis. Nature* 1997;**386**:44–51.

174. Schubert D, Clarenz O, Goodrich J. Epigenetic control of plant development by Polycomb-group proteins. *Curr Opin Plant Biol* 2005;**8**:553–61.

175. Makarevich G, Leroy O, Akinci U, Schubert D, Clarenz O, Goodrich J, et al. Different Polycomb group complexes regulate common target genes in *Arabidopsis. EMBO Rep* 2006;**7**:947–52.

176. Alvarez-Venegas R, Pien S, Sadder M, Witmer X, Grossniklaus U, Avramova Z. ATX1, an *Arabidopsis* homolog of *trithorax*, activates flower homeotic genes. *Curr Biol* 2003;**13**:627–37.

177. Pien S, Grossniklaus U. *Polycomb* group and *trithorax* group proteins in *Arabidopsis. Biochim Biophys Acta* 2007;**1769**:375–82.

178. Avramova Z. Evolution and pleiotropy of TRITHORAX function in *Arabidopsis. Int J Dev Biol* 2009;**53**:371–81.

179. Weake VM, Workman JL. Histone ubiquitination: triggering gene activity. *Mol Cell* 2008;**29**:653–63.

180. Chanvivattana Y, Bishop A, Schubert D, Stock C, Moon YH, Sung Z, et al. Interaction of Polycomb-group proteins controlling flowering in *Arabidopsis. Development* 2004;**131**:5263–76.

181. Xu L, Shen WH. Polycomb silencing of KNOX genes confines shoot stem cell niches in *Arabidopsis. Curr Biol* 2008;**18**:1966–71.

182. Calonje M, Sanchez R, Chen L, Sung ZR. EMBRYONIC FLOWER1 participates in polycomb group-mediated AG gene silencing in *Arabidopsis. Plant Cell* 2008;**20**:277–91.

183. Shilatifard A. Chromatin modifications by methylation and ubiquitination: implications in the regulation of gene expression. *Annu Rev Biochem* 2006;**75**:243–69.

184. Jiang D, Gu X, He Y. Establishment of the Winter-Annual Growth Habit via FRIGIDA-Mediated Histone Methylation at FLOWERING LOCUS C in *Arabidopsis. Plant Cell* 2009;**21**:1733–46.

185. Xia Z-B, Anderson M, Diaz MO, Zeleznik-Le N. MLL repression domain interacts with histone deacetylases, the polycomb group proteins HPC2 and BMI-1, and the co-repressor C-terminal-binding protein. *Proc Natl Acad Sci USA* 2003;**100**:8342–7.

186. Bernstein BE, Mikkelsen TS, Xie X, Kamal M, Huebert DJ, Cuff J, et al. A bivalent chromatin structure marks key developmental genes in embryonic stem cells. *Cell* 2006;**125**:315–26.

187. Sharov AA, Ko MS. Human ES cell profiling broadens the reach of bivalent domains. *Cell Stem Cell* 2007;**1**:237–8.

188. Jullien PE, Katz. A, Oliva M, Ohad N, Berger F. Polycomb group complexes self-regulate imprinting of the Polycomb group gene MEDEA in *Arabidopsis. Curr Biol* 2006;**16**:486–92.

189. Kinoshita R, Yadegari R, Harada JJ, Goldberg RB, Fischer RL. Imprinting of the *MEDEA* polycomb gene in the *Arabidopsis* endosperm. *Plant Cell* 1999;**11**:1945–52.

190. Saha A, Wittmeyer J, Cairns BR. Chromatin remodeling: the industrial revolution of DNA around histones. *Nat Rev Mol Cell Biol* 2006;**7**:437–47.

191. Kaya H, Shibahara K, Taoka K, Iwabuchi M, Stillman B, Araki T. FASCIATA genes for chromatin assembly factor-1 in *Arabidopsis* maintain the cellular organization of apical meristems. *Cell* 2001;**104**:131–42.

192. Zhu Y, Dong A, Meyer D, Pichon O, Renou J, Cao K, et al. *Arabidopsis* NRP1 and NRP2 encode histone chaperones and are required for maintaining postembryonic root growth. *Plant Cell* 2006;**18**:2879–92.

193. Liu Z, Zhu Y, Gao J, Yu F, Dong A, Shen WH. Molecular and reverse genetic characterization of NUCLEOSOME ASSEMBLY PROTEIN1 (NAP1) genes unravels their function in transcription and nucleotide excision repair in *Arabidopsis thaliana. Plant J* 2009;**59**:27–38.

194. Phelps-Durr TL, Thomas J, Vahab P, Timmermans MCP. Maize rough sheath2 and its *Arabidopsis* orthologue ASYMMETRIC LEAVES1 interact with HIRA, a predicted histone chaperone, to maintain knox gene silencing and determinacy during organogenesis. *Plant Cell* 2005;**17**:2886–98.

195. Ogas J, Kaufmann S, Henderson J, Somerville C. PICKLE is a CHD3 chromatin-remodeling factor that regulates the transition from embryonic to vegetative development in *Arabidopsis. Proc Natl Acad Sci USA* 1999;**96**:13839–44.

196. Caikovski M, Yokthongwattana C, Habu Y, Nishimura T, Mathieu O, Paszkowski J. Divergent evolution of CHD3 proteins resulted in MOM1 refining epigenetic control in vascular plants. *PLoS Genet* 2008;**4**(8):e1000165.

197. Sarnowski TJ, Ríos G, Jásik J, Swiezewski S, Kaczanowski S, Li Y, et al. SWI3 subunits of putative SWI/ SNF chromatin-remodeling complexes play distinct roles during *Arabidopsis* development. *Plant Cell* 2005;**17**:2454–72.

198. Bezhani S, Winter C, Hershman S, Wagner JD, Kennedy JF, Kwon CS, et al. Unique, shared, and redundant roles for the *Arabidopsis* SWI/SNF chromatin remodeling ATPases BRAHMA and SPLAYED. *Plant Cell* 2007;**19**:403–16.

199. Hurtado L, Farrona S, Reyes JC. The putative SWI/SNF complex subunit BRAHMA activates flower homeotic genes in *Arabidopsis thaliana*. *Plant Mol Biol* 2006;**62**:291–304.

200. Jerzmanowski A. SWI/SNF chromatin remodeling and linker histones in plants. *Biochim Biophys Acta* 2007;**1769**:330–45.

201. Tang X, Hou A, Babu M, Nguyen V, Hurtado L, Lu Q, et al. The *Arabidopsis* BRAHMA chromatin-remodeling ATPase is involved in repression of seed maturation genes in leaves. *Plant Physiol* 2008;**147**:1143–57.

202. Reyes JC. Chromatin modifiers that control plant development. *Curr Opin Plant Biol* 2006;**9**:21–7.

203. Kwon CS, Wagner D. Unwinding chromatin for development and growth: a few genes at a time. *Trends Genet* 2007;**23**:403–12.

204. Knizewski L, Ginalski K, Jerzmanowski A. Snf2 proteins in plants: gene silencing and beyond. *Trends Plant Sci* 2008;**13**:557–65.

205. March-Díaz R, García-Domínguez M, Lozano-Juste J, León J, Florencio FJ, Reyes J,C. Histone H2A.Z and homologues of components of the SWR1 complex are required to control immunity in *Arabidopsis*. *Plant J* 2008;**53**:475–87.

206. March-Diaz. R, García-Domínguez M, Florencio FJ, Reyes JC. SEF, a new protein required for flowering repression in *Arabidopsis*, interacts with PIE1 and ARP6. *Plant Physiol* 2007;**143**:893–901.

207. Barlow DP. Methylation and imprinting: from host defense to gene regulation? *Science* 1993;**260**:309–10.

208. Lawrence RJ, Earley K, Pontes O, Silva M, Chen ZJ, Neves N, et al. A concerted DNA methylation/histone methylation switch regulates rRNA gene dosage control and nucleolar dominance. *Mol Cell* 2004;**13**:599–609.

Metabolism and Epigenetics

CHAPTER 17

Metabolic Regulation of DNA Methylation in Mammals

Ji-Hoon E. Joo, Roberta H. Andronikos, and Richard Saffery
Developmental Epigenetics, Murdoch Children's Research Institute, and Department of Paediatrics, University of Melbourne, Parkville, Victoria, Australia

THE DYNAMIC METHYLOME

Over 100 methyltransferase enzymes (including four DNA methyltransferases) have been described in mammals, involved in the transfer of methyl groups to a large array of proteins, phospholipids, and nucleotides. These reactions are fundamental to many different cellular functions, and it is therefore not surprising that insufficient methyl donor availability has the potential to disrupt a wide variety of biological processes, including DNA nucleotide synthesis and methylation, and gene expression (among others). The production of sufficient methyl-donors is therefore of critical importance for faithful cell division and development.

The methylated maternal and paternal genomes are de-methylated at fertilization and specific patterns of methylation are then re-established progressively starting in the early post-conception period [1]. The *de novo* establishment of DNA methylation is carried out by DNMT3A and -3B methyltransferases and is modulated by DNMT3L, lacking direct catalytic activity [2]. Recent work with human cell lines has also shown dynamic remodeling of epigenetic markings during the cell cycle [3–5].

In addition to the *de novo* establishment and removal of DNA methylation markings during early development, genome-wide methylation profile of dividing cells is faithfully copied in newly synthesized DNA strands in daughter cells following cell division. This is carried out by the maintenance DNA methyltransferase, DNMT1 [6]. As around 4% of all cytosine nucleotides within genomic DNA of mammalian somatic cells are methylated [7], this necessitates the constant availability of a pool of methyl donors in dividing cells if the methylome is to be faithfully replicated. The focus of this chapter is dedicated to a discussion of the metabolic processes involved in one-carbon (methyl) donor production and their impact on DNA methylation.

ONE-CARBON METABOLISM AND METHYL DONOR PRODUCTION

The methyl groups required for establishment and maintenance of DNA methylation are derived solely from dietary methyl donors in association with specific enzymes and

281

Handbook of Epigenetics: The New Molecular and Medical Genetics. DOI: 10.1016/B978-0-12-375709-8.00017-4

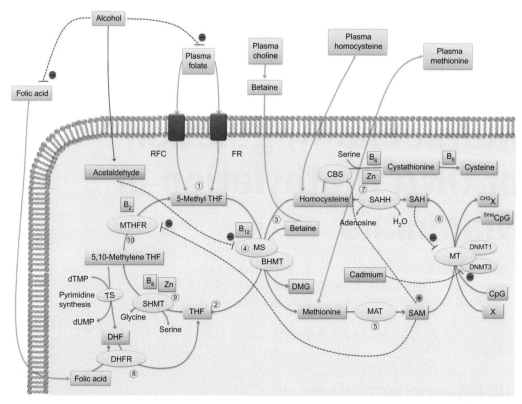

FIGURE 17.1

Schematic illustration of one-carbon metabolic pathway. Precursors and cofactors obtained exclusively from the diet are highlighted in orange. Red dashed lines indicate inhibition of enzymatic reactions: acetylaldehyde directly inhibits folate and absorption of folate. It also down-regulates MS [45,53]. High levels of SAH down-regulate the activity of MT. Increased concentration of SAM inhibits the activity of MTHFR, limiting the bioavailability of 5-methyl THF [8]. Cadmium is an inhibitor of DNMT activity [96]. Green dashed line indicates up-regulation of specific pathways. All mammalian tissues express MAT and MS, whereas BHMT is found only in the liver and kidney. SAM inhibits MTHFR and MS and activates CBS leading to homocysteine channeling down the trans-sulfuration pathway. **Abbreviations:** RFC, reduced folate carrier; FR, folate receptor; THF, tetrahydrofolate; MS, methionine synthase;, BHMT, betaine homocysteine methyltransferase; DMG, dimethylglycine; MAT, methionine adenosyltransferase; SAM, S-adenosylmethionine; MT, methyltransferase; X, substrates for methylation; SAH, S-adenosylhomocysteine; CBS, cystathionine beta-synthase; SHMT, serine hydroxymethyltransferase; MTHFR; methylene THF reductase; DHF, dihydrofolate; TS, thymidylate synthase. (Please refer to color plate section)

associated cofactors [8]. The addition of a methyl group to the 5th carbon of cytosine within CpG dinucleotides is catalyzed solely by DNMTs and utilizes the final methyl donor produced by one-carbon metabolism, S-adenosylmethionine (also known as SAM, SAMe, adomethionine, adoMet). This molecule represents the universal methyl donor in all cells [9].

Folates are the primary methyl donors and key mediators of one-carbon metabolic pathways along with choline and other cofactors such as B-group vitamins B_2, B_6, and B_{12}. Folate in blood plasma exists predominantly as polyglutamated methyl-tetrahydrofolate (methyl-THF) (① in Fig. 17.1). Following transport into cytoplasm, primarily by the reduced folate carrier (RFC), methyl-THF acts as a methyl group donor for the production of tetrahydrofolate (THF; ② in Fig. 17.1) and a precursor for homocysteine conversion to methionine (③ in Fig. 17.1). The process is catalysed by methionine synthase (MS; ④ in Fig. 17.1), and requires cofactor vitamin B_{12}. Since MS is the only reaction that utilizes methyltetrahydrofolate, folates remain "trapped" in this form by any impairment of the MS-catalyzed reaction [10,11].

Methionine is further activated to SAM by methionine adenosyltransferase (MAT; ⑤ in Fig. 17.1). SAM synthesis represents the preferred catabolic pathway for methionine in the liver, where up to half of the daily intake of methionine is converted to SAM and the majority of all subsequent methylation reactions take place [12]. Such reactions invariably result in conversion of SAM to S-adenosylhomocysteine (SAH; ⑥ in Fig. 17.1), which is further hydrolysed back to homocysteine by SAH hydrolase (SAHH; ⑦ in Fig. 17.1). SAH is a potent competitive inhibitor of transmethylation reactions; disruption of the SAH:SAM ratio, through an increase in SAH or a decrease in SAM, leads to inhibition of transmethylation reactions [13,14]. For this reason, SAH removal is essential and is carried out by SAH hydroxylase, but only in the absence of downstream products adenosine and homocysteine [15].

Folic acid is a synthetic form of folate which is widely used in food fortification and diet supplements. Unlike folate, it must be reduced, prior to entering the one-carbon cycle, to DHF, dihydrofolate, and then to THF (⑧ in Fig. 17.1). Both processes are mediated by dihydrofolate reductase (DHFR). THF formed in this manner (③, ④ in Fig. 17.1) is further converted and recycled to 5,10-methyl THF with the aid of vitamin B_6 (⑨ in Fig. 17.1), and to 5-methyl THF by methylene THF reductase (MTHFR) with the aid of vitamin B_2 (⑩ in Fig. 17.1). The former reaction is associated with conversion of serine to glycine, catalyzed by serine hydroxymethyltransferase (SHMT). Folic acid may have a negative effect on the absorption of naturally-occurring folates as it binds endogenous receptors with a higher affinity than THF [16]. Chemical structural formulas for the key compounds involved in this pathway are shown in Table 17.1.

S-ADENOSYLMETHIONINE (SAM): MASTER OF THE METHYLOME

SAM is an essential precursor molecule, found in all living organisms. It is the principal methyl donor in all tissues (transmethylation) and also the precursor of aminopropyl groups (aminopropylation) and of glutathione (transsulfuration) in the liver [15].

Second only to ATP in its utility, in terms of the number of reactions it facilitates, SAM is critical not only for DNA methylation, but also for other metabolic reactions such as nucleic acid synthesis and histone methylation (an important epigenetic regulatory mechanism in its own right). In addition, SAM has an important role in the regulation of activity of several enzymes (discussed above) and evidence is mounting for independent actions on cell growth, apoptosis, and differentiation, independent of its role as a methyl donor [17]. The structure of SAM was first reported by Catoni in 1951 [18]. This revealed the presence of a high energy sulphonium ion that activates each of the attached carbon atoms towards nucleophilic attack, making it an excellent substrate for distinct biochemical reactions (Fig. 17.2). SAM is produced following transfer of the adenosyl moiety of ATP to methionine by the MAT enzyme.

In the majority of instances SAM reacts by transfer of the S-associated methyl group to acceptor molecules as part of transmethylation reactions [15]. The common product for all reactions is S-adenosylhomocysteine (SAH). Most transmethylation reactions are inhibited by SAH, whereas decreasing SAM is stimulatory to enzyme activity [14]. In this way, SAM inhibits formation of THF, involved in transmethylation, while at the same time stimulating the transsulfuration pathway that converts homocysteine (a byproduct of transmethylation reactions) to cystathione and cysteine [19] (Fig. 17.1). Interestingly, oxidative stress may play an important role in determining the balance between transsulfuration and transmethylation reactions. The MS enzyme is inactivated by oxidation and requires reductive methylation for reactivation [20].

REGULATION OF SAM LEVELS BY PRECURSOR AND COFACTOR BIOAVAILABILITY

The production of SAM as a methyl donor for DNA methylation involves a complex interplay between metabolic substrate and cofactor bioavailability and enzymatic activity, the latter being also influenced by genetic variability.

TABLE 17.1 Chemical Structures of the Key Compounds Involved in the One-carbon Metabolic Pathway

Compound Name	Chemical Structure
Folate	
Tetrahydrofolate (THF)	
Dihydrofolate (DHF)	
5-methyl THF	
5,10-methyleneTHF	
Homocysteine	
Methionine	

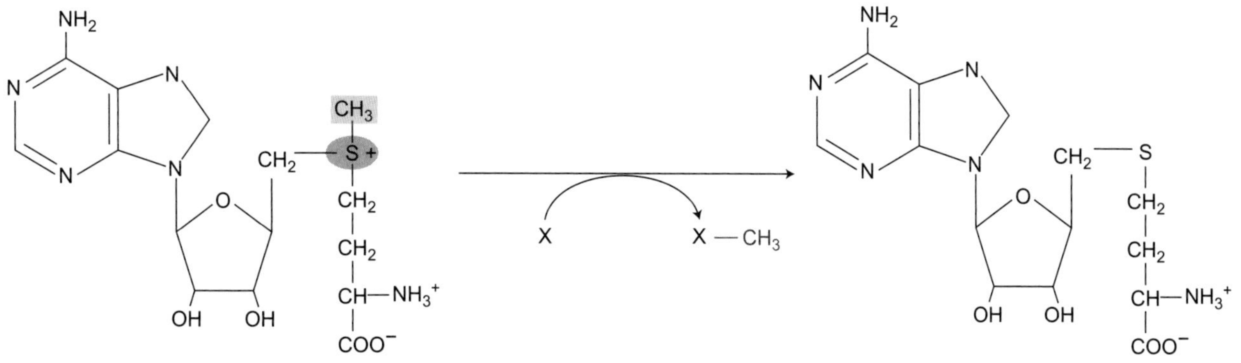

FIGURE 17.2
Conversion of S-adenosyl methionine to S-adenosyl homocysteine. The high energy sulfonium ion (orange shading) of SAM activates each of the attached carbons towards nucleophilic attack facilitating methyl donor transfer. (Please refer to color plate section)

As mentioned above, SAM is the major methyl donor and is derived solely from dietary methionine, choline, and folates [21]. Transfer of a single methyl group from SAM to a cytosine residue in DNA occurs exclusively at CpG dinucleotides in mammals. As SAH, produced during DNA methylation, functions as an inhibitor of SAM-dependent methyltransferases (Fig. 17.1), removal of this byproduct is a prerequisite if methylation demand within the cell is to be met [8].

Homocysteine Metabolism

Homocysteine (2-amino-4-mercaptobutyric acid) represents a point of intersection between the methionine cycle and transsulfuration pathway. Homocysteine is a non-essential amino acid, derived from metabolism of dietary methionine [8], which causes toxicity upon accumulation [22]. Homocysteine is produced following hydrolysis of SAH by the enzyme S-adenosylhomocysteine hydrolase [23]. Homocysteine produced via hydrolysis of SAH can either undergo re-methylation, yielding methionine, or transsulfuration, yielding cysteine [8,23]. It can also be exported into extracellular fluids [23]. The liver is the tissue thought to make the greatest contribution to the plasma homocysteine level via homocysteine export [24]. Regulation of homocysteine distribution between methylation and transsulfuration by SAM is speculated to occur primarily in the liver, which is unique in its ability to synthesize excess SAM in response to methionine availability. In non-hepatic tissues the level of SAM is tightly regulated and SAH may be the dominant effector [13].

The basic methionine cycle occurs in all mammalian cells. In contrast, transsulfuration has a limited distribution in mammalian cells, and those lacking this pathway require an exogenous source of cysteine [13]. Remethylation of homocysteine to methionine can be catalyzed by methionine synthase and requires cobolamin or vitamin B_{12} as a cofactor [8]. This is recycled via the action of MTHFR, an important reaction in one-carbon metabolism with the potential to influence DNA methylation [8]. In the liver, homocysteine can also be remethylated to methionine by betaine homocysteine methyltransferase (BHMT) [23].

Because the pathways for the metabolism of folate and choline intersect at the conversion of homocysteine to methionine, folate and choline act together to decrease homocysteine concentration. Impairment of these pathways, as a result of dietary deficiency or genetic polymorphisms reducing enzyme activity (see below), leads to elevated plasma homocysteine concentrations.

Hyperhomocysteinemia is an independent risk factor for several diseases including cancer, mental health disorders and vascular disease [25–29]; however, it remains unclear whether this is a causative agent or a marker of specific pathologies.

Glycine-N-methyltransferase

Glycine N-methyltransferase (GNMT) is a SAM-dependent enzyme present in liver, kidney, and pancreas. It plays an important role in regulating the ratio of SAM to SAH within the cell [23], which can be considered a "methylation potential" or indicator of capacity for functional SAM activity [30]. GNMT activity is inhibited by allosteric binding of 5-methyl THF, the synthesis of which is inhibited by SAM [8]. The activity of GNMT can therefore be regulated by the intracellular concentrations of SAM and 5-methyl THF. The SAM/SAH ratio is regulated via inhibition of SAM by 5,10-methylene-THF reductase (TMFR), and of GNMT by folate compounds [23]. Folate deficiency in rats leads to increased GNMT activity and a decrease in the SAM/SAH ratio [31].

Dietary Folate, SAM and DNA Methylation

Available evidence from both animal and human studies suggests that the effects of folate deficiency on DNA methylation are very complex, being cell type and target organ dependent, and are gene- and site-specific [32,33]. Both circumstantial and direct evidence exists for a link between disruption in the folate pathway (Fig. 17.1), changes in DNA methylation, altered gene expression, and disease predisposition. Animal studies have shown that prenatal feeding of a methyl-supplemented diet can increase DNA methylation and decrease expression of genes in offspring [34–36], while limiting folate supply in humans results in increasing levels of homocysteine and reduced DNA methylation [29,37–39]. For example, colonic DNA methylation in humans has been positively correlated with serum and RBC folate concentrations and negatively correlated with plasma homocysteine concentrations [40]. Conversely, folate supplementation at 2.5–25 times the daily requirement for 3–12 months significantly increases genomic DNA methylation in subjects with resected colorectal adenoma or cancer [40–43].

Dietary folates (tetrahydrofolates or THFs) are required as cofactors for reactions involved in one-carbon metabolism. 5,10-Methylene THF is reduced to 5-methyl THF by the enzyme methylene THF reductase (MTHFR). 5-Methyl THF is required for conversion of homocysteine to methionine, and therefore influences availability of methyl donor SAM [22]. Dietary choline is oxidized to betaine, which can be utilized in an alternative pathway for the conversion of homocysteine to methionine, also influencing availability of SAM [44].

Low dietary intake of folate and choline decreases concentrations of SAM, resulting in hypomethylation of DNA.

Alcohol Antagonism of One-carbon Metabolism

Numerous studies have demonstrated that additional dietary or behavioral factors, such as alcohol consumption, can affect the bioavailability of SAM, and therefore DNA methylation levels [reviewed in Refs 45–48]. Alcohol antagonizes one carbon metabolism by preventing the absorption of folate in the body [49]. In addition, alcohol has been reported to adversely affect both the availability and metabolism of folate through diminished or inadequate dietary intake, impaired intestinal absorption, and increased loss via renal excretion [50].

Direct effects of alcohol on folate metabolism occur primarily via inhibition of the enzyme methionine synthase (MS), resulting in decreased concentrations of downstream products, methionine and SAM (Fig. 17.1), and increased concentrations of precursors, homocysteine and SAH [50]. A study utilizing a rodent model of alcoholism found that prolonged heavy alcohol consumption resulted in decreased tissue levels of SAM, increased tissue levels of SAH, and substantial global hypomethylation of DNA in the colonic mucosa [51]. In another study, inhibition of MS by alcohol in a rodent model also resulted in an increase in BHMT activity and a decrease in betaine level [52]. Alcoholism is associated with alterations in DNA methylation at both global and gene-specific level [53]. It is also linked to increased risk of cancer, especially in colon and liver. Paternal chronic alcohol consumption has been linked to impaired DNMT function in sperm. This further leads to the disruption of genomic imprinting and altered fetal growth [54].

GENETIC VARIANTS AND REGULATION OF SAM LEVELS

The rate of passage through the one-carbon cycle can be influenced by genetic polymorphisms in genes encoding the enzymes involved in this pathway [55]. The C to T substitution at nucleotide 677 of the methylene tetrahydrofolate reductase (MTHFR) gene (677C > T; rs1801133) has been most widely studied in this regard. This common SNP results in a more thermo-labile enzyme with lower catalytic activity, lower levels of 5-methyltetrahydrofolate, and increasing levels of 5,10-methylenetetrahydrofolate and plasma homocysteine. TT homozygous individuals have generally lower levels of DNA methylation relative to CC homozygotes [55,56]. This variant of MTHFR has been linked to an increasing risk of adverse pregnancy outcomes, various cancers, coronary disease, atherosclerosis, alcoholic liver disease, Down Syndrome, and neuropsychiatric conditions in humans, in both case-control studies and meta analyses [57–60]. A second common variant of MTHFR (1298 C > A; rs1801131) has also been widely studied and shown to reduce enzymatic activity, without altering thermolability [61,62]. Interestingly, opposite effects on red blood cell folate levels have been reported for each of the common MTHFR variants [63] and protective effects against disease have also been reported for these variants in some studies. A recent genome-wide association study confirmed genetic variation in the MTHFR gene as a major determinant of serum homocysteine levels [64].

Other previously described genetic variants with the potential to disrupt one-carbon metabolic include MTHFD1 (rs2236225, rs1950902), MTHFD2 (rs1667627), MTRR (rs1532268), MTR (rs1805087), BHMT (rs3733890), RFC-1 (rs1051266), and SHMT (rs1979277) [65–68] (reviewed in Refs 69,70).

In addition to the proposed link between genetic variants of enzymes regulating one-carbon metabolism, variants in DNMTs have also been identified as risk factors for disease, including DNMT1 in systemic lupus erythematosus [71]. Genetic deficiency of DNMT3B causes a recessive human disorder characterized by immunodeficiency, centromere instability, and facial anomalies [72]. Variants in other DNMTs (i.e. DNMT3L, DNMT1) have been associated with increased cancer risk [73–76].

MECHANISMS OF DNA DEMETHYLATION

The DNA methylation state is highly dynamic, and therefore represents a novel target for potential therapeutic intervention. Although the mechanism by which methyl groups are added to CpG dinucleotides is now well defined, the mechanism for DNA demethylation remains the subject of intense debate. Three candidate mechanisms have been proposed in mammals. The first involves the removal of methyl groups from 5-methyl cytosine by a DNA (base-excision) repair-based system [77–79]. In this model, 5-methyl cytosine is first deaminated to thymine, resulting in a T-G base pair mismatch. Deamination can be triggered either by a DNA methyltransferase, or DNA deaminases, such as the AID (Activation-induced deaminase)/APOBEC1 (apolipoprotein B mRNA editing enzyme, catalytic polypeptide 1) family in the zygote. Those deaminases are potential candidates for a global demethylating mechanism in early development [5,77,78]. There are also conflicting data related to the potential demethylating role of Dnmt3a/b [77]. T-G mismatches are recognized by thymine DNA glycosylase (TDG) and the base is removed, leaving an abasic site. Methyl Binding Domain protein-4 (MBD4) is suggested to play role in this process [80]. The DNA repair system then repairs the abasic site by replacement with a non-methylated cytosine.

The second proposed mechanism of DNA demethylation involves the passive loss of methylation in the absence of the maintenance methyltransferase, DNMT1 [79]. During

DNA replication, methylation of the newly synthesized DNA strand is established by the action of DNMT1 in conjunction with SAM. However, a decrease in enzyme level or activity can lead to passive DNA demethylation as cells undergo subsequent rounds of division and DNA replication. 5-Aza-2′-deoxycytidine (AzdC) is a well known DNMT inhibitor, associated with passive DNA demethylation [81].

The third, and most controversial, mechanism for DNA demethylation involves the direct removal of the methyl group from 5-methyl cytosine, via hydrolytic attack, oxidation, or a DNA demethylase enzyme (MBD2 has been suggested as a candidate) [82–84]. A recent study in rats has linked a specific environmental trigger (maternal behavior) to recruitment of MBD2 at a specific gene promoter (glucocorticoid receptor gene). It has been proposed that the candidate DNA demethylase favors the up-regulation of such genes, and that the activity of the demethylase can be inhibited by methionine supplementation (producing methyl donors) [83]. The absence of any *in vitro* studies replicating these findings suggests that there are likely to be other enzymes or cofactors involved. Future studies should therefore focus on the identification of these components.

DISRUPTION OF ONE-CARBON METABOLISM, DNA METHYLATION, AND DISEASE: WHAT IS THE LINK?

Mounting evidence suggests a direct link between disruption of one-carbon metabolism and disease pathogenesis. This can arise either through genetic variability (described above) or environmental factors (dietary, lifestyle, or exposures), or a combination of both. There is little doubt that the downstream consequences of this disruption are manifold and include changes to genomic DNA methylation and an imbalance in homocysteine levels, each of which has been associated with increasing disease risk in numerous studies.

Cancer and One-carbon Deficiency

Perhaps the most compelling evidence for a link between altered one-carbon metabolism and disease comes from the success attributed to folic acid supplementation in reducing the rate of neural tube defects [50]. However, given the interplay between one-carbon production, DNA methylation, and other fundamental cellular activities, the direct contribution of DNA methylation to human disease risk remains controversial.

Irrespective of this, animal studies have proven to be invaluable in confirming much of the circumstantial data obtained in human studies linking insufficient methyl donors, DNA hypomethylation, and increasing disease risk. For example, insufficient zinc intake is associated with decreasing levels of SAM in the liver, DNA hypomethylation, low birth weight, and reduced growth in rats [45,85]. Similarly, another co-substrate of one-carbon metabolism, selenium, is also shown to induce hypomethylation in rat colon and liver with increased homocysteine level in the plasma [86]. A similar finding has been reported in a study using a human colon cancer cell line (Caco-2) [87]. An interesting study also has found that the group of rats fed with selenium were less susceptible to colon cancer than the control group, suggesting adequate selenium intake may have a protective effect against cancer [88]. Despite these findings, the mechanism(s) by which selenium interacts with other enzymes and substrates in one-carbon metabolism still remains to be clarified.

There is compelling evidence for folate deficiency increasing disease risk. In general, most studies have found that folate deficiency is directly associated with low SAM level. This limits the availability of the universal methyl donor and induces global hypomethylation, in addition to specific promoter hypomethylation of certain genes associated with tumorigenesis associated with some cancers [11,89–91].

Interestingly, some suggest that the link between folate intake and risk of cancer may be dosage- and time-dependent [11,90,91]. Inadequate folate may contribute to cancer progression but high folate levels may also promote cancer growth once established [90,91]. This may be due to the role of folate in synthetizing nucleotides which are required for rapid DNA replication and cell division in cancer; thus high folate levels may enhance tumor progression [11,92]. Therefore, it has been suggested that there exists an optimal folate intake for cancer prevention [90].

Additionally, the current evidence suggests that inadequate intake of vitamins B_6 or B_{12}, along with low folate, decreases the SAM levels which, in turn, causes the increased level of homocysteine in blood plasma due to a block in the one-carbon cycle. Both choline and methionine deficiency have also been linked to increased risk of cancer with a further elevated risk when the deficiencies occur in combination with each other or with folate deficiency [45]. The relationship between deficiencies in one-carbon metabolites/cofactors and cancer development/progression seems to be complex and requires further investigation [90].

Alcohol Related Disease

Given the accumulated data linking alcohol and folate antagonism through reduced folate absorption, increased folate excretion, and direct alteration of methionine synthase (MS) activity [45,53], it is not surprising that supplementing one-carbon metabolism has been investigated as a treatment for diseases associated with chronic alcohol consumption. Whereas inhibition of MS activity by chronic over-consumption of alcohol creates a condition that cannot be overcome by dietary folate supplementation alone, betaine supplementation can alleviate alcohol-induced alterations to one-carbon metabolism in the liver by restoring levels of SAM and preventing increased export of homocysteine [23]. Direct administration of SAM to subjects with alcoholic liver disease, has been associated with improvements in survival [93], and can prevent the development of liver cancer in at least some rodent models [94], but does not appear to reverse cancer progression once established [95].

CONCLUDING REMARKS

The metabolic pathways that generate the primary methyl donors needed for the *de novo* establishment and maintenance of the DNA methylation profile are complicated and do not exist in isolation from other essential biochemical reactions necessary for cell survival, division, and differentiation. However, despite the "mountains" of conflicting experimental data, the significance of methyl donor production and insufficiency is unequivocal. To date, unraveling the mechanisms underlying disease risk associated with deficiencies in these pathways has proven problematic, and in many cases, controversial. The recent rise in reports providing novel insights into the interaction of environmental, genetic, and epigenetic determinants suggest that an understanding of these processes is likely to be achieved in the not too distant future. Application of emergent technologies which can be applied to high-resolution studies of DNA methylation across the genome (e.g. next-generation sequencing), will further speed progress towards this important goal.

References

1. Kelsey G. Genomic imprinting – roles and regulation in development. *Endocr Dev* 2007;**12**:99–112.

2. Aapola U, Kawasaki K, Scott HS, Ollila J, Vihinen M, Heino M, et al. Isolation and initial characterization of a novel zinc finger gene, DNMT3L, on 21q22.3, related to the cytosine-5-methyltransferase 3 gene family. *Genomics* 2000;**65**:293–8.

3. Brown SE, Fraga MF, Weaver IC, Berdasco M, Szyf M. Variations in DNA methylation patterns during the cell cycle of HeLa cells. *Epigenetics* 2007;**2**:54–65.

4. Kangaspeska S, Stride B, Metivier R, Polycarpou-Schwarz M, Ibberson D, Carmouche RP, et al. Transient cyclical methylation of promoter DNA. *Nature* 2008;**452**:112–15.

5. Metivier R, Gallais R, Tiffoche C, Le Peron C, Jurkowska RZ, Carmouche RP, et al. Cyclical DNA methylation of a transcriptionally active promoter. *Nature* 2008;**452**:45–50.

6. Bestor TH. The DNA methyltransferases of mammals. *Hum Mol Genet* 2000;**9**:2395–402.

7. Gama-Sosa MA, Midgett RM, Slagel VA, Githens S, Kuo KC, Gehrke CW, et al. Tissue-specific differences in DNA methylation in various mammals. *Biochim Biophys Acta* 1983;**740**:212–19.

8. Ulrey CL, Liu L, Andrews LG, Tollefsbol TO. The impact of metabolism on DNA methylation. *Hum Mol Genet* 2005;**14**(Spec No 1):R139–R47.

9. Stanger O. Physiology of folic acid in health and disease. *Curr Drug Metab* 2002;**3**:211–23.

10. Ulrich CM, Reed MC, Nijhout HF. Modeling folate, one-carbon metabolism, and DNA methylation. *Nutr Rev* 2008;**66**(Suppl. 1):S27–S30.

11. Smith AD, Kim YI, Refsum H. Is folic acid good for everyone? *Am J Clin Nutr* 2008;**87**:517–33.

12. Mudd SH, Poole JR. Labile methyl balances for normal humans on various dietary regimens. *Metabolism* 1975;**24**:721–35.

13. Finkelstein JD. Methionine metabolism in mammals. *J Nutr Biochem* 1990;**1**:228–37.

14. Mato JM, Alvarez L, Ortiz P, Pajares MA. S-adenosylmethionine synthesis: molecular mechanisms and clinical implications. *Pharmacol Ther* 1997;**73**:265–80.

15. Lu SC. S-Adenosylmethionine. *Int J Biochem Cell Biol* 2000;**32**:391–5.

16. Qiu A, Jansen M, Sakaris A, Min SH, Chattopadhyay S, Tsai E, et al. Identification of an intestinal folate transporter and the molecular basis for hereditary folate malabsorption. *Cell* 2006;**127**:917–28.

17. Lu SC, Mato JM. S-Adenosylmethionine in cell growth, apoptosis and liver cancer. *J Gastroenterol Hepatol* 2008;**23**(Suppl. 1):S73–S7.

18. Catoni GL. Methylation of nicotinamide with a soluble enzyme system from rat liver. *J Biol Chem* 1951;**189**:203–16.

19. Selhub J. Homocysteine metabolism. *Annu Rev Nutr* 1999;**19**:217–46.

20. Chen Z, Banerjee R. Purification of soluble cytochrome b5 as a component of the reductive activation of porcine methionine synthase. *J Biol Chem* 1998;**273**:26248–55.

21. Zeisel SH. Importance of methyl donors during reproduction. *Am J Clin Nutr* 2009;**89**:673S–7S.

22. Beaudin AE, Stover PJ. Folate-mediated one-carbon metabolism and neural tube defects: balancing genome synthesis and gene expression. *Birth Defects Res C Embryo Today* 2007;**81**:183–203.

23. Grillo MA, Colombatto S. S-adenosylmethionine and its products. *Amino Acids* 2008;**34**:187–193.

24. Stead LM, Brosnan ME, Brosnan JT. Characterization of homocysteine metabolism in the rat liver. *Biochem J* 2000;**350**(Pt 3):685–92.

25. Boushey CJ, Beresford SA, Omenn GS, Motulsky AG. A quantitative assessment of plasma homocysteine as a risk factor for vascular disease. Probable benefits of increasing folic acid intakes. *JAMA* 1995;**274**:1049–57.

26. Fanapour PC, Yug B, Kochar MS. Hyperhomocysteinemia: an additional cardiovascular risk factor. *Wmj* 1999;**98**:51–4.

27. Herrmann W. The importance of hyperhomocysteinemia as a risk factor for diseases: an overview. *Clin Chem Lab Med* 2001;**39**:666–74.

28. Wu LL, Wu JT. Hyperhomocysteinemia is a risk factor for cancer and a new potential tumor marker. *Clin Chim Acta* 2002;**322**:21–8.

29. Muskiet FA. The importance of (early) folate status to primary and secondary coronary artery disease prevention. *Reprod Toxicol* 2005;**20**:403–10.

30. Mato JM, Alvarez L, Ortiz P, Mingorance J, Duran C, Pajares MA. S-adenosyl-L-methionine synthetase and methionine metabolism deficiencies in cirrhosis. *Adv Exp Med Biol* 1994;**368**:113–17.

31. Ogawa H, Gomi T, Takusagawa F, Fujioka M. Structure, function and physiological role of glycine N-methyltransferase. *Int J Biochem Cell Biol* 1998;**30**:13–26.

32. Kim YI. Folate and DNA methylation: a mechanistic link between folate deficiency and colorectal cancer? *Cancer Epidemiol Biomarkers Prev* 2004;**13**:511–19.

33. Jhaveri MS, Wagner C, Trepel JB. Impact of extracellular folate levels on global gene expression. *Mol Pharmacol* 2001;**60**:1288–95.

34. Cooney CA, Dave AA, Wolff GL. Maternal methyl supplements in mice affect epigenetic variation and DNA methylation of offspring. *J Nutr* 2002;**132**:2393S–400S.

35. Lillycrop KA, Phillips ES, Jackson AA, Hanson MA, Burdge GC. Dietary protein restriction of pregnant rats induces and folic acid supplementation prevents epigenetic modification of hepatic gene expression in the offspring. *J Nutr* 2005;**135**:1382–6.

36. Cropley JE, Suter CM, Beckman KB, Martin DI. Germ-line epigenetic modification of the murine A vy allele by nutritional supplementation. *Proc Natl Acad Sci USA* 2006;**103**:17308–12.

37. Singh SM, Murphy B, O'Reilly R. Epigenetic contributors to the discordance of monozygotic twins. *Clin Genet* 2002;**62**:97–103.

38. Jacob RA, Gretz DM, Taylor PC, James SJ, Pogribny IP, Miller BJ, et al. Moderate folate depletion increases plasma homocysteine and decreases lymphocyte DNA methylation in postmenopausal women. *J Nutr* 1998;**128**:1204–12.

39. Rampersaud GC, Kauwell GP, Hutson AD, Cerda JJ, Bailey LB. Genomic DNA methylation decreases in response to moderate folate depletion in elderly women. *Am J Clin Nutr* 2000;**72**:998–1003.

40. Pufulete M, Al-Ghnaniem R, Khushal A, Appleby P, Harris N, Gout S, et al. Effect of folic acid supplementation on genomic DNA methylation in patients with colorectal adenoma. *Gut* 2005;**54**:648–53.

41. Cravo M, Fidalgo P, Pereira AD, Gouveia-Oliveira A, Chaves P, Selhub J, et al. DNA methylation as an intermediate biomarker in colorectal cancer: modulation by folic acid supplementation. *Eur J Cancer Prev* 1994;**3**:473–9.

42. Cravo ML, Pinto AG, Chaves P, Cruz JA, Lage P, Nobre Leitao C, et al. Effect of folate supplementation on DNA methylation of rectal mucosa in patients with colonic adenomas: correlation with nutrient intake. *Clin Nutr* 1998;**17**:45–9.

43. Kim YI, Baik HW, Fawaz K, Knox T, Lee YM, Norton R, et al. Effects of folate supplementation on two provisional molecular markers of colon cancer: a prospective, randomized trial. *Am J Gastroenterol* 2001;**96**:184–95.

44. Niculescu MD, Zeisel SH. Diet, methyl donors and DNA methylation: interactions between dietary folate, methionine and choline. *J Nutr* 2002;**132**:2333S–5S.

45. Davis CD, Uthus EO. DNA methylation, cancer susceptibility, and nutrient interactions. *Exp Biol Med (Maywood)* 2004;**229**:988–95.

46. McGowan PO, Meaney MJ, Szyf M. Diet and the epigenetic (re)programming of phenotypic differences in behavior. *Brain Res* 2008;**1237**:12–24.

47. Ross SA. Diet and DNA methylation interactions in cancer prevention. *Ann N Y Acad Sci* 2003;**983**:197–207.

48. Ross SA, Milner JA. Epigenetic modulation and cancer: effect of metabolic syndrome? *Am J Clin Nutr* 2007;**86**:s872–7.

49. Halsted CH, Villanueva JA, Devlin AM, Chandler CJ. Metabolic interactions of alcohol and folate. *J Nutr* 2002;**132**:2367S–72S.

50. Mason JB, Choi SW. Effects of alcohol on folate metabolism: implications for carcinogenesis. *Alcohol* 2005;**35**:235–41.

51. Choi SW, Stickel F, Baik HW, Kim YI, Seitz HK, Mason JB. Chronic alcohol consumption induces genomic but not p53-specific DNA hypomethylation in rat colon. *J Nutr* 1999;**129**:1945–50.

52. Barak AJ, Beckenhauer HC, Tuma DJ. Hepatic transmethylation and blood alcohol levels. *Alcohol Alcohol* 1991;**26**:125–8.

53. Bleich S, Hillemacher T. Homocysteine, alcoholism and its molecular networks. *Pharmacopsychiatry* 2009;**42**(Suppl. 1):S102–S9.

54. Bielawski DM, Zaher FM, Svinarich DM, Abel EL. Paternal alcohol exposure affects sperm cytosine methyltransferase messenger RNA levels. *Alcohol Clin Exp Res* 2002;**26**:347–51.

55. Friso S, Choi SW, Girelli D, Mason JB, Dolnikowski GG, Bagley PJ, et al. A common mutation in the 5,10-methylenetetrahydrofolate reductase gene affects genomic DNA methylation through an interaction with folate status. *Proc Natl Acad Sci USA* 2002;**99**:5606–11.

56. Castro R, Rivera I, Ravasco P, Camilo ME, Jakobs C, Blom HJ, et al. 5,10-methylenetetrahydrofolate reductase (MTHFR) 677C→T and 1298A→C mutations are associated with DNA hypomethylation. *J Med Genet* 2004;**41**:454–8.

57. Loenen WA. S-adenosylmethionine: jack of all trades and master of everything? *Biochem Soc Trans* 2006;**34**:330–3.

58. Ueland PM, Hustad S, Schneede J, Refsum H, Vollset SE. Biological and clinical implications of the MTHFR C677T polymorphism. *Trends Pharmacol Sci* 2001;**22**:195–201.

59. Kim YI. 5,10-Methylenetetrahydrofolate reductase polymorphisms and pharmacogenetics: a new role of single nucleotide polymorphisms in the folate metabolic pathway in human health and disease. *Nutr Rev* 2005;**63**:398–407.

60. Shi J, Gershon ES, Liu C. Genetic associations with schizophrenia: meta-analyses of 12 candidate genes. *Schizophr Res* 2008;**104**:96–107.

61. van der Put NM, Gabreels F, Stevens EM, Smeitink JA, Trijbels FJ, Eskes TK, et al. A second common mutation in the methylene tetrahydrofolate reductase gene: an additional risk factor for neural-tube defects? *Am J Hum Genet* 1998;**62**:1044–51.

62. Weisberg I, Tran P, Christensen B, Sibani S, Rozen R. A second genetic polymorphism in methylene tetrahydrofolate reductase (MTHFR) associated with decreased enzyme activity. *Mol Genet Metab* 1998;**64**:169–72.

63. Parle-McDermott A, Mills JL, Molloy AM, Carroll N, Kirke PN, Cox C, et al. The MTHFR 1298CC and 677TT genotypes have opposite associations with red cell folate levels. *Mol Genet Metab* 2006;**88**:290–4.

64. Tanaka T, Scheet P, Giusti B, Bandinelli S, Piras MG, Usala G, et al. Genome-wide association study of vitamin B6, vitamin B12, folate, and homocysteine blood concentrations. *Am J Hum Genet* 2009;**84**:477–82.

65. Fredriksen A, Meyer K, Ueland PM, Vollset SE, Grotmol T, Schneede J. Large-scale population-based metabolic phenotyping of thirteen genetic polymorphisms related to one-carbon metabolism. *Hum Mutat* 2007;**28**:856–65.

66. Gast A, Bermejo JL, Flohr T, Stanulla M, Burwinkel B, Schrappe M, et al. Folate metabolic gene polymorphisms and childhood acute lymphoblastic leukemia: a case-control study. *Leukemia* 2007;**21**:320–5.

67. Koushik A, Kraft P, Fuchs CS, Hankinson SE, Willett WC, Giovannucci EL, et al. Nonsynonymous polymorphisms in genes in the one-carbon metabolism pathway and associations with colorectal cancer. *Cancer Epidemiol Biomarkers Prev* 2006;**15**:2408–17.

68. Ulvik A, Ueland PM, Fredriksen A, Meyer K, Vollset SE, Hoff G, et al. Functional inference of the methylene tetrahydrofolate reductase 677C→T and 1298A→C polymorphisms from a large-scale epidemiological study. *Hum Genet* 2007;**121**:57–64.

69. Molloy AM. Folate and homocysteine interrelationships including genetics of the relevant enzymes. *Curr Opin Lipidol* 2004;**15**:49–57.

70. Lucock M, Yates Z. Folic acid – vitamin and panacea or genetic time bomb? *Nat Rev Genet* 2005;**6**:235–40.

71. Park BL, Kim LH, Shin HD, Park YW, Uhm WS, Bae SC. Association analyses of DNA methyltransferase-1 (DNMT1) polymorphisms with systemic lupus erythematosus. *J Hum Genet* 2004;**49**:642–6.

72. Hansen RS, Wijmenga C, Luo P, Stanek AM, Canfield TK, Weemaes CM, et al. The DNMT3B DNA methyltransferase gene is mutated in the ICF immunodeficiency syndrome. *Proc Natl Acad Sci USA* 1999;**96**:14412–17.

73. Montgomery KG, Liu MC, Eccles DM, Campbell IG. The DNMT3B C→T promoter polymorphism and risk of breast cancer in a British population: a case-control study. *Breast Cancer Res* 2004;**6**:R390–R4.

74. Lee SJ, Jeon HS, Jang JS, Park SH, Lee GY, Lee BH, et al. DNMT3B polymorphisms and risk of primary lung cancer. *Carcinogenesis* 2005;**26**:403–9.

75. Kelemen LE, Sellers TA, Schildkraut JM, Cunningham JM, Vierkant RA, Pankratz VS, et al. Genetic variation in the one-carbon transfer pathway and ovarian cancer risk. *Cancer Res* 2008;**68**:2498–506.

76. Cebrian A, Pharoah PD, Ahmed S, Ropero S, Fraga MF, Smith PL, et al. Genetic variants in epigenetic genes and breast cancer risk. *Carcinogenesis* 2006;**27**:1661–9.

77. Gehring M, Reik W, Henikoff S. DNA demethylation by DNA repair. *Trends Genet* 2009;**25**:82–90.

78. Morgan HD, Dean W, Coker HA, Reik W, Petersen-Mahrt SK. Activation-induced cytidine deaminase deaminates 5-methylcytosine in DNA and is expressed in pluripotent tissues: implications for epigenetic reprogramming. *J Biol Chem* 2004;**279**:52353–60.

79. Patra SK, Patra A, Rizzi F, Ghosh TC, Bettuzzi S. Demethylation of (Cytosine-5-C-methyl) DNA and regulation of transcription in the epigenetic pathways of cancer development. *Cancer Metastasis Rev* 2008;**27**:315–34.

80. Zhu B, Zheng Y, Angliker H, Schwarz S, Thiry S, Siegmann M, et al. 5-Methylcytosine DNA glycosylase activity is also present in the human MBD4 (G/T mismatch glycosylase) and in a related avian sequence. *Nucleic Acids Res* 2000;**28**:4157–65.

81. Yoo CB, Jones PA. Epigenetic therapy of cancer: past, present and future. *Nat Rev Drug Discov* 2006;**5**:37–50.

82. Reik W. Stability and flexibility of epigenetic gene regulation in mammalian development. *Nature* 2007;**447**:425–32.

83. Szyf M. The early life environment and the epigenome. *Biochim Biophys Acta* 2009;**1790**:878–85.

84. Morgan HD, Santos F, Green K, Dean W, Reik W. Epigenetic reprogramming in mammals. *Hum Mol Genet* 2005;**14**(Spec No 1):R47–R58.

85. Wallwork JC, Duerre JA. Effect of zinc deficiency on methionine metabolism, methylation reactions and protein synthesis in isolated perfused rat liver. *J Nutr* 1985;**115**:252–62.

86. Uthus EO, Ross SA, Davis CD. Differential effects of dietary selenium (se) and folate on methyl metabolism in liver and colon of rats. *Biol Trace Elem Res* 2006;**109**:201–14.

87. Davis CD, Uthus EO, Finley JW. Dietary selenium and arsenic affect DNA methylation in vitro in Caco-2 cells and in vivo in rat liver and colon. *J Nutr* 2000;**130**:2903–9.

88. Finley JW, Davis CD, Feng Y. Selenium from high selenium broccoli protects rats from colon cancer. *J Nutr* 2000;**130**:2384–9.

89. Mason JB, Cole BF, Baron JA, Kim YI, Smith AD. Folic acid fortification and cancer risk. *Lancet* 2008;**371**:1335–6.

90. Ulrich CM. Folate and cancer prevention: a closer look at a complex picture. *Am J Clin Nutr* 2007;**86**:271–3.

91. Ulrich CM, Potter JD. Folate and cancer – timing is everything. *JAMA* 2007;**297**:2408–9.

92. Kim YI. Folate: a magic bullet or a double edged sword for colorectal cancer prevention? *Gut* 2006;**55**:1387–9.

93. Mato JM, Camara J, Fernandez de Paz J, Caballeria L, Coll S, Caballero A, et al. S-adenosylmethionine in alcoholic liver cirrhosis: a randomized, placebo-controlled, double-blind, multicenter clinical trial. *J Hepatol* 1999;**30**:1081–9.

94. Pascale RM, Marras V, Simile MM, Daino L, Pinna G, Bennati S, et al. Chemoprevention of rat liver carcinogenesis by S-adenosyl-L-methionine: a long-term study. *Cancer Res* 1992;**52**:4979–86.

95. Lu SC, Ramani K, Ou X, Lin M, Yu V, Ko K, et al. S-adenosylmethionine in the chemoprevention and treatment of hepatocellular carcinoma in a rat model. *Hepatology* 2009;**50**:462–71.

96. Takiguchi M, Achanzar WE, Qu W, Li G, Waalkes MP. Effects of cadmium on DNA-(Cytosine-5) methyltransferase activity and DNA methylation status during cadmium-induced cellular transformation. *Exp Cell Res* 2003;**286**:355–65.

CHAPTER **18**

Dietary and Metabolic Compounds Affecting Chromatin Dynamics/ Remodeling

F.I. Milagro, J. Campión, and J.A. Martinez
Department of Nutrition and Food Sciences, Physiology and Toxicology,
University of Navarra, Spain

INTRODUCTION

Eukaryotic DNA is intimately associated with histones forming chromatin. Two different forms of chromatin have been described: heterochromatin is a tightly compacted form of DNA usually associated with transcriptionally silent genomic regions, whereas euchromatin is a lightly packed form of chromatin usually under active transcription. The term chromatin remodeling encompasses a wide variety of changes in chromatin structure but can be defined as a discernible change in histone–DNA contacts. This molecular mechanism is regulated by epigenetic processes, for example covalent modification of histones, DNA methylation, non-coding RNA (microRNA and others), and Polycomb-group proteins [1].

The fundamental unit of the chromatin is the nucleosome, which consists of approximately 147 base pairs of DNA wrapped around a histone octamer containing two copies of each of the four conserved core histones: H2A, H2B, H3, and H4 [2]. Chromatin is further compacted by the incorporation of the linker histone H1, which has been reported to have eight isoforms in higher eukaryotes [3]. There are also variant histone subspecies that are recognized by differences in their amino acid sequence relative to the major histone species [4]. Each protein has both a histone fold domain, which mediates the histone–histone and histone–DNA interactions that are crucial for the assembly of the nucleosome core particle, and a flexible amino-terminal tail domain, which protrudes from the nucleosome core particle [5]. As shown in Figure 18.1, there are various histone post-transcriptional modifications that decorate the canonical histones (H2A, H2B, H3, and H4), as well as variant histones (such as H3.1, H3.3, and HTZ.1). The combination of modifications ('marks') produced by specific enzymes has been proposed to constitute a code that regulates downstream processes such as gene transcription, DNA repair, and apoptosis [6,7].

Altering chromatin structure at the level of histone modifications involves the participation of multiple enzymes that induce biochemical changes on precise amino acids in a post-transcriptional manner. Thus, enzymes have been identified for lysine acetylation, lysine and arginine methylation, serine and threonine phosphorylation, and lysine biotinylation (Fig. 18.2),

295

Handbook of Epigenetics: The New Molecular and Medical Genetics. DOI: 10.1016/B978-0-12-375709-8.00018-6

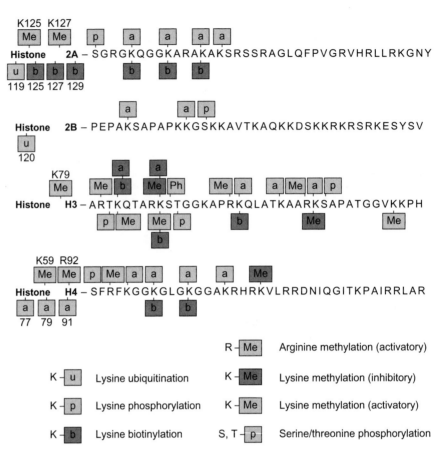

FIGURE 18.1

The histone code, including the main histone modifications and the amino acids affected.

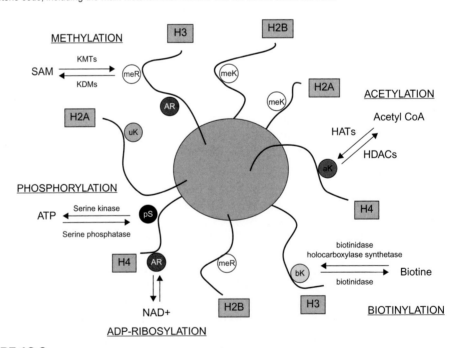

FIGURE 18.2

Main histone modifications, enzymes involved, and donors. Modifications: aK, acetyl lysine; meR, methyl arginine; meK, methyl lysine; uK, ubiquitinated lysine; pS, phosphorylated serine; bK, biotinylated lysine; AR, ADP-ribosylation. Enzymes and substrates: histone methyltransferases (KMTs) and histone lysine demethylases (KDMs); histone acetyltransferases (HATs) and histone deacetylases (HDACs); SAM, S-adenosyl methionine.

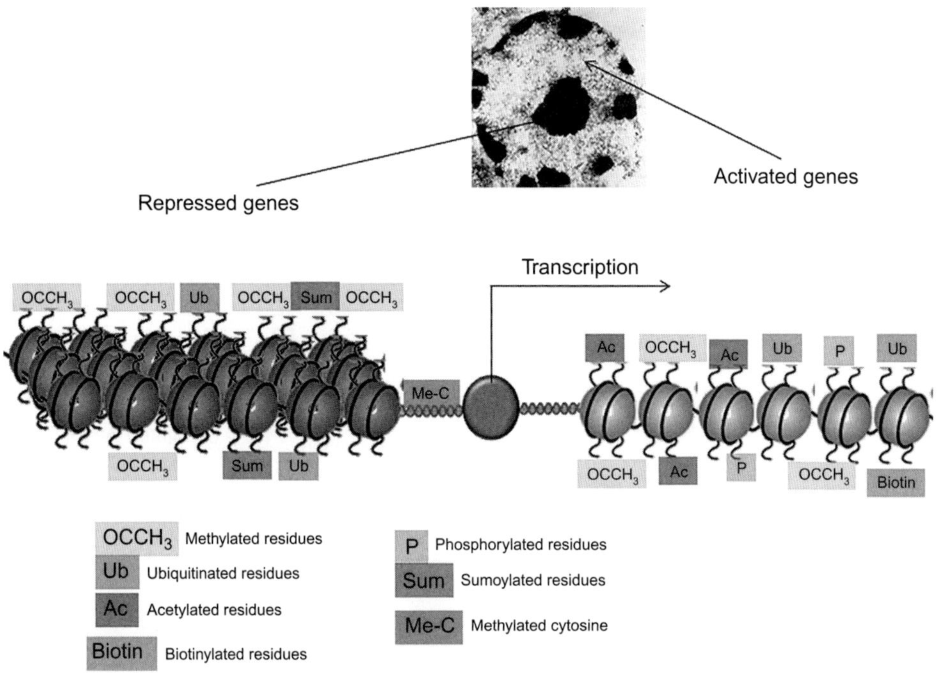

FIGURE 18.3
Histone modifications involved in the repression of gene transcription (left) or gene transcription activation (right). (Please refer to color plate section)

but also for ubiquitination, sumoylation, ADP-ribosylation, deimination, and proline isomerization [8]. Functions of these enzymes depend on cofactors, such as acetyl CoA, biotin, NAD, and S-adenosyl methionine (SAM), with the intranuclear levels of them depending on the metabolic, inflammatory, or redox state of the cell (Fig. 18.2). More subtle effects can be exerted by naturally occurring inhibitors, such as short chain fatty acids or nicotinamide. In any case, aging, environmental exposure, and lifestyle changes (including the diet) are closely associated with an epigenetic drift [9,10]. There are even transient, reversible circadian epigenetic patterns controlled by chromatin remodeling that are sensitive to environmental cues [11]. Especially important in epigenetic programming is the nutritional status of the mother during pregnancy and lactation, as well as the maternal behavior during interactions with pups [12].

Most modifications are localized to the amino- and carboxy-terminal histone tails, and a few are focused into the histone globular domains. There are over 60 different residues on histones where modifications have been detected, either by specific antibodies or by mass spectrometry. Extra complexity comes partly from the fact that methylation at lysines or arginines may be one of three different forms: mono-, di-, or trimethyl for lysines and mono- or di- (asymmetric or symmetric) for arginines. This vast array of modifications gives enormous potential for functional adaptive responses, but it has to be remembered that not all these modifications will be on the same histone at the same time. The majority of these histone changes regulate gene transcription, which has been explained by two different mechanisms. The first one proposes that chromatin packing is directly altered (either by a change in electrostatic charge or through internucleosomal contacts) to open or close the DNA polymer, thus controlling access of DNA-binding proteins such as transcription factors. The other one postulates that the attached chemical moieties alter the nucleosome surface to promote the association of chromatin binding proteins.

The effects of chromatin modifications on transcription regulation of the protein-encoding genome have been broadly classified into repressing and activating (Fig. 18.3). In other words, they correlate with, and perhaps directly regulate, gene repression and induction.

297

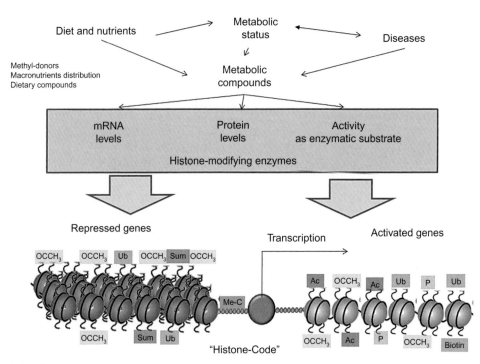

FIGURE 18.4
Influence of metabolic compounds and diseases on the activity of histone-modifying enzymes and gene transcription regulation. (Please refer to color plate section)

298

In general, histone acetylation and phosphorylation act as activators of gene expression, whereas histone deacetylation, biotinylation, and sumoylation inhibit gene expression. Methylation and ubiquitination can act as repressor or activator of gene expression depending on the histone residue being targeted [2,13,14].

The involvement of histone modifications in transcription factor binding processes provides a potentially enormous repertoire of signals. In fact, more than 100 possible chemical modifications of histones are known (acetylation, monomethylation, dimethylation, trimethylation, etc.), performed by different families of modifying and demodifying enzymes. Many of these enzymes require common metabolites as substrates or cofactors for regulating the "histone code" and the way the genes are regulated. Furthermore, changes in cellular metabolic compounds can affect enzyme function, acting as substrates or inhibitors of the modification of the N-terminal tail domains of the core histones. Cellular metabolic compounds can be affected in general by diet and nutrients, metabolic status of the body (hypoxia, hyperglycemia, redox status, inflammation), and also by endocrine unbalance and diseases that, in turn, could alter mRNA and protein levels of histone-modifying enzymes (Fig. 18.4).

This chapter does not intend to give an exhaustive review of the role of histones in epigenetics, but it is focused on the effect of metabolic and dietary compounds, as well as the metabolic state, on relevant post-translational histone modifications, with emphasis on acetylation and methylation of histones H3 and H4.

HISTONE ACETYLATION

Histones suffer acetylation and deacetylation, especially on lysine residues in the N-terminal tail. This regulatory mechanism is catalyzed by two types of enzyme, histone acetyltransferases (HATs) and histone deacetylases (HDACs), which not only act on histone substrates but also

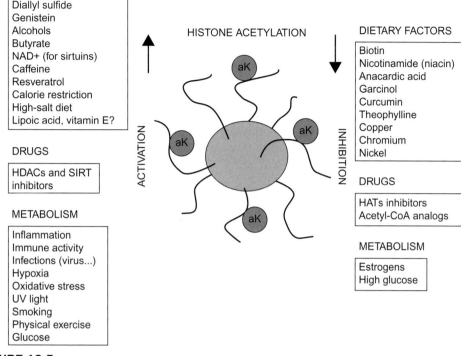

DIETARY FACTORS

Isothiocyanates
Diallyl sulfide
Genistein
Alcohols
Butyrate
NAD+ (for sirtuins)
Caffeine
Resveratrol
Calorie restriction
High-salt diet
Lipoic acid, vitamin E?

DRUGS

HDACs and SIRT
inhibitors

METABOLISM

Inflammation
Immune activity
Infections (virus...)
Hypoxia
Oxidative stress
UV light
Smoking
Physical exercise
Glucose

HISTONE ACETYLATION

ACTIVATION

INHIBITION

DIETARY FACTORS

Biotin
Nicotinamide (niacin)
Anacardic acid
Garcinol
Curcumin
Theophylline
Copper
Chromium
Nickel

DRUGS

HATs inhibitors
Acetyl-CoA analogs

METABOLISM

Estrogens
High glucose

FIGURE 18.5
Dietary and metabolic factors involved in histone acetylation and deacetylation.

299

on nonhistone proteins. In the acetylation process, acetyl-coenzyme A is the donor of the acetyl group, whereas this group is transferred to coenzyme A [15]. In fact, nuclear acetyl-CoA synthesis is a rate limiting step for histone acetylation. Thus, acetyl-CoA metabolism is directly linked to chromatin regulation and may affect diverse cellular processes in which acetylation and metabolism intersect, such as disease states and aging [16].

Many stimuli can activate histone acetylation (Fig. 18.5), such as inflammation (including IL-1β, TNFα, LPS), oxidative stress, UV light, bacteria, viruses, cigarette smoke [17], aging [18], hypoxia [19], cell adhesion-mediated signaling [20], biotin deficiency [21], quercetin [22], caffeine [23], or glucose availability [24]. Indeed, histone acetylation in mammalian cells is dependent on ATP-citrate lyase, the enzyme that converts glucose-derived citrate into acetyl CoA [25]. Regarding hormones, estrogens (i.e. 17beta-E2) attenuate H3 and H4 histone acetylation [26]. Finally, different metals are involved in histone acetylation, such as zinc, nickel, arsenite, chromium, and copper [27].

Histone Deacetylases (HDACs)

HDACs are specific deacetylases that remove the acetyl groups of lysine residues of histone tails leading to chromatin compaction and transcriptional repression [17,28,29]. HDACs are divided into four main classes based on their sequence homology and expression patterns. Class I, II, and IV HDACs are Zn-dependent deacetylases, whereas the Class III HDACs, also called sirtuins, are NAD-dependent deacetylases [30].

Different HDAC inhibitors have been clinically used for cancer therapy and psychiatry, and proposed for the treatment of metabolic diseases, such as stroke and cardiovascular diseases, asthma, arthritis, type 2 diabetes, and neurodegenerative processes [29,31,32]. These inhibitors include hydroxamates, such as trichostatin A (TSA) and vorinostat (SAHA); short chain fatty acids, such as butyrate and valproic acid; cyclic tetrapeptides, such as apicidin or depsipeptide, benzamides, epoxides, amides, ketones, and lactones [29,30,32].

Different dietary agents, such as biotin, lipoic acid, garlic organosulfur compounds (i.e. diallyl disulfide), and vitamin E metabolites show structural features compatible with HDAC inhibition [33]. Thus, diallyl disulfide is an organosulfur compound found in garlic that increases histone H3 and H4 acetylation in human colon tumor cell lines [34] and, such as allyl mercaptan and butanethiol, in liver and Morris hepatoma cells [35]. Sulforaphane (SFN) is an isothiocyanate found in cruciferous vegetables that weakly inhibits HDAC activity in human cell lines [36]. Isothiocyanates with a similar spacer length to that of sulforaphane, such as sulforaphene, erucin, and phenylbutyl isothiocyanate, exhibit comparable HDAC inhibitory activities. Even compounds with a longer or shorter spacer length, such as SFN[6], SFN[9], 6-erucin, and phenethyl isothiocyanate, display a similar effect [37]. Similarly, other natural organoselenium compounds found in yeasts (methylselenocysteine and selenomethionine) also act as HDAC inhibitors with potential anticancer properties [38]. The isoflavone genistein, with antiestrogenic activities, is able to inhibit HDAC6, which is the main cytoplasmic deacetylase in mammalian cells [39]. Other plant compounds with effects on the histone acetylation process are flavone [40], isoliquiritigenin, dihydrocoumarin, quercetin, and psammaplin A [5]. Finally, an activatory role of green tea's theophylline on HDAC activity has been reported [41]. However, despite the ability of these substances to induce punctual epigenetic changes, it has to be clarified whether they should be considered as real epigenetic modifiers [42].

Regarding metabolism, oxidative stress inhibits HDAC activity [43,44]. Function of Class IIa HDAC, at least in mouse models, is regulated by neuronal stimuli, immune activity, physical exercise, and perhaps fasting. Smoking alters the expression of Class I/II HDACs and a high-salt diet induces the expression of a kinase that can block the functioning of Class IIa HDACs in rats [45]. Finally, as all the above HDACs are zinc-dependent hydrolases [46], zinc levels in the diet could affect their activity.

Concerning Class III HDACs or sirtuins, resveratrol is a polyphenolic activator of SIRT1 found in red wine and different vegetable products that shows neuroprotective, antioxidant, and anti-aging effects [47]. Other SIRT1-activating compounds are the plant polyphenols butein and fisetin [48], the resveratrol metabolite piceatannol, the liquorice flavonoid isoliquiritigenin, and the flavonoid luteolin [49]. Similarly, calorie restriction upregulates SIRT1 in human mononuclear cells [50], fat, muscle, and liver [51], contributing thus to extend lifespan and ameliorate insulin resistance. However, as sirtuins need to bind to NAD^+, low levels of NAD^+ inhibit their activity [52]. Other sirtuin inhibitors are high nicotinamide levels and the naphthol-derived lactone sirtinol [47], as well as high glucose levels [53].

Histone Acetyltransferases (HATs)

HATs are acetyltransferases that acetylate lysines within lysine-rich amino-terminal tails of histone proteins, resulting in charge neutralization and a more relaxed, open, and transcriptionally active chromatin structure. They are broadly classified into two different classes, based on their functional localization: type A HATs, in the nucleus, and cytoplasmic type B HATs, that modify the newly synthesized histones before their assembly [54].

HAT inhibitors seem promising for the treatment of human diseases such as Alzheimer's disease, hyperlipidemias, and diabetes [55]. Despite that, few natural inhibitors of HATs have been reported. Thus, anacardic acid, isolated from cashew nut shell liquid, has been identified as a potent noncompetitive inhibitor of both p300 and PCAF HAT activity *in vitro* [56]. A polyprenylated benzophenone known as garcinol, isolated from *Garcinia indica* fruit rind, has been also identified as an inhibitor of p300 and PCAF HAT activity [57]. The polyphenol curcumin is a p300/CREB-binding protein-specific inhibitor of acetyltransferase that exhibits a variety of pharmacological effects including anti-tumor, anti-inflammatory, and anti-infectious activities [58,59]. The catechin epigallocatechin-3-gallate, abundant in tea and with potent antioxidant properties, acts also as a histone acetyltransferase inhibitor [60]. Although not natural products, several isothiazolone-based HAT inhibitors have been also identified showing inhibitory effects on PCAF and p300 [61], in a similar way to some alpha-methylene butyrolactones, benzylidene acetones, and alkylidene malonates [62]. Finally, there are different synthetic HAT inhibitors that are specific for p300 (i.e. lysyl CoA) and for PCAF (i.e. H3-CoA-20) [63].

Concerning HAT activators, glucose induces overall acetylation of H3K9, K18, and K27 and H4K5, K8, and K12 in yeasts, probably by regulating HAT activity [64]. Retinoic acid, an oxidized form of vitamin A, is also able to acetylate H3K9 [55]. Ethanol has been reported to acetylate histone H3 at lysine 9 (H3K9) in rat hepatocytes, as well as other hepatic nuclear and non-nuclear proteins, which probably contributes to alcohol-induced hepatotoxicity [65]. Surrogate alcohols, such as 1-propanol, 1-butanol, and isopentanol, modulate H3K9 via increasing HAT activity [66]. A high carbohydrate/fat ratio diet induces histone H3 acetylation at lysine 9 on the SGLT1 gene and its expression in the jejunum [67].

Other physiological factors that could participate in the acetylation process are cold exposure [68] and aging [69,70], at least for H4K16, whose acetylation plays a critical role in lifespan regulation, transcription activation, and protein interactions [71,72]. Chronic hypoxia induces H3K9 acetylation in prostate cancer cells, with several histone acetyltransferase enzymes (i.e. CBP and p300) being components of the HIF-1α complex and therefore potentiating the HIF-1α gene expression response to hypoxia [73].

Finally, apart from HATs and HDACs, other post-translational modifications regulate cellular acetylation. Thus, phosphorylation has been shown to activate HAT function and repress HDACs and, recently, methylation has also been shown to control HAT function [54].

HISTONE METHYLATION

Methylation of lysine or arginine residues can occur in several modification states. Thus, lysine residues can house either mono-, di- or trimethyl moieties on their amine group, whereas arginine residues can carry mono- or dimethyl groups on their guanidinyl group [74,75]. The dimethyl arginine state is further defined by whether the modification exists in the symmetric (me2s) or the asymmetric (me2a) configuration [74]. These modifications in histone methylation states can have different and profound implications for the function of chromatin, conforming *per se* an internal cryptogram within the histone code. Histone arginine methylation is more dynamic, correlating well with gene activation, whereas its loss from target arginines in H3 and H4 correlates with gene inactivation [76]. In contrast, lysine methylation seems to be a more stable mark, but with a more complicated readout [77]. Thus, although methylation of H3K4 and H3K79 correlates with gene activation, methylation of H3K27 correlates with repression [76].

Changes at any particular amino acid are determined by the dynamic equilibrium between the activities of modifying and de-modifying enzymes. Although histone methylation was discovered in the 1960s, histone methyltransferases (KMTs) and histone lysine demethylases (KDMs) have only recently been identified [75]. The most thoroughly studied lysine methylation marks in histones are found on H3K4, K9, K27, K36, K79, and H4K20. In general, H3K4, K36, and K79 methylation are localized near active or poised transcriptional units, and H3K9 and H4K20 modifications are hallmarks of silenced or heterochromatic regions. KMTs use SAM as methyl donor, while KDMs require either flavine adenine dinucleotide (FAD) or α-ketoglutarate, oxygen and Fe^{2+}, depending on whether they act on mono- and dimethyl lysine (e.g. KDME1/LSD1) or trimethyl lysine (e.g. jumonji domain enzymes) [78].

Histone arginine methylation occurs on H3R2, R8, R17 and R26 and H4R3. As lysine methylation, it can be either activatory or repressive for transcription, and the enzymes (protein arginine methyltransferases, PRMTs) are recruited to promoters by transcription factors [74]. The arginine demethylation process has recently been described as performed by deiminases and jumonji-group demethylases [79]. Thus, recently, a member of the JmjC family (JMJD6) was shown to apparently reverse histone methylation.

As different authors have described [80–83], epigenetic mechanisms could be directly controlled by metabolic and dietary compounds, metabolic state, endocrine unbalances and diseases (Fig. 18.4). Deepening into histone methylation, these studies early postulated possible effects of one-carbon and redox metabolism on methyltransferases action in addition to effects of oxygen on demethylases. Thus, DNA and histone methyltransferases (KMT and PRMTs) all use SAM as methyl donor [82], which can directly induce epigenetic marks. The bioavailability of SAM is directly influenced by the diet. Thus, SAM is formed from methyl groups derived from choline, methionine, or methyl-tetrahydrofolate. Betaine, choline, methionine, zinc, and folate are necessary for homocysteine conversion into methionine [1]. Thus, the effects of these nutrients on epigenetic marks are interrelated, interacting among them to alter epigenetic DNA and histone methylation processes. In fact, the perturbation of the metabolism of one of the methyl donors results in compensatory changes in the other methyl donors because of the intermingling of these metabolic pathways [84].

Several examples have been published in the literature showing a direct effect of metabolic compounds on methyltransferase activity. Thus, Clarke et al. [85] early described methyltransferases as being inhibited by the cellular ratio of SAM and S-adenosyl-homocysteine (SAH), especially by the inhibitory capacity of the latter. More recently, it has been observed that a methyl-deficient diet reduces protein and mRNA levels of some lysine and arginine methyltransferases in mice [85–88]. And a diet low in methionine and cysteine increases histone H1 relative to the total content of histones in rat liver [89]. On the other hand, gestational choline supplementation [90] and the *in vitro* experiments carried out by Ara et al. [see Ref.91] increase lysine methyltransferase mRNA levels. Finally, it is interesting to remark on the clinical relevance in cardiovascular diseases of the asymmetric dimethylarginine, a product of PRMTs [92,93].

Sometimes, changes in protein/mRNA levels of histone modifying enzymes are not experimentally related to changes in posttranslational modifications of histones. Thus, changes in the levels of methylated histones (lysines and arginines), in a global manner or on a specific locus, are more relevant (Table 18.1) in order to discriminate the direct effect of a metabolic compound. Changes in the methylation levels in lysine and arginine residues of histone 3 and 4 have been described as a result of changes in methyl donors in the diet. Thus, the levels of H3K9Me2 and H3K27Me3, tags of transcriptionally repressed chromatin, were up-regulated by choline supplementation, whereas the levels of H3K4Me2, associated with active promoters, were highest in choline-deficient rats [90]. A methyl-deficient diet increases considerably the level of histone macroH2A and H3K9me3 in mice, and reduces the levels of H3K27me3, H3K4me2, and H4K20me3, being sometimes related to changes in enzymes or not [88,94]. Similarly, it led to progressive loss of H4K20me3, H3K9me2, and H3K9me3 [94,95]. Food restriction and/or protein restriction increases H3K27me3, H3K9me3, and H3K9me1, and reduces H3K4me3, being related to changes in glucocorticoid receptor and IGf2 mRNA gene expression [96,97].

Interestingly, most of the JmjC domain-containing proteins are upregulated by hypoxia [98], and hypoxia induces an increase in methylated histones [99,100]. On the other hand, demethylase levels are regulated by oxygen tensions [100–102]. These findings suggest that an overexpression of jumonji histone demethylases helps to compensate for decreased levels of molecular oxygen in maintaining H3K4 methylation as a target of HIF-1 transactivation [100].

TABLE 18.1 Examples of Experimental Changes in the Level of Histone Methylations (lysines, K and arginines, R) by Different Metabolic Models

Effect	Residue/Methylation	Experimental Model	Reference
Increase	H3K4me1	Insulin via ROS, high glucose, choline deficiency	[90,104,111]
	H3K4me2	Methyl deficiency, hypoxia, hexavalent chromium, aging, glucose	[94,99,100,105,117]
	H3K4me3	Hexavalent chromium, arsenite, nickel, LPS, aging, hypoxia, glucose, ischemia, TNF-alpha	[91,99,100,105,106, 110,115,116,117]
	H3K9me1	Protein restriction, nickel, insulin via ROS with hyperglycemia, glucose deprivation, gestational choline supply, hypoxia, chromium, high glucose	[90,96,100,103,104, 105,107,113]
	H3K9me3	Food restriction, methyl deficient model, hexavalent chromium	[88,95,97,101]
	H3K27me3	Food restriction, hypoxia, gestational choline supply	[90,97,99]
	H3K36me3	Hypoxia	[100]
	H3K79me2	Hypoxia	[99]
	H4R3me2	Hypoxia	[99]
Reduction	H3K4me1	Insulin via ROS with hyperglycemia	[104]
	H3K4me2	High glucose, glucose induced cAMP, diabetic state	[110,113,114]
	H3K4me3	Food restriction, chromium, SAM, LPS	[91,97,101]
	H3K9me1	Insulin via ROS	[104]
	H3K9me2	Protein restriction, choline/methionine restriction, high glucose	[94,96,111,113]
	H3K9me3	db/db mice, high glucose	[111,112]
	H3K27me3	Hypoxia, hexavalent chromium, methyl deficient diet	[88,99,105]
	H3R17	Insulin	[109]
	H3R2me2	Hexavalent chromium	[105]
	H4K20me3	Food restriction, methyl deficient diet, glucose-induced cAMP	[88,95,97,110]

Also, glucose deprivation induces a higher NAD^+/NAD ratio that is related to a redox-induced increase in the histone methyltransferase Suv39H1 and in H3K9me2 [103]. Finally, recent results from Kabra et al. [104] have shown that insulin, via ROS, increases H3K4me1 and H3K9me1, demonstrating that theoretical previous studies were right concerning the control of epigenetic mechanisms by redox status.

These results open the door to the possibility of the control of histone modifications by other metabolic compounds. Different minerals, such as arsenite, nickel, selenite, or chromium, also induce changes in histone methylations (H3K4, H3K9, and H3K27), although most of the studies are related to toxicity and teratogenic effects [105–107]. In a similar way, high doses of some polyphenols, such as genistein [108] have been related to histone methylation. Insulin reduces the methylation levels at H3R17 in relation to a down-regulation of PEPCK and G6Pase [109]. Glucose and cAMP produce opposite effects on the methylation status of histone H3 associated with the L-PK promoter and coding regions [110]. The sustained up-regulation of the NFkappaB-p65 gene as a result of ambient or prior hyperglycemia was associated with increased H3K4m1 and the suppression of H3K9m2 and H3K9m3 methylation on the p65 promoter [111]. Remarkably, the diabetic state induces a decrease of H3K9me3 in VSMC cells from db/db mice [112] and a decrease of H3K9me2 and an increase of H3K4me2 in human peripheral blood cells [113]. In this context, the same group [114] described an epigenetic role for the increase of histone H3K9m2 in the lymphocytes of type 1 diabetic patients.

TABLE 18.2 Inductors of the Different Histone Modifications, Enzymes Implicated and Amino Acidic Residues Affected

	Phosphorylation	ADP-Ribosylation	Sumoylation	Ubiquitination	Deimination	Biotinylation	Proline Isomerization
Enzymes	Serine kinases Serine phosphatases	Poly-ADPribose polymerase Mono-ADP-ribosyltransferases ADP-ribosyl cyclases Sirtuin SirT4	Sumo-protein ligases	Ubiquitin ligase	Peptidylarginine deiminases (PAD 1–6)	Biotinidase Holocarboxylase synthetase	Cyclophilins FK506-binding proteins Pin1
Amino acids	Serine, threonine, tyrosine	Arginine, glutamic acid, phosphoserine		Lysine	Arginine	Lysine	Proline
Inductors	DNA breaks Benzene Bisphenol A UVA Oxidative stress Hypoxia Butyrate (inhibitor)	Oxidative stress Genotoxics DNA breaks Tryptophan NAD$^+$ Nicotinamide (niacin)	Ginkgolic acid Anacardic acid Kerriamycin B	PYR-41	Inflammation Oxidative stress	Biotin Cell proliferation Oxidative stress	Juglone

TABLE 18.3 Examples of Different Metabolic and Dietary Conditions, as Well as Several Plant Compounds, in the Level of Histone Methylation and Acetylation and the Activities of Related Enzymes

	Effector	Effect
Metabolism	Insulin	Increases H3K4me1 and H3K9me1 Reduces H3R17
	High glucose	Inhibits sirtuins and induce acetylation Increases H3K4me2 Decreases H3K9me2 and H3K9me3
	Glucose deprivation	Increases Suv39H1 and H3K9me2 Increases demethylase activities
	Diabetes	Decreases H3K9me3 and H3K4me2 Increases H3K9me2
	Hypoxia	Activates histone acetylation (i.e. H3K9) Activates JmjC-containing proteins Increases H3K9me2 and G9a
	Oxygen	Activates KDMs
	Aging	Acetylates H4K16 Increases H3K4me2 and H3K4me3
	Inflammation	Activates histone acetylation Increases H3K4me3 and H3K4 Recruits 65, p300, and SET7/9 Increases H2A.Z and SET1 MT
	Oxidative stress	Inhibits HDAC activity
	Estrogens	Decreases H3 and H4 acetylation
	Cold exposure	Acetylates H4K16
Diet	Methyl deficient diet	Down-regulates methyltransferases Loss of H4K20me3 and H3K9me3 Increases macroH2A and H3K9me3 Reduces H3K27me3 and H4K20me3 Increases histone H1 proportion
	Choline deficiency	Increases H3K4Me2
	Choline supplementation	Up-regulates lysine methyltransferase Up-regulates H3K9Me2, H3K27Me3
	Calorie restriction	Up-regulates SIRT1
	Food/protein restriction	Increases H3K27me3, H3K9me1/me3 Reduces H3K4me3
	High carbohydrate/fat ratio	Acetylates H3K9
	Biotin deficiency	Activates histone acetylation
	High-salt diet	Inhibits Class IIa HDACs
	Zinc deficiency	Inhibits Class I and II HDACs
Plant compounds	Genistein	Activates histone methylation Inhibits HDAC6
	Diallyl disulfide	Increases H3 and H4 acetylation
	Sulforaphane	Inhibits HDAC activity
	Theophylline	Increases HDAC activity
	Quercetin	Activates histone acetylation

(*Continued*)

TABLE 18.3 *(Continued)*

Effector	Effect
Resveratrol	Activates SIRT1
Luteolin	Activates SIRT1
Butein and fisetin	Activates SIRT1
Anacardic acid	Inhibits HAT activity
Garcinol	Inhibits HAT activity
Curcumin	Inhibits HAT activity
Epigallocatechin-3-gallate	Inhibits HAT activity
Retinoic acid	Acetylates H3K9
Ethanol	Acetylates H3K9
Surrogate alcohols	Acetylates H3K9

Inflammation, as a complex biological response of vascular tissues to harmful stimuli, such as pathogens, damaged cells, or irritants, has recently been related to multiple diseases such as obesity and cardiovascular diseases. There is evidence regarding the epigenetic control of inflammation. Thus, LPS increases H3K4me3 and H3K4 levels [91] in relation to iNOS and TNF-alpha genes and SET1MLL methyltransferase. TNF-alpha induces the recruitment of p65, p300, and SET7/9 (H3K4 methyltransferase) at the MCP-1 and TNF-alpha promoters, along with increased H3K4me levels (especially trimethylation) [115]. In this sense, a recent study reports that ischemic renal injury activates proinflammatory genes and progressive elevations of H3K4me3, H2A.Z, and SET1 methyltransferase in mice, suggesting that inflammation could be a trigger of methyl-histone modifications [116]. Epigenetic and transcriptional mechanisms also contribute to dysregulated inflammatory and autoimmune responses associated with aging, especially in changes associated to H3K4me2 and H3K4me3 [117].

OTHER HISTONE MODIFICATIONS

Other important post-translational histone modifications are led by enzymatic reactions. Thus, phosphorylation, usually an activator of gene expression, can be regulated by different dietary compounds, including genistein [118], sulforaphane [119], or the fungal estrogenic contaminant of moldy feeds, zearalenone [120].

Other histone modifications influenced by dietary compounds and metabolic processes are carbonylation, reduced by aging and calorie restriction [121], biotinylation, influenced by dietary biotin levels [122], and ubiquitination, regulated by nickel [123]. Finally, other less known histone modifications are ADP-ribosylation [124], sumoylation [125], deimination or citrullination [126], proline isomerization [127], and lysine propionylation and butyrylation [128]. Although they have been less exhaustively analyzed, some metabolic and dietary activators and inhibitors have been described (Table 18.2) and they are new promising target for studies concerning histone modifications by the metabolic state.

CONCLUSIONS

As a conclusion, it is well established that epigenetic processes allow plasticity of phenotype in a fixed genotype [129]. Thus, only environmental (including nutritional) factors are able to explain the phenotypical and epigenetic differences reported in monozygotic twins, which increase over the years [9]. These altered epigenetic marks are implicated in the etiology of several diseases including cancer and chronic metabolic diseases [130]. The metabolic compounds and physiological situations that regulate these enzymatic modifications are being thoroughly studied in order to understand the intrinsic mechanisms that are involved in histone action in human diseases (Table 18.3). Although research in nutritional epigenomics is just beginning, the number of histone modifications whose transcriptional effects are known is increasing every year. Thus, the coming years will undoubtedly show flourishing research in related fields and important advances in this topic.

References

1. Campion J, Milagro FI, Martinez JA. Individuality and epigenetics in obesity. *Obes Rev* 2009;**10**:383–92.

2. Hake SB, Xiao A, Allis CD. Linking the epigenetic 'language' of covalent histone modifications to cancer. *Br J Cancer* 2007:96 (Suppl):R31–R9.

3. Kimmins S, Sassone-Corsi P. Chromatin remodelling and epigenetic features of germ cells. *Nature* 2005;**434**:583–9.

4. Sarma K, Reinberg D. Histone variants meet their match. *Nat Rev Mol Cell Biol* 2005;**6**:139–49.

5. Delage B, Dashwood RH. Dietary manipulation of histone structure and function. *Annu Rev Nutr* 2008;**28**:347–66.

6. Ballestar E, Esteller M. Epigenetic gene regulation in cancer. *Adv Genet* 2008;**61**:247–67.

7. Strahl BD, Allis CD. The language of covalent histone modifications. *Nature* 2000;**403**:41–5.

8. Cosgrove MS, Wolberger C. How does the histone code work? *Biochem Cell Biol* 2005;**83**:468–76.

9. Fraga MF, Ballestar E, Paz MF, Ropero S, Setien F, Ballestar ML, et al. Epigenetic differences arise during the lifetime of monozygotic twins. *Proc Natl Acad Sci USA* 2005;**102**:10604–9.

10. Wang SC, Oelze B, Schumacher A. Age-specific epigenetic drift in late-onset Alzheimer's disease. *PLoS ONE* 2008;**3**:e2698.

11. Gallou-Kabani C, Vige A, Junien C. Lifelong circadian and epigenetic drifts in metabolic syndrome. *Epigenetics* 2007;**2**:137–46.

12. Gallou-Kabani C, Junien C. Nutritional epigenomics of metabolic syndrome: new perspective against the epidemic. *Diabetes* 2005;**54**:1899–906.

13. Ito T. Role of histone modification in chromatin dynamics. *J Biochem* 2007;**141**:609–14.

14. Weake VM, Workman JL. Histone ubiquitination: triggering gene activity. *Mol Cell* 2008;**29**:653–63.

15. Sadoul K, Boyault C, Pabion M, Khochbin S. Regulation of protein turnover by acetyltransferases and deacetylases. *Biochimie* 2008;**90**:306–12.

16. Takahashi H, McCaffery JM, Irizarry RA, Boeke JD. Nucleocytosolic acetyl-coenzyme A synthetase is required for histone acetylation and global transcription. *Mol Cell* 2006;**23**:207–17.

17. Adcock IM, Lee KY. Abnormal histone acetylase and deacetylase expression and function in lung inflammation. *Inflamm Res* 2006;**55**:311–21.

18. Suo L, Meng QG, Pei Y, Yan CL, Fu XW, Bunch TD, et al. Changes in acetylation on lysine 12 of histone H4 (acH4K12) of murine oocytes during maternal aging may affect fertilization and subsequent embryo development. *Fertil Steril* 2010;**93**:945–51.

19. Taylor PB, McDonald GW. Hypoxia stimulates myocardial RNA synthesis and histone acetylation. *Res Commun Chem Pathol Pharmacol* 1983;**40**:349–52.

20. Kim YB, Yu J, Lee SY, Lee MS, Ko SG, Ye SK, et al. Cell adhesion status-dependent histone acetylation is regulated through intracellular contractility-related signaling activities. *J Biol Chem* 2005;**280**:28357–64.

21. Petrelli F, Coderoni S, Moretti P, Paparelli M. Effect of biotin on phosphorylation, acetylation, methylation of rat liver histones. *Mol Biol Rep* 1978;**4**:87–92.

22. Jia J, Chen J. Histone hyperacetylation is involved in the quercetin-induced human leukemia cell death. *Pharmazie* 2008;**63**:379–83.

23. Mukwevho E, Kohn TA, Lang D, Nyatia E, Smith J, Ojuka EO. Caffeine induces hyperacetylation of histones at the MEF2 site on the Glut4 promoter and increases MEF2A binding to the site via a CaMK-dependent mechanism. *Am J Physiol Endocrinol Metab* 2008;**294**:E582–E8.

24. Mosley AL, Ozcan S. Glucose regulates insulin gene transcription by hyperacetylation of histone h4. *J Biol Chem* 2003;**278**:19660–6.

25. Wellen KE, Hatzivassiliou G, Sachdeva UM, Bui TV, Cross JR, Thompson CB. ATP-citrate lyase links cellular metabolism to histone acetylation. *Science* 2009;**324**:1076–80.

26. Adamski J, Ma Z, Nozell S, Benveniste EN. 17beta-Estradiol inhibits class II major histocompatibility complex (MHC) expression: influence on histone modifications and cbp recruitment to the class II MHC promoter. *Mol Endocrinol* 2004;**18**:1963–74.

27. Baccarelli A, Bollati V. Epigenetics and environmental chemicals. *Curr Opin Pediatr* 2009;**21**:243–51.

28. Fraga MF, Agrelo R, Esteller M. Cross-talk between aging and cancer: the epigenetic language. *Ann N Y Acad Sci* 2007;**1100**:60–74.

29. Langley B, Brochier C, Rivieccio MA. Targeting histone deacetylases as a multifaceted approach to treat the diverse outcomes of stroke. *Stroke* 2009;**40**:2899–905.

30. Smith KT, Workman JL. Histone deacetylase inhibitors: anticancer compounds. *Int J Biochem Cell Biol* 2009;**41**:21–5.

31. Gluckman PD, Hanson MA, Buklijas T, Low FM, Beedle AS. Epigenetic mechanisms that underpin metabolic and cardiovascular diseases. *Nat Rev Endocrinol* 2009;**5**:401–8.

32. Halili MA, Andrews MR, Sweet MJ, Fairlie DP. Histone deacetylase inhibitors in inflammatory disease. *Curr Top Med Chem* 2009;**9**:309–19.

33. Myzak MC, Dashwood RH. Histone deacetylases as targets for dietary cancer preventive agents: lessons learned with butyrate, diallyl disulfide, and sulforaphane. *Curr Drug Targets* 2006;**7**:443–52.

34. Druesne-Pecollo N, Chaumontet C, Pagniez A, Vaugelade P, Bruneau A, Thomas M, et al. In vivo treatment by diallyl disulfide increases histone acetylation in rat colonocytes. *Biochem Biophys Res Commun* 2007;**354**:140–7.

35. Lea MA, Randolph VM. Induction of histone acetylation in rat liver and hepatoma by organosulfur compounds including diallyl disulfide. *Anticancer Res* 2001;**21**:2841–5.

36. Myzak MC, Dashwood WM, Orner GA, Ho E, Dashwood RH. Sulforaphane inhibits histone deacetylase in vivo and suppresses tumorigenesis in Apc-minus mice. *FASEB J* 2006;**20**:506–8.

37. Myzak MC, Ho E, Dashwood RH. Dietary agents as histone deacetylase inhibitors. *Mol Carcinog* 2006;**45**:443–6.

38. Lee SA, Kim YM, Kwak TK, Kim HJ, Kim S, Ko W, et al. The extracellular loop 2 of TM4SF5 inhibits integrin {alpha}2 on hepatocytes under collagen type I environment. *Carcinogenesis* 2009;**30**:1872–9.

39. Basak S, Pookot D, Noonan EJ, Dahiya R. Genistein down-regulates androgen receptor by modulating HDAC6-Hsp90 chaperone function. *Mol Cancer Ther* 2008;**7**:3195–202.

40. Bontempo P, Mita L, Miceli M, Doto A, Nebbioso A, De Bellis F, et al. Feijoa sellowiana derived natural Flavone exerts anti-cancer action displaying HDAC inhibitory activities. *Int J Biochem Cell Biol* 2007;**39**:1902–14.

41. Ito K, Lim S, Caramori G, Cosio B, Chung KF, Adcock IM, et al. A molecular mechanism of action of theophylline: induction of histone deacetylase activity to decrease inflammatory gene expression. *Proc Natl Acad Sci USA* 2002;**99**:8921–6.

42. Calvanese V, Lara E, Kahn A, Fraga MF. The role of epigenetics in aging and age-related diseases. *Ageing Res Rev* 2009;**8**:268–76.

43. Bartling TR, Drumm ML. Oxidative stress causes IL8 promoter hyperacetylation in cystic fibrosis airway cell models. *Am J Respir Cell Mol Biol* 2009;**40**:58–65.

44. Rahman I, Marwick J, Kirkham P. Redox modulation of chromatin remodeling: impact on histone acetylation and deacetylation, NF-kappaB and pro-inflammatory gene expression. *Biochem Pharmacol* 2004;**68**:1255–67.

45. Yang XJ, Seto E. HATs and HDACs: from structure, function and regulation to novel strategies for therapy and prevention. *Oncogene* 2007;**26**:5310–18.

46. Attenni B, Ontoria JM, Cruz JC, Rowley M, Schultz-Fademrecht C, Steinkuhler C, et al. Histone deacetylase inhibitors with a primary amide zinc binding group display antitumor activity in xenograft model. *Bioorg Med Chem Lett* 2009;**19**:3081–4.

47. Pallas M, Verdaguer E, Tajes M, Gutierrez-Cuesta J, Camins A. Modulation of sirtuins: new targets for antiageing. *Recent Pat CNS Drug Discov* 2008;**3**:61–9.

48. Yang J, Kong X, Martins-Santos ME, Aleman G, Chaco E, Liu GE, et al. Activation of SIRT1 by resveratrol represses transcription of the gene for the cytosolic form of phosphoenolpyruvate carboxykinase (GTP) by deacetylating hepatic nuclear factor 4alpha. *J Biol Chem* 2009;**284**:27042–53.

49. Adams J, Klaidman L. Sirtuins, nicotinamide and aging: a critical review. *Lett Drug Des Discov* 2007;**4**:44–8.

50. Crujeiras AB, Parra D, Goyenechea E, Martinez JA. Sirtuin gene expression in human mononuclear cells is modulated by caloric restriction. *Eur J Clin Invest* 2008;**38**:672–8.

51. Chen D, Bruno J, Easlon E, Lin SJ, Cheng HL, Alt FW, et al. Tissue-specific regulation of SIRT1 by calorie restriction. *Genes Dev* 2008;**22**:1753–7.

52. Green KN, Steffan JS, Martinez-Coria H, Sun X, Schreiber SS, Thompson LM, et al. Nicotinamide restores cognition in Alzheimer's disease transgenic mice via a mechanism involving sirtuin inhibition and selective reduction of Thr231-phosphotau. *J Neurosci* 2008;**28**:11500–10.

53. Balestrieri ML, Rienzo M, Felice F, Rossiello R, Grimaldi V, Milone L, et al. High glucose downregulates endothelial progenitor cell number via SIRT1. *Biochim Biophys Acta* 2008;**1784**:936–45.

54. Selvi RB, Kundu TK. Reversible acetylation of chromatin: implication in regulation of gene expression, disease and therapeutics. *Biotechnol J* 2009;**4**:375–90.

55. Manzo F, Tambaro FP, Mai A, Altucci L. Histone acetyltransferase inhibitors and preclinical studies. *Expert Opin Ther Pat* 2009;**19**:761–74.

56. Balasubramanyam K, Swaminathan V, Ranganathan A, Kundu TK. Small molecule modulators of histone acetyltransferase p300. *J Biol Chem* 2003;**278**:19134–40.

57. Balasubramanyam K, Altaf M, Varier RA, Swaminathan V, Ravindran A, Sadhale PP, et al. Polyisoprenylated benzophenone, garcinol, a natural histone acetyltransferase inhibitor, represses chromatin transcription and alters global gene expression. *J Biol Chem* 2004;**279**:33716–26.

58. Balasubramanyam K, Varier RA, Altaf M, Swaminathan V, Siddappa NB, Ranga U, et al. Curcumin, a novel p300/CREB-binding protein-specific inhibitor of acetyltransferase, represses the acetylation of histone/nonhistone proteins and histone acetyltransferase-dependent chromatin transcription. *J Biol Chem* 2004;**279**:51163–71.

59. Morimoto T, Sunagawa Y, Kawamura T, Takaya T, Wada H, Nagasawa A, et al. The dietary compound curcumin inhibits p300 histone acetyltransferase activity and prevents heart failure in rats. *J Clin Invest* 2008;**118**:868–78.

60. Choi KC, Jung MG, Lee YH, Yoon JC, Kwon SH, Kang HB, et al. Epigallocatechin-3-gallate, a histone acetyltransferase inhibitor, inhibits EBV-induced B lymphocyte transformation via suppression of RelA acetylation. *Cancer Res* 2009;**69**:583–92.

61. Stimson L, Rowlands MG, Newbatt YM, Smith NF, Raynaud FI, Rogers P, et al. Isothiazolones as inhibitors of PCAF and p300 histone acetyltransferase activity. *Mol Cancer Ther* 2005;**4**:1521–32.

62. Gorsuch S, Bavetsias V, Rowlands MG, Aherne GW, Workman P, Jarman M, et al. Synthesis of isothiazol-3-one derivatives as inhibitors of histone acetyltransferases (HATs). *Bioorg Med Chem* 2009;**17**:467–74.

63. Lau OD, Kundu TK, Soccio RE, Ait-Si-Ali S, Khalil EM, Vassilev A, et al. HATs off: selective synthetic inhibitors of the histone acetyltransferases p300 and PCAF. *Mol Cell* 2000;**5**:589–95.

64. Friis RM, Wu BP, Reinke SN, Hockman DJ, Sykes BD, Schultz MC. A glycolytic burst drives glucose induction of global histone acetylation by picNuA4 and SAGA. *Nucleic Acids Res* 2009;**37**:3969–80.

65. Shepard BD, Tuma PL. Alcohol-induced protein hyperacetylation: mechanisms and consequences. *World J Gastroenterol* 2009;**15**:1219–30.

66. Choudhury M, Shukla SD. Surrogate alcohols and their metabolites modify histone H3 acetylation: involvement of histone acetyl transferase and histone deacetylase. *Alcohol Clin Exp Res* 2008;**32**:829–39.

67. Honma K, Mochizuki K, Goda T. Inductions of histone H3 acetylation at lysine 9 on SGLT1 gene and its expression by feeding mice a high carbohydrate/fat ratio diet. *Nutrition* 2009;**25**:40–4.

68. Kulkarni-Shukla S, Barge AP, Vartak RS, Kar A. Cold-induced alteration in the global structure of the male sex chromosome of In1BM2(reinverted) of *Drosophila melanogaster* is associated with increased acetylation of histone 4 at lysine 16. *J Genet* 2008;**87**:235–40.

69. Dang W, Steffen KK, Perry R, Dorsey JA, Johnson FB, Shilatifard A, et al. Histone H4 lysine 16 acetylation regulates cellular lifespan. *Nature* 2009;**459**:802–7.

70. Manosalva I, Gonzalez A. Aging alters histone H4 acetylation and CDC2A in mouse germinal vesicle stage oocytes. *Biol Reprod* 2009;**81**:1164–71.

71. Dou Y, Milne TA, Tackett AJ, Smith ER, Fukuda A, Wysocka J, et al. Physical association and coordinate function of the H3 K4 methyltransferase MLL1 and the H4 K16 acetyltransferase MOF. *Cell* 2005;**121**:873–85.

72. Shogren-Knaak M, Ishii H, Sun JM, Pazin MJ, Davie JR, Peterson CL. Histone H4-K16 acetylation controls chromatin structure and protein interactions. *Science* 2006;**311**:844–7.

73. Watson JA, Watson CJ, McCrohan AM, Woodfine K, Tosetto M, McDaid J, et al. Generation of an epigenetic signature by chronic hypoxia in prostate cells. *Hum Mol Genet* 2009;**18**:3594–604.

74. Bedford MT, Clarke SG. Protein arginine methylation in mammals: who, what, and why. *Mol Cell* 2009;**33**:1–13.

75. Nottke A, Colaiacovo MP, Shi Y. Developmental roles of the histone lysine demethylases. *Development* 2009;**136**:879–89.

76. Bannister AJ, Schneider R, Kouzarides T. Histone methylation: dynamic or static? *Cell* 2002;**109**:801–6.

77. Zhang Y, Reinberg D. Transcription regulation by histone methylation: interplay between different covalent modifications of the core histone tails. *Genes Dev* 2001;**15**:2343–60.

78. Shi Y. Histone lysine demethylases: emerging roles in development, physiology and disease. *Nat Rev Genet* 2007;**8**:829–33.

79. Ng SS, Yue WW, Oppermann U, Klose RJ. Dynamic protein methylation in chromatin biology. *Cell Mol Life Sci* 2009;**66**:407–22.

80. Dashwood RH, Ho E. Dietary agents as histone deacetylase inhibitors: sulforaphane and structurally related isothiocyanates. *Nutr Rev* 2008;**66**(Suppl 1):S36–S8.

81. Hitchler MJ, Domann FE. Metabolic defects provide a spark for the epigenetic switch in cancer. *Free Radic Biol Med* 2009;**47**:115–27.

82. Huang S. Histone methyltransferases, diet nutrients and tumour suppressors. *Nat Rev Cancer* 2002;**2**:469–76.

83. Junien C, Gallou-Kabani C, Vige A, Gross MS. [Nutritional epigenomics of metabolic syndrome]. *Med Sci (Paris)* 2005;**21**:396–404.

84. Zeisel SH. Epigenetic mechanisms for nutrition determinants of later health outcomes. *Am J Clin Nutr* 2009;**89**:1488S–93S.

85. Clarke S, Banfield K. *Homocysteine in Health and Disease*. Cambridge: Cambridge University Press; 2001.

86. Horiguchi K, Yamada M, Satoh T, Hashimoto K, Hirato J, Tosaka M, et al. Transcriptional activation of the mixed lineage leukemia-p27Kip1 pathway by a somatostatin analogue. *Clin Cancer Res* 2009;**15**:2620–9.

87. Osborne TC, Obianyo O, Zhang X, Cheng X, Thompson PR. Protein arginine methyltransferase 1: positively charged residues in substrate peptides distal to the site of methylation are important for substrate binding and catalysis. *Biochemistry* 2007;**46**:13370–81.

88. Pogribny IP, Tryndyak VP, Bagnyukova TV, Melnyk S, Montgomery B, Ross SA, et al. Hepatic epigenetic phenotype predetermines individual susceptibility to hepatic steatosis in mice fed a lipogenic methyl-deficient diet. *J Hepatol* 2009;**51**:176–86.

89. Norell M, von der Decken A. Effect of a diet low in methionine-cysteine on rat liver chromatin and nuclear proteins. *Cell Mol Biol* 1989;**35**:63–74.

90. Davison JM, Mellott TJ, Kovacheva VP, Blusztajn JK. Gestational choline supply regulates methylation of histone H3, expression of histone methyltransferases G9a (Kmt1c) and Suv39h1 (Kmt1a), and DNA methylation of their genes in rat fetal liver and brain. *J Biol Chem* 2009;**284**:1982–9.

91. Ara AI, Xia M, Ramani K, Mato JM, Lu SC. S-adenosylmethionine inhibits lipopolysaccharide-induced gene expression via modulation of histone methylation. *Hepatology* 2008;**47**:1655–66.

92. Nicholson TB, Chen T, Richard S. The physiological and pathophysiological role of PRMT1-mediated protein arginine methylation. *Pharmacol Res* 2009;**60**:466–74.

93. Puchau B, Zulet MA, Urtiaga G, Navarro-Blasco I, Martinez JA. Asymmetric dimethylarginine association with antioxidants intake in healthy young adults: a role as an indicator of metabolic syndrome features. *Metabolism* 2009;**58**:1483–8.

94. Dobosy JR, Fu VX, Desotelle JA, Srinivasan R, Kenowski ML, Almassi N, et al. A methyl-deficient diet modifies histone methylation and alters Igf2 and H19 repression in the prostate. *Prostate* 2008;**68**:1187–95.

95. Pogribny IP, Tryndyak VP, Muskhelishvili L, Rusyn I, Ross SA. Methyl deficiency, alterations in global histone modifications, and carcinogenesis. *J Nutr* 2007;**137**:216S–22S.

96. Lillycrop KA, Slater-Jefferies JL, Hanson MA, Godfrey KM, Jackson AA, Burdge GC. Induction of altered epigenetic regulation of the hepatic glucocorticoid receptor in the offspring of rats fed a protein-restricted diet during pregnancy suggests that reduced DNA methyltransferase-1 expression is involved in impaired DNA methylation and changes in histone modifications. *Br J Nutr* 2007;**97**:1064–73.

97. Sharif J, Nakamura M, Ito T, Kimura Y, Nagamune T, Mitsuya K, et al. Food restriction in pregnant mice can induce changes in histone modifications and suppress gene expression in fetus. *Nucleic Acids Symp Ser (Oxf)* 2007;**51**:125–6.

98. Yang J, Ledaki I, Turley H, Gatter KC, Montero JC, Li JL, et al. Role of hypoxia-inducible factors in epigenetic regulation via histone demethylases. *Ann N Y Acad Sci* 2009;**1177**:185–97.

99. Johnson AB, Denko N, Barton MC. Hypoxia induces a novel signature of chromatin modifications and global repression of transcription. *Mutat Res* 2008;**640**:174–9.

100. Xia X, Lemieux ME, Li W, Carroll JS, Brown M, Liu XS, et al. Integrative analysis of HIF binding and transactivation reveals its role in maintaining histone methylation homeostasis. *Proc Natl Acad Sci USA* 2009;**106**:4260–5.

101. Pollard PJ, Loenarz C, Mole DR, McDonough MA, Gleadle JM, Schofield CJ, et al. Regulation of Jumonji-domain-containing histone demethylases by hypoxia-inducible factor (HIF)-1alpha. *Biochem J* 2008;**416**:387–94.

102. Wellmann S, Bettkober M, Zelmer A, Seeger K, Faigle M, Eltzschig HK, et al. Hypoxia upregulates the histone demethylase JMJD1A via HIF-1. *Biochem Biophys Res Commun* 2008;**372**:892–7.

103. Murayama A, Ohmori K, Fujimura A, Minami H, Yasuzawa-Tanaka K, Kuroda T, et al. Epigenetic control of rDNA loci in response to intracellular energy status. *Cell* 2008;**133**:627–39.

104. Kabra DG, Gupta J, Tikoo K. Insulin induced alteration in post-translational modifications of histone H3 under a hyperglycemic condition in L6 skeletal muscle myoblasts. *Biochim Biophys Acta* 2009;**1792**:574–83.

105. Sun H, Zhou X, Chen H, Li Q, Costa M. Modulation of histone methylation and MLH1 gene silencing by hexavalent chromium. *Toxicol Appl Pharmacol* 2009;**237**:258–66.

106. Zhou X, Li Q, Arita A, Sun H, Costa M. Effects of nickel, chromate, and arsenite on histone 3 lysine methylation. *Toxicol Appl Pharmacol* 2009;**236**:78–84.

107. Chen H, Ke Q, Kluz T, Yan Y, Costa M. Nickel ions increase histone H3 lysine 9 dimethylation and induce transgene silencing. *Mol Cell Biol* 2006;**26**:3728–37.

108. Kikuno N, Shiina H, Urakami S, Kawamoto K, Hirata H, Tanaka Y, et al. Genistein mediated histone acetylation and demethylation activates tumor suppressor genes in prostate cancer cells. *Int J Cancer* 2008;**123**:552–60.

109. Hall RK, Wang XL, George L, Koch SR, Granner DK. Insulin represses phosphoenolpyruvate carboxykinase gene transcription by causing the rapid disruption of an active transcription complex: a potential epigenetic effect. *Mol Endocrinol* 2007;**21**:550–63.

110. Burke SJ, Collier JJ, Scott DK. cAMP prevents glucose-mediated modifications of histone H3 and recruitment of the RNA polymerase II holoenzyme to the L-PK gene promoter. *J Mol Biol* 2009;**392**:578–88.

111. Brasacchio D, Okabe J, Tikellis C, Balcerczyk A, George P, Baker EK, et al. Hyperglycemia induces a dynamic cooperativity of histone methylase and demethylase enzymes associated with gene-activating epigenetic marks that coexist on the lysine tail. *Diabetes* 2009;**58**:1229–36.

112. Villeneuve LM, Reddy MA, Lanting LL, Wang M, Meng L, Natarajan R. Epigenetic histone H3 lysine 9 methylation in metabolic memory and inflammatory phenotype of vascular smooth muscle cells in diabetes. *Proc Natl Acad Sci USA* 2008;**105**:9047–52.

113. Miao F, Wu X, Zhang L, Yuan YC, Riggs AD, Natarajan R. Genome-wide analysis of histone lysine methylation variations caused by diabetic conditions in human monocytes. *J Biol Chem* 2007;**282**:13854–63.

114. Miao F, Smith DD, Zhang L, Min A, Feng W, Natarajan R. Lymphocytes from patients with type 1 diabetes display a distinct profile of chromatin histone H3 lysine 9 dimethylation: an epigenetic study in diabetes. *Diabetes* 2008;**57**:3189–98.

115. Li Y, Reddy MA, Miao F, Shanmugam N, Yee JK, Hawkins D, et al. Role of the histone H3 lysine 4 methyltransferase, SET7/9, in the regulation of NF-kappaB-dependent inflammatory genes. Relevance to diabetes and inflammation. *J Biol Chem* 2008;**283**:26771–81.

116. Zager RA, Johnson AC. Renal ischemia-reperfusion injury upregulates histone-modifying enzyme systems and alters histone expression at proinflammatory/profibrotic genes. *Am J Physiol Renal Physiol* 2009;**296**:F1032–41.

117. El Mezayen R, El Gazzar M, Myer R, High KP. Aging-dependent upregulation of IL-23p19 gene expression in dendritic cells is associated with differential transcription factor binding and histone modifications. *Aging Cell* 2009;**8**:553–65.

118. Cappelletti V, Fioravanti L, Miodini P, Di Fronzo G. Genistein blocks breast cancer cells in the G(2)M phase of the cell cycle. *J Cell Biochem* 2000;**79**:594–600.

119. Davie JR, Chadee DN. Regulation and regulatory parameters of histone modifications. *J Cell Biochem Suppl* 1998;**30-31**:203–13.

120. Ahamed S, Foster JS, Bukovsky A, Wimalasena J. Signal transduction through the Ras/Erk pathway is essential for the mycoestrogen zearalenone-induced cell-cycle progression in MCF-7 cells. *Mol Carcinog* 2001;**30**:88–98.

121. Sharma R, Nakamura A, Takahashi R, Nakamoto H, Goto S. Carbonyl modification in rat liver histones: decrease with age and increase by dietary restriction. *Free Radic Biol Med* 2006;**40**:1179–84.

122. Hassan YI, Zempleni J. Epigenetic regulation of chromatin structure and gene function by biotin. *J Nutr* 2006;**136**:1763–5.

123. Ke Q, Davidson T, Chen H, Kluz T, Costa M. Alterations of histone modifications and transgene silencing by nickel chloride. *Carcinogenesis* 2006;**27**:1481–8.

124. Hassa PO, Haenni SS, Elser M, Hottiger MO. Nuclear ADP-ribosylation reactions in mammalian cells: where are we today and where are we going? *Microbiol Mol Biol Rev* 2006;**70**:789–829.

125. Fukuda I, Ito A, Hirai G, Nishimura S, Kawasaki H, Saitoh H, et al. Ginkgolic acid inhibits protein SUMOylation by blocking formation of the E1-SUMO intermediate. *Chem Biol* 2009;**16**:133–40.

126. Neeli I, Khan SN, Radic M. Histone deimination as a response to inflammatory stimuli in neutrophils. *J Immunol* 2008;**180**:1895–902.

127. Lu KP, Finn G, Lee TH, Nicholson LK. Prolyl cis-trans isomerization as a molecular timer. *Nat Chem Biol* 2007;**3**:619–29.

128. Chen Y, Sprung R, Tang Y, Ball H, Sangras B, Kim SC, et al. Lysine propionylation and butyrylation are novel post-translational modifications in histones. *Mol Cell Proteomics* 2007;**6**:812–19.

129. Mathers JC. Session 2: personalised nutrition. Epigenomics: a basis for understanding individual differences? *Proc Nutr Soc* 2008;**67**:390–4.

130. Feinberg AP. Phenotypic plasticity and the epigenetics of human disease. *Nature* 2007;**447**:433–40.

Functions of Epigenetics

CHAPTER 19

Epigenetics, Stem Cells, and Cellular Differentiation

Berry Juliandi[1,2], Masahiko Abematsu[1,3] and Kinichi Nakashima[1]
[1]Laboratory of Molecular Neuroscience, Graduate School of Biological Sciences, Nara Institute of Science and Technology, Ikoma, Nara 630-0192, Japan
[2]Department of Biology, Bogor Agricultural University (IPB), Bogor 16680, Indonesia
[3]Department of Orthopaedic Surgery, Graduate School of Medical and Dental Sciences, Kagoshima University, Kagoshima 890-8520, Japan

INTRODUCTION

Two cardinal features characterize stem cells: their ability to undergo unlimited self-renewal by division and their potential to generate at least two different cell types. Progenitor cells, which possess a limited capacity for self-renewal, are the immediate progeny of stem cells, and behave as transit amplifying cells that can expand the number of new differentiated cells owing to their higher rate of proliferation than the more quiescent stem cells. It can be difficult to distinguish these two cell types unambiguously, and they are sometimes referred to by the collective term "precursor cell".

Numerous studies have indicated that stem cells respond to a combination of intrinsic programs and extracellular cues from the environment that determines which types of progeny they will produce. One of these intrinsic programs is epigenetic modification, which encompasses DNA methylation, chromatin modification, and non-coding RNA-mediated processes. Epigenetic modifications are temporally regulated and reversible, thereby ensuring that stem cells can generate different types of cell from a fixed DNA sequence.

The excitement generated by recent vigorous research on stem cell epigenetic modification reflects the prospect that this new knowledge may enable us to reprogram or modulate the fate of stem cells, using treatments with defined components and at specific time points to alter the epigenetic status of the treated cell and thereby produce a desired cell phenotype. In this review, we discuss recent progress in the study of epigenetic modifications that regulate stem cell differentiation.

STEM CELLS

Animal stem cell research began in the fields of embryology and of the biology of organs with inherent regenerative ability [1]. Other organs with presumptive non-regenerative behavior, such as brain, heart, and lung, were thought to lack stem cells. However, there is increasing evidence that stem cells occur ubiquitously, from embryo to adult and in many organs of the body.

Embryogenesis in multicellular organisms starts with the fertilization of an ovum by a sperm to make a zygote. The zygote is totipotent: it has the potential to develop into a complete

Handbook of Epigenetics: The New Molecular and Medical Genetics. DOI: 10.1016/B978-0-12-375709-8.00019-8

organism and also to make a trophoblast, a structure that will form the placenta. Initial divisions of the zygote yield the morulla and later on the blastocyst, which is composed of the trophoblast, inner cell mass, and blastocyst cavity [2]. The inner cell mass can be isolated and cultured under specific conditions *in vitro* to generate embryonic stem cells (ESCs). ESCs are categorized as pluripotent, since they can generate cells of all body tissues except the trophoblast. This deficiency makes ESCs incapable of forming a complete organism upon implantation into the uterus. Nevertheless, ESCs have the capacity to generate somatic stem cells and subsequently differentiated cells of all three germ layers, ectoderm, mesoderm, and endoderm, if they are pre-treated under optimal *in vitro* culture conditions (Fig. 19.1).

During subsequent developmental stages, each germ layer retains cells that possess stem cell features. These cells are described as being multipotent, because they can generate all progenitor and differentiated cell types within their particular restricted lineage. Neural stem cells (NSCs) and hematopoietic stem cells (HSCs) are examples of such multipotent cells. NSCs can differentiate into neural progenitor cells, neurons, and glial cells (astrocytes and oligodendrocytes). The neuroepithelial cells lining the neural tube are considered as the primary NSCs. From this cell type, the central nervous system develops in a sophisticated temporal and spatial sequence, governed in part by epigenetic mechanisms [3–5]. Likewise, HSCs can give rise to all lineages of the blood, including T and B cells (the lymphoid lineage) and neutrophils, eosinophils, basophils, monocytes, macrophages, megakaryocytes, platelets, and erythrocytes (the myeloid lineage) [6].

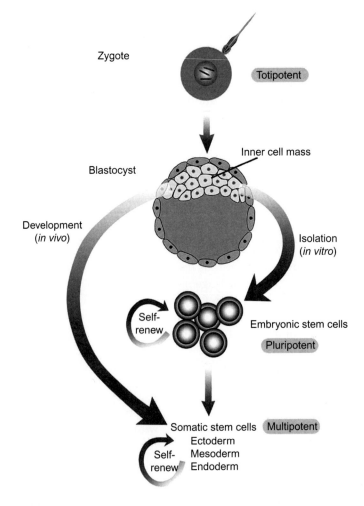

FIGURE 19.1

Developmental potential of stem cells. The totipotent level exists after the egg is fertilized. After several mitotic divisions that lead to the blastocyst, the inner cell mass can be isolated *in vitro*, yielding pluripotent embryonic stem cells (ESCs). ESCs can self-renew and differentiate into multipotent somatic stem cells specific to each of the three germ layers. (Please refer to color plate section)

EPIGENETICS OF STEM CELLS

There are two pathways in the developmental course of stem cells: either to self-renew while retaining pluripotency or multipotency, or to differentiate into other cell types. During this course, genes that are active at an earlier stage or in maintaining the potency gradually become silenced, and subsets of later-stage genes or cell type-specific genes responsible for the cells' differentiation are then turned on. This reduction in potency over time by progressive gene silencing can be achieved by epigenetic mechanisms in concert with differential expression of certain transcription factors (TFs).

DNA Methylation

Several studies have indicated that DNA methylation regulates the timing of differentiation and maintenance of cell type identity [7–9]. The DNA methylation pattern in the genome is established by a family of DNA methyltransferases (DNMT). Maintenance of methylation patterns is achieved by a function of DNMT1 during DNA replication, while new or *de novo* methylation is primarily catalyzed by DNMT3a and DNMT3b.

In the developing embryonic brain, neurons are generated first and glial cells (astrocytes and oligodendrocytes) afterward. At mid-gestation, embryonic day (E) 11.5 mouse (m) NSCs can only differentiate into neurons, not astrocytes. The glycoprotein 130 and signal transducer and activator of transcription 3 (gp130-STAT3) signaling is a well known pathway to induce astrocytogenesis. Although this pathway can be activated in primary culture of E11.5 mNSCs, its astrocytic target genes are not yet competent to respond to this signal [10,11]. This prevention of premature astrocytogenesis is established in NSCs through methylation in the promoter regions of astrocytic genes, such as glial fibrillary acidic protein (*gfap*). Even in the presence of interleukin-6 (IL-6) family cytokines such as leukemia inhibitory factor (LIF), which can activate the gp130-STAT3 signaling pathway, E11.5 mNSCs do not differentiate into astrocytes because STAT3 cannot bind to the methylated promoter region. The same promoter region becomes hypomethylated as gestation proceeds, which allows the binding of STAT3 and the expression of astrocytic genes in later-stage NSCs [10,12,13], leading them toward astrocytic lineages (Fig. 19.2A).

The astrocyte gene specific demethylation is not just confined to the *gfap* promoter, but is rather common among astrocytic genes. For example, an earlier astrocytic marker *S100β* also possesses a particular cytosine residue in its promoter which is highly methylated in mESCs but becomes lower methylated as mESCs differentiate into neural progenitors [13]. Demethylation of *S100β* promoter also occurs at mid-gestation coinciding with the onset of its expression in the mouse brain [12]. Furthermore, a genome-wide DNA methylation status of E11.5 and E14.5 mNSCs has been recently compared by the profiling method using microarrays [14], confirming that many astrocytic genes become demethylated in late-stage mNSCs.

The *gfap* promoter was also found to be hypomethylated in neurons derived from primary culture of later-stage mNSCs, raising the question of why these cells had not differentiated into astrocytes. Setoguchi et al. [15] showed that even though STAT3 can bind to the hypomethylated *gfap* promoter, *gfap* expression is blocked in neurons due to the association of methyl-CpG-binding protein 2 (MeCP2) with hypermethylated exon 1 (Fig. 19.2A). MeCP2 is a member of the methyl-CpG-binding domain (MBD) proteins, which is highly expressed in neurons. Although recent studies showed that MeCP2 can be found in astrocytes, its expression is very low [16,17]. Indeed, ectopic expression of MeCP2 *in vitro* directs mNSCs to become neurons and inhibits astrocytic differentiation even in the presence of astrocyte-inducing cytokines [18]. Moreover, Tsujimura et al. [18] showed that these astrocytic cytokines actually induce mNSCs to produce more neurons with ectopic MeCP2 expression, by as-yet-unknown mechanisms. It will be intriguing to study how the

317

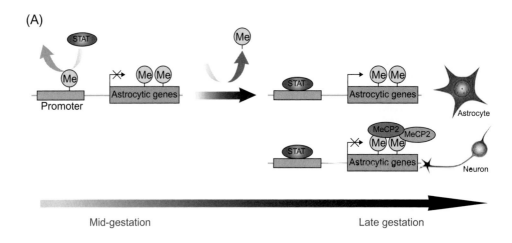

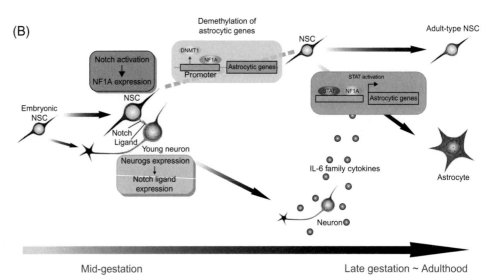

FIGURE 19.2

(A) Astrocytic gene methylation status during NSCs development. Although STAT3 can be activated in mid-gestational NSCs, it cannot bind to astrocytic gene promoters such as *gfap* due to promoter hypermethylation (left). As gestation proceeds, these promoters become demethylated, allowing STAT3 to bind and activate astrocytic genes, resulting in the differentiation of NSCs into astrocytes (upper right). Methyl-CpG binding protein 2 (MeCP2) blocks this activation in neurons (lower right). (B) Notch-induced demethylation of astrocytic genes. Activation of Notch signaling in residual NSCs by young neurons induces demethylation of astrocytic gene promoters by up-regulation of NFIA and release of DNMT1 from astrocytic gene promoters. In turn, at late gestation, IL-6 family cytokines activate the STAT3 pathway and induce NSCs to differentiate into astrocytic lineages. (Please refer to color plate section)

level of MeCP2 expression and its binding to the astrocytic genes such as *gfap*, influence NSC differentiation in the future. Nevertheless, recent studies show that oligodendrocytes which also have a hypomethylated *gfap* promoter, but unlike neurons which possess MeCP2 at the exon 1 region, can change their fate and become astrocytes by the stimulation of astrocytic cytokines *in vitro* and *in vivo* [17].

A further intriguing question is how astrocytic gene promoters become demethylated in later-stage NSCs. Several reports have indicated that Notch ligands are expressed in neuronally committed precursor cells and young neurons generated from NSCs [19–21]. Notch signaling is a conserved pathway from insects to mammals which contributes to cell-to-cell communication [22–25] and controls cell fate determination in the central nervous system (CNS) [26]. Namihira et al. [11] showed that in the cortex of mouse embryo, Notch ligands are expressed in neuronally committed precursor cells and young neurons, and that Notch signaling is activated in neighboring NSCs (Fig. 19.2B). During this activation, the Notch intracellular domain (NICD) is released from the plasma membrane and translocated into the nucleus, where it converts the CBF1(RBP-J)/Su(H)/LAG1 repressor complex into an activator complex [27,28]. Forced expression of NICD in primary culture of E11.5 mNSCs induced the up-regulation of nuclear factor IA (NFIA), which in turn accelerated demethylation of astrocytic gene promoters by preventing DNMT1 binding, thus allowing precocious astrocytic differentiation in response to LIF [11] (Fig. 19.2B).

The chicken ovalbumin upstream promoter transcription factors I and II (COUP-TFI/II) were also found to be important for unlocking the silencing epigenetic marks of astrocytic genes [29]. Expression of COUP-TFI/II, which is transiently up-regulated in the early neurogenic period, markedly decreased before the onset of astrocytogenesis. Using mESC-derived NSCs that recapitulate *in vivo* mouse CNS development [30], Naka et al. [29] showed that the CpG methylation status of the *gfap* promoter remained high after COUP-TFI/II knockdown. Moreover, COUP-TFI/II knockdown inhibited the switch from neurogenesis to gliogenesis in this culture system and at developing mouse forebrain. Taken together, these results indicate that COUP-TFI/II are important factors for *gfap* promoter demethylation, although it is not yet clear how demethylation occurs.

The maturation of erythrocytes during hematopoiesis is associated with increased expression of α- and β-globin genes. The β-globin locus consists of five genes, ε, Gγ, Aγ, δ, and β, which are under the regulation of the locus control region located 6–22 kb upstream of the ε-globin gene [31]. In non-erythroid cells, all of these genes are methylated and transcriptionally silent. During erythroid differentiation, individual genes corresponding to embryonic (ε), fetal (Gγ, Aγ), and adult (δ, β) stages of erythropoiesis are expressed in a sequential fashion, such that when the adult genes are activated the embryonic and fetal genes become silenced. Initial activation of embryonic/fetal genes is thought to be caused by demethylation of their promoters, since *in vitro* differentiation of baboon HSCs derived from fetal liver and adult bone marrow into mature erythroblasts is accompanied by a progressive decrease in γ-globin promoter methylation and an activation of transcription [32]. Promoter methylation also occurs in other hematopoietic lineages to regulate differentiation of their precursors. For example, the Ets family transcription factor PU.1 (SPI1) is highly expressed in human (h) HSCs and differentiated B cells, but not in T cells. Hypomethylation of *PU.1* can lead to diffuse large B-cell lymphoma, indicating a requirement for tight DNA methylation control of this gene for normal hematopoiesis [33]. *In vitro* study using murine-derived cell lines showed that differentiation of common progenitors into myeloid and erythroid lineages is also regulated by PU.1 which can recruit SUV39H1, HP1, and retinoblastoma (Rb) proteins when they associate with GATA-1 on its target genes, thereby inhibiting erythroid differentiation [34]. Differential methylation of other genes such as *GATA3*, *TCF-7*, *c-maf*, *TBX21*, and *Etv5* has also been observed to control lineage-commitment in human hematopoiesis [33].

Histone Modification

ACETYLATION

Histone modification processes are at least partially involved in the differentiation of ESCs into NSCs and neurons. mESCs appear to have higher global levels of histone acetylation than lineage-restricted stem cells and differentiated cells, which is consistent with their higher level of transcription and more open chromatin configuration [35]. In fact, heterochromatin markers such as HP1 are highly dynamic and dispersed in the nuclei of mESCs, and then become more concentrated at specific loci as differentiation proceeds [36,37].

Neuron-specific genes are repressed in mESCs by the binding of RE-1 silencing transcription factor (NRSF/REST) to its conserved 21–23 bp DNA response elements (RE-1) which forms a repressor complex by recruiting HDAC1/2 and Sin3A [38–41]. As the cells differentiate into neural progenitors and neurons, this HDAC containing-repressor complex is released from neuron-specific genes due to the degradation of REST/NRSF [38].

Adult rat (r) hippocampus-derived NSCs differentiate predominantly into neurons, at the expense of astrocytes and oligodendrocytes, if treated by the antiepileptic and HDAC inhibitor valproic acid (VPA) *in vitro*, even in conditions that favor glia-specific differentiation [42]. This HDAC inhibition up-regulates the neuron-specific gene *NeuroD*, a neurogenic basic helix-loop-helix transcription factor, resulting in the induction and suppression of neuronal and glial differentiation, respectively. In the developing rat brain and in cultured E14 rNSCs, VPA treatment can also promote neurogenesis by activating the Ras-ERK pathway [43].

Progression of the oligodendrocyte lineage in rat is also dependent on HDAC activity [44]. Postnatal administration of VPA was shown to delay the timing of NSC differentiation into myelin-forming oligodendrocytes in the developing rat brain [45], and significant hypomyelination in the developing corpus callosum together with sustained expression of progenitor markers and delayed expression of late differentiation markers were observed in this study. However, HDAC inhibition by VPA after the onset of myelination resulted in comparable myelin gene expression with control, attributed to further changes of nucleosomal histones from a state of reversible deacetylation to a more stably repressed state by histone methylation. It has recently been shown that HDAC1/2 contribute to the progression of murine oligodendrocyte differentiation by disrupting the β-catenin-TCF activator complex at inhibitor of differentiation genes *id2/4*, thereby preventing the synthesis of Id2/4 proteins to inhibit myelin gene expression [46].

During hematopoiesis, lineage-restricted TFs can also regulate specific gene expression by recruiting HAT or HDAC complexes to its promoter region [47]. *In vitro* study using erythroid cell line G1E showed that erythroid-specific TFs such as GATA1, which is necessary for red blood cell survival and maturation, recruit HAT-containing complexes to the β-globin locus, inducing acetylation of histones H3 and H4 thus stimulating globin gene expression [48]. Some co-activators such as p300/CBP can also be recruited by TFs to catalyze acetylation of histones, correlating with transcriptional activation of hematopoietic genes [49]. However, p300 can also repress transcription as in the case of acetylation of the promyelocytic leukemia zinc-finger (PLZF) protein [50].

METHYLATION

The methylation of histones on lysine and arginine residues by histone methyl transferase (HMT) represents another level of histone modification. The mixed-lineage leukemia (MLL), which belongs to the trithorax group (trxG) gene, can specifically methylate H3K4 for gene activation by recruiting HATs such as MOF and CBP in various cell lines [51–53], or it can repress target genes through the recruitment of polycomb group (PcG) proteins, HDACs and/ or SUV39H1 [54]. In the postnatal mouse brain, MLL1 is required for neurogenesis, and its deficiency in NSCs at the subventricular zone leads to a glial lineage preference [55]. MLL also plays a critical role in the proliferation and lineage determination of hematopoietic progenitors derived from RW4 mESC line, by maintaining the expression of HOX genes, such as *Hoxa7* and *Hoxa9* [56,57], whose up-regulation can confer leukemogenic potential [58].

Stem cell chromatin can be maintained in the bivalent state by PcG proteins [59,60]. The bivalent state is characterized by the existence of both activating and repressive histone methylation marks. In mESCs, while pluripotency-related genes are marked by active histone H3K4 trimethylation (H3K4me3), those that are necessary for differentiation are marked by both activating H3K4me3 and repressive histone H3K27 trimethylation (H3K27me3) chromatin marks [61,62] (Fig. 19.3). Moreover, H3K4me3 is found at nearly 70% of all gene promoters in hESCs [63], while the level of H3K27me3 is only around 10%

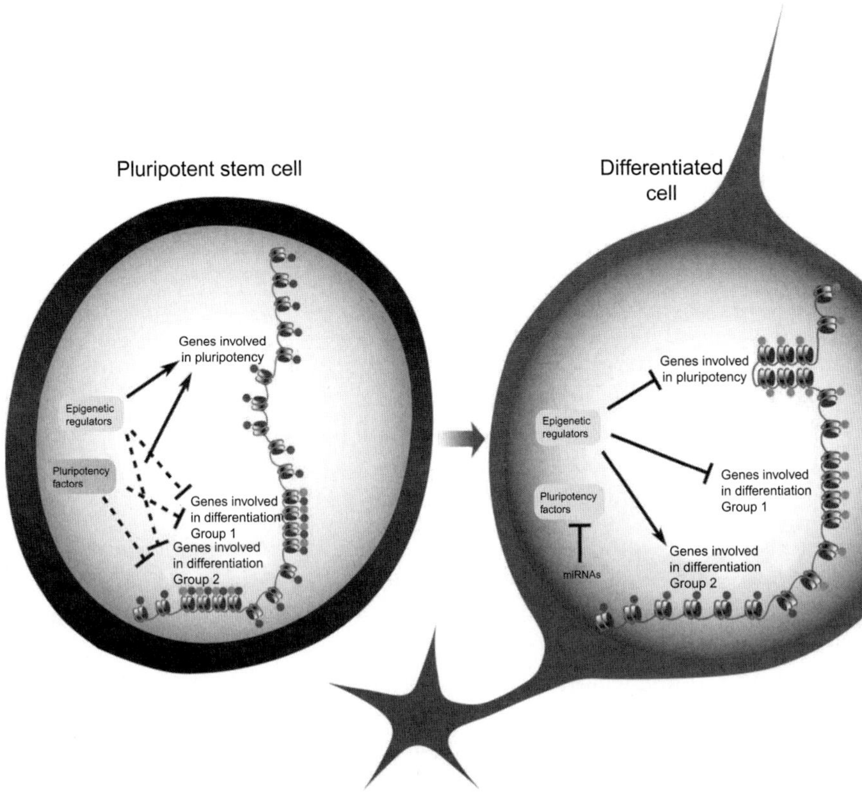

FIGURE 19.3
Regulatory mechanisms of pluripotency. Genes associated with pluripotency are actively transcribed in pluripotent cells (left), while differentiation-associated genes are kept in a silent poised state. Several epigenetic regulators and pluripotency factors regulate this state, in part by a combination of the activating methylation H3K4me3 (red circles) and the repressive methylation H3K27me3 (green circles). In pluripotent cells, only H3K4me3 is present at pluripotency-associated genes, but both H3K4me3 and H3K27me3 at differentiation-associated genes. Upon differentiation, miRNAs down-regulate pluripotency factors and differential repression of differentiation-associated gene groups is sustained only by epigenetic regulators. Pluripotency-associated genes and silenced differentiation-associated gene groups retain the H3K27me3 mark, while activated differentiation-associated gene groups retain the H3K4me3 mark. Chromatin status also changes from hyper to less dynamic during differentiation [36,37]. (Please refer to color plate section)

[59,60,64]. Large blocks of the silent chromatin mark H3K9 methylation also accumulate in differentiated cells rather than in mESCs [65]. In mESC derived-NSCs, on the other hand, the ESC pluripotency-related genes are repressed by methylated H3K9 and the bivalent state exists on neuronal and glial differentiation genes. Thus, it is conceivable that the bivalent state produced by PcG proteins is a common mechanism for maintaining the differentiation potential of many stem/progenitor cell types [64].

Pluripotency is also maintained in human and mouse ESCs by regulatory networks of several TFs, which in some cases, such as Oct3/4, Sox2, and Nanog, are believed to be the main controller [66,67]. Interestingly, these TFs' main direct targets are also transcriptional regulators that might extend the regulatory effects of the network to numerous subsequent targets [68]. Moreover, most of the differentiation regulatory genes are located at the chromatin domains with bivalent state modifications H3K4me3 and H3K27me3 [62] and they are enriched with binding sites for these TFs (Fig. 19.3). This suggests that bivalent-marked differentiation-related genes are kept in a poised state, ready for rapid transcriptional activation upon differentiation under the control of pluripotency TFs, a mechanism that might be responsible for the balance between maintenance of ESC pluripotency and differentiation.

Multipotent stem cells such as NSCs and HSCs, CD4$^+$ T cells and embryonic fibroblasts also possess bivalent histone modification, although less than that in ESCs [64,69–72]. Their lower amount of bivalent histone marks is a result of selective retention: genes whose expression levels are induced upon ESC differentiation retain the H3K4me3 mark, while genes that are silenced keep the H3K27me3 mark [62,64,71] (Fig. 19.3). The persistence of bivalent histone marks at some genes in these cells may ensure the genes' plasticity at later stages of differentiation.

MIcro RNA

MicroRNA (miRNA) is one of many types of non-coding RNA, and is typically a 20–25-nucleotide length that can bind to the 3'-untranslated region (UTR) of target mRNAs through an imperfect sequence match to repress their translation and stability [73]. Repression is achieved by the formation of a structure called the RNA-induced silencing complex (RISC). Interestingly, some human miRNAs also have been reported to function as activators of target mRNA translation [74].

Several lines of study using various stem cells have indicated the importance of miRNA in stem cell regulation, and especially in fate specification. The lack of miRNA's maturation processing machinery can resulted in differentiation deficiencies. For example, in Dicer-null mESCs, differentiation marker expression is not present even after the induction of differentiation [75]. The coding regions of pluripotency-markers Nanog, Oct4, and Sox2 in mESCs are also targets of differentiation-related miRNAs such as miR-134, miR-296, and miR-470 [76], and they fail to be silenced in DGCR8-null mESCs [77]. On the other hand, a subset of the miR-290 cluster, called the ES cell-specific cell cycle (ESCC) regulating miRNA, for example miR-291-3p, miR-294, and miR-295 is known to promote proliferation of mESC [78]. Interestingly, these miRNA gene promoters are targets of pluripotency-associated factors Nanog, Oct4, and Sox2 [79]. These observations indicate that the existence of miRNA is both important for ESC differentiation and proliferation, and that miRNAs can promote differentiation by reducing pluripotency-associated protein levels (Fig. 19.3).

In neural tissues, miR-124a is expressed predominantly and has been shown to participate in the *in vitro* differentiation of mNSCs into neurons by mediating degradation of non-neuronal gene transcripts [80]. miR-124a expression is regulated by REST/NRSF, which is expressed only in NSCs and non-neuronal cells including ESCs (see above). In NSCs, therefore, since the expression of the *miR-124a* gene is suppressed by REST/NRSF, the stability of non-neuronal gene transcripts can be increased, thus limiting NSCs to differentiate into neurons. When REST/NRSF is absent, the expression of *miR-124a* and neuronal genes is up-regulated, leading to a preference for neuronal lineage differentiation. miR-124 can also target small carboxy-terminal domain phosphatase 1 (SCP1), which, like REST/NRSF, is an anti-neuronal factor in non-neural tissues and is recruited to RE1-containing gene promoters by REST/NRSF [81], thus providing another mechanism to induce neurogenesis [82]. miR-124a and miR-9 were also found to promote neurogenesis *via* inhibition of STAT3 activation [83]; STAT3 activation induces rNSCs to differentiate into astrocytes while also inhibiting neuronal differentiation [84]. miR-124 and miR-128 are found exclusively in the neuronal lineage, while miR-23, miR-26, and miR-29 are expressed in the astrocytic lineage [85].

Differentiation of early progenitors of the hematopoietic lineage is prevented by miR-128 and miR-181. In addition, another set of miRNAs such as miR-16, miR-103, and miR-107 prevents proliferation of later progenitor cells, whereas miR-221, miR-222, and miR-223 control the terminal differentiation pathways [86]. Mouse hematopoietic progenitor cells can differentiate into lymphoid and myeloid progenitors by selective expression of miR-181 and miR-223, respectively [87]. Within the mouse lymphoid lineage, the differential expression of miR-150 regulates the lineage decision between T- and B-cells [88], while in the myeloid lineage miR-150, miR-155, miR-221, and miR-222 are progressively down-regulated, with up-regulation of miR-451 and miR-16 occurring during the late phase of human erythropoiesis in *in vitro* study [89].

REPROGRAMMING FOR PLURIPOTENCY

Cell differentiation has been depicted as a ball rolling down an epigenetic landscape [90], starting from totipotency, moving through pluripotency, and finally reaching lineage-committed states. In the last three years, multiple studies have reported that the "ball" can actually be pushed back up the hill (Fig. 19.4): several types of differentiated cells have been shown to be reprogrammable back to the pluripotent state under the influence of a few factors such as Oct4, Sox2, Klf4, Myc, Nanog, and Lin-28 [91–101]. Such reprogrammed cells, which have been called induced pluripotent stem (iPS) cells [91], are similar to ESCs in terms of their morphology, expression of major ESC marker genes, and capacity to self-renew and to differentiate into various cell types of the three germ layers.

Induced pluripotency in iPS cells was shown to be caused by changes in epigenetic modification of the treated cells. The promoter regions of various pluripotency-associated genes are hypermethylated in differentiated cells, but in iPS cells these genes are hypomethylated, resembling their state in ESCs [102]. How the above-mentioned inducing factors can trigger the demethylation of pluripotency genes remains elusive, because it is unclear whether they possess direct or indirect DNA demethylation activity. Interestingly, nevertheless, generation of iPS cells can be promoted by demethylating agents such as 5-azacytidine [103], underlining the importance of the DNA demethylation process in mediating induced pluripotency.

Bivalent methylation marks on histone H3 are also re-established at the promoter regions in iPS cells [103,104]. Both ESCs and iPS cells have H3K4me3 in the promoter regions of pluripotency-associated genes, while both active H3K4me3 and repressive H3K27me3 are present at their differentiation-associated genes [62] (Fig. 19.3). Because differentiated cells such as mouse fibroblasts have the opposite histone methylation pattern [104], it is very likely that histone methylation also plays a role in reprogramming.

323

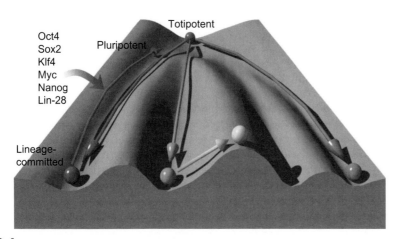

FIGURE 19.4
Epigenetic landscape. The totipotent fertilized egg can be depicted as a red ball that can roll down one of several possible valleys, passing through the pluripotent state and then differentiating into a particular tissue-lineage cell (blue, green, and pink balls). Reprogramming factors can push the ball back up the hill, enabling it to re-acquire pluripotency features. Lineage-committed cells can also trans-differentiate into cells of another lineage (yellow ball) by other epigenetic reprogrammings. The diagram is modified from Waddington [90]. (Please refer to color plate section)

In both ESCs and iPS cells, histones H3 and H4 at the promoter regions of pluripotency-associated genes are hyperacetylated, whereas these promoters in differentiated cells display hypoacetylated H3 and H4. Addition of the HDAC inhibitor VPA can lead to a hyperacetylated histone status, and has proved to be efficient in inducing pluripotency of human fibroblasts, with only Oct4 and Sox2 required as necessary supplemental factors [105]. Other factors may be dispensable under these conditions if any role they play in increasing acetylation by recruiting HATs, a known function in the case of c-Myc [106], can be substituted by VPA.

CONCLUDING REMARKS

Revealing the epigenetic mechanisms that contribute to stem cell potency and differentiation has been an exciting journey. Still, many avenues remain to be explored. For example, several HDAC inhibitors and DNA demethylating agents such as VPA and 5-azacytidine are now in clinical trials for therapeutic application to several disorders and diseases [107,108]. While HDAC inhibitors or DNA demethylating agents might be expected to affect a broad range of genes or their activator/repressor complexes, they actually do not. This differential effect must be attributable to characteristics that the affected genes alone possess. Therefore, the knowledge of how these compounds relieve or cure disorder and disease by changing epigenetic marks is highly important. It will be also interesting to explore the effect of disruption of certain miRNAs in order to generate iPS cells, because recent findings show that RNA binding protein Lin28 can promote reprogramming by selective inhibition of miRNA maturation machinery [109,110]. We also have to evaluate carefully the various origins of cells and induction methods available for stem cell differentiation in order to generate specific cell phenotypes that can be used for clinical applications. To develop an optimal method to generate specific cell types, we must urgently learn more about the precise mechanisms of stem cell fate specification.

ACKNOWLEDGEMENTS

We apologize to colleagues whose work, although relevant, we may not have included in this review due to space constraints. We thank our laboratory members for useful discussions on this topic, and Drs. Ian Smith and Siripong Thitamadee for critical reading of the manuscript. We have been supported by a Grant-in-Aid for Scientific Research in priority areas, the NAIST Global COE Program (Frontier Biosciences: Strategies for Survival and Adaptation in a Changing Global Environment), the Nakajima Foundation, and the Uehara Memorial Foundation.

References

1. Kempermann G. Adult neurogenesis. New York: Oxford University Press; 2006.

2. Keller G. Embryonic stem cell differentiation: emergence of a new era in biology and medicine. *Genes Dev* 2005;**19**:1129–55.

3. Namihira M, Kohyama J, Abematsu M, Nakashima K. Epigenetic mechanisms regulating fate specification of neural stem cells. *Phil Trans R Soc B* 2008;**363**:2099–109.

4. Allen ND. Temporal and epigenetic regulation of neurodevelopmental plasticity. *Phil Trans R Soc B* 2008;**363**:23–8.

5. Okano H, Temple S. Cell types to order: temporal specification of CNS stem cells. *Curr Opin Neurobiol* 2009;**19**:1–8.

6. Zhu J, Emerson SG. Hematopoietic cytokines, transcription factors and lineage commitment. *Oncogene* 2002;**21**:3295–313.

7. Liang G, Chan MF, Tomigahara Y, Tsai YC, Gonzales FA, Li E, et al. Cooperativity between DNA methyltransferases in the maintenance methylation of repetitive elements. *Mol Cell Biol* 2002;**22**:480–91.

8. Shiota K, Kogo Y, Ohgane J, Imamura T, Urano A, Nishino K, et al. Epigenetic marks by DNA methylation specific to stem, germ and somatic cells in mice. *Genes Cells* 2002;**7**:961–9.

9. Bibikova M, Chudin E, Wu B, Zhou L, Garcia EW, Liu Y, et al. Human embryonic stem cells have a unique epigenetic signature. *Genome Res* 2006;**16**:1075–83.

10. Takizawa T, Nakashima K, Namihira M, Ochiai W, Uemura A, Yanagisawa M, et al. DNA methylation is a critical cell-intrinsic determinant of astrocyte differentiation in the fetal brain. *Dev Cell* 2001;**1**:749–58.

11. Namihira M, Kohyama J, Semi K, Sanosaka T, Deneen B, Taga T, et al. Committed neuronal precursors confer astrocytic potential on residual neural precursor cells. *Dev Cell* 2009;**16**:245–55.

12. Namihira M, Nakashima K, Taga T. Developmental stage dependent regulation of DNA methylation and chromatin modification in a immature astrocyte specific gene promoter. *FEBS Lett* 2004;**572**:184–8.

13. Shimozaki K, Namihira M, Nakashima K, Taga T. Stage- and site-specific DNA demethylation during neural cell development from embryonic stem cells. *J Neurochem* 2005;**93**:432–9.

14. Hatada I, Namihira M, Morita S, Kimura M, Horii T, Nakashima K. Astrocyte-specific genes are generally demethylated in neural precursor cells prior to astrocytic differentiation. *PLoS One* 2008;**3**:e3189.

15. Setoguchi H, Namihira M, Kohyama J, Asano H, Sanosaka T, Nakashima K. Methyl-CpG binding proteins are involved in restricting differentiation plasticity in neurons. *J Neurosci Res* 2006;**84**:969–79.

16. Ballas N, Lioy DT, Grunseich C, Mandel G. Non-cell autonomous influence of MeCP2-deficient glia on neuronal dendritic morphology. *Nat Neurosci* 2009;**12**:311–17.

17. Kohyama J, Kojima T, Takatsuka E, Yamashita T, Namiki J, Hsieh J, et al. Epigenetic regulation of neural cell differentiation plasticity in the adult mammalian brain. *Proc Natl Acad Sci USA* 2008;**105**:18012–17.

18. Tsujimura K, Abematsu M, Kohyama J, Namihira M, Nakashima K. Neuronal differentiation of neural precursor cells is promoted by the methyl-CpG binding protein MeCP2. *Exp Neurol* 2009;**219**:104–11.

19. Campos LS, Duarte AJ, Branco T, Henrique D. mDll1 and mDll3 expression in the developing mouse brain: role in the establishment of the early cortex. *J Neurosci Res* 2001;**64**:590–8.

20. Kawaguchi A, Ikawa T, Kasukawa T, Ueda HR, Kurimoto K, Saitou M, et al. Single-cell gene profiling defines differential progenitor subclasses in mammalian neurogenesis. *Development* 2008;**135**:3113–24.

21. Yoon KJ, Koo BK, Im SK, Jeong HW, Ghim J, Kwon MC, et al. Mind bomb 1-expressing intermediate progenitors generate notch signaling to maintain radial glial cells. *Neuron* 2008;**58**:519–31.

22. Simpson P. Developmental genetics. The Notch connection. *Nature* 1995;**375**:736–7.

23. Nye JS, Kopan R. Developmental signaling. Vertebrate ligands for Notch. *Curr Biol* 1995;**5**:966–9.

24. Bray S. A Notch affair. *Cell* 1998;**93**:499–503.

25. Louvi A, Artavanis-Tsakonas S. Notch signaling in vertebrate neural development. *Nat Rev Neurosci* 2006;**7**:93–102.

26. Lundkvist J, Lendahl U. Notch and the birth of glial cells. *Trends Neurosci* 2001;**24**:492–4.

27. Nakayama K, Nagase H, Hiratochi M, Koh CS, Ohkawara T. Similar mechanisms regulated by gamma-secretase are involved in both directions of the bi-directional Notch-Delta signaling pathway as well as play a potential role in signaling events involving type 1 transmembrane proteins. *Curr Stem Cell Res Ther* 2008;**3**:288–302.

28. Wallberg AE, Pedersen K, Lendahl U, Roeder RG. p300 and PCAF act cooperatively to mediate transcriptional activation from chromatin templates by notch intracellular domains in vitro. *Mol Cell Biol* 2002;**22**:7812–19.

29. Naka H, Nakamura S, Shimazaki T, Okano H. Requirement for COUP-TFI and II in the temporal specification of neural stem cells in central nervous system development. *Nat Neurosci* 2008;**11**:1014–23.

30. Okada Y, Matsumoto A, Shimazaki T, Enoki R, Koizumi A, Ishii S, et al. Spatio-temporal recapitulation of central nervous system development by murine ES cell-derived neural stem/progenitor cells. *Stem Cells* 2008;**26**:3086–98.

31. Levings PP, Bungert J. The human beta-globin locus control region. *Eur J Biochem* 2002;**269**:1589–99.

32. Singh M, Lavelle D, Vaitkus K, Mahmud N, Hankewych M, DeSimone J. The gamma-globin gene promoter progressively demethylates as the hematopoietic stem progenitor cells differentiate along the erythroid lineage in baboon fetal liver and adult bone marrow. *Exp Hematol* 2007;**35**:48–55.

33. Ivascu C, Wasserkort R, Lesche R, Dong J, Stein H, Thiel A, et al. DNA methylation profiling of transcription factor genes in normal lymphocyte development and lymphomas. *Int J Biochem Cell Biol* 2007;**39**:1523–38.

34. Stopka T, Amanatullah DF, Papetti M, Skoultchi AI. PU.1 inhibits the erythroid program by binding to GATA-1 on DNA and creating a repressive chromatin structure. *EMBO J* 2005;**24**:3712–23.

35. Efroni S, Duttagupta R, Cheng J, Dehghani H, Hoeppner DJ, Dash C, et al. Global transcription in pluripotent embryonic stem cells. *Cell Stem Cell* 2008;**2**:437–47.

36. Meshorer E, Yellajoshula D, George E, Scambler PJ, Brown DT, Misteli T. Hyperdynamic plasticity of chromatin proteins in pluripotent embryonic stem cells. *Dev Cell* 2006;**10**:105–16.

37. Meshorer E, Misteli T. Chromatin in pluripotent embryonic stem cells and differentiation. *Nat Rev Mol Cell Biol* 2006;**7**:540–6.

38. Ballas N, Grunseich C, Lu DD, Speh JC, Mandel G. REST and its corepressors mediate plasticity of neuronal gene chromatin throughout neurogenesis. *Cell* 2005;**121**:645–57.

39. Lunyak VV, Burgess R, Prefontaine GG, Nelson C, Sze SH, Chenoweth J, et al. Co-repressor-dependent silencing of chromosomal regions encoding neuronal genes. *Science* 2002;**298**:1747–52.

40. Lunyak VV, Rosenfeld MG. No rest for REST: REST/NRSF regulation of neurogenesis. *Cell* 2005;**121**:499–501.

41. Rice JC, Allis CD. Histone methylation versus histone acetylation: new insights into epigenetic regulation. *Curr Opin Cell Biol* 2001;**13**:263–73.

42. Hsieh J, Nakashima K, Kuwabara T, Mejia E, Gage FH. Histone deacetylase inhibition-mediated neuronal differentiation of multipotent adult neural progenitor cells. *Proc Natl Acad Sci USA* 2004;**101**:16659–64.

43. Jung GA, Yoon JY, Moon BS, Yang DH, Kim HY, Lee SH, et al. Valproic acid induces differentiation and inhibition of proliferation in neural progenitor cells via the beta-catenin-Ras-ERK-p21$^{Cip/WAF1}$ pathway. *BMC Cell Biol* 2008;**9**:66.

44. Marin-Husstege M, Muggironi M, Liu A, Casaccia-Bonnefil P. Histone deacetylase activity is necessary for oligodendrocyte lineage progression. *J Neurosci* 2002;**22**:10333–45.

45. Shen S, Li J, Casaccia-Bonnefil P. Histone modifications affect timing of oligodendrocyte progenitor differentiation in the developing rat brain. *J Cell Biol* 2005;**169**:577–89.

46. Ye F, Chen Y, Hoang TN, Montgomery RL, Zhao X, Bu H, et al. HDAC1 and HDAC2 regulate oligodendrocyte differentiation by disrupting the β-catenin-TCF interaction. *Nat Neurosci* 2009;**12**:829–38.

47. Huo X, Zhang J. Important roles of reversible acetylation in the function of hematopoietic transcription factors. *J Cell Mol Med* 2005;**9**:103–12.

48. Letting DL, Rakowski C, Weiss MJ, Blobel GA. Formation of a tissue-specific histone acetylation pattern by the hematopoietic transcription factor GATA-1. *Mol Cell Biol* 2003;**23**:1334–40.

49. Blobel GA. CBP and p300: versatile coregulators with important roles in hematopoietic gene expression. *J Leukoc Biol* 2002;**71**:545–56.

50. Guidez F, Howell L, Isalan M, Cebrat M, Alani RM, Ivins S, et al. Histone acetyltransferase activity of p300 is required for transcriptional repression by the promyelocytic leukemia zinc finger protein. *Mol Cell Biol* 2005;**25**:5552–66.

51. Ernst P, Wang J, Huang M, Goodman RH, Korsmeyer SJ. MLL and CREB bind cooperatively to the nuclear coactivator CREB-binding protein. *Mol Cell Biol* 2001;**21**:2249–58.

52. Milne TA, Briggs SD, Brock HW, Martin ME, Gibbs D, Allis CD, et al. MLL targets SET domain methyltransferase activity to Hox gene promoters. *Mol Cell* 2002;**10**:1107–17.

53. Dou Y, Milne TA, Tackett AJ, Smith ER, Fukuda A, Wysocka J, et al. Physical association and coordinate function of the H3 K4 methyltransferase MLL1 and the H4 K16 acetyltransferase MOF. *Cell* 2005;**121**:873–85.

54. Xia ZB, Anderson M, Diaz MO, Zeleznik-Le NJ. MLL repression domain interacts with histone deacetylases, the polycomb group proteins HPC2 and BMI-1, and the corepressor C-terminal-binding protein. *Proc Natl Acad Sci USA* 2003;**100**:8342–7.

55. Lim DA, Huang YC, Swigut T, Mirick AL, Garcia-Verdugo JM, Wysocka J, et al. Chromatin remodelling factor Mll1 is essential for neurogenesis from postnatal neural stem cells. *Nature* 2009;**458**:529–33.

56. Ernst P, Wang J, Korsmeyer SJ. The role of MLL in hematopoiesis and leukemia. *Curr Opin Hematol* 2002;**9**:282–7.

57. Ernst P, Mabon M, Davidson AJ, Zon LI, Korsmeyer SJ. An Mll-dependent Hox program drives hematopoietic progenitor expansion. *Curr Biol* 2004;**14**:2063–9.

58. Ayton PM, Cleary ML. Transformation of myeloid progenitors by MLL oncoproteins is dependent on Hoxa7 and Hoxa9. *Genes Dev* 2003;**17**:2298–307.

59. Boyer LA, Plath K, Zeitlinger J, Brambrink T, Medeiros LA, Lee TI, et al. Polycomb complexes repress developmental regulators in murine embryonic stem cells. *Nature* 2006;**441**:349–53.

60. Lee TI, Jenner RG, Boyer LA, Guenther MG, Levine SS, Kumar RM, et al. Control of developmental regulators by Polycomb in human embryonic stem cells. *Cell* 2006;**125**:301–13.

61. Azuara V, Perry P, Sauer S, Spivakov M, Jørgensen HF, John RM, et al. Chromatin signatures of pluripotent cell lines. *Nat Cell Biol* 2006;**8**:532–8.

62. Bernstein BE, Mikkelsen TS, Xie X, Kamal M, Huebert DJ, Cuff J, et al. A bivalent chromatin structure marks key developmental genes in embryonic stem cells. *Cell* 2006;**125**:315–26.

63. Guenther MG, Levine SS, Boyer LA, Jaenisch R, Young RA. A chromatin landmark and transcription initiation at most promoters in human cells. *Cell* 2007;**130**:77–88.

64. Mikkelsen TS, Ku M, Jaffe DB, Issac B, Lieberman E, Giannoukos G, et al. Genome-wide maps of chromatin state in pluripotent and lineage-committed cells. *Nature* 2007;**448**:553–60.

65. Wen B, Wu H, Shinkai Y, Irizarry RA, Feinberg AP. Large histone H3 lysine 9 dimethylated chromatin blocks distinguish differentiated from embryonic stem cells. *Nat Genet* 2009;**41**:246–50.

66. Boyer LA, Lee TI, Cole MF, Johnstone SE, Levine SS, Zucker JP, et al. Core transcriptional regulatory circuitry in human embryonic stem cells. *Cell* 2005;**122**:947–56.

67. Loh YH, Wu Q, Chew JL, Vega VB, Zhang W, Chen X, et al. The Oct4 and Nanog transcription network regulates pluripotency in mouse embryonic stem cells. *Nat Genet* 2006;**38**:431–40.

68. Zhou Q, Chipperfield H, Melton DA, Wong WH. A gene regulatory network in mouse embryonic stem cells. *Proc Natl Acad Sci USA* 2007;**104**:16438–43.

69. Roh TY, Cuddapah S, Cui K, Zhao K. The genomic landscape of histone modifications in human T cells. *Proc Natl Acad Sci USA* 2006;**103**:15782–7.

70. Barski A, Cuddapah S, Cui K, Roh TY, Schones DE, Wang Z, et al. High-resolution profiling of histone methylations in the human genome. *Cell* 2007;**129**:823–37.

71. Pan G, Tian S, Nie J, Yang C, Ruotti V, Wei H, et al. Whole-genome analysis of histone H3 lysine 4 and lysine 27 methylation in human embryonic stem cells. *Cell Stem Cell* 2007;**1**:299–312.

72. Cui K, Zang C, Roh TY, Schones DE, Childs RW, Peng W, et al. Chromatin signatures in multipotent human hematopoeitic stem cells indicate the fate of bivalent genes during differentiation. *Cell Stem Cell* 2009;**4**:80–93.

73. Rana TM. Illuminating the silence: understanding the structure and function of small RNAs. *Nat Rev Mol Cell Biol* 2007;**8**:23–36.

74. Vasudevan S, Tong Y, Steitz JA. A switching from repression to activation: microRNAs can up-regulate translation. *Science* 2007;**318**:1931–4.

75. Kanellopoulou C, Muljo SA, Kung AL, Ganesan S, Drapkin R, Jenuwein T, et al. Dicer-deficient mouse embryonic stem cells are defective in differentiation and centromeric silencing. *Genes Dev* 2005;**19**:489–501.

76. Tay Y, Zhang J, Thomson AM, Lim B, Rigoutsos I. MicroRNAs to Nanog, Oct4, and Sox2 coding regions modulate embryonic stem cell differentiation. *Nature* 2008;**455**:1124–8.

77. Wang Y, Medvid R, Melton C, Jaenisch R, Blelloch R. DGCR8 is essential for microRNA biogenesis and silencing of embryonic stem cell self-renewal. *Nat Genet* 2007;**39**:380–5.

78. Wang Y, Baskerville S, Shenoy A, Babiarz JE, Baehner L, Blelloch R. Embryonic stem cell-specific microRNAs regulate the G1-S transition and promote rapid proliferation. *Nat Genet* 2008;**40**:1478–83.

79. Marson A, Levine SS, Cole MF, Frampton GM, Brambrink T, Johnstone S, et al. Connecting microRNA genes to the core transcriptional regulatory circuitry of embryonic stem cells. *Cell* 2008;**134**:521–33.

80. Conaco C, Otto S, Han JJ, Mandel G. Reciprocal actions of REST and a microRNA promote neuronal identity. *Proc Natl Acad Sci USA* 2006;**103**:2422–7.

81. Yeo M, Lee SK, Lee B, Ruiz EC, Pfaff SL, Gill GN. Small CTD phosphatases function in silencing neuronal gene expression. *Science* 2005;**307**:596–600.

82. Visvanathan J, Lee S, Lee B, Lee JW, Lee SK. The microRNA miR-124 antagonizes the anti-neural REST/SCP1 pathway during embryonic CNS development. *Genes Dev* 2007;**21**:744–9.

83. Krichevsky AM, Sonntag KC, Isacson O, Kosik KS. Specific microRNAs modulate embryonic stem cell-derived neurogenesis. *Stem Cells* 2006;**24**:857–64.

84. Gu F, Hata R, Ma YJ, Tanaka J, Mitsuda N, Kumon Y, et al. Suppression of Stat3 promotes neurogenesis in cultured neural stem cells. *J Neurosci Res* 2005;**81**:163–71.

85. Smirnova L, Gräfe A, Seiler A, Schumacher S, Nitsch R, Wulczyn FG. Regulation of miRNA expression during neural cell specification. *Eur J Neurosci* 2005;**21**:1469–77.

86. Georgantas RW 3rd, Hildreth R, Morisot S, Alder J, Liu CG, Heimfeld S, et al. CD34+ hematopoietic stem-progenitor cell microRNA expression and function: a circuit diagram of differentiation control. *Proc Natl Acad Sci USA* 2007;**104**:2750–5.

87. Chen CZ, Li L, Lodish HF, Bartel DP. MicroRNAs modulate hematopoietic lineage differentiation. *Science* 2004;**303**:83–6.

88. Zhou B, Wang S, Mayr C, Bartel DP, Lodish HF. miRNA-150, a microRNA expressed in mature B and T cells, blocks early B cell development when expressed prematurely. *Proc Natl Acad Sci USA* 2007;**104**:7080–5.

89. Bruchova H, Yoon D, Agarwal AM, Mendell J, Prchal JT. Regulated expression of microRNAs in normal and polycythemia vera erythropoiesis. *Exp Hematol* 2007;**35**:1657–67.

90. Waddington CH. *The Strategy of the Genes. A Discussion of Some Aspects of Theoretical Biology*. London: Allen & Unwin; 1957.

91. Takahashi K, Yamanaka S. Induction of pluripotent stem cells from mouse embryonic and adult fibroblast cultures by defined factors. *Cell* 2006;**126**:663–76.

92. Takahashi K, Okita K, Nakagawa M, Yamanaka S. Induction of pluripotent stem cells from fibroblast cultures. *Nat Protoc* 2007;**2**:3081–9.

93. Takahashi K, Tanabe K, Ohnuki M, Narita M, Ichisaka T, Tomoda K, et al. Induction of pluripotent stem cells from adult human fibroblasts by defined factors. *Cell* 2007;**131**:861–72.

94. Meissner A, Wernig M, Jaenisch R. Direct reprogramming of genetically unmodified fibroblasts into pluripotent stem cells. *Nat Biotechnol* 2007;**25**:1177–81.

95. Okita K, Ichisaka T, Yamanaka S. Generation of germline-competent induced pluripotent stem cells. *Nature* 2007;**448**:313–17.

96. Hanna J, Wernig M, Markoulaki S, Sun. CW, Meissner A, Cassady JP, et al. Treatment of sickle cell anemia mouse model with iPS cells generated from autologous skin. *Science* 2007;**318**:1920–3.

97. Aoi T, Yae K, Nakagawa M, Ichisaka T, Okita K, Takahashi K, et al. Generation of pluripotent stem cells from adult mouse liver and stomach cells. *Science* 2008;**321**:699–702.

98. Lowry WE, Richter L, Yachechko R, Pyle AD, Tchieu J, Sridharan R, et al. Generation of human induced pluripotent stem cells from dermal fibroblasts. *Proc Natl Acad Sci USA* 2008;**105**:2883–8.

99. Nakagawa M, Koyanagi M, Tanabe K, Takahashi K, Ichisaka T, Aoi T, et al. Generation of induced pluripotent stem cells without Myc from mouse and human fibroblasts. *Nat Biotechnol* 2008;**26**:101–6.

100. Park IH, Zhao R, West JA, Yabuuchi A, Huo H, Ince TA, et al. Reprogramming of human somatic cells to pluripotency with defined factors. *Nature* 2008;**451**:141–6.

101. Wernig M, Meissner A, Cassady JP, Jaenisch R. c-Myc is dispensable for direct reprogramming of mouse fibroblasts. *Cell Stem Cell* 2008;**2**:10–12.

102. Imamura M, Miura K, Iwabuchi K, Ichisaka T, Nakagawa M, Lee J, et al. Transcriptional repression and DNA hypermethylation of a small set of ES cell marker genes in male germline stem cells. *BMC Dev Biol* 2006;**6**:34.

103. Mikkelsen TS, Hanna J, Zhang X, Ku M, Wernig M, Schorderet P, et al. Dissecting direct reprogramming through integrative genomic analysis. *Nature* 2008;**454**:49–55.

104. Maherali N, Sridharan R, Xie W, Utikal J, Eminli S, Arnold K, et al. Directly reprogrammed fibroblasts show global epigenetic remodeling and widespread tissue contribution. *Cell Stem Cell* 2007;**1**:55–70.

105. Huangfu D, Osafune K, Maehr R, Guo W, Eijkelenboom A, Chen S, et al. Induction of pluripotent stem cells from primary human fibroblasts with only Oct4 and Sox2. *Nat Biotech* 2008;**26**:1269–75.

106. Knoepfler PS, Zhang XY, Cheng PF, Gafken PR, McMahon SB, Eisenman RN. Myc influences global chromatin structure. *EMBO J* 2006;**25**:2723–34.

107. Kazantsev AG, Thompson LM. Therapeutic application of histone deacetylase inhibitors for central nervous system disorders. *Nat Rev Drug Discov* 2008;**7**:854–68.

108. Yoo CB, Jones PA. Epigenetic therapy of cancer: past, present and future. *Nat Rev Drug Discov* 2006;**5**:37–50.

109. Viswanathan SR, Daley GQ, Gregory RI. Selective blockade of microRNA processing by Lin28. *Science* 2008;**320**:97–100.

110. Rybak A, Fuchs H, Smirnova L, Brandt C, Pohl EE, Nitsch R, et al. A feedback loop comprising lin-28 and let-7 controls pre-let-7 maturation during neural stem-cell commitment. *Nat Cell Biol* 2008;**10**:987–93.

CHAPTER 20

Epigenetic Basis of Skeletal Muscle Regeneration

Pier Lorenzo Puri[1,2] and Vittorio Sartorelli[3]
[1]Dulbecco Telethon Institute at Fondazione S. Lucia, Rome, Italy
[2]The Burnham Institute for Medical Research, San Diego, CA, USA
[3]Laboratory of Muscle Stem Cells and Gene Regulation, National Institute of Arthritis, Musculoskeletal and Skin Diseases, National Institutes of Health, Bethesda, MD, USA

INTRODUCTION

Therapeutic repopulation of diseased organs and tissues by endogenous progenitor cells is one of the most challenging tasks in regenerative medicine. Tissue and organ precursors are often referred to as adult "somatic stem cells" (SSCs) because of their functional analogies with the embryonic stem cells (ESCs), including the ability for long-term self-renewal and the potential to commit into multiple lineages. However, while ESCs are totipotent and can adopt virtually all lineages, SSCs are located within differentiated tissues and organs, have restricted "potency", and provide an immediate reservoir for repair upon injury or disease-associated events [1].

Lineage commitment, migration, proliferation, and differentiation of SSCs are regulated by the coordinated activation and repression of distinct subsets of genes in response to cues released within the regenerative environment [2]. Therefore, understanding how extrinsic signals are converted into the epigenetic information that controls gene expression at different regeneration stages is critical to devise strategies aimed at manipulating SSCs for therapeutic regeneration of diseased tissues and organs [3].

The extensive knowledge gained on muscle stem cells (MSCs) makes muscle regeneration an interesting paradigm to unveil general principles of epigenetic regulation of tissue regeneration and to investigate strategies for regenerative medicine using SSCs. Because of the extraordinary potential of muscle regeneration in the treatment of currently fatal genetic diseases (e.g. muscular dystrophies) or widespread muscular disorders, such as muscle atrophies, cachexia, and sarcopenia, the molecular and epigenetic basis of skeletal myogenesis has been the object of intense investigation [4]. As such, this chapter will focus on the epigenetic regulation of muscle regeneration, as a paradigm for other tissues.

EPIGENETIC PROFILE OF MUSCLE STEM CELLS

The epigenetic profile of MSCs consists of a variety of chromatin modifications and the expression pattern of a variety of microRNAs (miRNAs) that are transmitted along the transition through sequential stages of the regeneration program, and establish the "memory" of an active and a repressive gene state. These modifications contribute to sequentially reprogram the MSC genome toward a differentiated phenotype [3].

Handbook of Epigenetics: The New Molecular and Medical Genetics. DOI: 10.1016/B978-0-12-375709-8.00020-4

Satellite muscle cells are one typical example of adult MSCs that can regenerate injured or diseased skeletal myofibers [5]. Recent studies have reported on the heterogeneous nature of satellite cells and on the myogenic potential of additional subpopulations of putative MSCs that are distinct from satellite cells [6]. Other cell types that participate in muscle regeneration are resident fibroblasts and the inflammatory and hematopoietic cells recruited in the regenerative environment [7]. These cells contribute to regeneration of skeletal muscles directly, via cell-to-cell interactions, and indirectly, by releasing paracrine/autocrine cues. Independent studies have reported on the ability of MSCs and other regeneration-associated cell types to adopt the myogenic, adipogenic and, possibly, other lineages [8]. Although the actual impact of such "plasticity" is controversial, it suggests that signal-dependent regulation of cell fate might influence muscle regeneration and regulate the relative abundance of muscle and fat in adult organisms. Thus, the complete elucidation of the epigenetic network (epigenome) that regulates gene expression in distinct cell types within the regenerative environment is important to identify new targets for selective interventions that promote muscle regeneration by manipulating the expression of key genes, and to conceive strategies toward achieving favorable metabolic responses, via modulation of the relative proportion of skeletal muscles and adipose tissue.

GENOME REPROGRAMMING OF MUSCLE STEM CELLS

During skeletal myogenesis the nucleus of the multipotent muscle progenitor cell is sequentially reprogrammed to adopt and maintain the new pattern of gene expression [9]. This process is achieved epigenetically, via the engagement of chromatin modifying enzymes, which are recruited by tissue-specific transcription factors on target promoters in response to the activation of signaling cascades [10]. This global genome reprogramming allows the acquisition of the myogenic identity, and the proliferation of muscle progenitors and their subsequent differentiation into multinucleated myofibers. For instance, the myogenic lineage is determined by the selective activation of genes that establish the myogenic identity (e.g. Pax3 and Pax7, MyoD and Myf5) and the repression of genes associated with the acquisition of nonmuscle lineages – a process termed lineage-commitment [10]. One feature of SSCs is the asymmetric division [11–13], which predicts that the epigenetic information stored in one cell has to be segregated into two daughter cells that are committed toward distinct fates. One cell returns to quiescence and replenishes the pool of reserve satellite cells; another cell enters the differentiation program. Thus asymmetric division is deputed to regulate the proportion of cells that repopulate injured muscles, while maintaining the integrity of satellite cell potential to sustain repeated cycles of regeneration. Recent studies have identified putative molecular markers that correlate with distinct fates of the progenies generated by asymmetric division. In particular, Pax7 expression appears to co-segregate with the fraction of satellite cells that do not enter the differentiation program, while the expression of MyoD, Myf5, and Numb reflects the commitment to differentiation [11–13]. Pax7 is a paired box transcription factor required for satellite cells to generate skeletal muscle-committed progenitor cells [14,15]. It is expressed in quiescent satellite cells and the expression persists during the first stages of regeneration, when it promotes proliferation and survival, and induces the expression of MyoD and Myf5 [16,17]. Despite the established role of Pax7 in satellite cell lineage acquisition [14,15], recent studies indicate that Pax7 is dispensable for adult muscle regeneration [18]. Because of the asymmetric division of satellite cells, Pax7 is likely to drive two distinct transcription networks that determine the fate of satellite cell progeny. Pax7-mediated activation of MyoD and Myf5 specifies the population of MSCs that enter the differentiation program [13,19]. Studies from the Rudnicki lab have elucidated the mechanism underlying Pax7-mediated activation of Myf5 in satellite cells [20]. Pax7, which is by itself a weak transcriptional activator, associates with the Wdr5-Ash2LMLL2 histone methyl-transferase (HMT) complex, which directs the methylation of H3K4 on the chromatin of the Myf5 locus. By contrast, in the fraction of MSCs that returns to quiescence, MyoD and Myf5 loci is refractory to Pax7-mediated activation.

The different response of these loci to Pax7 might depend on the epigenetic memory, which is determined by the presence of different types of histone and by the methylation profile of the DNA at the regulatory elements of the MyoD locus. On the MyoD promoter, the presence of the histone H3.3 variant establishes the epigenetic memory conducive for transcription in differentiation-committed MSCs [21]. By contrast, the presence of the H1b isoform bound to the homeoprotein Msx1 induces repressive chromatin on the regulatory elements of MyoD in MSCs that re-enter quiescence [22]. Since the expression of MyoD is promoted by the cooperative activity of Pax7 and FoxO3 [17], it is likely that these two proteins establish the chromatin conformation permissive for MyoD expression in differentiation committed satellite cells – or vice versa; in these fractions of cells the particular histone composition might facilitate the access of Pax7 and FoxO3 to MyoD regulatory elements. Finally, MyoD expression is regulated by DNA methylation, which precludes the ectopic activation of MyoD in non-muscle cells [23]. Thus, the expression of MyoD is regulated by distinct epigenetic events, including histone exchange, binding of transcriptional activators and repressors, and DNA methylation. The precise relationship between these events is currently unclear.

TRANSCRIPTIONAL NETWORK THAT REGULATES ADULT SKELETAL MYOGENESIS

Genome wide-based approaches indicate that the progression of muscle progenitors through sequential stages of skeletal myogenesis is underlined by epigenetic changes that permit a coordinated expression of specific subsets of genes [24,25]. Sequence-specific transcription factors and chromatin-modifying enzymes form the transcriptional network that establishes a feed-forward circuit, which drives the genome reprogramming toward terminal differentiation.

The bHLH muscle-specific transcriptional activators – MyoD, Myf5, myogenin, and MRF4 – initiate and perpetuate the differentiation program in collaboration with the ubiquitously expressed E2A gene products (E12, E47, and HEB) and MEF2 proteins [3]. When ectopically introduced into somatic cells, MyoD reprograms the host genome toward the skeletal muscle lineage – a process referred to as "myogenic conversion" [26]. This potential depends on MyoD's ability to penetrate and remodel the chromatin at previously silent muscle loci [27]. Subsequently, a feed-forward circuit involving other bHLH muscle proteins, their hetero-dimerization partners (the products of the E2A gene), MEF2 proteins, and other downstream genes, amplifies the process of skeletal myogenesis by recruiting a variety of chromatin-modifying enzymes, which catalyze the deposition of epigenetic marks conducive for muscle gene expression [28]. Thus, the balance between transcriptional co-activators and co-repressors, and the consequent exchange of post-transcriptional modifications of histone tails, such as acetylation, methylation, phosphorylation, and ubiquitination, imparts to the chromatin at muscle loci the epigenetic profile that coordinates gene expression in muscle cells. In undifferentiated myoblasts, the premature activation of the differentiation program is precluded by the presence of histone deacetylases (HDACs), which prevent local hyperacetylation (Fig. 20.1A). Two additional events involved in the formation of the heterochromatin on promoters of muscle genes in myoblasts are Suv39h1-mediated dimethylation of H3-K9, which mediates the interaction with the chromodomain containing Heterochromatin Protein 1 (HP1) and Polycomb-mediated tri-methylation of H3-K27 [3] (Fig. 20.1A). Removal of these epigenetic marks by differentiation-related cues are likely to involve specific dimethylases and histone exchange, and permits the recruitment of the acetyltransferases p300/CBP, PCAF, the arginine-methyltransferases CARM1 and PRMT5, the ATPase-dependent SWI/SNF chromatin-remodeling complexes, and the MLL/TrxG-associated lysine methyl transferases [3]. These enzymatic complexes endow the myogenic

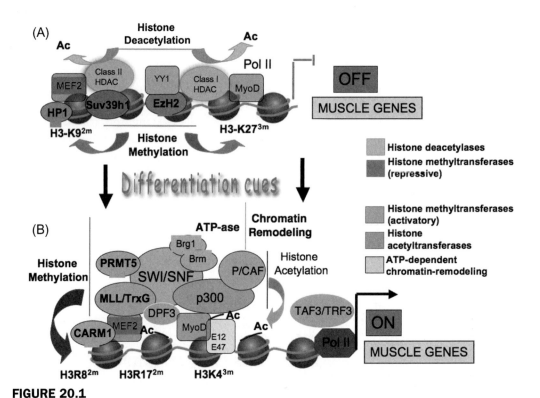

FIGURE 20.1

Schematic representation of the dynamic exchange in the chromatin-associated machinery that controls the epigenetic changes at the regulatory regions (promoter/enhancer elements) of muscle genes. (Please refer to color plate section)

transcriptosome with the enzymatic activities necessary to initiate the transcription of target genes (Fig. 20.1B). A further level of control of muscle gene expression is achieved by the switch of the core promoter recognition complex that confers specificity on muscle gene transcription. The prototypic core promoter recognition complex, TFIID, which initiates the preinitiation complex (PIC) on target sequences, is present in myoblasts, but is replaced by the TAF3/TRF3 complex in myotubes [29].

CHROMATIN-ASSOCIATED KINASES: REGULATORS OF THE EPIGENOME IN MYOGENIC CELLS IN RESPONSE TO REGENERATION CUES

Extrinsic cues in the regeneration environment (the stem cell niche) are converted into epigenetic modifications by signal-activated intracellular cascades, leading to the chromatin recruitment of kinases, which phosphorylate histones and chromatin-associated proteins [30,31].

Recent studies have identified p38 kinases as important regulators of chromatin associated proteins. p38 kinases direct the assembly of the chromatin-remodeling machinery by targeting distinct enzymatic complexes. Phosphorylation of the structural BAF60 subunits of the SWI/SNF complex mediates the SWI/SNF recruitment of on the chromatin of muscle genes [32]. Phosphorylation of MEF2D by p38 alpha/beta kinases promotes the recruitment of the histone methyltransferase Ash2L – the enzymatic subunit of the Trithorax group (TrxG) that catalyzes H3K4 tri-methylation [33], which favors gene transcription. Since the catalytic subunits of SWI/SNF and TrxG are functionally linked, p38 kinases appear to integrate at the chromatin level two functionally-related chromatin-remodeling complexes, via distinct biochemical events – e.g. direct phosphorylation of BAF60 and MEF2D. Thus, p38 signaling couples SWI/SNF-dependent remodeling of nucleosomes and TrxG-mediated H3K4me3 at the regulatory elements of muscle genes. Because p38-mediated phosphorylation of E47 induces the formation of MyoD-E47 heterodimers [34], which promotes optimal binding of MyoD to the DNA recognition sites (the Ebox), this event is likely to contribute to the stable recruitment of SWI/SNF and TrxG complexes to MyoD target promoters [3].

The IGF1-activated PI3K signaling proceeds as a pathway parallel to p38 and regulates the recruitment of p300 HAT. The terminal effectors of IGF1 signaling to the chromatin of muscle genes are AKT1 and −2 kinases, which phosphorylate the C-terminal region of p300 on serine 1834 at the onset of differentiation. This AKT-mediated phosphorylation is required for productive interactions between p300 and MyoD [35].

Chromatin re-setting during the transition from undifferentiated myoblasts to terminally-differentiated myotubes is essential for the exchange of transcriptional co-repressors with co-activators.

Indeed, recruitment of transcriptional co-activators to muscle genes must be preceded, or take place simultaneously, by the displacement of corepressor enzymes and the erasure of pre-existing epigenetic modifications generated by these enzymes. Displacement of histone deacetylases by calcium/calmodulin-dependent kinase (CaMK) and cell cycle events are necessary for local hyperacetylation [36,37]. Another mechanism that contributes to impart to the chromatin of undifferentiated myoblasts the conformation repressive for transcription relates to the recruitment of the Polycomb complex, which silences transcription by trimethylation of lysine 27 of histone 3 (H3K27me3), which is catalyzed by the enzymatic subunit – the histone–lysine methyltransferase Ezh2 [38]. The Ezh2 is recruited to the chromatin of muscle regulatory regions via interaction with YY1. Further association with HDAC1 forms a repressive complex. At the onset of differentiation, the down-regulation of Ezh2 and HDAC1 proteins, and the replacement of YY1 with SRF, allows the binding of MyoD and the recruitment of the positive co-activators, to form a productive transcriptosome [38]. Although the intracellular signaling that regulates these events on muscle regulatory sequences is still unknown, recent evidence indicates that interactions with specific miRNAs establish a regulatory circuit (see next section). The involvement of histone methylation in the dynamic regulation of muscle gene expression implicates the role of histone demethylases in the control of myogenesis. This model postulates that muscle gene expression in myoblasts is controlled by dynamic histone methylation/demethylation exchange [3]. Muscle-specific histone demethylases have not been discovered yet; their identification will complete our knowledge on the epigenetic network that controls chromatin structure and accessibility.

Current knowledge on the molecular mechanism that regulates muscle gene transcription has already inspired pharmacological interventions to boost muscle regeneration. Histone acetyltransferases and deacetylases regulate the acetylation status of target genes and are a target of epigenetic drugs. Inhibition of histone deacetylase in MSCs by drugs currently used in clinical practice (deacetylase inhibitors) implements muscle regeneration and counters the progression of muscular dystrophy in dystrophic mice [39–40].

The application of genome-wide technologies to complex systems, such as muscle regeneration, promises to elucidate the regulatory mechanism underlying signal-dependent distribution of epigenetic marks in the genome of muscle progenitors. This technology has revealed the existence of particular combinations of epigenetic modifications, such as the simultaneous H3K4 and H3K27 tri-methylation, that define a "poised" chromatin conformation typical of developmental genes in stem cells. It will be interesting to know if similar epigenetic marks regulate the temporal gene expression in MSCs during muscle regeneration.

EPIGENETIC REGULATION OF SKELETAL MYOGENESIS BY miRNA

MicroRNA (miRNAs) are short (20–24 nt-long) non-coding RNAs that post-transcriptionally regulate gene expression in both animals and plants [41–43]. It is estimated that more than

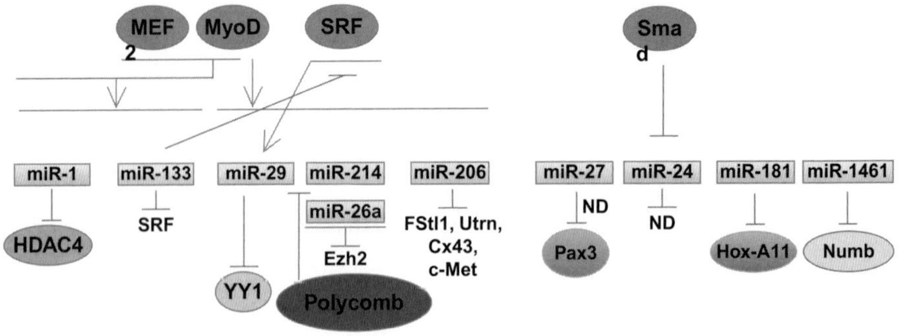

FIGURE 20.2

Regulatory circuits linking transcription factors and chromatin-modifying complexes to miRNA and their relative targets. (Please refer to color plate section)

50% of mammalian messenger RNAs (mRNAs) may be regulated by miRNAs, making it by far the most prevalent regulatory mechanism of mRNA availability. miRNAs are transcribed by RNA polymerase II as long primiRNAs, often encompassing more than one miRNA. In the nucleus, pri-miRNAs are cleaved into ~70 nucleotide hairpin RNA by the Drosha protein complex to generate pre-miRNA [44,45], which are subsequently exported to the cytoplasm via an actively regulated process controlled by Exportin-5 [46]. Dicer cleaves miRNAs into their mature forms so that they can be incorporated into the RNA-induced silencing complexes (RISC) [47].

miRNAs utilize base-pairing to target RISC to specific mRNAs with fully or partly complementary sequences located, in the majority of the cases, in the 3′ untranslated regions (UTRs) [43]. The most common outcome of RISC-recruitment is repression of the targeted mRNA via translational inhibition, site-specific endonucleolytic cleavage, or accelerated exonucleolytic mRNA decay [48]. However, RISC-mediated microRNA targeting can, upon cell cycle arrest, mediate translational activation of the target mRNA [49]. The repressive effect of an individual miRNA on the accumulation of a given protein is relatively small [50], rarely exceeding 2-fold [51]. However, an individual miRNA can target hundreds of targets. In addition, miRNAs often act in concert with other regulatory processes. For instance, an upstream event may promote miRNA transcription and concomitantly repress expression of its target mRNAs [52,53]. Thus, the final effect of miRNA on protein output is the summation of several and independent events.

miRNAs AND SKELETAL MYOGENESIS (see Fig. 20.2)

The fundamental role played by miRNAs during mouse development is underlined by the finding that experimental deletion of the processing enzyme Dicer results in embryonic lethality by day 7.5 [54]. In an attempt to overcome lethality and to specifically inactivate Dicer in developing skeletal muscle, O'Rourke and colleagues crossed a MyoD-Cre recombinase transgenic mouse with a floxed Dicer mouse. The MyoD-Cre:floxed Dicer animals have reduced muscle miRNAs, and die perinatally due to skeletal muscle hypoplasia accompanied by abnormal myofiber morphology. Moreover, Dicer mutant mice showed increased muscle apoptosis [55].

Role of Individual miRNAs in Muscle Development

Individual miRNAs have been shown to regulate skeletal myogenesis in cell culture system as well as in developing and adult embryos. In some instances, the target mRNAs have been identified as known important regulators of muscle gene expression [56]. However, unbiased approaches based on miRNA overexpression and underexpression coupled to mRNA and protein output [50,51] will be required to systematically describe the constellation of mRNA targets. A non-exhaustive discussion of some of the miRNAs regulating skeletal muscle differentiation is reported here below.

miR-1/206 and miR-133a/133b

miR-1-2 and miR-133a-1 (located on chromosome 18), miR-206 and miR-133b (chromosome 1), and miR1-1 and miR133a-2 (chromosome 2) are co-expressed in skeletal muscle and in the differentiating C2C12 cell line as a single bicistronic transcript [57] regulated by upstream regions bound by MyoD and myogenin [58,59].

Employing a MyoD-inducible system, MyoD was shown to directly promote miR-206/133b transcription [60]. MEF2 is also involved in regulating expression of miR-1 and miR-133a through binding to an intronic enhancer located between the two microRNA-expressing regions [61]. miR-1 and miR-133a regulate proliferation and differentiation of C2C12 skeletal muscle cells [57]. miR-1 exerts its effects, at least in part, by reducing the levels of the histone deacetylase HDAC4 protein. Whether the HDAC4 mRNA levels are also affected has not been investigated.

Since HDAC4 negatively regulates MEF2 activity, miR-1 establishes a positive feed-forward loop favoring cell differentiation. On the contrary, miR-133a promotes cell proliferation, thus counteracting differentiation, by repressing SRF. Therefore, two co-regulated miRNAs (miR-1 and miR-133a) exert opposing effects on muscle cell differentiation. In a negative feedback loop, SRF regulates miR-133a expression, thus providing a mechanism to finely regulate the relative abundance of the two molecules. Indicating a conserved evolutionary function, both miR-1 and miR-133 control muscle gene expression and sarcomeric actin organization in zebrafish [62]. Mir-206 expression promotes C2C12 cell differentiation [60,63] and targets several mRNAs, including follistatin-like 1 (Fstl1), utrophin (Utrn) [60], and connexin 43 (Cx43) [64]. Recently, miR-206 has been shown to block human rhabdomyosarcoma growth in xenotransplants by regulating the MET proto-oncogene [65].

miR-24

TGF-1 signaling inhibits skeletal muscle differentiation. miR-24 expression is up-regulated during myoblast differentiation. TGF-was found to inhibit the expression of miR24 in a Smad-dependent manner. While direct targets for miR-24 in skeletal muscle cells have not been identified yet, the results of overexpression and blocking experiments are consistent with a promyogenic role exerted by miR-24 and suggest that the inhibitory effects exerted by TGF-1 on muscle differentiation may be mediated, in part, by miR-24 [66].

miR-26a

In C2C12 skeletal muscle cells, miR-26a targets the Polycomb group (PcG) Ezh2 methyltransferase, a negative regulator of muscle differentiation [38], thus favoring myogenesis [67,68]. Up-regulation of miR-26a is evident at the latest stages of C2C12 cells differentiation, appearing only after several days of cell culture in differentiation medium. Initial reduction of the Ezh2 protein level at the earlier differentiation step coincides with activation of miR-214, a microRNA that also targets Ezh2 (see below). Thus, miR-214 and miR-26a may target Ezh2 at distinct developmental myogenic stages.

miR-27

miR-27b is expressed in several anatomical structures, including the somites. It is present in somitic regions from where Pax3 expression is absent. miR-27b (and a) directly target Pax3 3′UTR. Transgenic animals expressing miR-27b in Pax3-positive cells display a shift from Pax3/7-positive progenitor cells to cells that are myogenin-positive and have entered myogenic differentiation. Thus, miR-27b favors *in vivo* differentiation of muscle progenitor cells by reducing Pax3 [69].

335

miR-29

During myogenesis, miR-29 expression is induced by MEF2 and SRF. In undifferentiated myoblasts, miR-29 expression is silenced by the transcription factor YY1 and Polycomb proteins. In turn, YY1 is a primary target of miR-29. In rhabdomyosarcomas (RDs), elevated levels of YY1 recruit Polycomb proteins to miR16 29 regulatory regions, resulting in its transcriptional silencing and maintenance of the undifferentiated state of RD cells [70].

miR-146a

Expression of miR-146a increases when C2C12 cells undergo cyclical mechanical stretching – an intervention that prohibits differentiation. miR-146a targets the Notch-1 inhibitor Numb and miR-146a anatgomirs reverse the effect of mechanical stretching on C2C12 cell differentiation [71]. The causative role of Numb in C2C12 mechanical stretching remains to be demonstrated.

miR-181

Mir-181 is strongly up-regulated during differentiation of C2C12 skeletal muscle cells and in ES-derived embryonic bodies [72]. miR-181 – barely detectable in skeletal muscle of adult mice – is strongly up-regulated in regenerating myofibers. Blocking miR-181 with specific antagomirs interferes with C2C12 cell differentiation. This phenotypic effect may be mediated by miR-181 targeting of the homeobox Hox-A11, which represses MyoD through a mechanism that has not been fully explored yet.

miR-214

Mir-214 is co-transcribed with miR-199a from a conserved antisense intronic transcript at the dynamin3 (*Dnm3*) locus. In myoblasts, the Polycomb group (PcG) proteins Suz12 and Ezh2 occupy and repressed miR-214 transcription [73]. Transcriptional down-regulation of Ezh2 and concomitant recruitment of MyoD and myogenin at the *Dnm3* locus during the initial phases of skeletal muscle cell differentiation allow for miR-214 expression. miR-214 targets the 3′ UTR of Ezh2, thus contributing to further reducing the Ezh2 protein levels and activating by derepression its own expression in differentiating myocytes [73]. Since PcG proteins restrain cell differentiation [57], miR-214-mediated regulation of Ezh2 acts as a pro-myogenic switch. A similar miR-214-dependent regulation of PcG proteins operates in mouse embryonic stem cells [73]. Mice subjected to genetic ablation of the miR-199a/214 regions within the *Dnm3* locus die within a month of birth and displayed several abnormalities, including skeletal and muscle defects [74]. In zebrafish, miR-214 positively regulates the slow muscle phenotype by targeting the 3′ UTR of suppressor of fused Su(fu) [75]. Zebrafish and mammalian Su(fu) 3′UTRs have not been conserved throughout evolution, and thus miR-214 does not target mammalian Su(fu).

miRNAs Regulating Myosins

Exploiting an elegant regulatory mechanism, miRNAs embedded in myosin-encoding genes have been shown to target myosins themselves in heart and regulate stress-dependent cardiac remodeling [56,76].

miRNAs and Muscular Dystrophies

In both limb girdle myopathies and Duchenne muscular dystrophy (DMD), expression of subsets of miRNAs is modified [77]. For instance, miR-299-5p, miR-487b, and miR-362 are up-regulated in DMD but not in the milder Becker muscular dystrophy. Specific miRNA signatures could aid in distinguishing facioscapulohumeral muscular dystrophy (FSHD) from DMD, and other rare degenerative myopathies from inflammatory myopathies.

In addition to the diagnostic value, functional validation of the miRNA predicted targets should help in further identifying molecular pathways disregulated in muscular dystrophies.

ACKNOWLEDGEMENTS

P.L.P. is an associate scientist of Sanford Children's Health Center and of the Telethon Dulbecco Institute, and is supported by NIAMS (RO1AR052779) and AIRC. Research in V.S.'s laboratory is supported by the Intramural Research Program of the National Institute of Arthritis, Musculoskeletal, and Skin Diseases of the National Institutes of Health.

References

1. Singec I, Jandial R, Crain A, Nikkhah G, Snyder EY. The leading edge of stem cell therapeutics. *Annu Rev Med* 2007;**58**:313–28.

2. Palacios D, Puri PL. The epigenetic network regulating muscle development and regeneration. *J Cell Physiol* 2006;**207**(1):1–11.

3. Guasconi V, Puri PL. Chromatin: the interface between extrinsic cues and the epigenetic regulation of muscle regeneration. *Trends Cell Biol* 2009;**19**(6):286–94.

4. Kuang S, Rudnicki MA. The emerging biology of satellite cells and their therapeutic potential. *Trends Mol Med* 2008;**14**(2):82–91. Epub, Jan 22 2008.

5. Morgan JE, Partridge TA. Muscle satellite cells. *Int J Biochem Cell Biol* 2003;**35**(8):1151–6.

6. Péault B, Rudnicki M, Torrente Y, Cossu G, Tremblay JP, Partridge T, et al. Stem and progenitor cells in skeletal muscle development, maintenance, and therapy. *Mol Ther* 2007;**5**:867–77.

7. Charge SB, Rudnicki MA. Cellular and molecular regulation of muscle regeneration. *Physiol Rev* 2004;**84**:209–38.

8. Shi X, Garry DJ. Muscle stem cells in development, regeneration, and disease. *Genes Dev* 2006;**20**(13):1692–708.

9. Pomerantz J, Blau HM. Nuclear reprogramming: a key to stem cell function in regenerative medicine. *Nat Cell Biol* 2004;**6**(9):810–16.

10. Forcales SV, Puri PL. Signaling to the chromatin during skeletal myogenesis: novel targets for pharmacological modulation of gene expression. *Semin Cell Dev Biol* 2005;**16**(4–5):596–611.

11. Conboy IM, Rando. TA. The regulation of Notch signaling controls satellite cell activation and cell fate determination in postnatal myogenesis. *Dev Cell* 2002;**3**:397–409.

12. Shinin V, Gayraud-Morel B, Gomès D, Tajbakhsh S. Asymmetric division and cosegregation of template DNA strands in adult muscle satellite cells. *Nat Cell Biol* 2006;**7**:677–8.

13. Kuang S, Kuroda K, Le Grand F, Rudnicki MA. Asymmetric self-renewal and commitment of satellite stem cells in muscle. *Cell* 2007;**129**(5):999–1010.

14. Seale P, Sabourin LA, Girgis-Gabardo A, Mansouri A, Gruss P, Rudnicki MA. Pax7 is required for the specification of myogenic satellite cells. *Cell* 2000;**102**(6):777–86.

15. Oustanina S, Hause G, Braun T. Pax7 directs postnatal renewal and propagation of myogenic satellite cells but not their specification. *EMBO J* 2004;**23**(16):3430–9.

16. Buckingham M. Skeletal muscle progenitor cells and the role of Pax genes. *C R Biol* 2007;**330**(6–7):530–3.

17. Hu P, Geles KG, Paik JH, DePinho RA, Tjian R. Codependent activators direct myoblast-specific MyoD transcription. *Dev Cell* 2008;**15**(4):534–46.

18. Lepper C, Conway SJ, Fan CM. Adult satellite cells and embryonic muscle progenitors have distinct genetic requirements. *Nature* 2009;**460**(7255):627–31.

19. Zammit PS, Relaix F, Nagata Y, Ruiz AP, Collins CA, Partridge TA, et al. Pax7 and myogenic progression in skeletal muscle satellite 21 cells. *J Cell Sci* 2006;**119**(9):1824–32.

20. McKinnell IW, Ishibashi J, Le Grand F, Punch VG, Addicks GC, Greenblatt JF, et al. Pax7 activates myogenic genes by recruitment of a histone methyltransferase complex. *Nat Cell Biol Jan* 2008;**10**(1):77–84.

21. Ng RK, Gurdon JB. Epigenetic memory of an active gene state depends on histone H3.3 incorporation into chromatin in the absence of transcription. *Nat Cell Biol* 2008;**10**(1):102–9.

22. Lee H, Habas R, Abate-Shen C. MSX1 cooperates with histone H1b for inhibition of transcription and myogenesis. *Science* 2004;**304**(5677):1607–9.

23. Brunk BP, Goldhamer DJ, Emerson CP Jr. Regulated demethylation of the myoD distal enhancer during skeletal myogenesis. *Dev Biol* 1996;**177**(2):490–503.

24. Cao Y, Kumar RM, Penn BH, Berkes CA, Kooperberg C, Boyer LA, et al. Global and gene-specific analyses show distinct roles for Myod and Myog at a common set of promoters. *EMBO J* 2006;**25**(3):502–11.

337

25. Blais A, Tsikitis M, Acosta-Alvear D, Sharan R, Kluger Y, Dynlacht BD. An initial blueprint for myogenic differentiation. *Genes Dev* 2005;**19**(5):553–6.

26. Weintraub H, Tapscott SJ, Davis RL, Thayer MJ, Adam MA, Lassar AB, et al. Activation of muscle-specific genes in pigment, nerve, fat, liver, and fibroblast cell lines by forced expression of MyoD. *Proc Natl Acad Sci USA* 1989;**86**:5434–8.

27. Gerber AN, Klesert TR, Bergstrom DA, Tapscott SJ. Two domains of MyoD mediate transcriptional activation of genes in repressive chromatin: a mechanism for lineage determination in myogenesis. *Genes Dev* 1997;**11**:436–50.

28. Penn BH, Bergstrom DA, Dilworth FJ, Bengal E, Tapscott SJ. A MyoDegenerated feed-forward circuit temporally patterns gene expression during skeletal muscle differentiation. *Genes Dev* 2004;**18**(19):2348–53.

29. Deato MD, Marr MT, Sottero T, Inouye C, Hu P, Tjian R. MyoD targets TAF3/TRF3 to activate myogenin transcription. *Mol Cell* 2008 Oct 10;**32**(1):96–105.

30. Chow CW, Davis RJ. Proteins kinases: chromatin-associated enzymes. *Cell* 2006;**127**(5):887–90.

31. Edmunds JW, Mahadevan LC. Cell signaling. Protein kinases seek close encounters with active genes. *Science* 2006;**313**(5786):449–51.

32. Simone C, Forcales SV, Hill DA, Imbalzano AN, Latella L, Puri PL. p38 pathway targets SWI/SNF chromatin remodeling complex to muscle-specific loci. *Nat Genet* 2004;**36**:738–43.

33. Rampalli S, Li L, Mak E, Ge K, Brand M, Tapscott SJ, et al. p38 MAPK signaling regulates recruitment of Ash2L-containing methyltransferase complexes to specific genes during differentiation. *Nat Struct Mol Biol* 2007;**14**(12):1150–6.

34. Lluís F, Ballestar E, Suelves M, Esteller M, Muñoz-Cánoves P. E47 phosphorylation by p38 MAPK promotes MyoD/E47 association and muscle-specific gene transcription. *EMBO J* 2005;**24**(5):974–84.

35. Serra C, Palacios D, Mozzetta C, Forcales SV, Morantte I, Ripani M, et al. Functional interdependence at the chromatin level between the MKK6/p38 and IGF1/PI3K/AKT pathways during muscle differentiation. *Mol Cell* 2007;**28**(2):200–13.

36. McKinsey TA, Zhang CL, Olson EN. Signaling chromatin to make muscle. *Curr Opin Cell Biol* 2002;**14**:763–72.

37. Puri PL, Sartorelli V. Regulation of muscle regulatory factors by DNA-binding, interacting proteins, and post-transcriptional modifications. *J Cell Physiol* 2000;**185**:155–73.

38. Caretti G, Di Padova M, Micales B, Lyons GE, Sartorelli V. The Polycomb Ezh2 methyltransferase regulates muscle gene expression and skeletal muscle differentiation. *Genes Dev* 2004;**18**(21):2627–38.

39. Iezzi S, Di Padova M, Serra C, Caretti G, Simone C, Maklan E, et al. Deacetylase inhibitors increase muscle cell size by promoting myoblast recruitment and fusion through induction of follistatin. *Dev Cell* 2004;**6**:673–84.

40. Minetti GC, Colussi C, Adami R, Serra C, Mozzetta C, Parente V, et al. Functional and morphological recovery of systrophic muscles in mice treated with deacetylase inhibitors. *Nat Medi* 2006;**12**(10):1147–50.

41. Ambros V. microRNAs: tiny regulators with great potential. *Cell* 2001;**107**:823–6.

42. Pasquinelli AE, Ruvkun G. Control of developmental timing by micrornas and their targets. *Annu Rev Cell Dev Biol* 2002;**18**:495–513.

43. Bartel DP. MicroRNAs: genomics, biogenesis, mechanism, and function. *Cell* 2004;**116**:281–97.

44. Grishok A, Pasquinelli AE, Conte D, Li N, Parrish S, Ha I, et al. Genes and mechanisms related to RNA interference regulate expression of the small temporal RNAs that control *C. elegans* developmental timing. *Cell* 2001;**106**:23–34.

45. Lee Y, Ahn C, Han J, Choi H, Kim J, Yim J, et al. The nuclear RNase III Drosha initiates microRNA processing. *Nature* 2003;**425**:415–19.

46. Lund E, Guttinger S, Calado A, Dahlberg JE, Kutay U. Nuclear export of microRNA precursors. *Science* 2004;**303**:95–8.

47. Hutvagner G, Zamore PD. A microRNA in a multiple-turnover RNAi enzyme complex. *Science* 2002;**297**:2056–60.

48. Bartel DP. MicroRNAs: target recognition and regulatory functions. *Cell* 2009;**136**:215–33.

49. Vasudevan S, Tong Y, Steitz JA. Switching from repression to activation: microRNAs can up-regulate translation. *Science* 2007;**318**:1931–4.

50. Selbach M, Schwanhausser B, Thierfelder N, Fang Z, Khanin R, Rajewsky N. Widespread changes in protein synthesis induced by microRNAs. *Nature* 2008;**455**:58–63.

51. Baek D, Villen J, Shin C, Camargo FD, Gygi SP, Bartel DP. The impact of microRNAs on protein output. *Nature* 2008;**455**:64–71.

52. Alon U. Network motifs: theory and experimental approaches. *Nat Rev Genet* 2007;**8**:450–61.

53. Tsang J, Zhu J, van Oudenaarden A. MicroRNA-mediated feedback and feedforward loops are recurrent network motifs in mammals. *Mol Cell* 2007;**26**:753–67.

54. Bernstein E, Kim SY, Carmell MA, Murchison EP, Alcorn H, Li MZ, et al. Dicer is essential for mouse development. *Nat Genet* 2003;**35**:215–17.

55. O'Rourke JR, Georges SA, Seay HR, Tapscott SJ, McManus MT, Goldhamer DJ, et al. Essential role for Dicer during skeletal muscle development. *Dev Biol* 2007;**311**:359–68.

56. Williams AH, Liu N, van Rooij E, Olson EN. MicroRNA control of muscle development and disease. *Curr Opin Cell Biol* 2009;**21**:461–9.

57. Chen JF, Mandel EM, Thomson JM, Wu Q, Callis TE, Hammond SM, et al. The role of microRNA-1 and microRNA-133 in skeletal muscle proliferation and differentiation. *Nat Genet* 2006;**38**:228–33.

58. Rao PK, Kumar RM, Farkhondeh M, Baskerville S, Lodish HF. Myogenic factors that regulate expression of muscle-specific microRNAs. *Proc Natl Acad Sci USA* 2006;**103**:8721–6.

59. Zhao Y, Samal E, Srivastava D. Serum response factor regulates a muscle-specific microRNA that targets Hand2 during cardiogenesis. *Nature* 2005;**436**:214–20.

60. Rosenberg MI, Georges SA, Asawachaicharn A, Analau E, Tapscott SJ. MyoD inhibits Fstl1 and Utrn expression by inducing transcription of miR-206. *J Cell Biol* 2006;**175**:77–85.

61. Liu N, Williams AH, Kim Y, McAnally J, Bezprozvannaya S, Sutherland LB, et al. An intragenic MEF2-dependent enhancer directs muscle-specific expression of microRNAs 1 and 133. *Proc Natl Acad Sci USA* 2007;**104**:20844–9.

62. Mishima Y, Abreu-Goodger C, Staton AA, Stahlhut C, Shou C, Cheng C, et al. Zebrafish miR-1 and miR-133 shape muscle gene expression and regulate sarcomeric actin organization. *Genes Dev* 2009;**23**:619–32.

63. Kim HK, Lee YS, Sivaprasad U, Malhotra A, Dutta A. Muscle-specific microRNA miR-206 promotes muscle differentiation. *J Cell Biol* 2006;**174**:677–87.

64. Anderson C, Catoe H, Werner R. MIR-206 regulates connexin43 expression during skeletal muscle development. *Nucleic Acids Res* 2006;**34**:5863–71.

65. Taulli R, Bersani F, Foglizzo V, Linari A, Vigna E, Ladanyi M, et al. The muscle-specific microRNA miR-206 blocks human rhabdomyosarcoma growth in xenotransplanted mice by promoting myogenic differentiation. *J Clin Invest* 2009;**119**(8):2366–78.

66. Sun Q, Zhang Y, Yang G, Chen X, Cao G, Wang J, et al. Transforming growth factor-beta-regulated miR-24 promotes skeletal muscle differentiation. *Nucleic Acids Res* 2008;**36**:2690–9.

67. Sartorelli V, Caretti G. Mechanisms underlying the transcriptional regulation of skeletal myogenesis. *Curr Opin Genet Dev* 2005;**15**(5):528–35.

68. Wong CF, Tellam RL. microRNA-26a targets the histone methyltransferase Enhancer of Zeste homolog 2 during myogenesis. *J Biol Chem* 2008;**283**:9836–43.

69. Crist CG, Montarras D, Pallafacchina G, Rocancourt D, Cumano A, Conway SJ, et al. Muscle stem cell behavior is modified by microRNA-27 regulation of Pax3 expression. *Proc Natl Acad Sci USA* 2009;**106**:13383–7.

70. Wang H, Garzon R, Sun H, Ladner KJ, Singh R, Dahlman J, et al. NF-kappaB-YY1-miR-29 regulatory circuitry in skeletal myogenesis and rhabdomyosarcoma. *Cancer Cell* 2008;**14**:369–81.

71. Kuang W, Tan J, Duan Y, Duan J, Wang W, Jin F, et al. Cyclic stretch induced miR-146a upregulation delays C2C12 myogenic differentiation through inhibition of Numb. *Biochem Biophys Res Commun* 2009;**378**:259–63.

72. Naguibneva I, Ameyar-Zazoua M, Polesskaya A, Ait-Si-Ali S, Groisman R, Souidi M, et al. The microRNA miR-181 targets the homeobox protein Hox-A11 during mammalian myoblast differentiation. *Nat Cell Biol* 2006;**8**:278–84.

73. Juan AH, Kumar RM, Marx JG, Young RA, Sartonelli V. Mir-214-dependent regulation of the polycomb protein Ezh2 in skeletal muscle and embryonic stem cells. *Mol Cell* 2009;**36**:61–74.

74. Watanabe T, Sato T, Amano T, Kawamura Y, Kawamura N, Kawaguchi H, et al. Dnm3os, a non-coding RNA, is required for normal growth and skeletal development in mice. *Dev Dyn* 2008;**237**:3738–48.

75. Flynt AS, Li N, Thatcher EJ, Solnica-Krezel L, Patton JG. Zebrafish miR-214 modulates Hedgehog signaling to specify muscle cell fate. *Nat Genet* 2007;**39**:259–63.

76. van Rooij E, Sutherland LB, Qi X, Richardson JA, Hill J, Olson EN. Control of stress-dependent cardiac growth and gene expression by a microRNA. *Science* 2007;**316**:575–9.

77. Eisenberg I, Eran A, Nishino I, Moggio M, Lamperti C, Amato AA, et al. Distinctive patterns of microRNA expression in primary muscular disorders. *Proc Natl Acad Sci USA* 2007;**104**:17016–21.

339

CHAPTER 21

Epigenetics of X Chromosome Inactivation

Tamar Dvash and Guoping Fan

Department of Human Genetics and The Eli and Edythe Broad Center for Regenerative Medicine and Stem Cell Research, David Geffen School of Medicine, University of California Los Angeles, Los Angeles, CA 90095, USA

INTRODUCTION

During embryonic development, mammalian female cells have one of the two X chromosomes silenced through the process of X chromosome inactivation (XCI). The XCI phenomenon is regarded as a classic paradigm of epigenetic gene regulation and, indeed, it has attracted great interest in the scientific community over the past 60 years. The first hint of the difference between the two X chromosomes in mammalian female cells was given in 1949 by Barr and Bertram [1]. They discovered that one of the X chromosomes is comprised of facultative heterochromatin. This seminal work denoted the inactive X chromosome characterized by facultative heterochromatin as the "Barr body". In 1961 Mary Lyon [2] discovered the underlying process for the formation of the facultative heterochromatin on one of the X chromosomes. The milestone work of Lyon put forward her hypothesis that the "Barr body" is an inactive X chromosome (Xi) that appears in mammalian cells with more than one X chromosome (Fig. 21.1). The inactivation allows dosage compensation in females as compared to males who carry only one X chromosome. In 1962, asynchronous replication was discovered as another feature characterizing the Xi [3]. These studies opened up an entire research field that is still active today. Although many features of this process were discovered over the past 50 years, there are still more questions to be asked. In this chapter we will describe the key regulatory events in XCI.

341

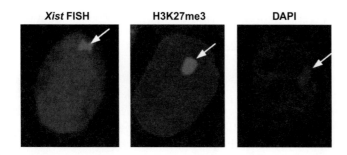

FIGURE 21.1

Visualization of major marks of XCI. Shown here are the three different visualizations of XCI: *Xist* RNA coating of the inactive X chromosome by fluorescence *in situ* hybridization (FISH); punctate staining of H3K27me3 as a major mark for XCI by immunostaining; and the Barr body detected as a dense DAPI staining within the nucleus. (Please refer to color plate section)

Handbook of Epigenetics: The New Molecular and Medical Genetics. DOI: 10.1016/B978-0-12-375709-8.00021-6

XCI REGULATION DURING DEVELOPMENT

XCI is a developmentally-regulated process that involves sequential acquisition of silencing markers on the X chromosome to be inactivated. Two different patterns of XCI exist: imprinted and random. The majority of XCI properties are shared between the two different patterns, yet some differences exist that reflect the nature and the degree of stability of inactivation. Most of the research concerning XCI in mammals has been conducted with the mouse model system. At the fertilization stage, the female mouse zygote has both X chromosomes active. The first inactivation during development occurs upon the first cleavage. This inactivation is imprinted and therefore only the paternal X chromosome is inactivated [4,5]. Later on, after the blastocyst has formed, cells from the inner cell mass (ICM) reactivate the inactive X [5,6]. At this stage the embryo has two types of XCI status; the ICM cells have both active X chromosomes while the trophectoderm and the primitive endoderm still retain their imprinted paternal XCI since the first cleavage. Then, only upon differentiation will the ICM cells again inactivate one of their X chromosomes but this time stochastically, in contrast to the first cleavage event [5,6]. Since the ICM cells are the origin of the embryo proper, the second round of inactivation will result in random XCI in each cell and throughout development its progenies will maintain that particular Xi. The primordial germ cells (PGC) are an exception in this regard since these cells again reactivate their Xi later on in mouse development (E11.5–E13.5) and this status is maintained in the female germ cells [7].

Both random and imprinted XCI are initiated by monoallelic *Xist* gene expression. This expression leads to a series of epigenetic modifications such as depletion of RNA polymerase II, transcription factors, and euchromatic markers (see Fig. 21.3). Imprinted XCI is temporary compared to the random XCI that remains stable from the moment of establishment throughout many cell divisions and across the entire lifespan. Therefore in order to establish stable random XCI, the mechanisms for CpG island methylation are employed [8]. This modification is considered to be more stable than histone modifications which are characteristic of imprinted XCI and early epigenetic events of random inactivation [9]. Although XCI occurs in a narrow time window during mouse development it is suggested that the kinetics of gene silencing varies. Existing evidence shows that genes located in the vicinity of the X chromosome inactivation center (XIC) are first silenced during differentiation [10].

Another interesting phenomenon in XCI is the "escape" from inactivation; although the majority of the genes on the Xi are subjected to complete silencing, some are able to express from both active and inactive X chromosomes. The exact mechanism for genes escaping XCI is not fully understood but a recent study using the transgene approach revealed that it is probably an intrinsic property of a specific locus. Random integration of BAC clones carrying normally silenced or escaped gene (*Jarid1c*) loci into the X chromosome of female ESC lines was able to recapitulate the endogenous expression pattern. The authors concluded that the DNA sequence itself is sufficient to determine whether a locus will be subjected to XCI [11].

Xist RNA AS A KEY PLAYER IN XCI

XCI occurs in three steps: initiation, spreading, and maintenance. The *Xist* gene is thought to be the major regulator of the XCI process and is the key component of the initiation and spreading stages. Interestingly, this gene lies within a specific region in the X chromosome designated as the XIC (Fig. 21.2) [12]. This center is believed to be important for the initiation of XCI and is involved in the process of counting and choice in random XCI. The counting and choice process achieves a ratio of one active X chromosome per diploid set of autosomes. In this process the number of sex chromosomes compared to autosomes is being "counted" and the number of X chromosomes to be inactivated is designated. Each cell needs to choose which X chromosome will be transcriptionally silenced. Random XCI is

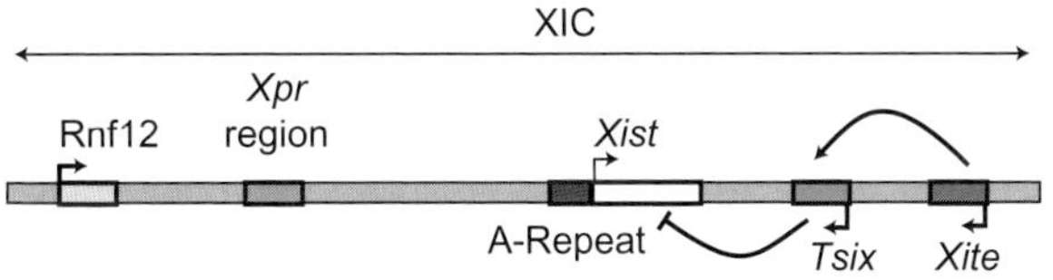

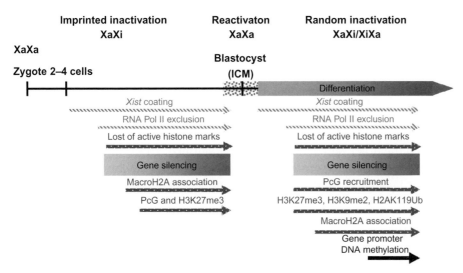

FIGURE 21.2

The X chromosome inactivation center (XIC) gene structure and interactions. The XIC contains three main genes that are involved in XCI. *Xist* is the main regulator of the initiation of XCI. *Tsix* is transcribed antisense to *Xist* and inhibits *Xist* expression and consequently XCI. *Xite* is a *Tsix* enhancer and therefore when it is expressed it prevents the expression of *Xist*. In addition, three more regulatory components are shown. The A-repeat located on the 5′ of *Xist* plays a major regulatory role in XCI. *Xpr* region plays a crucial role in X chromosome pairing, and RNF12 has been recently found to be a dose-dependent activator of XCI.

FIGURE 21.3

The layers of epigenetic marks characterizing imprinted and random XCI. Shown are the sequential events for the establishment of XCI. *Xist* coating and RNA polymerase II exclusion (striped light gray) are responsible for the initiation of the cascade. Later on the loss of active histone marks, the recruitment of the PcG proteins, and the induction of repressive histone marks such as H3K27me3 (dark dotted gray) allow for the maintenance of the inactive state to occur. Finally DNA methylation on the gene promoters of the Xi locks the inactive state in somatic cells (black).

343

known to be coupled with loss of pluripotency [13] and the involvement of the pluripotency factor Oct4 in counting and choice was recently demonstrated [14]. However, different models for the counting and choice mechanism exist and need to be further examined [15].

A specific portion of the XIC was shown to be the origin of inactivation spreading; this region contains three main genes: *Xist*, *Tsix*, and *Xite* [16]. *Xist* is a non-coding gene located within the XIC; it was shown to be transcribed solely from the X chromosome that will be inactivated [17–19]. *Xist* RNA coating on the *cis* X chromosome is the initial event in the cascade of silencing. Interestingly its antisense overlapping non-coding gene, *Tsix*, is thought to repress *Xist* expression and therefore it is highly expressed from the active X chromosome [20]. The balance between these two gene expressions allows inactivation of one X chromosome while the other remains active. Possible mechanisms for the regulation of *Xist* expression by *Tsix* and other components will be discussed below. Recently, the X-linked *RNF12* gene was discovered to regulate XCI initiation and counting. The *RNF12* gene lies upstream to the *Xist* gene (Fig. 21.2) and over-expression of this gene results in XCI initiation in both male female mouse ES cells. Additionally, XCI initiation is reduced in female ES cells that are heterozygous for this gene. This work identifies the first X-linked XCI activator [21].

Xist RNA coating the soon-to-be Xi is the initial step in silencing (Fig. 21.3). It is believed that *Xist* RNA is responsible for the induction of asynchronous DNA replication that is characteristic of the Xi [22]. Moreover, *Xist* RNA coating is known to induce the recruitment of histone variant macro H2A [23], a variety of histone H3 and H4 modification such as H3K9 [24] and H4 hypoacetylation [25], H4K20-monomethylation [26], H3K9 di-methylation [27,28], and H3K27 tri-methylation [29]. In addition, promoter CpG island methylation occurs after *Xist* coating [30]. Therefore the presence of *Xist* RNA appears to be the trigger for recruitment of epigenetic markers associated with gene silencing in the initiation step and the maintenance of the Xi. However, it was recently discovered that *Xist* coating is not necessary for establishment of imprinted XCI. In mouse embryos that carry a paternal X chromosome lacking the *Xist* gene, the imprinted XCI can still be initiated. However, *Xist* transcript is still important for stable silencing of the Xi [31].

Xist A-Repeat Role in Silencing

The *Xist* gene consists of several conserved repeat regions and areas of unique sequences. The A-Repeat is the most conserved sequence located in the 5′ region of the *Xist* gene and was shown to be essential for XCI. Its role was demonstrated by a series of deletion/ mutation analysis in an exogenous *Xist* cDNA transgene (Fig. 21.2). Interestingly the deletion of a 5′ 0.9 kb region resulted in complete abolition of silencing activity of the transgene although the transcript was still capable of associating with the chromatin and spreading throughout the X chromosome [32]. Recently, a non-coding RNA transcript generated from the A-Repeat region, RepA, was identified. This transcript was shown to recruit the PRC2 Polycomb group (PcG) proteins that are involved in chromatin remodeling for the maintenance of gene silencing. For instance the PcG protein Eed (embryonic ectoderm development) was shown to play a crucial role in preventing transcription activation of the inactive X in the extraembryonic cells upon their differentiation [33]. It was also suggested that the coupling of the RepA and the PRC2 results in tri-methylation of H3K27 and expression of the full length *Xist* transcript that consequently allows inactivation to occur. Interestingly, this group also showed that *Tsix* inhibits the RepA–PRC2 interaction and propose a regulatory mechanism for the initiation of XCI [34]. However, this mechanism seems to have certain redundancies. PcG proteins and H3K27me3 can also be recruited without a functional RepA in the mutated A-Repeat study [32]. Overall it seems that the A-Repeat is involved in gene silencing; however, its role in recruitment of H3K27me3 and PRC2 might be dispensable.

Xist GENE REGULATION

As described above *Xist* is the master regulator of both random and imprinted XCI process. As such, an interesting question is what regulates the master regulator and what promotes or stalls the initiation of this process. Interestingly the main negative regulator known for *Xist* expression is its antisense transcribed gene *Tsix*. This property of the antisense *Tsix* gene was first shown by mutating the *Tsix* gene on one allele of female embryos and embryonic stem (ES) cells. The result of these mutation experiments revealed preferential inactivation on the mutated X chromosome due to the uninhibited *Xist* expression on the mutated X chromosome [20,35]. However, the question of how *Tsix* modulates *Xist* activity mechanistically remains. Generally it is believed that *Tsix* affects the chromatin of the XIC locus. The proposed models for the repression mechanism of *Tsix* on the *Xist* promoter come from two different cellular systems and give diverse explanation for the repression mechanism. One group reported that mutating *Tsix* results in accumulation of repressive chromatin markers on *Xist* promoter such as H3K27me3 (Fig. 21.1) [36], while the other reported the opposite – that the ablation of *Tsix* expression leads to an increase in active chromatin markers such as H3K4me2, H3K4me3, and H3K9 acetylation at the XCI locus [37,38]. These seemingly contradictory results are not surprising since it had been described before that *Tsix* is not the exclusive regulator of *Xist* expression and that other mechanisms might exist. This notion was inspired by the study of XY and X0 ES cells that carry a *Tsix* mutation. Some of these cells, but not all, have up-regulation of *Xist* upon differentiation [39,40]. The variability effect of *Tsix* mutation in XY/X0 and between different cellular systems as shown by Sun et al. [36] and Navarro et al. [38], strongly suggests that there is a complementary mechanism to regulate *Xist* expression.

Xist Expression Regulation by *Tsix*

Tsix is transcribed antisense to *Xist* and regulates *Xist* expression. Several research groups have investigated the mechanism of *Xist* expression regulation. Since *Tsix* is also a noncoding RNA gene, it was suggested that the transcript itself is responsible for the repression of *Xist* expression. In an experiment where the *Tsix* transcript was truncated before it reached the *Xist* gene body, *Tsix* repressive activity was completely abolished [41]. Another group tried to further pinpoint the exact critical region required for *Tsix* activity. It was found that *Tsix* transcript that has less than 93% of its full size and in which the transcription does not go through the *Xist* promoter region would fail to repress the *Xist* expression [42]. The *Xist* promoter region is therefore a critical region of the *Tsix* transcript that allows the repression of *Xist* activity to occur.

The expression balance between the *Xist* and *Tsix* transcripts determines which X chromosome is active or inactive. An active X chromosome is characterized by high *Tsix* expression level and low/none *Xist* expression, and vice versa for the inactive X chromosome [20]. One research group claims that the coordinated action from both *Xist* and *Tsix* transcripts is needed to establish XCI. According to their model *Tsix* and *Xist* form a double stranded RNA and it then gets cleaved by the RNAi enzyme Dicer during XCI. The products from the cleavage are ~30 nt RNAs designated as xiRNAs, bigger than the expected fragment size cleaved by Dicer [20–24 nt]. Also, these fragments are claimed to be involved in *Xist* repression on the active X chromosome [43]. Their proposed model showing the involvement of RNAi machinery in *Xist* repression is elegant, but it is in conflict with another study showing that knockout Dicer female embryos have intact XCI [44]. The mechanism through which *Tsix* represses *Xist* during XCI still needs to be elucidated.

Regulation of *Xist* by Pluripotency Related Factors

During female mammalian embryogenesis, random XCI is coupled with transition from the undifferentiated to differentiation state. Interestingly, this feature is maintained even in female ES cells where XCI is initiated only upon differentiation *in vitro* or *in vivo* whereas the undifferentiated female ES cells carry two active X chromosomes. This correlation was also shown to exist in reprogramming of the mouse somatic cell. Reprogramming of a somatic cell line carrying one inactive X chromosome resulted in full reactivation of that Xi in the induced pluripotent stem cell (iPS) line that resulted in two active X chromosomes. Moreover, differentiation of these newly generated iPS cell line results in random XCI [45]. Thus induction of pluripotency can directly affect XCI status. Lately it has been shown that the pluripotency transcription factor Oct4 is involved in the process of pairing and counting. By protein–protein interactions with the transactivators, Ctcf and Yy1, Oct4 binds to the *Tsix* and *Xite* loci and regulates the pairing and counting [14]. This interesting coupling between differentiation and XCI commencement might suggest a tight regulation of *Xist* by factors which are also engaged in the transition from pluripotency to differentiation.

Indeed it was discovered recently that the three transcription factors that are known to be essential for initiation and maintenance of pluripotency–Oct4, Nanog, and Sox2 [46–48]–directly bind to the chromatin of the *Xist* gene in pluripotent cells [13]. Using chromatin immuno-precipitation (ChIP) analysis the researchers were able to show that all three pluripotency factors bind to *Xist* intron 1 in undifferentiated ES cells. However, upon differentiation the binding of these factors to *Xist* intron 1 is dramatically reduced. In addition, binding of the pluripotency transcription factors to the *Xist* intron 1 is a *Tsix*-independent mechanism. This supports the previous hypothesis that the *Tsix* regulation of *Xist* expression is complemented by another mechanism. Furthermore, by applying an inducible down-regulation system the authors were able to show that Oct4 is the main regulator of *Xist* expression. By down-regulating Oct4 all three pluripotency transcription factors lost their binding capacity to *Xist* intron 1. Consequently this led to comparable levels of *Xist* expression between female ES cells and differentiated female ES cells [13]. Hence the involvement of pluripotency factors in *Xist* regulation is direct and can explain the temporal nature of the XCI process.

The interplay between pluripotency factors and XCI in human ESCs (hESCs) is currently unknown. We assume that the relationships in the human cells won't be as simple as in mouse cells. It is known that hESCs can exist in three different XCI states [49–51] [reviewed in Ref. 52]. This requires re-assessment of the relationships between XCI and pluripotency factors in hESCs. It is possible that the different levels of pluripotency factors distinguish the three different XCI statuses or different binding capacities of the *Xist* promoter. It is also possible that the main pluripotency factor that regulates XCI in human cells is different than Oct4 and it has yet to be discovered.

Regulation of XCI by the Pairing of Sister X Chromosomes

Regulatory interactions between *Xist* and *Tsix* were clearly demonstrated in several different ways. Higher order chromatin structure of the region surrounding the XIC revealed interaction between *Xist* and *Tsix*. By using hypersensitive site mapping and chromosome conformation capture (3C) several domains of interactions between *Xist*, *Tsix*, and *Xite* were discovered. These interactions were shown to be regulated in a developmental manner, indicating a direct regulatory role of three dimensional organization in XCI [53]. It was demonstrated that an interaction between the two X chromosomes occurs even prior to the initiation of XCI in mouse. A region on the X chromosomes, an X-pairing region (*Xpr*) (Fig. 21.2), is responsible for homologous pairing of the two X chromosomes. This pairing indicates whether there is more than one X chromosome and signals for XCI to begin. Moreover it is suggested that the pairing of the X chromosomes also regulates complementary expression of *Xist*/*Tsix* [54]. These studies indicate a crucial role of the structural interactions in the initiation and developmental regulation of XCI.

CHROMATIN MODIFICATIONS CHARACTERIZING THE XCI

The initiation of XCI by *Xist* expression is followed by a distinct pattern of chromatin modifications that support silencing and the stabilization of the Xi in a timely manner. Histone modifications are known to be highly involved in the control of gene expression. These modifications allow easy modulation of gene expression that is stable upon cell division and yet can be reversible. Indeed it was shown that after initiation of XCI, *Xist* is dispensable while other repressive epigenetic modifications maintain the inactive state [55]. The combination of histone modifications provides a signature that is indicative for an either repressive or active chromatin region. It was shown by ChIP studies that the Xi is characterized by heterochromatic histone modifications such as H3K27me3, H3K9me2, H2AK119Ub, H4K20me1, and macroH2A. Conversely, the Xi is depleted of euchromatic histone modifications such as H3K4me2/3 and H3, H4 acetylated lysines [27,56,57]. It is believed that the active histone marks are the first to be lost upon *Xist* RNA coating of the Xi [57]. Then genes begin to be silenced with the appearance of histone modifications such as H3K9me2, H3K27me3. Later macroH2A accumulation occurs subsequent to asynchronous DNA replication [23,57]. In the later stages of XCI, Ezh2 (PRC2) mediates the trimethylation of H3K27 [29] (as reviewed in Ref. 58]. More careful examination in human somatic cells revealed that the heterochromatin marks can be divided into two spatially distinct groups. The first group is the chromatin modifications that are associated with *Xist* RNA such as macroH2A, H3K27me3, H2AK119Ub, and H4K20me1. The second group is the constitutive heterochromatic marks such as H3K9me3, H4K20me, and HP1 [27,56,57].

High throughput profiling studies [59,60] were able to provide a more comprehensive picture of the characteristics of the Xi and the changes occurring during XCI. It was found that the Xi contains approximately 1.5-fold more macroH2A1 than the autosomes. The distribution of macroH2A1 along the X chromosome is homogeneous along the entire Xi. However, the pseudoautosomal region is characterized by higher enrichment [59]. This wide distribution of macroH2A can indicate a role in chromosome structure and stability of the Xi. A ChIP-chip analysis combined with expression analysis during ESC differentiation was able to provide an insight into the dynamic chromatin changes in XCI. *Tsix* repression is characterized by H3K27me3 binding to its promoter region in female and male cells. However, the *Xist* locus in female cells shows active epigenetic marks in agreement with its high expression upon differentiation. Generally, an increase in H3K27me3 is widely observed on the Xi indicating a tight association of H3K27me3 with gene silencing on the Xi [60].

CpG island promoter methylation is another feature associated with the Xi. However, this epigenetic marker is not associated with the early stages of XCI. CpG island methylation is thought to be a more permanent lock of the inactive state for random XCI [56]. In addition, DNA methylation is required for the maintenance of random XCI. DNA demethylation induced by either mutagenesis of DNA methyl transferase 1 (Dnmt1) or exposure of 5′ azacytidine resulted in the reactivation of the Xi [61]. SmcHD1 is protein associated with structural maintenance of chromosomes and was localized to the Xi. This protein has a critical role in maintenance of XCI and the hypermethylation of CpG islands on the Xi [62]. SmcHD1 might be an example of the structural maintenance protein involved in epigenetic gene regulation.

DNA methylation also directly regulates *Xist* expression. Differentiating ES cells and developing mouse embryos with *Dnmt1* mutation have ectopic expression of *Xist*, even in male cells. This indicates that DNA methylation is required for repression of *Xist* on the active X chromosome [63,64]. Finally, it was found that the differentiated somatic cells have gene body methylation on the active X chromosome whereas promoter region methylation was on the Xi [65]. However, the role of gene body methylation in regulating gene expression is still unknown.

ROLE OF SPATIAL ORGANIZATION WITHIN THE NUCLEUS IN X INACTIVATION

As described above, *Xist* RNA coating of the X chromosome is the decisive event in the expression silencing of the inactive X chromosome. However, the mechanism of transcriptional silencing cannot be explained solely by the coating and chromatin modification accompanying it. It was discovered that the position of a gene within a distinct nuclear compartment indicates the activity level of transcription. More particularly, gene-rich regions might loop out of their chromosomal territory when they are transcriptionally active [66]. In somatic cells *Xist* RNA is not involved in gene silencing. X-linked genes are located in the internal region of the chromosomal territory of the Xi. However, these X-linked genes are found in the periphery of the chromosomal territory of the active X chromosome [67]. A later study shows that non-coding regions of the X chromosome are located within the *Xist* coated region and are involved in XCI. In addition it is suggested that *Xist* RNA interacts with the structural elements of the X chromosome to condense and silence it rather than interacting directly with the silenced genes [68]. In addition, many of the X-linked genes are located at the border of the Xi territory, not in the internal region as previously discovered. However, another research group seems to have a more precise view of the silencing compartment. In their 3D analysis of the X chromosome within the nucleus, *Xist* RNA accumulates and generates a silent compartment. This compartment mainly consists of repetitive sequences of the X chromosome and lacks RNA polymerase II and transcription factors. Also the compartment is independent of the *Xist* A-Repeat, which is involved in a later stage of gene repression and shifting of the genes into the silencing compartment [69]. Recently, SATB1 and SATB2 were shown to play a role in the gene silencing function of *Xist*. Interestingly, the previous work on these proteins during T lymphocyte development [70] has led the investigators to propose that SATB1 and SATB2 may be involved in the relocation of genes into the silent compartment during XCI [71]. Another interesting structural player is SAF-A (scaffold attachment factor A) that is abundant in territories of Xi [72]. SAF-A protein contains RNA and DNA binding domains and was found to be involved in gene expression regulation. It was shown that its binding to the Xi territories occurs via its RNA binding domain. Some evidence for interactions between SAF-A protein and *Xist* RNA on the Xi were provided. It is speculated that SAF-A plays a role in structural stabilization of inactive territories by interacting with *Xist* RNA [73].

Localization within the nucleus might complement the structural organization of the X chromosome itself. The Xi is known to be located at the periphery of the nucleus or at the nucleolus [74]. As these areas are associated with heterochromatin it was not surprising to find the Xi there. It was proposed that the localization of the Xi in the vicinity of the nucleolus during the S phase of the cell cycle correlates with maintenance of the inactivation markers. Interestingly, deletion of *Xist* resulted in the loss of peri-nucleolar localization and the heterochromatin characteristic of the Xi, which then consequently reactivate Xi. These authors propose that the separation of Xi from the nucleolus during the S phase allows the inactivated state to be maintained during cell cycles [75].

CONCLUDING REMARKS

As described here, XCI is characterized by a cascade of events that allows the silencing of one of the two X chromosomes during female mammalian embryogenesis. While it is believed that *Xist* coating is the initial event in XCI, the gradual progress of other epigenetic components is crucial for allowing stability and flexibility of the inactivated state to occur (as shown in Fig. 21.3). The transition from a transcriptional active to a silent chromosome requires the engagement of cellular relocalization as well as distinguished chromatin code. Nevertheless, the complexity and the multiple components involved in XCI continue to invite many researchers to discover more.

RESEARCH SUPPORT AND ACKNOWLEDGEMENTS

The research was supported by the California Institute of Regenerative Medicine training grant (TG2-01169) to T.D. and a comprehensive research grant RC1-0111 to G.F. We would like to thank Thuc Le for proofreading this chapter.

References

1. Barr ML, Bertram EG. A morphological distinction between neurones of the male and female, and the behaviour of the nucleolar satellite during accelerated nucleoprotein synthesis. *Nature* 1949;**163**:676.

2. Lyon MF. Gene action in the X-chromosome of the mouse (*Mus musculus L.*). *Nature* 1961;**190**:372–3.

3. Morishima A, Grumbach MM, Taylor JH. Asynchronous duplication of human chromosomes and the origin of sex chromatin. *Proc Natl Acad Sci USA* 1962;**48**:756–63.

4. Huynh KD, Lee JT. Inheritance of a pre-inactivated paternal X chromosome in early mouse embryos. *Nature* 2003;**426**:857–62.

5. Okamoto I, Otte AP, Allis CD, Reinberg D, Heard E. Epigenetic dynamics of imprinted X inactivation during early mouse development. *Science* 2004;**303**:644–9.

6. Mak W, Nesterova TB, de Napoles M, Appanah R, Yamanaka S, Otte AP, et al. Reactivation of the paternal X chromosome in early mouse embryos. *Science* 2004;**303**:666–9.

7. Nesterova TB, Mermoud JE, Hilton K, Pehrson J, Surani MA, McLaren A, et al. Xist expression and macroH2A1.2 localisation in mouse primordial and pluripotent embryonic germ cells. *Differentiation* 2002;**69**:216–25.

8. Sado T, Okano M, Li E, Sasaki H. De novo DNA methylation is dispensable for the initiation and propagation of X chromosome inactivation. *Development* 2004;**131**:975–82.

9. Wang J, Mager J, Chen Y, Schneider E, Cross JC, Nagy A, et al. Imprinted X inactivation maintained by a mouse Polycomb group gene. *Nat Genet* 2001;**28**:371–5.

10. Lin H, Gupta V, Vermilyea MD, Falciani F, Lee JT, O'Neill LP, et al. Dosage compensation in the mouse balances up-regulation and silencing of X-linked genes. *PLoS Biol* 2007;**5**:e326.

11. Li N, Carrel L. Escape from X chromosome inactivation is an intrinsic property of the Jarid1c locus. *Proc Natl Acad Sci USA* 2008;**105**:17055–60.

12. Rastan S. Non-random X-chromosome inactivation in mouse X-autosome translocation embryos–location of the inactivation centre. *J Embryol Exp Morphol* 1983;**78**:1–22.

13. Navarro P, Chambers I, Karwacki-Neisius V, Chureau C, Morey C, Rougeulle C, et al. Molecular coupling of Xist regulation and pluripotency. *Science* 2008;**321**:1693–5.

14. Donohoe ME, Silva SS, Pinter SF, Xu N, Lee JT. The pluripotency factor Oct4 interacts with Ctcf and also controls X-chromosome pairing and counting. *Nature* 2009;**460**:128–32.

15. Starmer J, Magnuson T. A new model for random X chromosome inactivation. *Development* 2009;**136**:1–10.

16. Lee JT, Lu N, Han Y. Genetic analysis of the mouse X inactivation center defines an 80-kb multifunction domain. *Proc Natl Acad Sci USA* 1999;**96**:3836–41.

17. Borsani G, Tonlorenzi R, Simmler MC, Dandolo L, Arnaud D, Capra V, et al. Characterization of a murine gene expressed from the inactive X chromosome. *Nature* 1991;**351**:325–9.

18. Brown CJ, Ballabio A, Rupert JL, Lafreniere RG, Grompe M, Tonlorenzi R, et al. A gene from the region of the human X inactivation centre is expressed exclusively from the inactive X chromosome. *Nature* 1991;**349**:38–44.

19. Brockdorff N, Ashworth A, Kay GF, Cooper P, Smith S, McCabe VM, et al. Conservation of position and exclusive expression of mouse Xist from the inactive X chromosome. *Nature* 1991;**351**:329–31.

20. Lee JT, Davidow LS, Warshawsky D. *Tsix*, a gene antisense to *Xist* at the X-inactivation centre. *Nat Genet* 1999;**21**:400–4.

21. Jonkers I, Barakat TS, Achame EM, Monkhorst K, Kenter A, Rentmeester E, et al. RNF12 is an X-Encoded dose-dependent activator of X chromosome inactivation. *Cell* 2009;**139**:999–1011.

22. Takagi N, Sugawara O, Sasaki M. Regional and temporal changes in the pattern of X-chromosome replication during the early post-implantation development of the female mouse. *Chromosoma* 1982;**85**:275–86.

23. Mermoud JE, Costanzi C, Pehrson JR, Brockdorff N. Histone macroH2A1.2 relocates to the inactive X chromosome after initiation and propagation of X-inactivation. *J Cell Biol* 1999;**147**:1399–408.

24. Boggs BA, Connors B, Sobel RE, Chinault AC, Allis CD. Reduced levels of histone H3 acetylation on the inactive X chromosome in human females. *Chromosoma* 1996;**105**:303–9.

25. Jeppesen P, Turner BM. The inactive X chromosome in female mammals is distinguished by a lack of histone H4 acetylation, a cytogenetic marker for gene expression. *Cell* 1993;**74**:281–9.

26. Kohlmaier A, Savarese F, Lachner M, Martens J, Jenuwein T, Wutz A. A chromosomal memory triggered by Xist regulates histone methylation in X inactivation. *PLoS Biol* 2004;**2**:E171.

27. Heard E, Rougeulle C, Arnaud D, Avner P, Allis CD, Spector DL. Methylation of histone H3 at Lys-9 is an early mark on the X chromosome during X inactivation. *Cell* 2001;**107**:727–38.

28. Mermoud JE, Popova B, Peters AH, Jenuwein T, Brockdorff N. Histone H3 lysine 9 methylation occurs rapidly at the onset of random X chromosome inactivation. *Curr Biol* 2002;**12**:247–51.

29. Plath K, Fang J, Mlynarczyk-Evans SK, Cao R, Worringer KA, Wang H, et al. Role of histone H3 lysine 27 methylation in X inactivation. *Science* 2003;**300**:131–5.

30. Norris DP, Brockdorff N, Rastan S. Methylation status of CpG-rich islands on active and inactive mouse X chromosomes. *Mamm Genome* 1991;**1**:78–83.

31. Kalantry S, Purushothaman S, Bowen RB, Starmer J, Magnuson T. Evidence of *Xist* RNA-independent initiation of mouse imprinted X-chromosome inactivation. *Nature* 2009;**460**:647–51.

32. Wutz A, Rasmussen TP, Jaenisch R. Chromosomal silencing and localization are mediated by different domains of *Xist* RNA. *Nat Genet* 2002;**30**:167–74.

33. Kalantry S, Mills KC, Yee D, Otte AP, Panning B, Magnuson T. The Polycomb group protein Eed protects the inactive X-chromosome from differentiation-induced reactivation. *Nat Cell Biol* 2006;**8**:195–202.

349

34. Zhao J, Sun BK, Erwin JA, Song JJ, Lee JT. Polycomb proteins targeted by a short repeat RNA to the mouse X chromosome. *Science* 2008;**322**:750–6.

35. Sado T, Wang Z, Sasaki H, Li E. Regulation of imprinted X-chromosome inactivation in mice by Tsix. *Development* 2001;**128**:1275–86.

36. Sun BK, Deaton AM, Lee JT. A transient heterochromatic state in *Xist* preempts X inactivation choice without RNA stabilization. *Mol Cell* 2006;**21**:617–28.

37. Navarro P, Pichard S, Ciaudo C, Avner P, Rougeulle C. Tsix transcription across the *Xist* gene alters chromatin conformation without affecting *Xist* transcription: implications for X-chromosome inactivation. *Genes Dev* 2005;**19**:1474–84.

38. Navarro P, Page DR, Avner P, Rougeulle C. Tsix-mediated epigenetic switch of a CTCF-flanked region of the *Xist* promoter determines the *Xist* transcription program. *Genes Dev* 2006;**20**:2787–92.

39. Clerc P, Avner P. Role of the region 3' to Xist exon 6 in the counting process of X-chromosome inactivation. *Nat Genet* 1998;**19**:249–53.

40. Morey C, Navarro P, Debrand E, Avner P, Rougeulle C, Clerc P. The region 3' to *Xist* mediates X chromosome counting and H3 Lys-4 dimethylation within the *Xist* gene. *EMBO J* 2004;**23**:594–604.

41. Shibata S, Lee JT. Tsix transcription-versus RNA-based mechanisms in *Xist* repression and epigenetic choice. *Curr Biol* 2004;**14**:1747–54.

42. Ohhata T, Hoki Y, Sasaki H, Sado T. Crucial role of antisense transcription across the *Xist* promoter in Tsix-mediated *Xist* chromatin modification. *Development* 2008;**135**:227–35.

43. Ogawa Y, Sun BK, Lee JT. Intersection of the RNA interference and X-inactivation pathways. *Science* 2008;**320**:1336–41.

44. Nesterova TB, Popova BC, Cobb BS, Norton S, Senner CE, Tang YA, et al. Dicer regulates *Xist* promoter methylation in ES cells indirectly through transcriptional control of Dnmt3a. *Epigenetics Chromatin* 2008;**1**:2.

45. Maherali N, Sridharan R, Xie W, Utikal J, Eminli S, Arnold K, et al. Directly reprogrammed fibroblasts show global epigenetic remodeling and widespread tissue contribution. *Cell Stem Cell* 2007;**1**:55–70.

46. Nichols J, Zevnik B, Anastassiadis K, Niwa H, Klewe-Nebenius D, Chambers I, et al. Formation of pluripotent stem cells in the mammalian embryo depends on the POU transcription factor Oct4. *Cell* 1998;**95**:379–91.

47. Chambers I, Colby D, Robertson M, Nichols J, Lee S, Tweedie S, et al. Functional expression cloning of Nanog, a pluripotency sustaining factor in embryonic stem cells. *Cell* 2003;**113**:643–55.

48. Mitsui K, Tokuzawa Y, Itoh H, Segawa K, Murakami M, Takahashi K, et al. The homeoprotein Nanog is required for maintenance of pluripotency in mouse epiblast and ES cells. *Cell* 2003;**113**:631–42.

49. Hall LL, Byron M, Butler J, Becker KA, Nelson A, Amit M, et al. X-inactivation reveals epigenetic anomalies in most hESC but identifies sublines that initiate as expected. *J Cell Physiol* 2008;**216**:445–52.

50. Shen Y, Matsuno Y, Fouse SD, Rao N, Root S, Xu R, et al. X-inactivation in female human embryonic stem cells is in a nonrandom pattern and prone to epigenetic alterations. *Proc Natl Acad Sci USA* 2008;**105**:4709–14.

51. Silva SS, Rowntree RK, Mekhoubad S, Lee JT. X-chromosome inactivation and epigenetic fluidity in human embryonic stem cells. *Proc Natl Acad Sci USA* 2008;**105**:4820–5.

52. Dvash T, Fan G. Epigenetic regulation of X-inactivation in human embryonic stem cells. *Epigenetics* 2009;**4**:19–22.

53. Tsai CL, Rowntree RK, Cohen DE, Lee JT. Higher order chromatin structure at the X-inactivation center via looping DNA. *Dev Biol* 2008;**319**:416–25.

54. Augui S, Filion GJ, Huart S, Nora E, Guggiari M, Maresca M, et al. Sensing X chromosome pairs before X inactivation via a novel X-pairing region of the Xic. *Science* 2007;**318**:1632–6.

55. Csankovszki G, Panning B, Bates B, Pehrson JR, Jaenisch R. Conditional deletion of Xist disrupts histone macroH2A localization but not maintenance of X inactivation. *Nat Genet* 1999;**22**:323–4.

56. Keohane AM, O'Neill LP, Belyaev ND, Lavender JS, Turner BM. X-Inactivation and histone H4 acetylation in embryonic stem cells. *Dev Biol* 1996;**180**:618–30.

57. Chaumeil J, Okamoto I, Guggiari M, Heard E. Integrated kinetics of X chromosome inactivation in differentiating embryonic stem cells. *Cytogenet Genome Res* 2002;**99**:75–84.

58. Chow J, Heard E. X inactivation and the complexities of silencing a sex chromosome. *Curr Opin Cell Biol* 2009;**21**:359–66.

59. Mietton F, Sengupta AK, Molla A, Picchi G, Barral S, Heliot L, et al. Weak but uniform enrichment of the histone variant macroH2A1 along the inactive X chromosome. *Mol Cell Biol* 2009;**29**:150–6.

60. Marks H, Chow JC, Denissov S, Francoijs KJ, Brockdorff N, Heard E, et al. High-resolution analysis of epigenetic changes associated with X inactivation. *Genome Res* 2009;**19**:1361–73.

61. Csankovszki G, Nagy A, Jaenisch R. Synergism of *Xist* RNA, DNA methylation, and histone hypoacetylation in maintaining X chromosome inactivation. *J Cell Biol* 2001;**153**:773–84.

62. Blewitt ME, Gendrel AV, Pang Z, Sparrow DB, Whitelaw N, Craig JM, et al. SmcHD1, containing a structural-maintenance-of-chromosomes hinge domain, has a critical role in X inactivation. *Nat Genet* 2008;**40**:663–9.

63. Beard C, Li E, Jaenisch R. Loss of methylation activates *Xist* in somatic but not in embryonic cells. *Genes Dev* 1995;**9**:2325–34.

64. Panning B, Jaenisch R. DNA hypomethylation can activate *Xist* expression and silence X-linked genes. *Genes Dev* 1996;**10**:1991–2002.

65. Hellman A, Chess A. Gene body-specific methylation on the active X chromosome. *Science* 2007;**315**:1141–3.

66. Chambeyron S, Bickmore WA. Does looping and clustering in the nucleus regulate gene expression? *Curr Opin Cell Biol* 2004;**16**:256–62.

67. Dietzel S, Schiebel K, Little G, Edelmann P, Rappold GA, Eils R, et al. The 3D positioning of ANT2 and ANT3 genes within female X chromosome territories correlates with gene activity. *Exp Cell Res* 1999;**252**:363–75.

68. Clemson CM, Hall LL, Byron M, McNeil J, Lawrence JB. The X chromosome is organized into a gene-rich outer rim and an internal core containing silenced nongenic sequences. *Proc Natl Acad Sci USA* 2006;**103**:7688–93.

69. Chaumeil J, Le Baccon P, Wutz A, Heard E. A novel role for *Xist* RNA in the formation of a repressive nuclear compartment into which genes are recruited when silenced. *Genes Dev* 2006;**20**:2223–37.

70. Cai S, Lee CC, Kohwi-Shigematsu T. SATB1 packages densely looped, transcriptionally active chromatin for coordinated expression of cytokine genes. *Nat Genet* 2006;**38**:1278–88.

71. Agrelo R, Souabni A, Novatchkova M, Haslinger C, Leeb M, Komnenovic V, et al. SATB1 defines the developmental context for gene silencing by *Xist* in lymphoma and embryonic cells. *Dev Cell* 2009;**16**:507–16.

72. Helbig R, Fackelmayer FO. Scaffold attachment factor A (SAF-A) is concentrated in inactive X chromosome territories through its RGG domain. *Chromosoma* 2003;**112**:173–82.

73. Fackelmayer FO. A stable proteinaceous structure in the territory of inactive X chromosomes. *J Biol Chem* 2005;**280**:1720–3.

74. Rego A, Sinclair PB, Tao W, Kireev I, Belmont AS. The facultative heterochromatin of the inactive X chromosome has a distinctive condensed ultrastructure. *J Cell Sci* 2008;**121**:1119–27.

75. Zhang LF, Huynh KD, Lee JT. Perinucleolar targeting of the inactive X during S phase: evidence for a role in the maintenance of silencing. *Cell* 2007;**129**:693–706.

CHAPTER 22

Genomic Imprinting

Wendy Chao
Department of Ophthalmology, Schepens Eye Research Institute, Harvard Medical School,
20 Staniford Street, Boston, MA 02114, USA

INTRODUCTION

When eukaryotic organisms reproduce sexually, each parent contributes a haploid set of chromosomes to create diploid progeny. Two copies, or *alleles*, exist for most gene loci. According to classical Mendelian genetics, both copies are expressed, and certain variations in DNA sequence may allow the phenotype of a dominant allele to prevail over a recessive one. However, some genes carry epigenetic marks that distinguish between maternally and paternally inherited alleles. These genomic "imprints" can dramatically alter gene expression depending on parent of origin – even if the two alleles are otherwise identical.

In mammals, genomic imprinting manifests in monoallelic silencing according to parental lineage. Because a second allele may provide genetic diversity and mask undesirable traits [1], it is somewhat counterintuitive to find functionally haploid genes in complex diploid species. However, genomic imprinting arose in mammalian evolution over 150 million years ago [2], which implies that monoallelic expression is not necessarily detrimental to genetic fitness. Rather, it is imperative for several loci to maintain imprinted monoallelic expression. In humans, aberrant imprinting underlies numerous developmental and neurological disorders [reviewed in Ref. 3], and loss of imprinting is common in cancer [reviewed in Ref. 4].

Even though it is highly relevant to human development and disease, genomic imprinting went undiscovered in mammals until relatively recently. Imprinted (as opposed to random) X chromosome inactivation has been a known phenomenon since the early 1970s [5,6], but it was twenty more years before imprinted autosomal genes were discovered in mammals [7–9]. The existence of these genes was predicted by earlier nuclear transplantation experiments, that produced mouse embryos with both sets of chromosomes derived from one parent. Not only were these uniparental embryos abnormal, but gynogenetic (female-derived) and androgenetic (male-derived) embryos displayed contrasting phenotypes [10–13]. These studies demonstrated the nonequivalence of maternal and paternal genomes – even after accounting for sex chromosome differences. Subsequent complementation studies narrowed these parental effects to discrete autosomal regions [14].

In 1991, three imprinted genes in mice were characterized: insulin-like growth factor 2 receptor (*Igf2r*), which is maternally expressed [7]; its ligand, insulin-like growth factor 2 (*Igf2*), a paternally-expressed regulator of growth and development [9]; and *H19*, a maternally-expressed noncoding RNA [8] that is physically linked to *Igf2* [15] and regulated by shared elements [16,17]. These archetypes of genomic imprinting have yielded much insight into various epigenetic regulatory mechanisms. The *Igf2* gene is particularly

353

Handbook of Epigenetics: The New Molecular and Medical Genetics. DOI: 10.1016/B978-0-12-375709-8.00022-8

interesting because its complex regulation involves both the *H19* gene at the transcriptional level and the Igf2r protein at the post-translational level [reviewed in Ref. 18]. *Igf2* is also highly conserved among vertebrates [2,19], but its imprinting status is not. The divergence of *Igf2* imprinting in the phylogenetic tree has fueled many theories on vertebrate evolution and the origin of genomic imprinting.

Genomic imprinting is often described as an exclusively mammalian phenomenon, yet parental effects on gene expression were documented in insects and plants long before the discovery of imprinted mammalian genes. The term "imprint" was actually used as early as 1960 to describe epigenetic parental effects in fungus gnats of the genus *Sciara*. At various stages of sciarid development, certain paternally-derived chromosomes are heterochromatized and eliminated from cells independently of genomic constitution and "determined only by the sex of the germ line through which the chromosome has been inherited" [20]. Allele-specific silencing in *Drosophila* was recorded in the mid-1930s, when vague reports noted the preferential silencing of the X-linked *scute-8* gene when paternally inherited [21,22]. The earliest (and perhaps most extreme) example of genomic imprinting in insects can be traced to a 1931 report, which described sex determination in the family Pseudococcidae [23]. Coccids (commonly known as mealybugs) represent a striking example of haplodiploidy, a system of sex determination commonly employed by insects, in which females are diploid but males are haploid [reviewed in Ref. 24]. In males, all paternally derived chromosomes are either silenced by heterochromatin or completely eliminated; thus, all male coccids are functionally haploid [reviewed in Ref. 25].

Epigenetic parent-specific effects were demonstrated even earlier in plants. In 1918 and 1919, two independent studies demonstrated parent-specific effects at the maize *R* locus, which controls anthocyanin pigment expression in the aleurone endosperm [26,27]. When the female gamete transmits the dominant *R* allele in *RR* (pigmented) × *rr* (colorless) crosses, the aleurone seed covering is solidly pigmented; conversely, if *R* originates from the paternal (pollen) parent in a reciprocal cross, the endosperm is lightly pigmented, and mottled or spotted in appearance (Fig. 22.1A). Although the endosperm of flowering plants is usually

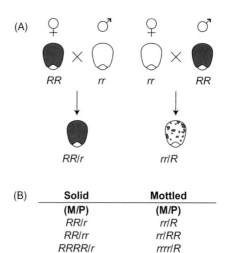

FIGURE 22.1
Influence of parental lineage on the *R*-mottled phenotype in maize *RR* (pigmented) × *rr* (colorless) crosses. (A) When *R* is transmitted by the maternal gamete (left), the *RR/r* triploid aleurone of the progeny is solidly pigmented. In the reciprocal cross (right), the resulting *rr/R* aleurone is lightly pigmented and mottled [26,27]. (B) Maize genotypes that yield solid or mottled phenotypes. M and P designate maternal and paternal origin, respectively. The solid phenotype occurs only when *R* is maternally derived, irrespective of gene dosage. A paternally transmitted *R* allele yields a mottled phenotype if the female gamete is homozygous for the *r* allele, and even when the maternal gamete is deficient (–) for the *R* region [28].

polyploid, later experiments [28] confirmed that the mottling effect is indeed dependent on parental origin rather than differences in gene dosage (Fig. 22.1B).

Definitions of "genomic imprinting" do not always include parent-of-origin effects in insects and plants, but are instead limited to those observed in mammals. However, syntactic differences aside, these processes use conserved regulatory mechanisms to achieve a common purpose: to epigenetically distinguish maternal and paternal genomes. Though genomic imprinting may not be a uniquely mammalian phenomenon, its discovery in mice did uncover an epigenetic basis of human disease, and catalyzed a field of research devoted to parent-specific gene expression. As with many genetic processes, the epigenetic phenomena in other organisms have helped elucidate the mechanisms of mammalian imprinting and its evolutionary origin.

MECHANISMS OF GENOMIC IMPRINTING

What distinguishes an imprinted gene from its non-imprinted counterparts, and destines it for allele-specific expression? This question has puzzled scientists for decades, and is now beginning to be understood. Genomic imprinting is currently known to involve numerous epigenetic processes – many of which are conserved among diverse species. While its exact catalyst is not entirely clear, the primary imprint in many organisms involves the classic epigenetic mark: DNA methylation. This primary mark is propagated by *cis* and *trans* factors that trigger additional modifications and culminate in allele-specific gene expression.

DNA Methylation

Many organisms – from primitive bacteria to complex eukaryotes – use methyl groups to distinguish DNAs of different origins [reviewed in Ref. 29]. In higher eukaryotes, methylation primarily occurs on the cytosine residue of CpG dinucleotides, which tend to cluster around promoter regions as "CpG islands" but appear sparsely in the rest of the genome due to spontaneous deamination of 5-methylcytosine to thymine [reviewed in Ref. 30]. In the mammalian genome, much of the DNA methylation targets transposable elements, which illustrates a role in host defense mechanisms that silence invasive DNAs [31]. Methylation is also a common denominator of differential DNA regulation; before the first imprinted genes were discovered in mammals, studies demonstrated that transgenes could acquire allele-specific methylation patterns depending on the transmitting parent [32,33]. Differential methylation was thus identified as a heritable epigenetic feature that distinguishes maternal and paternal alleles – and a central mechanism in genomic imprinting.

IMPRINT ESTABLISHMENT AND MAINTENANCE IN MAMMALS

Mammalian imprinted genes are often (if not always) situated near differentially-methylated regions (DMRs), also known as differentially-methylated domains (DMDs), which are believed to be the primary targets of epigenetic modifications. DMRs may in turn direct other *cis* and *trans* elements to achieve stable allele-specific gene expression [reviewed in Ref. 34]. Thus, DMRs may serve as imprinting control regions (ICRs), also known as imprinting control elements (ICEs) or imprinting centers (ICs). Aside from the high frequency of CpG dinucleotides, DMRs share little sequence homology; instead, they are characterized by tandemly repeated elements [35–37]. These repetitive structures are believed to trigger *de novo* differential methylation [35–37], similarly to how retrotransposon-derived repetitive sequences – such as short interspersed nuclear elements (SINEs) and CpG-rich Alu repeats – may acquire germline-specific differential methylation [38,39]. Primary imprints occur in the germline, where the prospective parent's existing imprints can be erased and reestablished in the haploid gametes. This occurs through global demethylation in germ cells, followed by differential methylation by the *de novo* DNA methyltransferase Dnmt3a [40] and its cofactor Dnmt3L [41].

355

It is not entirely clear how the *de novo* methyltransferases differentiate between maternal and paternal DMRs in the germline, though this seems to be partially determined by DMR location. Maternally-methylated DMRs coincide with transcription units, whereas the few known paternally-methylated DMRs occur within intergenic regions (Table 22.1). Paternal-specific germline methylation appears to target tandem repeat sequences, as evidenced by the *H19* [42] and *Rasgrf1* [43] loci, which contain two of the three known paternal germline DMRs. On the other hand, the act of transcription may dictate maternal-specific methylation in oocytes, as demonstrated by the *Gnas/Nesp* locus (Fig. 22.2). This complex imprinted domain [reviewed in Ref. 44] includes *Gnas*, which exhibits maternal-specific expression in some tissues. *Gnas* encodes the highly conserved signaling protein Gsaα; alternative promoters give rise to two paternal-specific transcripts, *Gnasxl* (which encodes the variant protein XLas) and the noncoding *IA*. The upstream *Nesp* locus encodes the Nesp55 protein, which is involved in the secretory pathway. This maternally expressed *Nesp* transcript also appears to have a major functional role in imprinting the entire locus. There are two DMRs in this domain: one encompassing the promoters for the paternally-expressed *Gnasxl* and the noncoding *Nespas*, and another at the *1A* promoter. Truncating *Nesp* transcription disrupts methylation of both DMRs in the female germline, suggesting that the act of transcription facilitates *de novo* methylation [45]. This model is supported by the fact that all known maternal DMRs occur in transcribed regions – either in introns or near promoters that are downstream of alternate transcription start sites (Table 22.1). It is hypothesized that oocyte-specific transcription facilitates germline DMR methylation by favorably altering chromatin structure; alternatively, the RNA itself might recruit *de novo* methyltransferases or other *trans* regulatory factors that promote germline methylation [45].

Allele-specific methylation also involves germline-specific timing of Dnmt3L expression. This protein lacks *in vitro* methyltransferase activity, but is required for *de novo* methylation by Dnmt3a [40,41]. Additionally, alternate germline-specific promoters lead to differential Dnmt3L expression in oocytes and spermatocytes [46]. Oocytes express Dnmt3L for only

TABLE 22.1 Known Germline DMRs and Their Locations

DMR/location	Locus		
Maternal/intron	Gnas (1A)	Inpp5f	Peg3
	Gnas (Nespas/Gnasxl)	Kcnq1 (KvDMR)	Peg13
	Grb10	Mcts2	Snrpn
	Igf2r (Air)	Nap1l5	U2af1-rs1
	Impact	Peg1	Zac1
Maternal/promoter	Peg10	Slc38a4	
Paternal/intergenic	Dlk1-Gtl2 (IG-DMR)	H19	Rasgrf1

Adapted from Ref. 45

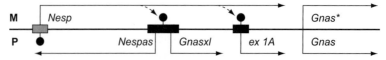

FIGURE 22.2
The *Gnas* imprinted domain (not drawn to scale), including the protein-coding transcripts (*Nesp*, *Gnasxl*, and *Gnas*) and noncoding transcripts (*Nespas* and *1A*). Maternal (M) transcripts are indicated by arrows above the line, while paternal (P) transcripts are below the line. Arrowheads indicate the direction of transcription. *Gnas* shows maternal-specific expression in some tissues, as indicated by the asterisk (*). Two germline DMRs (black boxes) acquire maternal methylation imprints (black circles) in oocytes. *Nesp* transcription is believed to facilitate germline methylation, as indicated by dashed arrows. A somatic DMR (gray box), which covers the *Nesp* promoter, is methylated on the paternal allele after fertilization. Adapted from Ref.45.

a few days before ovulation, and primary methylation imprints are established during this short time frame [41]. On the other hand, Dnmt3L expression begins prenatally in embryonic prospermatogonia, and continues until a few days after birth; paternal-specific DMR methylation then persists in the male germline well into adulthood [47]. Because methylated cytosine residues spontaneously deaminate to thymine, the prolonged methylation time may explain why paternal DMRs tend to have far fewer CpGs than their maternal counterparts; moreover, this gradual sequence degeneration may explain why so few paternal DMRs have been identified [reviewed in Ref. 48]. Coincidentally, retrotransposon silencing depends heavily on Dnmt3L in male germ cells [49] but to a lesser extent in the female germline [41,50]. Since paternal-specific DMR methylation requires tandemly-repeated elements [42,43], the primary paternal imprint likely targets the retrotransposon-like nature of DMRs.

After the maternal and paternal genomes join during fertilization, the primary germline imprints persist while the rest of the zygotic genome is demethylated [reviewed in Ref. 51]. It is not entirely clear how these primary parental imprints survive this early embryonic demethylation; however, methyl-CpG-binding proteins (MBDs), which regulate transcription by binding methylated DNA and recruiting additional silencing factors [52], are required to maintain differential methylation of imprinted mouse genes in somatic cells [53]. The murine maintenance methyltransferase Dnmt1 is also pivotal for preserving parental methylation patterns during zygotic demethylation and subsequent somatic cell divisions [54]. PGC7/Stella and ZFP57 are additional factors that appear to protect imprinted DMRs from zygotic demethylation. PCG7/Stella is a nuclear protein that is highly expressed in both male and female primordial germ cells (PGCs); maternal-specific expression of PCG7/Stella continues in the early embryo, and is essential for maintaining methylation at maternal DMRs [55]. ZFP57 is a Kruppel-associated box (KRAB) zinc finger protein that is expressed in oocytes and in certain somatic tissues; it is required for maternal imprint establishment (and for both maternal and paternal imprint maintenance) at some DMRs [56].

Many imprinted genes occur in clusters in the genome, which can be several megabases (Mb) in length and contain multiple differentially-expressed genes under the control of one or two DMRs. Such is the case with an approximately 3-Mb region on human chromosome 15q11-13 that is implicated in Prader–Willi syndrome (PWS) and Angelman syndrome (AS) [reviewed in Ref. 57]. At least five paternally expressed genes (*MKRN3, MAGEL2, NDN, SNURF,* and *SNRPN*) and two maternally expressed genes (*UBE3A* and *ATP-10A*) lie in this region (Fig. 22.3). On the homologous region on mouse chromosome 7c, *Atp10a* is not imprinted [58,59]; however, an additional paternally-expressed gene, *Peg12/Frat3*, lies distal to the *Mkrn3/Magel2/Ndn* cluster [60,61]. A bipartite ICR for this region was defined by observed microdeletions in PWS and AS patients; it encompasses the maternally-methylated *SNURF/SNRPN* promoter and a region 35 kb upstream that also has ICR function. Deletions in this upstream region (known as AS-IC) cause AS when maternally inherited, whereas

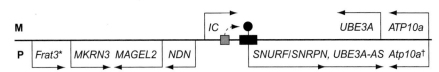

FIGURE 22.3

The PWS/AS imprinted domain on human 15q11-13 (not drawn to scale). Maternal transcripts are above the line; paternal transcripts are below the line. An additional paternal-specific gene, *Peg12/Frat3*, lies distal to *Mkrn3* on mouse chromosome 7c (*); however, murine *Atp10a* is not imprinted (†). The ICR near the *SNURF/SNRPN* promoter (PWS-IC, black box) is methylated in the female germline (black circle) and implicated in PWS. The upstream ICR (AS-IC, gray box) is implicated in Angelman syndrome, and facilitates germline methylation of PWS-IC (indicated by the curved dashed arrow). *IC* transcripts that arise from oocyte-specific alternate promoters may facilitate methylation of PWC-IC [64].

deletions in the *SNURF/SNRPN* promoter (called PWS-IC) cause PWS when paternally inherited [reviewed in Ref. 62]. The upstream ICR is believed to help establish the primary maternal imprint on PWS-IC [reviewed in Ref. 63]. In mouse and human oocytes, alternative promoters upstream of AS-IC give rise to maternal-specific *IC* transcripts that may facilitate PWS-IC methylation [64]. This is consistent with the hypothesis that maternal germline imprints are dependent on transcription [45].

The 5' DMR of *H19* is one of three known paternal germline DMRs (Table 22.1) and displays persistent paternal methylation in both mice [65] and humans [66]. It was initially identified as an ICR when its deletion in mice abolished imprinting for both *Igf2* and *H19* [67]. CpG mutations that prevent methylation also disrupt imprinting in this region [68]. In humans, *IGF2* overexpression is implicated in Wilms' tumor and in the overgrowth disease Beckwith-Wiedemann syndrome (BWS) [69]. Loss of imprinting (resulting in biallelic expression) is one of several known mechanisms that cause *IGF2* overexpression [70,71]. Not surprisingly, many BWS or Wilms' tumor cases involve mutations or deletions in the ICR [72–76]. Because *IGF2* overexpression is a common denominator in carcinogenesis, it is also not surprising that ICR dysfunction has been noted in numerous cancers [reviewed in Refs 77,78].

METHYLATION AND GENOMIC IMPRINTING IN PLANTS AND INVERTEBRATES

Some plants and insects also use DNA methylation to achieve parent-specific gene expression, albeit with several distinct features. In plants, the major role of methylation appears to be in maintaining differential expression [79,80], whereas the primary imprint is established by a DNA glycosylase rather than a *de novo* methyltransferase [81]. In mealybugs, parent-specific genome silencing also involves differential methylation, though the heterochromatic paternal DNA is associated with hypomethylation rather than hypermethylation [82]. In *Drosophila*, parent-specific genomic imprints appear to involve chromatin-modifying proteins rather than DNA methylation [83,84]. It was once believed that DNA methylation does not occur at all in *Drosophila* [85,86]; however, it is now known that methylation does in fact occur – though only at early stages of development at non-CpG dinucleotides (most often CpT and CpA) [87,88]. It remains unclear whether methylation is involved in *Drosophila* imprinting, yet homologs of *Drosophila* Polycomb group (PcG) proteins appear to coordinate differential methylation of imprinted genes in mice [89] and plants [reviewed in Ref. 90]. Another model organism, *Caenorhabditis elegans*, exhibits imprinted X chromosome inactivation [91] and also imprints exogenous transgenes [92], despite the long-held notion that *C. elegans* does not genomically imprint [93]. However, these parent-specific effects in *C. elegans* apparently do not involve methylation, as attempts to detect 5-methylcytosine have failed [94]. Because the *C. elegans* genome is considerably less complex than higher eukaryotic genomes, constitutive silencing mechanisms such as DNA methylation are thought to be less critical [95].

Replication Timing

Asynchronous DNA replication is a curious hallmark of imprinted alleles and other monoallelically-expressed genes, including those on the active and inactive X chromosomes [5]; actively transcribed genes tend to replicate early, while late replication is characteristic of repressed genes and transcriptionally silent heterochromatin [reviewed in Ref. 96]. Differential methylation correlates with replication asynchrony on human chromosomal region 15q11-q13, which contains multiple imprinted genes and is associated with PWS and AS (see Fig. 22.3). This implies that replication and methylation are coordinately regulated [97]. While differential replication could simply be a consequence of genomic imprinting, there is some evidence that it may occur independently of methylation imprints and might even play a regulatory role. In some cases of aberrant human *IGF2/H19* imprinting, loss of differential methylation does not disrupt asynchronous replication [98]. Moreover, in mouse embryonic stem (ES) cells that lack *de novo* methylation machinery (Dnmt1 and Dnmt3L), the *Igf2/H19* locus continues to replicate asynchronously despite loss of imprinting [99]. On the other hand, methylation imprints may also be established at the *Igf2/H19* locus without affecting replication timing [100]. Nonetheless, these results suggest that replication asynchrony does not necessarily occur secondary to genomic imprinting. Of note, asynchronous replication is reset during gametogenesis and maintained throughout zygotic development, which coincides temporally with imprint erasure, reestablishment, and maintenance [101]. Thus, a component of the primary epigenetic imprint may indeed involve replication timing; however, this remains uncertain.

Chromatin Modifications

Gene expression is not only affected by covalent DNA modifications such as methylation, but also by higher-order changes in chromatin structure that involve DNA-protein interactions. Conformational changes in chromatin (revealed by differential DNAseI hypersensitivity) may determine whether genes are accessible to transcription factors and other regulatory proteins [reviewed in Ref. 102]. Thus, differential modifications of key chromatin structures, such as core histones, are epigenetic events that can contribute to allele-specific gene expression.

HISTONE MODIFICATIONS

Differential histone modifications occur in many examples of genomic imprinting, including paternal genome silencing/elimination in insects [reviewed in Ref. 103] and imprinted X chromosome inactivation in mammals [104]. Modifications (generally to the lysine residues in histone N-terminal tails) may result in either transcriptional activation or silence, and often coordinate with DNA methylation status [reviewed in Ref. 105]. Acetylation is a well-known histone modification that generally associates with active transcription; deacetylation may require DNA methylation, as certain methyl-CpG binding proteins (such as MECP2) can recruit histone deacetylases that repress transcription in mice [106] and frogs [107].

Methylation not only takes place on CpG residues of DNA, but also on lysine residues of histones; these may be either transcriptionally repressing or activating, depending on lysine position and level of methylation [reviewed in Ref. 108]. Methylation on lysine 9 of histone H3 (H3K9) is required for DNA methylation in *Neurospora* [109] and *Arabidopsis* [110]; it also coincides with methylation of pericentric heterochromatin [111] and the inactive X chromosome in mammals [reviewed in Ref. 108]. Differential H3K9 methylation corresponds with imprinting in the PWS/AS domain (see Fig. 22.3). In human cells, both H3K9 methylation and CpG methylation occur at the maternal PWS-IC [112]. In mouse ES cells, deleting the gene that encodes G9a (the H3K9 methyltransferase) reduces PWS-IC methylation and disrupts imprinting at the PWS/AC locus, suggesting that H3K9 methylation regulates allele-specific ICR methylation [113].

While H3K9 methylation is associated with transcriptional repression, methylation of histone H3 lysine 4 (H3K4) is associated with transcriptional activation [reviewed in Ref. 108]. On the unmethylated and transcriptionally active paternal PWS-IC (see Fig. 22.3), H3K4 is methylated and H3K9 is unmethylated [112]. H3K4 has also emerged as an important variable in primary imprint establishment. Dnmt3L, the critical cofactor of the *de novo* DNA methylase Dnmt3a [41], functions in part by binding H3 and recruiting Dnmt3a to DNA; this complex cannot occur with methylated H3K4, which essentially prevents *de novo* CpG methylation [114]. H3K4 demethylation by the lysine demethylase KDM1B, which is highly expressed during late oogenesis, is required for *de novo* methylation of some (but not all) maternal DMRs [115]. Because different maternal DMRs acquire methylation imprints at specific stages of oocyte development [116], and only DMRs imprinted during late oogenesis are associated with H3K4 demethylation, KDM1B expression timing is likely to be a factor in maternal DMR specificity [115]. At the three known paternal DMRs (see Table 22.1), histone methylation may also direct primary imprint establishment. In sperm, H3K4 methylation occurs specifically at unmethylated maternal DMRs; in somatic tissues, the methylated paternal DMRs coincide with H3K9 and H4K20 methylation [117]. This concurs with the DNA/histone methylation patterns of the heterochromatin in pericentric satellite repeats [111]. Thus, histone methylation appears to be a critical precursor to primary imprinting of both maternal and paternal DMRs.

POLYCOMB AND TRITHORAX GROUP PROTEINS

Polycomb group (PcG) and trithorax group (TrxG) proteins, which have reciprocal functions in maintaining chromatin stability, may also regulate differential expression at imprinted regions. Both PcG and TrxG proteins were originally identified as modifiers of *Drosophila* position effect variegation, but are now known to control gene expression in mammals and many other species [reviewed in Ref. 118]. In *Drosophila*, TrxG and Su(var) (suppressor of variegation) proteins mediate imprinting of the mini-X chromosome [119]. *Drosophila* PcG proteins are also able to recognize the murine *Igf2/H19* ICR [120], and the mouse PcG protein Eed apparently mediates histone methylation and initiation of imprinted (but not random) X chromosome inactivation [104,121]. Murine Eed may also regulate autosomal imprinted loci, where it appears to modulate differential methylation of ICRs [89] and histones [122]. PcG proteins appear to coordinately regulate DNA and histone methylation to achieve tissue-specific gene expression, as evidenced by the differential methylation patterns associated with tissue-specific *Grb10* imprinting [123]. Interestingly, genes that encode PcG proteins may themselves be imprinted; in mice, the PcG gene *Sfmbt2* has paternal-specific expression in extraembryonic tissues [124], and several PcG genes that control endosperm development are imprinted in *Arabidopsis* and maize [reviewed in Ref. 90].

Chromosomal Position Effects

When inserted into imprinted chromosomal regions, transgenes acquire allele-specific methylation patterns that are determined by the transmitting parent [32,33]. This phenomenon bears resemblance to *Drosophila* position effect variegation, in which gene expression patterns may change upon transposition to other chromosomal locations – either by juxtaposition to enhancing elements, or insertion into heterochromatic DNA [reviewed in Ref. 125]. Thus, it has been proposed that position effects, similar to those observed in *Drosophila* variegation, may occur at imprinted domains. In this model, primary imprints established at ICRs lead to secondary methylation and heterochromatization, which can in turn spread to surrounding regions and silence distant genes [126]. In *Drosophila*, classical modifiers of position-effect variegation (such as chemicals and PcG proteins) mediate paternal-specific silencing of three closely linked genes on a mini-X chromosome [84,119]. The site of the primary imprint is a *cis* regulatory ICR; its effects extend to silence a distal gene within a 1.2-Mb region, as well as the entire 1.5-Mb mini-X chromosome, which is distinguished by reduced transcription and late replication [126]. Imprinting of a novel gene in the mouse PWS/AS cluster, *Peg12/Frat3* (Fig. 22.3), is believed to be a product of analogous position effects; this gene lacks a human homolog, and was likely retrotransposed during species divergence into the mouse PWS/AS locus, where it acquired the imprinted status of the surrounding genes [60,61]. Interestingly, ICRs themselves remain faithfully imprinted when inserted into non-imprinted regions, and can imprint hybrid transgenes [127,128]. Even more remarkably, mammalian ICRs function as silencers in *Drosophila*; although imprinting is not established, this demonstrates that genomic imprinting involves highly conserved silencing mechanisms [129,130].

Chromatin Insulators

Chromatin insulators establish boundaries between different DNA regulatory domains, an demarcate transcriptionally inactive heterochromatin from euchromatin that is conducive

Handbook of Epigenetics (978-0-12-375709-8)

ERRATA:

Figure 22.5 should appear as below:

FIGURE 22.5

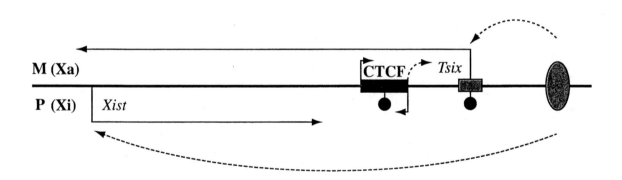

We apologize for this oversight. Elsevier is working with our suppliers to ensure this change is made to all newly-produced electronic and print versions of the book.

to transcription. Thus, insulators can protect against chromosomal positional effects and *cis* regulatory elements such as enhancing and silencing elements [reviewed in Ref. 131]. Insulators feature at several imprinted domains, most notably at the *Igf2/H19* locus (Fig. 22.4), which has become a paradigm of imprinted gene regulation. The ICR between *Igf2* and *H19* contains binding sites for CCCTC binding factor (CTCF), a multifunctional transcription factor; at the *Igf2/H19* locus, CTCF acts as a methylation-sensitive insulator protein that dictates whether a shared downstream enhancer can activate either promoter [132,133]. On the maternal chromosome, CTCF binds the ICR and forms a chromatin barrier between *Igf2* and the downstream enhancer, resulting in *Igf2* silence and *H19* expression. On the paternal chromosome, methylation of the ICR prevents CTCF binding – thus allowing the enhancer to activate the *Igf2* promoter (Fig. 22.4). This likely occurs through CTCF-mediated chromatin arrangements, that allow the enhancer to differentially activate the *Igf2* and *H19* promoters [134]. ICR methylation then leads to secondary *H19* methylation and transcriptional repression on the paternal allele [135].

Chromatin insulators have been proposed to regulate several other imprinted domains. At the *Xist/Tsix* locus, the identification of functional CTCF binding sites [136,137] has led to a similar model for X chromosome inactivation (Fig. 22.5). The ICR for this region exhibits differential methylation [138] and bidirectional promoter activity [139]; it is also part of a bipartite enhancer that is believed to activate *Tsix* expression on the active X chromosome [140,141]. CTCF also binds at the ICRs of several other imprinted loci [142], and is strongly correlated with imprinting in cross-species comparisons [143]. Another multifunctional

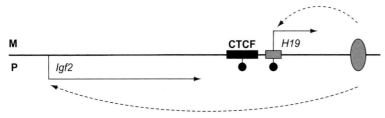

FIGURE 22.4
The insulator model of genomic imprinting at the *Igf2/H19* locus. This region (not drawn to scale) contains a downstream enhancer (gray oval) that may activate either *Igf2* or *H19* (curved dashed arrows), likely by chromatin looping events that allow the enhancer to contact either promoter [134]. The intergenic ICR (black box) is one of three known paternal germline DMRs (Table 22.1). When this DMR is paternally methylated (black circle), the enhancer activates *Igf2* expression on that allele, as indicated by the curved dashed arrow. The paternal *H19* promoter is then silenced by secondary methylation (gray box with black circle). On the maternal allele, the unmethylated DMR binds CTCF, which prevents the enhancer from accessing *Igf2*. The downstream enhancer thus activates maternal-specific *H19* expression (dashed curved arrow).

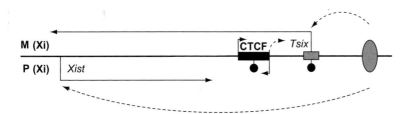

FIGURE 22.5
The *Xist/Tsix* region (not drawn to scale), which controls both random and imprinted X chromosome inactivation. The *Xist* ncRNA mediates silencing of the inactive X chromosome (Xi), and is antagonized by the *Tsix* antisense ncRNA on the active X chromosome (Xa) [reviewed in Ref. 152]. The ICR for this region exhibits paternal-specific methylation (black circles) and contains multiple CTCF binding sites; it also shows bidirectional promoter activity (arrows). The ICR and a downstream element (gray oval) both have enhancing activity on the *Tsix* promoter (dashed curved arrows). The *Tsix* promoter is believed to acquire biallelic methylation (gray box) as a secondary event after X inactivation takes place [138].

transcription factor, yin yang 1 (YY1), similarly functions as a methylation-sensitive insulator that mediates parent-specific expression at several imprinted loci [144–146]. Interestingly, YY1 associates with CTCF through protein-protein interactions and serves as a cofactor in X chromosome inactivation; because both proteins are ubiquitously expressed, they are hypothesized to alternately regulate (or co-regulate) developmentally or tissue-specific imprinting [147].

Noncoding RNAs

While only a fraction of the mammalian genome is actively transcribed in differentiated cells [148], much of the transcriptional activity produces non-coding RNAs (ncRNAs) [reviewed in Ref. 149]. Several of these ncRNAs – ranging from ~21 nucleotides to several kilobases in length – are now known to serve regulatory functions [reviewed in Refs 150,151]. Perhaps the most well known ncRNA in mammalian gene regulation is *Xist*, which mediates long-range silencing of the X chromosome to achieve dosage compensation [reviewed in Ref. 152]. Many imprinted loci studied to date contain ncRNAs [reviewed in Ref. 153], which form the basis of another genomic imprinting paradigm.

The maternally-expressed *Igf2r* was the first of three imprinted mouse genes identified in 1991 [7]. It encodes a receptor for the Igf2 protein, and primarily serves as a negative regulator by internalizing Igf2 and targeting it for degradation [reviewed in Ref. 154]. The *Igf2r* domain includes several nonimprinted genes (*Slc22a1*, *Mas1*, and *Plg*), two additional maternally-expressed genes (*Slc22a2* and *Slc22a3*), and a single paternally-expressed gene that encodes the antisense *Igf2r* RNA (*Air*) (Fig. 22.6). The first intron of the *Igf2r* gene contains a DMR that coordinates the maternal-specific expression of *Igf2r* [128], *Slc22a2*, and *Slc22a3* [155]. *Air* transcription originates within this DMR, and proceeds in an antisense orientation to *Igf2r* [156]. When *Air* transcription is prematurely terminated, the remaining promoter retains its imprint; however, all of the paternally-silenced genes (*Igf2r*, *Slc22a2*,

FIGURE 22.6
The *Igf2r/Air* region (not drawn to scale), which demonstrates ncRNA-dependent imprinting. Maternally expressed genes (*Slc22a2* and *Slc22a3*) are represented by solid arrows above the line, and the single paternally expressed antisense *Igf2r* RNA (*Air*) is indicated below the line. Biallelic genes (*Plg*, *Slc22a2*, and *Mas1*) are also shown. The DMR for this region resides in the first intron of the *Igf2r* gene, and is methylated on the maternal allele (black circle); it also serves as the origin of *Air* transcription on the paternal allele. The *Air* ncRNA overlaps the reciprocally imprinted *Igf2r*, and also mediates silencing at *Slc22a2* and *Slc22a3* [157]. The *Air* transcript overlaps (but does not silence) *Mas1* [156].

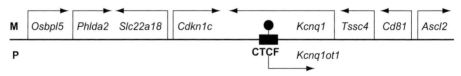

FIGURE 22.7
The *Kcnq1* locus (not drawn to scale) contains several maternally expressed genes (indicated above the line); a single paternally expressed *Kcnq1ot1* ncRNA is transcribed antisense to the *Kcnq1*. The ICR lies within a *Kcnq1* intron and is methylated on the maternal allele; on the paternal allele, the unmethylated ICR binds CTCF and also serves as the origin of *Kcnq1ot1* transcription. While the *Kcnq1ot1* ncRNA is required to imprint the maternally expressed genes [159], CTCF has been proposed to regulate imprinting of *Cdkn1c* in some tissues [161].

and *Slc22a3*) become active, which indicates that the full-length *Air* transcript is required to silence the maternal-specific genes in this region [157].

The *Kcnq1* locus also supports an ncRNA-dependent model of genomic imprinting. It contains several maternally expressed genes and one paternally expressed ncRNA, *Kcnq1ot1*, which is transcribed antisense to the *Kcnq1* potassium channel gene (Fig. 22.7). The ICR for this domain, which is also the origin of *Kcnq1ot1* transcription, lies within a *Kcnq1* intron (Table 22.1) and is methylated on the maternal allele [158]. Premature termination of the *Kcnq1ot1* ncRNA disrupts imprinting in the locus, but the truncated *Kcnq1ot1* retains its imprint [159]. This is consistent with the observations at the *Igf2r-Air* locus [157].

Interestingly, the *Kcnq1* locus may also utilize the insulator model of genomic imprinting, as the unmethylated paternal ICR not only serves as the origin of *Kcnq1ot1* transcription but also binds CTCF [160]. Truncating *Kcnq1ot1* transcription does not affect *Cdkn1c* imprinting in some tissues, which implicates ncRNA-independent mechanisms that are perhaps mediated by CTCF [161]. Likewise, the *Igf2/H19* domain (Fig. 22.4), which is believed to follow a strict CTCF-dependent insulator model [162], may also use multiple imprinting mechanisms – some of which may require the *H19* ncRNA. In a targeted disruption of the *H19* transcriptional unit, the DNA cassette inserted in its place becomes imprinted – yet *Igf2* becomes biallelic; this indicates that full-length *H19* is required for imprinting *Igf2* [163], though its precise function remains unclear.

These ncRNAs are distinct from the protein-coding transcripts that contribute to maternal germline imprints [45], and the antisense orientations of *Air* and *Kcnq1ot1* evoke a possible dsRNA-based mechanism for silencing *Igf2r* and *Kcnq1*, respectively. This model would emulate RNAi, a system that likely evolved to silence transposable elements, viral DNAs, and other parasitic nucleic acids [reviewed in Ref. 150]. However, these antisense ncRNAs do not overlap all oppositely imprinted genes in their domains; moreover, non-imprinted genes may be overlapped, such as *Mas1* in the *Igf2r/Air* locus [156]. These observations argue against the likelihood of a homology-dependent silencing mechanism.

Imprinted silencing by ncRNAs may be similar to *Xist*-mediated X chromosome inactivation. In this scenario, transcripts coat the DNA and recruit chromatin modifying proteins and silencing factors [reviewed in Ref. 152]. Consistent with this hypothesis, *Kcnq1ot1* associates with the PcG proteins at the chromatin level [164], and both *Air* and *Kcnq1ot1* recruit repressive histone methyltransferases to their target promoters – an effect that requires the full-length ncRNAs [165,166]. The silent genes in both the *Kcnq1* and *Igf2r* loci also become contracted into repressive nuclear compartments that exclude RNA polymerase II [164], which mirrors the transcriptionally silent nuclear compartment formed by the repressive *Xist* RNA [167]. Thus, imprinted silencing by ncRNA has been proposed to be mechanistically similar to *Xist*-mediated gene silencing [164].

However, there is some indication that *Xist* antagonism by the complementary *Tsix* does involve RNAi, whereas RNAi does not appear to be sufficient for silencing at the *Kcnq1* locus. The *Xist* and *Tsix* ncRNAs have been shown to form double-stranded duplexes processed by Dicer, a central protein in the RNAi pathway [168]. Though this study showed that Dicer deficiency abolishes *Tsix*-dependent *Xist* repression, other studies have shown Dicer-deficient ES cells to have normal *Xist* expression patterns and X chromosome inactivation [169,170]. Likewise, abolishing Dicer function does not affect *Kcnq1ot1*-mediated gene silencing, which suggests that RNAi pathways are not involved [171]. Interestingly, in the Dicer-deficient embryos that exhibit normal *Xist* expression, the *Xist* promoter is hypomethylated – suggesting that Dicer affects promoter methylation indirectly by regulating Dnmt3a [170], the *de novo* methylase involved in primary imprint establishment [40]. These studies demonstrate an interesting intersection of various epigenetic mechanisms, which may all contribute to genomic imprinting.

363

Transposable Elements

As evidenced by the incorporation of *Peg12/Frat3* into the murine PWS/AS domain (Fig. 22.3), previously non-imprinted genes may acquire allele-specific expression when transposed next to other imprinted genes [60,61]. However, there is increasing evidence that transposition not only adds genes to existing imprinted domains, but may also establish genomic imprinting *de novo*. This is conceptually similar to position effect variegation in *Drosophila*, in which genes are silenced when placed adjacent to heterochromatin [reviewed in Ref. 125]. Interestingly, genomic imprinting in *Drosophila* is usually confined to heterochromatin [reviewed in Ref. 172], which is characterized by repetitive sequences and transposable elements [reviewed in Ref. 173]. Due to its ability to silence genes by juxtaposition and its correlation with genomic imprinting in *Drosophila*, heterochromatin has been hypothesized to establish imprinted domains when transposed to euchromatin [172].

The concept of transposable "controlling elements" originated with Barbara McClintock's seminal discovery of transposons in maize [174], and transposable elements are now believed to serve major regulatory functions in genomic imprinting. Retrotransposons, which are transposons that replicate via RNA intermediates, are especially abundant in eukaryotes; much of the eukaryotic 5-methylcytosine is targeted to these elements [reviewed in Ref. 31]. In plants, transposable elements are differentially methylated and maintained by DDM1 [175], a homolog of the yeast SWI2/SNF2 chromatin-remodeling complex [176]. Differential methylation of the imprinted *Arabidopsis FWA* promoter is targeted to a retrotransposon-derived SINE element, which in itself is sufficient for imprinted silencing of *FWA* [177]. This SINE element of *FWA* also corresponds to small interfering RNAs (siRNAs), which supports a role for RNAi in genomic imprinting in plants [175].

Retrotransposons compose about 50% of the human genome [reviewed in Ref. 178], which also contains several hundred coding sequences for reverse transcriptases that facilitate retrotransposon replication [31]. While mammalian imprinted domains tend to have an overall lower frequency of retrotransposon-derived SINEs [143,179], DMRs and ICRs themselves are highly enriched with repeats [35–37], which may be remnants of transposition. Indeed, the repeat-rich ICR of the *Xist/Tsix* locus (Fig. 22.5) bears a striking resemblance to the ERV family of endogenous retrovirus-like transposons [139]. Known paternal DMRs are also correlated with tandemly repeated sequences [42,43], which are likely to be vestiges of retrotransposons that are targeted for allele-specific methylation by Dnmt3L [49].

Retrotransposons have been shown to act as novel promoters [180], which may also be imprinted. In mice, a retrotransposon inserted upstream of the *agouti* gene drives ectopic expression in a parent-specific manner [181,182]. Several imprinted retrotransposons have been found within introns, which are sometimes called "microimprinted" domains; these may serve as maternal germline DMRs that give rise to paternal-specific transcripts on the opposite allele [183]. Retrotransposons also serve as oocyte-specific promoters, which produce abundant transcripts (over 10% of the mRNA pool) that persist into the early embryo [184]. It has been proposed that transcripts from oocyte-specific alternative promoters facilitate maternal DMR methylation [45]; given the abundance of retrotransposon-derived transcripts in the oocyte [184], it is very likely that these alternative promoters are of retrotransposon origin.

ON THE ORIGIN OF GENOMIC IMPRINTING

The underlying mechanisms of genomic imprinting – such as DNA methylation, chromatin modification, and ncRNAs – are well conserved across diverse taxa and may theoretically be traced to common origins. It is likely that diverging species independently recruited these mechanisms, and evolved modes of imprinting that fundamentally differ while retaining striking similarities. For example, entire chromosomes are often silenced or eliminated in insects [reviewed in Refs 25,185] whereas in mammals (with the exception of the X chromosome), imprinting generally targets discrete loci [14]. However, when transgenically introduced into *Drosophila* DNA, mammalian ICRs can silence neighboring genes [129,130], which demonstrates remarkable mechanistic conservation. The most highly conserved silencing mechanism (at least among higher eukaryotes) is methylation, which is used by both animals and plants to maintain differential gene expression [reviewed in Ref. 186]. In mammals, *de novo* methylases establish primary imprints [reviewed in Ref. 187]; in contrast, imprinted alleles in plants are selectively activated from a default silent state by DNA glycosylases [81]. Yet in both phyla, imprinted genes feature prominently in tissues that support embryonic development – such as the mammalian placenta and the seed endosperm of flowering plants [reviewed in Ref. 186]. Selective forces in these tissues form the basis of prominent evolutionary models of genomic imprinting.

Though ongoing research is steadily increasing our understanding of genomic imprinting, the phenomenon remains puzzling in many ways. Opinions differ as to how genomic imprinting originated, how parent-specific expression is selected, why imprinting affects only some genes, and why only certain taxa exhibit genomic imprinting. As expected, the field is rife with theories and models – the most prominent of which consider the phylogenetic distribution of genomic imprinting in vertebrates. The *Igf2* gene is of particular interest because it is well conserved among vertebrates but not universally imprinted [reviewed in Refs 188,189]; thus, studies have examined *Igf2* in various vertebrate classes [2,19,143, 190–194] in an effort to pinpoint the evolutionary origin of genomic imprinting. Only mammals in the subclass Theria, which includes metatherians (marsupials) and eutherians (mice, humans, and most other contemporary mammals), are known to genomically imprint; therefore, established theories are largely based on both physiological and genomic differences between therian and prototherian (monotreme) mammals [reviewed in Ref. 195].

According to philosophical tradition, scientific theories may be classified as *organismic*, *mechanistic*, or *reductionist* [196]. Organismic theories center on interactions between individual organisms; in terms of genomic imprinting, these interactions may favor allele-specific expression, rather than biallelic or stochastic (random) monoallelic expression, for certain genes. These models consider the physiological functions of imprinted genes, and attribute their parent-specific expression to genetic conflicts or co-adaptive interactions within populations. In contrast, mechanistic models focus mainly on the fundamental processes of genomic imprinting, such as silencing mechanisms that target certain genomic elements. According to the established mechanistic theories, genomic imprinting evolved from host defense mechanisms against invasive genetic elements, which reflect increasing genome complexity. Finally, reductionist theories dissect complex systems into interactions between individual parts; in the case of genomic imprinting, this may focus on the interaction between two alleles. Reductionist theories of genomic imprinting examine the possible benefits of monoallelic gene expression in diploid organisms. Various organismic, mechanistic, and reductionist theories have been presented as opposing viewpoints; however, they address different levels of hierarchy in imprinted gene regulation, and may actually form complementary models of genomic imprinting.

Organismic Models of Genomic Imprinting

The majority of known imprinted genes have established roles in growth, metabolism, or behavior during mammalian development [reviewed in Refs 188,189]. Because resources must be carefully allocated between the mother, offspring, and siblings during this time, it has been proposed that competition for resources imposes selective pressure on these genes. Prominent evolutionary models of genomic imprinting are based on these organismic interactions between related individuals. The *parent–offspring conflict theory* was originally formulated by Trivers [197], before imprinted genes were characterized, and proposed that genes (such as those governing altruistic behavior and maternal instinct) would be positively or negatively selected to optimize resource allocation during maternal care. A variation on this concept, known as the *kinship theory*, includes adaptations driven by additional familial or social interactions – such as in litters or social insects [24]. Thus, these selective pressures optimize survival not just for individuals, but also for populations. After the discovery of genomic imprinting, genes were increasingly viewed as modular alleles as opposed to singular units; thus, these concepts were adopted by Haig to explain parent-specific monoallelic expression in angiosperm (flowering) plants [198] and mammals [199].

Therian mammals are distinguished from the extant egg-laying prototherians by viviparity (live birth) and the presence of a highly-developed placenta, which is the site of nutrient transfer between mother and fetus [reviewed in Ref. 200]. Therefore, many organismic models of genomic imprinting correlate the known functions of imprinted genes with prolonged gestation and postnatal care. The parent–offspring conflict theory proposes that growth-promoting genes, such as *Igf2*, favor paternal expression – particularly in the placenta – to maximize resource transfer to the embryo, which represents the paternal genetic contribution; in contrast, maternal expression of growth-suppressing genes (such as *H19* and *Igf2r*, which both negatively regulate *Igf2*) would optimize maternal health [reviewed in Ref. 188]. Imprinted genes are not only common in the placenta but also in the brain [reviewed in Ref. 201], where they may contribute to cognitive processes [202], postnatal adaptation to feeding and novel environments [203,204], and other neurological processes that may also be relevant to parent-offspring interactions.

Because imprinted genes were first identified based on developmental phenotypes, this may have created a sampling bias for genes involved in embryonic or postnatal development; thus, it is not surprising that most imprinted genes conform to the parent-offspring conflict theory. However, not all imprinted expression patterns are so easily predicted by this theory. *Mash2*, a gene required for trophoblast development in mice [205], is a notable example [reviewed in Ref. 206]. The trophoblast is one of the more critical placental tissues for embryonic growth, as it promotes nutrient transfer to the embryo [207]. The conflict theory predicts a paternal expression pattern, yet *Mash2* is biallelic in the early embryo then maternally expressed by 8.5 days past coitum [208]. Complex trophoblast-mediated processes involving placental hormones may justify this paradoxical expression pattern [reviewed in Ref. 206]. Imprinted genes, such as *Rasgrf1*, may also indirectly control growth by regulating non-imprinted growth factors such as insulin-like growth factor I (*Igf1*) [209]. Other genes may display complex imprinting patterns that manifest more strongly in adulthood, after maternal contribution has ceased [210]; these expression patterns may not be obviously consistent with the simple parent-offspring conflict theory, but the more intricate aspects of the kinship theory may apply.

In social animals, survival is not restricted to maternal-offspring conflict, but also involves interactions between related individuals and other members of society [reviewed in Refs 206,211]. Hence, the more inclusive kinship theory may apply to imprinted genes without obvious relevance to simple maternal-offspring conflict. For animals that group together for warmth, such as emperor penguins and species with large litters, this might include genes that govern "huddling" behavior in addition to metabolic processes, such as thermogenesis [reviewed in Ref. 212]. Furthermore, parental investment in higher mammals is not limited to perinatal development, but includes courtship and mating; these complex interactions may explain why some genes that affect sexual behavior may also be imprinted. One such gene is the paternally expressed *Peg3*, which not only regulates suckling behavior in mouse pups but also olfactory-dependent maternal instincts (such as licking and grooming of pups) and male sexual behavior [213–216]. Because *Peg3* is involved in complex co-adaptive interactions that are beyond the scope of maternal–offspring conflict, the kinship theory has been challenged with the *co-adaptation theory* [217,218]. The co-adaptation theory has also been proposed for unexpected patterns of imprinted expression, such as the apparent preponderance of maternally-expressed genes [219]. However, since the kinship theory is not limited to maternal–offspring interactions but involves all interactions between related individuals [206], the co-adaptation theory may not be that dissimilar to the kinship theory.

Theories based on organismic interactions also predict genomic imprinting to occur in species that make significant maternal contributions to their young during gestation and postnatal development, such as the placental (therian) mammals. Indeed, genomic imprinting is characteristic of viviparous therian mammals [reviewed in Ref. 200] and apparently lacking in egg-laying (oviparous) animals, including monotreme mammals [191] and birds [19]. However, viviparity is also characteristic of certain placental fish and reptiles [reviewed in Ref. 220], which so far have not been associated with genomic imprinting [reviewed in Refs 188,189]. Comparative analyses in teleost fishes (which include both oviparous and viviparous species) have linked placental development with positive selection at the *Igf2* locus, which supports the notion that genomic imprinting is coincident with placental evolution [192]. However, the apparent lack of genomic imprinting in monotreme mammals – which have primordial placentas yet lay eggs – suggests that a stronger link may exist between genomic imprinting and viviparity [reviewed in Ref. 200].

Oviparity has also been linked to primordial imprinting mechanisms. Orthologs to imprinted mammalian genes (complete with CpG islands) occur in many egg-laying fish, such as zebrafish [221], puffer fish [222,223], and goldfish [194]; however, conservation is rather poor in terms of synteny (chromosomal position) [190,222] and differential methylation [194,223]. The biallelic methylation patterns in fish orthologs may be explained in part by Dnmt3L, the cofactor in primary imprint establishment in mouse [41]. Dnmt3L is conserved among therian mammals [224] but lacking in animals that do not genomically imprint, such as monotremes [225], fish, and birds [224]. Interestingly, though zebrafish lack Dnmt3L, they can differentially methylate exogenous transgenes according to parent of origin [226]. Furthermore, the CpG island near the goldfish *Igf2* gene is hypermethylated in goldfish sperm but not in eggs; this mirrors the methylation patterns of mammalian orthologs, though differential methylation is not maintained zygotically in goldfish [194]. Together with the reduced viability of uniparental zebrafish [227] and goldfish [228], both of which are oviparous, these data suggest that a primordial form of genomic imprinting exists in fish – and is not strongly correlated with viviparity or placentation. In addition to fish, other invertebrate genomes (such as chicken and frog) contain orthologous arrays with varying degrees of synteny with mammalian imprinted loci; this suggests that primordial imprinting mechanisms existed in a common vertebrate ancestor prior to mammalian divergence [190].

Mechanistic Models of Genomic Imprinting

It is important to note that the organismic conflict-based theories seek to explain the parent-specific expression patterns of imprinted genes, as well as to justify their natural selection during evolution; they do not explain how or why the mechanisms of genomic imprinting arose. This aspect of the kinship theory is sometimes interpreted as a weakness. However, the kinship theory assumes preexisting mechanisms and proposes selective pressures to impart allele specificity on these mechanisms; it does not actually attempt to explain their origins [206]. Mechanistic models of genomic imprinting address these fundamental processes.

One such model proposes that genomic imprinting evolved from a primitive host-defense mechanism [229,230]. In many species, this manifests as methylation and silencing of foreign nucleic acids, endogenous transposed DNAs, and repetitive elements [229]. Unlike conflict-based theories, the host defense hypothesis does not consider the biological functions of imprinted genes or species-specific reproductive features. Rather, it correlates imprinting with distinguishing genomic features and mammalian divergence [reviewed in Ref. 195].

Molecular and genomic evidence for this hypothesis again centers on Dnmt3L. This component of the *de novo* methylation machinery not only establishes germline-specific imprints in mammals [41], but also silences retrotransposons and repetitive DNA sequences [49]. Cross-species genome comparisons also support the link between retrotransposon silencing and imprinting, as Dnmt3L is present in therian mammals [224] but lacking in monotremes, fish, and birds [224,225]. Interestingly, most *de novo* methyltransferases are highly conserved between mice and humans (at least 80% identical), yet the Dnmt3L protein sequence is highly divergent (<60% identical); this rapid rate of evolution is consistent with a role in host defense [231].

Recent comparative genome analyses also support the notion that genomic imprinting is an incarnation of a primordial host defense mechanism against retrotransposons. *Paternally expressed 10 (Peg10)* belongs to the sushi class of retrotransposon-derived genes that have lost

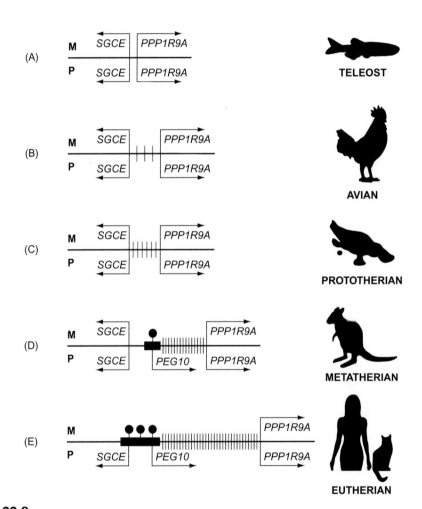

FIGURE 22.8

The *Peg10* locus (not drawn to scale), which correlates genomic imprinting with retrotransposition and genome expansion during vertebrate evolution. (A) In teleost fish, *SGCE* lies adjacent to the *PPP1R9A* gene. (B,C) This intergenic region is expanded in avian (bird) and prototherian (monotreme) genomes by retrotransposon-derived short and long interspersed elements (SINES and LINES), indicated by the short gray vertical lines. (D,E) In metatherian (marsupial) and eutherian mammals, the acquisition of the *Peg10* gene is coincident with imprinted paternal expression (solid arrow below the line) and maternal methylation (black circle). This novel imprinted gene is associated with increased long terminal repeat (LTR)-like sequences (short black lines), which are also indicative of retrotransposition. (E) In eutherians (humans and most contemporary mammals), *SGCE* is also imprinted; this is coincident with further genome expansion. [Adapted from Ref. 234].

the ability to transpose in mammals, which presents a link between genomic imprinting and retrotransposon silencing [232,233]. *Peg10* and the neighboring sarcoglycan epsilon (*Sgce*) gene are imprinted with paternal-specific expression in eutherians (mice and humans); however, the *Peg10* gene is conspicuously absent in the non-imprinted homologous region in platypus (Fig. 22.8), which correlates its retrotransposition into the therian genome with genomic imprinting [234]. Moreover, *Peg10* is imprinted in a marsupial (the tammar wallaby) but *Scge* is not (Fig. 22.8). This suggests that *Peg10* acquired differential methylation and primordial imprinting upon retrotransposition [234]. Compared to imprinted domains in therians, orthologous monotreme regions also have fewer repeat elements; thus, genomic imprinting coincides evolutionarily with retrotransposition and the accumulation of repetitive sequences in therian genomes [235].

The genomic evidence strongly correlates imprinting with increased retrotransposition during therian divergence. This implicates a host-defense origin for genomic imprinting mechanisms, yet does not address why certain alleles are preferentially expressed based on parent of origin. This may be better resolved with organismic models, such as the kinship theory; however, the host defense hypothesis may also provide underlying mechanisms for allele-specific silencing. The primary imprint in the male germline requires repeat elements in paternal DMRs [42,43] and is coincident with retrotransposon silencing by Dnmt3L [49]; thus, it is possible that Dnmt3L expression in the male germline targets the retrotransposon-like characteristics of DMRs. Conversely, maternal germline imprints appear to require transcription from oocyte-specific alternative promoters [45], which may also be of retrotransposon origin. This is supported by the regulatory role of retrotransposons in oocyte-specific transcription [184]. Since Dnmt3L-mediated retrotransposon silencing appears to be limited to the male germline [50], this may explain the abundance of retrotransposon-derived transcripts in the oocyte. Nonetheless, it remains unanswered why genes should be monoallelically expressed at all – whether they are silenced randomly or in a parent-dependent fashion. This question is addressed with reductionist theories of genomic imprinting.

Reductionist Models of Genomic Imprinting

Considering the purported genetic advantages of diploidy [1], it is perplexing for autosomal genes to be hemizygous – or expressed from just one allele – in diploid organisms. Reductionist theories of genomic imprinting address this paradoxical nature of genomic imprinting. These hypotheses not only apply to imprinted genes, which number around 125 according to several online databases (Table 22.2), but also to autosomal genes that are subject to stochastic (random) monoallelic expression. This includes genes in the immune and odorant systems that are randomly silenced via allelic exclusion [reviewed in Refs 96,236,237], as opposed to genes on the X chromosome that are silenced to achieve dosage compensation. Random monoallelic expression is a widespread phenomenon; according to a recent genome survey, 5–10% of autosomal genes – or well over 1000 – may be randomly expressed from only one allele at any given time [238]. While genomic imprints are set in the germline, random monoallelic expression is a zygotic process; however, similar mechanisms are used in either imprinted or random allelic silencing, which suggests that the two processes are related [reviewed in Ref. 96]. The prevalence of monoallelic expression – imprinted or random – implies that it may serve some sort of evolutionary purpose.

It is generally agreed that hemizygosity increases the evolvability of a particular locus, and hence the adaptability of the overall population [217,239,240]. Because diploidy may mask both deleterious and beneficial mutations, functional haploidy may quickly eliminate undesirable recessive traits while simultaneously promoting beneficial mutations. In complex multicellular organisms, monoallelic expression of multiple loci may combinatorially increase phenotypic variability and facilitate adaptive responses.

TABLE 22.2 Select Online Databases of Imprinted Genes

Database	Taxonomic Group(s)
Brain Imprinted Source Tables | Cardiff University http://www.bgg.cf.ac.uk/imprinted_tables	Mouse (brain)
Catalog of Parent of Origin Effects | Otago University http://igc.otago.ac.nz	Human, mouse, rat, cow, pig, sheep, marsupial, monotreme
Geneimprint | Duke University http://www.geneimprint.com	Human, mouse, rat
Genomic Imprinting | MRC Harwell http://www.har.mrc.ac.uk/research/genomic_imprinting	Mouse
WAMIDEX | King's College London https://atlas.genetics.kcl.ac.uk/atlas.php	Mouse

This is plainly illustrated by the broad range of highly specific receptors generated via allelic exclusion in the immune and olfactory systems [reviewed in Ref. 241]. Monoallelic expression is even characteristic of *Drosophila* odorant receptor genes, which are unrelated to vertebrate olfactory genes [reviewed in Ref. 242]. Thus, monoallelic expression is fundamental to adaptive processes, and may in fact be evolutionarily advantageous.

Because hemizygosity is so widespread, one might speculate that it reflects the economic costs associated with maintaining diploid genomes. Heterozygosity may be genetically advantageous; however, for a multicellular organism, the benefits of diploidy must outweigh the costs of replicating or expressing two copies of a locus in every single cell. An analogy can be made using one of the simplest model organisms, the bacterium *Escherichia coli*. It is a widely known fact that *E. coli* cultures only retain plasmids if they confer some selective advantage, such as antibiotic resistance; if there is no selective pressure, then cells without plasmids are at a growth advantage because they do not expend resources replicating extraneous plasmid DNA [243]. More primitive forms of genomic imprinting, such as paternal genome elimination in mealybugs and fungus gnats [reviewed in Refs 25,185], might illustrate this extraneous nature of duplicate genomes. In higher eukaryotes, the need to repress superfluous DNA is demonstrated by the targeted methylation and silencing of duplicated genes [244,245]. Gene silencing is also a fundamental property of complex multicellular organisms [95], and it is estimated that less than 10% of mammalian genes are transcribed at any given time in differentiated cell lineages [148]. Thus, it should not come as a surprise that monoallelic silencing is a widespread phenomenon. In the most basic model of genomic imprinting, monoallelic expression may simply represent an economical means of maintaining a diploid genome. Whatever the rationale, monoallelic expression may be subject to additional selective pressures that determine random or allele-specific silence.

One common limitation of reductionism is that it may not adequately accommodate complex systems or concepts [246]. Indeed, neither the mechanistic nor reductionist models presented here can fully account for the parent-specific expression that is central to genomic imprinting, which is best explained by the organismic kinship theory. Nonetheless, the mechanistic theories provide insight into the origin of genomic imprinting processes, and, as with many reductionist theories, those presented here may serve to facilitate the understanding of this complex phenomenon. Most importantly, none of these theories are mutually exclusive, and may serve complementary functions in deciphering the complex phenomenon of genomic imprinting.

CONCLUSION

Genomic imprinting was once perceived as a bizarre characteristic of plants, insects, and a handful of mammalian genes; it has since become the focus of intensive research, which has produced numerous implications for development, disease, and evolution. Genomic imprinting may also explain the long-standing mystery of reciprocal interspecies hybrids, which produce progeny with dramatically different phenotypes. For example, crossing a male tiger with a female lion produces a tigon, which is about the same size as either parent; however, the reciprocal cross produces a liger, which is known for its great size [247]. Another notable example is the mule, a hybrid of a male donkey and a female horse. It has been known for millennia that the hinny, which is the lesser-known reciprocal hybrid, differs remarkably in appearance from the mule [248]. Genomic imprinting is now known to underlie the inequality of reciprocal hybrid crosses [reviewed in Ref. 249].

In just three decades, the number of known imprinted genes in mice and humans has grown from three to over 100. In the past, most imprinted genes were recognized based on measurable phenotypes or by association with previously known imprinted genes; however, newer genome-wide analyses may predict novel imprinted genes based on expression profiles or even DNA sequence characteristics. Conservative estimates suggest that up to 150 or so additional imprinted genes exist [250,251]. Recent genome-wide analyses have also revealed complex patterns of imprinting that manifest over multiple generations, and depend not just on parent of origin but also on other imprinted alleles [210]. Thus, more finely tuned approaches may identify additional candidates – particularly those with expression profiles that are more complicated than the traditional binary (on-or-off) definition of genomic imprinting.

References

1. Otto SP, Goldstein DB. Recombination and the evolution of diploidy. *Genetics* 1992;**131**:745–51.
2. Smits G, Mungall AJ, Griffiths-Jones S, Smith P, Beury D, Matthews L, et al. Conservation of the H19 noncoding RNA and H19-IGF2 imprinting mechanism in therians. *Nat Genet* 2008;**40**:971–6.
3. Amor DJ, Halliday J. A review of known imprinting syndromes and their association with assisted reproduction technologies. *Hum Reprod* 2008;**23**:2826–34.
4. Feinberg AP, Ohlsson R, Henikoff S. The epigenetic progenitor origin of human cancer. *Nat Rev Genet* 2006;**7**:21–33.
5. Sharman GB. Late DNA replication in the paternally derived X chromosome of female kangaroos. *Nature* 1971;**230**:231–2.
6. Takagi N, Sasaki M. Preferential inactivation of the paternally derived X chromosome in the extraembryonic membranes of the mouse. *Nature* 1975;**256**:640–2.
7. Barlow DP, Stoger R, Herrmann BG, Saito K, Schweifer N. The mouse insulin-like growth factor type-2 receptor is imprinted and closely linked to the Tme locus. *Nature* 1991;**349**:84–7.
8. Bartolomei MS, Zemel S, Tilghman SM. Parental imprinting of the mouse H19 gene. *Nature* 1991;**351**:153–5.
9. DeChiara TM, Robertson EJ, Efstratiadis A. Parental imprinting of the mouse insulin-like growth factor II gene. *Cell* 1991;**64**:849–59.
10. McGrath J, Solter D. Completion of mouse embryogenesis requires both the maternal and paternal genomes. *Cell* 1984;**37**:179–83.
11. Surani MA, Barton SC, Norris ML. Development of reconstituted mouse eggs suggests imprinting of the genome during gametogenesis. *Nature* 1984;**308**:548–50.
12. Mann JR, Lovell-Badge RH. Inviability of parthenogenones is determined by pronuclei, not egg cytoplasm. *Nature* 1984;**310**:66–7.
13. Barton SC, Surani MA, Norris ML. Role of paternal and maternal genomes in mouse development. *Nature* 1984;**311**:374–6.
14. Cattanach BM, Kirk M. Differential activity of maternally and paternally derived chromosome regions in mice. *Nature* 1985;**315**:496–8.
15. Zemel S, Bartolomei MS, Tilghman SM. Physical linkage of two mammalian imprinted genes, H19 and insulin-like growth factor 2. *Nat Genet* 1992;**2**:61–5.
16. Leighton PA, Saam JR, Ingram RS, Stewart CL, Tilghman SM. An enhancer deletion affects both H19 and Igf2 expression. *Genes Dev* 1995;**9**:2079–89.
17. Leighton PA, Ingram RS, Eggenschwiler J, Efstratiadis A, Tilghman SM. Disruption of imprinting caused by deletion of the H19 gene region in mice. *Nature* 1995;**375**:34–9.
18. Chao W, D'Amore PA. IGF2: epigenetic regulation and role in development and disease. *Cytokine Growth Factor Rev* 2008;**19**:111–20.

19. O'Neill MJ, Ingram RS, Vrana PB, Tilghman SM. Allelic expression of IGF2 in marsupials and birds. *Dev Genes Evol* 2000;**210**:18–20.

20. Crouse HV. The controlling element in sex chromosome behavior in sciara. *Genetics* 1960;**45**:1429–43.

21. Noujdin NI. An investigation of an unstable chromosome in *Drosophila melanogaster* and the mosaicism caused by it. *Zoologishcheski Zh* 1935;**14**:317–52.

22. Noujdin NI. Influence of the Y chromosome and of the homologous region of the X on mosaicism in *Drosophila*. *Nature* 1936;**137**:319–20.

23. Schrader FaH-SS. Haploidy in metazoa. *Q Rev Biol* 1931;**6**:411–38.

24. Trivers RL, Hare H. Haploidploidy and the evolution of the social insect. *Science* 1976;**191**:249–63.

25. Khosla S, Mendiratta G, Brahmachari V. Genomic imprinting in the mealybugs. *Cytogenet Genome Res* 2006;**113**:41–52.

26. Emerson RA. A fifth pair of factors, *Aa*, for aleurone color in maize, and its relation to the *Cc* and *Rr* pairs. *Cornell Univ Agric Esp Station Memoir* 1918;**16**:231–89.

27. Kempton JH. Inheritance of spotted aleurone color in hybrids of chinese maize. *Genetics* 1919;**4**:261–74.

28. Kermicle JL. Dependence of the R-mottled aleurone phenotype in maize on mode of sexual transmission. *Genetics* 1970;**66**:69–85.

29. Doerfler W. In pursuit of the first recognized epigenetic signal–DNA methylation: a 1976 to 2008 synopsis. *Epigenetics* 2008;**3**:125–33.

30. Illingworth RS, Bird AP. CpG islands–"a rough guide". *FEBS Lett* 2009;**583**:1713–20.

31. Yoder JA, Walsh CP, Bestor TH. Cytosine methylation and the ecology of intragenomic parasites. *Trends Genet* 1997;**13**:335–40.

32. Swain JL, Stewart TA, Leder P. Parental legacy determines methylation and expression of an autosomal transgene: a molecular mechanism for parental imprinting. *Cell* 1987;**50**:719–27.

33. Reik W, Collick A, Norris ML, Barton SC, Surani MA. Genomic imprinting determines methylation of parental alleles in transgenic mice. *Nature* 1987;**328**:248–51.

34. Lewis A, Reik W. How imprinting centres work. *Cytogenet Genome Res* 2006;**113**:81–9.

35. Reinhart B, Paoloni-Giacobino A, Chaillet JR. Specific differentially methylated domain sequences direct the maintenance of methylation at imprinted genes. *Mol Cell Biol* 2006;**26**:8347–56.

36. Paoloni-Giacobino A, D'Aiuto L, Cirio MC, Reinhart B, Chaillet JR. Conserved features of imprinted differentially methylated domains. *Gene* 2007;**399**:33–45.

37. Hutter B, Helms V, Paulsen M. Tandem repeats in the CpG islands of imprinted genes. *Genomics* 2006;**88**:323–32.

38. Hellmann-Blumberg U, Hintz MF, Gatewood JM, Schmid CW. Developmental differences in methylation of human Alu repeats. *Mol Cell Biol* 1993;**13**:4523–30.

39. Rubin CM, VandeVoort CA, Teplitz RL, Schmid CW. Alu repeated DNAs are differentially methylated in primate germ cells. *Nucleic Acids Res* 1994;**22**:5121–7.

40. Kaneda M, Okano M, Hata K, Sado T, Tsujimoto N, Li E, et al. Essential role for de novo DNA methyltransferase Dnmt3a in paternal and maternal imprinting. *Nature* 2004;**429**:900–3.

41. Bourc'his D, Xu GL, Lin CS, Bollman B, Bestor TH. Dnmt3L and the establishment of maternal genomic imprints. *Science* 2001;**294**:2536–9.

42. Lewis A, Mitsuya K, Constancia M, Reik W. Tandem repeat hypothesis in imprinting: deletion of a conserved direct repeat element upstream of H19 has no effect on imprinting in the Igf2-H19 region. *Mol Cell Biol* 2004;**24**:5650–6.

43. Pearsall RS, Plass C, Romano MA, Garrick MD, Shibata H, Hayashizaki Y, et al. A direct repeat sequence at the Rasgrf1 locus and imprinted expression. *Genomics* 1999;**55**:194–201.

44. Plagge A, Kelsey G. Imprinting the Gnas locus. *Cytogenet Genome Res* 2006;**113**:178–87.

45. Chotalia M, Smallwood SA, Ruf N, Dawson C, Lucifero D, Frontera M, et al. Transcription is required for establishment of germline methylation marks at imprinted genes. *Genes Dev* 2009;**23**:105–17.

46. Shovlin TC, Bourc'his D, La Salle S, O'Doherty A, Trasler JM, Bestor TH, et al. Sex-specific promoters regulate Dnmt3L expression in mouse germ cells. *Hum Reprod* 2007;**22**:457–67.

47. Li JY, Lees-Murdock DJ, Xu GL, Walsh CP. Timing of establishment of paternal methylation imprints in the mouse. *Genomics* 2004;**84**:952–60.

48. Bourc'his D, Proudhon C. Sexual dimorphism in parental imprint ontogeny and contribution to embryonic development. *Mol Cell Endocrinol* 2008;**282**:87–94.

49. Bourc'his D, Bestor TH. Meiotic catastrophe and retrotransposon reactivation in male germ cells lacking Dnmt3L. *Nature* 2004;**431**:96–9.

50. Lucifero D, La Salle S, Bourc'his D, Martel J, Bestor TH, Trasler JM. Coordinate regulation of DNA methyltransferase expression during oogenesis. *BMC Dev Biol* 2007;**7**:36.

51. Reik W, Dean W, Walter J. Epigenetic reprogramming in mammalian development. *Science* 2001;**293**:1089–93.

52. Ohki I, Shimotake N, Fujita N, Jee J, Ikegami T, Nakao M, et al. Solution structure of the methyl-CpG binding domain of human MBD1 in complex with methylated DNA. *Cell* 2001;**105**:487–97.

53. Reese KJ, Lin S, Verona RI, Schultz RM, Bartolomei MS. Maintenance of paternal methylation and repression of the imprinted H19 gene requires MBD3. *PLoS Genet* 2007;**3**:e137.

54. Hirasawa R, Chiba H, Kaneda M, Tajima S, Li E, Jaenisch R, et al. Maternal and zygotic Dnmt1 are necessary and sufficient for the maintenance of DNA methylation imprints during preimplantation development. *Genes Dev* 2008;**22**:1607–16.

55. Nakamura T, Arai Y, Umehara H, Masuhara M, Kimura T, Taniguchi H, et al. PGC7/Stella protects against DNA demethylation in early embryogenesis. *Nat Cell Biol* 2007;**9**:64–71.

56. Li X, Ito M, Zhou F, Youngson N, Zuo X, Leder P, et al. A maternal-zygotic effect gene, Zfp57, maintains both maternal and paternal imprints. *Dev Cell* 2008;**15**:547–57.

57. Horsthemke B, Wagstaff J. Mechanisms of imprinting of the Prader–Willi/Angelman region. *Am J Med Genet A* 2008;**146A**:2041–52.

58. Kayashima T, Yamasaki K, Joh K, Yamada T, Ohta T, Yoshiura K, et al. Atp10a, the mouse ortholog of the human imprinted ATP10A gene, escapes genomic imprinting. *Genomics* 2003;**81**:644–7.

59. Dubose AJ, Johnstone KA, Smith EY, Hallett RA, Resnick JL. Atp10a, a gene adjacent to the PWS/AS gene cluster, is not imprinted in mouse and is insensitive to the PWS-IC. *Neurogenetics* 2009.

60. Kobayashi S, Kohda T, Ichikawa H, Ogura A, Ohki M, Kaneko-Ishino T, et al. Paternal expression of a novel imprinted gene, Peg12/Frat3, in the mouse 7C region homologous to the Prader–Willi syndrome region. *Biochem Biophys Res Commun* 2002;**290**:403–8.

61. Chai JH, Locke DP, Ohta T, Greally JM, Nicholls RD. Retrotransposed genes such as Frat3 in the mouse Chromosome 7C Prader–Willi syndrome region acquire the imprinted status of their insertion site. *Mamm Genome* 2001;**12**:813–21.

62. Nicholls RD, Knepper JL. Genome organization, function, and imprinting in Prader–Willi and Angelman syndromes. *Annu Rev Genomics Hum Genet* 2001;**2**:153–75.

63. Kantor B, Shemer R, Razin A. The Prader–Willi/Angelman imprinted domain and its control center. *Cytogenet Genome Res* 2006;**113**:300–5.

64. Mapendano CK, Kishino T, Miyazaki K, Kondo S, Yoshiura K, Hishikawa Y, et al. Expression of the Snurf-Snrpn IC transcript in the oocyte and its putative role in the imprinting establishment of the mouse 7C imprinting domain. *J Hum Genet* 2006;**51**:236–43.

65. Tremblay KD, Duran KL, Bartolomei MS. A 5′ 2-kilobase-pair region of the imprinted mouse H19 gene exhibits exclusive paternal methylation throughout development. *Mol Cell Biol* 1997;**17**:4322–9.

66. Jinno Y, Sengoku K, Nakao M, Tamate K, Miyamoto T, Matsuzaka T, et al. Mouse/human sequence divergence in a region with a paternal-specific methylation imprint at the human H19 locus. *Hum Mol Genet* 1996;**5**:1155–61.

67. Thorvaldsen JL, Duran KL, Bartolomei MS. Deletion of the H19 differentially methylated domain results in loss of imprinted expression of H19 and Igf2. *Genes Dev* 1998;**12**:3693–702.

68. Engel N, West AG, Felsenfeld G, Bartolomei MS. Antagonism between DNA hypermethylation and enhancer-blocking activity at the H19 DMD is uncovered by CpG mutations. *Nat Genet* 2004;**36**:883–8.

69. Scott J, Cowell J, Robertson ME, Priestley LM, Wadey R, Hopkins B, et al. Insulin-like growth factor-II gene expression in Wilms' tumour and embryonic tissues. *Nature* 1985;**317**:260–2.

70. Ogawa O, Eccles MR, Szeto J, McNoe LA, Yun K, Maw MA, et al. Relaxation of insulin-like growth factor II gene imprinting implicated in Wilms' tumour. *Nature* 1993;**362**:749–51.

71. Weksberg R, Shen DR, Fei YL, Song QL, Squire J. Disruption of insulin-like growth factor 2 imprinting in Beckwith-Wiedemann syndrome. *Nat Genet* 1993;**5**:143–50.

72. Sparago A, Cerrato F, Vernucci M, Ferrero GB, Silengo MC, Riccio A. Microdeletions in the human H19 DMR result in loss of IGF2 imprinting and Beckwith-Wiedemann syndrome. *Nat Genet* 2004;**36**:958–60.

73. Prawitt D, Enklaar T, Gartner-Rupprecht B, Spangenberg C, Lausch E, Reutzel D, et al. Microdeletion and IGF2 loss of imprinting in a cascade causing Beckwith-Wiedemann syndrome with Wilms' tumor. *Nat Genet* 2005;**37**:786–7.

74. Prawitt D, Enklaar T, Gartner-Rupprecht B, Spangenberg C, Oswald M, Lausch E, et al. Microdeletion of target sites for insulator protein CTCF in a chromosome 11p15 imprinting center in Beckwith-Wiedemann syndrome and Wilms' tumor. *Proc Natl Acad Sci USA* 2005;**102**:4085–90.

75. Ravenel JD, Broman KW, Perlman EJ, Niemitz EL, Jayawardena TM, Bell DW, et al. Loss of imprinting of insulin-like growth factor-II (IGF2) gene in distinguishing specific biologic subtypes of Wilms tumor. *J Natl Cancer Inst* 2001;**93**:1698–703.

76. Cui H, Niemitz EL, Ravenel JD, Onyango P, Brandenburg SA, Lobanenkov VV, et al. Loss of imprinting of insulin-like growth factor-II in Wilms' tumor commonly involves altered methylation but not mutations of CTCF or its binding site. *Cancer Res* 2001;**61**:4947–50.

77. Yun K. Genomic imprinting and carcinogenesis. *Histol Histopathol* 1998;**13**:425–35.

78. Ohlsson R. Loss of IGF2 imprinting: mechanisms and consequences. *Novartis Found Symp* 2004;**262**:108–21. discussion 121-104, 265-108.

79. Vielle-Calzada JP, Thomas J, Spillane C, Coluccio A, Hoeppner MA, Grossniklaus U. Maintenance of genomic imprinting at the *Arabidopsis medea* locus requires zygotic DDM1 activity. *Genes Dev* 1999;**13**:2971–82.

80. Jullien PE, Kinoshita T, Ohad N, Berger F. Maintenance of DNA methylation during the *Arabidopsis* life cycle is essential for parental imprinting. *Plant Cell* 2006;**18**:1360–72.

81. Kinoshita T, Miura A, Choi Y, Kinoshita Y, Cao X, Jacobsen SE, et al. One-way control of FWA imprinting in Arabidopsis endosperm by DNA methylation. *Science* 2004;**303**:521–3.

82. Bongiorni S, Cintio O, Prantera G. The relationship between DNA methylation and chromosome imprinting in the coccid *Planococcus citri. Genetics* 1999;**151**:1471–8.

83. Golic KG, Golic MM, Pimpinelli S. Imprinted control of gene activity in *Drosophila. Curr Biol* 1998;**8**:1273–6.

84. Lloyd VK, Sinclair DA, Grigliatti TA. Genomic imprinting and position-effect variegation in *Drosophila melanogaster. Genetics* 1999;**151**:1503–16.

85. Patel CV, Gopinathan KP. Determination of trace amounts of 5-methylcytosine in DNA by reverse-phase high-performance liquid chromatography. *Anal Biochem* 1987;**164**:164–9.

86. Urieli-Shoval S, Gruenbaum Y, Sedat J, Razin A. The absence of detectable methylated bases in *Drosophila melanogaster* DNA. *FEBS Lett* 1982;**146**:148–52.

87. Lyko F, Ramsahoye BH, Jaenisch R. DNA methylation in *Drosophila melanogaster. Nature* 2000;**408**:538–40.

88. Ramsahoye BH, Biniszkiewicz D, Lyko F, Clark V, Bird AP, Jaenisch R. Non-CpG methylation is prevalent in embryonic stem cells and may be mediated by DNA methyltransferase 3a. *Proc Natl Acad Sci USA* 2000;**97**:5237–42.

89. Mager J, Montgomery ND, de Villena FP, Magnuson T. Genome imprinting regulated by the mouse Polycomb group protein Eed. *Nat Genet* 2003;**33**:502–7.

90. Huh JH, Bauer MJ, Hsieh TF, Fischer R. Endosperm gene imprinting and seed development. *Curr Opin Genet Dev* 2007;**17**:480–5.

91. Bean CJ, Schaner CE, Kelly WG. Meiotic pairing and imprinted X chromatin assembly in *Caenorhabditis elegans. Nat Genet* 2004;**36**:100–5.

92. Sha K, Fire A. Imprinting capacity of gamete lineages in *Caenorhabditis elegans. Genetics* 2005;**170**:1633–52.

93. Hodgkin J. Epigenetics and the maintenance of gene activity states in *Caenorhabditis elegans. Dev Genet* 1994;**15**:471–7.

94. Simpson VJ, Johnson TE, Hammen RF. *Caenorhabditis elegans* DNA does not contain 5-methylcytosine at any time during development or aging. *Nucleic Acids Res* 1986;**14**:6711–19.

95. Bird AP. Gene number, noise reduction and biological complexity. *Trends Genet* 1995;**11**:94–100.

96. Goldmit M, Bergman Y. Monoallelic gene expression: a repertoire of recurrent themes. *Immunol Rev* 2004;**200**:197–214.

97. Shemer R, Hershko AY, Perk J, Mostoslavsky R, Tsuberi B, Cedar H, et al. The imprinting box of the Prader–Willi/Angelman syndrome domain. *Nat Genet* 2000;**26**:440–3.

98. Reik W, Brown KW, Schneid H, Le Bouc Y, Bickmore W, Maher ER. Imprinting mutations in the Beckwith-Wiedemann syndrome suggested by altered imprinting pattern in the IGF2-H19 domain. *Hum Mol Genet* 1995;**4**:2379–85.

99. Gribnau J, Hochedlinger K, Hata K, Li E, Jaenisch R. Asynchronous replication timing of imprinted loci is independent of DNA methylation, but consistent with differential subnuclear localization. *Genes Dev* 2003;**17**:759–73.

100. Cerrato F, Dean W, Davies K, Kagotani K, Mitsuya K, Okumura K, et al. Paternal imprints can be established on the maternal Igf2-H19 locus without altering replication timing of DNA. *Hum Mol Genet* 2003;**12**:3123–32.

101. Simon I, Tenzen T, Reubinoff BE, Hillman D, McCarrey JR, Cedar H. Asynchronous replication of imprinted genes is established in the gametes and maintained during development. *Nature* 1999;**401**:929–32.

102. Feil R, Khosla S. Genomic imprinting in mammals: an interplay between chromatin and DNA methylation? *Trends Genet* 1999;**15**:431–5.

103. Sha K. A mechanistic view of genomic imprinting. *Annu Rev Genomics Hum Genet* 2008;**9**:197–216.

104. Plath K, Fang J, Mlynarczyk-Evans SK, Cao R, Worringer KA, Wang H, et al. Role of histone H3 lysine 27 methylation in X inactivation. *Science* 2003;**300**:131–5.

105. Cedar H, Bergman Y. Linking DNA methylation and histone modification: patterns and paradigms. *Nat Rev Genet* 2009;**10**:295–304.

106. Nan X, Ng HH, Johnson CA, Laherty CD, Turner BM, Eisenman RN, et al. Transcriptional repression by the methyl-CpG-binding protein MeCP2 involves a histone deacetylase complex. *Nature* 1998;**393**:386–9.

107. Jones PL, Veenstra GJ, Wade PA, Vermaak D, Kass SU, Landsberger N, et al. Methylated DNA and MeCP2 recruit histone deacetylase to repress transcription. *Nat Genet* 1998;**19**:187–91.

108. Sims RJ 3rd, Nishioka K, Reinberg D. Histone lysine methylation: a signature for chromatin function. *Trends Genet* 2003;**19**:629–39.

109. Tamaru H, Selker EU. A histone H3 methyltransferase controls DNA methylation in *Neurospora crassa*. *Nature* 2001;**414**:277–83.

110. Jackson JP, Lindroth AM, Cao X, Jacobsen SE. Control of CpNpG DNA methylation by the KRYPTONITE histone H3 methyltransferase. *Nature* 2002;**416**:556–60.

111. Lehnertz B, Ueda Y, Derijck AA, Braunschweig U, Perez-Burgos L, Kubicek S, et al. Suv39h-mediated histone H3 lysine 9 methylation directs DNA methylation to major satellite repeats at pericentric heterochromatin. *Curr Biol* 2003;**13**:1192–200.

112. Xin Z, Allis CD, Wagstaff J. Parent-specific complementary patterns of histone H3 lysine 9 and H3 lysine 4 methylation at the Prader–Willi syndrome imprinting center. *Am J Hum Genet* 2001;**69**:1389–94.

113. Xin Z, Tachibana M, Guggiari M, Heard E, Shinkai Y, Wagstaff J. Role of histone methyltransferase G9a in CpG methylation of the Prader–Willi syndrome imprinting center. *J Biol Chem* 2003;**278**:14996–5000.

114. Ooi SK, Qiu C, Bernstein E, Li K, Jia D, Yang Z, et al. DNMT3L connects unmethylated lysine 4 of histone H3 to de novo methylation of DNA. *Nature* 2007;**448**:714–17.

115. Ciccone DN, Su H, Hevi S, Gay F, Lei H, Bajko J, et al. KDM1B is a histone H3K4 demethylase required to establish maternal genomic imprints. *Nature* 2009;**461**:415–18.

116. Obata Y, Kono T. Maternal primary imprinting is established at a specific time for each gene throughout oocyte growth. *J Biol Chem* 2002;**277**:5285–9.

117. Delaval K, Govin J, Cerqueira F, Rousseaux S, Khochbin S, Feil R. Differential histone modifications mark mouse imprinting control regions during spermatogenesis. *EMBO J* 2007;**26**:720–9.

118. Schwartz YB, Pirrotta V. Polycomb silencing mechanisms and the management of genomic programmes. *Nat Rev Genet* 2007;**8**:9–22.

119. Joanis V, Lloyd VK. Genomic imprinting in *Drosophila* is maintained by the products of Suppressor of variegation and Trithorax group, but not Polycomb group, genes. *Mol Genet Genomics* 2002;**268**:103–12.

120. Erhardt S, Lyko F, Ainscough JF, Surani MA, Paro R. Polycomb-group proteins are involved in silencing processes caused by a transgenic element from the murine imprinted H19/Igf2 region in *Drosophila*. *Dev Genes Evol* 2003;**213**:336–44.

121. Kalantry S, Magnuson T. The Polycomb group protein EED is dispensable for the initiation of random X-chromosome inactivation. *PLoS Genet* 2006;**2**:e66.

122. Umlauf D, Goto Y, Cao R, Cerqueira F, Wagschal A, Zhang Y, et al. Imprinting along the Kcnq1 domain on mouse chromosome 7 involves repressive histone methylation and recruitment of Polycomb group complexes. *Nat Genet* 2004;**36**:1296–300.

123. Yamasaki-Ishizaki Y, Kayashima T, Mapendano CK, Soejima H, Ohta T, Masuzaki H, et al. Role of DNA methylation and histone H3 lysine 27 methylation in tissue-specific imprinting of mouse Grb10. *Mol Cell Biol* 2007;**27**:732–42.

124. Kuzmin A, Han Z, Golding MC, Mann MR, Latham KE, Varmuza S. The PcG gene Sfmbt2 is paternally expressed in extraembryonic tissues. *Gene Expr Patterns* 2008;**8**:107–16.

125. Weiler KS, Wakimoto BT. Heterochromatin and gene expression in *Drosophila*. *Annu Rev Genet* 1995;**29**:577–605.

126. Anaka M, Lynn A, McGinn P, Lloyd VK. Genomic imprinting in *Drosophila* has properties of both mammalian and insect imprinting. *Dev Genes Evol* 2009;**219**:59–66.

127. Reinhart B, Eljanne M, Chaillet JR. Shared role for differentially methylated domains of imprinted genes. *Mol Cell Biol* 2002;**22**:2089–98.

128. Wutz A, Smrzka OW, Schweifer N, Schellander K, Wagner EF, Barlow DP. Imprinted expression of the Igf2r gene depends on an intronic CpG island. *Nature* 1997;**389**:745–9.

129. Lyko F, Brenton JD, Surani MA, Paro R. An imprinting element from the mouse H19 locus functions as a silencer in *Drosophila*. *Nat Genet* 1997;**16**:171–3.

130. Lyko F, Buiting K, Horsthemke B, Paro R. Identification of a silencing element in the human 15q11-q13 imprinting center by using transgenic *Drosophila*. *Proc Natl Acad Sci USA* 1998;**95**:1698–702.

131. Geyer PK. The role of insulator elements in defining domains of gene expression. *Curr Opin Genet Dev* 1997;**7**:242–8.

132. Bell AC, Felsenfeld G. Methylation of a CTCF-dependent boundary controls imprinted expression of the Igf2 gene. *Nature* 2000;**405**:482–5.

133. Hark AT, Schoenherr CJ, Katz DJ, Ingram RS, Levorse JM, Tilghman SM. CTCF mediates methylation-sensitive enhancer-blocking activity at the H19/Igf2 locus. *Nature* 2000;**405**:486–9.

375

134. Murrell A, Heeson S, Reik W. Interaction between differentially methylated regions partitions the imprinted genes Igf2 and H19 into parent-specific chromatin loops. *Nat Genet* 2004;**36**:889–93.

135. Srivastava M, Frolova E, Rottinghaus B, Boe SP, Grinberg A, Lee E, et al. Imprint control element-mediated secondary methylation imprints at the Igf2/H19 locus. *J Biol Chem* 2003;**278**:5977–83.

136. Chao W, Huynh KD, Spencer RJ, Davidow LS, Lee JT. CTCF, a candidate trans-acting factor for X-inactivation choice. *Science* 2002;**295**:345–7.

137. Xu N, Donohoe ME, Silva SS, Lee JT. Evidence that homologous X-chromosome pairing requires transcription and Ctcf protein. *Nat Genet* 2007;**39**:1390–6.

138. Boumil RM, Ogawa Y, Sun BK, Huynh KD, Lee JT. Differential methylation of Xite and CTCF sites in Tsix mirrors the pattern of X-inactivation choice in mice. *Mol Cell Biol* 2006;**26**:2109–17.

139. Cohen DE, Davidow LS, Erwin JA, Xu N, Warshawsky D, Lee JT. The DXPas34 repeat regulates random and imprinted X inactivation. *Dev Cell* 2007;**12**:57–71.

140. Stavropoulos N, Rowntree RK, Lee JT. Identification of developmentally specific enhancers for Tsix in the regulation of X chromosome inactivation. *Mol Cell Biol* 2005;**25**:2757–69.

141. Ogawa Y, Lee JT. Xite, X-inactivation intergenic transcription elements that regulate the probability of choice. *Mol Cell* 2003;**11**:731–43.

142. Kim TH, Abdullaev ZK, Smith AD, Ching KA, Loukinov DI, Green RD, et al. Analysis of the vertebrate insulator protein CTCF-binding sites in the human genome. *Cell* 2007;**128**:1231–45.

143. Weidman JR, Murphy SK, Nolan CM, Dietrich FS, Jirtle RL. Phylogenetic footprint analysis of IGF2 in extant mammals. *Genome Res* 2004;**14**:1726–32.

144. Kim J, Kim JD. In vivo YY1 knockdown effects on genomic imprinting. *Hum Mol Genet* 2008;**17**:391–401.

145. Kim J, Kollhoff A, Bergmann A, Stubbs L. Methylation-sensitive binding of transcription factor YY1 to an insulator sequence within the paternally expressed imprinted gene, Peg3. *Hum Mol Genet* 2003;**12**:233–45.

146. Kim JD, Hinz AK, Choo JH, Stubbs L, Kim J. YY1 as a controlling factor for the Peg3 and Gnas imprinted domains. *Genomics* 2007;**89**:262–9.

147. Donohoe ME, Zhang LF, Xu N, Shi Y, Lee JT. Identification of a Ctcf cofactor, Yy1, for the X chromosome binary switch. *Mol Cell* 2007;**25**:43–56.

148. Allis CD, Gasser SM. Chromosomes and expression mechanisms New excitement over an old word: 'chromatin'. *Curr Opin Genet Dev* 1998;**8**:137–9.

149. Carninci P, Kasukawa T, Katayama S, Gough J, Frith MC, Maeda N, et al. The transcriptional landscape of the mammalian genome. *Science* 2005;**309**:1559–63.

150. Ghildiyal M, Zamore PD. Small silencing RNAs: an expanding universe. *Nat Rev Genet* 2009;**10**:94–108.

151. Mercer TR, Dinger ME, Mattick JS. Long non-coding RNAs: insights into functions. *Nat Rev Genet* 2009;**10**:155–9.

152. Lee JT. Lessons from X-chromosome inactivation: long ncRNA as guides and tethers to the epigenome. *Genes Dev* 2009;**23**:1831–42.

153. Koerner MV, Pauler FM, Huang R, Barlow DP. The function of non-coding RNAs in genomic imprinting. *Development* 2009;**136**:1771–83.

154. Hebert E. Mannose-6-phosphate/insulin-like growth factor II receptor expression and tumor development. *Biosci Rep* 2006;**26**:7–17.

155. Zwart R, Sleutels F, Wutz A, Schinkel AH, Barlow DP. Bidirectional action of the Igf2r imprint control element on upstream and downstream imprinted genes. *Genes Dev* 2001;**15**:2361–6.

156. Lyle R, Watanabe D, te Vruchte D, Lerchner W, Smrzka OW, Wutz A, et al. The imprinted antisense RNA at the Igf2r locus overlaps but does not imprint Mas1. *Nat Genet* 2000;**25**:19–21.

157. Sleutels F, Zwart R, Barlow DP. The non-coding Air RNA is required for silencing autosomal imprinted genes. *Nature* 2002;**415**:810–13.

158. Fitzpatrick GV, Soloway PD, Higgins MJ. Regional loss of imprinting and growth deficiency in mice with a targeted deletion of KvDMR1. *Nat Genet* 2002;**32**:426–31.

159. Mancini-Dinardo D, Steele SJ, Levorse JM, Ingram RS, Tilghman SM. Elongation of the Kcnq1ot1 transcript is required for genomic imprinting of neighboring genes. *Genes Dev* 2006;**20**:1268–82.

160. Fitzpatrick GV, Pugacheva EM, Shin JY, Abdullaev Z, Yang Y, Khatod K, et al. Allele-specific binding of CTCF to the multipartite imprinting control region KvDMR1. *Mol Cell Biol* 2007;**27**:2636–47.

161. Shin JY, Fitzpatrick GV, Higgins MJ. Two distinct mechanisms of silencing by the KvDMR1 imprinting control region. *EMBO J* 2008;**27**:168–78.

162. Ideraabdullah FY, Vigneau S, Bartolomei MS. Genomic imprinting mechanisms in mammals. *Mutat Res* 2008;**647**:77–85.

163. Ripoche MA, Kress C, Poirier F, Dandolo L. Deletion of the H19 transcription unit reveals the existence of a putative imprinting control element. *Genes Dev* 1997;**11**:1596–604.

376

164. Terranova R, Yokobayashi S, Stadler MB, Otte AP, van Lohuizen M, Orkin SH, et al. Polycomb group proteins Ezh2 and Rnf2 direct genomic contraction and imprinted repression in early mouse embryos. *Dev Cell* 2008;**15**:668–79.

165. Pandey RR, Mondal T, Mohammad F, Enroth S, Redrup L, Komorowski J, et al. Kcnq1ot1 antisense noncoding RNA mediates lineage-specific transcriptional silencing through chromatin-level regulation. *Mol Cell* 2008;**32**:232–46.

166. Nagano T, Mitchell JA, Sanz LA, Pauler FM, Ferguson-Smith AC, Feil R, et al. The Air noncoding RNA epigenetically silences transcription by targeting G9a to chromatin. *Science* 2008;**322**:1717–20.

167. Chaumeil J, Le Baccon P, Wutz A, Heard E. A novel role for Xist RNA in the formation of a repressive nuclear compartment into which genes are recruited when silenced. *Genes Dev* 2006;**20**:2223–37.

168. Ogawa Y, Sun BK, Lee JT. Intersection of the RNA interference and X-inactivation pathways. *Science* 2008;**320**:1336–41.

169. Kanellopoulou C, Muljo SA, Dimitrov SD, Chen X, Colin C, Plath K, et al. X chromosome inactivation in the absence of Dicer. *Proc Natl Acad Sci USA* 2009;**106**:1122–7.

170. Nesterova TB, Popova BC, Cobb BS, Norton S, Senner CE, Tang YA, et al. Dicer regulates Xist promoter methylation in ES cells indirectly through transcriptional control of Dnmt3a. *Epigenetics Chromatin* 2008;**1**:2.

171. Redrup L, Branco MR, Perdeaux ER, Krueger C, Lewis A, Santos F, et al. The long noncoding RNA Kcnq1ot1 organises a lineage-specific nuclear domain for epigenetic gene silencing. *Development* 2009;**136**:525–30.

172. Lloyd V. Parental imprinting in *Drosophila*. *Genetica* 2000;**109**:35–44.

173. Dimitri P. Constitutive heterochromatin and transposable elements in *Drosophila melanogaster*. *Genetica* 1997;**100**:85–93.

174. McClintock B. The origin and behavior of mutable loci in maize. *Proc Natl Acad Sci USA* 1950;**36**:344–55.

175. Lippman Z, Gendrel AV, Black M, Vaughn MW, Dedhia N, McCombie WR, et al. Role of transposable elements in heterochromatin and epigenetic control. *Nature* 2004;**430**:471–6.

176. Jeddeloh JA, Stokes TL, Richards EJ. Maintenance of genomic methylation requires a SWI2/SNF2-like protein. *Nat Genet* 1999;**22**:94–7.

177. Fujimoto R, Kinoshita Y, Kawabe A, Kinoshita T, Takashima K, Nordborg M, et al. Evolution and control of imprinted FWA genes in the genus *Arabidopsis*. *PLoS Genet* 2008;**4**:e1000048.

178. Rafalski A, Morgante M. Corn and humans: recombination and linkage disequilibrium in two genomes of similar size. *Trends Genet* 2004;**20**:103–11.

179. Greally JM. Short interspersed transposable elements (SINEs) are excluded from imprinted regions in the human genome. *Proc Natl Acad Sci USA* 2002;**99**:327–32.

180. Ferrigno O, Virolle T, Djabari Z, Ortonne JP, White RJ, Aberdam D. Transposable B2 SINE elements can provide mobile RNA polymerase II promoters. *Nat Genet* 2001;**28**:77–81.

181. Morgan HD, Sutherland HG, Martin DI, Whitelaw E. Epigenetic inheritance at the agouti locus in the mouse. *Nat Genet* 1999;**23**:314–18.

182. Michaud EJ, van Vugt MJ, Bultman SJ, Sweet HO, Davisson MT, Woychik RP. Differential expression of a new dominant agouti allele (Aiapy) is correlated with methylation state and is influenced by parental lineage. *Genes Dev* 1994;**8**:1463–72.

183. Wood AJ, Roberts RG, Monk D, Moore GE, Schulz R, Oakey RJ. A screen for retrotransposed imprinted genes reveals an association between X chromosome homology and maternal germ-line methylation. *PLoS Genet* 2007;**3**:e20.

184. Peaston AE, Evsikov AV, Graber JH, de Vries WN, Holbrook AE, Solter D, et al. Retrotransposons regulate host genes in mouse oocytes and preimplantation embryos. *Dev Cell* 2004;**7**:597–606.

185. Goday C, Esteban MR. Chromosome elimination in sciarid flies. *Bioessays* 2001;**23**:242–50.

186. Feil R, Berger F. Convergent evolution of genomic imprinting in plants and mammals. *Trends Genet* 2007;**23**:192–9.

187. Scott RJ, Spielman M. Epigenetics: imprinting in plants and mammals–the same but different? *Curr Biol* 2004;**14**:R201–R3.

188. Wilkins JF, Haig D. What good is genomic imprinting: the function of parent-specific gene expression. *Nat Rev Genet* 2003;**4**:359–68.

189. Reik W, Lewis A. Co-evolution of X-chromosome inactivation and imprinting in mammals. *Nat Rev Genet* 2005;**6**:403–10.

190. Dunzinger U, Haaf T, Zechner U. Conserved synteny of mammalian imprinted genes in chicken, frog, and fish genomes. *Cytogenet Genome Res* 2007;**117**:78–85.

191. Killian JK, Nolan CM, Stewart N, Munday BL, Andersen NA, Nicol S, et al. Monotreme IGF2 expression and ancestral origin of genomic imprinting. *J Exp Zool* 2001;**291**:205–12.

192. O'Neill MJ, Lawton BR, Mateos M, Carone DM, Ferreri GC, Hrbek T, et al. Ancient and continuing Darwinian selection on insulin-like growth factor II in placental fishes. *Proc Natl Acad Sci USA* 2007;**104**:12404–9.

377

193. Suzuki S, Renfree MB, Pask AJ, Shaw G, Kobayashi S, Kohda T, et al. Genomic imprinting of IGF2, p57(KIP2) and PEG1/MEST in a marsupial, the tammar wallaby. *Mech Dev* 2005;**122**:213–22.

194. Xie B, Zhang L, Zheng K, Luo CT. The evolutionary foundation of genomic imprinting in lower vertebrates. *Chinese Sci Bull* 2009;**54**:1354–60.

195. Renfree MB, Hore TA, Shaw G, Marshall Graves JA, Pask AJ. Evolution of genomic imprinting: Insights from marsupials and monotremes. *Annu Rev Genomics Hum Genet* 2009;**10**:241–62.

196. Nagel E. *The Structure of Science: Problems in the Logic of Scientific Explanation*. New York: Harcourt, Brace & World;1961.

197. Trivers RL. Parent-offspring conflict. *Am Zool* 1974;**14**:249–64.

198. Haig D, Westoby M. Parent-specific gene expression and the triploid endosperm. *Am Nat* 1989;**134**:147–55.

199. Haig D. The kinship theory of genomic imprinting. *Annu Rev Ecol Sys* 2000;**31**:9–32.

200. Renfree MB, Ager EI, Shaw G, Pask AJ. Genomic imprinting in marsupial placentation. *Reproduction* 2008;**136**:523–31.

201. Davies W, Isles AR, Humby T, Wilkinson LS. What are imprinted genes doing in the brain? *Adv Exp Med Biol* 2008;**626**:62–70.

202. Davies W, Isles A, Smith R, Karunadasa D, Burrmann D, Humby T, et al. Xlr3b is a new imprinted candidate for X-linked parent-of-origin effects on cognitive function in mice. *Nat Genet* 2005;**37**:625–9.

203. Plagge A, Gordon E, Dean W, Boiani R, Cinti S, Peters J, et al. The imprinted signaling protein XL alpha s is required for postnatal adaptation to feeding. *Nat Genet* 2004;**36**:818–26.

204. Plagge A, Isles AR, Gordon E, Humby T, Dean W, Gritsch S, et al. Imprinted Nesp55 influences behavioral reactivity to novel environments. *Mol Cell Biol* 2005;**25**:3019–26.

205. Guillemot F, Nagy A, Auerbach A, Rossant J, Joyner AL. Essential role of Mash-2 in extraembryonic development. *Nature* 1994;**371**:333–6.

206. Haig D. Genomic imprinting and kinship: how good is the evidence? *Annu Rev Genet* 2004;**38**:553–85.

207. Constancia M, Hemberger M, Hughes J, Dean W, Ferguson-Smith A, Fundele R, et al. Placental-specific IGF-II is a major modulator of placental and fetal growth. *Nature* 2002;**417**:945–8.

208. Guillemot F, Caspary T, Tilghman SM, Copeland NG, Gilbert DJ, Jenkins NA, et al. Genomic imprinting of Mash2, a mouse gene required for trophoblast development. *Nat Genet* 1995;**9**:235–42.

209. Drake NM, Park YJ, Shirali AS, Cleland TA, Soloway PD. Imprint switch mutations at Rasgrf1 support conflict hypothesis of imprinting and define a growth control mechanism upstream of IGF1. *Mamm Genome* 2009;**20**:654–63.

210. Wolf JB, Cheverud JM, Roseman C, Hager R. Genome-wide analysis reveals a complex pattern of genomic imprinting in mice. *PLoS Genet* 2008;**4**:e1000091.

211. Constancia M, Kelsey G, Reik W. Resourceful imprinting. *Nature* 2004;**432**:53–7.

212. Haig D. Huddling: brown fat, genomic imprinting and the warm inner glow. *Curr Biol* 2008;**18**:R172–R4.

213. Champagne FA, Curley JP, Swaney WT, Hasen NS, Keverne EB. Paternal influence on female behavior: the role of Peg3 in exploration, olfaction, and neuroendocrine regulation of maternal behavior of female mice. *Behav Neurosci* 2009;**123**:469–80.

214. Swaney WT, Curley JP, Champagne FA, Keverne EB. Genomic imprinting mediates sexual experience-dependent olfactory learning in male mice. *Proc Natl Acad Sci USA* 2007;**104**:6084–9.

215. Curley JP, Barton S, Surani A, Keverne EB. Coadaptation in mother and infant regulated by a paternally expressed imprinted gene. *Proc Biol Sci* 2004;**271**:1303–9.

216. Curley JP, Pinnock SB, Dickson SL, Thresher R, Miyoshi N, Surani MA, et al. Increased body fat in mice with a targeted mutation of the paternally expressed imprinted gene Peg3. *FASEB J* 2005;**19**:1302–4.

217. Keverne EB, Curley JP. Epigenetics, brain evolution and behaviour. *Front Neuroendocrinol* 2008;**29**:398–412.

218. Hager R, Johnstone RA. The genetic basis of family conflict resolution in mice. *Nature* 2003;**421**:533–5.

219. Wolf JB, Hager R. A maternal-offspring coadaptation theory for the evolution of genomic imprinting. *PLoS Biol* 2006;**4**:e380.

220. Zeh JA, Zeh DW. Viviparity-driven conflict: more to speciation than meets the fly. *Ann N Y Acad Sci* 2008;**1133**:126–48.

221. Hahn Y, Yang SK, Chung JH. Structure and expression of the zebrafish mest gene, an ortholog of mammalian imprinted gene PEG1/MEST. *Biochim Biophys Acta* 2005;**1731**:125–32.

222. Paulsen M, Khare T, Burgard C, Tierling S, Walter J. Evolution of the Beckwith-Wiedemann syndrome region in vertebrates. *Genome Res* 2005;**15**:146–53.

223. Brunner B, Grutzner F, Yaspo ML, Ropers HH, Haaf T, Kalscheue VM. Molecular cloning and characterization of the *Fugu rubripes* MEST/COPG2 imprinting cluster and chromosomal localization in *Fugu* and *Tetraodon nigroviridis*. *Chromosome Res* 2000;**8**:465–76.

224. Yokomine T, Hata K, Tsudzuki M, Sasaki H. Evolution of the vertebrate DNMT3 gene family: a possible link between existence of DNMT3L and genomic imprinting. *Cytogenet Genome Res* 2006;**113**:75–80.

225. Warren WC, Hillier LW, Marshall Graves JA, Birney E, Ponting CP, Grutzner F, et al. Genome analysis of the platypus reveals unique signatures of evolution. *Nature* 2008;**453**:175–83.

226. Martin CC, McGowan R. Genotype-specific modifiers of transgene methylation and expression in the zebrafish, *Danio rerio*. *Genet Res* 1995;**65**:21–8.

227. Corley-Smith GE, Lim CJ, Brandhorst BP. Production of androgenetic zebrafish (*Danio rerio*). *Genetics* 1996;**142**:1265–76.

228. Luo C, Li B. Diploid-dependent regulation of gene expression: a genetic cause of abnormal development in fish haploid embryos. *Heredity* 2003;**90**:405–9.

229. Barlow DP. Methylation and imprinting: from host defense to gene regulation? *Science* 1993;**260**:309–10.

230. McDonald JF, Matzke MA, Matzke AJ. Host defenses to transposable elements and the evolution of genomic imprinting. *Cytogenet Genome Res* 2005;**110**:242–9.

231. Bestor TH, Bourc'his D. Transposon silencing and imprint establishment in mammalian germ cells. *Cold Spring Harb Symp Quant Biol* 2004;**69**:381–7.

232. Youngson NA, Kocialkowski S, Peel N, Ferguson-Smith AC. A small family of sushi-class retrotransposon-derived genes in mammals and their relation to genomic imprinting. *J Mol Evol* 2005;**61**:481–90.

233. Ono R, Nakamura K, Inoue K, Naruse M, Usami T, Wakisaka-Saito N, et al. Deletion of Peg10, an imprinted gene acquired from a retrotransposon, causes early embryonic lethality. *Nat Genet* 2006;**38**:101–6.

234. Suzuki S, Ono R, Narita T, Pask AJ, Shaw G, Wang C, et al. Retrotransposon silencing by DNA methylation can drive mammalian genomic imprinting. *PLoS Genet* 2007;**3**:e55.

235. Pask AJ, Papenfuss AT, Ager EI, McColl KA, Speed TP, Renfree MB. Analysis of the platypus genome suggests a transposon origin for mammalian imprinting. *Genome Biol* 2009;**10**:R1.

236. Zakharova IS, Shevchenko AI, Zakian SM. Monoallelic gene expression in mammals. *Chromosoma* 2009;**118**:279–90.

237. Olender T, Lancet D, Nebert DW. Update on the olfactory receptor (OR) gene superfamily. *Hum Genomics* 2008;**3**:87–97.

238. Gimelbrant A, Hutchinson JN, Thompson BR, Chess A. Widespread monoallelic expression on human autosomes. *Science* 2007;**318**:1136–40.

239. Beaudet AL, Jiang YH. A rheostat model for a rapid and reversible form of imprinting-dependent evolution. *Am J Hum Genet* 2002;**70**:1389–97.

240. McGowan RA, Martin CC. DNA methylation and genome imprinting in the zebrafish, *Danio rerio*: some evolutionary ramifications. *Biochem Cell Biol* 1997;**75**:499–506.

241. Cedar H, Bergman Y. Choreography of Ig allelic exclusion. *Curr Opin Immunol* 2008;**20**:308–17.

242. Fuss SH, Ray A. Mechanisms of odorant receptor gene choice in *Drosophila* and vertebrates. *Mol Cell Neurosci* 2009;**41**:101–12.

243. Casali N, Preston AE. coli plasmid vectors: methods and applications. Totowa, NJ: Humana Press; 2003.

244. Kricker MC, Drake JW, Radman M. Duplication-targeted DNA methylation and mutagenesis in the evolution of eukaryotic chromosomes. *Proc Natl Acad Sci USA* 1992;**89**:1075–9.

245. Lynch M, Conery JS. The evolutionary fate and consequences of duplicate genes. *Science* 2000;**290**:1151–5.

246. Anderson PW. More is different. *Science* 1972;**177**:393–6.

247. Gray AP. Mammalian hybrids: a checklist with bibliography. Farnham Royal, Bucks: Commonwealth Agricultural Bureaux; 1972.

248. Savory TH. The mule. *Sci Am* 1970;**223**:102–9.

249. Glaser RL, Morison IM. Equality of the sexes? parent-of-origin effects on transcription and de novo mutations. *Bioinform Syst Biol* 2009:485–513.

250. Luedi PP, Dietrich FS, Weidman JR, Bosko JM, Jirtle RL, Hartemink AJ. Computational and experimental identification of novel human imprinted genes. *Genome Res* 2007;**17**:1723–30.

251. Pollard KS, Serre D, Wang X, Tao H, Grundberg E, Hudson TJ, et al. A genome-wide approach to identifying novel-imprinted genes. *Hum Genet* 2008;**122**:625–34.

CHAPTER 23

Epigenetics of Memory Processes

Tania L. Roth, Eric D. Roth, and J. David Sweatt
Department of Neurobiology and Evelyn F. McKnight Brain Institute,
University of Alabama at Birmingham, Birmingham, AL 35294, USA

INTRODUCTION

Seminal studies continue to demonstrate that the histone proteins and DNA that comprise chromatin are targets of neuronal signaling pathways involved in CNS plasticity and memory formation. As it applies to cognition, we define *epigenetics* as the covalent modification of chromatin that influences activity-dependent changes in gene expression. These changes can be transient, underlying the dynamic regulation of gene activity states, or they can be long-term and responsible for lasting alterations in gene activity states. The combination of dynamic and stable components renders chromatin an ideal substrate for signal integration and storage of cellular information in the CNS. Indeed, studies are being published at a rapid pace demonstrating that epigenetic mechanisms mediate experience-driven changes in the CNS. Concerning the field of learning and memory, there are two basic molecular epigenetic mechanisms that are currently studied – post-translational modifications of histone proteins and direct covalent methylation of cytosines. In the following sections, we first briefly introduce the reader to these mechanisms, and then summarize the current body of literature pertaining to their role in learning and memory.

381

EPIGENETIC MODIFICATION OF HISTONES UNDERLYING MEMORY

Histones are proteins that organize DNA in the nucleus. There are eight histone proteins (histones 2A, 2B, 3, and 4, with two copies of each molecule) at the heart of the chromatin core. DNA either coils or uncoils around this core, a process that is mediated in part by the post-translational modifications of the *N-terminal tail* of the histone proteins. Modifications of the tails include acetylation, phosphorylation, methylation, ubiquitination, and sumoylation [1,2]. These modifications, along with DNA methylation, are the principal epigenetic mechanisms that help govern the activity of genes in the CNS.

Histone modifications have different effects on gene activity. For example, histone acetylation is coupled to gene activation, while sumoylation is coupled to gene repression. Histone methylation on the other hand is a more complex process, and can be associated with either gene activation or repression dependent upon the particular amino acid residue modified. Since histone acetylation has been the most extensively studied epigenetic modification in the field of learning and memory, we will briefly review the enzymes catalyzing this modification and how this reaction promotes gene transcription.

Handbook of Epigenetics: The New Molecular and Medical Genetics. DOI: 10.1016/B978-0-12-375709-8.00023-X

Enzymes known as histone acetyltransferases (HATs) catalyze the direct transfer of an acetyl group from acetyl-CoA to the ε-NH$^+$ group of the lysine residues within a histone [3]. The addition of an acetyl group decreases the affinity between the protein tail and DNA, thus relaxing the chromatin structure and providing access for transcriptional machinery. Acetylated histone tails at the same time also provide a substrate for the binding of additional co-activators with domains that recognize acetylated lysines. Thus, histone acetylation is generally associated with transcriptional activation and is widely regarded as one of the epigenetic marks associated with active chromatin, often referred to as euchromatin.

Histone acetylation is also reversible, and the enzymes that catalyze the reversal of histone acetylation are known as histone deacetylases (HDACs). There are a total of eleven classic HDAC isoforms, most of which are expressed in the CNS. HDACs remove the acetyl groups from lysine residues, a reaction that promotes DNA condensation around the histone core. Trichostatin A, sodium butyrate, valproic acid, and suberoylanilide hydroxamic acid (SAHA) are the most widely used HDAC inhibitors, each having varying degrees of selectivity for the classical HDAC isoforms. The use of these inhibitors has been instrumental in helping define a role for histone modifications in adult memory formation. Furthermore, these inhibitors have potential therapeutic value in alleviating cognitive deficits.

Some of the earliest evidence for the role of histone modifications in adult cognition came from studies that investigated the role of a particular HAT, cAMP Response Element Binding Protein (CBP), in long-term memory formation. Long-term memory formation is the stabilization of recently learned information, a process that evokes gene expression and structural synaptic changes in restricted regions of the brain. Mice with a truncated form of CBP were found to have significant deficits in long-term memory following several tasks, including step-through passive avoidance (a paradigm in which animals will learn to avoid the dark box of the apparatus, although they have a natural preference for dark), novel object recognition (a test based on the premise that rodents will explore a novel object more than a familiar one, but only if they remember the familiar one) and cued-fear conditioning (an associative learning paradigm in which an association is made between a neutral stimulus such as an odor and an aversive stimulus such as foot-shock) [4,5]. However, since CREB and CBP govern developmental processes, these animals also had developmental abnormalities. Thus, straightforward interpretation of data pertaining to memory is difficult. Three laboratories later developed CBP-deficient mice that were void of the effects of CBP on development [6–8]. Similar to results from previous studies, acquisition of new information (learning) and short-term memory were spared in these mice, but these mice exhibited significant impairments in novel object recognition, spatial and fear memory, and also had significant deficits in hippocampal long-term potentiation (LTP) [6–8]. In a recent report, Marcelo Wood's group has shown that the HDAC inhibitor sodium butyrate can establish and generate more persistent forms of long-term novel object recognition memory in both CBP mutant and wildtype mice [9].

Evidence continues to mount in support of the hypothesis that specific histone modifications are involved in long-term memory formation. In our initial studies, we investigated whether there were hippocampal histone modifications following contextual-fear conditioning in mice. Commonly used to assess hippocampal-dependent learning and memory function, this is a behavioral paradigm in which animals learn to associate a novel context (conditioned stimulus) with a mildly aversive unconditioned stimulus (foot-shock) that naturally elicits a freezing response (unconditioned response). After a few presentations of the two stimuli, animals readily learn this association and freeze in response to presentation of the context in the absence of the foot-shock (conditioned response). Their memory of this association can be assessed by returning the animal to the same context (typically 24 hours following the training) and measuring their freezing behavior in the absence of the foot-shock. Freezing behavior in then used as an index to illustrate whether the animals have successfully learned and maintained memory of the association. We observed significant increases in both acetylation and phosphorylation of histone 3 (H3), but not histone 4 (H4), in mice that had learned fear, and found that these chromatin remodeling events were regulated by the extracellular signal-regulated kinase/mitogen-activated protein kinase (ERK/MAPK) pathway [10,11]. Furthermore, HDAC inhibitors (sodium butyrate and trichostatin A) were shown to enhance both LTP and the fear memory itself [11]. These observations were the first to indicate that epigenetic marking of the genome occurs in long-term memory formation, and that manipulation of epigenetic processes is a viable way to alter memory capacity.

In a landmark study in 2007, Li-Huei Tsai's group showed that the beneficial effects of environmental enrichment on restoring learning and memory in neurodegenerative mice involves increased hippocampal and cortical H3 acetylation [12]. Also in that study, they demonstrated that the use of the HDAC inhibitor sodium butyrate is sufficient to restore learning and memory in these impaired mice. This beneficial effect of various HDAC inhibitors in improving learning and memory in non-diseased rodents and in other models of neurodegeneration and brain injury has continued to be replicated [13–16]. Tsai's group has now been working to delineate the functions of particular HDACs in learning and memory. In a recent report, they presented data that indicate HDAC2 negatively regulates memory by suppressing the activity of genes that are necessary for synaptic plasticity, such as the brain-derived neurotrophic factor gene (*BDNF*) and glutamate receptor 1 gene (*GLUR1*) [17].

Several studies have also implicated the involvement of histone modifications at specific gene loci during fear memory formation. The gene that has undoubtedly received the most attention is *BDNF*, as it has been established to have an essential role in regulating neuronal structure and neural function, and is critical for the synaptic plasticity underlying long-term memory formation. This gene has also been the focus in several DNA methylation studies. In a recent report, we showed that acetylation of H3 increases at promoter IV of *BDNF*, and that these modifications parallel changes in expression of *BDNF* mRNA during the consolidation period (Fig. 23.1) [18]. Other reports also indicate that neural plasticity and long-term memory requires specific histone modifications at *BDNF* loci. For example, there are increased H3 acetylation levels within several regions of promoter I of *BDNF* following NMDA treatment in cultures of hippocampal neurons [19]. The extinction of conditioned fear in mice induces H4 acetylation around exon IV in the prefrontal cortex [20]. Additionally, chronic social defeat stress in adult mice produces a lasting down-regulation of hippocampal *BDNF* transcripts III and IV that are associated with increased H3 lysine methylation at the particular promoters [21]. Investigators continue to show that other stressors and memories of stressful events evoke similar changes in hippocampal H3 phosphorylation and acetylation [22–25].

In summary, there are now varied but extensive data indicating that an adult animal's capacity to form and consolidate memories depends in part on specific histone modifications in the CNS. Data also highlight the therapeutic value of HDAC inhibitors in restoring memory capacity. Despite this emerging role of histone modifications in learning and memory, several questions remain to be addressed. For example, though the current body of literature suggests that histone acetylation and phosphorylation are the major histone modifications supporting memory, this may be attributable to the fact that these are the two modifications that have received the most attention. Whether there is a role for histone methylation and sumoylation in memory has been largely unexplored. An additional caveat of the data is that the currently available HDAC inhibitors can also target non-histone substrates [26]. Thus, it is possible that some of the beneficial effects of HDAC therapy on memory capacity is through non-histone effects. Thus, to gain a better appreciation of the role of histone modifications in memory and how they can mediate memory capacity, it will be crucial to resolve whether histone-modifying enzymes have any non-histone mediated functions.

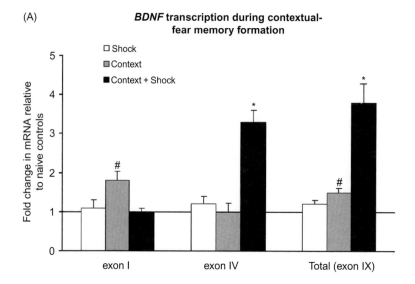

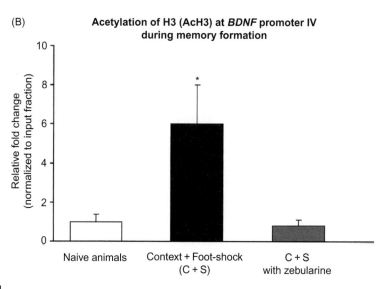

FIGURE 23.1

Contextual-fear memory formation elicits hippocampal *BDNF* transcription that is associated with histone modifications at exon-specific *BDNF* promoters. (A) Quantitative real-time PCR data indicate that 2 hr following contextual-fear conditioning there are significant increases in *BDNF* exons IV and IX mRNA in area CA1 of Context + Foot-shock animals (*p-value significant versus Naive, Shock, and Context controls). Context exposure alone also elicits an increase in *BDNF* exon I and IX mRNA (#p-value significant versus Naive controls). (B) At promoter IV of the *BDNF* gene, there is also a significant increase in acetylation of histone 3 (H3) in Context + Foot-shock animals. This increase is blocked if DNA methylation is inhibited (by pre-training treatment with zebularine). *BDNF* = brain-derived neurotrophic factor gene.

COVALENT MODIFICATION OF DNA UNDERLYING MEMORY

In addition to histone modifications, DNA methylation appears to have some active role in regulating synaptic plasticity and memory. DNA methylation is a direct chemical modification that adds a $-CH_3$ group through a covalent bond. This modification occurs at cytosine–guanine dinucleotide (CpG) sequences that occur in clusters in and around gene regulatory regions as well as within intragenic regions. DNA methylation is catalyzed by a class of enzymes known as DNA methyltransferases (DNMTs) [27,28]. DNMT3a and DNMT3b, the *de novo* DNMTs, methylate previously unmethylated CpG sites in DNA – sites which have no methyl-cytosine on either DNA strand. The maintenance DNMT isoform, DNMT1, perpetuates methylation marks after cell division, regenerating the methyl-cytosine marks on the newly synthesized complementary DNA strand that arises from DNA replication.

DNA methylation is a process that is generally associated with suppression of gene transcription. In essence, methylation of cytosines at CpG dinucleotides recruits methyl-DNA binding proteins and HDACs at specific sites in the genome. This triggers localized regulation of the three-dimensional structure of DNA and its associated histone proteins, resulting in a higher-affinity interaction between DNA and the histone core [27,28]. In all, this ultimately suppresses transcription. However, while DNA methylation is usually associated with transcriptional suppression, it is important to point out that recent studies provide evidence that DNA methylation status alone does not always reflect that a gene is repressed, as MeCP2 can also be associated with active genes [29–31].

The idea of whether there is active DNA demethylation in post-mitotic cells, such as neurons, remains a controversial topic in the field. DNA demethylation has been historically viewed as a passive and largely irreversible process in differentiated cells, whereby multiple rounds of cell division without DNMT-mediated remethylation is necessary to erase epigenetic marks. However, recent evidence argues that there is indeed active methylation and demethylation in mature cells, mediated by either DNMT3a and 3b or Gadd45b through a glycosylase reaction/DNA repair mechanism, though it should be noted that Gadd45-mediated demethylation remains controversial [32–37]. Furthermore, an active demethylase that can remove methyl groups has not been identified.

Regardless of the exact mechanism involved in the mature CNS, results continue to highlight the capacity of DNA methylation to regulate synaptic plasticity and memory [18,38–42]. Initial studies in 2003 indicated that DNA methylation might play a key role in activity-dependent neural plasticity, as the methylation status of *BDNF* was shown to undergo dynamic changes in response to stimulation [43,44]. For example, Martinowich and colleagues demonstrated in neurons that transcription of *BDNF* exon IV is suppressed by methyl-CpG binding protein 2 (MeCP2), and upon depolarization, MeCP2 is released along with HDAC1 [44]. Since these studies, 5-azadeoxycytidine (5-aza-C), which disrupts DNA methylation, and the HDAC inhibitor trichostatin A, have been shown to up-regulate specific *BDNF* transcripts, including exons I and IV [18,45].

Work from our own laboratory provided the first demonstration that the acute application of 5-aza-C or zebularine (a drug that also disrupts DNA methylation) to mouse hippocampal slices not only influences the methylation status of *BDNF* and *RELN* DNA, but also blocks hippocampal LTP [39,40]. The effects of their ability to disrupt synaptic plasticity have since then been confirmed and extended by Lisa Monteggia's group [42]. Using *in vivo* approaches we have begun to more definitively link DNA methylation with fear memory capacity. Our first efforts showed that *DNMT3a* and *DNMT3b* were up-regulated in the hippocampus following contextual-fear conditioning in adult rats [41]. In those same rats, we observed a decrease in methylation (demethylation) and transcriptional activation of *RELN*, and an increase in methylation and transcriptional silencing of the memory suppressor gene protein phosphatase 1 (*PP1*) [41]. Our recent efforts have focused on epigenetic regulation of the *BDNF* gene during fear memory formation, and have shown that contextual-fear conditioning also evokes *BDNF* DNA demethylation (Fig. 23.2A) [18]. These modifications are associated with localized histone modifications at specific *BDNF* promoters and up-regulation of *BDNF* transcription (Fig. 23.1) [18]. Furthermore, we have shown that disrupting DNA methylation with various agents blocks these epigenetic changes, as well as the fear memory (Fig. 23.2B) [18]. As a final point, a recent report has shown that disrupting the function of MeCP2 is sufficient to impair the ability to form a fear memory (amygdala-dependent cued-fear conditioning) [38]. All together, the available data indicate that both DNA methylation and demethylation are key components in adult memory formation.

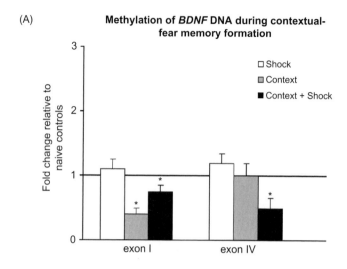

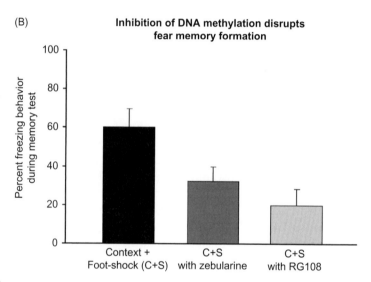

FIGURE 23.2

Contextual-fear memory formation also elicits changes in DNA methylation at specific *BDNF* loci within the hippocampus. (A) Methyl-specific quantitative real-time PCR data indicate that 2 hr following contextual-fear conditioning there is significant demethylation of DNA within the regulatory regions of exons I and IV in area CA1 of Context + Foot-shock animals (*p-value significant versus Naive controls). Context exposure alone also elicits demethylation of exon I. (B) Animals that receive contextual-fear conditioning demonstrate significant freezing behavior, which is indicative of learned fear. Pharmacological agents that interfere with DNA methylation, such as zebularine or RG108, disrupt fear memory capacity in these animals (*p-value significant versus C + S animals). *BDNF* = brain-derived neurotrophic factor gene.

As DNA methylation presents itself as a new mechanism underlying memory, questions remain to be answered to overcome limitations of the current data. For example, the bulk of what we know regarding the role of DNA methylation in synaptic physiology and memory is from pharmacological studies using drugs whose mechanisms we do not fully understand. Because both 5-aza and zebularine are nucleoside analogs that need to be incorporated into DNA to trap DNMT and block DNA methylation, the mechanism of how these drugs are able to alter methylation in post-mitotic neurons, or glia for that matter, is not clear. They may do so by actively demethylating DNA in non-dividing cells through a replication-independent event, such as a DNA repair process. There is only one behavior study to date that has used a non-nucleoside compound (RG108) to directly inhibit DNMT enzyme activity. Results indicated that RG108 had similar effects on memory as that of zebularine and 5-aza-C [18]. We also do not know whether these methylation changes are

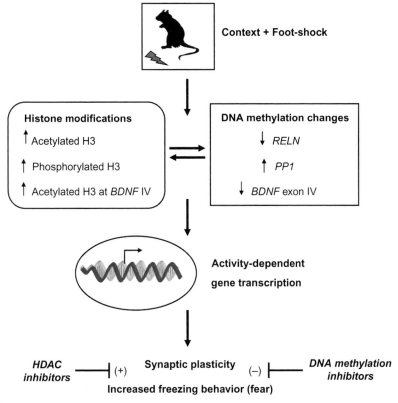

**Epigenetic changes occurring in the adult hippocampus
in response to contextual-fear conditioning**

Context + Foot-shock

Histone modifications

↑ Acetylated H3

↑ Phosphorylated H3

↑ Acetylated H3 at *BDNF* IV

DNA methylation changes

↓ *RELN*

↑ *PP1*

↓ *BDNF* exon IV

Activity-dependent
gene transcription

HDAC
inhibitors ———| (+) **Synaptic plasticity** (−) |——— *DNA methylation
inhibitors*

Increased freezing behavior (fear)

FIGURE 23.3
Overview of the epigenetic changes that we have shown to date are responsible for supporting adult hippocampal plasticity and fear memory formation. Contiguous presentations of Context + Foot-shock elicit several histone modifications and DNA methylation changes at several gene loci. This evokes activity-dependent changes in gene transcription necessary for synaptic plasticity and behavior modifications underlying fear memory. HDAC inhibitors (e.g. trichostatin A, sodium butyrate) enhance contextual fear memory capacity, while DNA methylation inhibitors (e.g. 5-azadeoxycytidine (5-aza-C), zebularine, RG108) decrease contextual fear memory capacity. *RELN* = Reelin gene; *PP1* = protein phosphatase 1 gene; *BDNF* = brain-derived neurotrophic factor gene.

occurring genome-wide, or only at a handful of gene loci. Moreover, all studies to date have investigated epigenetic mechanisms in memory using a target-gene approach to assessing changes at specific candidate gene loci. Finally, we do not know yet whether epigenetic changes contribute to the maintenance and persistence of memory. It has only been shown that they are involved in the early stages of memory formation and consolidation. The best evidence that they might subserve memory persistence is in the studies that demonstrate that there are epigenetic changes that underlie the persisting effects, or "memory", of early-life experiences [46–51].

SUMMARY

It is becoming increasingly clear that epigenetic mechanisms have some necessary role in the dynamic nature of the adult CNS in response to the environment, and that epigenetic regulation of gene transcription facilitates memory formation (Fig. 23.3). Studies also indicate that manipulation of such mechanisms can restore learning and memory deficits in several rodent models of neurodegeneration and brain injury. Thus, one of the most important questions that future studies will continue to address is whether epigenetic drugs can alleviate cognitive deficits in humans, particularly those associated with age-related neurological disorders (i.e. Alzheimer's) and neuropsychiatric disorders (such as schizophrenia). Indeed studies continue to demonstrate that epigenetic alterations occur in these patients [52–55]. The continued study of epigenetic marks in the regulation of cognition and in postmortem tissue promises a future where we will fully appreciate the role of epigenetic molecular mechanisms in CNS control, and importantly, the viability of epigenetic therapy treatment to alleviate the growing prevalence of memory dysfunction in society.

ACKNOWLEDGEMENTS

This work was funded by grants from the National Institutes of Health, the National Alliance for Research on Schizophrenia and Depression, Civitan International, the Rotary Clubs CART fund, and the Evelyn F. McKnight Brain Research Foundation.

References

1. Luger K, Mader AW, Richmond RK, Sargent DF, Richmond TJ. Crystal structure of the nucleosome core particle at 2.8 A resolution. *Nature* 1997;**389**:251–60.

2. Strahl BD, Allis CD. The language of covalent histone modifications. *Nature* 2000;**403**:41–5.

3. Marmorstein R. Structure and function of histone acetyltransferases. *Cell Mol Life Sci* 2001;**58**:693–703.

4. Bourtchouladze R, Lidge R, Catapano R, Stanley J, Gossweiler S, Romashko D. A mouse model of Rubinstein–Taybi syndrome: defective long-term memory is ameliorated by inhibitors of phosphodiesterase 4. *Proc Natl Acad Sci USA* 2003;**100**:10518–22.

5. Oike Y, Hata A, Mamiya T, Kaname T, Noda Y, Suzuki M. Truncated CBP protein leads to classical Rubinstein–Taybi syndrome phenotypes in mice: implications for a dominant-negative mechanism. *Hum Mol Genet* 1999;**8**:387–96.

6. Alarcon JM, Malleret G, Touzani K, Vronskaya S, Ishii S, Kandel ER, et al. Chromatin acetylation, memory, and LTP are impaired in CBP+/– mice: a model for the cognitive deficit in Rubinstein–Taybi syndrome and its amelioration. *Neuron* 2004;**42**:947–59.

7. Korzus E, Rosenfeld MG, Mayford M. CBP histone acetyltransferase activity is a critical component on memory consolidation. *Neuron* 2004;**42**:961–72.

8. Wood MA, Kaplan MP, Park A, Blanchard EJ, Oliviera AM, Lombardi TL, et al. Transgenic mice expressing a truncated form of CREB-binding protein (CBP) exhibit deficits in hippocampal synaptic plasticity and memory storage. *Learn Mem* 2005;**12**:111–19.

9. Stefanko DP, Barrett RM, Ly AR, Reolon GK, Wood MA. Modulation of long-term memory for object recognition via HDAC inhibition. *Proc Natl Acad Sci USA* 2009;**106**:9447–52.

10. Chwang WB, O'Riordan KJ, Levenson JM, Sweatt JD. ERK/MAPK regulates hippocampal histone phosphorylation following contextual fear conditioning. *Learn Mem* 2006;**13**:322–8.

11. Levenson JM, O'Riordan KJ, Brown KD, Trinh MA, Molfese DL, Sweatt JD. Regulation of histone acetylation during memory formation in the hippocampus. *J Biol Chem* 2004;**279**:40545–59.

12. Fischer A, Sananbenesi F, Wang X, Dobbin M, Tsai L-H. Recovery of learning and memory is associated with chromatin remodeling. *Nature* 2007;**447**:178–82.

13. Bredy TW, Barad M. The histone deacetylase inhibitor valproic acid enhances acquisition, extinction, and reconsolidation of conditioned fear. *Learn Mem* 2008;**15**:39–45.

14. Dash PK, Orsi SA, Moore AN. HDAC inhibition combined with behavioral therapy enhances learning and memory following traumatic brain injury. *Neurosci* 2009;**163**:1–8.

15. Fontán-Lozano Á, Romero-Granados R, Troncoso J, Múnera A, Delgado-García JM, Carrión ÁM. Histone deacetylase inhibitors improve learning consolidation in young and in KA-induced-neurodegeneration and SAMP-8-mutant mice. *Mol Cell Neurosci* 2008;**39**:193–201.

16. Ricobaraza A, Cuadrado-Tejedor M, Perez-Mediavilla A, Frechilla D, Del Rio J, Garcia-Osta A. Phenylbutyrate ameliorates cognitive deficit and reduces tau pathology in an Alzheimer's Disease mouse model. *Neuropsychopharmacol* 2009;**34**:1721–32.

17. Guan J-S, Haggarty SJ, Giacometti E, Dannenberg J-H, Joseph N, Gao J, et al. HDAC2 negatively regulates memory formation and synaptic plasticity. *Nature* 2009;**459**:55–60.

18. Lubin FD, Roth TL, Sweatt JD. Epigenetic regulation of *bdnf* gene transcription in the consolidation of fear memory. *J Neurosci* 2008;**28**:10576–86.

19. Tian F, Hu X-Z, Wu X, Jiang H, Pan H, Marini AM, et al. Dynamic chromatin remodeling events in hippocampal neurons are associated with NMDA receptor-mediated activation of *Bdnf* gene promoter 1. *J Neurochem* 2009;**109**:1375–88.

20. Bredy TW, Wu H, Crego C, Zellhoefer J, Sun YE, Barad M. Histone modifications around individual *BDNF* gene promoters in prefrontal cortex are associated with extinction of conditioned fear. *Learn Mem* 2007;**14**:268–76.

21. Tsankova N, Berton O, Renthal W, Kumar A, Neve R, Nestler E. Sustained hippocampal chromatin regulation in a mouse model of depression and antidepressant action. *Nature Neurosci* 2006;**9**:519–25.

22. Collins A, Hill LE, Chandramohan Y, Whitcomb D, Droste SK, Reul JMHM. Exercise improves cognitive responses to psychological stress through enhancement of epigenetic mechanisms and gene expression in the dentate gyrus. *PLoS ONE* 2009;**4**:e4330.

23. Fuchikami M, Morinobu S, Kurata A, Yamamoto S, Yamawaki S. Single immobilization stress differentially alters the expression profile of transcripts of the brain-derived neurotrophic factor (*BDNF*) gene and histone acetylation at its promoters in the rat hippocampus. *Int J Neuropsychopharmacol* 2009;**12**:73–82.

24. Jakobsson J, Cordero MI, Bisaz R, Groner AC, Busskamp V, Bensadoun J-C, et al. KAP1-mediated epigenetic repression in the forebrain modulates behavioral vulnerability to stress. *Neuron* 2008;**60**:818–31.

25. Reul JMHM, Chandramohan Y. Epigenetic mechanisms in stress-related memory formation. *Psychoneuroendocrinology* 2007;**32**:S21–S5.

26. Yuan Z-l, Guan Y-j, Chatterjee D, Chin YE. Stat3 dimerization regulated by reversible acetylation of a single lysine residue. *Science* 2005;**307**:269–73.

27. Bird A. DNA methylation patterns and epigenetic memory. *Genes Dev* 2002;**16**:6–21.

28. Miranda TB, Jones PA. DNA methylation: the nuts and bolts of repression. *J Cell Physiol* 2007;**213**:384–90.

29. Chahrour M, Jung SY, Shaw C, Zhou X, Wong STC, Qin J, et al. MeCP2, a key contributor to neurological disease, activates and represses transcription. *Science* 2008;**320**:1224–9.

30. Cohen S, Zhou Z, Greenberg ME. Activating a repressor. *Science* 2008;**320**:1172–3.

31. Yasui DH, Peddada S, Bieda MC, Vallero RO, Hogart A, Nagarajan RP, et al. Integrated epigenomic analyses of neuronal MeCP2 reveal a role for long-range interaction with active genes. *PNAS* 2007;**104**:19416–21.

32. Brown SE, Weaver ICG, Meaney MJ, Szyf M. Regional-specific global cytosine methylation and DNA methyltransferase expression in the adult rat hippocampus. *Neurosci Lett* 2008;**440**:49–53.

33. Kangaspeska S, Stride B, Metivier R, Polycarpou-Schwarz M, Ibberson D, Carmouche RP, et al. Transient cyclical methylation of promoter DNA. *Nature* 2008;**452**:112–15.

34. Kriaucionis S, Heintz N. The nuclear DNA base 5-hydroxymethylcytosine is present in Purkinje neurons and the brain. *Science* 2009;**324**:929–30.

35. Ma DK, Jang M-H, Guo JU, Kitabatake Y, Chang M-l, Pow-anpongkul N, et al. Neuronal activity-induced Gadd45b promotes epigenetic DNA demethylation and adult neurogenesis. *Science* 2009;**323**:1074–7.

36. Metivier R, Gallais R, Tiffoche C, Le Peron C, Jurkowska RZ, Carmouche RP, et al. Cyclical DNA methylation of a transcriptionally active promoter. *Nature* 2008;**452**:45–50.

37. Tahiliani M, Koh KP, Shen Y, Pastor WA, Bandukwala H, Brudno Y, et al. Conversion of 5-methylcytosine to 5-hydroxymethylcytosine in mammalian DNA by the MLL fusion partner TET1. *Science* 2009;**324**:930–5.

38. Adachi M, Autry AE, Covington HE III, Monteggia LM. MeCP2-mediated transcription repression in the basolateral amygdala may underlie heightened anxiety in a mouse model of Rett Syndrome. *J Neurosci* 2009;**29**:4218–27.

39. Levenson JM, Roth TL, Lubin FD, Miller CA, Huang IC, Desai P, et al. Evidence that DNA (cytosine-5) methyltransferase regulates synaptic plasticity in the hippocampus. *J Biol Chem* 2006;**281**:15763–73.

40. Miller CA, Campbell SL, Sweatt JD. DNA methylation and histone acetylation work in concert to regulate memory formation and synaptic plasticity. *Neurobiol Learn Mem* 2008;**89**:599–603.

41. Miller CA, Sweatt JD. Covalent modification of DNA regulates memory formation. *Neuron* 2007;**53**:857–69.

42. Nelson ED, Kavalali ET, Monteggia LM. Activity-dependent suppression of miniature neurotransmission through the regulation of DNA methylation. *J Neurosci* 2008;**28**:395–406.

43. Chen WG, Chang Q, Lin Y, Meissner A, West AE, Griffith EC, et al. Depression of BDNF transcription involves calcium-dependent phosphorylation of MeCP2. *Science* 2003;**302**:885–9.

44. Martinowich K, Hattori D, Wu H, Fouse S, He F, Hu Y, et al. DNA methylation-related chromatin remodeling in activity-dependent *Bdnf* gene regulation. *Science* 2003;**302**:890–3.

45. Aid T, Kazantseva A, Piirsoo M, Palm K, Timmusk T. Mouse and rat BDNF gene structure and expression revisited. *J Neurosci Res* 2007;**85**:525–35.

46. Champagne FA, Curley JP. Epigenetic mechanisms mediating the long-term effects of maternal care on development. *Neurosci Biobehav Rev* 2008;**33**:593–600.

47. McGowan PO, Sasaki A, D'Alessio AC, Dymov S, Labonte B, Szyf M, et al. Epigenetic regulation of the glucocorticoid receptor in human brain associates with childhood abuse. *Nat Neurosci* 2009;**12**:342–8.

48. McGowan PO, Sasaki A, Huang TCT, Unterberger A, Suderman M, Ernst C, et al. Promoter-wide hypermethylation of the ribosomal RNA gene promoter in the suicide brain. *PLoS ONE* 2008;**3**:e2085.

49. Mueller BR, Bale TL. Sex-specific programming of offspring emotionality after stress early in pregnancy. *J Neurosci* 2008;**28**:9055–65.

50. Roth TL, Lubin FD, Funk AJ, Sweatt JD. Lasting epigenetic influence of early-life adversity on the *BDNF* gene. *Biol Psychiatry* 2009;**65**:760–9.

51. Weaver ICG, Cervoni N, Champagne FA, D'Alessio AC, Sharma S, Seckl JR, et al. Epigenetic programming by maternal behavior. *Nat Neurosci* 2004;**7**:847–54.

52. Abdolmaleky HM, Cheng KH, Russo A, Smith CL, Faraone SV, Wilcox M, et al. Hypermethylation of the reelin (RELN) promoter in the brain of schizophrenic patients: a preliminary report. *Am J Med Genet B Neuropsychiatr Genet* 2005;**134**:60–6.

53. Akbarian S, Huang H-S. Epigenetic regulation in human brain – focus on histone lysine methylation. *Biol Psychiatry* 2009;**65**:198–203.

54. Costa E, Chen Y, Dong E, Grayson DR, Kundakovic M, Maloku E, et al. GABAergic promoter hypermethylation as a model to study the neurochemistry of schizophrenia vulnerability. *Expert Rev Neurother* 2009;**9**:87–98.

55. Wang S-C, Oelze B, Schumacher A. Age-specific epigenetic drift in late-onset Alzheimer's Disease. *PLoS ONE* 2008;**3**:e2698.

CHAPTER 24

Transgenerational Epigenetics

James P. Curley, Rahia Mashoodh, and Frances A. Champagne
Columbia University, Department of Psychology, New York NY 10027, USA

INTRODUCTION

The regulation of gene expression through epigenetic modifications provides a dynamic route through which environmental experiences can lead to persistent changes in cellular phenotype. This plasticity plays an important role in mediating cellular differentiation and the potential stability of these modifications can lead to persistent and heritable variations in gene expression. Though there are numerous types of epigenetic mechanisms, studies of environmentally-induced changes in the epigenome have focused primarily on DNA methylation and post-translational modification of histone proteins. The process of DNA methylation whereby cytosine is converted to 5-methylcytosine is mediated by methyltransferases which either promote maintenance (i.e. DNMT1) or *de novo* DNA methylation (i.e. DNMT3) [1–3]. The process of methylation is dependent on the presence of methyl donors (provided by nutrients such as folic acid, methionine, and choline) and the transcriptional repression associated with DNA methylation is sustained through methyl-binding proteins such as MeCP2 [4]. Histone proteins, which form the core of the nucleosome, also significantly alter gene expression through their interactions with DNA. Histones can undergo multiple post-translational modifications, including methylation, acetylation, and ubiquitination, which can alter the accessibility of DNA and the density of chromatin structure. In particular, histone acetylation is associated with increased transcriptional activity whereas histone deacetylation is associated with transcriptional repression [1,5].

The role of epigenetic mechanisms in mediating the long-term effects of environmental experiences is a rapidly expanding field of study, and it has become evident that experiences across the lifespan can induce modifications to the epigenome. Moreover, these epigenetic effects can have implications for neurobiology, physiology, and behavior of an organism leading to divergent developmental outcomes. Thus the molecular mechanisms that regulate gene expression can contribute to the "epigenesis" of phenotype as described by Waddington in the 1940s, in which the term "epigenetics" has its roots [6]. Within the study of mammalian development, the quality of interactions between parents and offspring is a particularly salient aspect of the early environment and there is converging evidence from numerous experimental paradigms for parental influences on the regulation of gene expression and behavior [7–10]. Though maternal effects have been well established in the literature, there is increasing evidence for paternal regulation of offspring development

Handbook of Epigenetics: The New Molecular and Medical Genetics. DOI: 10.1016/B978-0-12-375709-8.00024-1

which may provide important insights into the role of epigenetic mechanisms in mediating the transmission of environmental experiences across generations. In this review, we will discuss evidence of maternal and paternal epigenetic influence on offspring development, with particular focus on studies indicating an association between parental experiences/ environmental exposures and epigenetic alterations in offspring. An emerging theme within these studies is the transgenerational implications of these environmentally-induced effects (i.e. effects observed in grand-offspring generations or later) and here we will explore the pathways through which parental influences may persist across multiple generations leading to the stable inheritance of an epigenetically-mediated phenotype.

Epigenetic Consequences of Prenatal Maternal Exposures

The quality of the maternal nutritional environment during pregnancy can have a significant impact on the growth and development of the fetus, with long-term consequences for brain development and metabolism [11–13]. Epidemiological studies of cohorts exposed prenatally to conditions of famine, as in the Dutch Hunger Winter, suggest a heightened risk of schizophrenia and other neurodevelopmental abnormalities – with the specific consequences dependent on the timing of exposure to maternal undernutrition [14,15]. Analysis of blood samples from siblings gestated during periods with or without maternal famine indicates that there is decreased DNA methylation of the insulin-like growth factor II (*IGF2*) gene as a consequence of maternal periconceptual exposure to famine [16]. Laboratory studies in rodents have subsequently identified specific nutritional deficits, such as prenatal protein restriction or folic acid/choline deficiency, as having similar epigenetic consequences. Offspring of female rats placed on a protein deficient diet throughout gestation were found to have elevated hepatic glucocorticoid receptor (*GR*) and peroxisomal proliferator-activated receptor (*PPAR*) gene expression associated with decreased DNA methylation of these genes [17,18]. Moreover, these epigenetic effects are not observed when gestational protein restriction is accompanied by folic acid supplementation [17]. Dietary effects on levels of DNMT1 may account for these observed modifications in global and gene-specific methylation, as *DNMT1* expression is increased in hepatic [19] and brain tissue [20] as a function of protein/choline restriction. The impact of dietary supplementation with methyl-donors during fetal development is also clearly demonstrated by the consequences for phenotype among mice with the A^{vy} allele of the Agouti gene or $Axin^{Fu}$ epiallele of the Axin gene. The expression of these alleles is epigenetically regulated through levels of DNA methylation, with decreased methylation associated with yellow coat color and obesity among A^{vy} mice or a "kinky" tail phenotype among $Axin^{Fu}$ mice [21,22]. Though there is typically an epigenetic inheritance of these phenotypes, gestational exposure to methyl donors through dietary supplementation of the mother can effectively silence the expression of these alleles with the consequence of inducing a pseudo wild-type phenotype [23,24]. Thus the maternal nutritional environment can have a sustained impact on development through alterations in gene expression that are maintained through DNA methylation. Though the focus of these nutritional studies (as well as the majority of studies to be described further in this chapter) has been on epigenetic modifications within candidate genes implicated in the outcome of interest, it is likely that the transcriptional activity of multiple genes is altered by these experiences with the role of DNA methylation in these experience-dependent genome-wide changes yet to be determined.

The rapid period of cellular proliferation and differentiation that occurs during fetal development provides a critical window during which maternal gestational exposure to toxins may lead to long-term disruptions in offspring and there is increasing evidence for the epigenetic basis of these effects. *In utero* methyl mercury exposure in mice has been shown to lead to DNA hypermethylation, increased histone tri-methylation and decreased histone acetylation within the IV promoter of the brain derived neurotrophic factor (*BDNF*) gene in the hippocampus of offspring and is associated with depressive-like behaviors [25]. Exposure of pregnant mice to inhaled diesel exhaust particles combined with an allergen results in altered offspring immunoglobulin (IgE) levels associated with hypermethylation of the interferon (*IFN*)-gamma promoter and hypomethylation of the interleukin (*IL*)-4 promoter [26]. Altered DNA methylation within these immune pathways may account for observed maternal effects of prenatal smoking on offspring asthma risk [27]. In rats, prenatal exposure to the anti-androgenic fungicide vinclozolin or the estrogenic pesticide methoxychlor results in increased rates of prostate disease, kidney disease, immune system abnormalities, testis abnormalities, and tumor development [28]. Though the molecular pathways through which these endocrine disrupting chemicals exert epigenetic modifications has yet to be determined, this exposure is associated with altered DNA methylation patterns in sperm and impairments in reproduction in male offspring [29]. *In utero* exposure to the endocrine disruptor bisphenol-A (BPA) has been demonstrated to induce widespread changes in promoter methylation in the fetal mouse brain, with consequences for neural development [30]. BPA-induced hypomethylation of the A^{vy} allele in mice leads to metabolic abnormality and obesity in adulthood. Interestingly, these toxin induced effects can be reversed through folate supplementation in the mother's diet [31], suggesting that abnormalities in DNA methylation can be ameliorated through exposure to increased levels of methyl-donors.

Evidence for the epigenetic influence of antenatal maternal mood has emerged from human cohort studies and animal models – providing further support for the role of epigenetic mechanisms in mediating developmental outcomes. Analysis of cord blood samples from infants born to mothers with elevated ratings of depression (using the Hamilton Depression Scale) during the 3rd trimester of pregnancy indicates elevated GR 1_F promoter DNA methylation levels associated with maternal depressed mood [32]. Moreover, the level of methylation within the neonatal GR 1_F promoter predicts increased salivary cortisol levels of infants at 3 months of age, and these effects are independent of exposure to selective serotonin reuptake inhibitors during pregnancy. This study provides preliminary evidence for the utility of using epigenetic markers within blood samples to predict developmental outcomes; however, the relationship between these markers and changes in the brain in human cohorts remains an issue of debate. In rodents, the long-term consequences of prenatal stress for brain and behavior have been explored with recent evidence of altered gene expression and DNA methylation within the placenta and hypothalamus as possible mediators of these maternal effects. In mice, chronic variable stress during the 1st trimester is associated with decreased DNA methylation of the corticotrophin-releasing-factor (*CRF*) gene promoter and increased methylation of the GR exon 1_7 promoter region in hypothalamic tissue of adult male offspring [33]. Gestational stress within these experiments was found to exert sex-specific effects on the expression of *DNMT1* in the placenta which may induce disruption of the epigenetic status of genes within this critical interface between mother and fetus. Imprinted genes, such as *IGF2*, may be particularly sensitive to this disruption, leading to impairments in placental growth and function with subsequent consequences for offspring growth and neurodevelopment [34].

Postnatal Maternal Regulation of the Epigenome

Though dynamic epigenetic modifications were once thought to be limited to the very early stages of development, evidence for continued parental influence on DNA methylation beyond the prenatal period has challenged this view. Studies of the effects of natural variations in postnatal care in rodents have established the mediating role of epigenetic factors in shaping individual differences in brain and behavior [9,35]. Postnatal maternal licking/grooming (LG) behavior, in particular, has been found to induce increased hippocampal *GR* expression leading to more efficient negative feedback of the stress response, and cross-fostering studies have confirmed that these effects are mediated by the level of maternal care received during postnatal development [36,37]. Analysis of the GR 1_7 promoter region suggests that variations in *GR* expression associated with differential levels of maternal care are maintained though altered DNA methylation [38]. Thus, offspring who receive high levels of maternal LG during the early postnatal period have decreased hippocampal $GR\ 1_7$ promotor methylation, increased GR expression, and decreased stress responsivity whereas low levels of LG are associated with increased $GR\ 1_7$ methylation, decreased GR expression, and an increased HPA response to stress. Time course analysis has indicated that these maternally-induced epigenetic profiles emerge during the postnatal period and are sustained into adulthood [38]. The pathways through which these effects are achieved are currently being elucidated and it appears likely that maternal LG mediated up-regulation of nerve growth factor inducible protein A (NGFI-A) in infancy may be critical to activating GR transcription and maintaining low levels of DNA methylation within the $GR\ 1_7$ promoter. Though the exploration of these brain region-specific maternal effects in humans is limited by the inaccessibility of brain tissue, recent studies have illustrated the long-term effects of childhood abuse on hippocampal DNA methylation patterns of suicide victims [39,40]. Analysis of hippocampal tissue from suicide victims with a history of childhood abuse indicates decreased GR expression and elevated $GR\ 1_F$ promoter methylation associated with disruptions of the early environment and confirms the potential role of NGFI-A as a mediator of differential GR promoter methylation. Early life effects on GR signaling pathways in humans are further illustrated by a recent genome-wide analysis of gene expression of peripheral blood mononuclear cells from healthy adults who had experienced conditions of low vs. high socioeconomic (SES) status during childhood, with low childhood SES associated with a down-regulation of genes containing GR response elements [41].

Paternal Influence on Offspring Development

Mammalian development is characterized by intense prenatal and postnatal mother–infant interactions and thus studies of parental influence have primarily focused on maternal rather than paternal effects. However, even among species in which biparental care is not typical, significant paternal modulation of offspring development has been observed. In rodents, pre-mating exposure of males to alcohol is associated with reduced offspring litter size, reduced birth weight, increased mortality, and numerous cognitive and behavioral abnormalities [42–47]. Likewise, offspring of cocaine-exposed males perform poorly on tests of visuo-spatial attention, spatial working memory, and spontaneous alternation and have a reduced cerebral volume [48,49]. When pre-mating housing conditions of male mice lead to reduced oxygen and increased carbon dioxide, female offspring are found to have elevated blood hemoglobin [50]. Among isogenic Balb/c mice, offspring anxiety-like behavior can be predicted based on paternal levels of open-field exploration, even when offspring have had no interaction with their fathers [51]. Significantly, these effects persist when factors such as maternal care, litter characteristics, and duration of time the male was housed with the mother during the mating period are statistically controlled. Moreover, variation in the dietary environment of fathers appears to be transmissible to offspring. For instance, reduced serum glucose and altered levels of corticosterone and IGF1 are found among offspring of male mice that undergo a 24-hour complete fast two weeks before mating [52]. Finally, epidemiological studies in humans have demonstrated increased risk of autism and schizophrenia that emerge as a function of increased paternal age [53–55]. Laboratory studies of paternal age effects in genetically-identical rodents also indicate that offspring of "old" fathers have reduced longevity and perform more poorly on learning and memory tasks [56–58]. The transmission of these paternal effects to offspring in the absence of any postnatal contact with fathers suggests that these exposures may lead to alterations in the male germline with consequences for early embryonic development.

Investigation of the role of epigenetic mechanisms in mediating these paternal effects suggests that environmentally-induced changes in DNA methylation within sperm cells may be transmitted to offspring with implications for development. In the case of paternal age, hypermethylation of ribosomal DNA has been found in the sperm and liver cells of "old"

(21–28 months) compared to "young/adult" (6 months) male rats [59], and twin studies suggest that a drift in epigenetic patterns of various cell types occurs with age, such that "old" twins have relatively divergent DNA methylation patterns compared to "young" twins [60]. Though there are many genetic and morphological abnormalities in sperm associated with aging, these epigenetic modifications may contribute to the aberrant developmental outcomes associated with increasing paternal age. In males, chronic exposure to alcohol or cocaine can induce chromatin remodeling and changes in DNA methylation within numerous genes in both the brain and periphery [61–63]. In particular, alcohol exposure has been shown to decrease *DNMT* mRNA levels in the sperm cells of adult male rats [64] and chronic cocaine exposure in adult male mice has been shown to decrease *DNMT1* while increasing *DNMT3* mRNA expression in the germ cell-rich cells of the seminiferous tubules of the testes [49]. Altered DNMT levels may have particular implications for imprinted regions within the genome as analysis of sperm DNA methylation levels in heavy drinkers indicates reduced methylation in the normally hypermethylated *H19* and *IG* regulatory regions [62]. Thus, environmental exposures in males may lead to altered levels of enzymes involved in maintenance of epigenetic marks, with possible paternal transmission of the epigenetic abnormalities to offspring.

Transgenerational Effects of Parental Influence

The stability of epigenetic modifications within an individual's own development and evidence supporting a transmission of parental epigenetic changes to offspring provide a new perspective on the stable inheritance of traits. Moreover, there is increasing evidence that this non-genomic inheritance can be maintained over multiple generations, such that in addition to the developmental effects of parental experiences on offspring, there may be observed influences of parental (F0) experiences on grand-offspring (F2) and possibly great-grand-offspring (F3). In general, there may be two distinct routes through which these types of epigenetic inheritance patterns can occur: germline-mediated vs. experience-dependent/non-germline-mediated (Fig. 24.1). Within germline-mediated transgenerational effects, grandparental environmental exposures are thought to induce epigenetic alterations within the developing gametes that persist in the absence of continued exposure with consequences for F1, F2, and F3 generations. In contrast, experience-dependent/non-germline mediated epigenetic transmission requires that a particular experience or environmental exposure be repeated in each generation to re-establish the epigenetic modifications which permit the trait to persist in subsequent generations. The distinction between these two routes can be difficult to establish experimentally, particularly in the case of prenatal exposures in which F1 offspring and the F1 offspring's germline, which will give rise to the F2 generation, are exposed to the inducing environmental factor. Though both of these processes can lead to the stable inheritance of phenotype, there is certainly divergence in the routes through which this is achieved.

Germline-mediated Transgenerational Inheritance

Evidence for the transgenerational impact of early life nutrition or prenatal exposure to toxins/chemicals provides support for an inheritance pattern that is likely germline-mediated, though in many cases, the specific effect on the germline has yet to be elucidated and the experimental design may not conclusively identify the effect as being independent from experiences occurring during formation of the germline which persist through developmental effects. Analysis of archival records from Sweden in which crop success (used as a proxy for food intake) and longevity can be determined in multiple generations, suggests that in humans, a high level of nutrition during the slow growth period that precedes puberty is associated with diabetes and cardiovascular disease mortality of grand-offspring [65,66]. Interestingly, these effects are sex-specific, with paternal grandfather nutrition predicting grandson mortality and paternal grandmother nutrition predicting grand-daughter

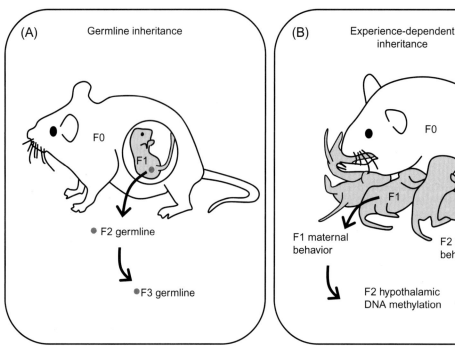

FIGURE 24.1

Illustration of the distinction between a germline epigenetic inheritance (A) and an experience-dependent inheritance of an epigenetic effect (B). In an example of a germline inheritance, an environmental exposure occurring during prenatal development results in an epigenetic alteration within the F1 germline that is transmitted to F2 and F3 generation offspring. In contrast, experience-dependent inheritance, such as the transmission of maternal behavior across generations, requires that each generation is exposed to differential maternal care in infancy.

longevity [67]. Laboratory studies in rodents have confirmed the transgenerational impact of nutrition and indicate that prenatal protein restriction can exert effects on growth and metabolism of offspring and grand-offspring through changes in methylation status of *GR* [68]. When F0 female mice are exposed to caloric restriction during late gestation, F2 grand-offspring are found to have impaired glucose tolerance and this effect is maintained even when the F1 generation is provided with *ad libitum* food throughout their lifetime. In human cohort studies, paternal consumption of betel nuts (which contain nitrosamines) leads to dose-dependent increases in offspring risk of metabolic syndrome [69] and in transgenerational studies of mice, 2–6 days of betel nut consumption by F0 generation males was found associated with increased glucose intolerance amongst F1, F2, and F3 generation offspring [70]. Similar metabolic effects are observed when males are exposed *in utero* to dexamethasone, with increased glucose intolerance observed among the offspring of these males when mated with non-exposed females [71]. However, in the case of prenatal dexamethasone exposure, these metabolic phenotypes do not persist beyond the F2 generation indicating that there is either compensation for the germline effects or that the effect is mediated by experience-dependent transmission.

The consequences of *in utero* exposure to endocrine-disrupting compounds has also been explored within a transgenerational model and provides evidence for the pervasive effects on epigenetic profiles of these early life exposures. In humans, matrilineal transmission of the effects of diethylstilbestrol (DES)-induced hypomethylation and increased cancer risk has been observed in daughters and granddaughters [72]. *In utero* exposure to vinclozolin in rats has been demonstrated to disrupt DNA methylation in sperm and increase rates of infertility and risk of prostrate and kidney disease in F1, F2, and F3 offspring with the transmission though the patriline [29]. Vinclozolin-induced alterations in gene expression within the hippocampus and amygdala have also been observed for up to three generations post-exposure with sex-specific effects on anxiety-like behavior [73]. Interestingly, mate-choice studies suggest that females presented with F3 vinclozolin-exposed or non-exposed males show a significant partner preference for non-exposed males, indicating an additional measure of decreased reproductive success as a consequence of treatment with endocrine disruptors [74]. The persistence of these disruptions beyond the F2 generation suggests that the effects of these exposures have become incorporated into the germline and there is incomplete erasure of the associated epigenetic marks during the process of gametogenesis, fertilization, and embryogenesis [75]. The sensitivity of sperm and oocytes to epigenetic disruption is further illustrated in findings of increased incidence of imprinting disorders, such as Angelmann Syndrome and Beckwith–Wiedemann Syndrome, occurring following *in vitro* conception using assisted reproductive technology (ART) [76]. Ovarian stimulation with gonadotropins and the quality of the embryo culture medium have been found to alter DNA methylation and gene expression [77–79], particularly within imprinted genes, and there is recent evidence that *in vitro* conception is associated with decreased DNA methylation within the placenta and increased methylation within cord blood samples [80]. Thus, understanding the mechanisms through which these environmental effects lead to alterations in the epigenome will have significant implications for offspring disease risk.

Experience-dependent Transgenerational Inheritance

Across species, there is evidence for the transmission of individual differences in maternal behavior from mother to offspring and grand-offspring. In humans, mother–infant attachment classifications (secure, anxious/resistant, avoidant, disorganized) are similar across generations of female offspring [81,82] as are levels of parental bonding [83]. In rhesus and pigtail macaques, the frequency of postpartum maternal behavior has been observed to be transmitted across matrilines as are rates of maternal rejection and infant abuse [84–86]. Cross-fostering studies conducted between abusive and non-abusive macaques females indicates that the transmission of abusive behavior from mother to daughter is dependent on the experience of abuse during the postnatal period [87]. As such, females born to abusive mothers who are then fostered to a non-abusive mother do not show increased rates of infant abuse. This matrilineal transmission is also also evident in laboratory rodents. Natural variations in maternal LG observed in the F0 generation are associated with similar levels of LG in F1 and F2 generation females [88,89]. As such, under stable environmental conditions, offspring and grand-offspring of Low LG females display low levels of LG whereas offspring and grand-offspring of High LG females display high levels of LG. Similar to the transgenerational effects of abuse in macaques, cross-fostering studies have demonstrated that the transmission of maternal LG from mother to female offspring is dependent on the level of maternal LG received in infancy [36,88]. Further evidence for the experience-dependent nature of these effects comes from studies in which maternal LG is altered, through chronic exposure to stress [90] or manipulation of the juvenile environment [89], leading to a disruption of the inheritance of the predicted maternal phenotype.

There is evidence that epigenetic mechanisms may be critical in mediating the transmission of maternal behavior across generations. Female offspring of low LG mothers exhibit a reduced sensitivity to estrogen and have reduced levels of estrogen receptor α (*ERα*) mRNA in the medial preoptic area (MPOA) of the hypothalamus [91,92]. Analysis of methylation within the *ERα* 1_B promoter region indicates that the experience of low levels of LG in infancy is associated with increased methylation whereas high levels of LG in infancy are associated with low levels of methylation at several sites within the promoter [93]. This differential methylation results in reduced binding of signal transducer and activator of transcription (Stat)5 to the *ERα* promoter with consequences for the transcriptional activity of this gene. Thus, epigenetic modifications to a gene that regulates several aspects of reproduction, including postpartum maternal behavior, results in differential levels of expression of *ERα* in adulthood, which alters estrogen sensitivity and consequently leads to variations in the level of maternal care that these females provide to their own offspring. The transmission from mother to daughter of variations in maternal LG within this transgenerational framework is mediated by the stability of brain region-specific epigenetic modifications that occur in infancy and influence behavior in adulthood [94]. Similar experience-dependent effects of the postnatal environment in rats have been induced through exposure to abuse. Increase in methylation of exon IV of the *BDNF* promoter and consequent decrease in *BDNF* mRNA in the prefrontal cortex has been found in association with exposure to periods of abusive maternal care (dragging, burying etc.) [95]. Moreover, these effects on exon 1_V methylation are perpetuated to the F1 offspring of abused females suggesting a role for epigenetic mechanisms in this transgenerational effect. Enrichment of the postnatal environment in mice through use of communal nursing (multiple mothers and litters housed together) has also been found to alter F1 and F2 offspring brain and behavior [96], though the role of epigenetic effects has not yet been explored within this model. Overall, these studies highlight the stable inheritance of traits that can be achieved through a behavioral transmission of epigenetic modifications.

Epigenetics, Plasticity and Evolving Concepts of Inheritance

Though the study of mechanisms of inheritance and the origins of individual differences has traditionally been the domain of the field of genetics, there is increasing evidence for the role of epigenetic modifications in maintaining environmentally-induced variations in phenotype both within and across generations. The dynamic nature of these epigenetic effects provides a mechanism through which a single genotype can give rise to multiple phenotypic outcomes conferring a heightened level of developmental plasticity to an organism. In contrast to environmentally-induced genetic alterations/mutations, which are thought to be non-directed, there may be adaptive consequences associated with experience-dependent epigenetic modifications. For example, nutritional "programming" of fetal metabolism has been explored as an adaptive consequence of early life experience [97,98] and, as has been described in previous sections, there is clearly a role for epigenetic mechanisms in mediating the effects of variations in prenatal food intake. When the prenatal period is characterized by undernutrition, a "thrifty phenotype" may result which allows an individual to be conservative with regard to energy output and which promotes storage of glucose [99,100] – with adverse health consequences associated with a mismatch between the quality of the prenatal and postnatal nutritional environment [101,102], Similar adaptive consequences may be relevant to the development of heightened HPA reactivity. Though elevated stress responses are typically considered to be a negative outcome and associated with increased susceptibility to physical and psychiatric disease, within an evolutionary perspective, the ability to respond rapidly to threat would be particularly advantageous under conditions of high predation/low resource availability [103]. Laboratory studies of maternal care in rodents suggest that chronic stress and social impoverishment can lead to reduced LG with consequences for the increased stress response of offspring *via* differential methylation of hippocampal *GR* [37,38,89,90]. Though this environmentally-induced phenotype is associated with impaired cognitive performance under standard testing conditions [104], recent evidence suggests that synaptic plasticity is enhanced in offspring of low LG mothers when corticosterone levels are elevated [105]. Thus, the consequences of early life experience can be considered as adaptive or maladaptive dependent on the consistency or "match" between early and later environmental conditions, and epigenetic mechanisms play a critical role in shaping these phenotypic adaptations.

The concept that experience-induced characteristics can be transmitted across generations is reminiscent of Lamarckian theories of use/disuse and the inheritance of acquired characteristics [106]. Though the role of heritable epigenetic modifications in evolutionary processes is still questionable [107], the plasticity that these modifications confer certainly has implications for our understanding of the developmental origins of health and disease. Importantly, there is growing support for transgenerational epigenetic consequences of environmental exposures, though our understanding of the molecular, cellular, and behavioral pathways through which these outcomes are achieved is still in its infancy. Though these epigenetic effects have often been explored from the perspective of pathology, recent evidence suggests that genetically-induced impairments in learning and memory can be overcome through exposure to environmental enrichment – with improvements in cognition persisting across generations [108]. Thus, broadening our concept of inheritance to include both genetic and epigenetic mechanisms may provide insights into effective therapeutic approaches and lead to a greater appreciation of the benefits that can be achieved through intervention in parental and grand-parental generations.

ACKNOWLEDGEMENTS

The authors wish to acknowledge funding received from Grant Number DP2OD001674 from the Office of the Director, National Institutes of Health.

References

1. Feng J, Fouse S, Fan G. Epigenetic regulation of neural gene expression and neuronal function. *Pediatr Res* 2007;**61**:58R–63R.

2. Razin A. CpG methylation, chromatin structure and gene silencing – a three-way connection. *EMBO J* 1998;**17**:4905–8.

3. Turner B. *Chromatin and Gene Regulation*. Oxford: Blackwell Science Ltd; 2001.

4. Fan G, Hutnick L. Methyl-CpG binding proteins in the nervous system. *Cell Res* 2005;**15**:255–61.

5. Strathdee G, Brown R. Aberrant DNA methylation in cancer: potential clinical interventions. *Expert Rev Mol Med* 2002;**2002**:1–17.

6. Goldberg AD, Allis CD, Bernstein E. Epigenetics: a landscape takes shape. *Cell.* 2007;**128**:635–8.

7. Hane AA, Fox NA. Ordinary variations in maternal caregiving influence human infants' stress reactivity. *Psychol Sci* 2006;**17**:550–6.

8. Lippmann M, Bress A, Nemeroff CB, Plotsky PM, Monteggia LM. Long-term behavioural and molecular alterations associated with maternal separation in rats. *Eur J Neurosci* 2007;**25**:3091–8.

9. Meaney MJ. Maternal care, gene expression, and the transmission of individual differences in stress reactivity across generations. *Annu Rev Neurosci* 2001;**24**:1161–92.

10. Sanchez MM, Ladd CO, Plotsky PM. Early adverse experience as a developmental risk factor for later psychopathology: evidence from rodent and primate models. *Dev Psychopathol* 2001;**13**:419–49.

11. Godfrey KM, Barker DJ. Fetal programming and adult health. *Public Health Nutr* 2001;**4**:611–24.

12. Symonds ME, Stephenson T, Gardner DS, Budge H. Long-term effects of nutritional programming of the embryo and fetus: mechanisms and critical windows. *Reprod Fertil Dev* 2007;**19**:53–63.

13. Zeisel SH. Importance of methyl donors during reproduction. *Am J Clin Nutr* 2009;**89**:673S–7S.

14. Hulshoff Pol HE, Hoek HW, Susser E, Brown AS, Dingemans A, Schnack HG, et al. Prenatal exposure to famine and brain morphology in schizophrenia. *Am J Psychiatry* 2000;**157**:1170–2.

15. Susser ES, Lin SP. Schizophrenia after prenatal exposure to the Dutch Hunger Winter of 1944–1945. *Arch Gen Psychiatry* 1992;**49**:983–8.

16. Heijmans BT, Tobi EW, Stein AD, Putter H, Blauw GJ, Susser ES, et al. Persistent epigenetic differences associated with prenatal exposure to famine in humans. *Proc Natl Acad Sci USA* 2008;**105**:17046–9.

17. Lillycrop KA, Phillips ES, Jackson AA, Hanson MA, Burdge GC. Dietary protein restriction of pregnant rats induces and folic acid supplementation prevents epigenetic modification of hepatic gene expression in the offspring. *J Nutr* 2005;**135**:1382–6.

18. Lillycrop KA, Phillips ES, Torrens C, Hanson MA, Jackson AA, Burdge GC. Feeding pregnant rats a protein-restricted diet persistently alters the methylation of specific cytosines in the hepatic PPAR alpha promoter of the offspring. *Br J Nutr* 2008;**100**:278–82.

19. Lillycrop KA, Slater-Jefferies JL, Hanson MA, Godfrey KM, Jackson AA, Burdge GC. Induction of altered epigenetic regulation of the hepatic glucocorticoid receptor in the offspring of rats fed a protein-restricted diet

during pregnancy suggests that reduced DNA methyltransferase-1 expression is involved in impaired DNA methylation and changes in histone modifications. *Br J Nutr* 2007;**97**:1064–73.

20. Kovacheva VP, Mellott TJ, Davison JM, Wagner N, Lopez-Coviella I, Schnitzler AC, et al. Gestational choline deficiency causes global and Igf2 gene DNA hypermethylation by upregulation of Dnmt1 expression. *J Biol Chem* 2007;**282**:31777–88.

21. Morgan HD, Sutherland HG, Martin DI, Whitelaw E. Epigenetic inheritance at the agouti locus in the mouse. *Nat Genet* 1999;**23**:314–18.

22. Rakyan VK, Chong S, Champ ME, Cuthbert PC, Morgan HD, Luu KV, et al. Transgenerational inheritance of epigenetic states at the murine Axin(Fu) allele occurs after maternal and paternal transmission. *Proc Natl Acad Sci USA* 2003;**100**:2538–43.

23. Waterland RA, Dolinoy DC, Lin JR, Smith CA, Shi X, Tahiliani KG. Maternal methyl supplements increase offspring DNA methylation at Axin Fused. *Genesis* 2006;**44**:401–6.

24. Wolff GL, Kodell RL, Moore SR, Cooney CA. Maternal epigenetics and methyl supplements affect agouti gene expression in Avy/a mice. *FASEB J* 1998;**12**:949–57.

25. Onishchenko N, Karpova N, Sabri F, Castren E, Ceccatelli S. Long-lasting depression-like behavior and epigenetic changes of BDNF gene expression induced by perinatal exposure to methylmercury. *J Neurochem* 2008;**106**:1378–87.

26. Liu J, Ballaney M, Al-alem U, Quan C, Jin X, Perera F, et al. Combined inhaled diesel exhaust particles and allergen exposure alter methylation of T helper genes and IgE production in vivo. *Toxicol Sci* 2008;**102**:76–81.

27. Li Y-F, Langholz B, Salam MT, Gilliland FD. Maternal and grandmaternal smoking patterns are associated with early childhood asthma. *Chest* 2005;**127**:1232–41.

28. Anway MD, Leathers C, Skinner MK. Endocrine disruptor vinclozolin induced epigenetic transgenerational adult-onset disease. *Endocrinology* 2006;**147**:5515–23.

29. Anway MD, Cupp AS, Uzumcu M, Skinner MK. Epigenetic transgenerational actions of endocrine disruptors and male fertility. *Science* 2005;**308**:1466–9.

30. Yaoi T, Itoh K, Nakamura K, Ogi H, Fujiwara Y, Fushiki S. Genome-wide analysis of epigenomic alterations in fetal mouse forebrain after exposure to low doses of bisphenol A. *Biochem Biophys Res Commun* 2008;**376**:563–7.

31. Dolinoy DC, Huang D, Jirtle RL. Maternal nutrient supplementation counteracts bisphenol A-induced DNA hypomethylation in early development. *Proc Natl Acad Sci USA* 2007;**104**:13056–61.

32. Oberlander TF, Weinberg J, Papsdorf M, Grunau R, Misri S, Devlin AM. Prenatal exposure to maternal depression, neonatal methylation of human glucocorticoid receptor gene (NR3C1) and infant cortisol stress responses. *Epigenetics* 2008;**3**:97–106.

33. Mueller BR, Bale TL. Sex-specific programming of offspring emotionality after stress early in pregnancy. *J Neurosci* 2008;**28**:9055–65.

34. Reik W, Constancia M, Fowden A, Anderson N, Dean W, Ferguson-Smith A, et al. Regulation of supply and demand for maternal nutrients in mammals by imprinted genes. *J Physiol* 2003;**547**:35–44.

35. Szyf M, Weaver IC, Champagne FA, Diorio J, Meaney MJ. Maternal programming of steroid receptor expression and phenotype through DNA methylation in the rat. *Front Neuroendocrinol* 2005;**26**:139–62.

36. Francis D, Diorio J, Liu D, Meaney MJ. Nongenomic transmission across generations of maternal behavior and stress responses in the rat. *Science* 1999;**286**:1155–8.

37. Liu D, Diorio J, Tannenbaum B, Caldji C, Francis D, Freedman A, et al. Maternal care, hippocampal glucocorticoid receptors, and hypothalamic-pituitary-adrenal responses to stress. *Science* 1997;**277**:1659–62.

38. Weaver IC, Cervoni N, Champagne FA, D'Alessio AC, Sharma S, Seckl JR, et al. Epigenetic programming by maternal behavior. *Nat Neurosci* 2004;**7**:847–54.

39. McGowan PO, Sasaki A, D'Alessio AC, Dymov S, Labonte B, Szyf M, et al. Epigenetic regulation of the glucocorticoid receptor in human brain associates with childhood abuse. *Nat Neurosci* 2009;**12**:342–8.

40. McGowan PO, Sasaki A, Huang TC, Unterberger A, Suderman M, Ernst C, et al. Promoter-wide hypermethylation of the ribosomal RNA gene promoter in the suicide brain. *PLoS ONE* 2008;**3**:e2085.

41. Miller GE, Chen E, Fok AK, Walker H, Lim A, Nicholls EF, et al. Low early-life social class leaves a biological residue manifested by decreased glucocorticoid and increased proinflammatory signaling. *Proc Natl Acad Sci USA* 2009;**106**:14716–21.

42. Abel E. Paternal contribution to fetal alcohol syndrome. *Addict Biol* 2004;**9**:127–33. Discussion 135–26.

43. Bielawski DM, Abel EL. Acute treatment of paternal alcohol exposure produces malformations in offspring. *Alcohol (Fayetteville, N.Y.)* 1997;**14**:397–401.

44. Cicero TJ, Nock B, O'Connor LH, Sewing BN, Adams ML, Meyer ER. Acute paternal alcohol exposure impairs fertility and fetal outcome. *Life Sci* 1994;**55**:PL33–6.

45. Ledig M, Misslin R, Vogel E, Holownia A, Copin JC, Tholey G. Paternal alcohol exposure: developmental and behavioral effects on the offspring of rats. *Neuropharmacology* 1998;**37**:57–66.

46. Meek LR, Myren K, Sturm J, Burau D. Acute paternal alcohol use affects offspring development and adult behavior. *Physiol Behav* 2007;**91**:154–60.

47. Wozniak DF, Cicero TJ, Kettinger L 3rd, Meyer ER. Paternal alcohol consumption in the rat impairs spatial learning performance in male offspring. *Psychopharmacology* 1991;**105**:289–302.

48. Abel EL, Moore C, Waselewsky D, Zajac C, Russell LD. Effects of cocaine hydrochloride on reproductive function and sexual behavior of male rats and on the behavior of their offspring. *J Androl* 1989;**10**:17–27.

49. He F, Lidow IA, Lidow MS. Consequences of paternal cocaine exposure in mice. *Neurotoxicol Teratol* 2006;**28**:198–209.

50. Kahn AJ. Alteration of paternal environment prior to mating: effect on hemoglobin concentration in offspring of CF1 mice. *Growth* 1970;**34**:215–20.

51. Alter MD, Gilani AI, Champagne FA, Curley JP, Turner JB, Hen R. Paternal transmission of complex phenotypes in inbred mice. *Biol Psychiatry* 2009;**66**:1061–6.

52. Anderson LM, Riffle L, Wilson R, Travlos GS, Lubomirski MS, Alvord WG. Preconceptional fasting of fathers alters serum glucose in offspring of mice. *Nutrition* 2006;**22**:327–31.

53. Brown AS, Schaefer CA, Wyatt RJ, Begg MD, Goetz R, Bresnahan MA, et al. Paternal age and risk of schizophrenia in adult offspring. *Am J Psychiatry* 2002;**159**:1528–33.

54. Malaspina D, Harlap S, Fennig S, Heiman D, Nahon D, Feldman D, et al. Advancing paternal age and the risk of schizophrenia. *Arch Gen Psychiatry* 2001;**58**:361–7.

55. Reichenberg A, Gross R, Weiser M, Bresnahan M, Silverman J, Harlap S, et al. Advancing paternal age and autism. *Arch Gen Psychiatry* 2006;**63**:1026–32.

56. Auroux M. [Age of the father and development]. *Contracep, Fertilite, Sexualite (1992)* 1993;**21**:382–5.

57. Garcia-Palomares S, Navarro S, Pertusa JF, Hermenegildo C, Garcia-Perez MA, Rausell F, et al. Delayed fatherhood in mice decreases reproductive fitness and longevity of offspring. *Biol Reprod* 2009;**80**:343–9.

58. Garcia-Palomares S, Pertusa JF, Minarro J, Garcia-Perez MA, Hermenegildo C, Rausell F, et al. Long-term effects of delayed fatherhood in mice on postnatal development and behavioral traits of offspring. *Biol Reprod* 2009;**80**:337–42.

59. Oakes CC, Smiraglia DJ, Plass C, Trasler JM, Robaire B. Aging results in hypermethylation of ribosomal DNA in sperm and liver of male rats. *Proc Natl Acad Sci USA* 2003;**100**:1775–80.

60. Fraga MF, Ballestar E, Paz MF, Ropero S, Setien F, Ballestar ML, et al. Epigenetic differences arise during the lifetime of monozygotic twins. *Proc Natl Acad Sci USA* 2005;**102**:10604–9.

61. Novikova SI, He F, Bai J, Cutrufello NJ, Lidow MS, Undieh AS. Maternal cocaine administration in mice alters DNA methylation and gene expression in hippocampal neurons of neonatal and prepubertal offspring. *PloS ONE* 2008;**3**:e1919.

62. Ouko LA, Shantikumar K, Knezovich J, Haycock P, Schnugh DJ, Ramsay M. Effect of alcohol consumption on CpG methylation in the differentially methylated regions of H19 and IG-DMR in male gametes: implications for fetal alcohol spectrum disorders. *Alcohol Clin Exp Res* 2009;**33**:1615–27.

63. Pandey SC, Ugale R, Zhang H, Tang L, Prakash A. Brain chromatin remodeling: a novel mechanism of alcoholism. *J Neurosci* 2008;**28**:3729–37.

64. Bielawski DM, Zaher FM, Svinarich DM, Abel EL. Paternal alcohol exposure affects sperm cytosine methyltransferase messenger RNA levels. *Alcohol Clin Exp Res* 2002;**26**:347–51.

65. Kaati G, Bygren LO, Edvinsson S. Cardiovascular and diabetes mortality determined by nutrition during parents' and grandparents' slow growth period. *Eur J Hum Genet* 2002;**10**:682–8.

66. Kaati G, Bygren LO, Pembrey M, Sjostrom M. Transgenerational response to nutrition, early life circumstances and longevity. *Eur J Hum Genet* 2007;**15**:784–90.

67. Pembrey ME, Bygren LO, Kaati G, Edvinsson S, Northstone K, Sjostrom M, et al. Sex-specific, male-line transgenerational responses in humans. *Eur J Hum Genet* 2006;**14**:159–66.

68. Zambrano E, Martinez-Samayoa PM, Bautista CJ, Deas M, Guillen L, Rodriguez-Gonzalez GL, et al. Sex differences in transgenerational alterations of growth and metabolism in progeny (F2) of female offspring (F1) of rats fed a low protein diet during pregnancy and lactation. *J Physiol* 2005;**566**:225–36.

69. Chen TH, Chiu YH, Boucher BJ. Transgenerational effects of betel-quid chewing on the development of the metabolic syndrome in the Keelung community-based integrated screening program. *Am J Clin Nutr* 2006;**83**:688–92.

70. Boucher BJ, Ewen SW, Stowers JM. Betel nut (*Areca catechu*) consumption and the induction of glucose intolerance in adult CD1 mice and in their F1 and F2 offspring. *Diabetologia* 1994;**37**:49–55.

71. Drake AJ, Walker BR, Seckl JR. Intergenerational consequences of fetal programming by in utero exposure to glucocorticoids in rats. *Am J Physiol* 2005;**288**:R34–R8.

72. Newbold RR, Padilla-Banks E, Jefferson WN. Adverse effects of the model environmental estrogen diethylstilbestrol are transmitted to subsequent generations. *Endocrinology* 2006;**147**:S11–S7.

73. Skinner MK, Anway MD, Savenkova MI, Gore AC, Crews D. Transgenerational epigenetic programming of the brain transcriptome and anxiety behavior. *PloS ONE* 2008;**3**:e3745.

74. Crews D, Gore AC, Hsu TS, Dangleben NL, Spinetta M, Schallert T, et al. Transgenerational epigenetic imprints on mate preference. *Proc Natl Acad Sci USA* 2007;**104**:5942–6.

75. Skinner MK. What is an epigenetic transgenerational phenotype? F3 or F2. *Reprod Toxicol* 2008;**25**:2–6.

76. Allen C, Reardon W. Assisted reproduction technology and defects of genomic imprinting. *BJOG.* 2005;**112**:1589–94.

77. Fortier AL, Lopes FL, Darricarrere N, Martel J, Trasler JM. Superovulation alters the expression of imprinted genes in the midgestation mouse placenta. *Hum Mol Genet* 2008;**17**:1653–65.

78. Sato A, Otsu E, Negishi H, Utsunomiya T, Arima T. Aberrant DNA methylation of imprinted loci in superovulated oocytes. *Hum Reprod* 2007;**22**:26–35.

79. Khosla S, Dean W, Reik W, Feil R. Culture of preimplantation embryos and its long-term effects on gene expression and phenotype. *Hum Reprod Update* 2001;**7**:419–27.

80. Katari S, Turan N, Bibikova M, Erinle O, Chalian R, Foster M, et al. DNA methylation and gene expression differences in children conceived in vitro or in vivo. *Hum Mol Genet* 2009;**18**:3769–78.

81. Benoit D, Parker KC. Stability and transmission of attachment across three generations. *Child Dev* 1994;**65**:1444–56.

82. Sroufe LA. Attachment and development: a prospective, longitudinal study from birth to adulthood. *Attach Hum Dev* 2005;**7**:349–67.

83. Miller L, Kramer R, Warner V, Wickramaratne P, Weissman M. Intergenerational transmission of parental bonding among women. *J Am Acad Child Adolesc Psychiatry* 1997;**36**:1134–9.

84. Berman C. Intergenerational transmission of maternal rejection rates among free-ranging rhesus monkeys on Cayo Santiago. *Anim Behav* 1990;**44**:247–58.

85. Maestripieri D, Tomaszycki M, Carroll KA. Consistency and change in the behavior of rhesus macaque abusive mothers with successive infants. *Dev Psychobiol* 1999;**34**:29–35.

86. Maestripieri D, Wallen K, Carroll KA. Infant abuse runs in families of group-living pigtail macaques. *Child Abuse & Neglect* 1997;**21**:465–71.

87. Maestripieri D. Early experience affects the intergenerational transmission of infant abuse in rhesus monkeys. *Proc Natl Acad Sci USA* 2005;**102**:9726–9.

88. Champagne FA, Francis DD, Mar A, Meaney MJ. Variations in maternal care in the rat as a mediating influence for the effects of environment on development. *Physiol Behav* 2003;**79**:359–71.

89. Champagne FA, Meaney MJ. Transgenerational effects of social environment on variations in maternal care and behavioral response to novelty. *Behav Neurosci* 2007;**121**:1353–63.

90. Champagne FA, Meaney MJ. Stress during gestation alters postpartum maternal care and the development of the offspring in a rodent model. *Biol Psychiatry* 2006;**59**:1227–35.

91. Champagne F, Diorio J, Sharma S, Meaney MJ. Naturally occurring variations in maternal behavior in the rat are associated with differences in estrogen-inducible central oxytocin receptors. *Proc Natl Acad Sci USA* 2001;**98**:12736–41.

92. Champagne FA, Weaver IC, Diorio J, Sharma S, Meaney MJ. Natural variations in maternal care are associated with estrogen receptor alpha expression and estrogen sensitivity in the medial preoptic area. *Endocrinology* 2003;**144**:4720–4.

93. Champagne FA, Weaver IC, Diorio J, Dymov S, Szyf M, Meaney MJ. Maternal care associated with methylation of the estrogen receptor-alpha1b promoter and estrogen receptor-alpha expression in the medial preoptic area of female offspring. *Endocrinology* 2006;**147**:2909–15.

94. Champagne FA. Epigenetic mechanisms and the transgenerational effects of maternal care. *Frontiers in Neuroendocrinol* 2008;**29**:386–97.

95. Roth TL, Lubin FD, Funk AJ, Sweatt JD. Lasting epigenetic influence of early-life adversity on the BDNF gene. *Biol Psychiatry* 2009;**65**:760–9.

96. Curley JP, Davidson S, Bateseon P, Champagne FA. Social enrichment during postnatal development induces transgenerational effects on emotional and reproductive behavior in mice. *Front Behav Neurosci* 2009;**3**:25.

97. Barker DJ. Fetal origins of coronary heart disease. *BMJ (Clinical Research edn)* 1995;**311**:171–4.

98. McMillen IC, Robinson JS. Developmental origins of the metabolic syndrome: prediction, plasticity, and programming. *Physiol Rev* 2005;**85**:571–633.

99. Wells JC. The thrifty phenotype as an adaptive maternal effect. *Biol Rev Camb Philos Soc* 2007;**82**:143–72.

100. Gluckman PD, Hanson MA. The developmental origins of the metabolic syndrome. *Trends Endocrinol Metab* 2004;**15**:183–7.

101. Godfrey KM, Lillycrop KA, Burdge GC, Gluckman PD, Hanson MA. Epigenetic mechanisms and the mismatch concept of the developmental origins of health and disease. *Pediatr Res* 2007;**61**:5R–10R.

102. Gluckman PD, Hanson MA, Pinal C. The developmental origins of adult disease. *Matern Child Nutr* 2005;**1**:130–41.

103. Cameron NM, Champagne FA, Parent C, Fish EW, Ozaki-Kuroda K, Meaney MJ. The programming of individual differences in defensive responses and reproductive strategies in the rat through variations in maternal care. *Neurosci Biobehav Rev* 2005;**29**:843–65.

104. Liu D, Diorio J, Day JC, Francis DD, Meaney MJ. Maternal care, hippocampal synaptogenesis and cognitive development in rats. *Nat Neurosci* 2000;**3**:799–806.

105. Champagne DL, Bagot RC, van Hasselt F, Ramakers G, Meaney MJ, de Kloet ER, et al. Maternal care and hippocampal plasticity: evidence for experience-dependent structural plasticity, altered synaptic functioning, and differential responsiveness to glucocorticoids and stress. *J Neurosci* 2008;**28**:6037–45.

106. Lamarck J.-B. *Philosophie Zoologique*. 1809.

107. Haig D. Weismann Rules! OK? Epigenetics and the Lamarckian temptation. *Biol Philos* 2006;**22**:415–28.

108. Arai JA, Li S, Hartley DM, Feig LA. Transgenerational rescue of a genetic defect in long-term potentiation and memory formation by juvenile enrichment. *J Neurosci* 2009;**29**:1496–502.

403

CHAPTER 25

Aging Epigenetics

Axel Schumacher
Centre for Addiction and Mental Health, Toronto, ON M5T 1R8, Canada

INTRODUCTION

Although there are many theories of aging and longevity, epigenetic hypotheses only constitute a recent addition to the well-established principles that have been developed to explain some aspects of aging. Overall, aging epigenetics is still a neglected field that may have a significant impact on biogerontology and other biomedical fields in the near future. While an endless supply of evidence supporting classical evolutionary theories of aging appears to exist [1], they are always subject to change as new knowledge becomes available. In this context, epigenetics is of special interest, as epigenetic events may play a role, directly or indirectly, in nearly all aging theories. One of the earliest theories of aging is the theory of programmed death, proposed by the German experimental biologist, August Weisman, in 1891 [2]; Weisman was heavily influenced by Darwin's work on evolutionary biology, published 32 years earlier. His initial idea was that a limited lifespan of an organism confers an evolutionary advantage to the species, but not the individual, mainly because older members of the population would no longer compete with younger generations for vital resources. His ideas sprouted numerous new theories, including the *theory of mutation accumulation*, which states that old age is not under selective pressure *per se*. This explains the fact that there is no evolutionary mechanism to rid a population of detrimental mutations in old animals [3]. A second popular theory, the *antagonistic pleiotropy hypothesis*, appeared shortly thereafter, and states that aging evolved due to the pleiotropic effects of genes that are beneficial early in life, but become harmful at late ages [4]. On the epigenetic level, these theories can be combined to form new explanations for observed aging phenomena. For example, epigenetic patterns that cause overproduction of sex hormones, thereby increasing reproductive output, may interfere with other cellular processes in old age after hormone levels in the body are dramatically altered, and may even cause sex-related cancers such as prostate or ovarian cancer. Aging is unlikely to have adaptive traits in humans, because natural selection does not occur in elderly individuals, hence, it does not provide any significant additional contribution to offspring numbers [1]. Since abnormal epigenetic patterns (epimutations) can be induced during aging by various processes and no selective pressure regulates these epimutations, it is possible for hazardous, late-acting epigenetic patterns to exist and form the basis for development of many age-related disorders. Aberrant epigenetic patterns have already been linked to a number of age-related disorders, including cancer, Alzheimer's disease (AD), and autoimmune disorders. Evidence is beginning to mount that the epigenetic code may be – at least in part – responsible for many other age-related diseases, including diabetes, osteoarthritis, chronic pancreatitis, and Parkinson's disease, just to name a few. As there is no longer any doubt that age-dependent epigenetic mechanisms exist, this chapter will focus on some of the components that are relevant to aging epigenetics.

405

Handbook of Epigenetics: The New Molecular and Medical Genetics. DOI: 10.1016/B978-0-12-375709-8.00025-3

DNA METHYLATION CHANGES DURING AGING

DNA methylation is a universal phenomenon observed in bacteria, plants, and animals. In mammals, DNA methylation usually refers to the methylation of cytosine bases at the 5 position (^mC), and is found almost exclusively as symmetrical methylation within CpG dinucleotides. DNA methylation is mediated by a family of conserved DNA methyltransferases (DNMTs). Maintenance methylation, which becomes necessary after each round of DNA replication, is carried out by DNMT1, the most abundant DNA methyltransferase in mammalian cells [5]. The transmission accuracy of epigenetic information to the next cellular generation critically depends on the degree of DNMT1 specificity for hemimethylated DNA. In contrast, DNMT3a and DNMT3b are responsible for *de novo* DNA methylation reactions of unmethylated DNA. Hence, they are able to influence numerous biological processes mediated by gene repression, as DNA methylation has been shown to silence gene expression in a heritable manner. Recently, 5-hydroxymethylcytosine (^{hm}C) was discovered to be an additional DNA modification in humans [6–8]. The formation of 5-hydroxymethylcytosine occurs by conversion of ^mC by the oxygenase, TET1. Overrepresentation of ^{hm}C occurs in regulatory regions, and the base appears to be highly tissue-specific, as it is only present in some tissues. However, since ^{hm}C is a recent discovery, no information exists on the roles of the ^{hm}C modification at molecular and cellular levels, nor is it known if this DNA modification plays an essential role in the aging process.

Dynamics of DNA Methylation During Aging

DNA methylation is a very dynamic process throughout life, especially during early embryogenesis, when methylation patterns change dynamically on larger scales to facilitate X-inactivation, genomic imprinting, or to initiate various differentiation processes. It was found that methylation patterns vary during a single cell cycle, with global levels of DNA methylation decreasing in G1 and increasing during S phase [9]. DNA methylation changes are even observed in fully differentiated, post-mitotic cells. Neurons, for example, show altered DNA methylation in response to environmental changes affecting behavioral tasks, such as learning and memory [10]. Gain or loss of methylation during aging, for example by deamination of methylcytosine, must be prevented by well-organized repair mechanisms. Efficient repair may be particularly critical for cells that are required to function normally for most of the lifetime of an individual, such as post-mitotic neurons.

In general, it can be assumed that a significant number of methylation errors during aging are due to replication errors, which are mainly stochastic in nature. Overall, maintenance methylation is a highly efficient process, with DNMT1 moving along the DNA in a random walk, methylating hemimethylated substrates with high processivity [11]. However, unlike DNA sequence, methylation patterns change noticeably with age in human cells at specific loci, suggesting that epigenetic replication error rates may be significantly greater than genetic replication errors rates [12,13]. Such epimutations seem to occur at dissimilar frequencies in different tissues, because somatic cells within an individual do not have the same mitotic ages (total number of cell divisions since the zygote); this is attributable to the fact that each cell type divides at its own rate, time-period, and developmental stage. Cells with greater mitotic ages (e.g. epithelial cells) would usually accumulate a higher number of epigenetic errors relative to cells with fewer divisions [12,14]. However, methylation changes also occur independent of cell division, since age-dependent DNA methylation changes can also be observed in post-mitotic tissues [15], although not much is known about DNA methylation dynamics in post-mitotic cells. Intriguingly, despite the high expression levels of DNMT1 in post-mitotic neurons [16], mouse models have shown that DNMT1 deficiency in post-mitotic neurons neither affects levels of global DNA methylation nor influences cell survival during postnatal life [17]. The absence of an obvious phenotype raises the question of "Why do neurons retain so much maintenance methyltransferase?". One reasonable role of DNMT1 in neurons might be the re-methylation of cytosine residues that have lost their regulative methyl-group, for example, due to environmental factors or spontaneous deamination of methylcytosine to thymine and subsequent DNA mismatch repair [18]. Alternatively, DNMT1 may still be essential for maintaining DNA methylation if methylation turnover occurs in post-mitotic neurons. In aging cells, striking changes in the gene expression of methyltransferases were observed, with the mRNA of DNMT1 and DNMT3a becoming reduced, while the production of DNMT3b increased steadily [19]. These studies indicate that changes in transcriptional control of the DNMTs are one potential cause for alterations in DNA methylation in aging cells. A study on 50 loci encompassing primarily CpG islands of genes related to CNS growth and development indicated that there is a robust and progressive rise in DNA methylation levels across an organism's lifespan for a variety of loci, such as *GABRA2*, GAD1, *HOXA1*, *NEUROD1*, or *SYK*, usually in conjunction with declining levels of the corresponding mRNAs [20]. Additionally, methylation studies on sorted nuclei have provided evidence for bidirectional methylation events in cortical neurons during the transition from childhood to advanced age, as reflected by a significant increase or decrease in DNA methylation of specific loci [20]. These methylation changes indicate that epigenetic repair mechanisms may not be restricted to maintenance methylation alone, but may also involve active demethylation. Indeed, Mu and colleagues demonstrated that the murine stress response gene *Gadd45b* can be transiently induced by neuronal activity and may promote adult neurogenesis through dynamic DNA demethylation of specific gene promoters, potentially regulating long-lasting changes in neural plasticity in mammalian brains [21].

Global Methylation Changes in Aging Cells

Several studies that examined age-dependent changes in total methylated cytosine content have demonstrated that a large amount of vertebrate tissues tend to demethylate with aging (overview in Richardson 2003, Ref. 22). The highest amount of methylated cytosines in vertebrates is most commonly observed in DNA isolated from embryos and newborn animals, and gradually decreases with aging. This phenomenon may be accelerated under certain conditions, as loss of global DNA methylation has been found in a variety of common human age-related diseases, namely cancer or disorders of the immune system. It is likely that global genomic DNA hypomethylation is partly attributable to demethylation in transposable repetitive elements, including the Alu and LINE-1 sequences, which play a crucial role in gene regulation and genomic stability. It is noteworthy that the majority of all methylated cytosines are found within CpG islands located in these and other repetitive elements. Data from the Boston Normative Aging Study indicated a progressive loss of DNA methylation in specific repetitive elements dispersed throughout the genome [23]. In total, the authors measured DNA methylation in 1097 blood DNA samples from 718 elderly subjects between 55 and 92 years of age, who had been repeatedly evaluated over an 8-year time span. The results demonstrated a significant decrease in average Alu methylation over time compared to blood samples collected up to 8 years earlier. The longitudinal decline in Alu methylation was linear and highly correlated with time since the first measurement, whereas average LINE-1 methylation did not vary over time. Unfortunately, it is difficult to estimate the impact of the observed genome-wide hypomethylation of Alu elements over time, as the changes are relatively small, particularly when compared with the wide interindividual variability in DNA methylation [24]. Nevertheless, it may be speculated that demethylation with age promotes chromosomal translocations by activating transposable elements, which are usually repressed by DNA methylation in humans [25]. Although future follow-up studies will be needed to determine the exact impact of age-dependent hypomethylation on human health, these findings suggest that genome-wide losses of methylation through aging may account, at least in part, for the increased rates of many complex diseases that are not explained by DNA sequence-based genetics [26,27].

The Boston Normative Aging Study is a good example of a prospective study design, demonstrating that a single age-related epigenetic mechanism that affects methylation patterns in the entire human genome is probably unlikely. Most aging studies are not prospective and merely compare different age groups based on average methylation values, rather than paired sampling in the same individuals over time. It is not surprising that the initial phase of the human epigenome project (HEP), which interrogated about 1.9 million CpG methylation values derived from 12 different tissues, failed to identify any significant age-attributable effects on DNA methylation [28], likely due to averaging of methylation levels across individuals for a given age group. To prove that longitudinal changes exist in the global epigenetic patterns of individuals, a prospective study design with direct examination of methylation in the same individuals is indispensable. Using a quantitative luminometric measurement of genome-wide DNA methylation, the effectiveness of this methodology was demonstrated in a landmark paper by Bjornsson and co-workers. The study utilized DNA samples in individuals from an Icelandic and a Utah cohort, in which the samples were collected several years apart [29]. The Icelandic set consisted of 111 individuals who donated DNA at two visits separated on average by 11 years [30]. Although the average methylation difference over this time-period was zero, a wide range of changes were observed in the majority of samples, including significant global methylation gains (up to 26%) and losses (−30%) [29]. In total, 29 percent of the individuals showed greater than 10% methylation change over time. The 126 three-generation family-based Utah samples, derived from the Salt Lake City CEPH pedigrees collected an average of 16 years apart, also showed intra-individual changes over time, and additionally demonstrated familial clustering of methylation change. This clustering occurred for both decreased and increased methylation, indicating that the stringency of global methylation pattern maintenance is itself a heritable trait. Intriguingly, a gene-specific analysis from the same DNA samples demonstrated that within the genes that showed the greatest change over time, there was a significant enrichment of imprinted genes. This is a notable observation, as imprinted genes seem to be very sensitive to environmental factors, such as diet or *in vitro* manipulation [31,32].

Gene-Specific Methylation Changes

Global methylation studies carry only limited information, as some parts of the genome can show an increase in DNA methylation, whereas other loci may exhibit an inverse age-dependent development, resulting in a net modification of zero over the whole epigenome (Fig. 25.1). In most organisms, it is likely that specific genes will experience age-related methylation abnormalities before larger, global changes occur; only in the last stages of an animal's life will variation of methylation levels change exponentially (unpublished data from our lab). Indeed, the list of genes that change their methylation during aging grows steadily, and many of these loci can be linked to human diseases, especially cancer. Such genes include the proto-oncogenes *c-fos* and *c-myc*, which were among the first genes determined to be epigenetically unstable during aging in the liver of aged mice [33]. Similarly, transcriptional silencing of tumor suppressor genes associated with age-dependent DNA methylation is a common epigenetic event in various malignancies. In particular, an age-dependent inactivation is frequently detected at the *RASSF1A* tumor suppressor genes [36] and CDKN2, which is probably the most commonly altered gene known in human cancers [34]. The loss of CDKN2 activity through promoter hypermethylation is a common step in tumor development and progression.

Most of the epigenetically unstable genes currently identified may underlie the age-related risk of cancer development. Cancer is one of the most rigorously studied diseases; hence, the observation of aberrant methylation in cancer-related genes is highly biased. The situation is further complicated by the fact that epigenetic changes are highly tissue-specific. For example, it could be demonstrated that in aged mice, the methylation status of *c-myc* tends to be unaffected in the brain, whereas it is prone to hypermethylation in the liver, and hypomethylation in the spleen at the same time [33].

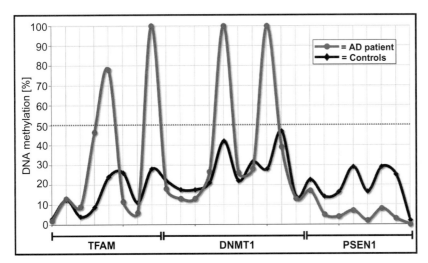

FIGURE 25.1
Gene-specific methylation changes during aging. The overall methylation content of DNA in very old individuals may be stable during aging, as was shown in the case of a 97-year-old Alzheimer's patient [15]. However, the post-mortem brain DNA of this individual (orange [upper] line) exhibited a hypomethylation of the PSEN1 promoter, whereas the promoters of DNMT1 and TFAM were significantly hypermethylated compared to the median brain methylation pattern observed in healthy individuals (black [lower] line). (Please refer to color plate section)

It should be mentioned that gene promoters are not always completely affected over their entire sequence. In a study of autopsied human cerebral cortex, it was found that the changes in methylation status of the *tau* gene differed among transcription factor binding sites with age [35]. CpG sites within AP2-binding sites were never methylated in any of the cases studied at any age, whereas methylation within binding sites for SP1, a transcriptional activator, significantly increased with age. Also, CpG methylation in the binding sites for GCF, a repressor of GC-rich promoters, significantly decreased with age. These findings suggest that the methylation status of the promoter region of the *tau* gene may have changed actively with age to adjust the transcriptional activity of this locus in the human cerebral cortex. In a recent study of 217 non-pathologic human DNA samples from 10 different tissues, methylation profiles were shown to be significantly associated with age and were efficient predictors of tissue origin [36]. In solid tissues, the authors found that loci in CpG islands gained methylation with age, whereas loci outside of CpG islands lost methylation. Furthermore, many of the methylation profiles showed association with environmental factors, such as asbestos, smoking, and alcohol consumption. Some genes may be generally unstable on the epigenetic level, resulting in frequent methylation switches. As a certain gene may be hypermethylated in one cell and demethylated in the next, these methylation changes can be difficult to discover, and may not be detected when analyzing whole tissues where the net effect is zero.

One-Carbon Metabolism in Aging

The presence of familial clustering of age-dependent DNA methylation changes [29] raises the possibility that epigenetic stability might be directly related to genetic and epigenetic variation, for instance, in genes controlling DNA methyltransferase activity or 1-carbon metabolism. Intriguingly, it was found that some genes that are essential components of the 1-carbon metabolism in humans, *MTHFR*, and *DNMT1*, show a notable interindividual variation in DNA methylation [15]. A significant interindividual variance may be an indicator of age-related instability of these genes that, in turn, could influence methylation homeostasis in cells. This can result in aberrant levels of important methylation metabolism components, such as homocysteine (Hcy) and folate (see Fig. 25.2). Many epidemiological

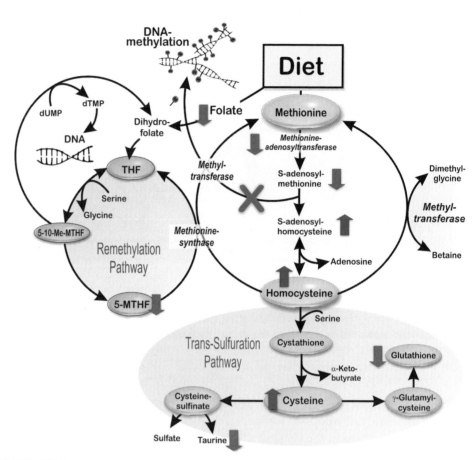

FIGURE 25.2

Aberrant methylation metabolism observed in aging. Several components of methylation pathways are frequently found to be abnormal (see arrows) in age-related disorders, either through environmental, hereditary, or nutritional factors. It was reported that S-adenosylmethionine (SAM), which is required for the methylation of DNA as well as methylation of histones, is severely decreased in the spinal fluid and brains of late-onset Alzheimer patients. A reduction of SAM can be the cause of a decreased uptake of folate and vitamin B12, which in turn induces the accumulation of homocysteine and S-adenosylhomocysteine (SAH). A number of steps in the metabolism of SAM, such as the transformation of Hcy to methionine or a methionine adenosyltransferase deficiency, which is frequently observed in dementia patients, could lead to altered methylation levels within a cell. The accumulation of Hcy reverts the methionine metabolism to SAH, which acts as a strong inhibitor of DNA methyltransferases, resulting ultimately in DNA hypomethylation. Clearance of Hcy by methionine synthase maintains a favorable SAM:SAH ratio, an index of cellular methylation potential. These age-dependent metabolic changes may be very important factors that play a paramount role in the genesis of age-related disorders. THF = tetrahydrofolate; 5-MTHF = 5-methyltetrahydrofolate. (Please refer to color plate section)

and experimental studies have linked elevated plasma Hcy and low serum folate concentrations with several age-related conditions, including stroke, AD, and Parkinson's disease [37].

Epidemiological studies have shown that elevations in plasma Hcy temporally precede the development of age-dependent dementia, and that there is an inverse linear relationship between plasma Hcy concentrations and cognitive performance in older persons [38–40]. These observations indicate that some epigenetic pathways may be disrupted before the main age-related phenotypes occur, such as formation of amyloid plaques in AD. In a healthy cell, low Hcy levels are maintained by the remethylation pathway that requires folate and vitamin B12, and by the activity of the enzyme cystathionine-beta-synthase (CBS) in the trans-sulfuration pathway that converts Hcy to cystathionine; Hcy induces oxidative stress in the cells and impairs DNA repair. Interestingly, this process results in the activation of poly-(ADP ribose) polymerase (PARP), another central component of the epigenetic machinery, which contributes to an epigenetic deregulation in aging cells.

CHANGES OF HIGHER ORDER CHROMATIN STRUCTURE IN AGING

Changes in the composition and higher order structure of chromatin with age represent a major causative mechanism for the deterioration of cellular and tissue functions. The major components of chromatin are DNA and histones, although numerous other chromosomal proteins also have important functions. The smallest unit of chromatin is the nucleosome, which consists of a histone octamer core that contains two copies of histones H2A, H2B, H3, and H4, around which a 147 base pair DNA segment is wrapped in a super-helical turn. To lock the DNA into place, the linker histone H1 binds to the nucleosome at the entry and exit sites of the DNA, therefore allowing the formation of higher order chromatin structures. While age-associated DNA methylation changes are well documented, information on chromatin structural rearrangements in aging has only just begun to accumulate.

Histone Modifications During Aging

Long N-terminal tails protrude from the nucleosome at histones H3 and H4, and several locations on these tails can undergo a number of covalent post-translational modifications, such as acetylation, methylation, sumoylation, phosphorylation, ubiquitination, ribosylation, biotinylation, deamination (citrullination), proline isomerization, and the lesser known carbonylation [41,42]. The different combinations of modifications are thought to constitute a histone code, which directs the binding of transcription factors to the DNA and changes the chromatin structure on a genome-wide scale. A 2005 study analyzed the major DNA methylation and histone-modification differences in aging in a large cohort of monozygotic (MZ) twins, revealing a significant epigenetic drift between siblings during aging [43]. At the time, these changes were associated with phenotypic discordance, which was attributed to an unshared environment, although several of the observed epigenetic changes are more likely to result from stochastic or other influences, rather than environmental effects. Overall, it could be shown that histone modifications change during aging and frequently occur in concert with changes in DNA methylation.

At present, only small subsets of histone modifications have been studied in aging, such as histone methylation at lysine residues. Lysines can be mono-, di-, or trimethylated, and this likely plays a role in transcriptional control and DNA repair [41]. Usually, sites showing trimethylation correlate with transcription starts, while those showing mainly dimethylation occur elsewhere near active genes [44]. One of the first studies of histone methylation detected a significant increase in trimethylated histone H4 (H4K20) in kidney and liver of rats older than 30 days, whereas the amounts of mono- and dimethylated forms did not essentially change in organs from young (10 days) or old animals (30 and 450 days) [45]. Using bone-marrow-derived dendritic cells (DCs) from C57BL/6 mice, it was shown that with age-dependent up-regulation of the newly-discovered cytokine, IL-23, gene expression is associated with chromatin remodeling, characterized by di- and trimethylation of histone H3K4 [46]. In contrast to methylation, it was found that histone carbonylation, which is specific for histone H1, H2A, H2B, and H3, is decreased with age in the rat organs [42]. Whereas a reduction of trimethylation potentially affects multiple processes, including gene activation, gene repression, chromatin condensation, S phase progression, mitosis, and DNA-damage checkpoint signaling [47], the role of carbonylation, a form of oxidative modification of proteins, is not yet understood. Since reduced carbonylation in basic amino acid residues would increase the positive charge in histones, these findings are consistent with the fact that chromatin is more condensed in old animals, consequently reducing gene activities [42].

The most investigated histone modification is histone acetylation, which changes dynamically with developmental age at individual gene regions, and is correlated with the state of transcription [48]. Histone acetylation occurs via histone acetyltransferases (HATs), which acetylate histones H3 and H4 at a number of residues. Acetylation is a reversible process, performed by histone deacetylases (HDACs) that actively deacetylate the histone tails. In a crosstalk with phosphorylation, methylation, ubiquitination, sumoylation, and DNA methylation, lysine acetylation is believed to be a major signal for an active chromatin configuration and dynamic control of cellular pathways [49]. A study of histone modification changes with age in rat liver showed that histone H3 Lys9 acetylation (H3K9ac) was significantly decreased, whereas H3 Ser10 phosphorylation (H3S10ph) was increased with age [50]. These findings may indicate that a decrease in acetylation and an increase in phosphorylation of H3 histones may inhibit gene activity.

HAT:HDAC Balance

Recent data indicate that one of the key determinants of aging is the HAT:HDAC balance, as transcriptional events involve alternate exchange of HDAC complexes with those containing acetylase activities [51]. Such switching in steady-state cells suggests a stoichiometrical balance in HAT and HDAC enzymatic activities, which in turn confers stability to the cellular homeostasis by coordinating time-dependent gene expression. In healthy cells, the HAT:HDAC equilibrium is maintained very stringently and any disruption of this equilibrium severely interferes with cellular homeostasis. Emerging evidence suggests that the disproportionate presence of HATs or HDACs and the resulting loss of acetylation homeostasis is a major reason for neuronal dysfunction, toxicity and, particularly, age-dependent neurodegeneration [51,52]. Severe disruptions of the HAT:HDAC balance can potentially collapse the chromatin structure, thereby obstructing the expression of critical genes. This mechanism was supported by the observation that, in a model of primary neurons, histone acetylation levels decreased at the onset of apoptosis. The hypoacetylation was induced by decreased activity of the CREB-binding protein (CBP), a HAT that also acts as a co-activator controlling CREB-dependent transcriptional activity, which usually exerts neuroprotective functions [53]. A CBP-dependent histone deacetylation was also observed in two different pathological contexts: amyloid precursor protein-dependent signaling and amyotrophic lateral sclerosis model mice, indicating that HAT down-regulation likely contributes to neurodegenerative diseases [53]. Similarly, treatment of neurons with HDAC inhibitors like trichostatin A (TSA) induces neuronal apoptosis [52]. Taken together, alteration of the HAT–HDAC regulatory system, with its involvement in regulation of chromatin structure and transcription machinery, may contribute significantly to the progression of aging.

Sirtuins

In humans, four classes of histone deacetylases have been recognized [54]. Classes I and II are zinc-dependent amidohydrolases, class IV are not well-characterized enzymes classified only by their DNA sequence similarity to other HDACs, whereas class III enzymes depend on their catalysis of NAD(+). Due to homology with the yeast histone deacetylase, SIR2, the seven known human NAD(+) deacetylases are also termed sirtuins. Sirtuins target a wide range of nuclear and cytoplasmic proteins by acetylation (SIRT1, 2, 3, 5) or ADP ribosylation (SIRT4, 6). Through these mechanisms, they regulate a variety of biological processes, such as gene expression, cell cycle, fat mobilization, insulin secretion, and apoptosis. Intriguingly, the sirtuin family members with histone deacetylase activity have also been implicated in modulating lifespan [55]. Of these, human SIRT1 and SIRT2 have gained a lot of attention, as evidence suggests that their expression might be age-dependent and that they are altered in cancer cells [54,56]. Their importance in the aging process may primarily depend on their ability to regulate the cellular responses to stress, which ensures that damaged DNA is repaired and that mutations do not accumulate during aging [57,58].

The first indications that sirtuins play an important role in aging came from studies on the silent information regulator 2 (SIR2) in *Saccharomyces cerevisiae*, where epigenetic changes are a primary cause of aged phenotypes. One function of the SIR2 protein is repression of the silent mating type loci, HML and HMR [59]. As yeast cells age, the SIR protein complex dissociates from the HM loci and moves to the nucleus in response to accumulation of toxic rDNA circles (ERCs) [60]. This redistribution of the chromatin-modifying complex results in loss of HM silencing, which causes sterility, a hallmark of aging in yeast [60,61]. Deletion of SIR2 shortened the lifespan of the cells, whereas an extra copy of this gene increased it [reviewed in Ref. 62]. It is probable that elongation of lifespan is mediated by caloric restrictions that decrease the carbon flow in glycolysis and the tricarboxylic acid (TCA) cycle. In turn, the caloric restriction limits the amount of NAD available for SIR2, directly affecting NAD-dependent gene regulation and chromatin remodeling [56,62].

The mammalian ortholog of SIR2, SIRT1, exhibits a similar NAD-dependent deacetylase activity, primarily modifying histones at H4K16 and H3K9 [63]. SIRT1 is widely expressed in most tissues and regulates the expression of individual genes and the formation of facultative (cell-type specific) heterochromatin. It was linked to DNA damage response mechanisms via p53, and also implicated in the regulation of DNA methylation patterns at damaged CpG-rich regions [64,65]. Using embryonic stem cells, it was shown that SIRT1 represses repetitive DNA, among other loci, across the mouse genome [57]. In response to DNA damage, it dissociates from these loci and relocalizes to the DNA breaks to initiate repair. Similarly, it was found that another member of the sirtuin family, SIRT6, promotes resistance to DNA damage and suppresses genomic instability in mouse cells, in association with a role in base excision repair [66]. Loss of SIRT6, which has a H3K9 deacetylase activity, leads to abnormalities in mice that overlap with aging-associated degenerative processes. Thus, it seems likely that DNA damage-induced redistribution of sirtuins may be a conserved mechanism of aging in eukaryotes.

Epigenetic Control of Telomeres in Aging

413

It is interesting to note that mammalian SIRT6 also has a role in modulation of telomere maintenance [67]. Telomeres are protective structures at eukaryotic chromosome ends that contain a number of hexanucleotide TTAGGG repeats. The protective function is due to the formation of a nucleoprotein complex called shelterin, which binds to the repeats [68], thereby preventing the end of the DNA molecules from activating DNA damage pathways. Usually, telomere length is maintained by telomerase, a reverse transcriptase that adds telomeric repeats *de novo* after each cell division, using an associated RNA molecule (TERC) as a template. This process counteracts incomplete DNA replication of telomeres, a common problem occurring in eukaryotic DNA strands with a 5′ end. If the shelterin complexes are disrupted, the chromosomes become highly unstable and the cellular DNA damage response is activated, resulting in senescence or apoptosis [69]. Nevertheless, the majority of adult cells progressively lose telomeres during cell division and tissue renewal, which is thought to result from limiting amounts of telomerase activity in aging organisms and results in a failure to compensate for the progressive telomere shortening [reviewed in Ref. 69]. These observations led to the proposal that telomere shortening is rate limiting for human lifespan and contributes to the development of age-related disorders [70]. This limited self-renewal capacity results in a finite division potential of human cells, a process that was first proposed nearly 120 years ago by the aforementioned August Weismann [71]. Weismann had the exceptional theoretical speculation that there is a specific limitation on the number of divisions that somatic cells might undergo in the course of an individual's life, and that this number is already determined in the embryo.

In recent years, it has become evident that chromatin modifications are important regulators of mammalian telomeres and adjacent subtelomeric regions, which possess the epigenetic marks of constitutive heterochromatin. The condensed chromatin formation is able to silence nearby genes, apparently due to spreading of silent heterochromatin, a phenomenon called "telomere position effect" [TPE, reviewed in Ref. 72]. TPE involves histone hypoacetylation, as it can be disrupted by treatment with the deacetylase inhibitor, TSA. Similarly to mammalian pericentromeric regions, telomeric regions are characterized by their extremely rich variety of epigenetic modifications, particularly trimethylation of H3K9 and H4K20, dimethylation of H3K79, and low levels of acetylated H3 and H4 [73]. Alterations of histone modifications in telomeric chromatin or of DNA methylation in subtelomeric regions correlate with telomere-length deregulation, indicating that a proper epigenetic regulation of higher-order chromatin structures at telomeres is necessary to control telomere length. For example, it could be demonstrated that telomerase-deficient Terc(−/−) mice with shortened telomeres show decreased trimethylation of histone positions H3K9 and H4K20 in telomeric and subtelomeric chromatin, as well as increased H3 and H4 acetylation at these regions [74]. Furthermore, the Terc(−/−) mice displayed a significant subtelomeric DNA methylation, which acts as a negative regulator of telomere length and telomere recombination independent of histone methylation [75]. The shortening of telomeres observed in normal aging and age-related pathologies is also associated with an increase in H3 and H4 acetylation. It is important to note that accelerated telomere shortening can also occur due to environmental influences, such as smoking, obesity, or stress [76–78]. These epigenetic changes might be an important link between telomere deregulation and typical aging phenotypes, particularly cancer development. Indeed, tumor formation often occurs in the context of altered DNA methylation, loss of H4K20me3, and altered expression of histone-methylases [79]. Dysfunctional telomeric components can also lead to other diseases, for example premature aging syndromes, such as aplastic anemia, dyskeratosis congenita, and idiopathic pulmonary fibrosis [reviewed in Refs 79,80]. Telomere homeostasis can also be affected by mutations in various DNA repair genes, causing several disorders, in particular ataxia telangiectasia, Werner- and Bloom-syndromes, Fanconi anemia, and Nijmegen breakage syndrome. These patients display a substantially increased risk of developing disease states characterized by a loss of tissue renewal, ultimately leading to premature death.

Other Epigenetic Mediators that Influence Longevity

Recently, telomeres have also been shown to generate long, *non-coding RNAs*, named telomeric repeat-containing RNA (TERRA or TelRNA) [81,82]. TERRA molecules are transcribed from a number of subtelomeric loci toward the chromosome ends and remain associated to the telomeric chromatin, a feature similar to the non-coding XIST RNA that controls mammalian dosage compensation. Telomeric RNAs have been proposed to orchestrate chromatin remodeling throughout development and cellular differentiation, and to control telomere elongation by telomerase [82,83]. Up-regulation of TERRA molecules, for example, by defective DNA methylation in the subtelomeric regions, can lead to interference with telomere replication. This can eventually result in a loss of telomere tracts – a phenotype that can be observed in cells from ICF (Immunodeficiency, Centromeric region instability, Facial anomalies) patients.

Overall, it seems that non-coding RNAs (ncRNA), ranging from 0.5 to over 100 kb, are general and widespread determinants that transmit information for the assembly and modification of local chromatin structures. A wide diversity of mechanisms already had been reported, by which ncRNAs can regulate chromatin over a single promoter, a gene cluster, or an entire chromosome, in order to activate or silence genes *in cis* or *in trans* [reviewed in Ref. 84]. RNAs are of special interest because they are likely to have trans-generational effects, as they are able to cross the germline to the offspring and mediate an epigenetic phenotype called paramutation [85]. Once transmitted to the next generation, these RNAs may distribute chromatin-modifying complexes to specific chromosomal loci, thus generating an epigenetic starting-pattern that might direct the development of the new organism. Importantly, RNA homeostasis seems to be altered during aging, as reflected in the tendency toward predominant transcriptional up-regulation of specific ncRNA species (i.e. miRNAs) during aging [86,87]. Studies on liver tissues indicated that miRNA specifically target genes that predominantly participate in oxidative defence, DNA repair, intermediate metabolism, cytoskeletal organization, cell cycle control, and apoptosis, all of which have central protective roles in the aging process [86]. Hence, it was suggested that, during aging, miRNA posttranscriptional regulation of genes involved in dynamic signaling may affect many intertwining networks, disrupting and deforming various regulatory networks.

Other epigenetic mediators may undergo changes with aging and influence longevity. The *Polycomb group genes* (PcGs), a family of proteins that can silence specific target genes by modifying chromatin organization [88], are one example. PcG complexes involve at least two kinds of large multimeric complexes: the initiation complex, Polycomb complex 2 (PRC2), which in humans consists of EZH2, EED, and SUZ12, and the maintenance complex, PRC1, with the core proteins RNF2, HPC, and BMI1. These epigenetic repressors are characterized by an intrinsic histone lysine methyltransferase activity and work jointly to regulate cellular senescence, apoptosis, X-inactivation, stem cell self-renewal, and aging, among other processes. For instance, it was shown that the PcG gene *Bmi1* is required in neurons to suppress apoptosis and the induction of a premature aging-like program characterized by reduced antioxidant defenses [89].

Initially discovered as epigenetic silencers during embryogenesis, it was demonstrated that PcGs control cellular lifespan through regulation of both the p16(Ink4a)/Rb and the Arf/p53 pathways, as well as by interaction with components of the Wnt signaling pathway [90,91]. The *Wnt pathway* describes a complex network of proteins that elicit intracellular signaling cascades with demonstrated roles in cell proliferation, cell fate determination, apoptosis, and axis polarity induction [92]. Emerging data indicate that canonical Wnt signaling also regulates stem and progenitor cell renewal and delays the onset of age-related changes [reviewed in Ref. 93]. Induction of the Wnt inhibitor, GSK-3b, promotes replicative senescence in mammalian brain cells, suggesting that inhibition of the Wnt signaling cascade might be accountable for the age-related decline in cell proliferation, thereby affecting memory functions that are progressively compromised in aging. While growing evidence supports the hypothesis that Wnt signal activation delays aging, others have surprisingly pointed in the opposite direction [93]. These conflicting data indicate that additional work is needed to further test the hypothesis that responses to Wnt signaling change in an age-dependent manner.

MODEL OF AGE-DEPENDENT EPIGENETIC DRIFT

The projected model of evolutionary epigenetics of aging offers a theoretical framework to explain many aging phenomena that are difficult to explain by classical aging theories alone. The theory suggests that, among other causes, aging results from progressive accumulation of epigenetic damage as a direct consequence of evolved limitations in the genetic and epigenetic settings of maintenance and repair functions. Age-dependent epigenetic drift is a natural phenomenon that is present in all healthy individuals, but may become hazardous with age, potentially playing a central role in many complex disorders. Indeed, epigenetic drift was reported to be present in tissues from healthy individuals, but also presented a notably increased drift in tissues derived from patients with late-onset AD [15,43].

Mammalian aging is a complex individual phenotype arising from a variety of risk factors, such as environmental effects, nutrition, or stochastic fluctuations, among others, which directly act on the epigenomic machinery and increase epigenetic variability with age. Whereas genetic mutations accumulate in a nearly linear fashion during aging, epimutations seem to increase exponentially once a certain threshold of cellular epigenetic deregulation is reached (unpublished results from our laboratory). Reaching the threshold may often cause a ripple effect that influences various other genetic and epigenetic maintenance processes,

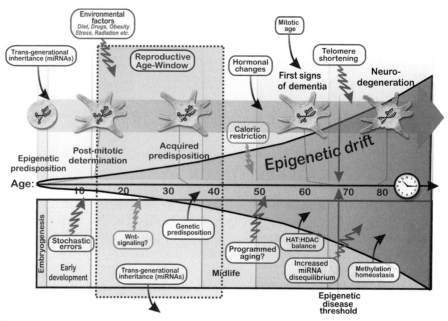

FIGURE 25.3

Age-dependent epigenetic drift. Shown here is the example of neuronal epigenetic drift that may lead to age-dependent neurodegenerative disorders, such as AD or Parkinson's. The relatively high frequency of *de novo* epimutations suggests that epigenetic alterations accumulate during aging. Small epimutations may be tolerated by the cells; however, once the epigenetic deregulation reaches a critical threshold, the cells no longer function properly. The phenotypic outcome depends on the overall effect of the series of pre- and postnatal impacts on the pre-epimutation. Only some predisposed individuals will reach the "threshold" of epigenetic deregulation that causes the phenotypic changes that meet the diagnostic criteria for a clinical disorder. Epigenetic drift may not only be affected by several internal and external factors – some individuals may already be epigenetically predisposed at birth, due to trans-generational epigenetic effects. Trans-generational epigenetic inheritance results either from incomplete erasure of parental epigenetic marks during phases of epigenetic reprogramming at fertilization, or may be established by small RNA species that pass through the germline. Since all organisms eventually die from different causes, epigenetic patterns beneficial in early life are favored by natural selection over patterns advantageous later in life. Deleterious epigenetic drift occurring after the reproductive phase is relatively neutral to selection, because its bearers have already transmitted their genes (and potentially epigenetic information) to the next generation. (Please refer to color plate section)

ultimately leading to a significantly quicker epigenetic drift (see Fig. 25.3). Compared to genetic mutations, it is likely that epigenetic drift is partially caused by the higher epigenetic turnover, as the replication-independent epigenetic maintenance in cells is very dynamic and inherently probabilistic. For example, it could be shown that one of the main genes maintaining a transcriptionally silent state, Hp1, which mediates communication between histone and DNA methyltransferases, transiently binds to target chromatin domains. Intriguingly, the turnover time for the entire cellular pool of HP1 at a given heterochromatic domain is a matter of seconds [94].

The rate of epigenetic drift may not only be affected by singular events in the genome, but may also be driven by genome-wide systemic mechanisms. This was shown in a methylome network analysis of post-mortem brain samples of schizophrenia cases and controls, which uncovered decreased epigenetic modularity (co-regulation) in both the brain and the germline of affected individuals, suggesting that systemic epigenetic dysfunction may be associated with complex disorders [95]. A systemic effect can be potentially positive or negative in relation to longevity. It seems reasonable that, in very old individuals, some favorable epigenotypes exist that act as buffers against the deleterious effects of age-related disease genes. As a result, one could expect that the frequency of deleterious epimutations may increase among individuals with extreme lifespan, because their protective epigenotype allows disease-related genes or epigenetic patterns to accumulate.

On the epigenetic level, some genes may not predispose to age-related disease *per se*, but only in connection with specific environmental triggers. Interactions between genes and putative risk environments may explain why association studies of complex disorders often cannot be replicated in studies from different geographical areas. The first direct evidence for such an interplay in the etiology of a complex disorder was reported by Caspi and co-workers, who found that a polymorphism in the serotonin transporter gene, *5-Htt*, regulates the effect of stressful life events on susceptibility to depression [96]. Generally, it can be assumed that environmental insults leave an "epigenetic footprint" of mammalian cells, especially in post-mitotic cells, such as neurons. These footprints may be harmful to an organism, even years after exposure in old age. Fittingly, it was reported that developmental exposure of rats to the xenobiotic metal, lead (Pb), resulted in delayed overexpression (20 months later) of APP, a gene with a critical role in AD [97]. Aged monkeys exposed to Pb as infants responded in a similar way, suggesting that environmental effects occurring during brain development predetermine the expression and regulation of APP later in life, potentially influencing the course of disease development. The occurrence of significant epidemiological differences in age-related disease prevalence is a strong sign that environmental triggers may drive disease development [98]. A more intense cross-cultural comparison of age related diseases is needed to understand how the environment (i.e. diet and behavioral circumstances) affects these disorders. Nevertheless, the "Western-lifestyle" may be an essential component in the development of late-onset disorders of the brain. A similar situation exists in other complex diseases, such as cancer, where many migration studies have shown that cancer incidence changes following migration, pointing to a predominant environmental contribution to disease causation [99].

Lessons From Twin Research

Traditionally, in twin studies, when variation due to environmental factors is ruled out, an influence at the genetic level is assumed. However, epigenetic factors are probably better suited to explain the observed anomalies in age-related disorders, as aberrant epigenetic patterns may be acquired during many developmental stages. In monozygotic (MZ) twins, it has been repeatedly demonstrated that an epigenetic drift occurs during aging that affects DNA methylation as well as histone-modification, and is associated with phenotypic discordance [43,100,101]. For example, in a recent genome-wide study on MZ and dizygotic (DZ) twins using a microarray-based DNA methylation profiling method [102], it was demonstrated that, although a large degree of MZ co-twin DNA methylation variation exists, the epigenetic difference in buccal cells of DZ co-twins was significantly higher. These data suggest that molecular mechanisms of heritability may not be limited to DNA sequence differences. In a landmark study by Fraga and colleagues, genome-wide differences in DNA methylation and histone acetylation with age were examined [43]. Consistent with the idea that MZ twins are epigenetically similar at birth, little or no epigenetic differences were detected in twins early in life. In contrast, the elderly MZ twin pairs showed substantial variations in several tissues. Notably, the extent of epigenetic variation was related to environmental differences (i.e. lifestyle and non-shared environment) between twins and correlated significantly with differences in expression patterns. These findings are of major importance for understanding the results of classic twin heritability and highlight a potential mechanism by which environmental factors can influence age-dependent expression patterns [reviewed in Ref. 103]. On the other hand, one of the first large-scale twin studies, the Minnesota Study of Twins Reared Apart, challenged the notion that environment plays a significant role in determining disease phenotypes [104], indicating that much more research on the dynamic interaction of environment and epigenotype is necessary.

In aging studies, rare sets of MZ twins discordant for disease phenotypes are of special interest, as they offer the opportunity to examine factors that may contribute to the probabilistic nature of the association between the epigenome and age-related disorder. Information as to whether a twin pair is concordant (both are affected) or discordant (only one twin has the disorder) relies on the rationale that MZ twins are genetically almost identical, whereas DZ twins, on average, share 50% of their segregating genes. In agreement with the theory of epigenetic drift, the degree of concordance in MZ twins is lower than 100% for nearly all complex diseases, but substantially higher in comparison to the concordance rate in DZ twins [105]. Nevertheless, data derived from aging twin studies have to be used cautiously. The disadvantage of aging twin studies is the late clinical diagnosis of the disease, and it is obvious that in some cases a concordance may be missed if one twin dies earlier than the other, often from causes unrelated to the studied disorder. For example, in a recent study on a twin pair discordant for AD, the authors found a significant global demethylation in the affected brain areas of the AD twin [106]. However, the unaffected twin also showed weak signs of AD pathology, indicating that he may have been predisposed to AD as well. It seems likely that, had he lived longer, he would have developed AD symptoms, too. In this context, it is important to note that the age of onset in MZ twins tends to differ considerably, often by more than two decades [107], further supporting the effect of either harmful or protective environmental factors playing a significant role in the occurrence of age-dependent disorders. Identification of such causative factors may help to develop preventive strategies to delay the onset of age-related pathologies significantly.

CONCLUSIONS AND FUTURE DIRECTIONS

It remains unknown whether aging is epigenetically programmed, but it is highly likely that aging partially results from the accumulation of epigenetic somatic damage, owing to organisms' limited investments in maintenance and repair of epigenetic patterns in old age. To understand these processes, it will be beneficial not only to focus on research of age-related disorders, but also to incorporate longevity studies, e.g. observing the epigenetic patterns in centenarians. If environment plays a critical role in epigenetic drift, it is to be expected that long-lived people can carry just as many deleterious aging epimutations as the rest of the population; however, their protective "longevity epigenotype" may work as a buffer to protect from the harmful effects of other epimutations. With an increasing population of over 65 years of age, a much stronger investment in the field of aging epigenetics seems advisable, as the risk of developing age related diseases will increase dramatically in the near future. The risk of developing a neurodegenerative disorder doubles every 5 years over age 65, and it is estimated that more than half of the individuals older than 85 have neurodegenerative phenotypes. These facts are significant, because the group with the highest risk, those older than 85, is the fastest growing population group in the Western world. If the threshold model of epigenetic drift holds true, it would be important to adjust an individual's epigenome before the critical threshold is reached, as epigenetic treatments become increasingly difficult once a subject enters a disease state. Reversal of complex epigenetic patterns is a challenging task; however, there are many lines of evidence that lifestyle may inhibit or at least postpone the onset of age-related disorders. It seems plausible that, with appropriate preventive strategies, the onset of disease may be delayed nearly indefinitely. This is the case for various cancers, where early diagnosis is essential, but treatment becomes very complicated after a certain threshold is crossed (i.e. metastasis). This problem underscores the need for early diagnostic markers in age-dependent disorders, and epigenetic markers may be excellent tools for development of an efficient treatment strategy [15]. Used in conjunction with other therapies, "epigenetic" drugs could provide a faster, safer, and more reliable treatment for patients. Already, several of these drugs, such as HDAC or DNA methyltransferase inhibitors, have been tested in clinical trials on age-related disease (i.e. cancer). However, all current drug candidates are very unspecific and, consequently, large-scale deregulation of the methylation and deacetylation machinery of the cell is unavoidable. In the future, it will be essential to identify technologies that target only specific elements of the epigenome, or even better, to find natural ways (e.g. lifestyle, nutrition) to protect us from aging phenotypes by increasing the body's tolerance for epigenetic insults.

References

1. Ljubuncic P, Reznick AZ. The evolutionary theories of aging revisited – a mini-review. *Gerontology* 2009;**55**:205–16.

2. Weismann A. *Essays Upon Heredity*. Oxford: Clarendon Press; 1891.

3. Medawar PB. *An Unsolved Problem of Biology*. London: HK Lewis; 1952.

4. Williams GC. Pleiotropy, natural selection, and the evolution of senescence. *Evolution* 1957;**11**:398–411.

5. Bestor T, Laudano A, Mattaliano R, Ingram V. Cloning and sequencing of a cDNA encoding DNA methyltransferase of mouse cells. The carboxyl-terminal domain of the mammalian enzymes is related to bacterial restriction methyltransferases. *J Mol Biol* 1988;**203**:971–83.

6. Kriaucionis S, Heintz N. The nuclear DNA base 5-hydroxymethylcytosine is present in Purkinje neurons and the brain. *Science* 2009;**324**:929–30.

7. Loenarz C, Schofield CJ. Oxygenase catalyzed 5-methylcytosine hydroxylation. *Chem Biol* 2009;**16**:580–3.

8. Tahiliani M, Koh KP, Shen Y, Pastor WA, Bandukwala H, Brudno Y, et al. Conversion of 5-methylcytosine to 5-hydroxymethylcytosine in mammalian DNA by MLL partner TET1. *Science* 2009;**324**:930–5.

9. Brown SE, Fraga MF, Weaver IC, Berdasco M, Szyf M. Variations in DNA methylation patterns during the cell cycle of HeLa cells. *Epigenetics* 2007;**2**:54–65.

10. Martinowich K, Hattori D, Wu H, Fouse S, He F, Hu Y, et al. DNA methylation-related chromatin remodeling in activity-dependent BDNF gene regulation. *Science* 2003;**302**:890–3.

11. Goyal R, Reinhardt R, Jeltsch A. Accuracy of DNA methylation pattern preservation by the Dnmt1 methyltransferase. *Nucleic Acids Res* 2006;**34**:1182–8.

12. Chu MW, Siegmund KD, Eckstam CL, Kim JY, Yang AS, Kanel GC, et al. Lack of increases in methylation at three CpG-rich genomic loci in non-mitotic adult tissues during aging. *BMC Med Genet* 2007;**8**:50.

13. Issa JP. CpG-island methylation in aging and cancer. *Curr Top Microbiol Immunol* 2000;**249**:101–18.

14. Kim JY, Tavare S, Shibata D. Counting human somatic cell replications: methylation mirrors endometrial stem cell divisions. *Proc Natl Acad Sci USA* 2005;**102**:17739–44.

15. Wang S-C, Oelze B, Schumacher A. Age-specific epigenetic drift in late-onset Alzheimer's disease. *PLoS ONE* 2008;**3**:e2698.

16. Inano K, Suetake I, Ueda T, Miyake Y, Nakamura M, Okada M, et al. Maintenance-type DNA methyltransferase is highly expressed in post-mitotic neurons and localized in the cytoplasmic compartment. *J Biochem* 2000;**128**:315–21.

17. Fan G, Beard C, Chen RZ, Csankovszki G, Sun Y, Siniaia M, et al. DNA hypomethylation perturbs the function and survival of CNS neurons in postnatal animals. *J Neurosci* 2001;**21**:788–97.

18. Brooks PJ, Marietta C, Goldman D. DNA mismatch repair and DNA methylation in adult brain neurons. *J Neurosci* 1996;**16**:939–45.

19. Casillas MA Jr., Lopatina N, Andrews LG, Tollefsbol TO. Transcriptional control of the DNA methyltransferases is altered in aging and neoplastically-transformed human fibroblasts. *Mol Cell Biochem* 2003;**252**:33–43.

20. Siegmund KD, Connor CM, Campan M, Long TI, Weisenberger DJ, Biniszkiewicz D, et al. DNA methylation in the human cerebral cortex is dynamically regulated throughout the life span and involves differentiated neurons. *PLoS ONE* 2007;**2**:e895.

21. Ma DK, Jang MH, Guo JU, Kitabatake Y, Chang ML, Pow-Anpongkul N, et al. Neuronal activity-induced Gadd45b promotes epigenetic DNA demethylation and adult neurogenesis. *Science* 2009;**323**:1074–7.

22. Richardson B. Impact of aging on DNA methylation. *Ageing Res Rev* 2003;**2**:245–61.

23. Bollati V, Schwartz J, Wright R, Litonjua A, Tarantini L, Suh H, et al. Decline in genomic DNA methylation through aging in a cohort of elderly subjects. *Mech Ageing Dev* 2009;**130**:234–9.

24. Flanagan JM, Popendikyte V, Pozdniakovaite N, Sobolev M, Assadzadeh A, Schumacher A, et al. Intra- and interindividual epigenetic variation in human germ cells. *Am J Hum Genet* 2006;**79**:67–84.

25. Barbot W, Dupressoir A, Lazar V, Heidmann T. Epigenetic regulation of an IAP retrotransposon in the aging mouse: progressive demethylation and de-silencing of the element by its repetitive induction. *Nucleic Acids Res* 2002;**30**:2365–73.

26. Petronis A. Human morbid genetics revisited: relevance of epigenetics. *Trends Genet* 2001;**17**:142–6.

27. Wang SC, Oelze B, Schumacher A. Age-specific epigenetic drift in late-onset Alzheimer's disease. *PLoS ONE* 2008;**3**:e2698.

28. Eckhardt F, Lewin J, Cortese R, Rakyan VK, Attwood J, Burger M, et al. DNA methylation profiling of human chromosomes 6, 20 and 22. *Nat Genet* 2006;**38**:1378–85.

29. Bjornsson HT, Sigurdsson MI, Fallin MD, Irizarry RA, Aspelund T, Cui H, et al. Intra-individual change over time in DNA methylation with familial clustering. *JAMA* 2008;**299**:2877–83.

30. Harris TB, Launer LJ, Eiriksdottir G, Kjartansson O, Jonsson PV, Sigurdsson G, et al. Age, gene/environment susceptibility – Reykjavik Study: multidisciplinary applied phenomics. *Am J Epidemiol* 2007;**165**:1076–87.

31. Schumacher A, Doerfler W. Influence of in vitro manipulation on the stability of methylation patterns in the Snurf/Snrpn-imprinting region in mouse embryonic stem cells. *Nucleic Acids Res* 2004;**32**:1566–76.

32. Waterland RA, Lin JR, Smith CA, Jirtle RL. Post-weaning diet affects genomic imprinting at the insulin-like growth factor 2 (Igf2) locus. *Hum Mol Genet* 2006;**15**:705–16.

33. Ono T, Tawa R, Shinya K, Hirose S, Okada S. Methylation of the c-myc gene changes during aging process of mice. *Biochem Biophys Res Commun* 1986;**139**:1299–304.

34. Hayslip J, Montero A. Tumor suppressor gene methylation in follicular lymphoma: a comprehensive review. *Mol Cancer* 2006;**5**:44.

35. Tohgi H, Utsugisawa K, Nagane Y, Yoshimura M, Ukitsu M, Genda Y. The methylation status of cytosines in a tau gene promoter region alters with age to downregulate transcriptional activity in human cerebral cortex. *Neurosci Lett* 1999;**275**:89–92.

36. Christensen BC, Houseman EA, Marsit CJ, Zheng S, Wrensch MR, Wiemels JL, et al. Aging and environmental exposures alter tissue-specific DNA methylation dependent upon CpG island context. *PLoS genet* 2009;**5**:e1000602.

37. Mattson MP, Shea TB. Folate and homocysteine metabolism in neural plasticity and neurodegenerative disorders. *Trends Neurosci* 2003;**26**:137–46.

38. Lehmann M, Gottfries CG, Regland B. Identification of cognitive impairment in the elderly: homocysteine is an early marker. *Dement Geriatr Cogn Disord* 1999;**10**:12–20.

39. Seshadri S, Beiser A, Selhub J, Jacques PF, Rosenberg IH, D'Agostino RB, et al. Plasma homocysteine as a risk factor for dementia and Alzheimer's disease. *N Engl J Med* 2002;**346**:476–83.

40. Quadri P, Fragiacomo C, Pezzati R, Zanda E, Forloni G, Tettamanti M, et al. Homocysteine, folate, and vitamin B-12 in mild cognitive impairment, Alzheimer disease, and vascular dementia. *Am J Clin Nutr* 2004;**80**:114–22.

41. Kouzarides T. Chromatin modifications and their function. *Cell* 2007;**128**:693–705.

42. Sharma R, Nakamura A, Takahashi R, Nakamoto H, Goto S. Carbonyl modification in rat liver histones: decrease with age and increase by dietary restriction. *Free Radic Biol Med* 2006;**40**:1179–84.

43. Fraga MF, Ballestar E, Paz MF, Ropero S, Setien F, Ballestar ML, et al. Epigenetic differences arise during the lifetime of monozygotic twins. *Proc Natl Acad Sci USA* 2005;**102**:10604–9.

44. Bernstein BE, Kamal M, Lindblad-Toh K, Bekiranov S, Bailey DK, Huebert DJ, et al. Genomic maps and comparative analysis of histone modifications in human and mouse. *Cell* 2005;**120**:169–81.

45. Sarg B, Koutzamani E, Helliger W, Rundquist I, Lindner HH. Postsynthetic trimethylation of histone H4 at lysine 20 in mammalian tissues is associated with aging. *J Biol Chem* 2002;**277**:39195–201.

46. El Mezayen R, El Gazzar M, Myer R, High KP. Aging-dependent upregulation of IL-23p19 gene expression in dendritic cells is associated with differential transcription factor binding and histone modifications. *Aging Cell* 2009;**8**:553–65.

47. Yang H, Mizzen CA. The multiple facets of histone H4-lysine 20 methylation. *Biochem Cell Biol* 2009;**87**:151–61.

48. Yin W, Barkess G, Fang X, Xiang P, Cao H, Stamatoyannopoulos G, et al. Histone acetylation at the human beta-globin locus changes with developmental age. *Blood* 2007;**110**:4101–7.

49. Yang XJ, Seto E. Lysine acetylation: codified crosstalk with other posttranslational modifications. *Mol Cell* 2008;**31**:449–61.

50. Kawakami K, Nakamura A, Ishigami A, Goto S, Takahashi R. Age-related difference of site-specific histone modifications in rat liver. *Biogerontology* 2009;**10**:415–21.

51. Saha RN, Pahan K. HATs and HDACs in neurodegeneration: a tale of disconcerted acetylation homeostasis. *Cell Death Differ* 2006;**13**:539–50.

52. Boutillier AL, Trinh E, Loeffler JP. Selective E2F-dependent gene transcription is controlled by histone deacetylase activity during neuronal apoptosis. *J Neurochem* 2003;**84**:814–28.

53. Rouaux C, Jokic N, Mbebi C, Boutillier S, Loeffler JP, Boutillier AL. Critical loss of CBP/p300 histone acetylase activity by caspase-6 during neurodegeneration. *EMBO J* 2003;**22**:6537–49.

54. Saunders LR, Verdin E. Sirtuins: critical regulators at the crossroads between cancer and aging. *Oncogene* 2007;**26**:5489–504.

55. Longo VD, Kennedy BK. Sirtuins in aging and age-related disease. *Cell* 2006;**126**:257–68.

56. Fraga MF, Esteller M. Epigenetics and aging: the targets and the marks. *Trends Genet* 2007;**23**:413–18.

57. Oberdoerffer P, Michan S, McVay M, Mostoslavsky R, Vann J, Park SK, et al. SIRT1 redistribution on chromatin promotes genomic stability but alters gene expression during aging. *Cell* 2008;**135**:907–18.

58. Vijg J, Maslov AY, Suh Y. Aging: a sirtuin shake-up? *Cell* 2008;**135**:797–8.

59. Rine J, Strathern JN, Hicks JB, Herskowitz I. A suppressor of mating-type locus mutations in *Saccharomyces cerevisiae*: evidence for and identification of cryptic mating-type loci. *Genetics* 1979;**93**:877–901.

60. Sinclair DA, Guarente L. Extrachromosomal rDNA circles – a cause of aging in yeast. *Cell* 1997;**91**:1033–42.

61. Smeal T, Claus J, Kennedy B, Cole F, Guarente L. Loss of transcriptional silencing causes sterility in old mother cells of *S. cerevisiae*. *Cell* 1996;**84**:633–42.

62. Guarente L, Picard F. Calorie restriction – the SIR2 connection. *Cell* 2005;**120**:473–82.

63. Vaquero A, Scher M, Lee D, Erdjument-Bromage H, Tempst P, Reinberg D. Human SirT1 interacts with histone H1 and promotes formation of facultative heterochromatin. *Mol Cell* 2004;**16**:93–105.

64. Luo J, Nikolaev AY, Imai S, Chen D, Su F, Shiloh A, et al. Negative control of p53 by Sir2alpha promotes cell survival under stress. *Cell* 2001;**107**:137–48.

65. O'Hagan HM, Mohammad HP, Baylin SB. Double strand breaks can initiate gene silencing and SIRT1-dependent onset of DNA methylation in an exogenous promoter CpG island. *PLoS Genet* 2008;**4**:e1000155.

66. Mostoslavsky R, Chua KF, Lombard DB, Pang WW, Fischer MR, Gellon L, et al. Genomic instability and aging-like phenotype in the absence of mammalian SIRT6. *Cell* 2006;**124**:315–29.

67. Michishita E, McCord RA, Berber E, Kioi M, Padilla-Nash H, Damian M, et al. SIRT6 is a histone H3 lysine 9 deacetylase that modulates telomeric chromatin. *Nature* 2008;**452**:492–6.

68. de Lange T. Shelterin: the protein complex that shapes and safeguards human telomeres. *Genes Dev* 2005;**19**:2100–10.

69. McCord RA, Broccoli D. Telomeric chromatin: roles in aging, cancer and hereditary disease. *Mutat Res* 2008;**647**:86–93.

70. Collins K, Mitchell JR. Telomerase in the human organism. *Oncogene* 2002;**21**:564–79.

71. Weismann A. *Über Leben und Tod*. Jena: Fisher; 1892.

72. Perrod S, Gasser SM. Long-range silencing and position effects at telomeres and centromeres: parallels and differences. *Cell Mol Life Sci* 2003;**60**:2303–18.

73. Blasco MA. The epigenetic regulation of mammalian telomeres. *Nat Rev Genet* 2007;**8**:299–309.

74. Benetti R, Garcia-Cao M, Blasco MA. Telomere length regulates the epigenetic status of mammalian telomeres and subtelomeres. *Nat Genet* 2007;**39**:243–50.

75. Gonzalo S, Jaco I, Fraga MF, Chen T, Li E, Esteller M, et al. DNA methyltransferases control telomere length and telomere recombination in mammalian cells. *Nat Cell Biol* 2006;**8**:416–24.

76. Cawthon RM, Smith KR, O'Brien E, Sivatchenko A, Kerber RA. Association between telomere length in blood and mortality in people aged 60 years or older. *Lancet* 2003;**361**:393–5.

77. Epel ES, Blackburn EH, Lin J, Dhabhar FS, Adler NE, Morrow JD, et al. Accelerated telomere shortening in response to life stress. *Proc Natl Acad Sci USA* 2004;**101**:17312–15.

78. Valdes AM, Andrew T, Gardner JP, Kimura M, Oelsner E, Cherkas LF, et al. Obesity, cigarette smoking, and telomere length in women. *Lancet* 2005;**366**:662–4.

79. Schoeftner S, Blasco MA. A "higher order" of telomere regulation: telomere heterochromatin and telomeric RNAs. *EMBO J* 2009;**28**:2323–36.

80. Blasco MA. Mice with bad ends: mouse models for the study of telomeres and telomerase in cancer and aging. *EMBO J* 2005;**24**:1095–103.

81. Azzalin CM, Reichenbach P, Khoriauli L, Giulotto E, Lingner J. Telomeric repeat containing RNA and RNA surveillance factors at mammalian chromosome ends. *Science* 2007;**318**:798–801.

82. Luke B, Lingner J. TERRA: telomeric repeat-containing RNA. *EMBO J* 2009;**28**:2503–10.

83. Schoeftner S, Blasco MA. Developmentally regulated transcription of mammalian telomeres by DNA-dependent RNA polymerase II. *Nat Cell Biol* 2008;**10**:228–36.

84. Whitehead J, Pandey GK, Kanduri C. Regulation of the mammalian epigenome by long noncoding RNAs. *Biochim Biophys Acta* 2009;**1790**:936–47.

85. Rassoulzadegan M, Grandjean V, Gounon P, Vincent S, Gillot I, Cuzin F. RNA-mediated non-mendelian inheritance of an epigenetic change in the mouse. *Nature* 2006;**441**:469–74.

86. Bates DJ, Liang R, Li N, Wang E. The impact of noncoding RNA on the biochemical and molecular mechanisms of aging. *Biochim Biophys Acta* 2009;**1790**:970–9.

87. Maes OC, An J, Sarojini H, Wang E. Murine microRNAs implicated in liver functions and aging process. *Mech Ageing Dev* 2008;**129**:534–41.

88. Valk-Lingbeek ME, Bruggeman SW, van Lohuizen M. Stem cells and cancer; the polycomb connection. *Cell* 2004;**118**:409–18.

89. Chatoo W, Abdouh M, David J, Champagne MP, Ferreira J, Rodier F, et al. The polycomb group gene Bmi1 regulates antioxidant defenses in neurons by repressing p53 pro-oxidant activity. *J Neurosci* 2009;**29**:529–42.

90. Gil J, Bernard D, Martinez D, Beach D. Polycomb CBX7 has a unifying role in cellular lifespan. *Nat Cell Biol* 2004;**6**:67–72.

91. Shi B, Liang J, Yang X, Wang Y, Zhao Y, Wu H, et al. Integration of estrogen and Wnt signaling circuits by the polycomb group protein EZH2 in breast cancer cells. *Mol Cell Biol* 2007;**27**:5105–19.

92. Clevers H. Wnt/beta-catenin signaling in development and disease. *Cell* 2006;**127**:469–80.

93. DeCarolis NA, Wharton KA Jr., Eisch AJ. Which way does the Wnt blow? Exploring the duality of canonical Wnt signaling on cellular aging. *Bioessays* 2008;**30**:102–6.

94. Cheutin T, McNairn AJ, Jenuwein T, Gilbert DM, Singh PB, Misteli T. Maintenance of stable heterochromatin domains by dynamic HP1 binding. *Science* 2003;**299**:721–5.

95. Mill J, Tang T, Kaminsky Z, Khare T, Yazdanpanah S, Bouchard L, et al. Epigenomic profiling reveals DNA-methylation changes associated with major psychosis. *Am J Hum Genet* 2008;**82**:696–711.

96. Caspi A, Sugden K, Moffitt TE, Taylor A, Craig IW, Harrington H, et al. Influence of life stress on depression: moderation by a polymorphism in the 5-HTT gene. *Science* 2003;**301**:386–9.

97. Zawia NH, Basha MR. Environmental risk factors and the developmental basis for Alzheimer's disease. *Rev Neurosci* 2005;**16**:325–37.

98. Ineichen B. The geography of dementia: an approach through epidemiology. *Health Place* 1998;**4**:383–94.

99. Hemminki K, Lorenzo Bermejo J, Forsti A. The balance between heritable and environmental aetiology of human disease. *Nat Rev Genet* 2006;**7**:958–65.

100. Kristiansen M, Knudsen GP, Bathum L, Naumova AK, Sorensen TI, Brix TH, et al. Twin study of genetic and aging effects on X chromosome inactivation. *Eur J Hum Genet* 2005;**13**:599–606.

101. Kaminsky ZA, Tang T, Wang SC, Ptak C, Oh GH, Wong AH, et al. DNA methylation profiles in monozygotic and dizygotic twins. *Nat Genet* 2009;**41**:240–5.

102. Schumacher A, Kapranov P, Kaminsky Z, Flanagan J, Assadzadeh A, Yau P, et al. Microarray-based DNA methylation profiling: technology and applications. *Nucleic Acids Res* 2006;**34**:528–42.

103. Poulsen P, Esteller M, Vaag A, Fraga MF. The epigenetic basis of twin discordance in age-related diseases. *Pediatr Res* 2007;**61**:38R–42R.

104. Bouchard TJ Jr., Lykken DT, McGue M, Segal NL, Tellegen A. Sources of human psychological differences: the Minnesota Study of Twins Reared Apart. *Science* 1990;**250**:223–8.

105. Schumacher A, Petronis A. Epigenetics of complex diseases: from general theory to laboratory experiments. *Curr Top Microbiol Immunol* 2006;**310**:81–115.

106. Mastroeni D, McKee A, Grover A, Rogers J, Coleman PD. Epigenetic differences in cortical neurons from a pair of monozygotic twins discordant for Alzheimer's disease. *PLoS ONE* 2009;**4**:e6617.

107. Nee LE, Lippa CF. Alzheimer's disease in 22 twin pairs – 13-year follow-up: hormonal, infectious and traumatic factors. *Dement Geriatr Cogn Disord* 1999;**10**:148–51.

Evolutionary Epigenetics

CHAPTER 26

Epigenetics in Adaptive Evolution and Development
The Interplay Between Evolving Species and Epigenetic Mechanisms

Simon H. House*
Cambridge, UK

PRESSURE FOR MOLECULAR EVIDENCE

As research advances rapidly the definition of "epigenetics" evolves, but a definition currently acknowledged widely is: "changes in phenotype (appearance) or gene expression caused by mechanisms other than changes in the underlying DNA sequence" [1].

425

Practical Indications of Epigenetics

Epigenetics is more than a fascinating and fast burgeoning field of biological research: it is of vital consequence to the human race as we have come into arguably the worst crises in new forms of disease the human race has encountered. These include the "non-communicable" diseases related to the metabolic syndrome. Obesity, diabetes, and cardiovascular and mental health disorders are increasingly recognized as connected with epigenetic changes of early origin.

One of the first scientists to prophesy specific permanent effects on health of nutrition from before conception and during development was Professor Michael A. Crawford, Director of the Institute of Brain Chemistry and Human Nutrition. In 1972 he related his and others' findings to the effects of current nutrition on our current evolution as well as health, with particular emphasis on preconception nutrition, and on marine omega-3 oils in brain and heart development [2–4]. Such lifelong effects of fetal and infant nutrition are better known by the term "The Barker Hypothesis" [5,6]. Crawford used the adjective "epigenetic" in a broad sense, but although the term "epigenetics" had already been coined by Conrad Waddington in 1937 to describe environmental effects on the phenotype, molecular comprehension was delayed for half a century.

Nutrition is one aspect of the environment readily studied objectively. Less amenable to objective study are stressors and stress, particularly the lasting impacts of maternal–fetal stress. Nonetheless remarkable evidence has come from work of pre- and perinatal societies,

*Declaration – I declare I have no conflict of interest.

Handbook of Epigenetics: The New Molecular and Medical Genetics. DOI: 10.1016/B978-0-12-375709-8.00026-5

and notably of Dr Frank Lake, between 1954 and 1982 [7–9], instanced in the section below, "Continuing evolutionary impacts, p.438". This evidence is becoming more intelligible through epigenetics.

Environment and Evolution

Crawford's proclaiming of the powerful effects of diet on development, culminated in his insistence that only at the waterside could the human species have achieved so large a brain, sustained by plentiful fish and shellfish with their essential marine oils, docosahexaenoic and eicosapentaenoic acids (DHA and EPA). The only other mammals to retain a large brain as they became at least as large were marine, such as dolphins and whales. Not until the 1990s was the power of Crawford's case acknowledged by leaders in paleontology, having re-dated remains with electron-spin technology, and related them to evidence of contemporary water levels [10,11]. David Marsh and Michael Crawford [12,13] had emphasized that Darwin, in *On The Origin of Species* [14], had attributed adaptation to "Conditions of Existence", as a higher law than "Unity of Type". The former was compatible with Lamarck's [15] first basic law that "organs, thus species, change in response to a need created by a changing environment". His second law that "such change was passed through the hereditary mechanism to the offspring" was intrinsic to Darwin's "Unity of Type", even if Darwin made more of random change. Sifted from the theistic and teleological elements, these two scientists' theories stand the light of current science [16].

The commonly held view that adaptation was by natural selection of merely random changes, the "neo-Darwinist" view, is attributable to "the Weismann barrier" [17] – there are even scientists who use the term "Darwinism" clearly meaning "neo-Darwinism". Although in his introduction Darwin writes, "I am convinced that Natural Selection has been the main, but not exclusive, means of modification", concluding Chapter 6 he affirms:

> "The expression of conditions of existence… is fully embraced by natural selection [which acts by] adapting the varying parts of each being to its organic and inorganic conditions of life". He adds "[Of the] two great laws – Unity of Type, and the Conditions of Existence… the law of the Conditions of Existence is the higher law; as it includes, through the inheritance of former variations and adaptations, that of Unity of Type."

Darwin goes on:

> "I can see no very great difficulty… in believing that *natural selection has converted the simple apparatus of an optic nerve* merely coated with pigment and invested by transparent membrane, into an optical instrument as perfect as is possessed by any member of the great Articulate class [nor] that *natural selection has actually converted a swimbladder into a lung*, or organ used exclusively for respiration [my italics]".

Darwin here seems to lean towards Lamarck. If he were right, that natural selection somehow caused the conversion, might it be through epigenetic switching prompted by conditions of existence, advantageous enough to survive as adaptation? Among others leaning towards Lamarck, Jablonka, Fox et al. [18] hold – controversially – that:

> "it is quite wrong to think of the environment as just a selector of heritable variation. *The environment has a dual role in evolution – it does not just select among heritable phenotypic variations, it also induces them*" [my italics].

Origin's final words reveal Darwin's sense of wonder that "from so simple a beginning endless forms most beautiful and most wonderful have been, and *are being evolved*." [14] Our growing awareness of mechanisms of epigenetics and genomic imprinting makes us all the more sensitive to the importance of keeping today's increasingly artificial conditions of existence as beneficial in relation to evolution as we possibly can.

EVOLUTIONARY CONCEPTS FROM THE 19TH CENTURY TO TODAY

David Marsh [12] well describes how Lamarck's concept of heritable acquired characteristics, modified by Darwin, had been flatly contradicted by Weismann. Weismann held a "germ plasm theory" of sealed germ-cells, insulated against all somatoplasmic influence, ruling out heritability of acquired characteristics. Weismann's ridiculous "disproof" of acquired characteristics – by cutting off mice-tails for 20 generations to no hereditary effect (despite centuries of docking lambs' tails, or indeed millennia of human male circumcision!) – led to a widespread view of mutations being purely random, advantages enduring through natural selection. Such a concept of evolution became inappropriately termed "neo-Darwinism", many scientists even referring to it mistakenly as "Darwinism". By the 1930s mathematical coordination of neo-Darwinism with Malthus's principle of population growth, Mendel's statistical approach, and human population genetics led to the "Modern Synthesis". Setting the seal on this rigid "primacy of DNA" was the 1950s discovery of the double helix by Watson, Crick et al. reinforcing the "blue-print" paradigm. All along, caught up in the controversies were the Church, Creationism and Intelligent Design; and politics, industrial revolution, and trade with religious expansionism [12]. In the turbulence, Darwin's image became stripped of his wavering religious propensity, to be branded agnostic, even atheist, and his image also became stripped of his profound environmentalism "Conditions of Existence".

The opinion of David Marsh and Michael Crawford has not wavered since their publication – over a decade before the human genome was mapped – of *The Driving Force: Food in Evolution and the Future*, 1989, that natural selection and environmental conditions work hand in hand, as current epigenetics research is now showing. To these two author friends I am indebted for their insights in *Nutrition and Evolution*, as their excellent book was renamed in 1995 [13].

For the controversial background I strongly recommend David E. Marsh's *The Origins of Diversity* [12], an excellent short history of evolutionary debate. It ranges from the late 18th century – Lamarck's theory [15] – through to the human genome sequencing and current surge in epigenetics research, which is beginning to clarify precise mechanisms of evolutionary change.

I pay tribute to Michael Crawford particularly for his globally outstanding research focused on maternal–child nutrition and brain development. Crawford, Marsh and I are among the increasing number who stress the value of young children's developing a taste for fish, sea foods, seaweeds, and beneficial algae like spirulina and chlorella. For the sake of future generations people need to eat more aquatic foods as the main and richest source of brain specific nutrients, docosahexaenoic and eicosapentaenoic acids (DHA & EPA), whose dietary depletion in many countries is threatening the brain [19–21]. Covering 70% of our planet's surface, water represents our greatest future food potential, while the demise of ocean plant-life can seriously aggravate climate change. At our peril do we continue wrecking our evolutionary habitats of oceans, rivers, and lakes. Their saving and enhancement, rooted in agriculture of the ocean beds, is a vital contribution to our future. Epigenetics makes even clearer the urgency of guarding aquatic habitats, if only for the sake of human health.

Many epidemiological and molecular researchers are exploring transgenerational effects, including genomic imprinting, opening windows on modifications in species. Among them are Barry Keverne [22–24], Marcus Pembrey [25–28], Michael Golubovsky [29], Marilyn Monk [30,31], and Jonathan Mill [33a,33b]. We seem on the verge of clarifying the relationship of heritability, yet reversibility, to more permanent changes in DNA sequence, namely, mutations. David Marsh [12] sees this as the threshold of a new paradigm. We are recognizing the molecular variations through which we are affected by our great-grandparents, and in turn affect our great-grandchildren-to-be, passing on generation by generation much of what we inherit, and a substantial part of how we live. How substantial? Despite Barry Keverne's description [22] of genomic imprinting as co-adaptive, the results of epigenetic transmission to the next generation are still controversial – shades of Weismann?

Keverne writes [22]:

> Genomic imprinting acts primarily through key regulatory genes which in turn have a cascade effect through other genes. Possible effects vary widely, for instance the mother's food intake and weight gain; maternal fat and blood glucose; letdown of milk and post-natal pup growth. Other effects include her maternal behavior, nest-building, and placental hormones, placental blood flow and nutrient transfer, fetal growth, and early weaning and puberty onset. In these ways the placenta enables the fetus to regulate its own destiny, mainly by genomic co-adaptation affecting hormonal action on receptors in the maternal hypothalamus. The two genomes, infant and maternal, are co-adaptive for infant wellbeing and reproductive success. Offspring that have extracted 'good' maternal nurturing will be genetically predisposed towards good mothering.

When David Barker describes the pregnant woman as reading her environment for her grandchildren, "adaptively" is the implication [34]. Marcus Pembrey's epidemiology indicates environmental effects in the gender-line and the trans-gender-line [25], opportunities for which are revealed in Golubovsky and Manton's three-generational physiology of female epigenetics [29]. Golubovsky and Manton indicate that establishment and maintenance of epigenetic states is a flexible and vulnerable process citing Singh et al. [35], that environmental factors (such as pollution, nutrition, and lifestyle) influence the epigenetic dynamics of the oocyte in the maternal grandmother and mother possibly causing genotype/phenotype changes in the grandchildren, citing Issa [36]. Boyano et al. report mammalian male imprinting at different stages over time [37], while Grandjean and Rassoulzadegan [38] identified epigenetic inheritance mediated by RNA and micro-RNA released by sperm. They consider RNA molecules present in the spermatozoon head may be possible vectors for the hereditary transfer of epigenetic modifications.

EVIDENCE – EPIDEMIOLOGICAL AND MOLECULAR
Evidence – Epidemiological, Transgenerational, and Molecular

Marcus Pembrey has provocatively referred to himself as "neo-Lamarckian" – in apparent support of adaptive evolution. He and Marilyn Monk, both of London's Institute of Child Health, are among those describing transgenerational transmission of environmentally induced epigenetic change. They include epidemiological and molecular evidence of parent to child transmission [25,26,30,39] with particular susceptibility during reproduction and development.

Epigenetics has qualified Mendelian theory in that randomization studies need not only assume a random distribution of alleles in the offspring, but also a random distribution of epigenetic changes at conception, in order for the core assumptions of the Mendelian randomization methodology to remain valid [32].

Although their effects were recognized early in the 20th century, epigenetic mechanisms were scarcely being revealed until its last decade. Then in 2001 the shock finding of the Human Genome Project completion – with less than a quarter the number of genes expected – turned minds to explore gene-expression to explain the immense variations in human beings, and their fine-tuning to their environment. The result is a reasonable explanation of two mysteries – (a) how environment can stamp its mark on an organism's phenotype and (b) how, within an organism, the position of each cell might induce its specific role-development – both by this single type of process. Yet neat as this explanation may be, it is in fact a complex range of processes, as will be summarized.

Inheritance and Genomic Imprinting

Genomic imprinting tends to stabilize transmission from parents to child with apparently rare changes to imprints. Imprints are usually sustained in somatic cells through life – see under "Human reproduction", "Physiological stages of imprinting processes, p. 437", below [40]. Flexibility comes through environmentally-caused epigenetic changes affecting only the individual, but including the fetus or infant indirectly through the mother. Highly-complex processes are involved in various reproductive processes: oogenesis and spermatogenesis, fertilization, and early embryonic development [2]. The most complex and seemingly most significant stage is periconceptional, from just before conception up to gastrulation – the beginning of the morula's nutritional opening of what is to become the alimentary canal [41]. Notably some specific gene-switch settings are inherited alleles – specifically as active or inactive. This process of genomic imprinting is accredited by Keverne [22] with the value of co-adaptive genetic behavior: paternal, maternal, embryonic, and placental. In contrast, genomic imprinting is open to danger. Being mono-allelic, an imprinted gene lacks the safeguard of an alternative copy of the gene similarly marked. Should one be flawed, it can result in a genetic disorder, for instance Prader–Willi or Angelman syndrome. Following gastrulation there is growing evidence of epigenetic effects from nutrition in gestation and infancy, [26, 42,43] and more strongly from stress, cortisol [7,8,44–49].

Tell-Tale Rapid Changes

Setting innovations in their evolutionary context provides an impression of the interplay between evolving organisms and evolving mechanisms. Significantly we recognize some of our conditions of existence to be changing now so fast and so adversely as to introduce some traits into our evolution that some hold to be seriously degenerative, such as the sudden escalation in obesity and diabetes, and cardiovascular and mental disorders. Technologies that contribute to this could also provide insights and tools to contend with the increasing complexities. Epigenetics highlights the need for farsightedness in matters of human health. To give some substance to this view, Crawford and Sinclair [50] prophesied in 1972 that whilst heart disease and cancer headed the burden of ill-health in the 20th century, mental ill health would replace them in the 21st century. A terrible confirmation of their prophecy is its fulfillment. In the UK the annual £77 billion cost of brain disorders is more than that of cancer and heart disease combined [51]. From the U.S. Department of Health and Human Services we read, "The cost of treating mental disorders rose sharply between 1996 and 2006, from $35 billion (in 2006 dollars) to almost $58 billion... 27 million people were using antidepressants in 2005, compared to 5.84%, or 13.3 million people, in 1996" [52]. For Europe, "… the true economic cost of disorders of the brain is substantially higher than our estimate of 386 billion Euros, perhaps in the range 500–700 billion Euros" [53]. The gene sequence has not changed [54]: conditions of existence are now constantly changing, and affecting epigenetic settings.

TIME PERSPECTIVE – THE BIG BANG – BEGINNING OF EVOLUTION – INTO THE FUTURE
Inorganic Matter to Cellular Life

The beginning of life and accelerating evolutionary processes need to be seen in perspective, against the Big Bang some 15 billion years ago (bya), the earth's formation some 4.5 bya and assembly of elements into molecules. In the submarine volcanic heat, fatty acid molecules may have formed on minerals, then conglomerated, drawn together by their oily end into spheres, their water-soluble ends outward in the ocean. Possibly RNA was established by this stage. A likely date-scale suggests amino-acids made feasible the protocell, "prokaryote", some 3.8 bya, then DNA and microbes by 3.5 bya, with powerfully photosynthetic cyanobacteria by 3 bya. By 2.8 bya cell nuclei and eukaryotic cells formed, the basis of today's immense variety of life, with cell memory, light-sensitivity and photosynthesis. By 2.5 bya the ocean may have been rich in omega-3s, docosahexaenoic acid (DHA), and eicosapentaenoic (EPA), significant particularly to the blue-green algae and phytoplankton, with the new process of photosynthesis, binding carbon. So successful were they that they threatened themselves with excess of their own oxygen, until rescued by the mitochondria at 2.3 bya. Burning oxygen back to carbon dioxide were the mitochondria, symbiotically with the blue-greens and phytoplankton on the one hand, and on the other hand energizing not just themselves but becoming the new range of oxygenated life-forms. This vast evolutionary potential remained poised on the threshold, inhibited by the deepest of all ice-ages, for some billion and a half years. Then, 670 million years ago (mya) a new warm climate brought on a burgeoning variety of new life, the Cambrian Explosion, evidence of which was first unearthed among the ancient rocks of Wales (Cambria).

Further Key Transitions in Evolution With Likely Changes in Epigenetic Processes

Million years ago (mya):

670   from single to multi-cell life
540   to shell-bearing life
500   vertebrate life
400   amphibians and vitamins for land-life
180   marsupials and mammals diverge
130   flowering plants, seeds with omega-6s
 50   sea mammals with omega-3s
  2   *Homo*'s brain enlargement with omega-3 docosahexaenoic acid (DHA).

Once this oxygenated life became multi-cellular, it sprang to highest complexity: shell-bearing invertebrates, vertebrates, and brain-powered life. Within 200 million years legs had appeared and amphibians. Vitamins, in evidence by 350 mya, seemed a requirement for land-life metabolism. A significant benefit featured in the transition between aquatic and terrestrial life, from a diet combining omega-3s with omega-6s, the very prescription for brain development. So the leap forward: reptiles 360 mya, mammals 210 mya, and *Homo habilis* 2 mya. This perspective spanning from the Big Bang to the present shows the long time-scale and widely varying pace of evolution with the sense of its continuity into the future. This is important in grasping the rapidity of a sequence of civilizational changes. One after another in a mere 400 generations, beginning with the hunter-gatherer, have fallen such impacts as agriculture, industrialization, intensification of farming, high-tech medicine, food manufacture, and marketing; and now the challenges of pandemic diseases, climate change, and ecological depletion, are calling for urgent ocean-bed agriculture. Now our myriad artificial environmental interventions come at a stage when our human sensitivity has never been higher in terms of epigenetics, genomic imprinting included. Through this young field of study can we divine any moments in this lay-out of evolution during which novel mechanisms were achieved?

Although we can tell many ways that environment affects cells and organisms, and how it does so, we can largely only speculate on when and how the changes came about to accumulate to the complex systems now regulating life. The most likely indications will come from the transitional stages in evolution, found by comparing fossil records across transitions with epigenetics of the descendants we know today.

MECHANISMS EMERGING IN EVOLUTION
Subcellular, Cellular, and Gene-Related
SWITCHING MECHANISMS IN EVOLUTION

There are many mechanisms affecting gene settings whose activity or silence depends on the spatial structure of molecules. A small proportion of these have been worked out. Discerning which species first achieved a mechanism, and at what stage in its evolution, reveals some of the major moments in evolution at which changes in organisms and mechanism took place. Some landmarks:

1. Chemical activation and inactivation (silencing) is not confined to genes but can operate in nucleosomes around which the chromosomes wrap themselves, affecting their spatial positioning and so the organism's life.
2. Even protocells and single-cell organisms – some 3.8 to 2.7 by a – needed components with some kind of marking for their self-arrangement and replication. With the assembly of DNA at 3.5 bya, and cell-memory at 2.8 bya, epigenetics would presumably have begun. Some such mechanisms might have been operating in RNA, which could possibly have preceded DNA. RNA is the less stable. It not only has a part in transcribing proteins but in transcribing back to the DNA as influenced by environment.
3. By 2.7 bya eukaryotes existed, with responses to environmental changes becoming increasingly elaborate. From this divergence of eukaryotes from prokaryotes, at 2.7 bya, springs sexual reproduction [55] with increasingly complex mechanisms.
4. In multi-cellular organisms, by 670 mya, each stem-cell needs markings not only for its internal components but to differentiate appropriately for its position. Cell-to-cell communication may help in this process [56]. Markings basically remain according to initial settings in the embryo except as changed by immediate environmental conditions.

TWO COMMON MECHANISMS – DNA METHYLATION AND HISTONE ACETYLATION

Two of the commonest mechanisms, and best understood, are DNA methylation and histone acetylation. Switching genes off is effected by attachment of methyl groups to cytosines, several in a cluster, or by detaching acetyl groups from lysines on histones. Often the two processes work together, tightening the way that DNA is coiled around the "nucleosomes" (spheres of histone protein), spatially changing the shape and physical forces of the DNA system. Histones also may be methylated, which may either turn a gene on (H3K4me2,3), or off (H3K9me2,3). Silencing genetic expression of is mostly through switching off controlling genes upstream. Changing gene expression substantially affects transcription. Some yeast cells, for example, can abruptly change from single sex to a capacity for switching between two sexual types. Most epigenetic changes are made through regulatory genes, having a cascade effect on many genes.

Yet methylation and acetylation are just the two best known mechanisms, and there are many others at play. Zuckerkandl, for instance, describes how "junk DNA" along the chromosome is probably influencing the extent to which methylation is happening. Influences such as this may be affecting how lasting the epigenetic change may become, and whether it may cause a change in DNA sequence, a mutation [57].

RELATIONSHIP IN EVOLUTION BETWEEN GENETIC CHANGES AND EPIGENETIC

Mutations, irreversible, are almost invariably passed down to half the children. Epigenetic settings, reversible, are mostly erased during early embryo implantation: generally imprints only are re-established. A key question is the relationship between epigenetic changes and

431

mutations. Current speculation is that changes in gene sequence – mutations – correlate with lasting composite epigenetic changes [61].

The power in evolution of various mechanisms of epigenetic transmission is emphasized by Jablonka and Lamb [18], who write of the combinations "…of different active/inactive loops, cellular architectures, chromatin marks, and RNA-mediated silencing patterns". The amount of cell variation, they say, is vast and the evolutionary potential of inheritance systems therefore considerable. They draw the general conclusions that when conditions change, epigenetic events can increase the rate of adaptive evolution, by activating silent genes and through heritable variants. They also quote Belyaev's [59] work in Russian with the silver morph of the red fox. He took two groups of silver foxes, farm-raised one and domesticated the other, observing the differences between each group and those remaining in the wild. Even over 10 generations remarkable differences emerged in coloring and behavior, leading him to conclude that "induced heritable epigenetic variations play an integral role in adaptive evolution". Although the epigenetic functional differences were not analyzed, Belyaev detected 40 gene differences between domesticated and farm-raised, and between them and wild foxes, 2700 different genes [60].

Belyaev's demonstration of how few generations of silver foxes it takes to induce heritable adaptive changes, combined with human transgenerational effects such as those found in Pembrey's Swedish study (see section below on "Transgenerational effects"), keeps open the search for mechanisms that can, so quickly, and often enough favorably, select from alternative settings those lasting composite epigenetic changes that correlate with mutations. Rando and Verstrepen [61] have attended to time-scales of variations and suggest that some organisms have evolved mechanisms with variability that relates to the variability of selective pressure encountered. The search for non-teleological adaptive mechanism continues.

Single Cell to Amphibians
EARLY PROCESSES IN PROTOCELLS, NUCLEI, AND MEMORY

Components and cells, and thereby organisms, are affected by factors including: component position and shaping; memory-responses; neighboring components; gravity and pressure; and environment – chemical, electromagnetic, and radiation – cosmic, geological, artificial. Some of these environmental forces can induce mutations. Since cells, or even organelles, require components to organize themselves appropriately to their position, we can reasonably assume markers or signals to be involved at a very basic level. We are at least clear that DNA strands have markers making fine distinctions [62]. Memory can lie latent yet ready to be reactivated [63]. DNA methylation can be altered by "activation-induced (cytidine) deaminase" (AID), an enzyme involved in the formation of the immune system, leaving cells with inaccurate memory. AID initiates immunoglobulin class switch recombination and somatic hypermutation by producing uridine:guanidine mismatches in DNA, which can also induce DNA damage including double-stranded breaks and chromosome translocations. Strict regulation of AID is vital for genomic stability [64]. Wolf Reik proposed that methyl-cytosine can be changed to thymine by deamination with AID, being repaired to normal cytosine [65]. Nat Heintz showed that methyl-cytosine can be converted to hydroxymethyl-cytosine by other enzymes and is similarly repaired [66]. See also AID in the paragraph after next, "Invertebrate to vertebrate transition".

BACTERIA AND MITOCHONDRIA

Bacteria make widespread use of DNA methylation, following replication, for epigenetic control of DNA–protein interactions. Bacteria methylate adenine in DNA, rather than cytosine, as an epigenetic signal [67]. Mitochondria evolved possibly within bacteria at the same time as the eukaryotic cell nucleus [68,69]. Yet they seem to have emerged independently and entered eukaryotes, and symbiotically have provided signaling, regulating nuclear gene expression [70] and oxygen-powered energy [13]. Mitochondrial DNA has a mutation rate 10 times that of eukaryotes. Somehow, with their signaling role, this has potential to speed up mutations to rates more closely compatible with the recognized pace of evolution [71]. The mitochondria are rewarded with access to plentiful oxygen. The radically diverse trends in mitochondrial genome evolution, recognized in different phylogenetic groups, has allowed pinpointing of specific protist relatives of multicellular lineages – animals, plants, and fungi. This research revealed "unique and fascinating aspects of mitochondrial gene expression, highlighting the mitochondrion as an evolutionary playground par excellence" [72].

INVERTEBRATE TO VERTEBRATE TRANSITION

DNA methylation is a mark associated with gene regulation and cell memory, silencing of transposable elements, genomic imprinting, and repression of spurious transcription of duplicated sequences. These roles have varied widely during animal evolution. DNA-methylation machinery includes three groups of enzymes (Dnmt methyltransferase), and five binding proteins (Mbd methyl-DNA).

Albalat [73] has identified changes in the presence of these Dnmt and Mbd gene families at the juncture between invertebrates and vertebrates (in the cephalochordate amphioxus, *Branchiostoma floridae*), a group closest to vertebrates. Whereas three major groups of Dnmt enzymes were found in the invertebrates, in the vertebrates only two Mbd members were found. Although during the invertebrate–vertebrate transition, methyltransferases were little changed, new Mbd proteins arose, which perhaps minimized certain collateral effects associated with the major genomic changes that occurred.

Between fish and tetrapods appeared class switch recombination, the last of the lymphocyte-specific DNA modification reactions to appear in the evolution of the adaptive immune system. Class switching is initiated by activation-induced cytidine deaminase (AID), which is also required for somatic hypermutation. Fish AID differs from orthologs found in tetrapods in several respects, including its catalytic domain and carboxy-terminal region, both of which are essential for the switching reaction. Fish AID was found to catalyze class switch recombination in mammalian B cells, and therefore had the potential to catalyze this reaction before the teleost and tetrapod lineages diverged [74]. See AID in "Early processes", above.

AMPHIBIANS – REVERSION POTENTIAL OF SOMATIC CELLS TO STEM CELLS

Salamander somatic cells have been remodeled to the stem cell state, clearly a "reverse" epigenetics change. The levels of signal intensity in differentiated cells that are then treated, resemble those detected in embryonic stem cells, which are unaffected by these extracts. Selectively somatic cells exposed to oocyte extracts undergo demethylation [75]. Plants, fish, and animals that abandoned the sea for the land and inland waters, lacked a wide range of nutrients, iodine, and other elements, also marine antioxidants, prompting improved production of various antioxidants which became essential vitamins to life on land.

Mammals

MAMMALS – TRANSITIONS TO PLACENTA, LIVE BIRTH, AND GENOMIC IMPRINTING

Genomic imprinting marks a gene active or inactive mono-allelically according to its being the maternal or paternal allele. It is mainly evident in some flowering plants and in placental mammals. Origins of mammalian genomic imprinting are emerging from studies of two transitions: (a) from egg-laying "prototherian" mammals such as the platypus, with only a short-lived placenta, to fully placental "therian" marsupials such as the kangaroo; (b) from therians on to fully placental "eutherian" mammals. Since even egg-laying prototherians have a short-lived placenta, imprinting appears to correlate with giving live birth (viviparity)

433

rather than with placentation. Marsupial live-birth follows a short gestation supported by a fully functional placenta [76,77].

MAMMALIAN ACQUISITION OF GENOMIC IMPRINTING – FROM PROTOTHERIAN TO THERIAN TO EUTHERIAN

The acquisition and evolution of genomic imprinting is among the most fundamental genetic questions. Genomic imprinting is an epigenetic phenomenon that regulates many aspects of growth and development. Apparently absent from the egg-laying prototherian mammals such as the platypus, genomic imprinting is widespread in (therian) marsupials, such as the kangaroo, as well as more advanced (eutherian) mammals. According to favored hypotheses, genomic imprinting evolved within the cell, benefiting it by silencing foreign DNA elements entering the genome, and by balancing maternal and fetal nutrient supply [77,78].

Pask et al. [78] showed that the platypus has significantly fewer repeats of certain classes in the regions of the genome that have become imprinted in therian mammals. They conclude that the accumulation of repeats in therian imprinted genes and gene clusters, especially long terminal repeats and DNA elements, may have been a driving force in the development of mammalian genomic imprinting. All orthologs of eutherian imprinted genes examined have a conserved expression in the marsupial placenta regardless of their imprint status.

In eutherian mammals the most common mechanisms controlling genomic imprinting are "differentially methylated regions" (DMRs), whereas in the marsupial the mechanism used to silence the equivalent genes appears to be histone modification.

> "At least three genes in marsupials have DMRs: H19, IGF2 and PEG10. PEG10 is particularly interesting as it is derived from a retrotransposon [a DNA sequence that can move its position within the genome], providing the first direct evidence that retrotransposon insertion can drive the evolution of an imprinted region and of a DMR in mammals. The insertion occurred after the prototherian-therian mammal divergence, suggesting that there may have been strong selection for the retention of imprinted regions that arose during the evolution of placentation" [77].

NON-CODING RNAs IN MAMMALIAN ACQUISITION OF GENOMIC IMPRINTING

Most imprinted genes, located in clusters and regulated by imprinting control regions, included non-coding RNAs. Some of these non-coding RNA-expressions were changed dramatically at this stage, and also many novel non-coding RNAs were added. A study of imprinted small nucleolar RNA genes from 15 vertebrates "suggests that the origination of imprinted snoRNAs occurred after the divergence between eutherians and marsupials". Subsequently rapid expansion led to the fixation of major gene families in the eutherian ancestor, and then the radiation of modern placental mammals. The non-coding RNAs' major roles during the acquisition of genomic imprinting in mammals seem to be the regulation of imprinting silence, and mediation of the chromatins' epigenetic modification [76].

IMPRINTING DIFFERENCES IN THE MAMMALIAN EMBRYO AND PLACENTA

The maternal genome in the zygote is highly methylated in both its DNA and its histones, its imprinted genes mostly having maternal germline methylation imprints. The paternal genome is rapidly remodeled by protamine removal, addition of acetylated histones, and rapid demethylation of DNA before replication. A minority of imprinted genes are silent, having paternal germline methylation imprints. Methylation and chromatin reprogramming continues during cleavage divisions. At the blastocyst stage, DNA and histone methylation increases dramatically.

This may set up major epigenetic differences between embryonic and extraembryonic tissues, placenta included. X-chromosome inactivation is involved and perhaps imprinting. Maintaining asymmetry appears important for development. In cloned embryos asymmetry is lost, most having developmental defects, particularly an imbalance between extraembryonic and embryonic tissue development [79].

Hominids
NEW SEXUAL AND PARENTING BEHAVIORAL IMPRINTS IN HOMINIDS

Two major developments have affected mammalian sex differences in behavior: the placenta's hormonal effects on the maternal brain, especially in small-brained rodents; and the brain's massive expansion in primates, especially hominids. In both developments genomic imprinting has been significant. "Most of the imprinted genes investigated to date are expressed [mono-allelicly] in the placenta and a subset are expressed in both placenta and hypothalamus." Recent knockout studies suggest the co-adaptive effect of imprinting may have been significant in imprinting's evolution, rather than parental conflict as commonly held, contributing to co-adaptation between male and female and offspring. Evidence supports a co-adaptive evolution of placenta and hypothalamus, particularly in neurohormonal regulation of maternalism. The neocortex and other parts of the brain which have expanded are undoubtedly under the influence of imprinted genes. In small-brained mammals, a female's short estrus demands greater olfactory powers from males, compensated for by a male accessory olfactory system. Evidently the same imprinted gene that regulates mammalian maternal care and offspring development also regulates male olfaction and sexual behavior. In hominids, humans particularly, differences in sexual behavior owe much to social structure and strategies of intelligence. Social learning has become as important as hormones in epigenetic effects on brain development [23,24].

STABILITY OF THE ORIGINAL CHROMATIN STRAND IN DIFFERENTIATING CELLS

From problems including cancer and atheromatous plaque increase with aging, we learn about protection of chromatin strands. In healthy cell division the stem cells retain the original chromatin strands while the differentiated daughter cell receives new chromatin strands, which are more susceptible to error in self-repair. Since a differentiated cell has a limited life before dying and being replaced, there is little chance of a problem. Occasionally, however, the differentiating daughter cell receives the original strands, leaving the stem cell with the newer strands in which any fault will continue to be replicated. Epidemiological time-patterns indicate a mechanism that almost always prevents "a final step" of mutation to cancer. The final step could be when a rare fully mutant cancerous stem cell produces a daughter cell that is freed from some nongenetic imperative to differentiate and die. This appears to be a mechanism, like the one that induces appropriate cell differentiation, which marks the strands as original or new so that persistent errors are effectively prevented [52].

According to D. Simmonds, although epigenetic changes do not alter the sequence of DNA, they can cause mutations. About half of the genes that cause familial or inherited forms of cancer are turned off by methylation. Most of these genes that normally suppress tumor formation and help repair DNA, including O^6-methylguanine-DNA methyltransferase (*MGMT*) are turned on. For example, hypermethylation of the promoter of *MGMT* causes the number of G-to-A mutations to increase.

Unlike DNA sequence mutations, which are irreversible, many diseases such as cancer involve epigenetic changes. Being reversible, these could be responsive to epigenetic treatment. The most popular of these treatments aim to alter either DNA methylation or histone acetylation [81].

435

GENDER EFFECTS AS WELL AS SEX-CHROMOSOME EFFECTS IN HEALTH AND DISEASE

The vast majority of common diseases, including atherosclerosis, diabetes, osteoporosis, asthma, and neuropsychological and autoimmune diseases, which often take root in early development, display some degree of sex bias, very marked in some cases. This bias could be explained by the role of sex chromosomes, the various regulatory pathways affecting sexual development of most organs and the continuing fluctuating impact of sex hormones.

In a gender-related manner, environmental factors such as social behavior, nutrition, or chemical compounds can influence these flexible marks during particular developmental stages and subsequent changes in life. Each developmental process may be more sensitive for one gender or the other during specific environmental challenges, particularly developmental programming and gametogenesis, but also throughout the individual's life as influenced by sex steroid hormones and/or sex chromosomes. An unfavorable programming could thus lead to defects and susceptibilities to diseases differing between males and females. Recent studies suggest that such programming can be sex-specifically transmitted to subsequent generations leading to transgenerational effects. Gabory et al.'s review highlights the importance of studying both sexes in epidemiological protocols or dietary interventions whether in humans or in experimental animal models [82].

In the mouse, gender-dependent genomic imprinting effects, not related to sex-chromosome effects, have been demonstrated. Thirteen loci on 11 chromosomes showed significant differences between the genders in imprinting effects. Most loci showed imprinting effects in only one sex, with eight imprinted effects found in males and six in females, but one locus showed sex-dependent imprinting effects in both sexes for different traits [83]. The degree of imprinting is often tissue-specific or developmental stage-specific, or both. In some diseases, cancer included, it may be altered. Some 1% of genes may be imprinted and the balance between alleles can vary from 100% one-way to 50–50 [84]. Wang et al. found in neonatal mouse brains that both known imprinted genes and novel genes were all close to differentially-methylated regions (DMRs).

Human Reproduction – Health of Future Generations

ORIGINS OF HEALTH AND DISEASE – PRIMACY OF EPIGENETICS

Epigenetics could contribute importantly to lifelong prevention of common chronic health conditions. The focus of the International Society for Developmental Origins of Health and Disease (DOHaD) [6] is on the earliest stages of human development. The Society's 3rd International Congress, 2005, added new perspectives, including developmental plasticity, influences of social hierarchies, effects of prematurity, and populations in transition. Emerging areas of science included:

- Infant weight gain and prediction of adult obesity, diabetes, and cardiovascular disease.
- The era of epidemic obesity – the over-nourished fetus and growth retarded fetus.
- Environmental toxins' broad range of long-lasting effects on the developing human.

The Society recognized that epigenetic mechanisms could unite several strands of human and animal observations. They could explain, for instance, how genetically identical individuals, even raised in similar postnatal environments, can nonetheless develop widely differing phenotypes [85–88]. Improving the individual's environment during development may be as important as any other public health effort to enhance population health world wide.

Nathanielsz uses animal models to evaluate specific exposures such as nutrient restriction, overfeeding during pregnancy, maternal stress, and exogenously administered substances such as glucocorticoids on developmental programming, revealing effects of hypertension, diabetes, obesity, and altered pituitary-adrenal function in offspring in later life. Although, for example, the fetus responds to challenges such as hypoxia and nutrient restriction in ways that help to ensure its survival, this "developmental plasticity" may have long-term consequences that may not be beneficial in adult life [89].

Recent Findings

During development, there are critical periods of vulnerability to suboptimal conditions when programming may permanently modify disease susceptibility. Programming involves structural changes in important organs: altered cell number, imbalance in distribution of different cell types within the organ, and altered blood supply or receptor numbers [90].

TRANSGENERATIONAL EFFECTS

Transgenerational sex-specific links were evident in a study of Swedish cohorts by Pembrey et al. [25]. The paternal grandfather's food supply was associated with the mortality risk ratio of grandsons only, and the paternal grandmother's with the granddaughters' only. The effects followed exposure during the slow growth period (both grandparents), or fetal/infant life (grandmothers), but not during either grandparent's puberty. These sex-specific, male-line effects in humans suggest mediation by the sex chromosomes, X and Y. The same study showed early paternal smoking, before 11 years old, to be associated with greater body mass index (BMI) at 9 years in sons, but not daughters. A subsequent study by Kaati et al. [26] confirmed these effects. It also showed that early social circumstances influenced longevity for sons. The main influence on longevity was transgenerational responses to ancestors' nutrition. But less expected was Pembrey's study [27] showing *scarcity* of food in the grandfather's slow growth period to be associated with a significantly *extended* survival of his grandchildren for many years, whilst food *abundance* was associated with a greatly *shortened* life span of the grandchildren. Even more surprising is that overall the results show that *cardiovascular mortality* was reduced with *poor* availability of food in the father's slow growth period, but also with *good* availability in the mother's slow growth period... "hinting at some 'see-saw' effect down the generations".

What is known about these mechanisms? Hung et al. including Pembrey [91] studied effects of a particular variant gene, the "small heterodimer partner" (SHP), which has repressive effects hormonally and transcriptively. They concluded that although mutations in SHP are not a common cause of severe human obesity, genetic variation in the SHP locus may influence birth weight and have effects on body-mass index (BMI), possibly through effects on insulin secretion. Even in 1996 Pembrey [28] was speculating on the beneficial potential of such knowledge. Poised between transcriptionally-active and silent states, imprinted genes seem good candidates for the evolution of transgenerational adaption systems, where coordinated changes in gene expression over the generations are a selective advantage.

PHYSIOLOGICAL STAGES OF IMPRINTING PROCESSES

Clearly the process of imprinting is complicated and Barry Keverne does not claim to know the full story yet, but I relay here the 4-stage cycle with two notes that he has kindly set out [40]:

1. Epigenetic reprogramming events occur in the primordial germ cells at embryonic day 8. This results in a global loss of methylation.
2. Following this erasure there is a resetting of the epigenome in the germline – this may account for those instances of transgenerational inheritance of epimutations.
3. Later in oogenesis and spermatogenesis, "*de novo*" methylation occurs in a sex-specific manner for imprinted genes.

437

4. There is a further wave of genome-wide reprogramming events immediately after fertilization, but imprinted genes are thought to be protected from demethylation in the zygote.

(a) The general rule is that the imprint is maintained throughout development, so all somatic cells that express this imprinted gene do so in a haploid dominant manner according to parent of origin, i.e. the memory is sustained through mitotic divisions.

(b) This imprint never switches in somatic cells, only in germline cells following reprogramming. For the most part, imprints retain their mono-allelic stability, except in tumors. It has also been shown that handling of fertilized embryos (e.g. *in vitro* fertilization/embryo transfer) can also influence imprints.

This sequence makes evident:

- Some reasons for high vulnerability of the fertilization/early-embryonic phase–in 1, 4, and (a).
- A likely stage at which occasional epimutations may be transmitted – in 2.
- The setting of sex-specific imprinting – in 3.
- The way in which imprint memory is sustained for transmission transgenerationally – in (b).

CONTINUING EVOLUTIONARY IMPACTS ON HEALTH AND DISEASE

Subtle environmental changes throughout early development affect later health and disease. Genetic evolutionary background contributes significantly to our susceptibility to perinatal imprinting. Epigenetic modulation, in reaction to a given environment, results in functional adaptation of the genomic response, more plastic than the genome sequence.

> "Evolutionarily acquired genomic susceptibilities, and environmentally induced epigenomic modulations occurring early in life, impact on later development of human diseases" [92].

Such evidence as amply supplied by Tremblay and Hamet [92] provides molecular understanding that might explain some of the extensive evidence gathered in *The Unborn Child* [21].

For half a billion years there has been no change in our inherited genetic mechanism for human blood clotting. Genetically it remains the same as in the puffer-fish, reports hematology professor Edward Tuddenham. Yet in a mere 150 years our cardiovascular problems include a huge thrombosis epidemic with high mortality. The cause is changes in blood-clotting due to smoking and modern diet. "The cure must lie in returning our diet towards its premodern state" [54]. Epigenetic implication is supported by studies including that by Sharma et al. [93].

Studies of genes in relation to violence so far have largely concerned the MAOA gene, which, in children, only after maltreatment, correlates with nine times the risk of violent behavior [94]. This seems ripe for epigenetic study. Professor Bruce Lipton perceives a process of "survival of the most loving." Scant nurture or other severe conditions, he writes, will program the new being for anxiety about survival: generous nurture will program a child for trust, love and creativity, for reproducing and nurturing [95].

Bridging between cytology and psychiatry between 1954 and 1982 was a scientist, Dr Frank Lake [7–9]. After 13 years of studying cell memory, Lake qualified in psychiatry. At first using lysergic acid diethylamide 25 (LSD25), but then finding therapy more effective through current feelings and memory evoked without any drug, he found patients were accessing early memories and also, through physical movement, very early cell memory. Lake's discoveries were contemporary with Arthur Janov's [96,97] yet independent, and had striking similarities, particularly in re-experience of birth. Since then this experience has become common to those involved in pre- and perinatal psychotherapy. Lake's insights have been incorporated into William Emerson's renowned work [98]. Lasting psychological effects of human gestation and birth have been set in an evolutionary context by Ludwig Janus [99]. Lake went on to train teams of psychotherapists. Patients and therapists, including medical people, became convinced that powerful body movements were expressing cell memory from early somatic or cellular memory from as far back as conception. I am not implying embryonic consciousness, or that such early events were "remembered". Karl Pribram held that a short-term memory could resonate through the brain's stored holograms until an association is triggered in long-term memory. According to Lake [3,100]:

> "The holograms of cellular memory are still broadcasting from infinitesimally small, but collectively audible transmitting stations. These minute radio stations belong to successive periods of development, from conception to implantation and the developmental stages of pregnancy. It seems they are still transmitting and it is possible to tune into them."

Among thousands of the Lake team's patients, for many the periconceptional and embryonic stages seem to have had the greatest permanent effects on the person's later feelings and behavior. Inevitably controversial, those of us with these experiences find them hard to dissociate from earliest cell memory. Evidence of epigenetic effects at stages round the lifecycle help to make such findings more intelligible, and may well cast new light on early life plasticity. Extensive records of pre- and perinatal psychotherapists witness to the impacts of early life circumstances, which help to unravel the effects of cultural traditions on the emotions and mind-set of generations [101,102].

MENTAL DISORDERS AND HERITABILITY, AND PSYCHOTIC DRUG EFFECTS

Alarming epigenetic evidence relating to mental disorders includes long-term antipsychotic drug use and transgenerational effects. These further strengthen the impression of our continuing evolutionary state. Mill et al. [33a] describe correlations of epigenetic misregulation with various non-Mendelian features of schizophrenia and bipolar disorder, including several affecting brain development, neurotransmission, and other processes. Epigenetic disruption was evident in mitochondrial function, brain development, and stress response.

Like the DNA sequence, the epigenetic profile of somatic cells is mitotically inherited, but unlike the DNA sequence, the signals are dynamic. The epigenetic status of the genome is tissue-specific, developmentally regulated, and influenced by both environmental and random factors. Genes must therefore be both in the right sequence and also expressed in the appropriate amount, at the correct time of the cell cycle, and in the correct compartment of the nucleus. During gametogenesis some epigenetic signals, rather than being erased and reset, can be transmitted meiotically across generations [103,104]. For both brain and germline DNA, evidence suggested that the epigenome is regionalized relating to distinct physical regions and/or functional pathways. Comparing the affected group to the control group:

(a) The number of interconnections between genomic regions is higher, resulting in more interference between regions in both brain and germline DNA, and therefore very possibly between specific functional tasks [105].
(b) In the major-psychosis samples, these regions' greater activation (lower degree of DNA-methylation), indicates some degree of systemic epigenetic dysfunction associated with major psychosis.

> "Our data are consistent with the epigenetic theory of major psychosis and suggest that DNA-methylation changes are important to the etiology of schizophrenia and bipolar disorder" [33a].

Mill and Petronis also propose that understanding the epigenetic processes involved in linking specific environmental pathogens to an increased risk for ADHD may offer new possibilities for preventative and therapeutic intervention [33b].

SAFEGUARDING HUMAN REPRODUCTIVE HEALTH WITH EPIGENETICS UNDERSTANDING

For decades findings in nutrition and psychotherapy have shown the impact of periconceptional as well as pre- and perinatal circumstances on development. In the authoritative words of E.L. Ford-Jones et al:

> "Diseases of modernism, rather than infectious diseases and chronic medical conditions, increasingly cause childhood morbidity and mortality. Thus, the goal of enhancing life outcomes for all children has become imperative... The new neuroscience of experience-based brain and biological development has caught up with the social epidemiology literature. It is now known from both domains that a child's poor developmental and health outcomes are a product of early and ongoing socioeconomic and psychological experiences. In the era of epigenetics, it is now understood that both nature and nurture control the genome... A challenge is to connect the traditional population health approach with traditional primary care responsibilities. New and enhanced collaborative interdisciplinary networks with, for example, public health, primary care, community resources, education and justice systems are required" [106]...

...required to benefit not merely our health but our continuing evolution.

SIGNIFICANCE OF EVOLUTIONARY DIFFERENCES TO ASSISTED REPRODUCTIVE TECHNOLOGY

Since imprinting is related to various processes in reproduction, it is naturally liable to impacts from assisted reproduction technologies (ARTs), especially during embryo and trophoblast development, as ARTs are used during these periods.

Wilkins-Haug et al. found that induced ovulation, and oocytes with potentially less stable imprints, may contribute to the higher rate of the maternal imprint disorders. Paternal imprinting abnormalities in oligospermatic men may indicate that subfertility itself is associated with epimutations – low sperm quality does correlate with low sperm count [2]. Higher rates of adverse outcomes that follow ARTs, such as growth restriction, may be found related to placental epimutations [107,108].

The association, particularly in assisted conception, of disease and genesis of tumors with perturbed imprinting means that monitoring for normal imprinted gene expression in human embryos is critical. Monk and Salpekar showed mono-allelic expression to be tissue-specific and time of onset to vary between different imprinted genes. Three of six genes analyzed were clearly expressed in human preimplantation embryos, and the expression of one, SNRPN, was mono-allelic from the paternal allele. This gene was also mono-allelically expressed in mouse preimplantation embryos, being "correlated with differential methylation of Xist promoter sites in egg and sperm, and specific binding of a protein only to the methylated maternal (egg) allele." But in human preimplantation embryos, "unlike the mouse, XIST is expressed from both parental alleles."

Monk and Salpekar's studies highlight not only a significant step in the evolution of an epigenetic mechanism, but also the criticality of checks that animal research findings are entirely appropriate to techniques in human medicine [31].

IN CONCLUSION

Darwin's final statement in *On the Origin of Species* [14] is: "There is grandeur in this view of life... that... from so simple a beginning endless forms most beautiful and most wonderful have been, and are being evolved." This ties in with Chapter VI's concluding climactic paragraph "…that all organic beings have been formed on two great laws – Unity of Type, and the Conditions of Existence, [which is] the higher law…". The term "Conditions of Existence" Darwin here introduces weightily as that of "the illustrious Cuvier". Although using it three times in this paragraph out of only four in the main text, he does not distinguish it from his own term "conditions of life" which, in his main text, he synonymously uses no less than 118 times.

Darwin did try to explain how conditions could affect change in organisms and their heritability, offering the hypothesis of "pangenesis" [109], but it failed to hold back the general acceptance that changes were random and subject to the slow work of natural selection. Today's awareness of mechanisms of epigenetics and genomic imprinting vindicate Darwin's search, explaining even the power of changing conditions on organisms that we are forcibly witnessing in our lifetime. Just as very few generations of changed conditions were needed for changes in Belyaev's silver foxes, very few generations have been needed to bring human heart disease from a rarity to top killer over the last century. Now brain disorders are overtaking the cost of all other burdens of ill health – at least in Europe – and spreading globally [51,52]. The need is urgent to address the "Conditions of Existence" law of Cuvier and Darwin, by which most serious epigenetic impacts on *Homo sapiens* are threatening subsequent generations.

Epigenetics could be our most powerful technological insight against the current crises, microscopically highlighting the dependence of human health on the biosphere. The cradle of biosystems has been the estuarial and inshore beds of photosynthetic systems, basic to the marine food chain, whose demise threatens both collapse of fisheries and aggravated climate change. Of the Philippines coral reefs, only 5% remains healthy. The fight is on to restore their thousands of island shores [110].

How can we reverse the drastic trend in brain and other disorders? Draconian steps have necessarily begun, to cut the industrial pollution destroying our biosystem, and to build systems of marine agriculture, include filtering of pollution with seaweed and shellfish, as in Sungo Bay, China, since 1991. Such measures are essential, if only to meet the human brain's prime need for the marine source of omega-3 oil docosahexaenoic acid and other essential nutrients, that is greatest in gestation and brain development [51,52]. Healing fertile marine regions is basic to restoring human brain health in today's mental health pandemic.

> "The concept of fetal programming is an area that is now under rigorous investigation in many laboratories throughout the world. We need to engender a fascination in all segments of society, not just pregnant women, about life in the womb. *Conclusion*: Everyone needs to understand that improving the condition of the fetus will have personal, social and economic benefits. The time has come to realize that, in a sense, it is not just women who are pregnant but it is the family and the whole of society" [111].

Our growing awareness of mechanisms of epigenetics and genomic imprinting should raise our sensitivity to the importance of keeping today's increasingly artificial conditions of existence as beneficial as possible to evolution and health.

ACKNOWLEDGEMENTS

For help with this text I profoundly thank, among others, particularly David Marsh on the history aspect, and Edward Tuddenham, Barry Keverne, and Michael Crawford, without shifting onto them any responsibilities for its inadequacies.

References

1. Bird A. Perceptions of epigenetics. *Nature* 2007;447(7143):396–8.
2. House S. Stages in reproduction particularly vulnerable to xenobiotic hazards and nutritional deficits. In: *Generating Healthy People: Nutrition and Health* 2000;14(3). (Medline and www.healthierbabies.org click "Healthy People").

3. Ridgway R, House SH. *The Unborn Child*. London: Karnac; 2006.

4. House SH. Nurturing the brain, nutritionally and emotionally, from before conception to late adolescence. In: Generating healthy brains. *Nutr Health* 2007;**19**(1–2). Bicester UK: AB Academic.

5. Barker DJP. *Mothers, Babies and Health in Later Life*. Edinburgh: Churchill Livingstone; 1998.

6. Gillman MW, Barker D, Bier D, Cagampang F, Challis J, Fall C, et al. Meeting report on the 3rd International Congress on Developmental Origins of Health and Disease (DOHaD). *Pediatr Res* 2007;**May**;61(5 Pt 1):625–9.

7. Maret SM. *The prenatal person: Frank lake's maternal-fetal distress syndrome*. Lanham: University Press of America; 1997 PhD thesis–Frank Lake's Maternal-Fetal Distress Syndrome. NJ: Drew University; 1992.

8. Lake F. *Tight corners in pastoral counselling*. Birmingham UK: Bridge Pastoral Foundation; 2005 London: Darton L.T. 1981.

9. House SH. Primal Integration Therapy–School of Lake–Dr Frank Lake, MB MRCPsych DPM. *Int J Prenat & Perinat Psychol & Med* 1999;**11**(4):437–57. Heidelberg: Mattes. (2000). and *J Prenatal Perinatal Psychol Health*, 14(3-4):195-352, Lawrence KA: Allen.

10. Tobias PV. Some aspects of the multifaceted dependence of early humanity on water. *Nutr Health* 2002;**16**(1):13–17.

11. Stringer C. Coasting out of Africa. *Nature* 2000;4(405(6782)):24–5, 27.

12. Marsh DE. The origins of diversity: Darwin's conditions and epigenetic variations. *Nutr Health* 2007;**19**(1-2):103–32.

13. Crawford M, Marsh D. *Nutrition and evolution*. New Canaan CT, USA: Keats; 1995 The Driving Force, London: Heinemann; 1989.

14. Darwin C. *On the origin of species*. London: John Murray; 1859.

15. Lamarck J-B. *Philosophie Zoologique*, editor Charles M (2 vols). Paris: Savy; 1873.

16. Koonin EV, Wolf YI. Is evolution Darwinian or/and Lamarckian? *Biol Direct* 2009;**4**(1):42.

17. Weismann A. *The effects of external influences on development*. Romanes lecture. London: Frowde; 1894.

18. Jablonka E, Lamb MJ. The epigenetic turn: the challenge of soft inheritance. www.mfo.ac.uk/files/images/Jablonka-ms_MPGM_EEEMclean.doc. In: Fox CW, Wolf JB, editors. *Evolutionary genetics: concepts and case studies*. UK: Oxford University Press; 1995.

19. Saugstad LF. From superior adaptation and function to brain dysfunction–the neglect of epigenetic factors. *Nutr Health* 2004;**18**(1):3–27.

20. Crawford MA, Bloom M, Linseisen F, et al. Evidence for the unique function of DHA during the evolution of the modern hominid brain. *Lipids* 1999;**34**:S39–47. See http://www.physics.ubc.ca/~flinseis/Lipids/HumanBrain.pdf

21. House SH. Schoolchildren, maternal nutrition and generating healthy brains: the importance of lifecycle education for fertility, health and peace. *Nutr Health* 2008;**19**(4). Bicester UK: AB Academic.

22. Keverne EB. The significance of genomic imprinting for brain development and behavior. In: Generating Healthy Brains. *Nutr Health* 2007;**19**:1–2.

23. Swaney WT, Curley JP, Champagne FA, Keverne EB. Genomic imprinting mediates sexual experience-dependent olfactory learning in male mice. *Proc Natl Acad Sci USA* 2007;**104**(14):6084–9.

24. Keverne EB. Genomic imprinting and the evolution of sex differences in mammalian reproductive strategies. *Adv Genet* 2007;**59**:217–43.

25. Pembrey ME, Bygren LO, Kaati G, Edvinsson S, Northstone K, Sjostrom M, et al. Sex-specific, male-line transgenerational responses in humans. *Eur J Hum Genet* 2006;**14**(2):159–66. Also Pembrey M (2006) *The Ghost in Your Genes* (2005). http://www.bbc.co.uk/sn/tvradio/programmes/horizon/ghostgenes.shtml.

26. Kaati G, Bygren LO, Pembrey M, Sjöström M. Transgenerational response to nutrition, early life circumstances and longevity. *Eur J Hum Genet* 2007;**15**(7):784–90.

27. Pembrey ME. Time to take epigenetic inheritance seriously. *Eur J Hum Genet* 2002;**10**(11):669–71.

28. Pembrey M. Imprinting and transgenerational modulation of gene expression; human growth as a model. *Acta Genet Med Gemellol (Roma)* 1996;**45**(1-2):111–25.

29. Golubovsky M, Manton KG. A three-generation approach in biodemography is based on the developmental profiles and the epigenetics of female gametes. *Front Biosci* 2005;**10**:187–91. http://www.bioscience.org/2005/v10/af/1520/pdf.pdf

30. Monk M. Epigenetic programming of differential gene expression in development and evolution. *Dev Genet* 1995;**17**(3):188–97.

31. Monk M, Salpekar A. Expression of imprinted genes in human preimplantation development. *Mol Cell Endocrinol* 2001;**183**(Suppl 1):S35–S40.

32. Ogbuanu IU, Zhang H, Karmaus W. Can we apply the Mendelian randomization methodology without considering epigenetic effects? *Emerg Themes Epidemiol* 2009;**6**:3.

33a. Mill J, Tang T, Kaminsky Z, Khare T, Yazdanpanah S, Bouchard L, et al. Epigenomic profiling reveals DNA-methylation changes associated with major psychosis. *Am J Hum Genet* 2008;**82**(3):696–711.

33b. Mill J, Petronis A. Pre- and peri-natal environmental risks for attention-deficit hyperactivity disorder (ADHD): the potential role of epigenetic processes in mediating susceptibility. *J Child Psychol Psychiatry* 2008;**49**(10):1020–30.

34. Barker DJP, in personal communication.

35. Singh SM, Murphy B, O'Reily R. Epigenetic contributors to the discordance of monozygotic twins. *Clin Genet* 2002;**62**:97–103.

36. Issa JP. Epigenetic variation and human disease. *J Nutr* 2002;**132**:2388S–92S.

37. Boyano MD, Andollo N, Zalduendo MM, Aréchaga J. Imprinting of mammalian male gametes is gene specific and does not occur at a single stage of differentiation. *Int J Dev Biol* 2008;**52**(8):1105–11.

38. Grandjean V, Rassoulzadegan M. Epigenetic inheritance of the sperm: an unexpected role of RNA. *Gynecol Obstet Fertil* 2009;**37**(6):558–61.

39. Monk M. Genomic imprinting. Memories of mother and father. *Nature* 1987;**328**(6127):203–4.

40. Keverne B. Personal communication, 2009.

41. Razin A. *Israel J Vet Med* 2001;**56**(2). http://www.isrvma.org/article/56_2_8.htm

42. Heijmans BT, Tobi EW, Stein AD, Putter H, Blauw GJ, Susser ES, et al. Persistent epigenetic differences associated with prenatal exposure to famine in humans. *Proc Natl Acad Sci USA* 2008;**105**(44):17046–9.

43. Cropley JE, Suter CM, Beckman KB, Martin DI. Germ-line epigenetic modification of the murine A vy allele by nutritional supplementation. *Proc Natl Acad Sci USA* 2006;**103**(46):17308–12.

44. Oberlander TF, Weinberg J, Papsdorf M, Grunau R, Misri S, Devlin AM. Prenatal exposure to maternal depression, neonatal methylation of human glucocorticoid receptor gene (NR3C1) and infant cortisol stress responses. *Epigenetics* 2008;**3**(2):97–106.

45. McGowan PO, Sasaki A, D'Alessio AC, Dymov S, Labonté B, Szyf M, et al. Epigenetic regulation of the glucocorticoid receptor in human brain associates with childhood abuse. *Nat Neurosci* 2009;**12**:342–8.

46. Glynn LM, Wadhwa PD, Dunkel-Schetter C, Chicz-Demet A, Sandman CA. When stress happens matters: effects of earthquake timing on stress responsivity in pregnancy. *Am J Obstet Gynecol* 2001;**185**(3):779–80.

47. Yehuda R, Engel SM, Brand SR, Seckl J, Marcus SM, Berkowitz GS. Transgenerational effects of posttraumatic stress disorder in babies of mothers exposed to the World Trade Center attacks during pregnancy. *J Clin Endocrinol Metab* 2005;**90**(7):4115–18.

48. Brand SR, Engel SM, Canfield RL, Yehuda R. The effect of maternal PTSD following in utero trauma exposure on behavior and temperament in the 9-month-old infant. *Ann N Y Acad Sci* 2006;**1071**:454–8.

49. Raine A, Brennan P, Mednick SA. Birth complication combined with early maternal rejection at age 1 year predispose to violent crimes at age 18 years. *Arch Gen Psychiatry* 1994;**51**:984–8.

50. Crawford M.A, Sinclair A.J. Nutritional influences in the evolution of the mammalian brain. In *Lipids, malnutrition and the developing brain*. Elliot K, Knight J, editors. A Ciba Foundation Symposium (19–21 October, 1971). Amsterdam: Elsevier; 1972. P. 267–92.

51. http://www.publications.parliament.uk/pa/ld200809/ldhansrd/text/91104-gc0003.htm

52. U.S. Department of Health and Human Services. http://news.health.com/2009/08/06/u-s-spending-mental-health-care-soaring/

53. Costs of Disorders of the Brain in Europe. *Eur J Neurol* 2005;**12**(Suppl. 1). http://www.europeanbraincouncil.org/pdfs/Publications_/EBC%20-%20Cost%20Doc%20-%20EN.pdf

54. Tuddenham E. Genome Evolution, Blood and Soil–a message from deep time. *Post-Genome: Health implications for research and food policy*; A McCarrison Society Conference, The Medical Society of London, September 19. europium.csc.mrc.ac.uk. 2001.

55. Holliday R. Meiosis and sex: potent weapons in the competition between early eukaryotes and prokaryotes. *Bioessays* 2006;**28**(11):1123–5.

56. Ruiz-Trillo I, Roger AJ, Burger G, Gray MW, Lang BF. A phylogenomic investigation into the origin of metazoa. *Mol Biol Evol* 2008;**25**(4):664–72.

57. Zuckerkandl E, Cavalli G. Combinatorial epigenetics, "junk DNA", and the evolution of complex organisms. *Gene* 2007;**390**(1–2):232–42.

58. Jablonka E, Raz G. Transgenerational epigenetic inheritance: prevalence, mechanisms, and implications for the study of heredity and evolution. *Q Rev Biol* 2009;**84**(2):131–76.

59. Belyaev DK, Ruvinsky AO, Trut LN. Inherited activation-inactivation of the star gene in foxes: its bearing on the problem of domestication. *J Hered* 1981;**72**(4):267–74.

60. Lindberg J, Björnerfeldt S, Saetre P, Svartberg K, Seehuus B, Bakken M, et al. Election for tameness has changed brain gene expression in silver foxes. *Curr Biol* 2005;**15**(22):R915–R16.

61. Rando OJ, Verstrepen KJ. Timescales of genetic and epigenetic inheritance. *Cell* 2007;**128**(4):655–68.

62. Burger G, Forget L, Zhu Y, Gray MW, Lang BF. Unique mitochondrial genome architecture in unicellular relatives of animals. *Proc Natl Acad Sci USA* 2003;**100**(3):892–7.

63. Dodd IB, Micheelsen MA, Sneppen K, Thon G. Theoretical analysis of Nucleosome mediated epigenetics. In *Theoretical Analysis of Epigenetic Cell Memory by Nucleosome Modification*. Cell 2007;**129**:813–22. http://cmol.nbi.dk/models/epigen/

64. McBride KM, Barreto V, Ramiro AR, Stavropoulos P, Nussenzweig MC. Somatic hypermutation is limited by CRM1-dependent nuclear export of activation-induced deaminase. *J Exp Med* 2004;**199**(9):1235–44.

65. Morgan HD, Dean W, Coker HA, Reik W, Petersen-Mahrt SK. Activation-induced cytidine deaminase deaminates 5-methylcytosine in DNA and is expressed in pluripotent tissues: implications for epigenetic reprogramming. *J Biol Chem* 2004;**10**(279(50)):52353–60.

66. Kriaucionis S, Heintz N. The nuclear DNA base 5-hydroxymethylcytosine is present in Purkinje neurons and the brain. *Science 15*, 2009;**329**(5929):929–930.

67. Casadesús J, Low D. Epigenetic gene regulation in the bacterial world. *Microbiol Mol Biol Rev* 2006; **70**(3):830–56.

68. Gray MW, Burger G, Lang BF. Mitochondrial evolution. *Science* 1999;**283**(5407):1476–81.

69. Gray MW, Burger G, Lang BF. The origin and early evolution of mitochondria. *Genome Biol* 2001;**2**(6). http://genomebiology.com/2001/2/6/reviews/1018

70. Coffman JA. Mitochondria and metazoan epigenesis. *Semin Cell Dev Biol* 2009;**20**(3):321–9.

71. Margulis L. *Symbiosis in cell evolution*. San Francisco: W. H. Freeman and Company; 1981. p. 206–227. http://cmol.nbi.dk/models/epigen/ building on concept of Finnan AP, (1870) in Crawford & Marsh (1989/1995) p.70-71.

72. Gray MW, Lang BF, Burger G. Mitochondria of protists. *Annu Rev Genet* 2004;**38**:477–524.

73. Albalat R. Evolution of DNA-methylation machinery: DNA methyltransferases and methyl-DNA binding proteins in the amphioxus *Branchiostoma floridae*. *Dev Genes Evol* 2008;**218**(11-12):691–701.

74. Barreto VM, Pan-Hammarstrom Q, Zhao Y, Hammarstrom L, Misulovin Z, Nussenzweig MC. AID from bony fish catalyzes class switch recombination. *J Exp Med* 2005;**202**(6):733–8.

75. Bian Y, Alberio R, Allegrucci C, Campbell KH, Johnson AD. Epigenetic marks in somatic chromatin are remodelled to resemble pluripotent nuclei by amphibian oocyte extracts. *Epigenetics* 2009:194–202.

76. Zhang Y, Qu L. Non-coding RNAs and the acquisition of genomic imprinting in mammals. *Sci China C Life Sci* 2009;**52**(3):195–204.

77. Renfree MB, Ager EI, Shaw G, Pask AJ. Genomic imprinting in marsupial placentation. *Reproduction* 2008;**136**(5):523–31.

78. Pask AJ, Papenfuss AT, Ager EI, McColl KA, Speed TP, Renfree MB. Analysis of the platypus genome suggests a transposon origin for mammalian imprinting. *Genome Biol* 2009;**10**(1):R1.

79. Reik W, Santos F, Mitsuya K, Morgan H, Dean W. Epigenetic asymmetry in the mammalian zygote and early embryo: relationship to lineage commitment? *Philos Trans R Soc Lond B Biol Sci* 2003;**358**:1403–9.

80. Prehn RT. Rate-limiting step in the progression of mouse breast tumors. *Int J Cancer* 1977;**19**(5):670–2.

81. Simmons D. Epigenetic influences and disease (*Write Science Right*). *Nat Educ* 2008;**1**(1). http://www.nature.com/scitable/topicpage/epigenetic-influences-and-disease-895

82. Gabory A, Attig L, Junien C. Sexual dimorphism in environmental epigenetic programming. *Mol Cell Endocrinol* 2009;**304**(1-2):8–18.

83. Hager R, Cheverud JM, Leamy LJ, Wolf JB. Sex dependent imprinting effects on complex traits in mice. *BMC Evol Biol* 2008;**8**:303.

84. Wang X, Sun Q, McGrath SD, Mardis ER, Soloway PD, Clark AG. Transcriptome-wide identification of novel imprinted genes in neonatal mouse brain. *PLoS ONE* 2008;**3**(12):e3839.

85. Mitchell SR, Reiss AL, Tatusko DH, Ikuta I, Kazmerski DB, Botti JA, et al. Neuroanatomic alterations and social and communication deficits in monozygotic twins discordant for autism disorder. *Am J Psychiatry* 2009;**166**(8):917–25.

86. Pietiläinen KH, Söderlund S, Rissanen A, Nakanishi S, Jauhiainen M, Taskinen MR, et al. HDL subspecies in young adult twins: heritability and impact of overweight. *Obesity (Silver Spring)* 2009;**17**(6):1208–14.

87. Mastroeni D, McKee A, Grover A, Rogers J, Coleman PD. Epigenetic differences in cortical neurons from a pair of monozygotic twins discordant for Alzheimer's disease. *PLoS ONE* 2009;**4**(8):e6617.

88. Haque FN, Gottesman II, Wong AH. Not really identical: epigenetic differences in monozygotic twins and implications for twin studies in psychiatry. *Am J Med Genet C Semin Med Genet* 2009;**15**;(151C(2)):136–41.

89. Nathanielsz PW. Animal models that elucidate basic principles of the developmental origins of adult diseases. *ILAR J* 2006;**47**(1):73–82.

90. Nijland MJ, Ford SP, Nathanielsz PW. Prenatal origins of adult disease. *Curr Opin Obstet Gynecol* 2008;**20**(2):132–8.

91. Hung CC, Farooqi IS, Ong K, Luan J, Keogh JM, Pembrey M, et al. Contribution of variants in the small heterodimer partner gene to birthweight, adiposity, and insulin levels: mutational analysis and association studies in multiple populations. *Diabetes* 2003;**52**(5):1288–91.

92. Tremblay J, Hamet P. Impact of genetic and epigenetic factors from early life to later disease. *Metabolism* 2008;**57**(Suppl 2):S27–S31.

93. Sharma P, Senthilkumar RD, Brahmachari V, Sundaramoorthy E, Mahajan A, Sharma A, et al. Mining literature for a comprehensive pathway analysis: a case study for retrieval of homocysteine related genes for genetic and epigenetic studies. *Lipids Health Dis* 2006;**23**(5):1.

94. Caspi A, Moffitt TE, et al. Role of genotype in the cycle of violence in maltreated children. *Science* 2002;**2**;(297(5582)):851–4.

95. Lipton BH. Nature, nurture and the power of love. *J Prenat Perinat Psychol Health* 1998;**13**(1):3–10.

96. Janov A. *The primal scream*. London: Abacus; 1973/1991.

97. Janov A. Life: How experience in the womb can affect our lives forever. *J Prenat & Perinat Psychol & Health* 2009;**23**(3):143–84.

98. Emerson W. http://www.emersonbirthrx.com/bio.html – and in 3. *The Unborn Child*.

99. Janus L. *The enduring effects of prenatal experience*. Northvale NJ: Aronson; 1997.

100. Pribram KH. *Languages of the brain*. Englewood Cliffs NJ: Prentice-Hall; 1971.

101. International Society of Prenatal and Perinatal Psychology and Medicine (ISPPM) http://www.isppm.de/

102. *Association For Pre- & Perinatal Psychology and Health*. http://www.birthpsychology.com/

103. Petronis A. The origin of schizophrenia: genetic thesis, epigenetic antithesis, and resolving synthesis. *Biol Psychiatry* 2004;**55**:965–970.

104. Richards EJ. Inherited epigenetic variation–revisiting soft inheritance. *Nat Rev Genet* 2006;**7**:395–401.

105. Newman ME. Modularity and community structure in networks. *Proc Natl Acad Sci USA* 2006;**103**:8577–82.

106. Ford-Jones EL, Williams R, Bertrand J. Social paediatrics and early child development: Part 1. *Paediatr Child Health* 2008;**13**(9):755–8.

107. Wilkins-Haug L. Epigenetics and assisted reproduction. *Curr Opin Obstet Gynecol* 2009;**21**(3):201–6.

108. Charalambous M, da Rocha ST, Ferguson-Smith AC. Genomic imprinting, growth control and the allocation of nutritional resources: consequences for postnatal life. *Curr Opin Endocrinol Diabetes Obes* 2007;**14**(1):3–12.

109. Darwin C. *The variation of animals and plants under domestication*. London: John Murray; 1868.

110. East Asian Seas Congress summary. November. 2009. *IISD Reporting Services* http://www.iisd.ca/ymb/easc2009/

111. Nathanielsz PW, Berghorn KA, Derks JB, Giussani DA, Docherty C, Unno N, et al. Life before birth: effects of cortisol on future cardiovascular and metabolic function. *Acta Paediatr* 2003;**92**(7):766–72.

Epigenetic Epidemiology

CHAPTER 27

The Effects of Diet on Epigenetic Processes[†]

Sheila C.S. Lima[1,2], Luis Felipe Ribeiro Pinto[2], and Zdenko Herceg[1]
[1]Epigenetics Group, International Agency for Research on Cancer (IARC), Lyon, France
[2]Instituto Nacional do Cancer (INCA), Rio de Janeiro, Brazil

INTRODUCTION

In the past few decades, there has been an increasing interest in understanding how the environment, diet, and lifestyle affect human health. A number of dietary compounds have been linked with either the induction or prevention of different diseases. For example, excessive salt intake has been related to hypertension, cardiovascular events, and gastric cancer [1,2]. On the other hand, it has been demonstrated that vitamin E and selenium may prevent prostate cancer [3], whereas nut consumption appears to reduce the risk of coronary heart disease [4]. Many other epidemiological studies showed an association between diet and human diseases; however, the exact mechanisms by which dietary compounds affect key cell functions and cause diseases are largely unknown.

Several studies have demonstrated that micronutrients can interact with the genome, modify gene expression, and alter protein and metabolite composition within the cell by affecting epigenetic states [5,6]. The term "epigenetics" is used to describe the stable and heritable changes in gene expression that are not due to alterations in the DNA sequence [7]. There are three main epigenetic mechanisms of gene regulation: DNA methylation, histone modifications, and non-coding RNAs. DNA methylation refers to the addition of a methyl group to cytosines located 5' to guanines in the so-called CpG dinucleotides. This reaction is catalyzed by DNA methyltransferase (DNMT) enzymes and depends on the concentration of S-adenosyl-L-methionine (SAM), the methyl donor, within the cell [8,9]. The methylation of DNA has multiple roles in normal cellular functions, and it has been shown to regulate gene expression [10]. Most notably, the genes that contain CpG islands in their promoters may be silenced by aberrant DNA methylation. Although it is well established that the predominant consequence of methylation is transcriptional repression, it is less clear if this is mediated directly or indirectly [11]. For example, the addition of a methyl group may inhibit the binding of transcription factors and accessory elements; however, it may also recruit other proteins, such as methyl binding proteins and histone deacetylases, leading to condensation of chromatin and ultimately to a partial or complete shutdown of gene expression [12,13].

Another important mechanism of epigenetic gene regulation involves modifications of histones. This refers to covalent modifications of N-terminal tails of histone proteins and includes phosphorylation, acetylation, methylation, and ubiquitination [9]. The set of these

[†]**Competing interests statement:** The authors declare no competing financial interests.

Handbook of Epigenetics: The New Molecular and Medical Genetics. DOI: 10.1016/B978-0-12-375709-8.00027-7

modifications is called the "histone code" and it determines the level of expression of the associated gene [14,15]. Nowadays, it is proposed that the "histone code" considerably extends the information potential of the genetic code. It is an important regulatory mechanism of all the processes involving chromatin remodeling and, therefore, plays an important role in cell fate [16].

Finally, RNA-mediated gene silencing is an important epigenetic mechanism that either alone or in collaboration with other epigenetic mechanisms, participates in stable propagation of gene activity states. MicroRNAs (miRNAs) are small [19–24] nucleotides non-coding RNAs that regulate the expression of their targets by mRNA cleavage or by inhibition of translation. These molecules have been shown to influence several cellular crucial pathways and their impaired expression has been linked to diseases, most notably cancer [17].

All these epigenetic mechanisms seem to reinforce each other in the regulation of gene expression and play critical roles in different steps of development. In this chapter, we will focus on the interplay between diet and DNA methylation patterns and how dietary factors may alter the DNA methylation status and phenotypic traits, and ultimately lead to diseases including cancer.

DIET IN EARLY LIFE DEVELOPMENT

During early development, both paternal and maternal DNA undergo a remarkable epigenetic reprogramming, most notably DNA demethylation. After implantation, methylation patterns are reestablished via *de novo* methylation [18]. This epigenetic reprogramming during development must be a well-tuned process since it is an attempt to establish a configuration of the genome that can respond to changing needs of the early life development [19]. Here we will discuss how maternal diet may affect the phenotype of the offspring by epigenetic mechanisms.

MATERNAL PROTEIN RESTRICTION MODEL

Poor nutrition during pregnancy and consequent low birth weight have been related to an increase in the incidence of metabolic syndromes in adulthood, such as type 2 diabetes, hypertension and cardiovascular diseases [20]. The availability of protein is a limiting factor of fetal growth. During fetal life, amino acids control insulin secretion by β-pancreatic cells. Since insulin is an important fetal growth hormone, the availability of amino acids influences the rate of fetal growth [20]. However, this is not the only mechanism by which maternal protein restriction affects fetal growth and induces metabolic syndromes in adulthood. Epigenetic deregulation also seems to be involved.

One of the best models to study human metabolic syndromes is the maternal protein restriction (PR) model in rats. In this model, pregnant rats are fed with a PR diet, which induces phenotypic alterations in the offspring that mimic human metabolic syndromes [21]. Lillycrop et al. have shown that a PR diet during pregnancy induces hypomethylation of the *peroxisomal proliferator-activated receptor* α (*Ppara*) and *glucocorticoid receptor* (*GR*) promoters and increases expression of the *GR* and *Ppara* in the liver of the recently-weaned offspring [22]. This was associated with an increased expression of target genes of these transcription factors, *acyl-CoA oxidase* (*AOX*) and *phosphoenolpyruvate carboxykinase* (*Pepck*) [22,23]. *AOX*, target gene of *Ppara*, is involved in peroxisomal β-oxidation, whereas *Pepck*, target gene of *GR*, is involved in gluconeogenesis. Therefore, a link between maternal protein restriction, epigenetic alterations, and metabolic effects in the offspring was established. The hypomethylation of the *GR* and *Ppara* persisted after weaning, despite direct influence of the maternal dietary restriction. This suggests that the expression of these transcription factors was regulated by stable epigenetic modifications [22].

In addition, it was shown that *Dnmt1* expression was significantly lower in the liver of the PR offspring compared to controls. Therefore, the induced expression of *GR* and *Ppara* may be a consequence of reduced capacity to methylate hemimethylated DNA during mitosis [23]. According to these findings, a PR diet during pregnancy induces in the liver of the offspring an increase in gluconeogenesis and peroxisomal fatty acid β-oxidation capacity [23]. As the alterations leading to this phenotype are stable, it lasts until adulthood, and may lead to metabolic syndromes, such as type 2 diabetes and obesity. It is interesting to note that the predisposition to these diseases results from a mismatch between the diet to which offspring are exposed early in development and nutrient availability later in development and adulthood [24]. In other words, the organisms that are programmed during early development to survive in an environment with nutrients restriction, may be prone to metabolic dysfunctions when they are offered a *normal* diet in adulthood.

NUTRIENTS INVOLVED IN ONE-CARBON METABOLISM

All developmental stages of an organism depend on precise patterns of gene expression triggered by epigenetic alterations. Since the process of DNA methylation requires dietary methyl donors and cofactors, dietary factors may directly influence different developmental stages [24]. Methylation reactions, catalyzed by methyltransferases, depend on the pool of the methyl donor S-adenosylmethionine (SAM) in the body. In these reactions, SAM is converted to S-adenosylhomocysteine (SAH), and the result is the methylation of a substrate (DNA, RNA, or proteins). SAH is then converted to homocysteine and adenosine by the enzyme SAH hydrolase. Homocysteine is methylated to methionine by transferring a methyl group from 5-methyl-tetrahydrofolate or betaine (choline metabolite). Methionine is then converted to SAM, which may be used again as a methyl donor for methylation reactions. This is the so-called one-carbon metabolism and several dietary factors may affect this process (Fig. 27.1). For example, folate, methionine, and choline from the diet are the main external sources of methyl groups [25]. Also, some cofactors are required for different reactions of the one-carbon metabolism, such as vitamins B6 and B12 [26]. Therefore, the intake of these sources of methyl groups and the intake of these cofactors directly influence one-carbon metabolism, and consequently, the methylation reactions.

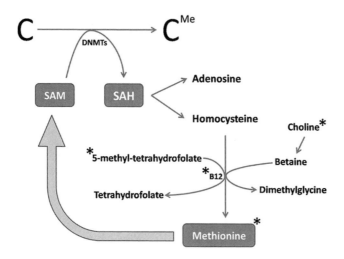

FIGURE 27.1

One-carbon metabolism. This figure illustrates how the methylation reactions occur. The transfer of a methyl group from SAM to the cytosine is catalyzed by DNMTs and the final products of this reaction are 5-methyl-cytosine and SAH. This product must then be recycled to enable another methylation reaction to take place. This recycling process involves the conversion of homocysteine to methionine, which is then converted to SAM. Many dietary components may affect this process (indicated by asterisks) and consequently affect the methylation reaction. (Please refer to color plate section)

One of the best models to study the methylation effects of dietary one-carbon metabolism intermediates is the viable yellow agouti mouse. In these mice, the coat color is determined by the expression of the *agouti* gene, which encodes a signaling molecule that stimulates the production of yellow pigment [27]. Usually, this gene is only expressed in the cells surrounding the hair follicle and only during specific stages of hair growth, and the wild-type mice present a brown agouti coat color [27]. However, mutations in the regulatory region of the *agouti* locus arose spontaneously by the insertion of an intracisternal A particle (IAP) retroviral element into the gene, generating several agouti viable yellow alleles (A^{vy}, A^{iapy}, or A^{hvy}) [27,28]. The mice expressing these dominant alleles present a yellow coat color since the *agouti* gene is under control of the IAP promoter and is continuously transcribed in almost every tissue [28].

The expression of the viable yellow alleles can be regulated by an epigenetic mechanism. When IAP is hypermethylated, the mice show a brown agouti coat color, identical to that of wild-type mice. On the other hand, when IAP is unmethylated, the gene is continuously expressed, leading to the characteristic phenotype: yellow coat color, obesity, and tumors [29]. Wolff et al. have investigated whether maternal diet may affect the phenotype of agouti offspring. In this study, pregnant mice were fed on a diet rich in methyl donors and cofactors of the methyl metabolism, such as choline, betaine, folic acid, and vitamin B12, and the offspring coat color was evaluated. The authors observed that the offspring of mothers fed on the supplemented diet showed a higher frequency of wild-type coat color in comparison with the offspring of mothers fed on the normal diet [28]. These results are in accordance with what is expected, since a higher offer of methyl donors and cofactors would enhance the methylation rates and this would result in the silencing of the IAP element.

Besides these effects of one-carbon metabolism intermediates on mouse metastable epialleles, other studies have demonstrated different mechanisms by which these intermediates may affect the offspring phenotype. As discussed above, choline is an important intermediate of methyl metabolism and is also required for normal brain development. The offspring of mice fed on a choline deficient diet has decreased memory and this is related to decreased cell proliferation in hippocampus. Niculescu et al. showed that these effects are, at least in part, attributable to DNA methylation. They found that the pups born of choline-deficient mothers show an increased expression of Kap, an important cyclin-dependent kinase inhibitor, in the neuroepithelial layer. In addition, this increased expression was correlated with a hypomethylation of the gene promoter [30]. These findings suggest that a choline-deficient diet during pregnancy may affect offspring neuronal proliferation by epigenetic mechanisms. The deficiency in this methyl donor leads to hypomethylation of the gene *Cdkn3* that encodes for Kap, leading to its over-expression and to cell cycle arrest.

Together these studies point to a fundamental role of one-carbon metabolism intermediates in epigenetic developmental programming. They also show that these epigenetic marks established during early life development are stable and carried through life. Therefore, it is important to establish what should be the optimal dietary intake of methyl donors during pregnancy. This information could then be used in the design of novel strategies for prevention.

OTHER NUTRIENTS

Not only nutrients deficiencies may influence epigenetic programming during early life development, but also over-nutrition and the consumption of isoflavones (a class of phytoestrogens), for example, may affect epigenetic patterns [31]. It has been reported that a maternal high-fat diet is correlated with an increased incidence of mammary cancer in the offspring. This might be explained by what has been observed in animal models. During aging, Sprague–Dawley rats show a progressive *estrogen receptor* (*ER*) silencing, which is mediated by promoter hypermethylation. However, when they are exposed to a high-fat diet during early life development, the *ER* promoter becomes hypomethylated and the consequent overexpression of the gene increases the risk for mammary cancer development [31].

Other dietary components that are capable of inducing epigenetic changes are the phytoestrogens. Environmental estrogens were shown to induce epigenetic shifts and alter gene expression [32]. It has been also shown that maternal exposure to genistein, the major isoflavone in soy, induces offspring coat color shifts in agouti mice. Genistein was capable of inducing hypermethylation of the IAP element and, as a consequence, the offspring phentotype was similar to that of the wild-type mice: brown agouti coat color and normal weight [33], although the underlying mechanism remains unclear.

Together these studies argue that dietary components, not only one-carbon metabolism intermediates, may affect epigenetic gene regulation, and that these effects seem to be even more striking during early development, during which epigenetic programming takes place. Therefore, future studies should investigate not only the effect of other diet components in early development, but also the mechanisms involved.

THE DUTCH HUNGER WINTER

Most of the studies correlating maternal diet and epigenetic changes in the offspring were carried out using animal models. This, however, raises the question of whether these findings can be extrapolated to humans. A recent study by Heijmans et al. investigated the effect of maternal diet in the offspring in humans using one of the few models available: people exposed to famine during the Dutch Hunger Winter. This period (winter of 1944–1945) was characterized by a food embargo in the western part of the Netherlands, which culminated in famine of the population. During this season, the food consumption was well documented and this enables comprehensive studies concerning the food deprivation effects. The authors investigated the epigenetic consequences of maternal famine in the offspring six decades later [34]. Interestingly, the results obtained were very similar to those reported in animal models. Preconceptional exposure to famine was found to be associated with *IGF2* DMR hypomethylation. The *IGF2* gene encodes the insulin growth factor 2 and is known to be an imprinted gene. This means that its expression is determined preferentially by one of the parental alleles and it is determined epigenetically. The *IGF2* gene is paternally expressed and shares regulatory elements with the maternally-expressed *H19* gene [35]. The expression of these two genes is regulated by the methylation status of the differentially-methylated regions (DMR). Hypomethylation of *IGF2* DMR leads to biallelic expression of the gene and to loss of imprinting (LOI) [34].

The *IGF2* DMR hypomethylation following preconceptional famine was similar to the hypomethylation of *GR* and *Ppara* promoters following maternal protein restriction in rats. However, this was not observed when the famine exposure occurred later in fetal development [34]. This study shows that in humans, as in other mammals, the maternal diet during critical periods of early life development may have an important influence on offspring methylation patterns, and consequently, on their phenotype (Fig. 27.2). However, many other dietary exposures, such as over-nutrition and specific nutrient deficiencies, should be investigated in humans. This would help us in understanding the origin of important diseases and how they could be prevented.

Besides, maternal diet and exposure to environmental factors may affect not only the health of their children but also of their grandchildren and even beyond. Better understanding of the mechanisms underlying these effects may unravel the inheritance of important human diseases with major impact on public health.

EFFECTS OF DIET ON EPIGENETIC STATES AND DISEASE IN ADULTHOOD

As discussed above, it is suspected that dietary factors have a major influence on extensive epigenetic reprogramming during early development. After birth, the individual is not submitted to such dramatic alterations and this could imply that the epigenetic profile would not be affected by environmental factors. However, in each cell division the epigenetic

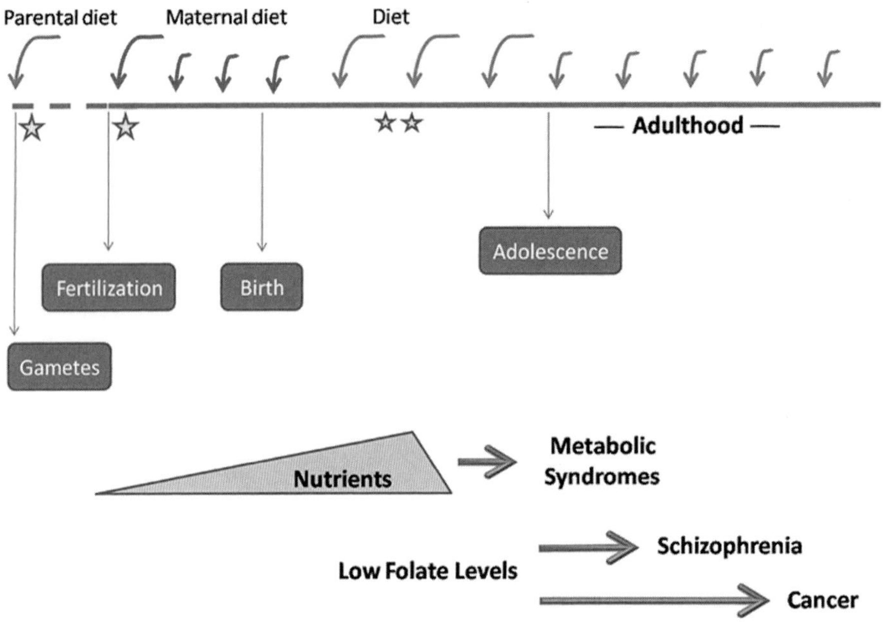

FIGURE 27.2
Diet influence on epigenetic traits during life time. During his life time, an individual faces periods of important epigenetic programming (indicated by stars). Dietary compounds may affect epigenetic patterns, especially during these critical periods. This dietary influence is not restricted to the individual's diet, but also the parental (during gamete formation) and maternal (during intrauterine life) diets exert striking effects. When an individual is subjected to a nutrient-deficient diet during early life development and faces a normal supply of nutrients after birth, the incidence of metabolic syndromes increases. Also, low folate levels during life are correlated to diseases such as schizophrenia and cancer. (Please refer to color plate section)

marks must be copied to the daughter cell in order to perpetuate the phenotype. Therefore, nutrition could have an impact on faithful transmission of epigenetic marks and gene activity states.

Some of the most puzzling suggestions of an environmental influence on epigenetic marks in adulthood come from identical twins studies. It has been observed that, immediately after birth, monozygotic twins exhibit virtually identical epigenetic patterns. However, with aging, these profiles seem to change and these individuals show major differences in DNA methylation and histone acetylation contents. Also, the biggest differences were observed between those twins who had spent less of their lives together, which supports the theory of an environmental influence [36].

Main evidences of dietary influence on the maintenance of epigenetic patterns during adulthood come from experimental models. It has been reported that agouti mice fed on a supplemented diet, containing extra folic acid, vitamin B12, choline, and betaine, show and increased level of IAP methylation and also show a corresponding shift in coat color [37]. On the other hand, when a diet deficient in methyl donors was used, the mice showed *Igf2* loss of imprinting [38]. This proves that one-carbon metabolism intermediates are determinants of methylation levels not only during early life development, but also in adulthood. Surprisingly, another study carried out in rats showed that the long-term administration of a diet lacking methionine, choline, and folic acid resulted in global DNA hypermethylation in the brain of these animals. This effect was associated with an increased expression of Dnmt3a and methyl-CpG-binding protein 2. This hypermethylation in the brain is in discordance with what was found in the liver of the same animals. In addition, the studies on transgenic mice indicated an altered expression of Dnmt3a associated with obese adipose tissues, suggesting the potential role of this enzyme in both developing embryo and adult tissue [39]. Therefore, not only do dietary factors have an influence on epigenetic patterns in adulthood, but also the effects seem to be tissue-specific [40].

In humans, the evidence for a nutrition role in epigenetic alterations in adulthood comes from a study carried out in patients with uremia. This disorder is frequently accompanied by hyperhomocysteinemia, characterized by an increased concentration of homocysteine and its precursor, SAH, in the organism. S-adenosylhomocysteine is a powerful competitive inhibitor of methyltransferases and may, therefore, impair DNA methylation and gene expression. Ingrosso et al. showed that patients with uremia have lower global levels of DNA methylation and it was correlated with defects in the expression of genes regulated by methylation. In addition, this effect was reversed by folate treatment. Because homocysteine may be methylated to methionine by transferring a methyl group from 5-methyl-tetrahydrofolate, folate could decrease homocysteine levels, resulting in the restoration of DNA methylation levels and corresponding gene expression [41].

Another disease shown to be characterized by high homocysteine blood levels is Alzheimer's disease (AD) [42]. Folate and vitamin B12 levels, which are essential for the transformation of homocysteine to methyonine, are also reduced in AD patients [43]. Therefore, it seems that an imbalance in one-carbon metabolism and subsequent alterations in DNA methylation could be involved in this disease. Scarpa et al. showed that *Presenilin1* (*PS1*) expression is regulated by DNA methylation and is reduced by exogenous administration of SAM. PS1 is a γ-secretase involved in amyloid precursor protein (APP) processing and B-amyloid (Aβ) production. Since the accumulation of Aβ in AD patients is largely documented, the overexpression of PS1 could be the underlying mechanism [44]. Furthermore, Fuso et al. showed that PS1 and BACE (a β-secretase) are induced by folate and vitamin B12 deprivation and this regulation involved DNA methylation [45]. Taken together, these results show that low levels of one-carbon metabolism intermediates may be involved in the genesis of AD. These findings not only show a strong link between dietary deficiencies, methylation, and AD, but could also have an impact in diagnosis and treatment of patients.

FOLATE, DISEASES, AND EPIGENETICS

Many studies looking into dietary influence on epigenetic patterns focused on folate effects. The deficiency of this important one-carbon metabolism intermediate has long been associated with birth defects and folate was found to be important for the maintenance of epigenetic patterns [46,47]. Folate is a major component of one-carbon metabolism and methylation reactions depend on this vitamin. When folate levels are below an optimal level, homocysteine accumulates and DNA methylation becomes impaired. Despite this important role, folates cannot be synthesized *de novo* by mammals and, therefore, their cellular levels depend on dietary intake [47].

Reinforcing the role of folate in health maintenance, one of the risk factors for schizophrenia is folate deficiency. It has been reported that patients with schizophrenia have low circulating folate levels and these levels are conversely correlated with the severity of the symptoms. Besides, patients affected by this disease show hypomethylation of *membrane-bound catechol-O-methyltransferase* (*MB-COMT*) promoter, resulting in an increased activity of the enzyme. COMT catalyzes the first step in the degradation of neurotransmitters and its hyperactivity has been associated with disturbances in attention, executive cognition, and working memory [48]. This shows a possible correlation between low dietary folate intake, DNA hypomethylation, and the symptoms observed in schizophrenia patients.

Furthermore, several epidemiologic studies have suggested an inverse association between folate status and the risk of cancer of lungs, oropharynx, esophagus, stomach, colorectum, pancreas, cervix, ovary, prostate, and breast, and risk of neuroblastoma and leukemia [49]. This is consistent with the idea that low folate levels induce DNA hypomethylation as a consequence of SAH accumulation. Folate may also modulate the expression of the DNA methylation machinery, including the DNMT enzymes and methyl CpG binding domain proteins [50,51]. Thus, availability of this compound may directly influence DNA methylation through these mechanisms and could explain some of the alterations found in tumors.

Taken together, these findings point to a fundamental role of folate in health and disease. This vitamin has been extensively studied and its effects seem to involve predominantly epigenetic alterations. This might be the first strong evidence for an epigenetic cause of several diseases, but further studies concerning other dietary components and other epigenetic mechanisms are necessary.

CONCLUSION AND FUTURE REMARKS

Here we have discussed how diet may influence epigenetic patterns and what the main consequences may be. However, the knowledge regarding this subject is still in its infancy. Most of the studies have focused on one-carbon metabolism intermediates and have been carried out in animal models. In addition, underlying mechanisms are largely unknown. Therefore, more comprehensive studies need to be carried out, and should focus on other dietary components and effects of dietary regimes in humans. Particular attention should be given to early life exposure and epigenetic reprogramming during development ("the window of vulnerability") and their effect on the susceptibility to diseases in later life.

Epigenetic research, a field in expansion, can help us understand how dietary, lifestyle and environmental factors can influence the phenotype. Intriguingly, this influence seems not only to affect the exposed individual, but also the future generations. Future studies may show that nutrition can be more important than we ever thought [52]. Therefore, the effect of dietary factors on epigenetic patterns and underlying mechanisms will be an important focus of nutritional epigenetics, and molecular epidemiology warrants further studies.

ACKNOWLEDGEMENTS

We apologize to colleagues whose relevant works were not cited owing to space limitations. We gratefully acknowledge Dr Anupam Paliwal for critical reading of the manuscript. The work in the IARC Epigenetics Group is supported by grants from the National Institutes of Health/National Cancer Institute (NIH/NCI), United States; the Association pour la Recherche sur le Cancer (ARC), France; la Ligue Nationale (Française) Contre le Cancer, France; the European Network of Excellence Environmental Cancer Risk, Nutrition and Individual Susceptibility (ECNIS); Agence Nationale de Recherche Contre le Sida et Hépatites Virales (ANRS, France); and the Swiss Bridge Award.

References

1. Shaldon S, Vienken J. Salt, the neglected silent killer. *Semin Dial* 2009;**22**(3):264–66.
2. Wang XQ, Terry PD, Yan H. Review of salt consumption and stomach cancer risk: epidemiological and biological evidence. *World J Gastroenterol* 2009;**15**(18):2204–13.
3. Jayachandran J, Freedland SJ. Prevention of prostate cancer: what we know and where we are going. *Am J Men's Health* 2008;**2**(2):178–89.
4. Sabaté J, Ang Y. Nuts and health outcomes: new epidemiologic evidence. *Am J Clin Nutr* 2009;**89**(5): 1643S–8S.
5. Subbiah MT. Understanding the nutrigenomic definitions and concepts at the food-genome junction. *OMICS* 2008;**12**(4):229–35.
6. Feil R. Environmental and nutritional effects on the epigenetic regulation of genes. *Mutat Res* 2006;**600** (1-2):46–57.
7. Holliday R. DNA methylation and epigenetic defects in carcinogenesis. *Mutat Res* 1987;**181**(2):215–17.
8. Adams RL, McKay EL, Craig LM, Burdon RH. Mouse DNA methylase: methylation of native DNA. *Biochim Biophys Acta* 1979;**561**(2):345–57.

9. Sawan C, Vaissière T, Murr R, Herceg Z. Epigenetic drivers and genetic passengers on the road to cancer. *Mutat Res* 2008;**642**(1-2):1–13.

10. Bird A. DNA methylation patterns and epigenetic memory. *Genes Dev* 2002;**16**(1):6–21.

11. Vaissière T, Sawan C, Herceg Z. Epigenetic interplay between histone modifications and DNA methylation in gene silencing. *Mutat Res* 2008;**659**(1-2):40–8.

12. Comb M, Goodman HM. CpG methylation inhibits proenkephalin gene expression and binding of the transcription factor AP-2. *Nucleic Acids Res* 1990;**18**(13):3975–82.

13. Fujita N, Takebayashi S, Okumura K, Kudo S, Chiba T, Saya H, et al. Methylation-mediated transcriptional silencing in euchromatin by methyl-CpG binding protein MBD1 isoforms. *Mol Cell Biol* 1999;**19**(9): 6415–26.

14. McGowan PO, Meaney MJ, Szyf M. Diet and the epigenetic (re)programming of phenotypic differences in behavior. *Brain Res* 2008;**1237**:12–24.

15. Shukla V, Vaissière T, Herceg Z. Histone acetylation and chromatin signature in stem cell identity and cancer. *Mutat Res* 2008;**637**(1-2):1–15.

16. Jenuwein T, Allis CD. Translating the histone code. *Science* 2009;**293**(5532):1074–80.

17. Garzon R, Calin GA, Croce CM. MicroRNAs in cancer. *Annu Rev Med* 2009;**60**:167–79.

18. Reik W, Dean W, Walter J. Epigenetic reprogramming in mammalian development. *Science* 2001;**293**(5532):1089–93.

19. Tang WY, Ho SM. Epigenetic reprogramming and imprinting in origins of disease. *Rev Endocr Metab Disord* 2007;**8**(2):173–82.

20. Bertram CE, Hanson MA. Animal models and programming of the metabolic syndrome. *Br Med Bull* 2001;**60**:103–21.

21. Burdge GC, Lillycrop KA, Jackson AA, Gluckman PD, Hanson MA. The nature of the growth pattern and of the metabolic response to fasting in the rat are dependent upon the dietary protein and folic acid intakes of their pregnant dams and post-weaning fat consumption. *Br J Nutr* 2008;**99**(3):540–9.

22. Lillycrop KA, Phillips ES, Jackson AA, Hanson MA, Burdge GC. Dietary protein restriction of pregnant rats induces and folic acid supplementation prevents epigenetic modification of hepatic gene expression in the offspring. *J Nutr* 2005;**135**(6):1382–6.

23. Lillycrop KA, Slater-Jefferies JL, Hanson MA, Godfrey KM, Jackson AA, Burdge GC. Induction of altered epigenetic regulation of the hepatic glucocorticoid receptor in the offspring of rats fed a protein-restricted diet during pregnancy suggests that reduced DNA methyltransferase-1 expression is involved in impaired DNA methylation and changes in histone modifications. *Br J Nutr* 2007;**97**(6):1064–73.

24. Junien C, Nathanielsz P. Report on the IASO Stock Conference 2006: early and lifelong environmental epigenomic programming of metabolic syndrome, obesity and type II diabetes.. *Obes Rev* 2007;**8**(6):487–502.

25. Niculescu MD, Zeisel SH. Diet, methyl donors and DNA methylation: interactions between dietary folate, methionine and choline. *J Nutr* 2002;**132**(8 Suppl):2333S–5S.

26. James SJ, Melnyk S, Pogribna M, Pogribny IP, Caudill MA. Elevation in S-adenosylhomocysteine and DNA hypomethylation: potential epigenetic mechanism for homocysteine-related pathology. *J Nutr* 2002;**132** (8 Suppl):2361S–6S.

27. Waterland RA. Assessing the effects of high methionine intake on DNA methylation. *J Nutr* 2006;**136**(6 Suppl):1706S–10S.

28. Wolff GL, Kodell RL, Moore SR, Cooney CA. Maternal epigenetics and methyl supplements affect *agouti* gene expression in A^{vy}/a mice. *FASEB J* 1998;**12**(11):949–57.

29. Jaenisch R, Bird A. Epigenetic regulation of gene expression: how the genome integrates intrinsic and environmental signals. *Nat Genet* 2003;**33**(Suppl):245–54.

30. Niculescu MD, Craciunescu CN, Zeisel SH. Dietary choline deficiency alters global and gene-specific DNA methylation in the developing hippocampus of mouse fetal brains. *FASEB J* 2006;**20**(1):43–9.

31. Tang WY, Ho SM. Epigenetic reprogramming and imprinting in origins of disease. *Rev Endocr Metab Disord* 2007;**8**(2):173–82.

32. Li S, Hansman R, Newbold R, Davis B, McLachlan JA, Barrett JC. Neonatal diethylstilbestrol exposure induces persistent elevation of c-fos expression and hypomethylation in its exon-4 in mouse uterus. *Mol Carcinog* 2003;**38**(2):78–84.

33. Dolinoy DC, Weidman JR, Waterland RA, Jirtle RL. Maternal genistein alters coat color and protects A^{vy} mouse offspring from obesity by modifying the fetal epigenome. *Environ Health Perspect* 2006;**114**(4):567–72.

34. Heijmans BT, Tobi EW, Stein AD, Putter H, Blauw GJ, Susser ES, et al. Persistent epigenetic differences associated with prenatal exposure to famine in humans. *Proc Natl Acad Sci USA* 2008;**105**(44):17046–9.

35. Waterland RA, Jirtle RL. Early nutrition, epigenetic changes at transposons and imprinted genes, and enhanced susceptibility to adult chronic diseases. *Nutrition* 2004;**20**(1):63–8.

36. Fraga MF, Ballestar E, Paz MF, Ropero S, Setien F, Ballestar ML, et al. Epigenetic differences arise during the lifetime of monozygotic twins. *Proc Natl Acad Sci USA* 2005;**102**(30):10604–9.

37. Waterland RA, Jirtle RL. Transposable elements: targets for early nutritional effects on epigenetic gene regulation. *Mol Cell Biol* 2003;**23**(15):5293–300.

38. Waterland RA, Lin JR, Smith CA, Jirtle RL. Post-weaning diet affects genomic imprinting at the insulin-like growth factor 2 (*Igf2*) locus. *Hum Mol Genet* 2006;**15**(5):705–16.

39. Kaimei Y, Suganami T, Ehara T, Kanai S, Hayashi K, Yamamoto Y, et al. Increased Expression of DNA Methyltransferase 3a in obese adipose tissue: studies with transgenic mice. *Obesity* 2009;**18**(2):314–21.

40. Pogribny IP, Karpf AR, James SR, Melnyk S, Han T, Tryndyak VP. Epigenetic alterations in the brains of Fisher 344 rats induced by long-term administration of folate/methyl-deficient diet. *Brain Res* 2008;**1237**:25–34.

41. Ingrosso D, Cimmino A, Perna AF, Masella L, De Santo NG, De Bonis ML, et al. Folate treatment and unbalanced methylation and changes of allelic expression induced by hyperhomocysteinaemia in patients with uraemia. *Lancet* 2003;**361**(9370):1693–9.

42. Seshadri S, Beiser A, Selhub J, Jacques PF, Rosenberg IH, D'Agostino RB, et al. Plasma homocysteine as a risk factor for dementia and Alzheimer's disease. *N Engl J Med* 2002;**346**:476–83.

43. Prodan CI, Cowan LD, Stoner JA, Ross ED. Cumulative incidence of vitamin B12 deficiency in patients with Alzheimer disease. *J Neurol Sci* 2009;**284**:144–8.

44. Scarpa S, Fuso A, D'Anselmi F, Cavallaro RA. Presenilin 1 gene silencing by S-adenosylmethionine: a treatment for Alzheimer disease? *FEBS Lett* 2003;**541**:145–8.

45. Fuso A, Seminara L, Cavallaro RA, D'Anselmi F, Scarpa S. S-adenosylmethionine/homocysteine cycle alterations modify DNA methylation status with consequent deregulation of PS1 and BACE and beta-amyloid production. *Mol Cell Neurosci* 2005;**28**:195–204.

46. Nelson MM, Asling CW, Evans HM. Production of multiple congenital abnormalities in young by maternal pteroylglutamic acid deficiency during gestation. *J Nutr* 1952;**48**(1):61–79.

47. Hamid A, Wani NA, Kaur J. New perspectives on folate transport in relation to alcoholism-induced folate malabsorption-association with epigenome stability and cancer development. *FEBS J* 2009;**276**(8):2175–91.

48. Abdolmaleky HM, Cheng KH, Faraone SV, Wilcox M, Glatt SJ, Gao F, et al. Hypomethylation of *MB-COMT* promoter is a major risk factor for schizophrenia and bipolar disorder. *Hum Mol Genet* 2006;**15**(21):3132–45.

49. Kim YI. Folate and cancer prevention: a new medical application of folate beyond hyperhomocysteinemia and neural tube defects. *Nutr Rev* 1999;**57**(10):314–21.

50. Esfandiari F, Green R, Cotterman RF, Pogribny IP, James SJ, Miller JW. Methyl deficiency causes reduction of the methyl-CpG-binding protein, MeCP2, in rat liver. *Carcinogenesis* 2003;**24**(12):1935–40.

51. Ghoshal K, Li X, Datta J, Bai S, Pogribny I, Pogribny M, et al. A folate- and methyl-deficient diet alters the expression of DNA methyltransferases and methyl CpG binding proteins involved in epigenetic gene silencing in livers of F344 rats. *J Nutr* 2006;**136**(6):1522–7.

52. Herceg Z. Epigenetics and cancer: towards an evaluation of the impact of environmental and dietary factors. *Mutagenesis* 2007;**22**(2):91–103.

CHAPTER 28

Environmental Agents and Epigenetics

Adriana Arita and Max Costa
New York University School of Medicine, Nelson Institute of Environmental Medicine,
NY 10987, USA

INTRODUCTION

Environmental agents to which humans are exposed include metals, ionizing radiation, tobacco smoke, ambient particulate matter, and endocrine disruptors, among others. Exposure to certain environmental agents influences the risk of developing various chronic diseases, such as cancer, cardiovascular and pulmonary disease, diabetes, obesity, and neurological and behavioral disorders. Although the majority of the aberrant changes in gene expression linked to the health effects of exposure to environmental agents have been associated with genotoxic mechanisms, non-genotoxic mechanisms may also play a role. *In vitro*, animal and human studies have identified several environmental agents that may mediate their toxic and carcinogenic properties through epigenetic mechanisms. Most studies conducted so far on the epigenetic effects induced by environmental agents have reported changes in global and gene specific DNA methylation and histone modification levels – changes that are the same or similar to the observed epigenetic changes found in patients with the disease or condition induced by that particular environmental agent. Here, we review the reports of experimental and epidemiological studies that have identified epigenetic effects induced by exposure to metals (nickel, arsenic, chromium, cadmium, and cobalt), methylmercury, the semi-metal selenium, peroxisome proliferators, radiation, particulate matter, tobacco smoke, benzene, endocrine disruptors, and polycyclic aromatic hydrocarbons.

459

METALS
Nickel

Nickel (II) is a toxic non-essential transition metal used in modern industry with other metals to form alloys to produce coins, jewelry, and stainless steel as well as for nickel plating and manufacturing of Ni-Cd batteries. Among new applications, it is important to note its role as a catalyst for the production of carbon nanoparticles. Non-occupational sources of Ni exposure include jewelry, nickel-plated tools and utensils, orthodontic and surgical prostheses, and coins. Occupational exposure to both soluble [i.e. nickel chloride ($NiCl_2$) and nickel sulfate ($NiSO_4$)] and insoluble [i.e. nickel sulfide (NiS), nickel subsulfide (Ni_3S_2), and nickel monoxide (NiO)] nickel compounds has been associated with increased risks for acute respiratory syndromes, ranging from mild irritation and inflammation of the respiratory system to bronchitis, pulmonary fibrosis, asthma, and pulmonary edema. Nickel compounds are of great environmental concern since epidemiological, animal, and cell

Handbook of Epigenetics: The New Molecular and Medical Genetics. DOI: 10.1016/B978-0-12-375709-8.00028-9

culture studies have found them to be carcinogenic. However, the precise mechanism(s) of nickel carcinogenesis remains unknown.

Since the mutagenic activity of nickel compounds in mutation assays from *Salmonella* to mammalian cells *in vitro* has been low it has been suggested that Ni-induced mutagenic activity is not the underlying mechanism in nickel-induced carcinogenesis. Instead, structural alterations in chromatin and epigenetic changes have been implicated as the primary events in nickel carcinogenesis. Phagocytosed nickel sulfide particles have been shown to selectively target heterochromatin. One reason why nickel ions target heterochromatin may be because heterochromatin forms the inside lining of the interface nucleus and with this location toxins entering the nucleus encounter heterochromatin before they reach euchromatin. The observed effects of nickel compounds on heterochromatin led to the discovery that nickel compounds could silence genes by inducing DNA methylation. In the Chinese hamster G12 cell line which possesses a copy of the bacterial *gpt* transgene near the telomere of chromosome I it was demonstrated that nickel induced DNA hypermethylation and gene silencing of the *gpt* transgene [1]. The promoter of the tumor suppressor gene *p16* was also found hypermethylated in nickel-induced tumors of wild type C57BL/6 mice and mice heterozygous for the *p53* tumor suppressor gene injected with nickel sulfide [2]. A proposed model for nickel-induced DNA hypermethylation includes the ability of nickel to substitute for magnesium in the phosphate backbone of DNA; Ni^{2+} may be better at condensing heterochromatin than Mg^{2+} ions, increasing chromatin condensation and triggering *de novo* DNA methylation of critical tumor suppressor or senescence genes that can become incorporated into heterochromatin because of their proximity to this type of chromatin [1]. An additional reason why nickel ions target heterochromatin may be because of the higher concentration of Mg^{2+} present in the phosphate backbone of DNA in heterochromatin than euchromatin.

Nickel-induced changes on histone posttranslational modifications have been reported and include the loss of global histone acetylation in H2A, H2B, H3, and H4, increase in H3K9 dimethylation, and an increase in the ubiquitination of H2A and H2B [3–7]. A decrease in histone acetylation and an increase in H3K9 dimethylation was also observed in the promoter of the *gpt* transgene silenced by nickel [1,4,8]. The suggested mechanism by which exposure to nickel decreases histone acetylation is through inhibition of histone acetyltransferase activity [9]. The observed global increase in H3K9 dimethylation induced by nickel compounds has been recently reported to be a result of nickel ion inhibition of a novel class of histone H3K9 demethylases that depend upon iron and 2-oxoglutarate for their enzymatic activity. Ni ions compete and displace the iron ions that bind the active site of these dioxygenases inhibiting their demethylase activity and resulting in increased H3K9 dimethylation levels [4]. Since H3K9 methylation is important for DNA methylation and long-term gene silencing it is possible that the observed increase in DNA methylation after exposure to nickel compounds is a result of the effect of nickel on H3K9 methylation levels [4,10]. It was also suggested that gene silencing mediated by histone acetylation may play a role in nickel-induced cell transformation [11].

Additionally, a role for chromatin damage in nickel-induced toxicity and carcinogenicity has been suggested [12]. The binding of Ni (II) to specific motifs on histone H3 and H2A could cause several lesions resulting in oxidative DNA damage via Fenton-like mechanisms, disrupting the structure and function of the nucleosome, and changing gene expression that may contribute to nickel-induced toxicity and carcinogenicity [12–17].

Arsenic

Arsenic is an environmental contaminant found in soil, water, and airborne particles. Chronic low dose exposure to arsenic has been associated with skin, bladder, lung, kidney, and liver cancers. The potential toxic effects of arsenic exposure are of great health concern

460

since in 2000 the Food and Drug Administration (FDA) approved arsenic trioxide for the treatment of acute promyeolocytic leukemia resistant to other treatments. Until the late 1950s arsenic trioxide was also prescribed as Fowler's solution for the treatment of various medical conditions including psoriasis. Exposure to arsenic occurs generally in the form of either arsenite (As^{III}) or arsenate (As^{V}). The increased cancer risk observed in arsenic carcinogenesis is attributed to exposure to arsenite rather than the less toxic arsenate. Inside the cell, As^{V} is reduced to As^{III}. As^{III} is then converted into the methylated arsenic metabolites, pentavalent or trivalent monomethylated (MMA) and dimethylated (DMA), using S-adenosyl-methionine (SAM) as a methyl donor. The methylated metabolites are excreted through urine at a much faster rate than are As^{III} and As^{V}. Therefore, the levels of S-adenosyl-methionine are important in arsenic metabolism since a low intake of methyl groups (dietary methionine or folate) results in lower arsenic methylation and excretion from the body.

Although many mechanisms for the carcinogenicity of arsenic have been proposed, the precise mechanism(s) remains unclear. The widespread disruption in global gene expression observed in spite of arsenic's low mutagenic activity suggests that the carcinogenicity of arsenic may be mediated through epigenetic mechanisms. Indeed, various studies have reported that exposure to arsenic induces both DNA hypo- and hypermethylation. Long-term exposure to arsenic *in vitro* resulted in malignant transformation associated with depletion of S-adenosyl-methionine, an increase in global DNA hypomethylation levels, and decreased DNA methyltransferase activity [18,19]. Tissue culture, and animal and human cancer studies have also associated exposure to arsenic with epigenetic silencing of tumor suppressor genes by gene promoter hypermethylation [20–24]. The co-existence of both DNA hypo- and hypermethylation after exposure to arsenite may be one mechanism by which appropriate gene expression may be disrupted in arsenite-exposed cells [25].

An association between DNA methylation levels and exposure to arsenite has also been reported in human population studies. A study of Bangladeshi adults chronically exposed to arsenic reported that genomic methylation of blood DNA is positively associated with plasma folate levels [26]. A more recent study by this same group in a related Bangladeshi population reported that folate deficiency, hyperhomocysteinemia, and low urinary creatinine, each associated with decreased arsenic methylation, give an elevated risk for skin lesions [27]. In another human study conducted in India, a dose–response relationship between the hypermethylation of the promoter regions of the *p53* and *p16* tumor suppressor genes and arsenic levels in the drinking water was found in blood DNA of individuals exposed to toxic levels of arsenic [28].

Recently it was reported that exposure to arsenite alters global histone methylation levels. Exposure to arsenite of the human lung carcinoma A549 cell line resulted in increased dimethylated H3K9 and H3K4 trimethylation and decreased trimethylated H3K27 levels. The increased levels of dimethylated H3K9 were attributed to increased histone methyltransferase G9a mRNA and protein levels [29].

Chromium

Hexavalent chromium (Cr (VI)) is a well known human carcinogen with exposures occurring in both occupational and environmental settings. Occupational exposure to Cr (VI) occurs in chromate manufacturing, chrome plating, ferrochrome production, and stainless steel welding. Environmental exposure to chromium likely impacts millions of people drinking Cr contaminated water and residing near toxic sites and chemicals manufacturers. Epidemiological risk-assessment studies revealed a high incidence of lung cancer due to occupational exposure to Cr (VI). Chromium exists in two major stable oxidation states, hexavalent chromium (Cr (VI)) and trivalent chromium (Cr (III)). These two forms of chromium have strikingly different toxicities because of differences in their uptake into cells. Cr (VI) readily enters cells through the anionic transport systems since at physiological pH Cr (VI) exists as an oxyanion, with an overall charge of minus 2 (CrO_4^{-2}), and in this form it resembles sulfate and phosphate. In biological systems, Cr (VI) undergoes a series of reduction reactions that yield the thermodynamically stable Cr (III). Exposure to Cr (III) is not believed to be toxic since Cr (III) does not readily enter cells. Cr (VI) reduction to Cr (III) is the activation event responsible for the generation of the Cr-induced cellular toxicity and genotoxic damage since Cr (III) readily interacts with proteins and nucleic acids in the cell. The reduction of Cr (VI) to Cr (III) results in oxidative stress, the formation of Cr (III)-DNA adducts, protein-DNA crosslinks, and DNA single- and double-strand breaks. Cr (III) induced cellular damage can therefore affects DNA replication, transcription, and translation resulting in altered gene expression.

In addition to the DNA damage induced by chromium, several studies have recently highlighted the potential epigenetic effects of Cr (VI), and how both genetic and epigenetic mechanisms may contribute to the changes in gene expression induced by chromium and play a role in its toxicity and carcinogenicity. Potassium dichromate was shown to silence the transgene expression of a cell line expressing a bacterial *gpt* reporter gene by inducing DNA methylation [30]. Potassium dichromate was also shown to induce genome-wide cytosine-hypermethylation in the *Brassica napus* L. plant [31]. DNA hypermethylation on the promoter region of the *p16* tumor suppressor gene and the DNA mismatch repair (*MLH1*) gene was reported in lung cancers of chromate-exposed workers [32,33]. Additionally, chromium was shown to inhibit the expression of *Cyp1a1* by crosslinking the histone deacetylase 1-DNA methyltransferase 1 complexes to the chromatin of its promoter and inhibiting the phosphorylation of histone H3 Ser-10, trimethylation of H3K4, and various histone acetylation marks on H3 and H4 histones [34]. A more recent study reported that exposure of the human adenocarcinoma A549 cell line to potassium chromate increased global levels of di- and trimethylated histone H3 lysine 9 (H3K9) and lysine 4 (H3K4) but decreased the levels of trimethylated histone H3 lysine 27 (H3K27) and dimethylated histone H3 arginine 2 (H3R2) with increased dimethylation of H3K9 on the promoter region of *MLH1* and a decrease in its mRNA expression [35].

Cadmium

Cadmium (Cd), a classified carcinogen, is a non-essential transition metal belonging to group IIB of the periodic table. Sources of human exposure to cadmium include employment in the production of certain batteries, metal industries, and electroplating processes and the consumption of tobacco products. Cadmium is absorbed in the body mainly through inhalation although some Cd can be ingested. The organs that store Cd include the liver, kidney, testis, spleen, heart, lungs, thymus, salivary glands, epididymis, and prostate. Approximately 50% of the Cd found in the body is stored in the liver and kidney due to their high metallothionein (MT) levels; MT exhibits high binding affinity to Cd and other metals. Occupational exposure to cadmium has been linked to pulmonary, liver, prostate, renal, urinary bladder, pancreatic, and stomach cancers, as well as cancers of the hematopoietic system. A significant decline in the production and use of Cd has been reported in the last few years. However, Cd continues to be a major health concern primarily because of its long half-life (15–20 years) and persistence in the environment and tissues.

Suggested mechanisms for cadmium carcinogenesis include suppressed apoptosis and disruption of e-cadherin mediated cell–cell adhesion, aberrant gene activation, and altered DNA repair. Because cadmium is a poor mutagen its carcinogenicity has been attributed to epigenetic and indirect genotoxic mechanisms. Several studies have reported changes in global and gene specific DNA methylation levels after exposure to cadmium. Acute exposure (1 week) of TRL 1215 rat liver cells to cadmium inhibited DNA methyltransferase activity and induced global DNA hypomethylation while prolonged exposure (10 week) resulted in global DNA hypermethylation and enhanced DNA methyltransferase activity [36].

Increased global DNA hypermethylation levels and DNA methyltransferase (DNMT) activity was observed in human lung fibroblast (HLF) cells after long-term (2 months) low-dose exposure to cadmium [37]. A 10-week exposure to cadmium induced malignant transformation associated with global DNA hypermethylation, overexpression of DNMT3b DNA methyltransferase and increased DNMT activity, and promoter hypermethylation and reduced expression of the *RASSF1A* and *p16* tumor suppressor genes [38]. Global DNA hypomethylation as a potential facilitator of Cd-stimulated cell proliferation in the chronic myelogenous leukemia K562 cell line was also reported [39].

Cobalt

Cobalt is an essential trace element and a central component of the vitamin cobalamin, vitamin B_{12}. Cobalt is found in various metallic ores and is universally used in the preparation of alloys in the steel industry. Although cobalt is an essential element for life in trace amounts, at higher levels cobalt shows mutagenic and carcinogenic effects similar to those of nickel. Occupational exposure to cobalt has been linked to various lung diseases, such as pneumonitis, fibrosis, and asthma. Cobalt has been shown to induce alterations in gene expression and interfere with the cellular homeostasis of reactive oxygen species, calcium, and iron. However, the precise mechanism(s) by which cobalt exerts its toxic and carcinogenic activity is not known. A recent study reported that cobalt ions alter the cell's epigenetic homeostasis. Exposure of human lung adenocarcinoma A549 and human bronchial epithelial Beas-2B cells to cobalt increased both gene repressive histone marks (trimethylation of H3K9, H3K27, H3K36, and ubiquitination of H2A) as well as gene activation marks (trimethylation of H3K4 and ubiquitination of H2B) and acetylation of H4. The increase in trimethylated H3K4 and H3K27 induced by cobalt was associated with activation of histone methyltransferase activity, whereas the increase in trimethylated H3K9 and H3K36 was attributed to the inhibition of the enzymatic activity of the JMJD2A histone demethylase. The increase in ubiquitination of H2A and H2B induced by cobalt was associated with the inhibition of histone deubiquinating enzymatic activity [40].

Methylmercury

Mercury is a toxic metal that occurs naturally at low levels in rocks, soil, water, and air from the burning of fossil fuels. Mercury pollution deposits in water bodies where elemental mercury (Hg^0) is biotransformed by bacteria and converted to methylmercury (meHg), the most toxic form of mercury in the environment. Fish that eat the bacteria accumulate high levels of methyl mercury in their body. The US population is primarily exposed to methylmercury by eating fish such as swordfish, shark, and tuna. Ingested methylmercury is readily absorbed by the gastrointestinal tract and is mostly found complexed with free cysteine and with proteins and peptides containing that amino acid. The methylmercuric-cysteinyl complex mimics the essential amino acid methionine and is recognized by amino acid transporting proteins in the body. Methylmercury can be transported across the blood–brain barrier and placenta where it can be absorbed by a developing fetus. Exposure to methylmercury *in utero* has been linked to developmental deficits in children such as loss of IQ points, decreased performance tests of language skills, memory function, and attention deficits. One study reported that the depression-like behavior in mice induced by perinatal exposure to low levels of methylmercury was associated with long lasting epigenetic suppression (decreased histone H3 acetylation, increased trimethylated H3K27 and promoter DNA hypermethylation) of the neurotrophin brain-derived neurotrophic factor (*BDNF*) gene in the hippocampus of exposed mice [41].

Selenium

Selenium (Se) occurs naturally in a number of inorganic forms, including selenide, selenate, and selenite, and organic compounds such as selenomethionine and selenocysteine. Se is

463

most commonly produced from selenide in sulfide ores, such as those of copper, silver, or lead. Sources of exposure to Se include the burning of coal and mining and smelting of sulfide ores. Isolated Se occurs in several different forms, the most stable being the semiconductor (semi-metal) form that conducts electricity and is used in photocells. Selenium also occurs in many non-conductive forms such as several red crystalline forms and a black glass-like allotrope. However, silicon semiconductor devices have now replaced the electronic uses of Se. Se is an essential element for human health that has received considerable attention for its potential role as a chemotherapeutic agent. The major sources of Se for humans include ingestion of meats, fish, cereals, dairy products, and plant foods. Selenomethionine, sodium selenate, sodium selenite, and selenized yeast are the forms of Se in dietary supplements. In the body, selenium is an important component of several selenoproteins such as the antioxidant family of glutathione peroxidases as well as other enzymes such as the iodothyronine deiododinases and thioredin reductases. Therefore, Se deficiency has been linked to a variety of human diseases, including cancer. An inverse relationship between serum Se levels and cancer risks has been reported.

Although ingestion of Se can be beneficial, ingestion of high levels of Se can be toxic. Studies have shown that selenium induces changes in DNA methylation and inhibits the expression and activity of DNMTs. Sodium selenite was found to induce DNA hypomethylation and inhibit DNA methyltransferase in Friend erythroleukemic cells (FELC) [42]. It was also suggested that inhibition of DNA methyltransferase may be a major mechanism of chemoprevention by selenium compounds at the post-initiation stage of carcinogenesis [43]. Additionally, selenium deficiency was found to decrease both global DNA methylation and methylation in the promoter region of the *p53* gene in Caco-2 cells and rat liver and colon cells [44]. Se has also been associated with epigenetically modulating DNA and histones to activate methylation-silenced genes. Treatment of LNCaP prostate cancer cells with sodium selenite resulted in partial promoter demethylation and re-expression of the π-class glutathione-S-transferase (*GSTP1*) in a dose- and time-dependent manner, decreased mRNA levels of DNA methyltransferases *DNMT1* and *DNMT3A* and protein levels of *DNMT1*, decreased histone deacetylase activity, increased levels of acetylated histone H3K9, and decreased levels of methylated histone H3K9. Sodium selenite treatment reduced levels of methylated histone H3K9 but increased levels of associated acetylated H3K9 in the *GSTP1* promoter [45]. Se treatment also resulted in global DNA hypomethylation and promoter demethylation and re-expression of the *APC* tumor suppressor gene and *CSR1*, a gene involved in tumor growth and metastasis [45].

PEROXISOME PROLIFERATORS

Peroxisome proliferators are a structurally-diverse group of agents that comprise a wide range of substances such as natural compounds (long-chain fatty acids and prostaglandins), synthetic drugs, including drugs used for the treatment for certain diseases (anti-inflammatory drugs, NSAIDs), and environmental contaminants (herbicides). Long-term administration of peroxisome proliferators results in liver cancer in rodents; however, the risk of their administration on human health remains inconclusive. Determining the possible risks of peroxisome proliferators to human health is crucial since some peroxisome proliferators (PPARα agonists) are currently being used for the treatment of hyperlipidemia. In rodents, peroxisome proliferators induce liver cancer via a mode of action that includes activation of the peroxisome proliferator-activated receptor α (PPARα) transcription factor, increased cell proliferation, decreased apoptosis, and secondary oxidative stress leading to DNA damage.

Evidence from several studies suggests that epigenetic events may play a role in the mechanism of carcinogenesis of certain peroxisome proliferators. Short-term treatment with 4-chloro-6-(2,3-xylidino)-pyrimidilnylthioacetic acid (WY-14,643), a model peroxisome proliferator, induced hypomethylation of the *c-myc* gene in the liver of exposed mice [46].

Decreased methylation in the promoter regions of the *c-jun* and *c-myc* genes and increased levels of their mRNAs and proteins, dependent on methionine supplementation, were also found in livers of mice exposed to trichloroethylene (TCE), dichloroacetic acid (DCA), and TCA [47]. Global hypomethylation of liver DNA, decreased trimethylation of H4K20 and H3K9, and loss of cytosine methylation in major and minor satellites and repetitive elements was observed in male SV129 mice exposed to WY-14,643. These epigenetic effects were PPARα dependent and correlated with cell proliferation rates [48]. The suggested mechanism of global DNA hypomethylation induced by the peroxisome proliferators WY-14,643 and di-(2-ethylhexyl)phthalate (DEHP) in the liver of male Fisher rats includes an accumulation of single-strand breaks associated with an increase in cell proliferation and decreased expression of DNA methyltransferase 1 [49].

RADIATION

Ionizing radiation (IR) is a well known cancer-inducing agent. Human exposure to ionizing radiation occurs through acute diagnostic and therapeutic medical radiation procedures and chronic exposure to background radiation, cosmic rays, radioactive waste, radon decay, nuclear tests and accidents at nuclear power plants, as well as ultraviolet A and B radiation. IR-induced DNA damage and disruption of a variety of cellular processes are the suggested mechanisms of carcinogenesis. Radiation-induced responses are observed in the irradiated cell, in naïve "bystander" cells that were in contact with irradiated cells or received signals from the irradiated cells, and in the progeny of the irradiated cell generations after exposure.

Several groups have reported changes in DNA methylation and histone post-translational modifications after direct exposure to radiation. A decrease in global methylation levels was reported after acute exposures to either gamma or X-ray radiation [50,51]. Dose-dependent and sex- and tissue specific global DNA hypomethylation was also observed after exposure to IR [52–54]. IR-induced global DNA hypomethylation was associated with DNA repair, alterations in the expression of DNA methyltransferases, and genome instability in the exposed tissue [52–55]. Phosphorylation of histone H2AX and changes in histone methylation, specifically loss of histone H4 lysine 20 trimethylation, were also detected with IR [53]. Supporting the idea that radiation can induce changes in DNA methylation, the lung tumors of workers of the MAYAK plutonium plant showed a higher risk for *p16* methylation than control tumors [56].

Epigenetic changes have also been associated with naïve "bystander" cells that had been in contact with irradiated cells or received signals from the irradiated cells. The alteration in DNA methylation in naïve cultured human keratinocytes exposed to the medium of irradiated cells persists over 20 passages [57]. A significant increase in the levels of phosphorylated H2AX in bystander tissues and loss of nuclear DNA methylation was observed in a normal human three-dimensional artificial tissue system after microbeam irradiation [58]. Other studies have also reported radiation-induced changes in H2AX phosphorylation in bystander cells [59–62]. A study in an *in vivo* model suggested that epigenetic transcriptional regulation may be involved in the radiation-induced bystander effect by altering the levels of key proteins that modulate methylation patterns and gene silencing in the bystander tissue. This study reported radiation suppressed global methylation levels in the directly irradiated mouse tissue but not in the bystander tissue. However, a decrease in the expression of the DNA methyltransferases DNMT3a and DNMT3b and an increase in the levels of DNMT1 were detected in the bystander tissue. Additionally, the levels of MeCP2 and MBD2, two methyl-binding proteins known to be involved in transcriptional silencing, were also found increased in the bystander tissue [63].

The occurrence of genome instability and elevated mutation rates in the progeny of exposed parents has also been attributed to a possible epigenetic mechanism. A decrease in global cytosine DNA methylation in the thymus tissue of offspring of irradiated mice was reported.

This same study reported a significant accumulation of DNA strand breaks, elevated levels of phosphorylated H2AX, and a decrease in the levels of DNA methyltransferases DNMT1 and DNMT3a and DNMT3b, and the MeCP2 protein, in the thymus of the progeny of irradiated parents [64]. In another study, elevated mutation rates in the tissue of the progeny of exposed mice were associated with elevated levels of phosphorylated H2AX [65].

Exposure to ultraviolet (UV) radiation has also been associated with epigenetic effects. A decrease in global DNA methylation was detected in the skin of mic chronically exposed to UVB (30 weeks) [66]. Additionally, skin cancer studies have demonstrated an increase in DNA methylation of many tumor suppressor genes induced by UV radiation [67].

PARTICULATE MATTER

Ambient particulate matter (PM) is composed of all the solid and liquid particles suspended in air including dust, pollen, soot, smoke, and liquid droplets. The particles may vary in size and composition. PM is associated with increased morbidity and mortality from cardiorespiratory disease and lung cancer risk. However, the specific particulate matter components responsible for the adverse health effects of PM have not been defined. Animal and human studies have shown that air particles or air particle components can induce changes in global and promoter specific DNA methylation levels. Methylation and inactivation of *p16* and ER genes was observed in lung tumors of rats induced by the particulate carcinogens carbon black, diesel exhaust, and beryllium metal [68]. A decrease in the long interspersed nucleotide element (*LINE-1*) and *Alu* repetitive elements and the promoter of the inducible nitric oxide synthase (*iNOS*) was found in blood DNA of Foundry workers [69]. A recent study reported an association between ambient particulate matter and a decrease in blood DNA methylation levels of the *LINE-1* and *Alu* repetitive elements in blood DNA samples of elderly men in the Boston MA area [70]. Changes in global DNA methylation induced by particulate matter may represent a mechanism by which human health is affected, as alterations in DNA methylation levels are related to the disease development of patients with cancer and cardiovascular disease [71]. Sperm DNA of mice exposed to ambient air near two integrated steel mills and a major highway was found hypermethylated compared to control mice and its persistence even after removal of the exposure suggests the possibility that air pollutants may produce DNA methylation changes transgenerationally [72].

TOBACCO SMOKE

Recent studies on tobacco smoke have indicated that among its 4800 identified compounds, as many as 69 may be carcinogens. Although extensive reports on the genotoxic effects of tobacco smoke exist, recent reports on the epigenetic events associated in tobacco-related cancers are beginning to emerge. Global DNA hypomethylation was reported in squamous cell carcinoma cancers of cigarette smokers [73]. Hypermethylation and silencing of the *p16*, *MGMT* (O(6)-methylguanine-DNA methyltransferase), and death-associated protein (*DAP*)-kinase genes in lung cancer due to tobacco smoke has been reported [74]. Methylation of the *p16* promoter region studied in 185 patients with primary non-small-cell lung cancer (NSCLC) was significantly associated with smoked pack/years and duration of smoking, and negatively with time since quitting smoking [75]. The promoter of the *TSLC1/IGSF4* tumor suppressor gene was found hypermethylated in 44% of the 103 patients with NSCLC and was significantly associated with smoking history, cigarette consumption per day, and shorter disease-free survival [76]. Hypermethylation of the promoters of the *RASSF1A* and *BLU* tumor suppressor genes was positively correlated to tobacco consumption, age, gender, histology, and starting smoking age under 18 years [77]. Additionally, hypermethylation of the *FHIT* promoter and *p16* methylation was associated with exposure to tobacco smoke in squamous cell carcinomas [78]. Promoter hypermethylation may be an early event in the development of lung cancer since aberrant promoter methylation of several genes such as *p16* and *MGMT* can be detected in DNA from sputum of patients with lung squamous cell carcinoma even up to 3 years before clinical manifestation of cancer [79]. Increased expression of DNMT1 has been linked to tobacco-induced hypermethylation of the *RASSF1A* tumor suppressor gene [80]. Interestingly, cigarette smoke has been shown to induce the demethylation and abnormal expression of *synuclein-γ*, a pro-metastatic oncongene, in lung cancer cells through down-regulation of *DNMT3B* [81]. Demethylation of the *CYP1A1* promoter was also associated with tobacco smoke. An inverse relationship between methylation and the number of cigarettes smoked daily was identified [82].

BENZENE

Benzene, an airborne pollutant emitted from traffic exhaust fumes and cigarette smoking, has been associated with acute myelogenous leukemia (AML). Low level exposure to benzene has been shown to induce altered DNA methylation levels. Significant reduction in DNA methylation of the repetitive elements *LINE-1* and *Alu* and hypermethylation of *p15* as well as hypomethylation of the melanoma antigen-1 (*MAGE-1*) cancer antigen gene was found in the peripheral blood DNA of normal subjects with well-characterized benzene exposure [83]. Interestingly, global DNA hypomethylation, and gene specific hypermethylation or hypomethylation are common in AML [84].

ENDOCRINE DISRUPTORS

Endocrine disrupting chemicals (EDCs) include numerous compounds found in the environment that act to mimic estrogens or act as antiestrogens or antiandrogens. Examples of environmental endocrine disrupters include pesticides (dicholorodiphenyltrichloroethane (DTT) and methoxychlor), fungicides (vinclozolin), insecticides (trichlorfon), herbicides (atrazine), plastics (phthalates), and a wide range of synthetic estrogens. Exposure to EDCs has been implicated with reproduction abnormalities in the F1 generation and subsequent generations. Numerous studies have suggested an epigenetic mechanism for the abnormal transgenerational health effects observed after exposure to certain endocrine disruptors.

A correlation between the exposure to diethylstilbestrol (DES), a nonsteroidal synthetic estrogen, in mothers and the occurrence of adenocarcinoma of the vagina in their daughters, along with a high incidence of anatomical abnormalities of the genital tract that adversely affect their reproductive capacity, has been reported in both humans and female mice [85,86]. Early exposure to DES has been reported to cause aberrant CpG methylation and silencing of key uterine cancer. Aberrant hypermethylation in the promoter of the *Hoxa10* gene upon exposure to DES *in utero* and increased mRNA expression of DNMT1 and DNMT3b was reported [87]. Neonatal diethylstilbestrol exposure induces persistent elevation of *c-fos* expression and hypomethylation in its exon-4 in mouse uterus [88]. DES exposure has been shown to induce global DNA methylation in mouse uterus [89]. Several studies have now shown marked effects of environmental toxicants on the F3 generation through germ-line alterations on the epigenome. It is important to note that when using animal data to study transgenerational effects only the F3 generation can demonstrate the first signs of transgenerational inheritance. Both the F1 generation embryo and F2 generation germline are directly exposed when an F0 generation mother is exposed. Therefore, for animal data to demonstrate transgenerational effects, they need to show that epigenetic changes are inherited at least up to the F3 generation. Remarkably, a recent human study reported that women who were not themselves exposed to DES *in utero* may have altered reproductive tract function if their mothers had been exposed *in utero* [90].

Another endocrine disruptor whose effects on the F2 generation have been associated with an epigenetic mechanism is bisphenol A (BPA), a high production chemical with estrogenic properties present in many commonly used products such as food and beverage containers, baby bottles, and dental components. *In utero* and neonatal exposure to BPA is associated with higher body weight, increased breast and prostate cancer, and altered reproductive function. Hypermethylation of the phosphodiesterase type 4 variant 4 gene at a CpG island after exposure to BPA was reported [91]. Recently, maternal exposure to BPA decreased methylation of the metastable loci Avy and Capb^{IAP}. This effect on methylation by BAP was counteracted by maternal dietary supplementation with folic acid or the genistein phytoestrogen [92]. Alterations in DNA methylation patterns as a result of phytoestrogen consumption have been reported [93–95].

In vitro exposure of preimplantation embryos to the contaminant 2,3,7,8-tetra-chlorodibenzo-*p*-dioxin (TCDD) can alter DNA methylation in the H19 and IGF2 imprinted genes [96]. DNA hypermethylation of the estrogen receptor alpha gene induced by phthalates was also reported [97]. Additionally, an inverse linear relationship was found between blood global DNA methylation and the persistent organic pollutants (POPs) dichloro-diphenyl-trichloroethane (DTT), dichloro-diphenyl-dichloroethylene (DDE), β-benzenehexachloride (β-BHC), oxychlordane, α-chlordane, and mirex in *Alu* repeated elements in the Greenlandic Iunit [98]. The Greenlandic Iunit population is highly exposed to POPs from their diet and contamination of the Arctic environment. Exposure to POPs, many of which are estrogen disruptors, is associated with autoimmune diseases, diabetes, developmental neurotoxicity, birth defects, impaired male fertility, and increased cancer risk.

Alterations in the methylation of specific genes associated with transgeneration disease were found in the F2 and F3 generations of mice exposed to the androgenic compound vinclozolin [99]. In Fisher rats, both vinclozolin and methoxychlor induced transgenerational defects (F1–F4 generations) in spermatogenic capacity and sperm viability. These effects on reproduction correlated with altered DNA methylation patterns in the germline [100,101].

POLYCYCLIC AROMATIC HYDROCARBONS

Polycyclic aromatic hydrocarbons (PAHs) are chemical compounds consisting of fused aromatic rings that do not contain heteroatoms or carry substituents. PAHs are found in oil, coal, and tar deposits, and are produced as a byproduct of fuel burning. Exposure to PAHs can also occur from breathing cigarette smoke, wood smoke, vehicle exhaust, eating grilled or charred meats, and coming into contact with air, water, or soil near hazardous wastes. PAHs are of environmental concern since some compounds have been identified as carcinogenic, mutagenic, and teratogenic. PAHs known for their carcinogenic, mutagenic, and teratogenic activities include benz[a]anthracene, chrysene, benzo[b]fluoranthene, benzo[j]fluoranthene, benzo[k]fluoranthene, benzo[a]pyrene, benzo[ghi]perylene, coronene, dibenz[a,h]anthracene ($C_{20}H_{14}$), indeno[1,2,3-cd]pyrene ($C_{22}H_{12}$), and ovalene.

Global DNA hypermethylation following *in vitro* chronic exposure to benzo[a]pyrene (B[a]P), the most well studied PAH carcinogen, was found in mouse embryonic fibroblasts [102]. In a recent human cohort study, methylation sensitive restriction fingerprinting was used to analyze umbilical cord white blood cell DNA of 20 cohort children in order to study the effect of transplacental exposure to high levels of airborne PAHs on DNA methylation levels and dysregulation of gene expression and childhood asthma. Thirty DNA sequences were identified whose methylation status was dependent on the level of maternal PAH exposure. Methylation of *acetyl-CoA synthetase long chain family member 3* (*ACSL3*) 5'CGI was found to be significantly associated with maternal airborne PAH exposure and with a parental report of asthma symptoms in children prior to age 5 [103]. Differences in DNA methylation states were found in peripheral blood lymphocyte DNA of PAH-chronic exposed polish male nonsmoking coke-oven workers and matched controls. Global (*Alu* and *LINE-1* repetitive elements) methylation level, and to a lesser extent *IL-6*, were higher in the peripheral blood lymphocytes of PAH-exposed workers. Conversely, DNA methylation levels were lower in gene specific promoters (*p53* and *HIC1*) of PAH-exposed workers [104].

468

CONCLUSION

Human exposure to various environmental chemicals can increase the risk for reproductive malformations, cardiovascular disease, cancers, and other disorders. In this chapter we have reviewed the evidence up to the present day that associates epigenetic mechanisms with the toxic and carcinogenic effects of certain environmental agents. Many of these chemicals have been shown to modify the same or similar epigenetic marks found in patients with disease states associated with that agent. For example, the increase in global DNA hypomethylation levels and promoter specific hypermethylation of tumor suppressor genes in response to arsenic exposure is consistent with the observed promoter specific hypermethylation and widespread loss in DNA methylation levels exhibited by cancer cells. Exposure to environmental factors such as metals, the semi-metal selenium, peroxisome proliferators, radiation, particulate matter, tobacco smoke, benzene, endocrine disruptors, and polycyclic aromatic hydrocarbons, perturb global and gene specific DNA methylation levels (Tables 28.1 and 28.2) as well as histone posttranslational modifications (Table 28.3). Although the

TABLE 28.1 Effect of Environmental Agents on Global DNA Methylation

Agent	Effect	Tissue	Reference
Metals			
Arsenic	Hypomethylation	Human HaCat cell line, Rat liver	[18,19]
Chromium	Hypermethylation	Brassica napus L. plant	[31]
Cadmium	Hypomethylation (short-term treatment)	Rat liver cells, K562 Cell line	[36,39]
	Hypermethylation (long-term treatment)	Various	[36–38]
Selenium	Hypomethylation	FELC, LNCAP, Caco-2, rat and liver cells	[42,44,45]
Peroxisome proliferators			
WY-14,643	Hypomethylation	Mouse and rat liver	[48,49]
DEHP	Hypomethylation	Rat liver	[49]
Radiation			
Direct exposure	Hypomethylation	Various	[50–54,63,66]
Bystander cells/tissue	Hypomethylation	Cultured human 3D artificial tissue system	[58]
Progeny	Hypomethylation	Mouse thymus	[64]
Particulate matter	Hypomethylation	Human buffy coat, mouse sperm DNA	[69,70,72]
Tobacco smoke	Hypomethylation	Squamous cell carcinoma cancers	[73]
Benzene	Hypomethylation	Peripheral blood DNA	[83]
Endocrine disruptors			
DES	Hypomethylation	Mouse uterus	[89]
POPs	Hypomethylation	Peripheral blood DNA	[98]
PAHs			
Benzo[a]pyrene (B[a]P)	Hypermethylation	Mouse embryonic fibroblasts	[102]
	Hypermethylation	Peripheral blood DNA	[104]

TABLE 28.2 Effect of Environmental Agents on Gene Specific DNA Methylation

Agent	Gene	Effect	Tissue	Reference
Metals				
Nickel	*gpt* transgene	Hypermethylation	Chinese hamster G12 cell line	[1]
	p16	Hypermethylation	Tumors of C57BL/6 mice	[2]
Arsenic	Various tumor suppressor genes	Hypermethylation	Various	[20–24]
	Various tumor suppressor genes	Hypo- and hypermethylation	A549 cells	[25]
	p16 and *p53*	Hypermethylation	Blood DNA of humans exposed	[28]
Chromium	*gpt* transgene	Hypermethylation	Chinese hamster G12 cell line	[30]
	p16, MLH1	Hypermethylation	Lung cancer of chromate workers	[32,33]
Cadmium	*p16* and *RASSF1A*	Hypermethylation	Human prostate cells	[38]
Methylmercury	*BDNF*	Hypermethylation	Hippocampus of exposed mice	[41]
Selenium	*GSTP1, APC, CSR1, p53*	Hypomethylation	LNCap prostate cancer cells, Caco-2, rat and liver cells	[44,45]
Peroxisome proliferators				
WY-14,643	*c-myc*	Hypomethylation	Mouse liver	[46]
TCE, DCA, TCA	*c-myc, c-jun*	Hypomethylation	Mouse liver	[47]
Radiation				
Direct exposure	*p16*	Hypermethylation	Human lung tumor	[56]
	Tumor suppressor genes	Hypermethylation	Human skin cancer	[67]
Particulate matter	*p16, ER* genes	Hypermethylation	Rat lung tumors	[68]
	iNOS	Hypomethylation	Human buffy coat	[69]
Tobacco smoke	*DAP*	Hypermethylation	Human lung cancer	[74]
	Various tumor suppressor genes	Hypermethylation	Non-small-cell lung cancer	[75–77,79,80]
	p16, FHIT	Hypermethylation	Squamous cell carcinoma	[78]
	Synuclein-γ	Hypomethylation	Lung cancer cells	[81]
	CYP1A1	Hypomethylation	Various cells	[82]
Benzene	*p15*	Hypermethylation	Peripheral blood DNA	[83]
	MAGE-1	Hypomethylation	Peripheral blood DNA	[83]
Endocrine disruptors				
DES	*Hox10*	Hypermethylation		[87]
	c-fos	Hypomethylation	Mouse uterus	[88]
BPA	*Phosphodiesterase type 4 variant 4*	Hypermethylation	Rat prostate	[91]
	Avy and *Capb*[IAP]	Hypomethylation	Mouse embryo	[92]
Phytoestrogens	Various	Various	Various	[93–95]
TCDD	*H19, IGF2*	Hypermethylation	Mouse embryo	[96]

470

TABLE 28.2 Effect of Environmental Agents on Gene Specific DNA Methylation (Continued)

Agent	Gene	Effect	Tissue	Reference
Phthalates	Estrogen receptor alpha	Hypermethylation	Human breast cancer (MCF7) cell line	[97]
Vinclozolin	Gene specific	Hypermethylation	Ray testis	[101]
PAHs	ACSL3	Hypermethylation	Umbilical cord white blood cell	[103]
	IL-6	Hypermethylation	Peripheral blood DNA	[104]
	p53 and HIC1	Hypomethylation	Peripheral blood DNA	[104]

TABLE 28.3 Effect of Environmental Agents on Histone Posttranslational Modifications

Agent	Modification	Effect	Reference
Metals			
Nickel	H2A, H2B, H3, H4 acetylation	Decrease	[3,5,6,8,9]
	H3K9 dimethylation	Increase	[4,10]
	H2A and H2B ubiquitination	Increase	[7]
Arsenite	H3K9 dimethylation, H3K4 trimethylation	Increase	[29]
	H3K27 trimethylation	Decrease	[29]
Chromium	H3K9 di- and trimethylation, H3K4 di- and trimethylation	Increase	[35]
	H3K27 dimethylation, H3R2 dimethylation	Decrease	[35]
Cobalt	H3K9, H3K27, H3K36, H3K4 trimethylation	Increase	[40]
	H2A and H2B ubiquitination	Increase	[40]
	H4 acetylation	Increase	[40]
Selenium	H3K9 acetylation	Increase	[45]
	H3K9 dimethylation	Decrease	[45]
Peroxisome proliferators			
WY-14,643	H4K20 and H3K9 trimethylation	Decrease	[48]
Radiation			
Direct exposure	H2AX phosphorylation	Increase	[53]
	H4K20 trimethylation	Decrease	[53]
Bystander cells	H2AX phosphorylation	Increase	[58–62]
Progeny	H2AX phosphorylation	Increase	[64,65]

471

mechanism(s) by which these chemicals perturb the epigenome is unknown many studies suggest an alteration in the expression and/or activity of enzymes that modify DNA and histone tails such as DNA methyltransferases and histone methyltransferases, deacetylases, and demethylases. For chemicals such as arsenite and selenium the mechanism(s) of epigenetic alteration may depend upon methyl group availability. Exposure to carcinogenic chemicals such as metals induce their effects by affecting the DNA and histone marks that activate and repress gene expression resulting in DNA hypermethylation of tumor suppressor genes and hypomethylation of genes that may promote tumorigenesis and tumor progression. Additionally, an association between epigenetic effects and transgenerational effects of exposure to endocrine disruptors has been identified.

Therefore, the evidence summarized in this chapter emphasizes the importance of applying to toxicological research the understanding that genotoxic mechanisms are not the sole mechanism underlying the changes in gene expression leading to cancer and other disease states. Instead, it is necessary to consider that heritable alterations in phenotype may have an epigenetic basis as well. Mapping changes in the epigenome induced by toxic environmental chemicals may be useful in the future in determining the epigenetic alterations that develop over time and increase the risk for disease in exposed individuals and their subsequent generations. The emerging field of environmental epigenetics promises to help map epigenetic changes associated with the disease induced by a particular agent and provide a better understanding of how both genetic and epigenetic mechanisms interact to confer susceptibility on a disease disorder. In order for the field of environmental epigenetics to grow both *in vivo* experimental animal models and collections of biological specimens over time, cohort studies of populations exposed to a particular agent will need to be conducted. Technologies such as methylation sensitive fingerprinting, restriction landmark genomic scanning, ChIP-on-ChIP (ChIP-ChIP), ChIP-sequencing (ChIP-Seq), and RNA-seq, among others, will need to be applied to epidemiological studies of specific exposures. Environmental epigenetics promises to help identify the agents that pose health risks to the human population, and the diseases associated with a specific exposure, establish better guidelines for the acceptable levels of exposure to a particular agent, and develop better methods for preventive and treatment medicine.

472

References

1. Lee YW, Klein CB, Kargacin B, Salnikow K, Kitahara J, Dowjat K, et al. Carcinogenic nickel silences gene expression by chromatin condensation and DNA methylation: a new model for epigenetic carcinogens. *Mol Cell Biol* 1995;**15**:2547–57.

2. Govindarajan B, Klafter R, Miller MS, Mansur C, Mizesko M, Bai X, et al. Reactive oxygen-induced carcinogenesis causes hypermethylation of *p16(Ink4a)* and activation of MAP kinase. *Mol Med* 2002;**8**:1–8.

3. Broday L, Peng W, Kuo MH, Salnikow K, Zoroddu M, Costa M. Nickel compounds are novel inhibitors of histone H4 acetylation. *Cancer Res* 2000;**60**:238–41.

4. Chen H, Ke Q, Kluz T, Yan Y, Costa M. Nickel ions increase histone H3 lysine 9 dimethylation and induce transgene silencing. *Mol Cell Biol* 2006;**26**:3728–37.

5. Ke Q, Davidson T, Chen H, Kluz T, Costa M. Alterations of histone modifications and transgene silencing by nickel chloride. *Carcinogenesis* 2006;**27**:1481–8.

6. Golebiowski F, Kasprzak KS. Inhibition of core histones acetylation by carcinogenic nickel(II). *Mol Cell Biochem* 2005;**279**:133–9.

7. Ke Q, Ellen TP, Costa M. Nickel compounds induce histone ubiquitination by inhibiting histone deubiquitinating enzyme activity. *Toxicol Appl Pharmacol* 2008;**228**:190–9.

8. Yan Y, Kluz T, Zhang P, Chen HB, Costa M. Analysis of specific lysine histone H3 and H4 acetylation and methylation status in clones of cells with a gene silenced by nickel exposure. *Toxicol Appl Pharmacol* 2003;**190**:272–7.

9. Kang J, Zhang Y, Chen J, Chen H, Lin C, Wang Q, et al. Nickel-induced histone hypoacetylation: the role of reactive oxygen species. *Toxicol Sci* 2003;**74**:279–86.

10. Jackson JP, Johnson L, Jasencakova Z, Zhang X, PerezBurgos L, Singh PB, et al. Dimethylation of histone H3 lysine 9 is a critical mark for DNA methylation and gene silencing in *Arabidopsis thaliana*. *Chromosoma* 2004;**112**:308–15.

11. Zhang QW, Salnikow K, Kluz T, Chen LC, Su WC, Costa M. Inhibition and reversal of nickel-induced transformation by the histone deacetylase inhibitor trichostatin A. *Toxicol Appl Pharmacol* 2003;**192**:201–11.

12. Kasprzak KS, Bal W, Karaczyn AA. The role of chromatin damage in nickel-induced carcinogenesis. A review of recent developments. *J Environ Monit* 2003;**5**:183–7.

13. Bal W, Lukszo J, Jezowskabojczuk M, Kasprzak KS. Interactions of nickel(II) with histones – stability and solution structure of complexes with Ch3CO-Cys-Ala-Ile-His-NH2, a putative metal-binding sequence of histone H3. *Chem Res Toxicol* 1995;**8**:683–92.

14. Bal W, Lukszo J, Kasprzak KS. Interactions of Nickel(II) with histones: enhancement of 2′-deoxyguanosine oxidation by Ni(II) complexes with CH3CO-Cys-Ala-Ile-His-NH2, a putative metal binding sequence of histone H3. *Chem Res Toxicol* 1996;**9**:535–40.

15. Bal W, Lukszo J, Bialkowski K, Kasprzak KS. Interactions of nickel(II) with histones: interactions of Nickel(II) with CH3CO-Thr-Glu-Ser-His-His-Lys-NH2, a peptide modeling the potential metal binding site in the "C-Tail" region of histone H2A. *Chem Res Toxicol* 1998;**11**:1014–23.

16. Bal W, Karantza V, Moudrianakis EN, Kasprzak KS. Interaction of nickel(II) with histones: in vitro binding of nickel(II) to the core histone tetramer. *Arch Biochem Biophys* 1999;**364**:161–6.

17. Bal W, Liang RT, Lukszo J, Lee SH, Dizdaroglu M, Kasprzak KS. Ni(II) specifically cleaves the C-terminal tail of the major variant of histone H2A and forms an oxidative damage mediating complex with the cleaved-off octapeptide. *Chem Res Toxicol* 2000;**13**:616–24.

18. Zhao CQ, Young MR, Diwan BA, Coogan TP, Waalkes MP. Association of arsenic-induced malignant transformation with DNA hypomethylation and aberrant gene expression. *Proc Natl Acad Sci USA* 1997;**94**:10907–12.

19. Reichard JF, Schnekenburger M, Puga A. Long-term low-dose arsenic exposure induces loss of DNA methylation. *Biochem Biophys Res Commun* 2007;**352**:188–92.

20. Marsit CJ, Karagas MR, Danaee H, Liu M, Andrew A, Schned A, et al. Carcinogen exposure and gene promoter hypermethylation in bladder cancer. *Carcinogenesis* 2006;**27**:112–16.

21. Cui X, Wakai T, Shirai Y, Hatakeyama K, Hirano S. Chronic oral exposure to inorganic arsenate interferes with methylation status of *p16INK4a* and *RASSF1A* and induces lung cancer in A/J mice. *Toxicol Sci* 2006;**91**:372–81.

22. Chai CY, Huang YC, Hung WC, Kang WY, Chen WT. Arsenic salts induced autophagic cell death and hypermethylation of *DAPK* promoter in SV-40 immortalized human uroepithelial cells. *Toxicol Lett* 2007;**173**:48–56.

23. Mass MJ, Wang L. Arsenic alters cytosine methylation patterns of the promoter of the tumor suppressor gene *p53* in human lung cells: a model for a mechanism of carcinogenesis. *Mutat Res* 1997;**386**:263–77.

24. Chanda S, Dasgupta UB, Guhamazumder D, Gupta M, Chaudhuri U, Lahiri S, et al. DNA hypermethylation of promoter of gene *p53* and *p16* in arsenic-exposed people with and without malignancy. *Toxicol Sci* 2006;**89**:431–7.

25. Zhong CXY, Mass MJ. Both hypomethylation and hypermethylation of DNA associated with arsenite exposure in cultures of human cells identified by methylation-sensitive arbitrarily-primed PCR. *Toxicol Lett* 2001;**122**:223–34.

26. Pilsner JR, Liu XH, Ahsan H, Ilievski V, Slavkovich V, Levy D, et al. Genomic methylation of peripheral blood leukocyte DNA: influences of arsenic and folate in Bangladeshi adults. *Am J Clin Nutr* 2007;**86**:1179–86.

27. Pilsner JR, Liu XH, Ahsan H, Ilievski V, Slavkovich V, Levy D, et al. Folate deficiency, hyperhomocysteinemia, low urinary creatinine, and hypomethylation of leukocyte DNA are risk factors for arsenic-induced skin lesions. *Environ Health Perspect* 2009;**117**:254–60.

28. Chanda S, Dasgupta UB, GuhaMazumder D, Gupta M, Chaudhuri U, Lahiri S, et al. DNA hypermethylation of promoter of gene *p53* and *p16* in arsenic-exposed people with and without malignancy. *Toxicol Sci* 2006;**89**:431–7.

29. Zhou X, Sun H, Ellen TP, Chen H, Costa M. Arsenite alters global histone H3 methylation. *Carcinogenesis* 2008;**29**:1831–6.

30. Klein CB, Su L, Bowser D, Leszczynska J. Chromate-induced epimutations in mammalian cells. *Environ Health Perspect* 2002;**110**(Suppl 5):739–43.

31. Labra M, Grassi F, Imazio S, Di Fabio T, Citterio S, Sgorbati S, et al. Genetic and DNA-methylation changes induced by potassium dichromate in *Brassica napus* L. *Chemosphere* 2004;**54**:1049–58.

32. Kondo K, Takahashi Y, Hirose Y, Nagao T, Tsuyuguchi M, Hashimoto M, et al. The reduced expression and aberrant methylation of *p16(INK4a)* in chromate workers with lung cancer. *Lung Cancer* 2006;**53**:295–302.

33. Takahashi Y, Kondo K, Hirose T, Nakagawa H, Tsuyuguchi M, Hashimoto M, et al. Microsatellite instability and protein expression of the DNA mismatch repair gene, *hMLH1*, of lung cancer in chromate-exposed workers. *Mol Carcinog* 2005;**42**:150–8.

34. Schnekenburger M, Talaska G, Puga A. Chromium cross-links histone deacetylase 1-DNA methyltransferase 1 complexes to chromatin, inhibiting histone-remodeling marks critical for transcriptional activation. *Mol Cell Biol* 2007;**27**:7089–101.

35. Sun H, Zhou X, Chen H, Li Q, Costa M. Modulation of histone methylation and *MLH1* gene silencing by hexavalent chromium. *Toxicol Appl Pharmacol* 2009;**237**:258–66.

473

36. Takiguchi M, Achanzar WE, Qu W, Li G, Waalkes MP. Effects of cadmium on DNA-(Cytosine-5) methyltransferase activity and DNA methylation status during cadmium-induced cellular transformation. *Exp Cell Res* 2003;**286**:355–65.

37. Jiang G, Xu L, Song S, Zhu C, Wu Q, Zhang L, et al. Effects of long-term low-dose cadmium exposure on genomic DNA methylation in human embryo lung fibroblast cells. *Toxicology* 2008;**244**:49–55.

38. Benbrahim-Tallaa L, Waterland RA, Dill AL, Webber MM, Waalkes MP. Tumor suppressor gene inactivation during cadmium-induced malignant transformation of human prostate cells correlates with overexpression of de novo DNA methyltransferase. *Environ Health Perspect* 2007;**115**:1454–9.

39. Huang D, Zhang Y, Qi Y, Chen C, Ji W. Global DNA hypomethylation, rather than reactive oxygen species (ROS), a potential facilitator of cadmium-stimulated K562 cell proliferation. *Toxicol Lett* 2008;**179**:43–7.

40. Li Q, Ke Q, Costa M. Alterations of histone modifications by cobalt compounds. *Carcinogenesis* 2009;**30**:1243–51.

41. Onishchenko N, Karpova N, Sabri F, Castren E, Ceccatelli S. Long-lasting depression-like behavior and epigenetic changes of *BDNF* gene expression induced by perinatal exposure to methylmercury. *J Neurochem* 2008;**106**:1378–87.

42. Cox R, Goorha S. A study of the mechanism of selenite-induced hypomethylated DNA and differentiation of Friend erythroleukemic cells. *Carcinogenesis* 1986;**7**:2015–18.

43. Fiala ES, Staretz ME, Pandya GA, El-Bayoumy K, Hamilton SR. Inhibition of DNA cytosine methyltransferase by chemopreventive selenium compounds, determined by an improved assay for DNA cytosine methyltransferase and DNA cytosine methylation. *Carcinogenesis* 1998;**19**:597–604.

44. Davis CD, Uthus EO, Finley JW. Dietary selenium and arsenic affect DNA methylation in vitro in Caco-2 cells and in vivo in rat liver and colon. *J Nutr* 2000;**130**:2903–9.

45. Xiang N, Zhao R, Song G, Zhong W. Selenite reactivates silenced genes by modifying DNA methylation and histones in prostate cancer cells. *Carcinogenesis* 2008;**29**:2175–81.

46. Ge R, Wang W, Kramer PM, Yang S, Tao L, Pereira MA. Wy-14,643-induced hypomethylation of the *c-myc* gene in mouse liver. *Toxicol Sci* 2001;**62**:28–35.

47. Tao L, Ge R, Xie M, Kramer PM, Pereira MA. Effect of trichloroethylene on DNA methylation and expression of early-intermediate protooncogenes in the liver of B6C3F1 mice. *J Biochem Mol Toxicol* 1999;**13**:231–7.

48. Pogribny IP, Tryndyak VP, Woods CG, Witt SE, Rusyn I. Epigenetic effects of the continuous exposure to peroxisome proliferator WY-14,643 in mouse liver are dependent upon peroxisome proliferator activated receptor alpha. *Mutat Res* 2007;**625**:62–71.

49. Pogribny IP, Tryndyak VP, Boureiko A, Melnyk S, Bagnyukova TV, Montgomery B, et al. Mechanisms of peroxisome proliferator-induced DNA hypomethylation in rat liver. *Mutat Res* 2008;**644**:17–23.

50. Kalinich JF, Catravas GN, Snyder SL. The effect of gamma radiation on DNA methylation. *Radiat Res* 1989;**117**:185–97.

51. Tawa R, Kimura Y, Komura J, Miyamura Y, Kurishita A, Sasaki MS, et al. Effects of X-ray irradiation on genomic DNA methylation levels in mouse tissues. *J Radiat Res (Tokyo)* 1998;**39**:271–8.

52. Loree J, Koturbash I, Kutanzi K, Baker M, Pogribny I, Kovalchuk O. Radiation-induced molecular changes in rat mammary tissue: Possible implications for radiation-induced carcinogenesis. *Int J Radiat Biol* 2006;**82**:805–15.

53. Pogribny I, Koturbash I, Tryndyak V, Hudson D, Stevenson SM, Sedelnikova O, et al. Fractionated low-dose radiation exposure leads to accumulation of DNA damage and profound alterations in DNA and histone methylation in the murine thymus. *Mol Cancer Res* 2005;**3**:553–61.

54. Raiche J, Rodriguez-Juarez R, Pogribny I, Kovalchuk O. Sex- and tissue-specific expression of maintenance and de novo DNA methyltransferases upon low dose X-irradiation in mice. *Biochem Biophys Res Commun* 2004;**325**:39–47.

55. Pogribny I, Raiche J, Slovack M, Kovalchuk O. Dose-dependence, sex- and tissue-specificity, and persistence of radiation-induced genomic DNA methylation changes. *Biochem Biophys Res Commun* 2004;**320**:1253–61.

56. Belinsky SA, Klinge DM, Liechty KC, March TH, Kang T, Gilliland FD, et al. Plutonium targets the *p16* gene for inactivation by promoter hypermethylation in human lung adenocarcinoma. *Carcinogenesis* 2004;**25**:1063–7.

57. Kaup S, Grandjean V, Mukherjee R, Kapoor A, Keyes E, Seymour CB, et al. Radiation-induced genomic instability is associated with DNA methylation changes in cultured human keratinocytes. *Mutation Research – Fundamental and Molecular Mechanisms of Mutagenesis* 2006;**597**:87–97.

58. Sedelnikova OA, Nakamura A, Kovalchuk O, Koturbash I, Mitchell SA, Marino SA, et al. DNA double-strand breaks form in bystander cells after microbeam irradiation of three-dimensional human tissue models. *Cancer Res* 2007;**67**:4295–302.

59. Smilenov LB, Hall EJ, Bonner WM, Sedelnikova OA. A microbeam study of DNA double-strand breaks in bystander primary human fibroblasts. *Radiat Prot Dosimetry* 2006;**122**:256–9.

60. Burdak-Rothkamm S, Short SC, Folkard M, Rothkamm K, Prise KM. ATR-dependent radiation-induced gamma H2AX foci in bystander primary human astrocytes and glioma cells. *Oncogene* 2007;**26**:993–1002.

61. Yang H, Anzenberg V, Held KD. The time dependence of bystander responses induced by iron-ion radiation in normal human skin fibroblasts. *Radiat Res* 2007;**168**:292–8.

62. Sokolov MV, Smilenov LB, Hall EJ, Panyutin IG, Bonner WM, Sedelnikova OA. Ionizing radiation induces DNA double-strand breaks in bystander primary human fibroblasts. *Oncogene* 2005;**24**:7257–65.

63. Koturbash I, Rugo RE, Hendricks CA, Loree J, Thibault B, Kutanzi K, et al. Irradiation induces DNA damage and modulates epigenetic effectors in distant bystander tissue in vivo. *Oncogene* 2006;**25**:4267–75.

64. Koturbash I, Baker M, Loree J, Kutanzi K, Hudson D, Pogribny I, et al. Epigenetic dysregulation underlies radiation-induced transgenerational genome instability in vivo. *Int J Radiat Oncol Biol Phys* 2006;**66**:327–30.

65. Barber RC, Hickenbotham P, Hatch T, Kelly D, Topchiy N, Almeida GM, et al. Radiation-induced transgenerational alterations in genome stability and DNA damage. *Oncogene* 2006;**25**:7336–42.

66. Mittal A, Piyathilake C, Hara Y, Katiyar SK. Exceptionally high protection of photocarcinogenesis by topical application of (–)-epigallocatechin-3-gallate in hydrophilic cream in SKH-1 hairless mouse model: relationship to inhibition of UVB-induced global DNA hypomethylation. *Neoplasia* 2003;**5**:555–65.

67. van Doorn R, Gruis NA, Willemze R, van der Velden PA, Tensen CP. Aberrant DNA methylation in cutaneous malignancies. *Semin Oncol* 2005;**32**:479–87.

68. Belinsky SA, Snow SS, Nikula KJ, Finch GL, Tellez CS, Palmisano WA. Aberrant CpG island methylation of the *p16(INK4a)* and estrogen receptor genes in rat lung tumors induced by particulate carcinogens. *Carcinogenesis* 2002;**23**:335–9.

69. Tarantini L, Bonzini M, Apostoli P, Pegoraro V, Bollati V, Marinelli B, et al. Effects of particulate matter on genomic DNA methylation content and *iNOS* promoter methylation. *Environ Health Perspect* 2009;**117**:217–22.

70. Baccarelli A, Wright RO, Bollati V, Tarantini L, Litonjua AA, Suh HH, et al. Rapid DNA methylation changes after exposure to traffic particles. *Am J Respir Crit Care Med* 2009;**179**:572–8.

71. Castro R, Rivera I, Struys EA, Jansen EE, Ravasco P, Camilo ME, et al. Increased homocysteine and S-adenosylhomocysteine concentrations and DNA hypomethylation in vascular disease. *Clin Chem* 2003;**49**:1292–6.

72. Yauk C, Polyzos A, Rowan-Carroll A, Somers CM, Godschalk RW, Van Schooten FJ, et al. Germ-line mutations, DNA damage, and global hypermethylation in mice exposed to particulate air pollution in an urban/industrial location. *Proc Natl Acad Sci USA* 2008;**105**:605–10.

73. Piyathilake CJ, Frost AR, Bell WC, Oelschlager D, Weiss H, Johanning GL, et al. Altered global methylation of DNA: an epigenetic difference in susceptibility for lung cancer is associated with its progression. *Hum Pathol* 2001;**32**:856–62.

74. Pulling LC, Vuillemenot BR, Hutt JA, Devereux TR, Belinsky SA. Aberrant promoter hypermethylation of the *death-associated protein kinase* gene is early and frequent in murine lung tumors induced by cigarette smoke and tobacco carcinogens. *Cancer Res* 2004;**64**:3844–8.

75. Kim DS, Cha SI, Lee JH, Lee YM, Choi JE, Kim MJ, et al. Aberrant DNA methylation profiles of non-small cell lung cancers in a Korean population. *Lung Cancer* 2007;**58**:1–6.

76. Kikuchi S, Yamada D, Fukami T, Maruyama T, Ito A, Asamura H, et al. Hypermethylation of the *TSLC1/IGSF4* promoter is associated with tobacco smoking and a poor prognosis in primary nonsmall cell lung carcinoma. *Cancer* 2006;**106**:1751–8.

77. Marsit CJ, Kim DH, Liu M, Hinds PW, Wiencke JK, Nelson HH, et al. Hypermethylation of *RASSF1A* and *BLU* tumor suppressor genes in non-small cell lung cancer: implications for tobacco smoking during adolescence. *Int J Cancer* 2005;**114**:219–23.

78. Kim JS, Kim H, Shim YM, Han J, Park J, Kim DH. Aberrant methylation of the *FHIT* gene in chronic smokers with early stage squamous cell carcinoma of the lung. *Carcinogenesis* 2004;**25**:2165–71.

79. Palmisano WA, Divine KK, Saccomanno G, Gilliland FD, Baylin SB, Herman JG, et al. Predicting lung cancer by detecting aberrant promoter methylation in sputum. *Cancer Res* 2000;**60**:5954–8.

80. Burbee DG, Forgacs E, Zochbauer-Muller S, Shivakumar L, Fong K, Gao B, et al. Epigenetic inactivation of *RASSF1A* in lung and breast cancers and malignant phenotype suppression. *J Natl Cancer Inst* 2001;**93**:691–9.

81. Liu H, Zhou Y, Boggs SE, Belinsky SA, Liu J. Cigarette smoke induces demethylation of prometastatic oncogene *synuclein-gamma* in lung cancer cells by downregulation of DNMT3B. *Oncogene* 2007;**26**:5900–10.

82. Anttila S, Hakkola J, Tuominen P, Elovaara E, Husgafvel-Pursiainen K, Karjalainen A, et al. Methylation of *cytochrome P4501A1* promoter in the lung is associated with tobacco smoking. *Cancer Res* 2003;**63**:8623–8.

83. Bollati V, Baccarelli A, Hou LF, Bonzini M, Fustinoni S, Cavallo D, et al. Changes in DNA methylation patterns in subjects exposed to low-dose benzene. *Cancer Res* 2007;**67**:876–80.

84. Lubbert M, Oster W, Ludwig WD, Ganser A, Mertelsmann R, Herrmann F. A switch toward demethylation is associated with the expression of myeloperoxidase in acute myeloblastic and promyelocytic leukemias. *Blood* 1992;**80**:2066–73.

85. Herbst AL, Ulfelder H, Poskanze. Dc. Adenocarcinoma of vagina – association of maternal stilbestrol therapy with tumor appearance in young women. *N Engl J Med* 1971;**284**:878.

86. Mclachlan JA, Newbold RR, Bullock BC. Long-term effects on the female mouse genital-tract associated with prenatal exposure to diethylstilbestrol. *Cancer Res* 1980;**40**:3988–99.

87. Bromer JG, Wu J, Zhou YP, Taylor HS. Hypermethylation of *homeobox A10* by in utero diethylstilbestrol exposure: an epigenetic mechanism for altered developmental programming. *Endocrinology* 2009;**150**:3376–82.

88. Li S, Hansman R, Newbold R, Davis B, McLachlan JA, Barrett JC. Neonatal diethylstilbestrol exposure induces persistent elevation of *c-fos* expression and hypomethylation in its exon-4 in mouse uterus. *Mol Carcinog* 2003;**38**:78–84.

89. Li S, Hursting SD, Davis BJ, McLachlan JA, Barrett JC. Environmental exposure, DNA methylation, and gene regulation: lessons from diethylstilbesterol-induced cancers. *Ann N Y Acad Sci* 2003;**983**:161–9.

90. Titus-Ernstoff L, Troisi R, Hatch EE, Wisei LA, Palmer J, Hyer M, et al. Menstrual and reproductive characteristics of women whose mothers were exposed in utero to diethylstilbestrol (DES). *Int J Epidemiol* 2006;**35**:862–8.

91. Ho SM, Tang WY, de Frausto JB, Prins GS. Developmental exposure to estradiol and bisphenol A increases susceptibility to prostate carcinogenesis and epigenetically regulates *phosphodiesterase type 4 variant 4*. *Cancer Res* 2006;**66**:5624–32.

92. Dolinoy DC, Huang D, Jirtle RL. Maternal nutrient supplementation counteracts bisphenol A-induced DNA hypomethylation in early development. *Proc Natl Acad Sci USA* 2007;**104**:13056–61.

93. Day JK, Bauer AM, desBordes C, Zhuang Y, Kim BE, Newton LG, et al. Genistein alters methylation patterns in mice. *J Nutr* 2002;**132**(8 Suppl):2419S–23S.

94. Guerrero-Bosagna CM, Sabat P, Valdovinos FS, Valladares LE, Clark SJ. Epigenetic and phenotypic changes result from a continuous pre and post natal dietary exposure to phytoestrogens in an experimental population of mice. *BMC Physiol* 2008;**8**:17.

95. Lyn-Cook BD, Blann E, Payne PW, Bo J, Sheehan D, Medlock K. Methylation profile and amplification of proto-oncogenes in rat pancreas induced with phytoestrogens. *Proc Soc Exp Biol Med* 1995;**208**:116–19.

96. Wu Q, Ohsako S, Ishimura R, Suzuki JS, Tohyama C. Exposure of mouse preimplantation embryos to 2,3,7,8-tetrachlorodibenzo-p-dioxin (TCDD) alters the methylation status of imprinted genes H19 and Igf2. *Biol Reprod* 2004;**70**:1790–7.

97. Kang SC, Lee BM. DNA methylation of *estrogen receptor alpha* gene by phthalates. *J Toxicol Environ Health A* 2005;**68**:1995–2003.

98. Rusiecki JA, Baccarelli A, Bollati V, Tarantini L, Moore LE, Bonefeld-Jorgensen EC. Global DNA hypomethylation is associated with high serum-persistent organic pollutants in Greenlandic Inuit. *Environ Health Perspect* 2008;**116**:1547–52.

99. Chang HS, Anway MD, Rekow SS, Skinner MK. Transgenerational epigenetic imprinting of the male germline by endocrine disruptor exposure during gonadal sex determination (Retracted article. See vol. 150, p. 2976, 2009). *Endocrinology* 2006;**147**:5524–41.

100. Anway MD, Leathers C, Skinner MK. Endocrine disruptor vinclozolin induced epigenetic transgenerational adult-onset disease. *Endocrinology* 2006;**147**:5515–23.

101. Anway MD, Cupp AS, Uzumcu M, Skinner MK. Epigenetic transgenerational actions of endocrine disruptors and male fertility. *Science* 2005;**308**:1466–9.

102. Yauk CL, Polyzos A, Rowan-Carroll A, Kortubash I, Williams A, Kovalchuk O. Tandem repeat mutation, global DNA methylation, and regulation of DNA methyltransferases in cultured mouse embryonic fibroblast cells chronically exposed to chemicals with different modes of action. *Environ Mol Mutagen* 2008;**49**:26–35.

103. Perera F, Tang WY, Herbstman J, Tang D, Levin L, Miller R, et al. Relation of DNA methylation of 5'-CpG island of *ACSL3* to transplacental exposure to airborne polycyclic aromatic hydrocarbons and childhood asthma. *PLoS ONE* 2009;**4**:e4488.

104. Pavanello S, Bollati V, Pesatori AC, Kapka L, Bolognesi C, Bertazzi PA, et al. Global and gene-specific promoter methylation changes are related to anti-B[a]PDE-DNA adduct levels and influence micronuclei levels in polycyclic aromatic hydrocarbon-exposed individuals. *Int J Cancer* 2009;**125**:1692–7.

CHAPTER 29

Impact of Microbial Infections on the Human Epigenome and Carcinogenesis

Kjetil Søreide
Department of Surgery, Stavanger University Hospital, Stavanger, Norway and Department of
Surgical Sciences, University of Bergen, Bergen, Norway

477

INTRODUCTION

As human beings we are influenced by the environment we live in. In an evolutionary perspective, the human being has long lived in, and been dependent upon, symbiosis with several microbes that confer a variety of physiologic benefits. Probably in no other place is the interaction between human cells and infective agents as important as in the gastrointestinal tract (Fig. 29.1), where microbes may influence both physiological and pathogenetic processes through various molecularly-regulated mechanisms [1]. For example, the microbes of the large bowel provide us with genetic and metabolic attributes we have not been required to evolve on our own, including the ability to harvest otherwise inaccessible nutrients [2]. Advanced understanding and abilities in investigating the microbes have enabled progress in characterizing the taxonomic composition, metabolic capacity, and immunomodulatory activity of the human gut microbiota, further establishing the role of microbiota in human health and disease [1–3]. The human host has co-evolved with normal bacteria over thousands of years and developed complex mechanisms that monitor and control this ecosystem. Such cellular mechanisms have homeostatic roles beyond the traditional concept of defense against potential pathogens, suggesting these pathways contribute directly to the well-being of the gut [1]. In fact, the bacterial microbiota has established multiple mechanisms to influence the eukaryotic host, generally in a beneficial fashion, and maintain their stable niche [1,2,4]. As the prokaryotic genomes of the human microbes enable involvement and facilitation of a number of metabolic processes beyond that of the host genome, the microbes are seen as an essential part of normal physiology in humans. Gaining a fuller understanding of both partners in the normal gut–microbiota interaction may thus shed light on how the relationship can go awry and contribute to a number of immune, inflammatory, and metabolic disorders. Further, increased understanding may reveal mechanisms by which this relationship could be manipulated toward therapeutic ends [1,2].

Handbook of Epigenetics: The New Molecular and Medical Genetics. DOI: 10.1016/B978-0-12-375709-8.00029-0

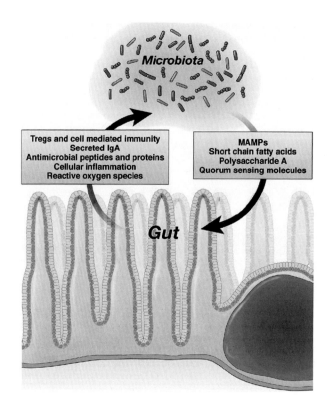

FIGURE 29.1

Mechanisms of microbes and crosstalk within the gut. The symbiosis between human cells of the large bowel and the intraluminal microbiotic flora (several strains of bacteria) is a dyadic relationship by which both parties alter and shape each other, resulting in a "negotiated settlement" at an equilibrium. A breakdown of this crosstalk may result in a "dysbiotic" microbiota and clinical consequences, resulting in diseases ranging from inflammatory bowel disease to cancer. Understanding of how and by which mechanisms microbes are involved in the genesis of human disease is evolving. MAMPs denote "microbial-associated molecular patterns". *(Reprinted from Neish, A.S. Microbes in gastrointestinal health and disease. Gastroenterology, 2009;136(1):65–80, Copyright (2009), with permission from Elsevier/AGA Institute.)* (Please refer to color plate section)

Of note, several viral and bacterial infections have been linked to different types of cancer. Infective agents are thought to be responsible for almost one-fifth of all cancers, with an estimated total of infection-attributable cancer at almost 2 million cases of the global cancer burden [5]. The principal agents involved have been estimated by Parkin [5] to be (in decreasing order): the bacterium *Helicobacter pylori* (5.5% of all cancer), the human papilloma viruses (5.2%), the hepatitis B and C viruses (4.9%), Epstein–Barr virus (1%), human immunodeficiency virus (HIV) together with the human herpes virus 8 (0.9%). Relatively less important causes of cancer are the schistosomes (0.1%), human T-cell lymphotropic virus type I (0.03%), and the liver flukes (0.02%). Estimates by Parkin suggest there would be about 26% fewer cancers in developing countries (1.5 million cases per year) and almost 8% fewer in developed countries (390,000 cases) if these infectious diseases were prevented. The fraction of infectious-induced neoplasia at the specific sites varies from 100% of cervix cancers attributable to the human papilloma viruses (HPV) to a tiny proportion (0.4%) of hepatocellular carcinomas caused by liver flukes on a global scale [5].

Cancer is both an epigenetic and a genetic disease, and epigenetic alterations in cancer are increasingly recognized [6]. Normal epigenetic modifications of DNA encompass three types of changes: chromatin modifications, DNA methylation, and genomic imprinting, each of which is altered in cancer cells, and typically characterized in colorectal cancer [7,8].

Minarovits has recently suggested that in addition to viruses and bacteria, other microparasites (protozoa) as well as macroparasites (helminths, arthropods, fungi) may induce pathological changes by epigenetic reprogramming of host cells they are interacting with [9]. Consequently, elucidation of the epigenetic consequences of microbe–host interactions may not only yield new insight into pathogenesis and understanding of diseases, but also have important therapeutic implications because epigenetic processes can be reverted and elimination of microbes inducing patho-epigenetic changes may prevent disease development.

This chapter will review some current known associations of infective agents (bacteria and viruses) and their (known or potential) influence on the human epigenome. Obviously, inclusion of all aspects, associations, and mechanisms of this emerging field is beyond the scope of this chapter. Thus, recent extensive reviews on related topics are referred to wherever applicable, and the cited work of several research groups are recognized and referenced at best ability. Consequently, interested readers are encouraged to seek further in-depth knowledge from these sources [1,2,9–15] as well as other references provided in the text. Owing to the vast field of potential anatomic locations (from immune system to parenchymal organs) and types of diseases (from autoimmune disease to neoplasia), the topic will mainly, but not exclusively, include examples from current aspects involving the gastrointestinal tract and development of neoplasia, with particular focus on colorectal cancer.

HUMANS AND INFECTIVE AGENTS AND EPIGENETICS

The human DNA is controlled not only by mechanisms regulating all aspects of the chromosomes to the base pair level [16], but also by epigenetic mechanisms. Epigenetic mechanisms are defined as all the meiotically and mitotically inherited changes in gene expression that are not encoded in the DNA sequence itself. Epigenetic modifications of chromatin and DNA have been recognized as important factors in controlling the expressed genome via gene transcription. Two major epigenetic mechanisms are: one, the post-translational modification of histone proteins in chromatin; and, two, the methylation of DNA itself. Obviously, the epigenetic state is a central regulator of cellular development and activation. Emerging evidence suggests a key role for epigenetics in human pathologies, including inflammatory and neoplastic disorders [17–20].

The epigenome is influenced by environmental factors throughout life. For example, nutritional factors can have profound effects on the expression of specific genes by epigenetic modification, and these may be passed on to subsequent generations with potentially detrimental effects. Many cancers are associated with altered epigenetic profiles [21–24], leading to altered expression of the genes involved in cell growth or differentiation. Autoimmune and neoplastic diseases increase in frequency with increasing age, with epigenetic dysregulation proposed as a potential explanation [19,20,25–27]. In support of this hypothesis, studies in monozygotic twins revealed increasing epigenetic differences with age. Differences in methylation status of CpG sites, monoallelic silencing, and other epigenetic regulatory mechanisms have been observed in key inflammatory response genes. The importance of the epigenome in the pathogenesis of common human diseases is likely to be as significant as that of traditional genetic mutations. With advances in technology, our understanding of this area of biology is likely to increase rapidly in the near future.

INFECTIONS AND EPIGENETIC MODIFICATIONS

The genomes of certain viruses and the proviral genomes of retroviruses are regularly targeted by epigenetic regulatory mechanisms (DNA methylation, histone modifications, binding of regulatory proteins) in infected cells [28]. In parallel, proteins encoded by viral genomes may affect the activity of a set of cellular promoters by interacting with the very same epigenetic regulatory machinery [9]. This may result in epigenetic dysregulation and subsequent cellular dysfunctions that may manifest in or contribute to the development of pathological changes (e.g. carcinogenesis or immunodeficiency). Bacteria infecting mammals may cause diseases in a similar manner, by causing hypermethylation of key cellular promoters at CpG dinucleotides (promoter silencing, e.g. by *Campylobacter rectus* in the placenta or by *Helicobacter pylori* in gastric mucosa).

Bacteria

Bacterial pathogens have evolved various strategies to avoid immune surveillance, depending on their *in vivo* "lifestyle" [29]. The identification of few bacterial effectors capable of entering the nucleus and modifying chromatin structure in host raises the questions of how pathogens modulate chromatin structure and why. Chromatin is a dynamic structure that maintains the stability and accessibility of the host DNA genome in relation to the transcription machinery. Arbibe [29] reviews the various strategies used by pathogens to interface with host chromatin. In some cases, chromatin injury can be a strategy to take control of major cellular functions, such as the cell cycle. In other cases, manipulation of chromatin structure at specific genomic locations by modulating epigenetic information provides a way for the pathogen to impose its own transcriptional signature onto host cells. This emerging field may influence our understanding of chromatin regulation at interphase nucleus and may provide invaluable openings to the control of immune gene expression in inflammatory and infectious diseases. Knowledge of how bacteria may influence the host environment though genetic and epigenetic manipulation is increasing [11,12,29,30]. Clearly, upon infection, pathogens reprogram host gene expression. In eukaryotic cells, genetic reprogramming is induced by the concerted activation/repression of transcription factors and various histone modifications that control DNA accessibility in chromatin [12]. One landmark study exploring the microbe–host interaction during infection reported that the bacterial pathogen *Listeria monocytogenes* induced a dramatic dephosphorylation of histone H3 as well as a deacetylation of histone H4 during early phases of infection. This effect was mediated by the major listerial toxin listeriolysin O in a pore-forming-independent manner. A similar effect was also observed with other toxins of the same family, such as *Clostridium perfringens* perfringolysin and *Streptococcus pneumoniae* pneumolysin. The decreased levels of histone modifications correlate with a reduced transcriptional activity of a subset of host genes, including key immunity genes. In their findings, control of epigenetic regulation emerged as an unsuspected function shared by several bacterial toxins, highlighting a common strategy used by intracellular and extracellular pathogens to modulate the host response early during infection [12].

Innate immunity is the first line of defense against a bacterial infection, and most organisms are able to mount an efficient early, nonspecific response leading to the recruitment of cellular effectors and inflammation. Microbial components that elicit an inflammatory response (Fig. 29.2) have been called microbial associated molecular patterns (MAMPs) and include LPS, bacterial flagellin, lipoteichoic acid, peptidoglycan, and nucleic acids [1,11]. Host cells recognize MAMPs through pattern recognition receptors (PRRs) present either at the cell surface and/or on endosomes, for Toll-like receptors (TLRs), or in the cytoplasm, for nucleotide-binding oligomerization domain proteins (NODs) and NOD-like receptors (NLRs). These receptors activate signaling cascades leading to transcriptional activation of immunity genes such as cytokine genes.

Virus

Viruses are capable of influencing the human genome as well, with proposed mechanisms for disease development. *Epstein–Barr virus* (EBV) is a human herpesvirus hiding in a latent form in memory B cells in the majority of the world population. Although primary EBV infection is asymptomatic or causes a self-limiting disease, infectious mononucleosis, the

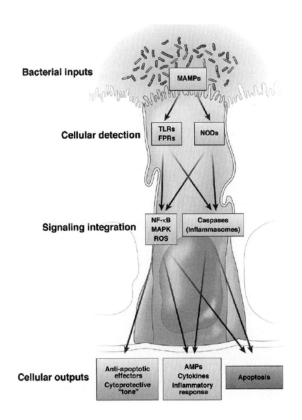

FIGURE 29.2

Cellular consequences to bacterial stimuli. Bacterial microbial-associated molecular patterns (MAMPs) may stimulate pattern recognition receptors (including extracellular Toll-like receptors (TLRs) and formylated peptide receptors (FPRs), or intracellular NODs). Intensity, duration, and spatial origin of the subsequent signaling responses are integrated by an intricate and interrelated network of transduction pathways that determine if MAMP perception warrants a "low gain" cytoprotective response, a "medium gain" inflammatory reaction, or "high gain" programmed cell death result. *(Reprinted from Neish, A.S. Microbes in gastrointestinal health and disease. Gastroenterology, 2009;136(1):65–80, Copyright (2009), with permission from Elsevier/AGA Institute.)* Abbreviations: NF-κB, nuclear factor κB; NLR, Nod-like receptor; PRR, pattern recognition receptor; ROS, reactive oxygen species; SCFA, short-chain fatty acid; TLR, Toll-like receptor. (Please refer to color plate section)

481

virus is associated with a wide variety of neoplasms developing in immunosuppressed or immunodeficient individuals, but also in patients with an apparently intact immune system [19]. In memory B cells, tumor cells, and lymphoblastoid cell lines (LCLs, transformed by EBV *in vitro*) the expression of the viral genes is highly restricted. There is no virus production (lytic viral replication associated with the expression of all viral genes) in tight latency. The expression of latent viral oncogenes and RNAs is under a strict epigenetic control via DNA methylation and histone modifications that results either in a complete silencing of the EBV genome in memory B cells, or in a cell-type dependent usage of latent promoters in tumor cells, germinal center B cells, and LCLs. Both the latent and lytic EBV proteins are potent immunogens and elicit vigorous B- and T-cell responses. In immunosuppressed and immunodeficient patients, or in individuals with a functional defect of EBV-specific T cells, lytic EBV replication is regularly activated and an increased viral load can be detected in the blood. Enhanced lytic replication results in new infection events and EBV-associated transformation events, and seems to be a risk factor for both malignant transformation and the development of autoimmune diseases [19]. As reviewed by Niller and colleagues, current speculation includes the idea that an increased load or altered presentation of a limited set of lytic or latent EBV proteins that cross-react with cellular antigens triggers and perpetuates the pathogenic processes that result in multiple sclerosis, systemic lupus erythematosus (SLE), and rheumatoid arthritis. In addition, in SLE patients EBV may cause defects of B-cell tolerance checkpoints because latent membrane protein 1, an EBV-encoded viral oncoprotein can induce BAFF, a B-cell activating factor that rescues self-reactive B cells and induces a lupus-like autoimmune disease in transgenic mice [19].

Chronic Inflammation

Chronic inflammatory disorders are often associated with an increased cancer risk [31]. A particularly striking example of the chronic inflammation-cancer link is seen in inflammatory bowel disease (IBD). The risk of developing CRC in patients affected with IBD is proportional to the number of years of active disease. Further, in a case-control study it was demonstrated that plasma CRP concentrations are elevated among average-risk individuals who subsequently developed colon (but not rectal) cancer [32]. These data support the hypothesis that inflammation is a risk factor for the development of CRC.

IBD results from a dysregulated immunologic response to commensal microbial flora residing in the intestinal lumen. Although this response is probably due at least in part to a genetic predisposition, patients with IBD have also been reported to house an abnormal intestinal microflora. Whether this altered flora is the cause or result of the associated chronic inflammation remains unclear. What appears important is the role of tumor necrosis factor (TNF)-α in the role of IBD development, as it may alter the microbial composition, enhance virulence, and increase adherence and invasion [33]. Of note, a single nucleotide polymorphism (SNP) in the promoter region of this proinflammatory cytokine has been found to be associated with an increased cancer-risk for patients with ulcerative colitis [34], and has been reported by others, although at other SNP sites [35]. Extrapolated from this is the role of inflammatory signals in alteration both of bacterial properties and of susceptibility to neoplasia, which may pave the way for new therapies and preventive strategies [36].

Further, animal models used for exploring the mechanisms by which inflammation increases the risk of CRC have shown that inflammatory cells, through the effects of the cytokines they produce, have a major role in promoting neoplastic transformation. However, the molecular mechanisms underlying these processes began only recently to be clarified. Indeed, from the initial concept that the release of free radicals during inflammation might induce the accumulation of genetic mutations thus leading to the onset of dysplastic cells [15,37], it is now becoming clear that the large amount of cytokines and growth factors released during inflammation may influence the carcinogenesis process [38]. While a detailed description is beyond the scope of this review, it includes immune system factors such as interleukins, cytokines, cyclo-oxygenases, activation of the transcription factor nuclear factor kappa B (NF-κB) [39–41], and more recently described molecules demonstrated as focal points of crosstalk between the signaling cascades such as the peroxisome proliferator-activated receptor delta (PPARdelta) [42], and the proteinase-activated receptor 2 (PAR-2) [43].

The role for *E. coli* in CRC carcinogenesis has been somewhat controversial, with studies indicating a relationship between localization [44], or the presence of certain strains of *E. coli*, as a risk factor for patients with CRC [41], but with only minor influence on initiation on chromosomal instability in a recently reported model [30]. *E. coli* is a normal inhabitant of the human intestine that becomes highly pathogenic following the acquisition of certain virulence factors, including a protein toxin named cytotoxic necrotizing factor 1 (CNF1) [41]. This toxin permanently activates the small GTP-binding proteins belonging to the Rho family, thus promoting changes in the cytoskeleton, protein expression, and cell physiology. CNF1 is receiving increased attention because of its ability to induce COX-2 expression, activate the transcription factor NF-κB, protect epithelial cells from apoptosis, release pro-inflammatory cytokines in epithelial and endothelial cells, and promote cellular motility [41]. As CRC may arise through dysfunction of the same regulatory systems, it seems possible that CNF1-producing *E. coli* infections can contribute to development of neoplasia. Taken together, several factors have been shown to promote the growth of colon tumors in experimental models.

VIRUS AND EPIGENETICS: ROLE IN CANCER DEVELOPMENT

Herpesviruses, papillomaviruses, and retroviruses are the three most important groups of infectious (viral) carcinogens [45,46]. In females, HPV infections on a global scale account for more than 50% of infection-linked cancers, in males for barely 5% [45]. Vaccines against the high-risk HPV types 16 and 18 represent the first preventive vaccines directly developed to protect against a major human cancer (cervical carcinoma) and its precursor steps (cervical intraepithelial neoplasia; CIN) [47–49].

Epstein–Barr virus (EBV), a human herpesvirus, is associated with a wide variety of malignant tumors [23,50]. The expression of the latent viral RNAs is under strict, host-cell dependent transcriptional control. This results in an almost complete transcriptional silencing of the EBV genome in memory B-cells. In tumor cells, germinal center B-cells, and lymphoblastoid cells, distinct viral latency promoters are active. Epigenetic mechanisms contribute to this strict control. In EBV-infected cells, epigenetic mechanisms also alter the expression of cellular genes, including tumor suppressor genes. In nasopharyngeal carcinoma, the hypermethylation of certain cellular promoters is attributed to the up-regulation of DNA methyltransferases by the viral oncoprotein LMP1 (latent membrane protein 1) via JNK/AP1-signaling. The role of other viral latency products in the epigenetic dysregulation of the cellular genome remains to be established. Analysis of epigenetic alterations in EBV-associated neoplasms may result in a better understanding of their pathogenesis and may facilitate the development of new therapies [23].

Proliferation is necessary for pretumor cells to accumulate genetic alterations and to acquire a transformed phenotype [51]. However, each cell division is associated with a progressive shortening of the telomeres, which can suppress tumor development by initiating senescence and irreversible cell cycle arrest. Therefore, the ability of virus-infected cells to circumvent the senescence program is essential for the long-term survival and proliferation of infected cells and the likelihood of transformation. Consequently, multiple strategies are being used by human DNA and RNA tumor viruses to subvert telomerase functions during cellular transformation and carcinogenesis [52]. Epstein–Barr virus, Kaposi sarcoma-associated herpesvirus, human papillomavirus, hepatitis B virus, hepatitis C virus, and human T-cell leukemia virus-1 each can increase transcription of the telomerase reverse transcriptase. Several viruses appear to mediate cis-activation or enhance epigenetic activation of telomerase transcription [52,53]. Epstein–Barr virus (EBV) and human papillomavirus (HPV) have each developed posttranscriptional mechanisms to regulate the telomerase protein. Finally, some tumor virus proteins can also negatively regulate telomerase transcription or activity. It is likely that, as future studies further expose the strategies used by viruses to deregulate telomerase activity and control of telomere length, novel mechanisms will emerge and underscore the importance of increased telomerase activity in sustaining virus-infected cells and its potential in therapeutic targeting.

Endogenous retrovirus-like elements, or ERVs, are an abundant component of all eukaryotic genomes. Their transcriptional and retrotranspositional activities have great potential for deleterious effects on gene expression. Consequences of such activity may include germline mutagenesis and cell transformation. As a result, mammalian genomes have evolved means of counteracting ERV transcription and mobilization. In a recent review by Maksakova et al. [54], the authors discuss epigenetic mechanisms of ERV and LTR retrotransposon control during mouse development, focusing on involvement of DNA methylation, histone modifications, and small RNAs and their interaction with one another; the relevance of research performed in the mouse system may be relevant for humans [54].

Boland et al. [55] have demonstrated a potential role for virus-induced carcinogenesis in CRC, in that most CRCs contain the DNA of JCV that encodes an oncogenic T-antigen, which is capable of interacting with key growth regulatory pathways (i.e. APC, p53, Wnt-signaling) in the colon, and has the potential to induce CIN. Thus, this suggests that JCV infection may be involved in the initiation of colorectal neoplasia. Apparently, JCV infection is ubiquitous and remains subclinical throughout the life of most individuals, but can cause disease when activated. Activation of the virus in the colon may lead to induction of the adenoma formation, CIN, and eventually CRC [56].

In a recent landmark study by Fernandez and colleagues [28] the authors investigated the complete DNA methylomes of the HPV16, HPV18, and HBV viruses and the DNA methylation analyses of all the transcription start sites of EBV obtained by bisulfite genomic sequencing of multiple clones. The dynamic changes in the viral DNA methylome and their functional relevance in the natural history of the disease were investigated. The researchers found that the DNA methylome of these viruses evolve from an unmethylated to a highly methylated genome in association with the progression of the disease, from asymptomatic healthy carriers, through tissues with chronical infection and pre-malignant lesions, to the full-blown invasive cancers [28]. One interpretation of this finding given by Fernandez et al. [28] is the possibility that DNA methylation might be a device to camouflage the virus from the human immune system. Further, the investigators suggest that the DNA methylomes found in the study could be used for further research in order to understand how the viral proteins themselves are able to use the human DNMTs to favor the establishment of persistent infection [28]. Also, the potential clinical applications of these findings include the non-invasive detection of methylated viral genomes in biological fluids, serum, and blood [28].

BACTERIAL INFLUENCE ON THE CELL CYCLE

The mammalian cell cycle is involved in many processes and, thus, it is not surprising that many bacterial pathogens manipulate the host cell cycle with respect to these functions. Cyclomodulins are a growing family of bacterial toxins and effectors that interfere with the eukaryotic cell cycle, and include cytolethal distending toxins (CDTs), vacuolating cytotoxin, the polyketide-derived macrolide mycolactone, cycle-inhibiting factor, cytotoxic necrotizing factors, dermonecrotic toxin, *Pasteurella multocida* toxin, and cytotoxin-associated antigen A [14].

Of particular interest are the CDTs [57]. These toxins are known to influence the control system of eukaryotic cells, with mechanisms depending on the cell type involved. For example, CDTs may initiate a eukaryotic cell cycle block at the G2 stage prior to mitosis – an effect which is produced by a number of bacterial pathogens [57,58]. The functional CDT is composed of three proteins; CdtA, B, and C. CdtB potentiates a cascade leading to cell cycle block, and CdtA and CdtC function as dimeric subunits, which bind CdtB and deliver it to the mammalian cell interior. Once inside the cell, CdtB enters the nucleus and exhibits a DNase I-like activity that results in DNA double-strand breaks. The eukaryotic cell responds to the DNA double-strand breaks by initiating a regulatory cascade that results in cell cycle arrest, cellular distension, and cell death. The result of CDT activity can differ somewhat depending on the eukaryotic cell types affected, but epithelial cells are arrested in the cell cycle at the G_2/M boundary. The affected cells enlarge until they finally undergo programmed cell death. Of notice, an enlarged, cigar-shaped, elongated cell-type within colorectal adenomatous epithelium has been recognized by histopathologists for decades, and has more recently been described by morphometry and linked to an increased long-term risk for developing CRC [59], together with several alterations in cell-cycle and apoptosis regulating proteins in the same adenomas [60,61]. While speculative at this stage, it should be of interest to further pursue a potential connection between any given bacterial infection with altered cell morphology, changes in intracellular signaling, and the development of neoplasia.

Several issues remain to be elucidated regarding CDT biology, including a molecular understanding of how CDT interacts with DNA [62]. Of note, other mechanisms such as *Cycle inhibiting factor* (Cif) act in a strikingly similar fashion to the CDTs [63]. However, while CDTs inhibit the G_2/M transition by activating the DNA-damage checkpoint pathway, Cif does not cause phosphorylation of histone H2AX, which is associated with DNA double-stranded breaks – thus, Cif works through a DNA damage-independent signaling pathway. Furthermore, is has been demonstrated that toxin capabilities may be transferred between bacteria [64].

Lastly, research has demonstrated that some bacterial toxins may have cancer protective effects [65], notably by pathogens commonly found in areas of low-incidence CRC, but where diarrheal diseases are prevalent [66]. The molecular mechanisms behind such enterotoxigenic *Escherichia coli* (ETEC) infections are developing, and may suggest a role for preventive measures or new therapeutic targets in the future [67].

BACTERIA AND EPIGENETIC MODIFICATION IN THE GUT

Humans live in direct and continuous interaction with a complex microbial environment. The total number of microbes (10^{14}) normally inhabiting our mucosal surfaces exceeds by an order of magnitude the quantity of cells (10^{13}) in our bodies. Perhaps the greatest habitat for microbes is the large bowel (Fig. 29.3) [2,4]. Scattered amidst the commensal microflora are potential pathogens – viruses, bacteria, or parasites – intrinsically capable of producing symptomatic infectious disease [1–4,55].

The long co-existence of bacterial pathogens with their eukaryotic hosts, and their co-evolution, have provided pathogens with an amazing capacity to exploit host cell functions for survival, replication inside or outside cells, and escape from early innate immune responses [11]. The fact that bacteria are so well adapted to their host has been of great benefit for cell biologists, who are increasingly using them to study fundamental cell processes. Similarly to viruses, bacteria provoke histone modifications and chromatin remodeling in infected cells, thereby altering the host's transcriptional program and, in most cases, dampening the host innate immune response.

The Human Intestine and Microflora

Many species of bacteria have evolved and adapted to live and grow in the human intestine. The intestinal habitat of an individual contains 300–500 different species of bacteria [3]. The stomach and small intestine contain only a few species of bacteria – for which the role of *Helicobacter* in the stomach is well know [4]. Contrasting this, the large intestine contains a complex and dynamic microbial ecosystem with high densities of living bacteria, which may achieve concentrations of >1000 cells per gram of luminal contents (Fig. 29.3). In fact, about 60% of fecal solids consist of bacteria. Several hundred grams of bacteria living within the colonic lumen affect host homoeostasis. Molecular analysis has demonstrated different communities of bacteria from patient to patient [68]. Obviously, some of these bacteria represent potential pathogens but can also confer important health benefits to the human host. Of note, the development of colorectal cancer has been related to infections of viruses [56,69,70], bacteria [71,72], and parasites [73].

The decline of gastric cancer in the Western world over the past few decades has been attributed (among other factors) to socioeconomic improvements, better hygiene, and thus less exposure to, for example, *H. pylori*. On the other hand, the incidence of CRC is still on the rise, and it is tempting to speculate that what we eat and new ways of processing food has provided new ways for different types and strains of bacteria to enter the gut epithelium [74,75], alter the bacterial milieu, and thus potentially induce carcinogenic effects at the molecular level [76].

Helicobacter pylori Infection and the Stomach

H. pylori are bacteria that colonize the stomach persistently in over half of the world's population. *H. pylori* is linked to various diseases of the stomach, such as peptic ulcer,

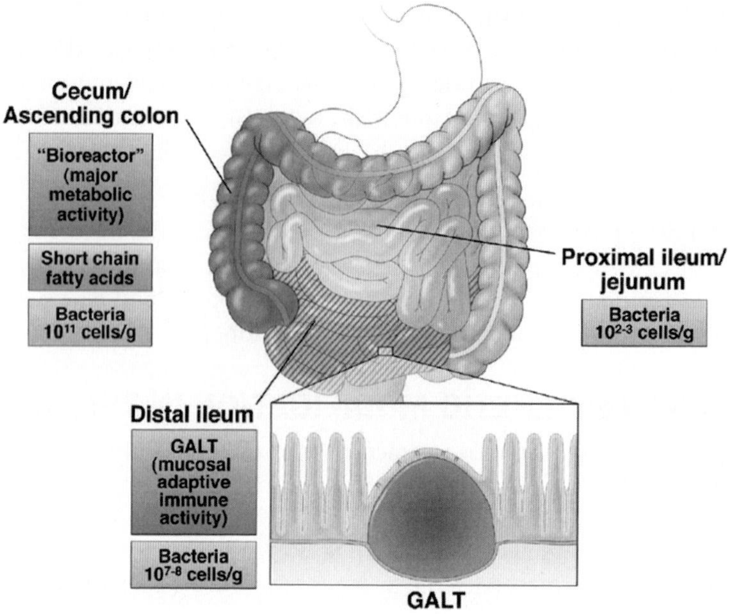

FIGURE 29.3
Preferred sites of commensal/probiotic interaction with the gut. Cecum/ascending colon is a "bioreactor" with the greatest amounts of bacteria, metabolic activity, and short-chain fatty acid (SCFA) fermentation. Concentration of SCFAs diminishes along the colon. The distal ileum is enriched in GALT (gut-associated lymphoid tissue; Peyer's patches) and is the dominant site of luminal sampling and mucosal adaptive immune activity. *(Reprinted from Neish, A.S. Microbes in gastrointestinal health and disease. Gastroenterology, 2009;136(1):65–80, Copyright (2009), with permission from Elsevier/AGA Institute.)* (Please refer to color plate section)

gastric lymphoma, and gastric neoplasia. In addition, *H. pylori* may also play a role in inflammatory bowel disease, irritable bowel disease, and, potentially, also in colorectal cancer [4,71,77].

Aberrant expression of cell cycle control proteins has been demonstrated in *H. pylori* infected gastric epithelial cells, suggesting that perturbation of the cell cycle plays a role in the pathogenesis of various *H. pylori* associated diseases. Down-regulation of E-cadherin (an adhesion molecule involved in tumour invasion and metastasis) in *H. pylori* associated gastric cancer has been known for more than a decade [78,79], and is caused by silencing of E-cadherin by promoter CpG methylation [79]. Further, in a study conducted by Xia and colleagues [80], the modulation of the cell cycle control protein p21(WAF1) by *H. pylori* in a gastric carcinoma cell line and in primary gastric cells derived from healthy tissue was investigated. In this study [80], the investigators observed an up-regulation of p21(WAF1) in both gastric cancer cells and primary cells. Analysis revealed that the increased expression of p21(WAF1) induced by *H. pylori* is associated with the release of HDAC-1 from the p21(WAF1) promoter and hyper-acetylation of histone H4 [80].

The epigenetic mechanisms involved are interesting in that they yield increased understanding of how bacteria may induce changes in cell cycle regulation, as similar effects may be extrapolated to other cancers [11], either by *H. pylori* itself [77], or by different strains of bacteria. Further, it has been demonstrated that *H. pylori* shows a predisposition to the proximal part of the colon [81], a site also known to harbor higher numbers of microsatellite instable cancers [82,83]. From a clinical point of view, it is also noted that the numbers of lymph nodes harvested are higher in the proximal colon, in association with MSI [84], something that speculatively might be attributed the infectious agents located there triggering an immune cell response in the lymph nodes. Further, MSI in sporadic colorectal cancer is usually associated with hypermethylation (most of the *MLH1*) in colorectal cancers [82,85] – and one may wonder if these are, in part, changes induced by bacterial strains located within the intestinal lumen.

Epigenetics and Microbial Infection in Colorectal Cancer

Colorectal cancer (CRC) represents an increasing health burden in the Western world [86]. Thus, new ways of diagnosing and preventing the disease are much welcomed. Central in this process is the understanding of the colorectal carcinogenesis and factors involved in the initiation, maintenance, and progression of neoplasia. While CRC is recognized as a heterogeneous disease with at least three modes of genetic instability involved [8,82,87], the prudent factors for initiation and maintenance of such genetic instability are less well understood. The association of inflammatory bowel disease (IBD) and the perceived increased risk of CRC has led researchers on the search for a possible role of infection and inflammation in CRC [55].

The thought that bacteria contribute to CRC development is not new [85,88]. However, evidence is now evolving from different disciplines and mechanistic insight is beginning to emerge [72]. Several mechanisms have been suggested [30,37,44,89,90]; some of these include the ability of the microbial flora of the colon to produce a state of continual low-grade inflammation with genotoxic stress that may contribute to colon carcinogenesis and also influence genetic stability.

Molecular pathways responsible for genomic instability in colorectal cancer include microsatellite instability (MSI), chromosomal instability (CIN), and epigenetics – referred to as CpG island methylator phenotype (CIMP) – as depicted in Figs 29.4 and 29.5 [8,83,91,92]. Approximately 15–20% of all CRCs demonstrate MSI, a reasonably well understood process caused by inactivation of the DNA mismatch repair (MMR) system either as an hereditary genetic trait, or through "epigenetic silencing" (hypermethylation of CpG sequences in the promoter region of some MMR genes, including MLH1, MLH2, and MSH6, for instance) in sporadic CRC [82]. Of note, most CRCs thought to develop through the epigenetic/microsatellite instable pathway are located right-sided in the colon (Fig. 29.6) – coinciding with the location of highest bacterial density in the colon (Fig. 29.3).

However, the majority of CRCs acquire genomic instability through the CIN pathway. CIN is characterized by aneuploidy and loss of heterozygosity (LOH), and leads to losses of tumor suppressor genes on chromosomes 5q, 18q, and 17p. It has long been appreciated that CRCs are commonly aneuploid (about 75–85%) and have CIN, but the molecular mechanisms causing CIN are poorly understood, and are currently an issue of controversy and intense investigation.

In a cellular model, it was demonstrated that macrophage cyclo-oxygenase-2 (COX-2) was induced by superoxide from *Enterococcus faecalis* and promoted CIN in mammalian cells through diffusible factors [30]. This mechanism links the oxidative physiology *of E. faecalis* to propagation of genomic instability through an indirect "bystander effect", and offers a novel theory for the role of commensal bacteria in the etiology of sporadic CRC. This notion of the important role of extracellular components, such as macrophages secreting COX-2, is in line with recent discoveries of other factors of the extracellular matrix which are acting as signal transmitters and initiating protease-cascades which again may affect epithelial cells in both inflammation and neoplasia development [93].

While bacteria may cause instability at the chromosomal level in colonic epithelium, it currently remains uncertain whether other genetic mechanisms are involved. However, recently it was demonstrated that antibiotic eradication of *H. pylori* in infected patients reversed the methylation pattern of important tumor promoters in the gastric mucosa [94]. Knowing that aberrant epigenetic changes may be induced by *H. pylori* infection, such as hypermethylation

487

FIGURE 29.4

Characteristics of the major pathways in colorectal cancer. APC, adenomatous polyposis coli; BAX, Bcl-2-associated X protein; CIMP, CpG island methylator phenotype; COX, cyclo-oxygenase; DCC, deleted in colorectal cancer; IGF-IIR, insulin-like growth factor II receptor; LOH, loss of heterozygosity; MLH, MutL homolog; MSH, MutB homolog; Smad, mothers against decapentaplegic homolog (*Drosophila*); TCF, T cell factor, TGF-βR, transforming growth factor β receptor. *(The figure is derived from Søreide et al. Copyright British Journal of Surgery Society Ltd. Reproduced with permission from John Wiley & Sons Ltd on behalf of the BJSS Ltd.)* (Please refer to color plate section)

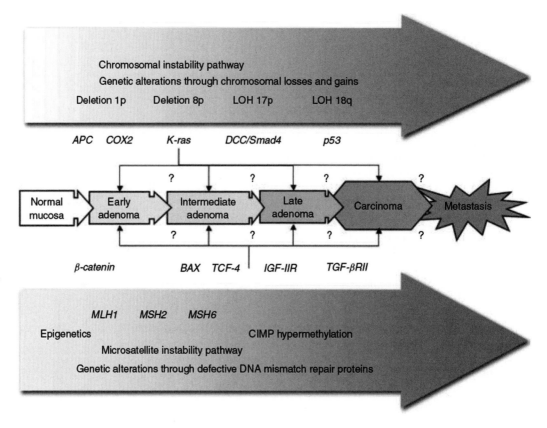

FIGURE 29.5

Molecular mechanisms leading to colorectal cancer. Depicted are chromosomal instability, epigenetic silencing, and microsatellite instability. The latter two mechanisms often coincide in sporadic CRC. *(Reproduced with permission from Søreide K, et al. Endoscopy, morphology, morphometry and molecular markers: predicting cancer risk in colorectal adenoma. Expert Rev Mol Diagn. 2009;9(2):125–37.)* (Please refer to color plate section)

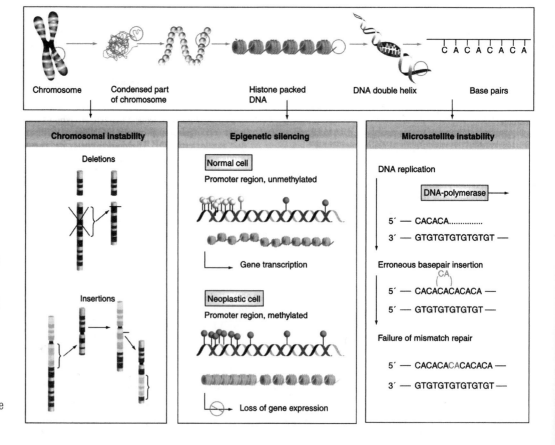

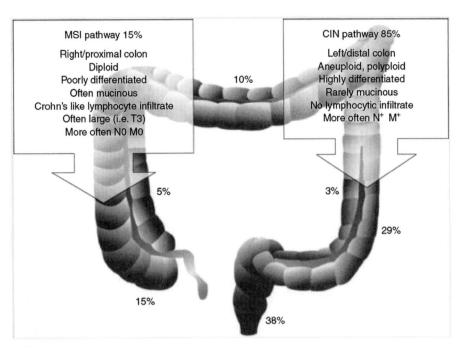

FIGURE 29.6
Pathological distinctions between tumors exhibiting microsatellite instability (MSI) and chromosomal instability (CIN). Percentages indicate the anatomical distribution of colorectal cancers (TNM refers to the tumor node metastasis staging system). *(The figure is derived from Søreide et al. Copyright British Journal of Surgery Society Ltd. Reproduced with permission from John Wiley & Sons Ltd on behalf of the BJSS Ltd.)*

and silencing of *hMLH1* [94], it can be postulated that the same effects may occur in the colorectal epithelium, caused either by *H. pylori*, or tentatively by other unrecognized bacterial species. Several bacterial toxins interfere with cellular signaling mechanisms in a way that is characteristic of tumor promoters [14]. Such toxins could play a direct, yet unappreciated, role in cancer causation and progression. Theoretically, the microflora within the colorectum may not only contribute to chromosomal instability but also to an "epigenetic field for cancerization". Since decreased transcription is involved in the specificity of methylated genes, it is likely that specific genes are methylated according to carcinogenic factors (i.e. bacteria), such as the DNA mismatch repair enzyme *hMLH1* causing widespread microsatellite instability.

POTENTIAL THERAPEUTIC IMPLICATIONS

As bacteria may cause cancer through hypermethylation of important tumor suppressor genes, they may at the same time hold a key for therapeutic interventions. The role of certain bacterial strains in specific cancers needs to be further investigated, but some experience has already been reported from both basic research and clinical investigations.

For one, eradication of *H. pylori* with antibiotics has demonstrated a reduction in gastric cancer, and extensive review of the literature suggests it can reverse many biochemical, genetic, and epigenetic changes that *H. pylori* infection induces in the stomach [95]. One such therapeutic effect is the reversal of methylation of E-cadherin induced by *H. pylori* in the stomach, and consequently reduction in the risk of cancer development [96,97]. A further option would be vaccines against the bacterial strain in question, such as investigated again for *H. pylori* [98], but not clinically effective as yet.

Epigenetic-silencing of aberrantly expressed tumor promoters is another therapeutic option. In fact, this has recently been demonstrated in a study on bacterial induced RNA-interference (RNAi) [99]. RNAi is a potent mechanism, conserved from plants to humans for specific silencing of genes, which holds promise for functional genomics and gene-targeted therapies. In their study, Xiang et al. [99], showed that bacteria engineered to produce a short hairpin RNA (shRNA) targeting a mammalian gene were able to induce trans-kingdom RNAi *in vitro* and *in vivo*. Nonpathogenic *E. coli* were engineered to transcribe shRNAs from a plasmid containing the invasin gene *Inv* and the listeriolysin O gene *HlyA*, which encode two bacterial factors needed for successful transfer of the shRNAs into mammalian cells. Upon oral or intravenous administration, *E. coli* encoding shRNA against β-catenin induced significant gene silencing in the intestinal epithelium and in human colon cancer xenografts in mice. These early results suggest the potential of bacteria-mediated RNAi for functional genomics, therapeutic target validation, and development of clinically-compatible RNAi-based therapies for molecular mechanisms known to be important for CRC development, such as the Wnt-pathway [100].

Lastly, and ending this chapter with a link to the introductory statement of our reliance on microbiota for normal human physiology and metabolism, the microflora of our intestines may be influenced by what we eat, and thus the epigenetic status of intestinal cell genes may be altered accordingly. Demonstrated in a recent clinical study investigating the effects of different nutrients on the microflora of the large bowel, significant differences in the types of bacteria prevalent in the feces were found, however with no associated epithelial or epigenetic changes [101]. However, it is likely that other positive results may develop from future studies in this regard.

CONCLUSIVE REMARKS

Modulation of host transcription by microbial pathogens is now a well-accepted concept, as demonstrated through several recent authoritative reviews [1,4,9,11,29,102]. How specific molecular programs are controlled by infective agents remains elusive [9,11]. For one, the fact that histones can be modified at specific promoters during infection starts to shed light on some of these important issues. There is a need to further determine the molecular mechanisms involved in epigenetic modifications induced by bacteria and viruses. Whether epigenetic changes are specifically induced by the microbiota to subvert normal host responses or are the normal host responses to these pathogens will have to be further investigated for specific diseases and conditions. Future work will determine how these epigenetic phenomena develop and influence disease processes, with potential new therapeutic implications evolving from this research field.

References

1. Neish AS. Microbes in gastrointestinal health and disease. *Gastroenterology* 2009;**136**:65–80.

2. Backhed F, Ley RE, Sonnenburg JL, Peterson DA, Gordon JI. Host-bacterial mutualism in the human intestine. *Science* 2005;**307**:1915–20.

3. Guarner F, Malagelada JR. Gut flora in health and disease. *Lancet* 2003;**361**:512–19.

4. Hecht G. In the beginning was *Helicobacter pylori*: roles for microbes in other intestinal disorders. *Gastroenterology* 2007;**132**:481–3.

5. Parkin DM. The global health burden of infection-associated cancers in the year 2002. *Int J Cancer* 2006;**118**:3030–44.

6. Iacobuzio-Donahue CA. Epigenetic changes in cancer. *Annu Rev Pathol* 2009;**4**:229–49.

7. Sanchez JA, Krumroy L, Plummer S, Aung P, Merkulova A, Skacel M, et al. Genetic and epigenetic classifications define clinical phenotypes and determine patient outcomes in colorectal cancer. *Br J Surg* 2009;**96**:1196–204.

8. Søreide K, Nedrebo BS, Knapp JC, Glomsaker TB, Soreide JA, Korner H. Evolving molecular classification by genomic and proteomic biomarkers in colorectal cancer: potential implications for the surgical oncologist. *Surg Oncol* 2009;**18**:31–50.

9. Minarovits J. Microbe-induced epigenetic alterations in host cells: the coming era of patho-epigenetics of microbial infections. A review. *Acta Microbiol Immunol Hung* 2009;**56**:1–19.

10. Selgrad M, Malfertheiner P, Fini L, Goel A, Boland CR, Ricciardiello L. The role of viral and bacterial pathogens in gastrointestinal cancer. *J Cell Physiol* 2008;**216**:378–88.

11. Hamon MA, Cossart P. Histone modifications and chromatin remodeling during bacterial infections. *Cell Host Microbe* 2008;**4**:100–9.

12. Hamon MA, Batsche E, Regnault B, Tham TN, Seveau S, Muchardt C, et al. Histone modifications induced by a family of bacterial toxins. *Proc Natl Acad Sci USA* 2007;**104**:13467–72.

13. Ge Z, Rogers AB, Feng Y, Lee A, Xu S, Taylor NS, et al. Bacterial cytolethal distending toxin promotes the development of dysplasia in a model of microbially induced hepatocarcinogenesis. *Cell Microbiol* 2007;**9**:2070–80.

14. Oswald E, Nougayrede JP, Taieb F, Sugai M. Bacterial toxins that modulate host cell-cycle progression. *Curr Opin Microbiol* 2005;**8**:83–91.

15. Hope ME, Hold GL, Kain R, El-Omar EM. Sporadic colorectal cancer–role of the commensal microbiota. *FEMS Microbiol Lett* 2005;**244**:1–7.

16. Felsenfeld G, Groudine M. Controlling the double helix. *Nature* 2003;**421**:448–53.

17. Yu W, Gius D, Onyango P, Muldoon-Jacobs K, Karp J, Feinberg AP, et al. Epigenetic silencing of tumour suppressor gene p15 by its antisense RNA. *Nature* 2008;**451**:202–6.

18. Egger G, Liang G, Aparicio A, Jones PA. Epigenetics in human disease and prospects for epigenetic therapy. *Nature* 2004;**429**:457–63.

19. Niller HH, Wolf H, Minarovits J. Regulation and dysregulation of Epstein–Barr virus latency: implications for the development of autoimmune diseases. *Autoimmunity* 2008;**41**:298–328.

20. Nagasaka T, Koi M, Kloor M, Gebert J, Vilkin A, Nishida N, et al. Mutations in both KRAS and BRAF may contribute to the methylator phenotype in colon cancer. *Gastroenterology* 2008;**134**:1950–60.

21. Timp W, Levchenko A, Feinberg AP. A new link between epigenetic progenitor lesions in cancer and the dynamics of signal transduction. *Cell Cycle* 2009;**8**:383–90.

22. Selaru FM, David S, Meltzer SJ, Hamilton JP. Epigenetic events in gastrointestinal cancer. *Am J Gastroenterol* 2009;**104**:1910–12.

23. Niller HH, Wolf H, Minarovits J. Epigenetic dysregulation of the host cell genome in Epstein–Barr virus-associated neoplasia. *Semin Cancer Biol* 2009;**19**:158–64.

24. Mathews LA, Crea F, Farrar WL. Epigenetic gene regulation in stem cells and correlation to cancer. *Differentiation* 2009;**78**:1–17.

25. Patra SK, Patra A, Rizzi F, Ghosh TC, Bettuzzi S. Demethylation of (cytosine-5-C-methyl) DNA and regulation of transcription in the epigenetic pathways of cancer development. *Cancer Metastasis Rev* 2008;**27**:315–34.

26. Haslberger A, Varga F, Karlic H. Recursive causality in evolution: a model for epigenetic mechanisms in cancer development. *Med Hypotheses* 2006;**67**:1448–54.

27. Muegge K, Young H, Ruscetti F, Mikovits J. Epigenetic control during lymphoid development and immune responses: aberrant regulation, viruses, and cancer. *Ann N Y Acad Sci* 2003;**983**:55–70.

28. Fernandez AF, Rosales C, Lopez-Nieva P, Grana O, Ballestar E, Ropero S, et al. The dynamic DNA methylomes of double-stranded DNA viruses associated with human cancer. *Genome Res* 2009;**19**:438–51.

29. Arbibe L. Immune subversion by chromatin manipulation: a 'new face' of host-bacterial pathogen interaction. *Cell Microbiol* 2008;**10**:1582–90.

30. Wang X, Huycke MM. Extracellular superoxide production by *Enterococcus faecalis* promotes chromosomal instability in mammalian cells. *Gastroenterology* 2007;**132**:551–61.

31. Karin M, Lawrence T, Nizet V. Innate immunity gone awry: linking microbial infections to chronic inflammation and cancer. *Cell* 2006;**124**:823–35.

32. Erlinger TP, Platz EA, Rifai N, Helzlsouer KJ. C-reactive protein and the risk of incident colorectal cancer. *JAMA* 2004;**291**:585–90.

33. Hecht GA. Inflammatory bowel disease–live transmission. *N Engl J Med* 2008;**358**:528–30.

34. Garrity-Park MM, Loftus EV Jr., Bryant SC, Sandborn WJ, Smyrk TC. Tumor necrosis factor-alpha polymorphisms in ulcerative colitis-associated colorectal cancer. *Am J Gastroenterol* 2008;**103**:407–15.

35. Suchy J, Klujszo-Grabowska E, Kladny J, Cybulski C, Wokolorczyk D, Szymanska-Pasternak J, et al. Inflammatory response gene polymorphisms and their relationship with colorectal cancer risk. *BMC Cancer* 2008;**8**:112.

36. Burstein E, Fearon ER. Colitis and cancer: a tale of inflammatory cells and their cytokines. *J Clin Invest* 2008;**118**:464–7.

37. Huycke MM, Gaskins HR. Commensal bacteria, redox stress, and colorectal cancer: mechanisms and models. *Exp Biol Med (Maywood)* 2004;**229**:586–97.

38. Itzkowitz SH. Molecular biology of dysplasia and cancer in inflammatory bowel disease. *Gastroenterol Clin North Am* 2006;**35**:553–71.

39. Fantini MC, Pallone F. Cytokines: from gut inflammation to colorectal cancer. *Curr Drug Targets* 2008;**9**:375–80.

40. Itzkowitz SH, Yio X. Inflammation and cancer IV. Colorectal cancer in inflammatory bowel disease: the role of inflammation. *Am J Physiol Gastrointest Liver Physiol* 2004;**287**:G7–G17.

41. Travaglione S, Fabbri A, Fiorentini C. The Rho-activating CNF1 toxin from pathogenic *E. coli*: a risk factor for human cancer development? *Infect Agent Cancer* 2008;**3**:4.

42. Wang D, DuBois RN. Inflammatory mediators and nuclear receptor signaling in colorectal cancer. *Cell Cycle* 2007;**6**:682–5.

43. Søreide K. Proteinase-activated receptor 2 (PAR-2) in gastrointestinal and pancreatic pathophysiology, inflammation and neoplasia. *Scand J Gastroenterol* 2008;**43**:902–9.

44. Swidsinski A, Khilkin M, Kerjaschki D, Schreiber S, Ortner M, Weber J, et al. Association between intraepithelial *Escherichia coli* and colorectal cancer. *Gastroenterology* 1998;**115**:281–6.

45. zur Hausen H. Papillomaviruses in the causation of human cancers–a brief historical account. *Virology* 2009;**384**:260–5.

46. Zur Hausen H. The search for infectious causes of human cancers: where and why. *Virology* 2009;**392**:1–10.

47. Quadrivalent vaccine against human papillomavirus to prevent high-grade cervical lesions. *N Engl J Med* 2007;**356**:1915–27.

48. Paavonen J, Jenkins D, Bosch FX, Naud P, Salmeron J, Wheeler C, et al. Efficacy of a prophylactic adjuvanted bivalent L1 virus-like-particle vaccine against infection with human papillomavirus types 16 and 18 in young women: an interim analysis of a phase III double-blind, randomised controlled trial. *Lancet* 2007;**369**: 2161–70.

49. Ault KA. Effect of prophylactic human papillomavirus L1 virus-like-particle vaccine on risk of cervical intraepithelial neoplasia grade 2, grade 3, and adenocarcinoma in situ: a combined analysis of four randomised clinical trials. *Lancet* 2007;**369**:1861–8.

50. Niller HH, Salamon D, Ilg K, Koroknai A, Banati F, Schwarzmann F, et al. EBV-associated neoplasms: alternative pathogenetic pathways. *Med Hypotheses* 2004;**62**:387–91.

51. Søreide K. Telomerase and survivin in colorectal and pancreatic cancer–biomarkers of life and death in the balance between proliferation and apoptosis. *Curr Cancer Ther Rev* 2008;**4**:253–61.

52. Bellon M, Nicot C. Regulation of telomerase and telomeres: human tumor viruses take control. *J Natl Cancer Inst* 2008;**100**:98–108.

53. Bellon M, Nicot C. Telomerase: a crucial player in HTLV-I-induced human T-cell leukemia. *Cancer Genomics Proteomics* 2007;**4**:21–5.

54. Maksakova IA, Mager DL, Reiss D. Keeping active endogenous retroviral-like elements in check: the epigenetic perspective. *Cell Mol Life Sci* 2008;**65**:3329–47.

55. Boland CR, Luciani MG, Gasche C, Goel A. Infection, inflammation, and gastrointestinal cancer. *Gut* 2005;**54**.1321–31.

56. Niv Y, Goel A, Boland CR. JC virus and colorectal cancer: a possible trigger in the chromosomal instability pathways. *Curr Opin Gastroenterol* 2005;**21**:85–9.

57. Smith JL, Bayles DO. The contribution of cytolethal distending toxin to bacterial pathogenesis. *Crit Rev Microbiol* 2006;**32**:227–48.

58. Nougayrede JP, Homburg S, Taieb F, Boury M, Brzuszkiewicz E, Gottschalk G, et al. *Escherichia coli* induces DNA double-strand breaks in eukaryotic cells. *Science* 2006;**313**:848–51.

59. Søreide K, Buter TC, Janssen EA, van Diermen B, Baak JP. A monotonous population of elongated cells (MPECs) in colorectal adenoma indicates a high risk of metachronous cancer. *Am J Surg Pathol* 2006;**30**:1120–9.

60. Søreide K, Buter TC, Janssen EA, Gudlaugsson E, Skaland I, Kørner H, et al. Cell-cycle and apoptosis regulators (p16INK4A, p21CIP1, beta-catenin, survivin, and hTERT) and morphometry-defined MPECs predict metachronous cancer development in colorectal adenoma patients. *Cell Oncol* 2007;**29**:301–13.

61. Søreide K, Gudlaugsson E, Skaland I, Janssen EA, Van Diermen B, Kørner H, et al. Metachronous cancer development in patients with sporadic colorectal adenomas–multivariate risk model with independent and combined value of hTERT and survivin. *Int J Colorectal Dis* 2008;**23**:389–400.

62. Nesic D, Hsu Y, Stebbins CE. Assembly and function of a bacterial genotoxin. *Nature* 2004;**429**:429–33.

63. Taieb F, Nougayrede JP, Watrin C, Samba-Louaka A, Oswald E. *Escherichia coli* cyclomodulin Cif induces G2 arrest of the host cell cycle without activation of the DNA-damage checkpoint-signalling pathway. *Cell Microbiol* 2006;**8**:1910–21.

64. Asakura M, Hinenoya A, Alam MS, Shima K, Zahid SH, Shi L, et al. An inducible lambdoid prophage encoding cytolethal distending toxin (Cdt-I) and a type III effector protein in enteropathogenic *Escherichia coli*. *Proc Natl Acad Sci USA* 2007;**104**:14483–8.

65. Pitari GM, Zingman LV, Hodgson DM, Alekseev AE, Kazerounian S, Bienengraeber M, et al. Bacterial enterotoxins are associated with resistance to colon cancer. *Proc Natl Acad Sci USA* 2003;**100**:2695–9.

66. Carrithers SL. Diarrhea or colorectal cancer: can bacterial toxins serve as a treatment for colon cancer? *Proc Natl Acad Sci USA* 2003;**100**:3018–20.

67. Li P, Schulz S, Bombonati A, Palazzo JP, Hyslop TM, Xu Y, et al. Guanylyl cyclase C suppresses intestinal tumorigenesis by restricting proliferation and maintaining genomic integrity. *Gastroenterology* 2007;**133**:599–607.

68. Green GL, Brostoff J, Hudspith B, Michael M, Mylonaki M, Rayment N, et al. Molecular characterization of the bacteria adherent to human colorectal mucosa. *J Appl Microbiol* 2006;**100**:460–9.

69. Damin DC, Caetano MB, Rosito MA, Schwartsmann G, Damin AS, Frazzon AP, et al. Evidence for an association of human papillomavirus infection and colorectal cancer. *Eur J Surg Oncol* 2007;**33**:569–74.

70. Bodaghi S, Yamanegi K, Xiao SY, Da Costa M, Palefsky JM, Zheng ZM. Colorectal papillomavirus infection in patients with colorectal cancer. *Clin Cancer Res* 2005;**11**:2862–7.

71. Zumkeller N, Brenner H, Zwahlen M, Rothenbacher D. *Helicobacter pylori* infection and colorectal cancer risk: a meta-analysis. *Helicobacter* 2006;**11**:75–80.

72. Lax AJ, Thomas W. How bacteria could cause cancer: one step at a time. *Trends Microbiol* 2002;**10**:293–9.

73. Zalata KR, Nasif WA, Ming SC, Lotfy M, Nada NA, El-Hak NG, et al. p53, Bcl-2 and C-Myc expressions in colorectal carcinoma associated with schistosomiasis in Egypt. *Cell Oncol* 2005;**27**:245–53.

74. Zheng J, Meng J, Zhao S, Singh R, Song W. Adherence to and invasion of human intestinal epithelial cells by *Campylobacter jejuni* and *Campylobacter coli* isolates from retail meat products. *J Food Prot* 2006;**69**:768–74.

75. Kadhum HJ, Ball HJ, Oswald E, Rowe MT. Characteristics of cytotoxic necrotizing factor and cytolethal distending toxin producing *Escherichia coli* strains isolated from meat samples in Northern Ireland. *Food Microbiol* 2006;**23**:491–7.

76. Dolara P, Luceri C, De Filippo C, Femia AP, Giovannelli L, Caderni G, et al. Red wine polyphenols influence carcinogenesis, intestinal microflora, oxidative damage and gene expression profiles of colonic mucosa in F344 rats. *Mutat Res* 2005;**591**:237–46.

77. Zhao YS, Wang F, Chang D, Han B, You DY. Meta-analysis of different test indicators: *Helicobacter pylori* infection and the risk of colorectal cancer. *Int J Colorectal Dis* 2008;**23**:875–82.

78. Terres AM, Pajares JM, O'Toole D, Ahern S, Kelleher D. H pylori infection is associated with downregulation of E-cadherin, a molecule involved in epithelial cell adhesion and proliferation control. *J Clin Pathol* 1998;**51**:410–12.

79. Chan AO, Lam SK, Wong BC, Wong WM, Yuen MF, Yeung YH, et al. Promoter methylation of E-cadherin gene in gastric mucosa associated with *Helicobacter pylori* infection and in gastric cancer. *Gut* 2003;**52**:502–6.

80. Xia G, Schneider-Stock R, Diestel A, Habold C, Krueger S, Roessner A, et al. *Helicobacter pylori* regulates p21(WAF1) by histone H4 acetylation. *Biochem Biophys Res Commun* 2008;**369**:526–31.

81. Maggio-Price L, Treuting P, Zeng W, Tsang M, Bielefeldt-Ohmann H, Iritani BM. *Helicobacter* infection is required for inflammation and colon cancer in SMAD3-deficient mice. *Cancer Res* 2006;**66**:828–38.

82. Søreide K, Janssen EA, Søiland H, Kørner H, Baak JP. Microsatellite instability in colorectal cancer. *Br J Surg* 2006;**93**:395–406.

83. Søreide K, Slewa A, Stokkeland PJ, van Diermen B, Janssen EA, Soreide JA, et al. Microsatellite instability and DNA ploidy in colorectal cancer: potential implications for patients undergoing systematic surveillance after resection. *Cancer* 2009;**115**:271–82.

84. Søreide K, Nedrebø BS, Søreide JA, Slewa A, Kørner H. Lymph node harvest in colon cancer: influence of microsatellite instability and proximal tumor location. *World J Surg* 2009;**33**:2695–703.

85. Sinicrope FA. Sporadic colorectal cancer: an infectious disease? *Gastroenterology* 2007;**132**:797–801.

86. Weitz J, Koch M, Debus J, Hohler T, Galle PR, Buchler MW. Colorectal cancer. *Lancet* 2005;**365**:153–65.

87. Goel A, Nagasaka T, Arnold CN, Inoue T, Hamilton C, Niedzwiecki D, et al. The CpG island methylator phenotype and chromosomal instability are inversely correlated in sporadic colorectal cancer. *Gastroenterology* 2007;**132**:127–38.

88. Hill MJ, Drasar BS, Hawksworth G, Aries V, Crowther JS, Williams RE. Bacteria and aetiology of cancer of large bowel. *Lancet* 1971;**1**:95–100.

89. Zumkeller N, Brenner H, Chang-Claude J, Hoffmeister M, Nieters A, Rothenbacher D. *Helicobacter pylori* infection, interleukin-1 gene polymorphisms and the risk of colorectal cancer: evidence from a case-control study in Germany. *Eur J Cancer* 2007;**43**:1283–9.

90. Yang L, Pei Z. Bacteria, inflammation, and colon cancer. *World J Gastroenterol* 2006;**12**:6741–6.

91. Søreide K. Genetics and molecular classification of colorectal cancer. *Tidsskr Nor Laegeforen* 2007;**127**:2818–23.

92. Søreide K, Nedrebo BS, Reite A, Thorsen K, Korner H. Endoscopy, morphology, morphometry and molecular markers: predicting cancer risk in colorectal adenoma. *Expert Rev Mol Diagn* 2009;**9**:125–37.

93. Søreide K. Proteinase-activated receptor 2 (PAR-2) in gastrointestinal and pancreatic pathophysiology, inflammation and neoplasia. *Scand J Gastroenterol* 2008;**43**:902–9.

493

94. Perri F, Cotugno R, Piepoli A, Merla A, Quitadamo M, Gentile A, et al. Aberrant DNA methylation in non-neoplastic gastric mucosa of *H. pylori* infected patients and effect of eradication. *Am J Gastroenterol* 2007;**102**:1361–71.

95. Kabir S. Effect of *Helicobacter pylori* eradication on incidence of gastric cancer in human and animal models: underlying biochemical and molecular events. *Helicobacter* 2009;**14**:159–71.

96. Chan AO, Peng JZ, Lam SK, Lai KC, Yuen MF, Cheung HK, et al. Eradication of *Helicobacter pylori* infection reverses E-cadherin promoter hypermethylation. *Gut* 2006;**55**:463–8.

97. Leung WK, Man EP, Yu J, Go MY, To KF, Yamaoka R, et al. Effects of *Helicobacter pylori* eradication on methylation status of E-cadherin gene in noncancerous stomach. *Clin Cancer Res* 2006;**12**:3216–21.

98. Kabir S. The current status of *Helicobacter pylori* vaccines: a review. *Helicobacter* 2007;**12**:89–102.

99. Xiang S, Fruehauf J, Li CJ. Short hairpin RNA-expressing bacteria elicit RNA interference in mammals. *Nat Biotechnol* 2006;**24**:697–702.

100. van de Wetering M, Sancho E, Verweij C, de Lau W, Oving I, Hurlstone A, et al. The beta-catenin/TCF-4 complex imposes a crypt progenitor phenotype on colorectal cancer cells. *Cell* 2002;**111**:241–50.

101. Worthley DL, Le Leu RK, Whitehall VL, Conlon M, Christophersen C, Belobrajdic D, et al. A human, double-blind, placebo-controlled, crossover trial of prebiotic, probiotic, and synbiotic supplementation: effects on luminal, inflammatory, epigenetic, and epithelial biomarkers of colorectal cancer. *Am J Clin Nutr* 2009;**90**:578–86.

102. Fabbri A, Travaglione S, Falzano L, Fiorentini C. Bacterial protein toxins: current and potential clinical use. *Curr Med Chem* 2008;**15**:1116–25.

CHAPTER 30

DNA Methylation Profiles in the 5′-Upstream Region of the Human *FMR1* Promoter and in an Adenovirus Transgenome

Walter Doerfler[1,2], Anja Naumann[1], Norbert Hochstein[1], and Stefanie Weber[1]
[1]Institute for Virology, Erlangen University Medical School Erlangen
[2]Institute of Genetics, University of Cologne, Cologne, Germany

495

INTRODUCTION

With the realization that the primary nucleotide sequences of genes in many mammalian species are very similar and in many genes even identical, the need to understand regulatory principles in mammalian genomes has become one of the exciting fields of research in molecular genetics. In the late 1970s and early 1980s, our laboratory demonstrated that the sequence-specific methylation of mammalian promoter and upstream regions frequently leads to promoter inactivation [1–6]. Moreover, we have recognized that foreign DNA integrated into an established mammalian genome frequently becomes *de novo* methylated [7–9]. This latter conclusion has been amply documented by our work with integrated adenovirus type 12 genomes and by that of others with retroviral DNA and with retrotransposons [10]. Conceivably, the site of foreign DNA insertion is an important factor to determine the degree of *de novo* methylation of integrated foreign DNA [8,11].

The insertion of foreign DNA into established eukaryotic genomes has become an important technique in experimental biology and medicine with the aim to generate transgenic cells or organisms. In spite of the generality of the application of these procedures, little attention has been focused on their consequences for the recipient cell or organism. It is often implicitly assumed that foreign DNA insertion into a genome might not have consequences other than the ones aspired to by the experimenter. Among the actually existing sequelae, the *de novo* methylation of the insert and changes in DNA methylation patterns in the recipient genome at sites adjacent to [12–14] and remote from the integration locus [15,16] have been studied to some degree [Ref. 17, for review].

Since in many experimental strategies the success of foreign DNA insertion is seen in the sustained genetic activity of the transgene or transgenome, *de novo* methylation and

Handbook of Epigenetics: The New Molecular and Medical Genetics. DOI: 10.1016/B978-0-12-375709-8.00030-7

the subsequent inactivation of transgenes or transgenomes is completely adverse to the intended experimental goal. An interesting example from plant molecular biology was the introduction of the *Zea mays A1* gene construct into a *Petunia hybrida* mutant with white flowers due to a block in its anthocyanin pathway. The *A1* gene was expected to lead to the production of pelargonidin which would bestow a red color to the *petunia* flowers. However, depending on the copy number introduced, on the *de novo* methylation of the *cauliflower mosaic virus* promoter used in the maize gene constructs, and other unknown factors, the flowers remained white, variegated, and only in some instances assumed the expected brick-red color [18].

It will be prudent to investigate systematically to what extent the trans-genes in genetically-modified organisms (GMOs), e.g. the toxin gene from *Bacillus thuringensis* in *Zea mays*, remain unmethylated and active under the variety of culture conditions in different countries and agricultural regimens employed. We suspect that temperature, soil conditions, humidity, intensity of UV irradiation, and additional parameters might influence the occurrence of *de novo* methylation of transgenes and their promoters. Without an improved understanding of the basic properties of GMOs, their acceptance in the public will remain controversial.

At present, we do not understand the rules by which foreign, chromosomally-inserted DNA becomes modified by *de novo* DNA methylation. There are many conditions that influence the fate of integrated foreign DNA in mammalian genomes: nucleotide sequence and size of the inserted DNA, structure and the sequence at site of insertion, strength of promoters attached to the transgene, and conditions for the cultivation of transgenic cells or organisms. Basic research on the mechanisms of *de novo* methylation of integrated foreign DNA and on the more general consequences of foreign DNA integration into established eukaryotic genomes constitutes one of the challenging topics in molecular biology.

In this report, we have summarized our former and more recent work on *de novo* methylation:

1. We present a synopsis of earlier studies from this laboratory.
2. We have identified a sharp DNA methylation boundary at a site between 650 and 800 nucleotides upstream of the CGG repeat in the first exon of the human *FMR1* gene. This boundary is present in all human cell lines and cell types, irrespective of age, gender, and developmental stage and also in different mouse tissues. In individuals with the fragile X syndrome (FRAXA), the methylation boundary is lost; methylation spreads into the *FMR1* promoter and inactivates the *FMR1* gene.
3. We have determined the epigenetic status of an adenovirus type 12 genome integrated into the hamster genome in an Ad12-transformed hamster cell line [9].

SYNOPSIS OF EARLIER WORK ON DNA METHYLATION AND GENE SILENCING
The Ad12-Hamster Cell System

This synopsis will be limited to work from our laboratory [5,19–21]. The biological system we have used in many studies has been hamster cells transformed in cell culture by human adenovirus type 12 (Ad12) or Ad12-induced hamster tumor cells. When Ad12 is injected into newborn Syrian hamsters (*Mesocricetus auratus*), undifferentiated tumors arise at the site of subcutaneous injection [22,23] or in the peritoneal cavity after intramuscular injection of Ad12 [24]. In a study on the mode of Ad12 integration into the hamster cell genome, it was observed that the methylation-sensitive restriction endonuclease *HpaII* did not cleave the integrated Ad12 DNA, whereas the control enzyme *MspI* did cut the integrated Ad12 DNA [1,2]. Ad12 DNA isolated from purified virus particles or from productively or abortively infected cells, which had been shown to be devoid of 5-methylcytosine (5-mC) [2,25], was cleaved by either enzyme. These studies [1,7] were among the first to document the *de novo* methylation of foreign integrated DNA. Further studies on the structure of integrated Ad12 DNA are described in Refs 26–31.

A Symmetric Recombinant Between Ad12 and Cellular DNA

During the infection of permissive human cells with Ad12, a symmetric recombinant (SYREC) between the 2081 left terminal nucleotide pairs of Ad12 and a huge palindrome of cellular DNA has been generated spontaneously [32,33]. The Ad12 SYREC molecule with two left terminal Ad12 ends is not methylated, whereas the cellular DNA in the SYREC DNA molecules, when part of the authentic human genome, is extensively methylated. Obviously, in SYREC DNA, viral or cellular DNA as part of an adenovirus genome, that replicates in the nucleus, escapes DNA methylation, probably because free intranuclear Ad12 DNA is capable of replicating in structures that are – to some degree – independent of the cellular DNA replication machinery.

Long-Term Silencing of Promoters by DNA Methylation

In Ad12-transformed cells and in Ad12-induced hamster tumor cells, some of the early viral genes are transcribed, whereas the late viral genes are almost completely silenced [9,24,34]. In this system, we detected one of the first examples of an inverse correlation between promoter methylation and promoter and gene activity in mammalian cells [1–3]. The data were subsequently refined to demonstrate that the *in vitro* pre-methylation of adenoviral promoters led to their inactivation in a transfection and transient expression system in mammalian cells in culture or upon microinjection into frog oocytes [3–6]. In cells expressing the E1A region of adenoviruses the inactivating effect of promoter methylation was weakened or abrogated [35,36]. The E1A genes of adenovirus have been shown to be transactivators of viral and cellular genes [37,38].

These early and numerous additional experimental contributions from many different biological systems and laboratories have confirmed that the sequence-specific methylation of eukaryotic and viral promoters leads to their transcriptional inactivation. The inactivation can be mitigated by the action of transactivators or by the presence of strong enhancers, like the *HCMV* enhancer, in the vicinity of the promoters [39]. In addition, the modification of histones in the promoter region of genes also contributes to the regulation of promoter activity [40–42].

An Interesting Renegade – Frog Virus 3

In contrast to the DNA of human adenoviruses, the DNA of the *Iridovirus Frog Virus 3 (FV3)* inside the virion and in infected cells is completely methylated in all CpG residues [43]. Even a completely CpG-methylated *FV3* promoter was active in transfection and transient expression assays in both fish and mammalian cells [44]. The results from this viral system, which deserves much more attention by researchers interested in DNA methylation, call for caution not to generalize concepts on the significance of DNA methylation too quickly.

De Novo Methylation – An Ancient Cellular Defense System?

The data on the *de novo* methylation of integrated foreign DNA [1,7–9] have led to the concept that *de novo* methylation might constitute an ancient cellular defense mechanism directed against the activity of foreign genes that had intruded into an established genome and sought access to its replication, transcription, and evolutionary potential [10,45].

497

The Fate of Food-Ingested Foreign DNA in the Mouse Organism

In this context, we asked the question of what major sources of foreign, other than viral, DNAs existed that obtain access to living organisms. One obvious, hitherto not investigated, possibility was readily apparent – the food supply of living organisms which provided an unlimited repertoire of foreign DNA with a wide variety and different provenance of nucleotide sequences. In proof-of-principle experiments, we demonstrated that foreign DNA of different sources survived the passage through the gastrointestinal tract in mice in minute amounts and in fragmented form. These DNA fragments were transiently found in different parts of the intestinal canal, in liver, spleen, and peripheral mononuclear cells. There was no evidence for the transcription of these DNA fragments in the mouse organism or for the intrusion of the food-ingested DNA into the mouse germ line. In rare instances, the foreign DNA taken up through the gastrointestinal tract was found in spleen cells to be integrated into the host cellular DNA [46–49]. The following DNA molecules were used in different experiments: *bacteriophage phage M13* DNA, adenovirus type 2 DNA, the cloned gene for the green fluorescent protein (*GFP*), and the cloned, strictly plant-specific *RUBISCO* (Ribulose-1, 5-Bisphosphate Carboxylase) gene.

Integration of Foreign DNA Can Alter Cellular Methylation Patterns at Sites Remote From the Insertion Locus

We also demonstrated that foreign DNA insertion into mammalian genomes can alter DNA methylation patterns in the recipient genome at sites remote from foreign DNA insertion [15,16,50]. In the Ad12-transformed hamster cell line T637, with 10 to 15 copies of integrated Ad12 DNA molecules, patterns of DNA methylation were substantially altered in the IAP (intracisternal A particles) genomes, and retroviral transposons, with about 900 copies per cell. In these studies, both restriction endonuclease analyses using the methylation sensitive enzymes *Hpa*II (CCGG) or *Hha*I (GCGC) and the bisulfite genomic sequencing method [51,52] were applied. Alterations in DNA methylation in this transgenomic cell line were not confined to the multi-copy *IAP* transposon sequences but were also seen in a number of single copy genes. Moreover, in a revertant of cell line T637, TR3, that had lost all of the integrated Ad12 DNA copies, the changes in DNA methylation in the *IAP* sequences were conserved. Obviously, the insertion of foreign DNA as such had somehow led to alterations in cellular methylation patterns, and these changes persisted irrespective of the continued presence of the transgenomes [15]. We also observed alterations in *IAP* DNA methylation in a number of hamster cell clones that had been rendered transgenic for the genome of *bacteriophage lambda* [16]. *Lambda* DNA, similarly to Ad12 DNA, was integrated in multiple copies at one chromosomal site of the hamster cells and had become *de novo* methylated like any integrated foreign DNA [15,16]. As in Ad12 DNA integration, the site of *lambda* DNA integration was different from clonal cell line to clonal cell line.

We further pursue the possibility that the integration of foreign DNA into mammalian, possibly in many eukaryotic, genomes can lead to alterations in cellular DNA methylation at sites remote from the integration locus and, as a consequence, to changes in cellular transcription patterns. It will be important to find out how general these alterations of DNA methylation patterns are, and to what extent these alterations are dependent on the site of foreign DNA integration and on the size and nature of the transgenome or transgene. We have recently extended these studies to human cells which have been rendered transgenic for a plasmid construct with the kanamycin gene under the control of an *SV40* promoter. Interchromosomal contacts might be important in transmitting signals from a chromosome, which had been the target of foreign DNA insertion, to other parts of the recipient genome. In recent years, evidence has been accumulating that DNA methylation patterns are subject to alterations in development, disease, particularly tumor diseases and depend on environmental conditions including the insertion of foreign genomes [53–60].

498

METHYLATION BOUNDARY IN THE 5′-UPSTREAM REGION OF THE HUMAN *FMR1* GENE: RELEVANCE FOR THE FRAGILE X SYNDROME

The Fragile X Syndrome (FRAXA)

This syndrome [OMIM 300624] is characterized by a fragile site on chromosome Xq27.3, by mental retardation, attention deficit/hyperactivity disorder, macroorchidism after puberty, and facial and skeletal dysmorphisms. At the molecular level, the expansion of a CGG repeat in the 5′-untranslated part of the first exon of the *FMR1* (fragile X mental retardation) gene [OMIM 309550] and the hypermethylation of its promoter region inactivate the *FMR1* gene early in human development [61–68]. In rare patients, amplification of the CGG repeat without *FMR1* promoter methylation does not result in the FRAXA syndrome [69,70].

The cause for the CGG repeat expansions to <200 repeats (pre-mutations) or >200 repeats (full mutations) is still unknown. Stable secondary DNA structures such as hairpins, and triplex and quadruplex DNA in the repeat have been discussed as problems for the replication, repair, or recombination machineries in this DNA segment [71,72]. In addition, the presence of an origin of DNA replication in the vicinity of the repeat might contribute to its instability [73–75]. Earlier analyses of DNA methylation in the *FMR1* promoter concentrated on a few methylation-sensitive restriction sites [67,68] or to bisulfite sequencing in a small segment of the promoter [76,77]. Here, we describe a detailed analysis of the methylation profile in a 5500 bp segment of the *FMR1* promoter and 5′-upstream sequences in DNA from numerous human cell lines, normal primary human cells from different tissues, and in DNA from FRAXA individuals.

A Sharp Boundary of DNA Methylation

In a genome segment of about 5500 bp 5′-upstream of the human *FMR1* gene and 65 to 70 CpG pairs upstream of its CGG repeat, a distinct DNA methylation boundary has been identified (Ref. 78 and Fig. 30.1D, E). On its downstream-side, a short segment with a mosaic methylation pattern is observed. This methylation boundary is consistently present in DNA both from human males and females, regardless of age, and in human cell lines. In females, the 5′-upstream and promoter regions of the *FMR1* gene contain DNA sequences from both X chromosomes. Hence, we observed DNA from one X chromosome with a more heavily methylated DNA sequence and without the methylation boundary plus DNA from the other X allele, with a methylation boundary at a location practically identical to that of male X chromosomes (data not shown).

In unrelated promoters in the human genome previously analyzed in our laboratory (*RET* [OMIM 164761], *CGGBP1* [OMIM 603363], genes of the erythrocyte membrane [OMIM 605331]), a comparable methylation boundary has not been seen [79–81]. However, at the 5′-end of a CpG island of the *glutathione S-transferase* (*GSTP1*) gene [OMIM 134660], a 5′-$(ATAAA)_{19-24}$-3′ repeat coincides with a distinct border between a methylated and an unmethylated segment in several human tissues [82]. In prostate cancer [OMIM 176807], this boundary is obliterated; the promoter is methylated and silenced [82]. The upstream methylation boundary of the *FMR1* gene does not contain a 5′-$(ATAAA)_{19-24}$-3′ repeat, but carries a 5′-$(CCAAA)_6$-3′ repeat downstream of the actual boundary [78].

Methylation Boundary at Equivalent Site in the Mouse Genome

Bisulfite sequencing of the 5′-upstream segment in the mouse genome also reveals a methylation boundary at the site equivalent to that in the human genome [78], although the nucleotide sequences of both genomes are only 46.7% identical in this genome segment. This conservation of the methylation boundary across mammalian species renders its structural and functional importance very likely.

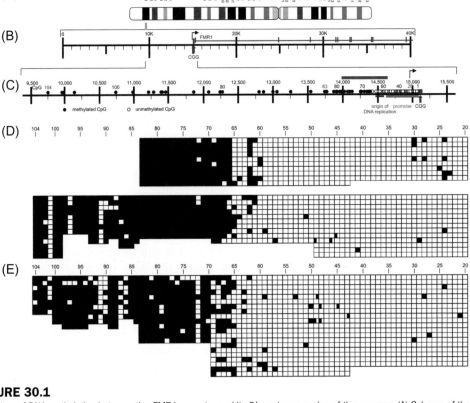

FIGURE 30.1
Boundary of DNA methylation between the *FMR1* promoter and its 5'-upstream region of the genome. (A) Scheme of the human X chromosome. (B) Map of the region 27.3 on the long arm of the X chromosome: *FMR1* gene, brown; CGG repeat blue, scale in 10,000 nucleotides (10 K). (C) Enlarged segment of the map in (B): Promoter, green. Arrow, site of initiation of transcription of the FMR1 gene; origin of DNA replication, gray. Each CG pair in the sequence as target for DNA methylation is marked by either ○ (unmethylated) or by ● (methylated). In (D) and (E) CG dinucleotides have been condensed (□ unmethylated; ■ methylated) in both DNA strands from the male human tumor cell line HCT116 (D) or for only one DNA strand from primary human fibroblasts (E). Numbers in (D) and (E) correspond to CG dinucleotide pairs counting upward from the CGG repeat (from Ref. 70). Each row of CG dinucleotide pairs corresponds to one DNA molecule, each column to the profile in one CG pair in different molecules. This figure was taken from Ref. 78. (Please refer to color plate section)

The Methylation Boundary is Lost in DNA From FRAXA Individuals

In FRAXA males, the boundary is completely lost, and almost all of the 88 CpG dinucleotides stretching down to the CGG repeat are methylated (Ref. 78 and Fig. 30.2A to C). In a genome segment far upstream in the boundary containing CpG dinucleotides 89 to 104, the degree of DNA methylation seems to be lower, particularly in the fibroblast sample GMO5848 (Fig. 30.2B), not however in DNA from PBMC sample 14,451 (Fig. 30.2C). This finding suggests a structural change of the methylation boundary in some FRAXA individuals.

Isolated Unmethylated CpG Dinucleotides in *De Novo* Methylated DNA

In FRAXA DNA, the completely *de novo* methylated promoter and 5'-upstream sequences of the *FMR1* gene contain isolated unmethylated CpGs (Fig. 30.2). Similar isolated unmethylated CpGs occur in the almost completely, also *de novo*, methylated adenovirus type 12 genome in a transformed hamster cell line (Ref. 9; see Fig. 30.4 below). Thus, mammalian and integrated adenoviral DNA segments with a high degree of *de novo* CpG methylation harbor one or several unmethylated CpG dinucleotides. Comparable

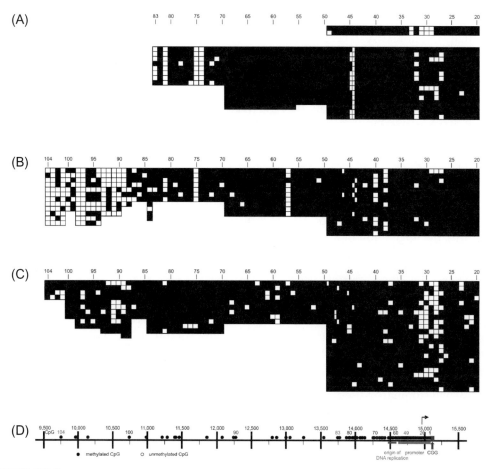

(A)

(B)

(C)

(D)

● methylated CpG   ○ unmethylated CpG

origin of   promoter CGG
DNA replication

FIGURE 30.2

Loss of the sharp DNA methylation boundary in the promoter region of individuals with the fragile X syndrome (FRAXA). Details and symbols are as described in the legend to Figure 30.1. (A,B) DNA from FRAXA fibroblasts; (C) DNA from FRAXA PMBCs. This figure was taken from Ref. 78.

isolated unmethylated CpGs are not apparent in *de novo* pre-methylated *FMR1* DNA when the bacterial M.Sss I (*Spiroplasma* species) DNA methyltransferase has been used (data not shown). Thus, the isolated unmethylated CpGs seem to arise only during *de novo* methylation in living mammalian cells and organisms [9,78].

Proteins Binding at the Boundary Sequence in the Human FMR1 5′-Upstream Region

We propose that the *FMR1* region carrying the methylation boundary is characterized by a specific chromatin structure and serves to demarcate the human *FMR1* and mouse *Fmr1* promoter regions from interspersed 5′-upstream located DNA sequences of as yet undetermined functions [78]. The analysis of specific DNA–protein complexes between the upstream *FMR1* boundary sequence and nuclear proteins from human cells might provide a guide to how this methylation boundary is preserved.

DNA in the transition zone from non-FRAXA individuals contains the methylated CpG dinucleotides 66 to 75 and the unmethylated CpGs 57 to 65. When the entire transition sequence of 630 bp was reacted with nuclear proteins, distinct DNA–protein complexes were observed both when unmethylated and the fully CpG M.Sss I methylated DNA fragment was incubated with nuclear protein extracts. The specificity of these complexes was ascertained by suitable competition experiments (for details see Ref. 78). We conclude that the DNA boundary sequence interacts specifically with as yet unidentified nuclear proteins from human cells. The methylation boundary is located in a functionally critical zone: a potentially unstable CGG repeat sequence, the promoter of a very important gene (*FMR1*), and an origin of DNA replication are in the vicinity of this boundary. Hence, this boundary deserves further studies.

EPIGENETIC PROFILE OF INTEGRATED AD12 DNA IN Ad12-TRANSFORMED HAMSTER CELLS

See Ref. 9.

De Novo Methylation of Integrated Foreign DNA

In molecular biology and medicine, in gene therapeutic experimental regimens of cells and organisms, in knock-in and knock-out experiments, and in the generation of transgenic organisms, the integration of foreign DNA into established eukaryotic genomes is an essential procedure. Other than the search for foreign DNA expression or the knock-out effect at the site of foreign DNA insertion, the consequences of integrating additional DNA into a genome have received limited attention. Among these sequelae, we have studied the *de novo* methylation of the integrate [1,7–9] and changes of DNA methylation in the recipient genome both at the site of insertion [14] and remote from it [15,16,50]. *De novo* methylation is, of course, of considerable interest beyond the context of integrated foreign genomes. During mammalian development, methylation patterns are erased and subsequently reestablished [57,83] by a mechanism akin to that of *de novo* methylation.

Here, we will summarize the results of an analysis of the mode of Ad12 DNA integration in the revertant cell line TR12. The Ad12-transformed hamster cell line T637 carries about 10 to 15 copies of integrated Ad12 DNA (Fig. 30.3A–C) which are all located at the same chromosomal location [9,31,84]. By the continuous cultivation of T637 cells, spontaneous revertants with an altered morphology arise [85]. One of these revertants, TR3, has lost all of the 10 to 15 copies of Ad12 DNA. In another revertant, TR12, only one copy of Ad12 DNA and a fragment of a second one (Fig. 30.3) persist in the cellular genome [9]. This revertant cell line appeared suitable for a precise analysis of the integration pattern and the epigenetic profile of the inserted Ad12 DNA.

Mode of Ad12 DNA Integration

In some of the Ad12-induced hamster tumor cell lines and in Ad12-transformed cell lines, Ad12 DNA is located in the vicinity of repetitive cellular DNA sequences, namely the *IAP* (internal A particle) retrotransposons [13]. Integration of Ad12 DNA into repetitive DNA might not be detrimental to cell survival, since essential cellular functions are less likely to be affected by this insertion, while sequences targeted to specific sites could lead to gene inactivation. A palindromic array is a likely arrangement both for the integrated Ad12 DNA and the recipient cellular DNA sequences in cell line T637 [9]. In the revertant cell line TR12, the cellular DNA at either terminus of the integrated Ad12 DNA represents a palindromic repeat DNA sequence of at least 5.2 kbp (Fig. 30.3, top panel). When cell line TR12 was generated from line T637 by the excision of a large part of the viral integrate, the abutting cellular DNA sequences apparently remained unaltered. The Ad12 transgenome is more extensively methylated in the revertant TR12 than in its parent T637 cell line [9]. Hypermethylated DNA, due to its altered degree of compaction, might have a better chance of resisting excision from the recipient genome.

The Epigenetic Profile of the Ad12 Integrates

The epigenetic status of integrated Ad12 DNA in hamster cell line TR12 has been determined by using the bisulfite sequencing technique [51,52]. Bisulfite sequencing reveals complete *de novo* methylation in most of the 1634 CpGs of the integrated viral DNA, except for its

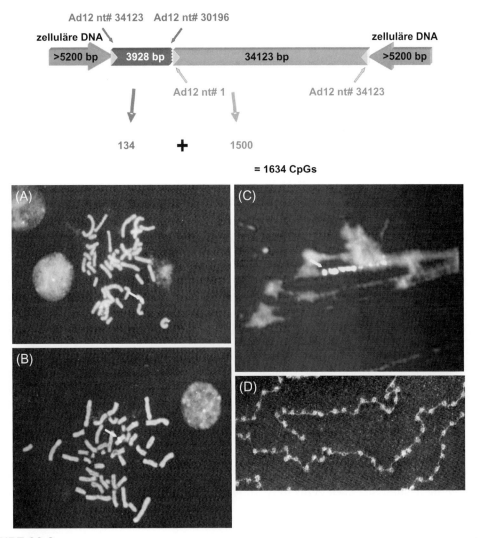

FIGURE 30.3

Integration of foreign, adenovirus type 12 (Ad12) DNA in the DNA of hamster cells. Upper panel: Scheme of an intact Ad12 DNA molecule (green) and of a fragment of a second Ad12 DNA molecule (red). Both genomes are integrated into hamster DNA (gray arrows). Lower panels: (A, B) Ad12 DNA integrated into one of the hamster chromosomes (red) are marked in yellow (arrows). (C) The chromosome in (cell line T637) (A) has been stretched and 10 to 15 copies of Ad12 DNA become apparent as individual yellow dots (arrow). In cell line TR12 (B) only one copy of Ad12 DNA and an Ad12 DNA fragment have remained integrated. The methylation profile of this integrate has been determined in Figure 30.4. (D) Nucleosome structure of an integrated Ad12 genome in hamster cells. Panels (A) and (B) were derived from Ref. 84, panel (C) from Ref. 31. (Please refer to color plate section)

termini (Fig. 30.4). Even in the most completely methylated Ad12 DNA segments in cell line TR12, isolated unmethylated CpGs have remained. The chromatin structure both of the cellular sites of Ad12 DNA integration and of the inserted viral genome(s) has been investigated. In the hamster cell lines T637 and TR12, the cellular DNA that abuts the integrated Ad12 DNA is more susceptible to micrococcal nuclease digestion than the same un-occupied site in non-transgenic hamster cells devoid of Ad12 genomes. Upon the insertion of viral DNA, the cellular chromatin structure likely becomes destabilized and opened up for the digestion with micrococcal nuclease. In addition, the micrococcal nuclease data [9] support the notion that the integrated Ad12 DNA has assumed nucleosome structure in cell lines T637 and TR12 (Fig. 30.3D).

FIGURE 30.4

Profile of DNA methylation in the Ad12 integrate in hamster cell line TR12. By using the bisulfite method, all 1634 CG dinucleotides in the Ad12 integrate in cell line TR12 (upper panel in Figure 30.3) have been interrogated for their methylation status [Ref. 9]. Symbols and presentation of methylation profiles have been explained in the legend to Figure 30.1. This figure was taken from Ref. 9.

Viral Gene Activities in the Integrates

The fully methylated Ad12 segments in cell lines T637 (not shown here) and TR12 are characterized by promoter silencing and histone H3 and H4 hypoacetylation. Hypermethylation of the E1A and E1B regions in the integrated Ad12 DNA in the revertant TR12 cells is associated with their transcriptional inactivation. Even under the control of the completely methylated major late promoter (MLP) in cell lines T637 and TR12, spurious amounts of Ad12 hexon, penton, and endoprotease transcripts are detected in cell lines T637 and TR12. Late Ad12 gene transcription from the hypermethylated MLP might be attributable to transient hemi-methylation of the MLP in the Ad12 integrates during DNA replication and cell division or to residual activities of fully methylated promoters detectable by the sensitive RT-PCR methods employed here [9].

The activity of the hypermethylated E1 genes in the Ad12 transgenome integrated in the revertant cell line TR12 has been reduced to a minimum, the E4 genes continue to be transcribed at about the same level as in the parent cell line, T637. The E4 sequence encodes up to seven different polypeptides. In adenovirus infections, the E1A functions activate the E4 promoter, which remains active also in the late phase of infection. E4 activation in the Ad12-transformed cell line TR12 seems to be under different control, since the Ad12 E1 genes are inactive in cell line TR12. The persistently active E4 functions may have played a role in stabilizing the residual Ad12 transgenome in cell line TR12. There is evidence that the E4 gene products from Ad12 have transforming potential [86]. Loss of the E1 genes from Ad12-induced tumor cells is compatible with the maintenance of the oncogenic phenotype in hamsters [23]. Thus, the continued transcription of the E4 functions might help in maintaining the transformed phenotype of cell line TR12 and keeping it viable in culture.

CONCLUSIONS

See Ref. 9.

1. Foreign (Ad12 viral) DNA is integrated into the hamster genome of the Ad12-transformed cell line T637 and of its revertant TR12 in an orientation co-linear with the virion genome. There is only one or a few nucleotides altered at the termini of the integrated Ad12 genome. The insert in cell line TR12 consists of one complete Ad12 genome and an additional 3.9 kb fragment of Ad12 DNA which has been derived from the right viral end and has been flip-flopped.

2. The recipient site in the cellular genome is an inverted repeat of cellular DNA. The sites of insertion in cell lines T637 and TR12 are identical.

3. The chromatin structure at the site of Ad12 DNA insertion has been decondensed as judged from the results of micrococcal nuclease analyses of the cellular integration site. The Ad12 integrate has assumed nucleosome structure.

4. The methylation status of all 1634 CpG dinucleotides in the trans-genome has been determined. Except for the two Ad12 E4 regions – one in the complete genome, a second one in the integrated 3.9 kb right-terminal fragment – most CpGs in the integrated Ad12 DNA are methylated.

5. There are, however, in a sea of methylated CpGs several isolated single CpGs which have remained unmethylated, possibly because these CpGs are part of binding sites for proteins. CpGs refractory to methylation in a background of hypermethylated DNA could be a characteristic of *de novo* methylated DNA. This finding parallels that of isolated similarly unmethylated CpG dinucleotides in the *de novo* methylated *FMR1* promoter in DNA from FRAXA patients (Fig. 30.2).

6. Transcription in the hypermethylated parts of the viral genome in cell lines T637 and TR12 has been silenced, and the H3 and H4 histones in these regions are hypoacetylated. The unmethylated E4 regions continue to be transcribed in cell lines T637 and TR12. There is an extremely low level of transcription even from the completely methylated Ad12 major late promoter. We interpret this low level of transient transcription to arise during the transiently hemi-methylated state of the MLP during DNA replication.

OUTLOOK

Our studies on the methylation profiles in an important regulatory region of the human genome, the promoter of the *FMR1* gene, and in an integrated adenoviral genome are considered a first step towards understanding the role of (epi)-genetic signals in the regulation of gene expression and the organization of genomes. The human epigenome project [87], which seeks to provide a complete map of all methylated CpGs, will help define new goals and initiate more precise functional approaches towards elucidating the significance of this genetic signal. With that information available some time in the future, however, researchers likely will find themselves again in a situation reminiscent of that after the human genome project was completed.

Viewing the complexity of genetic mechanisms, and with only 2% of the human nucleotide sequence coding for actual genes, we suggest that DNA can be regarded as a *molecule in search of additional functions*. There might be unknown codes hidden in DNA [5,88] that will be the challenge for future generations of geneticists. In this context, it will be interesting to recall that the physicists of the 1930s and 1940s were initially attracted to the study of biology because they surmised that biology might hold the clue to a *new type of elementary force.* Max Delbrück (1906–1981), a prominent member of this group of physicists, related to one of us (W.D.) in a discussion in the late 1970s [89] that the physicists then were disappointed to detect again "good old" physics and chemistry to be at work in biology. Perhaps the physicists with their speculations of those yonder years were on to something, and we might be well advised to at least think about their ideas of a new elementary force. The search for additional DNA coding principles [90] might initiate a novel and exciting research expedition.

ACKNOWLEDGEMENTS

We are indebted to the Institute for Virology, Erlangen University Medical School, Erlangen, Germany for their hospitality to W.D.'s senior investigator group. The experimental work summarized in this report was supported by amaxa GmbH, Cologne and by the Molecular Medicine Program of the Fritz Thyssen Foundation, Cologne. Earlier work on DNA methylation, then performed at the Institute of Genetics, University of Cologne, was aided by Sonderforschungsbereich 74 and 274 of the Deutsche Forschungsgemeinschaft in Bonn-Bad Godesberg, Germany, and by the Center for Molecular Medicine Cologne (CMMC) in Cologne, Germany.

References

1. Sutter D, Doerfler W. Methylation of integrated adenovirus type 12 DNA sequences in transformed cells is inversely correlated with viral gene expression. *Proc Natl Acad Sci USA* 1980;**77**:253–6.

2. Vardimon L, Neumann R, Kuhlmann I, Sutter D, Doerfler W. DNA methylation and viral gene expression in adenovirus-transformed and -infected cells. *Nucleic Acids Res* 1980;**8**:2461–73.

3. Vardimon L, Kressmann A, Cedar H, Maechler M, Doerfler W. Expression of a cloned adenovirus gene is inhibited by in vitro methylation. *Proc Natl Acad Sci USA* 1982;**79**:1073–7.

4. Kruczek I, Doerfler W. Expression of the chloramphenicol acetyltransferase gene in mammalian cells under the control of adenovirus type 12 promoters: effect of promoter methylation on gene expression. *Proc Natl Acad Sci USA* 1983;**80**:7586–90.

5. Doerfler W. DNA methylation and gene activity. *Annu Rev Biochem* 1983;**52**:93–124.

6. Langner K-D, Vardimon L, Renz D, Doerfler W. DNA methylation of three 5′ C-C-G-G 3′ sites in the promoter and 5′ region inactivates the E2a gene of adenovirus type 2. *Proc Natl Acad Sci USA* 1984;**81**:2950–4.

7. Sutter D, Westphal M, Doerfler W. Patterns of integration of viral DNA sequences in the genomes of adenovirus type 12-transformed hamster cells. *Cell* 1978;**14**:569–85.

8. Orend G, Knoblauch M, Kämmer C, Tjia ST, Schmitz B, Linkwitz A, et al. The initiation of de novo methylation of foreign DNA integrated into a mammalian genome is not exclusively targeted by nucleotide sequence. *J Virol* 1995;**69**:1226–42.

9. Hochstein N, Muiznieks I, Mangel L, Brondke H, Doerfler W. The epigenetic status of an adenovirus transgenome upon long-term cultivation in hamster cells. *J Virol* 2007;**81**:5349–61.

10. Walsh CP, Chaillet JR, Bestor TH. Transcription of IAP endogenous retroviruses is constrained by cytosine methylation. *Nat Genet* 1998;**20**:116–17.

11. Hertz J, Schell G, Doerfler W. Factors affecting *de novo* methylation of foreign DNA in mouse embryonic stem cells. *J Biol Chem* 1999;**274**:24232–40.

12. Jähner D, Jaenisch R. Retrovirus-induced de novo methylation of flanking host sequences correlates with gene inactivity. *Nature* 1985;**315**:594–7.

13. Lichtenberg U, Zock C, Doerfler W. Insertion of adenovirus type 12 DNA in the vicinity of an intracisternal A particle genome in Syrian hamster tumor cells. *J Virol* 1987;**61**:2719–26.

14. Lichtenberg U, Zock C, Doerfler W. Integration of foreign DNA into mammalian genome can be associated with hypomethylation at site of insertion. *Virus Res* 1988;**11**:335–42.

15. Heller H, Kämmer C, Wilgenbus P, Doerfler W. Chromosomal insertion of foreign (adenovirus type 12, plasmid, or bacteriophage λ) DNA is associated with enhanced methylation of cellular DNA segments. *Proc Natl Acad Sci USA* 1995;**92**:5515–19.

16. Remus R, Kämmer C, Heller H, Schmitz B, Schell G, Doerfler W. Insertion of foreign DNA into an established mammalian genome can alter the methylation of cellular DNA sequences. *J Virol* 1999;**73**:1010–22.

17. Doerfler W. *Foreign DNA in mammalian systems*. Weinheim, New York, Chichester, Brisbane, Singapore, Toronto: Wiley-VCH; 2000.

18. Linn F, Heidmann I, Saedler H, Meyer P. Epigenetic changes in the expression of the maize A1 gene in Petunia hybrida: role of numbers of integrated gene copies and state of methylation. *Mol Gen Genet* 1990;**222**:329–36.

19. Doerfler W. DNA methylation: *de novo* methylation, long-term promoter silencing, DNA methylation patterns and their changes. *Curr Top Microbiol Immunol* 2006;**301**:125–75.

20. Doerfler W. In pursuit of the first recognized epigenetic signal: DNA methylation. *Epigenetics* 2008;**3**:125–33.

21. Doerfler W. Epigenetic mechanisms in human adenovirus type 12 oncogenesis. *Semin Cancer Biol* 2009;**19**:136–43.

22. Trentin JJ, Yabe Y, Taylor G. The quest for human cancer viruses. *Science* 1962;**137**:835–41.

23. Kuhlmann I, Achten S, Rudolph R, Doerfler W. Tumor induction by human adenovirus type 12 in hamsters: loss of the viral genome from adenovirus type 12-induced tumor cells is compatible with tumor formation. *EMBO J* 1982;**1**:79–86.

24. Hohlweg U, Hösel M, Dorn A, Webb D, Hilger-Eversheim K, Remus R, et al. Intraperitoneal dissemination of Ad12-induced undifferentiated neuroectodermal hamster tumors: *de novo* methylation and transcription patterns of integrated viral and of cellular genes. *Virus Res* 2003;**98**:45–56.

25. Günthert U, Schweiger M, Stupp M, Doerfler W. DNA methylation in adenovirus, adenovirus-transformed cells, and host cells. *Proc Natl Acad Sci USA* 1976;**73**:3923–7.

26. Groneberg J, Chardonnet Y, Doerfler W. Integrated viral sequences in adenovirus type 12-transformed hamster cells. *Cell* 1977;**10**:101–11.

27. Doerfler W, Gahlmann R, Stabel S, Devring R, Lichtenberg U, Schulz M, et al. On the mechanism of recombination between adenoriral and cellular DNAs: the structure of junction sites. *Curr Top Microbiol Immunol* 1983;**109**:193–228.

28. Stabel S, Doerfler W, Friis RR. Integration sites of adenovirus type 12 DNA in transformed hamster cells and hamster tumor cells. *J Virol* 1980;**36**:22–40.

29. Knoblauch M, Schröer J, Schmitz B, Doerfler W. The structure of adenovirus type 12 DNA integration sites in the hamster cell genome. *J Virol* 1996;**70**:3788–96.

30. Deuring R, Winterhoff U, Tamanoi F, Stabel S, Doerfler W. Site of linkage between adenovirus type 12 and cell DNAs in hamster tumour line CLAC3. *Nature* 1981b;**293**:81–4.

31. Schröer J, Hölker I, Doerfler W. Adenovirus type 12 DNA firmly associates with mammalian chromosomes early after virus infection or after DNA transfer by the addition of DNA to the cell culture medium. *J Virol* 1997;**71**:7923–32.

32. Deuring R, Klotz G, Doerfler W. An unusual symmetric recombinant between adenovirus type 12 DNA and human cell DNA. *Proc Natl Acad Sci USA* 1981a;**78**:3142–6.

33. Deuring R, Doerfler W. Proof of recombination between viral and cellular genomes in human KB cells productively infected by adenovirus type 12: structure of the junction site in a symmetric recombinant (SYREC). *Gene* 1983;**26**:283–9.

34. Ortin J, Scheidtmann K-H, Greenberg R, Westphal M, Doerfler W. Transcription of the genome of adenovirus type 12. III. Maps of stable RNA from productively infected human cells and abortively infected and transformed hamster cells. *J Virol* 1976;**20**:355–72.

35. Langner K-D, Weyer U, Doerfler W. Trans effect of the E1 region of adenoviruses on the expression of a prokaryotic gene in mammalian cells: resistance to 5′-CCGG-3′ methylation. *Proc Natl Acad Sci USA* 1986;**83**:1598–602.

36. Weisshaar B, Langner K-D, Jüttermann R, Müller U, Zock C, Klimkait T, et al. Reactivation of the methylation-inactivated late E2A promoter of adenovirus type 2 by E1A (13S) functions. *J Mol Biol* 1988;**202**:255–70.

37. Jones N, Shenk T. An adenovirus type 5 early gene function regulates expression of other viral genes. *Proc Nat Acad Sci USA* 1979;**76**:3665–9.

38. Nevins JR. Mechanism of activation of early viral transcription by the early adenovirus E1A gene product. *Cell* 1981;**26**:213–20.

39. Knebel-Mörsdorf D, Achten S, Langner K-D, Rüger R, Fleckenstein B, Doerfler W. Reactivation of the methylation-inhibited late E2A promoter of adenovirus type 2 by a strong enhancer of human cytomegalovirus. *Virology* 1988;**166**:166–74.

40. Allfrey VG, Mirsky AE. Structural modifications of histones and their possible role in the regulation of RNA synthesis. *Science* 1964;**144**:559.

41. Jones PL, Wolffe AP. Relationships between chromatin organization and DNA methylation in determining gene expression. *Semin Cancer Biol* 1999;**9**:339–47.

42. Bernstein BE, Meissner A, Landers ES. The mammalian epigenome. *Cell* 2007;**128**:669–81.

43. Willis D, Granoff A. Frog virus 3 DNA is heavily methylated at CpG sequences. *Virology* 1980;**107**:250–7.

44. Munnes M, Schetter C, Hölker I, Doerfler W. A fully 5′-CG-3′ but not a 5′-CCGG-3′ methylated late frog virus 3 promoter retains activity. *J Virol* 1995;**69**:2240–7.

45. Doerfler W. Patterns of DNA methylation–evolutionary vestiges of foreign DNA inactivation as a host defense mechanism–A proposal. *Biol Chem Hoppe-Seyler* 1991;**372**:557–64.

46. Schubbert R, Lettmann C, Doerfler W. Ingested foreign (phage M13) DNA survives transiently in the gastrointestinal tract and enters the bloodstream of mice. *Mol Gen Genetics* 1994;**242**:495–504.

47. Schubbert R, Renz D, Schmitz B, Doerfler W. Foreign (M13) DNA ingested by mice reaches peripheral leukocytes, spleen and liver via the intestinal wall mucosa and can be covalently linked to mouse DNA. *Proc Natl Acad Sci USA* 1997;**94**:961–6.

48. Schubbert R, Hohlweg U, Doerfler W. On the fate of food-ingested foreign DNA in mice: chromosomal association and placental transmission to the fetus. *Mol Gen Genet* 1998;**259**:569–76.

49. Hohlweg U, Doerfler W. On the fate of plant or other foreign genes upon the uptake in food or after intramuscular injection in mice. *Mol Genet Genomics* 2001;**265**:225–33.

50. Müller K, Heller H, Doerfler W. Foreign DNA integration. Genome-wide perturbations of methylation and transcription in the recipient genomes. *J Biol Chem* 2001;**276**:14271–8.

51. Frommer M, McDonald LE, Millar DS, Collis CM, Watt F, Grigg GW, et al. A genomic sequencing protocol that yields a positive display of 5-methylcytosine residues in individual DNA strands. *Proc Natl Acad Sci USA* 1992;**89**:1827–31.

52. Clark SJ, Harrison J, Paul CL, Frommer M. High sensitivity mapping of methylated cytosines. *Nucleic Acids Res* 1994;**22**:2990–7.

53. Feinberg AP, Vogelstein B. Hypomethylation distinguishes genes of some human cancers from their normal counterparts. *Nature* 1983;**301**:89–92.

54. Heijmans BT, Tobi EW, Stein AD, Putter H, Blauw GJ, Susser ES, et al. Persistent epigenetic differences associated with prenatal exposure to famine in humans. *Proc Natl Acad Sci USA* 2008;**105**:17046–9.

55. McGowan PO, Sasaki A, D'Alessio AC, Dymov S, Labonté B, Szyf M, et al. Epigenetic regulation of the glucocorticoid receptor in human brain associates with childhood abuse. *Nat Neurosci* 2009;**12**:342–8.

56. Mill J, Tang T, Kaminsky Z, Khare T, Yazdanpanah S, Bouchard L, et al. Epigenomic profiling reveals DNA-methylation changes associated with major psychosis. *Am J Hum Genet* 2008;**82**:696–711.

57. Razin A, Shemer R. DNA. methylation in early development. *Hum Mol Geneti* 1995;**4**:1751–5.

58. Sharma S, Kelly TK, Jones PA. Epigenetics in cancer. *Carcinogenesis* 2010;**31**:27–36.

59. Szyf M, Weaver I, Meaney M. Maternal care, the epigenome and phenotypic differences in behavior. *Reprod Toxicol* 2007;**24**:9–19.

60. Timp W, Levchenko A, Feinberg AP. A new link between epigenetic progenitor lesions in cancer and the dynamics of signal transduction. *Cell Cycle* 2009;**8**:383–90.

61. O'Donnell WT, Warren ST. A decade of molecular studies of fragile X syndrome. *Annu Rev Neurosci* 2002;**25**:315–38.

62. Terracciano A, Chiurazzi P, Neri G. Fragile X syndrome. *Am J Med Genet C Semin Med Genet* 2005;**137C**:32–7.

63. Pieretti M, Zhang FP, Fu YH, Warren ST, Oostra BA, Caskey CT, et al. Absence of expression of the FMR-1 gene in fragile X syndrome. *Cell* 1991;**66**:817–22.

64. Fu YH, Kuhl DP, Pizzuti A, Pieretti M, Sutcliffe JS, Richards S, et al. Variation of the CGG repeat at the fragile X site results in genetic instability: resolution of the Sherman paradox. *Cell* 1991;**67**:1047–58.

65. Ashley Jr CT, Warren ST. Trinucleotide repeat expansion and human disease. *Annu Rev Genet* 1995;**29**:703–28.

66. Verkerk AJ, Pieretti M, Sutcliffe JS, Fu YH, Kuhl DP, Pizzuti A, et al. Identification of a gene (FMR-1) containing a CGG repeat coincident with a breakpoint cluster region exhibiting length variation in fragile X syndrome. *Cell* 1991;**65**:905–14.

67. Oberlé I, Rousseau F, Heitz D, Kretz C, Devys D, Hanauer A, et al. Instability of a 550-base pair DNA segment and abnormal methylation in fragile X syndrome. *Science* 1991;**252**:1097–102.

68. Hansen RS, Gartler SM, Scott CR, Chen SH, Laird CD. Methylation analysis of CGG sites in the CpG island of the human FMR1 gene. *Hum Mol Genet* 1992;**1**:571–8.

69. Smeets HJ, Smits AP, Verheij CE, Theelen JP, Willemsen R, van de Burgt I, et al. Normal phenotype in two brothers with a full FMR1 mutation. *Hum Mol Genet* 1995;**4**:2103–8.

70. Tabolacci E, Moscato U, Zalfa F, Bagni C, Chiurazzi P, Neri G. Epigenetic analysis reveals a euchromatic configuration in the FMR1 unmethylated full mutations. *Eur. J Hum Genet* 2008;**16**:1487–98.

71. Cleary JD, Nichol K, Wang YH, Pearson CE. Evidence of cis-acting factors in replication-mediated trinucleotide repeat instability in primate cells. *Nat Genet* 2002;**31**:37–46.

72. Nichol EK, Leonard MR, Pearson CE. Role of replication and CpG methylation in fragile X syndrome CGG deletions in primate cells. *Am J Hum Genet* 2005;**76**:302–11.

73. Brylawski BP, Chastain PD, Cohen SM, Cordeiro-Stone M, Kaufman DG. Mapping of an origin of DNA replication in the promoter of fragile X gene *FMR1*. *Exp Mol Pathol* 2007;**82**:190–6.

74. Gray SJ, Gerhardt J, Doerfler W, Small LE, Fanning E. An origin of DNA replication in the promoter region of the human fragile X mental retardation (*FMR1*) gene. *Mol Cell Biol* 2007;**27**:426–37.

75. Lucas I, Palakodeti A, Jiang Y, Young DJ, Jiang N, Fernald AA, et al. High-throughput mapping of origins of replication in human cells. *EMBO Rep* 2007;**8**:770–7.

76. Genç B, Müller-Hartmann H, Zeschnigk M, Deissler H, Schmitz B, Majewski F, et al. Methylation mosaicism of 5'-(CGG)$_n$-3' repeats in fragile X, premutation and normal individuals. *Nucleic Acids Res* 2000;**28**:2141–52.

77. Pietrobono R, Pomponi MG, Tabolacci E, Oostra B, Chiurazzi P, Neri G. Quantitative analysis of DNA demethylation and transcriptional reactivation of the *FMR1* gene in fragile X cells treated with 5-azadeoxycytidine. *Nucleic Acids Res* 2002;**30**:3278–85.

78. Naumann A, Hochstein N, Weber S, Fanning E, Doerfler. WA. Distinct DNA methylation border in the 5´-upstream sequence of the FMR1 promoter binds nuclear proteins and is lost in fragile X syndrome. *Am J Hum Genet* 2009;**86**:606–16.

79. Munnes M, Patrone G, Schmitz B, Romeo G, Doerfler W. A 5'-CG-3'-rich region in the promoter of the transcriptionally frequently silenced RET protooncogene lacks methylated cytidine residues. *Oncogene* 1998;**17**:2573–83.

80. Naumann F, Remus R, Schmitz B, Doerfler W. Gene structure and expression of the 5'-(CGG)$_n$-3'-binding protein (CGGBP1). *Genomics* 2004;**83**:106–18.

81. Remus R, Kanzaki A, Yawata A, Nakanishi H, Wada H, Sugihara T, et al. DNA methylation in promoter regions of red cell membrane protein genes in healthy individuals and patients with hereditary membrane disorders. *Int J Hematol* 2005;**81**:385–95.

82. Millar DS, Paul CL, Molloy PL, Clark SJ. A distinct sequence (ATAAA)$_n$ separates methylated and unmethylated domains at the 5'-end of the GSTP1 CpG island. *J Biol Chem* 2000;**275**:24893–9.

83. Frank D, Keshet I, Shani M, Levine A, Razin A, Cedar H. Demethylation of CpG islands in embryonic cells. *Nature* 1991;**351**:239–41.

84. Orend G, Knoblauch M, Doerfler W. Selective loss of unmethylated segments of integrated Ad12 genomes in revertants of the adenovirus type 12-transformed cell line T637. *Virus Res* 1995;**38**:261–7.

85. Groneberg J, Sutter D, Soboll H, Doerfler W. Morphological revertants of adenovirus type 12-transformed hamster cells. *J Gen Virol* 1978;**40**:635–45.

86. Shiroki K, Hashimoto S, Saito I, Fukui Y, Fukui Y, Kato H, et al. Expression of the E4 gene is required for establishment of soft-agar colony-forming rat cell lines transformed by the adenovirus 12 E1 gene. *J Virol* 1984;**50**:854–63.

87. Esteller M. The necessity of a human epigenome project. *Carcinogenesis* 2006;**27**:1121–5.

88. Doerfler W. In search of more complex genetic codes – can linguistics be a guide? *Med Hypotheses* 1982;563–79.

89. Doerfler W. Molecular Virology and Medical Genetics at the Institute of Genetics in Cologne, 1972 to 2002. In: Wenkel S, Deichmann U, editors. *Max Delbrück and Cologne: An early chapter of German molecular biology*. World Scientific Publishing Co. Pte. Ltd; 2007. pp. 159–77.

90. Doerfler W. DNA – a molecule in search of additional functions: recipient of *pool* wave emissions? – a hypothesis. *Med Hypotheses* 2010;**75**, Mar 29 [Epub ahead of print.].

CHAPTER 31

Population Pharmacoepigenomics

Jacob Peedicayil

Department of Pharmacology and Clinical Pharmacology, Christian Medical College, Vellore 632 002, India

INTRODUCTION

The extensive research work going on at present in epigenetics and epigenomics is impacting pharmacology, leading to new sub-disciplines in pharmacology, pharmacoepigenetics, and pharmacoepigenomics. Pharmacoepigenetics has been defined as the study of the epigenetic basis of variation in response to drugs [1]. Pharmacoepigenomics is the application of pharmacoepigenetics on a genome-wide basis. Together, these sub-disciplines of pharmacology involve the study of the role of epigenetics in the variations in response to drugs within and between individuals; in the effects of drugs on gene-expression profiles; in the mechanisms of action and adverse effects of drugs; and in the discovery of new drug targets [1,2]. It has been predicted that these sub-disciplines of pharmacology will have an ever-increasing role in pharmacology and clinical medicine [1,2].

Epigeneticsinvolves a number of molecular mechanisms including DNA methylation, histone modifications, small non-coding RNA-mediated regulation of gene expression, chromatin remodeling, and the use of histone variants [3]. Epigenetic mechanisms of gene expression are serving as new targets for the development of drugs, leading to the development of a new therapeutic option in pharmacology, epigenetic therapy, which involves the use of epigenetic drugs to treat or prevent disease [4,5]. Epigenetic mechanisms have been implicated in the pathogenesis of cancer for several years and, to date, most work on epigenetic therapy has focused on the use of epigenetic drugs in the management of cancer [6,7]. Population epigenetics and epigenomics, which address issues concerning the prevalence and importance of epigenetic variation within and between different populations, is a new and active area of research [8]. This chapter discusses some aspects of population pharmacoepigenomics, the study of how the epigenetic basis of variation in response to drugs varies within and between populations.

EPIGENETIC BIOMARKERS

Epigenetic changes may serve as biomarkers of disease and to date most work on epigenetic biomarkers in disease has focused on their use in the management of cancer [9,10]. At present DNA methylation markers appear to be the most promising epigenetic biomarkers in the management of patients with cancer [11–13]. DNA methylation markers are likely to be useful in the diagnosis and early detection of cancers, and in assessing prognosis of

511

Handbook of Epigenetics: The New Molecular and Medical Genetics. DOI: 10.1016/B978-0-12-375709-8.00031-9

TABLE 31.1 Racial/Ethnic Variations in DNA Methylation Patterns			
Diagnosis	**Phenotype Studied**	**Racial/Ethnic Difference**	**Reference**
Prostate Cancer	CD44 hypermethylation	43% in black men 25% in white men	[16]
Squamous cell carcinoma of lung (SCC)	5-methylcytosine immunostaining scores	Whites: 0.59 ± 0.06 in SCC 0.87 ± 0.07 in uninvolved bronchial mucosa African American: 0.55 ± 0.09 in SCC 0.60 ± 0.09 in uninvolved bronchial mucosa	[17]
Wilms' tumor	IGF2 loss of imprinting	31.7% of tumors in white children 0% of tumors in East-Asian children	[18]
Healthy women	p16 (INK4) tumor suppressor gene promoter hypermethylation	28% in African American 65% in European American	[19]
Colorectal cancer	hMLH 1 promoter hypermethylation	100% in Eskimos 42% in Aleut 50% in Native Americans	[20]

cancers, as well as in predicting response of cancers to treatment [14]. Although epigenetic biomarkers may not be a part of pharmacoepigenomics *per se*, in the broad sense they can be considered to be so, because they can assist in the pharmacological treatment of patients.

Epigenetic mechanisms are thought to play an important role in phenotypic differences in normal health and disease states between human races [15] and several studies suggest that DNA methylation patterns show racial and ethnic variations (16–20; Table 31.1). Hence, DNA methylation markers may also show racial and ethnic variations. Other epigenetic mechanisms in patients with cancer, like histone modifications [21] and microRNAs (miRNAs) [22], have also been shown to be of potential use as biomarkers in cancers. miRNAs may also show ethnic and racial variations in patients with tumors. For example, Wang and colleagues [23] examined the expression of 206 miRNAs in uterine leiomyomas in 15 black, 15 white, and 11 Asian/Hispanic women. A total of 31 miRNAs had expression levels that were significantly different among the three racial/ethnic groups. Hierarchical cluster analysis of the 31 miRNAs revealed a distinct expression profile between blacks and whites. Tumors from black women had a greater than 2-fold over-expression in miR-23 a/b, let-7s, miR-145, miR-197, miR-411, and miR-412 when compared to tumors from white women. The miRNA expression profile in tumors from Asian and Hispanic women was in between that of black and white women. At present, studies on histone modifications have not shown any clear ethnic and racial variations. The underlying mechanism of differences in epigenetic patterns between racial and ethnic groups is thought to be due to the differential epigenetic response of populations to varying environments [15].

Although epigenetic biomarkers have been most studied for their use in the management of cancer, there are many other clinical conditions where they may be useful like psychiatric disorders [24], obesity [25], and cardiovascular disease [26]. It is thought that the Human Epigenome Project, the ongoing international effort to map the human epigenome in various tissues in health and disease, is likely to greatly increase the number of available epigenetic biomarkers [27,28].

PATIENTS WITH DRUG-RESISTANT CANCER

Resistance of cancers to anticancer drugs is a common and important problem during the treatment of patients with cancer [29–31]. This phenomenon can be due to intrinsic

TABLE 31.2 Decrease of Resistance of Cancers to Conventional Anticancer Drugs by Combination with Epigenetic Drugs

Epigenetic Drug	Conventional Anticancer Drug	Diagnosis	Reference
Azacytidine	Cisplatin/Carboplatin/Temozolomide/Epirubicin	Ovarian/Colon cancer	[37]
Zebularine	Cisplatin	Ovarian cancer	[38]
Zebularine	Cisplatin	Oral squamous cell carcinoma	[39]
Trichostatin A	Etoposide	Small cell lung carcinoma	[40]
Valproate	Fludarabine, Cladribine	Chronic lymphocyte leukemia	[41]
Valproate	Pemetrexed/Cisplatin	Malignant mesothelioma	[42]

resistance of cancers before the start of treatment with anticancer drugs, or can be acquired during treatment [31]. Resistance of cancers to anticancer drugs is thought to be due to multiple types of genetic as well as epigenetic defects which act in concert to produce the drug-resistant phenotype [29–31]. Cellular mechanisms of resistance to anticancer drugs include the expression of energy-dependent transporters that detect and eject the drugs from cells, insensitivity to drug-induced apoptosis, alterations in membrane lipids, and induction of drug-detoxifying mechanisms [29]. An effective way of minimizing the chances of resistance of cancers to anticancer drugs is to concurrently use two or more drugs that act by different mechanisms [32–34].

Epigenetic drugs available at present have got their shortcomings such as the fact that they are nonspecific in their site of action, resulting in effects on many genes in different tissues, which could lead to adverse effects like mutagenicity and carcinogenicity [4–6]. Moreover, epigenetic defects, once corrected by the use of epigenetic drugs, may revert to the original state because of the reversible nature of epigenetic patterns [5,6]. Hence, it has been suggested that a combination of epigenetic drugs and conventional anticancer drugs may be ideal in the treatment of cancer [7,35,36], and such a combination may minimize the occurrence of drug resistance during the treatment of cancer. Several studies, in fact, have shown that epigenetic drugs, when combined with conventional anticancer drugs, decrease the development of resistance to conventional anticancer drugs in various types of cancers [37–42; Table 31.2].

NUTRITIONAL ASPECTS OF POPULATION PHARMACOEPIGENOMICS

An individual's nutrition can influence his or her epigenome causing changes in gene expression and modifications of health [43]. Nutritional epigenetics and epigenomics is a new area of research with most work done to date in this area having been on the effects on the body of the hematopoietic drugs, folic acid and vitamin B_{12} (cyanocobalamin), during the fetal and early postnatal periods of life [44,45]. These drugs are essential for the conversion of homocysteine to methionine [46]. For this conversion to occur, folic acid is metabolized in the body to methyltetrahydrofolate which acts as a methyl donor, and vitamin B_{12} acts as a cofactor. Methionine is metabolized to S-adenosyl-methionine, which is the primary methyl group donor for most biological methylation reactions, including that of DNA.

Abnormal amounts of intake of vitamin B_{12} and folic acid during prenatal and early postnatal periods have been implicated in the developmental origins of health and disease (DOHaD) hypothesis [47,48]. According to this hypothesis, during critical periods of prenatal and postnatal mammalian development, environmental factors, especially nutrition, influence developmental pathways and hence can induce permanent changes in metabolism and susceptibility to chronic disease. This hypothesis has been supported by a worldwide series of epidemiological studies which have provided evidence for a link between the perturbation of the early nutritional environment and health and disease status during adult life [47]. These data suggest that nutrient-based strategies will be of value in the prevention or delay of chronic diseases and that these strategies may require implementation during fetal life in order to achieve maximal benefit [49].

Nutritional influences on epigenetic mechanisms in early life were first suggested by studies of the viable yellow agouti (A^{vy}) metastable epiallele (a metastable epiallele is a locus that can be epigenetically modified in a variable and reversible manner, such that a distribution of phenotypes occurs from genetically-identical cells). The murine A^{vy} mutation resulted from transposition of a retrotransposon upstream of the agouti gene, the gene that regulates the production of yellow pigment in fur. Spontaneous variation in DNA methylation of the A^{vy} mutation causes dramatic variation in coat color among genetically identical A^{vy}/a mice [50]. Wolff et al. [51] found that supplementing the diet of mouse dams with the methyl donors and cofactors vitamin B_{12}, folic acid, betaine, and choline shifted the coat color distribution of their A^{vy}/a offspring from yellow to brown, suggesting hypermethylation at the A^{vy} mutation. It was later confirmed that maternal supplementation of mice with methyl donors and cofactors affects coat color of A^{vy}/a offspring by inducing hypermethylation at the A^{vy} mutation [52]. Hence, for the first time, it was shown that the effect of a mother's diet during pregnancy on the adult phenotype of her offspring was directly related to DNA methylation changes in the epigenome. This work marked the start of studies on the role of environmental effects on the epigenome during early development in relation to the pathogenesis of adult diseases [53].

More recently [54], it was shown in mature female sheep that restriction of the supply of vitamin B_{12}, folic acid and methionine within normal physiological doses during the periconceptual period had no effects on pregnancy establishment or birth weight of the offspring. However, this practice was found to result in adult offspring that were heavier and fatter, elicited altered immune responses to antigenic challenge, were insulin-resistant, and had elevated blood pressure–effects, especially in male offspring.

It was also found that there was altered DNA methylation in the livers of the sheep fetuses. In humans, epidemiological data has linked altered intake (ranging from constraint to abundance) of vitamin B_{12} and folic acid during fetal and early postnatal life with a number of clinical conditions such as cancer, Rett syndrome, cardiovascular disease, obesity, and type 2 diabetes [44,56,57]. Moreover, it has been shown that individuals who were prenatally exposed to famine during the Dutch Hunger Winter from 1944 to 1945, six decades later, had decreased DNA methylation of the imprinted gene *IGF2* compared to their unexposed, same-sex siblings [58]. One of the possible explanations that has been proposed for this finding is a deficiency of methyl donors in the diet during the famine [58].

EFFECTS OF AGE AND GENDER ON PHARMACOEPIGENOMICS

Aging is associated with epigenetic changes [59]. "Aging epigenetics" is an emerging discipline that promises exciting discoveries in the near future. Most epigenetic studies on aging to date have focused on DNA methylation. Epigenetic changes noted during aging include global DNA hypomethylation, promoter-specific DNA hypermethylation of a number of specific loci, and global histone acetylation [60].

Epigenetic changes are known to be associated with age-related diseases like cancer, type 2, diabetes and Alzheimer's disease [60]. These data have obvious implications for population pharmacoepigenomics since patient populations are likely to vary in response to drugs due to age-related differences in epigenetic patterns. Similarly, gender-specific diseases like diseases of the breast and the genitourinary tract which show epigenetic changes have obvious implications for population pharmacoepigenomics, since patient populations are likely to vary in response to drugs due to gender-related differences in epigenetic patterns. Diseases that are not gender-specific have also been found to show differences in DNA methylation [61] and histone modifications [62] between males and females, with implications for population pharmacoepigenomics.

OUTLOOK

Pharmacoepigenomics is a new sub-discipline in pharmacology dealing with the epigenetic basis of variation in response to drugs. At present, the knowledge of pharmacoepigenomics is limited with most work to date in this area having been done on DNA methylation. As knowledge advances, one can expect greater and better knowledge of pharmacoepigenomics, and how it varies within and between different populations.

References

1. Peedicayil J. Pharmacoepigenetics and pharmacoepigenomics. *Pharmacogenomics* 2008;**9**:1785–6.

2. Szyf M. Toward a discipline of pharmacoepigenomics. *Curr Pharmacogenomics* 2004;**2**:357–77.

3. Allis CD, Jenuwein T, Reinberg D. Overview and concepts. In: Allis CD, Jenuwein T, Reinberg D, editors. *Epigenetics*. New York: Cold Spring Harbor Laboratory Press; 2007, pp. 23–61.

4. Egger G, Liang G, Aparicio A, Jones PA. Epigenetics in human disease and prospects for epigenetic therapy. *Nature* 2004;**429**:457–63.

5. Peedicayil J. Epigenetic therapy–a new development in pharmacology. *Indian J Med Res* 2006;**123**:17–24.

6. Yoo CB, Jones PA. Epigenetic therapy of cancer: past, present and future. *Nat Rev Drug Discov* 2006;**5**:37–50.

7. Mai A, Altucci L. Epi-drugs to fight cancer: from chemistry to cancer treatment, the road ahead. *Int J Biochem Cell Biol* 2009;**41**:199–213.

8. Richards EJ. Population epigenetics. *Curr Opin Genet Dev* 2008;**18**:221–6.

9. Brena RM, Huang TH-M, Plass C. Quantitative assessment of DNA methylation: potential applications for disease diagnosis, classification, and prognosis in clinical settings. *J Mol Med* 2006;**84**:365–77.

10. Mulero-Navarro S, Esteller M. Epigenetic biomarkers for human cancer: the time is now. *Crit Rev Oncol Hematol* 2008;**68**:1–11.

11. Laird PW. The power and the promise of DNA methylation markers. *Nat Rev Cancer* 2003;**3**:253–66.

12. Paluszczak J, Baer-Dubowska W. Epigenetic diagnostics of cancer–the application of DNA methylation markers. *J Appl Genet* 2006;**47**:365–75.

13. Toyota M, Suzuki H, Yamashita T, Hirata K, Imai K, Tokino T, et al. Cancer epigenomics: implications of DNA methylation in personalized cancer therapy. *Cancer Sci* 2009;**100**:787–91.

14. McCabe MT, Brandes JC, Vertino PM. Cancer DNA methylation: molecular mechanisms and clinical implications. *Clin Cancer Res* 2009;**15**:3927–37.

15. Kuzawa CW, Sweet E. Epigenetics and the embodiment of race: developmental origins of US racial disparities in cardiovascular health. *Am J Hum Biol* 2009;**21**:2–15.

16. Woodson K, Hayes R, Wideroff L, Villaruz L, Tangrea J. Hypermethylation of *GSTP1*, *CD44*, and E-cadherin genes in prostate cancer among US blacks and whites. *Prostate* 2003;**55**:199–205.

17. Piyathilake CJ, Henao O, Frost AR, Macaluso M, Bell WC, Johanning GL, et al. Race- and age-dependent alterations in global methylation of DNA in squamous cell carcinoma of the lung (United States). *Cancer Causes Control* 2003;**14**:37–42.

18. Fukuzawa R, Breslow NE, Morison IM, Dwyer P, Kusafuka T, Kobayashi Y, et al. Epigenetic differences between Wilms' tumours in white and east-Asian children. *Lancet* 2004;**363**:446–51.

19. Dumitrescu RG, Marian C, Krishnan SS, Spear SL, Kallakury BV, Perry DJ, et al. Familial and racial determinants of tumor suppressor genes promoter hypermethylation in breast tissues from healthy women. *J Cell Mol Med* 2010. (in press).

20. Boardman LA, Lanier AP, French AJ, Schowalter KV, Burgart LJ, Koller KR, et al. Frequency of defective DNA mismatch repair in colorectal cancer among the Alaska native people. *Cancer Epidemiol Biomarkers Prev* 2007;**16**:2344–50.

21. Seligson DB, Horvath S, Shi T, Yu H, Tze S, Grunstein M, et al. Global histone modification patterns predict risk of prostate cancer recurrence. *Nature* 2005;**435**:1262–6.

22. Lu J, Getz G, Miska EA, Alvarez-Saavedra E, Lamb J, Peck D, et al. MicroRNA expression profiles classify human cancers. *Nature* 2005;**435**:834–8.

23. Wang T, Zhang X, Obijuru L, Laser J, Aris V, Lee P, et al. A micro-RNA signature associated with race, tumor size, and target gene activity in human uterine leiomyomas. *Genes Chromosomes Cancer* 2007;**46**:336–47.

24. Peedicayil J. Epigenetic biomarkers in psychiatric disorders. *Br J Pharmacol* 2008;**155**:795–6.

25. Campión J, Milagro FI, Martinez JA. Individuality and epigenetics in obesity. *Obesity Rev* 2009;**10**:383–92.

26. Stenvinkel P, Karimi M, Johansson S, Axelsson J, Suliman M, Lindholm B, et al. Impact of inflammation on epigenetic DNA methylation–a novel risk factor for cardiovascular disease? *J Intern Med* 2007;**261**:488–99.

27. Bradbury J. Human epigenome project–up and running. *Plos Biol* 2003;**1**:316–19.

28. American Association for Cancer Research Human Epigenome Task Force; European Union, Network of Excellence, Scientific Advisory Board. Moving AHEAD with an international human epigenome project. *Nature* 2008;**454**:711–15.

29. Gottesman MM. Mechanisms of cancer drug resistance. *Annu Rev Med* 2002;**53**:615–27.

30. Perez-Plasencia C, Duenas-Gonzalez A. Can the state of cancer chemotherapy resistance be reverted by epigenetic therapy? *Mol Cancer* 2006;**5**:27.

31. Humeniuk R, Mishra PJ, Bertino JR, Banerjee D. Molecular targets for epigenetic therapy of cancer. *Curr Pharm Biotechnol* 2009;**10**:161–5.

32. Sausville EA, Longo DL, et al. Principles of cancer treatment. In: Fauci. AS, Braunwald. E, Kasper. DL, Hauser. SL, Longo. DL, Jameson. JL, editors. *Harrison's principles of internal medicine*, Vol I. New York: McGraw Hill; 2008, pp. 514–33.

33. Brown R, Glasspool R. Epigenetic modulation of resistance to chemotherapy? *Ann Oncol* 2007;**18**:1429–30.

34. Epstein RJ, Leung TW. Reversing hepatocellular carcinoma progression by using networked biological therapies. *Clin Cancer Res* 2007;**13**:11–17.

35. Plimack ER, Stewart DJ, Issa J-PJ. Combining epigenetic and cytotoxic therapy in the treatment of solid tumors. *J Clin Oncol* 2007;**25**:4519–21.

36. Bots M, Johnstone RW. Rational combinations using HDAC inhibitors. *Clin Cancer Res* 2009;**15**:3970–7.

37. Plumb JA, Strathdee G, Sludden J, Kaye SB, Brown R. Reversal of drug resistance in human tumor xenografts by 2'-deoxy-5-azacytidine-induced demethylation of the *hMLH1*gene promoter. *Cancer Res* 2000;**60**:6039–44.

38. Balch C, Yan P, Craft T, Young S, Skalnik DG, Huang TH, et al. Antimitogenic and chemosensitizing effects of the methylation inhibitor zebularine in ovarian cancer. *Mol Cancer Ther* 2005;**4**:1505–14.

39. Suzuki M, Shinohara F, Nishimura K, Echigo S, Rikiishi H. Epigenetic regulation of chemosensitivity to 5-fluorouracil and cisplatin by zebularine in oral squamous cell carcinoma. *Int J Oncol* 2007;**31**:1449–56.

40. El-Khoury V, Breuzard G, Fourré N, Dufer J. The histone deacetylase inhibitor trichostatin A downregulates human MDR1 (*ABCB1*) gene expression by a transcription-dependent mechanism in a drug-resistant small cell lung carcinoma cell line model. *Br J Cancer* 2007;**97**:562–73.

41. Bouzar AB, Boxus M, Defoiche J, Berchem G, Macallan D, Pettengell R, et al. Valproate synergizes with purine nucleoside analogues to induce apoptosis of B-chronic lymphocyte leukaemia cells. *Br J Haematol* 2009;**144**:41–52.

42. Vandermeers F, Hubert P, Delvenne P, Mascaux C, Grigoriu B, Burny A, et al. Valproate, in combination with pemetrexed and cisplatin, provides additional efficacy to the treatment of malignant mesothelioma. *Clin Cancer Res* 2009;**15**:2818–28.

43. Cobiac L. Epigenomics and nutrition. *Forum Nutr* 2007;**60**:31–41.

44. Waterland RA, Jirtle RL. Early nutrition, epigenetic changes at transposons and imprinted genes, and enhanced susceptibility to adult chronic diseases. *Nutrition* 2004;**20**:63–8.

45. Mathers JC. Epigenomics: a basis for understanding individual differences? *Proc Nutr Soc* 2008;**67**:390–4.

46. Kaushansky K, Kipps TJ. Hematopoietic agents: Growth factors, minerals, and vitamins. In: Brunton LL, Lazo JS, Parker KL, editors. *The pharmacological basis of therapeutics*. New York: McGraw-Hill; 2006, pp. 1433–65.

47. McMillen IC, Robinson JS. Developmental origins of the metabolic syndrome: prediction, plasticity, and programming. *Physiol Rev* 2005;**85**:571–633.

48. Waterland RA, Michels KB. Epigenetic epidemiology of the developmental origins hypothesis. *Annu Rev Nutr* 2007;**27**:363–88.

49. Stover PJ, Garza C. Nutrition and developmental biology–implications for public health. *Nutr Rev* 2006;**64**:S60–71.

50. Morgan HD, Sutherland HG, Martin DI, Whitelaw E. Epigenetic inheritance at the agouti locus in the mouse. *Nat Genet* 1999;**23**:314–18.

51. Wolff GL, Kodell RL, Moore SR, Cooney CA. Maternal epigenetics and methyl supplements affect agouti gene expression in *Avy/a* mice. *FASEB J* 1998;**12**:949–57.

52. Waterland RA, Jirtle RL. Transposable elements: targets for early nutritional effects on epigenetic gene regulation. *Mol Cell Biol* 2003;**23**:5293–300.

53. Jirtle RL, Skinner MK. Environmental epigenomics and disease susceptibility. *Nat Rev Genet* 2007;**8**:253–62.

54. Sinclair KD, Allegrucci C, Singh R, Gardner DS, Sebastian S, Bispham J, et al. DNA methylation, insulin resistance, and blood pressure in offspring determined by maternal periconceptional B vitamin and methionine status. *Proc Natl Acad Sci USA* 2007;**104**:19351–6.

55. Burdge GC, Lillycrop KA, Jackson AA. Nutrition in early life, and risk of cancer and metabolic disease: alternative endings in an epigenetic tale? *Br J Nutr* 2009;**101**:619–30.

56. Zeisel SH. Epigenetic mechanisms for nutrition determinants of later health outcomes. *Am J Clin Nutr* 2009;**89**:1488S–93S.

57. Junien C, Nathanielsz P. Report on the IASO Stock Conference 2006: early and lifelong environmental epigenomic programming of metabolic syndrome, obesity and type II diabetes. *Obes Rev* 2007;**8**:487–502.

58. Heijmans BT, Tobi EW, Stein AD, Putter H, Blauw GJ, Susser ES, et al. Persistent epigenetic differences associated with prenatal exposure to famine in humans. *Proc Natl Acad Sci USA* 2008;**105**:17046–9.

59. Fraga MF, Esteller M. Epigenetics and aging: the targets and the marks. *Trends Genet* 2007;**23**:413–18.

60. Calvanese V, Lara E, Kahn A, Fraga MF. the role of epigenetics in aging and age-related diseases. *Ageing Res Rev* 2009;**8**:268–76.

61. Yu J, Zhang H, Gu J, Lin S, Li J, Lu W, et al. Methylation profiles of thirty four promoter-CpG islands and concordant methylation behaviours of sixteen genes that may contribute to carcinogenesis of astrocytoma. *BMC Cancer* 2004;**4**:65.

62. Wang A-G, Fang W, Han Y-H, Cho S-M, Choi JY, Lee KH, et al. Expression of the RERG gene is gender-dependent in hepatocellular carcinoma and regulated by histone deacetyltransferases. *J Korean Med Sci* 2006;**21**:891–6.

Epigenetics and Human Disease

CHAPTER 32

Cancer Epigenetics

Lee B. Riley and David W. Anderson
St. Luke's Hospital and Health Network, Bethlehem, PA 18015, USA

INTRODUCTION

Cancer is both a genetic and an epigenetic disease. However, because cancer represents a group of over three hundred specific diseases that share genetic, epigenetic, and pathologic features, it remains difficult to define. Currently, despite a vast array of information, physicians and scientists continue to rely on both pathologic and clinical features to define human cancers. Historically, in the early 1970s, three main theories addressed the origin of cancer [1,2]. One theory, based mostly on pathologic observations, posited that cancer was a disease of abnormal differentiation. This model held that malignant cells are the result of inappropriate epigenetic (non-mutational) changes that regulate the process of differentiation. A second model, advanced by virologists, argued that human cancers are caused by viruses similar to SV40 or the Rous sarcoma virus. The third model, supported by a strong correlation between the mutagenic and tumorigenic properties of chemical and physical agents, argued that cancer is a result of mutations. At that time the methods to test and promote this theory were limited. However, the discovery of the *v-src* gene in the oncogenic Rous sarcoma virus, followed by the discovery of the proto-oncogene *c-src* in human cells, focused the research efforts [3]. Shortly thereafter, several additional retroviruses were shown to carry oncogenes [4]. In 1982 a pivotal study solidified the mutational theory of cancer by demonstrating that a single point mutation in the *H-ras* oncogene is the causative event that converts the benign proto-oncogene into the oncogene *H-ras* [5]. The following year, Knudson's two-hit model for inactivation of tumor suppressor genes was clarified [6], and shortly thereafter numerous oncogenes and tumor suppressor genes were identified [7].

However, the activation of an oncogene and concomitant inactivation of a tumor suppressor gene does not account for all the alterations observed in pre-neoplastic tissues or in the spectrum of fully-metastatic cancers. Over the years further heritable changes in neoplastic cells were attributed to epigenetic mechanisms, including non-mutational activation and inactivation of oncogenes and tumor suppressor genes, respectively [8–10]. Additionally, in some circumstances epigenetic mechanisms are responsible for inducing downstream mutations in oncogenes [11–13]. Currently, epigenetic mechanisms appear to be responsible for a significant portion of the alterations in cancer, revitalizing one of the early theories of oncogenesis.

MECHANISMS OF EPIGENETIC CONTROL IN CANCER

Epigenetics represents the wide array of changes that regulate gene expression but that are not based on alterations in the primary base sequence of DNA. As such, epigenetic regulation

521

Handbook of Epigenetics: The New Molecular and Medical Genetics. DOI: 10.1016/B978-0-12-375709-8.00032-0

has a major role in the normal changes in gene expression required in embryogenesis and differentiation. Research regarding the mechanisms that effect these changes has focused predominantly on DNA methylation and modification of the histone proteins. In addition, several other mechanisms of epigenetic control including non-coding RNA and polycomb repressive complexes comprise additional important epigenetic mechanisms [14–16]. Each of these mechanisms has a significant role in the genesis and progression of cancer, though which mechanisms participate in different tumor histologies varies widely. These mechanisms are extensively reviewed in other chapters, but a brief summary of how they relate to cancer follows.

DNA Methylation

DNA methylation affects individual gene expression and, by interacting with nucleosomes that control DNA packaging, may also affect entire domains of DNA. In general, hypermethylation usually leads to silencing of gene activity. Methylation of DNA occurs on the cytosine in CpG dinucleotides. In normal DNA, approximately 4% of the cytosines is methylated [17]. In mammalian cells methylation is controlled by three main DNA methyltransferases. One of these, DNMT1, is the major maintenance enzyme that preserves existing methylation patterns following DNA replication by adding methyl groups to the hemi-methylated CpG sites [18]. DNMT3A and 3B methylate novel CpG sites are highly expressed during embryogenesis and minimally expressed in adult tissues [19]. Although DNMT3L lacks intrinsic methyltransferase activity, it interacts with DNMT3A and 3B to facilitate methylation of retrotransposons [20]. DNA methylation is not an irreversible step, and there are indications that under some conditions the methylated DNA pattern is a balance between active demethylation and re-methylation [21]. Demethylation in the absence of DNA replication normally occurs in the zygote and in the development of germ cells [22], but it also is active in malignant cells [23,24]. The molecular mechanisms for the demethylase activity remain unclear, though recent studies suggest that DNMT3A and 3B may be involved [25]. Therapeutic drugs that cause DNA demethylation currently are in use clinically and are discussed later in more detail [24,26,27].

Histone Modifications

In addition to the direct methylation of DNA, chromatin structure – influenced by various histone modifications – also plays an important role in gene expression and cancer [28]. Histone proteins coordinate the changes between tightly packaged DNA (heterochromatin), which is inaccessible to transcription, and exposed DNA (euchromatin), which is available for binding and regulation by various transcription factors. Two prominent mechanisms that regulate histone function are methylation and acetylation. Methylation of histone proteins is regulated by histone methyltransferases (HMTs) that transfer methyl groups from *S*-adenosylmethionine to the lysine or arginine residues in histones [9] and a family of demethylases that regulate histone demethylation [29]. Several HMTs have tumor suppressor functions (see section on "Epigenetic changes and malignant transformation" below) [9,30] and not surprisingly demethylases may have oncogenic activity [29,30]. Similarly, histone acetylation is regulated by histone acetyl-transferases (HATs) and deacetylases (HDACs) [31,32]. Additionally, other histone modifications including SUMOylation, ubiquitination, biotinylation, and poly-ADP ribosylation occur [33]. The balance of these chemical modifications and their effects on histone structure ultimately coordinate DNA exposure. This process is part of the "histone-code" theory, which posits that unique histone patterns guide recruitment of regulatory factors to the chromatin and result in conformational changes [34]. These structural changes allow or disallow transcription of single genes or of clusters of genes. Histone modification regulates many of the same pathways in oncogenesis as DNA methylation (see below).

Genomic Imprinting

Genomic imprinting is a phenomenon, controlled by DNA methylation, wherein unique genetic loci are expressed in a parent-specific manner; that is, only one allele is expressed. Only a small proportion of genes are imprinted (<1%), and the majority of imprinted genes are involved in the regulation of fetal and placental growth [35]. Interestingly, loss of imprinting (LOI), represented by biallelic expression or silencing of the imprinted allele, occurs in almost all tumor histologies. Wilms' tumors represent one of the best characterized scenarios of LOI, and data are mounting to suggest that LOI is the earliest lesion observed genetically or epigenetically in Wilms' tumors [36,37]. A fundamental role for LOI in oncogenesis was postulated by Feinberg, Ohlson and Henikoff, who suggested that LOI occurs in the stem cell population of a given organ/histology, ultimately leading to additional downstream genetic and epigenetic events [38].

Non-Coding RNA

The classic function of RNA is to translate DNA sequences into proteins; however, non-coding RNA (ncRNA) has numerous functions critical to cellular growth and maintenance [39]. Initially, ncRNAs were noted to have catalytic functions that facilitated RNA splicing [40] but later ncRNAs were noted to have a wide variety of functions, including: (i) regulation of DNA methylation [41,42]; (ii) chromatin remodeling [43]; (iii) regulation of transcription; and (iv) regulation of translation through microRNAs [44]. Recent studies have demonstrated that microRNAs play a major role in eukaryotic gene expression [45]. MicroRNAs have the potential to regulate global functions; a single microRNA can encode multiple mRNAs, each of which can then target numerous diverse pathways [45,46]. Although microRNAs appear to have a prominent role in development and differentiation [47], the observation that there are substantial differences in microRNAs between normal and malignant cells suggests that microRNAs play a significant role in cancer as well. The mechanisms by which microRNAs affect oncogenesis are not clearly defined. Typically, microRNAs are down-regulated during malignant transformation, which is consistent with their known role in the maintenance of the differentiated state [48]. However, their ability to regulate oncogenes and tumor suppressor genes also has been demonstrated [49]. In the future, understanding the role of microRNAs should significantly influence both the diagnosis and treatment of cancer.

EPIGENETICS AND ONCOGENESIS

Oncogenesis most likely occurs as a result of sequential heritable events. Early models of oncogenesis pointed to only a limited number of genetic alterations; however, most current models agree that oncogenesis is more complex than initially indicated and actually requires multiple events over many years. Additionally, given the observation that cancer is pathologically classified into over 300 different diseases, it is likely that oncogenesis occurs through a variety of different pathways. Consequently, the epigenetic alterations discussed below are arranged into loose pathologic groupings (Fig. 32.1) that can be characterized by routine histopathology as: (i) morphologically normal-appearing, though genetically or epigenetically abnormal; (ii) premalignant, which encompasses proliferative changes, dysplasia, atypical changes, and carcinoma *in situ*; (iii) malignant; and (iv) metastatic. This classification should not imply that a particular cancer needs to evolve through each successive group or that any particular genetic or epigenetic event that occurs at one stage in one cancer does not also occur in a different stage in a different cancer. For example, a mutation in a BRCA1 gene is an early event with no pathologic abnormities in familial breast cancer, but it also can be a late event with associated pathologic transformation in some sporadic breast cancers. This grouping simply hopes to accommodate the variability and plasticity observed in clinical cancers.

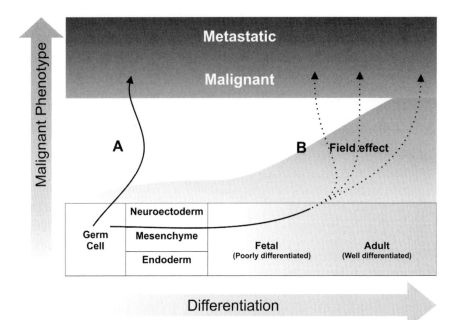

FIGURE 32.1

This figure depicts multiple potential pathways for oncogenesis as it mimics ontogeny. Cumulative epigenetic changes are represented by increasing shades of gray. Cell differentiation progresses from germ cells through fetal tissues and stem cells ultimately to become adult type with well-differentiated histology. Stem cells can follow this process early in embryogenesis or later as self-renewing stem cells. At any point along this normal process, a cell may acquire genetic and/or epigenetic alterations that initiate the transition to a malignant phenotype. If a malignant phenotype is achieved by a cell early in differentiation (pathway A), there may be few epigenetic changes (e.g. there is a rapid transformation to a malignant phenotype). Clinically, this is an uncommon event, with germ cell tumors comprising only about 1% of all cancers. In contrast, pathway B depicts how the vast majority (98–99%) of diagnosed human cancers acquire changes and progress to malignant phenotypes, whether they be poorly-differentiated (fetal-like with higher proliferation rates), well-differentiated (adult-like with lower proliferation rates), or of intermediate differentiation. In this scenario, cells may spend a long time evolving through a field effect. The field effect may be histologically indistinguishable from normal cells or may eventually manifest as an atypical, dysplastic or carcinoma *in situ* histology. Ultimately, accumulated genetic and epigenetic aberrations result in the evolution of a malignant cell of inversely varying proliferation rate and differentiation. In all scenarios there continue to be additional epigenetic changes as cells move from malignant to metastatic phenotypes.

Epigenetic Changes in "Normal" Stromal Cells of the Tumor Microenvironment

Although histopathology ultimately defines the malignant and premalignant phenotypes, advances in our ability to detect epigenetic changes have shown that routine histopathology cannot detect some of the earliest changes in oncogenesis. Rapidly accumulating evidence indicates that epigenetic changes are present in the normal-appearing tissue surrounding many tumors, including colon, breast, prostate, and lung cancer. It is likely that some of these changes represent a field effect that facilitates the ultimate development of cancer.

Numerous changes in promoter methylation are observed in the normal stroma surrounding primary cancers. In bladder cancer, approximately 70% of patients with transitional cell carcinoma develop recurrent disease. The recurrent tumors are thought to be of clonal origin since they have identical patterns of X chromosome inactivation [50] or identical p53 mutations [51]. While this result could be due to inadequate eradication of the primary tumor, it has been shown that histopathologically normal bladder mucosa from these patients also has similar loss of heterozygosity (LOH) patterns at microsatellite loci as well as similar methylation of the p16INK4 CpG island when compared to matched tumor specimens [52]. Similarly, in a series of 95 colon cancer patients and 33 normal subjects, Shen et al. [53] showed that 50% of the subset of patients with MGMT promoter methylation also exhibited MGMT promoter methylation in adjacent normal mucosa. Ramirez et al. [54] showed that the hMLH1 and MGMT genes are methylated in the normal mucosa surrounding primary colon cancers, and these methylated genes correlated with both microsatellite instability and K-ras activation in the primary tumors. In breast cancer, the RASSF1A promoter is frequently methylated, as is, though to a lesser degree, the normal tissue surrounding breast cancer; in contrast, methylation of this promoter is low in normal breast tissue from patients without cancer [55]. Likewise, the stromal cells that surround prostate cancer, while histologically normal, also have epigenetic modifications. The promoter of glutathione S-transferase P1 (GSTP1) is methylated in the tumor cells of more than 90% of prostate carcinomas [56] and is also methylated in tumor-associated stromal cells, but is not methylated in normal epithelium or normal stromal cells distant from the primary tumor [57]. The finding of an epigenetic field effect that precedes our ability to histologically detect these abnormalities has considerable implications for the diagnosis and treatment of cancer. Which epigenetic changes provide the earliest and most accurate diagnostic signal and which changes surgeons could use to monitor the margins of surgical resection will require further studies.

Premalignant Epigenetic Alterations

Pathologists have long recognized that for many tumor histologies there are non-malignant histologies, including proliferative changes, dysplasia, and atypia, that surround or are intertwined with malignant cells. Over the last several decades, the genetic characterization of some of these changes has greatly enhanced our understanding of oncogenesis. Accompanying these pathologic and genomic changes are alterations in numerous epigenetic pathways that eventuate in a multitude of epigenetic changes. Similar to epigenetic changes in histologically-normal cells, epigenetic changes that correlate with premalignant histologic features have been described for a variety of tumors, including carcinomas of esophageal [58], gastric [59], colon [59], breast [60,61] and lung origin [62]. Additionally, the mechanisms responsible for epigenetic changes associated with premalignant histologies include not only changes in promoter methylation [reviewed in Refs 21,63], but also histone modifications [64,65] and changes in microRNA expression [61,66].

While it is clear that some cancers can arise in the background of a "field-effect", it should be remembered that malignant cells also can influence their microenvironment by inducing histologic and epigenetic changes – potentially creating, rather than being a derivative of, the "field effect" [53–55,67,68]. It has been shown that normal human melanocytes suspended in a matrix preconditioned by metastatic melanoma cells assume an aggressive melanoma-like phenotype commensurate with increased migratory and invasive ability [69]. Expression profiling of the modified melanocytes showed similarities with the metastatic melanoma cells. Specifically, the melanocytes up-regulated genes associated with the malignant phenotype (VE-cadherin, VEGF-C, PAX8, kerain 7, CD13, laminin, urokinase, α3-integrin subunit, and c-met) [69]. Similar stromal-tumor interactions have been described in numerous histologies [reviewed in Refs 70,71]. Certainly there are several epigenetic changes associated with the premalignant cells; however, in the pathologic setting of premalignant cells interspersed with malignant cells, it remains difficult to discern which cells are responsible for a given event.

Epigenetic Changes and Malignant Transformation

Heritable epigenetic changes are involved in many of the pathways that lead to malignant transformation. The classic phenotype of a transformed cell is a combination of multiple features, including continual growth stimuli (e.g. oncogene activation), loss of checkpoint controls (e.g. silencing of tumor suppressor genes), and unlimited replication (e.g. prevention of telomere erosion). A detailed description of the epigenetic regulation of these events is beyond the scope of this section; consequently, only a brief description of key examples is provided.

Most oncogenes are activated by gene amplification or by mutation. However, there are some notable exceptions. While the majority of low-grade non-Hodgkin lymphomas have elevated expression of the bcl-2 gene as a result of translocation, elevated levels of bcl-2 in B-cell chronic lymphocytic leukemia appears to be the result of demethylation of both copies of bcl-2 [8]. Similarly, the paired-box (PAX) genes, which participate in epithelial proliferation and regulation of apoptosis [72], are expressed in a variety of tumors, including endometrial, breast, and ovarian cancer [73]. Wu et al. demonstrated in endometrial carcinomas that the pax2 gene, which is silent in normal tissues, is activated by estrogen and tamoxifen through hypomethylation of the pax2 promoter [74].

In contrast to oncogene activation, epigenetic mechanisms may play a more prominent role in the inactivation of tumor suppressor genes. Histone methyltransferases (HMT) can function as tumor suppressor genes [9,30]. The prototypical HMT, RIZ1 was originally identified bound to the tumor suppressor gene Rb [75]. However, RIZ1 also has intrinsic tumor suppressor activity, as evidenced by tumor formation in RIZ1 knock-out mice [76]. Inactivation of RIZ1 by hypermethylation occurs in many cancers, including the majority of thyroid, hepatocellular, and esophageal cancers [30,77]. Additional tumor suppressor genes (or putative tumor suppressor genes) also are silenced by hypermethylation, including Hint1 in hepatocellular cancers [78], RASSF1 in small-cell carcinoma of the urinary bladder [79], SLC5A8 in prostate cancer [80], and BRCA1 in breast cancer [81]. Additionally, some microRNAs have putative tumor suppressor activity and influence oncogenic target genes, such as C-MYC, E2F3, CDK6, and TGIF2. Several of these microRNAs are down-regulated by hypermethylation in tumor cells [14,82].

One prominent mechanism that controls cellular immortality is telomere erosion. With successive cellular division the telomeric ends of the chromosome progressively shorten, eventually leading to chromosomal fusion, karyotypic instability, and, ultimately, cell death [83]. The majority of invasive cancers circumvent telomere erosion by activating telomerase, an enzyme that maintains telomere length (hTERT) [2,84,85]. In most normal cells hTERT is silenced through deacetylation of the histone proteins that interact with the hTERT promoter. Reactivation of hTERT in cancer cells also occurs through complex epigenetic control involving hypo- and hypermethylation of the hTERT promoter as well as a variety of histone modifications [86–88].

Epigenetic Changes Associated with the Metastatic Phenotype

The metastatic phenotype is clearly regulated, in part, by epigenetic mechanisms. Metastatic behavior involves a multitude of biochemical pathways that facilitate invasion, migration, and angiogenesis. Epigenetic control of these events in malignant cells has been demonstrated globally by comparing CpG island hypermethylation patterns in selected genes between paired primary and metastatic cancers. Cavalli et al. evaluated a select set of genes (TP 16, THBS2, E-Cadherin, RARβ2, MINT1, MINT2, and MINT31) between primary breast cancers and their corresponding lymph node metastasis and showed differences in 29% of the methylation patterns [89]. Similar observations, though with different genes, have been reported for prostate and gastric cancers, as well [90,91].

A more specific way to demonstrate epigenetic control of the malignant phenotype is to focus on genes/proteins that participate in metastasis formation. The protease urokinase-type plasminogen activator (UPA) facilitates tissue invasion and is associated with a poor prognosis in both breast and prostate cancers [92,93]. Hypomethylation of the UPA promoter correlates with poor outcome in patients with breast [94] and prostate cancer [95]. Angiogenesis significantly contributes to the survival of cancer metastases, and the tissue inhibitor of metalloproteinases-3 (TIMP3) binds to the vascular endothelial growth factor-2 and inhibits angiogenesis. Hypermethylation of the TIMP3 promoter results in angiogenesis [96].

From a global perspective, metastatic cancer cells share a variety of features with normal, multi-step processes that occur during embryogenesis and differentiation, including migration and angiogenesis. Additionally, the invasive phenotype demonstrated in malignant cells also shares features with those of normal processes involved in early embryogenesis [97]. Specifically, during embryogenesis immobilized epithelial cells transition into cells with mesenchymal features, including motility and invasiveness. This regulated process is termed the "epithelial–mesenchymal transition" (EMT) [97]. Over the last several years, exciting work has demonstrated that both angiogenesis and EMT are globally regulated by microRNAs in both normal tissues and cancer [98–100]. Hopefully, in the near future the potential therapeutic advantages for controlling these pathways in cancer metastasis will be achieved.

EPIGENETIC DIAGNOSIS AND PROGNOSIS

Diagnosis of human cancer by the presence of epigenetic aberrations alone has yet to enjoy widespread acceptance in routine clinical care. Recent excellent reviews have addressed the diagnostic promise of epigenetic tests, including DNA methylation profiles [101], microRNA expression profiles [102], and histone modification profiling [67]. However, researchers are still in the validation phase of how best to integrate epigenetic test results into standard diagnostic practice. Early success has been achieved when the epigenetic test is employed as an aid to standard biopsy procedures, as in the case of the microRNA PCA3 [103]. This is a more sensitive and specific screening test (in urine) of prostate cancer than serum prostate specific antigen (PSA). When normalized to PSA mRNA copy numbers, PCA3 microRNA in prostate cancer tissue is up-regulated 34-fold (median value) relative to benign prostate tissue, allowing for efficient detection of cancer cells shed into urine. The differential signal strength of this epigenetic biomarker in lesional tissue is sufficient to reliably predict the presence of cancer when compared to non-lesional tissue from the same patient.

Epigenetic tests of several classes have the potential to redefine cancer types as well as to aid in their detection and diagnosis. However, epigenetic tests are several years behind protein-coding genomic RNA expression profiling [104] in their evolution to widespread clinical use. Breast cancer is an example. Diagnostic use of epigenetic biomarker profiles to predict tumor histotype or biologic behavior (e.g. grade) would be an important first step in this process. Profiles suggested today would employ between 40 and several hundred epigenetic biomarkers. The relative amounts of each biomarker are measured by chip-based hybridization technology and require cluster analysis to produce a similarity score. Examples for each class of epigenetic test are given below.

DNA methylation detection generally targets hypermethylation of CpG islands in promoter regions of silenced tumor suppressor genes (TSG). Of epigenetic tests, this class currently enjoys the most widespread application in translational research. Thus, diagnostic tests have been proposed that target a single methylated gene, such as glutathione S-transferase p1 (GSTP1) detected in urine of patients with prostate cancer [105]. Methylation profiles of multiple genes also offer promise, since multiple TSGs are rarely found to be silenced by promoter hypermethylation in non-cancerous human tissue [106]. Thus, TSG methylation profiling by nested methylation sensitive PCR (MS-PCR) methods has demonstrated accurate detection of lung, cervical, thyroid, and breast cancers. Although the specificity of these methods appears encouraging, the sensitivity is hampered by low-circulating copy number and potentially-interfering substances such as anticoagulants and uniform sets of positive and negative controls [101].

527

MicroRNA expression profiles [102] target up-regulated transcripts such as miR-127, a translational repressor of BCL-6 gene transcription, thus down-regulating this anti-apoptotic factor in lymphoma and in other hematological cancer. Expression profiles measuring 42 well-characterized microRNAs in tumors with non-diagnostic histology (tumors of unknown primary site) were found more effective at predicting the correct diagnosis than was protein-coding genomic RNA analysis [48].

Histone modification profiles target covalent modification of certain histone amino acids (lysine, arginine, serine), often found in the histone tail, and include phosphorylation, acetylation, methylation, ubiquitination, and sumoylation. Histone modification of specific genes is known to contribute to transcriptional activation (histone acetylation) or suppression (some histone methylation) of tumor suppressor genes as well as to affect DNA repair and chromosome organization. Gene-specific profiling of histones via ChIP on chip technology and mass spectroscopy are both highly specialized research techniques which await technologic innovation to become candidates for widespread clinical application.

Prognostic utility of global epigenetic aberrations is the focus of considerable recent activity in the literature, with particular focus on global hypomethylation and global histone modification. In general, the genetic instability within a tumor cell population is thought to reflect genome-wide epigenomic changes and is thus linked to prognosis [107,108]. Global hypomethylation has long been known to characterize human cancers but has not been as specific a predictor of cancer prognosis as are certain global histone modifications. Thus, certain global histone modifications determined by tumor immunohistochemistry are more readily measured in large specimen studies than are gene-specific histone modifications, which require chromatin immunoprecipitation. Recent studies [109–111] of global histone dimethylation (e.g. H3K4me2), trimethylation (e.g. H4K20me3), and acetylation (e.g. H2AK5ac and H3K18ac) have shown prognostic potential in invasive cancers (e.g. prostate, colon, breast, lung, and hematologic cancer). Conversely, loss of histone trimethylation in H4K20me3 was observed both in preneoplastic lesions and as an adverse outcome predictor in squamous cell carcinoma of lung [112].

The acceptance by Medicare and third party payers for use of epigenetic tests of cancer in routine clinical practice could follow a course similar to that of protein-coding genomic tests. By guiding personalized cancer management to avoid costs associated with chemotherapy choices of unlikely benefit, insurer acceptance can be justified, hopefully resulting in improved outcomes. Here, an important precedent has been set by various oncogene tests. Breast cancer cell expression of Her-2/neu by immunohistochemistry or fluorescence *in situ* hybridization (FISH), mRNA expression profiling in breast cancer by Oncotype DX (Genomic Health, Inc), and k-RAS mutational analysis in colon cancer by laboratory developed tests (LDTs) in specialty reference labs all enjoy national endorsement by NCCN Guidelines for cancer management. For many of these tests, large scale clinical trial validation has been documented and nationwide standardized criteria for quality control and quality assurance are already in place (e.g. Her-2/neu, k-RAS). A similar evolution awaits diagnostic epigenetic tests.

EPIGENETIC THERAPY OF CANCER

In comparison to normal cells, the genome from cancer cells is characterized by a decrease in histone modifications (e.g. acetylation) [32] and hypomethylation of CpG sites [113]. Pharmacologically correcting these potentially reversible changes is a rational therapeutic strategy. Specifically, inhibitors of DNMTs or HDACs could reactivate methylated or acetylated tumor suppressor genes and restore normal cell-cycle checkpoints.

Methylation of different genes in the hematologic malignancies correlates with worse prognosis. For example, methylation of p15-INK4B portends a worse prognosis in myelodysplastic syndromes (MDS) [114], acute myelogeneous leukemia (AML) [115], and high grade lymphomas [116]. Based on these and other findings, the hematologic malignancies provided the proving ground for the two FDA approved DNMT inhibitors: 5-azacytidine and 5-aza-2′deoxycytidine. In 2004, 5-azacytidine was approved for the treatment of both the low- and high-risk subtypes of myelodysplastic syndromes (MDS). In 2006, 5-aza-2′deoxycytidine, which is 10-fold more potent than 5-azacytidine, was approved for the treatment of MDS in previously treated and untreated, *de novo* and secondary MDS of all French-American-British subtypes. Unfortunately, while these drugs produce reasonable clinical responses in the hematologic malignancies, they have had limited benefit in solid tumors [117].

Post-translational acetylation of histone proteins facilitates transcription by opening the regional chromatin structures, which then allows access of regulatory proteins to the underlying DNA. In some cancers, genes that control cell-cycle checkpoints, apoptosis, or differentiation are silenced, thereby promoting the malignant phenotype. One potential therapeutic strategy has been to reactivate these suppressed genes by inhibiting the histone deacetylase enzymes. Re-expression of these genes was shown in culture for vorinostat and for several other histone deacetylase inhibitors (HDACi) [118]. Subsequent clinical studies demonstrated a clinical benefit of vorinostat, and in 2006 the FDA approved it for the treatment of refractory cutaneous T-cell lymphoma [119]. In addition to HDACi regulating histone proteins, a variety of non-histone proteins with diverse functions also are acetylated. Consequently, HDACi have multiple functions including deregulation of heat-shock proteins [120], alterations in microtubules [121], and NF-κB mediated transcriptional activation [122]. These associated functions of HDACi may account for some of the toxicities associated with higher doses but may also be responsible for the clinical efficacy of these agents.

MicroRNAs have a unique role in the anti-cancer armamentarium. Not only do microRNAs regulate many biologic processes, such as differentiation, proliferation, and apoptosis, but they also function as tumor suppressors and oncogenes [10]. The development of pharmacologic agents to modulate microRNAs is attractive because specific oligonucleotides can be synthesized and modified to specifically target a given microRNAs species [123]. Another potential benefit is that modulation of a single microRNAs species could simultaneously control multiple genes in a given pathway. Several experimentally-tested approaches to microRNA therapy have shown promise. One approach is to inhibit microRNAs that promote cancer growth and/or metastases. Liang et al. transfected breast cancer cells with a vector designed to block microRNA-155 in order to diminish CXCR4 levels [124]. The transfected cells exhibited reduced migration and invasion *in vitro* and formed fewer lung metastases *in vivo* compared to controls. A second approach is to replace microRNAs that are deficient in cancers. Murine hepatocellular carcinoma cells have low levels of MicroRNA-26a in comparison to most normal tissues. MicroRNA-26a induces cell-cycle arrest by targeting the cyclins D2 and E2. Kota et al. demonstrated that systemic administration of microRNAs-26a via an adeno-associated virus induced tumor-specific apoptosis and ablated tumor progression without toxicity [125]. The therapeutic potential for microRNAs-mediated therapy is applicable to a wide variety of tumor histologies and stages of disease and more clinical trials should be forthcoming.

CONCLUSION

The diverse field of epigenetics has rapidly added to our understanding of oncogenesis and has facilitated the development of novel therapeutic drugs, and will potentially generate new methods to prevent, diagnose, and treat cancers. Superimposing the field of epigenetics on the underpinnings of oncogenes and tumor suppressor genes has revealed a molecular framework that helps explain the observed pathologic transition from the normal to the malignant phenotype. The more familiar regulatory mechanisms like DNA-methylation and histone-acetylation in conjunction with additional pathways like microRNAs provide a diverse array of novel targets for therapeutic intervention. Additionally, because many of these changes are present prior to the malignant transformation, investigators may have the ability to create drugs or agents to diagnose or even prevent early cancers.

References

1. Potter VR. Initiation and promotion in cancer formation: the importance of studies on intercellular communication. *Yale J Biol Med* 1980;**53**:367–84.

2. Weinberg RA. Cancer: a genetic disorder. In: Mendelsohn J, Howley P, Israel M, Gray J, Thompson C, editors. *The molecular basis of cancer*. Philadelphia: Saunders Elsevier; 2008, pp. 3–16.

3. Stehelin D, Varmus HE, Bishop JM, Vogt PK. DNA related to the transforming gene(s) of avian sarcoma viruses is present in normal avian DNA. *Nature* 1976;**260**:170–3.

4. Bishop JM. Cellular oncogenes and retroviruses. *Annu Rev Biochem* 1983;**52**:301–54.

5. Reddy EP, Reynolds RK, Santos E, Barbacid M. A point mutation is responsible for the acquisition of transforming properties by the T24 human bladder carcinoma oncogene. *Nature* 1982;**300**:149–52.

6. Cavenee WK, Dryja TP, Phillips RA, Benedict WF, Godbout R, Gallie BL. Expression of recessive alleles by chromosomal mechanisms in retinoblastoma. *Nature*. 1983;**305**:779–84.

7. Weinberg RA. Oncogenes and tumor suppressor genes. *Trans Stud Coll Physicians Phila* 1988;**10**:83–94.

8. Hanada M, Delia D, Aiello A, Stadtmauer E, Reed JC. bcl-2 gene hypomethylation and high-level expression in B-cell chronic lymphocytic leukemia. *Blood* 1993;**82**:1820–8.

9. Kim KC, Huang S. Histone methyltransferases in tumor suppression. *Cancer Biol Ther* 2003;**2**:491–9.

10. Zhang B, Pan X, Cobb GP, Anderson TA. microRNAs as oncogenes and tumor suppressors. *Dev Biol* 2007;**302**:1–12.

11. Esteller M, Risques RA, Toyota M, Capella G, Moreno V, Peinado MA, et al. Promoter hypermethylation of the DNA repair gene O(6)-methylguanine-DNA methyltransferase is associated with the presence of G:C to A:T transition mutations in p53 in human colorectal tumorigenesis. *Cancer Res* 2001;**61**:4689–92.

12. Esteller M, Toyota M, Sanchez-Cespedes M, Capella G, Peinado MA, Watkins DN, et al. Inactivation of the DNA repair gene O_6-methylguanine-DNA methyltransferase by promoter hypermethylation is associated with G to A mutations in K-ras in colorectal tumorigenesis. *Cancer Res* 2000;**60**:2368–71.

13. Jones PA, Baylin SB. The fundamental role of epigenetic events in cancer. *Nat Rev Genet* 2002;**3**:415–28.

14. Lujambio A, Calin GA, Villanueva A, Ropero S, Sanchez-Cespedes M, Blanco D, et al. A microRNA DNA methylation signature for human cancer metastasis. *Proc Natl Acad Sci USA* 2008;**105**:13556–61.

15. Rajasekhar VK, Begemann M. Concise review: roles of polycomb group proteins in development and disease: a stem cell perspective. *Stem Cells* 2007;**25**:2498–510.

16. Tiwari VK, McGarvey KM, Licchesi JD, Ohm JE, Herman JG, Schubeler D, et al. PcG proteins, DNA methylation, and gene repression by chromatin looping. *PLoS Biol* 2008;**6**:2911–27.

17. Ehrlich M, Gama-Sosa MA, Huang LH, Midgett RM, Kuo KC, McCune RA, et al. Amount and distribution of 5-methylcytosine in human DNA from different types of tissues of cells. *Nucleic Acids Res* 1982;**10**:2709–21.

18. Goll MG, Bestor TH. Eukaryotic cytosine methyltransferases. *Annu Rev Biochem* 2005;**74**:481–514.

19. Okano M, Bell DW, Haber DA, Li E. DNA methyltransferases Dnmt3a and Dnmt3b are essential for de novo methylation and mammalian development. *Cell* 1999;**99**:247–57.

20. Hata K, Kusumi M, Yokomine T, Li E, Sasaki H. Meiotic and epigenetic aberrations in Dnmt3L-deficient male germ cells. *Mol Reprod Dev* 2006;**73**:116–22.

21. Molloy P. DNA Hypomethylation in cancer. In: Tollefsbol T, editor. *Cancer epigenetics*. Boca Raton: CRC Press; 2009, pp. 7–38.

22. Reik W. Stability and flexibility of epigenetic gene regulation in mammalian development. *Nature* 2007;**447**:425–32.

23. Patra SK, Patra A, Rizzi F, Ghosh TC, Bettuzzi S. Demethylation of (Cytosine-5-C-methyl) DNA and regulation of transcription in the epigenetic pathways of cancer development. *Cancer Metastasis Rev* 2008;**27**:315–34.

24. Szyf M. The role of DNA hypermethylation and demethylation in cancer and cancer therapy. *Curr Oncol* 2008;**15**:72–5.

25. Kangaspeska S, Stride B, Metivier R, Polycarpou-Schwarz M, Ibberson D, Carmouche RP, et al. Transient cyclical methylation of promoter DNA. *Nature* 2008;**452**:112–15.

26. Szyf M. DNA methylation and demethylation as targets for anticancer therapy. *Biochemistry (Mosc)* 2005;**70**:533–49.

27. Szyf M, Pakneshan P, Rabbani SA. DNA demethylation and cancer: therapeutic implications. *Cancer Lett* 2004;**211**:133–43.

28. Santos-Rosa H, Caldas C. Chromatin modifier enzymes, the histone code and cancer. *Eur J Cancer* 2005;**41**:2381–402.

29. Kampranis SC, Tsichlis PN. Histone demethylases and cancer. *Adv Cancer Res* 2009;**102**:103–69.

30. Huang S. Histone methylation and the initiation of cancer. In: Tollefsbol T, editor. *Cancer Epigenetics*. Boca Raton: CRC Press; 2009, pp. 109–50.

31. Orr JA, Hamilton PW. Histone acetylation and chromatin pattern in cancer. A review. *Anal Quant Cytol Histol* 2007;**29**:17–31.

32. Ropero S, Esteller M. The role of histone deacetylases (HDACs) in human cancer. *Mol Oncol* 2007;**1**:19–25.

33. Shukla A, Chaurasia P, Bhaumik SR. Histone methylation and ubiquitination with their cross-talk and roles in gene expression and stability. *Cell Mol Life Sci* 2009;**66**:1419–33.

34. Couture JF, Trievel RC. Histone-modifying enzymes: encrypting an enigmatic epigenetic code. *Curr Opin Struct Biol* 2006;**16**:753–60.

35. Morison IM, Ramsay JP, Spencer HG. A census of mammalian imprinting. *Trends Genet* 2005;**21**:457–65.

36. Ravenel JD, Broman KW, Perlman EJ, Niemitz EL, Jayawardena TM, Bell DW, et al. Loss of imprinting of insulin-like growth factor-II (IGF2) gene in distinguishing specific biologic subtypes of Wilms' tumor. *J Natl Cancer Inst* 2001;**93**:1698–703.

37. Yuan E, Li CM, Yamashiro DJ, Kandel J, Thaker H, Murty VV, et al. Genomic profiling maps loss of heterozygosity and defines the timing and stage dependence of epigenetic and genetic events in Wilms' tumors. *Mol Cancer Res* 2005;**3**:493–502.

38. Feinberg AP, Ohlsson R, Henikoff S. The epigenetic progenitor origin of human cancer. *Nat Rev Genet* 2006;**7**:21–33.

39. Pezer Z, Ugarkovic D. Role of non-coding RNA and heterochromatin in aneuploidy and cancer. *Semin Cancer Biol* 2008;**18**:123–30.

40. Kruger K, Grabowski PJ, Zaug AJ, Sands J, Gottschling DE, Cech TR. Self-splicing RNA: autoexcision and autocyclization of the ribosomal RNA intervening sequence of Tetrahymena. *Cell* 1982;**31**:147–57.

41. Weinberg MS, Villeneuve LM, Ehsani A, Amarzguioui M, Aagaard L, Chen ZX, et al. The antisense strand of small interfering RNAs directs histone methylation and transcriptional gene silencing in human cells. *RNA* 2006;**12**:256–62.

42. Morris KV, Chan SW, Jacobsen SE, Looney DJ. Small interfering RNA-induced transcriptional gene silencing in human cells. *Science* 2004;**305**:1289–92.

43. Brockdorff N, Ashworth A, Kay GF, McCabe VM, Norris DP, Cooper PJ, et al. The product of the mouse Xist gene is a 15 kb inactive X-specific transcript containing no conserved ORF and located in the nucleus. *Cell* 1992;**71**:515–26.

44. Zeng Y. Principles of micro-RNA production and maturation. *Oncogene* 2006;**25**:6156–62.

45. Chen K, Rajewsky N. The evolution of gene regulation by transcription factors and microRNAs. *Nat Rev Genet* 2007;**8**:93–103.

46. Hon LS, Zhang Z. The roles of binding site arrangement and combinatorial targeting in microRNA repression of gene expression. *Genome Biol* 2007;**8**:R166.

47. Song L, Tuan RS. MicroRNAs and cell differentiation in mammalian development. *Birth Defects Res C Embryo Today* 2006;**78**:140–9.

48. Lu J, Getz G, Miska EA, Alvarez-Saavedra E, Lamb J, Peck D, et al. MicroRNA expression profiles classify human cancers. *Nature* 2005;**435**:834–8.

49. Finoux AL, Chartrand P. Oncogenic and tumour suppressor microRNAs. *Med Sci (Paris)* 2008;**24**:1049–54.

50. Mao L, Lee D, Tcockman M, Erozan Y, Askin F, Sidransky D. Microsatellite alterations as clonal markers for the detection of human cancer. *Proc Natl Acad Sci USA* 1994;**91**:9871–5.

51. Habuchi T, Takahashi R, Yamada H, Kakehi Y, Sugiyama T, Yoshida O. Metachronous multifocal development of urothelial cancers by intraluminal seeding. *Lancet* 1993;**342**:1087–8.

52. Muto S, Horie S, Takahashi S, Tomita K, Kitamura T. Genetic and epigenetic alterations in normal bladder epithelium in patients with metachronous bladder cancer. *Cancer Res* 2000;**60**:4021–5.

53. Shen L, Kondo Y, Rosner GL, Xiao L, Hernandez NS, Vilaythong J, et al. MGMT promoter methylation and field defect in sporadic colorectal cancer. *J Natl Cancer Inst* 2005;**97**:1330–8.

54. Ramirez N, Bandres E, Navarro A, Pons A, Jansa S, Moreno I, et al. Epigenetic events in normal colonic mucosa surrounding colorectal cancer lesions. *Eur J Cancer* 2008;**44**:2689–95.

55. Yan PS, Venkataramu C, Ibrahim A, Liu JC, Shen RZ, Diaz NM, et al. Mapping geographic zones of cancer risk with epigenetic biomarkers in normal breast tissue. *Clin Cancer Res* 2006;**12**:6626–36.

56. Meiers I, Shanks JH, Bostwick DG. Glutathione S-transferase pi (GSTP1) hypermethylation in prostate cancer: review 2007. *Pathology* 2007;**39**:299–304.

57. Rodriguez-Canales J, Hanson JC, Tangrea MA, Erickson HS, Albert PS, Wallis BS, et al. Identification of a unique epigenetic sub-microenvironment in prostate cancer. *J Pathol* 2007;**211**:410–19.

58. Adams L, Roth MJ, Abnet CC, Dawsey SP, Qiao YL, Wang GQ, et al. Promoter methylation in cytology specimens as an early detection marker for esophageal squamous dysplasia and early esophageal squamous cell carcinoma. *Cancer Prev Res (Phila Pa)* 2008;**1**:357–61.

59. Chan AO, Rashid A. CpG island methylation in precursors of gastrointestinal malignancies. *Curr Mol Med* 2006;**6**:401–8.

60. Liu T, Niu Y, Feng Y, Niu R, Yu Y, Lv A, et al. Methylation of CpG islands of p16(INK4a) and cyclinD1 overexpression associated with progression of intraductal proliferative lesions of the breast. *Hum Pathol* 2008;**39**:1637–46.

61. Qi L, Bart J, Tan LP, Platteel I, Sluis T, Huitema S, et al. Expression of miR-21 and its targets (PTEN, PDCD4, TM1) in flat epithelial atypia of the breast in relation to ductal carcinoma in situ and invasive carcinoma. *BMC Cancer* 2009;**9**:163.

62. Breuer RH, Snijders PJ, Sutedja GT, Sewalt RG, Otte AP, Postmus PE, et al. Expression of the p16(INK4a) gene product, methylation of the p16(INK4a) promoter region and expression of the polycomb-group gene BMI-1 in squamous cell lung carcinoma and premalignant endobronchial lesions. *Lung Cancer* 2005;**48**:299–306.

63. DeAngelis J, Berletch B, Anderws L, Tollefsbol T. DNA hypermethylation and oncogenesis. In: Tollesfbol T, editor. *Cancer epigenetics*. Boca Raton: CRC Press; 2009, pp. 39–50.

64. Chang HH, Chiang CP, Hung HC, Lin CY, Deng YT, Kuo MY. Histone deacetylase 2 expression predicts poorer prognosis in oral cancer patients. *Oral Oncol* 2009;**45**:610–14.

65. Huang BH, Laban M, Leung CH, Lee L, Lee CK, Salto-Tellez M, et al. Inhibition of histone deacetylase 2 increases apoptosis and p21Cip1/WAF1 expression, independent of histone deacetylase 1. *Cell Death Differ* 2005;**12**:395–404.

66. Maru DM, Singh RR, Hannah C, Albarracin CT, Li YX, Abraham R, et al. MicroRNA-196a is a potential marker of progression during Barrett's metaplasia-dysplasia-invasive adenocarcinoma sequence in esophagus. *Am J Pathol* 2009;**174**:1940–8.

67. Esteller M. Cancer epigenomics: DNA methylomes and histone-modification maps. *Nat Rev Genet* 2007;**8**:286–98.

68. Hanson JA, Gillespie JW, Grover A, Tangrea MA, Chuaqui RF, Emmert-Buck MR, et al. Gene promoter methylation in prostate tumor-associated stromal cells. *J Natl Cancer Inst.* 2006;**98**:255–61.

69. Seftor EA, Brown KM, Chin L, Kirschmann DA, Wheaton WW, Protopopov A, et al. Epigenetic transdifferentiation of normal melanocytes by a metastatic melanoma microenvironment. *Cancer Res* 2005;**65**:10164–9.

70. Hu M, Polyak K. Microenvironmental regulation of cancer development. *Curr Opin Genet Dev* 2008;**18**:27–34.

71. Polyak K, Haviv I, Campbell IG. Co-evolution of tumor cells and their microenvironment. *Trends Genet* 2009;**25**:30–8.

72. Zhang SL, Chen YW, Tran S, Liu F, Nestoridi E, Hebert MJ, et al. Pax-2 and N-myc regulate epithelial cell proliferation and apoptosis in a positive autocrine feedback loop. *Pediatr Nephrol* 2007;**22**:813–24.

73. Muratovska A, Zhou C, He S, Goodyer P, Eccles MR. Paired-Box genes are frequently expressed in cancer and often required for cancer cell survival. *Oncogene* 2003;**22**:7989–97.

74. Wu H, Chen Y, Liang J, Shi B, Wu G, Zhang Y, et al. Hypomethylation-linked activation of PAX2 mediates tamoxifen-stimulated endometrial carcinogenesis. *Nature* 2005;**438**:981–7.

75. Buyse IM, Shao G, Huang S. The retinoblastoma protein binds to RIZ, a zinc-finger protein that shares an epitope with the adenovirus E1A protein. *Proc Natl Acad Sci USA* 1995;**92**:4467–71.

76. Steele-Perkins G, Fang W, Yang XH, Van Gele M, Carling T, Gu J, et al. Tumor formation and inactivation of RIZ1, an Rb-binding member of a nuclear protein-methyltransferase superfamily. *Genes Dev* 2001;**15**:2250–62.

77. Du Y, Carling T, Fang W, Piao Z, Sheu JC, Huang S. Hypermethylation in human cancers of the RIZ1 tumor suppressor gene, a member of a histone/protein methyltransferase superfamily. *Cancer Res* 2001;**61**:8094–9.

78. Zhang YJ, Li H, Wu HC, Shen J, Wang L, Yu MW, et al. Silencing of Hint1, a novel tumor suppressor gene, by promoter hypermethylation in hepatocellular carcinoma. *Cancer Lett* 2009;**275**:277–84.

79. Abbosh PH, Wang M, Eble JN, Lopez-Beltran A, Maclennan GT, Montironi R, et al. Hypermethylation of tumor-suppressor gene CpG islands in small-cell carcinoma of the urinary bladder. *Mod Pathol* 2008;**21**:355–62.

80. Park JY, Zheng W, Kim D, Cheng JQ, Kumar N, Ahmad N, et al. Candidate tumor suppressor gene SLC5A8 is frequently down-regulated by promoter hypermethylation in prostate tumor. *Cancer Detect Prev* 2007;**31**:359–65.

81. Esteller M, Silva JM, Dominguez G, Bonilla F, Matias-Guiu X, Lerma E, et al. Promoter hypermethylation and BRCA1 inactivation in sporadic breast and ovarian tumors. *J Natl Cancer Inst* 2000;**92**:564–9.

82. Lujambio A, Ropero S, Ballestar E, Fraga MF, Cerrato C, Setien F, et al. Genetic unmasking of an epigenetically silenced microRNA in human cancer cells. *Cancer Res* 2007;**67**:1424–9.

83. Shay JW, Zou Y, Hiyama E, Wright WE. Telomerase and cancer. *Hum Mol Genet* 2001;**10**:677–85.

84. Hahn WC. Telomere and telomerase dynamics in human cells. *Curr Mol Med* 2005;**5**:227–31.

85. Kelland L. Targeting the limitless replicative potential of cancer: the telomerase/telomere pathway. *Clin Cancer Res* 2007;**13**:4960–3.

86. Renaud S, Loukinov D, Abdullaev Z, Guilleret I, Bosman FT, Lobanenkov V, et al. Dual role of DNA methylation inside and outside of CTCF-binding regions in the transcriptional regulation of the telomerase hTERT gene. *Nucleic Acids Res* 2007;**35**:1245–56.

87. Xu D, Popov N, Hou M, Wang Q, Bjorkholm M, Gruber A, et al. Switch from Myc/Max to Mad1/Max binding and decrease in histone acetylation at the telomerase reverse transcriptase promoter during differentiation of HL60 cells. *Proc Natl Acad Sci USA* 2001;**98**:3826–31.

88. Zinn RL, Pruitt K, Eguchi S, Baylin SB, Herman JG. hTERT is expressed in cancer cell lines despite promoter DNA methylation by preservation of unmethylated DNA and active chromatin around the transcription start site. *Cancer Res* 2007;**67**:194–201.

89. Cavalli LR, Urban CA, Dai D, de Assis S, Tavares DC, Rone JD, et al. Genetic and epigenetic alterations in sentinel lymph nodes metastatic lesions compared to their corresponding primary breast tumors. *Cancer Genet Cytogenet* 2003;**146**:33–40.

90. Jarrard DF, Bova GS, Ewing CM, Pin SS, Nguyen SH, Baylin SB, et al. Deletional, mutational, and methylation analyses of CDKN2 (p16/MTS1) in primary and metastatic prostate cancer. *Genes Chromosomes Cancer* 1997;**19**:90–6.

91. Wang JF, Dai DQ. Metastatic suppressor genes inactivated by aberrant methylation in gastric cancer. *World J Gastroenterol* 2007;**13**:5692–8.

92. Foekens JA, Peters HA, Look MP, Portengen H, Schmitt M, Kramer MD. The urokinase system of plasminogen activation and prognosis in 2780 breast cancer patients. *Cancer Res* 2000;**60**:636–43.

93. Miyake H, Hara I, Yamanaka K, Gohji K, Arakawa S, Kamidono S. Elevation of serum levels of urokinase-type plasminogen activator and its receptor is associated with disease progression and prognosis in patients with prostate cancer. *Prostate* 1999;**39**:123–9.

94. Pakneshan P, Szyf M, Rabbani SA. Hypomethylation of urokinase (uPA) promoter in breast and prostate cancer: prognostic and therapeutic implications. *Curr Cancer Drug Targets* 2005;**5**:471–88.

95. Pulukuri SM, Estes N, Patel J, Rao JS. Demethylation-linked activation of urokinase plasminogen activator is involved in progression of prostate cancer. *Cancer Res* 2007;**67**:930–9.

96. Qi JH, Ebrahem Q, Moore N, Murphy G, Claesson-Welsh L, Bond M, et al. A novel function for tissue inhibitor of metalloproteinases-3 (TIMP3): inhibition of angiogenesis by blockage of VEGF binding to VEGF receptor-2. *Nat Med* 2003;**9**:407–15.

97. Kalluri R, Weinberg RA. The basics of epithelial-mesenchymal transition. *J Clin Invest* 2009;**119**:1420–8.

98. Fish JE, Srivastava D. MicroRNAs: opening a new vein in angiogenesis research. *Sci Signal* 2009;**2**:pe1.

99. Gregory PA, Bracken CP, Bert AG, Goodall GJ. MicroRNAs as regulators of epithelial-mesenchymal transition. *Cell Cycle* 2008;**7**:3112–18.

100. Suarez Y, Sessa WC. MicroRNAs as novel regulators of angiogenesis. *Circ Res* 2009;**104**:442–54.

101. Ramachandran K, Gordian E, Singal R. DNA methylation in tumor diagnosis. In: Tollefsbol T, editor. *Cancer Epigenetics*. New York: CRC Press; 2009, pp. 319–31.

102. Calin GA, Croce CM. MicroRNA signatures in human cancers. *Nat Rev Cancer* 2006;**6**:857–66.

103. Torres, A. and Marks, L. PCA3: a genetic marker of prostate cancer. *PCRI Insights*, **9**.

104. Paik S. Molecular assays to predict prognosis of breast cancer. *Clin Adv Hematol Oncol* 2007;**5**:681–2.

105. Hoque MO, Topaloglu O, Begum S, Henrique R, Rosenbaum E, Van Criekinge W, et al. Quantitative methylation-specific polymerase chain reaction gene patterns in urine sediment distinguish prostate cancer patients from control subjects. *J Clin Oncol* 2005;**23**:6569–75.

106. Herman JG, Baylin SB. Gene silencing in cancer in association with promoter hypermethylation. *N Engl J Med* 2003;**349**:2042–54.

107. Cadieux B, Ching TT, VandenBerg SR, Costello JF. Genome-wide hypomethylation in human glioblastomas associated with specific copy number alteration, methylenetetrahydrofolate reductase allele status, and increased proliferation. *Cancer Res* 2006;**66**:8469–76.

108. Jones PA, Baylin SB. The epigenomics of cancer. *Cell* 2007;**128**:683–92.

109. Barlesi F, Giaccone G, Gallegos-Ruiz MI, Loundou A, Span SW, Lefesvre P, et al. Global histone modifications predict prognosis of resected non small-cell lung cancer. *J Clin Oncol* 2007;**25**:4358–64.

110. Elsheikh SE, Green AR, Rakha EA, Powe DG, Ahmed RA, Collins HM, et al. Global histone modifications in breast cancer correlate with tumor phenotypes, prognostic factors, and patient outcome. *Cancer Res* 2009;**69**:3802–9.

533

111. Fraga MF, Ballestar E, Villar-Garea A, Boix-Chornet M, Espada J, Schotta G, et al. Loss of acetylation at Lys16 and trimethylation at Lys20 of histone H4 is a common hallmark of human cancer. *Nat Genet* 2005;**37**:391–400.

112. Van Den Broeck A, Brambilla E, Moro-Sibilot D, Lantuejoul S, Brambilla C, Eymin B. Loss of histone H4K20 trimethylation occurs in preneoplasia and influences prognosis of non-small cell lung cancer. *Clin Cancer Res* 2008;**14**:7237–45.

113. Feinberg AP, Vogelstein B. Hypomethylation distinguishes genes of some human cancers from their normal counterparts. *Nature* 1983;**301**:89–92.

114. Tien HF, Tang JH, Tsay W, Liu MC, Lee FY, Wang CH, et al. Methylation of the p15(INK4B) gene in myelodysplastic syndrome: it can be detected early at diagnosis or during disease progression and is highly associated with leukaemic transformation. *Br J Haematol* 2001;**112**:148–54.

115. Christiansen DH, Andersen MK, Pedersen-Bjergaard J. Methylation of p15INK4B is common, is associated with deletion of genes on chromosome arm 7q and predicts a poor prognosis in therapy-related myelodysplasia and acute myeloid leukemia. *Leukemia* 2003;**17**:1813–19.

116. Herman JG, Civin CI, Issa JP, Collector MI, Sharkis SJ, Baylin SB. Distinct patterns of inactivation of p15INK4B and p16INK4A characterize the major types of hematological malignancies. *Cancer Res* 1997;**57**:837–41.

117. Aparicio A, Weber JS. Review of the clinical experience with 5-azacytidine and 5-aza-2′-deoxycytidine in solid tumors. *Curr Opin Investig Drugs* 2002;**3**:627–33.

118. Glaser KB, Staver MJ, Waring JF, Stender J, Ulrich RG, Davidsen SK. Gene expression profiling of multiple histone deacetylase (HDAC) inhibitors: defining a common gene set produced by HDAC inhibition in T24 and MDA carcinoma cell lines. *Mol Cancer Ther* 2003;**2**:151–63.

119. Mann BS, Johnson JR, Cohen MH, Justice R, Pazdur R. FDA approval summary: vorinostat for treatment of advanced primary cutaneous T-cell lymphoma. *Oncologist* 2007;**12**:1247–52.

120. Kovacs JJ, Murphy PJ, Gaillard S, Zhao X, Wu JT, Nicchitta CV, et al. HDAC6 regulates Hsp90 acetylation and chaperone-dependent activation of glucocorticoid receptor. *Mol Cell* 2005;**18**:601–7.

121. Palazzo A, Ackerman B, Gundersen GG. Cell biology: tubulin acetylation and cell motility. *Nature* 2003;**421**:230.

122. Chen LF, Greene WC. Shaping the nuclear action of NF-kappaB. *Nat Rev Mol Cell Biol* 2004;**5**:392–401.

123. Krutzfeldt J, Rajewsky N, Braich R, Rajeev KG, Tuschl T, Manoharan M, et al. Silencing of microRNAs in vivo with 'antagomirs'. *Nature* 2005;**438**:685–9.

124. Liang Z, Wu H, Reddy S, Zhu A, Wang S, Blevins D, et al. Blockade of invasion and metastasis of breast cancer cells via targeting CXCR4 with an artificial microRNA. *Biochem Biophys Res Commun* 2007;**363**:542–6.

125. Kota J, Chivukula RR, O'Donnell KA, Wentzel EA, Montgomery CL, Hwang HW, et al. Therapeutic microRNA delivery suppresses tumorigenesis in a murine liver cancer model. *Cell* 2009;**137**:1005–17.

The Role of Epigenetics in Immune Disorders

Hanna Maciejewska-Rodrigues, Astrid Jüngel, and Steffen Gay
Center of Experimental Rheumatology, University Hospital and Zurich Center of Integrative
Human Physiology (ZIHP), University Hospital Zürich, Switzerland

IMMUNE SYSTEM AND AUTOIMMUNITY

The immune system enables the organism to resist infections and is composed of the innate
or non-specific system and the adaptive or specific system. Aberrations in any part of it can
lead to misdirected reactions resulting in autoimmunity where the immune system attack
turns against the host tissues.

535

Innate Immunity

Since the innate immune system constitutes the first defense against microorganisms such
as viruses, bacteria, and/or fungi its role is the nonspecific recognition of pathogens and
their removal without giving protective memory to the host. Upon recognition of foreign
substances, certain peptides derived from such pathogens, it is activated via pathogen
associated molecular patterns (PAMPs) ligating to pattern recognition receptors (PRRs)
such as Toll-like receptors (TLRs) expressed on the cells of the innate immune system [1,2].
Activation might also occur by ligation of diverse molecules expressed on stressed, injured,
infected, or transformed human cells that also act as PAMPs [3]. Therefore, stress factors
such as heat, fractures, necrosis, pressure, apoptosis, hypoxia, or carcinogenesis can also
result in the activation of the innate immunity. In addition to sensing and removing foreign
particles by phagocytosis or the production of nitric oxide (NO) the role of the innate
immune system is also the recruitment of immune cells to sites of infection or inflammation
by producing cytokines and chemokines [4]. Moreover, the function of the innate immune
system is the activation of the complement cascade resulting in massive amplification of
the response and activation of the cell-killing membrane attack complex and, last but not
least, the activation of the adaptive immune system [5,6]. The cells of the innate immune
system include phagocytic cells (monocyte/macrophages and neutrophils), natural killer
(NK) cells, dendritic cells (DCs), granulocytes, mast cells, and platelets. They act as antigen-
presenting cells (APCs) and express a cell surface molecule encoded by genes in the major
histocompatibility complex – class II MHC molecules, on which they present pathogenic
epitopes leading to the activation of the adaptive immune system.

Macrophages ($CD14^+/CD68^+$) play an important role in wound healing and in activation
of the immune responses [7]. They function in phagocytosis and intracellular killing of
microorganisms as well as in extracellular killing of altered self target cells. They are the

Handbook of Epigenetics: The New Molecular and Medical Genetics. DOI: 10.1016/B978-0-12-375709-8.00033-2

main source of tumor necrosis factor α (TNFα) and interleukin-1β (IL-1β) – two powerful proinflammatory cytokines [7,8].

NK cells (CD56$^+$, CD16$^+$) are capable of nonspecific killing of virus-infected and tumor cells by recognizing level changes of a surface molecule MHC class I. They release perforin, granzymes, and cytokines such as interferon-γ (IFN-γ), which has a protective role against many intracellular pathogens [9,10]. They also belong to the adaptive immune response since they act as effector cells in antibody-dependent cell-mediated cytotoxicity.

DCs are the most potent of the APCs [11]. After phagocytosing pathogens they up-regulate the expression of co-receptors, which enhance their ability to stimulate the cells of the adaptive immune system, such as T cells via CD40 co-receptors – CD40 ligand interaction [12]. Activated DCs also release IL-12, which primes and activates the adaptive immune system.

Adaptive Immunity

The adaptive immune system is composed of lymphocytes, T cells (CD3$^+$) and B cells (CD20$^+$), specialized immune cells that recognize specifically pathogenic epitopes and upon this recognition they proliferate resulting in multiple copies of a single specific cell clone [13]. Then, activated B cells differentiate to plasma cells and produce immunoglobulins (antibodies) binding specifically the corresponding antigen. The antigen–antibody complexes activate the complement system and can be cleared by phagocytosis via Fc receptors expressed on macrophages [14]. T cells specific for the antigen can stimulate a B-cell response (T helper; CD4$^+$) or directly kill the pathogen by releasing the cytotoxins perforin and granzymes (T cytotoxic; CD8$^+$) [15]. Whereas helper T 1 (Th1) cells are involved in cellular immunity against intracellular bacteria and viruses by expression of INF-γ and IL-2, Th2 cells mediate the humoral response to parasitic infection and are characterized by the production of IL-4. Th17 cells play a critical role in host defense against extracellular pathogens, particularly those colonizing exposed surfaces like skin, the mucosa of the airways, or lining of the intestines. They are the major source of IL-17 and induce the production of chemokine gradient leading to recruitment of the cells of the innate immunity [16]. Another subset of T cells, T regulatory cells (Treg; CD4$^+$, CD25$^+$, Foxp3$^+$) are responsible for limiting the immune response and for immune tolerance to self antigens and therefore they constitute defense against self-directed attack of the immune system [17]. The effector cytokine of Treg cells is anti-inflammatory IL-10.

Autoimmunity

In autoimmune disorders, such as systemic lupus erythematosus (SLE) or rheumatoid arthritis (RA), the cells of the adaptive immunity recognize self antigens and propagate a self-directed autoimmune reaction since the organism is unable to determine the difference between antigens coming from a pathogen such as virus or bacteria and the individual's cells and body tissues [18]. In consequence, the immune system creates autoantibodies, which are mostly directed against the organism's own tissues and, depending on the localization, results in diverse complications. Frequently, autoantibodies are present prior to diagnosis and even onset of the disease [19–21]. A minority of RA patients are seronegative, indicating that RA is a heterogeneous disease [22–25].

Most of the autoimmune diseases have an unclear pathogenesis. Nevertheless, the contribution of epigenetic changes is quite well recognized in SLE [26] and has recently emerged as being modified in RA as well as few other autoimmune disorders such as multiple sclerosis (MS), type 1 diabetes, systemic sclerosis (SSc), the immunodeficiency centromeric instability and facial anomalies (ICF) syndrome, and autoimmune thyroid disease, and even in experimental autoimmune encephalomyelitis [27,28]. In this chapter we focus on those of the immune disorders in which there is most evidence for the involvement of aberrant epigenetic modulations.

EPIGENETICS OF IMMUNE CELLS
CD4$^+$ T Cells
T REGULATORY CELLS

Recently, research has focused on elaborating the epigenetic regulation of the forkhead box P3 (Foxp3) – a transcriptional repressor and the master regulator of development and function of Treg cells [29]. It has been shown that CpG motifs in the locus of Foxp3 are demethylated in Tregs in contrast to naive T cells [30–33]. Furthermore, also histone modifications such as increased histone acetylation in the promoter of *Foxp3* have been described in natural Tregs [34,35]. Stimulation of Treg cells by ligation of TCR leads to the nuclear translocation of Foxp3 and its binding to promoters of proinflammatory IL-2 and INF-γ inducing their deacetylation and silencing, as well as to promoters of GITR, CD25, and CTLA-4 increasing their acetylation and expression which is characteristic for the functional Tregs [36,37]. It was shown that Foxp3 is optimally functional when in a complex with histone acetyltransferase Tip60 and histone deacetylase HDAC7 [38]. Furthermore, it was demonstrated that HDAC9 binds Foxp3 in resting, non-inhibitory Tregs, being a key repressor of Foxp3 function in Tregs, and upon ligation of TCR dissociates from the complex [38]. Therefore, it has been suggested that the selective blockade of HDAC9 could be used for treatment of autoimmune diseases and/or post-transplantation where Tregs with increased functionality could be of high relevance [35]. The use of HDAC inhibitors has already been shown to enhance the function of the Treg cells, and was beneficial in the treatment of homeostatic proliferation, inflammatory bowel disease, improved cardiac and islet survival, and allograft tolerance (in combination with rapamycin therapy) [29,34].

Th1 CELLS

Similarly to the naive T cells, Th1 cells have most of the CpG demethylated, including the promoter of *IFN-γ* [39]. Moreover, in Th1 cells also epigenetic changes in the histones of the IFN-γ locus such as increased histone acetylation and H3K4 dimethylation paralleled by complete loss of the repressive mark of H3K27 methylation were observed [40,41].

Th2 CELLS

It has been shown that chromatin remodeling involving changes in histone structure and DNA methylation surrounding the *IL-4* gene, encoding the effector cytokine of Th2 cells, lead to Th2-cell activation and differentiation [42]. In developing Th2 cells, the *IL-4* gene is positioned apart from heterochromatic regions and is demethylated [43,44], while silencing of Th1 specific *IFN-γ* occurs by nuclear reorganization to the heterochromatic domain, as well as increased CpG and H3K27 histone methylation [40,41,43]. Furthermore, histone acetylation and H3K4 dimethylation in the promoter of *IL-4* is increased while repressive H3K27 trimethylation is down-regulated [42,45]. Histone acetyltransferase CREB-binding protein (CBP) promotes expression of *IL-4* while HDAC1, 2, and 3 have the opposite effect. Aberrant expression of *IL-4* has been described in allergic and autoimmune disorders such as RA and diabetes type 1 [46–48].

Th17 CELLS

To date, there is not much data available on the epigenetic regulation in Th17 cells. Nevertheless, it was reported that the promoters of *IL17a* and *IL17f* genes are hyperacetylated in Th17 cells and this modification is dependent on *STAT3* [49,50].

B Cells

Many epigenetic changes occur during the development of B cells. At different stages of B-cell maturation different genes are expressed due to demethylation in their promoters. Thus, demethylation of *Pax-5*, *Pu-1*, and *Igα/mb-1* occurs in the early B-cell development [51–53].

Later on these DNA regions become methylated, and the transcription is switched off. At the pre-pro-B stage demethylation of DNA and histone trimethylation H3K4 in the promoter of B cell co-receptor *CD19* takes place [54]. As the cell matures demethylation of the *complement receptor II (CD21)* promoter occurs [55]. Upon stimulation of B cells via BCR the synthesis of DNMT1 increases leading to increased DNA and histone H3K9 methylation [56,57]. Also, acetylation of histones is involved in the development of B cells since it was shown to play a role in the generation of B-cell receptors (BCRs) [58] and in the regulation of B lymphocyte-induced maturation protein 1 (Blimp-1), the master regulator of plasma cell differentiation [59].

The use of HDAC inhibitors induces in B cells expression of genes such as *CMH-II, CD21,* and *IgM* while repressing *INF-γ, IL-12, IL-6* and *IL-10* [59–61]. Furthermore, it promotes B-cell terminal differentiation and apoptosis. It is suggested that this could be the mechanism of decreasing renal disease observed in MRL-lpr/lpr mice, a lupus-like animal model [62].

ICF syndrome caused by a mutation in the *de novo* DNA methyltransferase 3B (DNMT3B) [63] is characterized by B-cell immunodeficiency, in which cell hypomethylation of pericentromeric and subtelometic repeats is observed [64]. It is associated with blocked differentiation of naive B cells into memory/plasma cells, which leads to immunoglobulin deficiency [65].

Macrophages

In macrophages, epigenetic regulation has been described in the LPS tolerance [66]. Re-stimulation of macrophages by LPS leads to LPS tolerance, due to the loss of histone acetylation and H3K4me3 marks in the promoters of a set of genes including proinflammatory mediators [66].

Dendritic Cells

It has been reported that DC during acute sepsis undergo epigenetic modulations leading to subsequent impairment of IL-12 expression [67]. The down-regulation of IL-12 expression was connected with a decreased ratio between H3K4me3 and H3K27me2 at the IL-12 promoter. Taking into account that during severe sepsis there is an overwhelming pathological response and deregulation of cytokine expression, the authors hypothesized that TLR and cytokine signaling could be the potential contributors to the epigenetic changes in DC [68]. Therefore, it is also possible that in autoimmune diseases, where chronic inflammation occurs, DC could also undergo epigenetic modifications. Up to date, alterations in DC homeostasis have been described to be involved in the pathogenesis of various autoimmune diseases, such us SLE, RA, MS, and type I diabetes [69,70].

EPIGENETICS OF SLE

SLE is a chronic inflammatory connective tissue disease resulting in pain, inflammation, and, often, damage to organs. It affects different parts of the body, including the joints, skin, nervous system, blood, and kidneys [71]. Depending on the type of lupus, the symptoms can include rashes, hair loss, aching and swelling of joints, fever, anemia, and abnormal blood clotting. The disease can take a mild course when it affects only a few organs; nevertheless the severe form can be life-threatening. The main characteristic of lupus is the production of autoantibodies, such as anti-dsDNA antibodies, by self-reactive B cells. T-cell abnormalities and aberrant T helper cytokine profiles have been implicated in breaking immune tolerance to nuclear and cytoplasmic antigens [72,73].

Beside a strong genetic factor driving SLE [74], it is now clear that epigenetics plays a role in its development [26]. The first clue that the epigenetic alterations could participate in the pathogenesis of an autoimmune disease, or even be a cause for it, arose from the

538

experiments where T cells were treated with demethylating agents. Richardson in 1986 could show that CD4$^+$ T cells become autoreactive after treatment with the demethylator 5-azaC [75]. This work showed that the T cells with demethylated DNA lost the requirement of a specific antigen on the MHC II molecules and that only their exposure to the antigen-presenting cells was sufficient for inducing activation and proliferation. Consequently, it was reported that the DNA in T cells of both SLE and RA patients is hypomethylated [76].

Recent reports describe an increasing number of drugs to be implicated in drug-induced lupus with hydralazine, procainamide, and isoniazid accounting for the highest number of cases [77]. Approximately 12% of the patients under high dose hydralazine treatment over long periods of time developed lupus-like symptoms. The symptoms generally disappeared upon discontinuing the medication. Already in 1991 it had been reported that procainamide and hydralazine induced lupus-like disease also in mice [78]. Interestingly, hydralazine and procainamide are demethylating agents. Hydralazine decreases DNA methylation via inhibition of the ERK pathway [79] while the hypomethylation of the DNA in lupus T cells is due to decreased signaling via the ERK pathway [80]. Recently, it has been demonstrated in a mouse model that decreasing ERK signaling in T cells results in the down-regulation of DNA methyltransferase 1 (Dnmt1) and overexpression of the methylation-sensitive genes characteristic for T cells in human lupus. Moreover, these animals also developed anti-dsDNA antibodies [81]. Furthermore, the transfer of *in vitro* demethylated T cells to the syngeneic mice resulted in the development of lupus like disease [82].

Even though the mechanisms of the pathogenesis of SLE are not clear, it is well-accepted that epigenetic malfunctions are involved. It is not clear whether hypomethylation is the cause or result of inflammation in SLE; however, recent evidence suggest that inflammation may lead to epigenetic modifications [83].

As a result of promoter hypomethylation, genes including *perforin* [84], *CD70* [85], and *CD11a* are overexpressed in T cells [86]. Perforin is a cytotoxic effector molecule expressed in NK cells and a subset of T cells. The overexpression of perforin is thought to contribute to promiscuous monocyte killing by the CD4$^+$ T cells in lupus since the perforin inhibitor concanamycin A blocks autologous monocyte killing [84]. Cytotoxic T lymphocyte-mediated killing has been proposed to be a preferential and selective source of autoantigens in SLE [87]. Moreover, the increase in activated CD8$^+$ T lymphocytes expressing perforin and granzyme B was shown to correlate with disease activity in patients with SLE [88]. Overexpression of a B-cell co-stimulatory molecule CD70 on the other hand contributes to hyperstimulation of B cells [85]. CD11a (lymphocyte function-associated antigen 1) is an integrin overexpressed on an autoreactive subset of T cells from patients with active lupus which might also be contributing to the flares in this disease [86]. Furthermore, in women with lupus demethylation of the promoter of *CD40L* paralleled by the overexpression of *CD40L* was also observed [89]. This observation is of high interest since CD40L is encoded on the X chromosome and one copy of *CD40L* in women is normally silenced by hypermethylation of its promoter. Moreover, lupus is much more prevalent in women compared to men, where the ratio is 9:1 [90].

Besides the involvement of DNA hypomethylation in the development of lupus, impaired histone modifications have been recently described, in cells from lupus patients as well as from animal models of lupus-like disease. Hypoacetylation of H3 and H4 histones was first described in autoimmune field in splenocytes isolated from lupus prone mice [91]. Furthermore, treatment of these mice with HDAC inhibitors, TSA or SAHA, improved glomerulonephritis and splenomegaly [62,92]. In accordance with this, *in vitro* data also showed that splenocytes from lupus-prone mice treated with HDAC inhibitors had reduced expression of cytokines such as IL-6, IL-10, IL-12, INFγ, and TNFα as well as nitric oxide synthetase and NO itself. These data suggest that the activity of HDACs is increased and could contribute to the development of at least some of the symptoms in lupus. This idea is also supported by the data from a study wherein the presence of B cells with inactive histone acetyltransferase p300 leads to a development of severe lupus-like disease in mice [93]. These animals developed anti-dsDNA antibodies as well as glomerulonephritis and died prematurely. The animal model experiments are reflected by human data where, in CD4$^+$ T cells, global histone H3 and H4 hypomethylation was observed [94]. Most interestingly, the degree of hypomethylation correlated with the disease activity. Furthermore, it was also shown that in the CD4$^+$ T cells from lupus patients the global methylation of H3K9 is also decreased.

Also, the innate response is active in lupus patients, half of them carrying a characteristic type-I IFN signature in their PBMC [95]. Here, the modulation of miRNA has been recognized recently. In this regard, miR-146a targets several transcripts of this pathway, including *STAT-1* and *IRF5*. Interestingly, levels of miR-146a are significantly decreased in PBMC from lupus patients, pointing to a defect in the regulation of IFNα/β activation as an inductor mechanism of the disease [96]. Furthermore, autoantibodies in lupus and in several other connective tissue disorders target the microRNA pathway. It was recently found that several autoantibodies from patients with connective tissue diseases recognize some of the structures involved in the microRNA working machinery [97]. Anti-GWB antibodies (targeting Ge-1 and GWB182) have been identified in autoimmune PNP, Sjögren syndrome, SLE, RA, and PBC. In the same way, anti-Su autoantibodies have been shown in 20% of lupus patients as well as in scleroderma, and recognize the major components in the maturation of functional miRNA, including Dicer/TRBP and RISC, likely because of the presence of the regulating Argonaute protein Ago 2.

EPIGENETICS OF RHEUMATOID ARTHRITIS (RA)

RA is a chronic, systemic autoimmune disease of which the main characteristic is joint destruction [98]. Since the changes in joints lead to the loss of function, RA frequently results in disability and is associated with increased pain.

The pathogenic phenomena in RA occur mainly in the synovial tissues. The main function of the synovial tissue in a healthy joint is supplying the synovial fluid which lubricates the cartilage and minimizing the friction of the cartilage covering the bones. In RA, the synovial tissue overgrows as a result of infiltration by immune cells and the increased survival rate of the synovial resident cells [99]. Activation of the infiltrating immune cells leads to a self-directed attack localized to joints and is followed by joint inflammation (arthritis). Activation of resident synovial cells parallels the inflammation and results in an aggressive invasion of synovial fibroblasts and macrophages into cartilage and bone, leading to their damage.

Due to synovitis, the main symptoms of RA are swelling of multiple joints, which become tender, and their stiffness causes difficulties in movement [100]. Usually, RA affects multiple joints of the hands, feet, and cervical spine but also can involve larger joints like the wrist, elbow, shoulder, or knee. As the disease progresses, it can also involve other organs such as skin, where rheumatic nodules can be formed, and lungs. Moreover, in the severe course of RA, vasculitis can also be observed. Patients with RA are more prone to atherosclerosis and have generally an increased risk of cardiovascular events like heart attack and stroke [101]. RA affects about 1% of Western Europe's population, being the most common of the inflammatory joint diseases. Women have a three times increased risk of RA compared to men [102]. The disease onset is most commonly observed in individuals between 30 and 60 years old.

Recent studies on epigenetics have contributed to a better understanding of the pathogenesis of RA. The first evidence for epigenetic alterations describing impaired T-cell DNA methylation was published in 1990 [76]. Later on, our group demonstrated that RA synovial fibroblasts (RASFs) are intrinsically activated and resemble an activated phenotype [103]. The activated phenotype of synovial cells is characterized by increased production of proinflammatory cytokines that in turn attract more inflammatory cells to the joints [104]. Furthermore, RASFs express increased levels of adhesion molecules and matrix degrading enzymes. RASFs are also characterized by intrinsically high levels of the small ubiquitin modifier 1 (SUMO-1) which contributes to their resistance to apoptosis, especially at sites of invasion [105], as well as an increased expression of proto-oncogenes and cell cycle proteins and therefore have been suggested to resemble transformation [106]. Nevertheless, RASFs have not been shown to proliferate extensively like cancer cells; thus we refer to them as cells with an activated phenotype [104]. It needs to be investigated to which extent the epigenetic changes are responsible for this activated phenotype of RASFs.

Recently we have shown that RASFs have decreased levels of global DNA methylation [107], which could result in an increased expression of cell activating genes. Since unmethylated CpGs stimulate the innate immune response via TLR9 [108] the hypomethylated DNA from the apoptotic T cells or RASFs might trigger the innate immunity. We reported further that CpG islands in the promoter of the retrotransposable element *LINE-1* are hypomethylated in RA synovial cells which leads to the reactivation of transcription of *LINE-1* [109]. In another study, a single unmethylated CpG in a promoter of *IL-6* in monocytes from patients with RA was reported to render IL-6 more inducible by LPS stimulation [110]. On the other hand, the expression of a member of the apoptosis-inducing Fas gene family, *death receptor 3 (DR3)*, was shown to be down-regulated in synovial cells from patients with RA due to specific methylation of the CpG island in its promoter [111]. This down-regulation might render RA synovial cells resistant to apoptosis. These findings resemble certain aspects of malignant cells where methylation of specific promoters, for example in tumor suppressor genes, is followed by the hypomethylation of others, like proto-oncogenes.

Regarding histone modifications, our group described that in RA synovial tissue the balance of HAT/HDAC activity is strongly shifted towards histone acetylation [112]. Specifically, we observed a down-regulation of HDAC1 and 2 in the synovium of patients with RA.

In other reports, it has been shown by *in vivo* studies that the HDAC inhibitor FK228 inhibits joint swelling, synovial inflammation, and joint destruction in autoantibody-mediated arthritis in mice through the induction of p16INK4a and the up-regulation of p21(WAF1/Cip1) [113]. Furthermore, FK228 also suppressed the production of hypoxia-induced VEGF and blocked angiogenesis in synovial tissue in collagen-antibody-induced arthritis [114]. Accordingly, another HDAC inhibitor, TSA, was demonstrated to induce cell cycle arrest and apoptosis in RASF particularly in combination with ultrasound treatment [115]. TSA was also shown to render RASF sensitive for TRAIL-induced apoptosis; however, in our study TSA had no effect on apoptosis in the absence of TRAIL [116]. In summary, there is a significant down-regulation of specific HDACs observed in the tissue of RA patients while, on the contrary, in animal models inhibitors of HDACs ameliorate the disease. These data imply that HDACs might be less active, however, also localized aberrantly, at specific promoter sites, and hence participating in the pathogenesis of RA. On the other hand, the effect of HDAC inhibitors could be also mediated by non-histone substrates of HDACs. Therefore, it might be considered, as we predicted, whether the inhibition or rather relocalization of HDACs to a proper position is feasible for the treatment of RA [117].

In addition, modulations in the expression of miRNA have been detected in RA. The expression of miR-155, miR-146a, and miR-203 was shown to be higher in RASF compared to OASF [118–120] and miR-155 was inducible by TNFα, IL-1β, and the ligands of TLR2, 3, and 4 [118]. Furthermore, we have proven that miR-155 repressed the expression of matrix metalloproteinase (MMP)-1 and -3 in RASFs, implying that it counter-regulates inflammation and destruction. On the other hand, miRNA-155 has been suggested to be involved in B-cell malignancies [121], which is interesting in view of the fact that B cells play a major role in the pathogenesis of RA, as demonstrated by the successful application of anti-CD20 targeted therapies deleting B cells from the synovium [122].

Even though miR-146a was shown to be expressed at higher levels in RA, it was also reported to be unable to function properly by down-regulating the expression of TRAF6 and IRAK-1, two molecules involved into the pathway of TNFα synthesis [123]. In addition, miR-203 was shown to be involved in the up-regulation of IL-6 and MMP-1 in RASF indicating a role of miR-203 in the development of the activated phenotype of RASF [120]. Moreover, the expression of miR-203 seems to be regulated by DNA methylation, since it was increased upon treatment with 5-aza. Interestingly, microparticles which are small vesicles released from cells by exocytic budding during activation or death present in abundant quantities in the synovial fluid of patients with RA, contain miRNA, as shown recently by our laboratory [124,125]. This supports our notion that microparticles are inter-cellular mediators of inflammation participating in the pathogenesis of RA [126]. In summary, accumulating reports suggest that the epigenetic modulation could result in the intrinsic activation of RASF [99].

EPIGENETICS OF OTHER IMMUNE DISORDERS
Multiple Sclerosis

The recent data from the reports by Moscarello and co-authors strongly suggest that epigenetic aberrations could also lead to the development of MS [127–129]. MS refers to "multiple scars" that accumulate in the brain and spinal cord since it is a chronic inflammatory neurodegenerative autoimmune disease characterized by progressive demyelination of the neurons, leading to their destruction [130]. MS is characterized by the formation of focal demyelinated plaques in the white matter of the central nervous system [131]. Also, the atrophy of gray matter is observed with foci of demyelination, microglial activation, leptomeningeal inflammation, iron deposition, and neuronal loss. These changes result from the migration of lymphocytes into the brain early in the disease [132]. The majority of disability in MS however, relates to spinal cord dysfunction. Axonal and myelin loss are major pathological features of MS [133] which can result from attack of immune cells, such as CD8$^+$ cells damaging neurons or macrophages stripping myelin from the axon. Otherwise neurons can be damaged also by a release of toxic intermediates such as glutamate or nitric oxide [130]. There are different types and stages of MS, ranging from MS that progresses slowly to rare malignant forms leading to progressive disability [131].

Peptidyl argininedeiminase 2 (PAD2) is overexpressed in MS, which leads to increased deimination (citrullination) of myelin basic protein (MBP) in the white matter of brains of patients with MS [134]. Citrullination of myelin in turn results in the loss of myelin stability in the MS brain since it is accompanied by the loss of positive charge which compromises the ability of MBP to interact with the lipid bilayer. The increase of the PAD2 expression is most probably due to demethylation of a region in the *PAD2 promoter* [128]. Moreover, in the same report the DNA demethylase activity in the white matter of MS patients was also shown to be increased. Although no potential DNA demethylase was reported to be responsible for the decrease of DNA demethylase activity, the ratio of total cytidilic acid and 5-methyl cytidilic acid measured in the DNA demethylase assay was almost doubled in the white matter extracts from brains of patients with MS compared to controls [128].

Type 1 Diabetes Mellitus

Type 1 diabetes is an autoimmune disorder characterized by T-cell-mediated destruction of the insulin-secreting β cells of the Langerhans islets in the pancreas [135]. This in turn leads to an absence of insulin and hyperglycemia resulting from hepatic excessive production of glucose and decreased uptake of glucose from the circulation. The depletion of insulin results also in increase in fat breakdown and oxidation of fatty acids, delivering excess of ketones. Diabetes type 1 patients, if not supplemented with exogenous insulin, suffer from progressive central nervous depletion leading to coma and death. The early stages before the onset of symptoms of diabetes type 1 are characterized by insulitis, the infiltration of the pancreatic islets by immune cells such as macrophages, dendritic cells, and T cells [136]. The onset of type 1 diabetes usually occurs during childhood or adolescence. More than 80% of β cells have already been destroyed by the time that the first clinical symptoms become apparent, since the autoimmune processes destroying β cells remain subclinical for many years in most patients [137]. At the time of diagnosis, autoreactive T cells specific for β-cell proteins including insulin itself as well as autoantibodies against β-cell proteins are detected in the peripheral blood [138,139].

Leukocytes in type 1 diabetes patients have decreased synthesis of IL-2 and T-lymphocyte-associated antigen-4 (CTLA4), a major negative regulator of T-cell responses [140,141]. A defect in IL-2 signaling leads to a reduced function of Tregs and therefore promotes autoimmunity. The competency to produce IL-2 and/or the rate of transcription in type 1 diabetes mellitus depends on a segment of DNA that extends 10 kb upstream of the known *IL-2 promoter*. 33 of the 46 disease-associated SNPs are located in this 5′ region of the *IL-2* gene which has locus control region-like activity that determines the competency of a cell to express IL-2 mRNA [142]. This region undergoes specific epigenetic changes during both development and activation that regulate locus accessibility and recruitment of the transcription machinery [143–145]. After activation of T cells several areas of DNase hypersensitivity corresponding to open chromatin appear at different sites of this regulatory region (four between −2.5 and −3 kb and three at −7, −8, and −10 kb) [143]. Acetylation of histones in the promoter of *IL-2* has been found in naive T cells in contrast to non-T cells [144] and it is further increased upon activation of T cells [145,146]. Furthermore, upon activation of T cells (and potentiated by CD28 co-stimulation), CpG sites located within −600 bp of the IL-2 promoter become demethylated [146–148]. Also, methylation of arginine in proteins localizing to the *IL-2 promoter* has been shown to increase the expression of IL-2 [149].

Interestingly, non-obese diabetic mice with a defect in 10 kb upstream from *IL-2* develop type 1 diabetes, promoting the infiltration of autoreactive lymphocytes into the pancreas [150,151]. Even though there exists substantial evidence that IL-2 expression can be regulated by epigenetic modifications, more studies are needed to prove that epigenetic changes are responsible for the decreased production of IL-2 in patients with type 1 diabetes mellitus. On the other hand, it was shown recently that epigenetic modifications play a role in decreasing the expression of CLTA4 in type 1 diabetic patients [152]. Increased histone H3K9me2 repressive mark was observed in the *promoter of CLTA4* in blood lymphocytes of patients with diabetes type 1. Decrease in the levels of CLTA4 is thought to enhance T-cell activation and in consequence to lead to autoimmune response.

It has been described that administration of activators of histone deacetylase, sirt-1, could be beneficial for diabetes type 2 patients by preventing β-cell secretory failure and insulin resistance [153]. It is therefore possible that sirtuins could play a role also in the pathogenesis of diabetes type 1.

Systemic Sclerosis

Recently, it was demonstrated that epigenetic modifications play a key role in the pathogenesis of systemic sclerosis (SSc). Pathologic features of SSc include progressive tissue fibrosis and widespread vascular disorder [154,155]. Deposition of collagen occurs in the skin but can also occur in many internal organs, including the lungs, heart, gastrointestinal tract, and kidneys. Therefore, the clinical manifestations can differ broadly from a limited involvement of the skin and internal organs to diffuse skin involvement with fibrosis of internal organs, which in turn can lead to organ failure and death. The disease is driven by T cells, B cells, cytokines, and chemokines [156]. Activated antigen-specific T cells are the major cells infiltrating skin and lungs. It is believed that fibroblasts derived from involved

SSc skin mediate tissue fibrosis in SSc, since these cells synthesize excessive amounts of collagen and other components of the extracellular matrix together with reduced expression of matrix metalloproteinases, leading to excessive matrix accumulation [157,158]. Furthermore, SSc fibroblasts are characterized by hyper-responsiveness to cytokines and chemokines and their increased production.

It was reported in 2006 that the augmented collagen synthesis by SSc fibroblasts was linked to epigenetic repression of the collagen suppressor gene *FLI1* due to both hypermethylation and deacetylation of the *FLI1* promoter [159]. *FLI1* is a member of the Ets family of transcription factors which suppresses collagen production via a Sp-1-dependent pathway [160]. We could show that an HDAC inhibitor, TSA, prevented the induction of collagen type I and fibronectin in dermal fibroblasts from SSc patients and healthy controls. This beneficial effect of TSA was confirmed in the mouse model of bleomycin-induced dermal fibrosis, where TSA reduced the accumulation of collagen and the atrophy of the subcutis to levels comparable with those of control mice [161]. In our further studies we could show that TSA does not only block the enzymatic activity of HDAC, but additionally regulates the protein level of selective targets by influencing their transcription [162]. Subsequently, almost all transcripts of HDACs are reduced by TSA in SSc fibroblasts, and the expression of HDAC-7 is almost completely inhibited. Most interestingly, we demonstrated that silencing of HDAC-7 decreased the release of collagen types I and III in SSc fibroblasts. Silencing of HDAC-7 appears to be not only as effective as TSA, but also a more specific target for the treatment of SSc, because it does not up-regulate the expression of profibrotic molecules such as ICAM-1 and CTGF [162,163]. Moreover, it was reported just recently that inhibitors of DNA methyltransferases re-activate the expression of anti-fibrotic SOCS-3 and decrease the release of collagen in SSc fibroblasts [164]. Furthermore, the potent anti-fibrotic effect of 5-aza as well as of other inhibitors of DNMTs such as procainamide and hydralazine was also demonstrated *in vivo* using the mouse model of bleomycin-induced dermal fibrosis [164].

The epigenetic alterations were reported not only in SSc fibroblasts but also in T cells from patients with SSc [165]. Lei et al. described that the DNA in SSc CD4$^+$ T cells is hypomethylated and methylation-related genes were abnormally expressed [165].

ICF Syndrome

ICF syndrome is a rare autosomal disorder, caused by mutations of the *de novo* DNA methyltransferase 3B (DNMT3B) [63]. ICF syndrome is a recessive disease characterized by decreased levels of serum immunoglobulins in the presence of B cells, as well as chromatin decondensation and rearrangements, and satellite DNA hypomethylation specifically in the juxtacentromeric heterochromatin of chromosomes 1, 9, and 16. The hypomethylation of CpG sites in the pericentromeric satellite regions of these chromosomes was shown to be associated with centromeric instability.

USE OF EPIGENETIC MODIFIERS FOR POTENTIAL DIAGNOSIS AND THERAPY IN AUTOIMMUNE DISEASES

The knowledge base on epigenetic aberrations in autoimmune diseases is growing fast. Research in this field is important since the current diagnosis for most of these disorders is far from satisfactory. The diagnosis is possible often only after the appearance of symptoms, which also means long after the disease actually started, as is the case for RA. Moreover, the prediction of the progress of such diseases is also difficult. Similarly, even though there has been substantial progress in the treatment of some of these diseases, for example RA, there is no efficient therapy available for most of them and none of the autoimmune diseases described in this chapter can be cured.

The use of epigenetics for diagnostic purposes already has been shown in the field of cancer, where specific patterns of miRNA are of predictive value for the use of chemo- and radiotherapy in different tumor types [166]. Therefore, it is also possible that epigenetic differences could predict, for example, subgroups of patients with RA, such as destructive/nondestructive RA, or responders/nonresponders for the treatment available. Regarding potential therapy, there are accumulating data suggesting that HDAC inhibitors could be used for the treatment of RA to reduce joint swelling, synovial inflammation, and joint destruction [113].

Also, regarding lupus, the use of HDAC inhibitors appears to be promising for the development of future therapy via reduction of proteinuria, glomerulonephritis, and spleen weight [62]. Moreover, HDAC inhibitors could be also beneficial for the treatment of other autoimmune disorders since they have been shown to potentiate the suppressive function of Treg cells [35]. The specific inhibition of HDAC9 has been described as being particularly important in this context [35]. In general, the development of a potential drug targeting very specifically a distinct HDAC appears most promising since it would probably result in fewer side effects. Along this line, in SSc it was reported that both TSA treatment and specific inhibition of HDAC7 had a potential positive effect in inhibiting genes involved in pathogenesis, while the latter showed less adverse side effects [162].

On the other hand, it has been described that T cells which play a crucial role in the pathogenesis of RA and lupus have hypomethylated DNA [76]. Moreover, DNA methylation is also decreased in MS [128], and in the T cells of patients with SSc [165]. Therefore, developing a drug which could increase the activity of DNA methyltransferases could also be of therapeutic value. However, detailed analysis of drugs regulating DNA methylation have unfortunately revealed them to have numerous "off target" effects [167].

The use of demethylating drugs for treatment of other conditions such as cancer (where they reactivate the expression of tumor suppressor genes) could also have severe side effects such as inducing autoimmunity [77]. Targeting specific miRNA that are differentially expressed in autoimmune diseases, as in cancer [166], could be a potential promising treatment, but since one miRNA can have several targets, this approach needs careful examination. Most interestingly, there have also been distinct miRNAs regulating specific processes, such as miRNA-122 in the replication of hepatitic C [168] and miRNA-29 in the regulation of collagen synthesis in fibrosis [169]. Most impressively, with high speed plasma, miRNAs turn out to be sensitive and specific biomarkers of tissue injury [170]. Taking these facts together, it can be safely concluded that the identification and targeting of specific miRNAs will take place with accelerating speed in the near future.

References

1. Saito T, Gale Jr M, Principles of intracellular viral recognition. *Curr Opin Immunol* 2007;**19**:17–23.

2. Ospelt C, Gay S. TLRs and chronic inflammation. *Int J Biochem Cell Biol*, 2010;**42**:495–505.

3. Hietbrink F, Koenderman L, Rijkers G, Leenen L. Trauma: the role of the innate immune system. *World J Emerg Surg* 2006;**1**:15.

4. Castellheim A, Brekke OL, Espevik T, Harboe M, Mollnes TE. Innate immune responses to danger signals in systemic inflammatory response syndrome and sepsis. *Scand J Immunol* 2009;**69**:479–91.

5. Sjoberg AP, Trouw LA, Blom AM. Complement activation and inhibition: a delicate balance. *Trends Immunol* 2009;**30**:83–90.

6. Walzer T, Dalod M, Vivier E, Zitvogel L. Natural killer cell-dendritic cell crosstalk in the initiation of immune responses. *Expert Opin Biol Ther* 2005;**5**(Suppl 1):S49–S59.

7. Zhang X, Mosser DM. Macrophage activation by endogenous danger signals. *J Pathol* 2008;**214**:161–78.

8. Dinarello CA. Biologic basis for interleukin-1 in disease. *Blood* 1996;**87**:2095–147.

9. Biron CA, Brossay L. NK cells and NKT cells in innate defense against viral infections. *Curr Opin Immunol* 2001;**13**:458–64.

10. Chalifour A, Jeannin P, Gauchat JF, Blaecke A, Malissard M, N'Guyen T, et al. Direct bacterial protein PAMP recognition by human NK cells involves TLRs and triggers alpha-defensin production. *Blood* 2004;**104**:1778–83.

11. Banchereau J, Steinman RM. Dendritic cells and the control of immunity. *Nature* 1998;**392**:245–52.

12. Lee HK, Iwasaki A. Innate control of adaptive immunity: dendritic cells and beyond. *Semin Immunol* 2007;**19**:48–55.

13. Jiang H, Chess L. How the immune system achieves self-nonself discrimination during adaptive immunity. *Adv Immunol* 2009;**102**:95–133.

14. Matter MS, Ochsenbein AF. Natural antibodies target virus-antibody complexes to organized lymphoid tissue. *Autoimmun Rev* 2008;**7**:480–6.

15. McCullough KC, Summerfield A. Basic concepts of immune response and defense development. *ILAR J* 2005;**46**:230–40.

16. Peck A, Mellins ED. Precarious balance: Th17 cells in host defense. *Infect Immun* 2009;**78**:32–8.

17. Vignali DA, Collison LW, Workman CJ. How regulatory T cells work. *Nat Rev Immunol* 2008;**8**:523–32.

18. Harel M, Shoenfeld Y. Predicting and preventing autoimmunity, myth or reality? *Ann N Y Acad Sci* 2006;**1069**:322–45.

19. Kurki P, Aho K, Palosuo T, Heliovaara M. Immunopathology of rheumatoid arthritis. Antikeratin antibodies precede the clinical disease. *Arthritis Rheum* 1992;**35**:914–17.

20. Halldorsdottir HD, Jonsson T, Thorsteinsson J, Valdimarsson H. A prospective study on the incidence of rheumatoid arthritis among people with persistent increase of rheumatoid factor. *Ann Rheum Dis* 2000;**59**:149–51.

21. Nielen MM, van Schaardenburg D, Reesink HW, van de Stadt RJ, van der Horst-Bruinsma IE, et al. Specific autoantibodies precede the symptoms of rheumatoid arthritis: a study of serial measurements in blood donors. *Arthritis Rheum* 2004;**50**:380–6.

22. Calin A. The epidemiology of rheumatoid disease: past and present. *Dis Markers* 1986;**4**:1–6.

23. Sahatciu-Meka V, Izairi R, Rexhepi S, Manxhuka-Kerliu S. Comparative analysis of seronegative and seropositive rheumatoid arthritis regarding some epidemiological and anamnestic characteristics. *Reumatizam* 2007;**54**:5–11.

24. Alarcon GS, Koopman WJ, Acton RT, Barger BO. Seronegative rheumatoid arthritis. A distinct immunogenetic disease? *Arthritis Rheum* 1982;**25**:502–7.

25. Rantapaa-Dahlqvist S. What happens before the onset of rheumatoid arthritis? *Curr Opin Rheumatol* 2009;**21**:272–8.

26. Pan Y, Sawalha AH. Epigenetic regulation and the pathogenesis of systemic lupus erythematosus. *Transl Res* 2009;**153**:4–10.

27. Yin X, Latif R, Tomer Y, Davies TF. Thyroid epigenetics. *Ann N Y Acad Sci* 2007;**1110**:193–200.

28. Sobel RA. Genetic and epigenetic influence on EAE phenotypes induced with different encephalitogenic peptides. *J Neuroimmunol* 2000;**108**:45–52.

29. Tao R, de Zoeten EF, Ozkaynak E, Wang L, Li B, Greene MI, et al. Histone deacetylase inhibitors and transplantation. *Curr Opin Immunol* 2007;**19**:589–95.

30. Floess S, Freyer J, Siewert C, Baron U, Olek S, Polansky J, et al. Epigenetic control of the foxp3 locus in regulatory T cells. Epigenetic mechanisms of regulation of Foxp3 expression. *PLoS Biol* 2007;**5**:e38.

31. Kim HP, Leonard WJ. CREB/ATF-dependent T cell receptor-induced FoxP3 gene expression: a role for DNA methylation. *J Exp Med* 2007;**204**:1543–51.

32. Janson PC, Winerdal ME, Marits P, Thorn M, Ohlsson R, Winqvist O. FOXP3 promoter demethylation reveals the committed Treg population in humans. *PLoS ONE* 2008;**3**:e1612.

33. Baron U, Floess S, Wieczorek G, Baumann K, Grutzkau A, Dong J, et al. DNA demethylation in the human FOXP3 locus discriminates regulatory T cells from activated FOXP3(+) conventional T cells. *Eur J Immunol* 2007;**37**:2378–89.

34. Lal G, Zhang N, van der Touw W, Ding Y, Ju W, Bottinger EP, et al. Epigenetic regulation of Foxp3 expression in regulatory T cells by DNA methylation. *J Immunol* 2009;**182**:259–73.

35. Tao R, de Zoeten EF, Ozkaynak E, Chen C, Wang L, Porrett PM, et al. Deacetylase inhibition promotes the generation and function of regulatory T cells. *Nat Med* 2007;**13**:1299–307.

36. Chen C, Rowell EA, Thomas RM, Hancock WW, Wells AD. Transcriptional regulation by Foxp3 is associated with direct promoter occupancy and modulation of histone acetylation. *J Biol Chem* 2006;**281**:36828–34.

37. Ermann J, Fathman CG. Costimulatory signals controlling regulatory T cells. *Proc Natl Acad Sci USA* 2003;**100**:15292–3.

38. Li B, Samanta A, Song X, Iacono KT, Bembas K, Tao R, et al. FOXP3 interactions with histone acetyltransferase and class II histone deacetylases are required for repression. *Proc Natl Acad Sci USA* 2007;**104**:4571–6.

39. Jones B, Chen J. Inhibition of IFN-gamma transcription by site-specific methylation during T helper cell development. *EMBO J* 2006;**25**:2443–52.

40. Schoenborn JR, Dorschner MO, Sekimata M, Santer DM, Shnyreva M, Fitzpatrick DR, et al. Comprehensive epigenetic profiling identifies multiple distal regulatory elements directing transcription of the gene encoding interferon-gamma. *Nat Immunol* 2007;**8**:732–42.

41. Chang S, Aune TM. Dynamic changes in histone-methylation 'marks' across the locus encoding interferon-gamma during the differentiation of T helper type 2 cells. *Nat Immunol* 2007;**8**:723–31.

42. Valapour M, Guo J, Schroeder JT, Keen J, Cianferoni A, Casolaro V, et al. Histone deacetylation inhibits IL4 gene expression in T cells. *J Allergy Clin Immunol* 2002;**109**:238–45.

43. Grogan JL, Mohrs M, Harmon B, Lacy DA, Sedat JW, Locksley RM. Early transcription and silencing of cytokine genes underlie polarization of T helper cell subsets. *Immunity* 2001;**14**:205–15.

44. Bird JJ, Brown DR, Mullen AC, Moskowitz NH, Mahowald MA, Sider JR, et al. Helper T cell differentiation is controlled by the cell cycle. *Immunity* 1998;**9**:229–37.

45. Koyanagi M, Baguet A, Martens J, Margueron R, Jenuwein T, Bix M. EZH2 and histone 3 trimethyl lysine 27 associated with Il4 and Il13 gene silencing in Th1 cells. *J Biol Chem* 2005;**280**:31470–7.

46. Chan SC, Brown MA, Willcox TM, Li SH, Stevens SR, Tara D, et al. Abnormal IL-4 gene expression by atopic dermatitis T lymphocytes is reflected in altered nuclear protein interactions with IL-4 transcriptional regulatory element. *J Invest Dermatol* 1996;**106**:1131–6.

47. Rivas D, Mozo L, Zamorano J, Gayo A, Torre-Alonso JC, Rodriguez A, et al. Upregulated expression of IL-4 receptors and increased levels of IL-4 in rheumatoid arthritis patients. *J Autoimmun* 1995;**8**:587–600.

48. Wilson SB, Kent SC, Patton KT, Orban T, Jackson RA, Exley M, et al. Extreme Th1 bias of invariant Valpha24JalphaQ T cells in type 1 diabetes. *Nature* 1998;**391**:177–81.

49. Akimzhanov AM, Yang XO, Dong C. Chromatin remodeling of interleukin-17 (IL-17)-IL-17F cytokine gene locus during inflammatory helper T cell differentiation. *J Biol Chem* 2007;**282**:5969–72.

50. Wei L, Laurence A, Elias KM, O'Shea JJ. IL-21 is produced by Th17 cells and drives IL-17 production in a STAT3-dependent manner. *J Biol Chem* 2007;**282**:34605–10.

51. Danbara M, Kameyama K, Higashihara M, Takagaki Y. DNA methylation dominates transcriptional silencing of Pax5 in terminally differentiated B cell lines. *Mol Immunol* 2002;**38**:1161–6.

52. Amaravadi L, Klemsz MJ. DNA methylation and chromatin structure regulate PU.1 expression. *DNA Cell Biol* 1999;**18**:875–84.

53. Maier H, Colbert J, Fitzsimmons D, Clark DR, Hagman J. Activation of the early B-cell-specific mb-1 (Ig-alpha) gene by Pax-5 is dependent on an unmethylated Ets binding site. *Mol Cell Biol* 2003;**23**:1946–60.

54. Walter K, Bonifer C, Tagoh H. Stem cell-specific epigenetic priming and B cell-specific transcriptional activation at the mouse Cd19 locus. *Blood* 2008;**112**:1673–82.

55. Schwab J, Illges H. Regulation of CD21 expression by DNA methylation and histone deacetylation. *Int Immunol* 2001;**13**:705–10.

56. Baxter J, Sauer S, Peters A, John R, Williams R, Caparros ML, et al. Histone hypomethylation is an indicator of epigenetic plasticity in quiescent lymphocytes. *Embo J* 2004;**23**:4462–72.

57. Esteve PO, Chin HG, Smallwood A, Feehery GR, Gangisetty O, Karpf AR, et al. Direct interaction between DNMT1 and G9a coordinates DNA and histone methylation during replication. *Genes Dev* 2006;**20**:3089–103.

58. Xu CR, Feeney AJ. The epigenetic profile of Ig genes is dynamically regulated during B cell differentiation and is modulated by pre-B cell receptor signaling. *J Immunol* 2009;**182**:1362–9.

59. Lee SC, Bottaro A, Insel RA. Activation of terminal B cell differentiation by inhibition of histone deacetylation. *Mol Immunol* 2003;**39**:923–32.

60. Zabel MD, Weis JJ, Weis JH. Lymphoid transcription of the murine CD21 gene is positively regulated by histone acetylation. *J Immunol* 1999;**163**:2697–703.

61. Seo JS, Cho NY, Kim HR, Tsurumi T, Jang YS, Lee WK, et al. Cell cycle arrest and lytic induction of EBV-transformed B lymphoblastoid cells by a histone deacetylase inhibitor, Trichostatin A. *Oncol Rep* 2008;**19**:93–8.

62. Mishra N, Reilly CM, Brown DR, Ruiz P, Gilkeson GS. Histone deacetylase inhibitors modulate renal disease in the MRL-lpr/lpr mouse. *J Clin Invest* 2003;**111**:539–52.

63. Hansen RS, Wijmenga C, Luo P, Stanek AM, Canfield TK, Weemaes CM, et al. The DNMT3B DNA methyltransferase gene is mutated in the ICF immunodeficiency syndrome. *Proc Natl Acad Sci USA* 1999;**96**:14412–17.

64. Kondo T, Bobek MP, Kuick R, Lamb B, Zhu X, Narayan A, et al. Whole-genome methylation scan in ICF syndrome: hypomethylation of non-satellite DNA repeats D4Z4 and NBL2. *Hum Mol Genet* 2000;**9**:597–604.

65. Blanco-Betancourt CE, Moncla A, Milili M, Jiang YL, Viegas-Pequignot EM, Roquelaure B, et al. Defective B-cell-negative selection and terminal differentiation in the ICF syndrome. *Blood* 2004;**103**:2683–90.

66. Foster SL, Hargreaves DC, Medzhitov R. Gene-specific control of inflammation by TLR-induced chromatin modifications. *Nature* 2007;**447**:972–8.

67. Wen H, Dou Y, Hogaboam CM, Kunkel SL. Epigenetic regulation of dendritic cell-derived interleukin-12 facilitates immunosuppression after a severe innate immune response. *Blood* 2008;**111**:1797–804.

68. Wen H, Schaller MA, Dou Y, Hogaboam CM, Kunkel SL. Dendritic cells at the interface of innate and acquired immunity: the role for epigenetic changes. *J Leukoc Biol* 2008;**83**:439–46.

69. Banchereau J, Pascual V, Palucka AK. Autoimmunity through cytokine-induced dendritic cell activation. *Immunity* 2004;**20**:539–50.

70. Banchereau J, Pascual V. Type I interferon in systemic lupus erythematosus and other autoimmune diseases. *Immunity* 2006;**25**:383–92.

71. Namjou B, Kilpatrick J, Harley JB. Genetics of clinical expression in SLE. *Autoimmunity* 2007;**40**:602–12.

72. Paul E, Manheimer-Lory A, Livneh A, Solomon A, Aranow C, Ghossein C, et al. Pathogenic anti-DNA antibodies in SLE: idiotypic families and genetic origins. *Int Rev Immunol* 1990;**5**:295–313.

73. La Cava A. Lupus and T cells. *Lupus* 2009;**18**:196–201.

74. Moser KL, Kelly JA, Lessard CJ, Harley JB. Recent insights into the genetic basis of systemic lupus erythematosus. *Genes Immun* 2009;**10**:373–9.

75. Richardson B. Effect of an inhibitor of DNA methylation on T cells. II. 5-Azacytidine induces self-reactivity in antigen-specific T4+ cells. *Hum Immunol* 1986;**17**:456–70.

76. Richardson B, Scheinbart L, Strahler J, Gross L, Hanash S, Johnson M. Evidence for impaired T cell DNA methylation in systemic lupus erythematosus and rheumatoid arthritis. *Arthritis Rheum* 1990;**33**:1665–73.

77. Vedove CD, Del Giglio M, Schena D, Girolomoni G. Drug-induced lupus erythematosus. *Arch Dermatol Res* 2009;**301**:99–105.

78. Scheinbart LS, Johnson MA, Gross LA, Edelstein SR, Richardson BC. Procainamide inhibits DNA methyltransferase in a human T cell line. *J Rheumatol* 1991;**18**:530–4.

79. Deng C, Lu Q, Zhang Z, Rao T, Attwood J, Yung R, et al. Hydralazine may induce autoimmunity by inhibiting extracellular signal-regulated kinase pathway signaling. *Arthritis Rheum* 2003;**48**:746–56.

80. Oelke K, Richardson B. Decreased T cell ERK pathway signaling may contribute to the development of lupus through effects on DNA methylation and gene expression. *Int Rev Immunol* 2004;**23**:315–31.

81. Sawalha AH, Jeffries M, Webb R, Lu Q, Gorelik G, Ray D, et al. Defective T-cell ERK signaling induces interferon-regulated gene expression and overexpression of methylation-sensitive genes similar to lupus patients. *Genes Immun* 2008;**9**:368–78.

82. Richardson B, Ray D, Yung R. Murine models of lupus induced by hypomethylated T cells. *Methods Mol Med* 2004;**102**:285–94.

83. Sekigawa I, Kawasaki M, Ogasawara H, Kaneda K, Kaneko H, Takasaki Y, et al. DNA methylation: its contribution to systemic lupus erythematosus. *Clin Exp Med* 2006;**6**:99–106.

84. Kaplan MJ, Lu Q, Wu A, Attwood J, Richardson B. Demethylation of promoter regulatory elements contributes to perforin overexpression in CD4+ lupus T cells. *J Immunol* 2004;**172**:3652–61.

85. Lu Q, Wu A, Richardson BC. Demethylation of the same promoter sequence increases CD70 expression in lupus T cells and T cells treated with lupus-inducing drugs. *J Immunol* 2005;**174**:6212–19.

86. Lu Q, Kaplan M, Ray D, Zacharek S, Gutsch D, Richardson B. Demethylation of ITGAL (CD11a) regulatory sequences in systemic lupus erythematosus. *Arthritis Rheum* 2002;**46**:1282–91.

87. Casciola-Rosen L, Andrade F, Ulanet D, Wong WB, Rosen A. Cleavage by granzyme B is strongly predictive of autoantigen status: implications for initiation of autoimmunity. *J Exp Med* 1999;**190**:815–26.

88. Blanco P, Pitard V, Viallard JF, Taupin JL, Pellegrin JL, Moreau JF. Increase in activated CD8+ T lymphocytes expressing perforin and granzyme B correlates with disease activity in patients with systemic lupus erythematosus. *Arthritis Rheum* 2005;**52**:201–11.

89. Lu Q, Wu A, Tesmer L, Ray D, Yousif N, Richardson B. Demethylation of CD40LG on the inactive X in T cells from women with lupus. *J Immunol* 2007;**179**:6352–8.

90. Soto ME, Vallejo M, Guillen F, Simon JA, Arena E, Reyes PA. Gender impact in systemic lupus erythematosus. *Clin Exp Rheumatol* 2004;**22**:713–21.

91. Garcia BA, Busby SA, Shabanowitz J, Hunt DF, Mishra N. Resetting the epigenetic histone code in the MRL-lpr/lpr mouse model of lupus by histone deacetylase inhibition. *J Proteome Res* 2005;**4**:2032–42.

92. Reilly CM, Mishra N, Miller JM, Joshi D, Ruiz P, Richon VM, et al. Modulation of renal disease in MRL/lpr mice by suberoylanilide hydroxamic acid. *J Immunol* 2004;**173**:4171–8.

93. Forster N, Gallinat S, Jablonska J, Weiss S, Elsasser HP, Lutz W. p300 protein acetyltransferase activity suppresses systemic lupus erythematosus-like autoimmune disease in mice. *J Immunol* 2007;**178**:6941–8.

94. Hu N, Qiu X, Luo Y, Yuan J, Li Y, Lei W, et al. Abnormal histone modification patterns in lupus CD4+ T cells. *J Rheumatol* 2008;**35**:804–10.

95. Palucka AK, Blanck JP, Bennett L, Pascual V, Banchereau J. Cross-regulation of TNF and IFN-alpha in autoimmune diseases. *Proc Natl Acad Sci USA* 2005;**102**:3372–7.

96. Tang Y, Luo X, Cui H, Ni X, Yuan M, Guo Y, et al. MicroRNA-146A contributes to abnormal activation of the type I interferon pathway in human lupus by targeting the key signaling proteins. *Arthritis Rheum* 2009;**60**:1065–75.

97. Bhanji RA, Eystathioy T, Chan EK, Bloch DB, Fritzler MJ. Clinical and serological features of patients with autoantibodies to GW/P bodies. *Clin Immunol* 2007;**125**:247–56.

98. Harris ED Jr., DiBona DR, Krane SM. A mechanism for cartilage destruction in rheumatoid arthritis. *Trans Assoc Am Physicians* 1970;**83**:267–76.

99. Karouzakis E, Neidhart M, Gay RE, Gay S. Molecular and cellular basis of rheumatoid joint destruction. *Immunol Lett* 2006;**106**:8–13.

100. Grassi W, De Angelis R, Lamanna G, Cervini C. The clinical features of rheumatoid arthritis. *Eur J Radiol* 1998;**27**(Suppl 1):S18–S24.

101. Van Doornum S, McColl G, Wicks IP. Accelerated atherosclerosis: an extraarticular feature of rheumatoid arthritis? *Arthritis Rheum* 2002;**46**:862–73.

102. Symmons DP. Epidemiology of rheumatoid arthritis: determinants of onset, persistence and outcome. *Best Pract Res Clin Rheumatol* 2002;**16**:707–22.

103. Muller-Ladner U, Kriegsmann J, Franklin BN, Matsumoto S, Geiler T, Gay RE, et al. Synovial fibroblasts of patients with rheumatoid arthritis attach to and invade normal human cartilage when engrafted into SCID mice. *Am J Pathol* 1996;**149**:1607–15.

104. Ospelt C, Gay S. The role of resident synovial cells in destructive arthritis. *Best Pract Res Clin Rheumatol* 2008;**22**:239–52.

105. Meinecke I, Cinski A, Baier A, Peters MA, Dankbar B, Wille A, et al. Modification of nuclear PML protein by SUMO-1 regulates Fas-induced apoptosis in rheumatoid arthritis synovial fibroblasts. *Proc Natl Acad Sci USA* 2007;**104**:5073–8 [Epub 2007 Mar 5014].

106. Davis LS. A question of transformation: the synovial fibroblast in rheumatoid arthritis. *Am J Pathol* 2003;**162**:1399–402.

107. Karouzakis E, Gay RE, Michel BA, Gay S, Neidhart M. DNA hypomethylation in rheumatoid arthritis synovial fibroblasts. *Arthritis Rheum* 2009;**60**:3613–22.

108. Wen ZK, Xu W, Xu L, Cao QH, Wang Y, Chu YW, et al. DNA hypomethylation is crucial for apoptotic DNA to induce systemic lupus erythematosus-like autoimmune disease in SLE-non-susceptible mice. *Rheumatology (Oxford)* 2007;**46**:1796–803.

109. Neidhart M, Rethage J, Kuchen S, Kunzler P, Crowl RM, Billingham ME, et al. Retrotransposable L1 elements expressed in rheumatoid arthritis synovial tissue: association with genomic DNA hypomethylation and influence on gene expression. *Arthritis Rheum* 2000;**43**:2634–47.

110. Nile CJ, Read RC, Akil M, Duff GW, Wilson AG. Methylation status of a single CpG site in the IL6 promoter is related to IL6 messenger RNA levels and rheumatoid arthritis. *Arthritis Rheum* 2008;**58**:2686–93.

111. Takami N, Osawa K, Miura Y, Komai K, Taniguchi M, Shiraishi M, et al. Hypermethylated promoter region of DR3, the death receptor 3 gene, in rheumatoid arthritis synovial cells. *Arthritis Rheum* 2006;**54**:779–87.

112. Huber LC, Brock M, Hemmatazad H, Giger OT, Moritz F, Trenkmann M, et al. Histone deacetylase/acetylase activity in total synovial tissue derived from rheumatoid arthritis and osteoarthritis patients. *Arthritis Rheum* 2007;**56**:1087–93.

113. Nishida K, Komiyama T, Miyazawa S, Shen ZN, Furumatsu T, Doi H, et al. Histone deacetylase inhibitor suppression of autoantibody-mediated arthritis in mice via regulation of p16INK4a and p21(WAF1/Cip1) expression. *Arthritis Rheum* 2004;**50**:3365–76.

114. Manabe H, Nasu Y, Komiyama T, Furumatsu T, Kitamura A, Miyazawa S, et al. Inhibition of histone deacetylase down-regulates the expression of hypoxia-induced vascular endothelial growth factor by rheumatoid synovial fibroblasts. *Inflamm Res* 2008;**57**:4–10.

115. Nakamura C, Matsushita I, Kosaka E, Kondo T, Kimura T. Anti-arthritic effects of combined treatment with histone deacetylase inhibitor and low-intensity ultrasound in the presence of microbubbles in human rheumatoid synovial cells. *Rheumatology (Oxford)* 2008;**47**:418–24 [Epub 2008 Feb 2015].

116. Jungel A, Baresova V, Ospelt C, Simmen BR, Michel BA, Gay RE, et al. Trichostatin A sensitises rheumatoid arthritis synovial fibroblasts for TRAIL-induced apoptosis. *Ann Rheum Dis* 2006;**65**:910–2.

117. Maciejewska-Rodrigues H, Karouzakis E, Strietholt S, Hemmatazad H, Neidhart M, Ospelt C, et al. Epigenetics and rheumatoid arthritis: the role of SENP1 in the regulation of MMP-1 expression. *J Autoimmun* 2010;**35**:15–22.

118. Stanczyk J, Pedrioli DM, Brentano F, Sanchez-Pernaute O, Kolling C, Gay RE, et al. Altered expression of MicroRNA in synovial fibroblasts and synovial tissue in rheumatoid arthritis. *Arthritis Rheum* 2008;**58**:1001–9.

119. Nakasa T, Miyaki S, Okubo A, Hashimoto M, Nishida K, Ochi M, et al. Expression of microRNA-146 in rheumatoid arthritis synovial tissue. *Arthritis Rheum* 2008;**58**:1284–92.

120. Stanczyk Feldges J, Karouzakis E, Jüngel A, Ospelt C, Kolling C, Michel BA, et al. Mir-203 regulates the expression of IL-6 and matrix metalloproteinase (MMP)-1 in RA synovial fibroblasts. *Arthritis Rheum* 2009;**60**:S706.

121. Turner M, Vigorito E. Regulation of B- and T-cell differentiation by a single microRNA. *Biochem Soc Trans* 2008;**36**:531–3.

122. Perosa F, Favoino E, Caragnano MA, Prete M, Dammacco F. CD20: a target antigen for immunotherapy of autoimmune diseases. *Autoimmun Rev* 2005;**4**:526–31.

123. Pauley KM, Satoh M, Chan AL, Bubb MR, Reeves WH, Chan EK. Upregulated miR-146a expression in peripheral blood mononuclear cells from rheumatoid arthritis patients. *Arthritis Res Ther* 2008;**10**:R101.

124. Stanczyk J, Jungel A, Ospelt C, Gay RE, Distler O, Pisetsky D, et al. Detection of MicroRNA in the microparticles derived from monocytes. *Arthritis Rheum* 2008;**ACR2008**:S155.

125. Jungel A, Distler O, Schulze-Horsel U, Huber LC, Ha HR, Simmen B, et al. Microparticles stimulate the synthesis of prostaglandin E(2) via induction of cyclooxygenase 2 and microsomal prostaglandin E synthase 1. *Arthritis Rheum* 2007;**56**:3564–74.

126. Distler JH, Pisetsky DS, Huber LC, Kalden JR, Gay S, Distler O. Microparticles as regulators of inflammation: novel players of cellular crosstalk in the rheumatic diseases. *Arthritis Rheum* 2005;**52**:3337–48.

127. Moscarello MA, Mastronardi FG, Wood DD. The role of citrullinated proteins suggests a novel mechanism in the pathogenesis of multiple sclerosis. *Neurochem Res* 2007;**32**:251–6.

128. Mastronardi FG, Noor A, Wood DD, Paton T, Moscarello MA. Peptidyl argininedeiminase 2 CpG island in multiple sclerosis white matter is hypomethylated. *J Neurosci Res* 2007;**85**:2006–16.

129. Mastronardi FG, Tsui H, Winer S, Wood DD, Selvanantham T, Galligan C, et al. Synergy between paclitaxel plus an exogenous methyl donor in the suppression of murine demyelinating diseases. *Mult Scler* 2007;**13**:596–609.

130. Weiner HL. The challenge of multiple sclerosis: how do we cure a chronic heterogeneous disease? *Ann Neurol* 2009;**65**:239–48.

131. Lassmann H, Bruck W, Lucchinetti CF. The immunopathology of multiple sclerosis: an overview. *Brain Pathol* 2007;**17**:210–18.

132. Carrithers MD, Visintin I, Kang SJ, Janeway CA Jr.. Differential adhesion molecule requirements for immune surveillance and inflammatory recruitment. *Brain* 2000;**123**(6):1092–101.

133. Bjartmar C, Wujek JR, Trapp BD. Axonal loss in the pathology of MS: consequences for understanding the progressive phase of the disease. *J Neurol Sci* 2003;**206**:165–71.

134. Moscarello MA, Pritzker L, Mastronardi FG, Wood DD. Peptidylarginine deiminase: a candidate factor in demyelinating disease. *J Neurochem* 2002;**81**:335–43.

135. Kelly MA, Rayner ML, Mijovic CH, Barnett AH. Molecular aspects of type 1 diabetes. *Mol Pathol* 2003;**56**:1–10.

136. Gepts W, De Mey J. Islet cell survival determined by morphology. An immunocytochemical study of the islets of Langerhans in juvenile diabetes mellitus. *Diabetes* 1978;**27**(Suppl 1):251–61.

137. Reimann M, Bonifacio E, Solimena M, Schwarz PE, Ludwig B, Hanefeld M, et al. An update on preventive and regenerative therapies in diabetes mellitus. *Pharmacol Ther* 2009;**121**:317–31.

138. Naquet P, Ellis J, Tibensky D, Kenshole A, Singh B, Hodges R, et al. T cell autoreactivity to insulin in diabetic and related non-diabetic individuals. *J Immunol* 1988;**140**:2569–78.

139. Palmer JP, Asplin CM, Clemons P, Lyen K, Tatpati O, Raghu PK, et al. Insulin antibodies in insulin-dependent diabetics before insulin treatment. *Science* 1983;**222**:1337–9.

140. Zier KS, Leo MM, Spielman RS, Baker L. Decreased synthesis of interleukin-2 (IL-2) in insulin-dependent diabetes mellitus. *Diabetes* 1984;**33**:552–5.

141. Chistiakov DA, Voronova NV, Chistiakov PA. The crucial role of IL-2/IL-2RA-mediated immune regulation in the pathogenesis of type 1 diabetes, an evidence coming from genetic and animal model studies. *Immunol Lett* 2008;**118**:1–5.

142. Yamanouchi J, Rainbow D, Serra P, Howlett S, Hunter K, Garner VES, et al. Interleukin-2 gene variation impairs regulatory T cell function and causes autoimmunity. *Nat Genet* 2007;**39**:329–37.

143. Yui MA, Hernandez-Hoyos G, Rothenberg EV. A new regulatory region of the IL-2 locus that confers position-independent transgene expression. *J Immunol* 2001;**166**:1730–9.

144. Wang L, Kametani Y, Katano I, Habu S. T-cell specific enhancement of histone H3 acetylation in 5′ flanking region of the IL-2 gene. *Biochem Biophys Res Commun* 2005;**331**:589–94.

145. Adachi S, Rothenberg EV. Cell-type-specific epigenetic marking of the IL2 gene at a distal cis-regulatory region in competent, nontranscribing T-cells. *Nucleic Acids Res* 2005;**33**:3200–10.

146. Thomas RM, Gao L, Wells AD. Signals from CD28 induce stable epigenetic modification of the IL-2 promoter. *J Immunol* 2005;**174**:4639–46.

147. Northrop JK, Thomas RM, Wells AD, Shen H. Epigenetic remodeling of the IL-2 and IFN-gamma loci in memory CD8 T cells is influenced by CD4 T cells. *J Immunol* 2006;**177**:1062–9.

148. Murayama A, Sakura K, Nakama M, Yasuzawa-Tanaka K, Fujita E, Tateishi Y, et al. A specific CpG site demethylation in the human interleukin 2 gene promoter is an epigenetic memory. *Embo J* 2006;**25**:1081–92.

149. Richard S, Morel M, Cleroux P. Arginine methylation regulates IL-2 gene expression: a role for protein arginine methyltransferase 5 (PRMT5). *Biochem J* 2005;**388**:379–86.

150. Wicker LS, Todd JA, Prins JB, Podolin PL, Renjilian RJ, Peterson LB. Resistance alleles at two non-major histocompatibility complex-linked insulin-dependent diabetes loci on chromosome 3, Idd3 and Idd10, protect nonobese diabetic mice from diabetes. *J Exp Med* 1994;**180**:1705–13.

151. Lyons PA, Armitage N, Argentina F, Denny P, Hill NJ, Lord CJ, et al. Congenic mapping of the type 1 diabetes locus, Idd3, to a 780-kb region of mouse chromosome 3: identification of a candidate segment of ancestral DNA by haplotype mapping. *Genome Res* 2000;**10**:446–53.

152. Miao F, Smith DD, Zhang L, Min A, Feng W, Natarajan R. Lymphocytes from patients with type 1 diabetes display a distinct profile of chromatin histone H3 lysine 9 dimethylation: an epigenetic study in diabetes. *Diabetes* 2008;**57**:3189–98.

153. Leibiger IB, Berggren PO. A SIRTain role in pancreatic beta cell function. *Cell Metab* 2005;**2**:80–2.

154. Varga J, Abraham D. Systemic sclerosis: a prototypic multisystem fibrotic disorder. *J Clin Invest* 2007;**117**:557–67.

155. Abraham DJ, Krieg T, Distler J, Distler O. Overview of pathogenesis of systemic sclerosis. *Rheumatology (Oxford)* 2009;**48**(Suppl 3):iii3–7.

156. Gu YS, Kong J, Cheema GS, Keen CL, Wick G, Gershwin ME. The immunobiology of systemic sclerosis. *Semin Arthritis Rheum* 2008;**38**:132–60.

157. LeRoy EC. Increased collagen synthesis by scleroderma skin fibroblasts in vitro: a possible defect in the regulation or activation of the scleroderma fibroblast. *J Clin Invest* 1974;**54**:880–9.

158. Derk CT, Jimenez SA. Systemic sclerosis: current views of its pathogenesis. *Autoimmun Rev* 2003;**2**:181–91.

159. Wang Y, Fan PS, Kahaleh B. Association between enhanced type I collagen expression and epigenetic repression of the FLI1 gene in scleroderma fibroblasts. *Arthritis Rheum* 2006;**54**:2271–9.

160. Czuwara-Ladykowska J, Shirasaki F, Jackers P, Watson DK, Trojanowska M. Fli-1 inhibits collagen type I production in dermal fibroblasts via an Sp1-dependent pathway. *J Biol Chem* 2001;**276**:20839–48.

161. Huber LC, Distler JH, Moritz F, Hemmatazad H, Hauser T, Michel BA, et al. Trichostatin A prevents the accumulation of extracellular matrix in a mouse model of bleomycin-induced skin fibrosis. *Arthritis Rheum* 2007;**56**:2755–64.

162. Hemmatazad H, Maciejewska Rodrigues H, Maurer B, Brentano F, Pileckyte M, Distler JH, et al. Histone deacetylase 7 – A potential target for the anti-fibrotic treatment of systemic sclerosis. *Arthritis Rheum* 2009;**60**:1519–29.

163. Nassabeh N. Connective tissue diseases: is epigenetic modification the future of antifibrotic therapy? *Nat Rev Rheumatol* 2009;**5**:354–354.

164. Dees C, Akhmetshina A, Busch N, Horn A, Gusinde J, Nevskaya T, et al. Inhibitors of DNA methylotransferases exert potent anti-fibrotic effects via re-activation of SOCS-3. *Ann Rheum Dis* 2009;**68**(Suppl 3):94.

165. Lei W, Luo Y, Yan K, Zhao S, Li Y, Qiu X, et al. Abnormal DNA methylation in CD4 + T cells from patients with systemic lupus erythematosus, systemic sclerosis, and dermatomyositis. *Scand J Rheumatol* 2009;**38**:369–74.

166. Hummel R, Hussey DJ, Haier J. MicroRNAs: predictors and modifiers of chemo- and radiotherapy in different tumour types. *Eur J Cancer* 2010;**46**:298–311.

167. Flotho C, Claus R, Batz C, Schneider M, Sandrock I, Ihde S, et al. The DNA methyltransferase inhibitors azacitidine, decitabine and zebularine exert differential effects on cancer gene expression in acute myeloid leukemia cells. *Leukemia* 2009;**23**:1019–28.

168. Chang J, Guo JT, Jiang D, Guo H, Taylor JM, Block TM. Liver-specific microRNA miR-122 enhances the replication of hepatitis C virus in nonhepatic cells. *J Virol* 2008;**82**:8215–23.

169. van Rooij E, Sutherland LB, Thatcher JE, DiMaio JM, Naseem RH, Marshall WS, et al. Dysregulation of microRNAs after myocardial infarction reveals a role of miR-29 in cardiac fibrosis. *Proc Natl Acad Sci USA* 2008;**105**:13027–32.

170. Laterza OF, Lim L, Garrett-Engele PW, Vlasakova K, Muniappa N, Tanaka WK, et al. Plasma MicroRNAs as sensitive and specific biomarkers of tissue injury. *Clin Chem* 2009;**55**:1977–83.

CHAPTER 34

Epigenetics of Brain Disorders

Johannes Gräff, Tamara B. Franklin, and Isabelle M. Mansuy
Brain Research Institute, Medical Faculty of the University of Zürich and Department of Biology, Swiss Federal Institute of Technology, CH-8057 Zürich, Switzerland

INTRODUCTION

Epigenetics is most commonly defined as the study of alterations in gene function that are heritable through both mitosis and meiosis, but do not involve any change in the DNA sequence itself [1] [reviewed in Refs 2,3]. At a molecular level, epigenetic mechanisms comprise first and foremost chemical modifications of the DNA, and histone proteins, the major constituents of the chromatin. Additional mechanisms such as RNA interference, prion-like processes, and nucleosomes positioning also constitute epigenetic mechanisms [4], but are not covered in this chapter.

553

DNA methylation most commonly occurs at cytosine–guanine dinucleotides (CpG), and is generally associated with transcriptional silencing. Silencing takes place through direct inhibition of the binding of transcription factors, or the indirect recruitment of methyl-CpG binding proteins (MBPs) and associated repressive chromatin-remodeling activity [5,6]. Since DNA methylation also occurs at the promoter and coding region of actively transcribed genes, it has also been implicated in transcriptional activity but the underlying mechanisms remain unknown [7,8]. Epigenetic modifications at the chromatin refer essentially to covalent posttranslational modifications (PTMs) of histone proteins. Histones are basic proteins whose major roles are to shape the DNA and control the condensation and the accessibility of the chromatin. They are composed of a globular core, a C-terminus tail, and an N-terminus tail, which is a loosely structured sequence of amino acids protruding out of the core. PTMs of histones can be found on all portions of the histones but are most predominant on the N-terminus tail because they project through the major and minor grooves of the DNA helix, and are therefore accessible to listone-modifying enzymes [9]. These modifications primarily involve acetylation, methylation, phosphorylation, ubiquitination, and sumoylation, but also include ADP-ribosylation, glycosylation, and carbonylation [reviewed in Refs 10,11]. Owing to their specific chemical properties, PTMs dynamically modulate the structure and degree of compaction of the chromatin, and thereby control its accessibility to the transcriptional machinery. This largely depends on the panoply of PTMs present on individual genes at a given time point.

Both the acetylation and the phosphorylation of histones, which occur on lysine (K) and on serine (S), threonine (T), or tyrosine (Y) residues respectively, are associated with transcriptional activation [12]. For acetylation, this results in part from the neutralization

Handbook of Epigenetics: The New Molecular and Medical Genetics. DOI: 10.1016/B978-0-12-375709-8.00034-4

of the positive charge on lysine by the added acetyl group, while for phosphorylation, it involves the addition of a negative phospho group. This induces a repulsive force between the histone tail and the negatively-charged DNA, which relaxes the chromatin structure, and makes it more accessible to the transcriptional machinery. In contrast, histone methylation on lysine is associated with both actively transcribed and silenced genes [13,14], depending on the residue. It can occur as mono, di- or trimethylation, which modulate gene expression differently [15]. Histone methylation is also found on arginine (R) in both mono- or dimethylated form, but the impact on chromatin structure is not well understood [15]. Protein ubiquitination (also called ubiquitylation), which involves the attachment of the highly-conserved 76 amino acid polypeptide ubiquitin, is most commonly associated with the marking of proteins for degradation by the proteasome. However, ubiquitination also occurs on histone tails where its functions remain not fully characterized. It has been identified as a prerequisite for subsequent histone methylation, and correlates with both transcriptional activation and nucleosome loosening [9,15,16]. Finally, histone sumoylation is the least known histone PTM. In yeast, it has been shown to occur on all four core histones, and can negatively regulate transcription, possibly by interfering with histone acetylation and ubiquitination [17,18].

The repertoire of DNA and histone modifications is induced, maintained, and modulated, by specific enzymes that include DNA methyltransferases (DNMTs), histone acetyltransferases (HATs) and deacetylases (HDACs), histone methyltransferases (HMTs), protein kinases and phosphatases, and ubiquitin- and SUMO-associated enzymes [4,10,11,15,18,19]. These enzymes operate both independently and in synergy to establish an "epigenetic code". This code is highly dynamic and, in combination with chromatin-associated proteins, determines the pattern of gene expression in a gene-specific manner [20–22].

EPIGENETIC DYSREGULATION IN NEURODEVELOPMENTAL DISORDERS – THE EXAMPLE OF RETT SYNDROME

Neurodevelopmental disorders are characterized by impaired functions of the central nervous system that can appear early in development and persist into adulthood. The impairments can have their onset during initial phases of development, as in the case of fetal alcohol as syndrome when alcohol is ingested early in pregnancy, while others appear later and are due to a genetic predisposition, such as Down's syndrome. Several genetically-determined neurodevelopmental disorders have been documented to involve, at least in part, an epigenetic dysregulation of physiological functions. In this chapter, we will focus on Rett syndrome (RS), a disorder that implicates aberrant DNA methylation and histone PTMs. For a recent review of epigenetic dysregulation in other neurodevelopmental disorders such as Fragile X or Rubinstein–Taybi syndrome, the reader is referred to Refs 23 and 24.

RS is a relatively common and progressive neurological disorder characterized by an arrest of the development of the nervous system and mental retardation. It is classified as an autism spectrum disorder in DSM-IV (Diagnostic and Statistical Manual of Mental Disorders IV). In 80% of cases, RS is caused by a loss-of-function mutation in the X-linked *methyl-CpG-binding protein 2* (*MeCP2*) gene [25] (Table 34.1), which is lethal when hemizygous. The disease affects females exclusively and has a worldwide prevalence of 1:10,000. MeCP2 is a transcriptional regulator that binds specifically to methylated DNA. Its deficiency alters chromatin remodeling and induces a general dysregulation in gene transcription. Although MeCP2 is a member of the MBP family of transcriptional repressors (for a review see Ref. 5), it has recently also been implicated in transcriptional activation. It is thus both a transcriptional silencer and activator, and is associated with increased and decreased gene transcription [26].

Based on the fact that RS patients have a mutation in the *MeCP2* gene, RS was modeled in mice by *MeCP2* null mutation or neuron-specific deletion. Consistent with that observed in human, *MeCP2*-deficient mice have reduced brain weight and abnormal neuroanatomical

TABLE 34.1 Epigenetic Mechanisms in Selected Brain Disorders

Brain Disorder	Pathological Epigenetic Modification	Disrupted Enzymatic Machinery	Gene(s) Affected	Organism Studied	Potential "Epigenetic" Intervention	Reference(s)
Rett syndrome	DNA methylation	MeCP2	Bdnf	Young MeCP2 GOF transgenic mouse	Overexpression of MeCP2	[19,21–23,34]
	Histone acetylation	MeCP2	Bdnf	MeCP2 LOF transgenic mice	None suggested	[25,34]
	Histone methylation: H3K9/H3K4	MeCP2	Bdnf	Murine cell culture	None suggested	[25]
Alzheimer's Disease	DNA methylation	Not assessed	PS1	Human cell culture	Methyl-donor SAM	[44]
		Not assessed	PS1	Human post-mortem tissue	None suggested	[47]
		DNMT1	APP	Macaques	None suggested	[46]
	Histone acetylation	Not assessed	Not assessed	p25/Cdk5 transgenic mouse	HDAC SIRT1	[40]
		CBP	CREB-target genes	Human cell/murine neuronal culture	Substitution of PS1-mediated enzymatic activity	[39]
		Tip60	Not assessed	Human cell culture	None suggested	[38]
	Histone acetylation	Not assessed	Not assessed	p25/Cdk5 transgenic mouse	HDACi sodium butyrate	[43]
		CBP	Not assessed	PS1 transgenic mouse	None suggested	[42]
		CBP	c-fos, Bdnf	Murine neuronal culture	None suggested	[41]
		DNMT1	APP	Macaques	None suggested	[46]
Depression	DNA methylation	DNMT3b	GABA-A α1	Depressed suicide victims	None suggested	[58]
		MeCP2, MBD1	Not assessed	Rats	None suggested	[57]
	Histone methylation: H3K9/K27	Not assessed	Various, e.g. Bdnf	Chronic social defeat stress mouse model	Chronic administration of antidepressant fluoxetine	[54,55]
	Histone acetylation	HDAC5	Various, e.g. NK1R	Chronic social defeat stress mouse model	Chronic administration of antidepressant imipramine	[54,56]
Schizophrenia	DNA methylation	MeCP2	reelin	Human post-mortem tissue in vitro	None suggested	[63,64]
				Mice and in vitro	5-aza	[65]
				In vitro	HDACis valproic acid, TSA, MS-275	[65,66,80,81]
		Not assessed	GAD67	Human post-mortem tissue	None suggested	[68]
					HDACis TSA, MS-275	[80]
	DNA methylation	Not assessed	SOX10	Human post-mortem tissue	None suggested	[69]
Predisposition to stress	DNA methylation/ histone acetylation	Not assessed	COMT	Human post-mortem tissue	None suggested	[75]
	DNA methylation	Not assessed	GR1	Rats	TSA	[87]
	DNA methylation	Not assessed	GR1	Human post-mortem tissue	None suggested	[91]

structures resulting from smaller neurons [27]. They exhibit an overall decrease in exploratory activity [27,28], and have cognitive deficits and impaired synaptic plasticity [29]. Although severe, the cognitive deficits could be reversed by overexpression (by two fold) of wild-type human MeCP2 protein in young mutant animals (10-week-old) [29], confirming that they resulted from MeCP2 deficiency. However, intriguingly, in 20-week-old mice human MeCP2 overexpression induced seizures, suggesting that an excess in MeCP2 is deleterious to brain functions, and could also lead to neurological functions related to RS.

MeCP2 is a transcriptional regulator that controls multiple target genes by binding to their promoter in the chromatin. Its binding is regulated by calcium-dependent phosphorylation, and phosphorylation on S80 reduces this binding [30]. One known target of MeCP2 is the brain-derived neurotrophic factor (*Bdnf*) gene. MeCP2 binding to the *Bdnf* promoter derepresses *Bdnf* transcription and is associated with decreased methylation [31–33]. This suggests that MeCP2 deficiency should lead to increased *Bdnf* expression; however, this is not the case in MeCP2 mutant mice. These mice have reduced BDNF levels, a reduction that can be reversed by forebrain-specific overexpression of *Bdnf in vivo* [34]. This reversal is accompanied by reduced neuronal atrophy characteristic of RS, a beneficial effect recently confirmed in cultured hippocampal neurons [35]. These observations led to the hypothesis that a lack in MeCP2 binding *per se* is not the primary factor for determining the level of *Bdnf*, and that, perhaps an alteration in neuronal activity resulting from MeCP2 deficiency may be more important [36]. Alternatively, since MeCP2 is also expressed in astrocytes [37], it is conceivable that its mechanisms of *Bdnf* regulation are different in neuronal and glial cells. More refined analyses are needed to resolve this issue.

Further to *Bdnf*, seven other genes including myelin-associated proteins and dopamine decarboxylase, have recently been identified as direct binding targets of MeCP2 in the mouse brain [38]. Since MeCP2 phosphorylation is associated with dendritic growth and spine maturation [33], thought to occur through the derepression of a variety of target genes involved in development processes [26,39], many more genes are likely to be involved.

Further to regulating DNA methylation, MeCP2 also influences histone PTMs, in particular histone acetylation and methylation. At the promoter region of *Bdnf*, these changes are mediated by the formation of a complex between MeCP2 and HDAC1 [32], which reduces the acetylation of histone (H) 3 and H4. This effect is paralleled by an increased dimethylation of H3K9, which inhibits gene transcription, but decreased dimethylation of H3K4, which promotes gene transcription [40]. This suggests cooperation between epigenetic modifications on the DNA and histones for gene silencing that is mediated by MeCP2. This is consistent with the observation that mice deficient for MeCP2 have hyperacetylated H3 [41]. Taken together, these results would argue for an overall increase in gene transcription in RS, resulting from a loss in MeCP2-mediated gene silencing. Surprisingly however, several microarray analyses did not reveal any consistent change in gene expression in RS patients or in MeCP2 mouse models [reviewed in Ref. 42]. This may owe to the property of MeCP2 to act as both a transcriptional silencing and an activating factor [26], or to the fact that MeCP2 may require additional transcriptional regulators for functioning properly. Therefore, although it might be envisaged to attenuate the effect of MeCP2 deficiency by administring exogenous MeCP2, or with drugs targeting transcriptional and epigenetic processes, such approaches still need to be better evaluated.

EPIGENETIC DYSREGULATION IN NEURODEGENERATIVE DISORDERS – THE EXAMPLE OF ALZHEIMER'S DISEASE

Neurodegenerative diseases are pathological conditions characterized by a gradual loss of cells in the nervous system. In the central nervous system, such loss can have devastating consequences on cognition and locomotion since neurons are only marginally regenerated.

Alzheimer's disease (AD) and Huntington's disease are neurodegenerative diseases with an epigenetic basis. We will describe AD here, and refer the reader to Ref. 43 for an overview of the implications of epigenetic mechanisms in Huntington's disease.

AD is one of the most common neurodegenerative diseases worldwide, and has an estimated prevalence of 1:100 in the population over 65 years of age in Western countries. Despite years of intense research and multiple clinical trials, AD remains a non-curable brain disease [44]. The disease is primarily characterized by a progressive cognitive decline that gradually worsens with age. Its pathophysiology is manifested by two major hallmarks in the brain: extracellular amyloid plaques and intracellular neurofibrillary tangles (NFTs) [45]. Amyloid plaques are deposits of the amyloid β (Aβ) peptide, produced through enzymatic cleavage of the amyloid precursor protein (APP) by β and γ secretases. NFTs are intraneuronal aggregates of hyperphosphorylated tau, a microtubule-binding protein.

Several epigenetic modifications such as aberrant histone acetylation and DNA methylation are part of the APP-Aβ pathway, and are thought to contribute to AD. However, for histone acetylation, both hyper- and hypoacetylation have been reported. For hyperacetylation, a potential mechanism was proposed to involve the formation of a complex between the APP intracellular domain (AICD) produced from APP by γ secretase, the nuclear adaptor protein Fe65, and the HAT TIP60, a transcriptional activator [46]. Further, *presenilin 1* (*PS1*), a gene coding for the γ secretase complex, was also shown to contribute to histone hyperacetylation in AD pathology. Loss-of-function mutations in *PS1* or mutations associated with familial AD, an early onset form of AD affecting genetically-predisposed individuals, inhibit the proteasomal degradation of the HAT CREB binding protein (CBP), and result in increased CREB-mediated gene expression in cultured neurons [47]. Consistent with the involvement of hyperacetylation, overexpression of the HDAC SIRT1 (silent mating type information regulation 2 homolog) in a mouse model of AD (CK-p25 mice) overexpressing the cyclin-dependent kinase 5 (Cdk5), confers substantial protection against AD-related neurodegeneration and memory loss [48]. Although in this case it is not known whether SIRT1 acts via the epigenetic machinery in the nucleus and/or through cytoplasmic substrates, these findings suggest that pharmacological treatment activating SIRT1 such as the polyphenol resveratrol, might be beneficial for the treatment of AD.

In contrast, other lines of evidence have suggested that AD is associated with histone hypoacetylation. In cultured cortical neurons, the overexpression of APP was shown to lead to cell death, and to decrease H3 and H4 acetylation by reducing the level of CBP [49]. Similarly, loss-of-function mutations in *PS1* and *PS2* genes in mice were shown to reduce the expression of CBP, and of CBP/CREB target genes such as *c-fos* and *Bdnf*. They also impair synaptic plasticity, and spatial and contextual memory [50]. Moreover in CK-p25 mice, the intracerebroventricular injection of sodium butyrate, a potent class I/II HDAC inhibitor, rescues memory and synaptic connectivity [51], suggesting that HDAC inhibitors may have a potential for the treatment of AD-related pathologies. These findings, however, contrast with the observation that the Class III HDAC SIRT1 has similar beneficial effects in the same mouse model of AD. One possible explanation is that different HDACs have different functions and may affect distinct pools of substrates. Thus class I/II and III HDACs are structurally and functionally different. While class I/II HDACs are co-activated by zinc, SIRT1 requires NAD$^+$ as co-substrate [reviewed in Ref. 52]. Despite this, following the findings that AD is associated with an overall decrease in histone acetylation, several class I/II HDAC inhibitors are currently being tested in pre-clinical or phase I/II trials for the treatment of AD. Nonetheless, since histone acetylation appears to be dysregulated bidirectionally and possibly in a gene-specific manner, more research is required to fully evaluate the potential of HDAC inhibitors in AD treatment.

Besides pharmacological treatment, natural manipulations such as environmental enrichment may also represent promising means to alleviate AD symptoms. In CK-p25

557

mice, exposure to an enriched environment for four weeks was shown to improve synaptic connectivity and cognitive abilities [51]. The improvement was comparable to that achieved with the HDAC inhibitor sodium butyrate [51], which highlights the potential of environmental stimulation for the reversal of cognitive deficits in AD.

Further to histone acetylation, DNA methylation might also be involved in the pathology of AD. In cell culture, hypomethylation of the promoter region of *PS1* increases presenilin expression, which enhances β-amyloid formation [53]. This effect can be reversed by application of the methyl donor S-adenosylmethionine (SAM), which rescues methylation, decreases presenilin expression, and reduces β-amyloid formation. These observations suggest that methyl donors or drugs targeting the methyl metabolism may be potential therapeutic agents to treat AD [54]. The finding that DNA hypomethylation underlies some aspects of AD pathology was also recently confirmed in mouse and primate models of AD. In these models, exposure to lead (Pb) reduces the enzymatic activity of DNMT1 in cortical neurons and is associated with increased *APP* expression [55]. Further, a recent post mortem study in humans also reported hypomethylation of the presenilin promoter region in late-onset AD patients when compared to age-matched healthy subjects [56]. These findings therefore support the hypothesis that DNA *hypo*methylation, at least in the presenilin promoter, is causally associated with AD. In contrast, other AD-related susceptibility genes such as *BACE1*, which codes for β-secretase, or the gene coding for apolipoprotein E that facilitates amyloid plaque formation, are *hyper*methylated in late-onset AD [56]. This suggests that alterations in DNA methylation are presumably bidirectional and affect specific genes in AD, similarly to histone acetylation. Further studies are needed to identify the affected genes and determine the direction and extent of the changes in their expression.

EPIGENETIC DYSREGULATION IN PSYCHIATRIC DISORDERS – THE EXAMPLE OF DEPRESSION

Psychiatric disorders range from personality and anxiety disorders to addiction and depression. They are multi-faceted conditions with complex etiology and expression, and they are difficult to treat. One of the most prominent psychiatric disorders is depression. A substantial body of evidence has suggested that several depressive-like phenotypes are partly caused by epigenetic mechanisms.

According to DSM-IV, depression is a mental disease characterized by pessimistic thoughts, lack of enthusiasm and vitality, feelings of despair or guilt, and anhedonia. This chronic illness affects roughly five per cent of the population worldwide, although this number is likely to underestimate the actual prevalence. Depression is difficult to treat, and only half of depressed patients show complete remission [57,58]. One of the major issues with most treatments is that their benefit is often delayed, and symptoms are usually ameliorated only after a few weeks of treatment. The reasons for such delay are not known but could reflect the contribution of epigenetic mechanisms in the etiology of depression. Initial evidence for this hypothesis came from studies examining the effect of electroconvulsive therapy, a treatment that is effective only after repeated administration. Chronic electroconvulsive seizures (ECS) were shown to increase *Bdnf* mRNA and protein in the hippocampus, as well as *CREB* expression [59–61]. This increase was associated with H3 hyperacetylation at BDNF promoter 3 [62], suggesting that chromatin remodeling contributes to the beneficial effects of chronic ECS.

The importance of histone PTMs in depression was further demonstrated in a model of social defeat in rodents. This model of chronic stress induces symptoms of depression that can be reversed by repeated but not acute, antidepressant treatment, mimicking that observed in human patients [63]. In mice, chronic social defeat decreases the expression of two splice variants of *Bdnf* (*Bdnf III* and *Bdnf IV*) in the hippocampus, and is associated with increased dimethylation of H3K27, a mark of transcriptional repression [40], in their respective promoter regions P3 and P4 [63] (Fig. 34.1). While behavioral anomalies induced

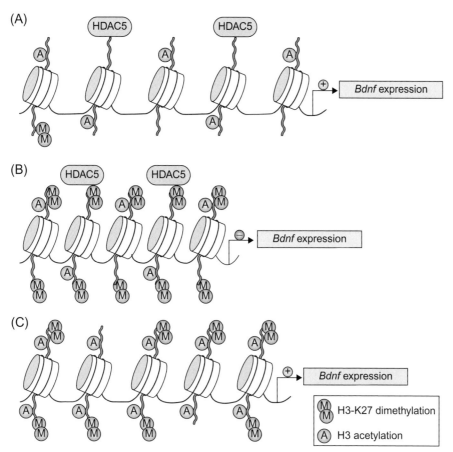

FIGURE 34.1

The importance of histone PTMs in a rodent model of depression. (A) Under physiological conditions, i.e. in the absence of stress, the promoter region of the *Bdnf* gene carries a moderate level of histone H3 acetylation and H3K27 dimethylation, and is bound by the histone deacetylase HDAC5. (B) Upon chronic social defeat stress, a rodent model of depression, H3K27 dimethylation is increased, leading to increased condensation of the chromatin, which shuts down *Bdnf* gene expression. (C) Upon chronic antidepressant (imipramine) treatment, HDAC5 levels are reduced, which leads to increased H3 acetylation, whereas H3K27 dimethylation remains unaffected. Nonetheless, the increase in H3 acetylation is sufficient to reinstate *Bdnf* gene expression. A, acetyl; Bdnf, brain-derived neurotrophic factor; HDAC, histone deacetylase; K, lysine; M, methyl. Figure reproduced, with permission, from Ref. 49. (Please refer to color plate section)

by social defeat can be reversed by chronic antidepressant treatment, the increase in H3K27-dimethylation can not [63]. Instead, antidepressant treatment appears to reverse the down-regulation of *Bdnf* expression by increasing H3 acetylation and H3K4 methylation, marks of transcriptional activation [40], at the same promoters [63,64]. Additionally, chronic antidepressant treatment down-regulates the expression of HDAC5, specifically in animals exposed to chronic stress [65]. Thus, histone modifications induced by chronic stress at the *Bdnf* gene in the hippocampus are likely to be an important mechanism for the development of depressive behaviors, and may also be a target for antidepressant treatments.

Recent evidence also showed that antidepressant treatments influence DNA methylation. Chronic antidepressant treatment was demonstrated to increase MeCP2 and methyl CpG-binding domain protein 1 (MBD1), two proteins that bind methylated CpG dinucleotides and act as transcriptional activator or repressor in the rodent brain [66]. The antidepressant-dependent increase in MeCP2 was specific to gamma aminobutyric acid (GABA)-ergic interneurons [66]. This finding is of particular interest, because abnormal GABAergic transmission and anomalies in GABA-related gene methylation have been linked to major depression and suicide. Post-mortem studies revealed that depressed patients who committed suicide have higher level of methylation in the GABA-A α1 receptor subunit promoter, and increased DNMT3b mRNA and protein in the prefrontal cortex when compared to control individuals who died of other causes [67]. This suggests the interesting possibility that antidepressant treatments can target the epigenetic machinery specifically in cell types affected by depression.

EPIGENETIC DYSREGULATION IN PSYCHOTIC DISORDERS – THE EXAMPLE OF SCHIZOPHRENIA

Psychotic disorders refer to mental illnesses characterized by a distorted perception of reality. The most common form of psychosis is schizophrenia, which has an approximate prevalence of one percent worldwide among people 18 years old or more [68]. Two main categories of symptoms characterize schizophrenia: positive symptoms such as delusions and hallucinations, and negative symptoms such as social withdrawal, lack of motivation, and overall apathy. While the causes of schizophrenia are not well understood, they are thought to involve the combination of a genetic predisposition and environmental factors during pre- and postnatal development. The importance of the environment is illustrated by the fact that the concordance rate for schizophrenia in monozygotic twins is only 50% [69]. However, the question of how environmental factors influence the development of schizophrenia is still open.

There is increasing evidence that schizophrenia is associated with an aberrant epigenetic profile, and with abnormal GABAergic neurotransmission in cortical areas [70]. The first line of evidence involves reelin, a glycoprotein expressed in GABAergic neurons during development and adulthood, required for neuronal migration [70,71]. Post-mortem analyses revealed that reelin mRNA and protein expression are significantly reduced in several brain regions in schizophrenic patients [72,73]. This reduction could result from an alteration in methylation, possibly hypermethylation, of the reelin promoter, which contains a large CpG island [58,74]. Importantly, reelin expression was found to be modulated by pharmacological manipulations of DNA methylation. *In vivo*, repeated methionine administration increased methylation of the reelin promoter, induced binding of MeCP2 to this promoter, and down-regulated *reelin* expression [75,76]. Consistently, *in vitro* administration of the DNMT inhibitor, 5-aza 2'-deoxycytidine, was observed to increase reelin expression [74].

Further evidence for the contribution of DNA hypermethylation in GABAergic dysfunctions observed in schizophrenic patients involves the glutamate decarboxylase GAD67, an enzyme that catalyzes the synthesis of GABA. GAD67 mRNA and protein expression are down-regulated in cortical structures of schizophrenic patients [72,73]. This correlates with increased methylation of the GAD67 promoter in prefrontal cortical areas in post-mortem brains from schizophrenic patients [77].

Further to the GABAergic system, oligodendrocytes also have increased DNA methylation. The CpG island of sex-determining region Y-box containing gene 10 (SOX10), an oligodendrocyte-specific transcription factor, is hypermethylated and SOX10 expression is decreased [78] in the brain of schizophrenic patients. This may provide a possible mechanism for the abnormalities in oligodendrocytes observed in schizophrenic patients.

The mechanisms that underlie the abnormal hypermethylation of promoters such as reelin, GAD67, and SOX10 in the brain of schizophrenic patients are not fully understood but are suggested to be partially accounted for by elevated levels of SAM [79,80]. Consistently, the administration of SAM can induce psychotic episodes in some schizophrenic patients [81]. An increased expression of DNMT1 mRNA was also shown [79,80], suggesting that DNMT inhibitors may be potential therapeutic agents in schizophrenia [4,74,82].

While in the GABAergic system and in oligodendrocytes, several instances of *hyper*methylation were observed in schizophrenic patients, *hypo*methylation was detected in the dopaminergic system. Increased activation of catechol-O-methyltransferase (COMT), an enzyme involved in the degradation of monoamine neurotransmitters such as dopamine, epinephrine, and norepinephrine, is associated with impairments in attention, executive cognition, and working memory, and has been linked with an increased risk to develop schizophrenia [83]. Reduced methylation of the COMT promoter was detected in the frontal lobe of schizophrenic patients, and is associated with increased activation of the gene [84]. Moreover, aberrant methylation of genes within the dopaminergic system has also been observed in monozygotic twins. In one study, methylation upstream of the locus for dopamine D2 receptor was investigated in two sets of twins, one concordant and one discordant for schizophrenia. In the discordant twins, the epigenetic profile of the affected twin was more similar to the profile in twins concordant for schizophrenia than to his own unaffected brother [85]. These observations corroborate the implication of DNA methylation in schizophrenia.

Finally, the brain of schizophrenic patients is also characterized by differential histone PTMs. Valproic acid (also known as valproate), a mood stabilizer commonly prescribed to alleviate some symptoms schizophrenia, is a potent HDAC inhibitor [86]. Its administration not only decreases HDAC activity, but also increases reelin expression both *in vitro* and *in vivo*. Valproate can further induce a decrease in DNA methylation at the reelin promoter [74,75]. *In vitro*, other HDAC inhibitors such as trichostatin A or MS-275 [for a review see Refs 52,87] have also been demonstrated to activate the expression of reelin and GAD67 [88]. While the mechanism for the link between HDAC inhibition and altered DNA methylation is still unclear, it is thought to operate by regulation of the accessibility of DNMTs to promoter regions, or by direct induction of DNA demethylase activity [58,88]. Valproate has further been demonstrated to reduce schizophrenia-like behaviors in a mouse model treated with methionine [89]. Two antipsychotic drugs acting as dopamine D2 receptor antagonists, haloperidol and raclopride, were also shown to induce phospho-acetylation of H3 in the mouse striatum [90]. Therefore, it appears that decreased acetylation contributes to the pathology of schizophrenia, which can be reversed by HDAC inhibitors. However, due to the non-specific effects of the antipsychotics described above, direct evidence for this hypothesis is still missing.

EPIGENETIC DYSREGULATION BY ENVIRONMENTAL STRESS – THE EXAMPLE OF EARLY LIFE STRESS

Increased stress vulnerability and other forms of pathological and inappropriate stress-coping behaviors are also common brain disorders. These disorders are complex and have intricate mechanisms, which are still poorly understood. Nonetheless, it is generally recognized that they are strongly influenced by environmental factors, in particular by detrimental early life experiences (pre- and/or postnatal). These experiences are thought to induce lasting epigenetic modifications resulting in persistent changes in gene expression in adult animals. In mammals, the quality of early life is primarily defined by maternal care and nutrition. In mice and rats, maternal care is provided by arched-back nursing (ABN) and licking/ grooming (LG), two behavioral traits exhibited by most rodent females but which vary greatly between species and strains [91]. These traits critically influence the offspring's behavior and determine their responsiveness to stress and their level of anxiety [92]. At the molecular level, this responsiveness is in part regulated by glucocorticoids and their receptors (GR). High levels of circulating glucocorticoids heighten the body's alertness and increase the stress response, while low levels result in a more relaxed behavior and attenuate the stress response. At the same time, high levels of GR in forebrain areas such as the hippocampus, provide a negative feedback that reduces the production of glucocorticoids and thereby dampens the stress response [reviewed in Ref. 93]. Intriguingly, the offspring of high ABN-LG mothers have increased GR expression and reduced reactivity to stress, whereas the offspring of low ABN-LG mothers have decreased GR expression and increased stress reactivity [94].

Both the reactivity to stress and the GR system are subject to epigenetic modifications in early life. In the offspring of high ABN-LG females, DNA methylation is reduced and H3K9 acetylation is increased in the promoter of the *GR* gene, a promoter that is, in part,

561

controlled by binding of the transcription factor NGFI-A (also known as Egr-1 or Zif268) [95]. In contrast, in the offspring of low ABN-LG females, promoter methylation is increased (but acetylation is not changed), suggesting that differential epigenetic marking underlies changes in GR expression. Further evidence indicates that NGFI-A itself may convey these epigenetic changes, since binding of NGFI-A to the GR promoter region is required for these changes to occur [96]. Importantly, although stable, these epigenetic changes can be reversed by environmental or pharmacological manipulations. Cross-fostering of pups, or treatment with the HDAC inhibitor trichostatin A, leads to hypomethylation of the GR promoter and histone hyperacetylation in low ABN-LG offspring [95]. Likewise, methyl supplementation via the administration of L-methionine, a SAM precursor, can reverse maternal programming of stress responses via GRs [97]. Both treatments have further been shown to modulate the transcriptome in the hippocampus in both high and low ABN-LG offspring [98], which suggests that the type of maternal care not only influences DNA methylation and histone acetylation at the GR promoter, but also on other genes.

The implication of epigenetic regulation of GR was recently confirmed in a study involving post-mortem human brain samples [99]. In this study, a correlated decrease in hippocampal *GR* expression and an increase in DNA methylation at the GR promoter were observed in suicide victims with a history of childhood abuse, but not in age-matched control subjects including suicide victims without such history, or in people who died from other causes. These findings strongly suggest that in rodents and humans, early life neglect can cause lifelong epigenetic alterations of gene expression in the brain's stress system. Remarkably, female rat pups raised by mothers with proficient nurturing become proficient nurturing dams themselves, suggesting that this behavioral trait is determined by early life and can be transmitted across generations [100].

Finally, a recent study extended the findings on *GR* to another gene implicated in stress-coping behaviors. Early life stress brought about by periodic mother–infant separation in mice led to DNA hypomethylation in the promoter region of the arginine vasopressin (avp) gene. This hypomethylation correlated with a concomitant increase in the expression of *avp*, which persisted in adulthood, and with increased helplessness in stressful situations [101]. Many more genes are likely to be involved and still need to be identified. Another recent study in mice further demonstrated that the negative effects of chronic postnatal stress in early life on behavior are associated with altered DNA methylation and can be transmitted across several generations [102].

CONCLUSIONS AND OUTLOOK

It is now clear that epigenetic mechanisms play a pivotal role in higher-order brain functions under physiological and pathological conditions. A deeper understanding of these mechanisms is of utmost importance for the development of potential treatments against brain disorders involving aberrant epigenetic modifications (for brain disorders not described here, the reader is referred to Refs 4, 22, 23, and 43). Promising results have already been obtained, in particular with the use of HDAC inhibitors [103]. However, most of these inhibitors are nonspecific, and more research is needed to better understand their mechanisms of action before any treatment can be safely administered. It is therefore important to not only identify more specific HDAC inhibitors, but also to determine which HDACs are involved in different disease states. In this direction, it was recently demonstrated that although HDAC1 and HDAC2 are structurally related class I HDACs that often form functional heterodimers, only HDAC2 is causally implicated in the negative regulation of memory formation and synaptic plasticity [104].

Finally, further to DNA methylation and histone PTMs, other forms of epigenetic regulation such as RNAi are likely to be involved in brain disorders and merit attention in the future. Thus although so far, no example of naturally-occurring RNAi-mediated epigenetic silencing of gene expression has been documented in diseases of the nervous system, this mechanism is quite common and may take place in the brain. In yeast, RNAi regulates the structure of heterochromatin around centromeres [105], while in mammals during development, it is implicated in X-chromosome inactivation [106]. RNAi has also proven to be an efficient means to artificially silence genes of interest, and has been used experimentally in the mouse brain [107]. Thus, RNAi-mediated gene silencing may emerge as a powerful tool against various brain disorders [108] including the neurodevelopmental and neurodegenerative disorders described here.

In summary, although the importance of epigenetic dysfunctions in brain diseases is now fully appreciated, a more thorough understanding of epigenetic processes is still required until safe and efficient treatments based on epigenetic drugs can be envisaged.

ACKNOWLEDGEMENTS

We thank the anonymous reviewers for their thorough reading of the manuscript and their useful comments. The lab of IMM is supported by the University of Zürich, the Swiss Federal Institute of Technology, the Swiss National Science Foundation, the National Center for competence in Research "Neural Plasticity and Repair", and the Human Frontier Science Program, EMBO, Roche.

References

1. Russo VEA, Martienssen RA, Riggs AD, editors. *Epigenetic mechanisms of gene regulation.* Woodbury: Cold Spring Harbour Laboratory Press; 1996.

2. Bird A. Perceptions of epigenetics. *Nature* 2007;**447**:396–8.

3. Jaenisch R, Bird A. Epigenetic regulation of gene expression: how the genome integrates intrinsic and environmental signals. *Nat Genet* 2003;**33**:245–54.

4. Levenson JM, Sweatt JD. Epigenetic mechanisms in memory formation. *Nat Rev Neurosci* 2005;**6**:108–18.

5. Bird A. DNA methylation patterns and epigenetic memory. *Genes Dev* 2002;**16**:6–21.

6. Klose RJ, Bird AP. Genomic DNA methylation: the mark and its mediators. *Trends Biochem Sci* 2006;**31**:89–97.

7. Fan M, Yan PS, Hartman-Frey C, Chen L, Paik H, Oyer SL, et al. Diverse gene expression and DNA methylation profiles correlate with differential adaptation of breast cancer cells to the antiestrogens tamoxifen and fulvestrant. *Cancer Res* 2006;**66**:11954–66.

8. Weber M, Hellmann I, Stadler MB, Ramos L, Paabo S, Rebhan M, et al. Distribution, silencing potential and evolutionary impact of promoter DNA methylation in the human genome. *Nat Genet* 2007;**39**:457–66.

9. Tweedie-Cullen RY, Reck JM, Mansuy IM. Comprehensive mapping of post-translational modifications on synaptic, nuclear, and histone proteins in the adult mouse brain. *J Proteome Res* 2009;**8**:4966–82.

10. Keppler BR, Archer TK. Chromatin-modifying enzymes as therapeutic targets–Part2. *Expert Opin Ther Targets* 2008;**12**:1457–67.

11. Keppler BR, Archer TK. Chromatin-modifying enzymes as therapeutic targets–Part1. *Expert Opin Ther Targets* 2008;**12**:1301–12.

12. Li B, Carey M, Workman JL. The role of chromatin during transcription. *Cell* 2007;**128**:707–19.

13. Klose RJ, Zhang Y. Regulation of histone methylation by demethylimination and demethylation. *Nat Rev Mol Cell Biol* 2007;**8**:307–18.

14. Peters AH, Schubeler D. Methylation of histones: playing memory with DNA. *Curr Opin Cell Biol* 2005;**17**:230–8.

15. Shilatifard A. Chromatin modifications by methylation and ubiquitination: implications in the regulation of gene expression. *Annu Rev Biochem* 2006;**75**:243–69.

16. He H, Lehming N. Global effects of histone modifications. *Brief Funct Genomic Proteomic* 2003;**2**:234–43.

17. Nathan D, Ingvarsdottir K, Sterner DE, Bylebyl GR, Dokmanovic M, Dorsey LA, et al. Histone sumoylation is a negative regulator in *Saccharomyces cerevisiae* and shows dynamic interplay with positive-acting histone modifications. *Genes Dev* 2006;**20**:966–76.

18. Shiio Y, Eisenman RN. Histone sumoylation is associated with transcriptional repression. *Proc Natl Acad Sci USA* 2003;**100**:13225–30.

19. Koshibu K, Gräff J, Beullens M, Heitz FD, Berchtold D, Russig H, et al. Protein phosphatase 1 regulates the histone code for long-term memory. *J Neurosci* 2009;**29**:13079–89.

20. Jenuwein T, Allis CD. Translating the histone code. *Science* 2001;**293**:1074–80.

21. Turner BM. Cellular memory and the histone code. *Cell* 2002;**111**:285–91.

22. Gräff J, Mansuy IM. Epigenetic codes in cognition and behavior. *Behav Brain Res* 2008;**192**:70–87.

23. Gräff J, Mansuy IM. Epigenetic dysregulation in cognitive disorders. *Eur J Neurosci* 2009;**30**:1–8.

24. Urdinguio RG, Sanchez-Mut JV, Esteller M. Epigenetic mechanisms in neurological diseases: genes, syndromes, and therapies. *Lancet Neurol* 2009;**8**:1056–72.

25. Amir RE, Van den Veyver IB, Wan M, Tran CQ, Francke U, Zoghbi HY. Rett syndrome is caused by mutations in X-linked MECP2, encoding methyl-CpG-binding protein 2. *Nat Genet* 1999;**23**:185–8.

26. Chahrour M, Jung SY, Shaw C, Zhou X, Wong ST, Qin J, Zoghbi HY. MeCP2, a key contributor to neurological disease, activates and represses transcription. *Science* 2008;**320**:1224–9.

27. Chen RZ, Akbarian S, Tudor M, Jaenisch R. Deficiency of methyl-CpG binding protein-2 in CNS neurons results in a Rett-like phenotype in mice. *Nat Genet* 2001;**27**:327–31.

28. Guy J, Hendrich B, Holmes M, Martin JE, Bird A. A mouse Mecp2-null mutation causes neurological symptoms that mimic Rett syndrome. *Nat Genet* 2001;**27**:322–6.

29. Collins AL, Levenson JM, Vilaythong AP, Richman R, Armstrong DL, Noebels JL, et al. Mild overexpression of MeCP2 causes a progressive neurological disorder in mice. *Hum Mol Genet* 2004;**13**:2679–89.

30. Tao J, Hu K, Chang Q, Wu H, Sherman NE, Martinowich K, et al. Phosphorylation of MeCP2 at Serine 80 regulates its chromatin association and neurological function. *Proc Natl Acad Sci USA* 2009;**106**:4882–7.

31. Chen WG, Chang Q, Lin Y, Meissner A, West AE, Griffith EC, Jaenisch R, Greenberg ME. Derepression of BDNF transcription involves calcium-dependent phosphorylation of MeCP2. *Science* 2003;**302**:885–9.

32. Martinowich K, Hattori D, Wu H, Fouse S, He F, Hu Y, et al. DNA methylation-related chromatin remodeling in activity-dependent BDNF gene regulation. *Science* 2003;**302**:890–3.

33. Zhou Z, Hong EJ, Cohen S, Zhao WN, Ho HY, Schmidt L, et al. Brain-specific phosphorylation of MeCP2 regulates activity-dependent Bdnf transcription, dendritic growth, and spine maturation. *Neuron* 2006;**52**:255–69.

34. Chang Q, Khare G, Dani V, Nelson S, Jaenisch R. The disease progression of Mecp2 mutant mice is affected by the level of BDNF expression. *Neuron* 2006;**49**:341–8.

35. Larimore JL, Chapleau CA, Kudo S, Theibert A, Percy AK, Pozzo-Miller L. Bdnf overexpression in hippocampal neurons prevents dendritic atrophy caused by Rett-associated MECP2 mutations. *Neurobiol Dis* 2009;**34**:199–211.

36. Sun YE, Wu H. The ups and downs of BDNF in Rett syndrome. *Neuron* 2006;**49**:321–3.

37. Maezawa I, Swanberg S, Harvey D, LaSalle JM, Jin LW. Rett syndrome astrocytes are abnormal and spread MeCP2 deficiency through gap junctions. *J Neurosci* 2009;**29**:5051–61.

38. Urdinguio RG, Lopez-Serra L, Lopez-Nieva P, Alaminos M, Diaz-Uriarte R, Fernandez AF, Esteller M. Mecp2-null mice provide new neuronal targets for Rett syndrome. *PLoS ONE* 2008;**3**:e3669.

39. Smrt RD, Eaves-Egenes J, Barkho BZ, Santistevan NJ, Zhao C, Aimone JB, et al. Mecp2 deficiency leads to delayed maturation and altered gene expression in hippocampal neurons. *Neurobiol Dis* 2007;**27**:77–89.

40. Kouzarides T. Chromatin modifications and their function. *Cell* 2007;**128**:693–705.

41. Shahbazian M, Young J, Yuva-Paylor L, Spencer C, Antalffy B, Noebels J, et al. Mice with truncated MeCP2 recapitulate many Rett syndrome features and display hyperacetylation of histone H3. *Neuron* 2002;**35**:243–54.

42. Monteggia LM, Kavalali ET. Rett syndrome and the impact of MeCP2 associated transcriptional mechanisms on neurotransmission. *Biol Psychiatry* 2009;**65**:204–10.

43. Deutsch SI, Rosse RB, Mastropaolo J, Long KD, Gaskins BL. Epigenetic therapeutic strategies for the treatment of neuropsychiatric disorders: ready for prime time? *Clin Neuropharmacol* 2008;**31**:104–19.

44. Cummings JL. Alzheimer's disease. *N Engl J Med* 2004;**351**:56–67.

45. LaFerla FM, Kitazawa M. Antipodal effects of p25 on synaptic plasticity, learning, and memory--too much of a good thing is bad. *Neuron* 2005;**48**:711–12.

46. Cao X, Sudhof TC. A transcriptionally (correction of transcriptively) active complex of APP with Fe65 and histone acetyltransferase Tip60. *Science* 2001;**293**:115–20.

47. Marambaud P, Wen PH, Dutt A, Shioi J, Takashima A, Siman R, et al. A CBP binding transcriptional repressor produced by the PS1/epsilon-cleavage of N-cadherin is inhibited by PS1 FAD mutations. *Cell* 2003;**114**:635–45.

48. Kim D, Nguyen MD, Dobbin MM, Fischer A, Sananbenesi F, Rodgers JT, et al. SIRT1 deacetylase protects against neurodegeneration in models for Alzheimer's disease and amyotrophic lateral sclerosis. *Embo J* 2007;**26**:3169–79.

49. Rouaux C, Jokic N, Mbebi C, Boutillier S, Loeffler JP, Boutillier AL. Critical loss of CBP/p300 histone acetylase activity by caspase-6 during neurodegeneration. *EMBO J* 2003;**22**:6537–49.

50. Saura CA, Choi SY, Beglopoulos V, Malkani S, Zhang D, Shankaranarayana Rao BS, et al. Loss of presenilin function causes impairments of memory and synaptic plasticity followed by age-dependent neurodegeneration. *Neuron* 2004;**42**:23–36.

51. Fischer A, Sananbenesi F, Wang XY, Dobbin M, Tsai LH. Recovery of learning and memory is associated with chromatin remodeling. *Nature* 2007;**447**:178–82.

52. Kazantsev AG, Thompson LM. Therapeutic application of histone deacetylase inhibitors for central nervous system disorders. *Nat Rev Drug Discov* 2008;**7**:854–68.

53. Scarpa S, Fuso A, D'Anselmi F, Cavallaro RA. Presenilin 1 gene silencing by S-adenosylmethionine: a treatment for Alzheimer disease? *FEBS Lett* 2003;**541**:145–8.

54. Scarpa S, Cavallaro RA, D'Anselmi F, Fuso A. Gene silencing through methylation: an epigenetic intervention on Alzheimer disease. *J Alzheimers Dis* 2006;**9**:407–14.

55. Wu J, Basha MR, Brock B, Cox DP, Cardozo-Pelaez F, McPherson CA, et al. Alzheimer's disease (AD)-like pathology in aged monkeys after infantile exposure to environmental metal lead (Pb): evidence for a developmental origin and environmental link for AD. *J Neurosci* 2008;**28**:3–9.

56. Wang SC, Oelze B, Schumacher A. Age-specific epigenetic drift in late-onset Alzheimer's disease. *PLoS ONE* 2008;**3**:e2698.

57. Berton O, Nestler EJ. New approaches to antidepressant drug discovery: beyond monoamines. *Nat Rev Neurosci* 2006;**7**:137–51.

58. Tsankova N, Renthal W, Kumar A, Nestler EJ. Epigenetic regulation in psychiatric disorders. *Nat Rev Neurosci* 2007;**8**:355–67.

59. Chen AC, Shin KH, Duman RS, Sanacora G. ECS-Induced mossy fiber sprouting and BDNF expression are attenuated by ketamine pretreatment. *J ECT* 2001;**17**:27–32.

60. Altar CA, Whitehead RE, Chen R, Wortwein G, Madsen TM. Effects of electroconvulsive seizures and antidepressant drugs on brain-derived neurotrophic factor protein in rat brain. *Biol Psychiatry* 2003;**54**:703–9.

61. Altar CA, Laeng P, Jurata LW, Brockman JA, Lemire A, Bullard J, et al. Electroconvulsive seizures regulate gene expression of distinct neurotrophic signaling pathways. *J Neurosci* 2004;**24**:2667–77.

62. Tsankova NM, Kumar A, Nestler EJ. Histone modifications at gene promoter regions in rat hippocampus after acute and chronic electroconvulsive seizures. *J Neurosci* 2004;**24**:5603–10.

63. Tsankova NM, Berton O, Renthal W, Kumar A, Neve RL, Nestler EJ. Sustained hippocampal chromatin regulation in a mouse model of depression and antidepressant action. *Nat Neurosci* 2006;**9**:519–25.

64. Wilkinson MB, Xiao G, Kumar A, LaPlant Q, Renthal W, Sikder D, et al. Imipramine treatment and resiliency exhibit similar chromatin regulation in the mouse nucleus accumbens in depression models. *J Neurosci* 2009;**29**:7820–32.

65. Renthal W, Maze I, Krishnan V, Covington HE 3rd, Xiao G, Kumar A, et al. Histone deacetylase 5 epigenetically controls behavioral adaptations to chronic emotional stimuli. *Neuron* 2007;**56**:517–29.

66. Cassel S, Carouge D, Gensburger C, Anglard P, Burgun C, Dietrich JB, Aunis D, et al. Fluoxetine and cocaine induce the epigenetic factors MeCP2 and MBD1 in adult rat brain. *Mol Pharmacol* 2006;**70**:487–92.

67. Poulter MO, Du L, Weaver IC, Palkovits M, Faludi G, Merali Z, et al. GABAA receptor promoter hypermethylation in suicide brain: implications for the involvement of epigenetic processes. *Biol Psychiatry* 2008;**64**:645–52.

68. Regier DA, Narrow WE, Rae DS, Manderscheid RW, Locke BZ, Goodwin FK. The de facto US mental and addictive disorders service system. Epidemiologic catchment area prospective 1-year prevalence rates of disorders and services. *Arch Gen Psychiatry* 1993;**50**:85–94.

69. Singh SM, Murphy B, O'Reilly R. Epigenetic contributors to the discordance of monozygotic twins. *Clin Genet* 2002;**62**:97–103.

70. Costa E, Chen Y, Dong E, Grayson DR, Kundakovic M, Maloku E, et al. GABAergic promoter hypermethylation as a model to study the neurochemistry of schizophrenia vulnerability. *Expert Rev Neurother* 2009;**9**:87–98.

71. Fatemi SH. Reelin glycoprotein: structure, biology and roles in health and disease. *Mol Psychiatry* 2005;**10**:251–7.

72. Impagnatiello F, Guidotti AR, Pesold C, Dwivedi Y, Caruncho H, Pisu MG, et al. A decrease of reelin expression as a putative vulnerability factor in schizophrenia. *Proc Natl Acad Sci USA* 1998;**95**:15718–23.

73. Guidotti A, Auta J, Davis JM, Di-Giorgi-Gerevini V, Dwivedi Y, Grayson DR, et al. Decrease in reelin and glutamic acid decarboxylase67 (GAD67) expression in schizophrenia and bipolar disorder: a postmortem brain study. *Arch Gen Psychiatry* 2000;**57**:1061–9.

74. Chen Y, Sharma RP, Costa RH, Costa E, Grayson DR. On the epigenetic regulation of the human reelin promoter. *Nucleic Acids Res* 2002;**30**:2930–9.

75. Tremolizzo L, Carboni G, Ruzicka WB, Mitchell CP, Sugaya I, Tueting P, et al. An epigenetic mouse model for molecular and behavioral neuropathologies related to schizophrenia vulnerability. *Proc Natl Acad Sci USA* 2002;**99**:17095–100.

76. Dong E, Agis-Balboa RC, Simonini MV, Grayson DR, Costa E, Guidotti A. Reelin and glutamic acid decarboxylase67 promoter remodeling in an epigenetic methionine-induced mouse model of schizophrenia. *Proc Natl Acad Sci USA* 2005;**102**:12578–83.

77. Huang HS, Akbarian S. GAD1 mRNA expression and DNA methylation in prefrontal cortex of subjects with schizophrenia. *PLoS ONE* 2007;**2**:e809.

78. Iwamoto K, Bundo M, Yamada K, Takao H, Iwayama-Shigeno Y, Yoshikawa T, et al. DNA methylation status of SOX10 correlates with its downregulation and oligodendrocyte dysfunction in schizophrenia. *J Neurosci* 2005;**25**:5376–81.

79. Veldic M, Guidotti A, Maloku E, Davis JM, Costa E. In psychosis, cortical interneurons overexpress DNA-methyltransferase 1. *Proc Natl Acad Sci USA* 2005;**102**:2152–7.

80. Guidotti A, Ruzicka W, Grayson DR, Veldic M, Pinna G, Davis JM, Costa E. S-adenosyl methionine and DNA methyltransferase-1 mRNA overexpression in psychosis. *Neuroreport* 2007;**18**:57–60.

81. Antun FT, Burnett GB, Cooper AJ, Daly RJ, Smythies JR, Zealley AK. The effects of L-methionine (without MAOI) in schizophrenia. *J Psychiatr Res* 1971;**8**:63–71.

82. Costa E, Chen Y, Davis J, Dong E, Noh JS, Tremolizzo L, et al. REELIN and schizophrenia: a disease at the interface of the genome and the epigenome. *Mol Interv* 2002;**2**:47–57.

83. van Vliet J, Oates NA, Whitelaw E. Epigenetic mechanisms in the context of complex diseases. *Cell Mol Life Sci* 2007;**64**:1531–8.

84. Abdolmaleky HM, Cheng KH, Faraone SV, Wilcox M, Glatt SJ, Gao F, et al. Hypomethylation of MB-COMT promoter is a major risk factor for schizophrenia and bipolar disorder. *Hum Mol Genet* 2006;**15**:3132–45.

85. Petronis A, Gottesman II, Kan P, Kennedy JL, Basile VS, Paterson AD, et al. Monozygotic twins exhibit numerous epigenetic differences: clues to twin discordance? *Schizophr Bull* 2003;**29**:169–78.

86. Gottlicher M. Valproic acid: an old drug newly discovered as inhibitor of histone deacetylases. *Ann Hematol* 2004;**83**(Suppl 1):S91–2.

87. Carew JS, Giles FJ, Nawrocki ST. Histone deacetylase inhibitors: mechanisms of cell death and promise in combination cancer therapy. *Cancer Lett* 2008;**269**:7–17.

88. Kundakovic M, Chen Y, Guidotti A, Grayson DR. The reelin and GAD67 promoters are activated by epigenetic drugs that facilitate the disruption of local repressor complexes. *Mol Pharmacol* 2009;**75**:342–54.

89. Tremolizzo L, Doueiri MS, Dong E, Grayson DR, Davis J, Pinna G, et al. Valproate corrects the schizophrenia-like epigenetic behavioral modifications induced by methionine in mice. *Biol Psychiatry* 2005;**57**:500–9.

90. Li J, Guo Y, Schroeder FA, Youngs RM, Schmidt TW, Ferris C, et al. Dopamine D2-like antagonists induce chromatin remodeling in striatal neurons through cyclic AMP-protein kinase A and NMDA receptor signaling. *J Neurochem* 2004;**90**:1117–31.

91. Champagne FA, Francis DD, Mar A, Meaney MJ. Variations in maternal care in the rat as a mediating influence for the effects of environment on development. *Physiol Behav* 2003;**79**:359–71.

92. Meaney MJ. Maternal care, gene expression, and the transmission of individual differences in stress reactivity across generations. *Annu Rev Neurosci* 2001;**24**:1161–92.

93. Seckl JR, Meaney MJ. Glucocorticoid "programming" and PTSD risk. *Ann N Y Acad Sci* 2006;**1071**:351–78.

94. Liu D, Diorio J, Tannenbaum B, Caldji C, Francis D, Freedman A, et al. Maternal care, hippocampal glucocorticoid receptors, and hypothalamic-pituitary-adrenal responses to stress. *Science* 1997;**277**:1659–62.

95. Weaver ICG, Cervoni N, Champagne FA, D'Alessio AC, Sharma S, Seckl JR, et al. Epigenetic programming by maternal behavior. *Nat Neurosci* 2004;**7**:847–54.

96. Weaver IC, D'Alessio AC, Brown SE, Hellstrom IC, Dymov S, Sharma S, et al. The transcription factor nerve growth factor-inducible protein a mediates epigenetic programming: altering epigenetic marks by immediate-early genes. *J Neurosci* 2007;**27**:1756–68.

97. Weaver ICG, Champagne FA, Brown SE, Dymov S, Sharma S, Meaney MJ, et al. Reversal of maternal programming of stress responses in adult offspring through methyl supplementation: altering epigenetic marking later in life. *J Neurosci* 2005;**25**:11045–54.

98. Weaver IC, Meaney MJ, Szyf M. Maternal care effects on the hippocampal transcriptome and anxiety-mediated behaviors in the offspring that are reversible in adulthood. *Proc Natl Acad Sci USA* 2006;**103**:3480–5.

99. McGowan PO, Sasaki A, D'Alessio AC, Dymov S, Labonte B, Szyf M, et al. Epigenetic regulation of the glucocorticoid receptor in human brain associates with childhood abuse. *Nat Neurosci* 2009;**12**:342–8.

100. Francis D, Diorio J, Liu D, Meaney MJ. Nongenomic transmission across generations of maternal behavior and stress responses in the rat. *Science* 1999;**286**:1155–8.

101. Murgatroyd C, Patchev AV, Wu Y, Micale V, Bockmuhl Y, Fischer D, et al. Dynamic DNA methylation programs persistent adverse effects of early-life stress. *Nat Neurosci* 2009;**12**:1559–66.

102. Franklin T, Russig H, Weiss IC, Gräff J, Linder N, Michalon A, et al. Non-genomic transmission of the impact of early stress across generations. *Biol Psych* 2010 (in press).

103. Abel T, Zukin RS. Epigenetic targets of HDAC inhibition in neurodegenerative and psychiatric disorders. *Curr Opin Pharmacol* 2008;**8**:57–64.

566

104. Guan JS, Haggarty SJ, Giacometti E, Dannenberg JH, Joseph N, Gao J, et al. HDAC2 negatively regulates memory formation and synaptic plasticity. *Nature* 2009;**459**:55–60.

105. Martienssen RA, Zaratiegui M, Goto DB. RNA interference and heterochromatin in the fission yeast *Schizosaccharomyces pombe*. *Trends Genet* 2005;**21**:450–6.

106. Chow JC, Yen Z, Ziesche SM, Brown CJ. Silencing of the mammalian X chromosome. *Annu Rev Genomics Hum Genet* 2005;**6**:69–92.

107. Peters M, Bletsch M, Catapano R, Zhang X, Tully T, Bourtchouladze R. RNA interference in hippocampus demonstrates opposing roles for CREB and PP1alpha in contextual and temporal long-term memory. *Genes Brain Behav* 2009;**8**:320–9.

108. Davidson BL, Boudreau RL. RNA interference: a tool for querying nervous system function and an emerging therapy. *Neuron* 2007;**53**:781–8.

CHAPTER 35

Complex Metabolic Syndromes and Epigenetics

Sally A. Litherland
Sanford-Burnham Medical Research Institute at Lake Nona, Orlando, Florida 32827, USA

METABOLIC DISEASE

Metabolic diseases are characterized by dysregulation of cellular biochemical processes that allow an individual to properly uptake and use basic nutrients, such as proteins, lipids and/or carbohydrates, in the process of energy metabolism, amino acid conversion for protein biosynthesis, fatty acid oxidation, and lipid storage [1]. The ability to sense and manage nutrient uptake and use is central to homeostasis and survival [2]. Pathological conditions stemming from dysfunction of these essential processes have both genetic and environmental components to their etiologies [3]. Several "in-born error" single gene mutations have been found which manifest in aberrant metabolic processes, such as storage diseases or malabsorption of proteins or lipids in the diet. However, these defects are relatively rare in the American population [1]. Here we will limit our discussion to epigenetic function in complex metabolic diseases, including Metabolic Syndrome (MetS), and Type 1 (T1D) and Type 2 (T2D) diabetes, which are more prevalent and alarmingly on the rise in most developed countries in the world.

The etiology of complex metabolic diseases conforms to the concepts outlined originally by the "threshold hypothesis" for multi-factorial genetic disease states, though it is acknowledged that these diseases have more environmental than inheritable components underlying their causes [2,4,5]. For the disease threshold to be met and overt pathology ensue, aberrations in multiple and interchangeable susceptibility factors from both inheritable ("nature") and environmental ("nurture") sources adversely affect a number of biochemical pathways. Though each metabolic function affected is on its own vital to life, each dysfunction alone cannot cause overt pathology. However, when multiple detrimental asymptomatic or subclinical characteristic changes occur, they add up over time to meet a "threshold" which tips the affected individual into disease. Thus, metabolic diseases can be characterized as insidious, progressive, and highly individualized in their deleterious effects on health and well-being.

Etiological Factors in Metabolic Syndrome, Type 1 and Type 2 Diabetes Susceptibility

"Metabolic Syndrome" (MetS) is a somewhat controversial but often used term that describes a complex of metabolic dysfunctions that include but is not limited to obesity, inflammation, abnormal glucose metabolism, insulin resistance, hypertension and dyslipidemia [6]. Individuals with the majority of these conditions are also at increased risk

569

of developing T2D and its associated cardiovascular complications. Several environmental components, including over-eating and being under-active, contribute to the etiology of MetS, but there is also an inheritable genetic susceptibility or tendency for this syndrome [1,3,6–9].

Hyperglycemia and resistance to insulin signaling are the hallmarks of T2D, formerly known as "Adult-onset diabetes" and "insulin independent diabetes". This devastating metabolic disease is on the rise in the developed and developing world populations. T2D and its complications have a significant impact on health costs and increase the risk for cardiovascular disease in America [10]. Genetic associations with risk for T2D include single nucleotide polymorphisms (SNP) in or near genes involved in energy metabolism have been associated with risk for disease. To date, 13 SNP affected genes have been associated with T2D; however, no single SNP site has been sufficiently correlative to be used as a biomarker for disease susceptibility [11]. As with MetS, lack of sufficient exercise and unbalanced nutrition leading to obesity and hypertension are major environmental factors contributing to T2D risk and the severity of its complications [10].

T1D, once called "juvenile diabetes" or "insulin dependent diabetes", has similar phenotypic characteristics and pathological outcomes as in T2D and MetS, but has a strong immune system dysfunction underlying its development. Risk for T1D is genetically distinct from the other two metabolic syndromes. Mouse genetic models for T1D and T2D show divergent genetic factors for each of these diseases, despite substantial overlap in pathophysiological effects and complications [12]. In addition, patients with T1D are at risk for development of other autoimmune syndromes, including Hashimoto's thyroiditis, Sjögren's syndrome, and gastrointestinal tract inflammatory diseases. The Major Histocompatibility Complex (MHC/HLA) locus and 17 to 21 other "susceptibility" loci as well as several "resistance" genetic loci have been implicated in the immunopathogenesis of T1D, but no one gene has been proven essential and sufficient for disease development [12–15].

The underlying causes of complex metabolic diseases involve interactions of multiple and varied genetic and environmental factors, most of which are still unknown. Moreover, the unique combination of these factors in each patient's risk for disease is individualized to the point of vexation in their study. The majority of studies to date have tried to link diverse environmental triggers with the myriad of polymorphisms and polygenic traits in the susceptibility and resistance loci discovered so far in mice and humans, but with relatively little success. Furthermore , the potential "event" that allows an environmental "trigger" to impact on genetic susceptibility locus can be both rapid and reversible, making the task of finding a general population-applicable linkage between nature (genetic) and nurture (environment) near impossible. However, if one approaches this problem from a mechanistic approach, i.e. not what and when do these factors interact but how do they do so, a potential solution emerges. Recent findings suggest that epigenetic dysregulation of gene expression can play a significant role in both the etiology and pathology of these complex metabolic diseases. Such a rapid, reversible, inheritable, and inducible mechanism for gene expression may be the common thread running through the regulatory signaling dysfunction in metabolic diseases [16–18].

Epigenetic Mechanism Underlying Genetic Susceptibility to Metabolic Diseases

The results of the Human Genome Project indicated the bulk of the human genome is involved in sterically and chemically protecting and regulating the expression of the ~10% of its DNA that encodes gene products [19]. The DNA–protein complexes that make up chromatin structure and shape can undergo changes in topological shape in non-transcribed regions to change DNA accessibility and regulation of multiple genes in a region. These epigenetic changes are often caused by chemical motifs added or subtracted from DNA and histones within chromatin structure. These modification patterns are inheritable and even resist reversal during recombination and DNA replication/repair. However, they also can be extremely rapidly reversible, depending on the type of modification made on DNA and the histones in the structure. In general, methylation of DNA and histone cores can "close" DNA in a susceptibility loci to prevent improper expression; whereas, other modifications of these molecules can reversibly "open" and "shut" chromosomal regions in response to environmentally-stimulated signal transduction, allowing gene expression, DNA repair, and DNA replication enzymes to have access to DNA within large regions of chromatin [16–18]. Such reversible changes in chromatin topology are actively promoted by reversible enzymatic modification of histones (primarily on specific lysine residues) and DNA with small molecules (e.g. acetylation, ubiquitination, sumoylation, and some sites of methylation) [3,20].

Epigenetic modifications can be inheritable as maternal/paternal imprinted methylation patterns of gene silencing. These changes can also be passed on through generations as a susceptibility sensitization, such as methylation patterns affected by *in utero* environmental factors. In addition, epigenetic drift or inducible changes can occur later in life in rapid response to environmental triggers such as infection, inflammation, and diet, promulgated through cytokine, hormonal, and antigen signaling [3,7,18,21–23]. MetS, with its associated angiogenesis, insulin resistance, and susceptibility to vascular disease, hypertension, and atherosclerosis, as well as frank T1D and T2D, incorporate changes by all three routes of epigenetic regulation – inherited imprinting, *in utero* induction, and environmentally activated – in their etiology and pathology (Fig. 35.1).

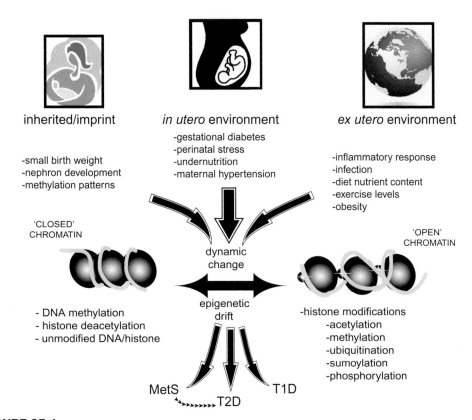

FIGURE 35.1

Routes of epigenetic regulation and dysregulation in metabolic diseases. Metabolic diseases such as Metabolic Syndrome (MetS), Type 1 diabetes (T1D), and Type 2 diabetes (T2D) all have potential epigenetic dysregulation underlying their pathologies [1–3,7–9,14]. *In utero* and inheritable epigenetic modifications can set up chromatin modification patterns to prejudice gene expression to be more conducive to disease development. Inducible epigenetic changes promote more dysregulation in response to aberrant environmental stimuli, adding to the underlying susceptibility set up during development, triggering overt disease. (Please refer to color plate section)

Inheritable Epigenetic Dysregulation in Metabolic Diseases

Epidemiological data suggest several co-morbidity factors associated with metabolic diseases correlate with changes in inheritable methylation patterns and the network of genes expressed in metabolic pathways [1]. Inherited imprinting of methylation patterns has been found on some genes known to contribute to T2D susceptibility in infants born of diabetic parents, including *LEP* (hunger and satiation), *GLUT4* (enzymatic activity in energy metabolism), *PPARalpha* and *gamma* (signaling in energy metabolism), and *PGC-1alpha* (regulation of mitochondrial function) [3,7,21,24]. Imprinting of methylation patterns and their control of gene silencing during development can influence gene expression and cellular function. The gene silencing patterns imposed by these inherited methylation patterns instilled *in utero* are thought to promote elevated risk for diabetes and cardiovascular disease later in life.

In studies of gestational diabetes, mothers with even mild hyperglycemia during pregnancy pass on altered DNA methylation modification levels and patterns. Recent evidence indicate that maternal inheritance of these traits may be more likely than paternal, suggesting that either maternal imprinting or *in utero* environmental effects on development are involved in the passing on of these characteristics [23,25–27]. Possible evidence for maternal imprinting as a factor in risk transmission comes from studies in rats where the link between inherited patterns of epigenetic modification patterns and diabetic hyperglycemic risk manifests through first, second, and third generation female offspring, but not in male offspring after the first generation [8,23,25–28].

Imprinting of DNA methylation and possibly histone modification (acetylation, phosphorylation, sumoylation, and ubiquitination) profiles may also be factors in regulation of transposons and in microRNA-mediated gene expression in diabetes [29–32]. MicroRNA sites within the genome have a strong correlation with epigenetic modification sites in non-coding chromatin [7,29–32]. MicroRNA regulation dysfunction also has been recently implicated in autoimmune disease development, especially in responsiveness to environmental triggers for susceptibility. CpG island motifs found within miRNA epigenetic regulatory sites are considered some of the most mutable DNA sequences in development, with a recombination rate 20 to 40 time higher than other nucleotide sequences during meiosis as well as mitosis [7,33]. The possibility of epigenetic drift in modification patterns at these highly mutable sites may contribute to misprogramming of gene expression during development that can have a persistent effect on later health in the offspring and their subsequent progeny.

INFLUENCE OF THE *IN UTERO* ENVIRONMENT ON EPIGENETIC MODIFICATION

Gene expression misprogramming would most likely require some extra-genetic "trigger" to initiate aberrant effects in later life. The question then arises, is it the developmental environment that triggers these epigenetic changes or do imprinted patterns persist because of *in utero* environmental factors? Two theories have arisen to answer this question, neither mutually exclusive: the "thrifty phenotype hypothesis" and the "fetal origins hypothesis" [22,28,34–37].

The "thrifty phenotype hypothesis" suggests that gestational or early postnatal life environmental factors, such as maternal malnutrition, gestational diabetes, and maternal hypertension, impose negative impact on an infant's inherited risk for later disease development [23,34,35]. One possible mechanism for such an effect would be maternally-inherited epigenetic modification patterns regulating susceptibility genes being re-programmed after exposure to such factors as maternal hyperglycemia during development [22,36–38].

The "fetal origins hypothesis" suggests that the passing on of risk traits for cardiovascular disease and T2D originates from "plasticity" within the developing genome of the infant [22,39]. This flexibility is needed to respond to the *in utero* environment changes, and allow for modifications that will modulate later response to the postnatal living and growth conditions [8,9,22,23,28,39–43]. For example, low birth weight predisposed infants to slower development and changes in DNA methylation patterns. Slower development caused by under-nourishment *in utero* leads to fewer nephrons developing in the kidney [37] and under-methylation of the angiotensin AT_{1b} receptor gene promoter in the adrenal gland [44]. Such changes promote increased renal stress, kidney dysfunction, and changes in vascular pressure. The kidney insufficiency caused by the increased renal retention, also contributes to a rise in systemic blood pressure. As a result, this vascular stress can increase risk of coronary heart disease, T2D, stroke and hypertension later in life. Thus, the decrease in plasticity of kidney development made these individuals less adaptable in their responses to environmental stress, through less flexibility in gene expression via epigenetic responsiveness [22,23,37,44].

Ultimately, these theories support each other in the importance of early gestational impacts on later disease susceptibility. It is most likely that combinations of these two mechanisms are involved in the unique individualized metabolic disease susceptibility patterns seen in the human population.

Several studies have suggested that there are defined windows of opportunity for the infant genome to establish set epigenetic modification patterns during development in both pre- and postnatal periods [22,38,41]. These windows allow for the proper amount and placement of methylation on DNA and small molecule modifications on histones to set the gene expression patterns dictated by inheritance and/or *in utero* environment [22,38,41]. A delay in gene access during these critical timeframes, due to slowed growth or metabolic stress, can set the gene expression pattern to a more pathology risky format [22,41]. Such temporal effects suggest that even inheritable changes in the epigenetic modification patterns may be influenced by the *in utero* environment.

For example, gestational diabetes exposes fetuses to hyperglycemia *in utero* and leads to impaired glucose tolerance, and impaired beta cell mass and function later in life [9,23,28,34]. This susceptibility was transmittable to later generations, suggesting that the *in utero* exposure led to an epigenetic change in the genome. Several studies in both T2D and T1D animal models suggest that proper epigenetic regulation of the key development and function control gene *Pdx-1 in utero* is crucial to not only proper beta cell development, but also controlling risk of beta cell dysfunction and diabetes susceptibility later in life [24,31,32,45,46].

In the T1D non-obese diabetic (NOD) mouse model, fostering of newborn NOD pups to nurse on non-diabetic prone mice decreased their diabetes incidence. Moreover, embryo transfer of NOD fetuses to non-diabetic pseudo-pregnant mice had an even greater effect on lowering their diabetes incidence. These findings suggest that both prenatal and immediate postnatal environments can contribute to autoimmune diabetes susceptibility [41].

To avoid auto-reactivity, immune antigen presenting cell (APC) regulation of T cell ontogeny instills self-tolerance in the immune system. These tolerizing events occur predominantly based on immune responsiveness to *in utero* and immediate perinatal period environment. Alterations in cytokine expression and function patterns during tolerance development can influence the epigenetic regulation of immune response gene expression and set the stage for autoimmune disease, including T1D, susceptibility later in life [21,38,47–49].

Cytokines, such as interleukin-3 (IL-3), granulocyte-macrophage colony stimulating factor (GM-CSF) and macrophage colony stimulating factor (M-CSF), comprise the growth factor milieu in bone marrow during hematopoietic differentiation. These factors rely heavily on

573

epigenetic gene expression regulatory mechanisms to exert their effects on immune cell maturation and functionality [47]. Temporal and sequential expression and reception of these immune hormones is vital to the development of self-tolerogenic APC within the myeloid lineage. The loss of this differentiation pattern of cytokine stimulation is seen in T1D mouse models, suggesting *in utero* as well as later in life changes in epigenetic gene regulation in response cytokine signaling can dramatically alter APC function in defining and maintaining self-tolerance [47–50].

Regulatory T cell lineages that are associated with control of inflammation and self-tolerance are dependent on the interplay of cytokines such as IL-2, IL-4, tumor growth factor beta (TGFbeta), and IL-17. These cytokines function through epigenetic adaptor proteins and transcriptional factors such as CREB, AP1, ATF, N-FAT, SMAD3, and STAT5, to promote epigenetic control of a key transcription factor, FOXP3, gene expression in their development [3,6,38,51,52]. Methylation state and later histone acetylation modulation promotes the stability of FOXP3 expression. In addition, microRNA regulation of genes involved in lineage differentiation genes is crucial to the regulation of FOXP3 [38]. Dysregulation of FOXP3 leads to loss of a critical class of regulatory T cells, which allows for the subsequent emergence of Th17 proinflammatory effector T cells. Both regulatory T cell loss and Th17 cell gain have been linked to the development of autoimmune diabetes in NOD mice [38,52]. Thus, epigenetic dysregulation of both APC and T-cell functions is a strong candidate mechanism in the etiology of autoimmune diseases such as T1D.

Inducible Epigenetic Dysregulation in Metabolic Diseases

The rapid response to nutrient and environmental stimuli in gene regulation is greatly dependent on the adequate and rapid response of the epigenetic modification machinery. Acetylation and deacetylation of histones provide rapidly reversible chromatin dynamics, allowing for precise and timely gene expression changes to adapt to an ever changing environment. Dysfunction in acetylation regulation seen in MetS, T1D, and T2D sets the stage for improper gene expression responses to environmental stimuli as well as to hormonal regulation of metabolic processes [3].

Susceptibility to MetS has been attributed to a complex of traits, exposures, and behaviors, including overeating and a sedentary lifestyle [7,10,42]. Circadian rhythm timing of activity and rest as well as hormonal signaling for hunger and satiation are governed by gene expression responses to environmental cues or stimuli (also called zeitgebers) [7]. A central regulatory gene in circadian timing regulation, CLOCK, is a histone acetyltransferase [7,53–55]. The CLOCK protein forms a dimer with BMAL1 and regulates the transcription of genes involved in energy metabolism, hormonal responsiveness, thermal control, and sleep induction/waking, including PPARgamma, PPAR alpha, PGC1alpha, Leptin, and other factors involved in lipid/carbohydrate energy utilization [7,53,54]. CLOCK dysfunction in regulation of PPAR molecules, PGC1alpha, and other genes central to lipid metabolism has been implicated in the development of MetS and T2D cardiovascular disease [7,55]. Environmental cues or triggers beyond the circadian zeitgebers have been implicated in the induction of T2D and T1D epigenetic dysregulation, through control of the metabolic pathways involved in nutrient sensing and the inflammatory responses [6].

EPIGENETIC DYSREGULATION OF ENERGY METABOLISM

Psychological, physiological, and pharmacological stress factors have all proven to influence energy metabolism, both by changes induced during development and by later induction by environment [56,57]. Epigenetic modification alterations can be induced in the perinatal period by environmental factors such as maternal obesity during gestation and lactation, increased caloric intake, high fat diets, and decreased energy expenditure [41,58]. The psychological and physiological stresses of birth, even including the decreased initial maternal contact after cesarean section, have been implicated in initial epigenetic modification pattern changes resulting in alter gene expression profiles leading to higher risk for MetS and T2D [58]. Studies now show that even genetically-predestined obesity and T2D induced in mice by leptin, agouti, or melanocortin receptor gene mutations can be reversed by severe caloric intake reduction, alteration of fatty acid diet content, or simply by voluntary exercise without restricted caloric intake [6,56,59]. These findings indicate that alterations in environment, both molecular and physical, can induce disease modulating changes in gene expression.

The gene responsiveness modulated by epigenetic control of genes involved in nutrient sensing and metabolism has multi-organ system effects [6]. Adipose tissue, liver, brain, and cardiovascular tissues are all adversely affected by dysregulated glucose and lipid nutrient sensing and metabolism. Adipocyte dysfunction in both fat and glucose metabolism and storage contributes to the development of insulin resistance in MetS and T2D and cardiovascular complications in MetS, T1D, and T2D [54]. Integration of the immune and metabolic systems on both a functional and molecular regulatory level provides a shared homeostatic mechanism dysregulated in both T1D and T2D.

Disruption of functional and biochemical interactions of immune and energy metabolism is emerging as a linked mechanism underlying many of the pathological events in common in chronic metabolic diseases such as T1D, T2D, and atherosclerosis [2]. Adipocytes, macrophages, and macrophage-derived foam cells share a gene expression signature which includes not only lipid and carbohydrate metabolic enzymes and cofactors (such as PPARgamma, pyruvate carboxylase, and carnitine acetyltransferase), but also genes involved in the induction and maintenance of inflammation (including IL-1beta, IL-6, tumor necrosis factor alpha (TNFalpha), GM-CSF, complement factors, COX2, iNOS, MCP-1and TGF beta) [6,16,17,49,60–69] (Fig. 35.2).

EPIGENETIC DYSREGULATION OF INFLAMMATION

Dysregulation of inflammation is a key component of many metabolic syndromes, including atherosclerosis, T1D, and T2D. Macrophage differentiation and function is integral to regulation of inflammation, and their dysfunction contributes to the pathology associated with metabolic disorders. Disruption of normal hormonal and cytokine signaling through NFkappaB, STAT proteins, and JNK/AP1 molecules affects the immune system's ability to control inflammation, exacerbating the pathology caused by chronic metabolic stresses such as obesity, dyslipidemia, and infection [6,69].

GM-CSF/M-CSF signaling disruptions promote inadequate tolerogenic myeloid APC development, macrophage foam cell deposition, and chronic inflammation in an aberrant response to lipid metabolism and environment stress factors. These dysregulated immune cell functions lead to increased vascular plaque formation, aberrant angiogenesis, and inflammatory tissue damage, which contribute to retinopathy and vascular complications in MetS, T1D, and T2D [16–18,50,54,57,61,67].

Epigenetic regulation of inflammation in T1D has been directly affected by persistent activation of the histone acetylase/deacetylase adaptor proteins, STAT5A and STAT5B, which are caused by and self-perpetuate through epigenetic dysregulation of cytokines such as GM-CSF and IL-10 in myeloid cells and IL-2, IL-17, and IL-10 in T cells [51–53,63,67]. STAT5 activation persists in myeloid cells due to its feedback promotion of histone acetylation at *CSF2* regulatory sequences and in T cells by *IL2* and *IL2Ra* dysregulation [63,64,67,68,70,71], and by apparent loss of mechanisms to deactivate or transport activated STAT5 proteins off of chromatin and out of the nucleus [63,67,71].

Dysregulation of *PTGS2* has been implicated in the pathogenesis of both T2D and T1D genome-wide screens for epigenetic sites [16,17]. *CSF2* and *IL10* were also found in the

	ENERGY METABOLISM	INFLAMMATION & IMMUNE RESPONSE	HUNGER & SATIATION	PANCREATIC DEVELOPMENT
METS/T2D	GLUT2 GLUT4 CLOCK		MC4R MC3R	ST1
OVERLAP	PPARalpha PPARgamma PGC-1 CREB NFκB	IL-6 NFκB P38 MAPK SAPK/JNK WNT/B-CAT TNF	CREB NFκB	PDX1 NOTCH
	INS IR IRK1 JUNB	PTGS2 MCP-1 TGFbeta IL-1beta STAT5		
T1D		IRAK1 CSF2 IL-10 IL-2 IL-2R IL-13 IL-18 LCK VDR TAB2 IFNgamma CXCL3 CD28 CTLA4 IL-4 CCR5		HOX4

FIGURE 35.2

Genome wide epigenetic modification screening reveals overlap in metabolic syndromes. Wren and Garner [16] and Maio et al. [17] conducted genome-wide data-mining analyses looking for genetic loci associated with T2D [16] or T1D [17]. Combining data from both analyses with epigenetic regulation data for gene candidates [46,52,53,59,63,64,68,69], reveals overlapping effects on genes in pathways essential to energy metabolism, inflammation, and macrophage function/differentiation, as well as signaling in hunger/satiation control and pancreatic development/differentiation. (Please refer to color plate section)

screening as strong candidates for T1D involvement [17], whereas *TNF* and *IL6* genes were identified as potential inflammation response genes associated with T2D [16] (Fig. 35.2).

These findings suggest that environmentally-induced hormone and cytokine signaling promotes dysregulation of epigenetic regulation seen in metabolic disorders, and genome-wide gene expression aberrations in pathways associated with inflammation and abnormal lipid metabolism.

POTENTIAL THERAPEUTIC INTERVENTIONS INVOLVING EPIGENETIC REGULATION

Understanding the convergence of nutrient-induced, inherent, and environment/pathogen-induced epigenetic modification signaling in metabolic diseases will not be an easy task, but could lead to the more comprehensive mechanistic approach to treatment of these diseases [3,6]. The genetic code alterations of SNP defects of T2D and the *IDD* susceptibility loci of T1D both incorporate the non-coding regulatory regions as the major contributing component of their genetic risk [1,15–17,72]. Genome wide screens in both diseases have revealed epigenetic sequences as having the highest potential impact on genetic susceptibility [16,17]. Modulation of epigenetic modification via substrate manipulation, environmental alteration, or pharmaceutical intervention may lead to reduction of disease-promoting gene expression.

Considering the tangled pathways involved in immune responsiveness, neurological signaling, and nutrient sensing and uptake in these metabolic disorders, selectively down-regulating epigenetic regulatory mechanisms will be difficult at best. Treatment through substrate control (e.g. lipid intake, insulin sensitivity/glucose tolerance, activity/energy intake balance) is the most promising approach, but also the most difficult to enforce or measure accurately.

Recent studies have focused on the potential therapeutic use of histone deacetylase (HDAC) inhibitors [1,11,57,73–76], and potential DNA methytransferase inhibitors [57,73,74], which have had some success as "epigenetic therapies" in cancers such as hematopoietic malignancies and breast/prostate cancers [74,75]. Unfortunately, there are severe side effects with these chemotherapeutic agents, including immunosuppression and dysregulation of regenerative processes. HDAC inhibitors affect pancreatic development and bone marrow immune cell differentiation [75,76], while MTDT inhibitors, such as 5-azacytidine and hydralazine, affect cell division and DNA repair/replication machinery [11,57,73–76]. Botanical derivatives such as bioflavones in green tea and poison substrates for lipid metabolism and storage have shown promise in cancer therapies, but their usefulness in reversing metabolic syndromes has yet to be fully explored [74].

As drug discovery searches continue to find molecular modulators of epigenetic modification, the application of such potential and less toxic chemotherapeutic agents to the correction of epigenetic dysregulation of gene expression could play a significant role in both treatment and prevention of complex metabolic diseases.

References

1. Lee DS, Park J, Kay KA, Christakis NA, Oltvai ZN, Barabasi AL. The implications of human metabolic network topology for disease comorbidity. *PNAS* 2008;**105**(29):9880–5.

2. McCarthy MI, Abecasis GR, Cardon LR, Goldstein DB, Little J, Ioannidis JPA, et al. Genome-wide association studies for complex traits: consensus, uncertainty and challenges. *Nat Rev Genet* 2008;**9**:356–69.

3. Liu L, Li Y, Tollefsbol TO. Gene-environment interactions and epigenetic basis of human diseases. *Curr Issues Mol Biol* 2008;**10**:25–36.

4. Galton F. The average contribution of each of several ancestors to the total heritage of the offspring. *Proc R Soc* 1897;**61**:401–13.

5. TEDDY Study Group. The environmental determinants of diabetes in the young (TEDDY) study. *Immunol Diabetes V. Ann NY Acad Sci* 2008;**V**(1150):1–13.

6. Hotamisligil GS, Erbay E. Nutrient sensing and inflammation in metabolic diseases. *Nat Rev Immunol* 2008;**8**:923–34.

7. Gallou-Kabani C, Vige A, Junien C. Lifelong circadian and epigenetic drifts in metabolic syndrome. *Epigenetics* 2007;**2**(3):137–46.

8. Dorner G, Mohnike A, Steindel E. On possible genetic and epigenetic modes of diabetes transmission. *Endokrinologie* 1975;**66**(2):225–7.

9. Pinney SE, Simmons RA. Epigenetic mechanisms in the development of type 2 diabetes. *Trends Endocrinol Metab* 2010;**21**(4):223–9.

10. Taube G. Prosperity's plague. *Science* 2009;**325**:256–60.

11. Salanti G, Southam L, Altshuler D, Ardlie K, Barroso I, Boehnke M, et al. Underlying genetic models of inheritance in established type 2 diabetes associations. *Am J Epidemiol* 2009;**170**(5):537–45.

12. Leiter EH, Lee CH. Mouse models and the genetics of diabetes: is there evidence for genetic overlap between type 1 and type 2 diabetes? *Diabetes* 2005;**54**(S2):S151–S8.

13. Hewagama A, Richardson B. The genetics and epigenetics of autoimmune diseases. *J Autoimmun* 2009;**33**(2009):3–11.

14. Huber A, Menconi F, Corathers S, Jacobson EM, Tomer Y. Joint genetic susceptibility to type 1 diabetes and autoimmune thyroiditis: from epidemiology to mechanisms. *Endocr Rev* 2008;**29**(6):697–725.

15. Morahan G, Morel L. Genetics of autoimmune diseases in humans and in animal models. *Curr Opin Immunol* 2002;**14**:803–11.

16. Wren JD, Garner HR. Data-mining analysis suggests an epigenetic pathogenesis for type 2 diabetes. *J Biomed Biotechnol* 2005;**2**(2005):104–12.

17. Maio F, Smith DD, Zhang L, Min A, Feng W, Natarajan R. Lymphocytes from patients with type 1 diabetes display a distinct profile of chromatin histone H3 lysine 9 dimethylation: an epigenetic study in diabetes. *Diabetes* 2008;**57**:3189–98.

18. Litherland SA. Immunopathogenic interaction of environmental triggers and genetic susceptibility in diabetes: is epigenetics the missing link? *Diabetes* 2008;**57**(12):3184–86.

19. Adams MD, Kelley JM, Gocayne JD, Dubnick M, Polymeropoulos MH, Xiao H, et al. Complementary DNA sequencing: expressed sequence tags and human genome project. *Science* 1991;**252**:1651–6.

20. Wilson AG. Epigenetic regulation of gene expression in the inflammatory response and relevance to common diseases. *J Periodontol* 2008;**79**:1514–19.

21. Mulder H, Ling C. Mitochondrial dysfunction in pancreatic beta-cells in type 2 diabetes. *Mol Cell Endocrinol* 2009;**297**(2009):34–40.

22. Barker DJP, Bagby SP, Hanson MA. Mechanisms of disease: *in utero* programming in the pathogenesis of hypertension. *Nat Clin Pract Nephrol* 2006;**2**(12):700–7.

23. Fetita LS, Sobngwi E, Serradas P, Calvo F, Gautier JF. Consequences of fetal exposure to maternal diabetes in offspring. *J Clin Endocrinol Metab* 2006;**91**(10):3718–24.

24. Lillycrop KA, Phillips ES, Torrens C, Hanson MA, Jackson AA, Burdge GC. Feeding pregnant rats a protein-restricted diet persistently alters the methylation of specific cytosine in the hepatic PPAR alpha promoter of the offspring. *Br J Nutr* 2008;**100**(2):278–82.

25. Aerts L, Holemans K, van Assche FA. Maternal diabetes during pregnancy: consequences for the offspring. *Diabetes Metab Rev* 1990;**6**:147–67.

26. Savana-Ventura C, Chircop M. Birth-weight influence on the subsequent development of gestational diabetes mellitus. *Acta Diabetalogia* 2003;**40**:101–4.

27. Aerts L, van Assche FA. Animal evidence for the transgenerational development of diabetes mellitus. *Int J Biochem Cell Biol* 2006;**38**:894–903.

28. Simmons RA, Templeton L, Gertz S, Niu H. Intrauterine growth retardation leads to type II diabetes in adulthood in the rat. *Diabetes* 2001;**50**:2279–86.

29. Muhonen P, Holthofer H. Epigenetic and microRNA-mediated regulation in diabetes. *Nephrol, Dial Transplant* 2009;**24**:1088–96.

30. Buysschaert I, Schmidt T, Roncal C, Carmeliet P, Lambrachts D. Genetics, epigenetics and pharmaco-(epi)genomics in angiogenesis. *J Cell Mol Med* 2008;**12**(6B):2533–51.

31. Waterland RA, Jirtle RL. Early nutrition, epigenetic changes at transposons and imprinted genes, and enhanced susceptibility to adult chronic diseases. *Nutrition* 2004;**20**:63–8.

32. Waterland RA, Jirtle RL. Transposable elements: targets for early nutritional effects. *Mol Cell Biol* 2003;**23**(15):5293–300.

33. Zhou X, Jeker LT, Fife BT, Zhu S, Anderson MS, McManus MT, et al. Selective miRNA disruption in T reg cells leads to uncontrolled autoimmunity. *J Exp Med* 2008;**205**(9):1983–91.

34. Tremblay J, Hamet P. Impact of genetic and epigenetic factors from early life to later disease. *Metab Clin Exp* 2008;**57**(S2):S27–S31.

35. Hales CN, Barker DJP. Type-2 (non-insulin-dependent) diabetes mellitus – the thrifty phenotype hypothesis. *Diabetologia* 1992;**35**:595–601.

36. Gluckman PD, Hanson MA, Buklijas T, Low FM, Beedle AS. Epigenetic mechanism that underpin metabolic and cardiovascular diseases. *Nat Rev Immunol* 2009;**5**:401–8.

37. Hinchliffe SA, Lynch MR, Sargent PH, Howard CV, van Velzen D. The effect of intrauterine growth retardation on the development of renal nephrons. *Br J Obstet Gynaecol* 1992;**99**(4):296–301.

38. Huehn J, Polansky JK, Hamann A. Epigenetic control of FOXP3 expression: the key to a stable regulatory T-cell lineage? *Nat Rev Immunol* 2009;**9**:83–9.

39. Reik W, Dean W, Walter J. Epigenetic reprogramming in mammalian development. *Science* 2001;**293**:1089–93.

40. Barker DJ, Forsen T, Eriksson JG, Osmond C. Growth and living conditions in childhood and hypertension in adult life: a longitudinal study. *J Hypertens* 2002;**20**:1951–6.

41. Washburn LR, Dang H, Tian J, Kaufman DL. The postnatal maternal environment influences diabetes development in nonobese diabetic mice. *J Autoimmun* 2007;**28**(2007):19–23.

42. Devaskar SU, Thamotharan M. Metabolic programming in the pathogenesis of insulin resistance. *Rev Endocrinol Metab Dis* 2007;**8**:105–13.

43. Freinkel N. Banting Lecture 1980: Of pregnancy and progeny. *Diabetes* 1980;**29**(12):1023–35.

44. Bogdarina I, Welham S, King PJ, Burns SP, Clark AJL. Epigenetic modification of the rennin-angiotensin system in the fetal programming of hypertension. *Circ Res* 2007;**100**:520–6.

45. Dongqi T, Burkhardt B, Cao L, Ahren K, Litherland SA, Aktinson M, et al. In vitro generation of functional insulin producing cells from human bone marrow-derived stem cells. *Diabetes* 2004;**53**(7):1721–32.

46. Park JH, Stoffers DA, Nicholls RD, Simmons RA. Development of type 2 diabetes following intrauterine growth retardation in rats is associated with progressive epigenetic silencing of Pdx1. *J Clin Invest* 2008;**118**(6):2316–24.

47. Hashimoto S, Komuro I, Yamada M, Akagawa KS. IL-10 inhibits granulocyte-macrophage colony-stimulating factor-dependent human monocyte survival at the early stage of the culture and inhibits the generation of macrophages. *J Immunol* 2001;**167**:3619–25.

48. Serreze DV, Gaskins HR, Leiter EH. Defects in the differentiation and function of antigen presenting cells in the NOD/Lt Mice. *J Immunol* 1993;**150**:2534–43.

49. Clare-Salzler MJ. The immunopathogenic roles of antigen presenting cells in the NOD mouse. In: Leiter EH, Atkinson MA, editors. *NOD mice and related strains: research applications in diabetes, AIDS, cancer, and other diseases*. Austin, TX: Landes Bioscience Publishers,; 1998, pp. 101–20.

50. Rumore-Maton B, Elf J, Belkin NS, Stutevoss B, Seydel F, Garrigan E, et al. M-CSF and GM-CSF regulation of STAT5 activation and DNA binding in myeloid cell differentiation is disrupted in nonobese diabetic mice. *Clin Dev Immunol* 2008;**2008**(76795):1–8.

51. Chen Z, O'Shea JJ. Regulation of IL-17 production in human lymphocytes. *Cytokine* 2008;**41**:71–8.

52. Murawski MR, Litherland SA, Clare-Salzler MJ, Davoodi-Semiromi A. Upregulation of FoxP3 expression in mouse and human Treg is IL-2/STAT5 dependent: implications for the NOD STAT5B mutation in diabetes. *Ann NY Acad Sci* 2006;**1079**:198–204.

53. Doi M, Hirayama J, Sassone-Cosi P. Circadian regulator CLOCK is a histone acetyltransferase. *Cell* 2006;**125**:497–508.

54. Canaple L, Rambaud J, Dkhissi-Benyahya O, Rayet B, Tan NS, Michalik L, et al. Reciprocal regulation of brain and muscle Arnt-like protein 1 and peroxisome proliferator-activated receptor alpha defines a novel positive feedback loop in the rodent liver circadian clock. *Mol Endocrinol* 2006;**20**:1715–27.

55. Staels B. When the clock stops ticking, metabolic syndrome explodes. *Nat Med* 2006;**12**:54–5.

56. Levin B. Epigenetic influences on food intake and physical activity level: review of animal studies. *Obesity* 2008;**16**(S3):S51–4.

57. Szyf M. Epigenetic therapeutics in autoimmune disease. *Clin Rev Allergy Immunol* 2009;**July 2009**. Humana Press online..

58. Schlinzig T, Johansson S, Gunnar A, Ekstrom TJ, Norman M. Epigenetic modulation at birth – altered DNA-methylation in white blood cells after caesarean section. *Acta Paediatr* 2009;**98**:1096–9.

59. Haskell-Luevano C, Schaub JW, Andreasen A, Haskell KR, Moore MC, Koerper LM, et al. Voluntary exercise prevents the obese and diabetic metabolic syndrome of the melanocortin-4 receptor knockout mouse. *FASEB J* 2008;**23**:642–55.

60. Hamilton JA. Colony-stimulating factors in inflammation and autoimmunity. *Nat Rev Immunol* 2008;**8**:533–44.

61. Mack CP. An epigenetic clue to diabetic vascular disease. *Circ Res* 2008;**103**(6):568–70.

62. Litherland SA, She J-X, Schatz D, Fuller K, Hutson AD, Li Y, et al. Aberrant monocyte prostaglandin synthase 2 (PGS2) in type 1 diabetes before and after disease onset. *Pediatr Diabetes* 2003;**4**:10–18.

63. Seydel F, Garrigan E, Stutevoss B, Belkin N, Makadia B, Carter J, et al. GM-CSF induces STAT5 binding at epigenetic regulatory sites within the Csf2 promoter of nonobese diabetic (NOD) mouse myeloid cells. *J Autoimmun* 2008;**31**(4):377–84.

64. Tsuji-Takayama K, Suzuki M, Yamamoto M, Harashima A, Okochi A, Otani T, et al. The production of IL-10 by human regulatory T cells is enhanced by IL-2 through a STAT5-responsive intronic enhancer in the IL10 locus. *J Immunol* 2008;**181**:3897–905.

65. Yamaoka K, Otsuka T, Niiro H, Arinobu Y, Nihho Y, Hamasaki N, et al. Activation of STAT5 by lipopolysaccharide through granulocyte-macrophage colony-stimulating factor production in human monocytes. *J Immunol* 1998;**160**:838–45.

66. Litherland SA, Xie XT, Li Y, Grebe KM, Reddy S, Moldawer LL, et al. IL10 Resistant PGS2 expression in NOD mouse macrophages and at-risk/Type 1 diabetic human monocytes. *J Autoimmun* 2004;**22**(3):227–33.

67. Litherland SA, Xie XT, Grebe KM, Davoodi-Semiromi A, Elf J, Belkin NS, et al. Signal transduction activator of transcription 5 (STAT5) proteins are dysfunctional in autoimmune monocytes and macrophages. *J Autoimmun* 2005;**24**:297–310.

68. Dendrou CA, Wicker LS. The IL-2/CD25 pathway determines susceptibility to T1D in humans and NOD mice. *J Clin Immunol* 2008;**28**(6):685–96.

69. Stephens JM, Morrison RF, Wu Z, Farmer SR. PPARgamma ligand-dependent induction of STAT1, STAT5A, and STAT5B during adipogenesis. *Biochem Biophys Res Commun* 1999;**262**:216–22.

70. Soldaini E, John S, Moro S, Bollenbacher J, Schindler U, Leonard WJ. DNA binding site selection of dimeric and tetrameric STAT5 proteins reveals a large repertoire of divergent tetrameric STAT5a binding sites. *Mol Cell Biol* 2000;**20**(1):389–401.

71. McBride KM, McDonald C, Reich MC. Nuclear export signal located within the DNA-binding domain of the STAT1 transcription factor. *EMBO J* 2000;**19**(22):6196–206.

579

72. Litherland SA, Grebe KM, Belkin NS, Paek E, Elf J, Atkinson M, et al. Nonobese diabetic mouse congenic analysis reveals chromosome 11 locus contributing to diabetes susceptibility, macrophage STAT5 dysfunction, and GM-CSF overproduction. *J Immunol* 2005;**175**(7):4561–5.

73. Khan ZA, Chakrabarti S. Cellular signaling and potential new treatment targets in diabetic retinopathy. *Exp Diabetes Res* 2007;**2007**(31867):1–12.

74. Kirk H, Cefalu WT, Ribnicky D, Liu Z, Eilertsen KJ. Botanicals as epigenetic modulators for mechanisms contributing to development of metabolic syndrome. *Metabol Clin Exp* 2008;**57**:S16–S23.

75. Javierre BM, Esteller M, Ballestar E. Epigenetic connections between autoimmune disorders and haematological malignancies. *Trends Immunol* 2008;**29**(12):616–23.

76. Haumaitre C, Lenoir O, Scharfmann R. Histone deacetylase inhibitors modify pancreatic cell fate determination and amplify endocrine progenitors. *Mol Cell Biol* 2008;**28**(20):6373–83.

CHAPTER 36

Imprinting Disorders in Humans

Thomas Eggermann
Institut für Humangenetik, RWTH Aachen, D-52074 Aachen, Germany

INTRODUCTION

More than 100 human genes are currently believed to be imprinted by epigenetic mechanisms that allow expression from only one of the two paternal alleles.

These parental imprints undergo a cycle during the life of an organism that allows their reprogramming at each generation. The imprinted marks are inherited from the parental gametes and are then maintained and realized in the somatic cells of an individual. During early development, methylation modification of the mammalian genome undergoes dramatic changes and is linked to the rapid differentiation and formation of various tissues and organs. The imprint marks are erased in the germline and re-established according to the sex of the contributing individual for the next generation. Among others, one fundamental molecular process in this imprinting cycle is DNA methylation. It is mainly catalyzed by DNA methyltransferases (DNMTs) and is generally associated with gene silencing.

The balanced expression of imprinted genes is needed for a regular development of an individual, and it is therefore not surprising that many imprinted genes are involved in human growth. According to their hypothesized biological function as mediators of the "battle-of-sexes" (conflict theory) in the fetal period, paternally- and maternally-imprinted genes have opposite functions: while paternally-expressed factors promote growth, maternally-expressed ones suppress it [1,2].

Indeed, the majority of the known imprinting syndromes are associated with disturbed growth (Table 36.1). The most prominent imprinting disorders (IDs) are Prader–Willi and Angelman syndromes (PWS, AS); in both entities similar but opposite genetic and epigenetic mutations of chromosome 15 are present. Further imprinting diseases are transient neonatal diabetes mellitus (TNDM, chromosome 6q24), Silver–Russell syndrome (SRS, chromosomes 7 and 11p15), Beckwith–Wiedemann syndrome (BWS, chromosome 11p15) and the "maternal/paternal UPD14" (UPD(14)mat/pat) syndromes. However, there are overlapping clinical findings in some of these disorders (i.e. TNDM and BWS). Recent studies identified an association between an increased risk for epigenetic defects resulting in IDs and assisted reproduction techniques (ART) [for review, see Ref. 3] but it is not yet clear whether these defects are linked to the subfertility of parents or the technique.

Handbook of Epigenetics: The New Molecular and Medical Genetics. DOI: 10.1016/B978-0-12-375709-8.00036-8

TABLE 36.1 Information on the Frequencies, Genetic Alterations, and the Main Clinical Features in the Currently Known Human IDs

Imprinting Disorder	Acronym/ Abbreviation	Frequency	OMIM	Affected Chromosomes/ Imprinted Regions	Types of Mutations/ Epimutations	MLH	Detection Rate	Main Clinical Features
Transient neonatal diabetes mellitus	TNDM	1/800.000	601410	6q24: ZAC1/HYMA1	upd(6)pat; Paternal duplications; Methylation defects	Yes	85%	Transient diabetes, IUGR, macroglossia
Silver-Russell syndrome	Russell–Silver syndrome, SRS, RSS	1/10.000	180860	7; 11p15:	upd(7)mat; upd(11p15)mat; Maternal duplication; Hypomethylation	Yes	~10%; Single cases single cases > 38%	Pre- and/or postnatal growth retardation, rel. macrocephaly, asymmetry, triangular face
Beckwith–Wiedemann syndrome	Wiedemann-syndrome, EMG syndrome, BWS	1/15.000	130650	IGF2/H19; 11p15:; ICR1: IGF2/H19; ICR2: KCNQ1; CDKN1C	upd(11p15)pat; Chromosomal aberrations; Hypermethylation; Hypomethylation; Point mutations	Yes	~20%; 2-4%; 5–10%; 40–50%; 5% in sporadic; 40–50% in familial cases	Pre- and postnatal overgrowth, organomegaly, omphalocele, neonatal hypoglycemia, hemihypertrophia, increased tumor risk in specific molecular subgroups
UPD(14)mat syndrome	Temple syndrome	Rare		14q32: DLK1/GTL2	upd(14)mat; Deletions of the paternal chromosome 14; Aberrant methylation	No	?	IUGR/PNGR, hypotonia, scoliosis, precocious puberty
UPD(14)pat syndrome	–	Rare	608149	14	upd(14)pat; Aberrant methylation	No	?	IUGR, polyhydramnios, abdominal and thoracal wall defects, bell-shaped thorax
Angelman syndrome	Happy Puppet syndrome, AS	1/20.000–1/12.000	105830	15q11-q13:; UBE3A	Maternal deletion; upd(15)pat; Aberrant methylation; Point mutations	No	70%; 1-3%; ~4%; 10–15%	Mental retardation, microcephaly, no speech, unmotivated laughing, ataxia, seizures, scoliosis
Prader-Willi syndrome	Prader–Labhart–Willi syndrome, PWS	1/25.000–1/10.000	176270	15q11-q13	Paternal deletion; upd(15)mat; Aberrant methylation	No	70%; <30%; ~1%	Mental retardation, neonatal hypotonia, growth retardation, hypogenitalism, hypopigmentation, adipositas/hyperphagia

TYPES OF MUTATIONS AND EPIMUTATIONS IN IDs

The regular expression of imprinted genes can be influenced by different types of mutations and epimutations. For nearly all known IDs, the same genetic and epigenetic alterations affecting imprinted genes/gene clusters and their expression and regulation have been reported (Table 36.1). They include genomic mutations (uniparental disomy (UPD), chromosomal imbalances, point mutations) and true epigenetic defects (abnormal DNA methylation but without a genomic alteration) which affect the expression and regulation of imprinted loci. The incidence of the different classes of mutations and epimutations are different in the known IDs (Table 36.1), suggesting that some loci like the ICR1 and the ICR2 in 11p15 are more vulnerable for disturbed methylation than others.

Uniparental Disomy (UPD)

Nearly all human imprinting disorders have been detected through the identification of Uniparental Disomy (UPD). UPD is defined as the abnormal inheritance of both copies of a chromosome or a chromosomal segment from only one parent. Several modes of UPD formation have meanwhile been described; one prominent mechanism is trisomic rescue [for review see Ref. 4] (Fig. 36.1). In the majority of cases, UPD affects the whole chromosome, but meanwhile several UPDs affecting only parts of a chromosome have been reported ("segmental UPD"; [5]). In particular in Beckwith–Wiedemann syndrome (BWS) segmental UPD of chromosome 11p15 accounts for up to 20% of cases.

Two types of UPD can be distinguished, uniparental heterodisomy (UPhD) and uniparental isodisomy (UPiD). UPhD means the presence of the two different homologous chromosomes

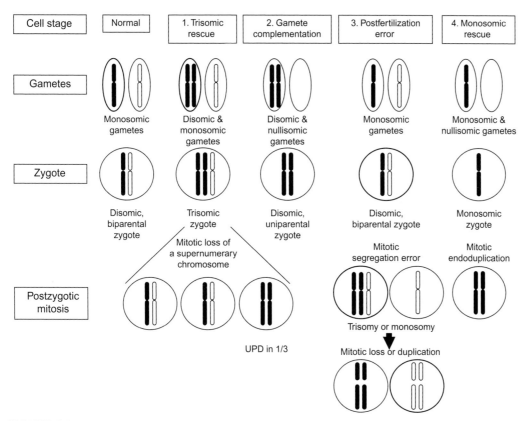

FIGURE 36.1

Schematic overview on UPD formation by trisomic rescue, gamete complementation, post-fertilization error, and monosomic rescue.

from the same parent, while in UPiD the same chromosome is duplicated. Up to now, three different possibilities as to how UPD might influence the phenotype have been reported:

1. If UPD affects imprinted genes it will lead to their imbalanced expression. UPD has been reported for nearly all human chromosomes but due to the tendency of imprinted genes to cluster, only some UPDs are associated with specific phenotypes (Table 36.2). Interestingly, on some chromosomes oppositely imprinted genes are clustered (chromosomes 11, 14, 15) whereas for others one parental UPD is risky while the other is not (chromosomes 6 and 7).

2. If one parent carries an autosomal recessive mutation and has a UPiD offspring, this child will be homozygous for the same mutated allele and will be affected by the respective disease. Indeed, the first reported case of UPD was a patient suffering from cystic fibrosis caused by maternal UPD of chromosome 7 [6]. Thus, despite the lack of imprinted genes on several human chromosomes, a reduction to homozygosity of recessive alleles has to be considered in any case of UPD.

3. If UPD is derived from a trisomic zygote, some trisomic cells can survive. The clinical outcome of this mosaic constitution can therefore be influenced by the UPD as well as by the trisomic cell line. Then, the phenotype is difficult to predict and a differentiation between symptoms caused by UPD or by the chromosomal mosaicism is often impossible.

The association between UPD and chromosomal disturbances has also to be borne in mind in prenatal diagnosis: in particular, in chorionic villous sampling trisomy mosaicism is a relatively-frequent finding and depending on the chromosome involved UPD testing should be considered after detection of an aneuploidy. Another group prone to UPD are carriers

TABLE 36.2 Overview on UPDs for the Single Human Chromosomes Reported so Far. So Far, Seven IDs are Known, While Numerous Paternal as well as Maternal UPDs have been Excluded to Cause Specific IDs ("−"). However, Some UPDs have not Yet been Reported ("?")

Chromosome	Maternal UPD	Paternal UPD
1	−	−
2	−	−
3	−	?
4	−	?
5	?	−
6	−	Transient diabetes melitus
7	Silver–Russell syndrome	−
8	−	−
9	−	−
10	−	−
11	Silver–Russell syndrome	Beckwith–Wiedemann syndrome
12	−	?
13	−	−
14	UPD(14)mat syndrome	UPD(14)pat syndrome
15	Prader–Willi syndrome	Angelman syndrome
16	−	−
17	−	?
18	?	?
19	?	?
20	−	−
21	−	−
22	−	−
X	−	−

of (Robertsonian) translocations involving chromosomes 6, 7, 11, 14, and 15. Indeed, the majority of chromosome 14 UPDs (UPD(14)mat, UPD(14)pat) has been detected among Robertsonian translocation carriers.

Chromosomal Deletions and Duplications

The functional relevance of chromosomal imbalances affecting imprinted genes is comparable to that of UPD. In PWS for example, a deletion of ~0.6 Mb in the paternal chromosome 15 accounts for nearly 70% of cases and affects the region 15q11-q13 while <30% of patients are UPD(15)mat carriers. Functionally, both (epi)mutations result in an inactivation of paternally-expressed genes [for review see Ref. 7]. Vice versa, ~70% of AS patients show a maternal deletion of the same chromosomal region in 15q, 1–3% a UPD(15)pat. The two chromosome 15 IDs impressively illustrate the profound role of chromosomal imbalances in the etiology of this group of diseases. The situation is similar for TNDM where 6q24 duplications account for approximately one third of cases. Single cases of duplications/deletions, maternal as well as paternal, have also been reported in SRS/BWS and UPD(14)mat/UPD(14)pat (Table 36.1).

Genomic Point Mutations in ID Patients

In some cases of IDs point mutations on genomic DNA level are responsible for the clinical course. These mutations might affect either genes underlying a regulation of imprinting centers, or genes involved in establishment or maintenance of methylation marks.

Point mutations in genes regulated by DMRs have been reported for BWS and AS and account for a significant number of cases. As aforementioned, *CDKN1C* mutations are responsible for a large proportion of familial BWS cases; *UBE3A1* mutations can be detected in 10–15% of AS patients (Table 36.1).

In principle, *genes responsible for the establishment or maintenance of methylation marks* are good candidates to carry mutations causing aberrant methylation. Indeed, a growing number of factors involved in DNA methylation and its regulation have been identified, but so far only two genes have been published to carry point mutations in ID patients. In TNDM patients with multilocus hypomethylation (MLH) autosomal-recessive mutations in the zinc-finger protein ZFP57 have been reported [see below; and Ref. 8]. In a BWS sibship a homozygous mutation in the *NLRP2* gene in the mother was described by Meyer et al. [9] providing evidence for a *trans* mechanism for the disturbed methylation pattern in 11p15 in the two children. However, screening studies in further enzymes involved in locus specific methylation and regulation such as *DNMT3L* and *PLAGL1* failed to identify pathogenic variants [10,11].

Imprinting Defects

Imprinting defects, or epimutations, describe altered DNA methylation patterns at specific differentially-methylated regions (DMRs) which regulate the expression of neighboring genes. The chromosomal region 11p15 represents a typical example of an imprinted region with complex regulation mechanisms (Fig. 36.2) and it is described below as an example.

The region 11p15 contains a number of imprinted genes, the expression of which is regulated by two different imprinting control regions (ICR1 and ICR2), also called H19 DMR and KvDMR1. The telomeric ICR1 confers a differential chromatin architecture to the two parental alleles leading to reciprocal expression of *H19* and *IGF2*. The two genes are co-expressed in endoderm- and mesoderm-derived tissues during embryonic development and compete for the same enhancers. The paternally-expressed *IGF2* is involved in fetal development and growth [12,13]. Although *H19* was one of the first non-coding transcripts identified, its function is still unknown. Knockout of *H19* removing the whole RNA coding

585

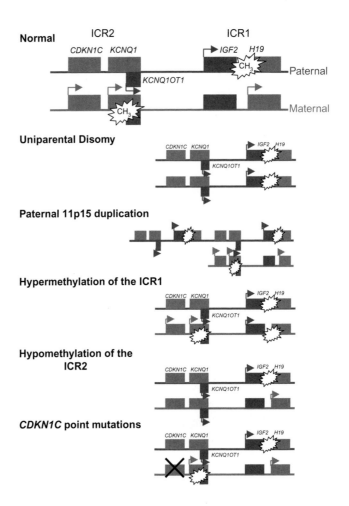

FIGURE 36.2

Epigenetic/genetic alterations in 11p15 which are currently known to be associated with BWS as an example for the complex molecular background of IDs. Only those genes in 11p15 are illustrated which are currently being discussed as being functionally involved in the pathogenesis of BWS (and SRS). (Please refer to color plate section)

sequence but leaving the promoter and surrounding transcription unit intact had no effect on the imprinted expression of *IGF2* [14]. These results indicate that the RNA itself might be non-functional; however, the fact that *H19* is a relatively highly conserved gene among mammals (77% identity between human and mouse) suggests a profound functional relevance. A recent study suggests that *H19* functions as a primary micro-RNA precursor involved in the post-transcriptional down-regulation of specific mRNAs during vertebrate development [15]. The ICR1 contains seven CTCF target sites in the DMR 2 kb upstream of *H19* and shows allele specific methylation. The Zinc-finger binding factor CTCF binds to the maternal unmethylated ICR1 copy and thereby forms a chromatin boundary. This CTCF binding mechanism blocks *IGF2* and promotes *H19* transcription of the maternal 11p15 copy. As a result of this complex and balanced regulation, hypo- as well as hypermethylation in the ICR1 in 11p15 cause IDs (SRS and BWS, respectively; Table 36.1).

The centromeric ICR2 in 11p15 regulates the expression of *CDKN1C*, *KCNQ1* (*potassium channel KQT-family member 1*) and further genes and is methylated only on the maternal allele. Mutations in the paternally-suppressed *CDKN1C* gene account for up to 40% of familial BWS cases and 5–10% of sporadic patients (Table 36.1). The gene encodes a cyclin dependent kinase inhibitor (p57^{KIP2}) and is part of the p21^{CIP2}Cdk inhibitor family. Functional analysis of *CDKN1C* germline mutations detected in two BWS patients showed the loss of cell-cycle inhibition [16]. The gene of another non-coding RNA in 11p15, *KCNQ1OT1* (*LIT1*), is localized in intron 10 of the *KCNQ1* gene. *KCNQ1OT1* is expressed by the paternal allele and probably represses the expression of the *CDKN1C* gene. Loss of methylation of the maternal ICR2 allele correlates with expression of *KCNQ1OT1*. In BWS, one central physiological change caused by ICR2 (epi)mutations (hypomethylation at ICR2 as well as *CDKN1C* point mutations) is the reduced expression of CDKN1C.

MULTILOCUS HYPOMETHYLATION (MLH) IN IDs

For all congenital IDs an association with aberrant methylation or mutations at specific loci has been well established. It is therefore amazing that patients with transient neonatal diabetes mellitus (TNDM) can exhibit hypomethylation at further imprinted loci in addition to the disease-specific in 6q24 [17,18]. Because all affected loci were maternally methylated, Mackay et al. [18] defined a maternal hypomethylation syndrome. In some of these patients, mutations in the *ZFP57* gene could recently be identified as the cause of the first heritable global human imprinting disorder [8]. *ZFP57* point mutations are autosomal recessive, a finding that is important for genetic counselling in TNDM. The association between mosaicism of hypomethylation and *ZFP57* mutations suggests that this factor is involved in the maintenance of imprinted DNA methylation in the early embryo. Recently, also for BWS and SRS patients with MLH in blood lymphocytes have been reported [19–22]. In these patients both paternally- and maternally-imprinted loci were affected in leukocytes. In all studies, a phenotypic difference between BWS or SRS patients with MLH and patients carrying isolated 11p15 aberrations was not obvious. A broad range of imprinted loci can be affected in the case of MHL; however, it seems that single loci are prone to demethylation, i.e. the ICR2 in 11p15 and the *DLK1/GTL2* locus in 14q32.

587

GENERAL ASPECTS OF IDs

At first glance the different IDs are clinically heterogeneous but due to the similar underlying molecular defects and the function of the affected genes they often show comparable clinical characteristics, i.e.:

- pre- and/or postnatal growth retardation;
- hypo- or hyperglycemia;
- failure to thrive in the newborn and early childhood period;
- neurological abnormalities in childhood.

UPD and epimutations mainly occur *denovo*; in the case of postzygotic origin, mosaicism for the disturbances can be observed. As a result, in IDs the following observations are common:

- There is asymmetry of body, head and/or limbs.
- Most cases are sporadic, and familial cases are rare.
- Discordant monozygotic twins occur.
- Genotype–phenotype correlations are difficult to delineate.

In family cases, deviations from Mendelian inheritance can be often observed due to the influence of the parental origin of the (epi)mutation and the sex of the contributing parent. For genetic counselling of ID families, the knowledge of the nature of molecular mutation or pimutation subtype is essential to delineate exact risk figures. Patients/carriers with deletions for example will have a 50% risk of conceiving a child with a UPD phenotype, depending on the sex of the patient. In the case of epimutations it might be more difficult to give risk figures, and genetic counsellors are therefore advised to continually update their knowledge for each disease.

TRANSIENT NEONATAL DIABETES MELLITUS/TNDM – CHROMOSOME 6

Transient neonatal diabetes mellitus (TNDM) is a rare disease; in addition to hypoglycemia IUGR and abdominal wall defects are common [23]. Insulin therapy is required for an average of 3 months; afterwards the diabetes resolves. However, the majority of TNDM patients develop type 2 diabetes.

TNDM is associated with an overexpression of the imprinted locus *PLAGL1/ZAC* in 6q24. As with the other IDs, three (epi)genetic causes of TNDM have been identified (Table 36.1): UPD(6)pat, paternal duplications of 6q24, and aberrant methylation at the *PLAGL1/ZAC* locus. The *PLAGL1/ZAC* gene is a maternally-imprinted gene and therefore only expressed from the paternal allele. It encodes a zinc-finger protein which binds DNA and hence influences the expression of other genes [for review see Ref. 24].

SILVER–RUSSELL SYNDROME/SRS – CHROMOSOMES 7 AND 11

Silver–Russell syndrome (SRS) is a congenital disorder mainly characterized by pre- and postnatal growth restriction. The children are relatively macrocephalic and their face is triangular-shaped with a broad forehead and a pointed, small chin. In many cases, asymmetry of limbs and body and clinodactyly V is present. Growth failure is often accompanied by severe failure to thrive, and feeding difficulties are reported. For those children without catch-up growth by the age of two, growth hormone therapy is encouraged.

The genetic basis of SRS is very heterogeneous. In approximately 10% of SRS patients a maternal uniparental disomy for chromosome 7 (UPD(7)mat) can be found [for review, see Ref. 25]. The majority of SRS patients (~40%) show a hypomethylation of the ICR1 in the imprinted region 11p15 (Fig. 36.2); in single cases maternal duplications of the whole chromosomal region in 11p15 have been reported ([26]; for review, see Ref. 25). Until now only one SRS patient with a maternal duplication restricted to the ICR2 has been identified [27]. With the recent identification of a patient with a UPD(11)mat the currently known spectrum of (epi)mutations has been accomplished in SRS [28]. Interestingly, the opposite 11p15 epigenetic and genetic findings can be observed in BWS (Table 36.1). Numerous (submicroscopic) chromosomal disturbances have been described in SRS patients, and thus screening for cryptic genomic imbalances is indicated after exclusion of UPD(7)mat and 11p15 epimutations [29,30]. Generally, the recurrence risk for SRS is low because the majority of patients are sporadic. Nevertheless, the situation changes in case of familial genetic or epigenetic alterations like chromosomal rearrangements or untypical aberrant methylation.

A genotype–phenotype correlation is difficult in SRS. In general, the 11p15 epimutation carriers show the more characteristic phenotype while UPD(7)mat patients are affected more mildly, but exceptions exist [31]. Interestingly, there is evidence for a correlation between the genotype/epigenotype and endocrinological parameters [32].

BECKWITH–WIEDEMANN SYNDROME/BWS – CHROMOSOME 11

Beckwith–Wiedemann syndrome (BWS) was initially called EMG syndrome, from its three main features of exomphalos, macroglossia, and (neonatal) gigantism. Additional features include neonatal hypoglycemia, hemihypertrophy, organomegaly, earlobe creases, polyhydramios, hemangioma, and cardiomyopathy. In 5–7% of children, embryonal tumors (most commonly Wilms' tumor) are diagnosed. The clinical diagnosis of BWS is often difficult due to its variable presentation and the phenotypic overlap with other overgrowth syndromes [for review, see Ref. 33].

The genetics of BWS are complex, but in the majority of cases an altered expression or mutations of several closely linked genes in 11p15 associated with cell cycle and growth control can be observed. As described above (Fig. 36.2), the region contains two imprinted gene clusters. In contrast to SRS, the ICR2 regulating *KCNQ1* and *CDKN1C* is the preponderant altered region in BWS; nearly 50% of patients carry a hypomethylation at this locus (Table 36.1). UPD(11p15)pat is the second important alteration, while ICR1 hypermethylation is rare. Most BWS cases are sporadic but familial inheritance is observed in 15% of all cases. In BWS families without aberrant 11p15 methylation, *CDKN1C* point mutations are frequent.

These BWS pedigrees resemble that of an autosomal dominant inheritance with incomplete penetrance. Interestingly, an increased frequency of monozygotic twinning has been reported in BWS families with epimutations (for review, see Ref. 22). It was therefore hypothesized that a methylation error precedes and possibly triggers twinning. BWS twins are nearly always female and discordant for the disorder. In some twin pairs, aberrant methylation was also detectable in leukocytes of the unaffected twin while it was not detectable in buccal swab DNA of the healthy child.

A rudimental genotype/epigenotype–phenotype correlation has recently been established for BWS [34]: hemihypertrophy is strongly associated with UPD(11)pat, exomphalos with ICR2 hypomethylation and *CDKN1C* mutations. Most importantly, the risk of neoplasias is significantly higher in ICR1 hypermethylation and UPD(11)pat than in the other molecular subgroups. In BWS the determination of the molecular subtype is therefore important for an individual prognosis and therapy. Nevertheless, the phenotypic transitions are fluid and testing for all molecular subtypes should be considered in patients with BWS features.

MATERNAL AND PATERNAL UPD(14) SYNDROMES/ UPD(14)mat/pat – CHROMOSOME 14

The so-far called UPD(14)mat and UPD(14)pat syndromes were first described in 1991 by Temple et al. and Wang et al. [35,36]; meanwhile distinct clinical phenotypes have been defined. However, the frequencies of both syndromes are currently unknown.

In both IDs, the original genetic alterations were UPDs associated with Robertsonian translocations. Considering the most important formation mechanism of UPD via trisomy rescue, this observation was consequent because Robertsonian translocations are prone to trisomic offspring. Meanwhile, several cases with isolated IDs and microdeletions affecting the *DLK1/GTL2* locus in 14q32 have been described [37,38], resulting in the same phenotypes, and thus new names for these two syndromes are necessary [39].

The UPD(14)mat phenotype is characterized by prenatal and postnatal growth retardation, muscular hypotonia, feeding difficulties, small hands and feet, recurrent otitis media, joint laxity, motor delay, truncal obesity, and early onset of puberty. The facial gestalt comprises a prominent forehead, a bulbous nasal tip and a short philtrum. Patients with UPD(14)-mat show clinical features overlapping with PWS, and thus screening for UPD(14)mat should be performed in patients with PWS-like phenotype after exclusion of the PWS specific (epi) mutations [40].

UPD(14)pat is associated with a severe clinical course with polyhydramios, a typical small, bell-shaped thorax, abdominal wall defects, and severe developmental delay. The majority of patients die *in utero* or in the first months of life. In addition to UPD(14)pat, isolated methylation defects at the *DLK1/GTL2* locus have meanwhile been identified in UPD(14)pat patients.

In all cases reported so far, the *DLK1/GTL2* locus is affected. The paternally-expressed gene *DLK1* (delta, *Drosophila* homolog-like 1) encodes a transmembrane signaling protein, the maternally-expressed *GTL2* (gene trap locus 2) is a microRNA which is involved in transcription regulation. However, the functional link between these genes and the phenotypes is currently unknown.

ANGELMAN AND PRADER–WILLI SYNDROMES – CHROMOSOME 15

Angelman (AS) and Prader–Willi syndrome (PWS) are currently the best-known IDs. Both disorders are associated with mental retardation but further clinical signs are different. Both neurodevelopmental diseases are caused by (epi)mutations in 15q11-q13 (Table 36.1). The lack of the paternal copy of this region results in PWS, while disturbances of the maternal copy of *UBE3A* lead to AS.

PWS is clinically characterized by neonatal hypotonia and failure to thrive, and then hyperphagia and obesity develop. Hypogonadism, short stature, behavior problems, and mild to moderate mental retardation are further characteristics [for review, see Ref. 41]. AS patients exhibit microcephaly, ataxia, seizures, absence of speech, and sleep disorder [42]. Due to the high percentage of microdeletions in 15q11-q13 in both syndromes, AS and PWS also belong to the so-called microdeletion syndromes. In AS, ~70% of patients have *de novo* deletions affecting the maternal chromosome. The same frequency can be observed in PWS where this region on the paternal chromosome 15 is deleted. Further AS specific (epi)mutations include *UBE3A1* mutations, imprinting defects and paternal UPD(15). In PWS, maternal UPD(15) is frequent and accounts for <30% of patients, while imprinting defects are rare (~1%).

CHROMOSOME 20 AND THE *GNAS* LOCUS

GNAS is a complex imprinted locus that encodes several transcripts by alternative promoters and splicing. Some loci are expressed biallelically, others exclusively either from the paternal or the maternal *GNAS* allele [43]. Mutations in the *GNAS1* gene are associated with pseudohypoparathyroidism Ia (PHP1A; Albright hereditary osteodystrophy with multiple hormone resistance), pseudopseudoparathyroidism (PPHP), and progressive osseous hypoplasia. PHP1A occurs only after maternal inheritance, whereas PHPP is only paternally inherited.

To date, only three patients with a UPD(20)mat have been reported (for review, see Ref. 4). All patients were characterized by severe pre- and postnatal growth retardation but none of them showed features belonging to the *GNAS* locus mutation spectrum.

GENETIC TESTING FOR IDs

With the growing knowledge on IDs and the rapid development of new molecular genetic technologies, genetic testing for epimutations and mutations in this group of diseases has been considerably improved. In particular, the development of bisulfite treatment assays to differentiate between methylated and unmethylated DNA was helpful to establish fast, reliable and low-cost strategies. Locus-specific Southern-blot assays have been replaced by different PCR-based methods. With the use of PCR only minimal amounts of patients' DNA are necessary, this aspect being particularly important for testing of neonates and deceased patients.

While techniques like methylation-specific PCRs and bisulfite sequencing allow the targeted analysis of single loci, methylation-specific *Multiplex Ligation Probe-dependent Amplification* (MS-MLPA) now allows the parallel characterization of different loci as well as of different types of (epi)mutations (duplications, UPD, aberrant methylation) in a one-tube reaction [for review, see Ref. 45]. The development of methylation-specific microarrays now helps us to gain insights into (aberrant) methylation of the total human genome.

The molecular confirmation of a clinical diagnosis is necessary not only for the patient but also for the family. As illustrated for BWS the molecular subtype of an ID allows a better prognosis, and therapeutic management (in the case of BWS the risk for neoplasias exists) but also helps to determine risk figures in the patients' families.

Nevertheless, prior to genetic testing the significance of genetic testing for the patient and his family should be critically discussed with the families. For each patient and his family, an individual strategy is necessary to avoid misleading and unclear results. Additionally, the putative predictive nature of a genetic test for affected as well as unaffected family members should be considered.

CONCLUDING REMARKS

In summary, there is a growing number of conditions where genomic imprinting effects are recognized to be associated with clinical disorders. Based on the observation that growth disturbance and behavior abnormalities are common features of IDs, genomic imprinting should be suspected in any disorder of unknown etiology characterized by these clinical signs. Furthermore, disorders with an unusual pattern of inheritance should be studied for the possibility that genomically-imprinted genes are involved. In view of the current lack of understanding of the functional basis of the known genetic/epigenetic alterations, identification of these (epi)mutations is of major importance in terms of recurrence risks, prediction of whether offspring will be affected, and risk of malignancies in the case of an ID.

New technologies such as microarrays in conjunction with computational approaches will help us to expand our knowledge and develop an imprinting map of the human genome. Thus, it is likely that many imprinted loci remain to be identified. As shown for the different IDs, epigenetics is a fascinating field of research and will provide us with profound insights into the etiology of many complex biological processes such as growth and we will be able to deduce the contribution of epigenetic changes to complex human disorders such as cancer and psychiatric diseases (see Chapters 33–36 of this book). Additionally, the prevalence and severity of these disorders are probably influenced by environmental factors (see Chapters 28–30 of this book). Thus, future epigenetic research will help us to discover the link between environment, genotype, and phenotype.

591

References

1. Haig D, Graham C. Genomic imprinting and the strange case of the insulin-like growth factor II receptor. *Cell* 1991;**64**:1045–6.

2. Moore T, Haig D. Genomic imprinting in mammalian development: a parental tug-of-war. *Trends Genet* 1991;**7**:45–9.

3. Horsthemke B, Ludwig M. Assisted reproduction: the epigenetic perspective. *Hum Reprod Update* 2005;**11**:473–82.

4. Kotzot D, Utermann G. Uniparental disomy (UPD) other than 15: phenotypes and bibliography updated. *Am J Med Genet* 2005;**136**:287–305.

5. Kotzot D. Complex and segmental uniparental disomy updated. *J Med Genet* 2008;**45**:545–56.

6. Spence JE, Perciaccante RG, Greig GM, Willard HF, Ledbetter DH, Helmancik JF, et al. Uniparental disomy as a mechanism for human genetic disease. *Am J Hum Genet* 1988;**42**:217–26.

7. Horsthemke B, Wagstaff J. Mechanisms of imprinting of the Prader–Willi/Angelman region. *Am J Med Genet* 2008;**146A**:2041–52.

8. Mackay JDG, Callaway JLA, Marks SM, White HE, Acerini CL, Boonen SE, et al. Hypomethylation of multiple imprinted loci in individuals with transient neonatal diabetes is associated with mutations in ZFP57. *Nat Genet* 2008;**40**:949–51.

9. Meyer E, Lim D, Pasha S, Tee LJ, Rahman F, Yates JRW, et al. Germline mutation in NLRP2 (NALP2) in a familial imprinting disorder (Beckwith–Wiedemann syndrome). *PLoS Genet* 2009;**5**:e1000423.

10. Bliek J, Verde G, Callaway J, Maas SM, De Crescenzo A, Sparago A, et al. Hypomethylation at multiple maternally methylated imprinted regions including PLAGL1 and GNAS loci in Beckwith–Wiedemann syndrome. *Eur J Hum Genet* 2008;**17**:611–19.

11. Jäger S, Schönherr N, Spengler S, Ranke MB, Wollmann HA, Binder G, et al. LOT1 (ZAC1/PLAGL1) as member of an imprinted gene network does not harbor Silver–Russell specific variants. *J Ped Endocrin Metabol* 2009;**22**:555–9.

12. DeChiara TM, Efstratiadis A, Robertson EJ. A growth-deficient phenotype in heterozygous mice carrying an insulin-like growth factor II gene disrupted by targeting. *Nature* 1990;**345**:78–80.

13. Constancia M, Hemberger M, Hughes J, Dean W, Ferguson-Smith A, Fundele R, et al. Placental specific IGFII is a major modulator of placental and fetal growth. *Nature* 2002;**417**:945–8.

14. Jones BK, Levorse JM, Tilghman SM. Igf2 imprinting does not require its own DNA methylation or H19 RNA. *Genes Dev* 1998;**12**:2200–7.

15. Cai X, Cullen BR. The imprinted H19 noncoding RNA is a primary micro RNA precursor. *RNA* 2007;**13**:313–16.

16. Bhuiyan ZA, Yatsuki H, Sasaguri T, Joh K, Soejima H, Zhu X, et al. Functional analysis of the p57KIP2 mutation in Beckwith-Wiedemann syndrome. *Hum Genet* 1999;**104**:205–10.

17. Arima T, Kamikihara T, Hayashi K. *ZAC, LIT1 (KCNQ1OT1)* and *p57^{KIP2}(CDKN1C)* are in an imprinted gene network that may play a role in Beckwith–Wiedemann syndrome. *Nucleic Acids Res* 2005;**33**:2650–60.

18. Mackay DJG, Bonnen SE, Clayton-Smith J, Goodship J, Hahnemann JM, Kant SG, et al. A maternal hypomethylation syndrome presenting as transient neonatal diabetes mellitus. *Hum Genet* 2006;**120**:262–9.

19. Rossignol S, Steunou V, Chalas C, Kerjean A, Rigolet M, Viegas-Pequignot E, et al. The epigenetic imprinting defect of patients with Beckwith–Wiedemann syndrome born after assisted reproduction technology is not restricted to the 11p15 region. *J Med Genet* 2006;**43**:902–7.

20. Bliek J, Verde G, Callaway J, Maas SM, De Crescenzo A, Sparago A, et al. Hyopmethylation at multiple maternally methylated imprinted regions including PLAGL1 and GNAS loci in Beckwith–Wiedemann syndrome. *Eur J Hum Genet* 2009;**17**:611–19.

21. Azzi S, Rossignol S, Steunou V, Sas T, Thibaud N, Danton F, et al. Multilocus methylation analysis in a large cohort of 11p15-related foetal growth disorders (Russell Silver and Beckwith Wiedemann syndromes) reveals simultaneous loss of methylation at paternal and maternal imprinted loci. *Hum Mol Genet* 2009;**18**:4724–33.

22. Bliek J, Alders M, Maas SM, Oostra RJ, Mackay DM, van der Lip K, et al. Lessons from BWS twins: complex maternal and paternal hypomethylation and a common source of haematopoietic stem cells. *Eur J Hum Genet* 2009;**17**:1625–34.

23. Temple K, Gardner RJ, Robinson DO, Kibirige MS, Ferguson AW, Baum JD, et al. Further evidence for an imprinted gene for neonatal diabetes localised to chromosome 6q22-q23. *Hum Mol Genet* 1996;**5**:1117–21.

24. Abdollahi A. LOT1 (ZAC1/PLAGL1) and its family members: mechanisms and functions. *J Cell Physiol* 2007;**210**:16–25.

25. Eggermann T, Eggermann K, Schönherr N. Growth retardation versus overgrowth: Silver–Russell syndrome is genetically opposite to Beckwith–Wiedemann syndrome. *Trends Genet* 2008;**24**:195–204.

26. Gicquel C, Rossignol S, Cabrol S, Houang M, Steunou V, Barbu V, et al. Epimutation of the telomeric imprinting center region on chromosome 11p15 in Silver–Russell syndrome. *Nat Genet* 2005;**37**:1003–7.

27. Schönherr N, Meyer E, Roos A, Schmidt A, Wollmann HA, Eggermann T. The centromeric 11p15 imprinting centre is also involved in Silver–Russell syndrome. *J Med Genet* 2007;**44**:59–63.

28. Bullman H, Lever M, Robinson DO, Mackay DJG, Holder SE, Wakeling EL. Mosaic uniparental disomy of chromosome 11 in patient with Silver–Russell syndrome. *J Med Genet* 2008;**45**:396–9.

29. Bruce S, Hannula-Jouppi K, Puoskari M, Fransson I, Simola KOJ, Lipsanen-Nyman M, et al. Submicroscopic genomic alterations in Silver–Russell syndrome and Silver-Russell like patients. *J Med Genet* 2009. Epub ahead of print.

30. Spengler S, Schonherr N, Binder G, Wollmann H, Fricke-Otto S, Muhlenberg R, et al. Submicroscopic chromosomal imbalances in idiopathic Silver–Russell syndrome (SRS): the SRS phenotype overlaps with the 12q14 microdeletion syndrome. *J Med Genet* 2009;**47**(5):356–60.

31. Eggermann T, Gonzalez D, Spengler S, Arslan-Kirchner M, Binder G, Schönherr N. Broad clinical spectrum in Silver–Russell syndrome and consequences for genetic testing in growth retardation. *Pediatr* 2009;**123**:e929.

32. Binder G, Seidel A-K, Martin DD, Ranke MB, Eggermann T, Wollmann HA. The endocrine phenotype in Silver-Russell syndrome is defined by the underlying epigenetic alteration. *J Clin Endocrin Metabol* 2008;**93**:1402–7.

33. Enklaar T, Zabel BU, Prawitt D. Beckwith–Wiedemann syndrome: multiple molecular mechanisms. *Expert Rev Mol Med* 2006;**8**:1–18.

34. Cooper WN, Luharia A, Evans GA, Raza H, Haire AC, Grundy R, et al. Molecular subtypes and phenotype expression of Beckwith–Wiedemann syndrome. *Eur J Hum Genet* 2005;**13**:1025–32.

35. Temple K, Cockwell A, Hassold T, Pettay D, Jacobs P. Maternal uniparental disomy for chromosome 14. *J Med Genet* 1991;**28**:511–14.

36. Wang JC, Passage MB, Yen PH, Shapiro LJ, Mohandas TK. Uniparental heterodisomy for chromosome 14 in a phenotypically abnormal familial balanced 13/14 translocation carrier. *Am J Hum Genet* 1991;**48**:1069–74.

37. Temple IK, Shrubb V, Lever M, Bullmann H, Mackay DJ. Isolated imprinting mutation of the DLK1/GTL2 locus associated with a clinical presentation of maternal uniparental disomy of chromosome 14. *J Med Genet* 2007;**44**:637–40.

38. Kagami M, Sekita Y, Nishimura G, Irie M, Kato F, Okada M, et al. Deletions and epimutations affecting the human 14q32.2 imprinted region in individuals with paternal and maternal upd(14)-like phenotypes. *Nat Genet* 2008;**40**:237–42.

39. Buiting K, Kanber D, Martin-Subero JI, Lieb W, Terhal P, Albrecht B, et al. Clinical features of maternal uniparental disomy 14 in patients with an epimutation and a deletion of the imprinted DLK1/GTL2 gene cluster. *Hum Mutat* 2008;**29**:1141.

40. Mitter D, Buiting K, von Eggeling F, Kuechler A, Liehr T, Mau-Holzmann UA, et al. Is there a higher incidence of maternal uniparental disomy 14 [upd(14)mat]? Detection of 10 new patients by methylation-specific PCR. *Am J Med Genet* 2006;**140A**:2039–49.

41. Goldstone AP. Prader-Willi syndrome: advances in genetics, pathophysiology and treatment. *Trends Endocrinol Metab* 2004;**15**:12–20.

42. Williams CA, Beaudet AL, Clayton-Smith J, Knoll JH, Kyllerman M, Laan LA, et al. Angelman 2005: updated consensus for diagnostic criteria. *Am J Med Genet* 2006;**140A**:413–18.

43. Bastepe M, Juppner H. GNAS locus and pseudoparathyroidism. *Hormone Res* 2005;**63**:65–74.

44. Geneviève D, Sanlaville D, Faivre L, Kottler M-L, Jambou M, Gosset P, et al. Paternal deletion of the GNAS imprinted locus (including Gnasx1) in two girls presenting with severe pre- and postnatal growth retardation and intractable feeding difficulties. *Eur J Hum Genet* 2005;**1**:1033–9.

45. Eggermann T, Schönherr N, Eggermann K, Buiting K, Ranke MB, Wollmann HA, et al. Use of multiplex ligation-dependent probe amplification increases the detection rate for 11p15 epigenetic alterations in Silver–Russell syndrome. *Clin Genet* 2008;**73**:79–84.

SECTION X

Epigenetic Therapy

CHAPTER 37

Clinical Applications of Histone Deacetylase Inhibitors

Scott Thomas[†], Ashleigh Miller[†], Kenneth T. Thurn[†], and Pamela Munster
Department of Medicine, Hematology/Oncology Division, University of California, San Francisco, San Francisco, CA 94143-1770, USA

INTRODUCTION

Histone acetyl-transferases (HATs) and histone deacetylases (HDAC) play a crucial role in gene regulation. HATs catalyze the transfer of acetyl groups to lysine residues of histones, resulting in the relaxation of chromosomal DNA [1]. In general, this promotes transcription at the affected chromosomal regions. On the other hand, HDACs function to reverse histone acetylation, causing chromosomal DNA to condense. Therefore, HDACs function to reduce transcription. HDACs have also been shown to directly modify a variety of non-histone substrates such as p53, Ku70, pRB, and E2F-1 [2–5]. There are currently 18 known human HDACs that are separated into four classes (see Table 37.1) [6]. Class I, II, and IV HDACs are zinc-dependent proteins, while class III HDACs require NAD^+ [7]. Most of the clinically-relevant work to date involves class I, II, and IV HDACs, and thus they will be the focus of this chapter.

HDAC inhibitors have shown great promise as a treatment option for a variety of cancer and non-cancer diseases. Despite the presumed global affect, pharmacological inhibition of HDACs alters the expression of only 2% of genes [8]. Preclinical work has demonstrated that HDAC inhibitors appear to have a stronger effect on transformed cells than normal cells [9]. Depending on the specific HDAC inhibitor, its concentration, and the cell type being treated, these drugs have demonstrated the ability to induce cell cycle arrest, increase the expression of pro-apoptotic genes, down-regulate anti-apoptotic genes, inhibit the expression of genes involved in angiogenesis, alter the expression of DNA repair genes, and down-regulate invasion and metastasis associated genes [1,6,10–13].

Clinically-relevant HDAC inhibitors are sub-divided by chemical structure and include the hydroxamic acids, short chain fatty acids, cyclic peptides, and benzamides. The potency of hydroxamic acids lies within the micro- to nano-molar range [6,14]. These drugs include vorinostat (suberoylanilide hydroxamic acid, SAHA), panobinostat (LBH589), belinostat (PXD101), trichostatin A (TSA), JNJ-26481585, and PCI-24781. Short chain fatty acids include butyrates and valproic acid, and typically have weaker potencies in the milli-molar

[†]These authors contributed equally to the writing of this chapter.

Handbook of Epigenetics: The New Molecular and Medical Genetics. DOI: 10.1016/B978-0-12-375709-8.00037-X

TABLE 37.1 Histone Deacetylases and their Inhibitors. Histone Deacetylases (HDACs) are Divided into Four Main Classes Based on their Similarity to Yeast Proteins [7]. HDACs have a Variety of Histone and Non-histone Substrates

	Class I	Class IIa	Class IIb	Class III	Class IV
Members	HDAC1 HDAC2 HDAC3 HDAC8	HDAC4 HDAC5 HDAC7 HDAC9	HDAC6 HDAC10	Sirt1-7	HDAC11
Location	Nuclear	Nuclear/ Cytoplasmic	Nuclear/ Cytoplasmic	Nuclear/ Cytoplasmic Mitochondrial	Nuclear
Substrates	Histones Stat3 Bcl-6 p53 E2F-1	Histones HP-1 GATA-1 Smad7	Histones alpha-Tubulin Hsp90 SMAD7	p53 NF-κB DNA Pol-B RNA-Pol I TAF Tubulin	Related to Rpd3 protein
Inhibitors	Belinostat Vorinostat Panobinostat Entinostat Valproic Acid Romidepsin MGDC0103 PCI-24781 JNJ-26481585	Belinostat Panobinostat Entinostat Valproic Acid JNJ-26481585	Belinostat Panobinostat Entinostat Valproic Acid PCI-24781 JNJ-26481585	Nicotinamides	Belinostat Vorinostat Panobinostat Entinostat Valproic Acid Romidepsin MGDC0103 JNJ-26481585

range [14]. They also generally have shorter half-lives. Valproic acid causes the specific degradation of HDAC2 [15]. Cyclic peptides are a group of potent HDAC inhibitors that function in the nano-molar range [14], the most notable being romidepsin (depsipeptide). Benzamides are generally effective inhibitors in the milli molar range, represented by entinostat (MS-275), which has received the most clinical evaluation [6]. The mechanism underlying the function of individual HDAC inhibitors remains largely unknown. Several co-crystal structures with HDAC8 suggest the hydroxamic acid moiety of TSA and vorinostat chelate with the zinc atom within the catalytic pocket of the protein [16,17]. The cyclic peptide romidepsin is thought to function in a similar fashion, perturbing the coordination of the zinc atom by HDAC8, and thus inhibiting its catalytic activity [18].

Despite the limited understanding of the mechanisms underpinning HDAC inhibitor action, there is mounting clinical evidence demonstrating an acceptable toxicity profile for most HDAC inhibitors [19]. Data compiled from phase I and II clinical trials have demonstrated HDAC inhibitors are clinically efficacious for the treatment of specific hematological malignancies. Further, HDAC inhibitors may potentiate the effects of existing treatments for solid and hematological malignancies. HDAC inhibitors are also being evaluated for the treatment of numerous non-cancer diseases, including neurological, viral, neuromuscular, pulmonary, and inflammatory. In this chapter, the clinical use of HDAC inhibitors will be discussed, with a strong emphasis on treatment options for solid and hematological malignancies.

HISTONE DEACETYLASE INHIBITORS FOR THE TREATMENT OF HEMATOLOGICAL TUMORS

HDAC inhibitors have demonstrated clinical efficacy in patients with hematological malignancies including cutaneous T-cell lymphoma (CTCL), acute myeloid leukemia (AML), and myelodysplastic syndromes (MDS). This is not surprising since HDACs have been shown to play a role in a variety of common translocations that occur in patients with hematological malignancies (e.g. AML/ETO fusion protein) [20]. There are currently two HDAC inhibitors specifically approved by the Food and Drug Administration for the treatment of hematological malignancies. Vorinostat was approved for the treatment of CTCL in 2006, followed by romidepsin in 2009. Despite promising results evaluating HDAC inhibitors in clinical trials for the treatment of other hematological malignancies, their efficacy as single agents has been modest. Numerous clinical trials are underway to evaluate their therapeutic potential when used in combination with other treatments (e.g. chemotherapeutics or immunological therapy). The relatively favorable toxicity profile of HDAC inhibitors in patients and their broad-ranging effects render these agents a promising novel addition to the current treatment of hematological malignancies.

Vorinostat (Suberoylanilide Hydroxamic Acid, SAHA)

Vorinostat is a hydroxymate HDAC inhibitor that targets class I, II, and IV HDACs. Vorinostat was approved as a therapeutic agent for patients who had undergone at least two systemic therapies for recurrent, progressive, or persistent cutaneous T-cell lymphoma (CTCL). CTCL, which include mycosis fungoides and Sézary syndrome, are non-Hodgkin's lymphomas with malignant T-lymphocytes present in the skin [21,22]. Early clinical trials in patients with various solid and hematological malignancies demonstrated that vorinostat was generally well tolerated through either intravenous or oral dosing [23,24]. The need, however, for frequent administration of the drug rendered the intravenous formulation more difficult. A phase II trial of daily oral administration of 400 mg of vorinostat in patients (n = 74) with progressive, persistent, or refractory mycosis fungoides or Sézary syndrome revealed a significant clinical benefit with an acceptable toxicity profile [25]. The study reported an objective response rate (a measurable response to treatment) in nearly 30% of patients, and relief from debilitating pruritus in 32% of the patients treated with vorinostat [25].

Treatment schedules appear to have a significant impact on vorinostat efficacy and their therapeutic window. In a phase II trial that examined various dosing regimens of vorinostat, 33 patients with CTCL were treated with 400 mg of vorinostat daily (group 1), 300 mg twice daily for two weeks with one week rest and then 200 mg daily (group 3), or 300 mg twice daily for 3 days followed by 4 days of rest (group 2) [26]. The percentage of patients who responded to treatment was higher in patients from groups 1 and 3 (31% and 33%, respectively) compared to patients from group 2 (9%) [26]. There was also significantly greater relief in pruritus in patients from group 1 compared to group 2 (73% versus 18%). This suggests continuous administration of vorinostat led to more beneficial results compared to higher-dose intermittent treatment.

Vorinostat has also demonstrated efficacy in patients with advanced acute myeloid leukemia (AML). A clinical trial evaluating vorinostat in 41 patients with advanced leukemias and myelodysplastic syndromes (MDS) reported a clinical benefit in 17% of patients [27]. All of the responses were seen in patients with AML [28]. The majority of non-responding AML patients had higher levels of 17 specific antioxidant genes in peripheral blood samples that could potentially contribute to the resistance of vorinostat therapy [27].

Data collected from multiple clinical trials evaluating vorinostat as a single therapy in patients with hematological and solid malignancies (n = 341) showed that vorinostat in general was well tolerated [19]. The most commonly observed low grade adverse affects included myelosuppression, anorexia, nausea, fatigue, vomiting, and diarrhea [19]. High-grade toxicities reported after vorinostat administration included dehydration, thrombocytopenia, and fatigue [19]. There were cardiac abnormalities reported after vorinostat treatment (e.g. non-specific ST changes) [29], however, no grade 3 or 4 drug-related acute cardiac toxicities were reported [23,25,26,29,30]. Overall, these clinical trials demonstrated that vorinostat caused a clinical response in patients with advanced and aggressive CTCL and some patients with AML with manageable toxicities.

Panobinostat (LBH589)

Panobinostat is a hydroxamic acid HDAC inhibitor with high potency and has exhibited promising results for the treatment of several hematological diseases including CTCL, Hodgkin's lymphomas, and leukemias. The reported mean terminal half-life of panobinostat is approximately 16 hours [31]. A subset of patients with CTCL that were part of a larger trial of patients (n = 32) with various solid tumors or CTCL, treated with panobinostat, showed a response in 8 of 10 patients (two complete responses, four partial responses, and two disease stabilizations; see Table 37.2) [31]. No objective responses were observed in patients with malignancies other than CTCL (as part of the larger trial with 32 patients), although five patients with other tumors did achieve disease stabilization [32]. Most patients with CTCL were treated with 20 mg of panobinostat on Mondays, Wednesdays, and Fridays (one patient was treated with 30 mg). The responses included one disease stabilization, four partial responses, and two complete responses [31]. Both complete responses occurred after cessation of panobinostat administration. The expression of 23 specific genes was altered in all patients tested (n = 6) after treatment with panobinostat [31]. Panobinostat administration was associated with cardiac toxicities, mainly electrocardiogram changes in a study of patients with hematological malignancies in which 13 of 15 patients had AML [33]. Extensive cardiac conduction evaluation is being undertaken in many studies without a clear association of cardiac toxicity to date. Other adverse effects associated with panobinostat treatment included fatigue, hypokalemia, anorexia, thrombocytopenia, nausea, vomiting, and fatigue [31,33]. Panobinostat is under further clinical investigation for patients with hematological malignancies as a single agent or in combination with a variety of treatments. Current combinational therapies include idarubicin, bortezomib, melphalan, and everolimus.

Belinostat (PXD101)

Numerous clinical trials are under way examining the efficacy of belinostat, a hydroxamic acid-type HDAC inhibitor, either alone or in conjunction with other modalities (e.g. idarubicin and bortezomib) in the treatment of patients with hematological malignancies (www.clinicaltrials.gov). Of the 16 patients enrolled in a phase I clinical trial, five patients demonstrated stable disease, yet no complete or partial responses were seen [34]. Patients were treated with 600 mg/m^2/d (n = 3), 900 mg/m^2/d (n = 3), or 1000 mg/m^2/d (n = 10). Belinostat was reported to be well-tolerated, and was not associated with significant cardiac toxicity [34]. There were two reported cases of acute renal failure in patients with multiple myeloma. Further, grade 3/4 toxicities observed included fatigue, lymphopenia, and neurological toxicities [34].

Valproic Acid

In 1962, valproic acid, used as a vehicle for drug dissolution for anticonvulsants, was serendipitously found to exhibit anticonvulsant activity by itself [35]. Since its approval for clinical use, valproic acid has been widely used for the treatment of a broad spectrum of epileptic conditions in both children and adults, with few contraindications and moderate adverse affects [35]. Valproic acid is an aliphatic short-branched chain fatty acid that is actively transported across the blood–brain barrier, and only recently was identified as an HDAC inhibitor [36]. As an HDAC inhibitor, valproic acid has been studied in patients with

TABLE 37.2 Clinical Trials of HDAC Inhibitors for the Treatment of Hematological Malignancies

Hematological Tumors	Select HDAC Inhibitor	Complete or Partial Response	Stable Disease	Hematological Improvement	Reference
CTCL	Vorinostat	22/74 (30%)			[25]
	Vorinostat	8/33 (24%)	1/33 (3%)		[26]
	Panobinostat	6/10 (60%)	1/10 (10%)		[31]
	Romidepsin	24/71 (34%)	26/71 (37%)		[45]
Relapsed/ Refractory acute leukemias	Entinostat	0/38 (0%)	3/38 (8%)		[53]
Leukemias or MDS (31 of 41 with AML)	Vorinostat	4/41 (10%)		7/41 (17%)	[27]
DLC, MM, CLL, FL	Belinostat	0/16 (0%)	5/16 (31%)		[34]
MM	Vorinostat	1/10 (10%)	9/10 (90%)		[126]
MDS and sAML/MDS	Valproic Acid	1/18 (6%)		7/18 (39%)	[38]
DLBCCL	Vorinostat	1/18 (6%)	1/18 (6%)		[125]
Hematological malignancies	Vorinostat	5/39 (13%)	3/39 (8%)		[30]
	Vorinostat	2/23 (9%)	4/23 (17%)		[23]
AML	Romidepsin	0/20 (0%)			[48]
AML or MDS	Romidepsin	1/11 (9%)	6/11 (55%)		[47]
CML	Romidepsin	0/10 (0%)			[46]
AML	Romidepsin	0/10 (0%)			[46]

CTCL = cutaneous T-cell lymphoma; MDS = myelodysplastic syndromes; AML = acute myeloid leukemia; DLC = diffuse large cell non-Hodgkin's lymphoma; CLL = chronic lymphocytic leukemia; MM = multiple myeloma; FL = follicular lymphoma; NHL = non-Hodgkin's leukemia; CML = chronic myelocytic leukemia; sAML/MDS = AML secondary to MDS; DLBCCL = diffuse large-B-cell lymphoma [125,126].

MDS and AML as a single therapy or in combination with other treatments. It has a half-life between 9 and 18 hours [37]. A phase II clinical trial, including 18 patients with MDS, treated with 50–100 μg/ml (serum concentration) of valproic acid, reported a response rate of nearly 44%. The responses consisted of a partial response (n = 1) and hematological improvements (n = 7) [38]. The addition of *all-trans* retinoic acid (ATRA) after disease progression on valproic acid alone resulted in an additional response in two of four patients [38]. Patients (n = 5) who received valproic acid and ATRA concurrently from the beginning of the trial, however, did not demonstrate a clinical response [38].

Treatment of poor-risk patients with MDS or AML with valproic acid (45–100 μg/ml serum concentrations) prior to ATRA administration resulted in significant hematological improvements in 30% of patients [39]. In this trial, however, the addition of ATRA after valproic acid administration did not improve the clinical efficacy. A similar study included patients with AML treated for one week with valproic acid before ATRA administration. Three of the 11 patients demonstrated a response. However, the study allowed the co-administration of theophylline (10–15 μg/ml serum concentration) to raise the levels of cAMP [40]. The role of valproic acid, ATRA, and theophylline was further evaluated in 22 patients with AML. The three-drug combination resulted in an improvement of the peripheral blood cells in nine of the patients (40%) [41]. The study suggested an increased survival time in the responding patients compared to non-responders (147 versus 48 days) [41].

These data are contrasted by another study in which patients with AML were treated with either valproic acid (n = 31) or ATRA (n = 27) [42]. Non-responding patients and those with disease progression after administration of valproic acid were allowed to cross over to ATRA, or to receive ATRA in addition to valproic acid (n = 13). No significant differences in improvement or disease stabilization between the groups were reported, and the response rates were low at 16% and 5% (criteria for MDS and AML, respectively) [42].

The addition of valproic acid to treatment with decitabine (5-aza-2′-deoxycytidine, DNA methyltransferase inhibitor) in patients with AML resulted in a minimal clinical benefit compared to treatment of patients with decitabine alone [43]. However, an increase in neurotoxicity was reported when administering both treatments concurrently [43].

Romidepsin (Istodax®, Depsipeptide, FK228)

Romidepsin was approved for cutaneous T-cell lymphoma after one line of therapy in 2009. In a case report of romidepsin, responses were noted in four patients with T-cell lymphomas (three cutaneous and 1 peripheral) after treatment with romidepsin (12.7 or 17.8 mg/m^2 weekly on days 1 and 5 of a three week cycle) [44]. The patient with peripheral T-cell lymphoma had a complete response, while the patients with CTCL had partial responses [44]. The study prompted interest in HDAC inhibitors as the Sézary cells showed an increase in the acetylation of histones. A larger phase II study that included 71 patients with CTCL demonstrated a response rate of 34% [45]. This included four complete responses, and 20 partial responses. There were also 26 disease stabilizations reported in the study [45].

In patients with chronic lymphocytic leukemia, AML, or MDS, romidepsin did not have significant clinical efficacy [46,47]. The drug did exhibit biological activity as measured by histone acetylation and p21 induction [46]. Romidepsin treatment was limited due to accumulated toxicities that included fatigue, nausea, and anorexia [46,47]. High-grade (3–4) toxicities including thrombocytopenia, were also associated with romidepsin treatment [47]. Romidepsin was reported to have a higher efficacy in a subset of patients with AML that had core binding factor dysregulation [48]. Core binding factor is a transcripton factor that is altered due to a chromosomal translocation in some patients with leukemia.

Preclinical studies raised the possibility of cardiac toxicity and abnormalities in electrocardiograms were noted in a phase I study [49]. A subsequent larger study (n = 42) investigating the effect of romidepsin on cardiac function in patients with T-cell lymphoma reported that there were no overt cardiac toxicities associated with romipedsin administration, despite some cardiac abnormalities [50].

Entinostat (MS-275, SND-275)

Entinostat is an HDAC inhibitor with a relatively long half-life (averaging between 33 and 52 hours) [51,52]. Trials have shown significant biological activity in patients with hematological malignancies receiving entinostat treatments [53]. However, the efficacy of entinostat as a single-agent therapy remains limited [51,53]. Reported dose-limiting toxicities associated with entinostat include neurotoxicity, fatigue, hypophosphatemia, anorexia, and vomiting [51–53].

The large number of clinical trials using HDAC inhibitors for the treatment of patients with hematological malignancies has demonstrated that these drugs are relatively well tolerated. Although the responses with the currently-available HDAC inhibitors are still limited, there are significant responses in some patients with advanced disease, where options are limited. With better patient selection and the development of more potent HDAC inhibitors, targeting HDACs for the treatment of hematological malignancies remains promising. Furthermore, a growing body of literature suggests that HDAC inhibitors may potentiate some of the currently used cytotoxic or biologic therapies based on their mechanism of action, with multiple trials currently ongoing. Promising combination partners include proteosome inhibitors, DNA demethylating agents, and anthracyclines.

HISTONE DEACETYLASE INHIBITORS IN THE TREATMENT OF SOLID TUMORS

HDAC inhibitors have had clinical success in treating Hodgkin's lymphoma, CTCL, and AML. Many studies are ongoing to define the optimal setting of HDAC inhibitors for the treatment of solid tumor malignancies. Preclinical experiments show HDAC inhibitors induce cell cycle arrest, differentiation, and apoptosis in solid tumor cancer cell lines [14]. In mouse xenograft models, treatment with HDAC inhibitors reduce tumor size and inhibit angiongenesis and metastasis [14]. Currently, there are at least 35 ongoing clinical trials testing HDAC inhibitors in a single-agent setting for treatment of solid tumors, and 92 trials testing HDAC inhibitors in combination with other treatments. While HDAC inhibitors have an acceptable toxicity profile, as single agents their anti-tumor activity has been relatively modest. Therefore, many studies investigating the benefit of HDAC inhibitors for the treatment of solid tumors use them in combination with other therapeutics.

HDAC Inhibitors as a Single-Agent Therapy of Solid Tumors

HDAC inhibitors have been tested as single-agents for the treatment of solid tumors in phase I and II clinical trials (Table 37.3). A phase I trial of belinostat in patients with advanced refractory solid tumors resulted in stable disease for 18/46 patients. The study included patients with colorectal tumors (n = 12), renal tumors (n = 6), melanoma (n = 6), and various other solid tumors. There was disease stabilization in 50% of patients at the highest tolerated dose [54]. A phase I trial of an oral formulation of MGCD-0103 in multiple advanced solid tumors saw 5 of 32 patients achieve stable disease [55]. Maximum tolerated doses and dose limiting toxicities were delineated in these studies; however, no objective response was noted in any of these trials. Recommendations were to combine HDAC inhibitors with other forms of treatment.

TABLE 37.3 Clinical Trials of Histone Deacetylase Inhibitor Treatment for Solid Tumors [127–130]

Solid Tumors	Select HDAC Inhibitor	Complete or Partial Response	Stable Disease	Reference
Melanoma	Entinostat	0	4/28 (14%)	[127]
Colorectal cancer	Romidepsin	0	4/28 (14%)	[61]
Thyroid cancer	Romidepsin + XRT	0	4/14 (29%)	[130]
Renal cell cancer	Romidepsin	2/29 (7%)	NR	[63]
Neuroendocrine	Romidepsin	0	2/14 (14%)	[65]
Head and Neck	Vorinostat	0	5/13 (38%)	[58]
Breast	Vorinostat	0	4/14 (29%)	[57]
Lung cancer	Pivanex	3/47 (6.4%)	10/47 (30%)	[129]
	Romidepsin	0/18	9/18 (50%)	[62]
Non-small cell lung cancer	Vorinostat	0/14	8/14 (57%)	[59]
Prostate	Romidepsin	1/31 (3%)	2/31 (6%)	[64]
	Vorinostat	0/27	2/27 (7%)	[60]
Ovarian	Vorinostat	1/27 (4%)	9/27 (33%)	[128]
Mesothelioma	Vorinostat			

Vorinostat is well tolerated at oral doses of 400 mg/d, leading to a substantial increase in acetylated biomarkers in peripheral blood mononuclear cells (PBMCs). A phase II trial of vorinostat to treat recurrent glioblastoma multiforme reported moderate single-agent efficacy, with 9/52 patients having stable disease at 6 months [56]. However, treatment with vorinostat in a single-agent setting did not achieve clinical responses in the treatment of solid tumors in phase II trials of breast [57], head and neck [58], non-small cell lung cancer [59], and prostate cancer [60]. Similarly, romidepsin was found to have a tolerable dosage of 13 mg/m^2, resulting in increased acetylation of biomarkers in PBMCs. However, romidepsin showed minimal clinical response when used as a single agent in phase II trials for metastatic colon cancer [61], lung cancer [62], renal cancer [63], and prostate cancer [64].

Similarly to their use in hematological malignancies, HDAC inhibitors are fairly well tolerated in patients with solid tumors. The most common toxicities include nausea, vomiting, anorexia, diarrhea, fatigue, neutropenia, and thrombocytopenia. Most toxicities are more pronounced with cumulative drug exposure. Cardiac irregularities have been reported with romidepsin [65] and other HDAC inhibitors; however, a recent study dedicated to determine the risk of QT-prolongation with supertherapeutic doses of vorinostat (800 mg) did not demonstrate an increased risk in cardiac toxicity including QT-prolongations or arrhythmias [62,66].

Unlike in the hematological malignancies, the efficacy in solid tumors is more modest. Data from phase I trials suggest treatment-induced acetylation in PBMCs occurs at pharmacological levels and acetylation in PBMCs correlates with acetylation in tumors [67,68]. However, no clear association between acetylation and tumor response has yet been established in these small early phase trials [69]. Further studies will be needed to determine whether acetylation is necessary and sufficient to induce target gene and protein modification required for tumor response.

The modest activity of HDAC inhibitors in solid tumors may be further explained by the frequent occurrence of cumulative toxicities, which limits prolonged exposure without interruptions and requires dose modifications in a number of patients. Additionally, most HDAC inhibitors exhibit a short pharmacological half-life, further influencing the therapeutic window of the currently available agents. Several preclinical *in vitro* and xenograft studies suggested that the effects of HDAC inhibitors on growth and differentiation were reversible upon drug withdrawal, and hence continuous drug exposure may indeed be necessary [70]. Some or all of these factors may contribute to the limited clinical efficacy of these drugs when given as single agents.

Combination Therapy with HDAC Inhibitors in the Treatment of Solid Tumors

Over 90 trials are currently underway examining HDAC inhibitors in combination with other therapies for the treatment of solid tumors (www.clinicaltrials.gov). Investigators are employing multiple strategies based on preclinical data to combine HDAC inhibitors in an effort to enhance efficacy of both treatments. Preclinical data demonstrated that HDAC inhibitors, specifically vorinostat, decondense chromatin, allowing for increased access to DNA-damaging chemotherapies such as topoisomerase inhibitors, nucleosides, cisplatin, and cisplatin derivatives [71]. Results from clinical trials with many of these combinations are promising. A phase I trial combining vorinostat with 5-fluorouracil, leucovorin, and oxaliplatin (FOLFOX) in refractory colon cancer patients established the maximum tolerated dose of vorinostat (300 mg oral, twice daily for 1 week every 2 weeks) in combination with FOLFOX [72]. Sequence-specific combination of the HDAC inhibitor valproic acid and the topoisomerase II inhibitor epirubicin in patients with advanced solid tumors resulted in a partial response in 22% of the patients (9/41), and stable disease in a further 39% of patients (16/41) [73]. A dose-expansion study in breast cancer showed a 50% response rate [67].

A follow-up phase II trial is currently recruiting patients. An ongoing phase I trial combining vorinostat and capecitabine (a 5-fluorouracil precursor) in patients with advanced solid tumors reported preliminary results consisting of partial responses in 4/28 of patients (14%), one unconfirmed complete response, and 18 patients (64%) with stable disease [74].

Some HDAC inhibitors act to deacetylate tubulin and thus sensitize cells to mitotic inhibitors such as paclitaxel [75]. Notably, a phase I study combining vorinostat with paclitaxel and carboplatin in patients with advanced solid tumors saw a partial response in 11 of 25 patients (44%), with 10 non-small cell lung cancers, and stable disease in seven patients [76]. A subsequent randomized, double-blind phase II trial with vorinostat and paclitaxel/carboplatin in non-small cell lung cancer reported a response of 34% of patients receiving the HDAC inhibitor combination, versus 12.5% receiving only the paclitaxel/carboplatin [77]. The addition of vorinostat increased toxicities including thrombocytopenia, fatigue, hyponatremia, and diarrhea, resulting in 27% of patients discontinuing with the study. Furthermore, the extended randomized phase III trial was halted, as the study did not meet its efficacy endpoint during a planned interim analysis (NCT 00473889).

Sequentially combining HDAC inhibitors with anti-estrogen therapies in hormone-refractive breast cancer cell lines re-sensitizes tumor cells to estrogen therapies, resulting in increased cell death, and reduction in tumor size in a mouse xenograft model [78]. Promising results are emerging from a phase II clinical trial with vorinostat and tamoxifen in breast cancer patients who progressed on aromatase inhibitors. The combination was well-tolerated and 8/43 patients had an objective response (19%), and 9/43 additional patients had stable disease for greater than 6 months (21%) [79]. Encouraging preliminary results are emerging from an ongoing phase I trial combining the HDAC inhibitor panobinostat with trastuzumab (a monoclonal antibody targeting the Her2/Neu receptor) in breast cancer patients who had prior progression under trastuzumab treatment. While the goal of this study was to determine the maximum tolerated dose, six patients experienced tumor reduction (33%) [80]. Additionally, minimal dose-limiting toxicities were observed. A follow-up phase II trial is currently underway (www.clinicaltrials.gov).

HDAC inhibitors also target Hypoxia-inducible Factor F-1 (HIF-1) transcription factors, and synergize with angiogenesis inhibitors such as bevacizumab (Avastin) and sorafenib [81]. In a phase I trial, the combination of vorinostat and bevacizumab in patients with advanced clear cell renal cancer was associated with increased acetylated biomarkers in PBMCs. Three patients (12%) had stable disease [82]. A phase II trial combining vorinostat with taxol and bevacizumab in patients with metastatic breast cancer report an overall response rate of 53%, with two complete responses, and 22 partial responses in 45 patients [83].

Preclinical experiments testing the combination of HDAC inhibitors and DNA methyltransferase inhibitors show synergistic anticancer activity in both *in vitro* and *in vivo* models. Treatment of hepatic cellular carcinoma cell lines with vorinostat and 5-aza-2'-deoxycitidine resulted in significant anti-proliferation and cell death, while having little to no effect on primary hepatocytes. Administering the drug combination in an *in vivo* murine xenograft model resulted in additive suppression of tumor growth [84]. Several phase I clinical trials incorporate HDAC inhibitors with 5-aza-2'-deoxycitidine. Valproic acid and 5-aza-2'-deoxycitidine in non-small cell lung cancer showed strong serum demethylation and saw one partial response in eight treated patients [85]. A subsequent phase I trial involving 66 patients with advanced cancers combining valproic acid with 5-aza-2'-deoxycitidine also showed strong histone H3 acetylation and global DNA methylation of PBMCs. In the trial, 14 patients achieved stable disease for 4–6 months (25%), suggesting this combination has therapeutic benefit on solid tumors [86]. Based on these initial findings, the combination of HDAC inhibitors and 5-aza-2'-deoxycitidine merits further exploration for the treatment of solid tumors.

605

CLINICAL APPLICATIONS OF HISTONE DEACETYLASE INHIBITORS FOR NON-CANCER DISEASES

As therapeutic targets, HDACs have elicited growing interest for the treatment of an array of diseases. Although recently much of this interest has been directed towards exploiting their anti-tumor activity [87], a substantial body of preclinical work demonstrates the promise of HDAC inhibitors for the treatment of non-cancer disorders, including neuromuscular, cardiovascular, hematological, pulmonary, infectious, and inflammatory diseases [88–91]. In this section, we will review results from clinical trials treating non-cancer diseases with HDAC inhibitors, focusing on valproic acid and phenylbutyrate, which have undergone extensive clinical testing.

Current Therapeutic Applications

VALPROIC ACID

For more than 30 years, valproic acid has been used to treat seizure disorders. In addition to its use as an anticonvulsant, valproic acid is commonly prescribed as a first line treatment for the prevention and maintenance of mania in patients with bipolar disorder and for migraine prevention [35,92,93]. A major factor limiting the use of valproic acid is its teratogenicity. Children of pregnant mothers treated with valproic acid are at greater risk for major congenital malformations such as neural tube defects, mid-line abnormalities, and the development of spina bifida, as well as neurodevelopmental delay [94].

Although well established in the clinical setting for the treatment of neurological disorders, the cellular mechanisms underpinning the therapeutic benefits of valproic acid are not clear, nor are the extent to which its HDAC inhibition activity contributes to these benefits. The effectiveness of valproic acid for the treatment of these varied neurological disorders argues for modulation of several target pathways. Furthermore, the response can vary significantly depending on the duration of treatment. For example, immediate short-term administration of valproic acid elicits rapid temporary control of seizure disorders, while to achieve full benefit for mood stabilization several days of treatment are required. Additionally, following the cessation of prolonged treatment, anti-seizure, -migraine, and -mania benefits have been found to persist in some patients [95]. One compelling hypothesis consistent with this behavior asserts that the rapid therapeutic actions of valproic acid are the result of direct interaction with components of key biochemical pathways, while the long-term and persistent effects are the result of HDAC inhibition and consequent epigenetic modulation and altered protein expression [95]. Immediate actions of valproic acid proposed for these disorders include altering function of several neurotransmitters including γ-aminobutyric acid, γ-hydroxybutyrate, glutamate, aspartate, serotonin, and dopamine as well as neuronal firing through voltage-regulated ion channel activity [35,96].

In an effort to identify valproic acid targets that contribute to its efficacy, rats were treated with valproic acid for 30 days and gene expression from brain homogenates was analyzed by microarray analysis. In this study, 121 genes were identified as up- or down-regulated, affecting a number of interesting pathways such as synaptic transmission, ion channels, heat shock proteins, apoptosis, sodium and potassium transport, and several protein kinases and phosphatases [97]. How these changes in expression translate to the therapeutic benefits achieved in the treatment of these neurological disorders is not yet fully understood.

PHENYLBUTYRATE

Phenylbutyrate is a fatty-acid HDAC inhibitor that targets class I and IIa HDACs, and for more than a decade has been applied in the clinic for the treatment of urea cycle disorders [98]. However, for this treatment, phenylbutyrate's efficacy does not lie with its HDAC inhibition activity. Rather, for patients unable to metabolize glutamine to urea, phenylbutyrate provides an alternate pathway for nitrogen excretion. Phenylbutyrate is metabolized to phenylacetyl-CoA and conjugated to glutamine in the liver, which can then be excreted in the urine [98].

Clinical Evaluation of Valproic Acid and Phenylbutyrate

The established use of valproic acid and phenylbutyrate in the clinic and their modest adverse effects in both children and adults has made them attractive HDAC inhibitors for the evaluative treatment of patients with disorders linked to the mis-expression of gene products or errors associated with acetylation regulation. Several afflictions have been evaluated in the clinic, including cystic fibrosis, spinal muscular atrophy, hemoglobinopathies, and human immunodeficiency viral infection.

Cystic fibrosis is an inherited disease with mutations in the chloride ion-channel protein, cystic fibrosis transmembrane conductance regulator (CFTR), which primarily manifests in mucus overproduction and impaired pulmonary function [99]. In two small randomized, double-blind, placebo-controlled studies, cystic fibrosis patients with homozygous deltaF508-CFTR mutations were treated with phenylbutyrate. In both studies, patients exhibited a modest, but statistically-significant improvement in chloride transport, indicating improved CFTR function [100,101].

Spinal muscular atrophy (SMA) is a severe hereditary neuromuscular disease caused by motor neuron degeneration, resulting in weakness and muscle wasting. Onset can occur in infancy or in adulthood. Depending on the progression of degeneration, SMA can be fatal in infancy or manifest as mild weakness. The most common form of SMA is due to mutations in the survival motor neuron 1 (SMN1) gene. A second gene, SMN2, is nearly identical to SMN1. However, a mutation in the coding sequence of SMN2 results primarily in the expression of a truncated, non-functional SMN product, with only a minor amount of full-length SMN protein produced [102]. It was hypothesized that increased expression of full length SMN from the SMN2 gene could be induced with HDAC inhibition, and thus compensate for mutant SMN1. In an initial pilot study, children with SMA treated with phenylbutyrate for 3 and 9 weeks demonstrated increased motor function [103]. Using leukocytes as a surrogate for motor neurons, phenylbutyrate was shown to elevate SMN mRNA levels [104]. However, in a randomized, double-blind, placebo-controlled trial enrolling 107 children with SMA, no significant improvement over placebo was observed with phenylbutyrate treatment [105]. With valproic acid treatment, increased levels of SMN mRNA and protein were observed in peripheral blood mononucleocytes in 7 of 10 SMA mutant carriers, and increased levels of SMN2 mRNA in 7 of 20 SMA patients [106]. In a small study of adults with milder SMA (type III/IV), increased muscle strength was observed with 8 months of valproic acid treatment [107]. In a recent phase II trial, tolerability and efficacy of valproic acid treatment was examined in a heterogeneous group of subjects ranging in age from 2 to 31 years with varying degrees of disease severity [108]. Of 42 patients, 27 showed an improvement in gross motor function. However, no significant increase in SMN mRNA was observed from whole blood purification. Benefits were primarily seen in non-ambulatory patients with type II SMA who were 5 years old or younger.

For the hemoglobinopathies, β-thalassemia, and sickle-cell anemia, anemia results from inherited defects in the production of adult hemoglobin [89]. It was hypothesized that treatment with an HDAC inhibitor would result in increased fetal hemoglobin (HbF) expression, which would decrease the severity of these disorders. In children treated for urea cycle disorder with phenylbutyrate, HbF was found to be elevated [109]. When patients with sickle-cell anemia or β-thalassemia were treated with phenylbutyrate, F-reticulocytes were induced and in a sub-group HbF expression was increased [110–113]. In a trial evaluating valproic acid for the treatment of sickle-cell anemia, a modest increase in HbF levels was observed; however, this increase did not translate into clinical benefit [114]. In two trials,

607

expression of HbF was measured in patients being treated for seizures with valproic acid. In one trial, the percentage of red blood cells expressing HbF was measured and compared to that of those of untreated normal subjects. The valproic acid treated patients exhibited a 3-fold increase in the number HbF-expressing red cells [115]. In the other trial, HbF levels in children were significantly increased, and correlated with valproic acid serum levels [116].

Substantial progress has been made in treating individuals infected with the human immunodeficiency virus (HIV). However, complete viral eradication is hampered by latency in infected $CD4^+$ T-cells. HDAC activity has been linked to viral latency and thus an HDAC inhibitor might help deplete this viral reservoir [117]. Keedy et al. implicated HDAC2 and 3, showing that they bind to the HIV-1 promoter, and when depleted by siRNA, viral protein expression is increased [118]. In a pilot study, valproic acid was administered with highly active anti-retroviral therapy in patients infected with HIV. Infected $CD4^+$ T cells were substantially reduced in three of four patients treated, with a mean decrease of 75%, and the treatment was well tolerated [119].

Current Clinical Trials Underway Evaluating HDAC Inhibitors for the Treatment of Non-Cancer Diseases

The therapeutic potential of valproic acid and phenylbutyrate continues to be evaluated for new and expanded applications for the treatment of non-cancer diseases. There are currently 205 trials involving valproic acid (www.clinicaltrials.gov), of which 166 are being evaluated as therapies for non-cancer diseases. Not surprisingly, most studies further examine its effectiveness for the treatment of mood disorder (89 trials), epilepsy (31 trials), and migraine/headache (eight trials). Four trials are currently evaluating valproic acid for the treatment of spinal muscular atrophy (SMA) in patients from infancy up to 17 years of age. Valproic acid also continues to be evaluated for the treatment of human immunodeficiency viral infection with five on-going trials. In three trials, valproic acid is being evaluated for its ability to promote the eradication of the latent HIV $CD4^+$ T-cell reservoir. A fourth trial is examining the ability of valproic acid to extend the half-life of the anti-viral drug zidovudine (AZT) in patients. Due to preclinical work suggesting valproic acid promotes neuronal protection [120], several trials have been initiated to evaluate its use for the treatment of neurodegenerative diseases such as Alzheimer's (three trials), Huntington's (one trial), and amyotrophic lateral sclerosis (one trial). These include a phase III double-blind, randomized, placebo-controlled trial for Alzheimer's and amyotrophic lateral sclerosis.

Currently, phenylbutyrate is being evaluated in 15 trials for the treatment of non-cancer diseases (www.clinicaltrials.gov). Three phase I/II trials are examining the efficacy of phenylbutyrate for the treatment of SMA in young infants and children. Two randomized, double-blind, placebo-controlled phase I/II trials are evaluating the use of phenylbutyrate for the treatment of cystic fibrosis, with one study enrolling patients who are homozygous for the delta-F508 CFTR mutation, and the other enrolling patients who are heterozygous. Two phase II trials are evaluating the use of combining the DNA methyltransferase 1 inhibitor, 5-azacytidine, with phenylbutyrate for the treatment of patients with β-thalassemia. As with valproic acid, phenylbutyrate is also being explored as a therapeutic agent for neurodegenerative disease in two trials. One trial is a phase II randomized, double-blind, placebo-controlled study treating adults with Huntington's disease. The second trial is a phase I/II study enrolling adults with amyotrophic lateral sclerosis.

Although most on-going trials are examining the use of vorinostat for the treatment of various malignancies (discussed earlier), a phase II trial is exploring its efficacy for treating graft versus host disease (GVHD) following bone marrow transplant in patients with blood cell cancers. GVHD can occur following a bone marrow transplant, where transplanted immune cells target host tissues. Transplanted immune cells also exhibit graft versus tumor activity, promotion of which has been hypothesized as a promising approach for the treatment of hematological malignancies [121]. In mouse bone marrow transplantation models, treatment with vorinostat reduces the pro-inflammatory activity associated with GVHD, but maintains the GVT activity [122]. This trial seeks to determine whether the addition of vorinostat to current GVHD therapy will ameliorate adverse GVHD effects, but preserve GVT activity.

A number of studies have demonstrated that HDAC inhibitors exhibit anti-inflammatory activity, including the hydroxamic acid derivative givinostat (ITF2357). In both cultured cells and mice treated with the pro-inflammatory agent lipopolysaccharide, administration of givinostat reduced pro-inflammatory cytokines, including tumor necrosis factor-α and interferon-γ [123]. Furthermore, using a head trauma mouse model, treatment with givinostat following injury attenuated inflammation and brain damage [124]. Based on anti-inflammatory pre-clinical findings, three clinical trials have been initiated that propose to evaluate givinostat for the treatment of inflammatory diseases. These include a phase II trial examining its use for patients with systemic onset juvenile idiopathic arthritis, a phase I/II random, double-blind, placebo-controlled trial evaluating mucosal healing in patients with Crohn's disease, and a phase II trial treating patients with auto-inflammatory syndromes.

CONCLUSION AND THE FUTURE DIRECTION OF THE CLINICAL APPLICATIONS OF HDAC INHIBITORS

HDAC inhibitors provide an exciting means to treat a variety of human diseases with acceptable side effects. The most advanced development of these agents is in the treatment of CTCL where several HDAC inhibitors have shown clinical efficacy in patients where other treatment options remain limited. Vorinostat and romidepsin have been approved for this indication. HDAC inhibitors have shown efficacy in other hematological malignancies including MDS, and leukemias and other lymphomas. However, as single agents for the treatment of solid tumor malignancies, the efficacy of HDAC inhibitors has been more modest.

Work to develop more specific HDAC inhibitors with an improved therapeutic window, and better target identification, is underway to increase their efficacy. Furthermore, multiple preclinical and clinical studies suggest that HDAC inhibitors may have a synergistic role in the target modulation of many novel biological therapeutics and chemotherapy in both hematological and solid tumor malignances. These strategies include the combined epigenetic modulation of targets by HDAC inhibitors and demethylation agents (e.g. decitabine), VEGF inhibitors, and anti-estrogens among many others. Promising clinical evidence using HDAC inhibitors also comes from clinical trials targeting non-cancer diseases such as spinal muscle atrophy, HIV infection, and cystic fibrosis. The established use of valproic acid and phenylbutyrate in the clinic has demonstrated their limited toxicity profile and thus they provide attractive options for evaluating the benefit of HDAC inhibition for the treatment of various diseases.

One of the limitations of current HDAC inhibitors is their relatively narrow therapeutic window. Many HDAC inhibitor effects on biological targets are reversible upon drug removal, and sustained effects often require continuous dosing. More specific patient selection will also have a significant impact on future clinical trials of HDAC inhibitors, especially as related to cancer treatment. Evidence is already being gathered on some of the potential markers that could correlate with resistance to HDAC inhibitors, including 17 specific antioxidant genes. Determining the specific type and grade of cancer most susceptible to HDAC inhibitor treatment will also aid in selecting patients most likely to respond.

Further insight into the molecular mechanisms that govern the inhibition of HDACs will help design more specific and potent drugs. The creation of more potent HDAC inhibitors has already shown great promise and will continue to increase their effectiveness in the future. Increasing target specificity will also have a profound effect on the creation of more effective treatment options. Recognizing the specific roles of select HDACs (e.g. HDAC1 or HDAC2) involved in exerting the desired effects of HDAC inhibitors and their synergy with other therapeutics will allow the development of more specific and potent drugs. The approval of two HDAC inhibitors has lent support to their use in the treatment of cancer, and their applicability as novel treatment strategies either alone or in combination with a large variety of other therapeutics will undoubtedly expand in the future.

References

1. Mottet D, Castronovo V. Histone deacetylases: target enzymes for cancer therapy. *Clin Exp Metastasis* 2008;**25**:183–9.

2. Chan HM, Krstic-Demonacos M, Smith L, Demonacos C, La Thangue NB. Acetylation control of the retinoblastoma tumour-suppressor protein. *Nat Cell Biol* 2001;**3**:667–74.

3. Kawaguchi Y, Kovacs JJ, McLaurin A, Vance JM, Ito A, Yao TP. The deacetylase HDAC6 regulates aggresome formation and cell viability in response to misfolded protein stress. *Cell* 2003;**115**:727–38.

4. Marzio G, Wagener C, Gutierrez MI, Cartwright P, Helin K, Giacca M. E2F family members are differentially regulated by reversible acetylation. *J Biol Chem* 2000;**275**:10887–92.

5. Chen CS, Wang YC, Yang HC, Huang PH, Kulp SK, Yang CC, et al. Histone deacetylase inhibitors sensitize prostate cancer cells to agents that produce DNA double-strand breaks by targeting Ku70 acetylation. *Cancer Res* 2007;**67**:5318–27.

6. Xu WS, Parmigiani RB, Marks PA. Histone deacetylase inhibitors: molecular mechanisms of action. *Oncogene* 2007;**26**:5541–52.

7. Carew JS, Giles FJ, Nawrocki ST. Histone deacetylase inhibitors: mechanisms of cell death and promise in combination cancer therapy. *Cancer Lett* 2008;**269**:7–17.

8. Van Lint C, Emiliani S, Verdin E. The expression of a small fraction of cellular genes is changed in response to histone hyperacetylation. *Gene Exp* 1996;**5**:245–53.

9. Ungerstedt JS, Sowa Y, Xu WS, Shao Y, Dokmanovic M, Perez G, et al. Role of thioredoxin in the response of normal and transformed cells to histone deacetylase inhibitors. *Proc Natl Acad Sci USA* 2005;**102**:673–8.

10. Wallace DM, Cotter TG. Histone deacetylase activity in conjunction with E2F-1 and p53 regulates Apaf-1 expression in 661W cells and the retina. *J Neurosci Res* 2009;**87**:887–905.

11. Yang QC, Zeng BF, Shi ZM, Dong Y, Jiang ZM, Huang J, et al. Inhibition of hypoxia-induced angiogenesis by trichostatin A via suppression of HIF-1a activity in human osteosarcoma. *J Exp Clin Cancer Res* 2006;**25**:593–9.

12. Marchion DC, Bicaku E, Daud AI, Richon V, Sullivan DM, Munster PN. Sequence-specific potentiation of topoisomerase II inhibitors by the histone deacetylase inhibitor suberoylanilide hydroxamic acid. *J Cell Biochem* 2004;**92**:223–37.

13. Zhang Y, Carr T, Dimtchev A, Zaer N, Dritschilo A, Jung M. Attenuated DNA damage repair by trichostatin A through BRCA1 suppression. *Radiat Res* 2007;**168**:115–24.

14. Dokmanovic M, Marks PA. Prospects: histone deacetylase inhibitors. *J Cell Biochem* 2005;**96**:293–304.

15. Kramer OH, Zhu P, Ostendorff HP, Golebiewski M, Tiefenbach J, Peters MA, et al. The histone deacetylase inhibitor valproic acid selectively induces proteasomal degradation of HDAC2. *EMBO J* 2003;**22**:3411–20.

16. Vannini A, Volpari C, Filocamo G, Casavola EC, Brunetti M, Renzoni D, et al. Crystal structure of a eukaryotic zinc-dependent histone deacetylase, human HDAC8, complexed with a hydroxamic acid inhibitor. *Proc Natl Acad Sci USA* 2004;**101**:15064–9.

17. Somoza JR, Skene RJ, Katz BA, Mol C, Ho JD, Jennings AJ, et al. Structural snapshots of human HDAC8 provide insights into the class I histone deacetylases. *Structure* 2004;**12**:1325–34.

18. Furumai R, Matsuyama A, Kobashi N, Lee KH, Nishiyama M, Nakajima H, et al. FK228 (depsipeptide) as a natural prodrug that inhibits class I histone deacetylases. *Cancer Res* 2002;**62**:4916–21.

19. Siegel D, Hussein M, Belani C, Robert F, Galanis E, Richon VM, et al. Vorinostat in solid and hematologic malignancies. *J Hematol Oncol* 2009;**2**:31.

20. Blum W, Marcucci G. Targeting epigenetic changes in acute myeloid leukemia. *Clin Adv Hematol Oncol* 2005;**3**(855–865):882.

21. Khan O, La Thangue NB. Drug insight: histone deacetylase inhibitor-based therapies for cutaneous T-cell lymphomas. *Nat Clin Pract Oncol* 2008;**5**:714–26.

22. Stimson L, Wood V, Khan O, Fotheringham S, La Thangue NB. HDAC inhibitor-based therapies and haematological malignancy. *Ann Oncol* 2009;**20**:1293–302.

23. Kelly WK, O'Connor OA, Krug LM, Chiao JH, Heaney M, Curley T, et al. Phase I study of an oral histone deacetylase inhibitor, suberoylanilide hydroxamic acid, in patients with advanced cancer. *J Clin Oncol* 2005;**23**:3923–31.

24. Kelly WK, Richon VM, O'Connor O, Curley T, MacGregor-Curtelli B, Tong W, et al. Phase I clinical trial of histone deacetylase inhibitor: suberoylanilide hydroxamic acid administered intravenously. *Clin Cancer Res* 2003;**9**:3578–88.

25. Olsen EA, Kim YH, Kuzel TM, Pacheco TR, Foss FM, Parker S, et al. Phase IIb multicenter trial of vorinostat in patients with persistent, progressive, or treatment refractory cutaneous T-cell lymphoma. *J Clin Oncol* 2007;**25**:3109–15.

26. Duvic M, Talpur R, Ni X, Zhang C, Hazarika P, Kelly C, et al. Phase II Trial of oral vorinostat (suberoylanilide hydroxamic acid, SAHA) for refractory cutaneous T-cell lymphoma (CTCL). *Blood* 2007;**109**:31–9.

27. Garcia-Manero G, Yang H, Bueso-Ramos C, Ferrajoli A, Cortes J, Wierda WG, et al. Phase 1 study of the histone deacetylase inhibitor vorinostat (suberoylanilide hydroxamic acid [SAHA]) in patients with advanced leukemias and myelodysplastic syndromes. *Blood* 2008;**111**:1060–6.

28. Garcia-Manero G, Assouline S, Cortes J, Estrov Z, Kantarjian H, Yang H, et al. Phase 1 study of the oral isotype specific histone deacetylase inhibitor MGCD0103 in leukemia. *Blood* 2008;**112**:981–9.

29. Kelly WK, O'Connor O, Richon V, Curley T, Richardson S, Chiao J, et al. Phase I clinical trial of an oral histone deacetylase inhibitor: suberoylanilide hydroxamic acid (SAHA). *Proc Molecular Targets Cancer Therapeutics* 2003:C174 [Abstract].

30. O'Connor OA, Heaney ML, Schwartz L, Richardson S, Willim R, MacGregor-Cortelli B, et al. Clinical experience with intravenous and oral formulations of the novel histone deacetylase inhibitor suberoylanilide hydroxamic acid in patients with advanced hematologic malignancies. *J Clin Oncol* 2006;**24**:166–73.

31. Ellis L, Pan Y, Smyth GK, George DJ, McCormack C, Williams-Truax R, et al. Histone deacetylase inhibitor panobinostat induces clinical responses with associated alterations in gene expression profiles in cutaneous T-cell lymphoma. *Clin Cancer Res* 2008;**14**:4500–10.

32. Prince HM, George DJ, Johnstone R, Williams-Truax R, Atadja P, Zhao C, et al. LBH589, a novel histone deacetylase inhibitor (HDACi), treatment of patients with cutaneous T-cell lymphoma (CTCL). Changes in skin gene expression profiles related to clinical response following therapy. *ASCO Annual Meeting Proceedings, Part I. Vol. 24 No 18S.* 2006.

33. Giles F, Fischer T, Cortes J, Garcia-Manero G, Beck J, Ravandi F, et al. A phase I study of intravenous LBH589, a novel cinnamic hydroxamic acid analogue histone deacetylase inhibitor, in patients with refractory hematologic malignancies. *Clin Cancer Res* 2006;**12**:4628–35.

34. Gimsing P, Hansen M, Knudsen LM, Knoblauch P, Christensen IJ, Ooi CE, et al. A phase I clinical trial of the histone deacetylase inhibitor belinostat in patients with advanced hematological neoplasia. *Eur J Haematol* 2008;**81**:170–6.

35. Loscher W. Basic pharmacology of valproate: a review after 35 years of clinical use for the treatment of epilepsy. *CNS Drugs* 2002;**16**:669–94.

36. Phiel CJ, Zhang F, Huang EY, Guenther MG, Lazar MA, Klein PS. Histone deacetylase is a direct target of valproic acid, a potent anticonvulsant, mood stabilizer, and teratogen. *J Biol Chem* 2001;**276**:36734–41.

37. Kuendgen A, Gattermann N. Valproic acid for the treatment of myeloid malignancies. *Cancer* 2007;**110**:943–54.

38. Kuendgen A, Strupp C, Aivado M, Bernhardt A, Hildebrandt B, Haas R, et al. Treatment of myelodysplastic syndromes with valproic acid alone or in combination with all-trans retinoic acid. *Blood* 2004;**104**:1266–9.

39. Pilatrino C, Cilloni D, Messa E, Morotti A, Giugliano E, Pautasso M, et al. Increase in platelet count in older, poor-risk patients with acute myeloid leukemia or myelodysplastic syndrome treated with valproic acid and all-trans retinoic acid. *Cancer* 2005;**104**:101–9.

40. Raffoux E, Chaibi P, Dombret H, Degos L. Valproic acid and all-trans retinoic acid for the treatment of elderly patients with acute myeloid leukemia. *Haematologica* 2005;**90**:986–8.

41. Ryningen A, Stapnes C, Lassalle P, Corbascio M, Gjertsen BT, Bruserud O. A subset of patients with high-risk acute myelogenous leukemia shows improved peripheral blood cell counts when treated with the combination of valproic acid, theophylline and all-trans retinoic acid. *Leuk Res* 2009;**33**:779–87.

42. Kuendgen A, Schmid M, Schlenk R, Knipp S, Hildebrandt B, Steidl C, et al. The histone deacetylase (HDAC) inhibitor valproic acid as monotherapy or in combination with all-trans retinoic acid in patients with acute myeloid leukemia. *Cancer* 2006;**106**:112–19.

43. Blum W, Klisovic RB, Hackanson B, Liu Z, Liu S, Devine H, et al. Phase I study of decitabine alone or in combination with valproic acid in acute myeloid leukemia. *J Clin Oncol* 2007;**25**:3884–91.

44. Piekarz RL, Robey R, Sandor V, Bakke S, Wilson WH, Dahmoush L, et al. Inhibitor of histone deacetylation, depsipeptide (FR901228), in the treatment of peripheral and cutaneous T-cell lymphoma: a case report. *Blood* 2001;**98**:2865–8.

611

45. Piekarz RL, Frye R, Turner M, Wright JJ, Allen SL, Kirschbaum MH, et al. Phase II multi-institutional trial of the histone deacetylase inhibitor romidepsin as monotherapy for patients with cutaneous T-cell lymphoma. *J Clin Oncol* 2009;**27**:5410–17.

46. Byrd JC, Marcucci G, Parthun MR, Xiao JJ, Klisovic RB, Moran M, et al. A phase 1 and pharmacodynamic study of depsipeptide (FK228) in chronic lymphocytic leukemia and acute myeloid leukemia. *Blood* 2005;**105**:959–67.

47. Klimek VM, Fircanis S, Maslak P, Guernah I, Baum M, Wu N, et al. Tolerability, pharmacodynamics, and pharmacokinetics studies of depsipeptide (romidepsin) in patients with acute myelogenous leukemia or advanced myelodysplastic syndromes. *Clin Cancer Res* 2008;**14**:826–32.

48. Odenike OM, Alkan S, Sher D, Godwin JE, Huo D, Brandt SJ, et al. Histone deacetylase inhibitor romidepsin has differential activity in core binding factor acute myeloid leukemia. *Clin Cancer Res* 2008;**14**:7095–101.

49. Sandor V, Bakke S, Robey RW, Kang MH, Blagosklonny MV, Bender J, et al. Phase I trial of the histone deacetylase inhibitor, depsipeptide (FR901228, NSC 630176), in patients with refractory neoplasms. *Clin Cancer Res* 2002;**8**:718–28.

50. Piekarz RL, Frye AR, Wright JJ, Steinberg SM, Liewehr DJ, Rosing DR, et al. Cardiac studies in patients treated with depsipeptide, FK228, in a phase II trial for T-cell lymphoma. *Clin Cancer Res* 2006;**12**:3762–73.

51. Ryan QC, Headlee D, Acharya M, Sparreboom A, Trepel JB, Ye J, et al. Phase I and pharmacokinetic study of MS-275, a histone deacetylase inhibitor, in patients with advanced and refractory solid tumors or lymphoma. *J Clin Oncol* 2005;**23**:3912–22.

52. Kummar S, Gutierrez M, Gardner ER, Donovan E, Hwang K, Chung EJ, et al. Phase I trial of MS-275, a histone deacetylase inhibitor, administered weekly in refractory solid tumors and lymphoid malignancies. *Clin Cancer Res* 2007;**13**:5411–17.

53. Gojo I, Jiemjit A, Trepel JB, Sparreboom A, Figg WD, Rollins S, et al. Phase 1 and pharmacologic study of MS-275, a histone deacetylase inhibitor, in adults with refractory and relapsed acute leukemias. *Blood* 2007;**109**:2781–90.

54. Steele NL, Plumb JA, Vidal L, Tjornelund J, Knoblauch P, Rasmussen A, et al. A phase 1 pharmacokinetic and pharmacodynamic study of the histone deacetylase inhibitor belinostat in patients with advanced solid tumors. *Clin Cancer Res* 2008;**14**:804–10.

55. Siu LL, Pili R, Duran I, Messersmith WA, Chen EX, Sullivan R, et al. Phase I study of MGCD0103 given as a three-times-per-week oral dose in patients with advanced solid tumors. *J Clin Oncol* 2008;**26**:1940–7.

56. Galanis E, Jaeckle KA, Maurer MJ, Reid JM, Ames MM, Hardwick JS, et al. Phase II trial of vorinostat in recurrent glioblastoma multiforme: a north central cancer treatment group study. *J Clin Oncol* 2009;**27**:2052–8.

57. Luu TH, Morgan RJ, Leong L, Lim D, McNamara M, Portnow J, et al. A phase II trial of vorinostat (suberoylanilide hydroxamic acid) in metastatic breast cancer: a California Cancer Consortium study. *Clin Cancer Res* 2008;**14**:7138–42.

58. Blumenschein GR Jr., Kies MS, Papadimitrakopoulou VA, Lu C, Kumar AJ, Ricker JL, et al. Phase II trial of the histone deacetylase inhibitor vorinostat (Zolinza, suberoylanilide hydroxamic acid, SAHA) in patients with recurrent and/or metastatic head and neck cancer. *Invest New Drugs* 2008;**26**:81–7.

59. Traynor AM, Dubey S, Eickhoff JC, Kolesar JM, Schell K, Huie MS, et al. Vorinostat (NSC# 701852) in patients with relapsed non-small cell lung cancer: a Wisconsin Oncology Network phase II study. *J Thorac Oncol* 2009;**4**:522–6.

60. Bradley D, Rathkopf D, Dunn R, Stadler WM, Liu G, Smith DC, et al. Vorinostat in advanced prostate cancer patients progressing on prior chemotherapy (National Cancer Institute Trial 6862): trial results and interleukin-6 analysis: A study by the Department of Defense Prostate Cancer Clinical Trial Consortium and University of Chicago Phase 2 Consortium. *Cancer* 2009.

61. Whitehead RP, Rankin C, Hoff PM, Gold PJ, Billingsley KG, Chapman RA, et al. Phase II trial of romidepsin (NSC-630176) in previously treated colorectal cancer patients with advanced disease: a Southwest Oncology Group study (S0336). *Invest New Drugs* 2008.

62. Schrump DS, Fischette MR, Nguyen DM, Zhao M, Li X, Kunst TF, et al. Clinical and molecular responses in lung cancer patients receiving Romidepsin. *Clin Cancer Res* 2008;**14**:188–98.

63. Stadler WM, Margolin K, Ferber S, McCulloch W, Thompson JA. A phase II study of depsipeptide in refractory metastatic renal cell cancer. *Clin Genitourin Cancer* 2006;**5**:57–60.

64. Molife LR, Attard G, Fong PC, Karavasilis V, Reid AH, Patterson S, et al. Phase II, two-stage, single-arm trial of the histone deacetylase inhibitor (HDACi) romidepsin in metastatic castration-resistant prostate cancer (CRPC). *Ann Oncol* 2010;**21**(1):109–13.

65. Shah MH, Binkley P, Chan K, Xiao J, Arbogast D, Collamore M, et al. Cardiotoxicity of histone deacetylase inhibitor depsipeptide in patients with metastatic neuroendocrine tumors. *Clin Cancer Res* 2006;**12**:3997–4003.

66. Munster PN, Rubin EH, Van Belle S, Friedman E, Patterson JK, Van Dyck K, et al. A single supratherapeutic dose of vorinostat does not prolong the QTc interval in patients with advanced cancer. *Clin Cancer Res* 2009;**15**:7077–84.

67. Munster P, Marchion D, Bicaku E, Lacevic M, Kim J, Centeno B, et al. Clinical and biological effects of valproic acid as a histone deacetylase inhibitor on tumor and surrogate tissues: phase I/II trial of valproic acid and epirubicin/FEC. *Clin Cancer Res* 2009;**15**:2488–96.

68. Munster P, Schmitt M, Marchion D, Bicaku E, Egorin M, Sullivan D, et al. Phase I trial of a sequence-specific combination of the HDAC inhibitor, vorinostat (SAHA) followed by doxorubicin in advanced solid tumor malignancies. *BMJ* 2009 in press.

69. Khan N, Jeffers M, Kumar S, Hackett C, Boldog F, Khramtsov N, et al. Determination of the class and isoform selectivity of small-molecule histone deacetylase inhibitors. *Biochem J* 2008;**409**:581–9.

70. Troso-Sandoval PN, Rosen T, Rifkind N, Marks R, Richon PA. The histone deacetylase inhibitor suberoylanilide hydroxamic acid induces differentiation of human breast cancer cells. *Cancer Res* 2001(61):8492–7.

71. Marchion DC, Bicaku E, Daud AI, Sullivan DM, Munster PN. Valproic acid alters chromatin structure by regulation of chromatin modulation proteins. *Cancer Res* 2005;**65**:3815–22.

72. Fakih MG, Pendyala L, Fetterly G, Toth K, Zwiebel JA, Espinoza-Delgado I, et al. A phase I, pharmacokinetic and pharmacodynamic study on vorinostat in combination with 5-fluorouracil, leucovorin, and oxaliplatin in patients with refractory colorectal cancer. *Clin Cancer Res* 2009;**15**:3189–95.

73. Munster P, Marchion D, Bicaku E, Schmitt M, Lee JH, DeConti R, et al. Phase I trial of histone deacetylase inhibition by valproic acid followed by the topoisomerase II inhibitor epirubicin in advanced solid tumors: a clinical and translational study. *J Clin Oncol* 2007;**25**:1979–85.

74. Townsley C, Oza AM, Tang P, Siu LL, Pond GR, Sarveswaran P, et al. Expanded phase I study of vorinostat (VOR) in combination with capecitabine (CAP) in patients (pts) with advanced solid tumors. *ASCO Meeting Abstracts* 2008;**26**:11096.

75. Dowdy SC, Jiang S, Zhou XC, Hou X, Jin F, Podratz KC, et al. Histone deacetylase inhibitors and paclitaxel cause synergistic effects on apoptosis and microtubule stabilization in papillary serous endometrial cancer cells. *Mol Cancer Ther* 2006;**5**:2767–76.

76. Ramalingam SS, Parise RA, Ramanathan RK, Lagattuta TF, Musguire LA, Stoller RG, et al. Phase I and pharmacokinetic study of vorinostat, a histone deacetylase inhibitor, in combination with carboplatin and paclitaxel for advanced solid malignancies. *Clin Cancer Res* 2007;**13**:3605–10.

77. Ramalingam SS, Maitland M, Frankel P, Argiris AE, Koczywas M, Gitlitz B, et al. Randomized, double-blind, placebo-controlled phase II study of carboplatin and paclitaxel with or without vorinostat, a histone deacetylase inhibitor (HDAC), for first-line therapy of advanced non-small cell lung cancer (NCI 7863). *ASCO Meeting Abstracts* 2009;**27**:8004.

78. Thomas S, Munster PN. Histone deacetylase inhibitor induced modulation of anti-estrogen therapy. *Cancer Lett* 2009;**280**:184–91.

79. Munster PN, Lacevic M, Thomas S, Christian C, Ismail-Khan R, Melisko M, et al. Phase II trial of the histone deacetylase inhibitor, vorinostat, to restore hormone sensitivity to the antiestrogen tamoxifen in patients with advanced breast cancer who progressed on prior hormone therapy. *ASCO Meeting Abstracts* 2009;**27**:1075.

80. Conte P, Campone M, Pronzato P, Amadori D, Frank R, Schuetz F, et al. Phase I trial of panobinostat (LBH589) in combination with trastuzumab in pretreated HER2-positive metastatic breast cancer (mBC): preliminary safety and tolerability results. *ASCO Meeting Abstracts* 2009;**27**:1081.

81. Ellis L, Hammers H, Pili R. Targeting tumor angiogenesis with histone deacetylase inhibitors. *Cancer Lett* 2009;**280**:145–53.

82. Hammers HJ, Verheul H, Wilky B, Salumbides B, Holleran J, Egorin MJ, et al. Phase I safety and pharmacokinetic/pharmacodynamic results of the histone deacetylase inhibitor vorinostat in combination with bevacizumab in patients with kidney cancer. *ASCO Meeting Abstracts* 2008;**26**:16094.

83. Ramaswamy B, Bhalla K, Cohen B, Pellegrino C, Hershman D, Chuang E, et al. Phase I-II study of the histone deacetylase inhibitor (HDACi) vorinostat plus paclitaxel and bevacizumab in metastatic breast cancer (MBC): New York Cancer Consortium trial P7703. *San Antonio Breast Cancer Symposium* 2008.

84. Venturelli S, Armeanu S, Pathil A, Hsieh CJ, Weiss TS, Vonthein R, et al. Epigenetic combination therapy as a tumor-selective treatment approach for hepatocellular carcinoma. *Cancer* 2007;**109**:2132–41.

85. Karpenko MJ, Liu Z, Aimiuwu J, Wang L, Wu X, Villalona-Calero MA, et al. Phase I study of 5-aza-2'-deoxycytidine in combination with valproic acid in patients with NSCLC. *ASCO Meeting Abstracts* 2008;**26**:3502.

86. Braiteh F, Soriano AO, Garcia-Manero G, Hong D, Johnson MM, Silva Lde P, et al. Phase I study of epigenetic modulation with 5-azacytidine and valproic acid in patients with advanced cancers. *Clin Cancer Res* 2008;**14**:6296–301.

87. Lee MJ, Kim YS, Kummar S, Giaccone G, Trepel JB. Histone deacetylase inhibitors in cancer therapy. *Curr Opin Oncol* 2008;**20**:639–49.

88. Kazantsev AG, Thompson LM. Therapeutic application of histone deacetylase inhibitors for central nervous system disorders. *Nat Rev Drug Discov* 2008;**7**:854–68.

89. Rotili D, Simonetti G, Savarino A, Palamara AT, Migliaccio AR, Mai A. Non-cancer uses of histone deacetylase inhibitors: effects on infectious diseases and beta-hemoglobinopathies. *Curr Top Med Chem* 2009;**9**:272–91.

90. Blanchard F, Chipoy C. Histone deacetylase inhibitors: new drugs for the treatment of inflammatory diseases? *Drug Discov Today* 2005;**10**:197–204.

91. Haberland M, Montgomery RL, Olson EN. The many roles of histone deacetylases in development and physiology: implications for disease and therapy. *Nat Rev Genet* 2009;**10**:32–42.

92. Haddad PM, Das A, Ashfaq M, Wieck A. A review of valproate in psychiatric practice. *Expert Opin Drug Metab Toxicol* 2009;**5**:539–51.

93. Mathew NT. Antiepileptic drugs in migraine prevention. *Headache* 2001;**41**(Suppl 1):S18–S24.

94. Duncan S. Teratogenesis of sodium valproate. *Curr Opin Neurol* 2007;**20**:175–80.

95. Rosenberg G. The mechanisms of action of valproate in neuropsychiatric disorders: can we see the forest for the trees? *Cell Mol Life Sci* 2007;**64**:2090–103.

96. Johannessen CU. Mechanisms of action of valproate: a commentatory. *Neurochem Int* 2000;**37**:103–10.

97. Bosetti F, Bell JM, Manickam P. Microarray analysis of rat brain gene expression after chronic administration of sodium valproate. *Brain Res Bull* 2005;**65**:331–8.

98. Feillet F, Leonard JV. Alternative pathway therapy for urea cycle disorders. *J Inherit Metab Dis* 1998;**21**(Suppl 1):101–11.

99. Rowe SM, Clancy JP. Advances in cystic fibrosis therapies. *Curr Opin Pediatr* 2006;**18**:604–13.

100. Rubenstein RC, Zeitlin PL. A pilot clinical trial of oral sodium 4-phenylbutyrate (Buphenyl) in deltaF508-homozygous cystic fibrosis patients: partial restoration of nasal epithelial CFTR function. *Am J Respir Crit Care Med* 1998;**157**:484–90.

101. Zeitlin PL, Diener-West M, Rubenstein RC, Boyle MP, Lee CK, Brass-Ernst L. Evidence of CFTR function in cystic fibrosis after systemic administration of 4-phenylbutyrate. *Mol Ther* 2002;**6**:119–26.

102. Lunn MR, Wang CH. Spinal muscular atrophy. *Lancet* 2008;**371**:2120–33.

103. Mercuri E, Bertini E, Messina S, Pelliccioni M, D'Amico A, Colitto F, et al. Pilot trial of phenylbutyrate in spinal muscular atrophy. *Neuromusc Disord* 2004;**14**:130–5.

104. Brahe C, Vitali T, Tiziano FD, Angelozzi C, Pinto AM, Borgo F, et al. Phenylbutyrate increases SMN gene expression in spinal muscular atrophy patients. *Eur J Hum Genet* 2005;**13**:256–9.

105. Mercuri E, Bertini E, Messina S, Solari A, D'Amico A, Angelozzi C, et al. Randomized, double-blind, placebo-controlled trial of phenylbutyrate in spinal muscular atrophy. *Neurology* 2007;**68**:51–5.

106. Brichta L, Holker I, Haug K, Klockgether T, Wirth B. In vivo activation of SMN in spinal muscular atrophy carriers and patients treated with valproate. *Ann Neurol* 2006;**59**:970–5.

107. Weihl CC, Connolly AM, Pestronk A. Valproate may improve strength and function in patients with type III/IV spinal muscle atrophy. *Neurology* 2006;**67**:500–1.

108. Swoboda KJ, Scott CB, Reyna SP, Prior TW, LaSalle B, Sorenson SL, et al. Phase II open label study of valproic acid in spinal muscular atrophy. *PLoS ONE* 2009;**4**:e5268.

109. Dover GJ, Brusilow S, Samid D. Increased fetal hemoglobin in patients receiving sodium 4-phenylbutyrate. *N Engl J Med* 1992;**327**:569–70.

110. Resar LM, Segal JB, Fitzpatric LK, Friedmann A, Brusilow SW, Dover GJ. Induction of fetal hemoglobin synthesis in children with sickle cell anemia on low-dose oral sodium phenylbutyrate therapy. *J Pediatr Hematol Oncol* 2002;**24**:737–41.

111. Hines P, Dover GJ, Resar LM. Pulsed-dosing with oral sodium phenylbutyrate increases hemoglobin F in a patient with sickle cell anemia. *Pediatr Blood Cancer* 2008;**50**:357–9.

112. Dover GJ, Brusilow S, Charache S. Induction of fetal hemoglobin production in subjects with sickle cell anemia by oral sodium phenylbutyrate. *Blood* 1994;**84**:339–43.

113. Collins AF, Pearson HA, Giardina P, McDonagh KT, Brusilow SW, Dover GJ. Oral sodium phenylbutyrate therapy in homozygous beta thalassemia: a clinical trial. *Blood* 1995;**85**:43–9.

114. Selby R, Nisbet-Brown E, Basran RK, Chang L, Olivieri NF. Valproic acid and augmentation of fetal hemoglobin in individuals with and without sickle cell disease. *Blood* 1997;**90**:891–3.

115. Collins AF, Dover GJ, Luban NL. Increased fetal hemoglobin production in patients receiving valproic acid for epilepsy. *Blood* 1994;**84**:1690–1.

116. Kieslich M, Schwabe D, Cinatl J Jr., Driever PH. Increase of fetal hemoglobin synthesis indicating differentiation induction in children receiving valproic acid. *Pediatr Hematol Oncol* 2003;**20**:15–22.

117. Demonte D, Quivy V, Colette Y, Van Lint C. Administration of HDAC inhibitors to reactivate HIV-1 expression in latent cellular reservoirs: implications for the development of therapeutic strategies. *Biochem Pharmacol* 2004;**68**:1231–8.

118. Keedy KS, Archin NM, Gates AT, Espeseth A, Hazuda DJ, Margolis DM. A limited group of class I histone deacetylases acts to repress human immunodeficiency virus type 1 expression. *J Virol* 2009;**83**:4749–56.

119. Lehrman G, Hogue IB, Palmer S, Jennings C, Spina CA, Wiegand A, et al. Depletion of latent HIV-1 infection in vivo: a proof-of-concept study. *Lancet* 2005;**366**:549–55.

120. Morrison BE, Majdzadeh N, D'Mello SR. Histone deacetylases: focus on the nervous system. *Cell Mol Life Sci* 2007;**64**:2258–69.

121. Rezvani AR, Storb RF. Separation of graft-vs.-tumor effects from graft-vs.-host disease in allogeneic hematopoietic cell transplantation. *J Autoimmun* 2008;**30**:172–9.

122. Reddy P, Maeda Y, Hotary K, Liu C, Reznikov LL, Dinarello CA, et al. Histone deacetylase inhibitor suberoylanilide hydroxamic acid reduces acute graft-versus-host disease and preserves graft-versus-leukemia effect. *Proc Natl Acad Sci USA* 2004;**101**:3921–6.

123. Leoni F, Fossati G, Lewis EC, Lee JK, Porro G, Pagani P, et al. The histone deacetylase inhibitor ITF2357 reduces production of pro-inflammatory cytokines in vitro and systemic inflammation in vivo. *Mol Med* 2005;**11**:1–15.

124. Shein NA, Grigoriadis N, Alexandrovich AG, Simeonidou C, Lourbopoulos A, Polyzoidou E, et al. Histone deacetylase inhibitor ITF2357 is neuroprotective, improves functional recovery, and induces glial apoptosis following experimental traumatic brain injury. *FASEB J* 2009;**23**:4266–75.

125. Crump M, Coiffier B, Jacobsen ED, Sun L, Ricker JL, Xie H, et al. Phase II trial of oral vorinostat (suberoylanilide hydroxamic acid) in relapsed diffuse large-B-cell lymphoma. *Ann Oncol* 2008;**19**:964–9.

126. Richardson P, Mitsiades C, Colson K, Reilly E, McBride L, Chiao J, et al. Phase I trial of oral vorinostat (suberoylanilide hydroxamic acid, SAHA) in patients with advanced multiple myeloma. *Leuk Lymphoma* 2008;**49**:502–7.

127. Hauschild A, Trefzer U, Garbe C, Kaehler KC, Ugurel S, Kiecker F, et al. Multicenter phase II trial of the histone deacetylase inhibitor pyridylmethyl-N-{4-[(2-aminophenyl)-carbamoyl]-benzyl}-carbamate in pretreated metastatic melanoma. *Melanoma Res* 2008;**18**:274–8.

128. Modesitt SC, Sill M, Hoffman JS, Bender DP. A phase II study of vorinostat in the treatment of persistent or recurrent epithelial ovarian or primary peritoneal carcinoma: a Gynecologic Oncology Group study. *Gynecol Oncol* 2008;**109**:182–6.

129. Reid T, Valone F, Lipera W, Irwin D, Paroly W, Natale R, et al. Phase II trial of the histone deacetylase inhibitor pivaloyloxymethyl butyrate (Pivanex, AN-9) in advanced non-small cell lung cancer. *Lung Cancer* 2004;**45**:381–6.

130. Su YB, Tuttle RM, Fury M, Ghossein R, Singh B, Herman K, et al. A phase II study of single agent depsipeptide (DEP) in patients (pts) with radioactive iodine (RAI)-refractory, metastatic, thyroid carcinoma: preliminary toxicity and efficacy experience. *ASCO Meeting Abstracts* 2006;**24**:5554.

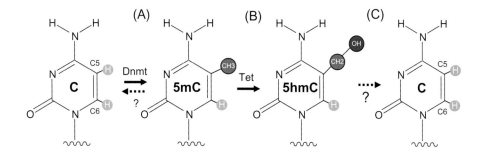

FIGURE 2.1

DNA cytosine methylation, hydroxylation, and demethylation. (A) The question mark indicates possible activity of DNA demethylases [150–155]. (B) Conversion of 5mC to 5hmC in mammalian DNA by the MLL fusion partner TET1 [129]. (C) It is currently unknown whether 5hmC is an end product or an intermediate in active DNA demethylation. The question mark indicates a possible MTase-assisted removal of the C5-bound hydroxymethyl group [131]. (Please refer to Chapter 2, page 10).

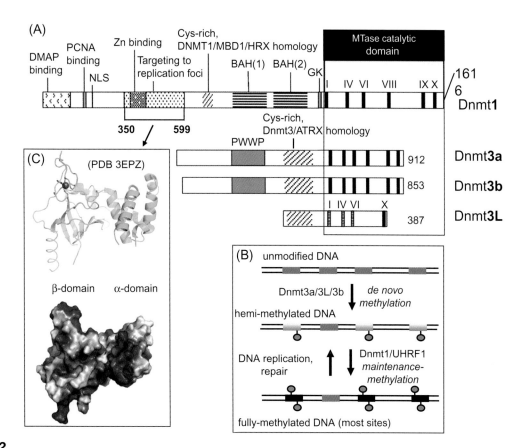

FIGURE 2.2

Schematic representation of Dnmt1 and Dnmt3. (A) Roman numerals refer to conserved motifs of DNA MTases [156]; motif IV includes the Cys nucleophile that forms a transient covalent bond to C6 of the target cytosine. (B) Maintenance vs. *de novo* methylation. The rectangular segments are substrate sequences (usually CpG), and the small ball shapes represent methyl groups on the cytosines. Following replication or repair, the duplex is methylated on one strand only. (C) The first domain structure of Dnmt1 (residues 350–599; PDB 3EPZ) [89] contains targeting sequence association with replication foci [88]. (Please refer to Chapter 2, page 11).

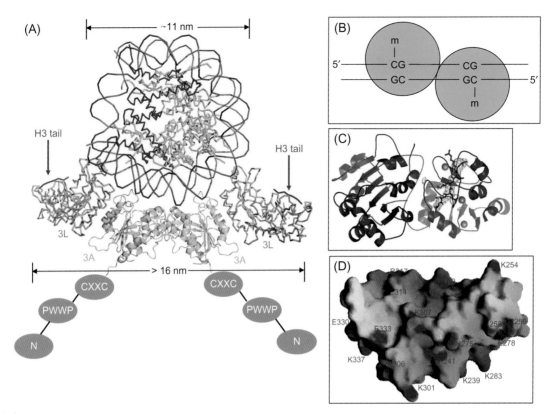

FIGURE 2.3

A model of interactions between Dnmt3a-3L tetramer and a nucleosome. (A) A nucleosome is shown, docked to a Dnmt3L-3a-3a-3L tetramer (3a-C in green; 3L full length in gray). The position of a peptide derived from the sequence of the histone H3 amino terminus (purple) is shown, taken from a co-crystal structure with this peptide bound to Dnmt3L [71]. Wrapping the tetramer around the nucleosome, the two Dnmt3L molecules could bind both histone tails from one nucleosome. The amino-proximal portion of Dnmt3a is labeled as N (for N-terminal domain), PWWP domain, and CXXC domain. By analogy to Dnmt3L, the CXXC domain of Dnmt3a might interact with histone tails from neighboring nucleosomes. (B) The Dnmt3a dimer could in theory methylate two CpGs separated by one helical turn in one binding event. (C) Structure of Dnmt3L with a bound histone H3 N-terminal tail (orange) [71]. (D) The PWWP domain structure of mouse Dnmt3b, rich in basic residues [42]. (Please refer to Chapter 2, page 12).

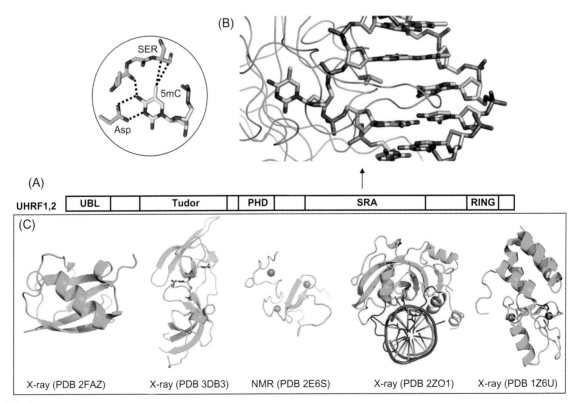

FIGURE 2.4

UHRF1 – a multi-domain protein. (A) Schematic representation of UHRF1 and its homolog UHRF2. (B) Structure of SRA-DNA complex. The 5mC flips out and is bound in a cage-like pocket. (C) Five domain structures are currently available. (Please refer to Chapter 2, page 15).

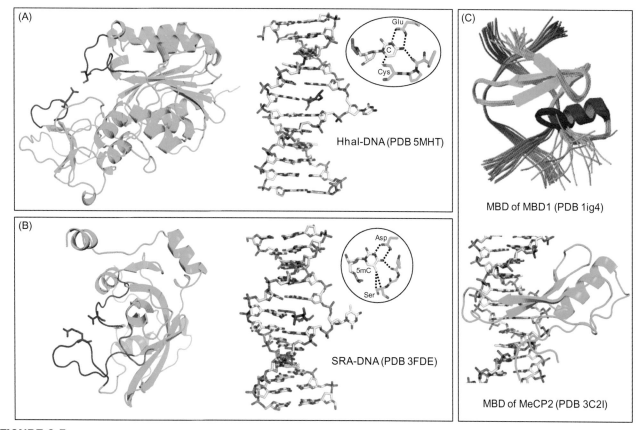

FIGURE 2.5

Comparison of base flipping by the SRA domain and HhaI methyltransferase. (A,B) DNA structures bound by HhaI (A) or SRA (B) show a flipped nucleotide. The intercalating amino acids are shown in each case. Structures of HhaI (A) and SRA (B) show the two opposite-side DNA-approaching loops. (C) NMR structure of MBD1-DNA (top) and X-ray structure of MeCP2-DNA (bottom) showed MBD domain inserts a beta-hairpin through the DNA major groove. The methyl-binding domains of MBD1 [112] and MeCP2 [113], instead of using a base-flipping mechanism, recognize changes in hydration of the major groove of a fully methylated CpG rather than detecting methyl groups directly. (Please refer to Chapter 2, page 16).

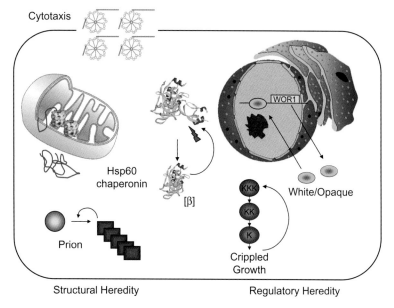

FIGURE 5.1

Schematic diagram of the prions and prion-like elements of eukaryotes discussed in this chapter. (Please refer to Chapter 5, page 72).

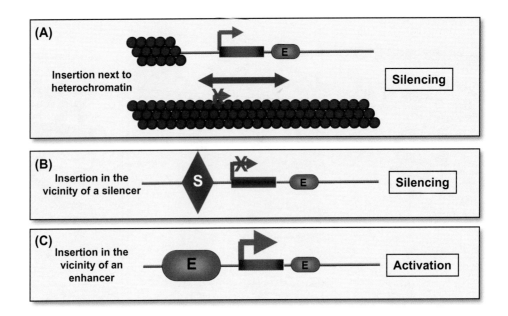

FIGURE 6.1

Chromatin configurations leading to the differential expression of an integrated transgene. Expression of integrated transgenes, even if they contain their own enhancers (E) can be modulated by the chromatin context at the site of integration. (A) Condensed chromatin (constitutive heterochromatin, repetitive DNA, telomeric DNA for instance) can induce the spreading of heterochromatin marks, the repositioning of nucleosomes, the recruitment of chromatin remodeling complexes, the silencing of the transgene, and the appearance of a variegated expression pattern from cell to cell. This phenomenon is called PEV (Position Effect Variegation) and has been observed in all eukaryotic cells. Alternatively, a transgene can be misregulated by the proximity of *cis*-regulating elements such as silencer (B) or enhancers (C) at the integration site, which might influence the expression of the gene of interest. (Please refer to Chapter 6, page 79).

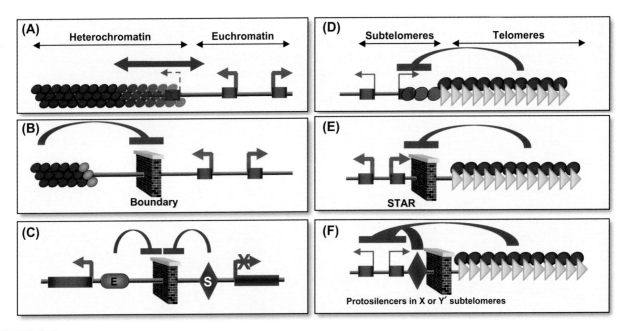

FIGURE 6.2

Modulation of position effect variegation by boundary elements. Two types of boundary element able to separate functional domains have been described in the literature [9,10]. (A) Regions named "fuzzy boundaries" depend on a dynamic equilibrium of chromatin proteins and are the more prone to variegation. (B) DNA sequences named "insulator elements" recruit specific factors to a precise location and define strict borders between chromatin regions. Two types of insulator element exist depending on their specificity (see text for details and [9,10]) and can be classified as boundary elements (B) or enhancer blocking insulators (C). (D) In *S. cerevisiae*, telomeres silence proximal gene through a mechanism named Telomeric Position Effect involving the nucleation of heterochromatin and the spreading of chromatin modifications from the telomere to the subtelomeric region [5,7]. (E) However, each of the 32 yeast chromosomes has a different composition in subtelomeric elements that can modulate TPE. Strong boundary elements dubbed STAR for SubTelomeric Antisilencing Regions consisting of binding sites for Tbf1p and Reb1p [78] are located in some subtelomeres and prevent the spreading of TPE and shelter gene expression. (F) Nevertheless, some elements in X or Y′ subtelomeric repeats can resume and reinforce TPE, bypassing the protective effect of STARs. The core X sequence behaves as a protosilencer, i.e. it does not act as a silencer by itself but reinforces silencing when located in the proximity of a master silencer [245,246]. Different combinations of STAR and protosilencer at native telomeres are likely to contribute to their respective behaviors with regard to TPE. (Please refer to Chapter 6, page 82).

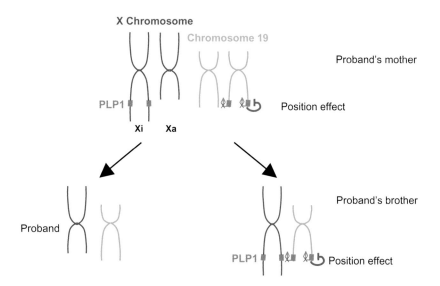

FIGURE 6.3

When the position effect rescues the phenotype. In the majority of patients with Pelizaeus–Merzbacher disease (PMD), duplication of the *PLP1* gene is responsible for the abnormal development of oligodendrocytes and the demyelinization of central nervous system neurons. In one family described in the literature [178], *PLP1* deletion resulted from a maternal balanced submicroscopic insertional translocation of the entire *PLP1* gene to the telomere of chromosome 19. The proband was a 10-year-old boy who showed early motor development and mental retardation. He had a healthy male sibling. The mother presented mild features of PMD such as difficulty in walking at the third decade, changes in her personality, and progressive mental deterioration. The patient carries a deletion of the *PLP1* gene on the X chromosome and further analysis of the family revealed that the translocation was inherited from his mother. The latter carried a translocation of the *PLP1* gene on the distal end of chromosome 19. The patient inherited the translocated X chromosome but did not inherit the derivative chromosome 19 containing *PLP1* while his healthy brother inherited this extra copy of the gene but does not carry the deleted X chromosome. However, the gene on chromosome 19 was silenced by the position effect and *PLP1* gene dosage was normal. Interestingly, in the mother, skewed X inactivation allowed the expression of the *PLP1* gene present on the inactive X chromosome suggesting that a low but detectable PLP1 level was associated with the late onset of the clinical signs. (Please refer to Chapter 6, page 89).

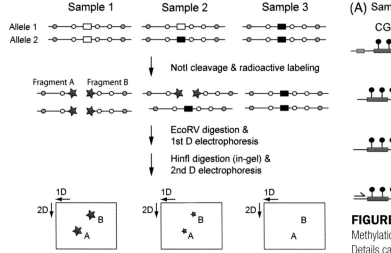

FIGURE 9.2

Restriction landmark genomic scanning (RLGS). A more detailed description of the main steps of RLGS is provided in the text. Open and filled rectangles label unmethylated and methylated NotI sites, respectively. Circles represent EcoRV and HinfI cleavage sites. (Please refer to Chapter 9, page 136).

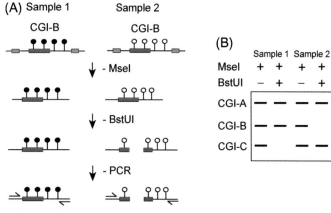

FIGURE 9.3

Methylation sensitive restriction fingerprinting (MSRF). (A) Principle of MSRF. Details can be found in the text. Filled and open lollipops mark methylated and unmethylated CpG sites, respectively. Rectangles refer to MseI and BstUI cleavage sites. Arrows show the locations of the arbitrary primers' annealing sites. (B) Theoretical outcome of an MSRF experiment (autoradiogram). *CGI-A*: there is no difference between samples 1 and 2. *CGI-B*: hypermethylation in sample 1 (this case is outlined in panel A). *CGI-C*: hypomethylation in sample 1. (Please refer to Chapter 9, page 137).

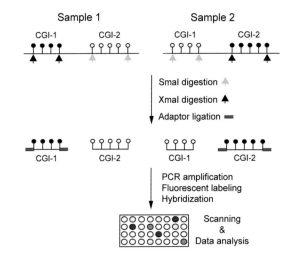

FIGURE 9.4

The main steps of methylated CpG island amplification coupled microarray (MCAM). Details are in the text. Arrows point to SmaI and XmaI cleavage sites. (Please refer to Chapter 9, page 138).

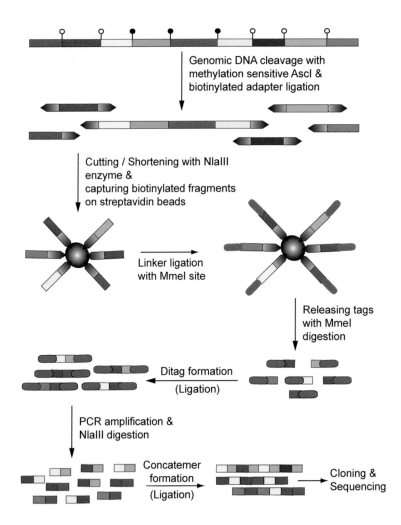

FIGURE 9.5

Methylation specific digital karyotyping (MSDK). A step-by-step introduction to the main steps can be found in the text. Filled and open lollipops mark methylated and unmethylated CpG sites, respectively. (Please refer to Chapter 9, page 139).

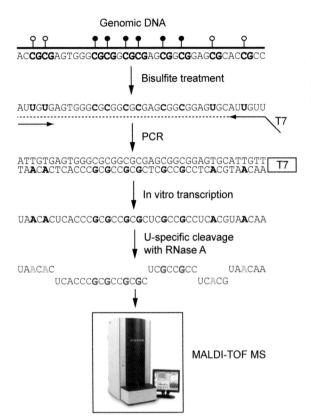

FIGURE 9.6

MALDI-TOF MS ("Sequenom" approach). The principle of MALDI-TOF MS methylation detection, which is based on bisulfite conversion, is outlined in the text. Filled and open lollipops mark methylated and unmethylated CpG sites, respectively. Boldfaced and colored letters help in following the nucleotide changes through the procedure. (Please refer to Chapter 9, page 141).

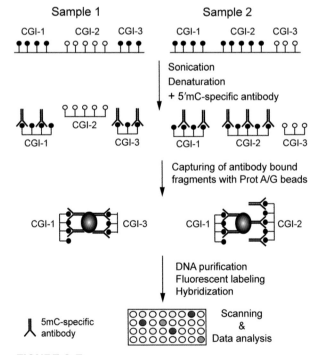

FIGURE 9.7

Methylated DNA immunoprecipitation (MeDIP). Delineation of MeDIP is given in the text. Filled and open lollipops mark methylated and unmethylated CpG sites, respectively. (Please refer to Chapter 9, page 142).

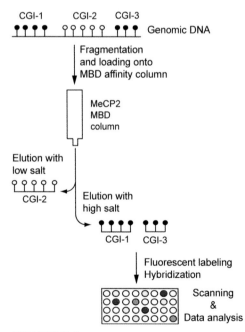

FIGURE 9.8

MBD-affinity column (MAC). Description of MAC is found in the text. Filled and open lollipops mark methylated and unmethylated CpG sites, respectively. (Please refer to Chapter 9, page 143).

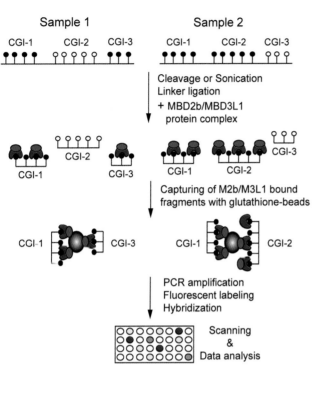

FIGURE 9.9
Methylated-CpG island recovery assay (MIRA). The steps of the MIRA procedure are described in the text. Filled and open lollipops mark methylated and unmethylated CpG sites, respectively. (Please refer to Chapter 9, page 144).

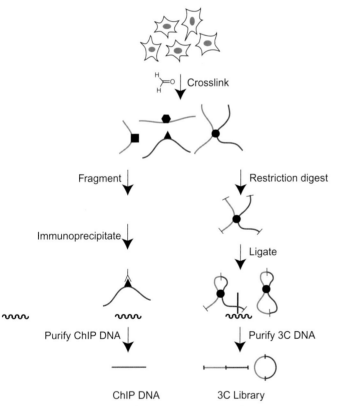

FIGURE 11.1
Overview of ChIP and 3C techniques. Cultured or primary cells are treated with formaldehyde to chemically crosslink DNA-binding proteins to their binding loci *in situ*. The crosslinked nuclei are processed for ChIP as outlined in the left panels or for 3C as outlined in right panels. Nuclear DNA is fragmented to yield DNA segments with associated proteins (squares, triangles, and circles, here) still intact. Proteins of interest (black triangle, here) are selected for by immuno-affinity precipitation, and crosslinks are reversed. The resulting co-precipitated DNA is then analyzed by whole-genome microarray (ChIP-chip) or high-throughput sequencing (ChIP-seq). For analyzing three-dimensional folding and looping, 3C employs crosslinked chromatin that has been restriction digested and ligated under dilute conditions for intramolecular ligation of DNA ends. The ligated DNA junctions are purified and analyzed by PCR, DNA microarrays, or sequencing. (Please refer to Chapter 11, page 160).

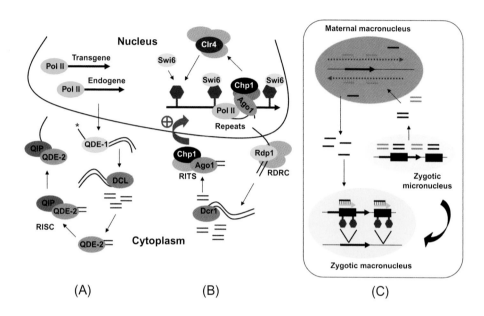

(A)   (B)   (C)

FIGURE 13.1

Models of RNAi in various eukaryotic microbes. (A) Quelling in *N. crassa*. Aberrant RNAs (*) are produced at loci that present repeats in large tandem arrays. Features of these aberrant RNAs are unknown, but they must be recognized by the RdRP QDE-1 and then convert into double stranded RNA molecules (dsRNA)[147]. dsRNA molecules are the typical substrate of the Dicer-like proteins DCL-1 and DCL-2 that chop them into siRNAs of 21–25 nucleotides. These siRNAs are integrated into the RISC complex, along with the Argonaute QDE-2 protein. They are then processed by the QIP nuclease and used as specific guides to target homologous mRNAs, which, once trapped, are most likely degraded by QDE-2. (B) RNAi silencing in *S. pombe*. The nascent transcript model proposes that RNA pol II continuously generates non-coding transcripts (*) from reverse promoter of heterochromatic repeats [36]. These aberrant RNAs are first cleaved by Ago1 and then recruited by the RNA-directed RNA polymerase complex (RDRC) to be converted into dsRNA by Rdp1 [148]. Using these dsRNAs as substrate, Dcr1 produces siRNA, which then bind to RNA-induced transcriptional silencing (RITS) complex, by means of Ago1 [149,150]. While RISC complexes target and degrade cytoplasmic mRNA, the RITS complex is tethered to chromatin through protein-protein interactions established between the chromodomain protein Chip1 and the H3K9me nucleosomes [35] (hexagons). The close association of the RITS complex and chromatin allows base-paring interactions between siRNA loaded on Ago1 and the nascent non-coding transcript soon to be cleaved by this protein. This amplification step of siRNA is likely to form a positive-feedback loop (plus arrow), which is believed to ensure the heterochromatin inheritance through cell divisions. As long as siRNA from a specific genomic region are produced, they continuously target the Clr4 histone methyltransferase complex (CLRC) to nucleosomes [151,152]. Thus, using H3K9me as signposts, heterochromatin spreads to large genomic territories in a sequence-independent but Swi6-dependent manner. As a result, transcription of the forward strand is silenced as in classical TGS systems. Gray ovals: known additional effectors. (C) Genome-scanning model in *Paramecium*. Because the micronucleus genome is unrearranged (rectangles represent IESs), it produces both IES-homologous (black) and non-IES-homologous (gray) scnRNAs. These diffusible molecules would enter and scan the IES-free maternal macronucleus. As a result of pairing with the maternal ncRNAs (dotted arrows), the non-IES-homologous scnRNAs would be sequestered. The remaining pool of scnRNA, highly enriched with IES-homologous scnRNAs, would be free to reach the developing zygotic macronucleus and pair with the nascent transcripts. At the IES targeted loci, chromatin shows H3K9 methylation [153] (hexagons), suggesting that this excision mechanism might have a TGS component. As for *S. pombe*, chromatin modifications could be used as signposts to direct an endonuclease towards the IESs to be excised. The curved arrow indicates that zygotic micronuclei develop into zygotic macronuclei throughout the course of the sexual phase. (Please refer to Chapter 13, page 187).

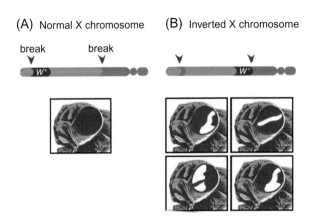

FIGURE 14.1

Diagram of the ln[1]w[m4] inversion that relocates the w[+] gene next to centromeric heterochromatin. The resulting clonal silencing of the w[+] gene is evidenced by pigmentless sectors in the inversion-bearing flies. (Please refer to Chapter 14, page 205).

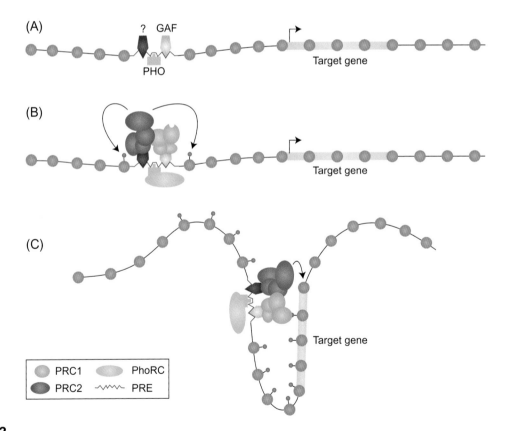

FIGURE 14.2

A model depicting the binding of PHO and GAF to a PRE, the recruitment of PRC1 and PRC2 resulting in methylation of nucleosomes, and the extension of the methylated domain by loop formation *(from [113], reprinted with permission from Macmillan Publishers Ltd).* (Please refer to Chapter 14, page 214).

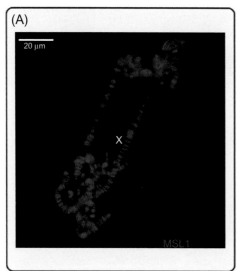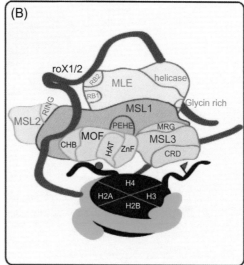

FIGURE 14.3

(A) The MSL complex is distributed along the male X chromosome as evidenced by indirect immunostaining of one of its protein components. (B) Subunit components of the MSL complex *(from Ref. 139, reprinted with permission from Springer Science + Business Media).* (Please refer to Chapter 14, page 216).

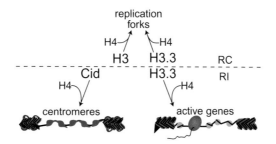

FIGURE 14.4

Diagram of the incorporation pathways of histones. H3 uses a replication-coupled (RC) mechanism and H3.3 and Cid a replication independent (RI) mechanism. Note that some degree of H3.3 incorporation can occur in an replication-coupled manner. *(from Ref. 217, reprinted with permission from The National Academy of Sciences, USA).* (Please refer to Chapter 14, page 219).

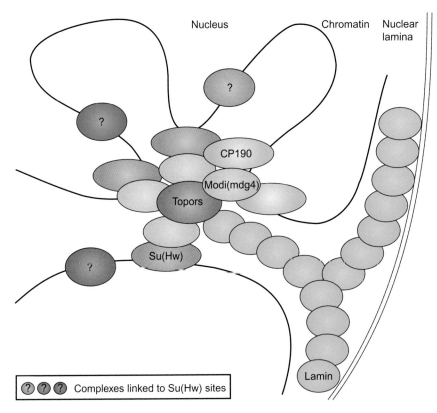

FIGURE 14.5
Detailed diagram of the association of Su(Hw) complexes with each other and with the nuclear matrix *(from Ref. 218, reprinted with permission from Macmillan Publishers Ltd).* (Please refer to Chapter 14, page 222).

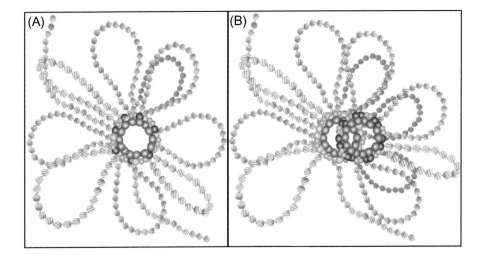

FIGURE 14.6
Models of the formation of DNA rosettes by the association of Su(Hw) (blue balls), Mod(mdg4) (green balls), CTCF (purple balls) and CP190 (pink balls). (A) Su(Hw) and CTCF insulators bind distinct but interspersed DNA sequences and, therefore, cluster together in the same insulator body. (B) Each insulator type forms its own insulator body but different bodies cluster together in particular nuclear regions *(from Ref. 211, reprinted with permission from Elsevier Limited).* (Please refer to Chapter 14, page 223).

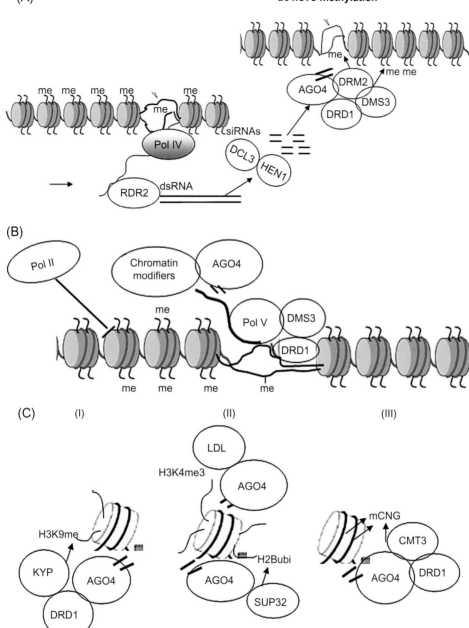

FIGURE 16.2

AGO4-siRNAs complexes involved in chromatin modifications. (A) A model for *de novo* DNA methylation involving Pol IV transcription as suggested by Matzke et al. [80]. The role of Pol IV is to produce single stranded RNA transcripts to be used as substrates by RDR2. Pol IV may transcribe from methylated DNA (as illustrated): DRM2 establishes new methyl groups at DNA sequences complementary to the small RNA loaded onto AGO4. The single stranded RNA produced from methylated DNA by Pol IV is used as a template for a dsRNA synthesis by RDR2 triggering the 24nt siRNA pathway. dsRNA is processed by DCL3 and HEN1 into small 3′-end methylated siRNAs. The 24nt siRNAs guide the AGO4 complex containing DRM2/DRD1/DMS3 to homologous genomic sequences. DRD1, a putative SNF2-like chromatin remodeler, and DMS3, an SMC-hinge domain-containing protein are accessory subunits of the complex [80]. (B) A model for spreading of silent chromatin and inhibiting Pol II activity through Pol V transcription, according to Wierzbicki et al. [85]. siRNAs and Pol V transcripts are produced by two independent pathways that collaborate to silence genes and to block Pol II activity. Pol V transcribes noncoding sequences enabled by DRD1 and DMS3. AGO4-siRNA complexes originated in a separate pathway recognize target loci by pairing with Pol V generated transcripts (see text). AGO4 recruits also DNA and histone modifiers (see panel C) to generate heterochromatin. The mechanism of recruiting chromatin modifiers is not clear. (C) AGO4-siRNA complexes in histone modifications and in DNA methylation establishing and propagating silenced chromatin. Once at a target locus, AGO4 and siRNA complexes might recruit several different chromatin-modifying enzymes to effect gene silencing. The order of action of these chromatin-modifying enzymes is not known, and their relative importance for gene silencing might be locus-specific. (I) Establishing the silencing H3K9me2 mark: SUVH4/KYP cooperates with the AGO4 complex to establish H3K9me2 according to [8–10,37,93]. (II) Removal of activating marks: LDL enzyme brought about by the AGO4-siRNA complex demethylates H3K4me3; de-ubiquitination of ubiquitinated H2B (H2Bubi) by the ubiquitinase SUP32 recruited and targeted by AGO4-siRNA [94]. (III) Establishing the CNG methylation: guided by homologous RNAs, AGO4 recruits the DNA methyltransferase CMT3 to produce CNG methylation at target loci [43,46,48]. (Please refer to Chapter 16, page 257).

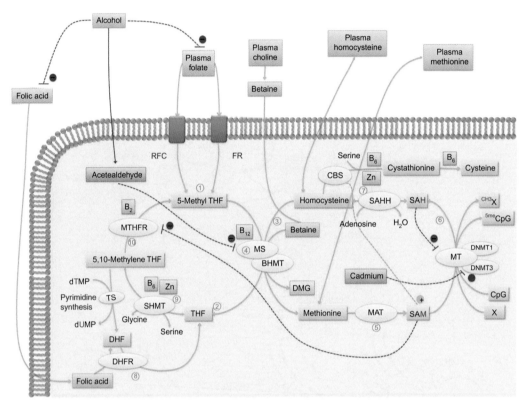

FIGURE 17.1

Schematic illustration of one carbon metabolic pathway. Precursors and cofactors obtained exclusively from the diet are highlighted in orange. Red dashed lines indicate inhibition of enzymatic reactions: Acetylaldehyde directly inhibits folate and absorption of folate. It also downregulate MS [45,53]. High level of SAH down-regulate the activity of MT. Increased concentration of SAM inhibits the activity of MTHFR, limiting the bioavailability of 5-methyl THF [8]. Cadmium is an inhibitor of DNMT activity [96]. Green dashed line indicates up-regulation of specific pathways. All mammalian tissues express MAT and MS, whereas BHMT is found only in the liver and kidney. SAM inhibits MTHFR and MS and activates CBS leading to homocysteine channeling down the transsulfuration pathway. **Abbreviations:** RFC, reduced folate carrier; FR, folate receptor; THF, tetrahydrofolate; MS, methionine synthase; BHMT, betaine homocysteine methyltransferase; DMG, dimethylglycine; MAT, methionine adonosyltransferase; SAM, S-adenosylmethionine; MT, methyltransferase; X, substrates for methylation; SAH, S-adenosylhomocysteine; CBS, cystathionine g-lyse; SHMT, serine hydroxymethyltransferase; MTHFR; methylene THF reductase; DHF, dihydrofolate; TS, thymidylate synthase. (Please refer to Chapter 17, page 282).

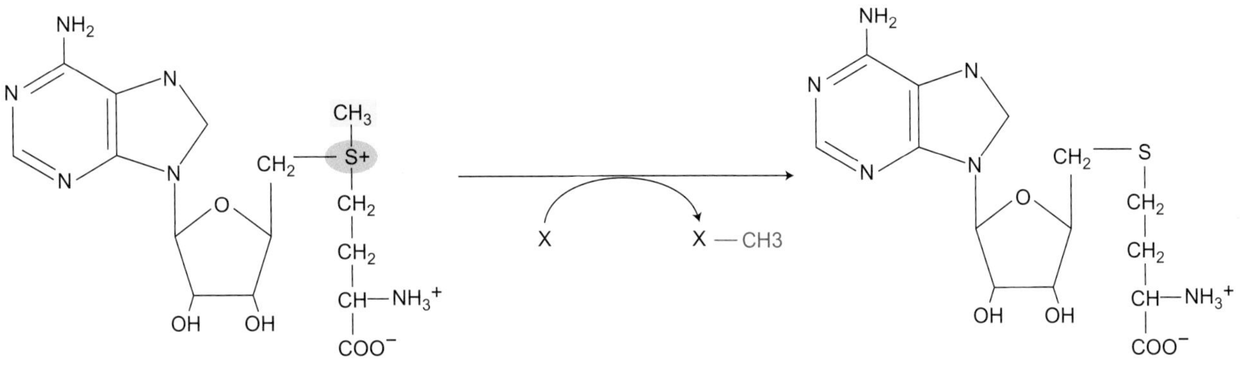

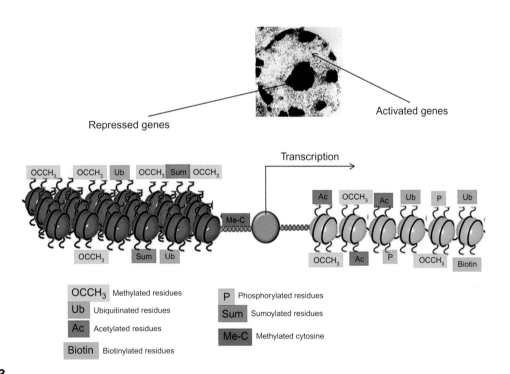

FIGURE 17.2
Conversion of S-adenosyl methionine to S-adenosyl homocysteine. The high energy sulfonium ion (orange shading) of SAM activates each of the attached carbons towards nucleophilic attack facilitating methyl donor transfer. (Please refer to Chapter 17, page 285).

Repressed genes

Activated genes

Transcription

OCCH₃	Methylated residues
Ub	Ubiquitinated residues
Ac	Acetylated residues
Biotin	Biotinylated residues
P	Phosphorylated residues
Sum	Sumoylated residues
Me-C	Methylated cytosine

FIGURE 18.3
Histone modifications involved in the repression of gene transcription (left) or gene transcription activation (right). (Please refer to Chapter 18, page 297).

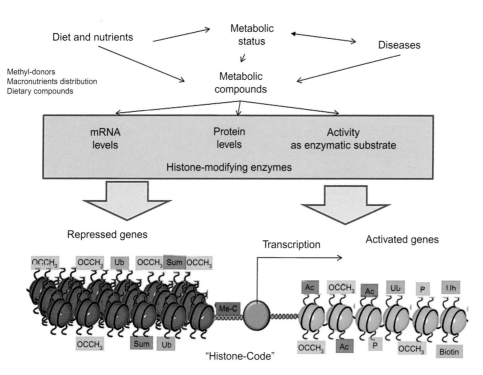

FIGURE 18.4
Influence of metabolic compounds and diseases on the activity of histone-modifying enzymes and gene transcription regulation. (Please refer to Chapter 18, page 298).

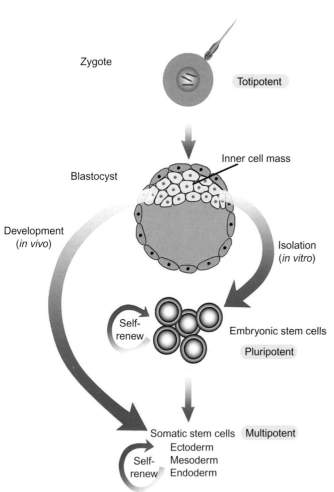

FIGURE 19.1
Developmental potential of stem cells. The totipotent level exists after the egg is fertilized. After several mitotic divisions that lead to the blastocyst, the inner cell mass can be isolated *in vitro*, yielding pluripotent embryonic stem cells (ESCs). ESCs can self-renew and differentiate into multipotent somatic stem cells specific to each of the three germ layers. (Please refer to Chapter 19, page 316).

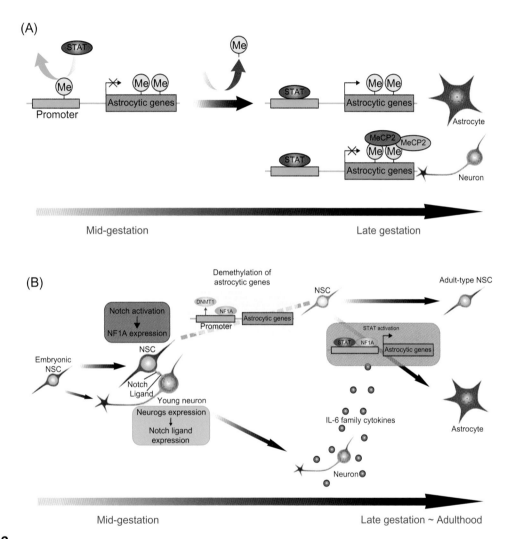

FIGURE 19.2

(A) Astrocytic gene methylation status during NSCs development. Although STAT3 can be activated in mid-gestational NSCs, it cannot bind to astrocytic gene promoters such as *gfap* due to promoter hypermethylation (left). As gestation proceeds, these promoters become demethylated, allowing STAT3 to bind and activate astrocytic genes, resulting in the differentiation of NSCs into astrocytes (upper right). Methyl-CpG binding protein 2 (MeCP2) blocks this activation in neurons (lower right). (B) Notch-induced demethylation of astrocytic genes. Activation of Notch signaling in residual NSCs by young neurons induces demethylation of astrocytic gene promoters by up-regulation of NFIA and release of DNMT1 from astrocytic gene promoters. In turn, at late gestation, IL-6 family cytokines activate the STAT3 pathway and induce NSCs to differentiate into astrocytic lineages. (Please refer to Chapter 19, page 318).

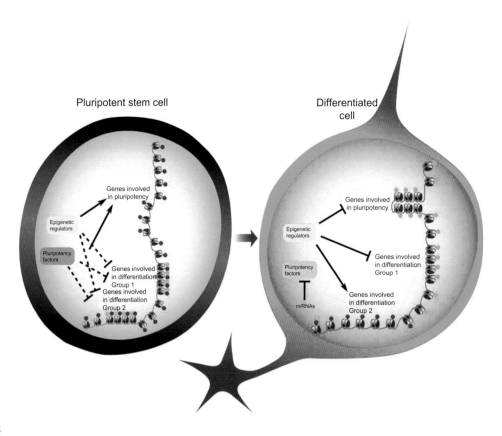

FIGURE 19.3

Regulatory mechanisms of pluripotency. Genes associated with pluripotency are actively transcribed in pluripotent cells (left), while differentiation-associated genes are kept in a silent poised state. Several epigenetic regulators and pluripotency factors regulate this state, in part by a combination of the activating methylation H3K4me3 (red circles) and the repressive methylation H3K27me3 (green circles). In pluripotent cells, only H3K4me3 is present at pluripotency-associated genes, but both H3K4me3 and H3K27me3 at differentiation-associated genes. Upon differentiation, miRNAs down-regulate pluripotency factors and differential repression of differentiation-associated gene groups is sustained only by epigenetic regulators. Pluripotency-associated genes and silenced differentiation-associated gene groups retain the H3K27me3 mark, while activated differentiation-associated gene groups retain the H3K4me3 mark. Chromatin status also changes from hyper to less dynamic during differentiation [36,37]. (Please refer to Chapter 19, page 321).

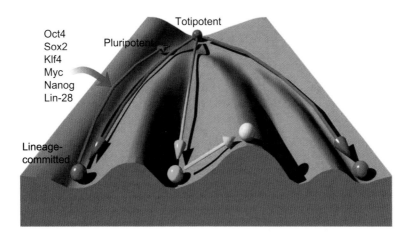

FIGURE 19.4

Epigenetic landscape. The totipotent fertilized egg can be depicted as a red ball that can roll down one of several possible valleys, passing through the pluripotent state and then differentiating into a particular tissue-lineage cell (blue, green, and pink balls). Reprogramming factors can push the ball back up the hill, enabling it to re-acquire pluripotency features. Lineage-committed cells can also trans-differentiate into cells of another lineage (yellow ball) by other epigenetic reprogrammings. The diagram is modified from Waddington [90]. (Please refer to Chapter 19, page 323).

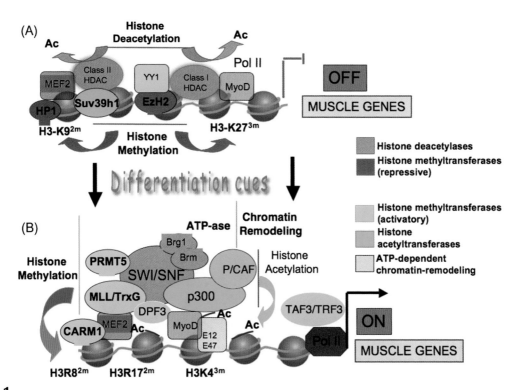

FIGURE 20.1

Schematic representation of the dynamic exchange in the chromatin-associated machinery that controls the epigenetic changes at the regulatory regions (promoter/enhancer elements) of muscle genes. (Please refer to Chapter 20, page 332).

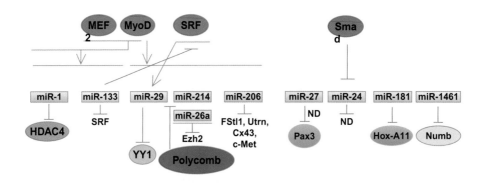

FIGURE 20.2

Regulatory circuits linking transcription factors and chromatin-modifying complexes to miRNA and their relative targets. (Please refer to Chapter 20, page 334).

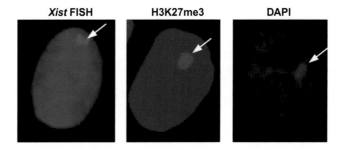

FIGURE 21.1

Visualization of major marks of XCI. Shown here are the three different visualizations of XCI: *Xist* RNA coating of the inactive X chromosome by fluorescence *in situ* hybridization (FISH); punctate staining of H3K27me3 as a major mark for XCI by immunostaining; and the Barr body detected as a dense DAPI staining within the nucleus. (Please refer to Chapter 21, page 341).

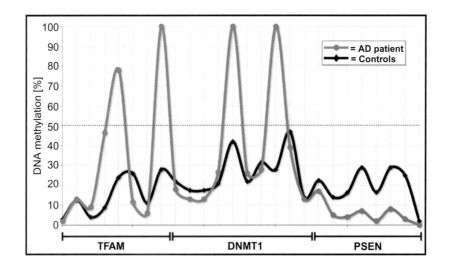

FIGURE 25.1

Gene-specific methylation changes during aging. The overall methylation content of DNA in very old individuals may be stable during aging, as was shown in the case of a 97-year-old Alzheimer's patient [15]. However, the post-mortem brain DNA of this individual (orange [upper] line) exhibited a hypomethylation of the PSEN1 promoter, whereas the promoters of DNMT1 and TFAM were significantly hypermethylated compared to the median brain methylation pattern observed in healthy individuals (black [lower] line). (Please refer to Chapter 25, page 409).

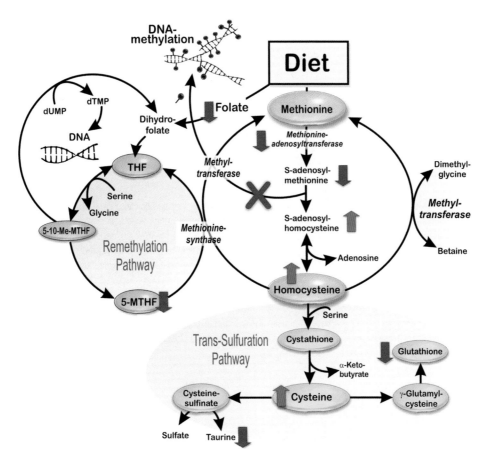

FIGURE 25.2

Aberrant methylation metabolism observed in aging. Several components of methylation pathways are frequently found to be abnormal (see arrows) in age-related disorders, either through environmental, hereditary, or nutritional factors. It was reported that S-adenosylmethionine (SAM), which is required for the methylation of DNA as well as methylation of histones, is severely decreased in the spinal fluid and brains of late-onset Alzheimer patients. A reduction of SAM can be the cause of a decreased uptake of folate and vitamin B12, which in turn induces the accumulation of homocysteine and S-adenosylhomocysteine (SAH). A number of steps in the metabolism of SAM, such as the transformation of Hcy to methionine or a methionine adenosyltransferase deficiency, which is frequently observed in dementia patients, could lead to altered methylation levels within a cell. The accumulation of Hcy reverts the methionine metabolism to SAH, which acts as a strong inhibitor of DNA methyltransferases, resulting ultimately in DNA hypomethylation. Clearance of Hcy by methionine synthase maintains a favorable SAM:SAH ratio, an index of cellular methylation potential. These age-dependent metabolic changes may be very important factors that play a paramount role in the genesis of age-related disorders. THF = tetrahydrofolate; 5-MTHF = 5-methyltetrahydrofolate. (Please refer to Chapter 25, page 410).

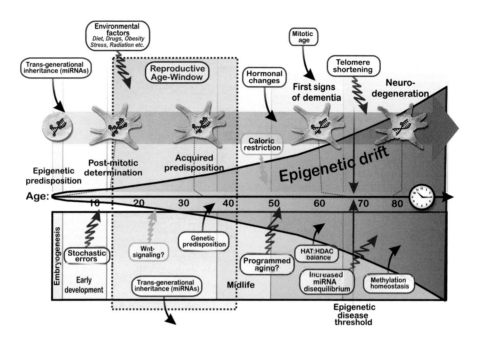

FIGURE 25.3

Age-dependent epigenetic drift. Shown here is the example of neuronal epigenetic drift that may lead to age-dependent neurodegenerative disorders, such as AD or Parkinson's. The relatively high frequency of *de novo* epimutations suggests that epigenetic alterations accumulate during aging. Small epimutations may be tolerated by the cells; however, once the epigenetic deregulation reaches a critical threshold, the cells no longer function properly. The phenotypic outcome depends on the overall effect of the series of pre- and postnatal impacts on the pre-epimutation. Only some predisposed individuals will reach the "threshold" of epigenetic deregulation that causes the phenotypic changes that meet the diagnostic criteria for a clinical disorder. Epigenetic drift may not only be affected by several internal and external factors – some individuals may already be epigenetically predisposed at birth, due to trans-generational epigenetic effects. Trans-generational epigenetic inheritance results either from incomplete erasure of parental epigenetic marks during phases of epigenetic reprogramming at fertilization, or may be established by small RNA species that pass through the germline. Since all organisms eventually die from different causes, epigenetic patterns beneficial in early life are favored by natural selection over patterns advantageous later in life. Deleterious epigenetic drift occurring after the reproductive phase is relatively neutral to selection, because its bearers have already transmitted their genes (and potentially epigenetic information) to the next generation. (Please refer to Chapter 25, page 416).

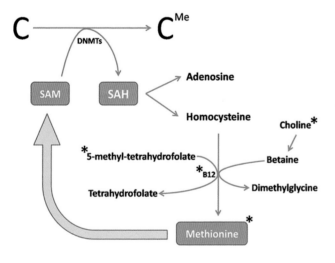

FIGURE 27.1

One-carbon metabolism. This figure illustrates how the methylation reactions occur. The transfer of a methyl group from SAM to the cytosine is catalyzed by DNMTs and the final products of this reaction are 5-methyl-cytosine and SAH. This product must then be recycled to enable another methylation reaction to take place. This recycling process involves the conversion of homocysteine to methionine, which is then converted to SAM. Many dietary components may affect this process (indicated by asterisks) and consequently affect the methylation reaction. (Please refer to Chapter 27, page 451).

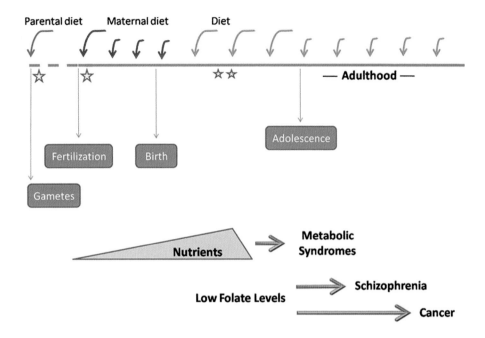

FIGURE 27.2

Diet influence on epigenetic traits during life time. During his life time, an individual faces periods of important epigenetic programming (indicated by stars). Dietary compounds may affect epigenetic patterns, especially during these critical periods. This dietary influence is not restricted to the individual's diet, but also the parental (during gamete formation) and maternal (during intrauterine life) diets exert striking effects. When an individual is subjected to a nutrient-deficient diet during early life development and faces a normal supply of nutrients after birth, the incidence of metabolic syndromes increases. Also, low folate levels during life are correlated to diseases such as schizophrenia and cancer. (Please refer to Chapter 27, page 454).

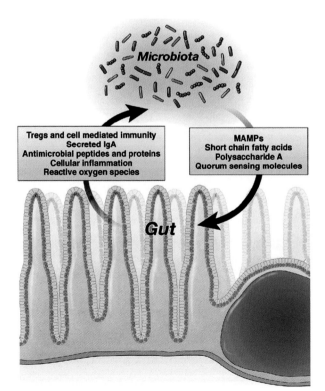

FIGURE 29.1

Mechanisms of microbes and crosstalk within the gut. The symbiosis between human cells of the large bowel and the intraluminal microbiotic flora (several strains of bacteria) is a dyadic relationship by which both parties alter and shape each other, resulting in a "negotiated settlement" at an equilibrium. A breakdown of this crosstalk may result in a "dysbiotic" microbiota and clinical consequences, resulting in diseases ranging from inflammatory bowel disease to cancer. Understanding of how and by which mechanisms microbes are involved in the genesis of human disease is evolving. MAMPs denote "microbial-associated molecular patterns". *(Reprinted from Neish, A.S. Microbes in gastrointestinal health and disease. Gastroenterology, 2009;136(1):65–80, Copyright (2009), with permission from Elsevier/AGA Institute.)* (Please refer to Chapter 29, page 478).

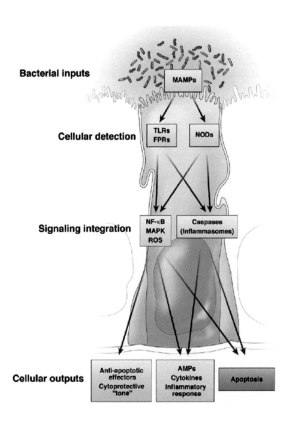

FIGURE 29.2

Cellular consequences to bacterial stimuli. Bacterial microbial-associated molecular patterns (MAMPs) may stimulate pattern recognition receptors (including extracellular Toll-like receptors (TLRs) and formylated peptide receptors FPRs, or intracellular NODs). Intensity, duration, and spatial origin of the subsequent signaling responses are integrated by an intricate and interrelated network of transduction pathways that determine if MAMP perception warrants a "low gain" cytoprotective response, a "medium gain" inflammatory reaction, or "high gain" programmed cell death result. *(Reprinted from Neish, A.S. Microbes in gastrointestinal health and disease. Gastroenterology, 2009;136(1):65–80, Copyright (2009), with permission from Elsevier/AGA Institute.)* Abbreviations: NF-κB, nuclear factor κB; NLR, Nod-like receptor; PRR, pattern recognition receptor; ROS, reactive oxygen species; SCFA, short-chain fatty acid; TLR, Toll-like receptor. (Please refer to Chapter 29, page 481).

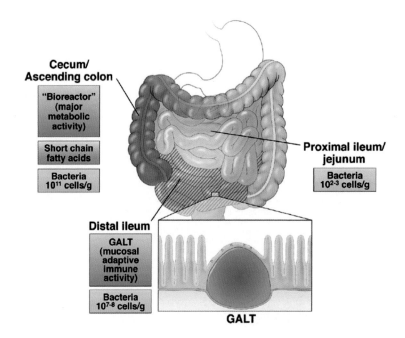

FIGURE 29.3

Preferred sites of commensal/probiotic interaction with the gut. Cecum/ascending colon is a "bioreactor" with the greatest amounts of bacteria, metabolic activity, and short-chain fatty acid (SCFAs) fermentation. Concentration of SCFA diminishes along the colon. The distal ileum is enriched in GALT (gut-associated lymphoid tissue; Peyer's patches) and is the dominant site of luminal sampling and mucosal adaptive immune activity. *(Reprinted from Neish, A.S. Microbes in gastrointestinal health and disease. Gastroenterology, 2009;136(1):65–80, Copyright (2009), with permission from Elsevier/AGA Institute.)* (Please refer to Chapter 29, page 486).

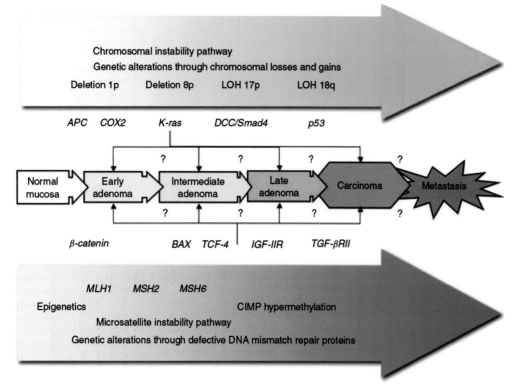

FIGURE 29.4

Characteristics of the major pathways in colorectal cancer. APC, adenomatous polyposis coli; BAX, Bcl-2-associated X protein; CIMP, CpG island methylator phenotype; COX, cyclo-oxygenase; DCC, deleted in colorectal cancer; IGF-IIR, insulin-like growth factor II receptor; LOH, loss of heterozygosity; MLH, MutL homolog; MSH, MutB homolog; Smad, mothers against decapentaplegic homolog (*Drosophila*); TCF, T cell factor, TGF-βR, transforming growth factor β receptor. *(The figure is derived from Søreide et al. Copyright British Journal of Surgery Society Ltd. Reproduced with permission from John Wiley & Sons Ltd on behalf of the BJSS Ltd.)* (Please refer to Chapter 29, page 488).

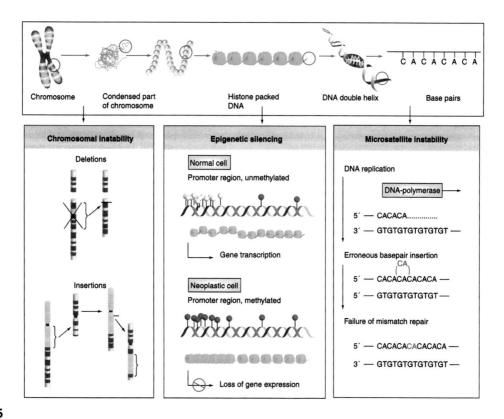

FIGURE 29.5

Molecular mechanisms leading to colorectal cancer. Depicted are chromosomal instability, epigenetic silencing, and microsatellite instability. The latter two mechanisms often coincide in sporadic CRC. *(Reproduced with permission from Søreide K, et al. Endoscopy, morphology, morphometry and molecular markers: predicting cancer risk in colorectal adenoma. Expert Rev Mol Diagn. 2009;9(2):125–37.)* (Please refer to Chapter 29, page 488).

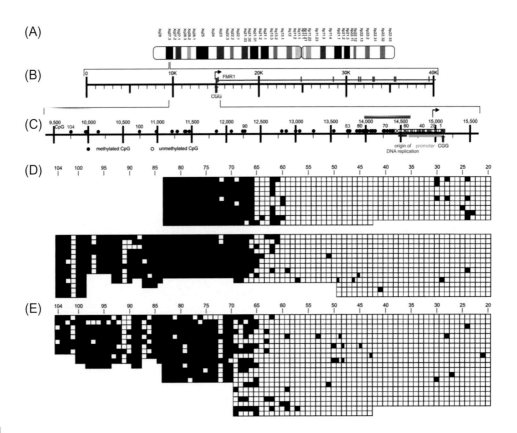

FIGURE 30.1

Boundary of DNA Methylation between the *FMR1* promoter and its 5′-upstream region of the genome. (A) Scheme of the human X chromosome. (B) Map of the region 27.3 on the long arm of the X chromosome: *FMR1* gene brown; CGG repeat blue, scale in 10,000 nucleotides (10K). (C) Enlarged segment of the map in (B): Promoter green. Arrow, site of initiation of transcription of the FMR1 gene; origin of DNA replication, gray. Each CG pair in the sequence as target for DNA methylation is marked by either ○ (unmethylated) or by ● (methylated). In (D) and (E) CG dinucleotides have been condensed (□ unmethylated; ■ methylated) in both DNA strands from the male human tumor cell line HCT116 (D) or for only one DNA strand from primary human fibroblasts (E). Numbers in (D) and (E) correspond to CG dinucleotide pairs counting upward from the CGG repeat (from Ref. 78). Each row of CG dinucleotide pairs corresponds to one DNA molecule, each column to the profile in one CG pair in different molecules. This figure was taken from Ref. 78. (Please refer to Chapter 30, page 500).

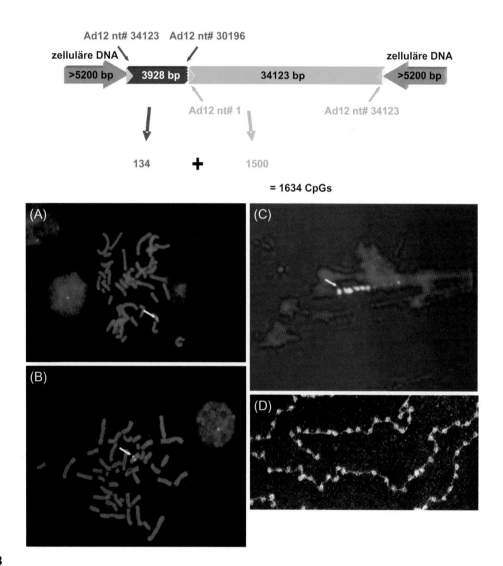

FIGURE 30.3

Integration of foreign, adenovirus type 12 (Ad12) DNA in the DNA of hamster cells. Upper panel: Scheme of an intact Ad12 DNA molecule (green) and of a fragment of a second Ad12 DNA molecule (red). Both genomes are integrated into hamster DNA (gray arrows). Lower panels: (A, B) Ad12 DNA integrated into one of the hamster chromosomes (red) are marked in yellow (arrows). (C) The chromosome in (cell line T637) (A) has been stretched and 10 to 15 copies of Ad12 DNA become apparent as individual yellow dots (arrow). In cell line TR12 (B) only one copy of Ad12 DNA and an Ad12 DNA fragment have remained integrated. The methylation profile of this integrate has been determined in Figure 30.4. (D) Nucleosome structure of an integrated Ad12 genome in hamster cells. Panels (A) and (B) were derived from Ref. 84, panel (C) from Ref. 31. (Please refer to Chapter 30, page 503).

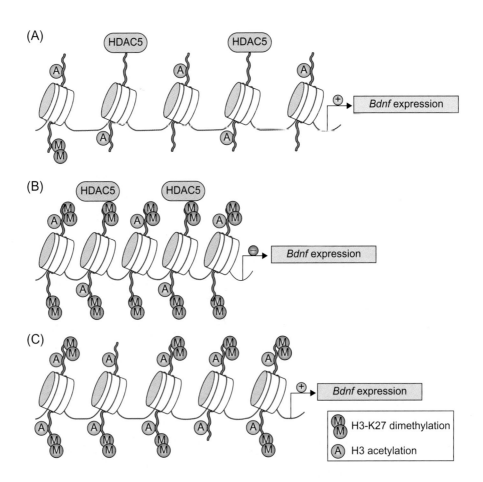

FIGURE 34.1

The importance of histone PTMs in a rodent model of depression. (A) Under physiological conditions, i.e. in the absence of stress, the promoter region of the *Bdnf* gene carries a moderate level of histone H3 acetylation and H3K27 dimethylation, and is bound by the histone deacetylase HDAC5. (B) Upon chronic social defeat stress, a rodent model of depression, H3K27 dimethylation is increased, leading to increased condensation of the Chromatin, which shuts down *Bdnf* gene expression. (C) Upon chronic antidepressant (imipramine) treatment, HDAC5 levels are reduced, which leads to increased H3 acetylation, whereas H3K27 dimethylation remains unaffected. Nonetheless, the increase in H3 acetylation is sufficient to reinstate *Bdnf* gene expression. A, acetyl; Bdnf, brain-derived neurotrophic factor; HDAC, histone deacetylase; K, lysine; M, methyl. Figure reproduced, with permission, from Ref. 49. (Please refer to Chapter 34, page 559).

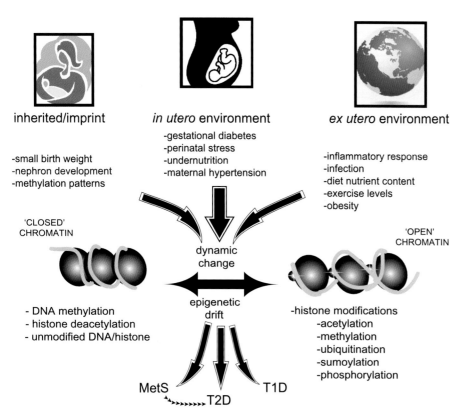

FIGURE 35.1

Routes of epigenetic regulation and dysregulation in metabolic diseases. Metabolic diseases such as Metabolic Syndrome (MetS), Type 1 diabetes (T1D), and Type 2 diabetes (T2D) all have potential epigenetic dysregulation underlying their pathologies [1–3,7–9,14]. *In utero* and inheritable epigenetic modifications can set up chromatin modification patterns to prejudice gene expression to be more conducive to disease development. Inducible epigenetic changes promote more dysregulation in response to aberrant environmental stimuli, adding to the underlying susceptibility set up during development, triggering overt disease. (Please refer to Chapter 35, page 571).

	ENERGY METABOLISM	INFLAMMATION & IMMUNE RESPONSE	HUNGER & SATIATION	PANCREATIC DEVELOPMENT
METS/T2D	GLUT2 GLUT4 CLOCK		MC4R MC3R	ST1
OVERLAP	PPARalpha PPARgamma PGC-1 CREB NFκB INS IR IRK1 JUNB	IL-6 NFκB P38 MAPK SAPK/JNK WNT/B-CAT TNF PTGS2 MCP-1 TGFbeta IL-1beta STAT5	CREB NFκB	PDX1 NOTCH
T1D		IRAK1 CSF2 IL-10 IL-2 IL-2R IL-13 IL-18 LCK VDR TAB2 IFNgamma CXCL3 CD28 CTLA4 IL-4 CCR5		HOX4

FIGURE 35.2

Genome wide epigenetic modification screening reveals overlap in metabolic syndromes. Wren and Garner [16] and Maio et al. [17] conducted genome-wide data-mining analyses looking for genetic loci associated with T2D [16] or T1D [17]. Combining data from both analyses with epigenetic regulation data for gene candidates [46,52,53,59,63,64,68,69], reveals overlapping effects on genes in pathways essential to energy metabolism, inflammation, and macrophage function/differentiation, as well as signaling in hunger/satiation control and pancreatic development/differentiation. (Please refer to Chapter 35, page 576).

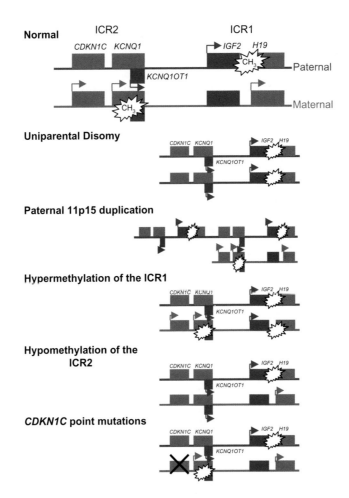

FIGURE 36.2

Epigenetic/genetic alterations in 11p15 which are currently known to be associated with BWS as an example for the complex molecular background of IDs. Only those genes in 11p15 are illustrated which are currently being discussed as being functionally involved in the pathogenesis of BWS (and SRS). (Please refer to Chapter 36, page 586).